Handbook of Thermal Conductivity

Volume 3

Organic Compounds C_8 to C_{28}

LIBRARY OF PHYSICO-CHEMICAL PROPERTY DATA

Handbook of Vapor Pressure

Volume 1: C_1 to C_4 Compounds (Product #5189)
Volume 2: C_5 to C_7 Compounds (Product #5190)
Volume 3: C_8 to C_{28} Compounds (Product #5191)
Volume 4: Inorganic Compounds and Elements (Product #5394)
Carl L. Yaws

Handbook of Viscosity

Volume 1: C_1 to C_4 Compounds (Product #5362)
Volume 2: C_5 to C_7 Compounds (Product #5364)
Volume 3: C_8 to C_{28} Compounds (Product #5368)
Carl L. Yaws

Handbook of Thermal Conductivity

Volume 1: C_1 to C_4 Compounds (Product #5382)
Volume 2: C_5 to C_7 Compounds (Product #5383)
Volume 3: C_8 to C_{28} Compounds (Product #5384)
Carl L. Yaws

Each of the above series contains data for more than 1,000 organic compounds, including hydrocarbons, oxygenates, halogenates, nitrogenates, sulfur compounds, and silicon compounds. The data are presented in graphs for *vapor pressure, viscosity,* or *thermal conductivity* as a function of temperature and are arranged by carbon number and chemical formula to enable the engineer to quickly determine values at the desired temperatures.

Handbook of Transport Property Data (Product #5392)

Carl L. Yaws

Comprehensive data on viscosity, thermal conductivity, and diffusion coefficients of gases and liquids are presented in convenient tabular format.

Physical Properties of Hydrocarbons

Volume 1, Second Edition (Product #5067)
Volume 2, Third Edition (Product #5175)
Volume 3 (Product #5176)
Volume 4 (Product #5272)
R. W. Gallant and Carl L. Yaws

The four-volume series provides chemical, environmental, and safety engineers with quick and easy access to vital physical property data needed for production and process design calculations.

Thermodynamic and Physical Property Data (Product #5031)

Carl L. Yaws

Property data for 700 major hydrocarbons and organic chemicals, including oxygen, nitrogen, fluorine, chlorine, bromine, iodine, and sulfur compounds, are provided.

LIBRARY OF PHYSICO-CHEMICAL PROPERTY DATA

Handbook of Thermal Conductivity

Volume 3

Organic Compounds C_8 to C_{28}

Carl L. Yaws

Gulf Publishing Company
Houston, London, Paris, Zurich, Tokyo

Handbook of Thermal Conductivity, Volume 3

Gulf Publishing Company
Book Division
P.O. Box 2608 □ Houston Texas 77252-2608

Printed in the United States of America

Transfered to Digital Printing, 2011

Printed on Acid-Free Paper (∞)

Library of Congress Cataloging-in-Publication Data

Yaws, Carl L.
Handbook of thermal conductivity / Carl L. Yaws.
p. cm. — (Library of physico-chemical property data)
Includes bibliographical references.
Contents: v. 1. Organic compounds, C_1 to C_4 — v. 2. Organic compounds, C_5 to C_7 — v. 3. Organic compounds, C_8 to C_{28}.
ISBN 0-88415-382-7 (v. 1). — ISBN 0-88415-383-5 (v. 2). — **ISBN 0-88415-384-3** (v. 3)
1. Heat—Conduction—Handbooks, manuals, etc.
I. Title. II. Series.
QD257.7.Y38 1995
661.8—dc20 95-5676
CIP

CONTENTS

CONTRIBUTORS

Li Bu	Graduate student, Chemical Engineering Department, Lamar University, Beaumont, Texas 77710, U.S.A.
Jack R. Hopper	Professor and Chair, Chemical Engineering Department, Lamar University, Beaumont, Texas 77710, U.S.A.
Xiaoyan Lin	Graduate student, Chemical Engineering Department, Lamar University, Beaumont, Texas 77710, U.S.A.
Sachin Nijhawan	Graduate student, Chemical Engineering Department, Lamar University, Beaumont, Texas 77710, U.S.A.
Carl L. Yaws	Professor, Chemical Engineering Department, Lamar University, Beaumont, Texas 77710, U.S.A.

ACKNOWLEDGMENTS

Many colleagues and students have made contributions and helpful comments over the years. The author is grateful to each: Jack R. Hopper, Joe W. Miller, Jr., C.S. Fang, K.Y. Li, Keith C. Hansen, Daniel H. Chen, P.Y. Chiang, H.C. Yang, Xiang Pan, Xiaoyan Lin, Duane G. Piper, Jr., Li Bu, and Sachin Nijhawan.

The author wishes to extend special appreciation to his wife (Annette) and family (Kent, Michele, Chelsea, and Brandon; Lindsay and Rebecca; and Matthew and Sarah).

The author wishes to acknowledge that the Gulf Coast Hazardous Substance Research Center provided partial support to this work.

DISCLAIMER

PREFACE

Thermal conductivity data are important in many engineering applications in the chemical processing and petroleum refining industries. The objective of this book is to provide the engineer with such data. The data are presented in graphs covering a wide temperature range to enable the engineer to quickly determine values at the temperatures of interest. The contents of the book are arranged in the following order: graphs, references, and appendixes.

The graphs for thermal conductivity as a function of temperature are arranged by carbon number and chemical formula to provide ease of use. Most of the graphs for the liquid cover a wide temperature range extending from melting point to boiling point to near the critical point. The graphs for gas cover a wide temperature range and are applicable at low pressure. Common units are used for thermal conductivity (W/m K). For those involved in English usage, each graph displays a conversion factor to provide the English units.

The coverage encompasses a wide range of organic compounds including hydrocarbons such as alkanes, olefins, acetylenes, and cycloalkanes; oxygenates such as alcohols, aldehydes, ketones, acids, ethers, glycols, and anhydrides; halogenates such as chlorinated, brominated, fluorinated, and iodinated compounds; nitrogenates such as nitriles, amines, cyanates, and amides; sulfur compounds such as mercaptans, sulfides, and sulfates; silicon compounds such as silanes and chlorosilanes; and many other chemical types.

The graphs for liquids are based on both experimental data and estimates. Estimates for liquid thermal conductivity were primarily based on modified Missenard and Pachaiyappan methods. Both experimental data and estimates were then regressed to provide the same equation with wide and extended temperature coverage for all compounds.

Very limited experimental data for liquid thermal conductivities are available at temperatures in the region of the melting point. Also, there are very few reliable data at temperatures above a reduced temperature of $T_f = 0.65$. Thus, the values in the regions of melting point and reduced temperatures above 0.65 should be considered rough approximations. The values in the intermediate region (above melting point and below reduced temperature of 0.65) are more accurate.

The graphs for gases are also based on both experimental data and estimates. In the absence of experimental data, estimates were primarily based on correlations of Roy and Thodos; Misic and Thodos; and Stiel and Thodos and modified Eucken models. Experimental data and estimates were then regressed to provide the same equation for all compounds. The graphs are applicable for low pressure gas. The presented values may be adjusted to provide values at higher pressure using the methods in Reid, Prausnitz, and Poling (25).

Very limited experimental data for gases are available for highly polar and high molecular weight compounds. Also, very few experimental data are available for organic compounds at high temperatures above 600 K. Thus, the values for these compounds and high temperatures should be considered rough approximations.

The literature has been carefully searched in construction of the graphs. The references are given in the section following the graphs near the end of the book. The references provide documentation for the original sources used in preparing the work.

The equations for liquid and gas thermal conductivities are:

$$\log_{10} \lambda_{liq} = A + B\,[1 - T/C]^{2/7} \qquad \text{(liquid)}$$
$$\lambda_{gas} = A + B\,T + C\,T^2 \qquad \text{(gas)}$$

The coefficients for the equations are provided in the appendix near the end of the book. The tabulated values are especially arranged for quick usage with hand calculator or computer. Computer programs, containing coefficients for all compounds, are available for liquid and gas.

Each program is in ASCII which can be accessed by other software. For information, contact Carl L. Yaws, Ph.D., P.O. Box 10053, Beaumont, Texas 77710, phone/fax (409) 880-8787.

A list of compounds is also given near the end of the book to aid the user in quickly locating the compound of interest by the chemical formula or name.

Handbook of Thermal Conductivity

Volume 3

Organic Compounds C_8 to C_{28}

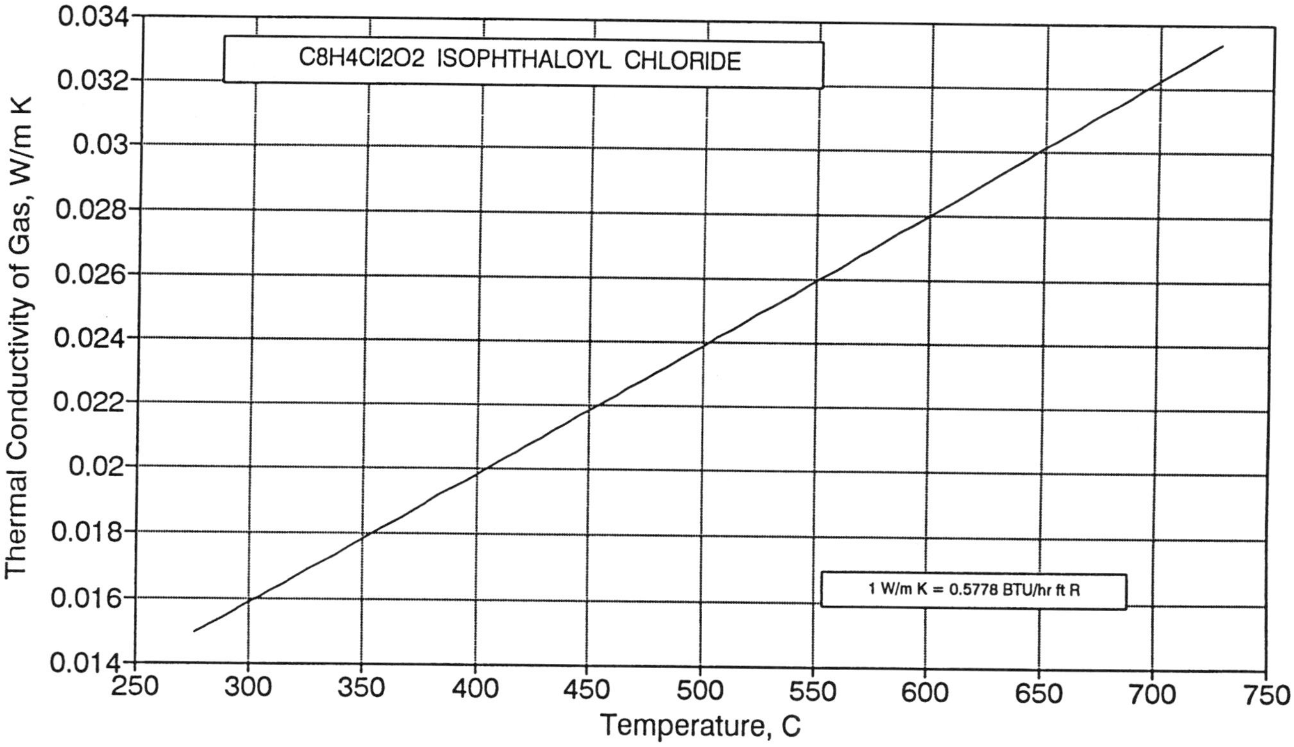
C8H4Cl2O2 ISOPHTHALOYL CHLORIDE
1 W/m K = 0.5778 BTU/hr ft R
Thermal Conductivity of Gas, W/m K
0.034
0.032
0.03
0.028
0.026
0.024
0.022
0.02
0.018
0.016
0.014
250
300
350
400
450
500
550
600
650
700
750
Temperature, C

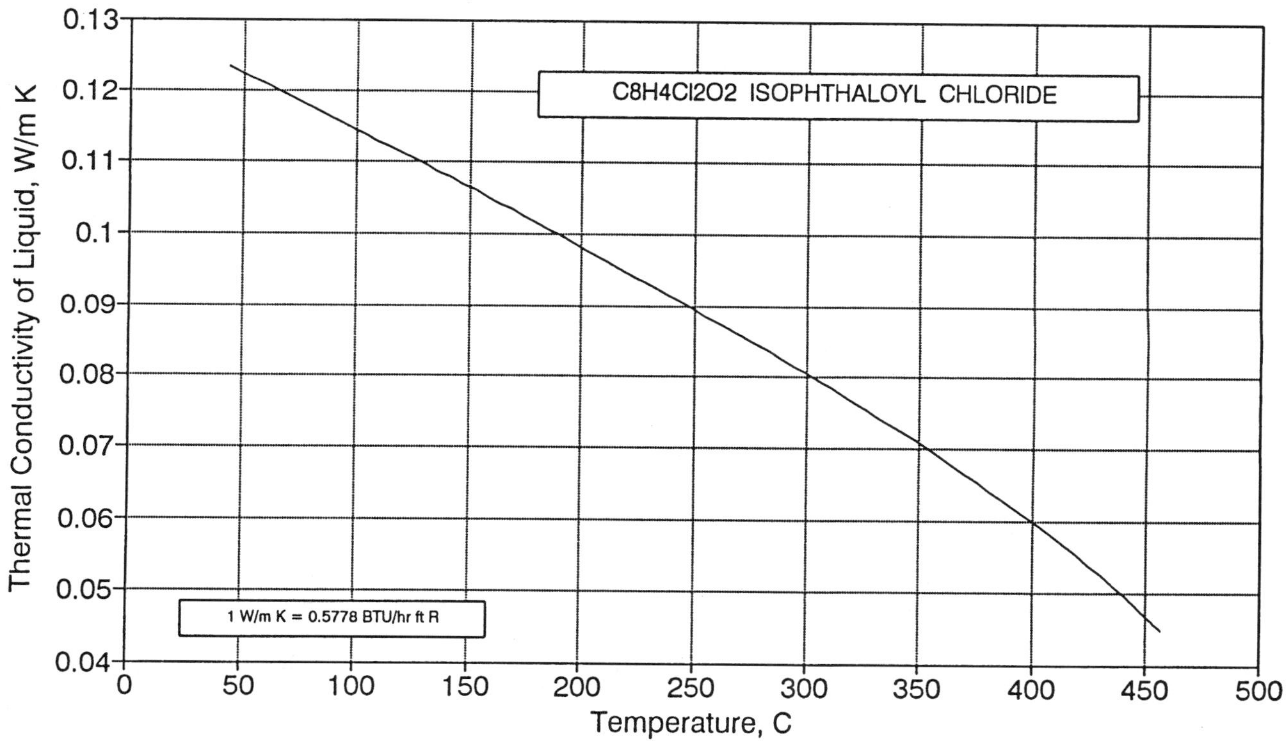
C8H4Cl2O2 ISOPHTHALOYL CHLORIDE
1 W/m K = 0.5778 BTU/hr ft R
Thermal Conductivity of Liquid, W/m K
0.13
0.12
0.11
0.1
0.09
0.08
0.07
0.06
0.05
0.04
0
50
100
150
200
250
300
350
400
450
500
Temperature, C

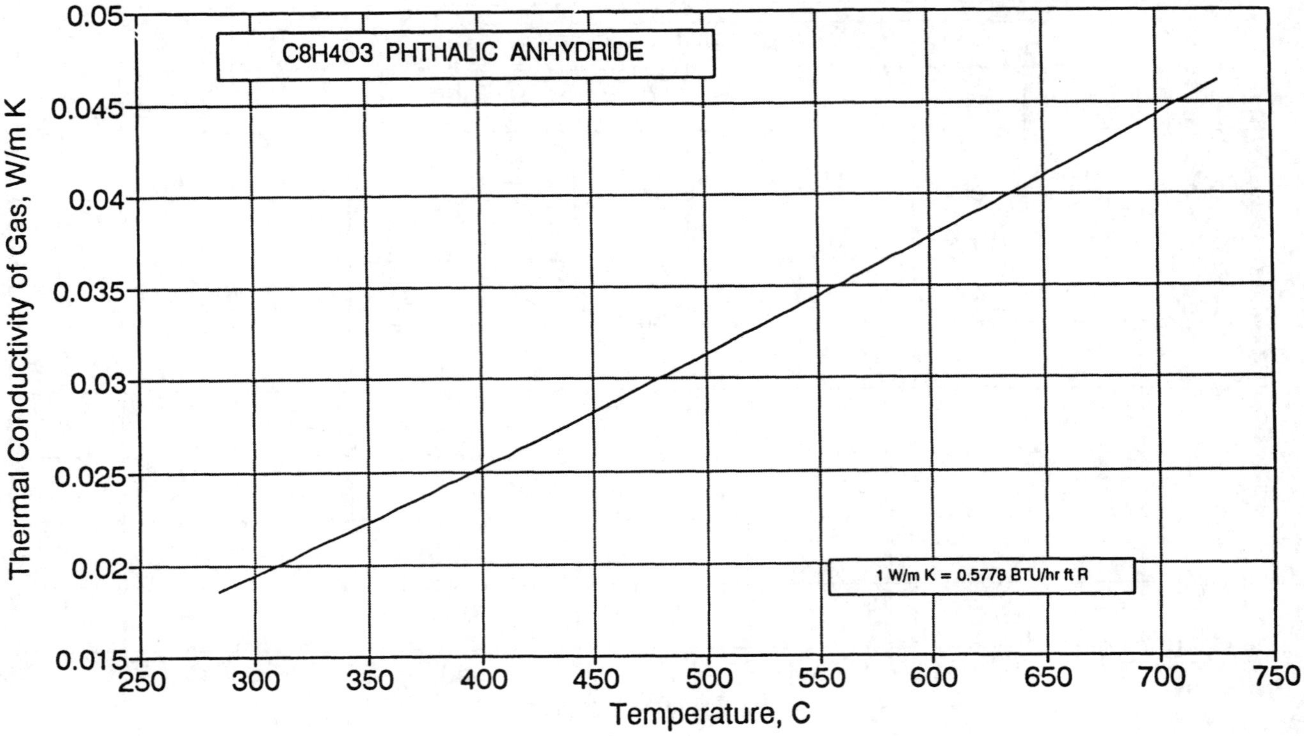
C8H4O3 PHTHALIC ANHYDRIDE
Thermal Conductivity of Gas, W/m K
0.05
0.045
0.04
0.035
0.03
0.025
0.02
0.015
1 W/m K = 0.5778 BTU/hr ft R
250
300
350
400
450
500
550
600
650
700
750
Temperature, C

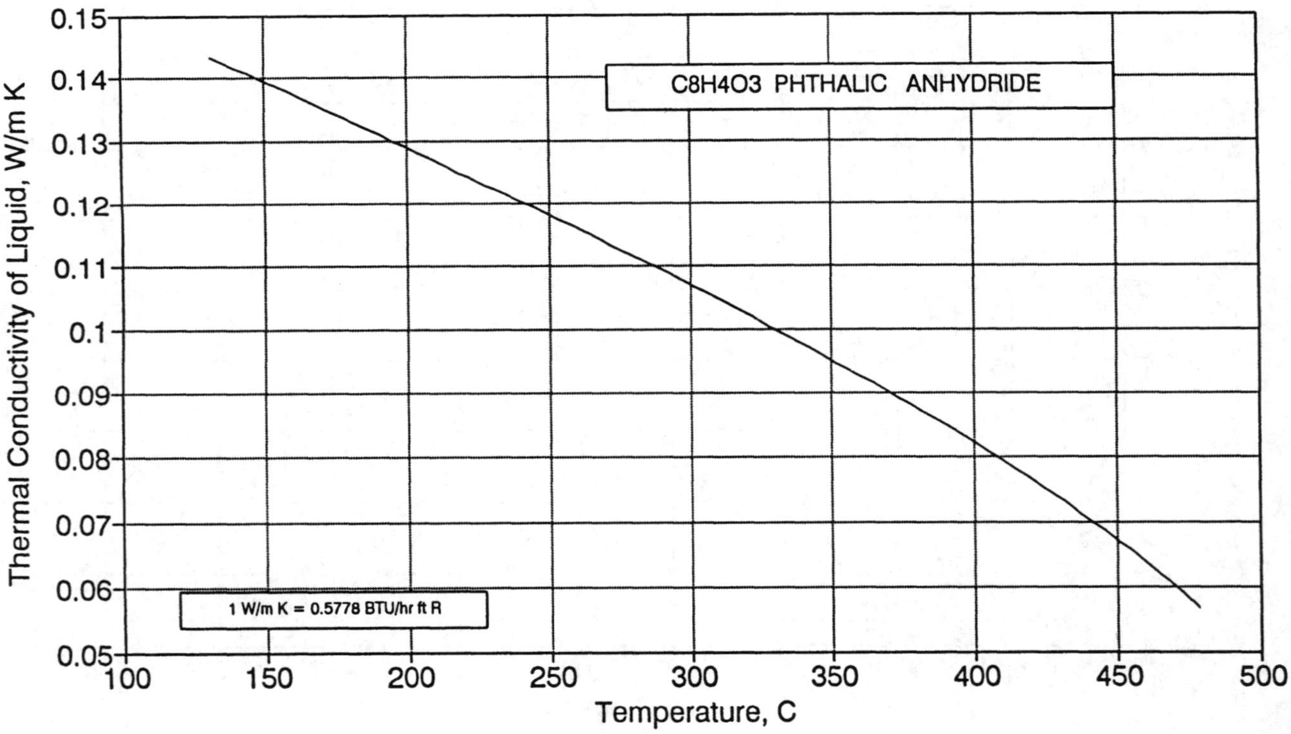
C8H4O3 PHTHALIC ANHYDRIDE
Thermal Conductivity of Liquid, W/m K
0.15
0.14
0.13
0.12
0.11
0.1
0.09
0.08
0.07
0.06
0.05
1 W/m K = 0.5778 BTU/hr ft R
100
150
200
250
300
350
400
450
500
Temperature, C

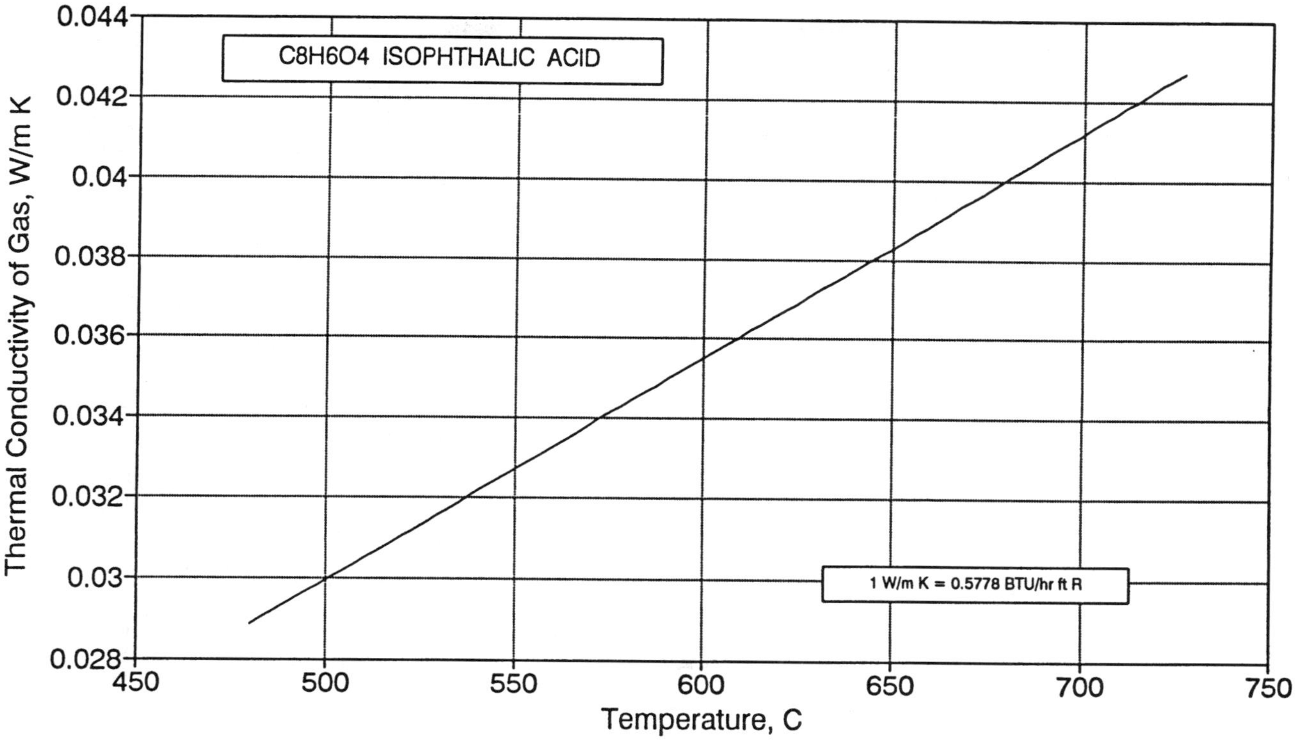

C8H6O4 ISOPHTHALIC ACID

Liquid thermal conductivity data are not available .

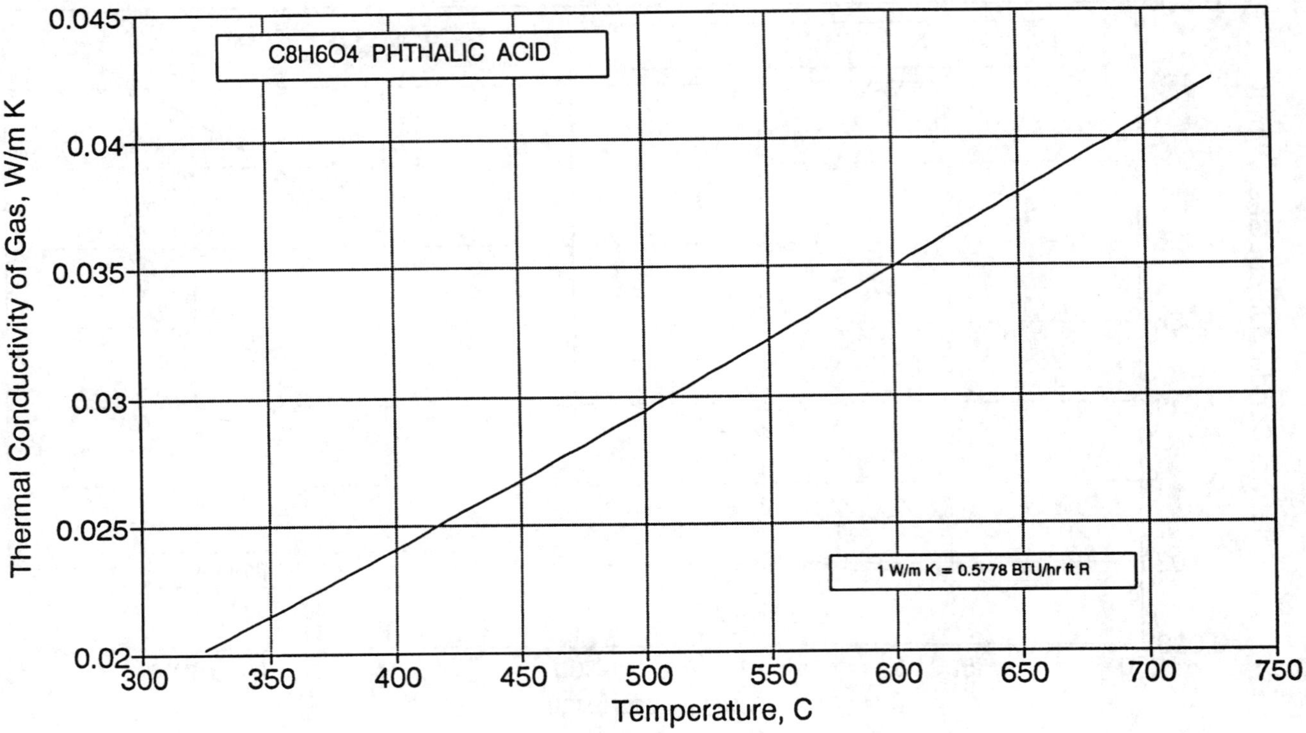

C8H6O4 PHTHALIC ACID

Liquid thermal conductivity data are not available .

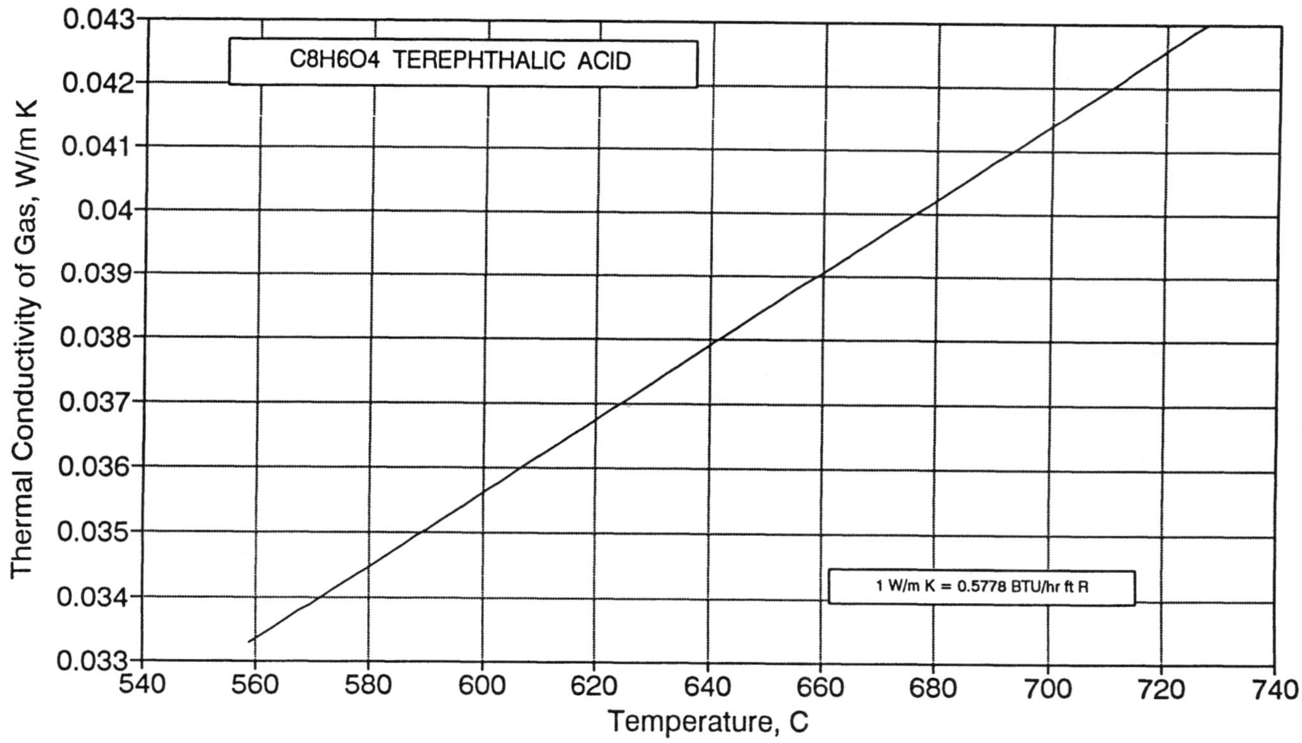

C8H6O4 TEREPHTHALIC ACID

Liquid thermal conductivity data are not available .

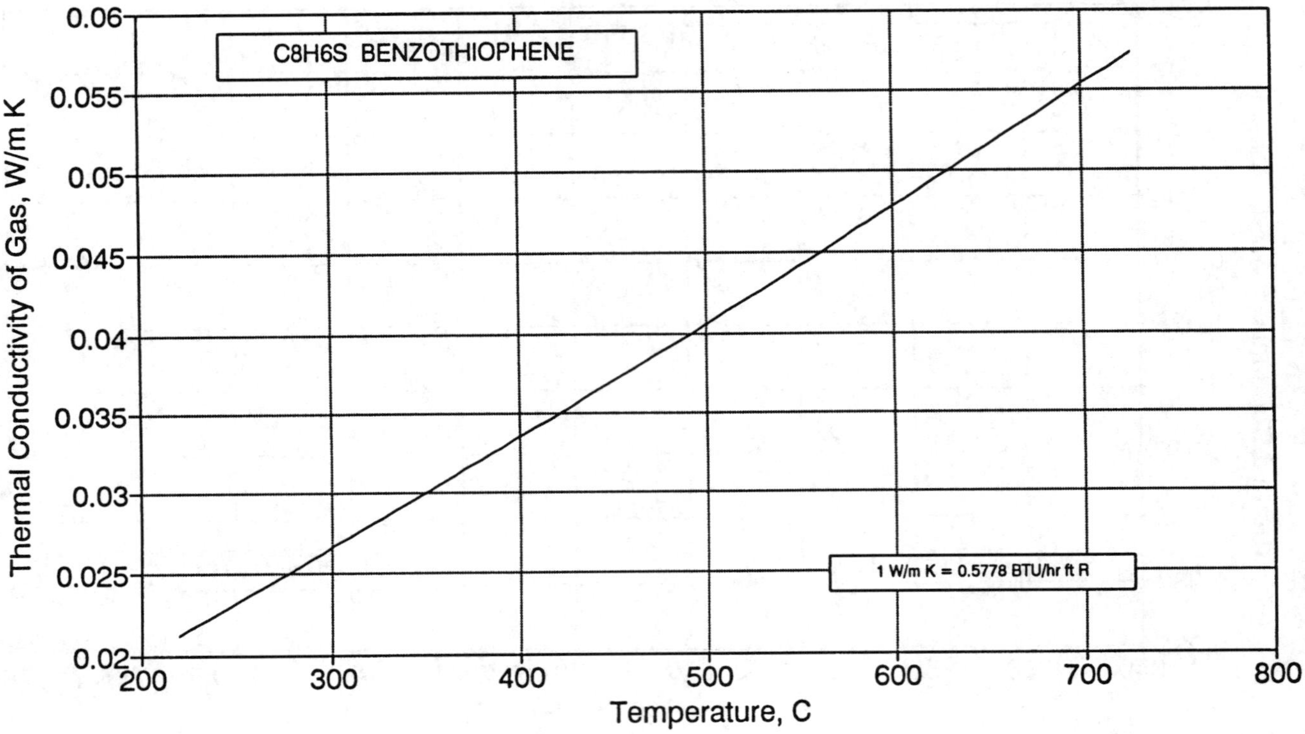

C8H6S BENZOTHIOPHENE
1 W/m K = 0.5778 BTU/hr ft R
Thermal Conductivity of Gas, W/m K
0.06
0.055
0.05
0.045
0.04
0.035
0.03
0.025
0.02
200
300
400
500
600
700
800
Temperature, C

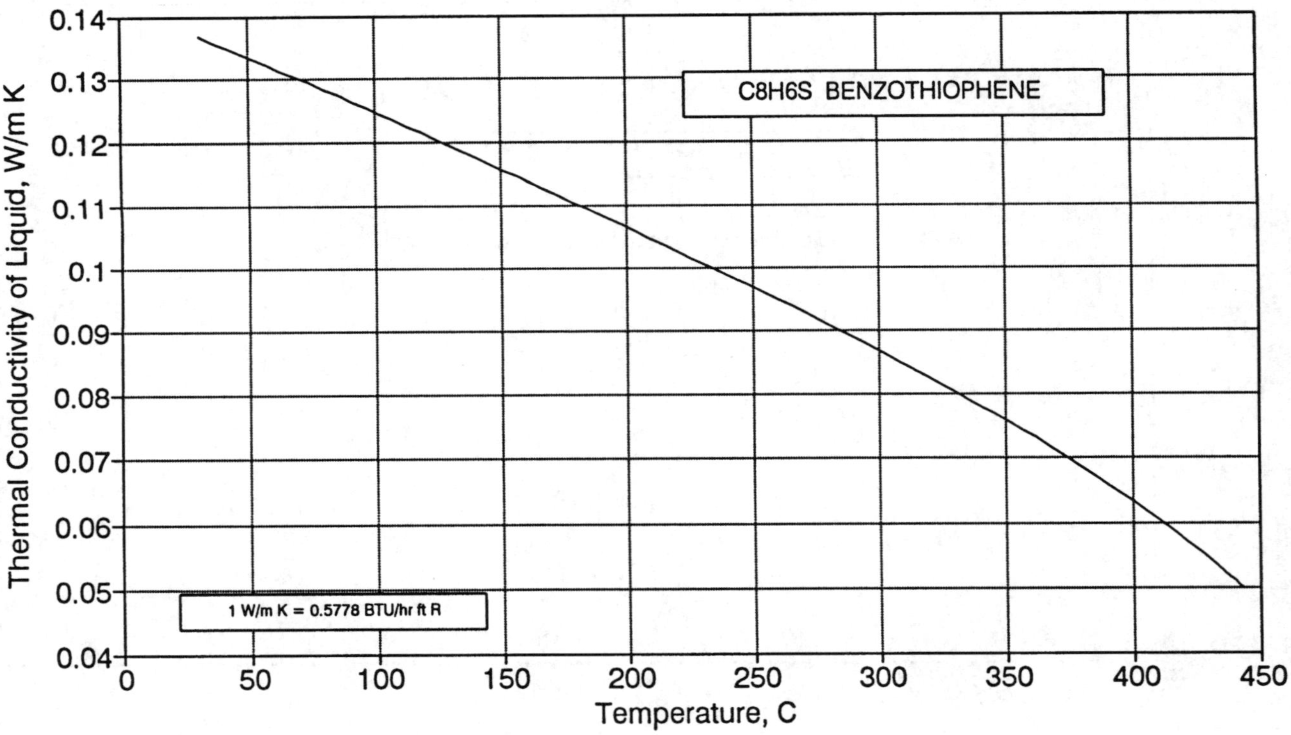

C8H6S BENZOTHIOPHENE
1 W/m K = 0.5778 BTU/hr ft R
Thermal Conductivity of Liquid, W/m K
0.14
0.13
0.12
0.11
0.1
0.09
0.08
0.07
0.06
0.05
0.04
0
50
100
150
200
250
300
350
400
450
Temperature, C

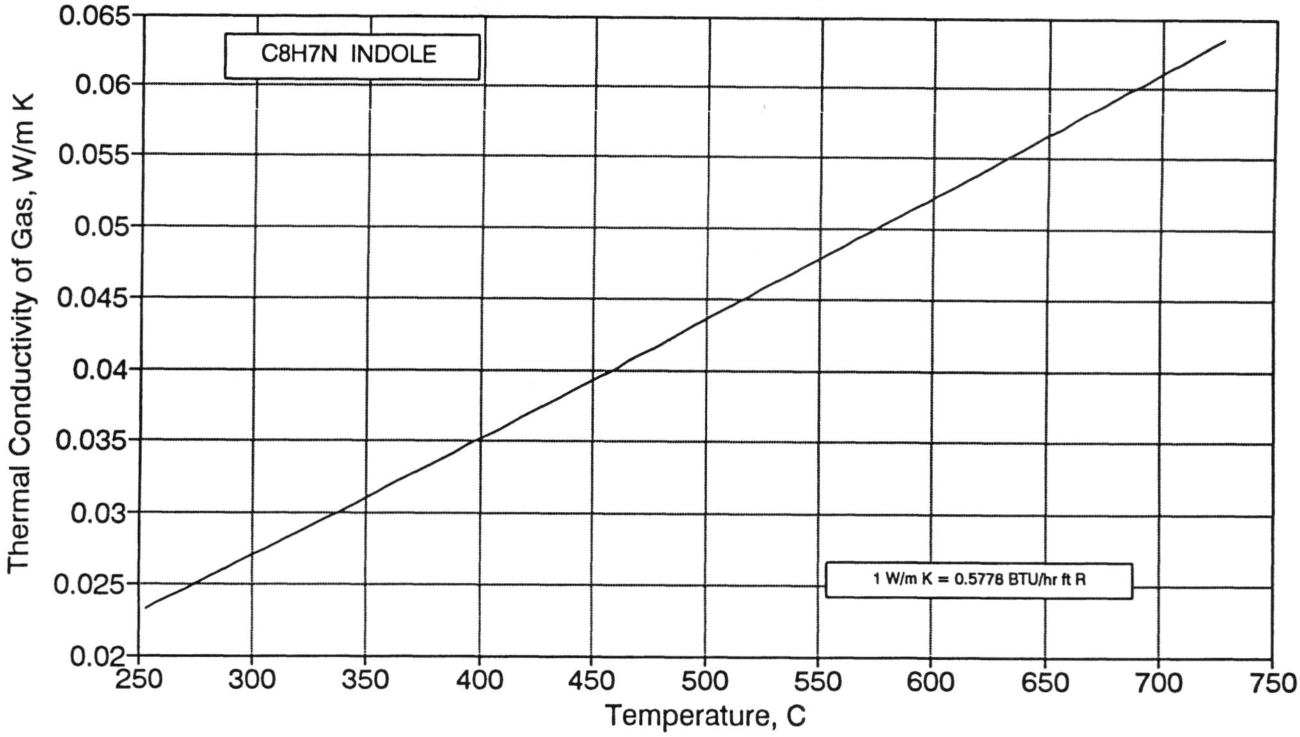
C8H7N INDOLE
Thermal Conductivity of Gas, W/m K
Temperature, C
1 W/m K = 0.5778 BTU/hr ft R
0.065
0.06
0.055
0.05
0.045
0.04
0.035
0.03
0.025
0.02
250
300
350
400
450
500
550
600
650
700
750

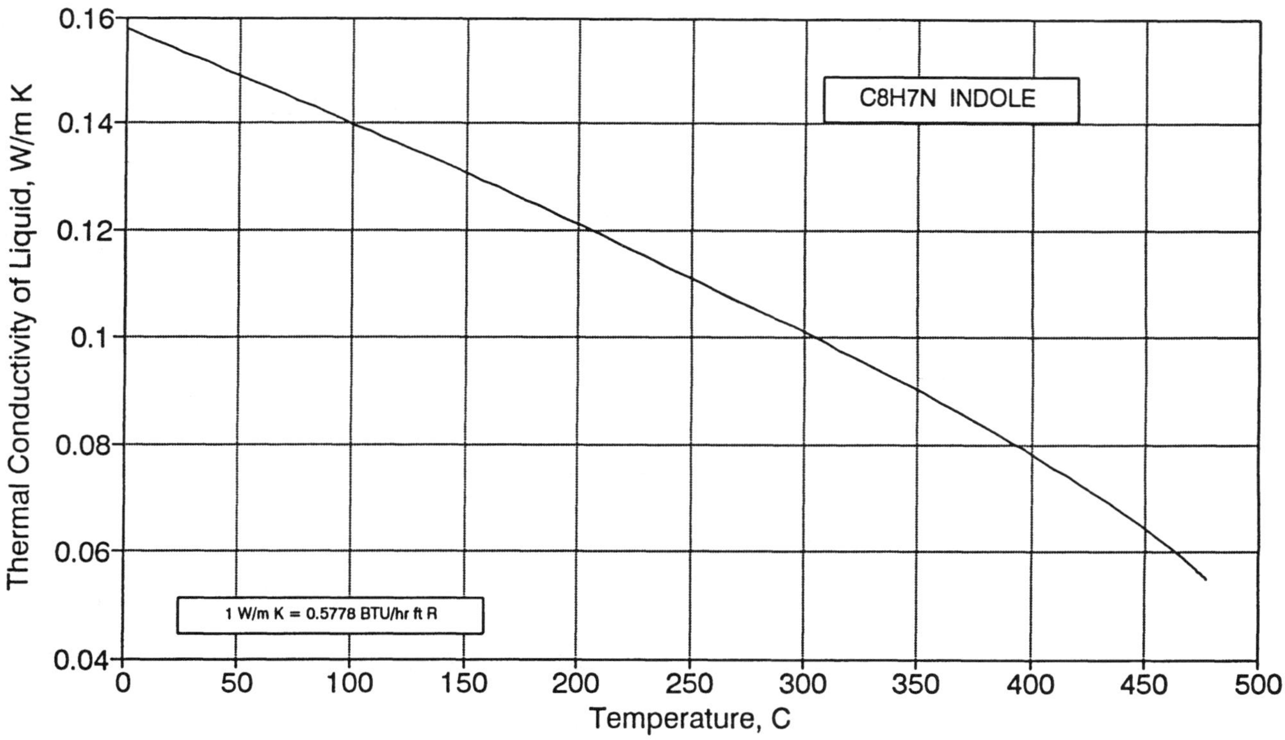
C8H7N INDOLE
Thermal Conductivity of Liquid, W/m K
Temperature, C
1 W/m K = 0.5778 BTU/hr ft R
0.16
0.14
0.12
0.1
0.08
0.06
0.04
0
50
100
150
200
250
300
350
400
450
500

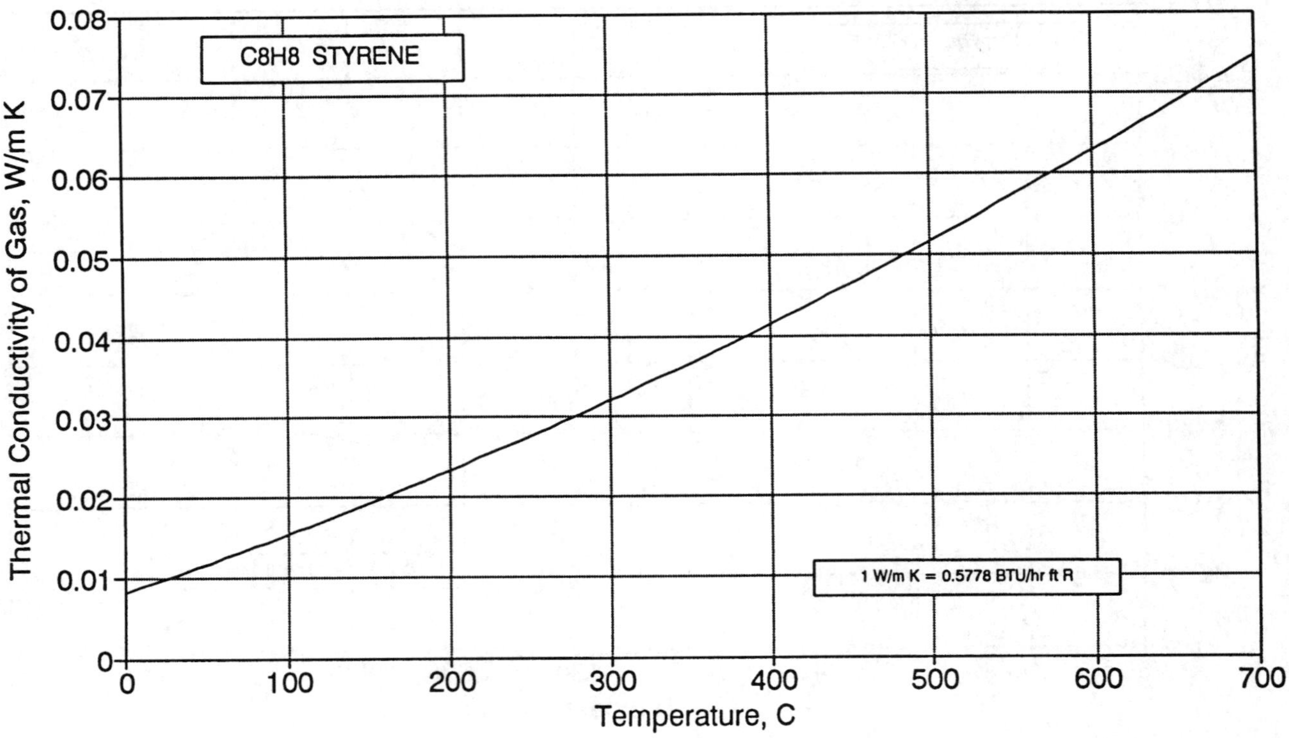
C8H8 STYRENE
1 W/m K = 0.5778 BTU/hr ft R
Thermal Conductivity of Gas, W/m K
Temperature, C

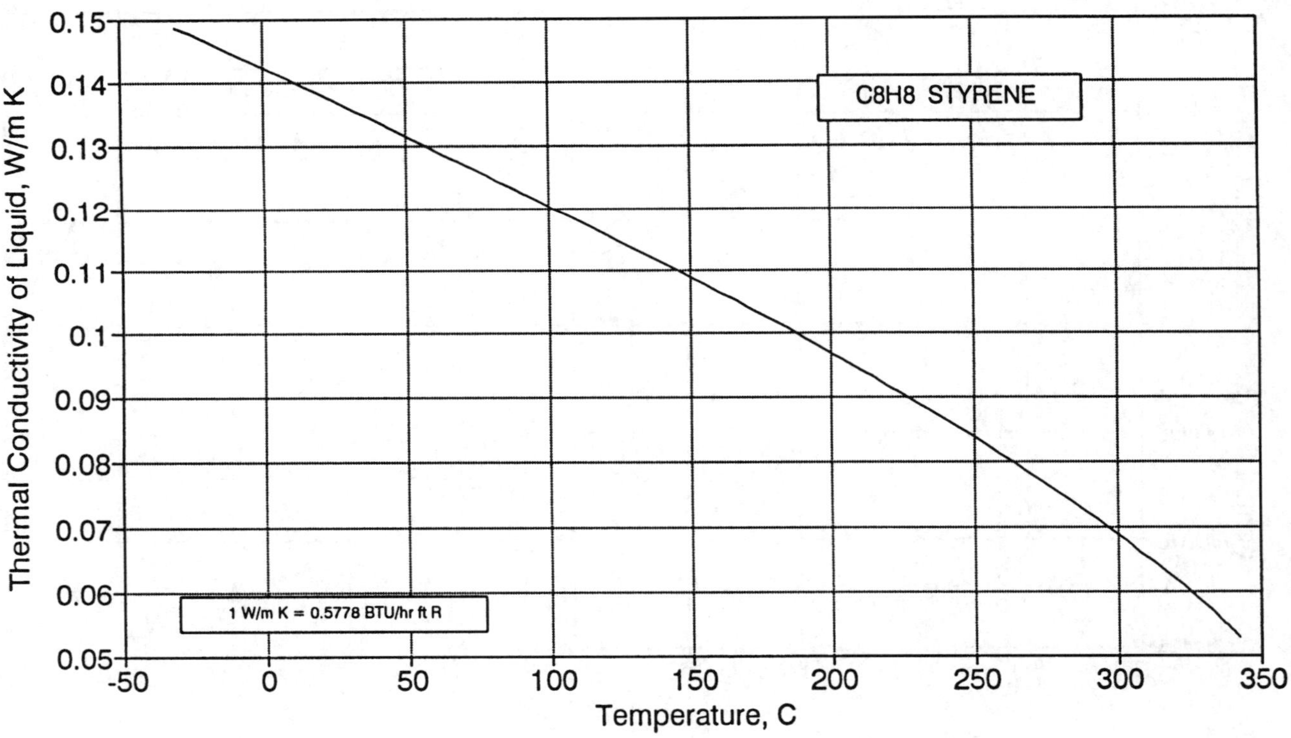
C8H8 STYRENE
1 W/m K = 0.5778 BTU/hr ft R
Thermal Conductivity of Liquid, W/m K
Temperature, C

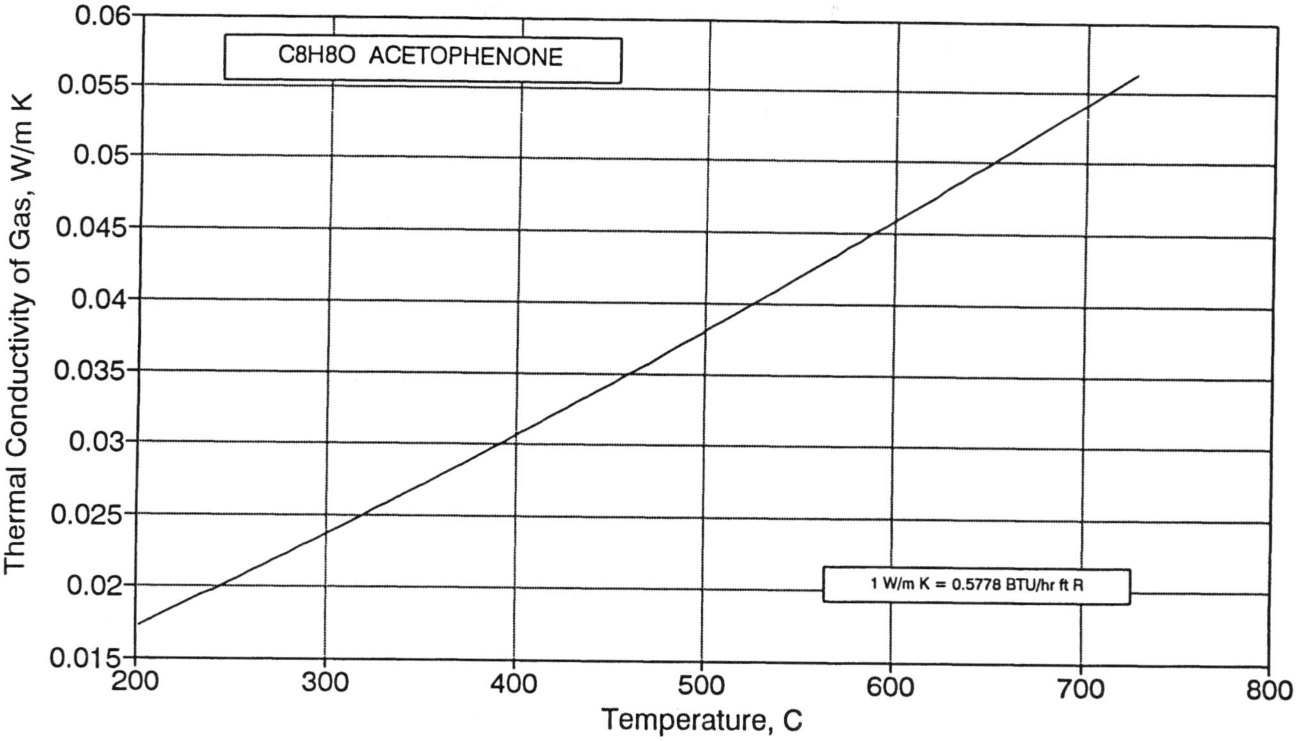
C8H8O ACETOPHENONE
1 W/m K = 0.5778 BTU/hr ft R
Thermal Conductivity of Gas, W/m K
Temperature, C
0.06
0.055
0.05
0.045
0.04
0.035
0.03
0.025
0.02
0.015
200
300
400
500
600
700
800

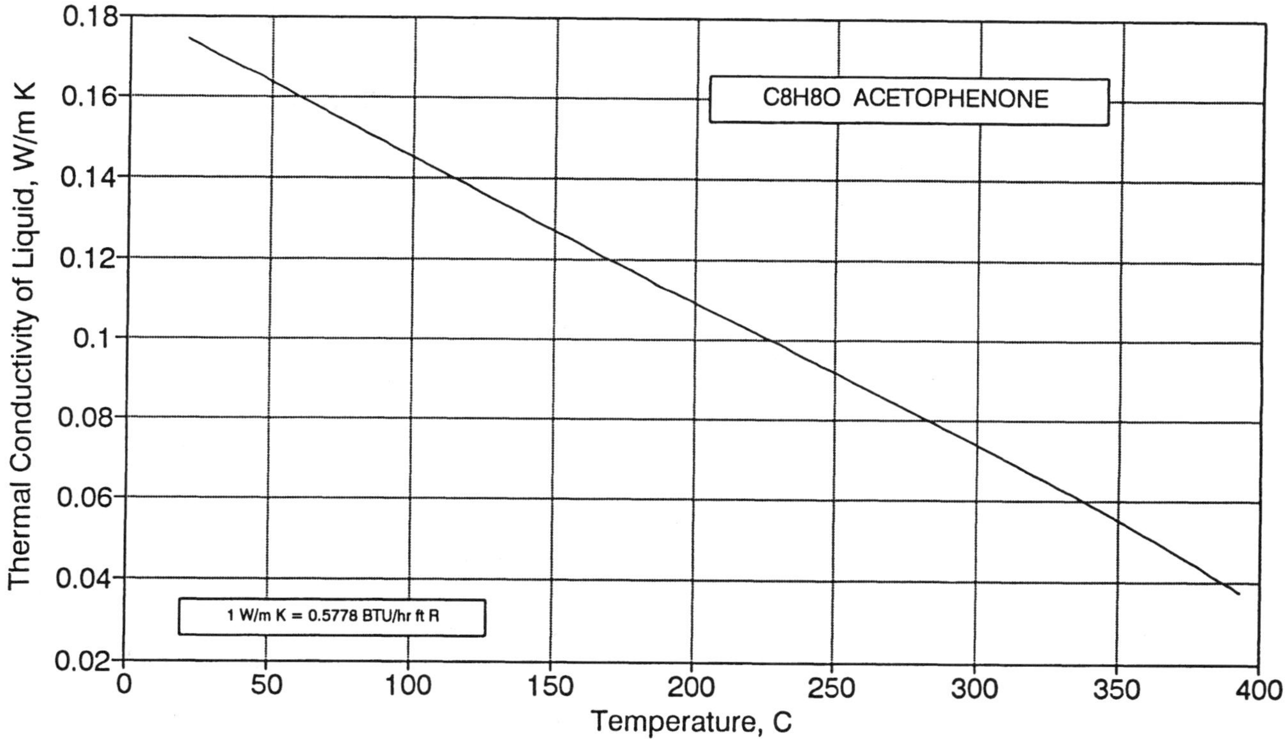
C8H8O ACETOPHENONE
1 W/m K = 0.5778 BTU/hr ft R
Thermal Conductivity of Liquid, W/m K
Temperature, C
0.18
0.16
0.14
0.12
0.1
0.08
0.06
0.04
0.02
0
50
100
150
200
250
300
350
400

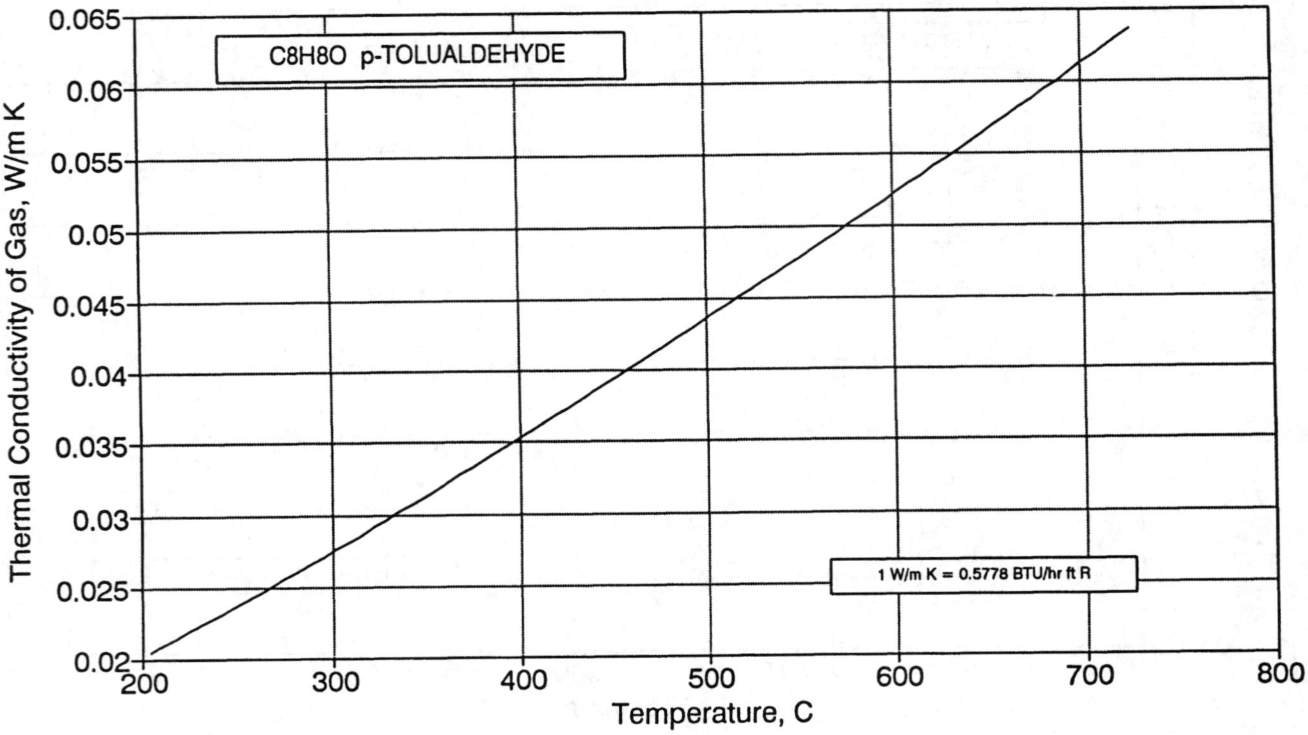
C8H8O p-TOLUALDEHYDE
1 W/m K = 0.5778 BTU/hr ft R
Thermal Conductivity of Gas, W/m K
0.065
0.06
0.055
0.05
0.045
0.04
0.035
0.03
0.025
0.02
200
300
400
500
600
700
800
Temperature, C

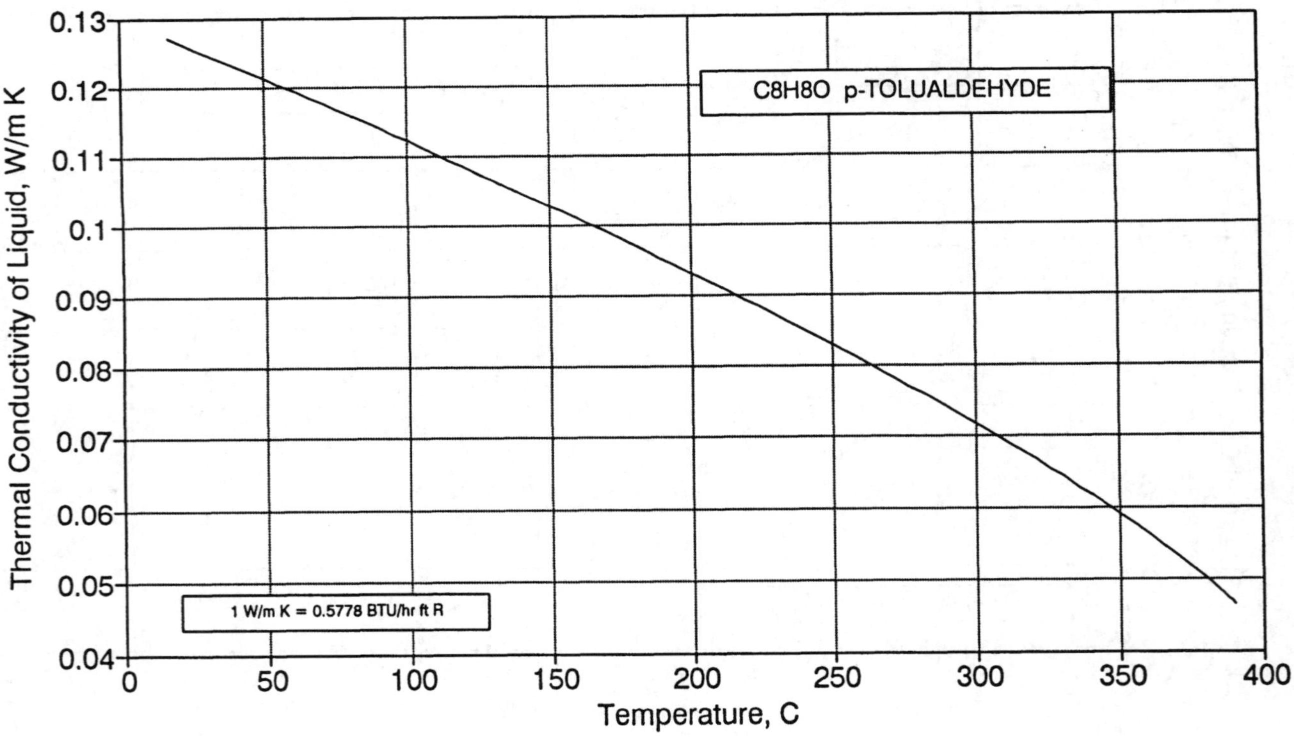
C8H8O p-TOLUALDEHYDE
1 W/m K = 0.5778 BTU/hr ft R
Thermal Conductivity of Liquid, W/m K
0.13
0.12
0.11
0.1
0.09
0.08
0.07
0.06
0.05
0.04
0
50
100
150
200
250
300
350
400
Temperature, C

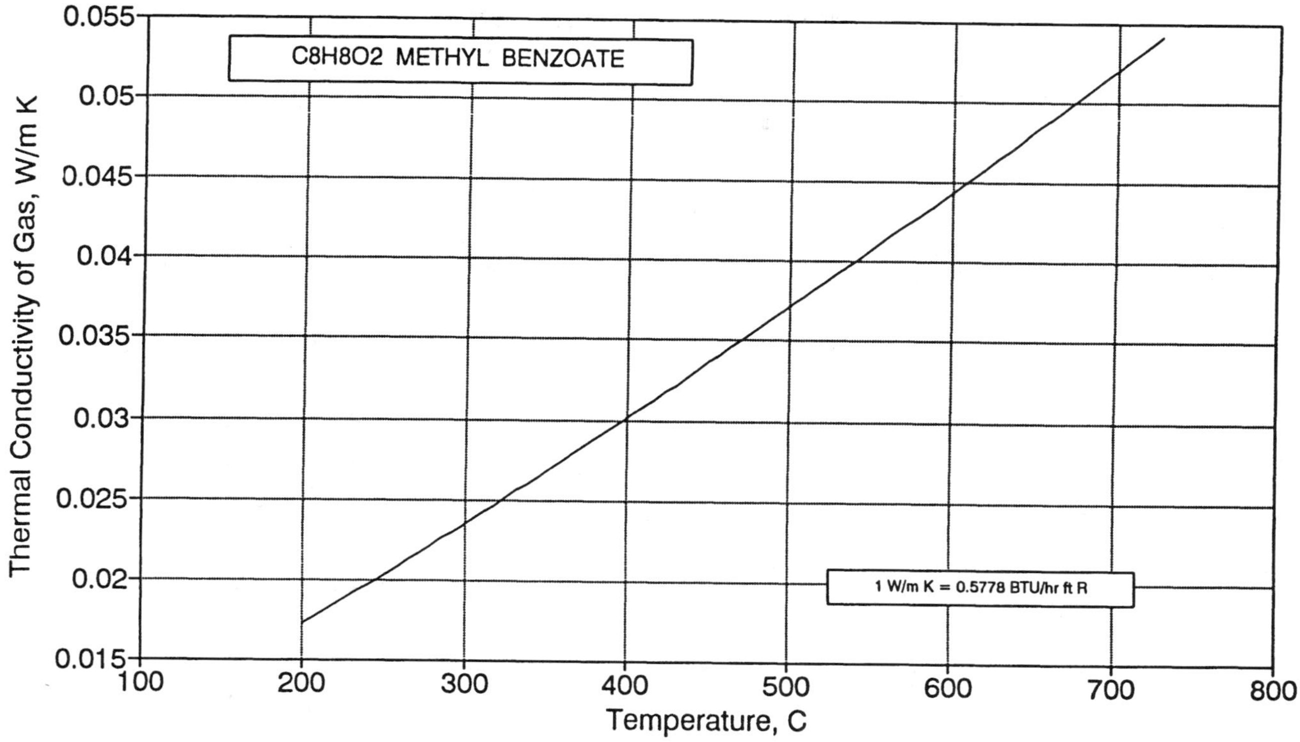
C8H8O2 METHYL BENZOATE
1 W/m K = 0.5778 BTU/hr ft R
Thermal Conductivity of Gas, W/m K
Temperature, C
0.055
0.05
0.045
0.04
0.035
0.03
0.025
0.02
0.015
100
200
300
400
500
600
700
800

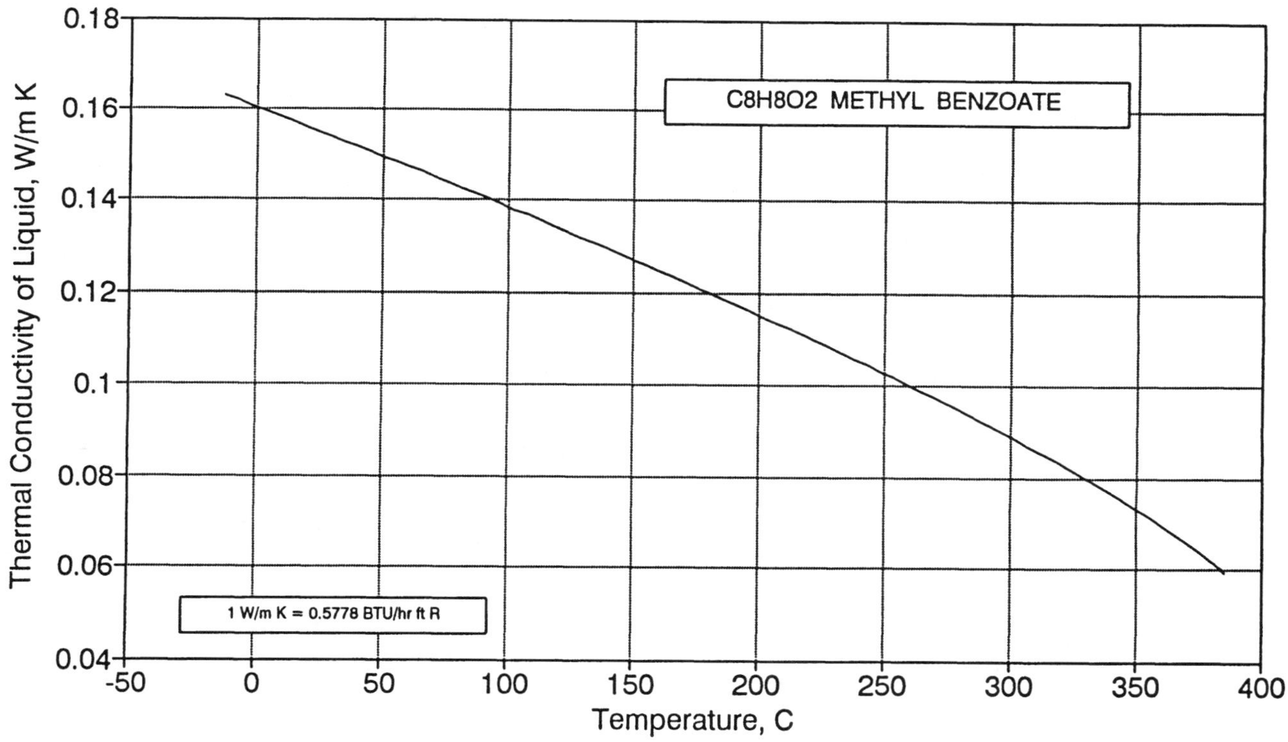
C8H8O2 METHYL BENZOATE
1 W/m K = 0.5778 BTU/hr ft R
Thermal Conductivity of Liquid, W/m K
Temperature, C
0.18
0.16
0.14
0.12
0.1
0.08
0.06
0.04
-50
0
50
100
150
200
250
300
350
400

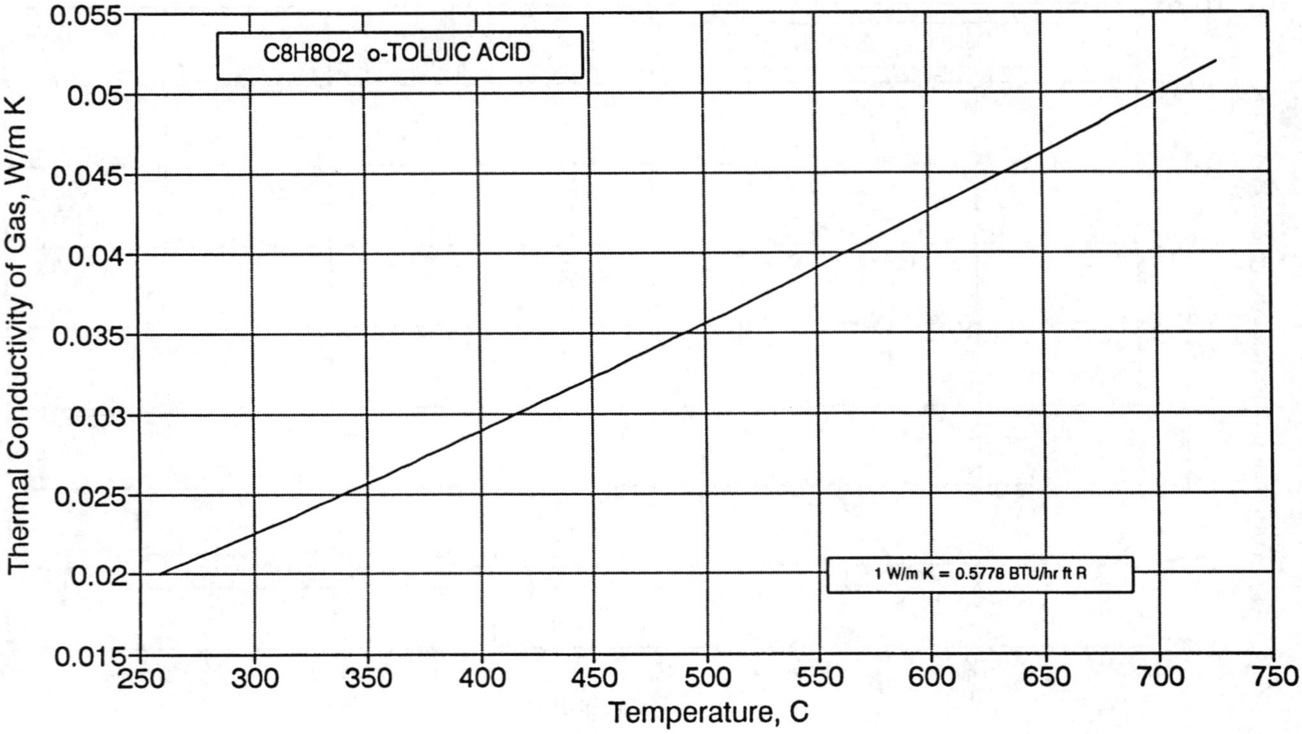
C8H8O2 o-TOLUIC ACID
1 W/m K = 0.5778 BTU/hr ft R
Thermal Conductivity of Gas, W/m K
Temperature, C
0.055
0.05
0.045
0.04
0.035
0.03
0.025
0.02
0.015
250
300
350
400
450
500
550
600
650
700
750

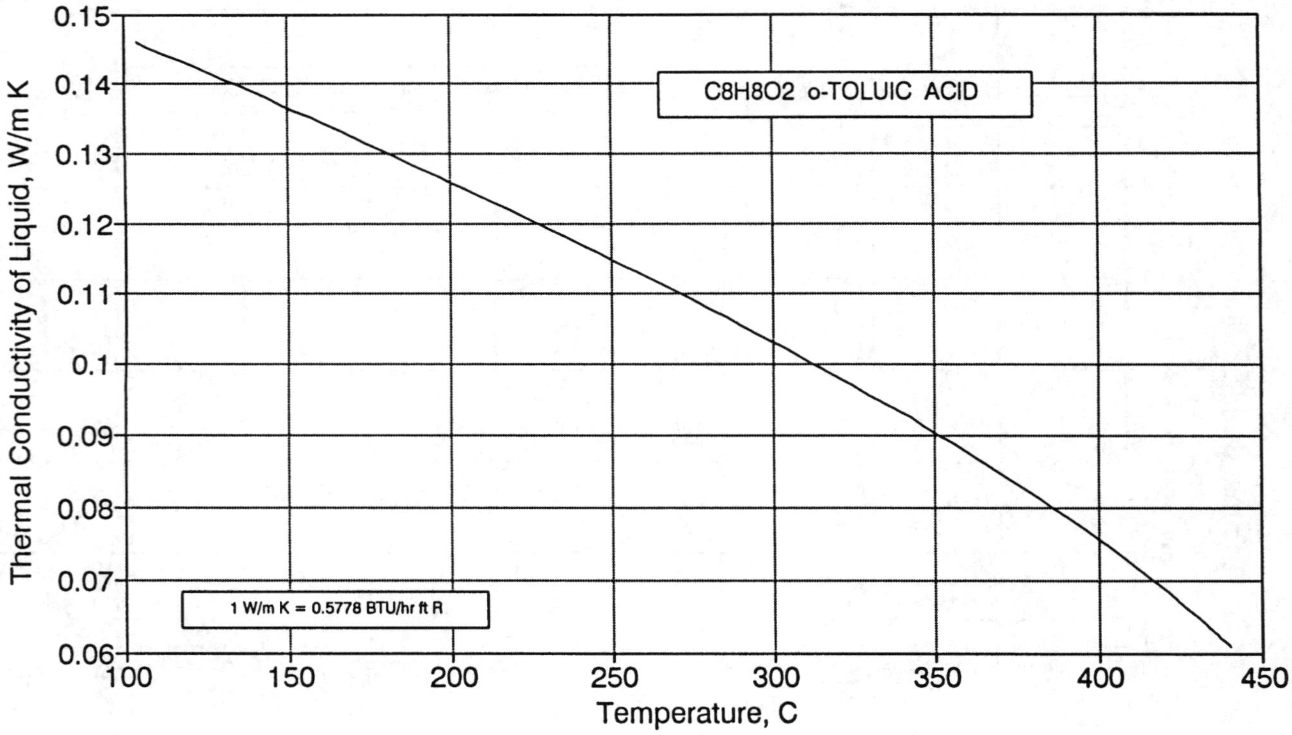
C8H8O2 o-TOLUIC ACID
1 W/m K = 0.5778 BTU/hr ft R
Thermal Conductivity of Liquid, W/m K
Temperature, C
0.15
0.14
0.13
0.12
0.11
0.1
0.09
0.08
0.07
0.06
100
150
200
250
300
350
400
450

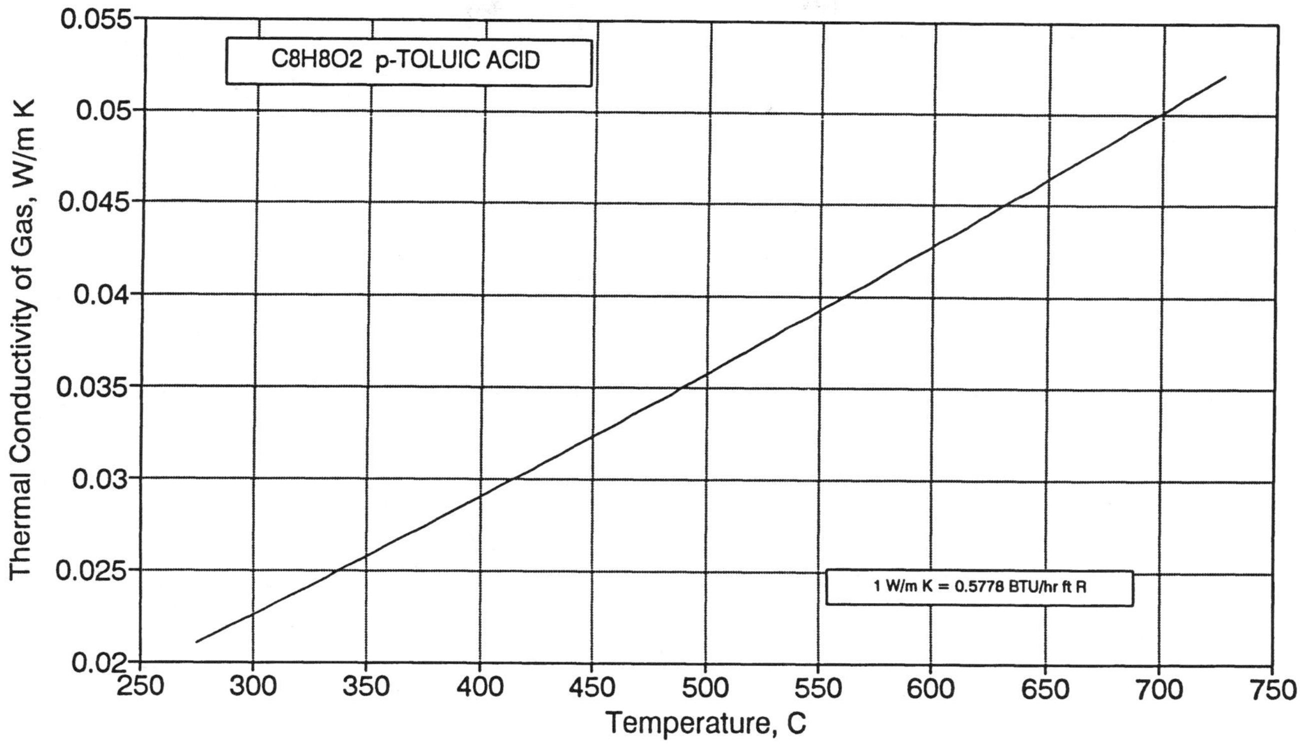
C8H8O2 p-TOLUIC ACID
1 W/m K = 0.5778 BTU/hr ft R
Thermal Conductivity of Gas, W/m K
Temperature, C
0.055
0.05
0.045
0.04
0.035
0.03
0.025
0.02
250
300
350
400
450
500
550
600
650
700
750

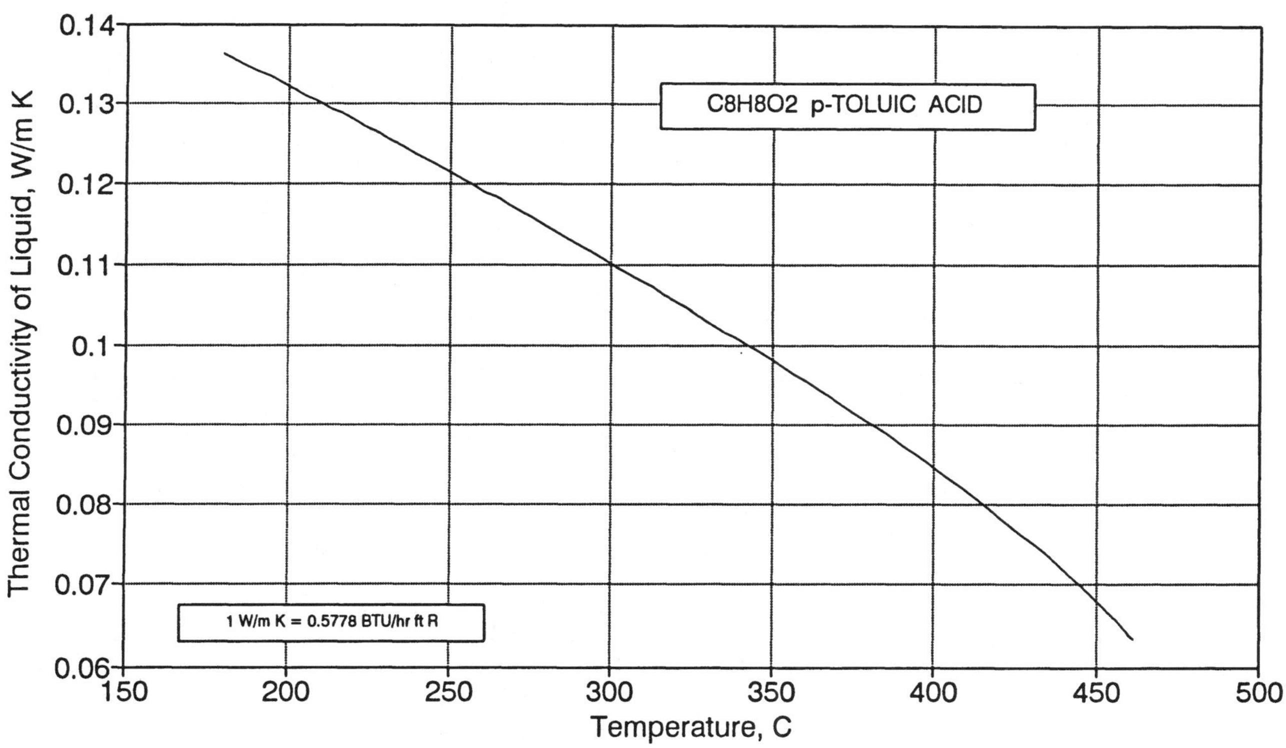
C8H8O2 p-TOLUIC ACID
1 W/m K = 0.5778 BTU/hr ft R
Thermal Conductivity of Liquid, W/m K
Temperature, C
0.14
0.13
0.12
0.11
0.1
0.09
0.08
0.07
0.06
150
200
250
300
350
400
450
500

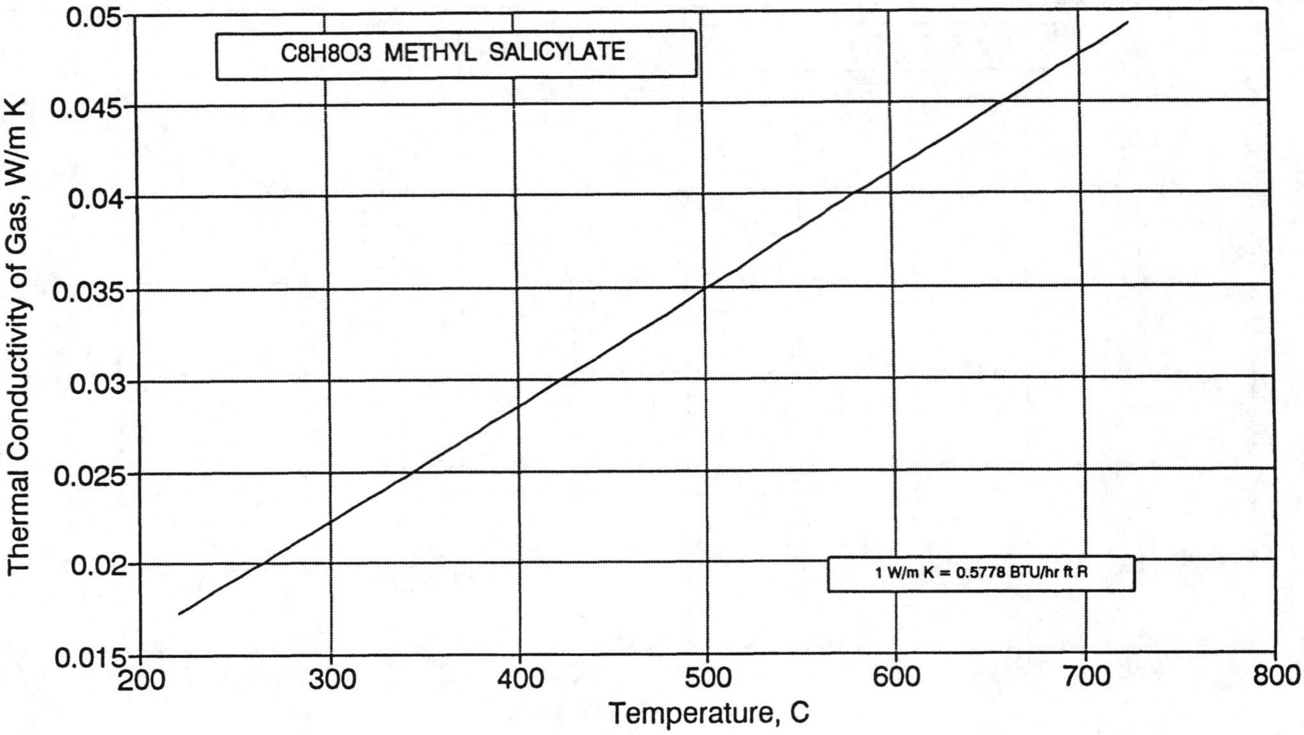
C8H8O3 METHYL SALICYLATE
1 W/m K = 0.5778 BTU/hr ft R
Thermal Conductivity of Gas, W/m K
Temperature, C
0.05
0.045
0.04
0.035
0.03
0.025
0.02
0.015
200
300
400
500
600
700
800

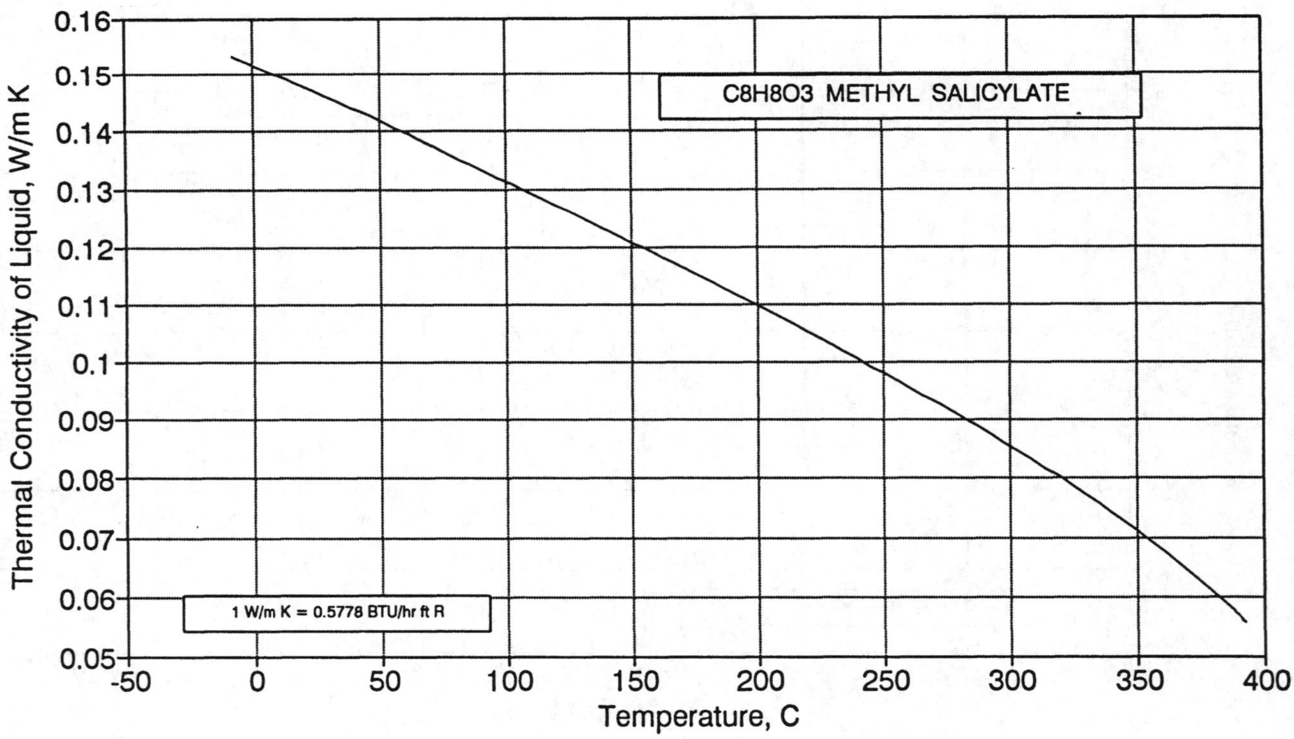
C8H8O3 METHYL SALICYLATE
1 W/m K = 0.5778 BTU/hr ft R
Thermal Conductivity of Liquid, W/m K
Temperature, C
0.16
0.15
0.14
0.13
0.12
0.11
0.1
0.09
0.08
0.07
0.06
0.05
-50
0
50
100
150
200
250
300
350
400

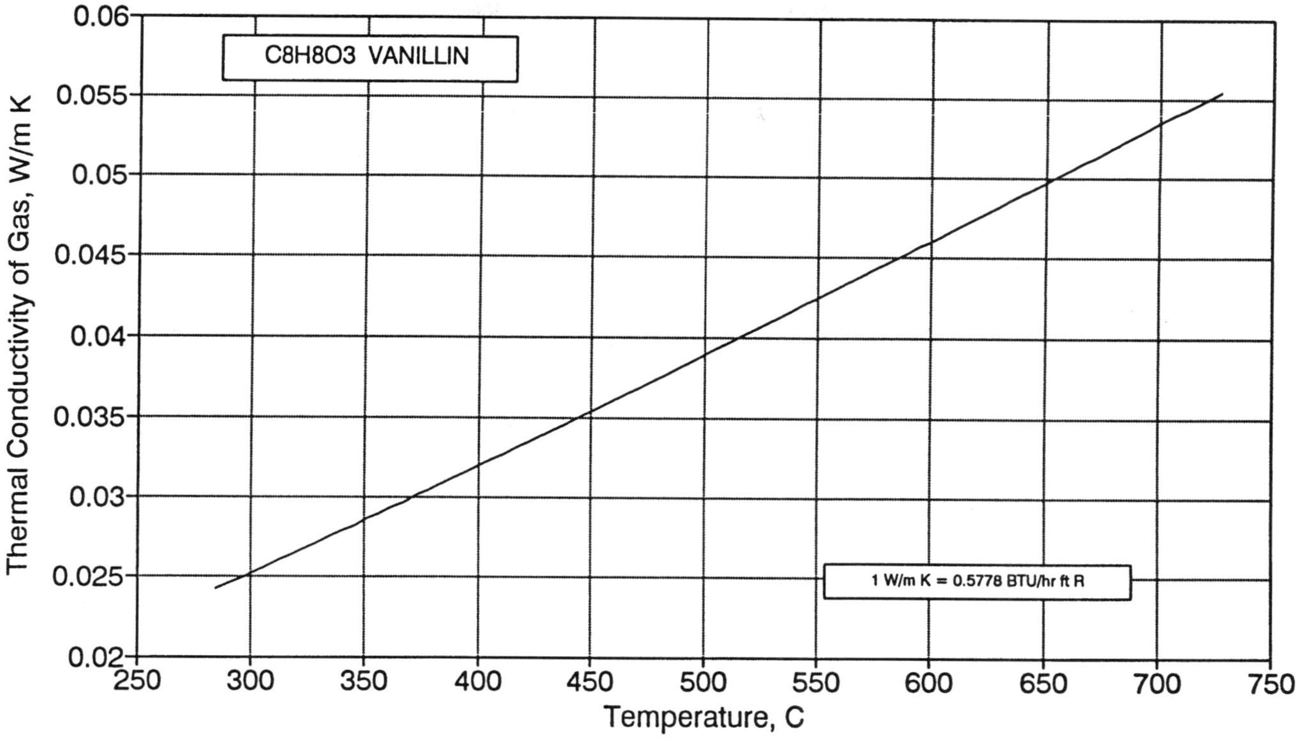
C8H8O3 VANILLIN
1 W/m K = 0.5778 BTU/hr ft R
Thermal Conductivity of Gas, W/m K
0.06
0.055
0.05
0.045
0.04
0.035
0.03
0.025
0.02
250
300
350
400
450
500
550
600
650
700
750
Temperature, C

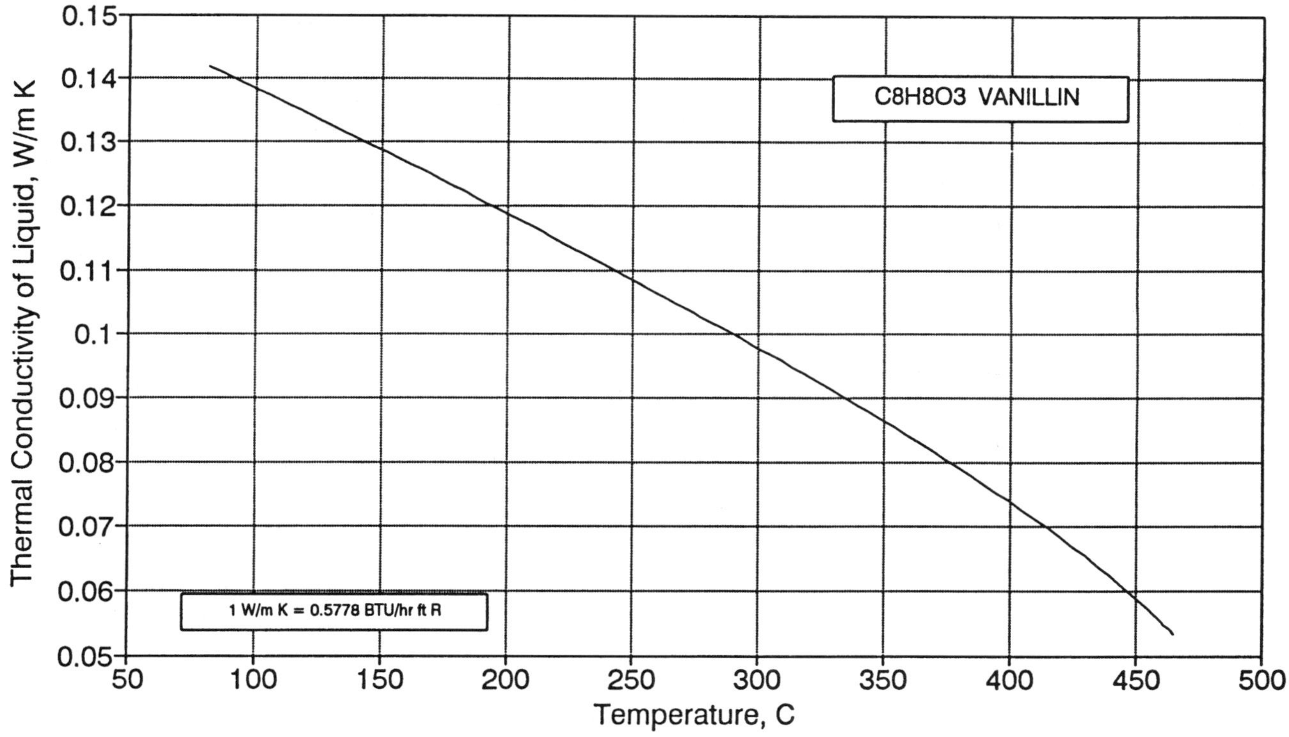
C8H8O3 VANILLIN
1 W/m K = 0.5778 BTU/hr ft R
Thermal Conductivity of Liquid, W/m K
0.15
0.14
0.13
0.12
0.11
0.1
0.09
0.08
0.07
0.06
0.05
50
100
150
200
250
300
350
400
450
500
Temperature, C

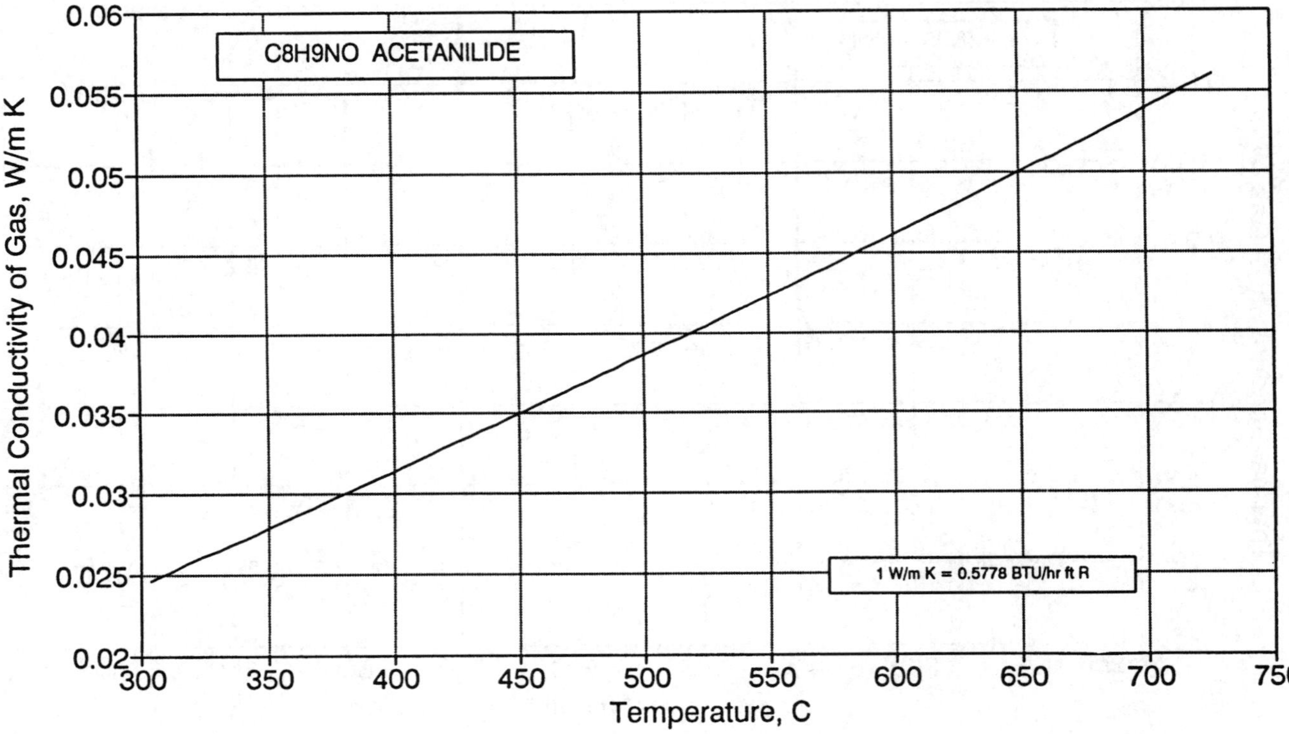

C8H9NO ACETANILIDE
1 W/m K = 0.5778 BTU/hr ft R
Thermal Conductivity of Gas, W/m K
Temperature, C
0.06
0.055
0.05
0.045
0.04
0.035
0.03
0.025
0.02
300
350
400
450
500
550
600
650
700
750

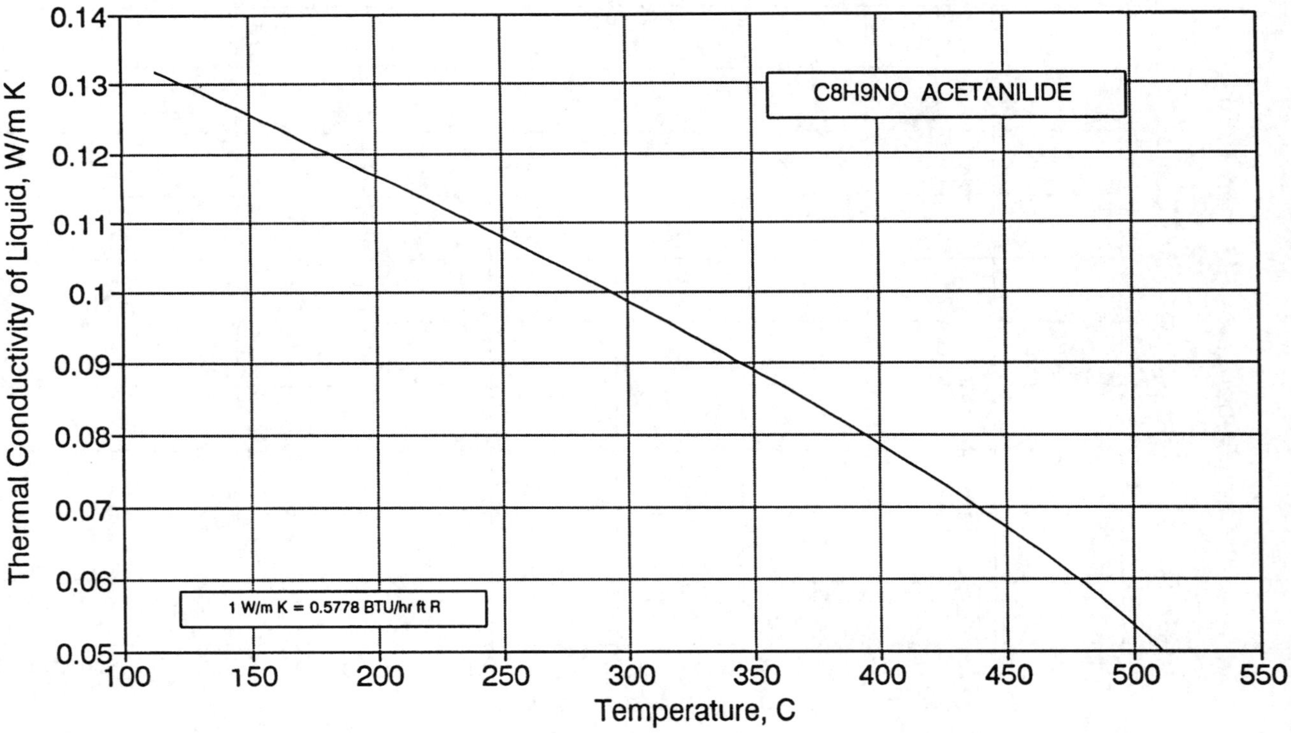

C8H9NO ACETANILIDE
1 W/m K = 0.5778 BTU/hr ft R
Thermal Conductivity of Liquid, W/m K
Temperature, C
0.14
0.13
0.12
0.11
0.1
0.09
0.08
0.07
0.06
0.05
100
150
200
250
300
350
400
450
500
550

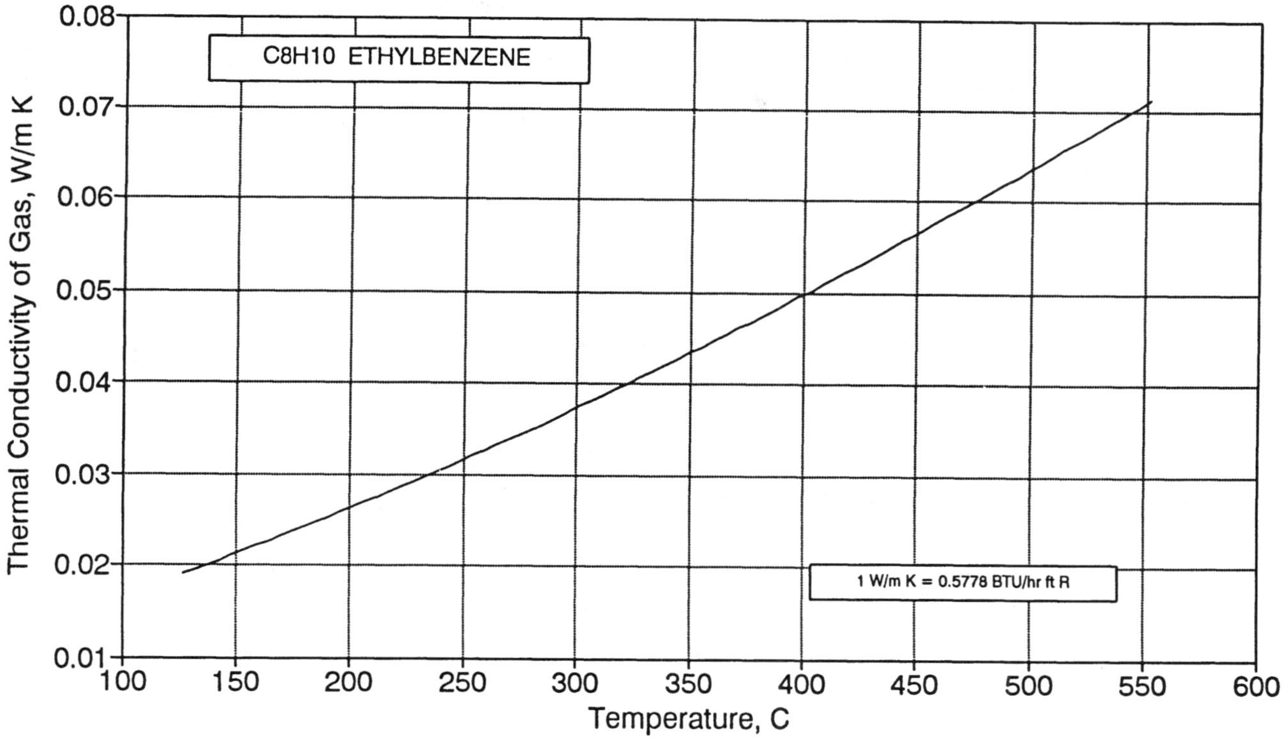
C8H10 ETHYLBENZENE
1 W/m K = 0.5778 BTU/hr ft R
Thermal Conductivity of Gas, W/m K
0.08
0.07
0.06
0.05
0.04
0.03
0.02
0.01
100
150
200
250
300
350
400
450
500
550
600
Temperature, C

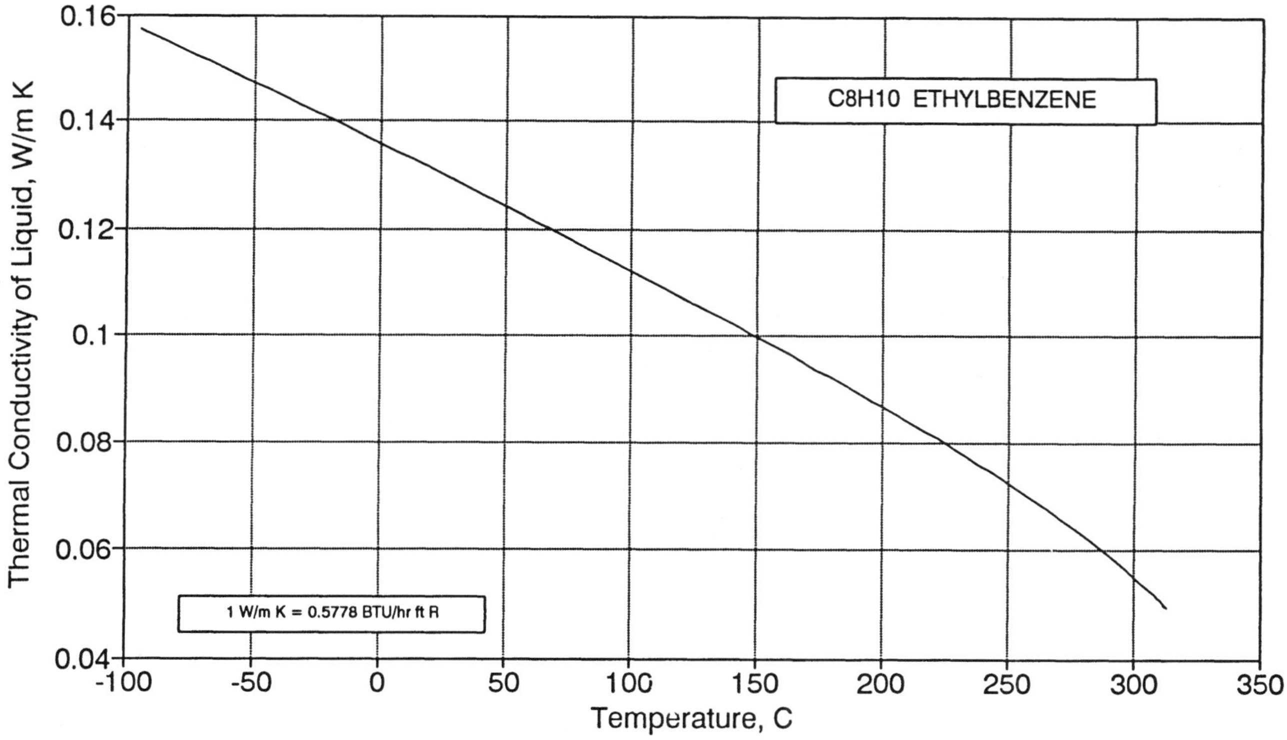
C8H10 ETHYLBENZENE
1 W/m K = 0.5778 BTU/hr ft R
Thermal Conductivity of Liquid, W/m K
0.16
0.14
0.12
0.1
0.08
0.06
0.04
-100
-50
0
50
100
150
200
250
300
350
Temperature, C

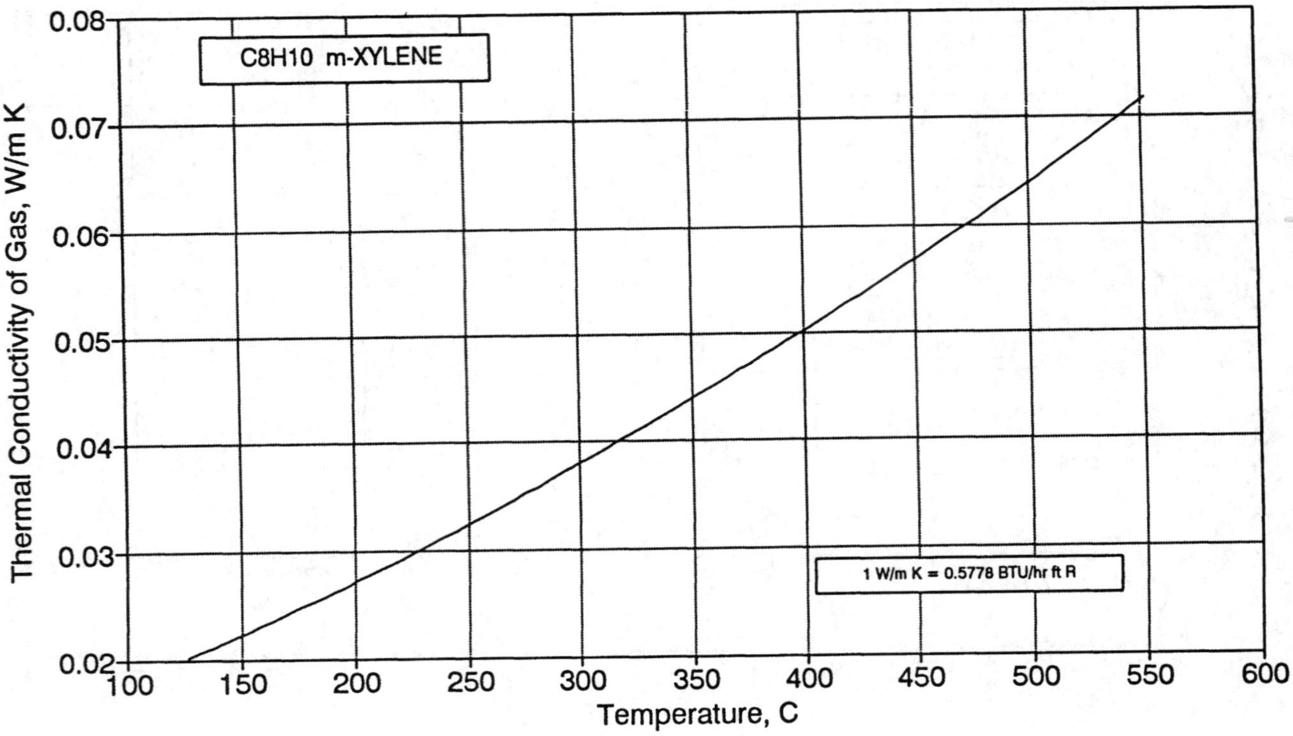
C8H10 m-XYLENE
1 W/m K = 0.5778 BTU/hr ft R
Thermal Conductivity of Gas, W/m K
0.08
0.07
0.06
0.05
0.04
0.03
0.02
100
150
200
250
300
350
400
450
500
550
600
Temperature, C

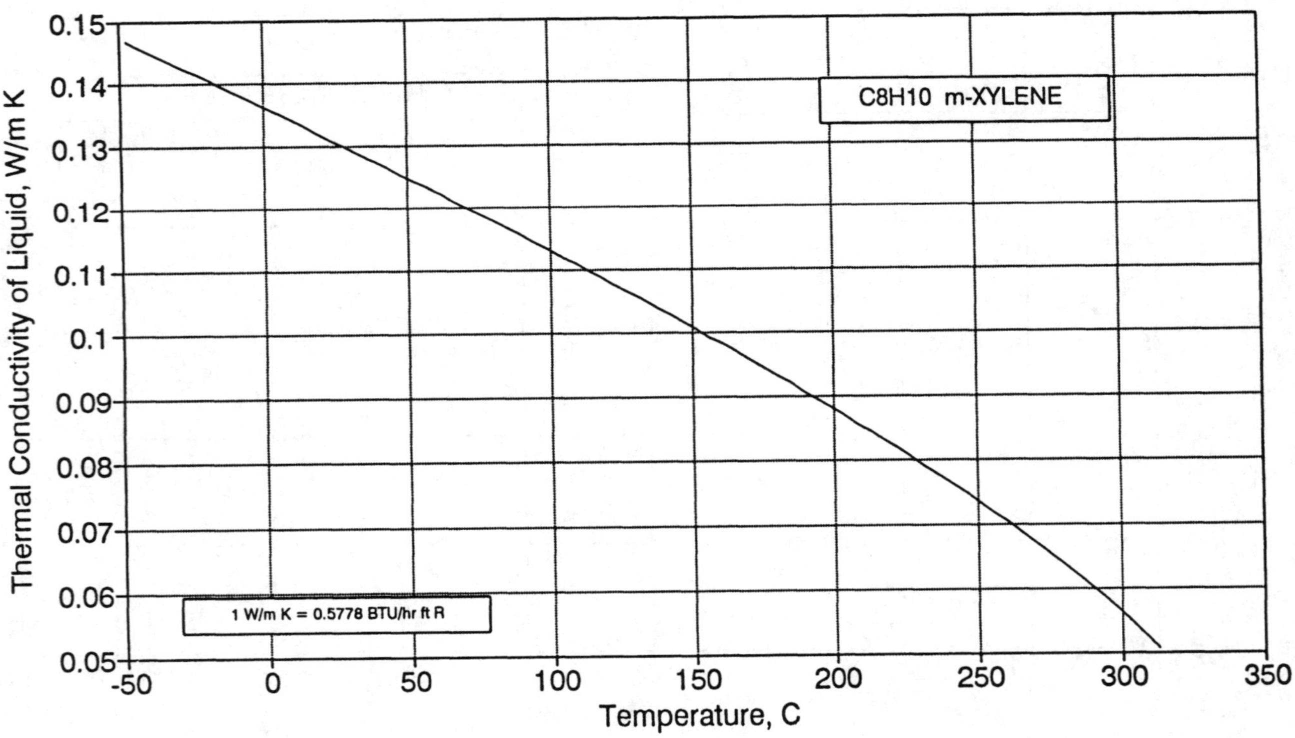
C8H10 m-XYLENE
1 W/m K = 0.5778 BTU/hr ft R
Thermal Conductivity of Liquid, W/m K
0.15
0.14
0.13
0.12
0.11
0.1
0.09
0.08
0.07
0.06
0.05
-50
0
50
100
150
200
250
300
350
Temperature, C

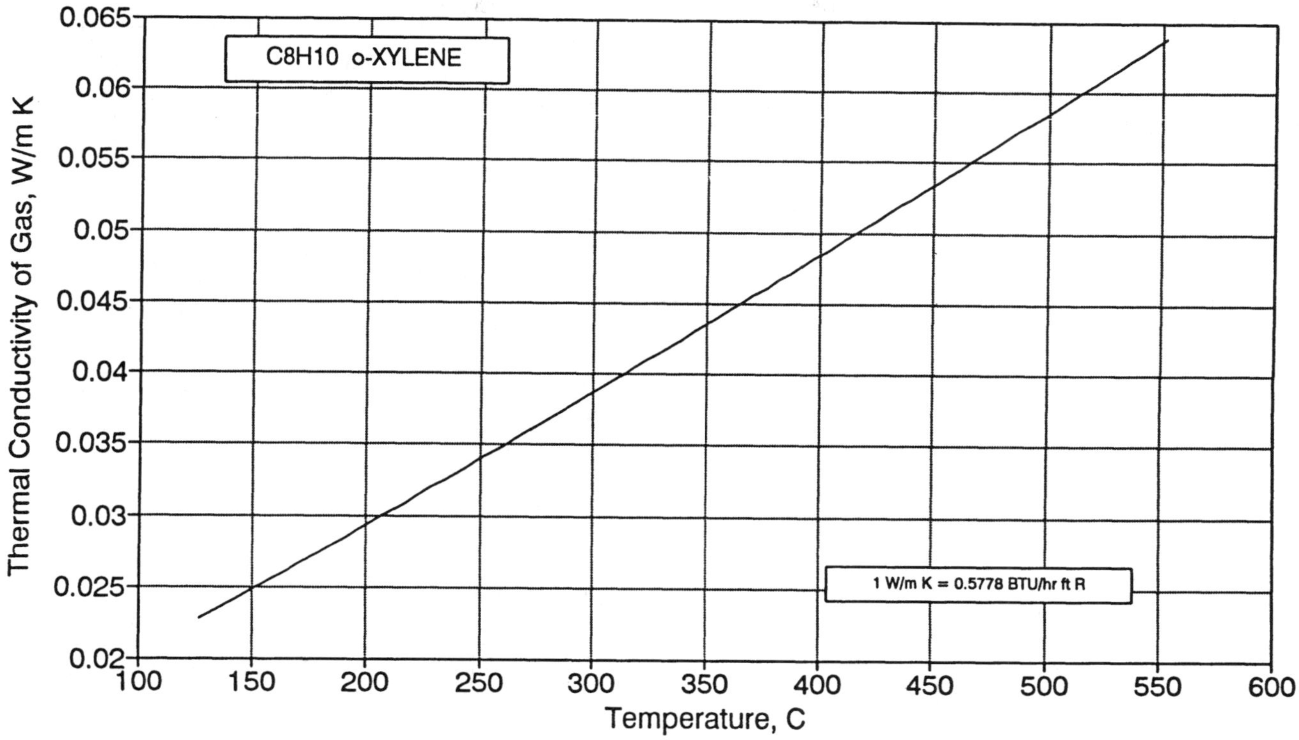

C8H10 o-XYLENE
1 W/m K = 0.5778 BTU/hr ft R
Thermal Conductivity of Gas, W/m K
0.065
0.06
0.055
0.05
0.045
0.04
0.035
0.03
0.025
0.02
100
150
200
250
300
350
400
450
500
550
600
Temperature, C

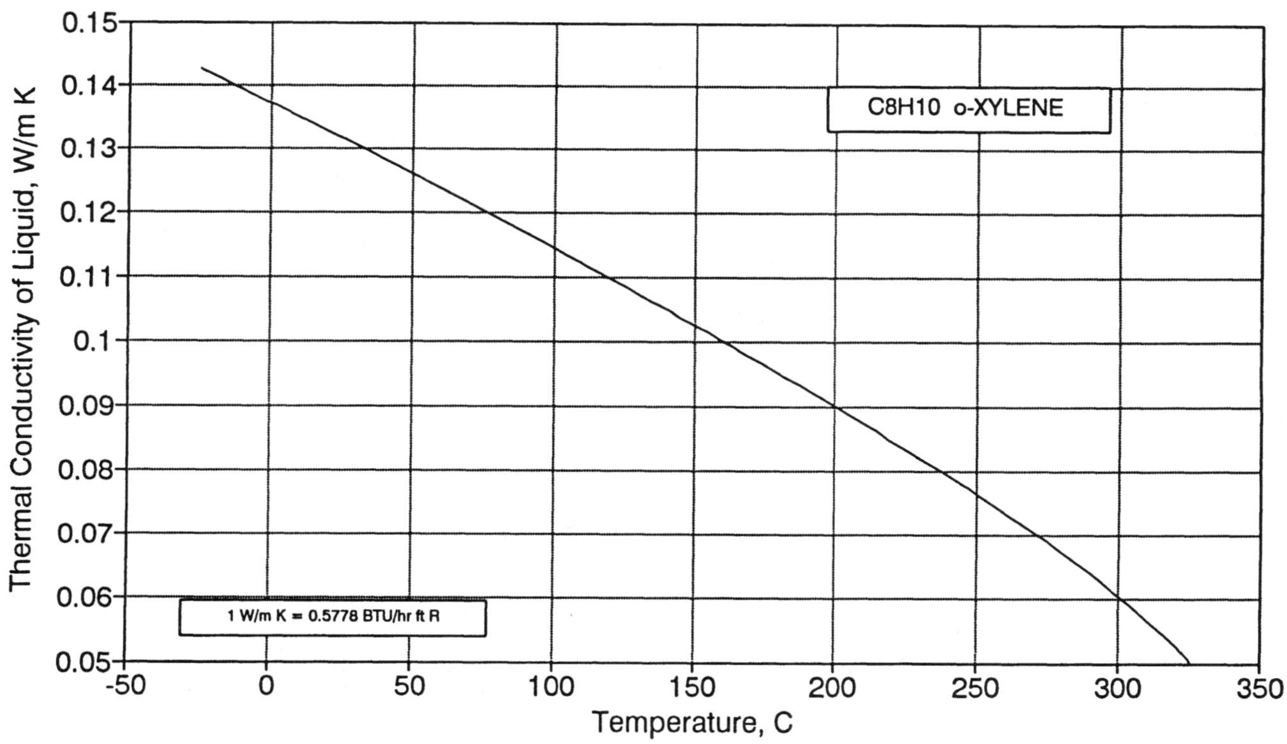

C8H10 o-XYLENE
1 W/m K = 0.5778 BTU/hr ft R
Thermal Conductivity of Liquid, W/m K
0.15
0.14
0.13
0.12
0.11
0.1
0.09
0.08
0.07
0.06
0.05
-50
0
50
100
150
200
250
300
350
Temperature, C

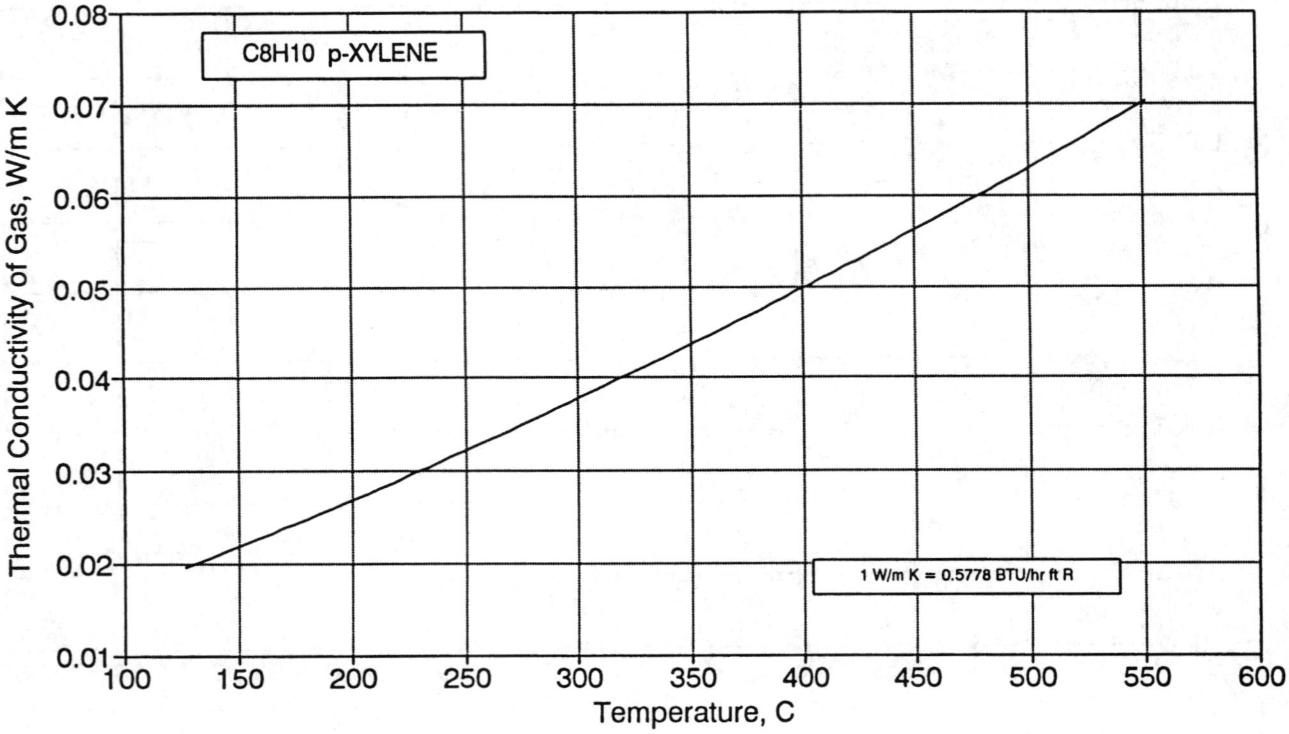
C8H10 p-XYLENE
Thermal Conductivity of Gas, W/m K
0.08
0.07
0.06
0.05
0.04
0.03
0.02
0.01
1 W/m K = 0.5778 BTU/hr ft R
100
150
200
250
300
350
400
450
500
550
600
Temperature, C

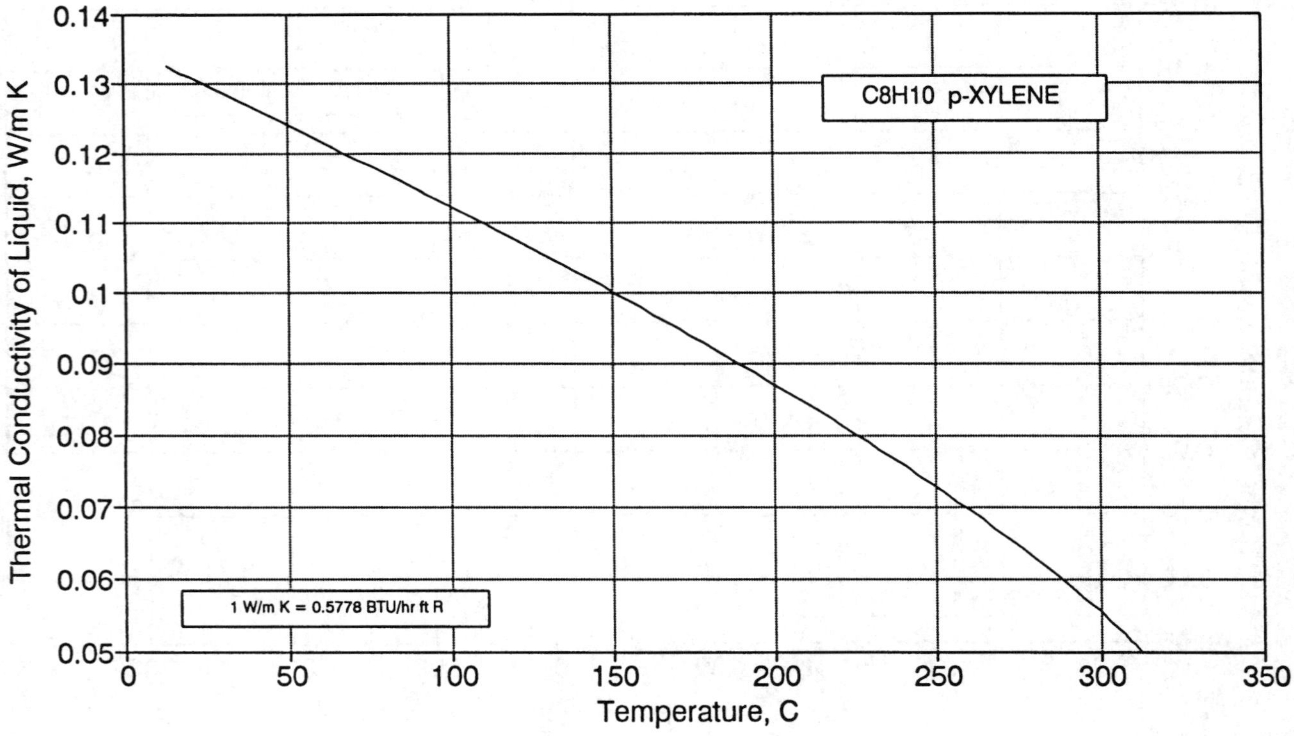
C8H10 p-XYLENE
Thermal Conductivity of Liquid, W/m K
0.14
0.13
0.12
0.11
0.1
0.09
0.08
0.07
0.06
0.05
1 W/m K = 0.5778 BTU/hr ft R
0
50
100
150
200
250
300
350
Temperature, C

C8H10O m-ETHYLPHENOL

Gas thermal conductivity data are not available .

TF , K = less than 255.15

TB , K = 480.15

C8H10O m-ETHYLPHENOL

Liquid thermal conductivity data are not available .

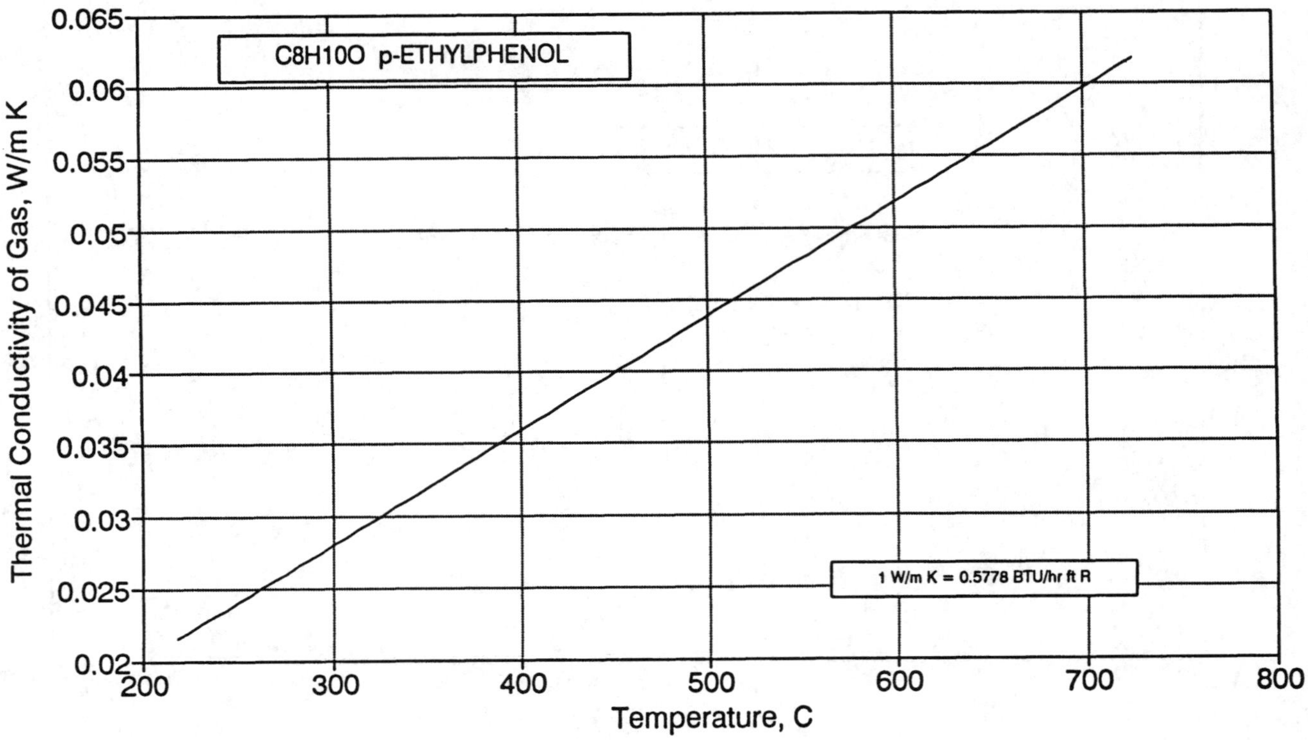
C8H10O p-ETHYLPHENOL
1 W/m K = 0.5778 BTU/hr ft R
Thermal Conductivity of Gas, W/m K
Temperature, C
0.065
0.06
0.055
0.05
0.045
0.04
0.035
0.03
0.025
0.02
200
300
400
500
600
700
800

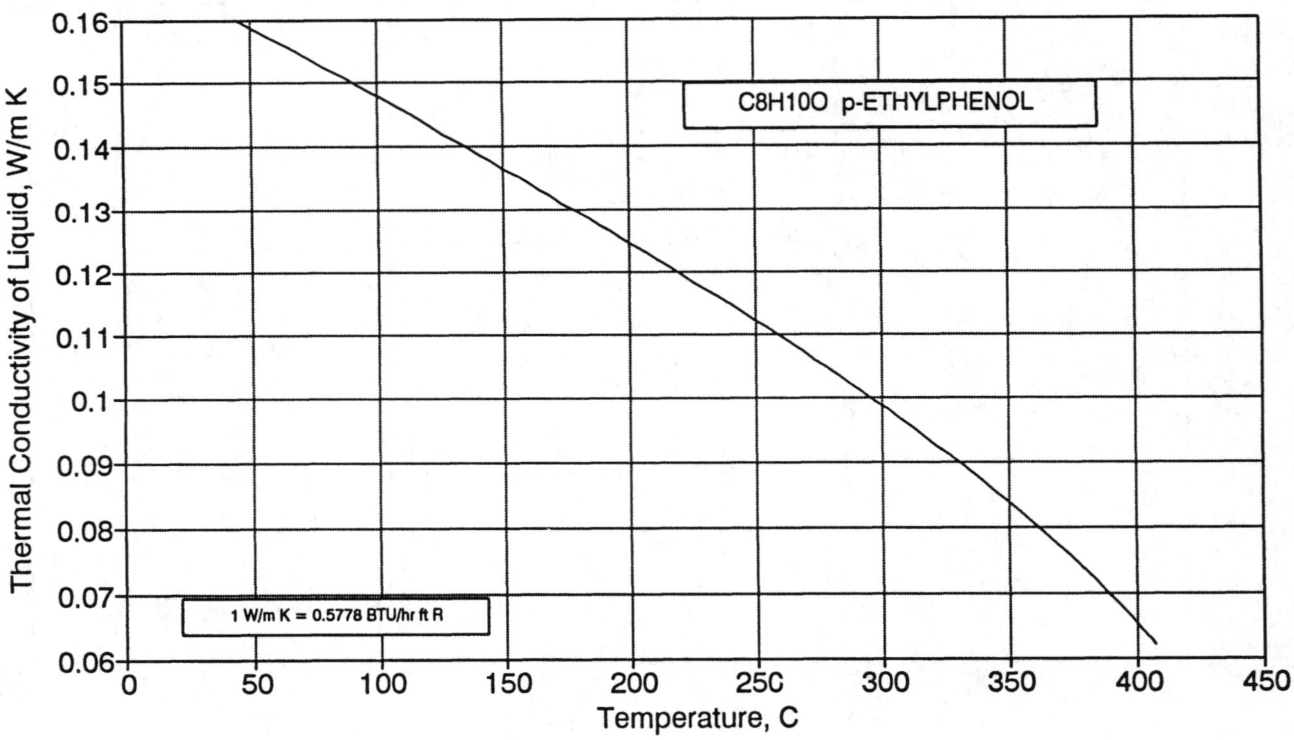
C8H10O p-ETHYLPHENOL
1 W/m K = 0.5778 BTU/hr ft R
Thermal Conductivity of Liquid, W/m K
Temperature, C
0.16
0.15
0.14
0.13
0.12
0.11
0.1
0.09
0.08
0.07
0.06
0
50
100
150
200
250
300
350
400
450

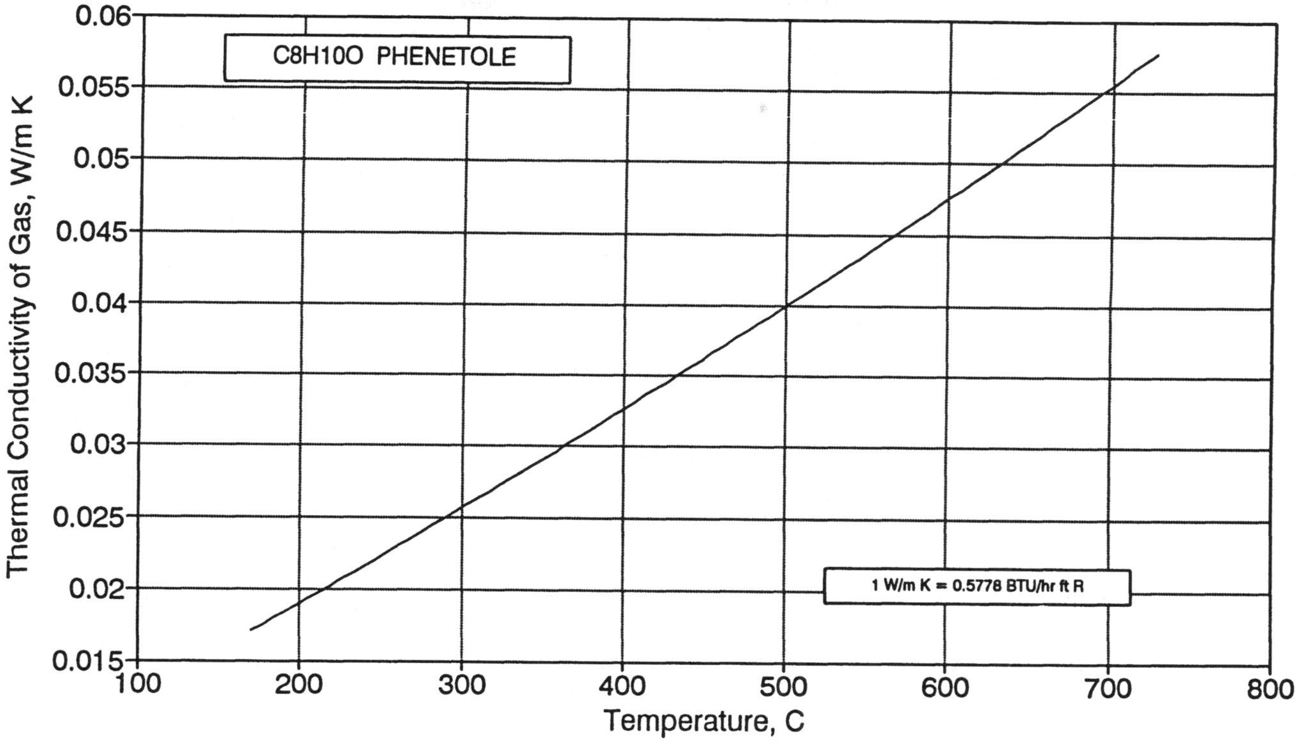
C8H10O PHENETOLE
1 W/m K = 0.5778 BTU/hr ft R
Thermal Conductivity of Gas, W/m K
Temperature, C
0.06
0.055
0.05
0.045
0.04
0.035
0.03
0.025
0.02
0.015
100
200
300
400
500
600
700
800

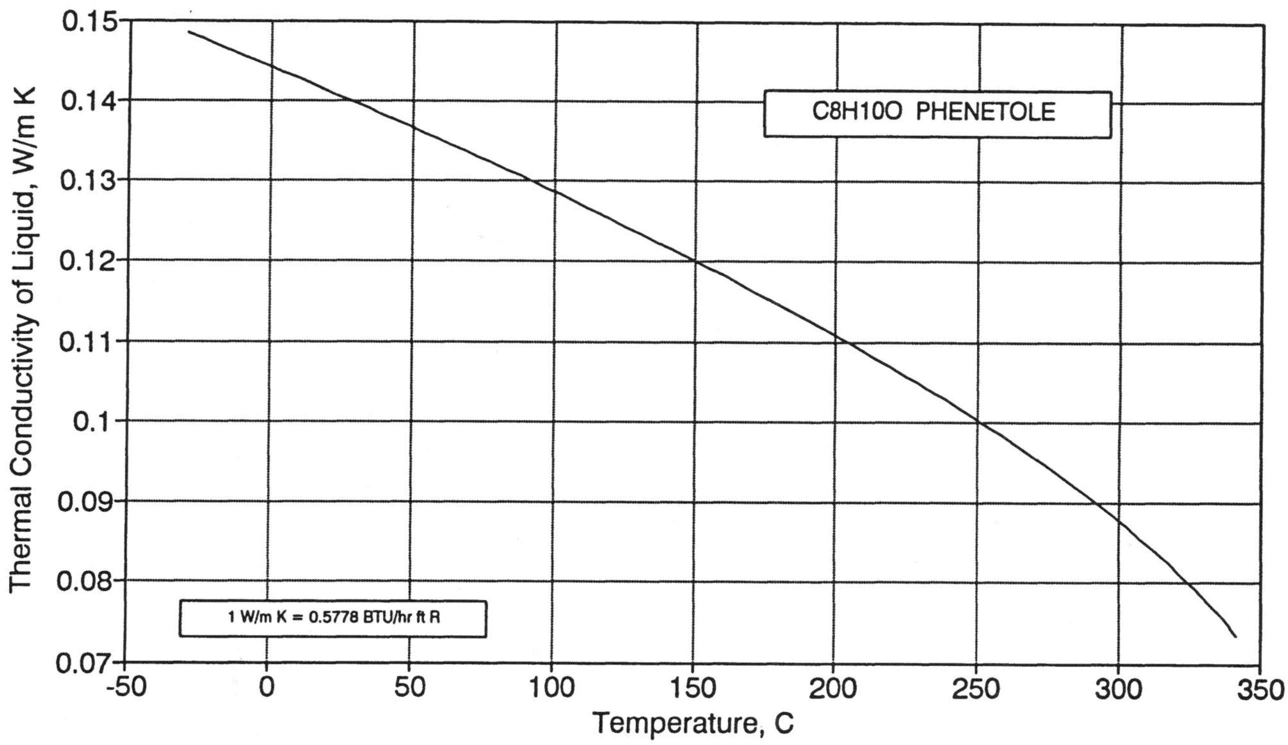
C8H10O PHENETOLE
1 W/m K = 0.5778 BTU/hr ft R
Thermal Conductivity of Liquid, W/m K
Temperature, C
0.15
0.14
0.13
0.12
0.11
0.1
0.09
0.08
0.07
-50
0
50
100
150
200
250
300
350

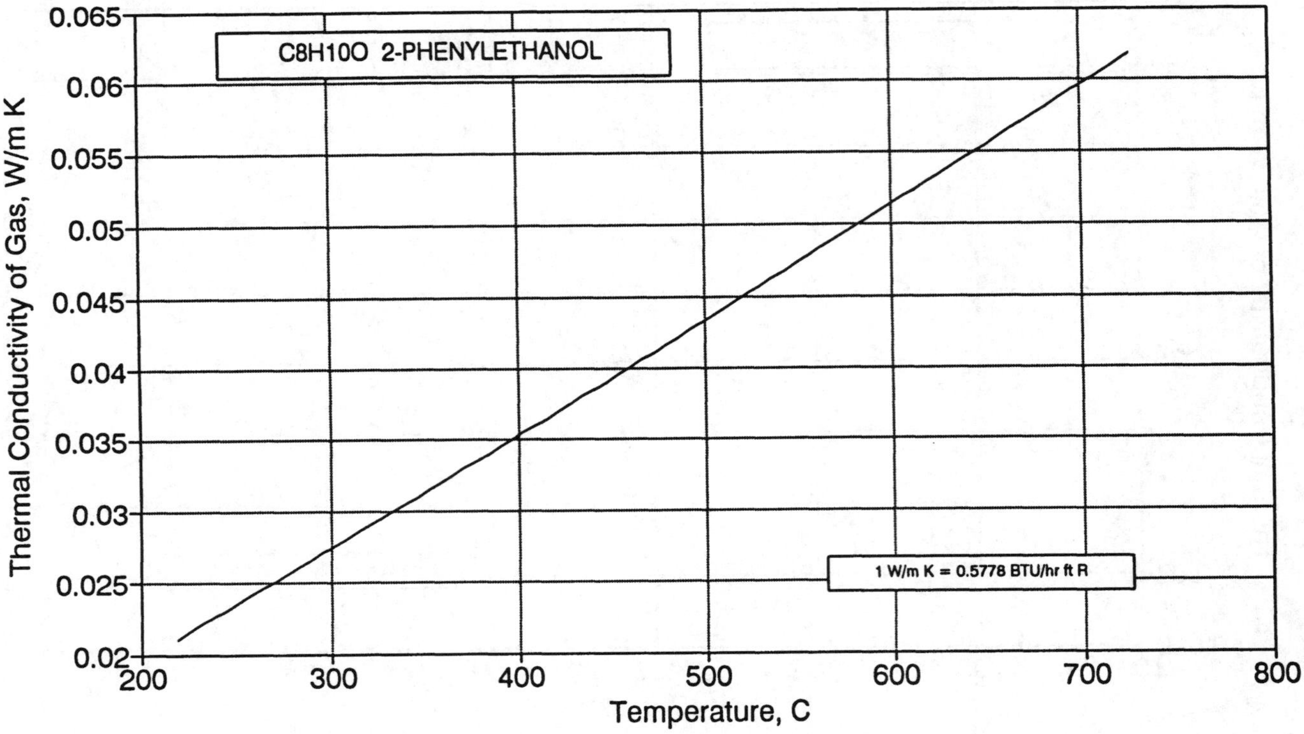

C8H10O 2-PHENYLETHANOL
Thermal Conductivity of Gas, W/m K
0.065
0.06
0.055
0.05
0.045
0.04
0.035
0.03
0.025
0.02
200
300
400
500
600
700
800
Temperature, C
1 W/m K = 0.5778 BTU/hr ft R

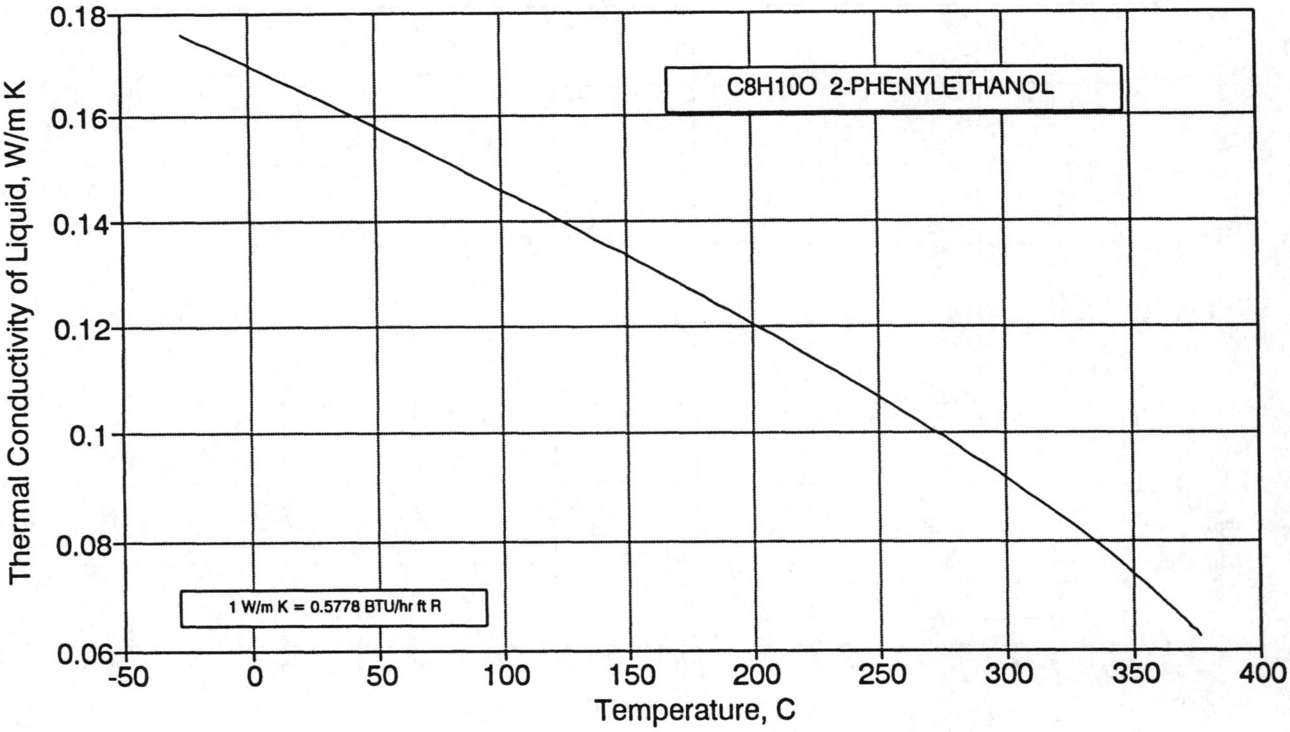

C8H10O 2-PHENYLETHANOL
Thermal Conductivity of Liquid, W/m K
0.18
0.16
0.14
0.12
0.1
0.08
0.06
-50
0
50
100
150
200
250
300
350
400
Temperature, C
1 W/m K = 0.5778 BTU/hr ft R

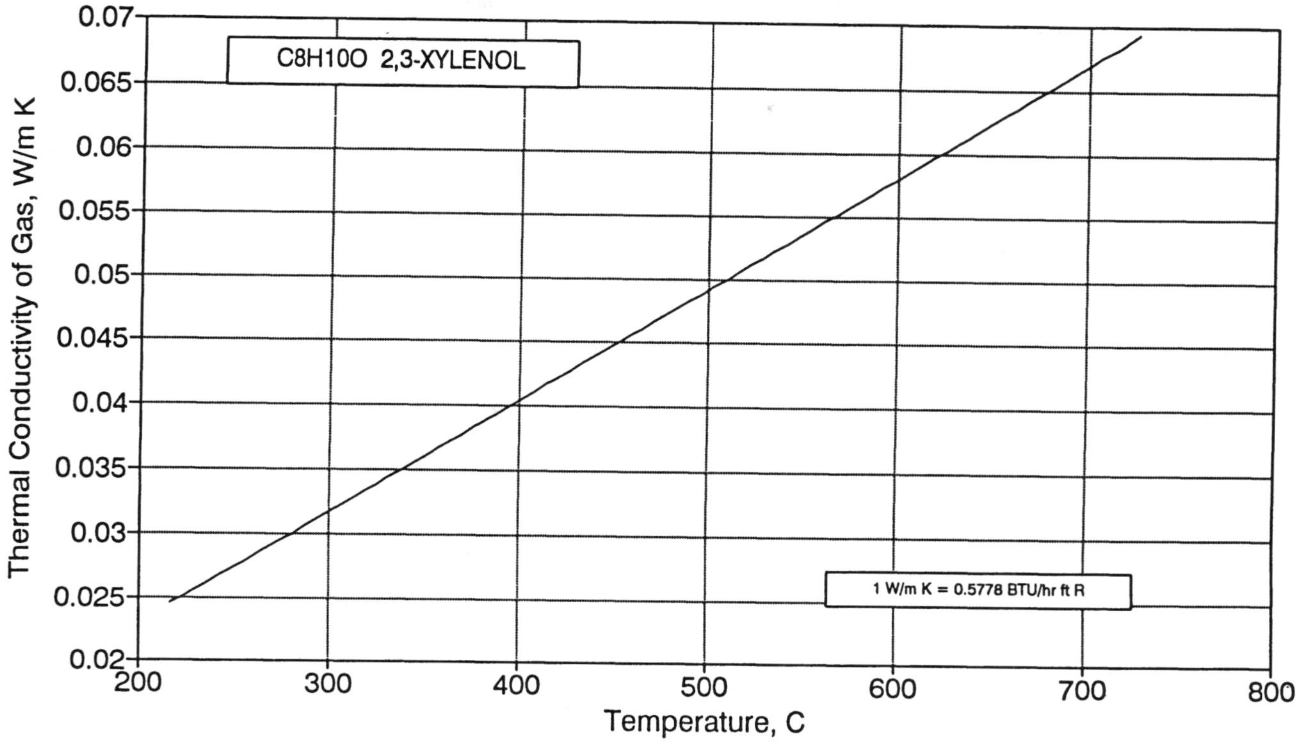
C8H10O 2,3-XYLENOL
1 W/m K = 0.5778 BTU/hr ft R
Thermal Conductivity of Gas, W/m K
0.07
0.065
0.06
0.055
0.05
0.045
0.04
0.035
0.03
0.025
0.02
200
300
400
500
600
700
800
Temperature, C

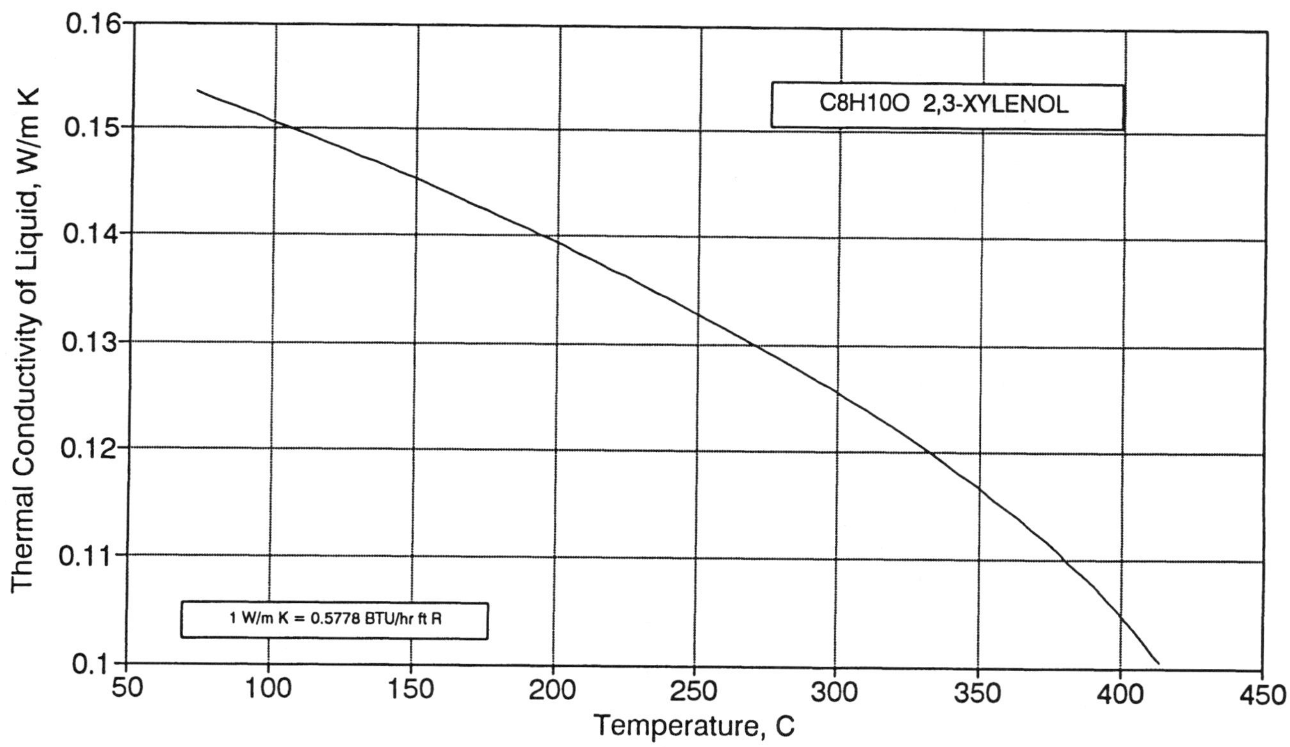
C8H10O 2,3-XYLENOL
1 W/m K = 0.5778 BTU/hr ft R
Thermal Conductivity of Liquid, W/m K
0.16
0.15
0.14
0.13
0.12
0.11
0.1
50
100
150
200
250
300
350
400
450
Temperature, C

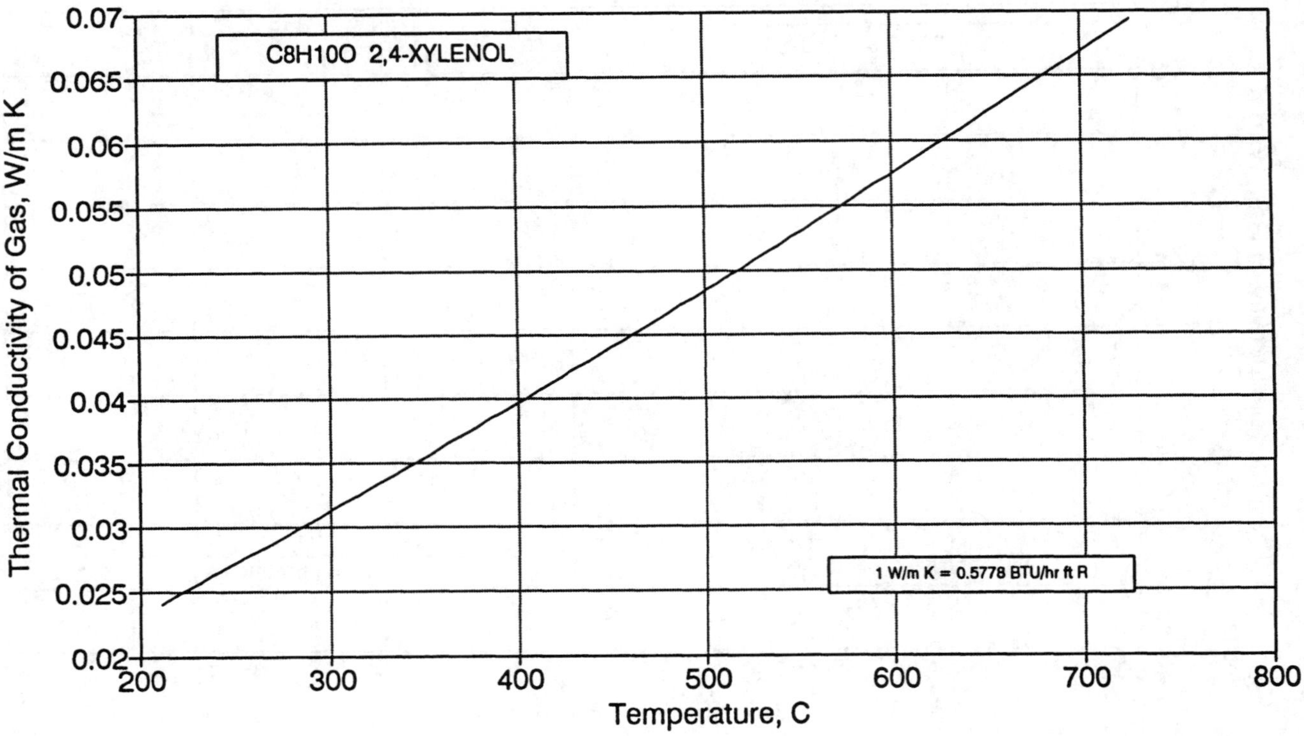
C8H10O 2,4-XYLENOL
Thermal Conductivity of Gas, W/m K
Temperature, C
1 W/m K = 0.5778 BTU/hr ft R
0.07
0.065
0.06
0.055
0.05
0.045
0.04
0.035
0.03
0.025
0.02
200
300
400
500
600
700
800

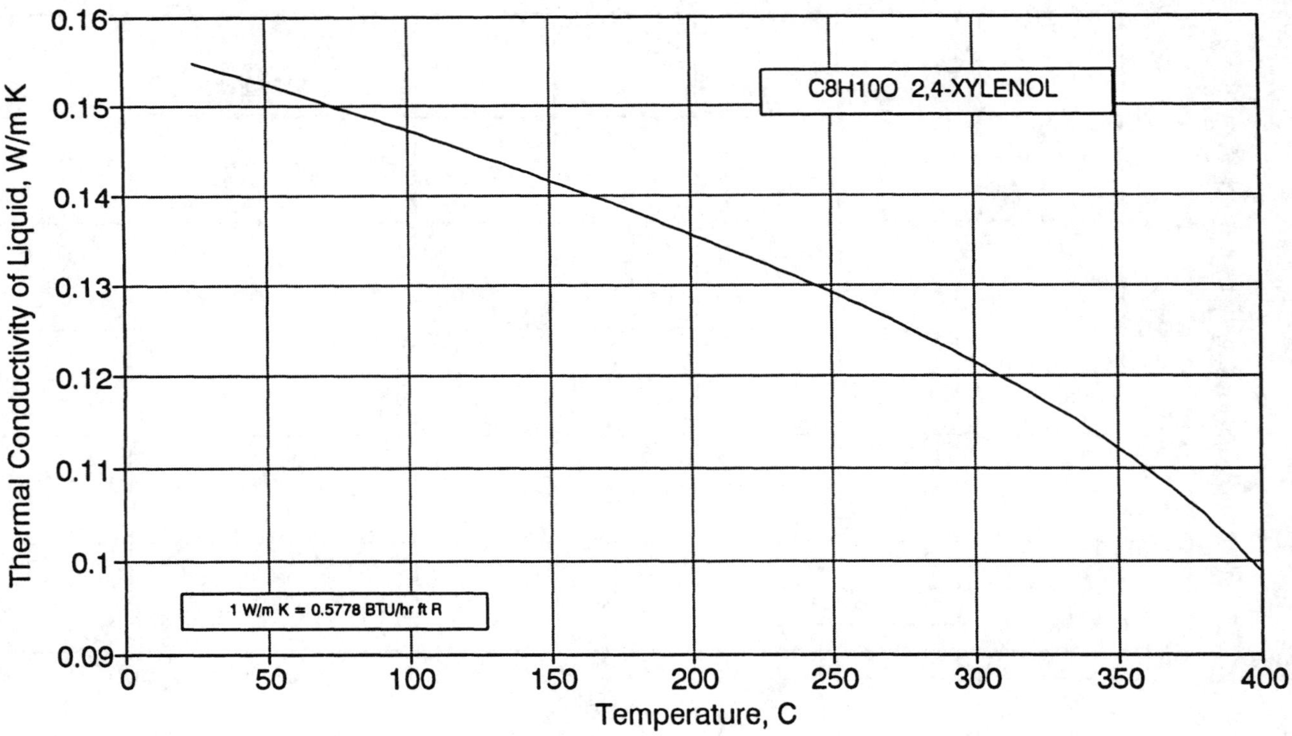
C8H10O 2,4-XYLENOL
Thermal Conductivity of Liquid, W/m K
Temperature, C
1 W/m K = 0.5778 BTU/hr ft R
0.16
0.15
0.14
0.13
0.12
0.11
0.1
0.09
0
50
100
150
200
250
300
350
400

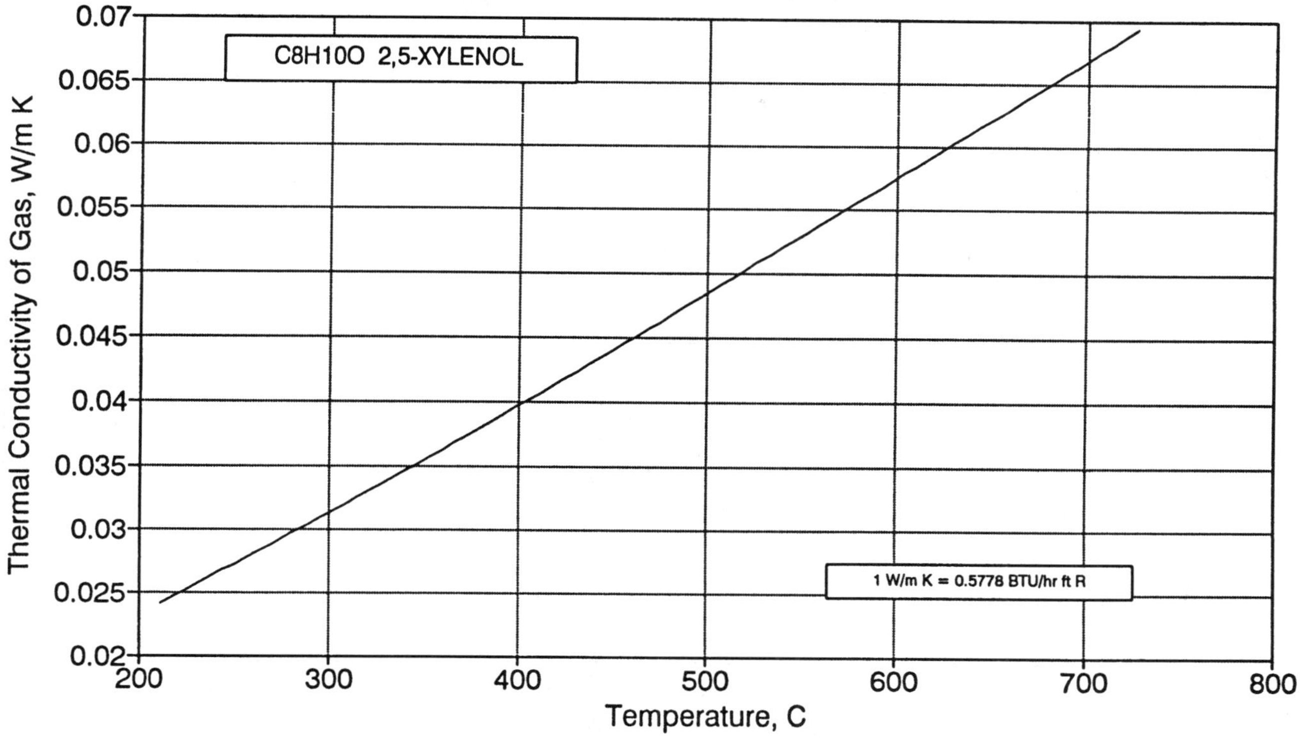
C8H10O 2,5-XYLENOL
1 W/m K = 0.5778 BTU/hr ft R
Thermal Conductivity of Gas, W/m K
0.07
0.065
0.06
0.055
0.05
0.045
0.04
0.035
0.03
0.025
0.02
200
300
400
500
600
700
800
Temperature, C

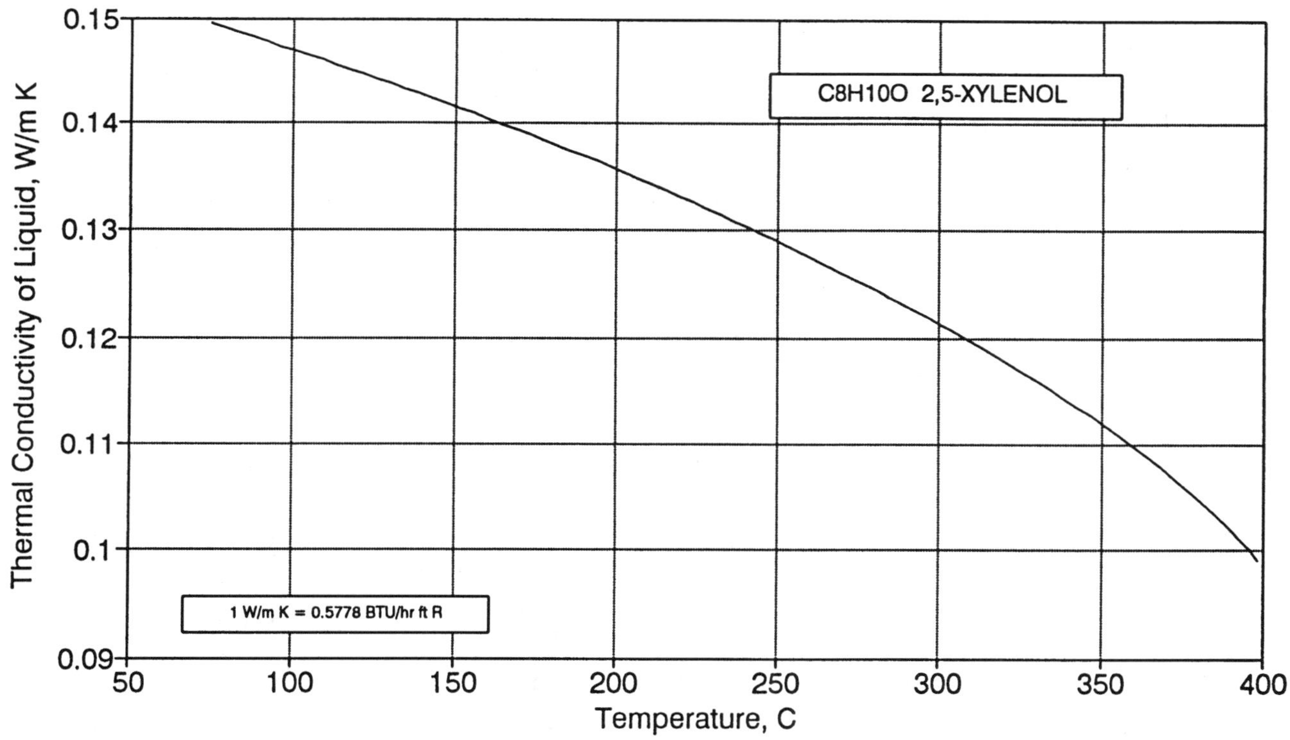
C8H10O 2,5-XYLENOL
1 W/m K = 0.5778 BTU/hr ft R
Thermal Conductivity of Liquid, W/m K
0.15
0.14
0.13
0.12
0.11
0.1
0.09
50
100
150
200
250
300
350
400
Temperature, C

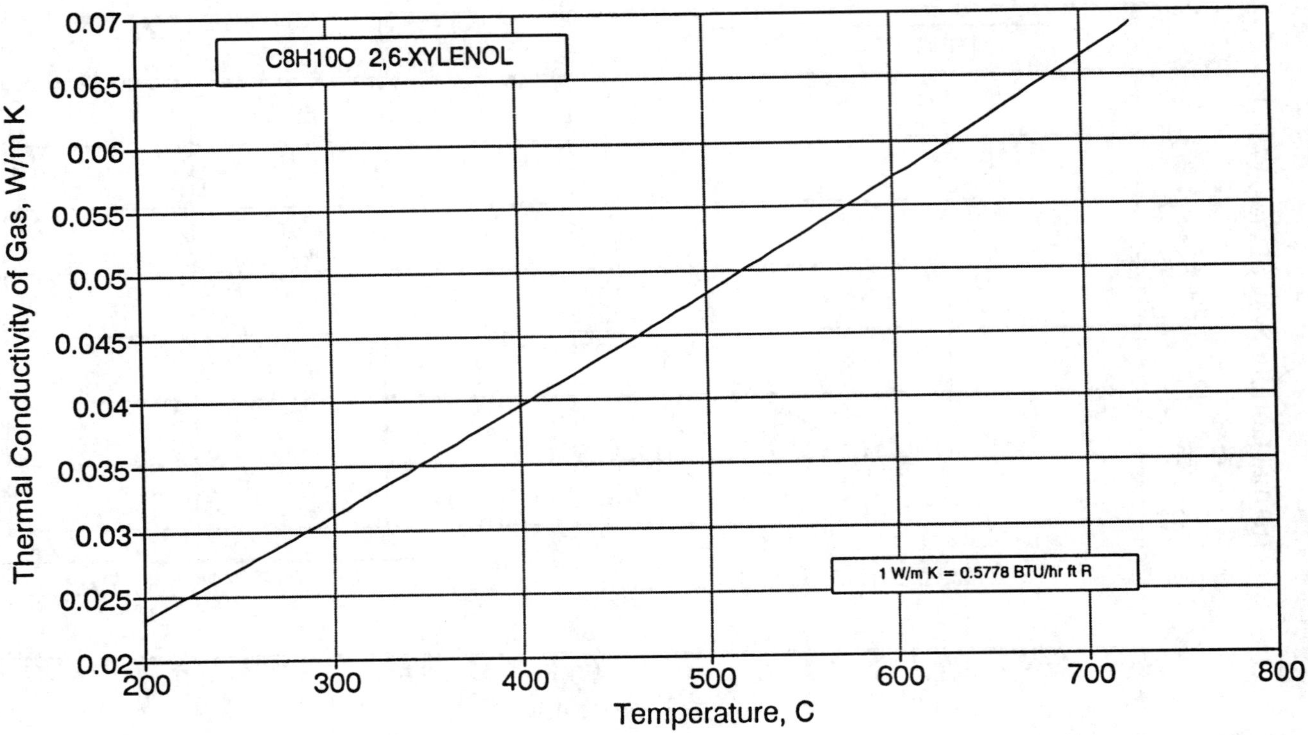
C8H10O 2,6-XYLENOL
1 W/m K = 0.5778 BTU/hr ft R
Thermal Conductivity of Gas, W/m K
0.07
0.065
0.06
0.055
0.05
0.045
0.04
0.035
0.03
0.025
0.02
200
300
400
500
600
700
800
Temperature, C

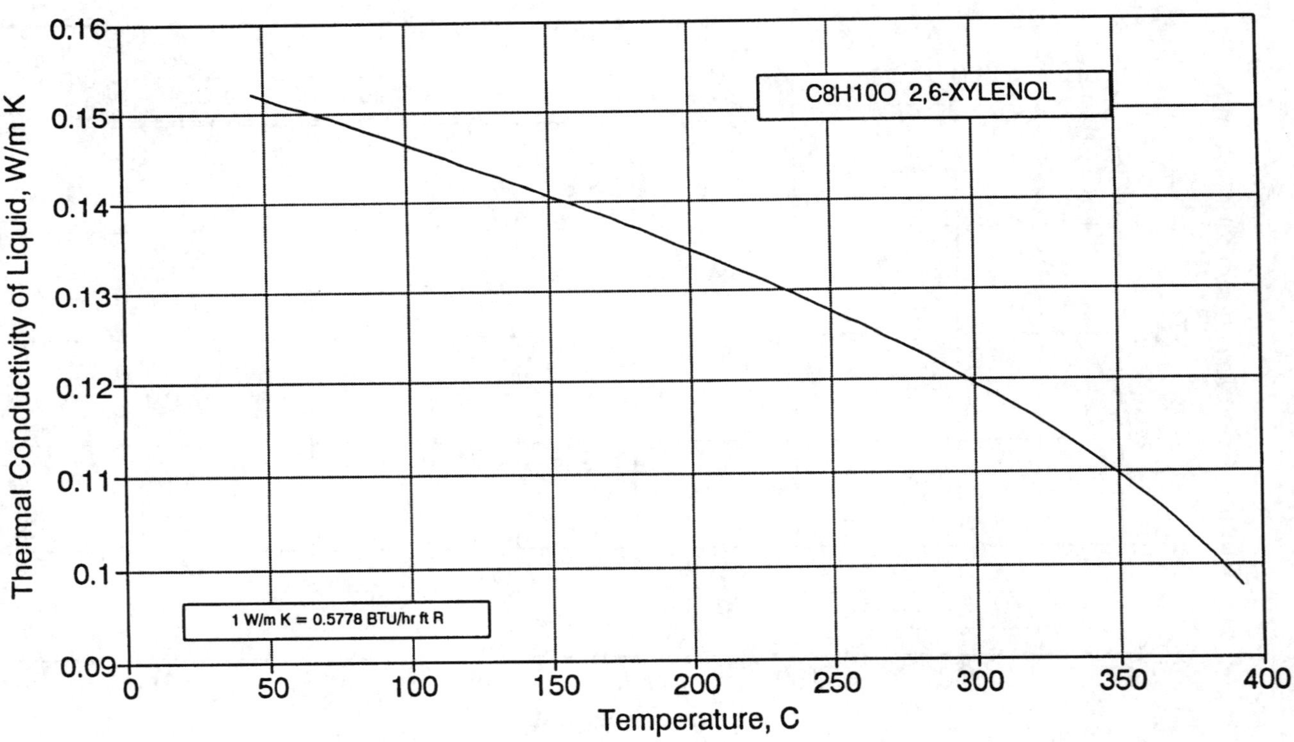
C8H10O 2,6-XYLENOL
1 W/m K = 0.5778 BTU/hr ft R
Thermal Conductivity of Liquid, W/m K
0.16
0.15
0.14
0.13
0.12
0.11
0.1
0.09
0
50
100
150
200
250
300
350
400
Temperature, C

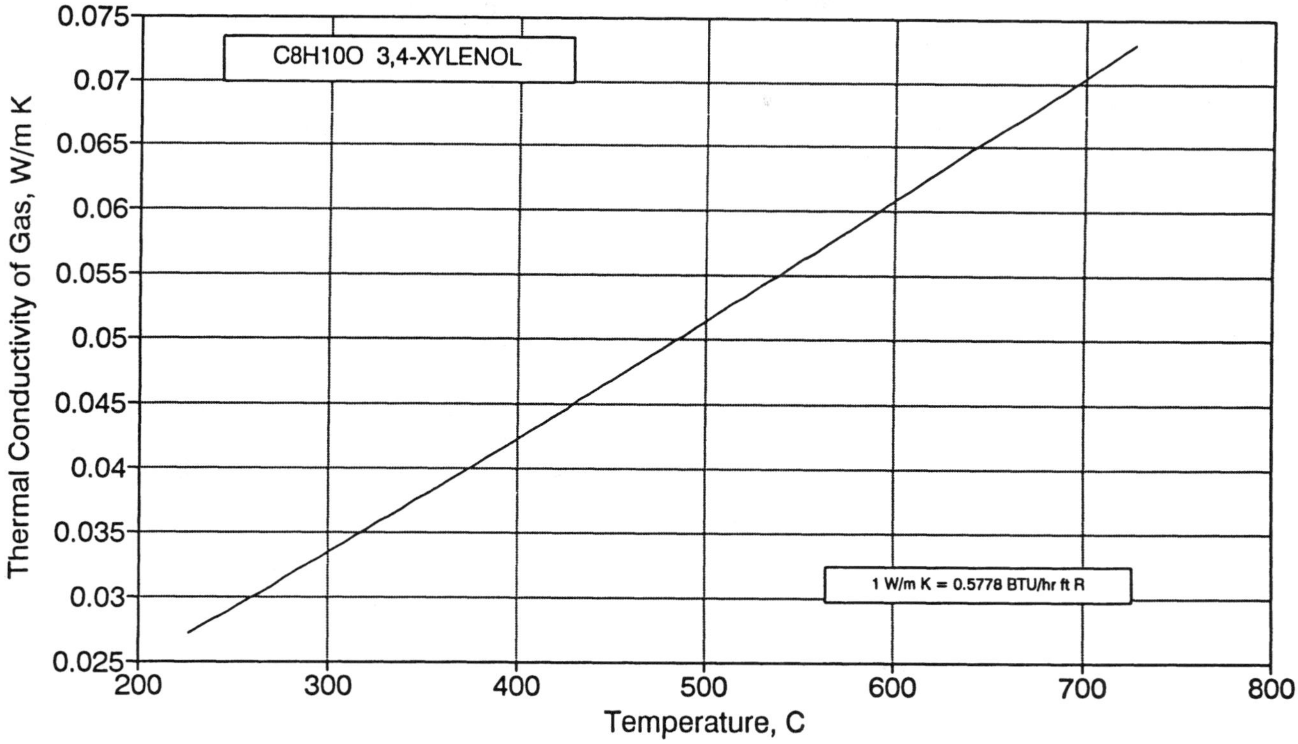
C8H10O 3,4-XYLENOL
1 W/m K = 0.5778 BTU/hr ft R
Thermal Conductivity of Gas, W/m K
0.075
0.07
0.065
0.06
0.055
0.05
0.045
0.04
0.035
0.03
0.025
200
300
400
500
600
700
800
Temperature, C

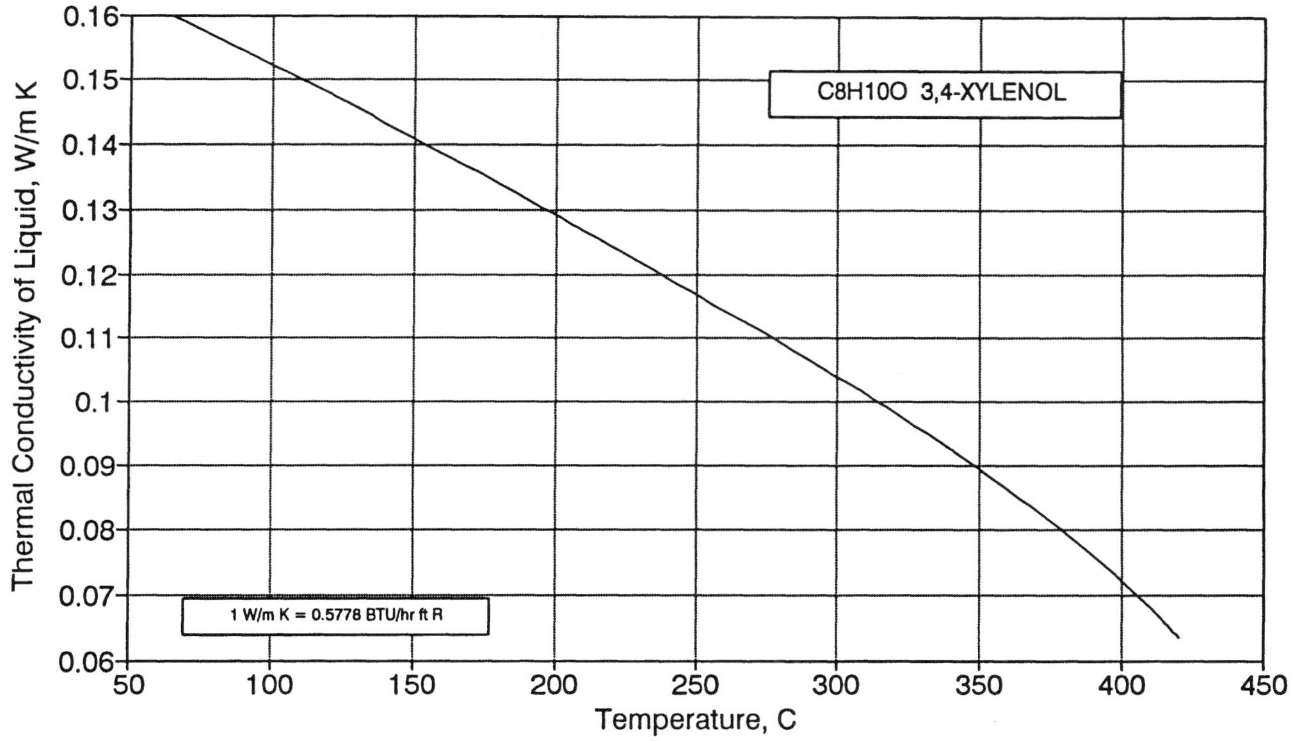
C8H10O 3,4-XYLENOL
1 W/m K = 0.5778 BTU/hr ft R
Thermal Conductivity of Liquid, W/m K
0.16
0.15
0.14
0.13
0.12
0.11
0.1
0.09
0.08
0.07
0.06
50
100
150
200
250
300
350
400
450
Temperature, C

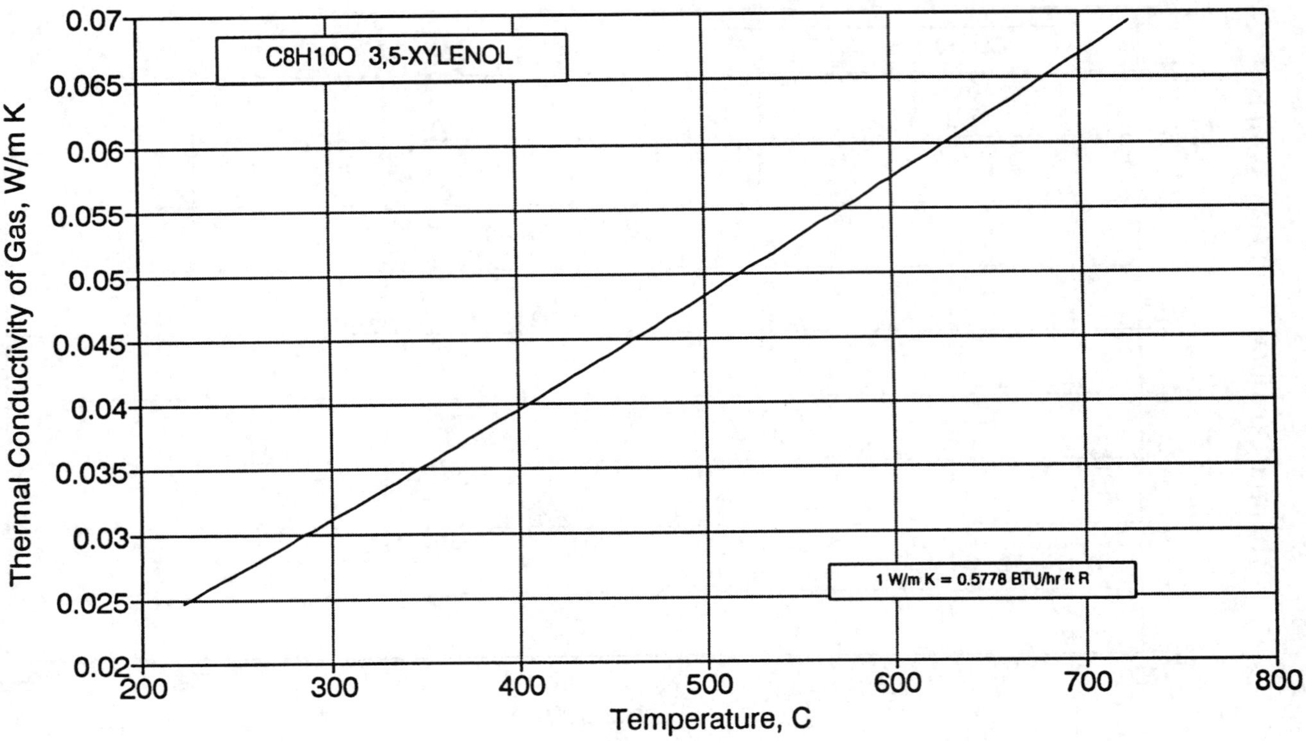
C8H10O 3,5-XYLENOL
1 W/m K = 0.5778 BTU/hr ft R
Thermal Conductivity of Gas, W/m K
0.07
0.065
0.06
0.055
0.05
0.045
0.04
0.035
0.03
0.025
0.02
200
300
400
500
600
700
800
Temperature, C

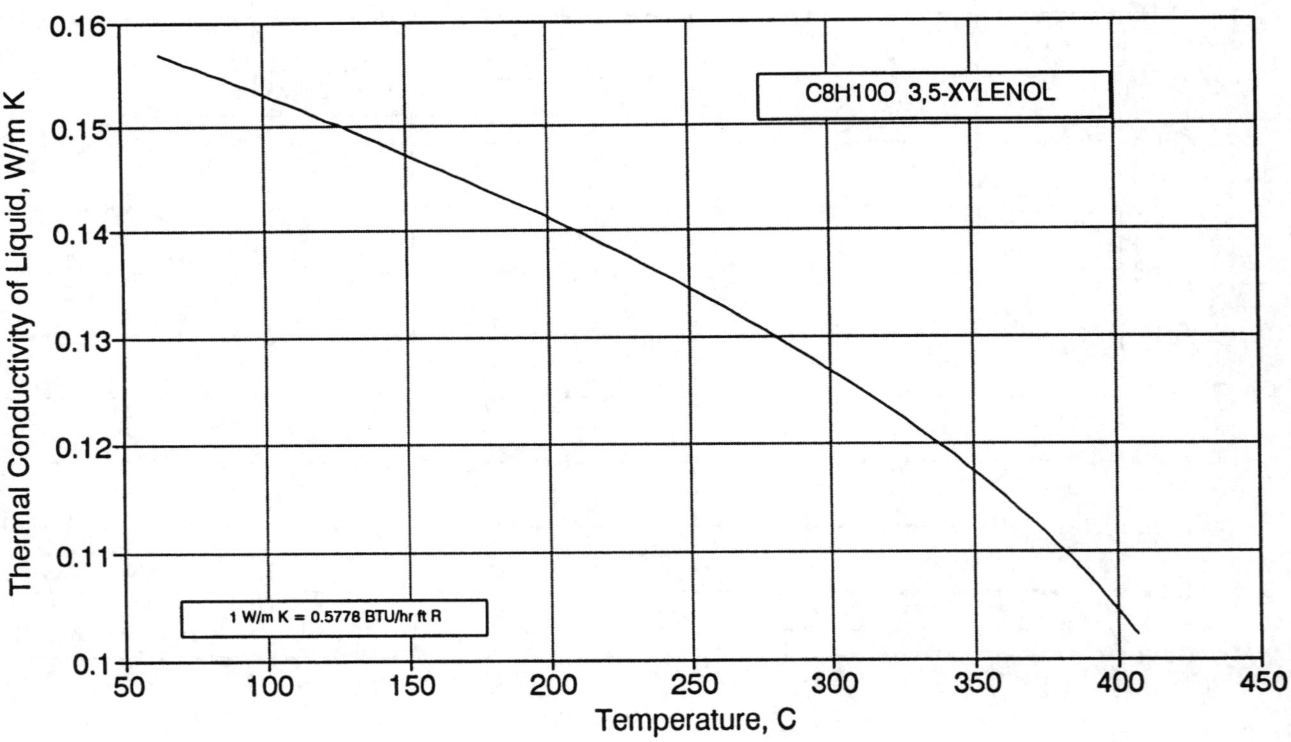
C8H10O 3,5-XYLENOL
1 W/m K = 0.5778 BTU/hr ft R
Thermal Conductivity of Liquid, W/m K
0.16
0.15
0.14
0.13
0.12
0.11
0.1
50
100
150
200
250
300
350
400
450
Temperature, C

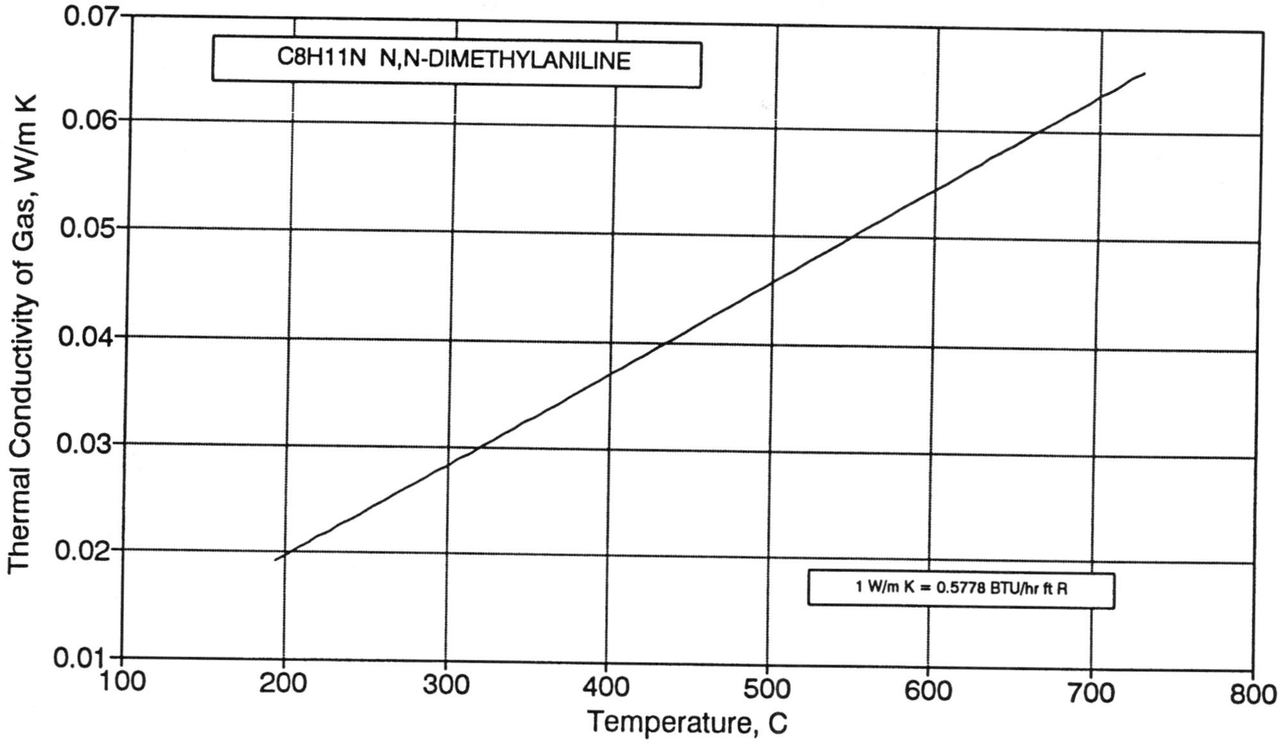

C8H11N N,N-DIMETHYLANILINE
1 W/m K = 0.5778 BTU/hr ft R
Thermal Conductivity of Gas, W/m K
0.07
0.06
0.05
0.04
0.03
0.02
0.01
100
200
300
400
500
600
700
800
Temperature, C

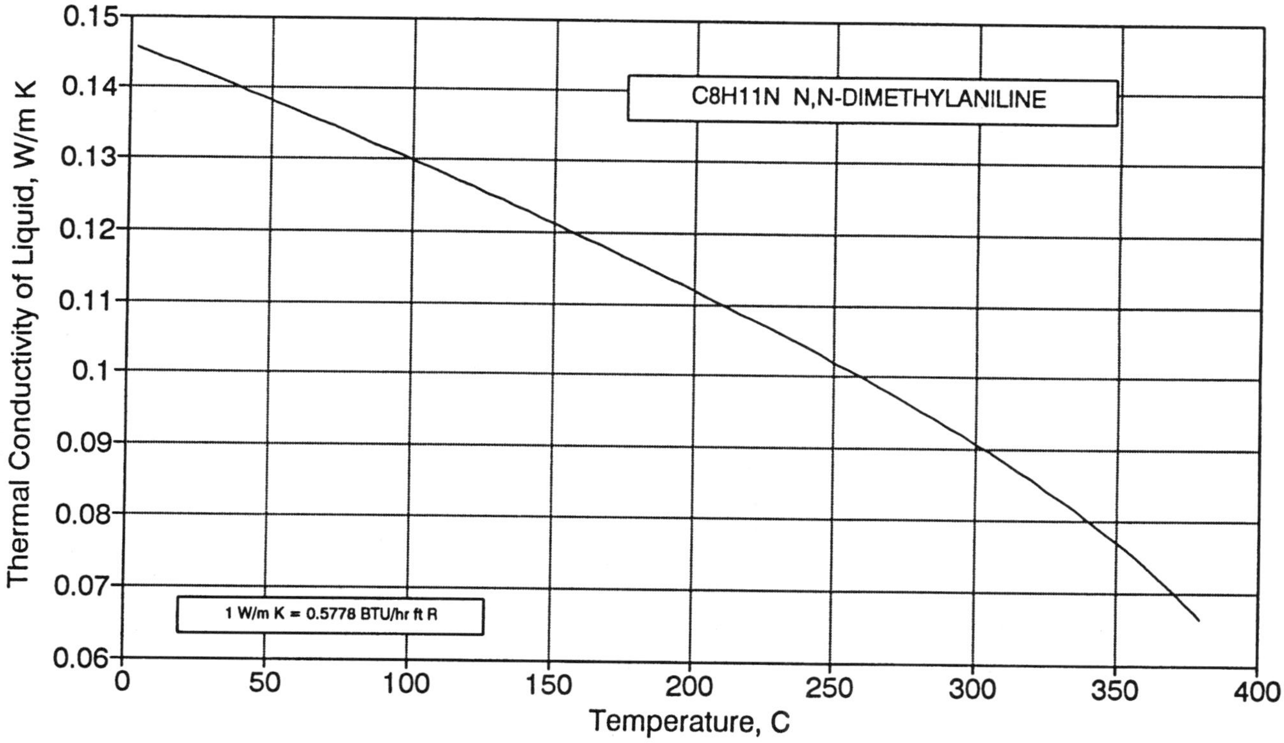

C8H11N N,N-DIMETHYLANILINE
1 W/m K = 0.5778 BTU/hr ft R
Thermal Conductivity of Liquid, W/m K
0.15
0.14
0.13
0.12
0.11
0.1
0.09
0.08
0.07
0.06
0
50
100
150
200
250
300
350
400
Temperature, C

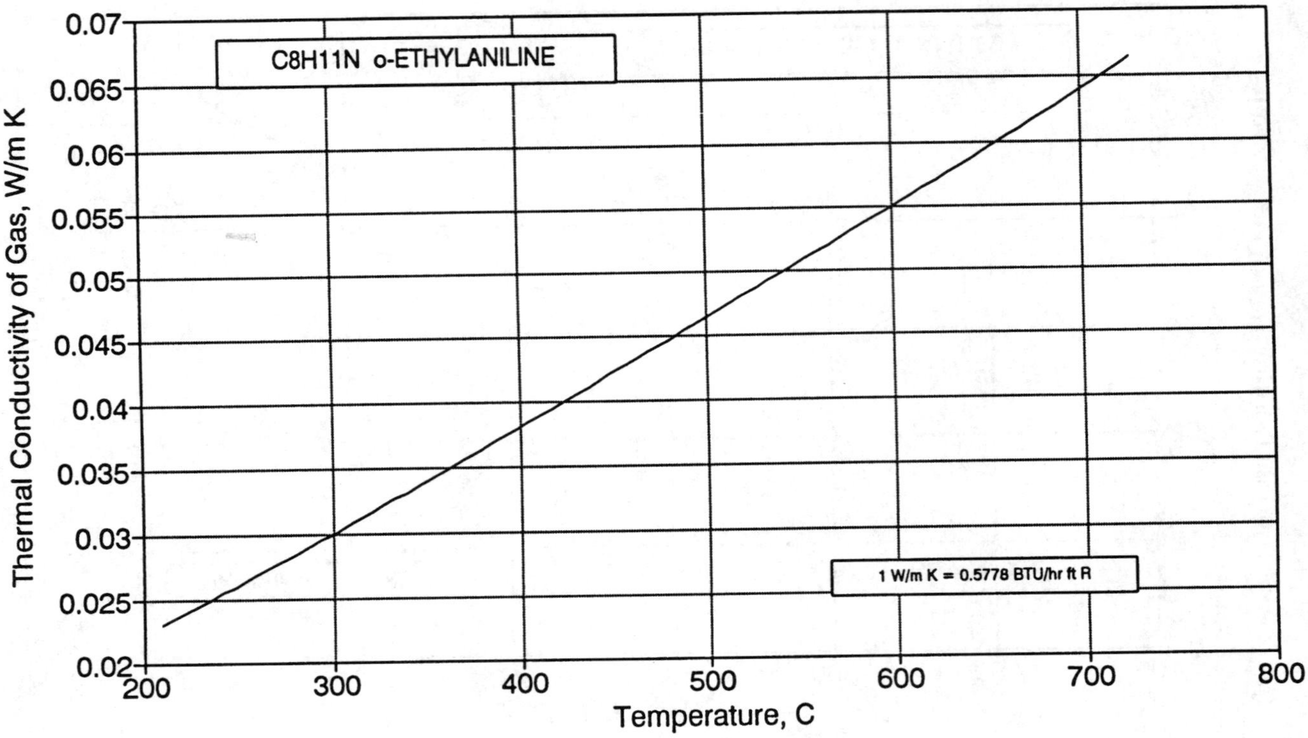
C8H11N o-ETHYLANILINE
1 W/m K = 0.5778 BTU/hr ft R
Thermal Conductivity of Gas, W/m K
0.07
0.065
0.06
0.055
0.05
0.045
0.04
0.035
0.03
0.025
0.02
200
300
400
500
600
700
800
Temperature, C

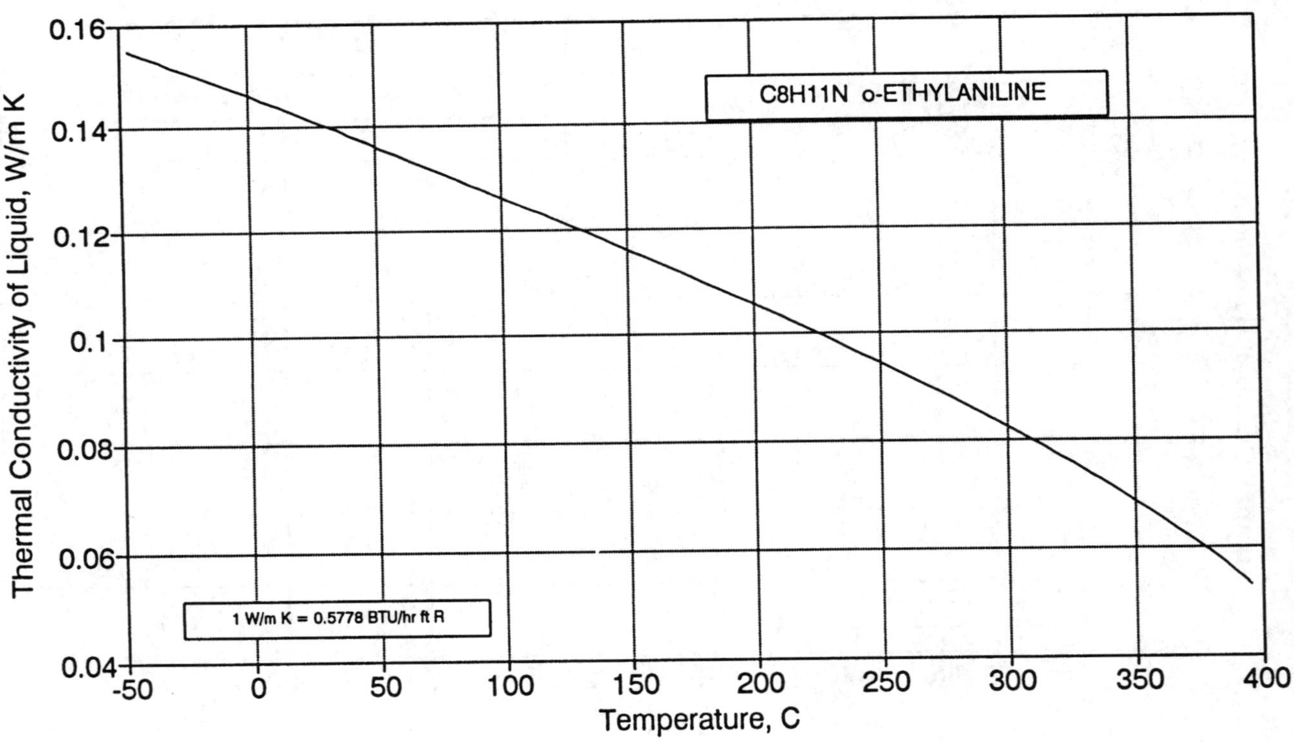
C8H11N o-ETHYLANILINE
1 W/m K = 0.5778 BTU/hr ft R
Thermal Conductivity of Liquid, W/m K
0.16
0.14
0.12
0.1
0.08
0.06
0.04
-50
0
50
100
150
200
250
300
350
400
Temperature, C

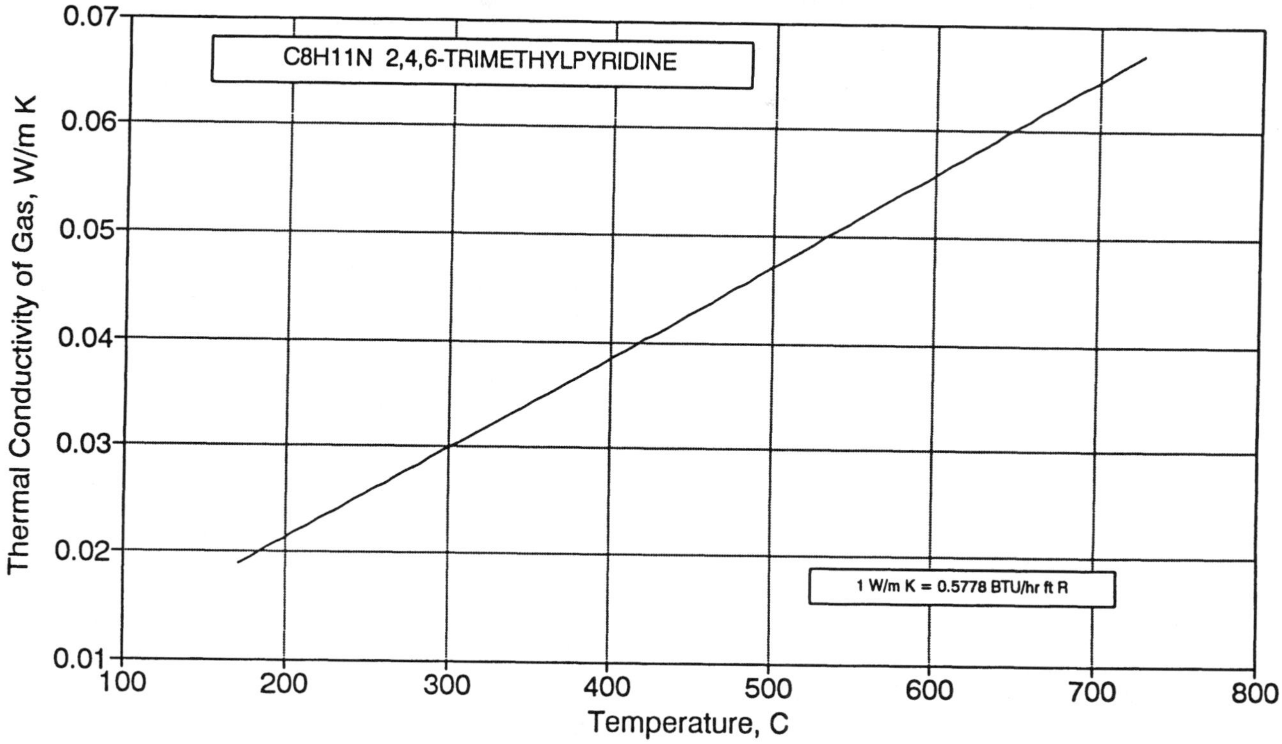
C8H11N 2,4,6-TRIMETHYLPYRIDINE
1 W/m K = 0.5778 BTU/hr ft R
Thermal Conductivity of Gas, W/m K
0.07
0.06
0.05
0.04
0.03
0.02
0.01
100
200
300
400
500
600
700
800
Temperature, C

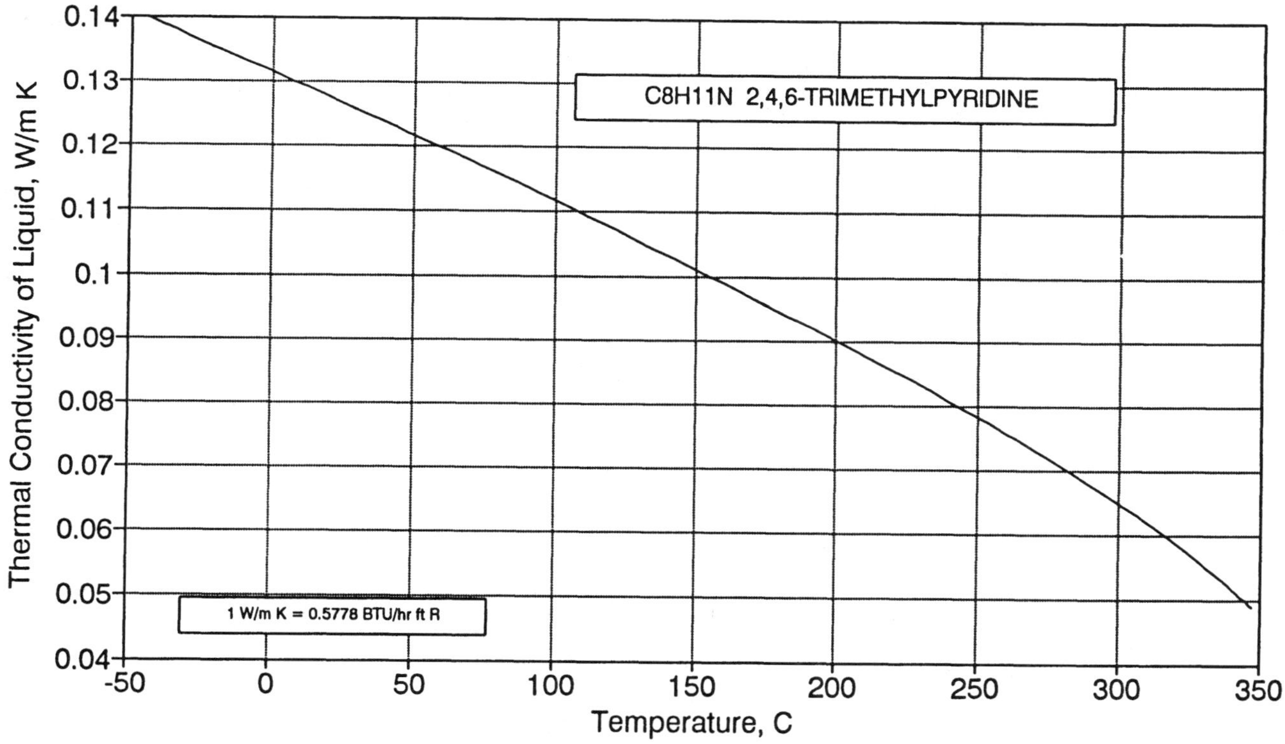
C8H11N 2,4,6-TRIMETHYLPYRIDINE
1 W/m K = 0.5778 BTU/hr ft R
Thermal Conductivity of Liquid, W/m K
0.14
0.13
0.12
0.11
0.1
0.09
0.08
0.07
0.06
0.05
0.04
-50
0
50
100
150
200
250
300
350
Temperature, C

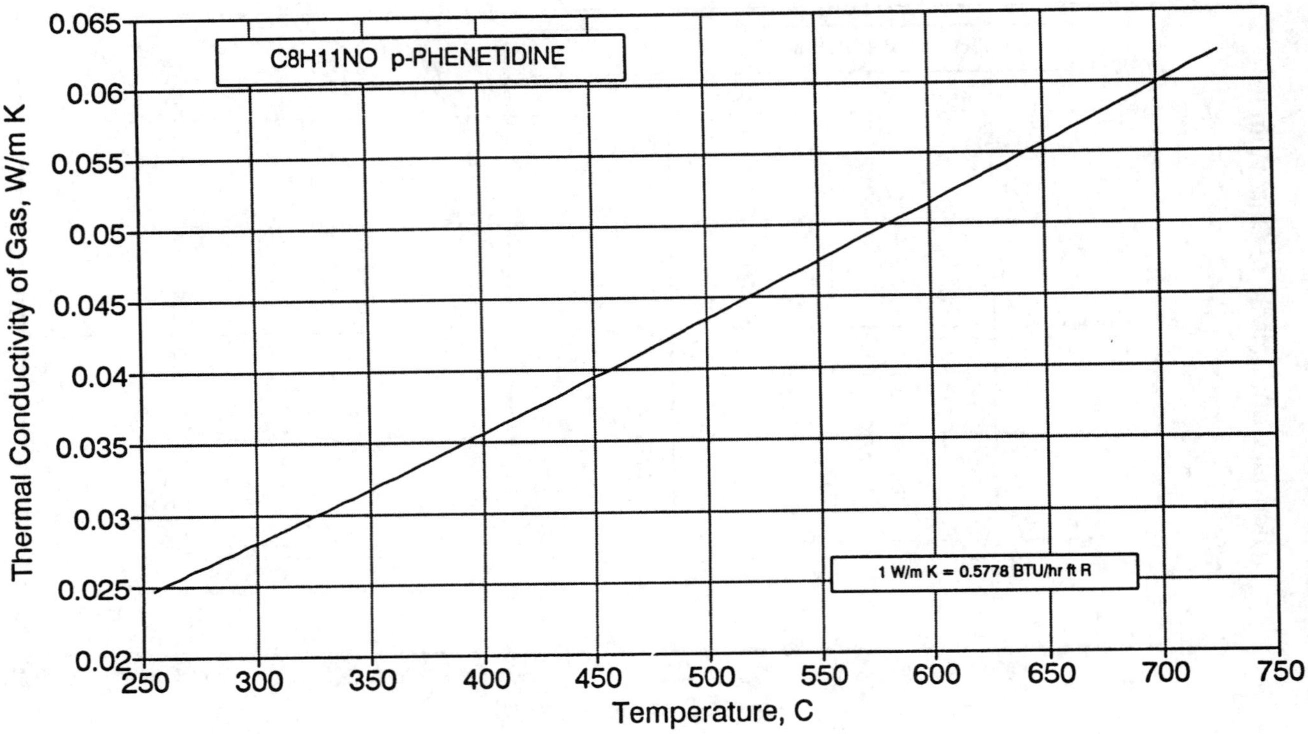
C8H11NO p-PHENETIDINE
1 W/m K = 0.5778 BTU/hr ft R
Thermal Conductivity of Gas, W/m K
0.065
0.06
0.055
0.05
0.045
0.04
0.035
0.03
0.025
0.02
250
300
350
400
450
500
550
600
650
700
750
Temperature, C

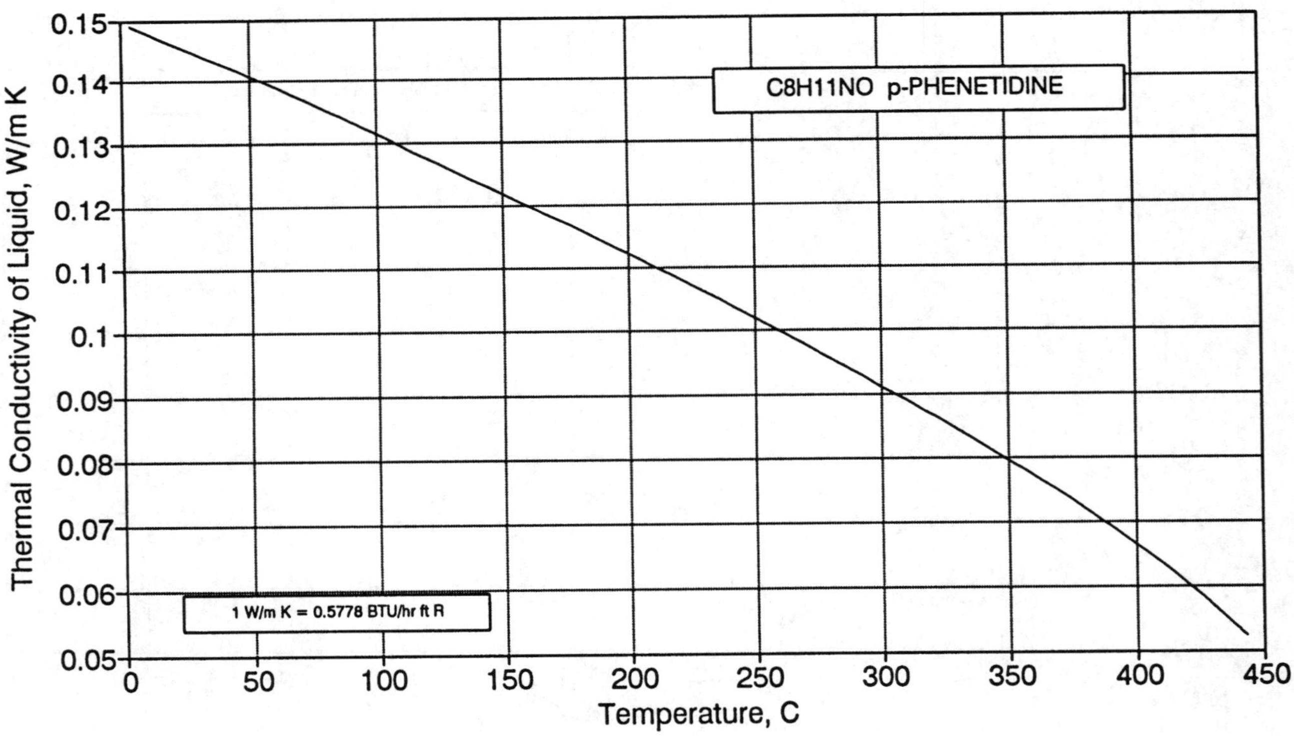
C8H11NO p-PHENETIDINE
1 W/m K = 0.5778 BTU/hr ft R
Thermal Conductivity of Liquid, W/m K
0.15
0.14
0.13
0.12
0.11
0.1
0.09
0.08
0.07
0.06
0.05
0
50
100
150
200
250
300
350
400
450
Temperature, C

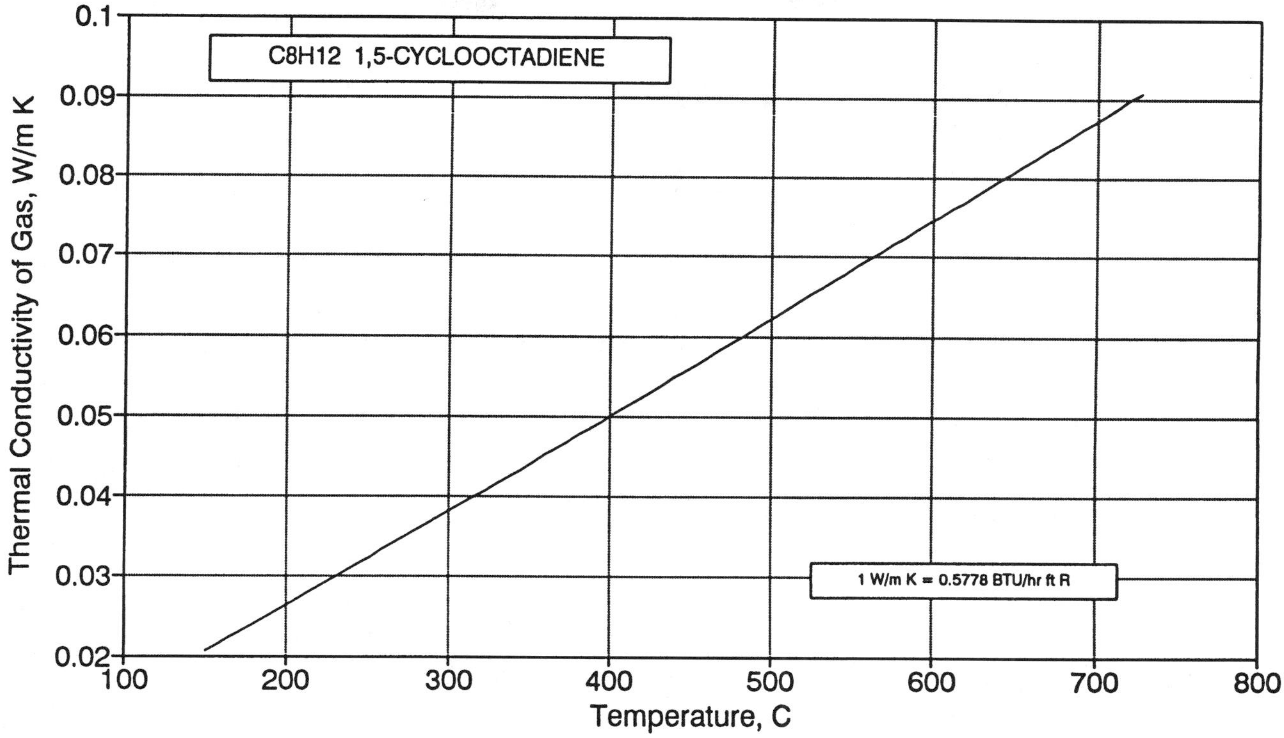
C8H12 1,5-CYCLOOCTADIENE
1 W/m K = 0.5778 BTU/hr ft R
Thermal Conductivity of Gas, W/m K
Temperature, C
0.1
0.09
0.08
0.07
0.06
0.05
0.04
0.03
0.02
100
200
300
400
500
600
700
800

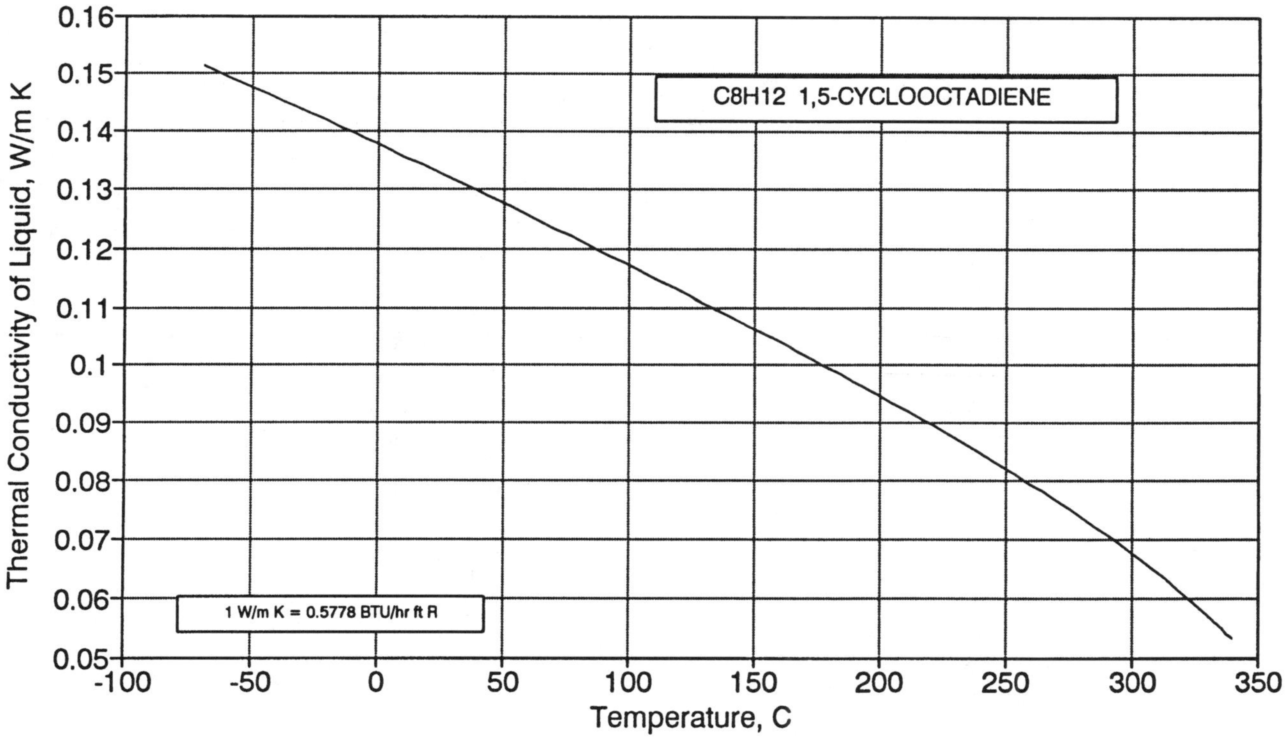
C8H12 1,5-CYCLOOCTADIENE
1 W/m K = 0.5778 BTU/hr ft R
Thermal Conductivity of Liquid, W/m K
Temperature, C
0.16
0.15
0.14
0.13
0.12
0.11
0.1
0.09
0.08
0.07
0.06
0.05
-100
-50
0
50
100
150
200
250
300
350

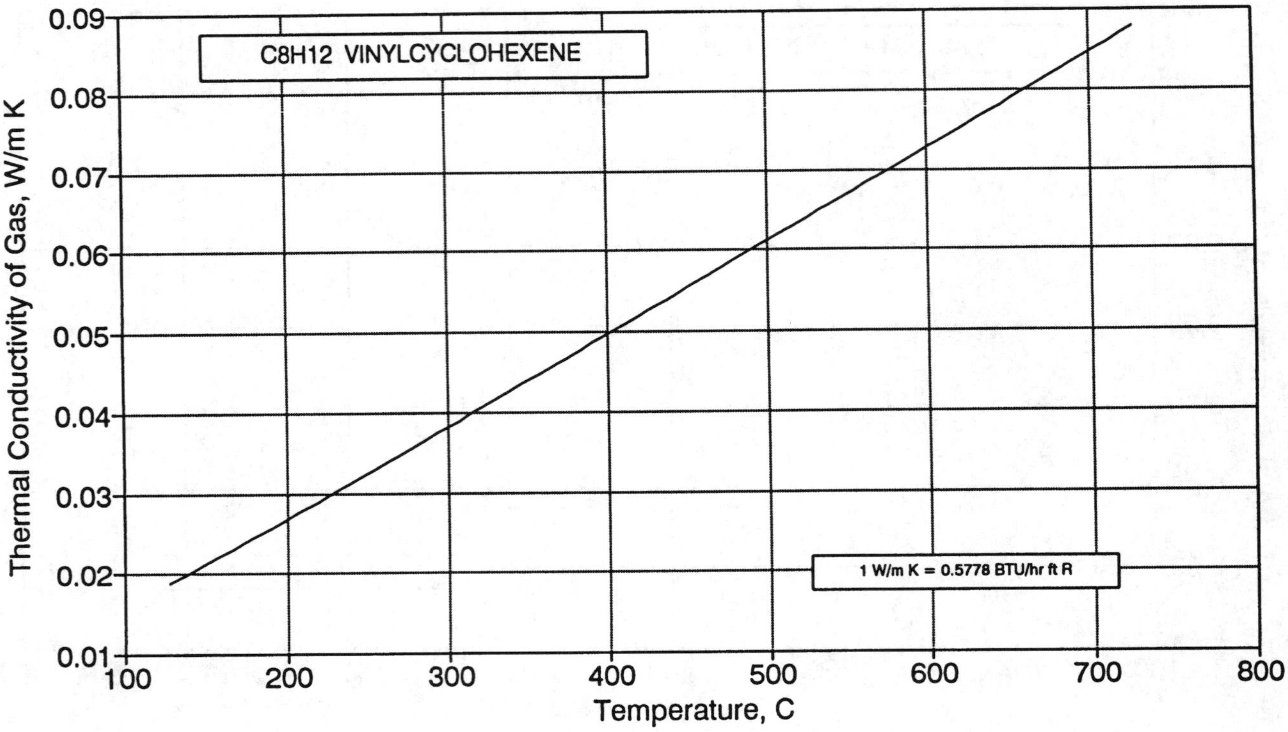
C8H12 VINYLCYCLOHEXENE
1 W/m K = 0.5778 BTU/hr ft R
Thermal Conductivity of Gas, W/m K
Temperature, C
0.09
0.08
0.07
0.06
0.05
0.04
0.03
0.02
0.01
100
200
300
400
500
600
700
800

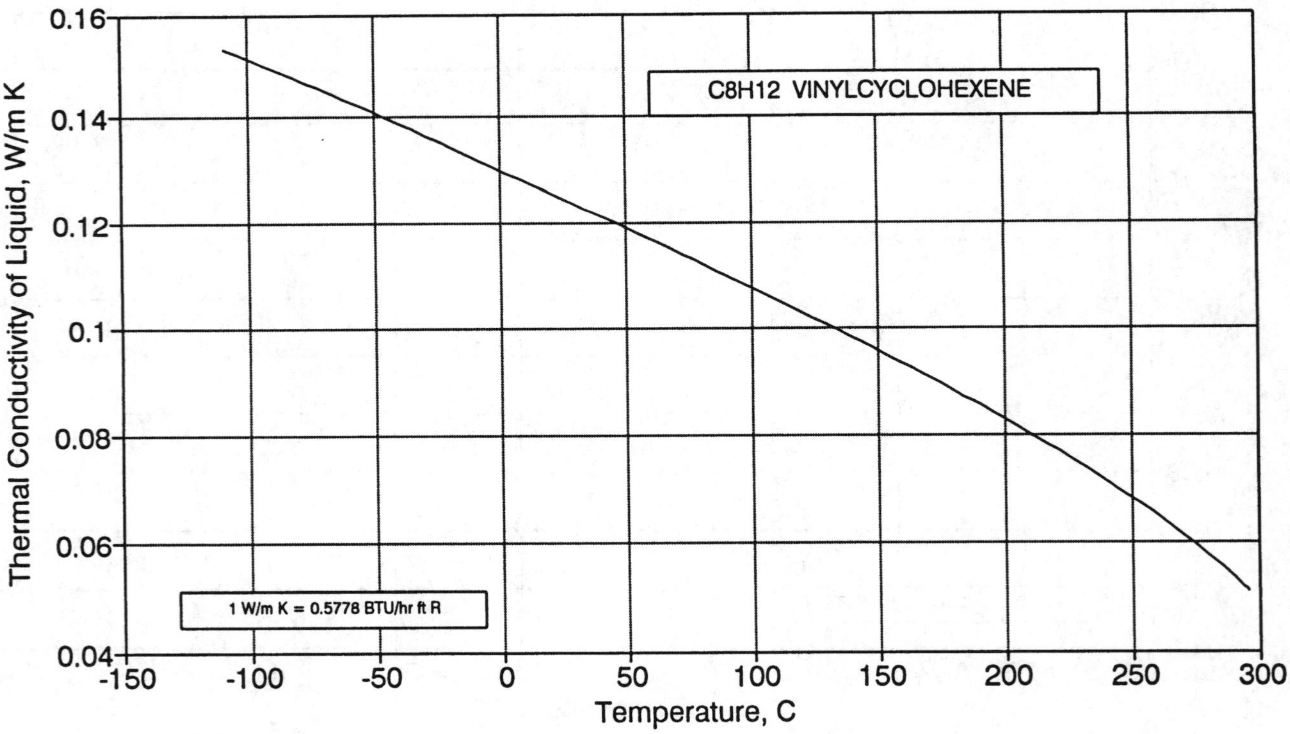
C8H12 VINYLCYCLOHEXENE
1 W/m K = 0.5778 BTU/hr ft R
Thermal Conductivity of Liquid, W/m K
Temperature, C
0.16
0.14
0.12
0.1
0.08
0.06
0.04
-150
-100
-50
0
50
100
150
200
250
300

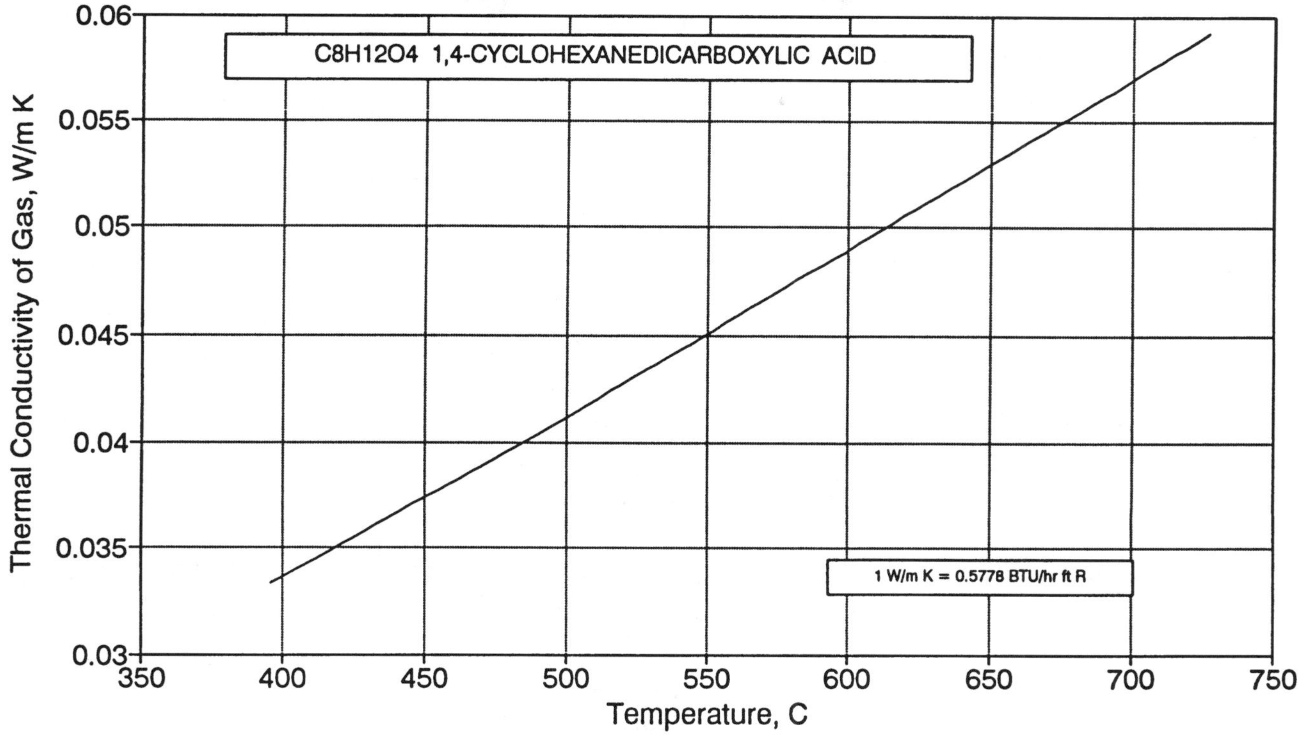
C8H12O4 1,4-CYCLOHEXANEDICARBOXYLIC ACID
1 W/m K = 0.5778 BTU/hr ft R
Thermal Conductivity of Gas, W/m K
Temperature, C
0.06
0.055
0.05
0.045
0.04
0.035
0.03
350
400
450
500
550
600
650
700
750

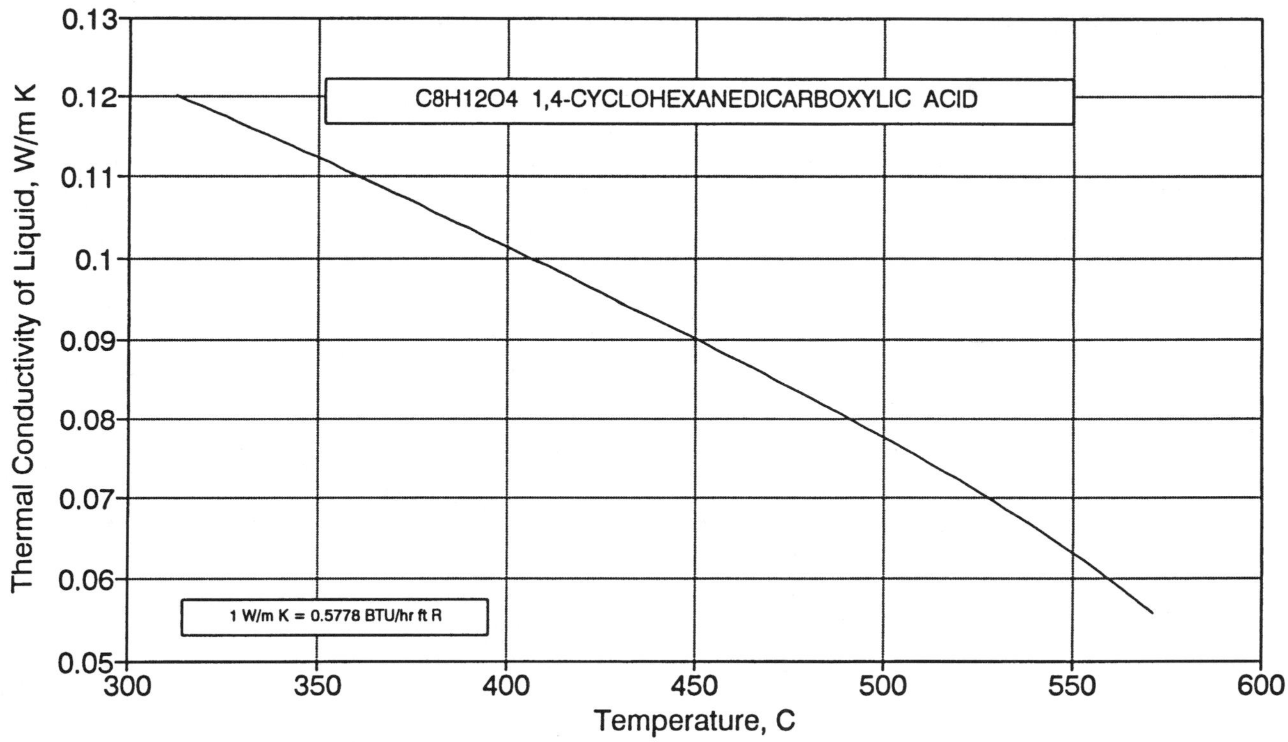
C8H12O4 1,4-CYCLOHEXANEDICARBOXYLIC ACID
1 W/m K = 0.5778 BTU/hr ft R
Thermal Conductivity of Liquid, W/m K
Temperature, C
0.13
0.12
0.11
0.1
0.09
0.08
0.07
0.06
0.05
300
350
400
450
500
550
600

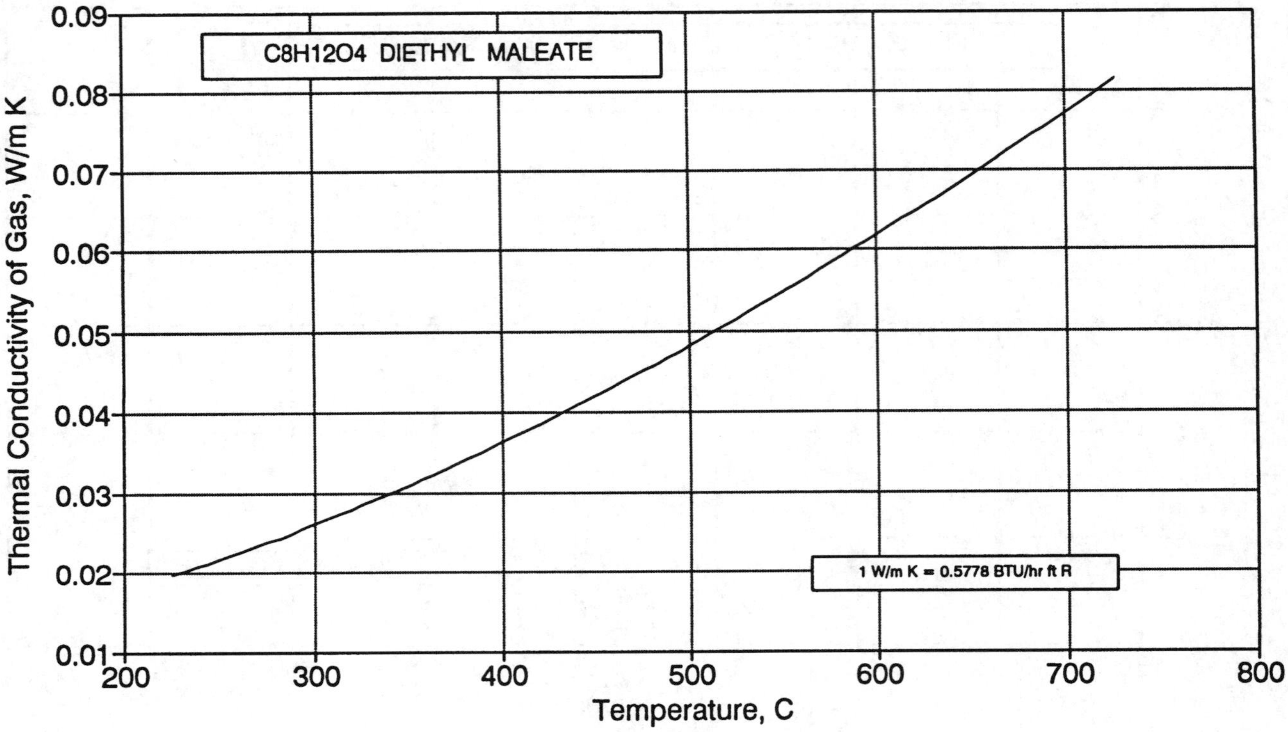
C8H12O4 DIETHYL MALEATE
1 W/m K = 0.5778 BTU/hr ft R
Thermal Conductivity of Gas, W/m K
Temperature, C
0.09
0.08
0.07
0.06
0.05
0.04
0.03
0.02
0.01
200
300
400
500
600
700
800

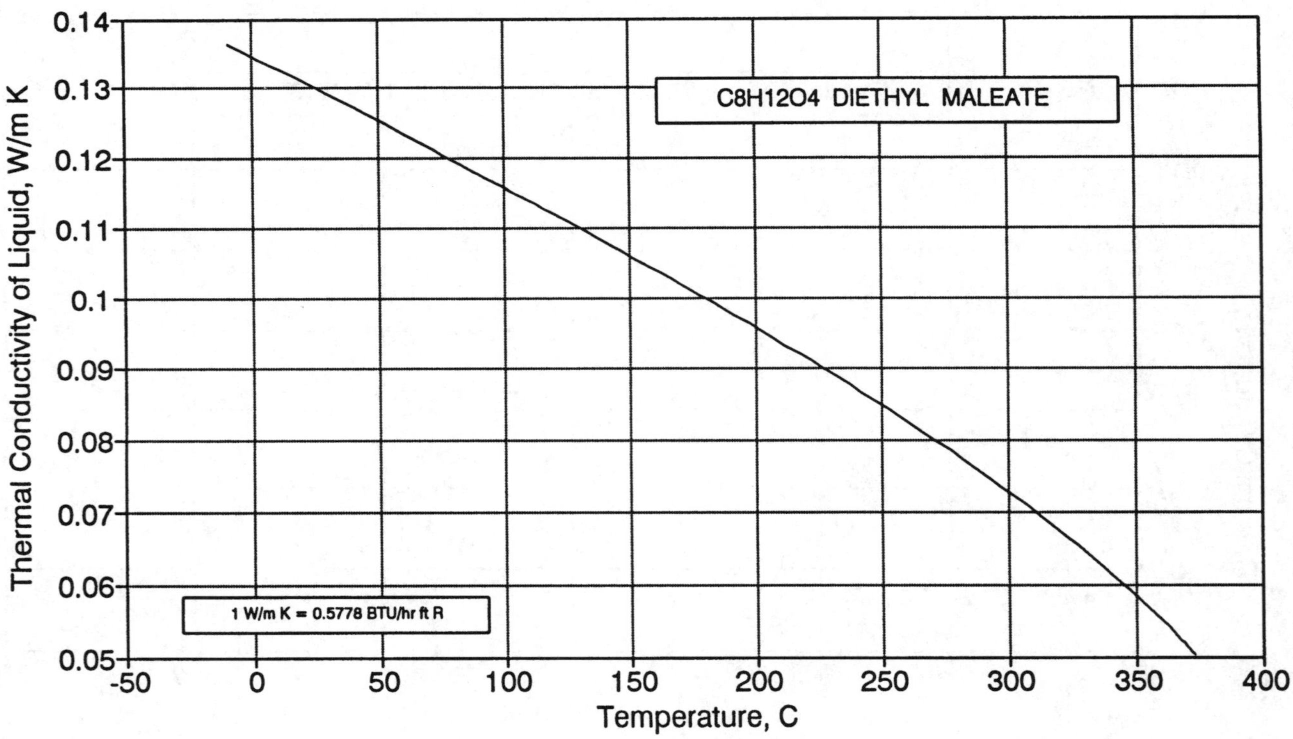
C8H12O4 DIETHYL MALEATE
1 W/m K = 0.5778 BTU/hr ft R
Thermal Conductivity of Liquid, W/m K
Temperature, C
0.14
0.13
0.12
0.11
0.1
0.09
0.08
0.07
0.06
0.05
-50
0
50
100
150
200
250
300
350
400

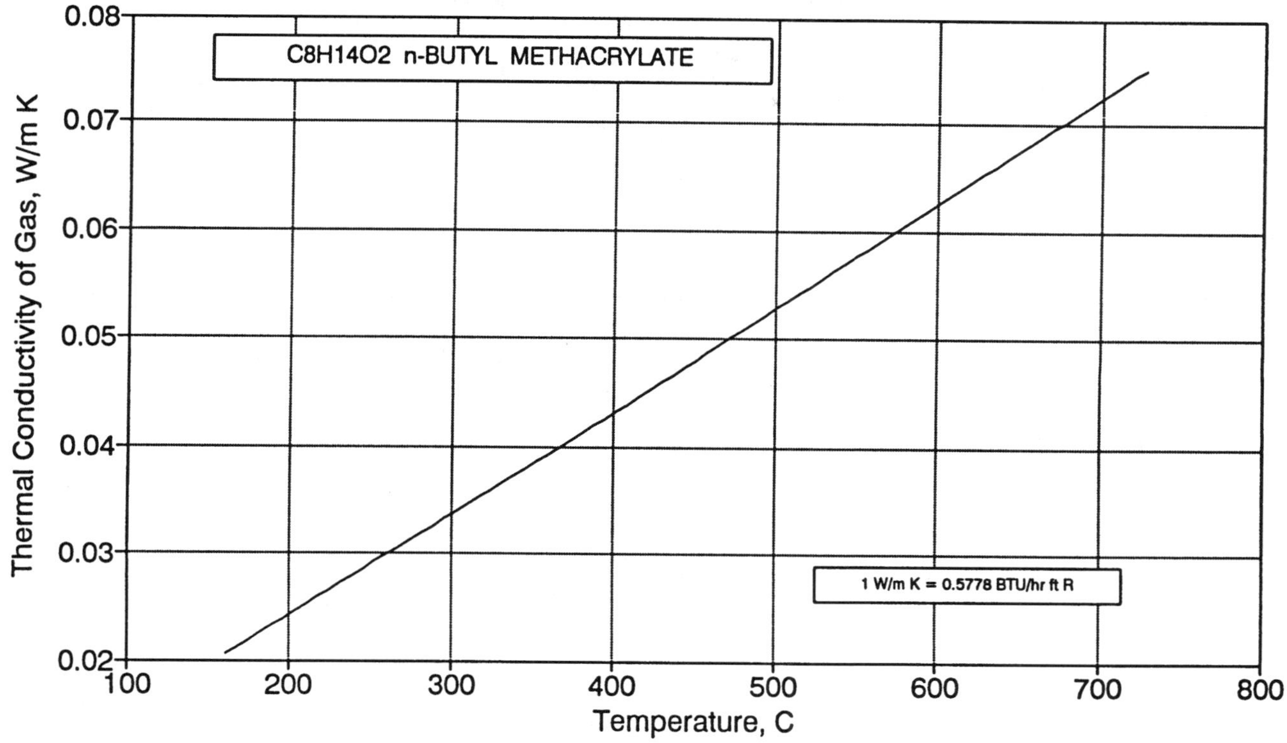
C8H14O2 n-BUTYL METHACRYLATE
1 W/m K = 0.5778 BTU/hr ft R
Thermal Conductivity of Gas, W/m K
Temperature, C
0.08
0.07
0.06
0.05
0.04
0.03
0.02
100
200
300
400
500
600
700
800

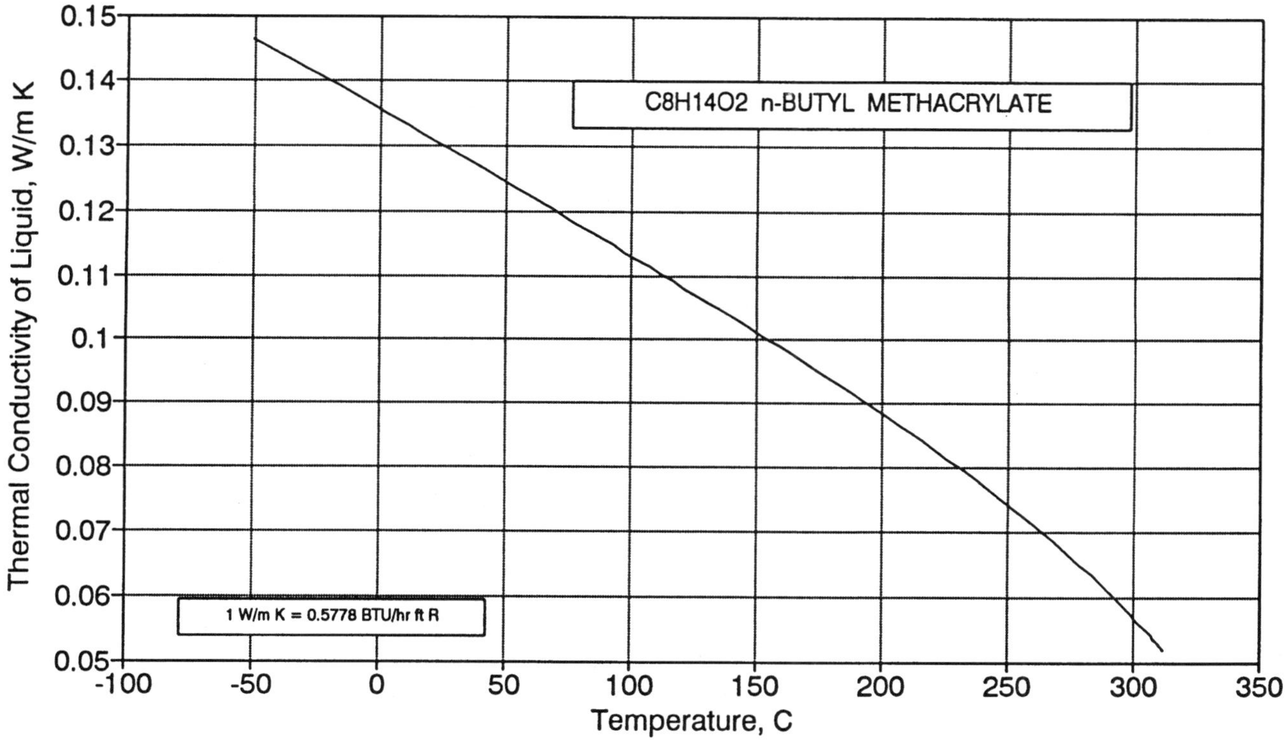
C8H14O2 n-BUTYL METHACRYLATE
1 W/m K = 0.5778 BTU/hr ft R
Thermal Conductivity of Liquid, W/m K
Temperature, C
0.15
0.14
0.13
0.12
0.11
0.1
0.09
0.08
0.07
0.06
0.05
-100
-50
0
50
100
150
200
250
300
350

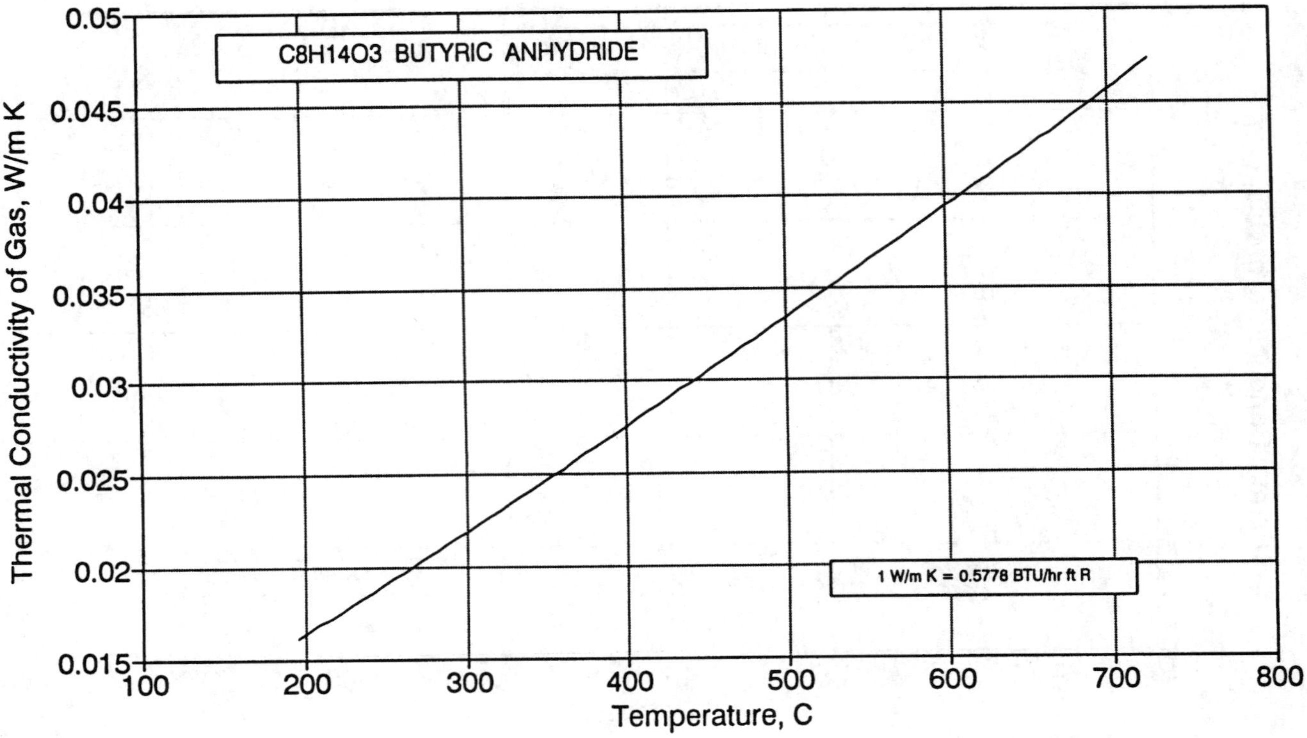
C8H14O3 BUTYRIC ANHYDRIDE
1 W/m K = 0.5778 BTU/hr ft R
Thermal Conductivity of Gas, W/m K
0.05
0.045
0.04
0.035
0.03
0.025
0.02
0.015
100
200
300
400
500
600
700
800
Temperature, C

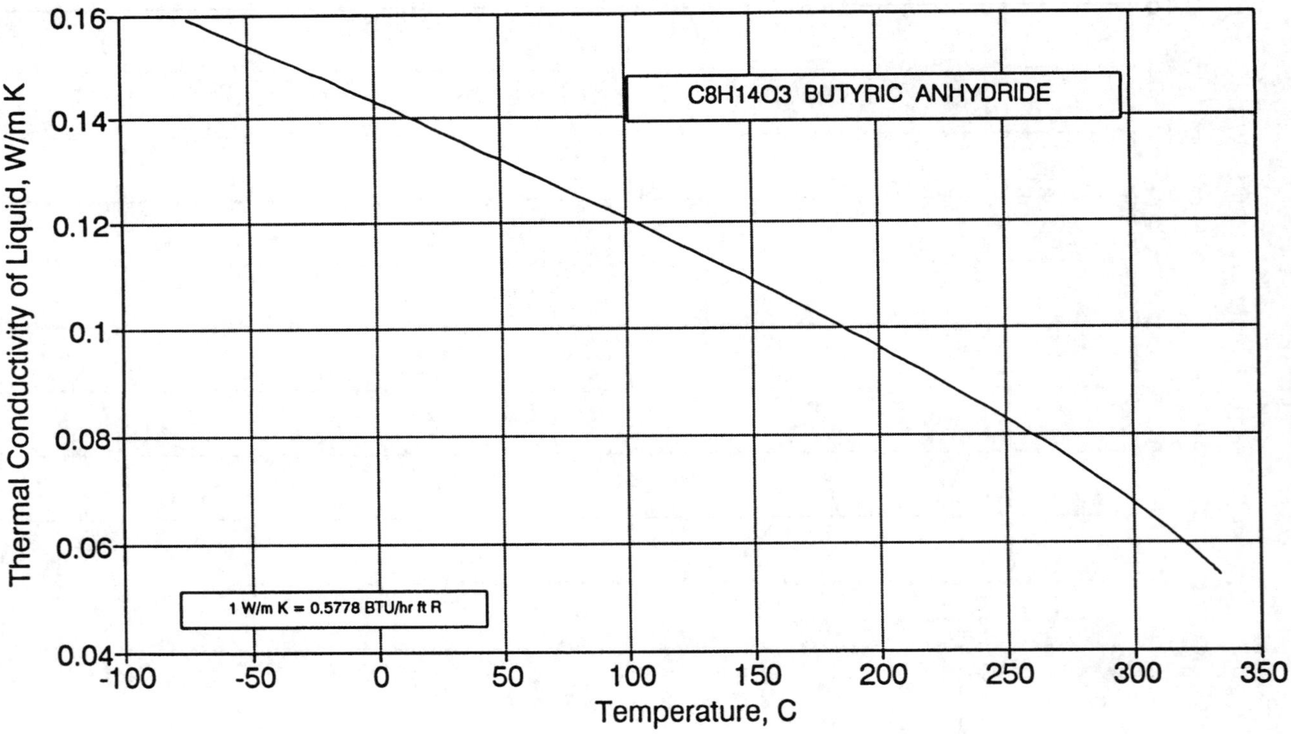
C8H14O3 BUTYRIC ANHYDRIDE
1 W/m K = 0.5778 BTU/hr ft R
Thermal Conductivity of Liquid, W/m K
0.16
0.14
0.12
0.1
0.08
0.06
0.04
-100
-50
0
50
100
150
200
250
300
350
Temperature, C

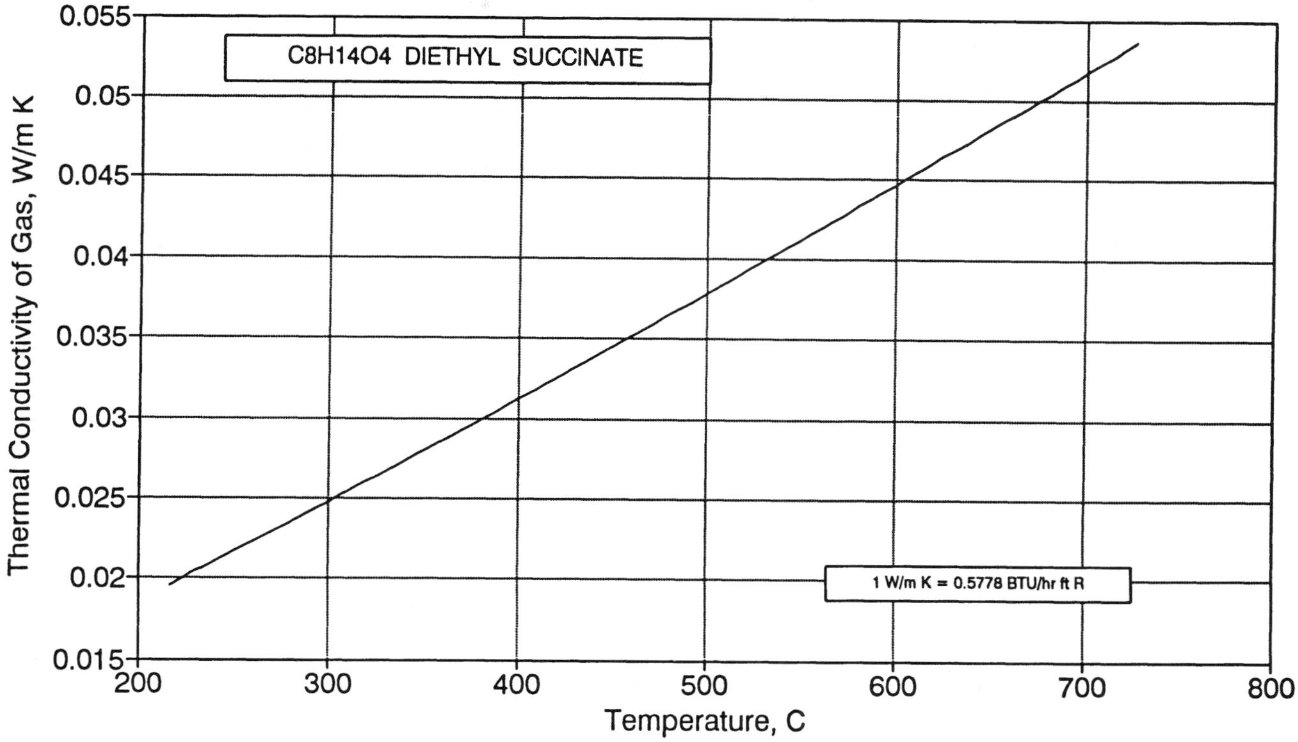
C8H14O4 DIETHYL SUCCINATE
1 W/m K = 0.5778 BTU/hr ft R
Thermal Conductivity of Gas, W/m K
0.055
0.05
0.045
0.04
0.035
0.03
0.025
0.02
0.015
200
300
400
500
600
700
800
Temperature, C

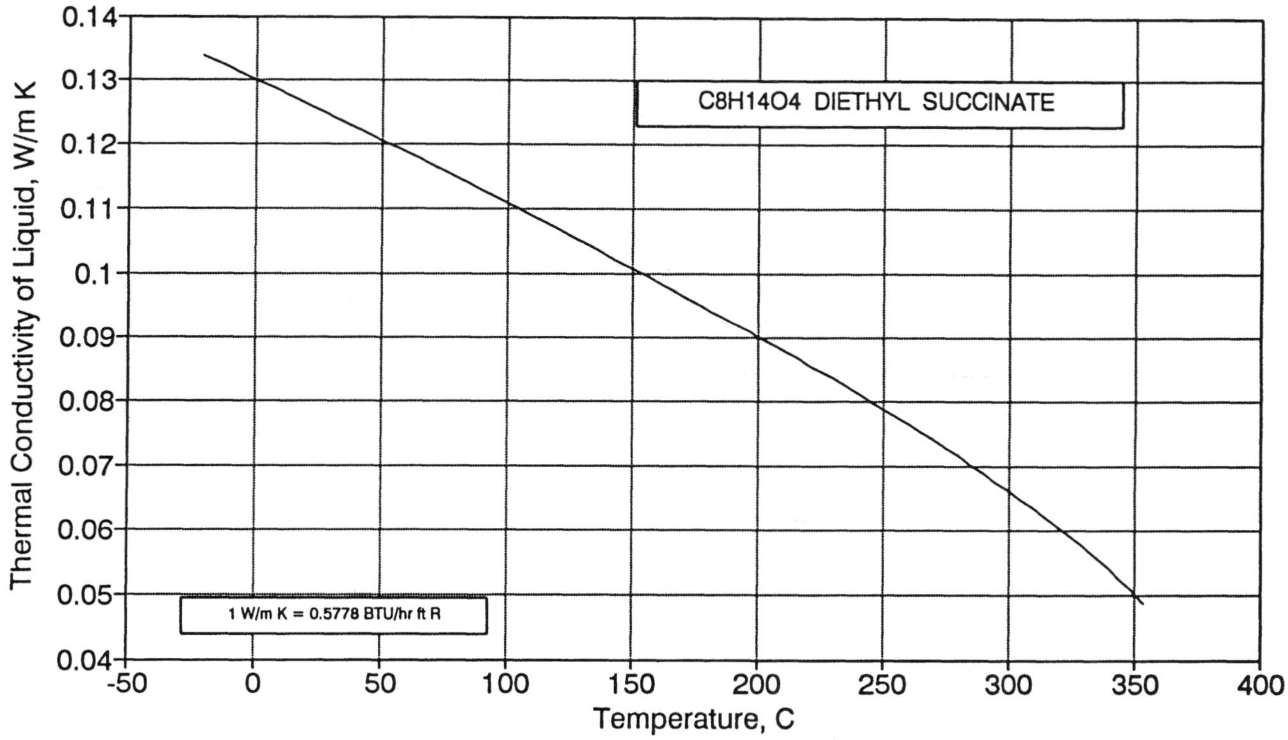
C8H14O4 DIETHYL SUCCINATE
1 W/m K = 0.5778 BTU/hr ft R
Thermal Conductivity of Liquid, W/m K
0.14
0.13
0.12
0.11
0.1
0.09
0.08
0.07
0.06
0.05
0.04
-50
0
50
100
150
200
250
300
350
400
Temperature, C

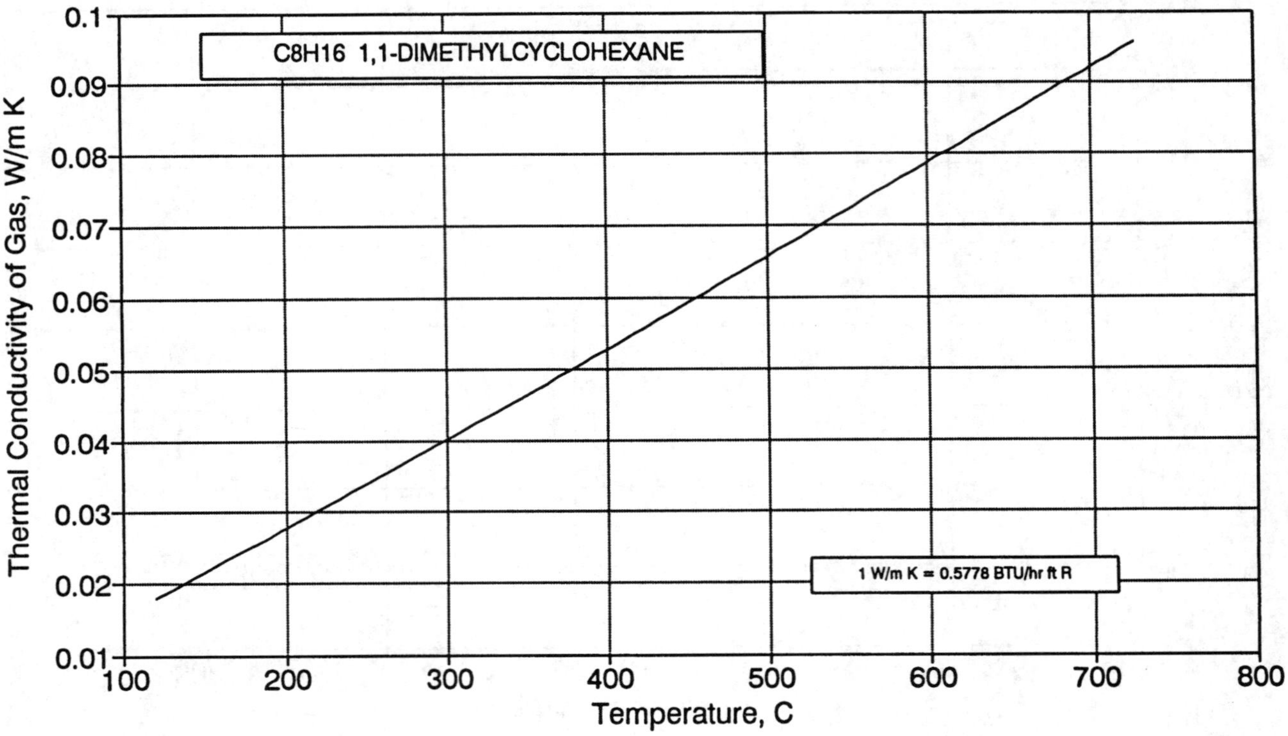
C8H16 1,1-DIMETHYLCYCLOHEXANE
1 W/m K = 0.5778 BTU/hr ft R
Thermal Conductivity of Gas, W/m K
0.1
0.09
0.08
0.07
0.06
0.05
0.04
0.03
0.02
0.01
100
200
300
400
500
600
700
800
Temperature, C

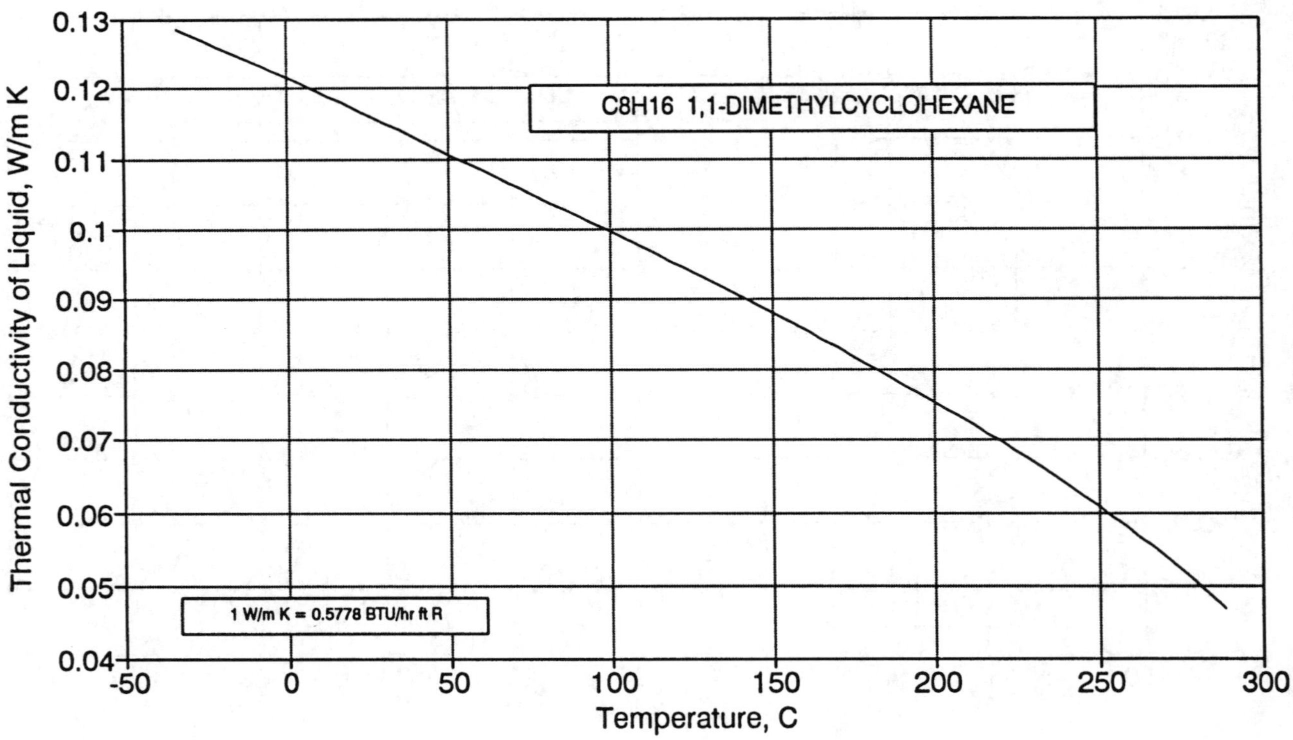
C8H16 1,1-DIMETHYLCYCLOHEXANE
1 W/m K = 0.5778 BTU/hr ft R
Thermal Conductivity of Liquid, W/m K
0.13
0.12
0.11
0.1
0.09
0.08
0.07
0.06
0.05
0.04
-50
0
50
100
150
200
250
300
Temperature, C

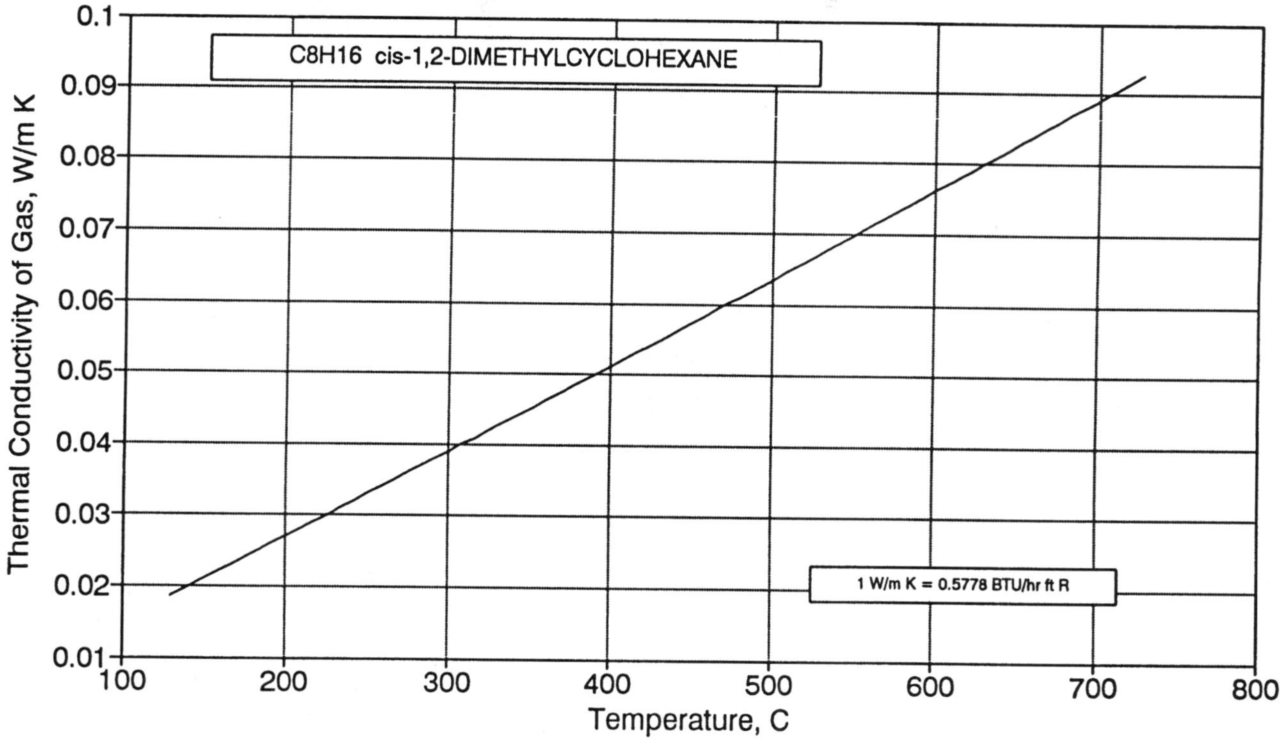
C8H16 cis-1,2-DIMETHYLCYCLOHEXANE
1 W/m K = 0.5778 BTU/hr ft R
Thermal Conductivity of Gas, W/m K
0.1
0.09
0.08
0.07
0.06
0.05
0.04
0.03
0.02
0.01
100
200
300
400
500
600
700
800
Temperature, C

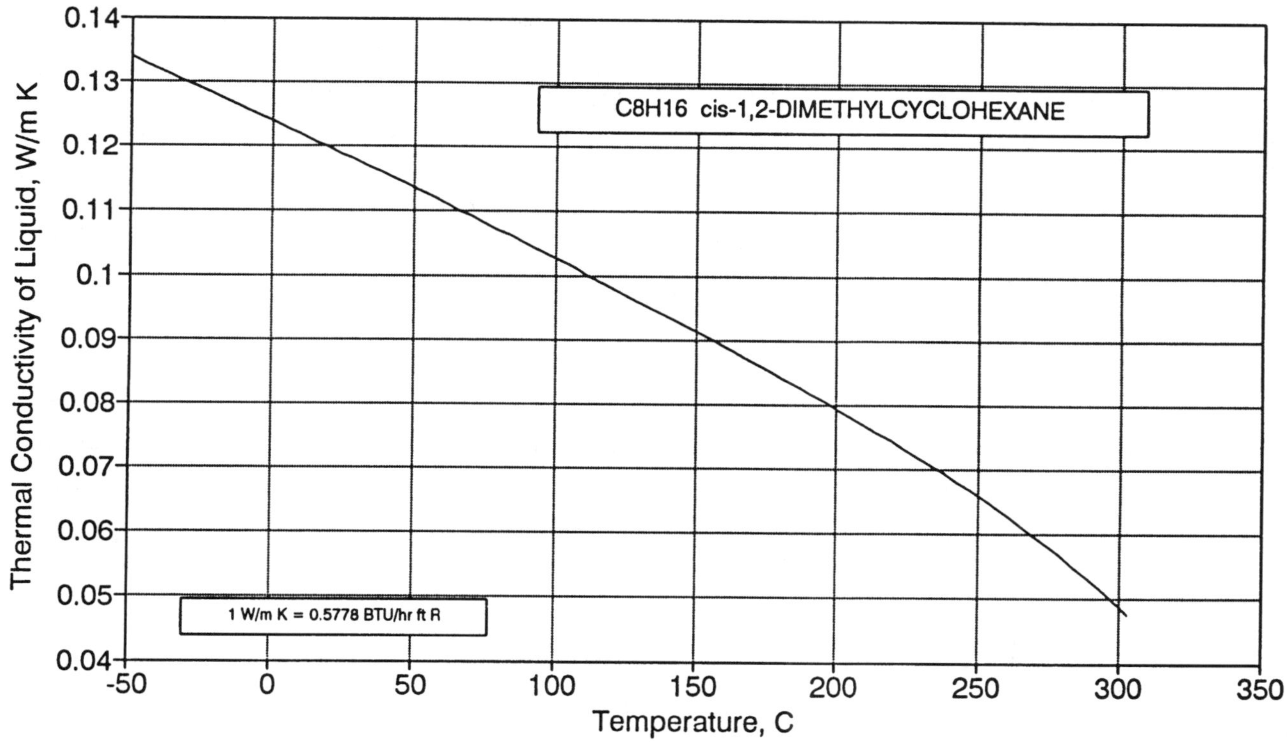
C8H16 cis-1,2-DIMETHYLCYCLOHEXANE
1 W/m K = 0.5778 BTU/hr ft R
Thermal Conductivity of Liquid, W/m K
0.14
0.13
0.12
0.11
0.1
0.09
0.08
0.07
0.06
0.05
0.04
-50
0
50
100
150
200
250
300
350
Temperature, C

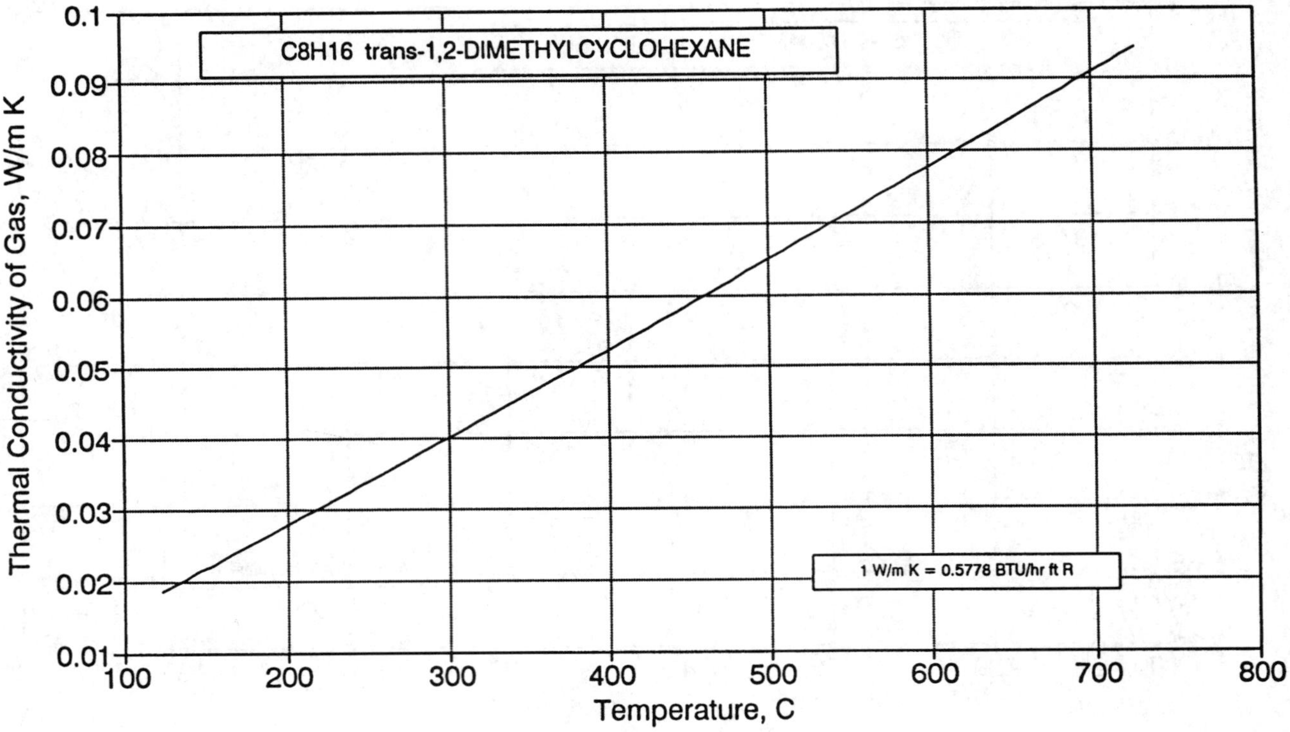
C8H16 trans-1,2-DIMETHYLCYCLOHEXANE
1 W/m K = 0.5778 BTU/hr ft R
Thermal Conductivity of Gas, W/m K
0.1
0.09
0.08
0.07
0.06
0.05
0.04
0.03
0.02
0.01
100
200
300
400
500
600
700
800
Temperature, C

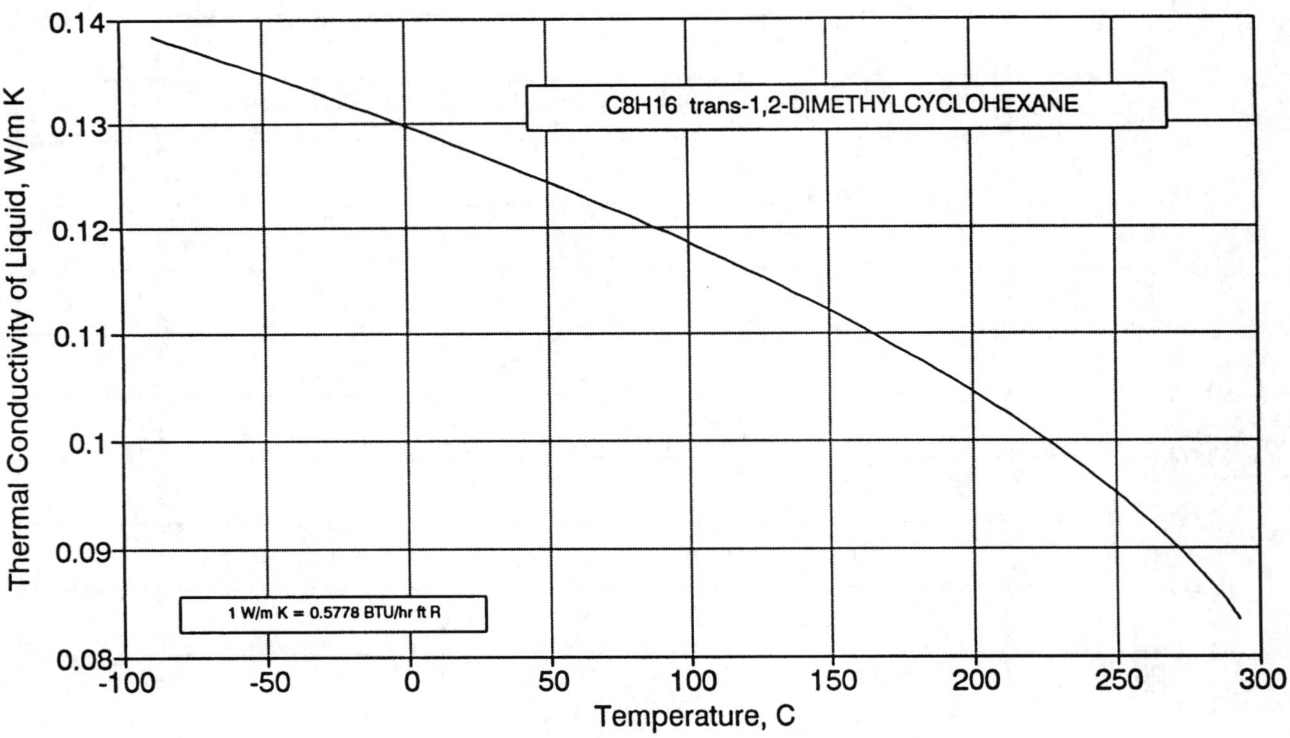
C8H16 trans-1,2-DIMETHYLCYCLOHEXANE
1 W/m K = 0.5778 BTU/hr ft R
Thermal Conductivity of Liquid, W/m K
0.14
0.13
0.12
0.11
0.1
0.09
0.08
-100
-50
0
50
100
150
200
250
300
Temperature, C

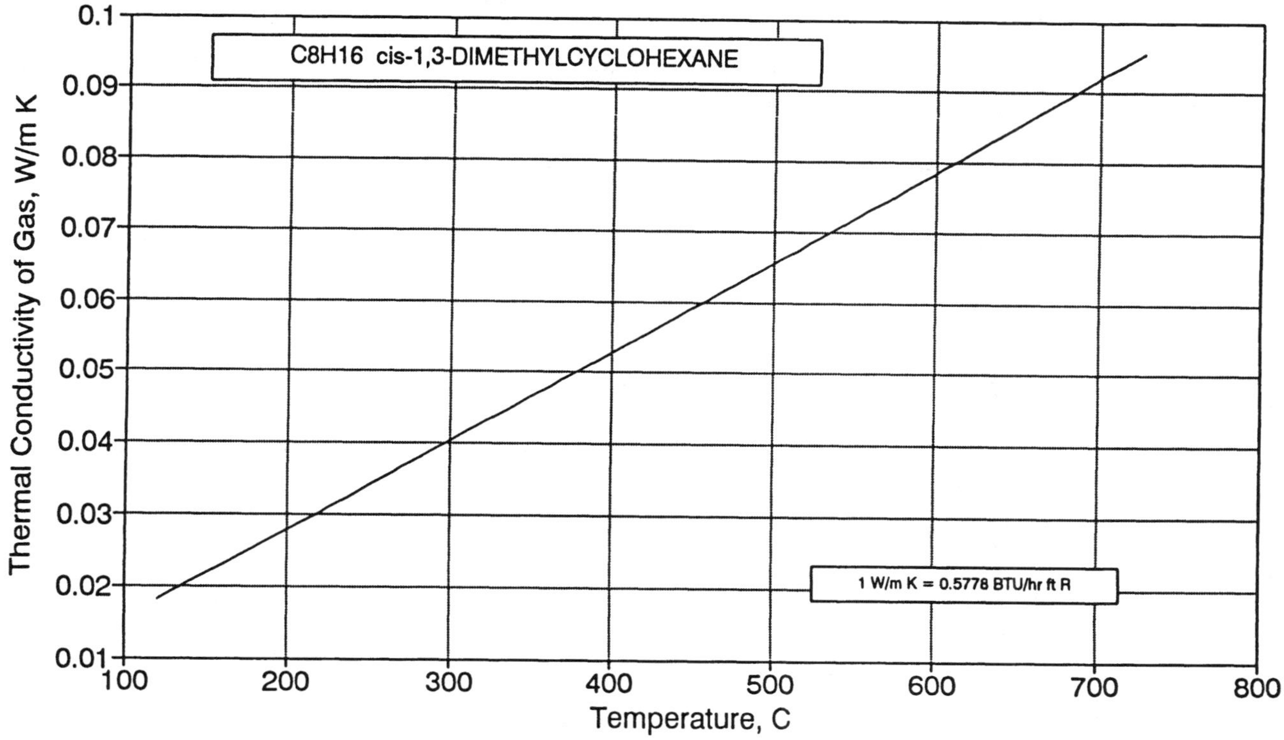

C8H16 cis-1,3-DIMETHYLCYCLOHEXANE
1 W/m K = 0.5778 BTU/hr ft R
Thermal Conductivity of Gas, W/m K
Temperature, C
0.1
0.09
0.08
0.07
0.06
0.05
0.04
0.03
0.02
0.01
100
200
300
400
500
600
700
800

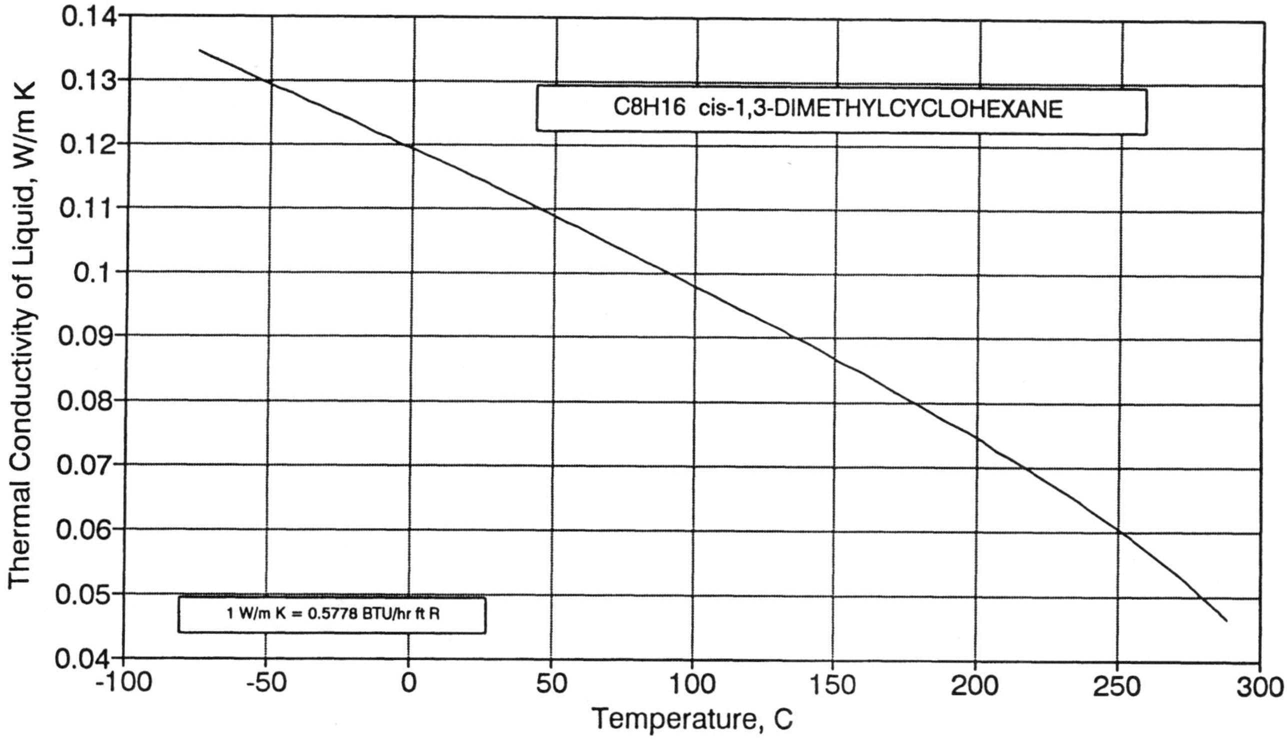

C8H16 cis-1,3-DIMETHYLCYCLOHEXANE
1 W/m K = 0.5778 BTU/hr ft R
Thermal Conductivity of Liquid, W/m K
Temperature, C
0.14
0.13
0.12
0.11
0.1
0.09
0.08
0.07
0.06
0.05
0.04
-100
-50
0
50
100
150
200
250
300

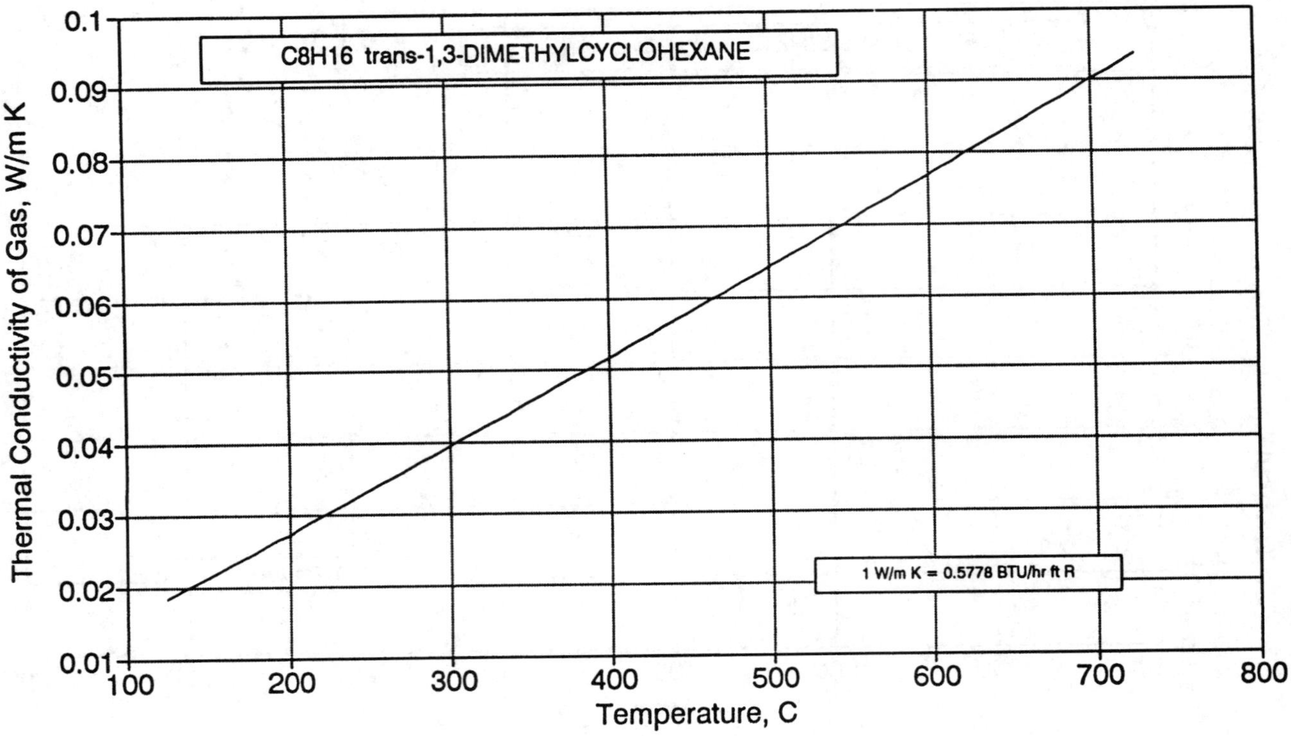
C8H16 trans-1,3-DIMETHYLCYCLOHEXANE
1 W/m K = 0.5778 BTU/hr ft R
Thermal Conductivity of Gas, W/m K
Temperature, C
0.1
0.09
0.08
0.07
0.06
0.05
0.04
0.03
0.02
0.01
100
200
300
400
500
600
700
800

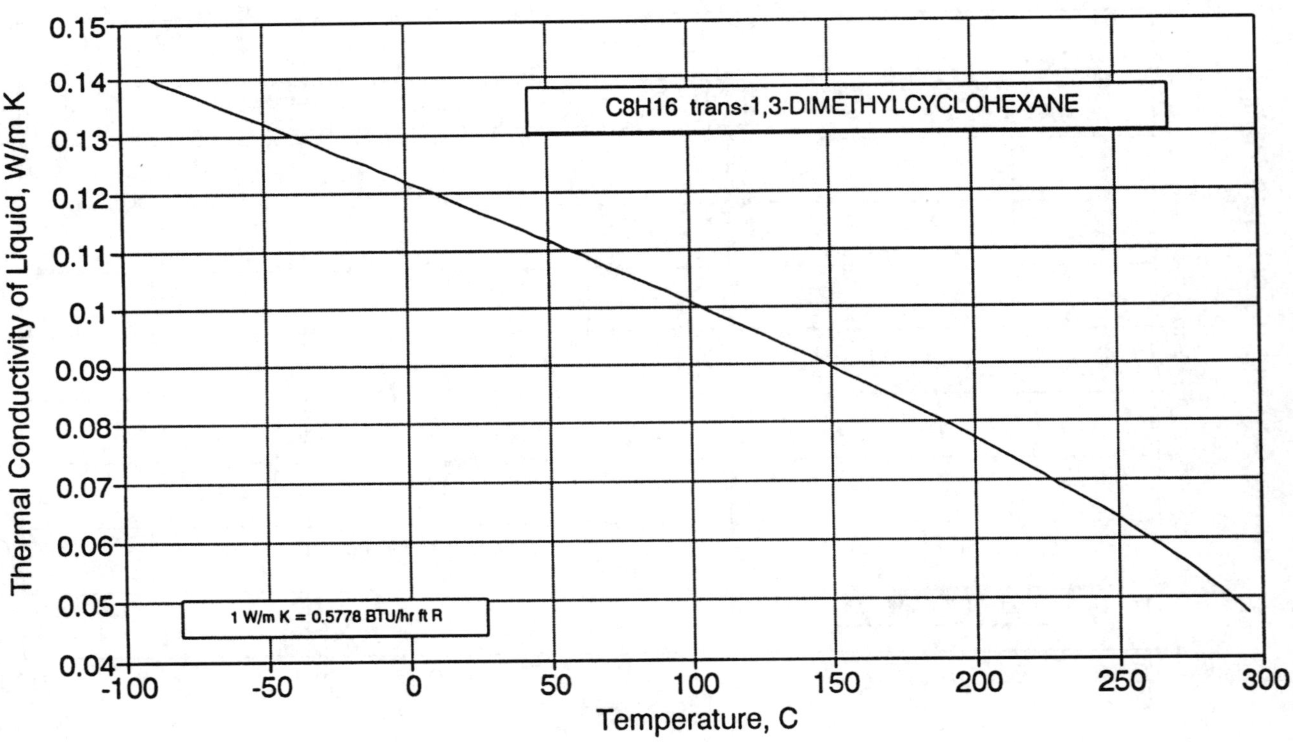
C8H16 trans-1,3-DIMETHYLCYCLOHEXANE
1 W/m K = 0.5778 BTU/hr ft R
Thermal Conductivity of Liquid, W/m K
Temperature, C
0.15
0.14
0.13
0.12
0.11
0.1
0.09
0.08
0.07
0.06
0.05
0.04
-100
-50
0
50
100
150
200
250
300

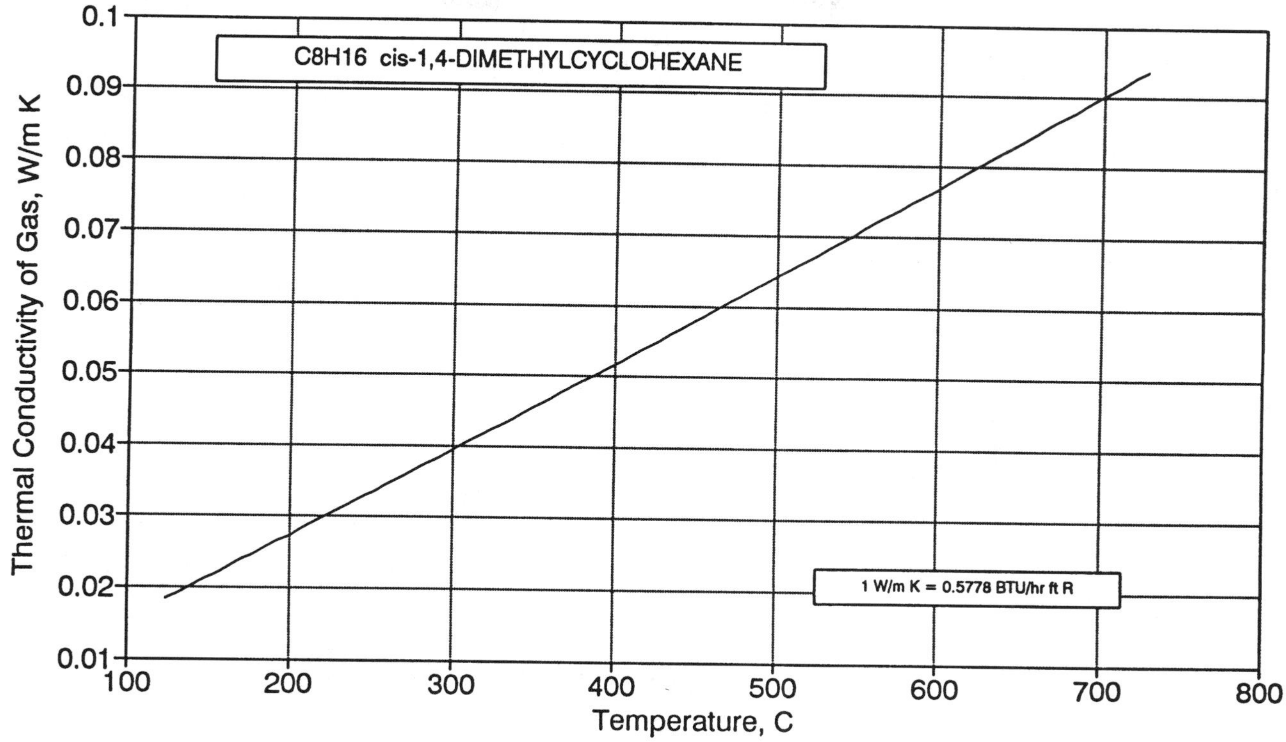
C8H16 cis-1,4-DIMETHYLCYCLOHEXANE
1 W/m K = 0.5778 BTU/hr ft R
Thermal Conductivity of Gas, W/m K
0.1
0.09
0.08
0.07
0.06
0.05
0.04
0.03
0.02
0.01
100
200
300
400
500
600
700
800
Temperature, C

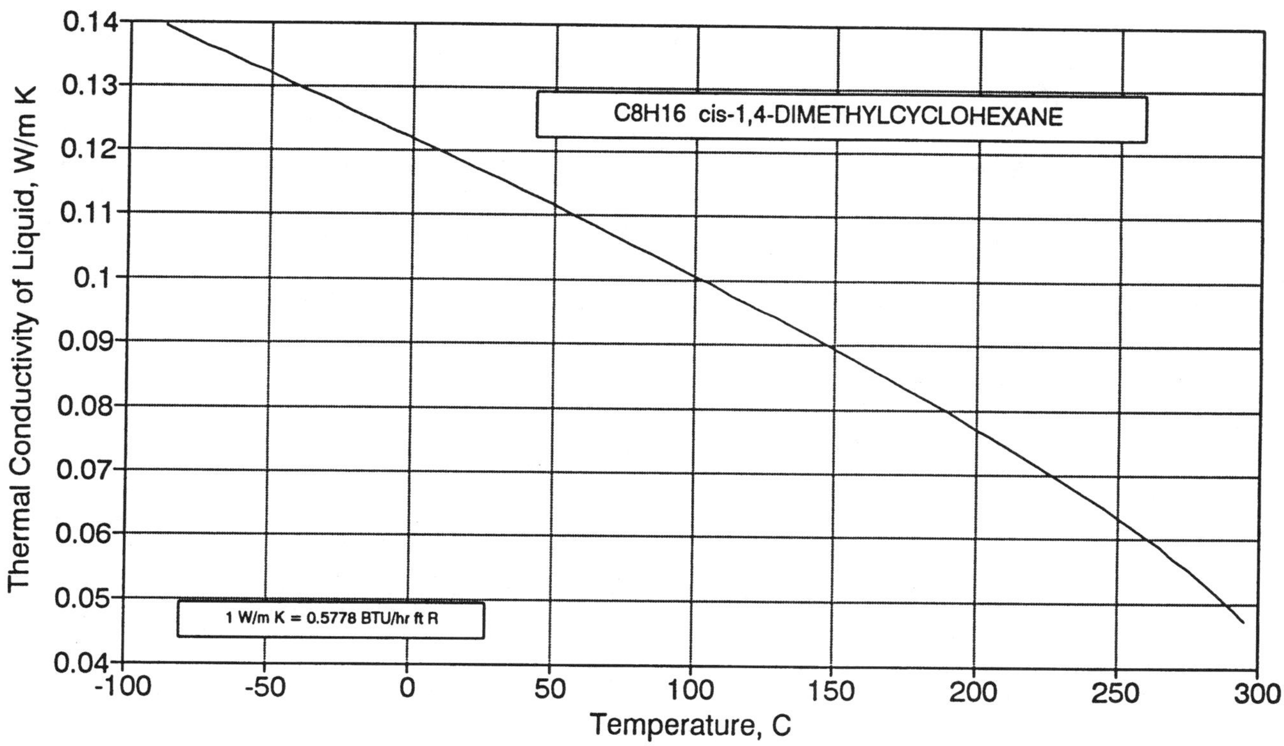
C8H16 cis-1,4-DIMETHYLCYCLOHEXANE
1 W/m K = 0.5778 BTU/hr ft R
Thermal Conductivity of Liquid, W/m K
0.14
0.13
0.12
0.11
0.1
0.09
0.08
0.07
0.06
0.05
0.04
-100
-50
0
50
100
150
200
250
300
Temperature, C

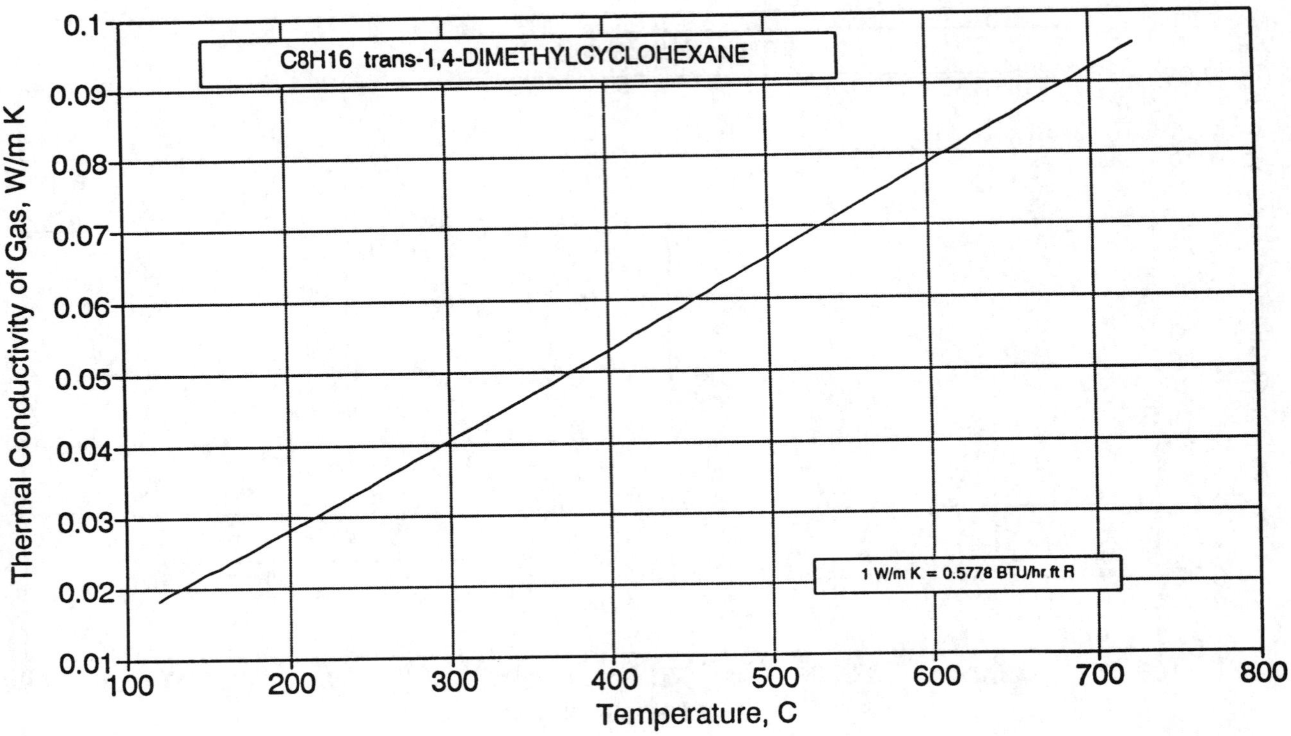
C8H16 trans-1,4-DIMETHYLCYCLOHEXANE
Thermal Conductivity of Gas, W/m K
0.1
0.09
0.08
0.07
0.06
0.05
0.04
0.03
0.02
0.01
1 W/m K = 0.5778 BTU/hr ft R
100
200
300
400
500
600
700
800
Temperature, C

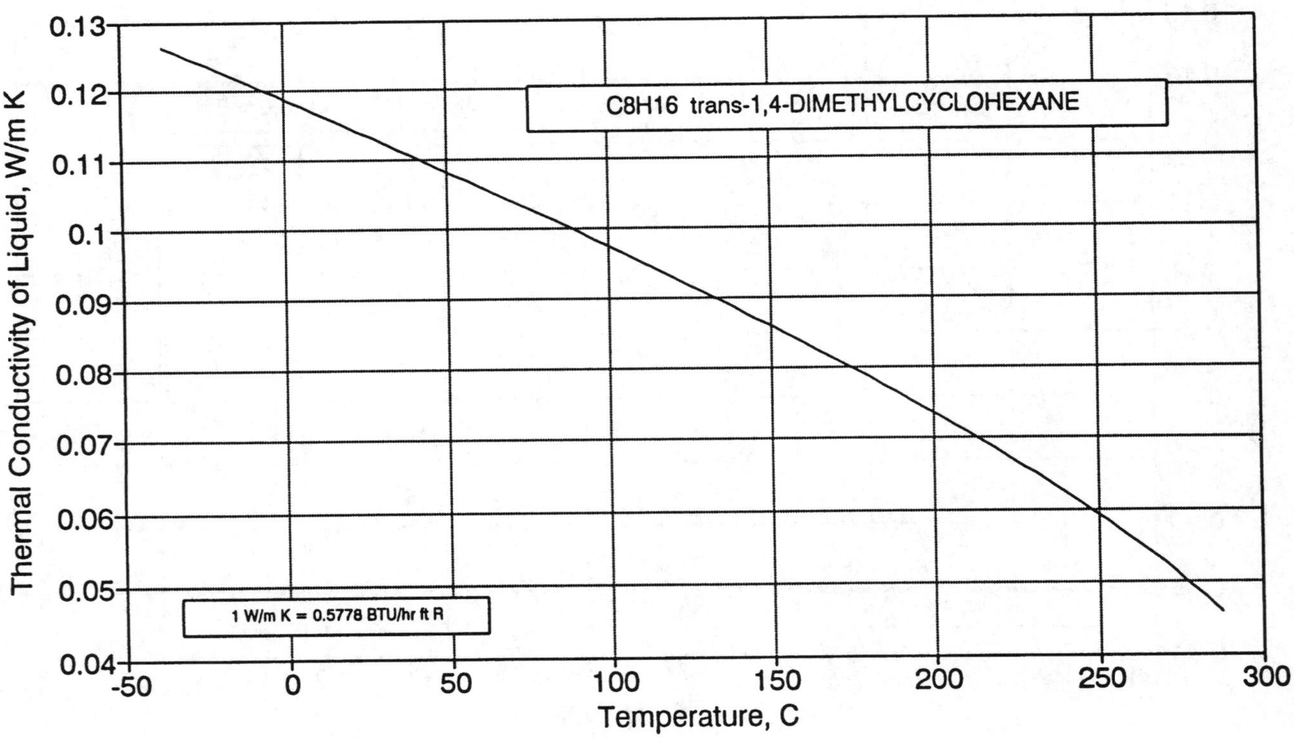
C8H16 trans-1,4-DIMETHYLCYCLOHEXANE
Thermal Conductivity of Liquid, W/m K
0.13
0.12
0.11
0.1
0.09
0.08
0.07
0.06
0.05
0.04
1 W/m K = 0.5778 BTU/hr ft R
-50
0
50
100
150
200
250
300
Temperature, C

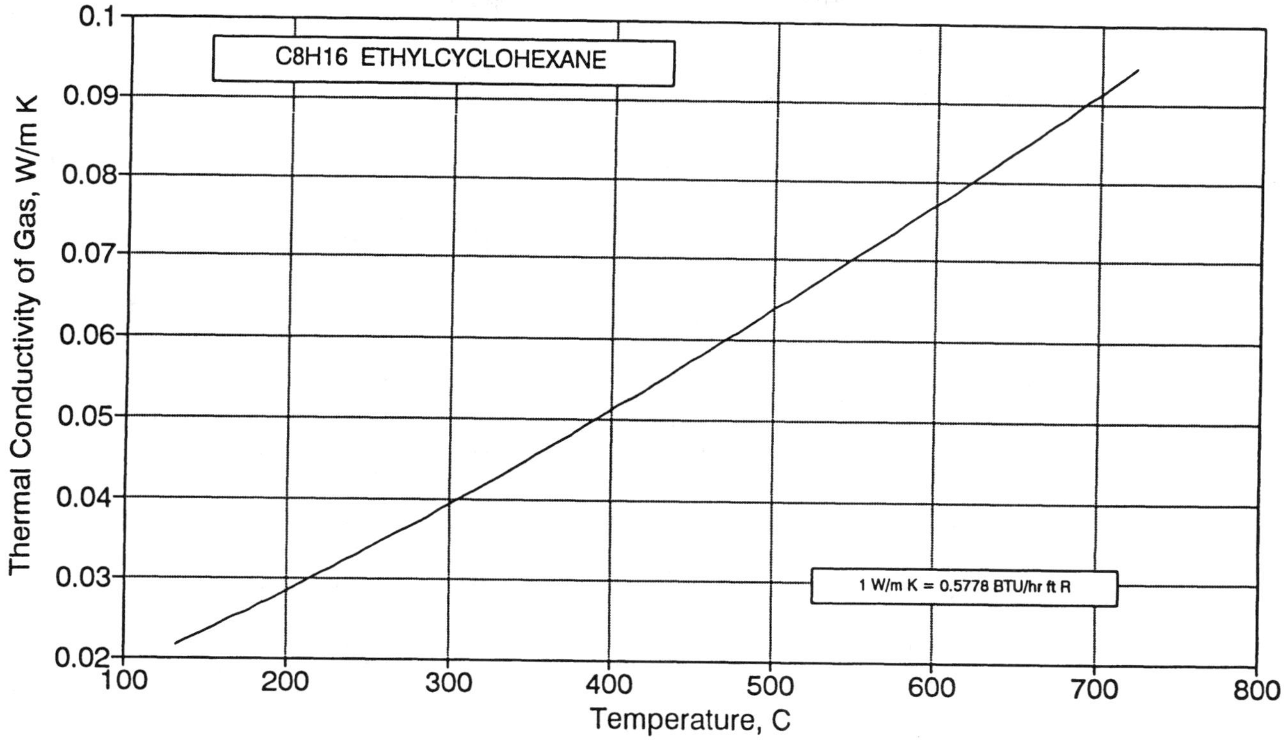

C8H16 ETHYLCYCLOHEXANE
1 W/m K = 0.5778 BTU/hr ft R
Thermal Conductivity of Gas, W/m K
0.1
0.09
0.08
0.07
0.06
0.05
0.04
0.03
0.02
100
200
300
400
500
600
700
800
Temperature, C

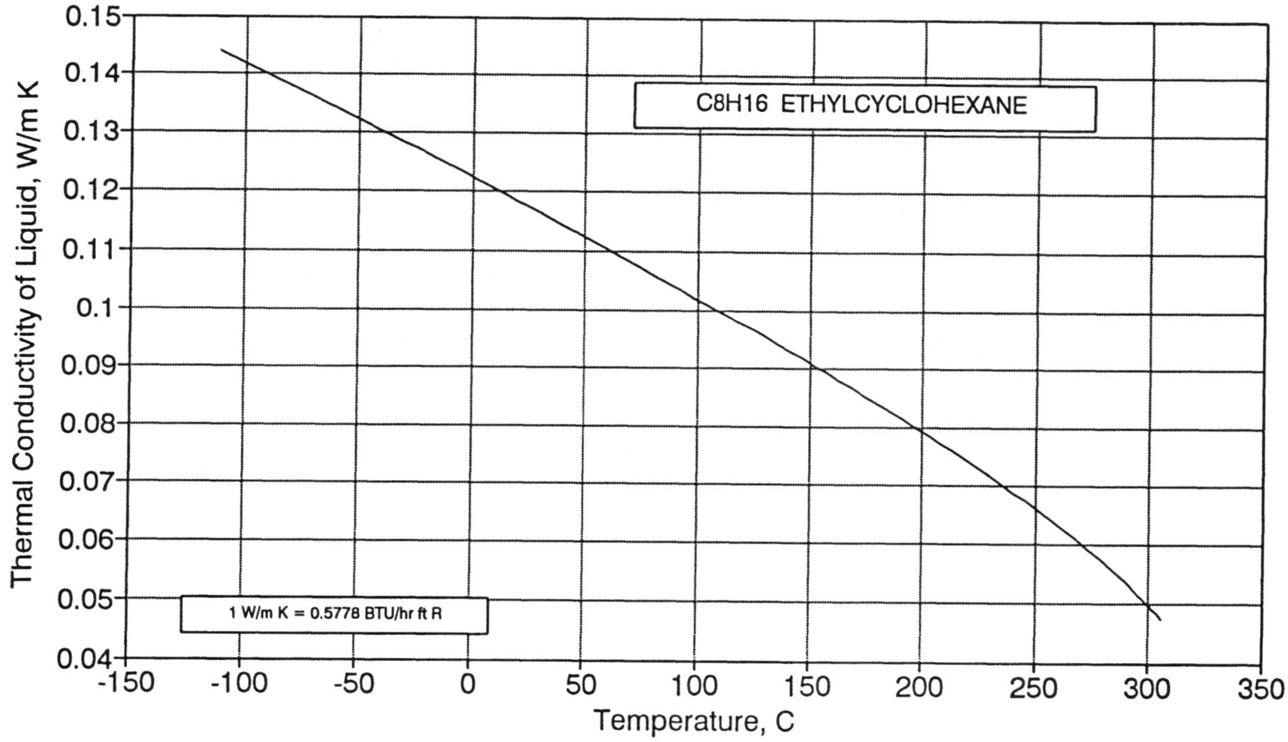

C8H16 ETHYLCYCLOHEXANE
1 W/m K = 0.5778 BTU/hr ft R
Thermal Conductivity of Liquid, W/m K
0.15
0.14
0.13
0.12
0.11
0.1
0.09
0.08
0.07
0.06
0.05
0.04
-150
-100
-50
0
50
100
150
200
250
300
350
Temperature, C

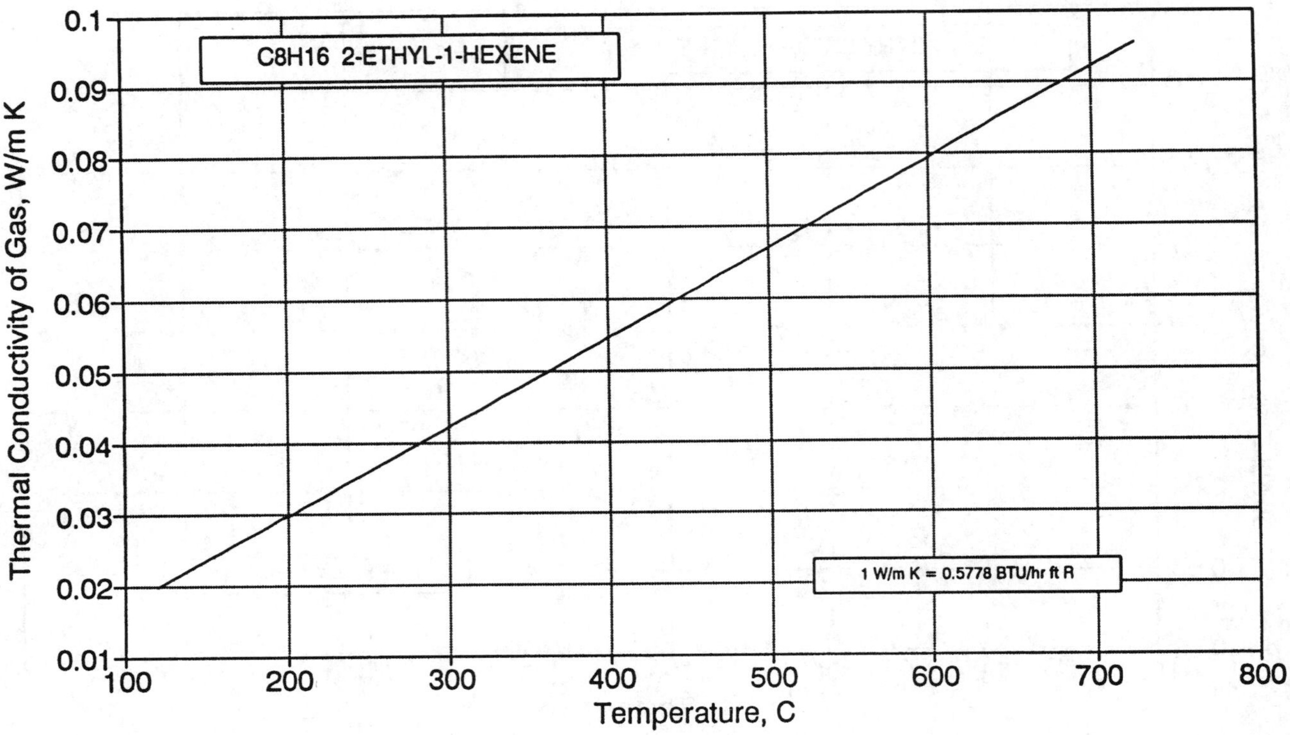
C8H16 2-ETHYL-1-HEXENE
1 W/m K = 0.5778 BTU/hr ft R
Thermal Conductivity of Gas, W/m K
Temperature, C
0.1
0.09
0.08
0.07
0.06
0.05
0.04
0.03
0.02
0.01
100
200
300
400
500
600
700
800

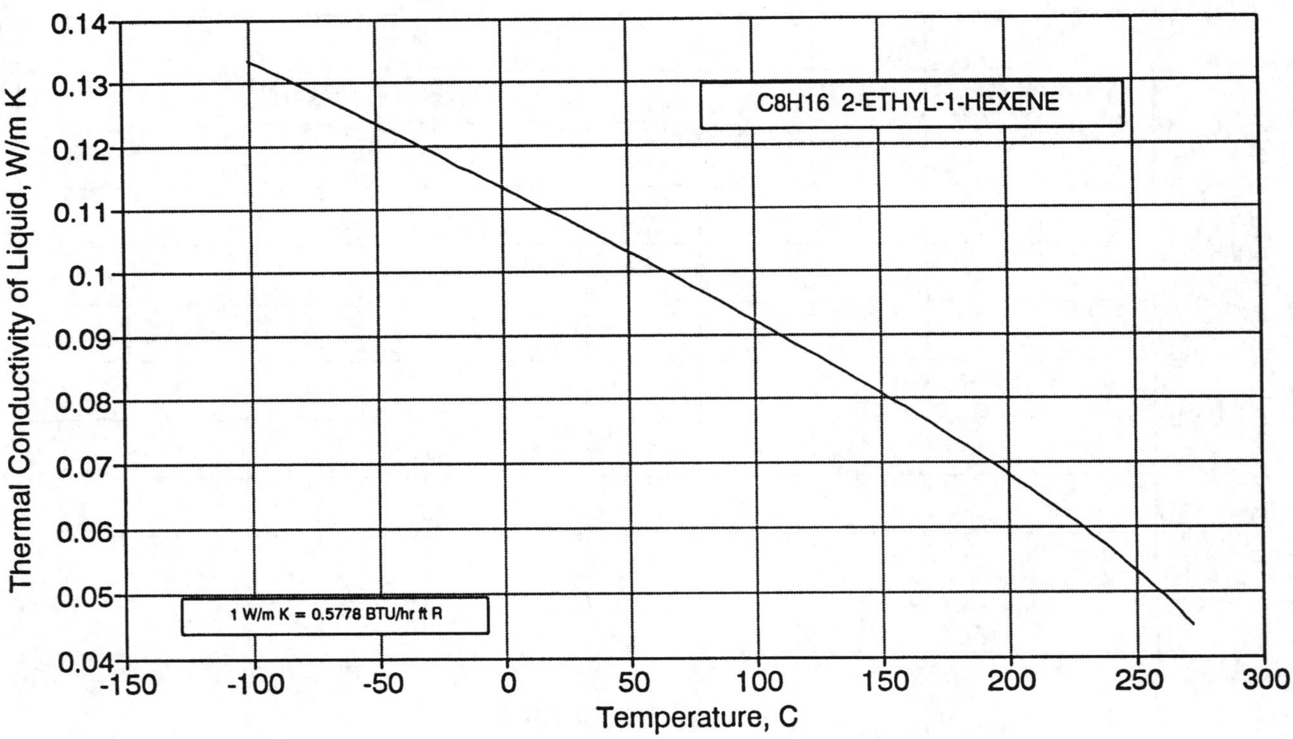
C8H16 2-ETHYL-1-HEXENE
1 W/m K = 0.5778 BTU/hr ft R
Thermal Conductivity of Liquid, W/m K
Temperature, C
0.14
0.13
0.12
0.11
0.1
0.09
0.08
0.07
0.06
0.05
0.04
-150
-100
-50
0
50
100
150
200
250
300

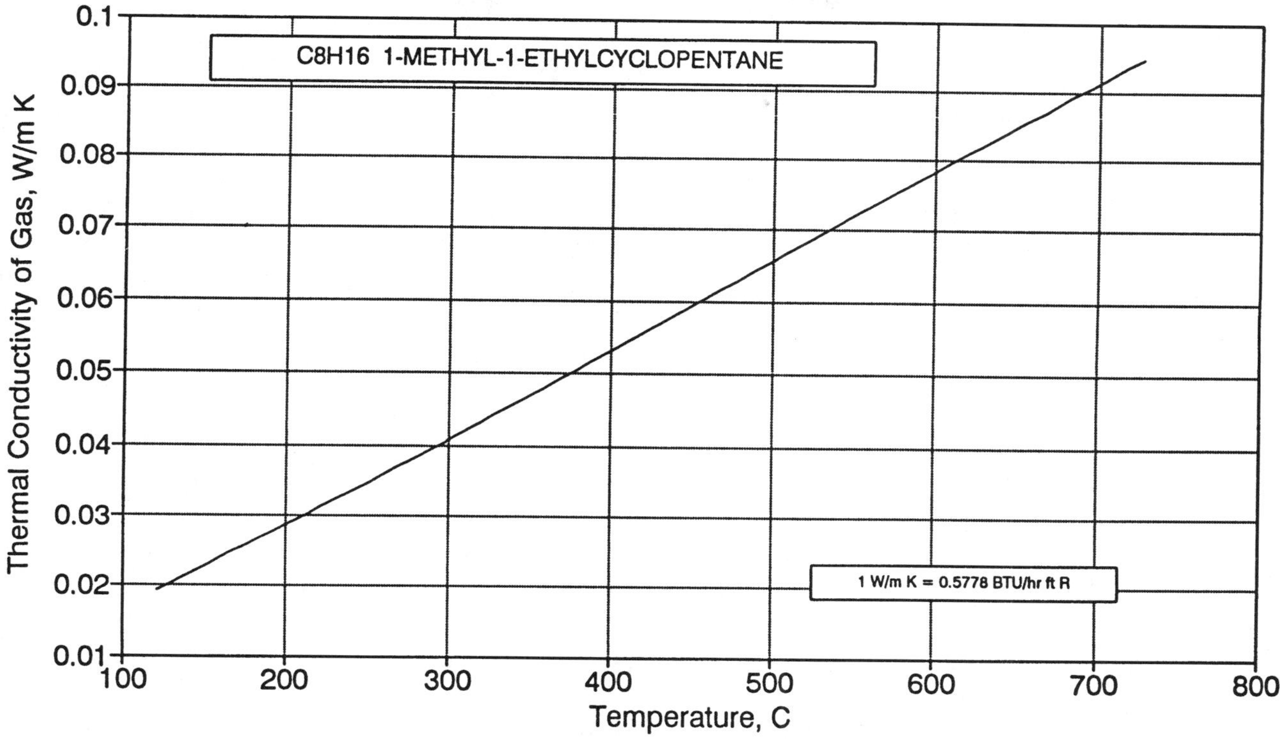

C8H16 1-METHYL-1-ETHYLCYCLOPENTANE
1 W/m K = 0.5778 BTU/hr ft R
Thermal Conductivity of Gas, W/m K
0.1
0.09
0.08
0.07
0.06
0.05
0.04
0.03
0.02
0.01
100
200
300
400
500
600
700
800
Temperature, C

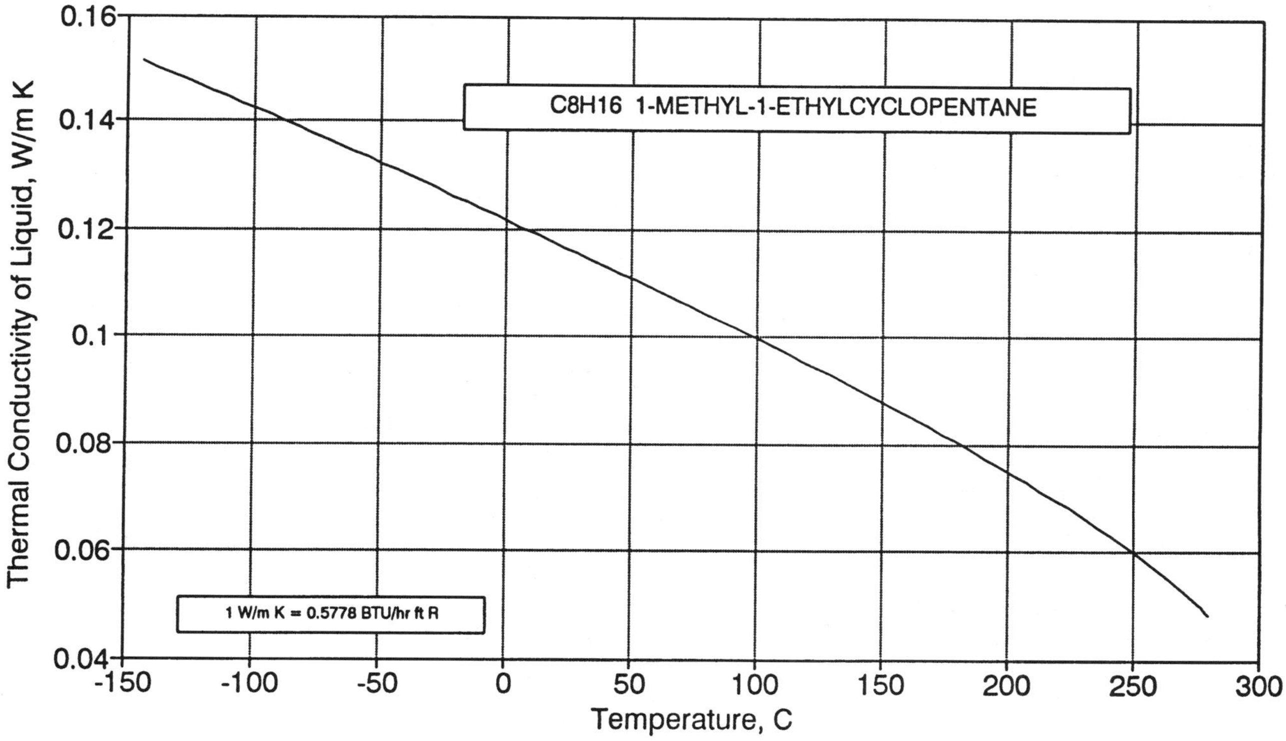

C8H16 1-METHYL-1-ETHYLCYCLOPENTANE
1 W/m K = 0.5778 BTU/hr ft R
Thermal Conductivity of Liquid, W/m K
0.16
0.14
0.12
0.1
0.08
0.06
0.04
-150
-100
-50
0
50
100
150
200
250
300
Temperature, C

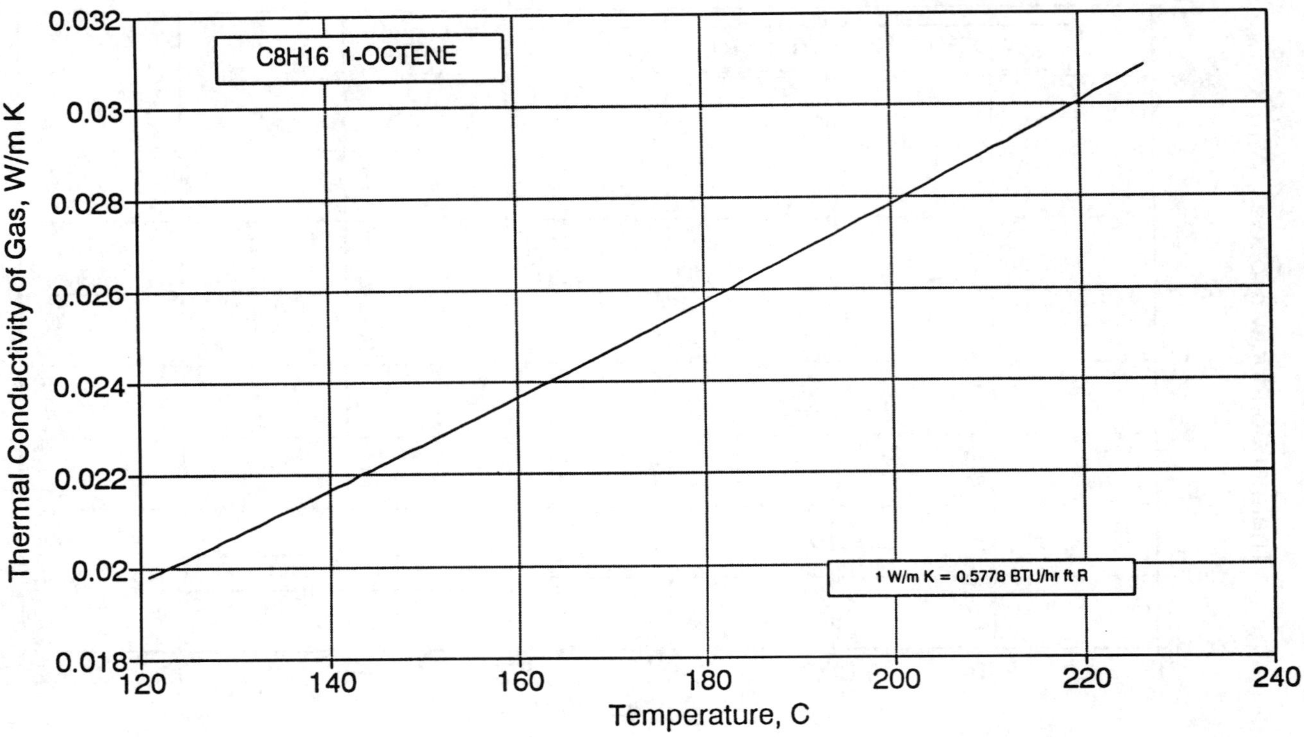

C8H16 1-OCTENE
1 W/m K = 0.5778 BTU/hr ft R
Thermal Conductivity of Gas, W/m K
Temperature, C
0.032
0.03
0.028
0.026
0.024
0.022
0.02
0.018
120
140
160
180
200
220
240

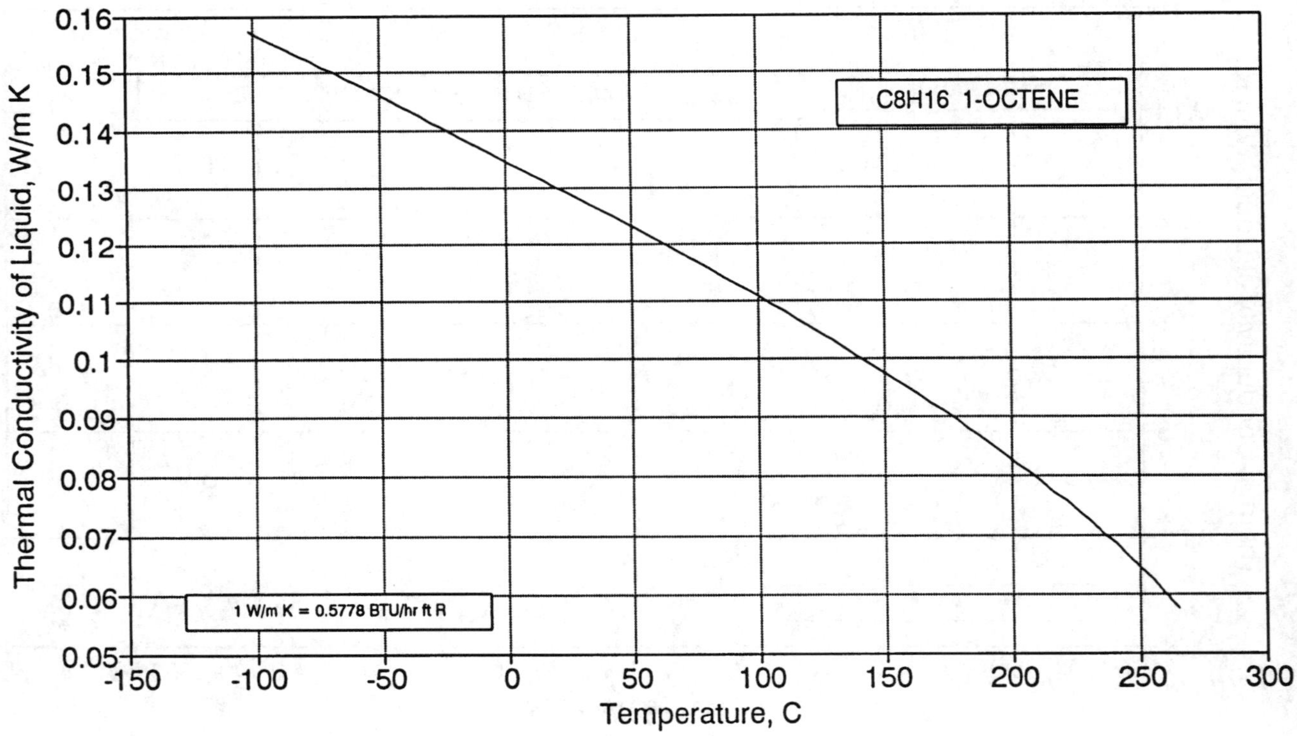

C8H16 1-OCTENE
1 W/m K = 0.5778 BTU/hr ft R
Thermal Conductivity of Liquid, W/m K
Temperature, C
0.16
0.15
0.14
0.13
0.12
0.11
0.1
0.09
0.08
0.07
0.06
0.05
-150
-100
-50
0
50
100
150
200
250
300

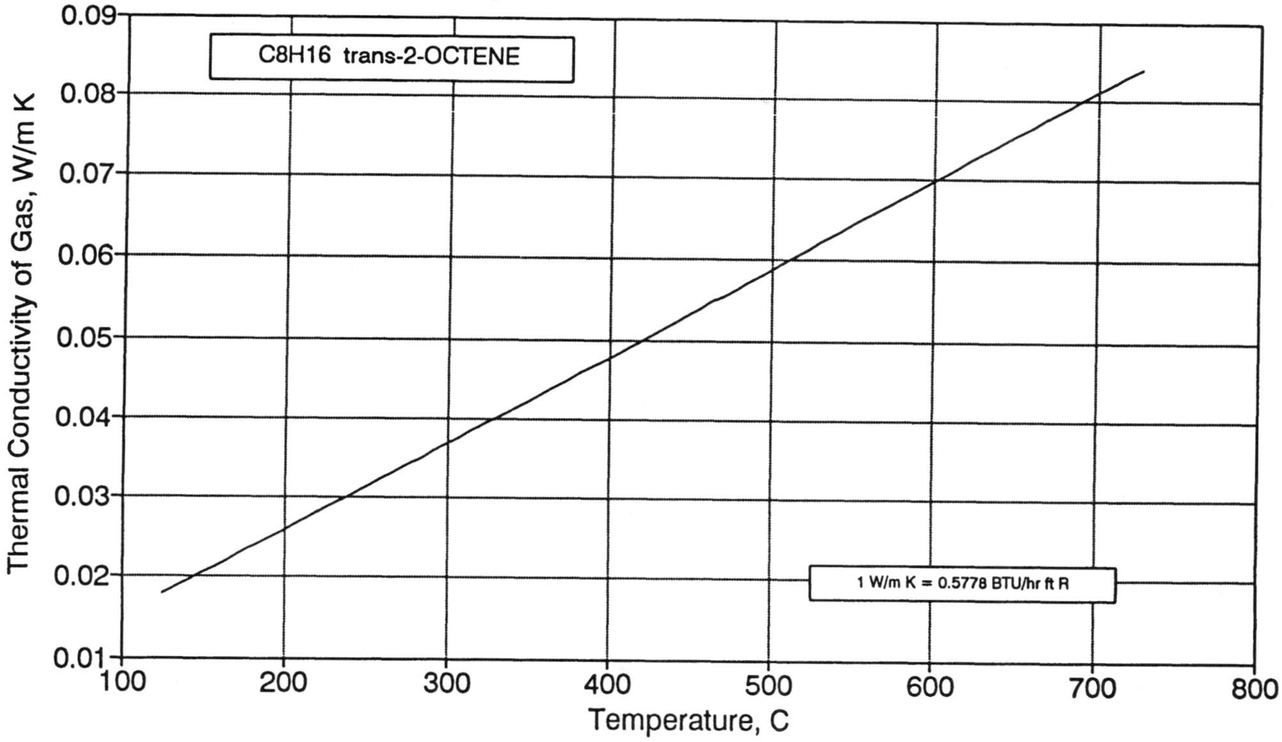
C8H16 trans-2-OCTENE
1 W/m K = 0.5778 BTU/hr ft R
Thermal Conductivity of Gas, W/m K
0.09
0.08
0.07
0.06
0.05
0.04
0.03
0.02
0.01
100
200
300
400
500
600
700
800
Temperature, C

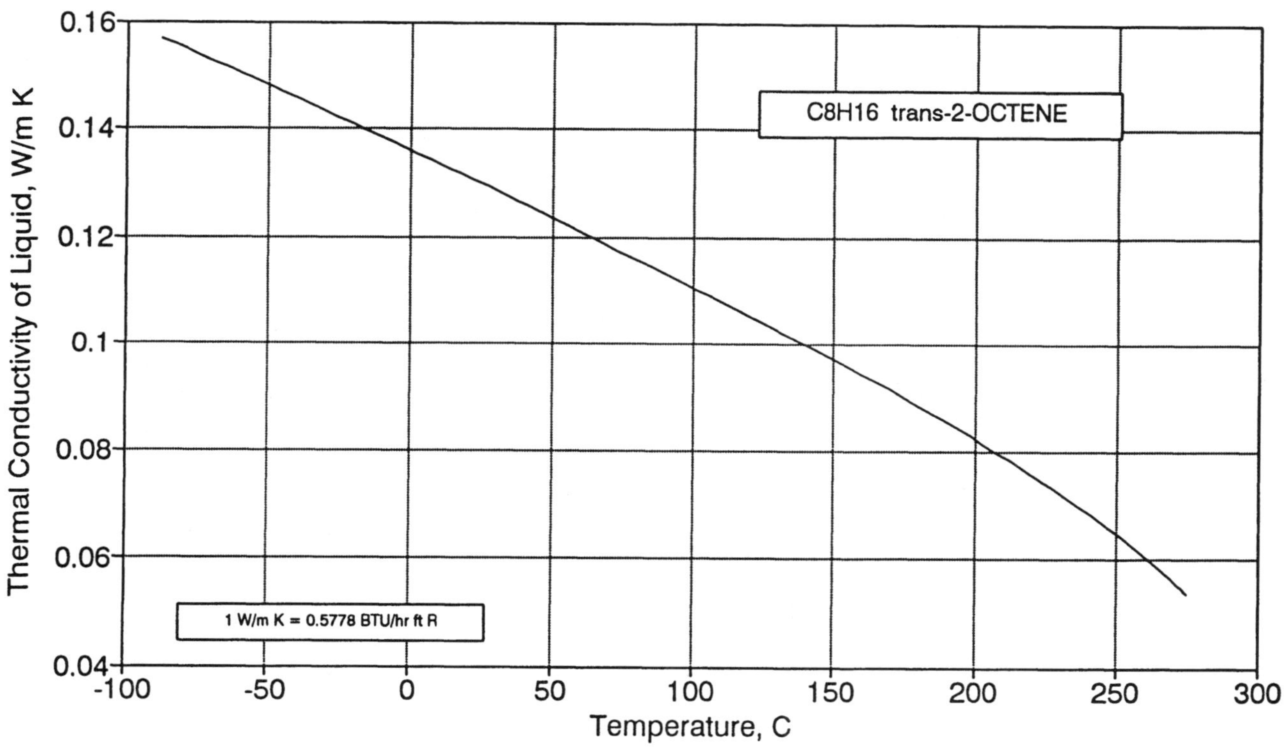
C8H16 trans-2-OCTENE
1 W/m K = 0.5778 BTU/hr ft R
Thermal Conductivity of Liquid, W/m K
0.16
0.14
0.12
0.1
0.08
0.06
0.04
-100
-50
0
50
100
150
200
250
300
Temperature, C

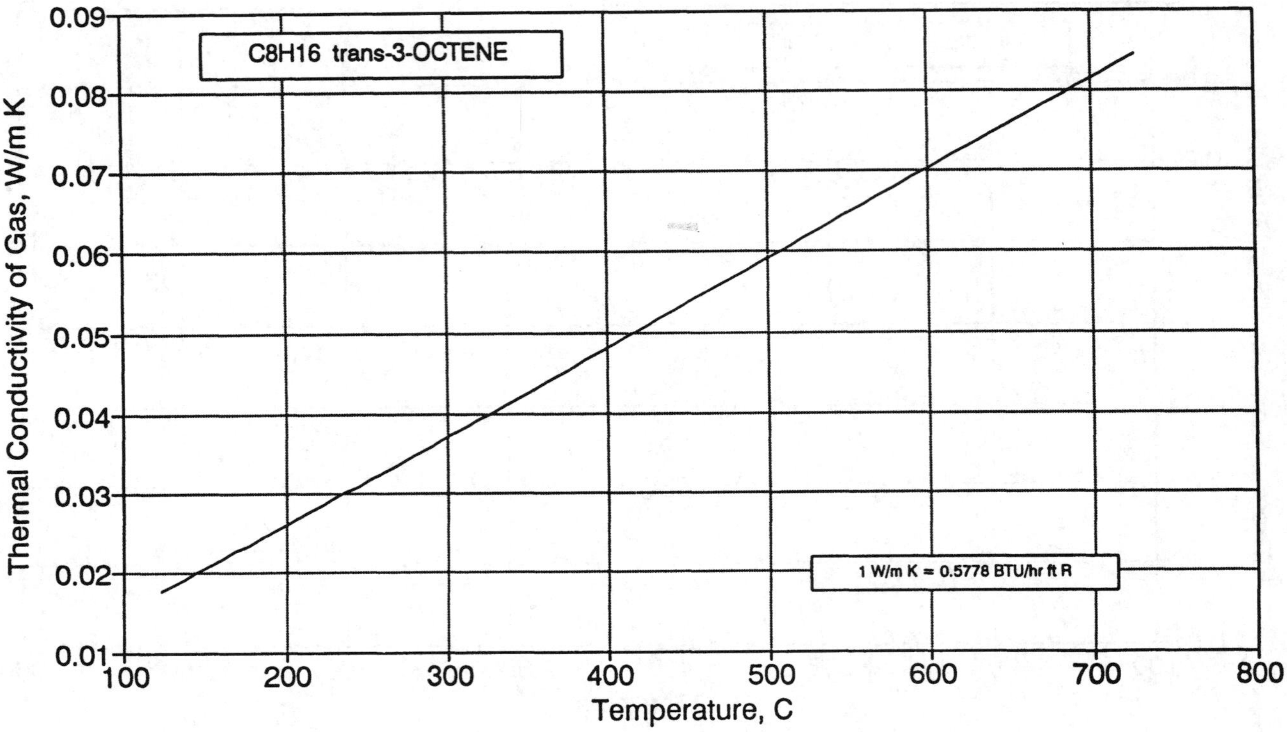
C8H16 trans-3-OCTENE
1 W/m K = 0.5778 BTU/hr ft R
Thermal Conductivity of Gas, W/m K
0.09
0.08
0.07
0.06
0.05
0.04
0.03
0.02
0.01
100
200
300
400
500
600
700
800
Temperature, C

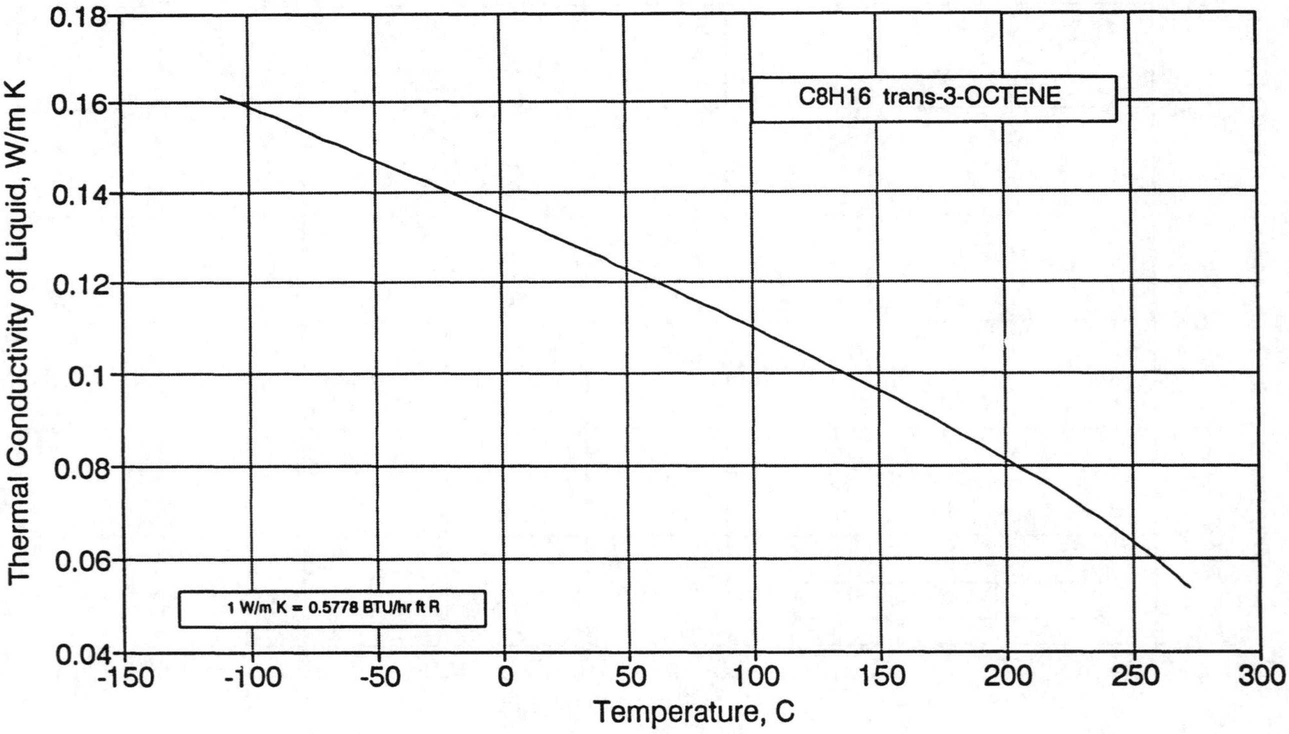
C8H16 trans-3-OCTENE
1 W/m K = 0.5778 BTU/hr ft R
Thermal Conductivity of Liquid, W/m K
0.18
0.16
0.14
0.12
0.1
0.08
0.06
0.04
-150
-100
-50
0
50
100
150
200
250
300
Temperature, C

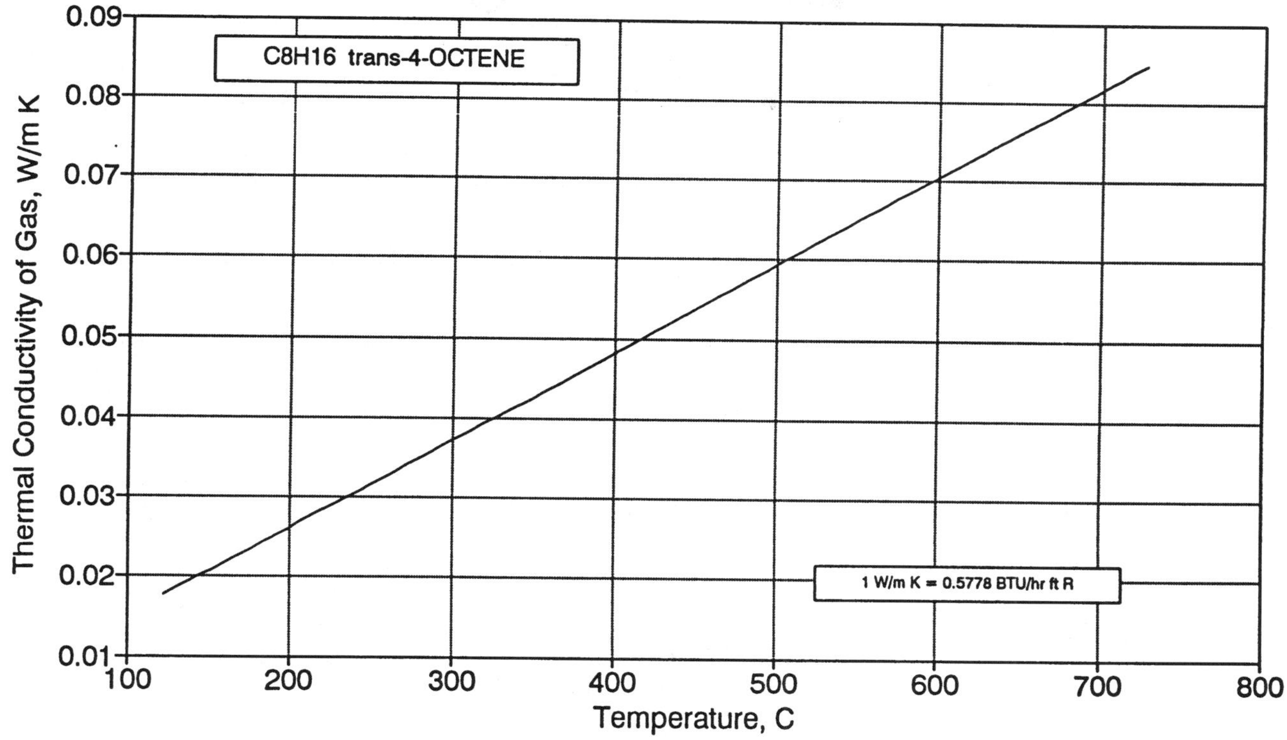
C8H16 trans-4-OCTENE
1 W/m K = 0.5778 BTU/hr ft R
Thermal Conductivity of Gas, W/m K
0.09
0.08
0.07
0.06
0.05
0.04
0.03
0.02
0.01
100
200
300
400
500
600
700
800
Temperature, C

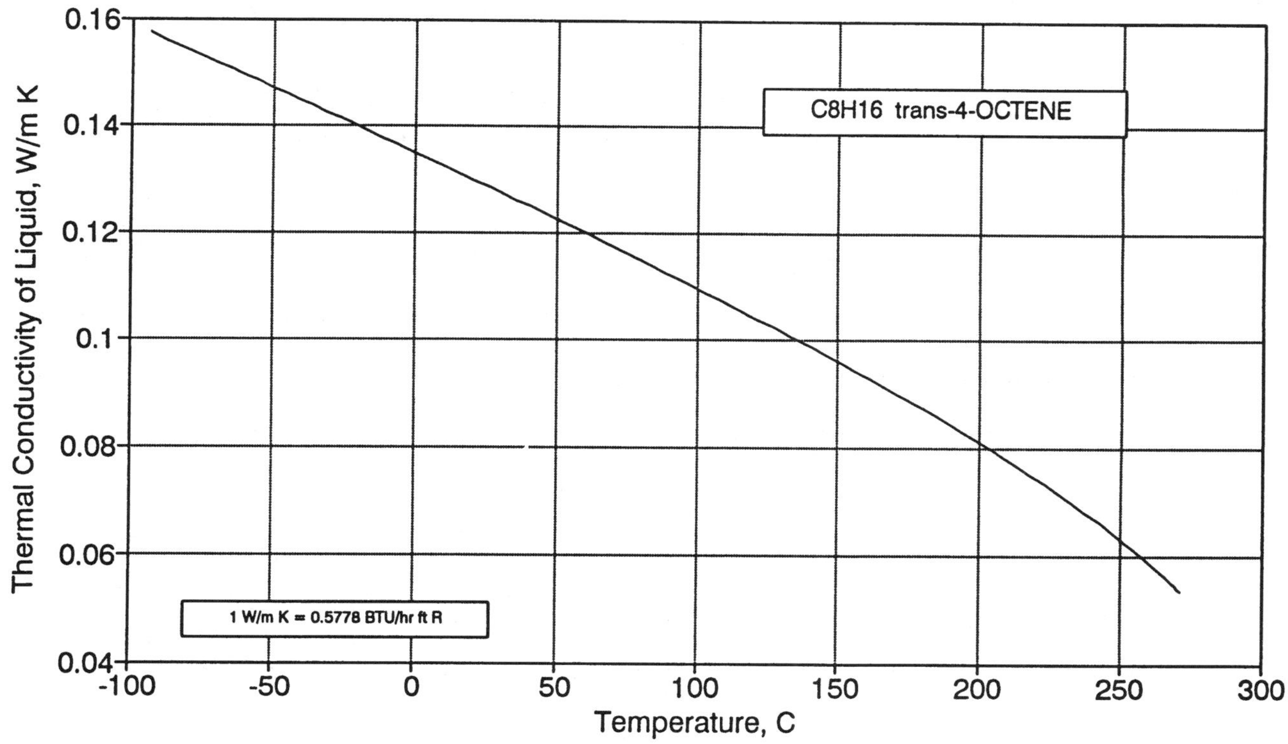
C8H16 trans-4-OCTENE
1 W/m K = 0.5778 BTU/hr ft R
Thermal Conductivity of Liquid, W/m K
0.16
0.14
0.12
0.1
0.08
0.06
0.04
-100
-50
0
50
100
150
200
250
300
Temperature, C

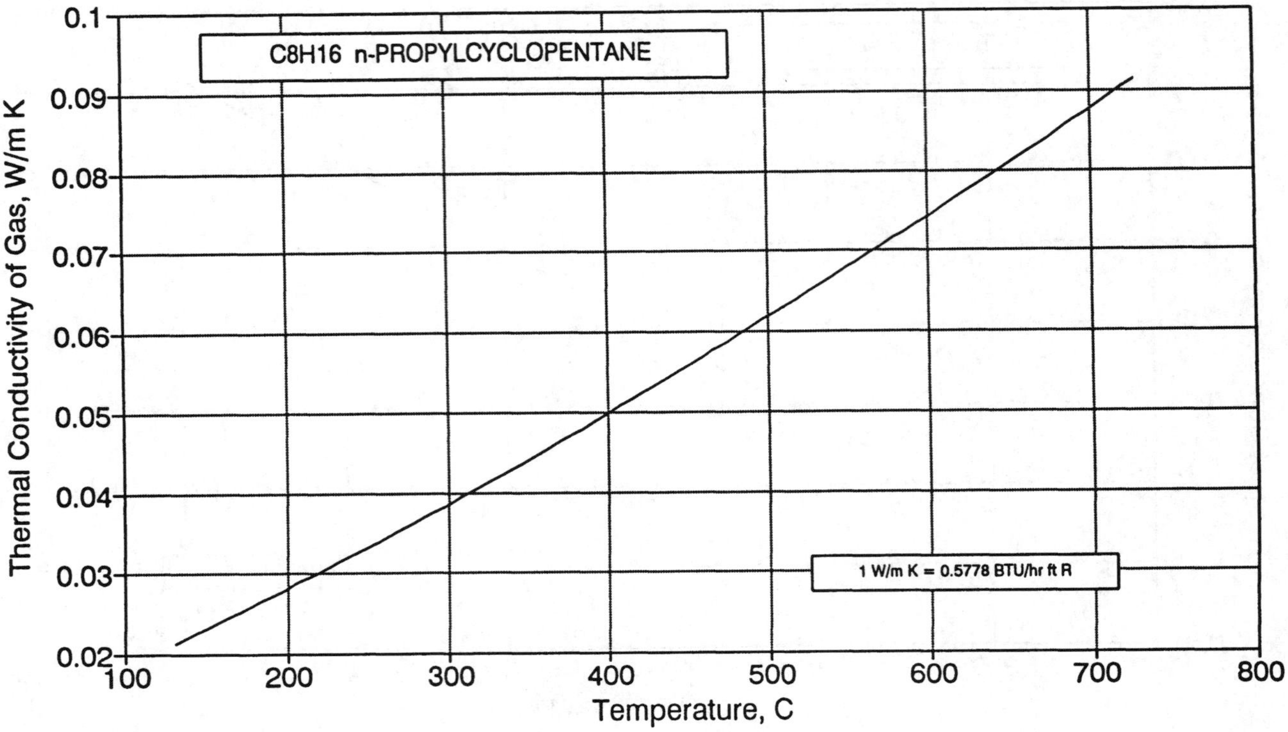
C8H16 n-PROPYLCYCLOPENTANE
1 W/m K = 0.5778 BTU/hr ft R
Thermal Conductivity of Gas, W/m K
Temperature, C
0.1
0.09
0.08
0.07
0.06
0.05
0.04
0.03
0.02
100
200
300
400
500
600
700
800

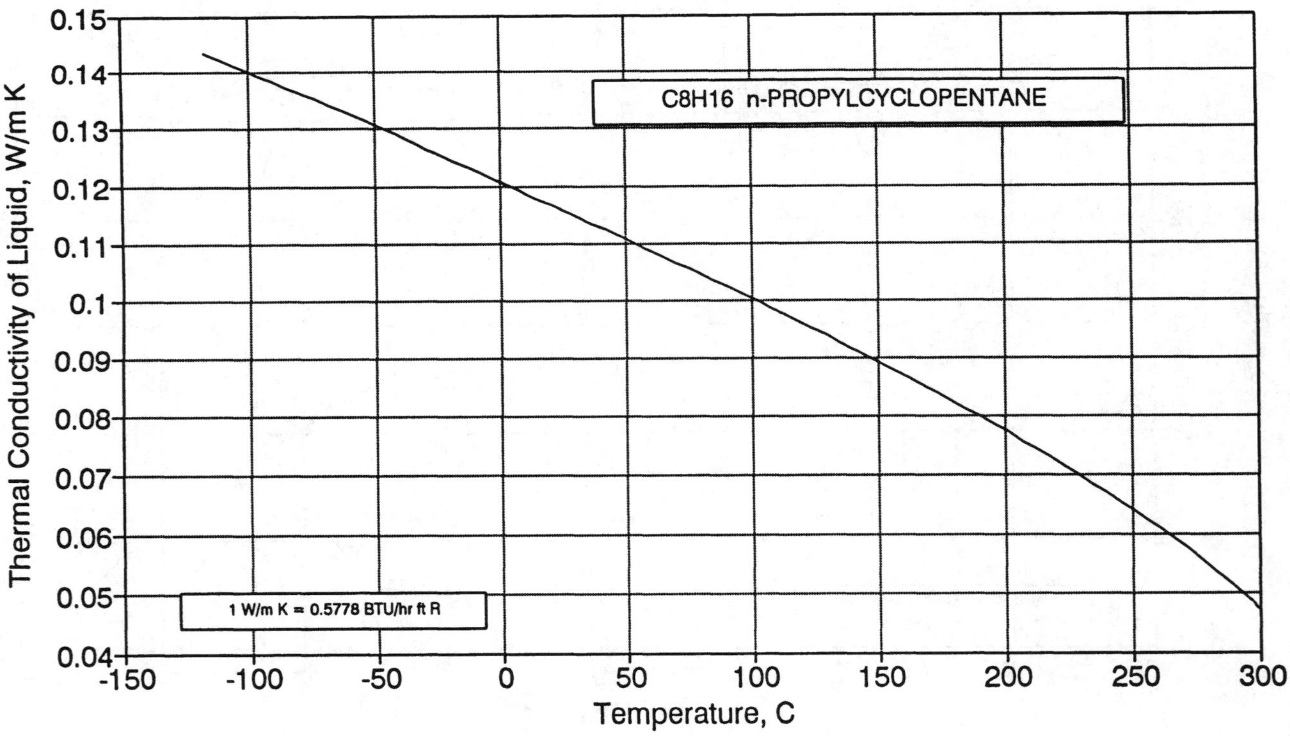
C8H16 n-PROPYLCYCLOPENTANE
1 W/m K = 0.5778 BTU/hr ft R
Thermal Conductivity of Liquid, W/m K
Temperature, C
0.15
0.14
0.13
0.12
0.11
0.1
0.09
0.08
0.07
0.06
0.05
0.04
-150
-100
-50
0
50
100
150
200
250
300

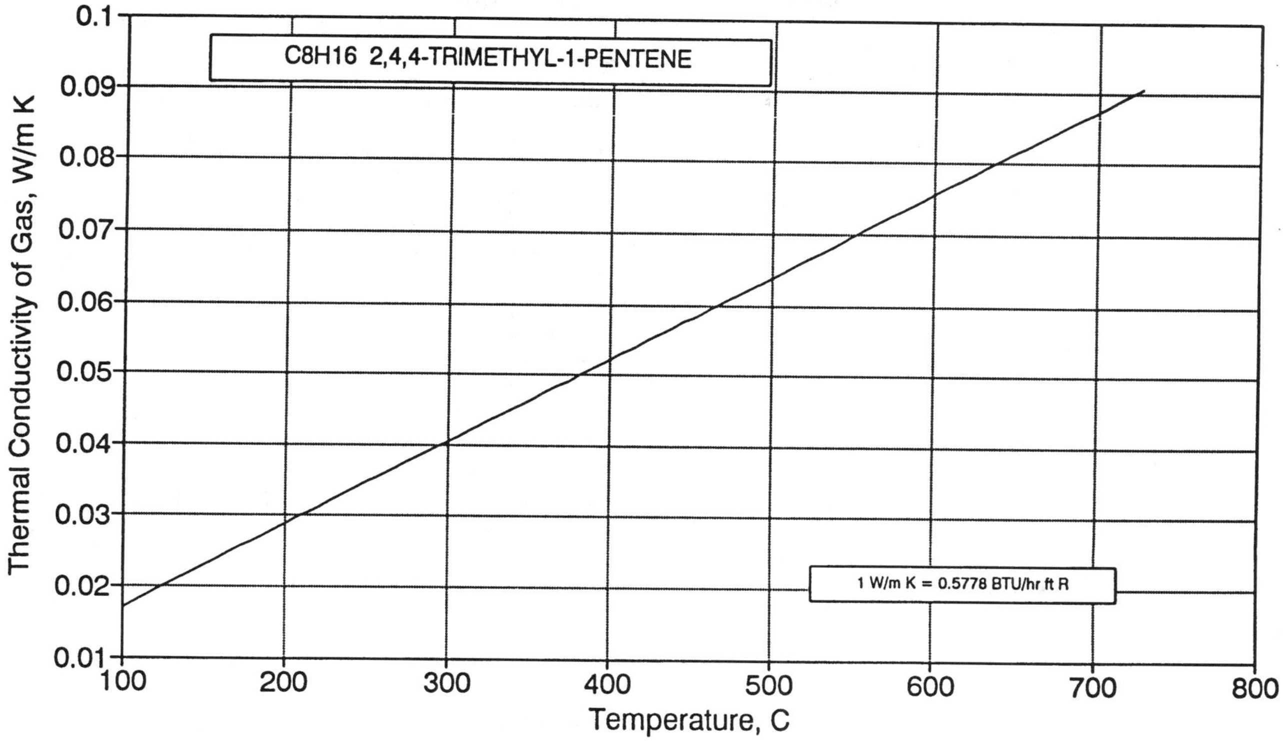
C8H16 2,4,4-TRIMETHYL-1-PENTENE
1 W/m K = 0.5778 BTU/hr ft R
Thermal Conductivity of Gas, W/m K
0.1
0.09
0.08
0.07
0.06
0.05
0.04
0.03
0.02
0.01
100
200
300
400
500
600
700
800
Temperature, C

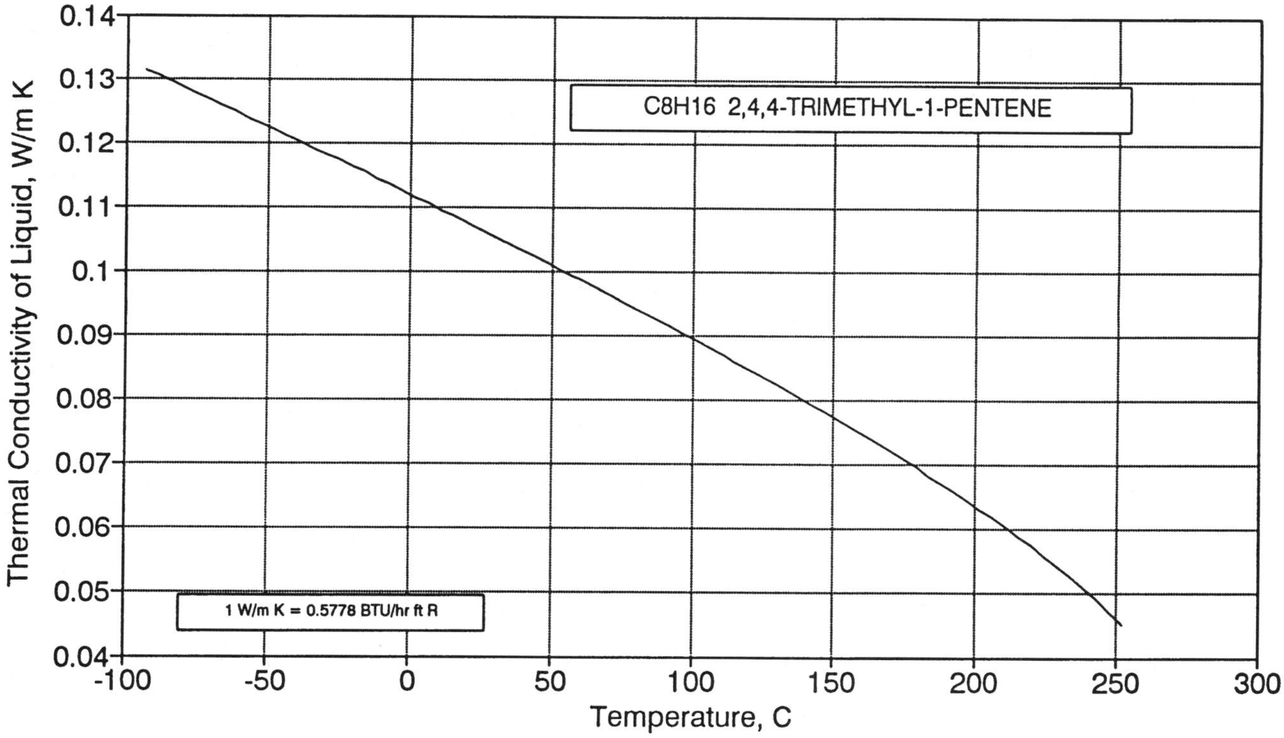
C8H16 2,4,4-TRIMETHYL-1-PENTENE
1 W/m K = 0.5778 BTU/hr ft R
Thermal Conductivity of Liquid, W/m K
0.14
0.13
0.12
0.11
0.1
0.09
0.08
0.07
0.06
0.05
0.04
-100
-50
0
50
100
150
200
250
300
Temperature, C

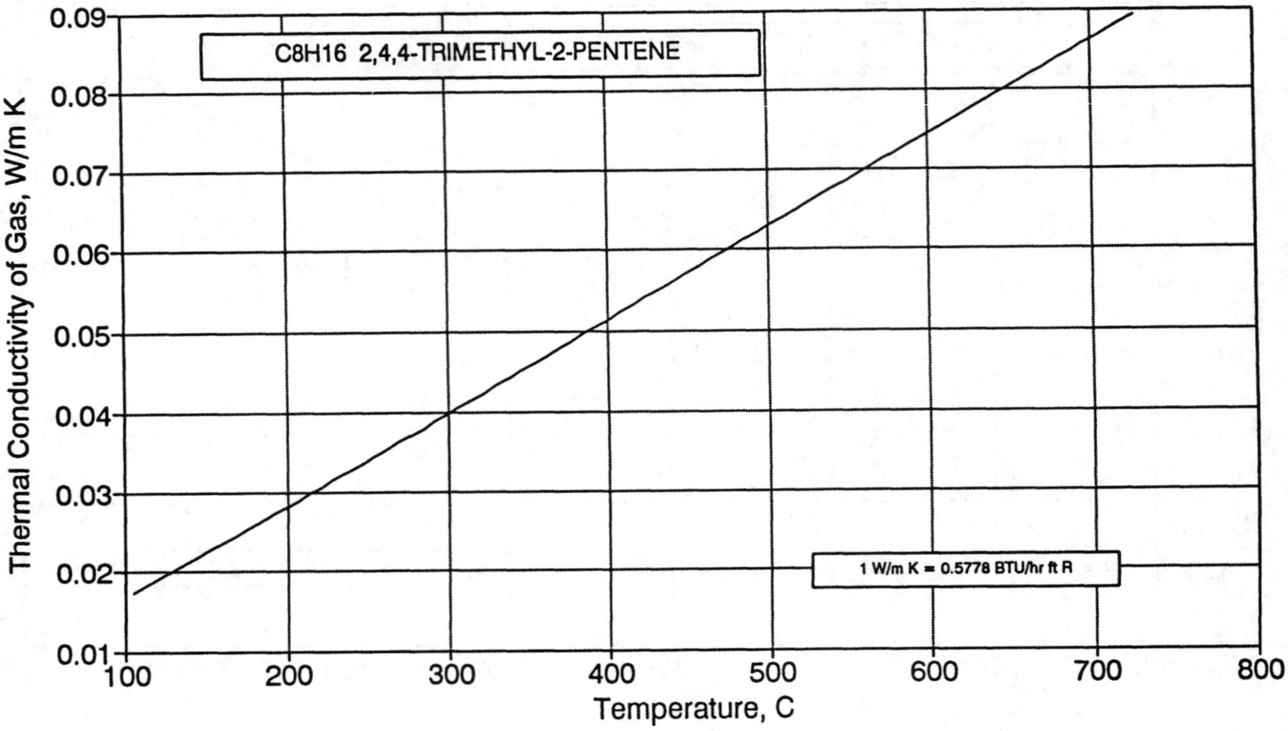
C8H16 2,4,4-TRIMETHYL-2-PENTENE
Thermal Conductivity of Gas, W/m K
0.09
0.08
0.07
0.06
0.05
0.04
0.03
0.02
0.01
100
200
300
400
500
600
700
800
Temperature, C
1 W/m K = 0.5778 BTU/hr ft R

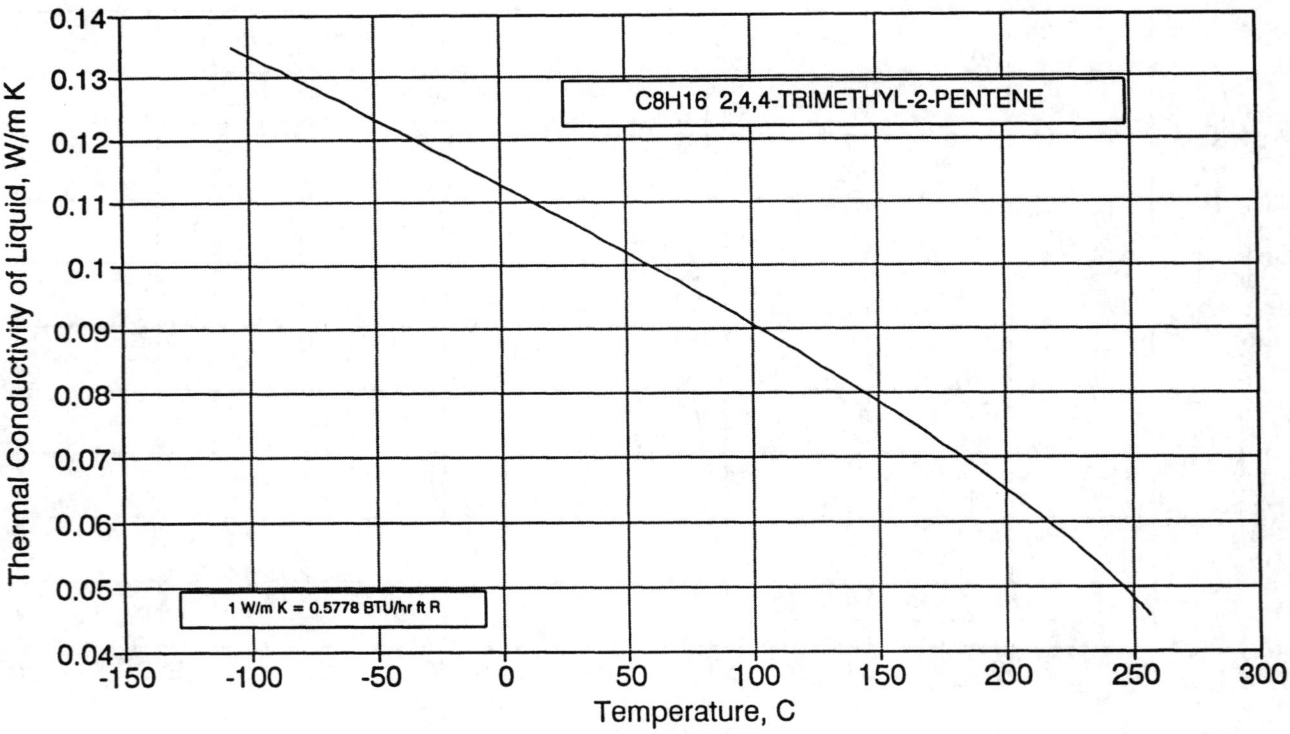
C8H16 2,4,4-TRIMETHYL-2-PENTENE
Thermal Conductivity of Liquid, W/m K
0.14
0.13
0.12
0.11
0.1
0.09
0.08
0.07
0.06
0.05
0.04
-150
-100
-50
0
50
100
150
200
250
300
Temperature, C
1 W/m K = 0.5778 BTU/hr ft R

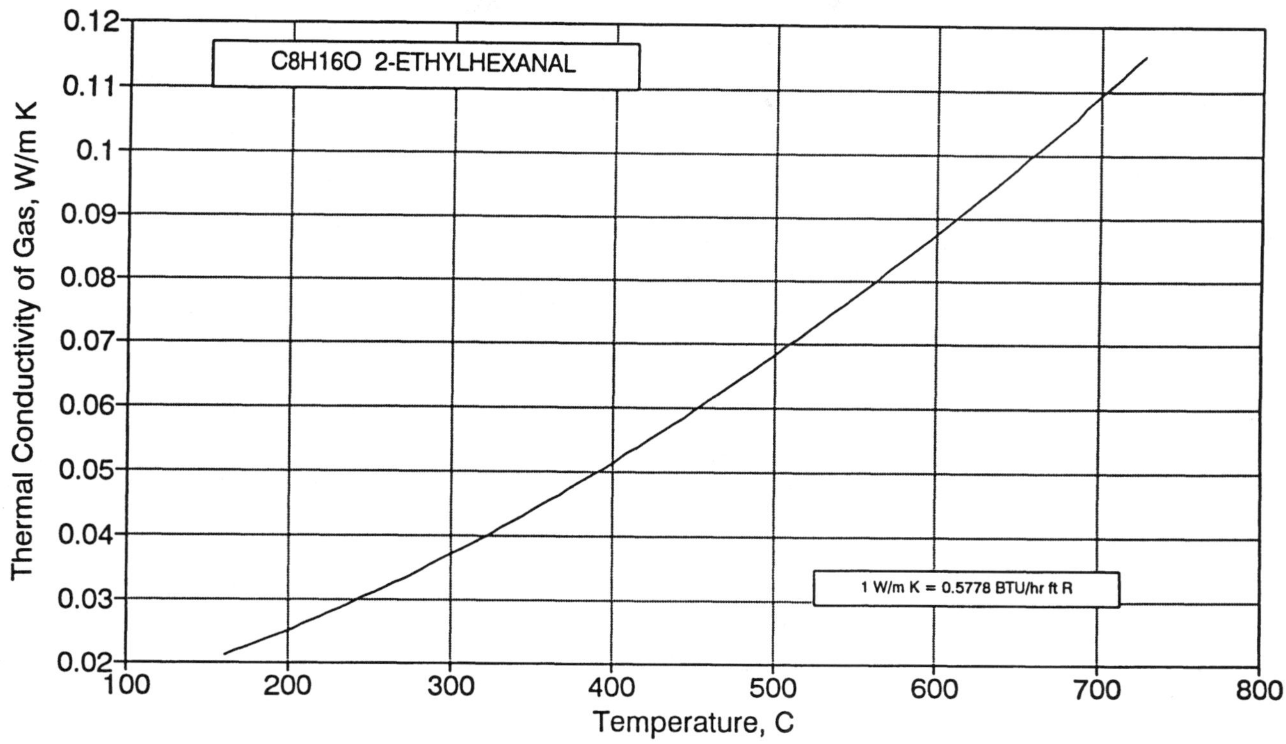
C8H16O 2-ETHYLHEXANAL
1 W/m K = 0.5778 BTU/hr ft R
Thermal Conductivity of Gas, W/m K
0.12
0.11
0.1
0.09
0.08
0.07
0.06
0.05
0.04
0.03
0.02
100
200
300
400
500
600
700
800
Temperature, C

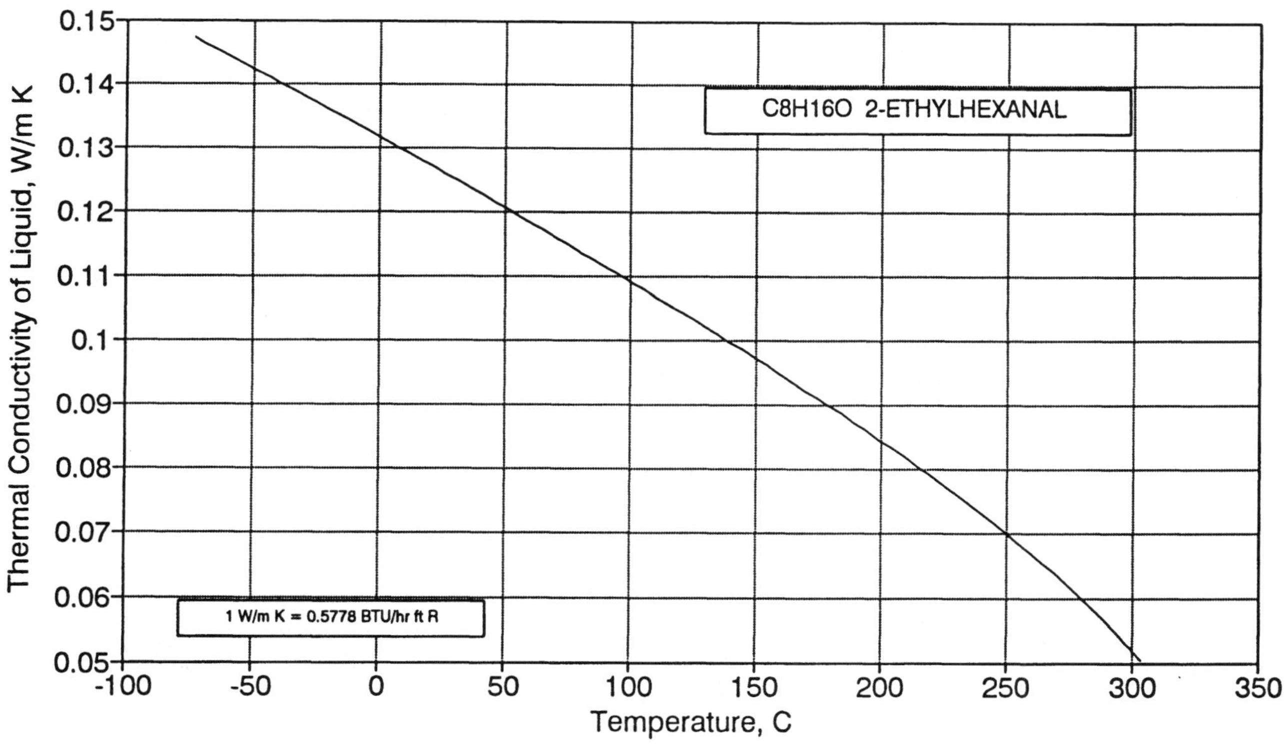
C8H16O 2-ETHYLHEXANAL
1 W/m K = 0.5778 BTU/hr ft R
Thermal Conductivity of Liquid, W/m K
0.15
0.14
0.13
0.12
0.11
0.1
0.09
0.08
0.07
0.06
0.05
-100
-50
0
50
100
150
200
250
300
350
Temperature, C

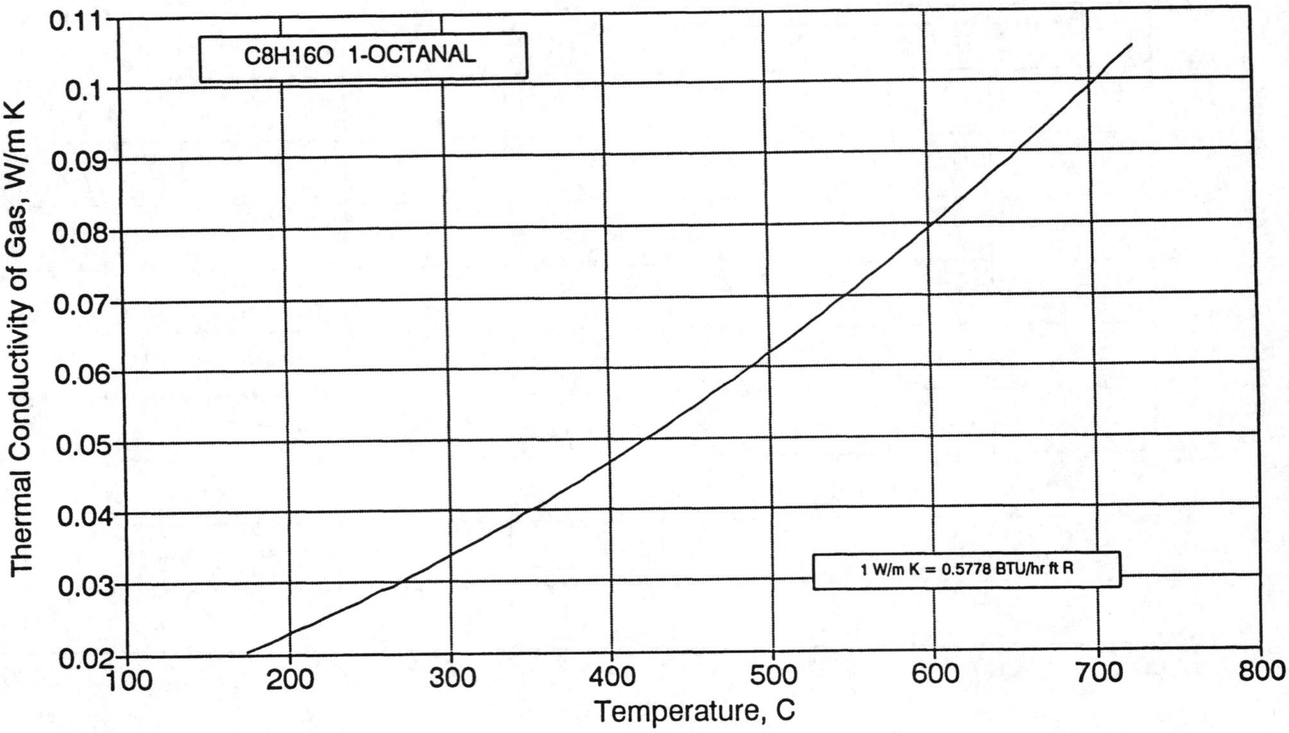
C8H16O 1-OCTANAL
1 W/m K = 0.5778 BTU/hr ft R
Thermal Conductivity of Gas, W/m K
0.11
0.1
0.09
0.08
0.07
0.06
0.05
0.04
0.03
0.02
100
200
300
400
500
600
700
800
Temperature, C

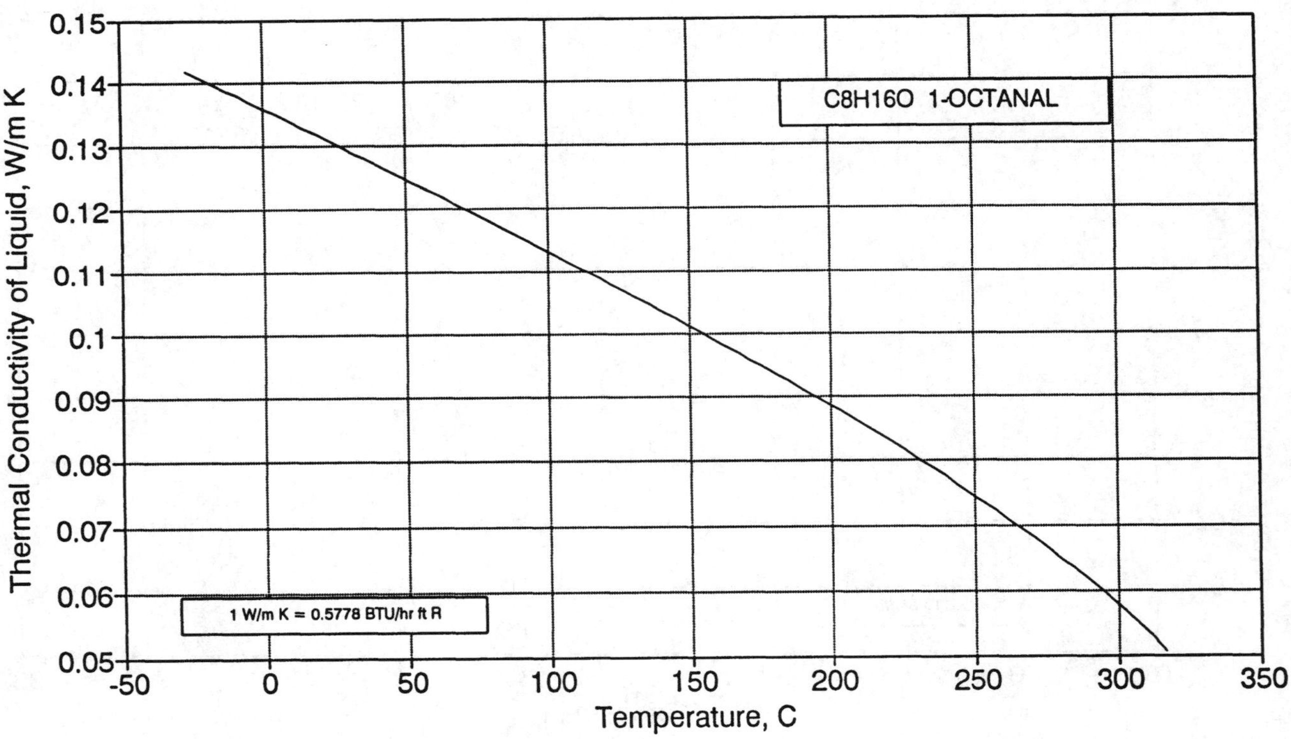
C8H16O 1-OCTANAL
1 W/m K = 0.5778 BTU/hr ft R
Thermal Conductivity of Liquid, W/m K
0.15
0.14
0.13
0.12
0.11
0.1
0.09
0.08
0.07
0.06
0.05
-50
0
50
100
150
200
250
300
350
Temperature, C

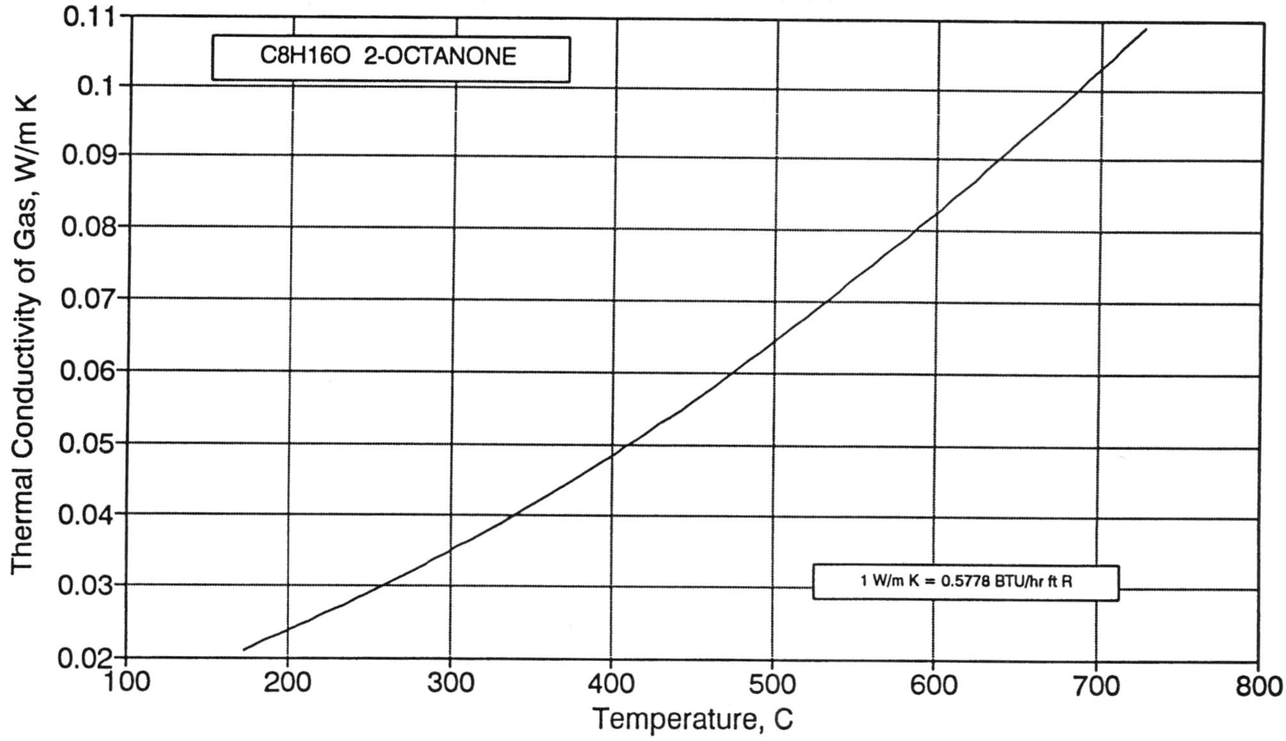
C8H16O 2-OCTANONE
1 W/m K = 0.5778 BTU/hr ft R
Thermal Conductivity of Gas, W/m K
0.11
0.1
0.09
0.08
0.07
0.06
0.05
0.04
0.03
0.02
100
200
300
400
500
600
700
800
Temperature, C

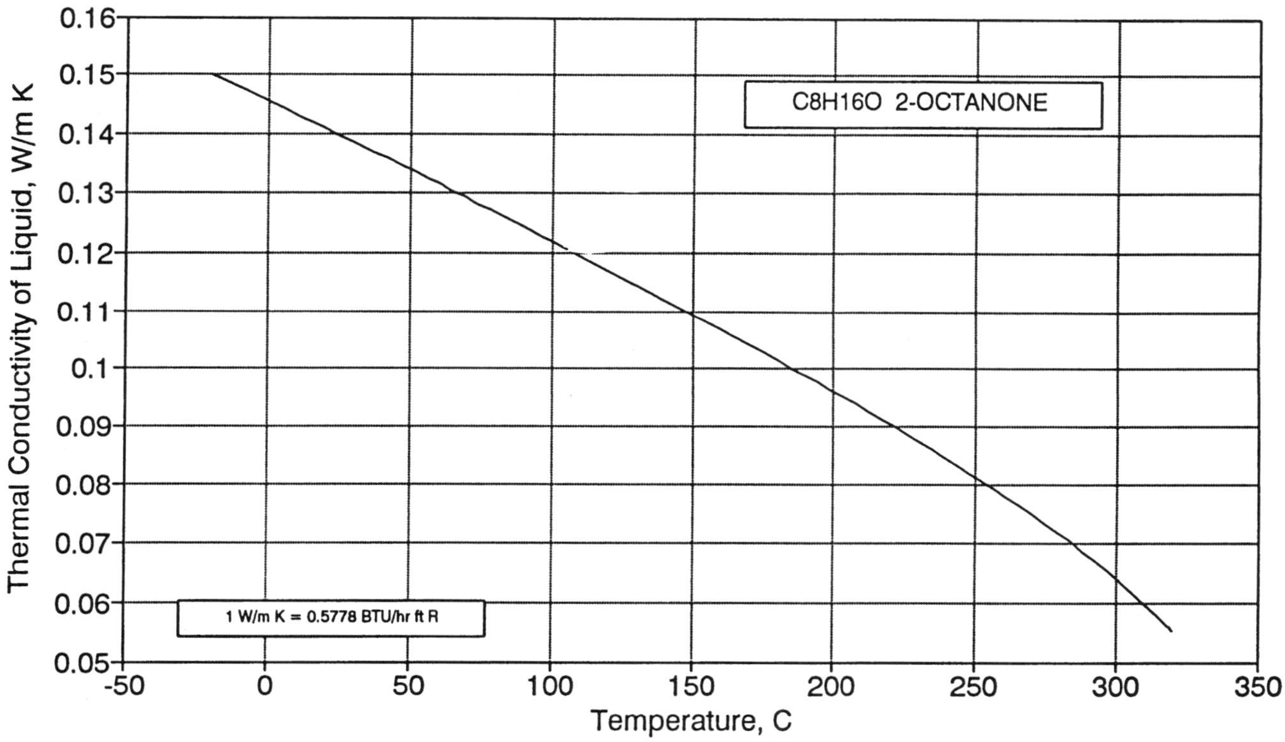
C8H16O 2-OCTANONE
1 W/m K = 0.5778 BTU/hr ft R
Thermal Conductivity of Liquid, W/m K
0.16
0.15
0.14
0.13
0.12
0.11
0.1
0.09
0.08
0.07
0.06
0.05
-50
0
50
100
150
200
250
300
350
Temperature, C

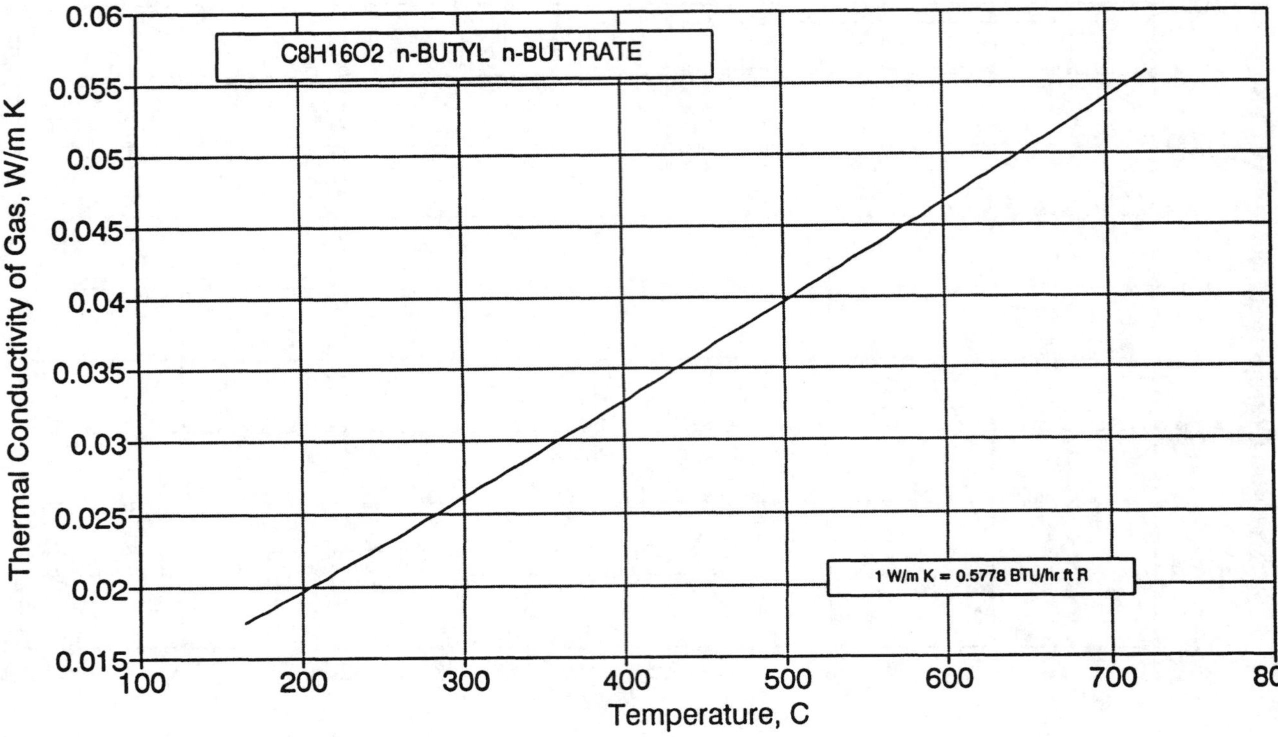
C8H16O2 n-BUTYL n-BUTYRATE
1 W/m K = 0.5778 BTU/hr ft R
Thermal Conductivity of Gas, W/m K
0.06
0.055
0.05
0.045
0.04
0.035
0.03
0.025
0.02
0.015
100
200
300
400
500
600
700
800
Temperature, C

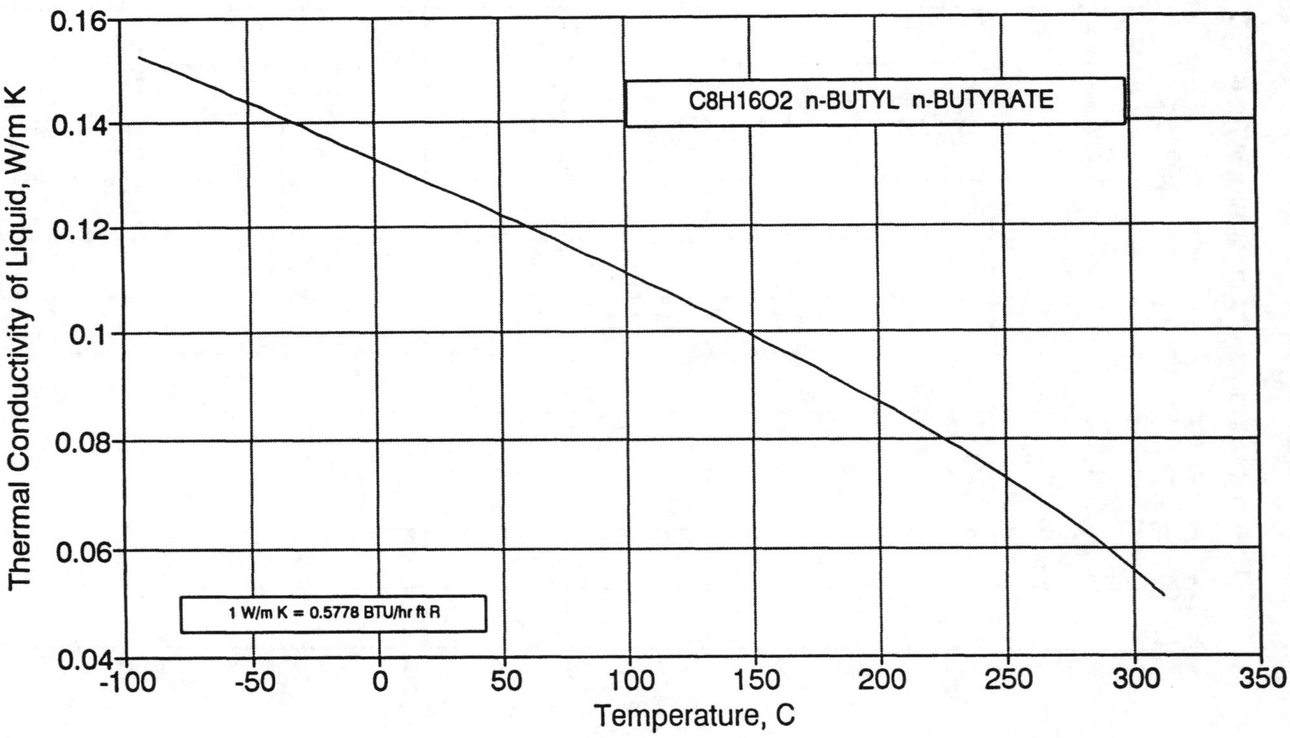
C8H16O2 n-BUTYL n-BUTYRATE
1 W/m K = 0.5778 BTU/hr ft R
Thermal Conductivity of Liquid, W/m K
0.16
0.14
0.12
0.1
0.08
0.06
0.04
-100
-50
0
50
100
150
200
250
300
350
Temperature, C

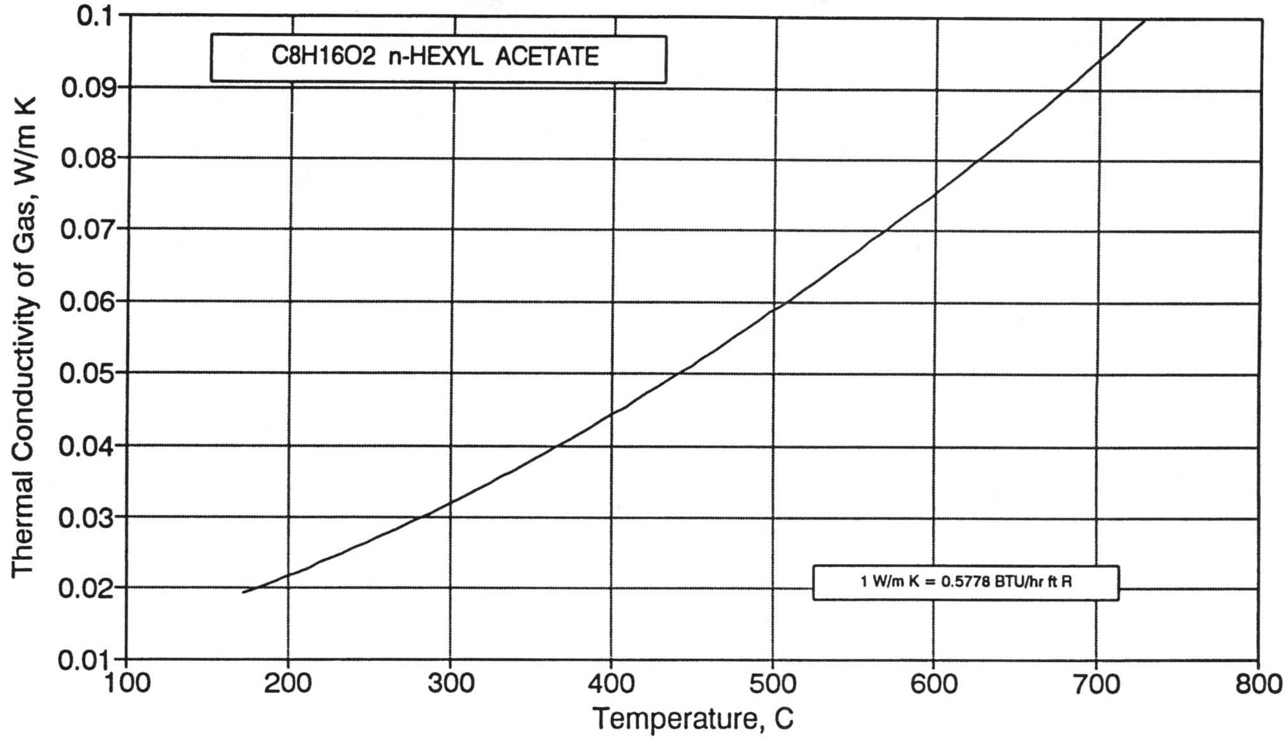
C8H16O2 n-HEXYL ACETATE
1 W/m K = 0.5778 BTU/hr ft R
Thermal Conductivity of Gas, W/m K
Temperature, C
0.1
0.09
0.08
0.07
0.06
0.05
0.04
0.03
0.02
0.01
100
200
300
400
500
600
700
800

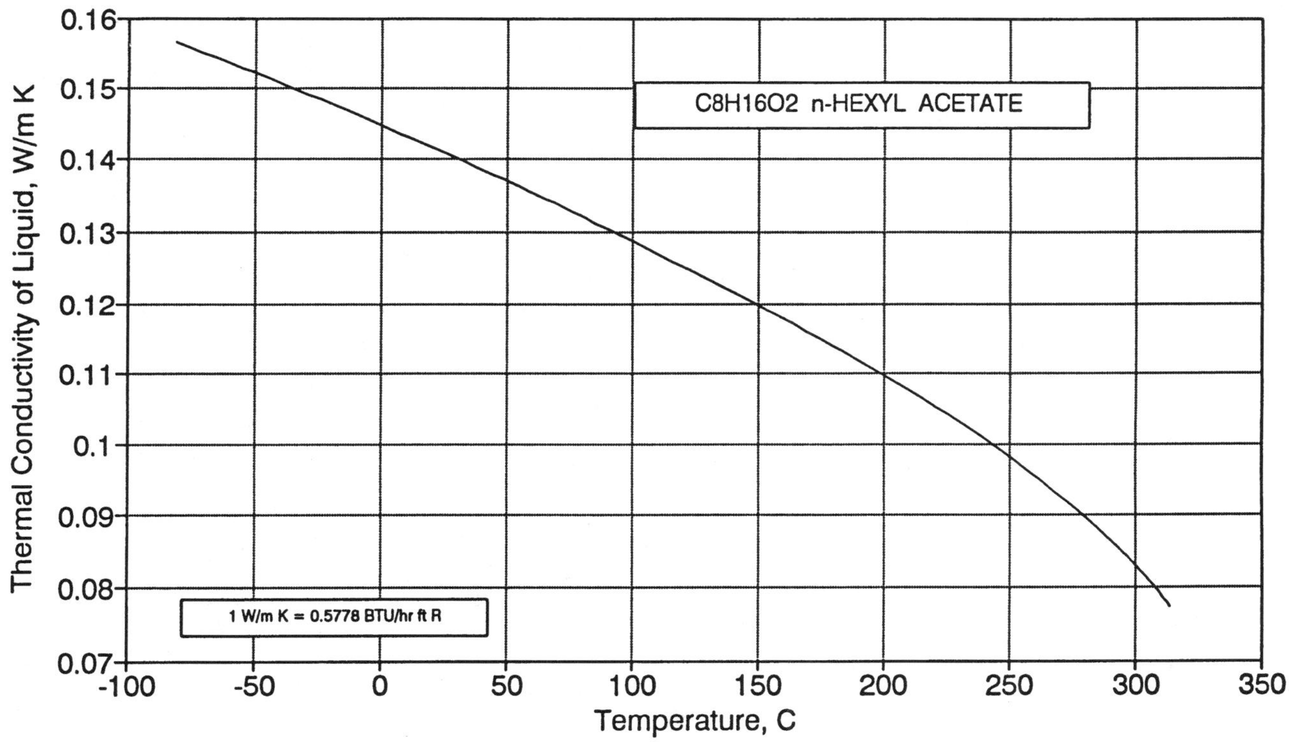
C8H16O2 n-HEXYL ACETATE
1 W/m K = 0.5778 BTU/hr ft R
Thermal Conductivity of Liquid, W/m K
Temperature, C
0.16
0.15
0.14
0.13
0.12
0.11
0.1
0.09
0.08
0.07
-100
-50
0
50
100
150
200
250
300
350

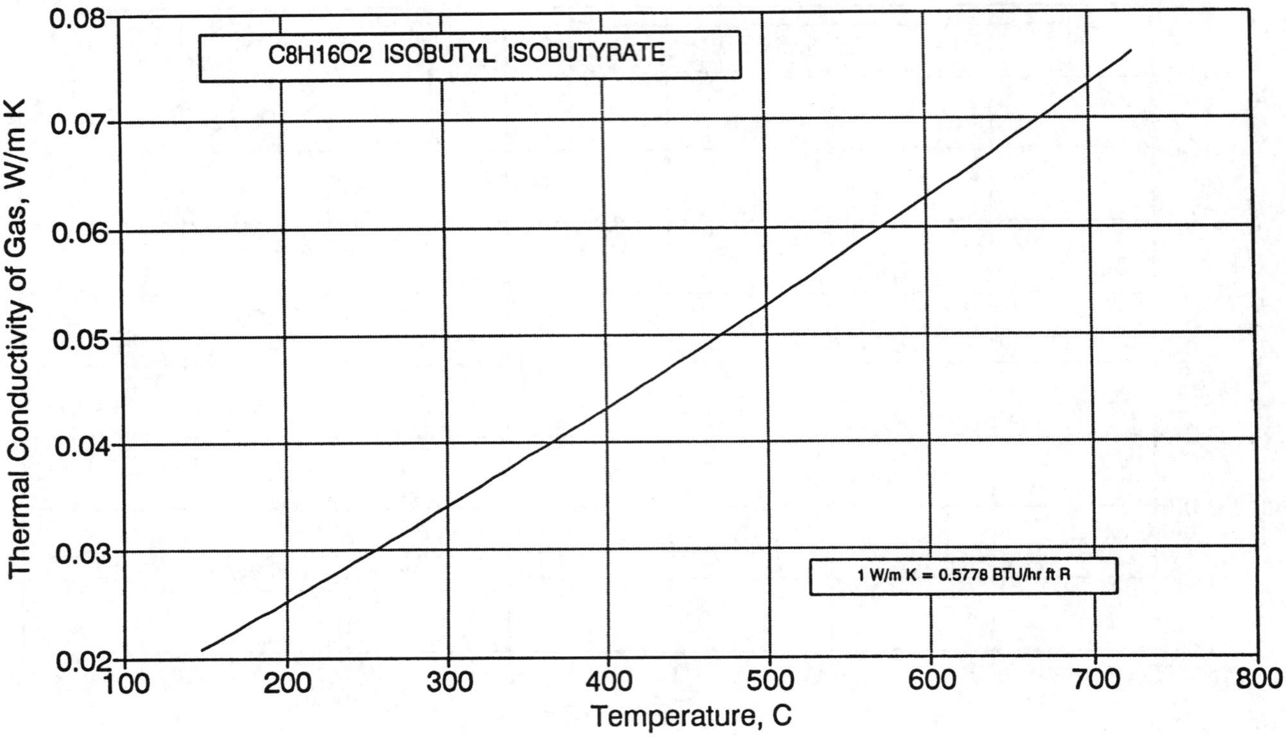
C8H16O2 ISOBUTYL ISOBUTYRATE
1 W/m K = 0.5778 BTU/hr ft R
Thermal Conductivity of Gas, W/m K
Temperature, C
0.08
0.07
0.06
0.05
0.04
0.03
0.02
100
200
300
400
500
600
700
800

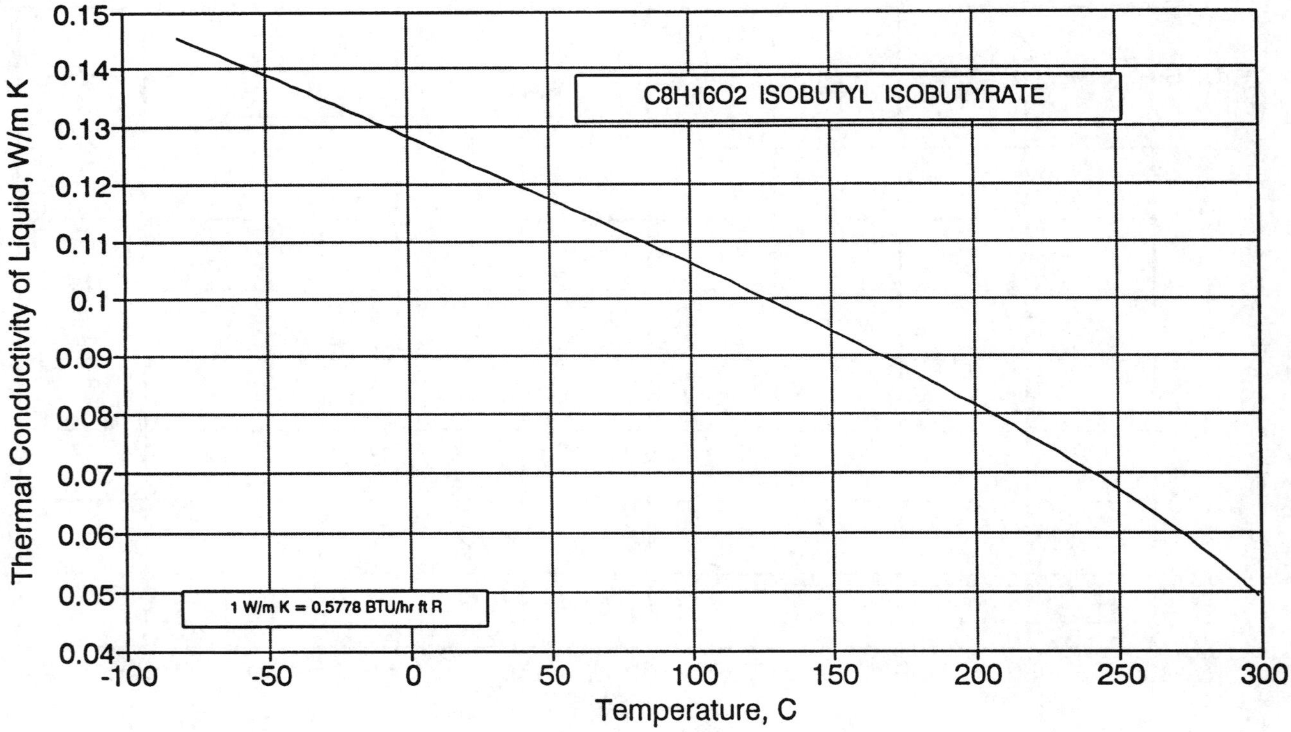
C8H16O2 ISOBUTYL ISOBUTYRATE
1 W/m K = 0.5778 BTU/hr ft R
Thermal Conductivity of Liquid, W/m K
Temperature, C
0.15
0.14
0.13
0.12
0.11
0.1
0.09
0.08
0.07
0.06
0.05
0.04
-100
-50
0
50
100
150
200
250
300

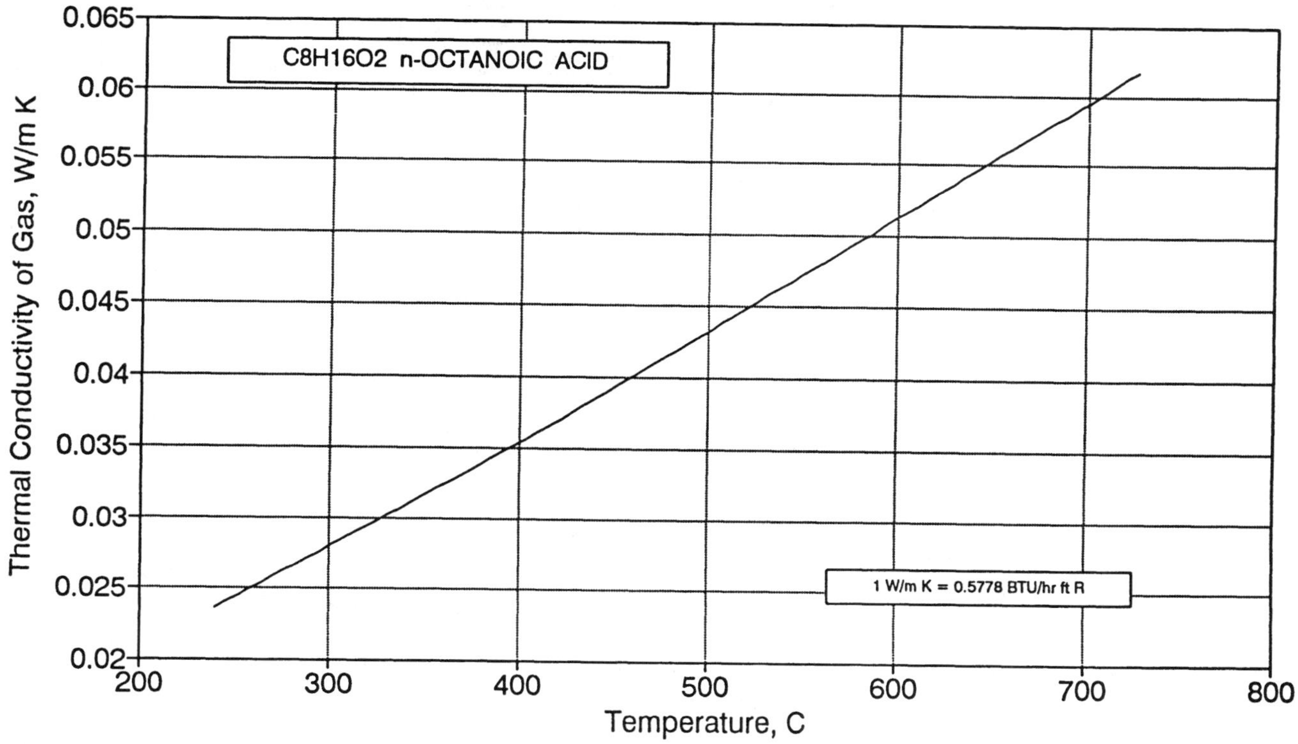

C8H16O2 n-OCTANOIC ACID
1 W/m K = 0.5778 BTU/hr ft R
Thermal Conductivity of Gas, W/m K
Temperature, C
0.065
0.06
0.055
0.05
0.045
0.04
0.035
0.03
0.025
0.02
200
300
400
500
600
700
800

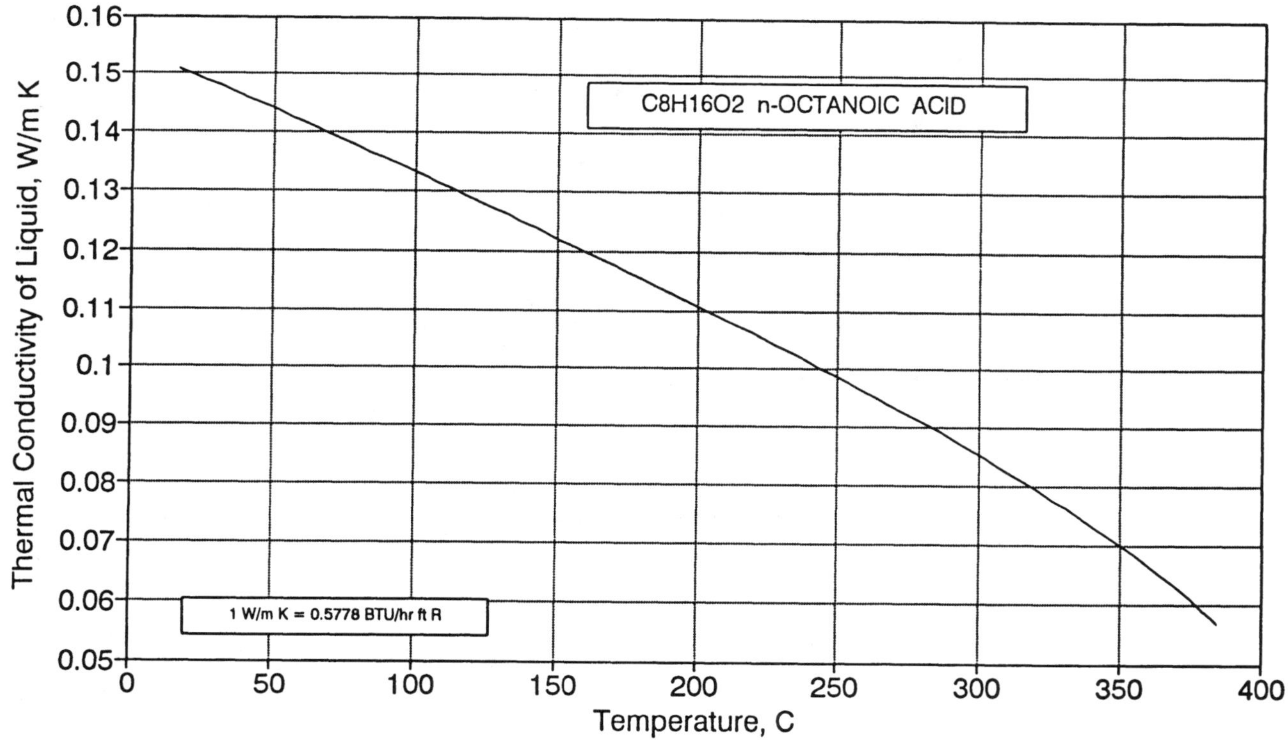

C8H16O2 n-OCTANOIC ACID
1 W/m K = 0.5778 BTU/hr ft R
Thermal Conductivity of Liquid, W/m K
Temperature, C
0.16
0.15
0.14
0.13
0.12
0.11
0.1
0.09
0.08
0.07
0.06
0.05
0
50
100
150
200
250
300
350
400

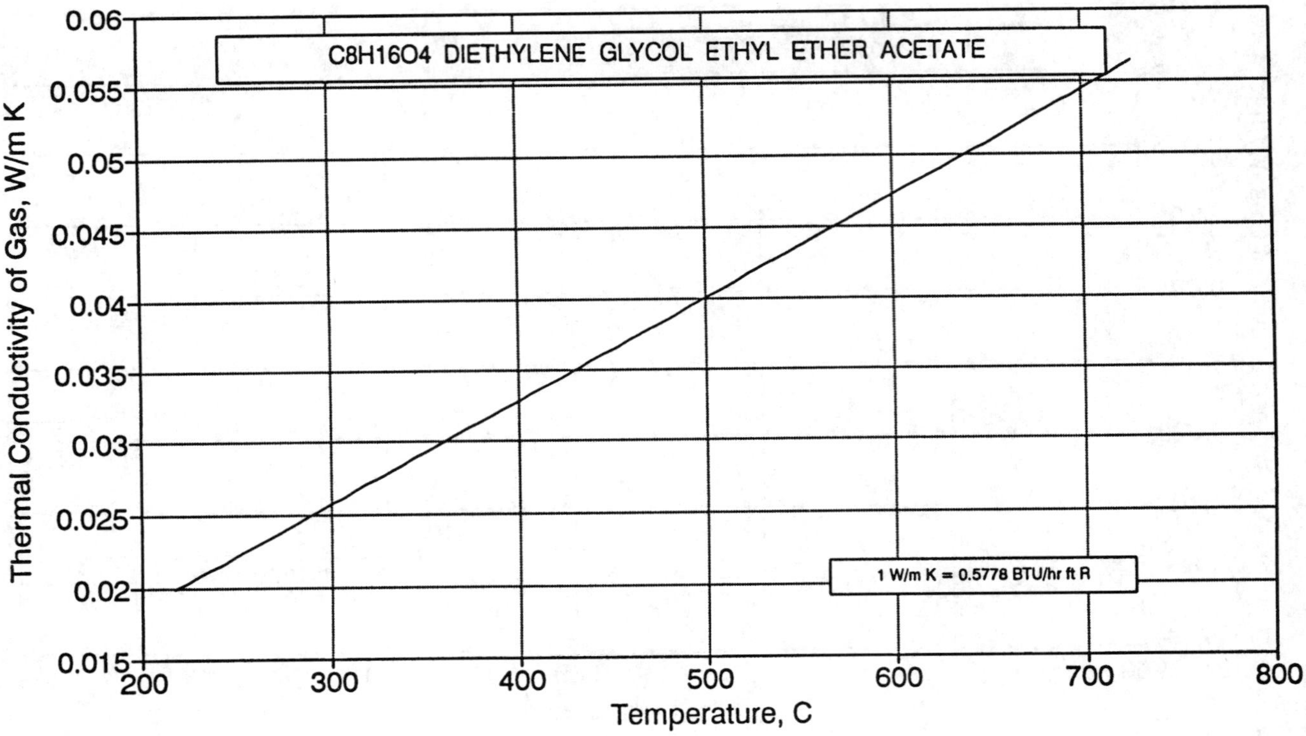
C8H16O4 DIETHYLENE GLYCOL ETHYL ETHER ACETATE
1 W/m K = 0.5778 BTU/hr ft R
Thermal Conductivity of Gas, W/m K
0.06
0.055
0.05
0.045
0.04
0.035
0.03
0.025
0.02
0.015
200
300
400
500
600
700
800
Temperature, C

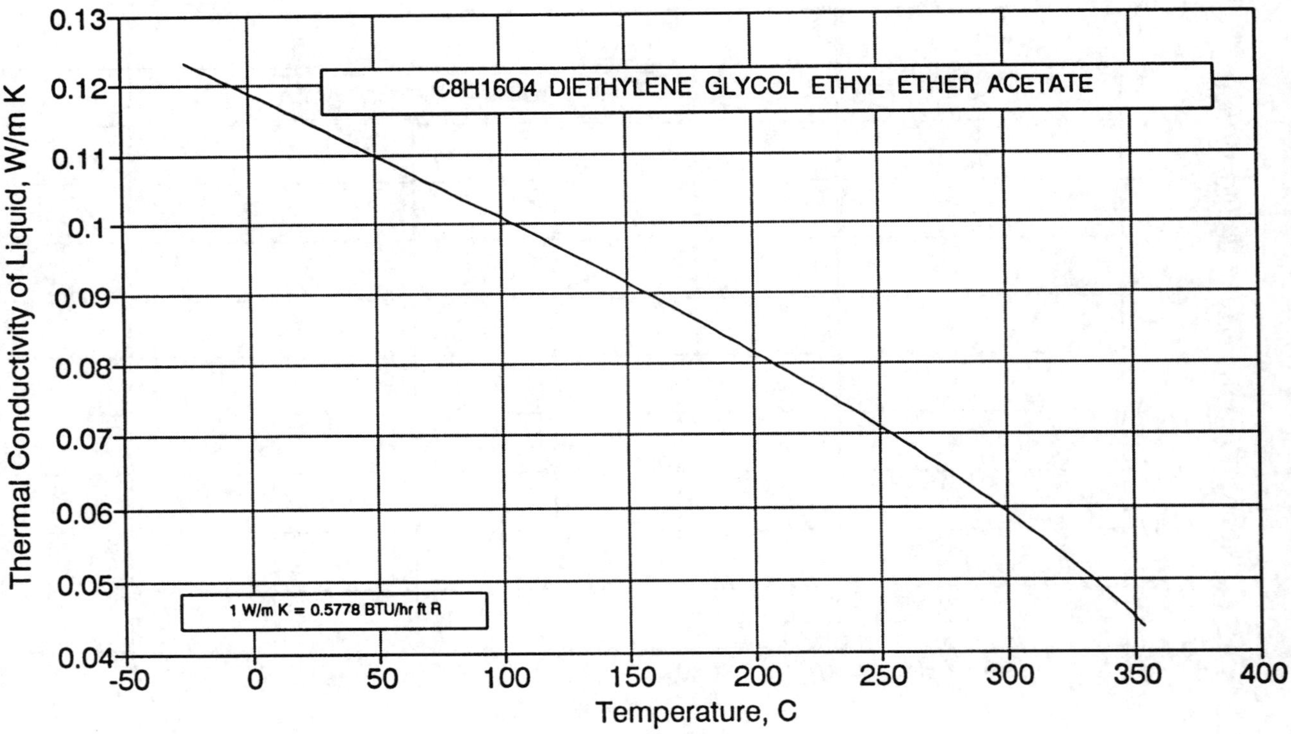
C8H16O4 DIETHYLENE GLYCOL ETHYL ETHER ACETATE
1 W/m K = 0.5778 BTU/hr ft R
Thermal Conductivity of Liquid, W/m K
0.13
0.12
0.11
0.1
0.09
0.08
0.07
0.06
0.05
0.04
-50
0
50
100
150
200
250
300
350
400
Temperature, C

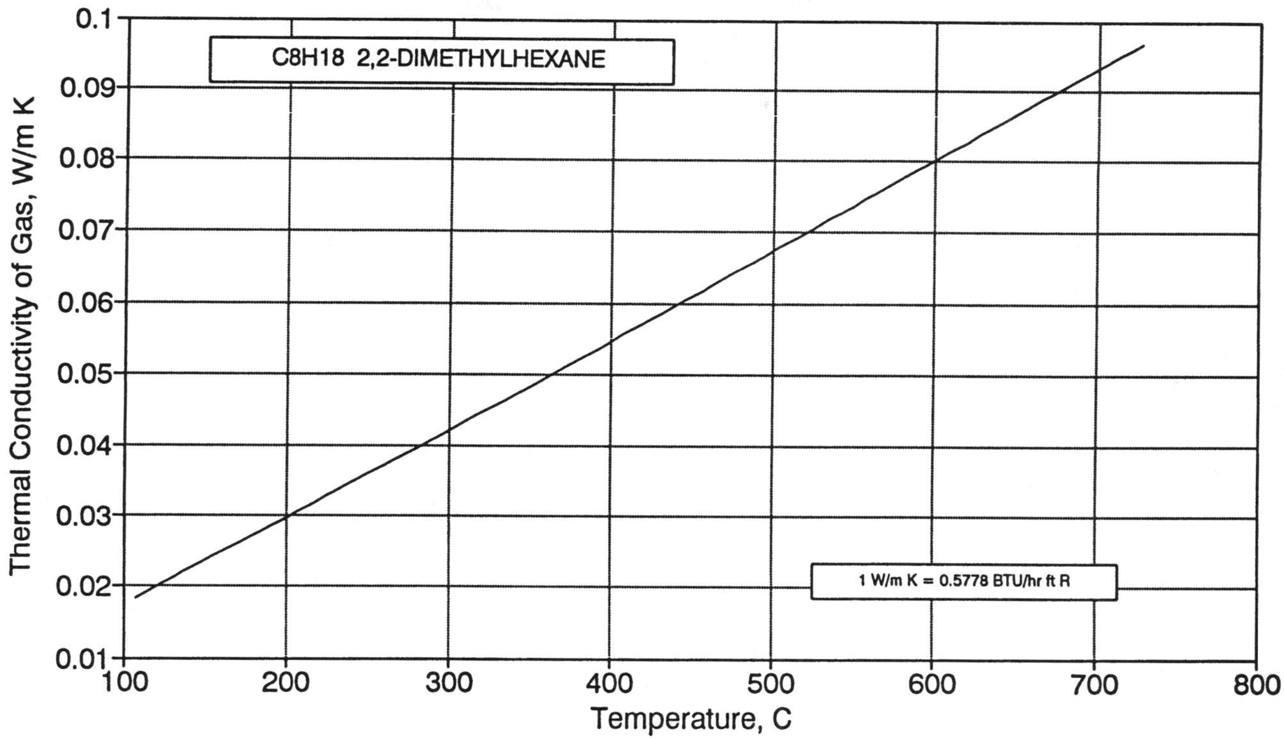
C8H18 2,2-DIMETHYLHEXANE
Thermal Conductivity of Gas, W/m K
0.1
0.09
0.08
0.07
0.06
0.05
0.04
0.03
0.02
0.01
100
200
300
400
500
600
700
800
Temperature, C
1 W/m K = 0.5778 BTU/hr ft R

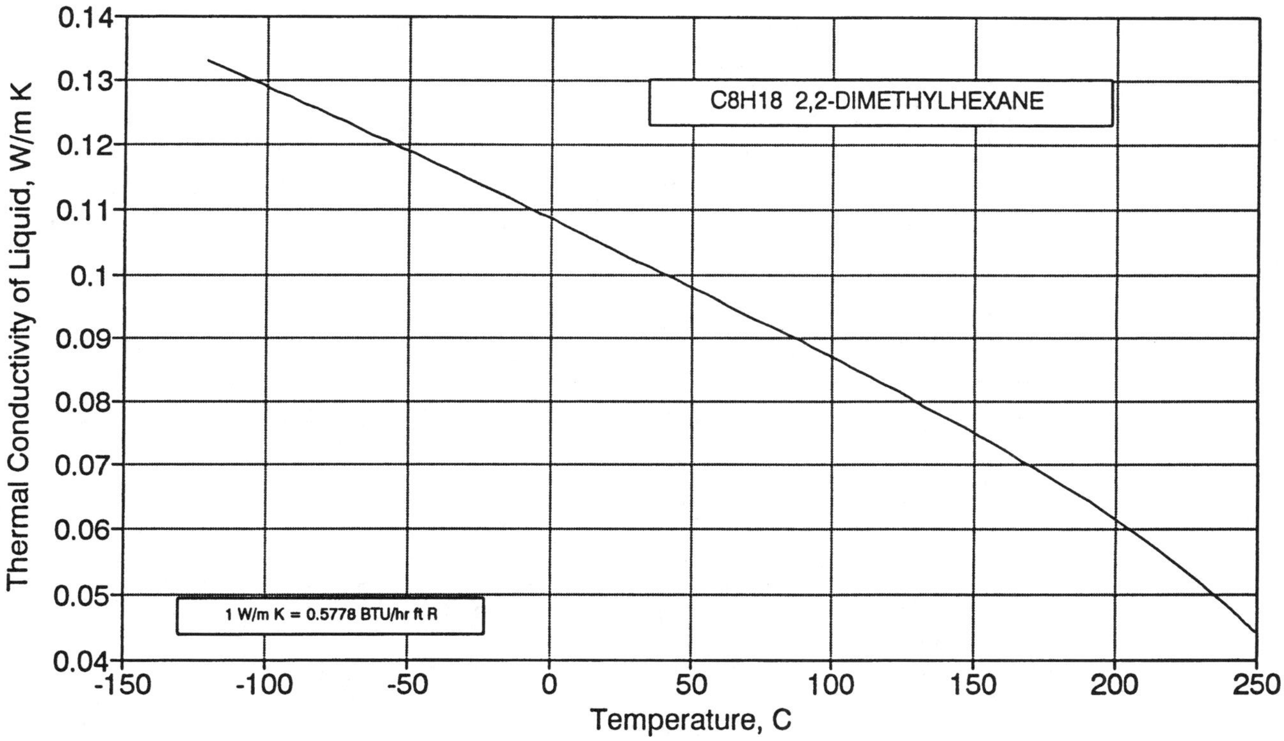
C8H18 2,2-DIMETHYLHEXANE
Thermal Conductivity of Liquid, W/m K
0.14
0.13
0.12
0.11
0.1
0.09
0.08
0.07
0.06
0.05
0.04
-150
-100
-50
0
50
100
150
200
250
Temperature, C
1 W/m K = 0.5778 BTU/hr ft R

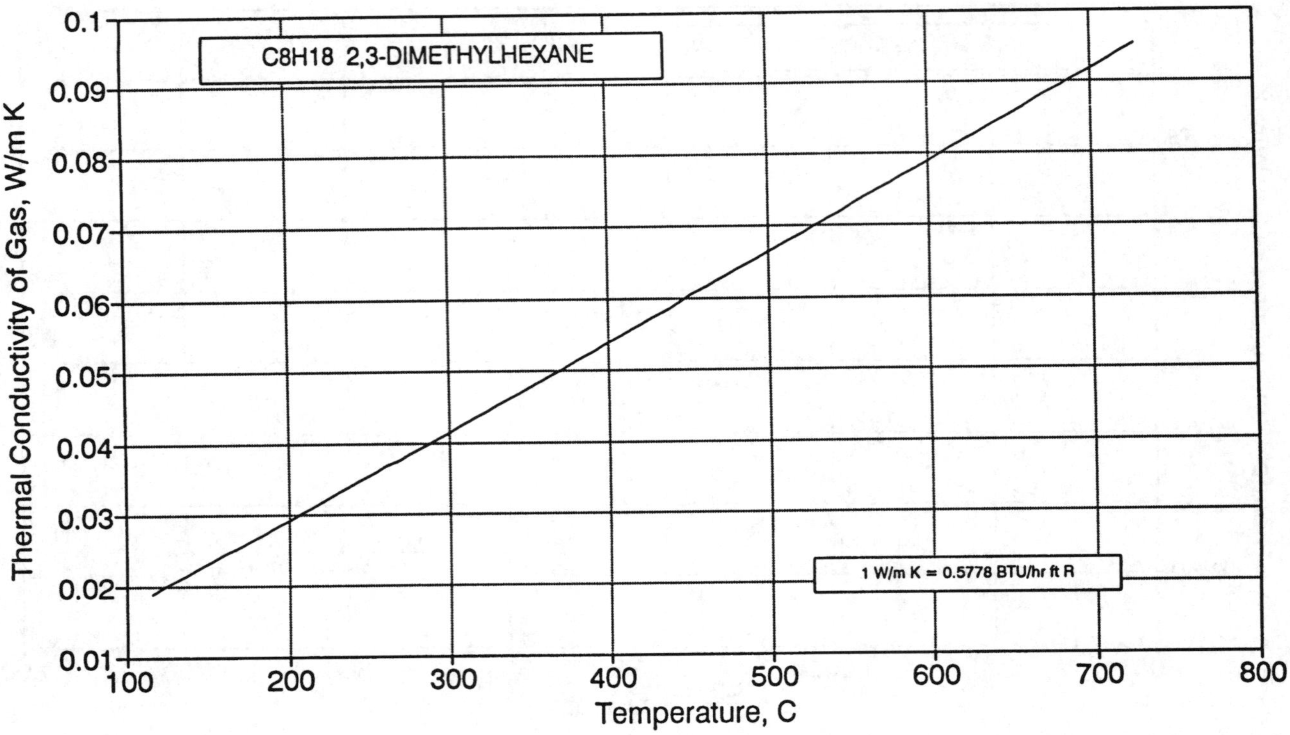
C8H18 2,3-DIMETHYLHEXANE
1 W/m K = 0.5778 BTU/hr ft R
Thermal Conductivity of Gas, W/m K
0.1
0.09
0.08
0.07
0.06
0.05
0.04
0.03
0.02
0.01
100
200
300
400
500
600
700
800
Temperature, C

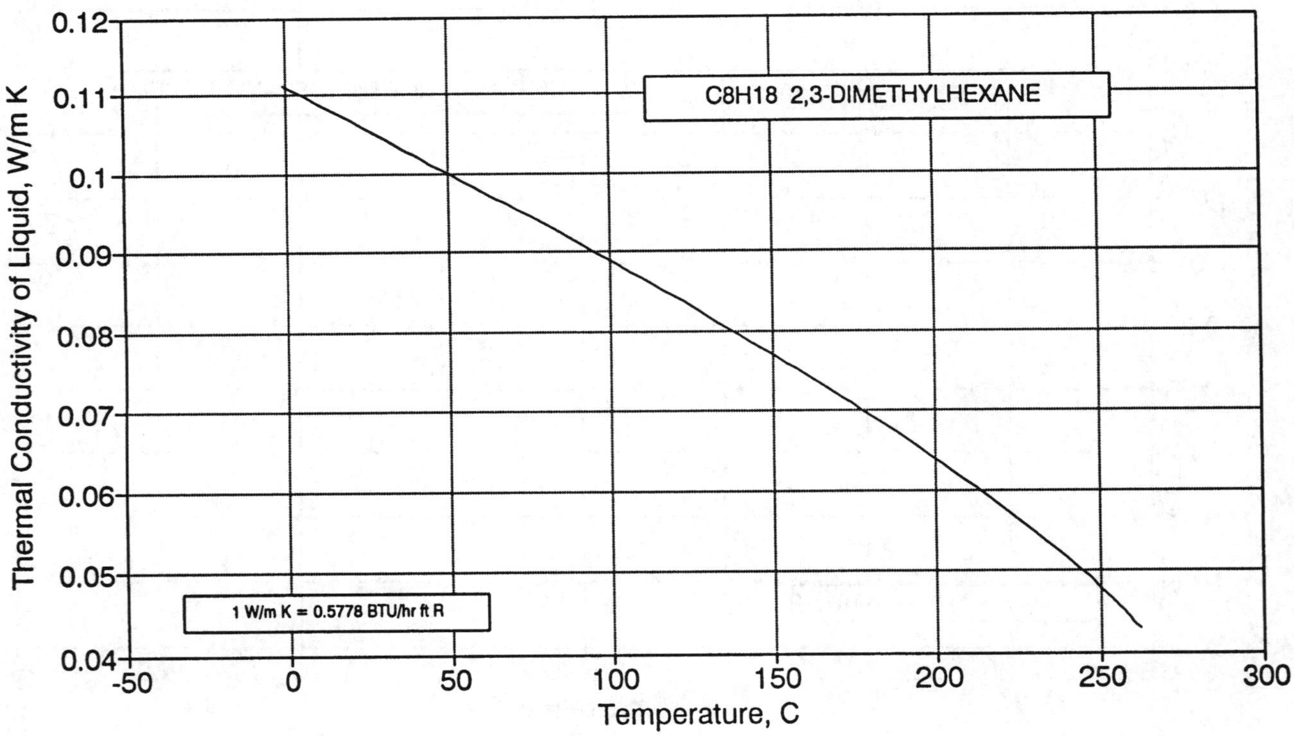
C8H18 2,3-DIMETHYLHEXANE
1 W/m K = 0.5778 BTU/hr ft R
Thermal Conductivity of Liquid, W/m K
0.12
0.11
0.1
0.09
0.08
0.07
0.06
0.05
0.04
-50
0
50
100
150
200
250
300
Temperature, C

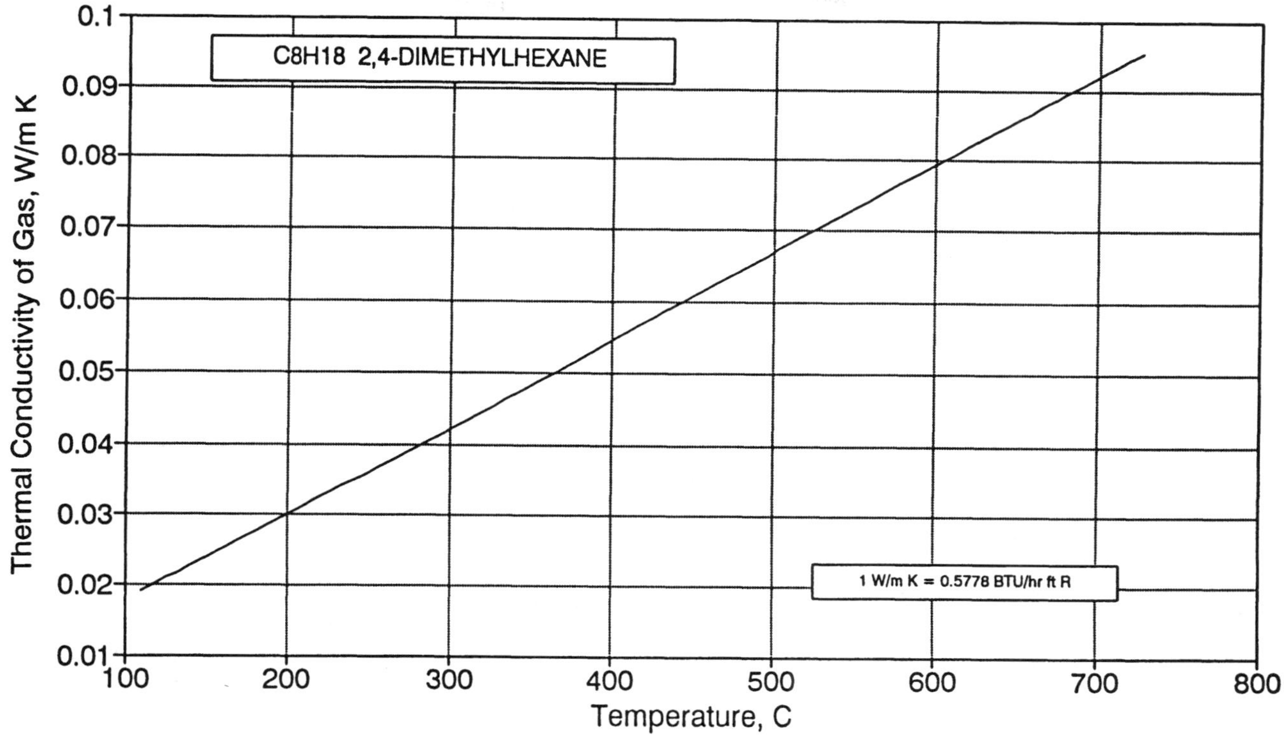
C8H18 2,4-DIMETHYLHEXANE
1 W/m K = 0.5778 BTU/hr ft R
Thermal Conductivity of Gas, W/m K
Temperature, C
0.1
0.09
0.08
0.07
0.06
0.05
0.04
0.03
0.02
0.01
100
200
300
400
500
600
700
800

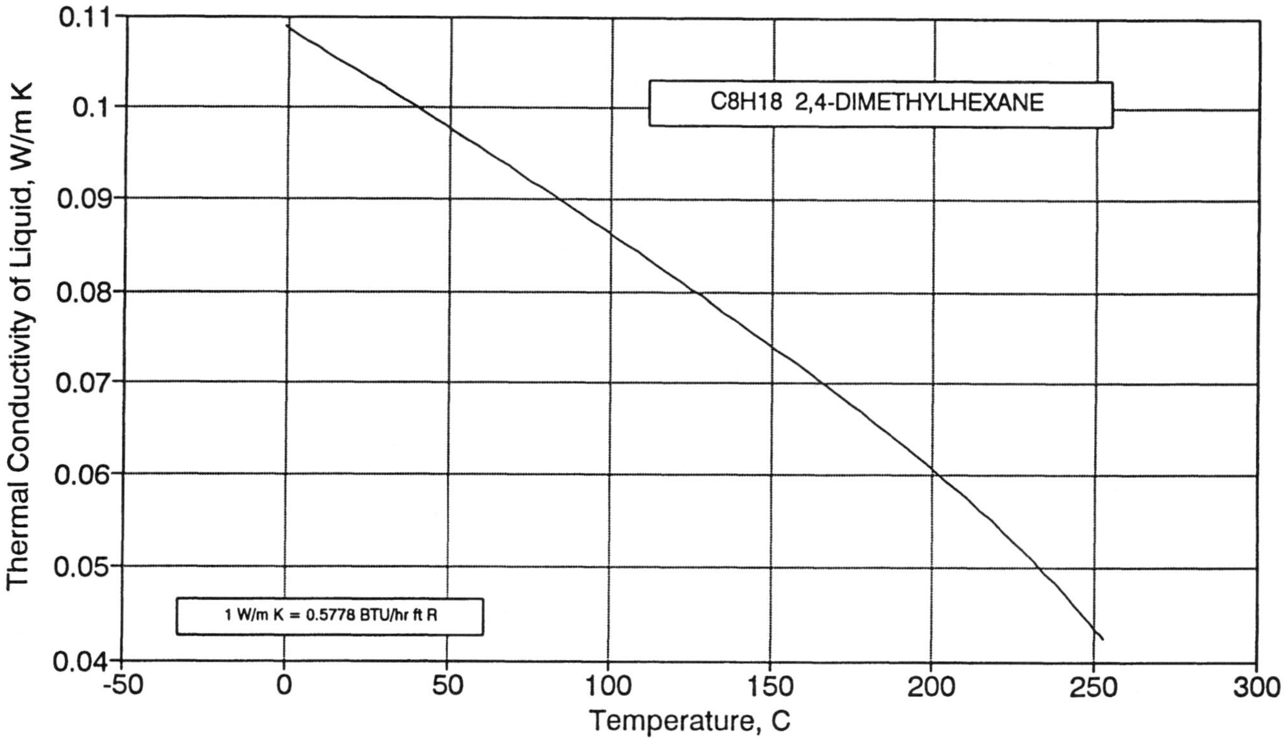
C8H18 2,4-DIMETHYLHEXANE
1 W/m K = 0.5778 BTU/hr ft R
Thermal Conductivity of Liquid, W/m K
Temperature, C
0.11
0.1
0.09
0.08
0.07
0.06
0.05
0.04
-50
0
50
100
150
200
250
300

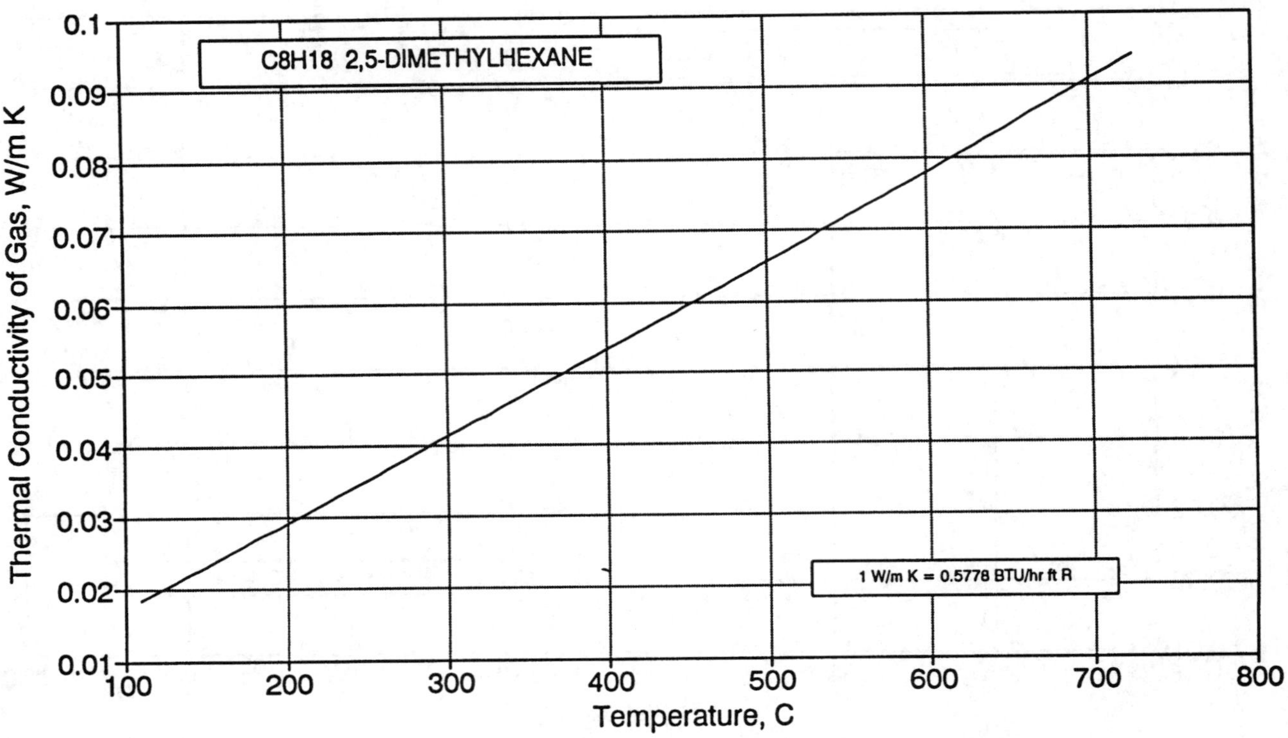
C8H18 2,5-DIMETHYLHEXANE
1 W/m K = 0.5778 BTU/hr ft R
Thermal Conductivity of Gas, W/m K
0.1
0.09
0.08
0.07
0.06
0.05
0.04
0.03
0.02
0.01
100
200
300
400
500
600
700
800
Temperature, C

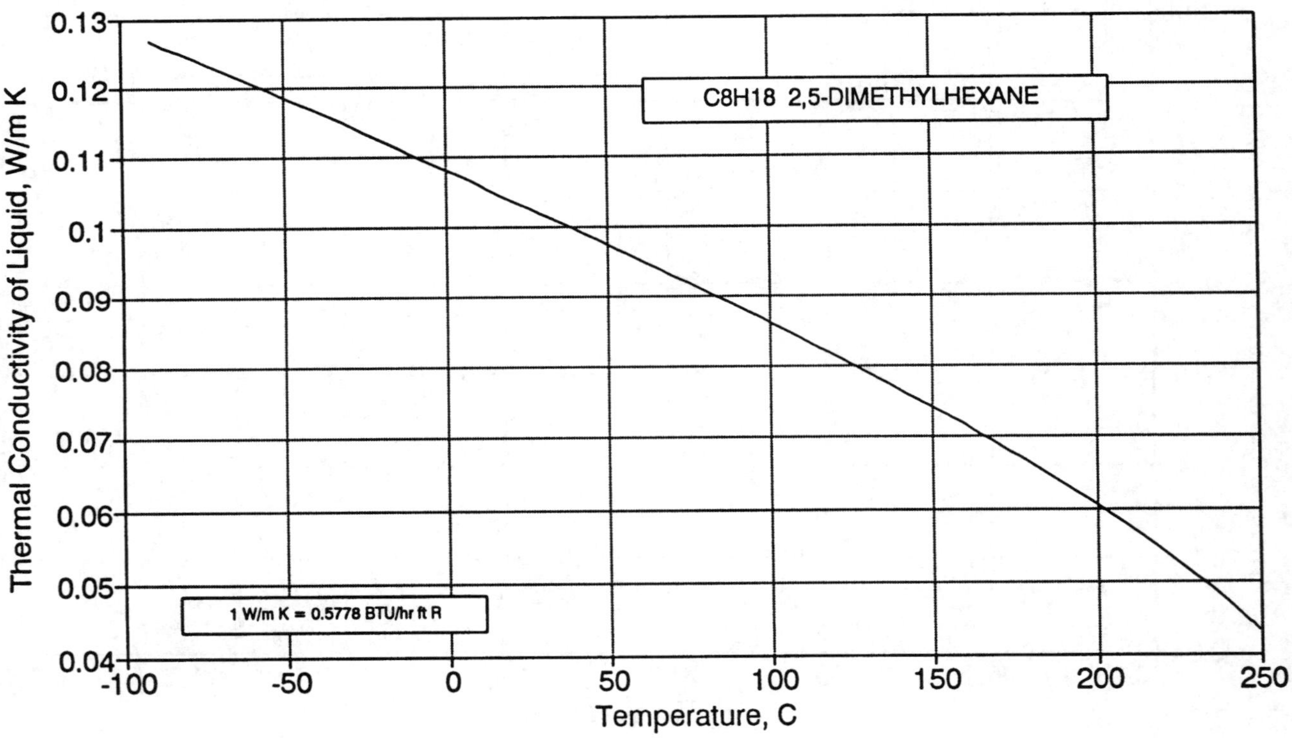
C8H18 2,5-DIMETHYLHEXANE
1 W/m K = 0.5778 BTU/hr ft R
Thermal Conductivity of Liquid, W/m K
0.13
0.12
0.11
0.1
0.09
0.08
0.07
0.06
0.05
0.04
-100
-50
0
50
100
150
200
250
Temperature, C

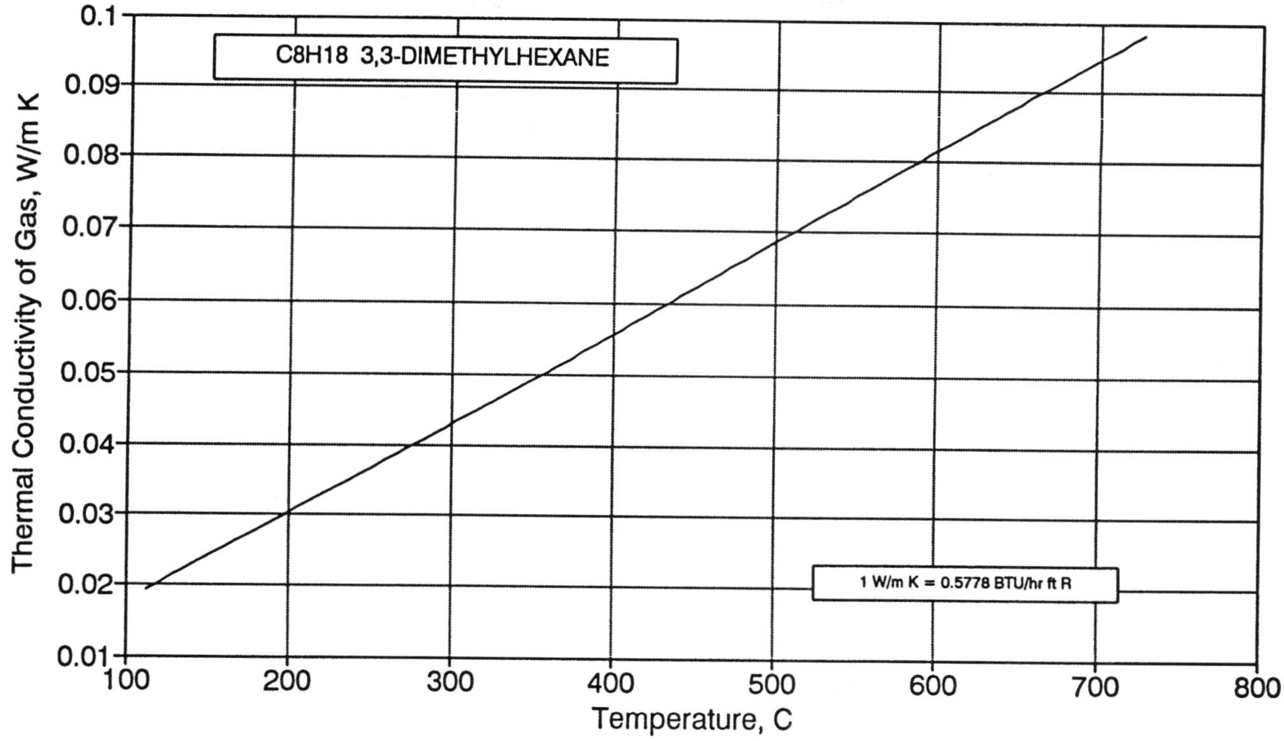
C8H18 3,3-DIMETHYLHEXANE
1 W/m K = 0.5778 BTU/hr ft R
Thermal Conductivity of Gas, W/m K
Temperature, C
0.1
0.09
0.08
0.07
0.06
0.05
0.04
0.03
0.02
0.01
100
200
300
400
500
600
700
800

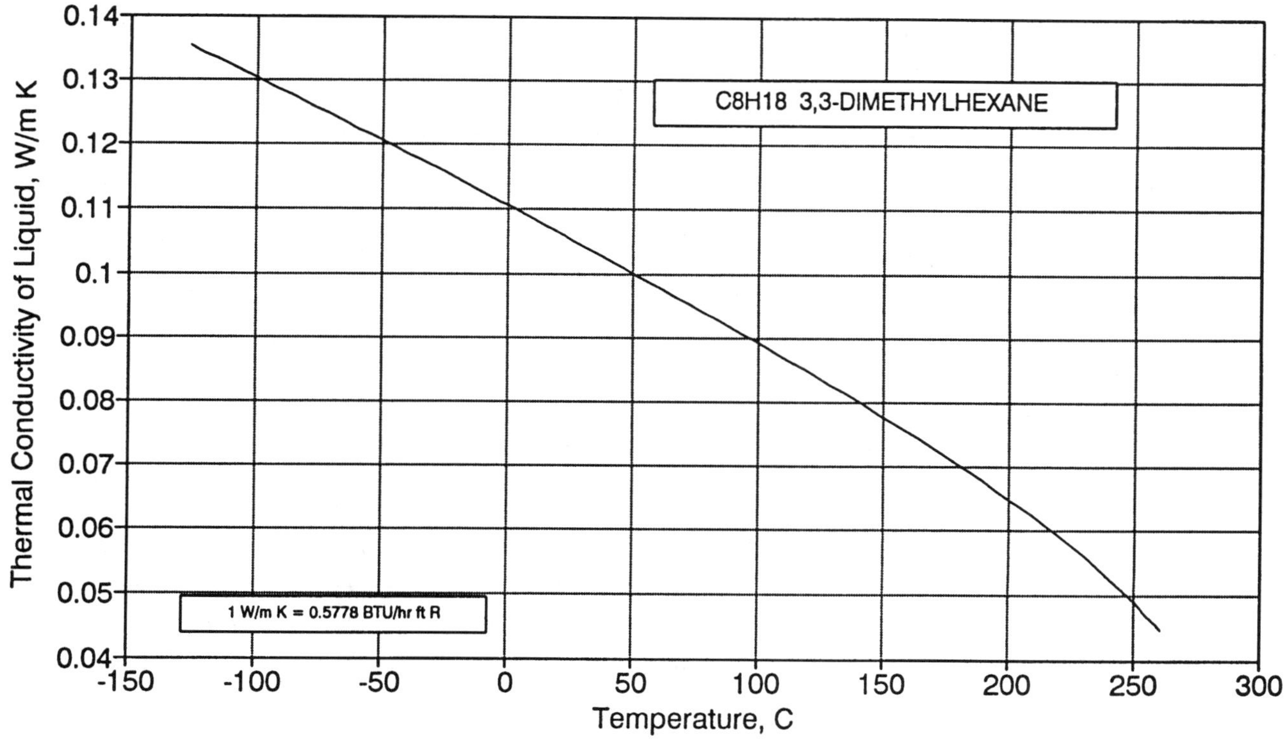
C8H18 3,3-DIMETHYLHEXANE
1 W/m K = 0.5778 BTU/hr ft R
Thermal Conductivity of Liquid, W/m K
Temperature, C
0.14
0.13
0.12
0.11
0.1
0.09
0.08
0.07
0.06
0.05
0.04
-150
-100
-50
0
50
100
150
200
250
300

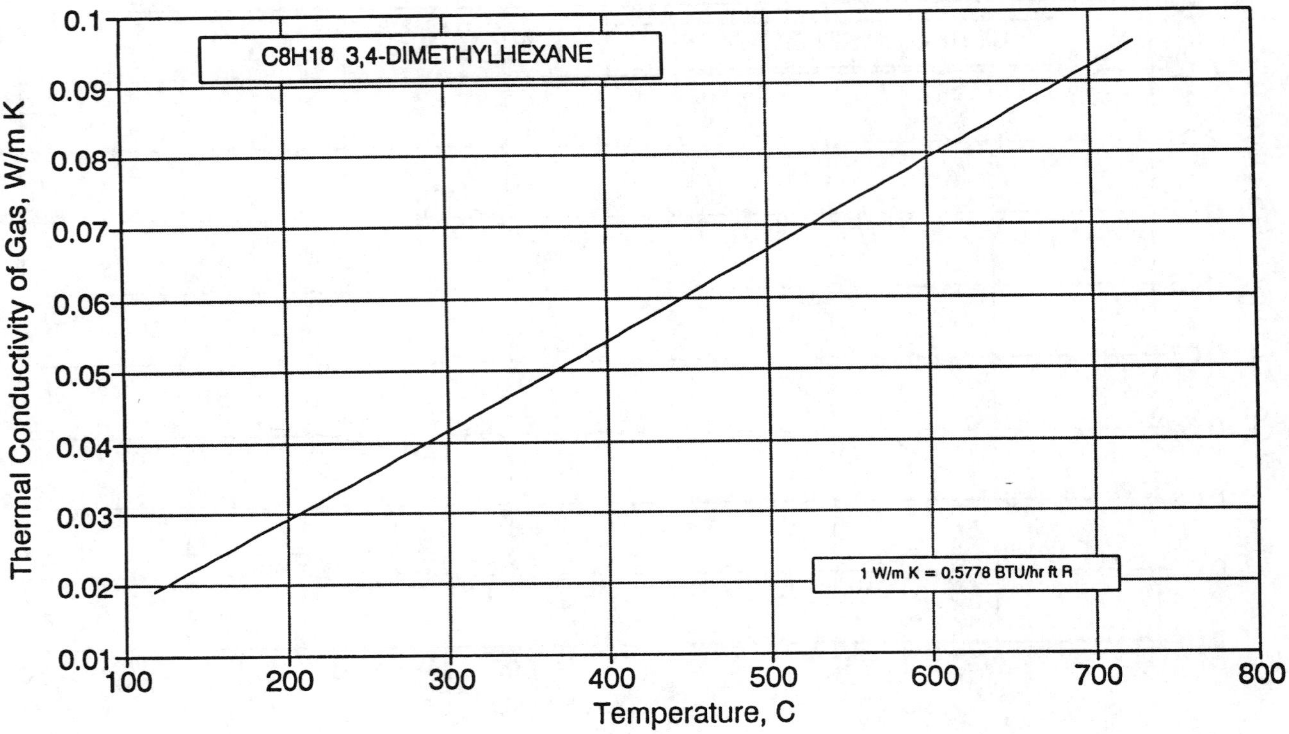

C8H18 3,4-DIMETHYLHEXANE
1 W/m K = 0.5778 BTU/hr ft R
Thermal Conductivity of Gas, W/m K
0.1
0.09
0.08
0.07
0.06
0.05
0.04
0.03
0.02
0.01
100
200
300
400
500
600
700
800
Temperature, C

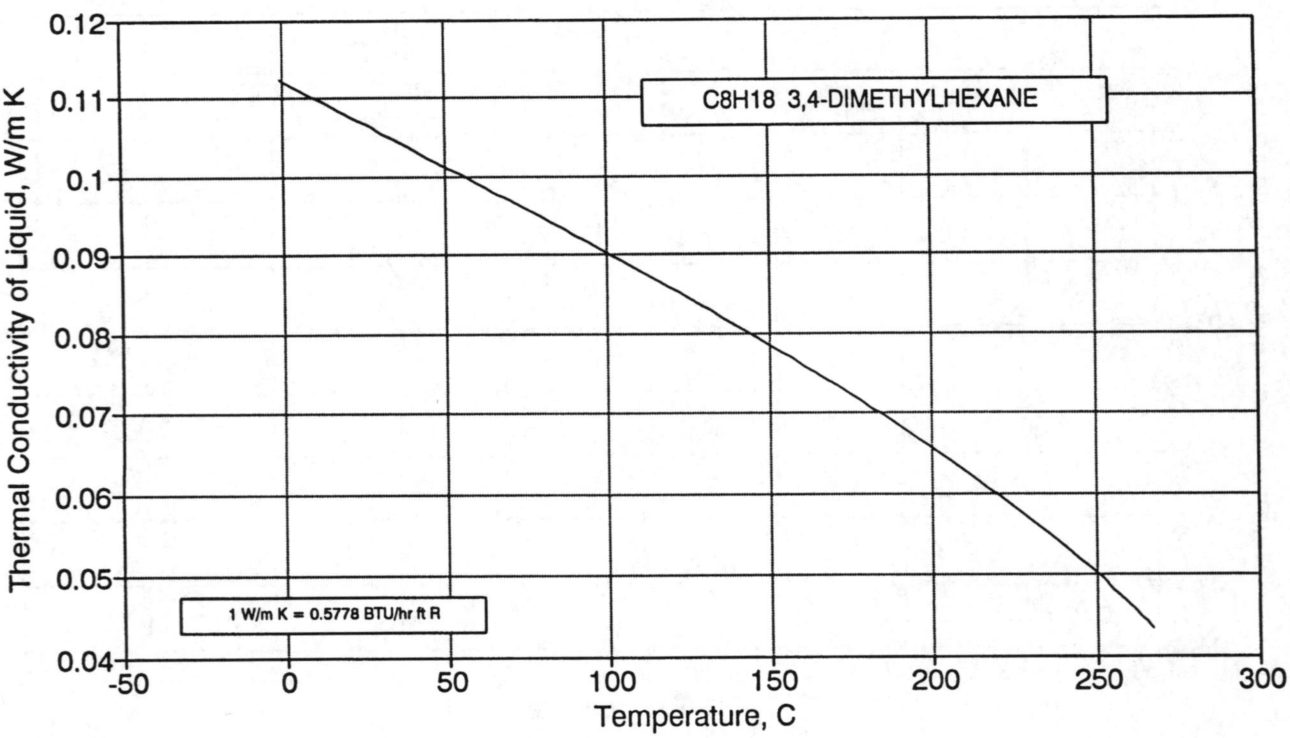

C8H18 3,4-DIMETHYLHEXANE
1 W/m K = 0.5778 BTU/hr ft R
Thermal Conductivity of Liquid, W/m K
0.12
0.11
0.1
0.09
0.08
0.07
0.06
0.05
0.04
-50
0
50
100
150
200
250
300
Temperature, C

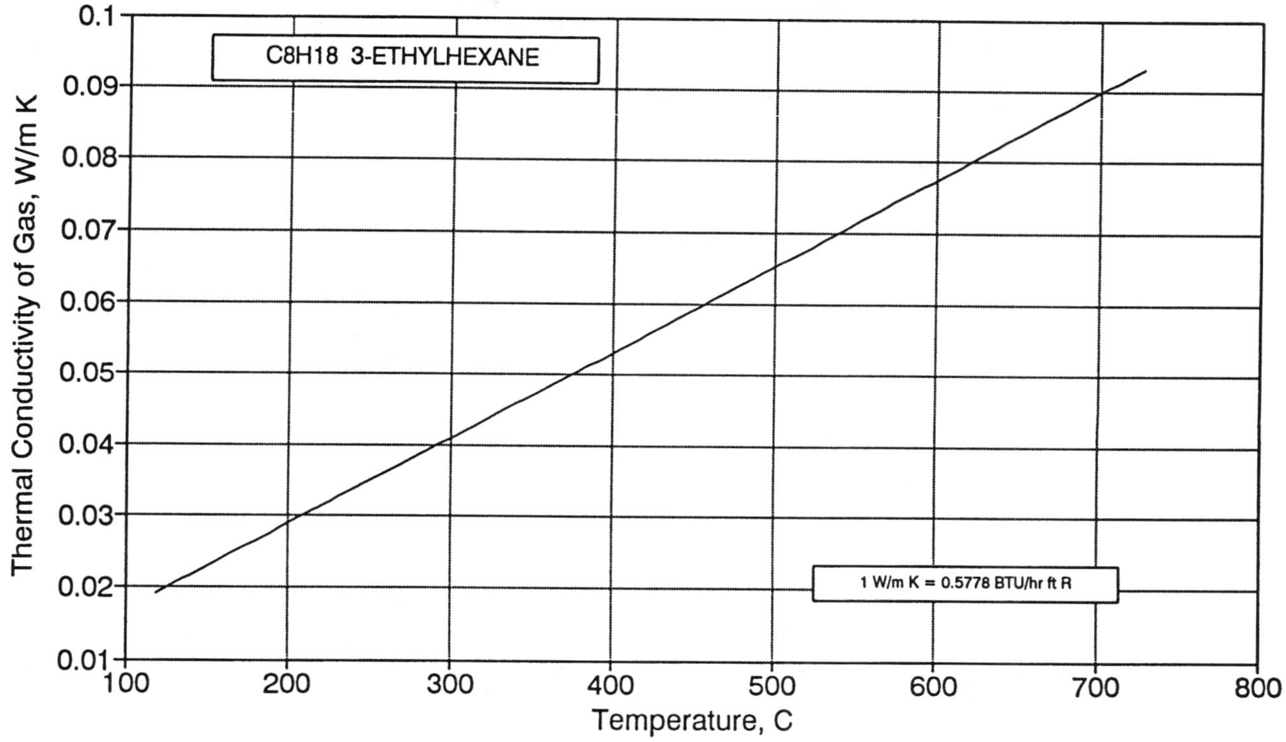
C8H18 3-ETHYLHEXANE
1 W/m K = 0.5778 BTU/hr ft R
Thermal Conductivity of Gas, W/m K
0.1
0.09
0.08
0.07
0.06
0.05
0.04
0.03
0.02
0.01
100
200
300
400
500
600
700
800
Temperature, C

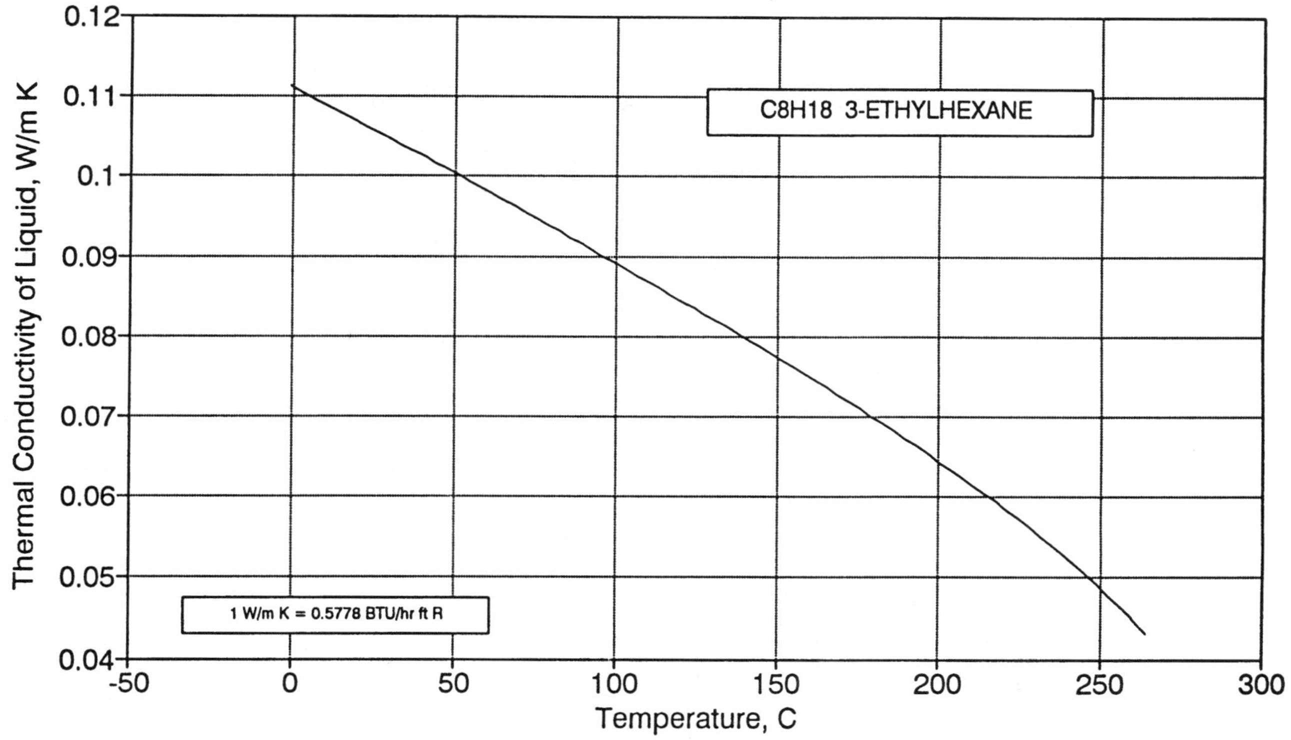
C8H18 3-ETHYLHEXANE
1 W/m K = 0.5778 BTU/hr ft R
Thermal Conductivity of Liquid, W/m K
0.12
0.11
0.1
0.09
0.08
0.07
0.06
0.05
0.04
-50
0
50
100
150
200
250
300
Temperature, C

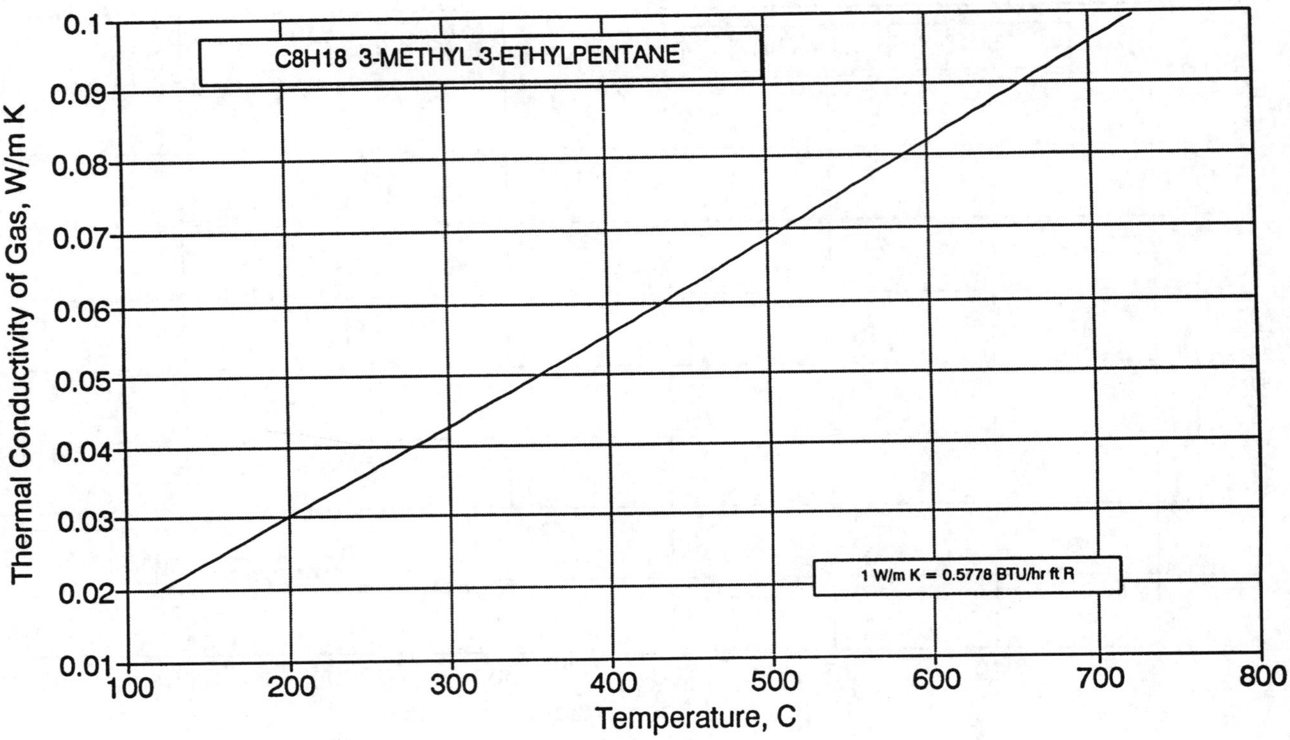
C8H18 3-METHYL-3-ETHYLPENTANE
1 W/m K = 0.5778 BTU/hr ft R
Thermal Conductivity of Gas, W/m K
0.1
0.09
0.08
0.07
0.06
0.05
0.04
0.03
0.02
0.01
100
200
300
400
500
600
700
800
Temperature, C

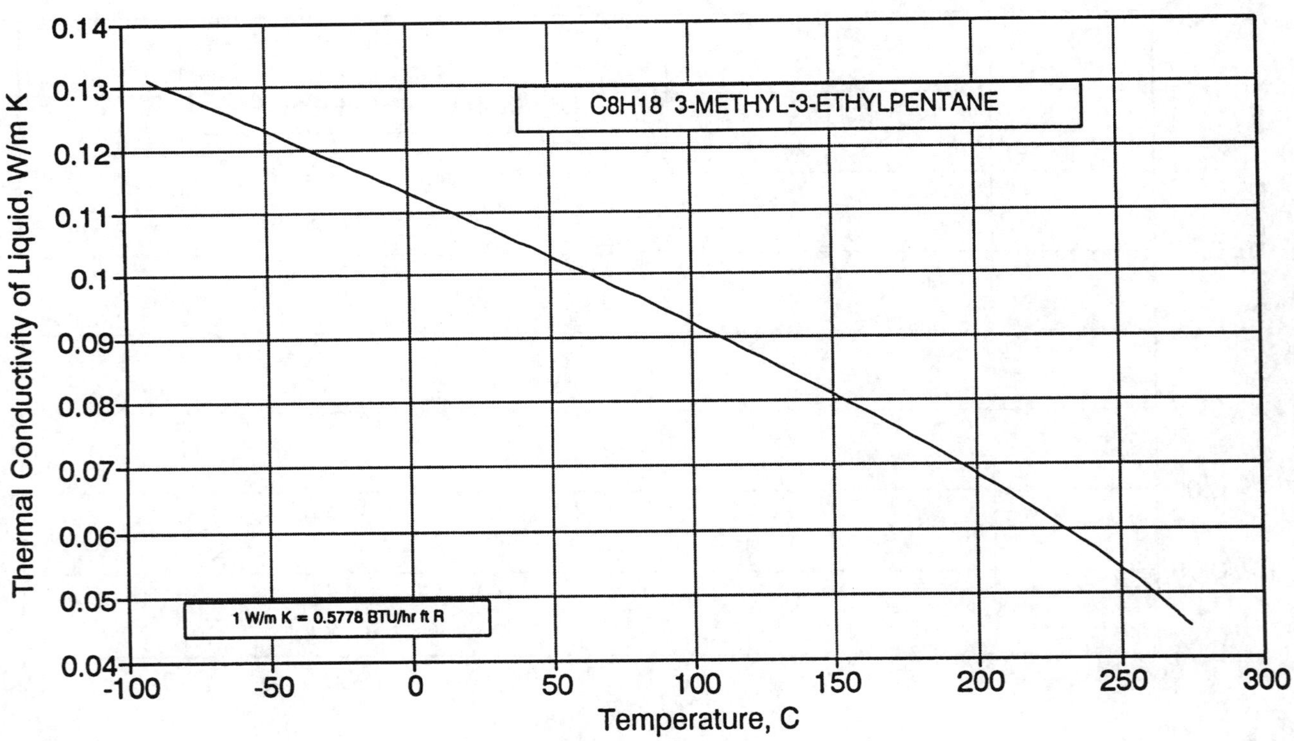
C8H18 3-METHYL-3-ETHYLPENTANE
1 W/m K = 0.5778 BTU/hr ft R
Thermal Conductivity of Liquid, W/m K
0.14
0.13
0.12
0.11
0.1
0.09
0.08
0.07
0.06
0.05
0.04
-100
-50
0
50
100
150
200
250
300
Temperature, C

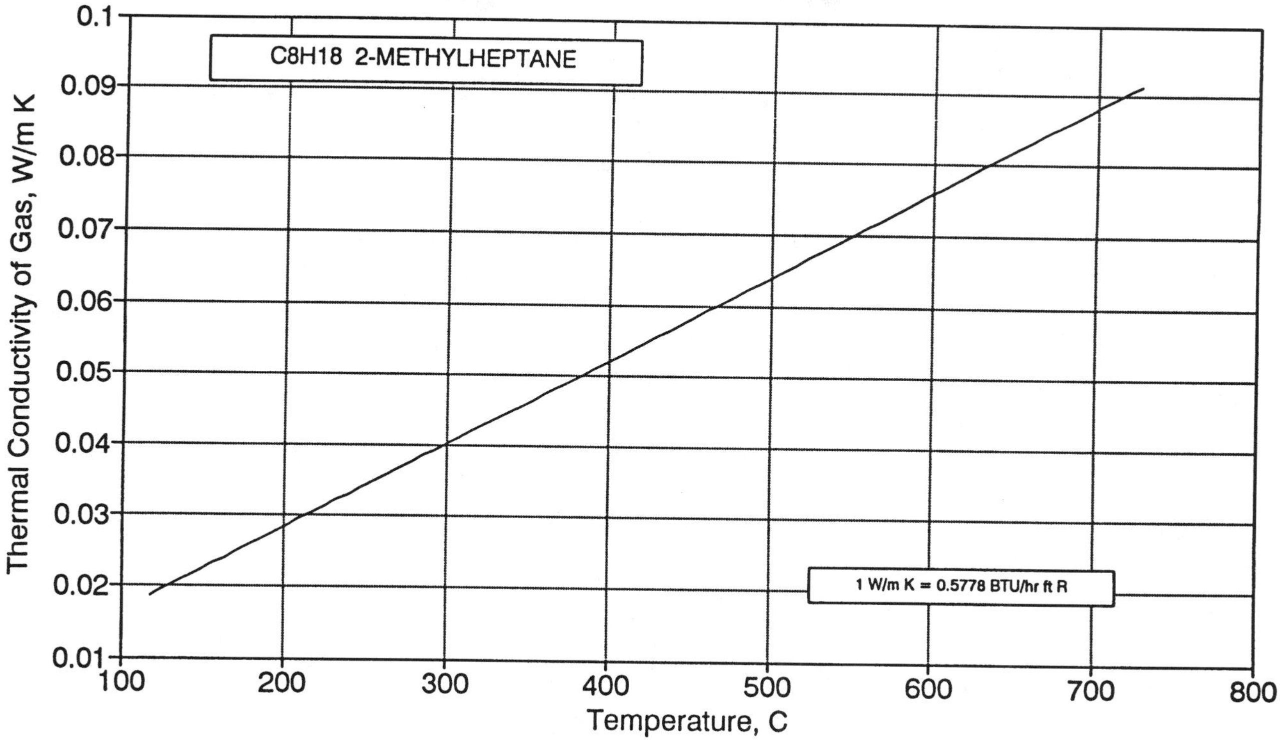
C8H18 2-METHYLHEPTANE
1 W/m K = 0.5778 BTU/hr ft R
Thermal Conductivity of Gas, W/m K
Temperature, C
0.1
0.09
0.08
0.07
0.06
0.05
0.04
0.03
0.02
0.01
100
200
300
400
500
600
700
800

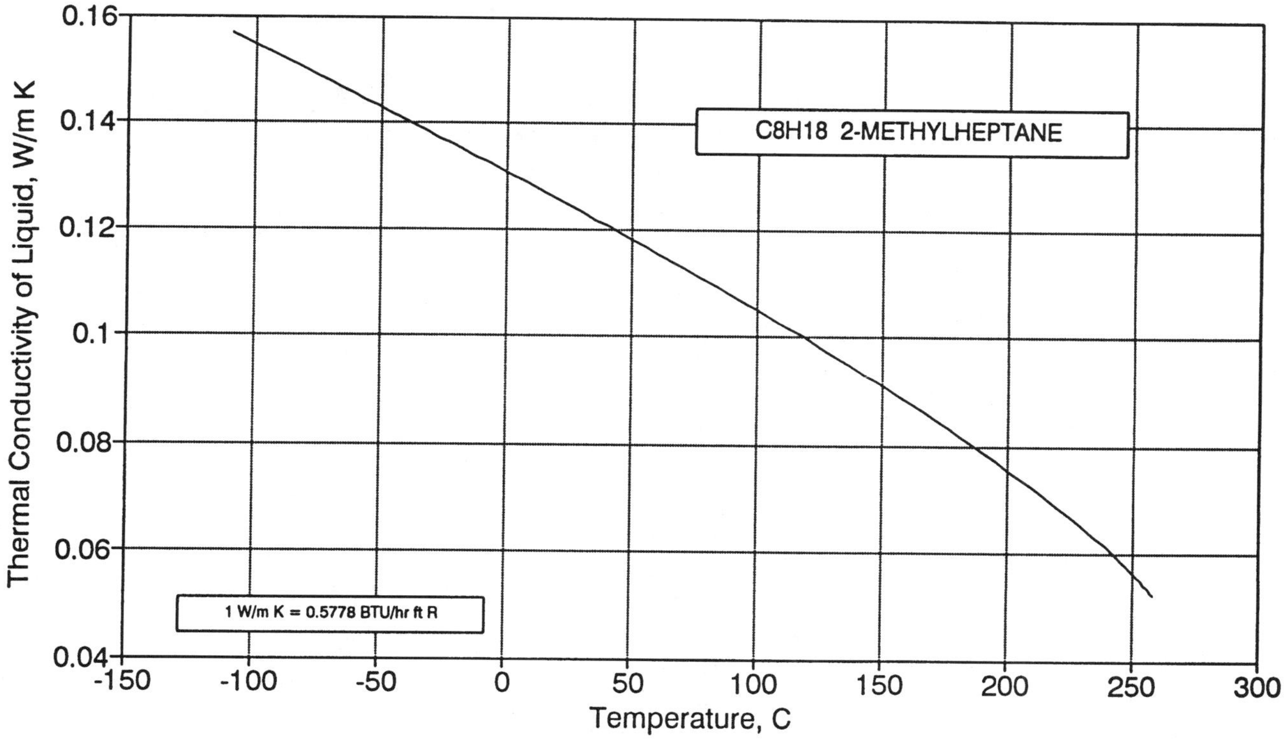
C8H18 2-METHYLHEPTANE
1 W/m K = 0.5778 BTU/hr ft R
Thermal Conductivity of Liquid, W/m K
Temperature, C
0.16
0.14
0.12
0.1
0.08
0.06
0.04
-150
-100
-50
0
50
100
150
200
250
300

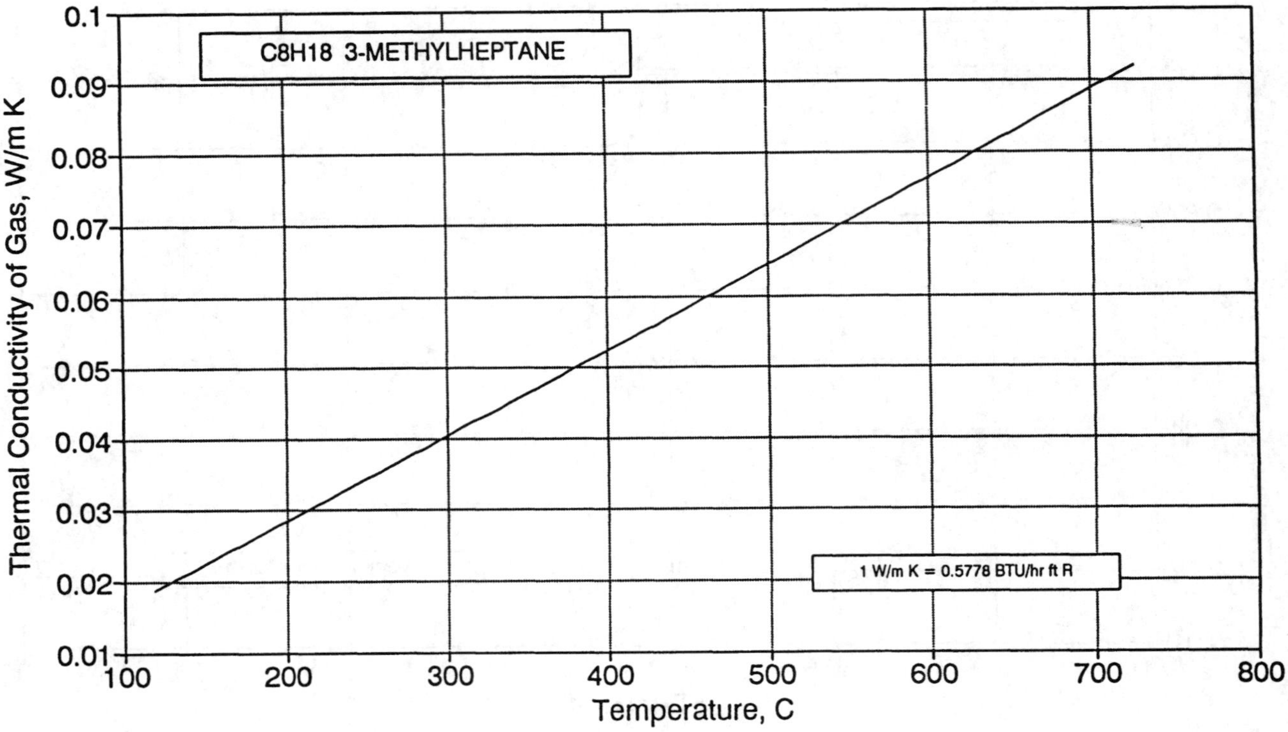
C8H18 3-METHYLHEPTANE
1 W/m K = 0.5778 BTU/hr ft R
Thermal Conductivity of Gas, W/m K
0.1
0.09
0.08
0.07
0.06
0.05
0.04
0.03
0.02
0.01
100
200
300
400
500
600
700
800
Temperature, C

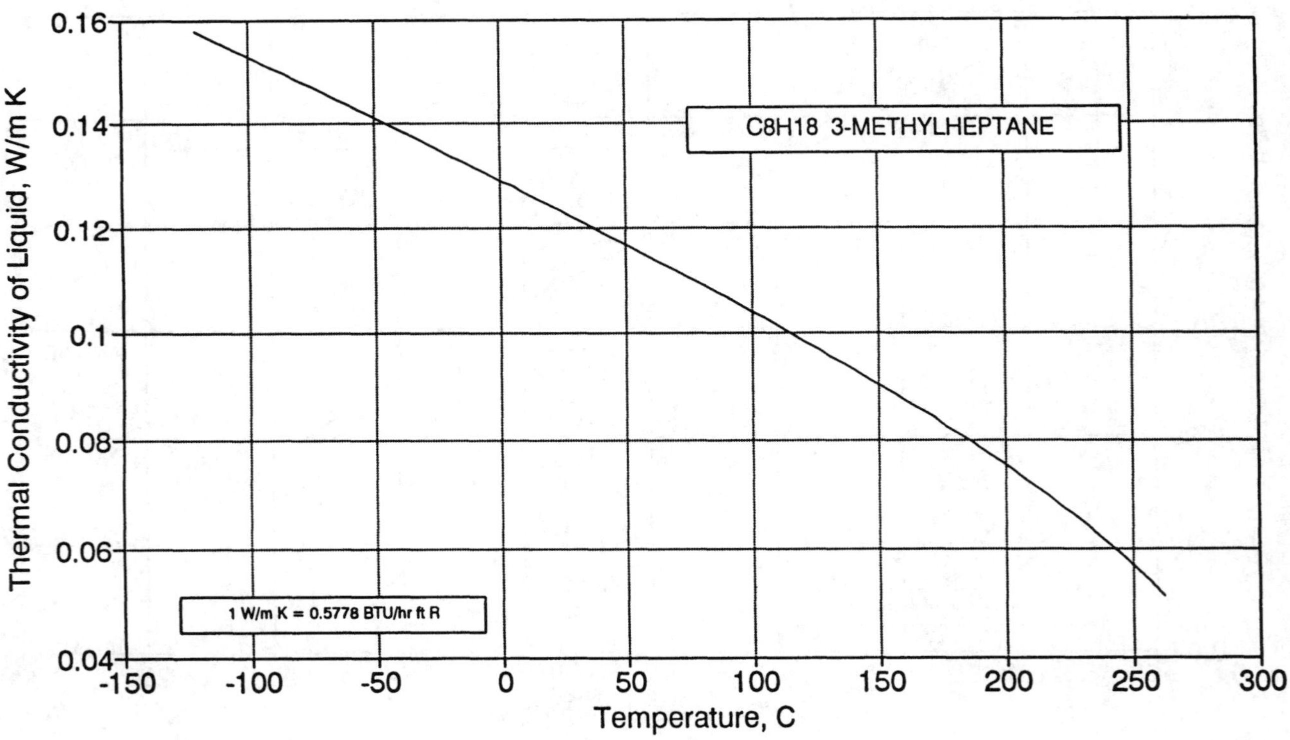
C8H18 3-METHYLHEPTANE
1 W/m K = 0.5778 BTU/hr ft R
Thermal Conductivity of Liquid, W/m K
0.16
0.14
0.12
0.1
0.08
0.06
0.04
-150
-100
-50
0
50
100
150
200
250
300
Temperature, C

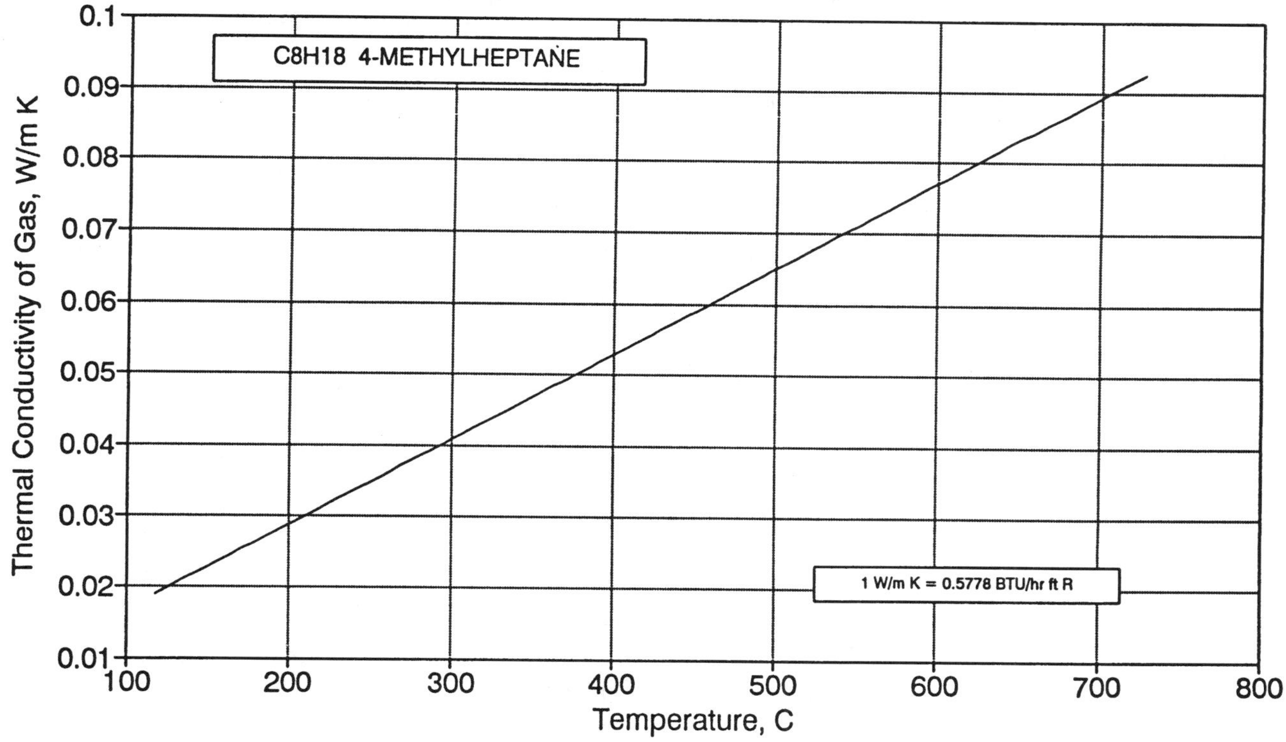
C8H18 4-METHYLHEPTANE
1 W/m K = 0.5778 BTU/hr ft R
Thermal Conductivity of Gas, W/m K
Temperature, C
0.1
0.09
0.08
0.07
0.06
0.05
0.04
0.03
0.02
0.01
100
200
300
400
500
600
700
800

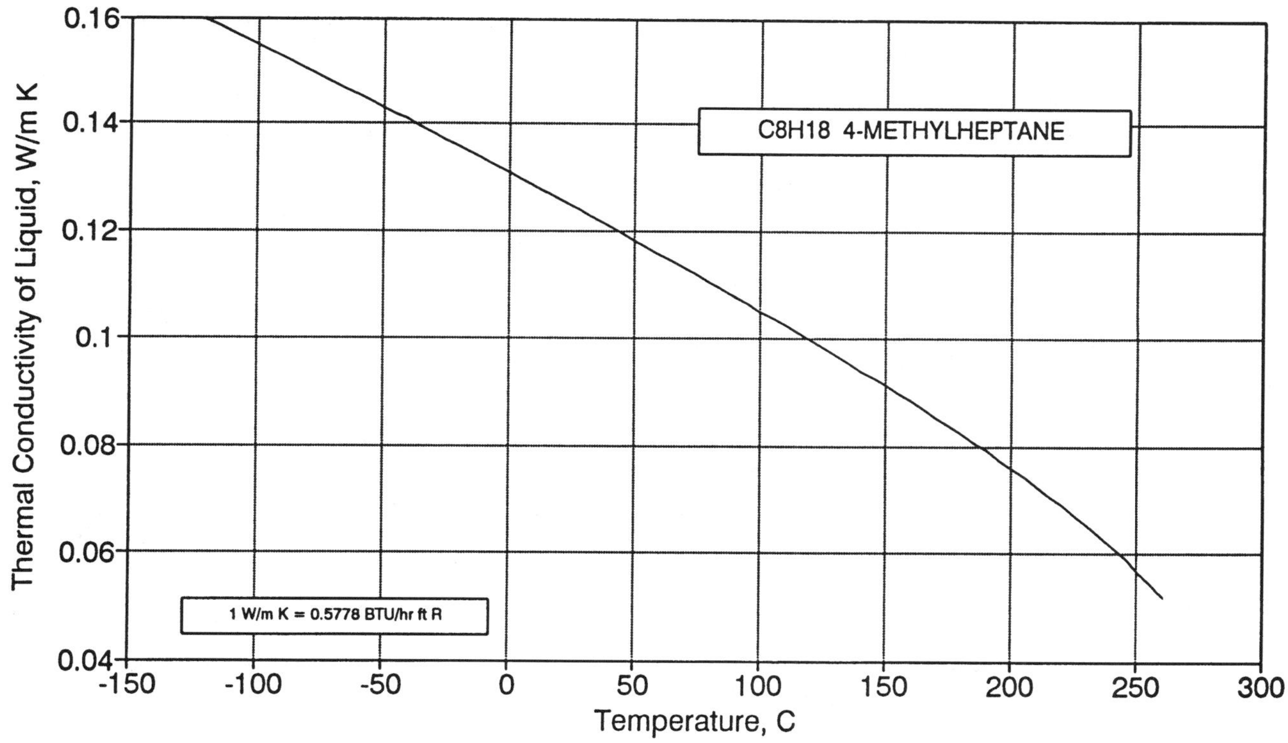
C8H18 4-METHYLHEPTANE
1 W/m K = 0.5778 BTU/hr ft R
Thermal Conductivity of Liquid, W/m K
Temperature, C
0.16
0.14
0.12
0.1
0.08
0.06
0.04
-150
-100
-50
0
50
100
150
200
250
300

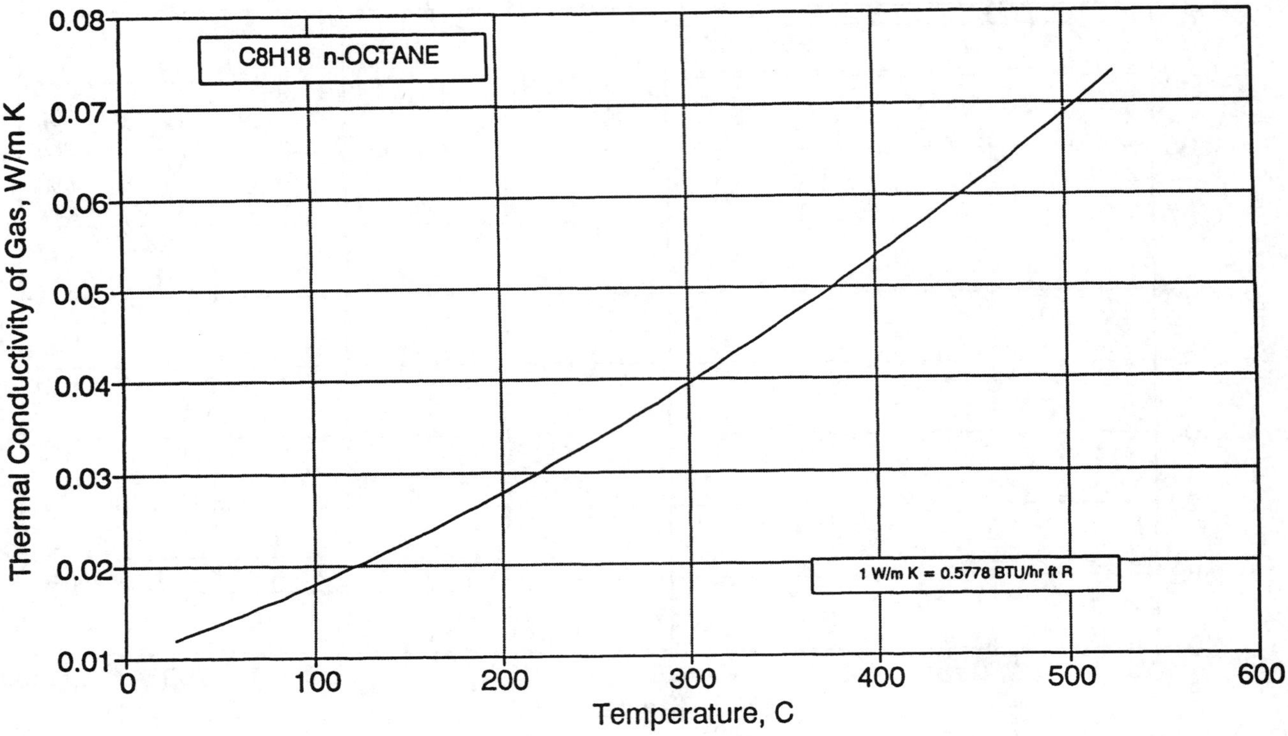
C8H18 n-OCTANE
Thermal Conductivity of Gas, W/m K
0.08
0.07
0.06
0.05
0.04
0.03
0.02
0.01
1 W/m K = 0.5778 BTU/hr ft R
0
100
200
300
400
500
600
Temperature, C

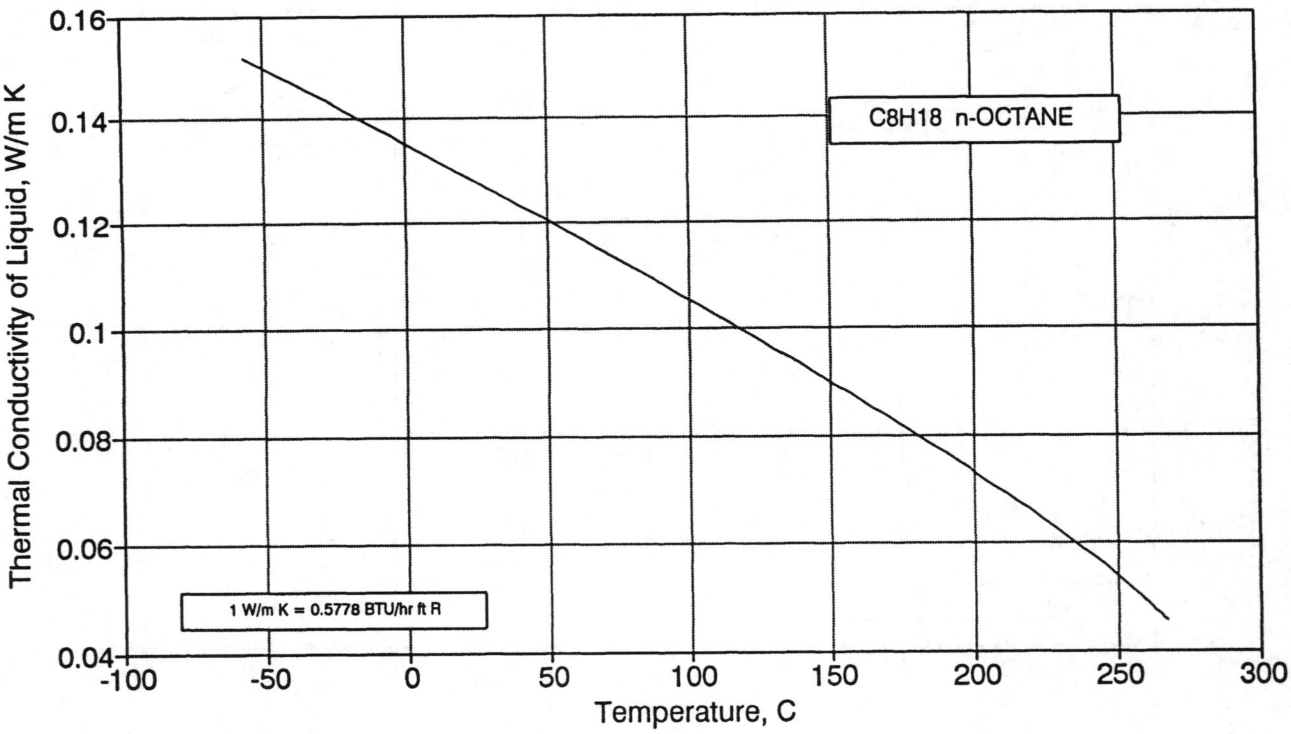
C8H18 n-OCTANE
Thermal Conductivity of Liquid, W/m K
0.16
0.14
0.12
0.1
0.08
0.06
0.04
1 W/m K = 0.5778 BTU/hr ft R
-100
-50
0
50
100
150
200
250
300
Temperature, C

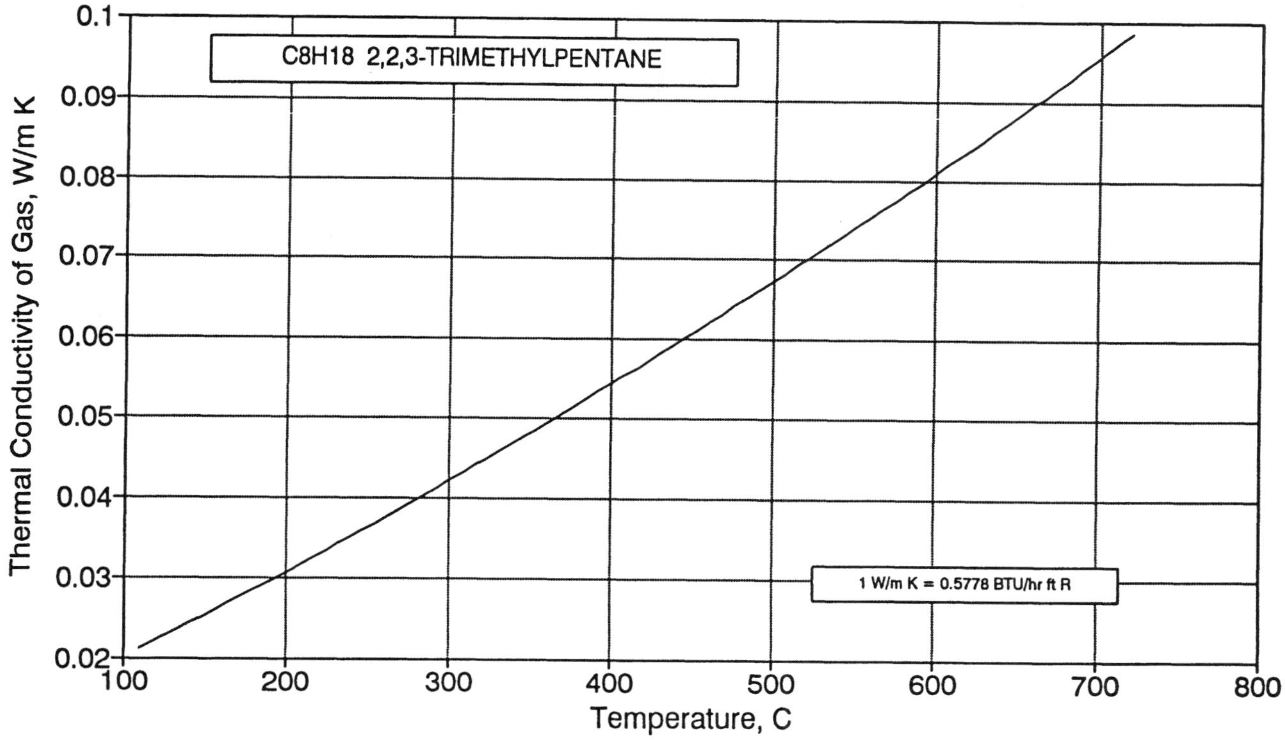

C8H18 2,2,3-TRIMETHYLPENTANE
1 W/m K = 0.5778 BTU/hr ft R
Thermal Conductivity of Gas, W/m K
0.1
0.09
0.08
0.07
0.06
0.05
0.04
0.03
0.02
100
200
300
400
500
600
700
800
Temperature, C

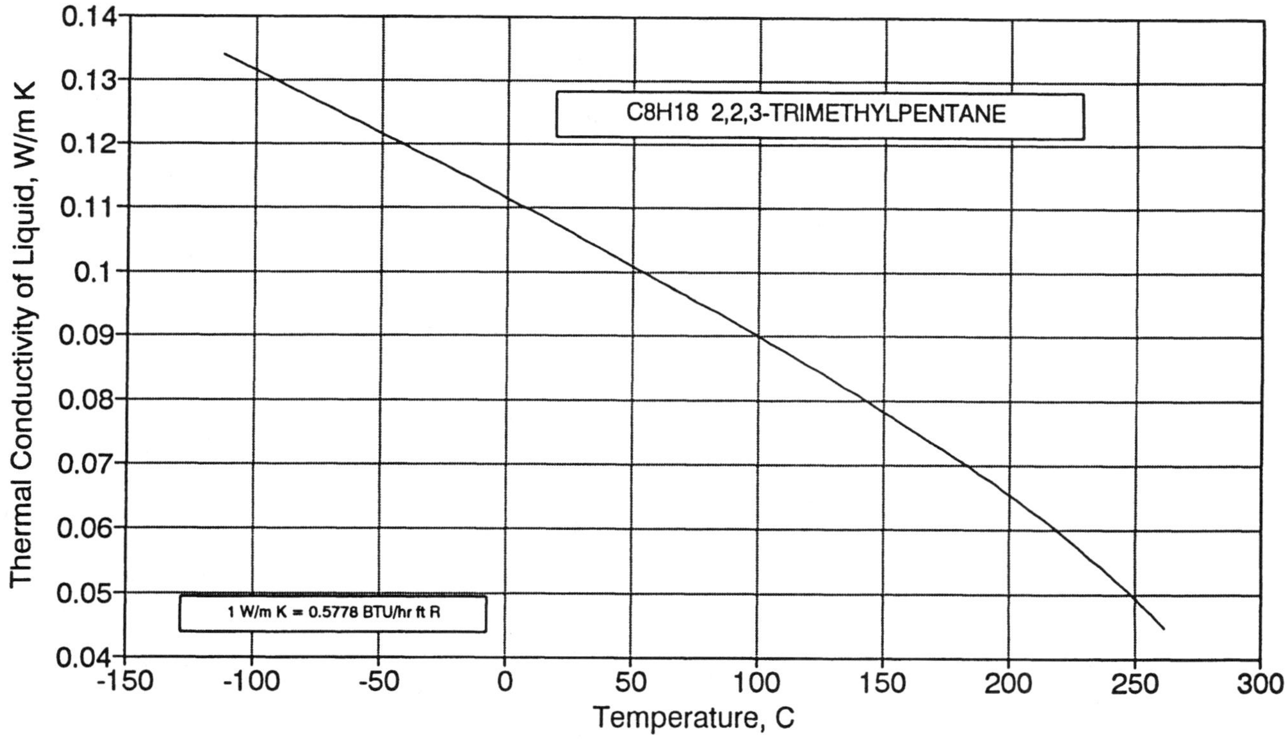

C8H18 2,2,3-TRIMETHYLPENTANE
1 W/m K = 0.5778 BTU/hr ft R
Thermal Conductivity of Liquid, W/m K
0.14
0.13
0.12
0.11
0.1
0.09
0.08
0.07
0.06
0.05
0.04
-150
-100
-50
0
50
100
150
200
250
300
Temperature, C

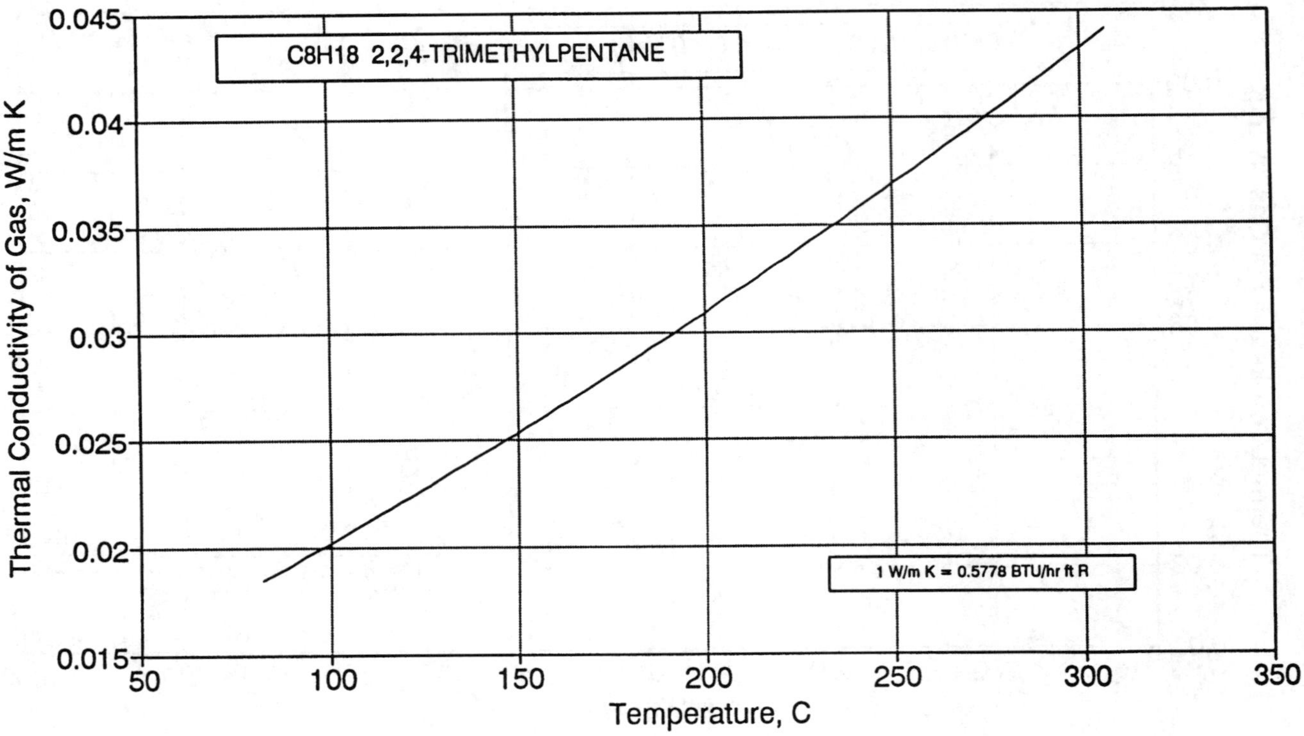
C8H18 2,2,4-TRIMETHYLPENTANE
1 W/m K = 0.5778 BTU/hr ft R
Thermal Conductivity of Gas, W/m K
0.045
0.04
0.035
0.03
0.025
0.02
0.015
50
100
150
200
250
300
350
Temperature, C

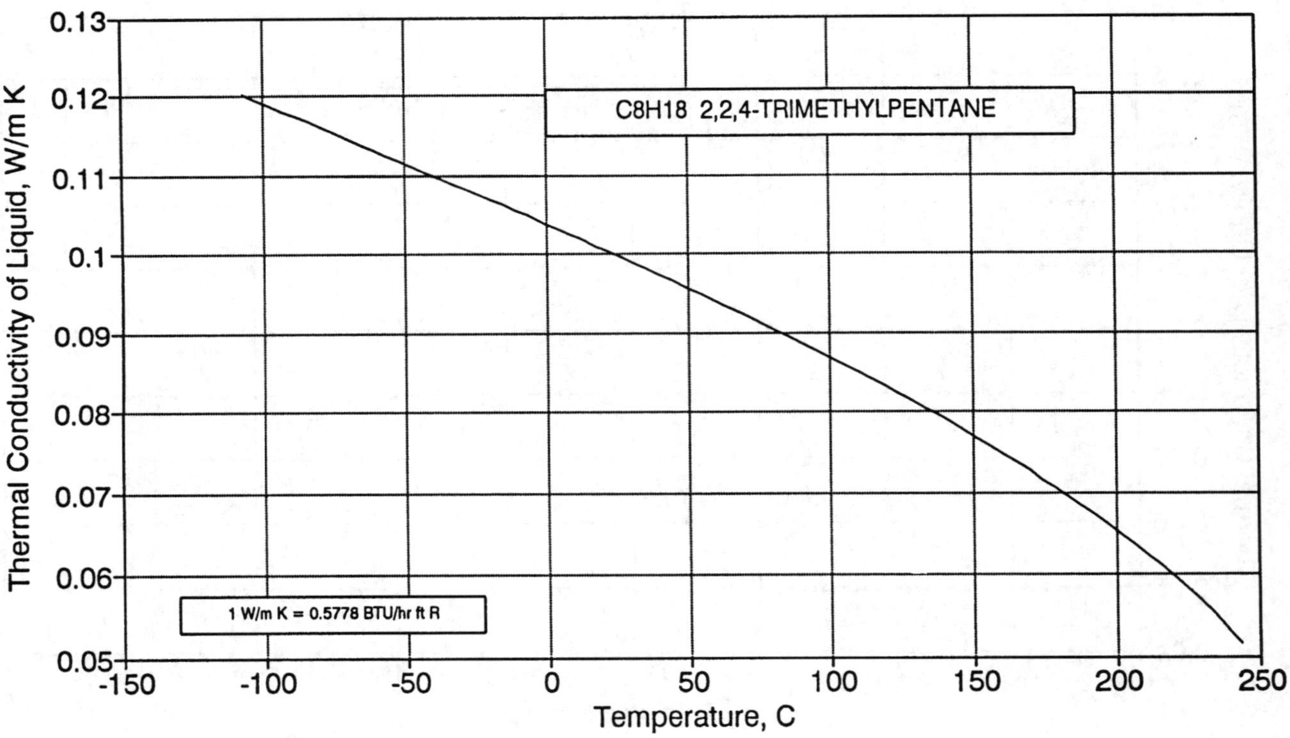
C8H18 2,2,4-TRIMETHYLPENTANE
1 W/m K = 0.5778 BTU/hr ft R
Thermal Conductivity of Liquid, W/m K
0.13
0.12
0.11
0.1
0.09
0.08
0.07
0.06
0.05
-150
-100
-50
0
50
100
150
200
250
Temperature, C

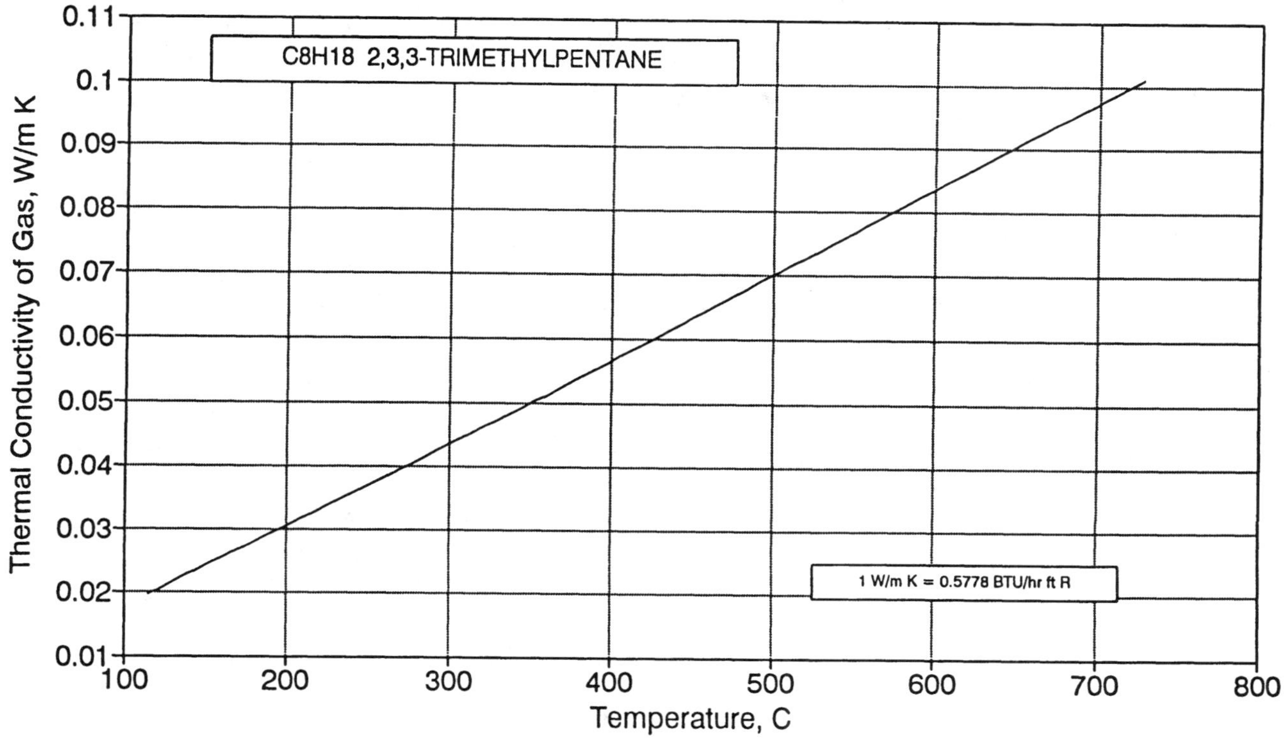
C8H18 2,3,3-TRIMETHYLPENTANE
1 W/m K = 0.5778 BTU/hr ft R
Thermal Conductivity of Gas, W/m K
0.11
0.1
0.09
0.08
0.07
0.06
0.05
0.04
0.03
0.02
0.01
100
200
300
400
500
600
700
800
Temperature, C

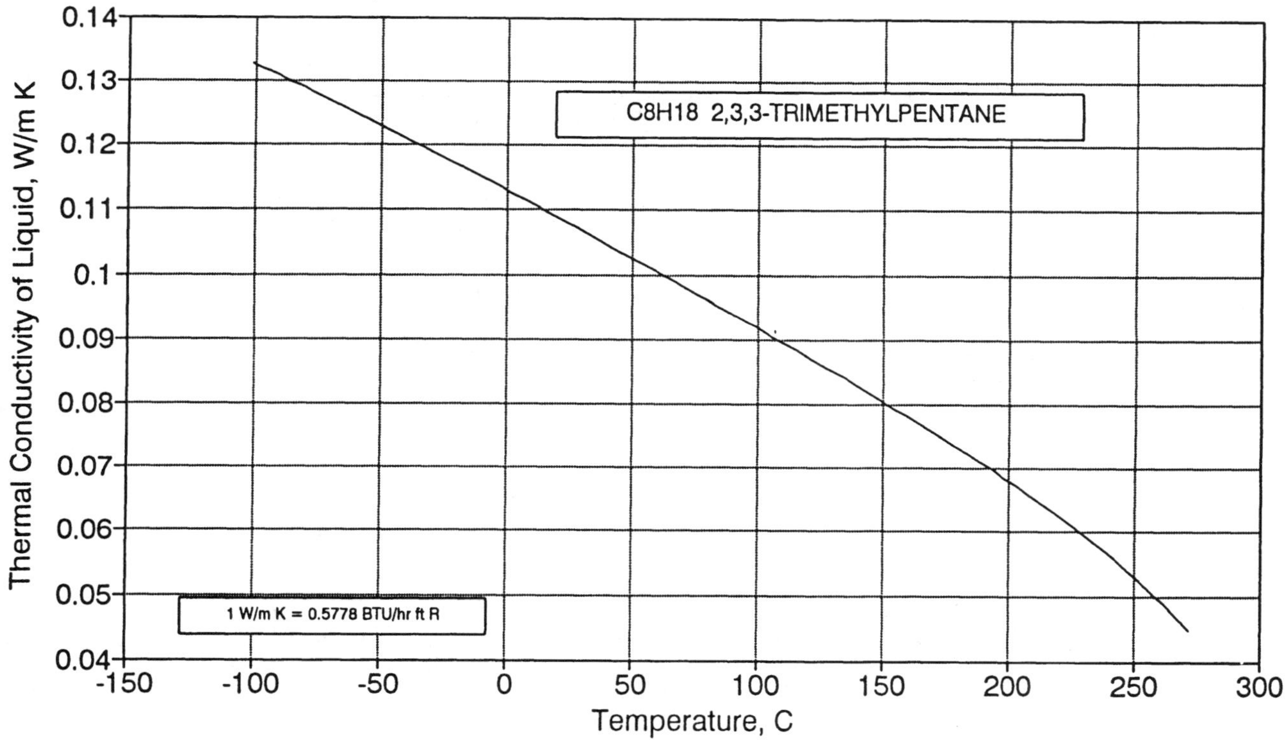
C8H18 2,3,3-TRIMETHYLPENTANE
1 W/m K = 0.5778 BTU/hr ft R
Thermal Conductivity of Liquid, W/m K
0.14
0.13
0.12
0.11
0.1
0.09
0.08
0.07
0.06
0.05
0.04
-150
-100
-50
0
50
100
150
200
250
300
Temperature, C

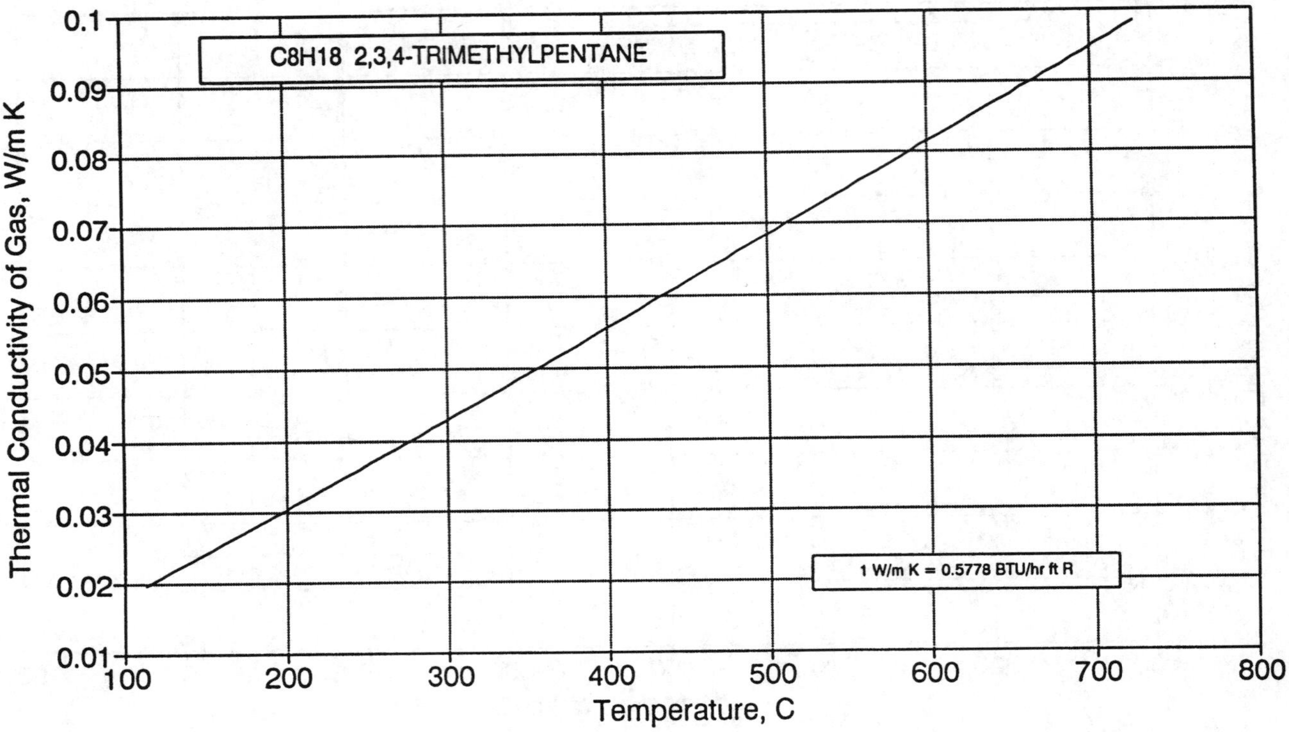

C8H18 2,3,4-TRIMETHYLPENTANE
1 W/m K = 0.5778 BTU/hr ft R
Thermal Conductivity of Gas, W/m K
0.01
0.02
0.03
0.04
0.05
0.06
0.07
0.08
0.09
0.1
100
200
300
400
500
600
700
800
Temperature, C

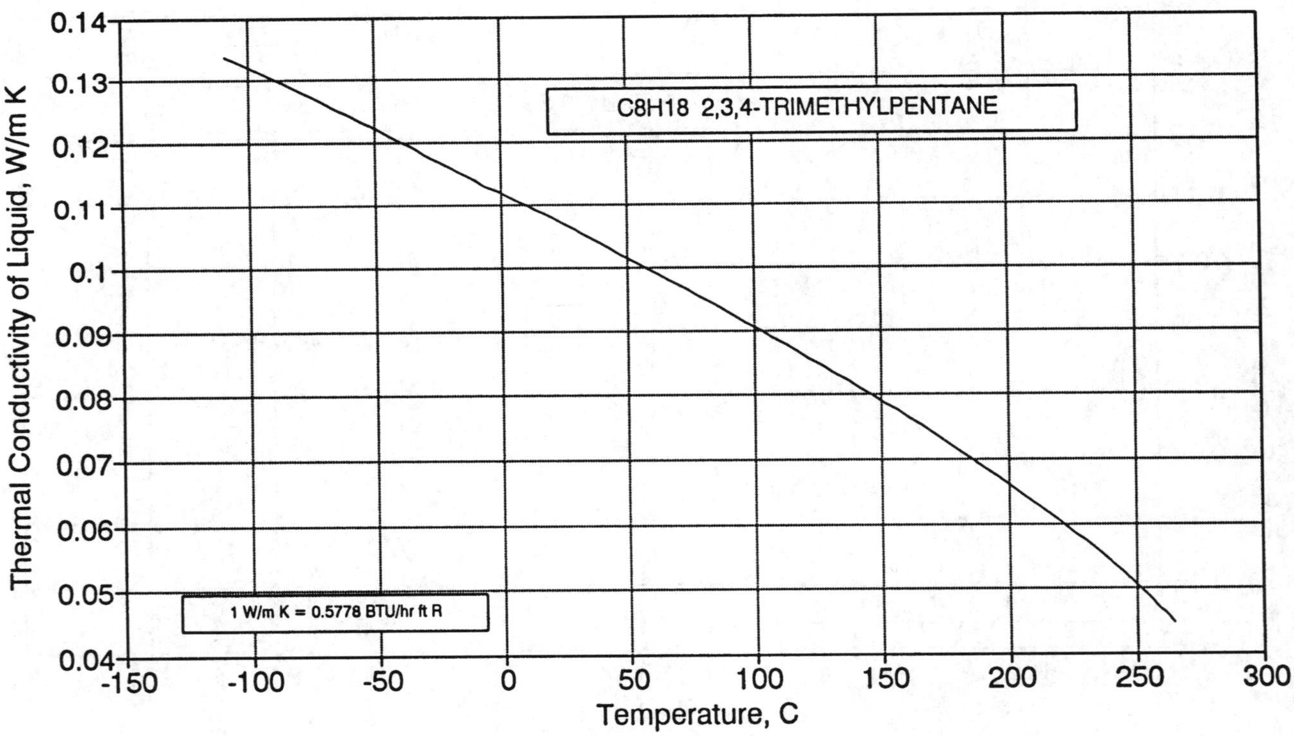

C8H18 2,3,4-TRIMETHYLPENTANE
1 W/m K = 0.5778 BTU/hr ft R
Thermal Conductivity of Liquid, W/m K
0.04
0.05
0.06
0.07
0.08
0.09
0.1
0.11
0.12
0.13
0.14
-150
-100
-50
0
50
100
150
200
250
300
Temperature, C

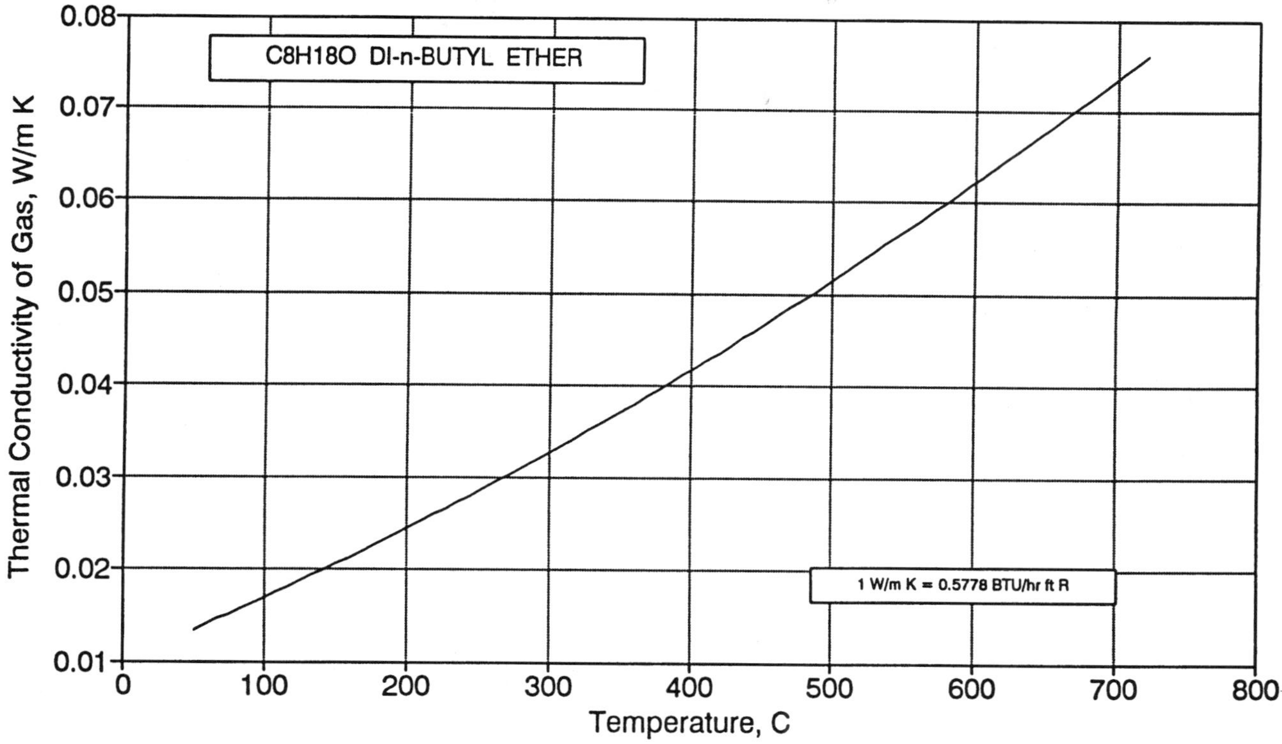
C8H18O DI-n-BUTYL ETHER
1 W/m K = 0.5778 BTU/hr ft R
Thermal Conductivity of Gas, W/m K
0.08
0.07
0.06
0.05
0.04
0.03
0.02
0.01
0
100
200
300
400
500
600
700
800
Temperature, C

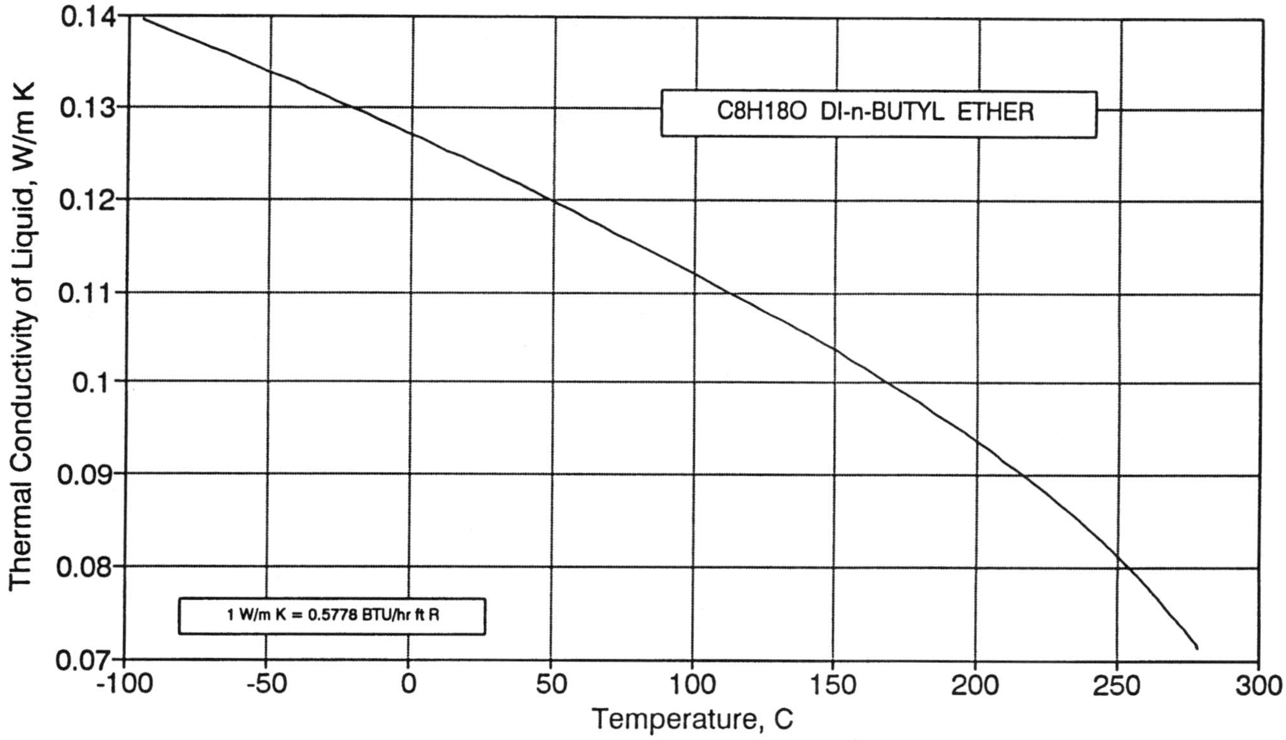
C8H18O DI-n-BUTYL ETHER
1 W/m K = 0.5778 BTU/hr ft R
Thermal Conductivity of Liquid, W/m K
0.14
0.13
0.12
0.11
0.1
0.09
0.08
0.07
-100
-50
0
50
100
150
200
250
300
Temperature, C

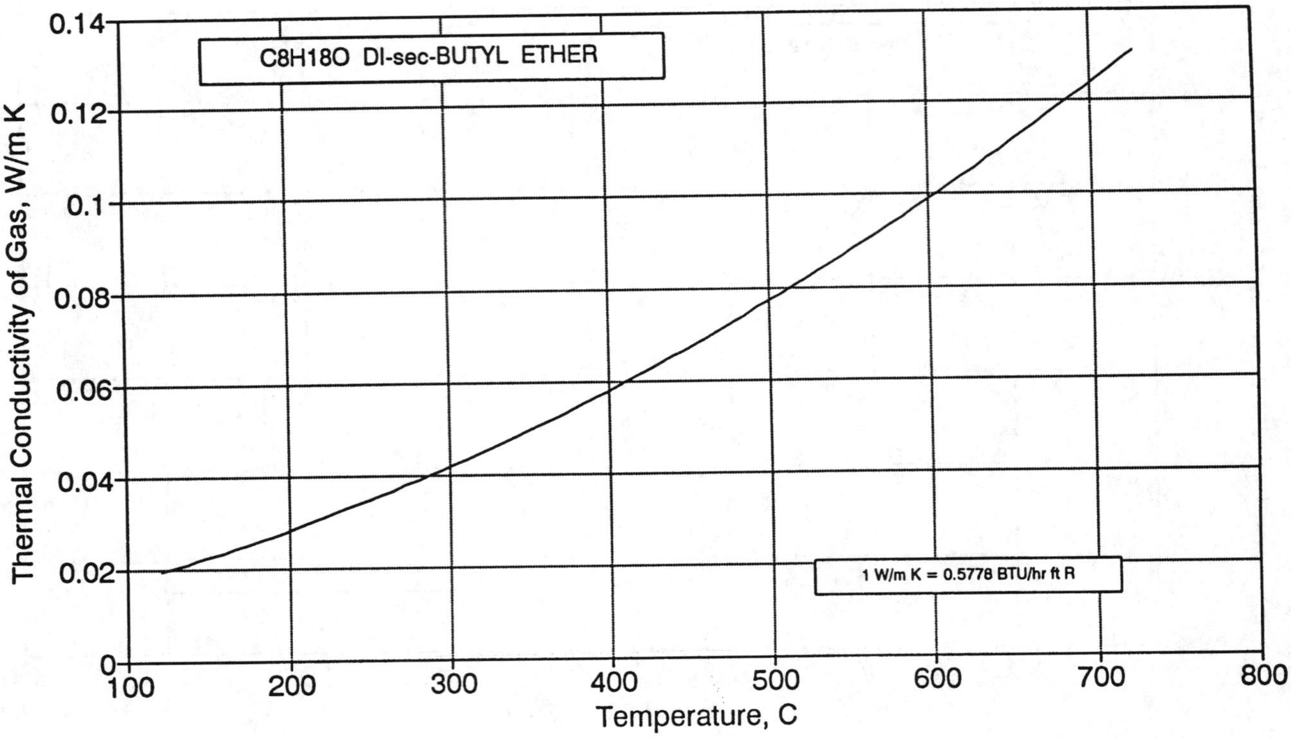
C8H18O DI-sec-BUTYL ETHER
1 W/m K = 0.5778 BTU/hr ft R
Thermal Conductivity of Gas, W/m K
Temperature, C
0.14
0.12
0.1
0.08
0.06
0.04
0.02
0
100
200
300
400
500
600
700
800

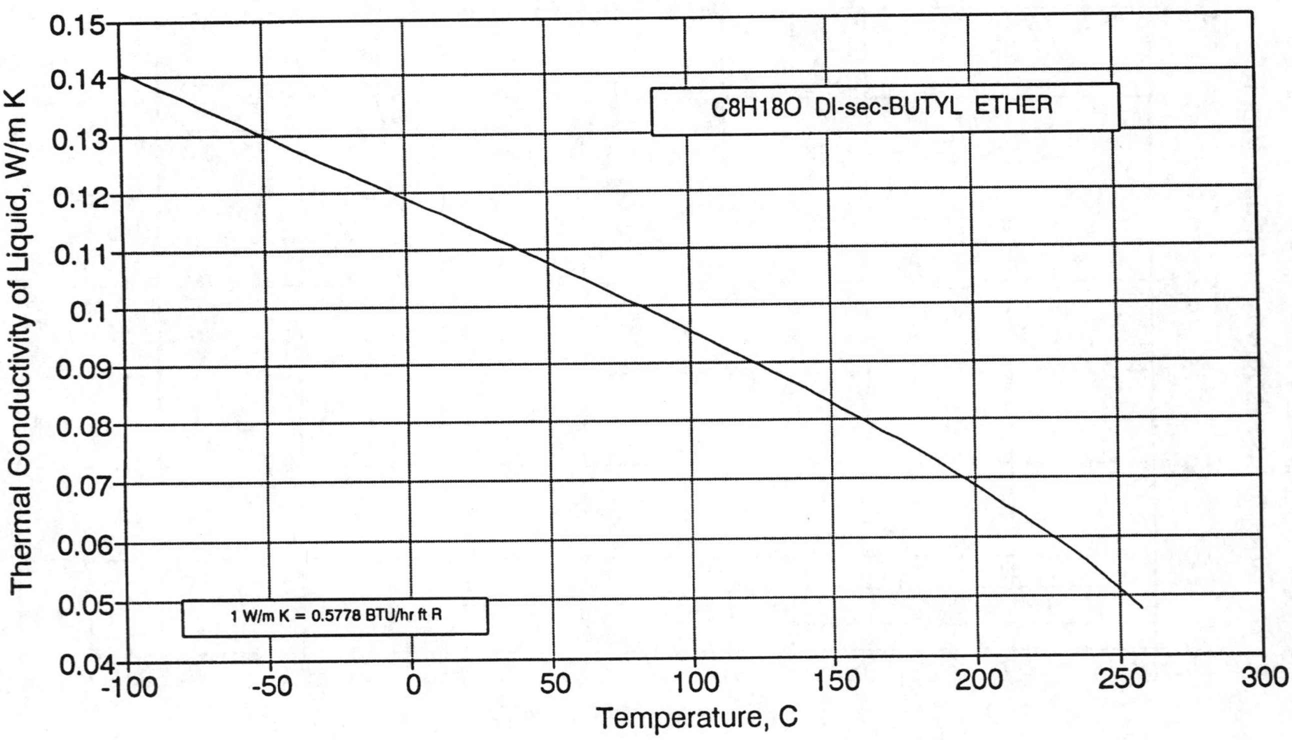
C8H18O DI-sec-BUTYL ETHER
1 W/m K = 0.5778 BTU/hr ft R
Thermal Conductivity of Liquid, W/m K
Temperature, C
0.15
0.14
0.13
0.12
0.11
0.1
0.09
0.08
0.07
0.06
0.05
0.04
-100
-50
0
50
100
150
200
250
300

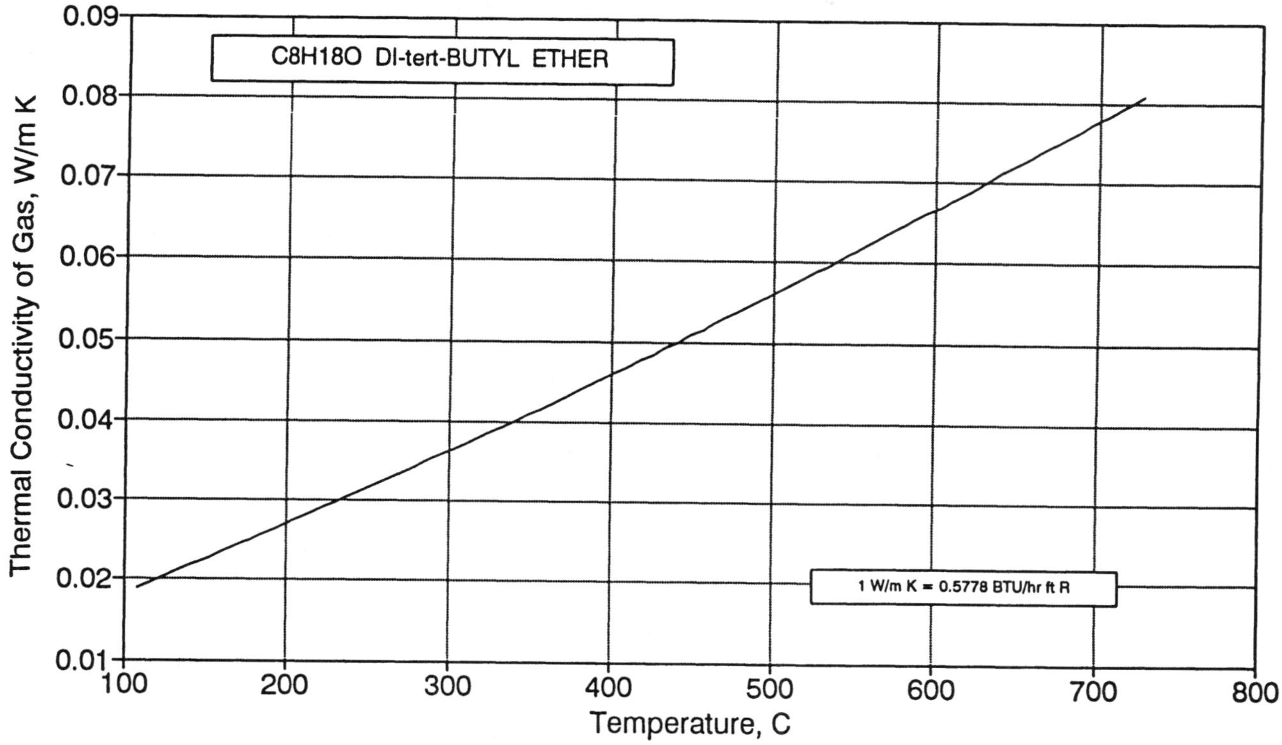
C8H18O DI-tert-BUTYL ETHER
1 W/m K = 0.5778 BTU/hr ft R
Thermal Conductivity of Gas, W/m K
0.09
0.08
0.07
0.06
0.05
0.04
0.03
0.02
0.01
100
200
300
400
500
600
700
800
Temperature, C

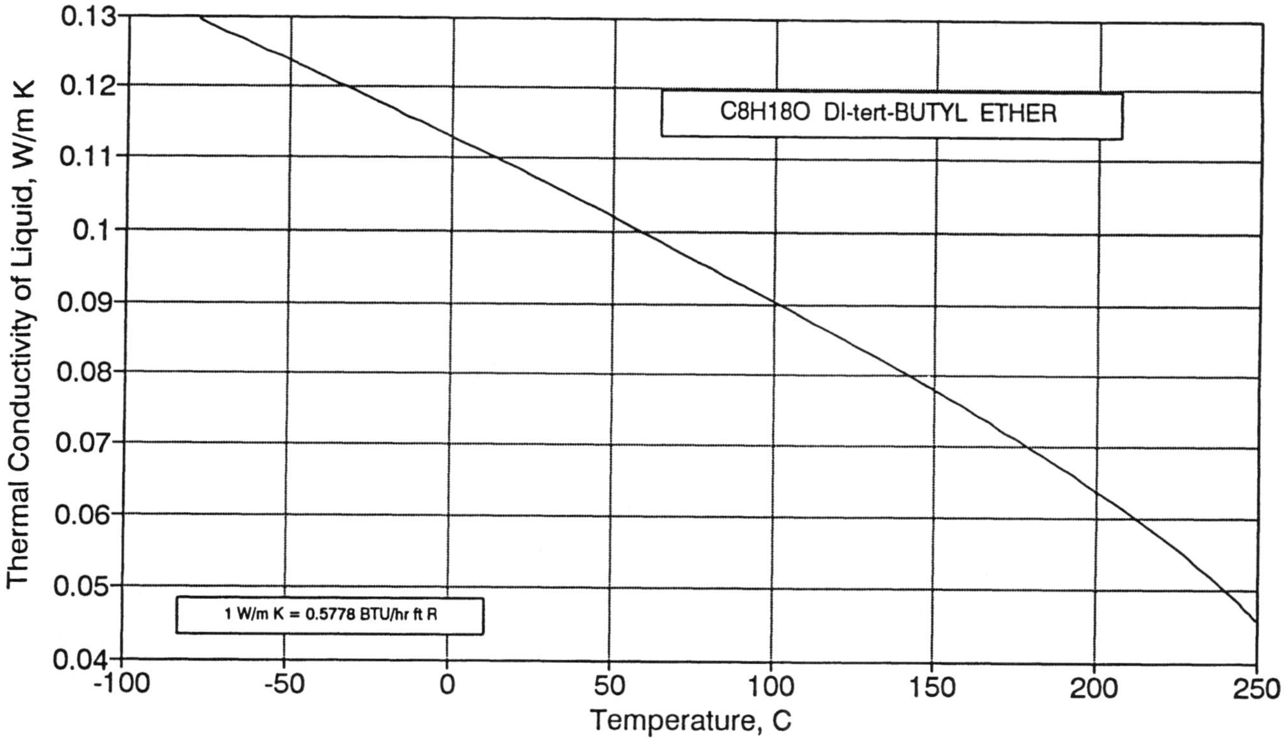
C8H18O DI-tert-BUTYL ETHER
1 W/m K = 0.5778 BTU/hr ft R
Thermal Conductivity of Liquid, W/m K
0.13
0.12
0.11
0.1
0.09
0.08
0.07
0.06
0.05
0.04
-100
-50
0
50
100
150
200
250
Temperature, C

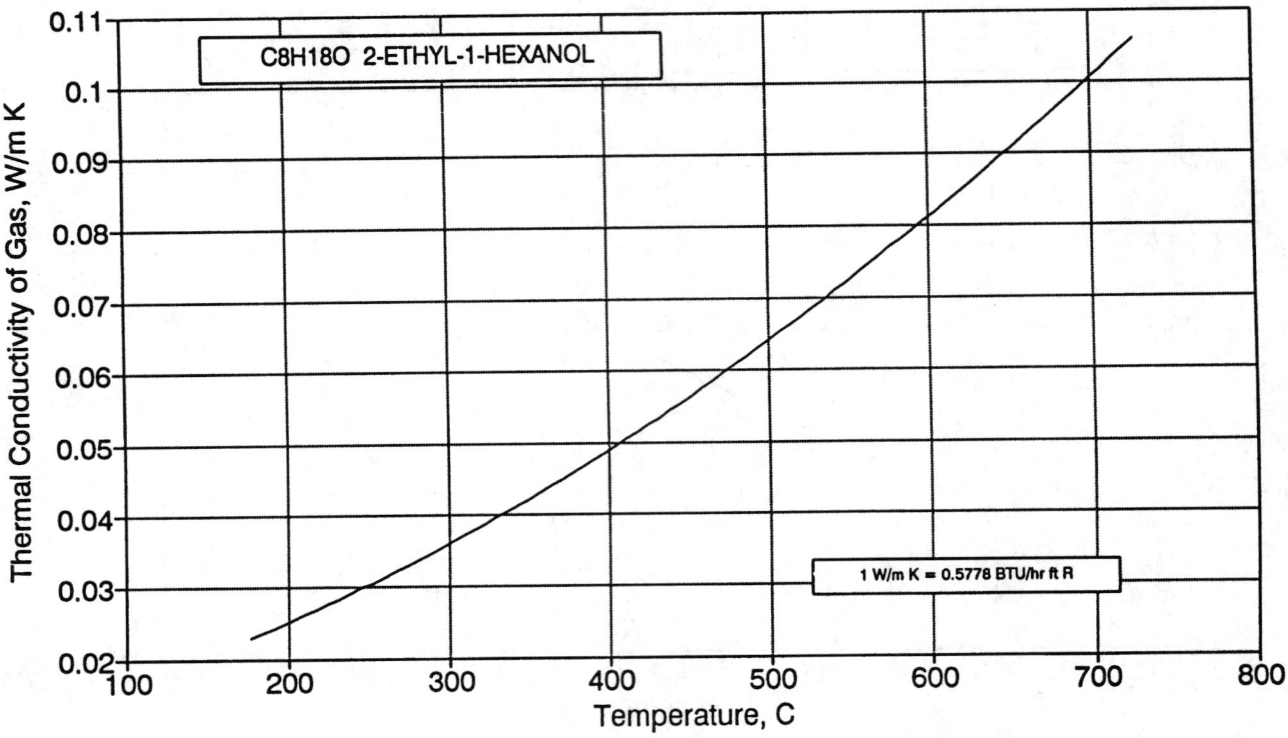
C8H18O 2-ETHYL-1-HEXANOL
1 W/m K = 0.5778 BTU/hr ft R
Thermal Conductivity of Gas, W/m K
0.11
0.1
0.09
0.08
0.07
0.06
0.05
0.04
0.03
0.02
100
200
300
400
500
600
700
800
Temperature, C

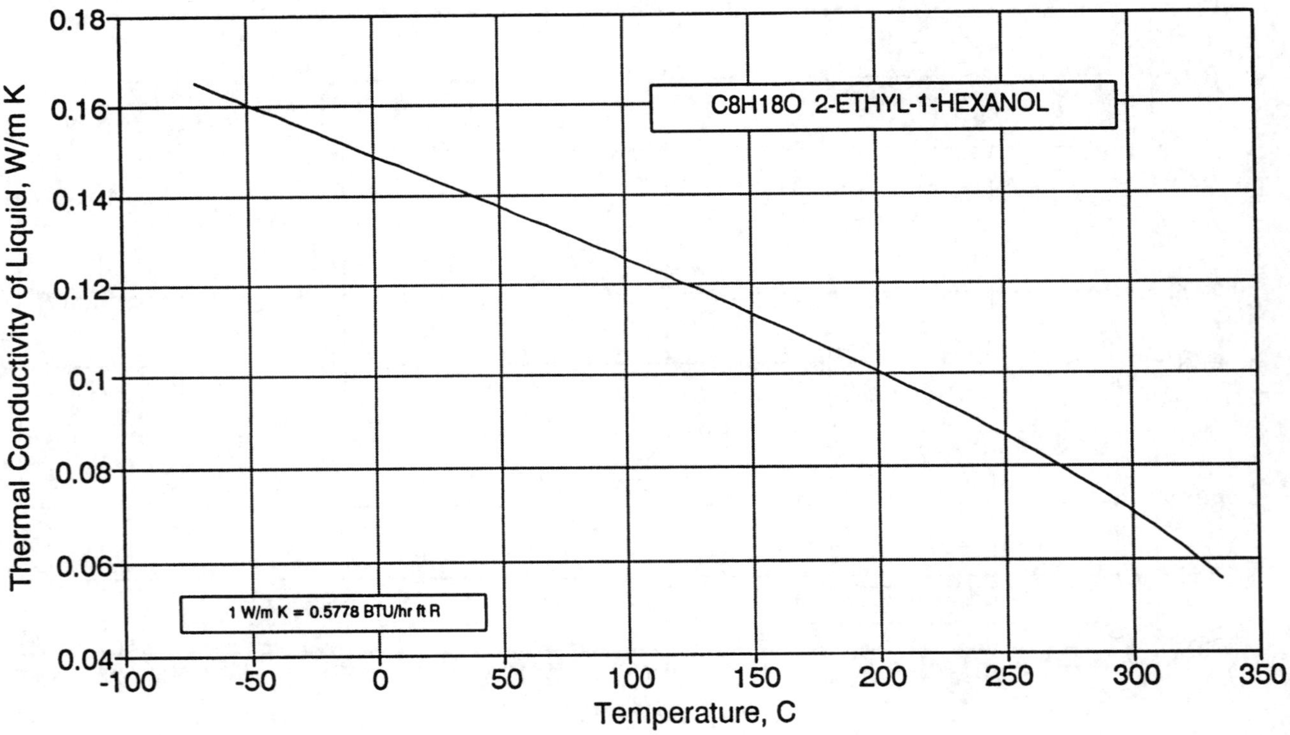
C8H18O 2-ETHYL-1-HEXANOL
1 W/m K = 0.5778 BTU/hr ft R
Thermal Conductivity of Liquid, W/m K
0.18
0.16
0.14
0.12
0.1
0.08
0.06
0.04
-100
-50
0
50
100
150
200
250
300
350
Temperature, C

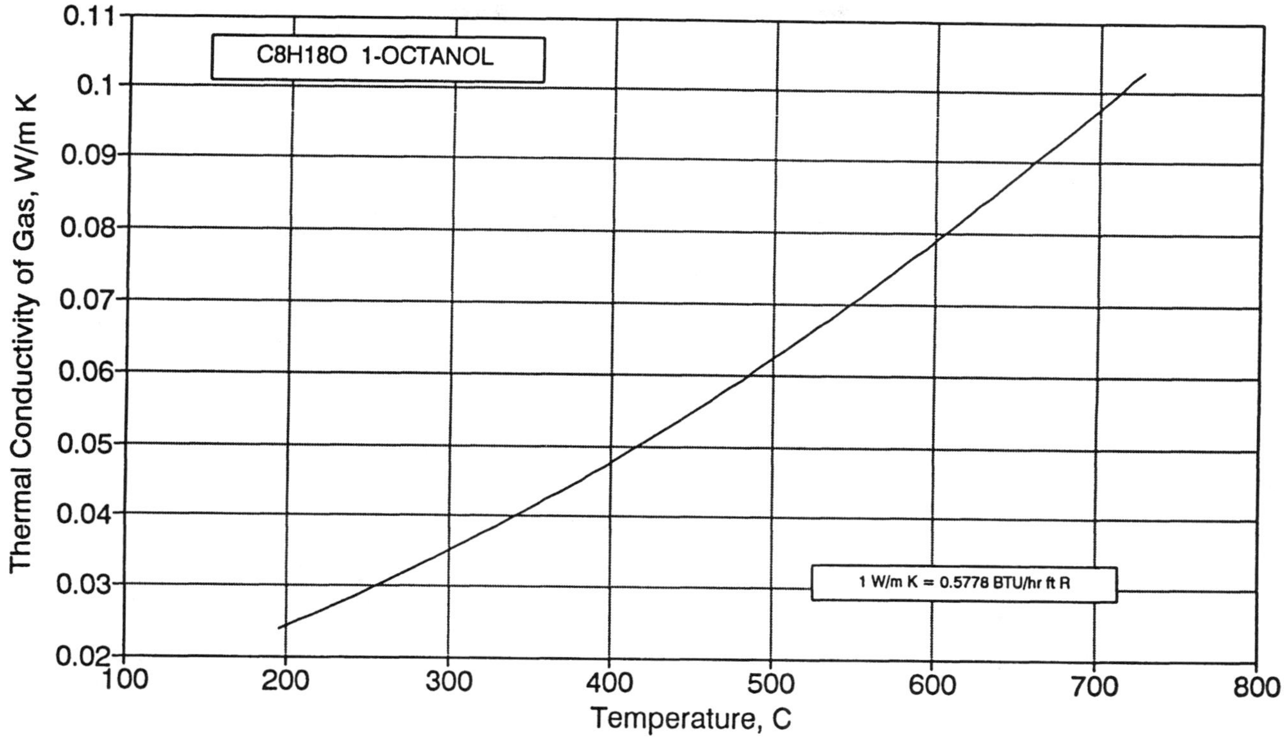
C8H18O 1-OCTANOL
1 W/m K = 0.5778 BTU/hr ft R
Thermal Conductivity of Gas, W/m K
0.11
0.1
0.09
0.08
0.07
0.06
0.05
0.04
0.03
0.02
100
200
300
400
500
600
700
800
Temperature, C

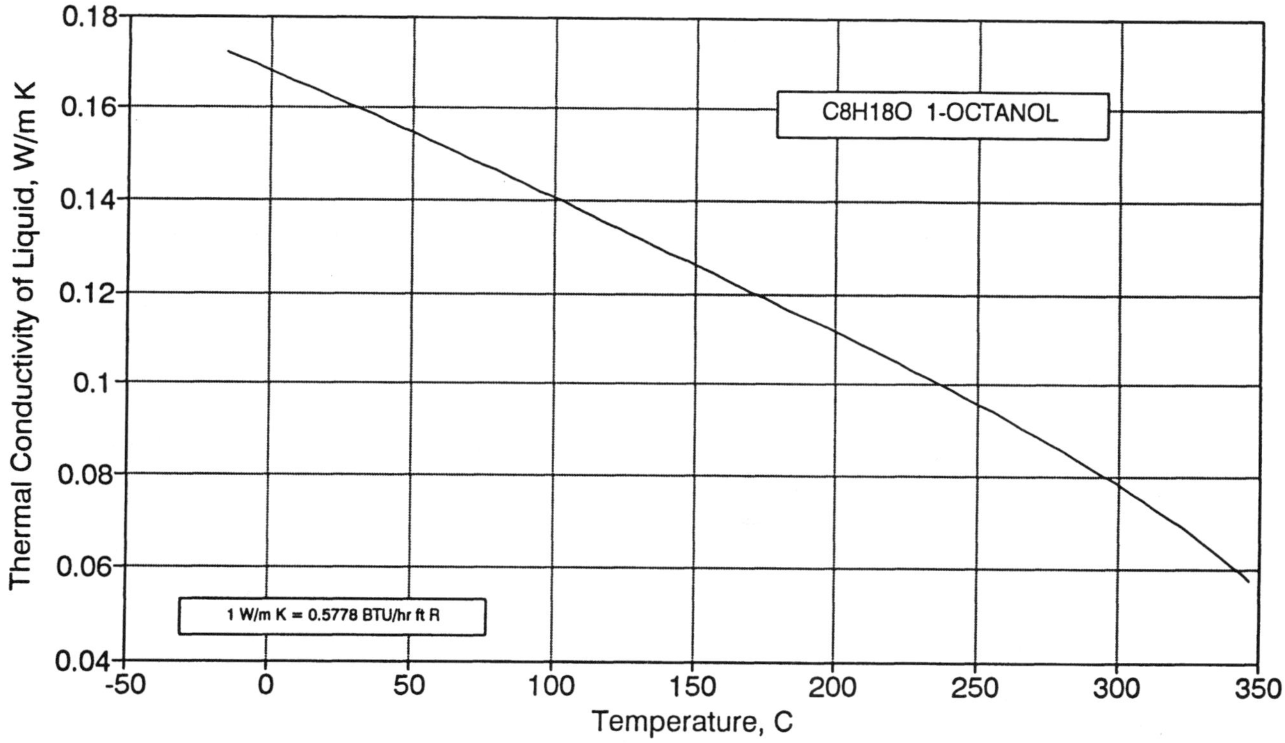
C8H18O 1-OCTANOL
1 W/m K = 0.5778 BTU/hr ft R
Thermal Conductivity of Liquid, W/m K
0.18
0.16
0.14
0.12
0.1
0.08
0.06
0.04
-50
0
50
100
150
200
250
300
350
Temperature, C

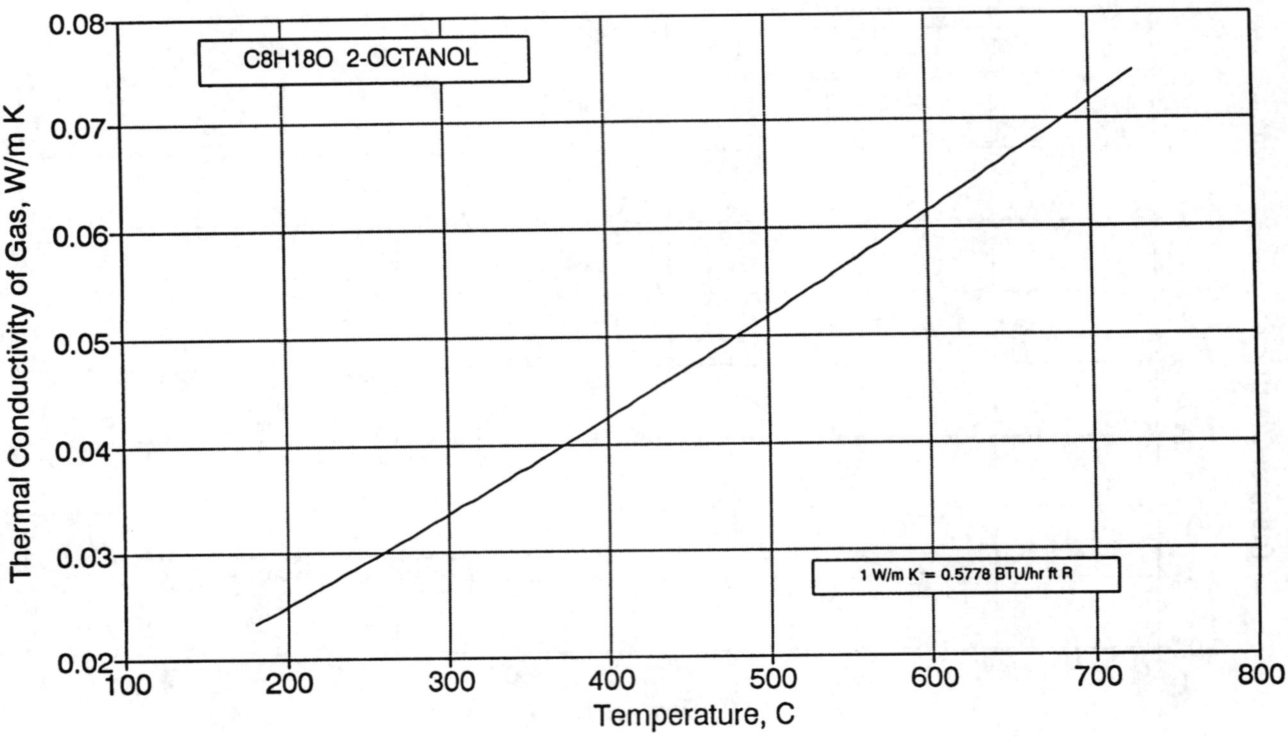
C8H18O 2-OCTANOL
1 W/m K = 0.5778 BTU/hr ft R
Thermal Conductivity of Gas, W/m K
0.08
0.07
0.06
0.05
0.04
0.03
0.02
100
200
300
400
500
600
700
800
Temperature, C

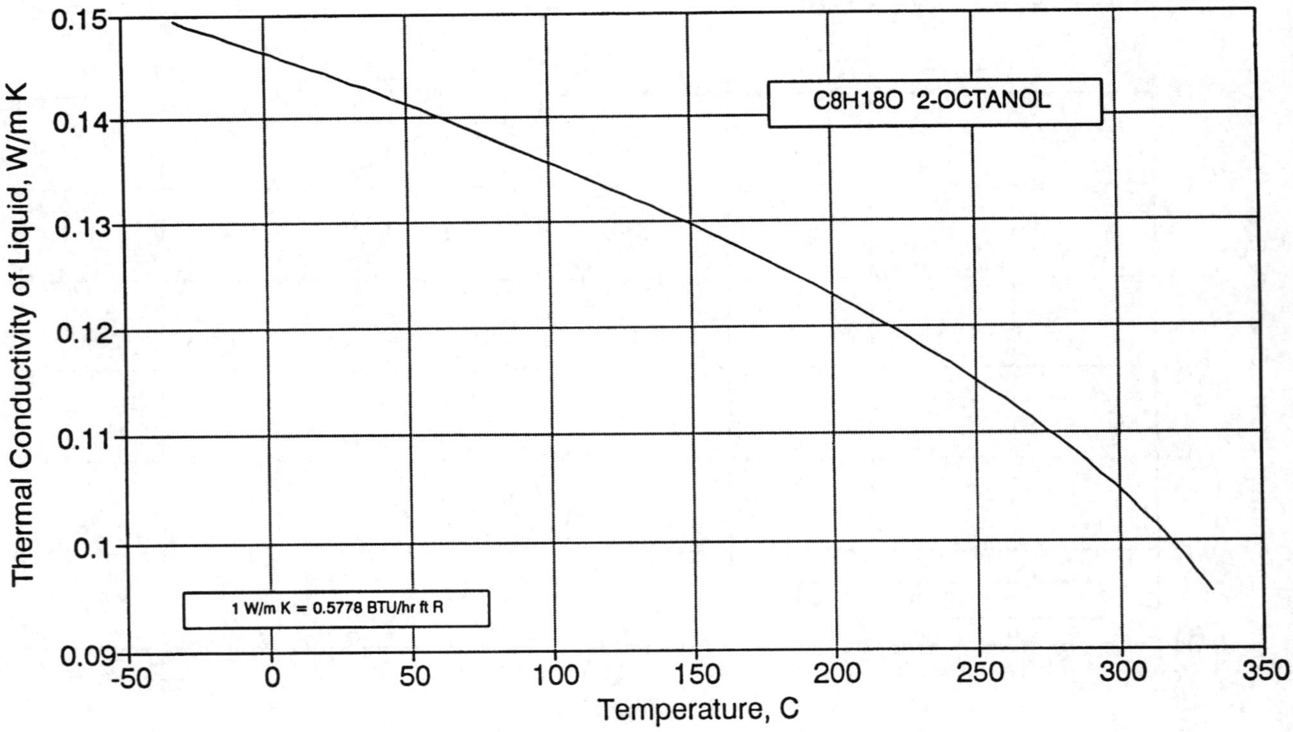
C8H18O 2-OCTANOL
1 W/m K = 0.5778 BTU/hr ft R
Thermal Conductivity of Liquid, W/m K
0.15
0.14
0.13
0.12
0.11
0.1
0.09
-50
0
50
100
150
200
250
300
350
Temperature, C

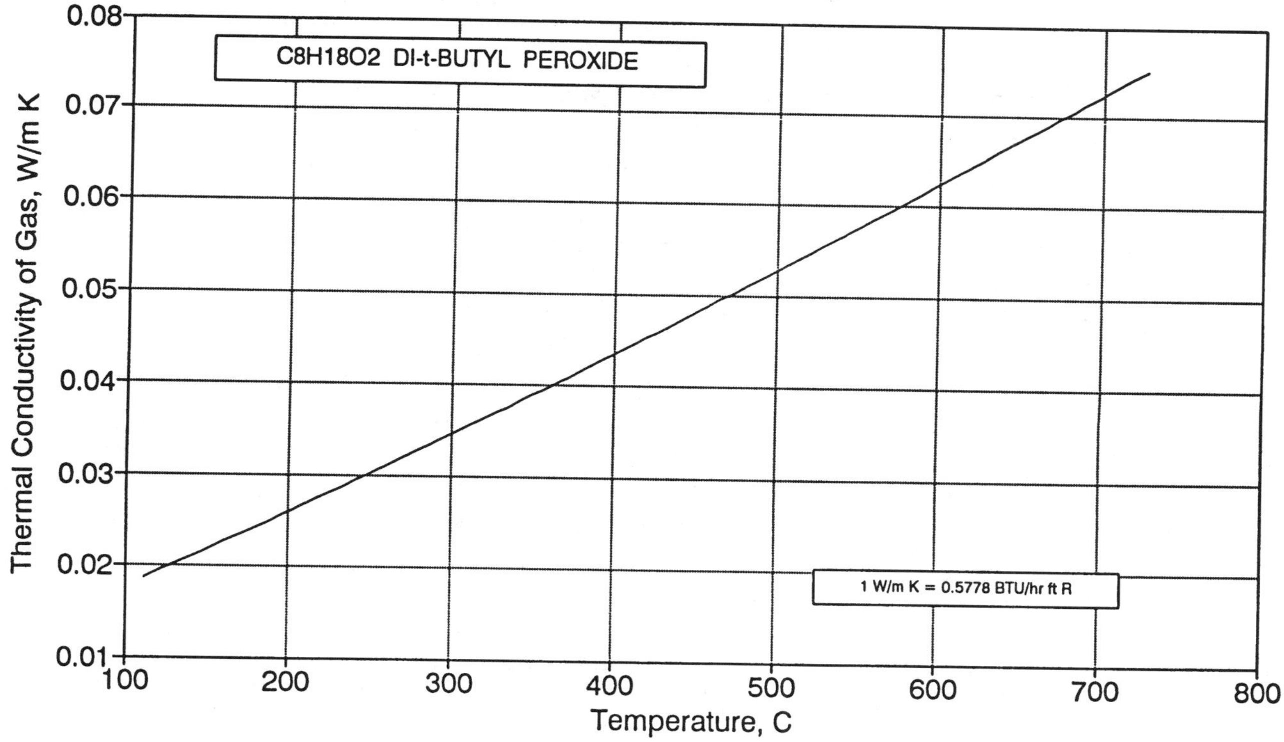
C8H18O2 DI-t-BUTYL PEROXIDE
1 W/m K = 0.5778 BTU/hr ft R
Thermal Conductivity of Gas, W/m K
0.08
0.07
0.06
0.05
0.04
0.03
0.02
0.01
100
200
300
400
500
600
700
800
Temperature, C

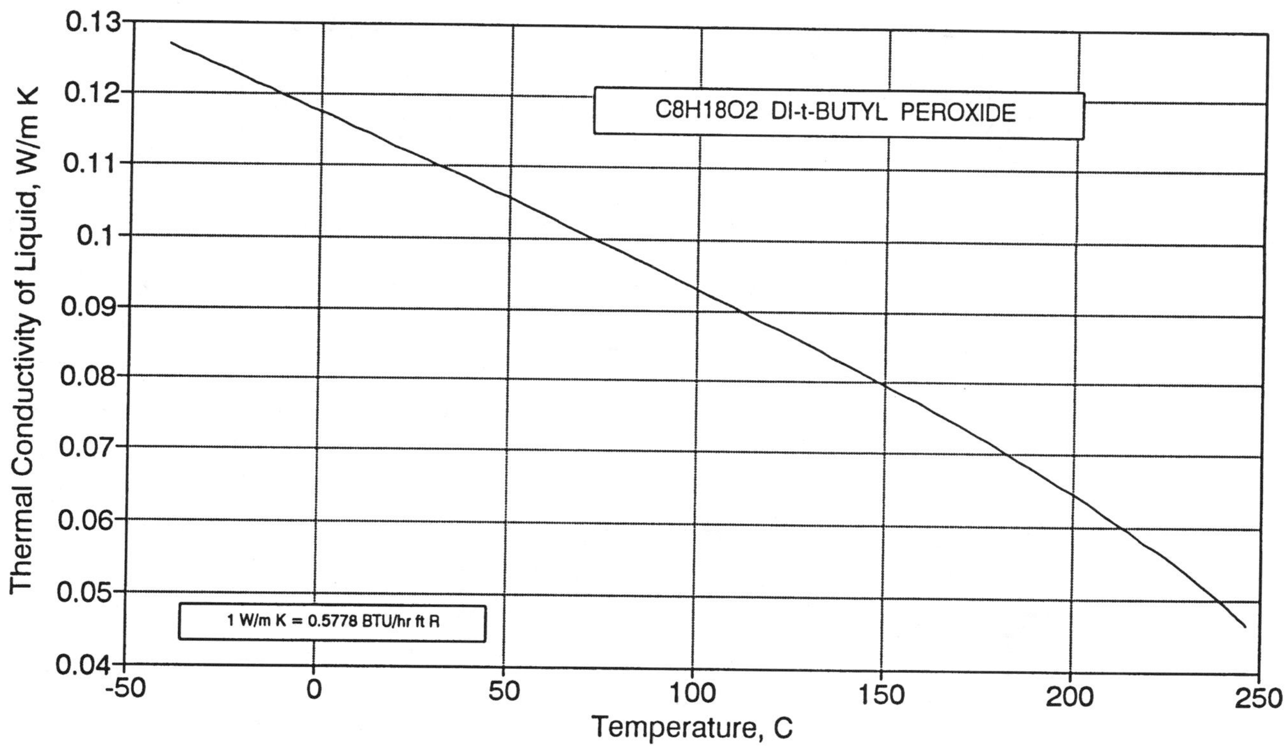
C8H18O2 DI-t-BUTYL PEROXIDE
1 W/m K = 0.5778 BTU/hr ft R
Thermal Conductivity of Liquid, W/m K
0.13
0.12
0.11
0.1
0.09
0.08
0.07
0.06
0.05
0.04
-50
0
50
100
150
200
250
Temperature, C

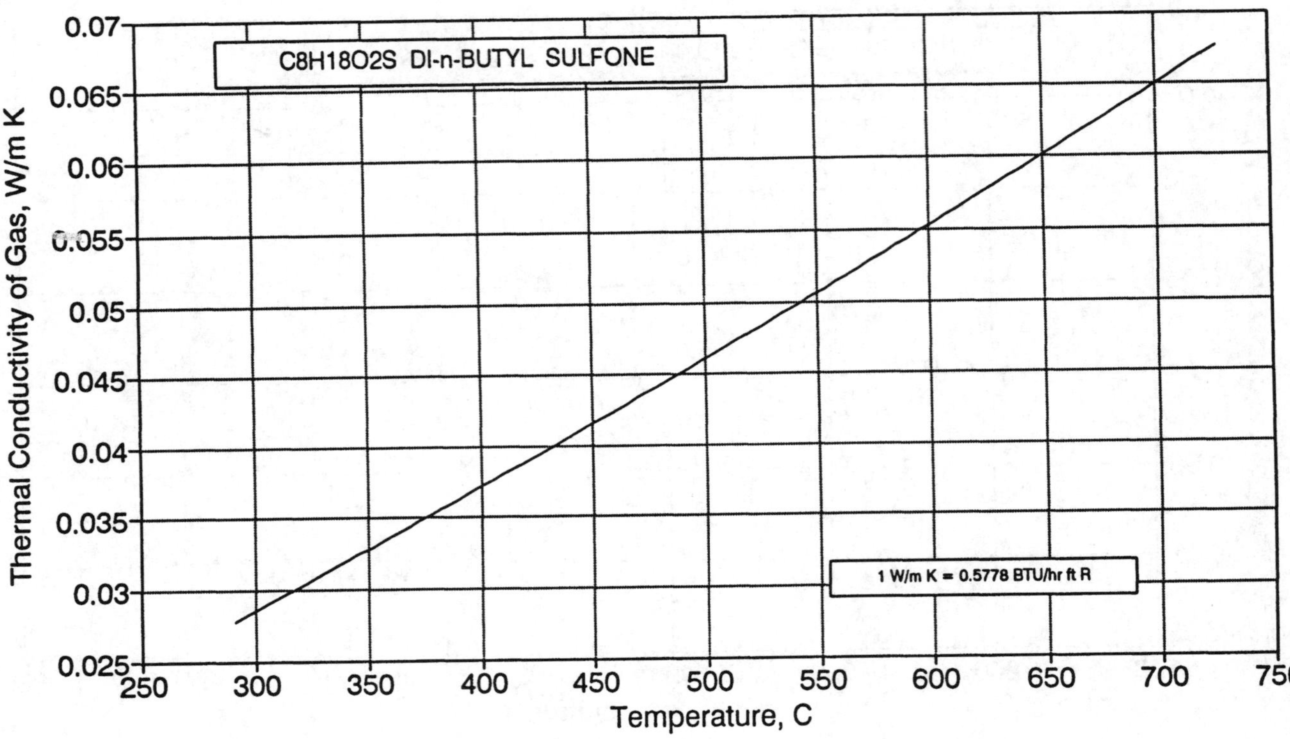
C8H18O2S DI-n-BUTYL SULFONE
1 W/m K = 0.5778 BTU/hr ft R
Thermal Conductivity of Gas, W/m K
0.07
0.065
0.06
0.055
0.05
0.045
0.04
0.035
0.03
0.025
250
300
350
400
450
500
550
600
650
700
750
Temperature, C

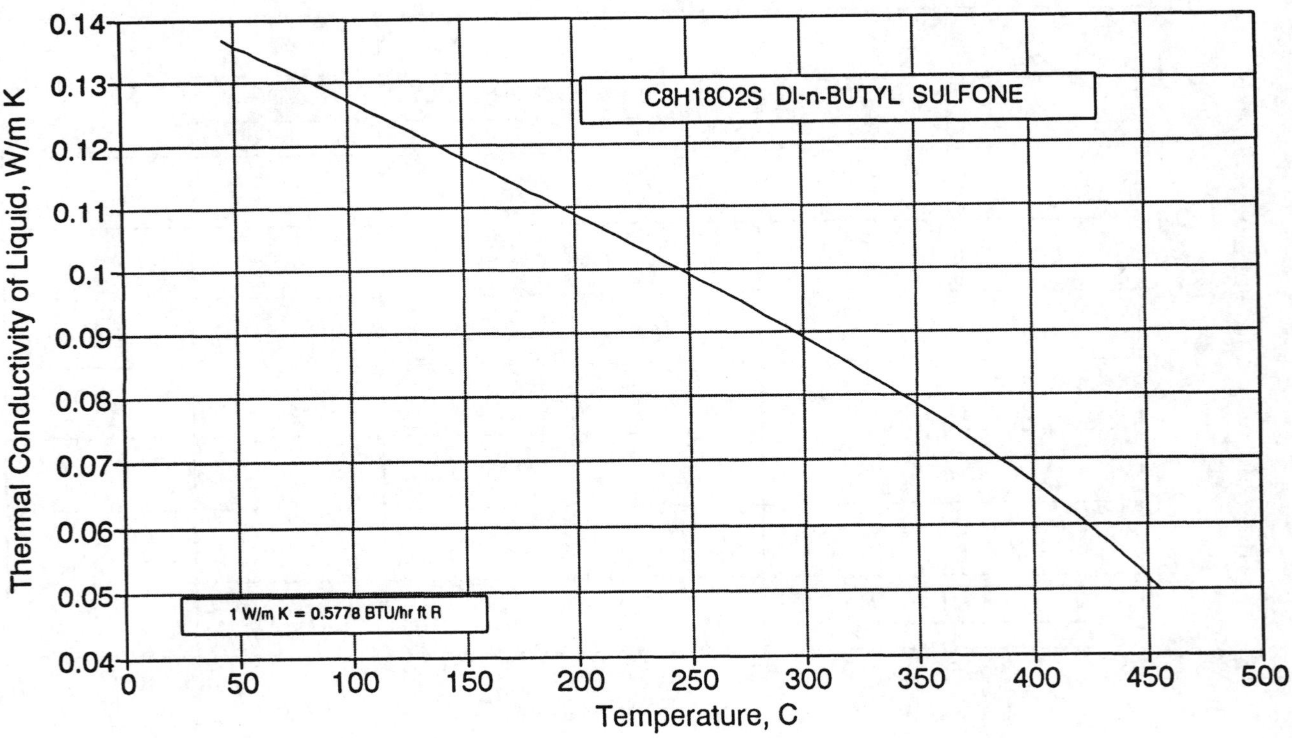
C8H18O2S DI-n-BUTYL SULFONE
1 W/m K = 0.5778 BTU/hr ft R
Thermal Conductivity of Liquid, W/m K
0.14
0.13
0.12
0.11
0.1
0.09
0.08
0.07
0.06
0.05
0.04
0
50
100
150
200
250
300
350
400
450
500
Temperature, C

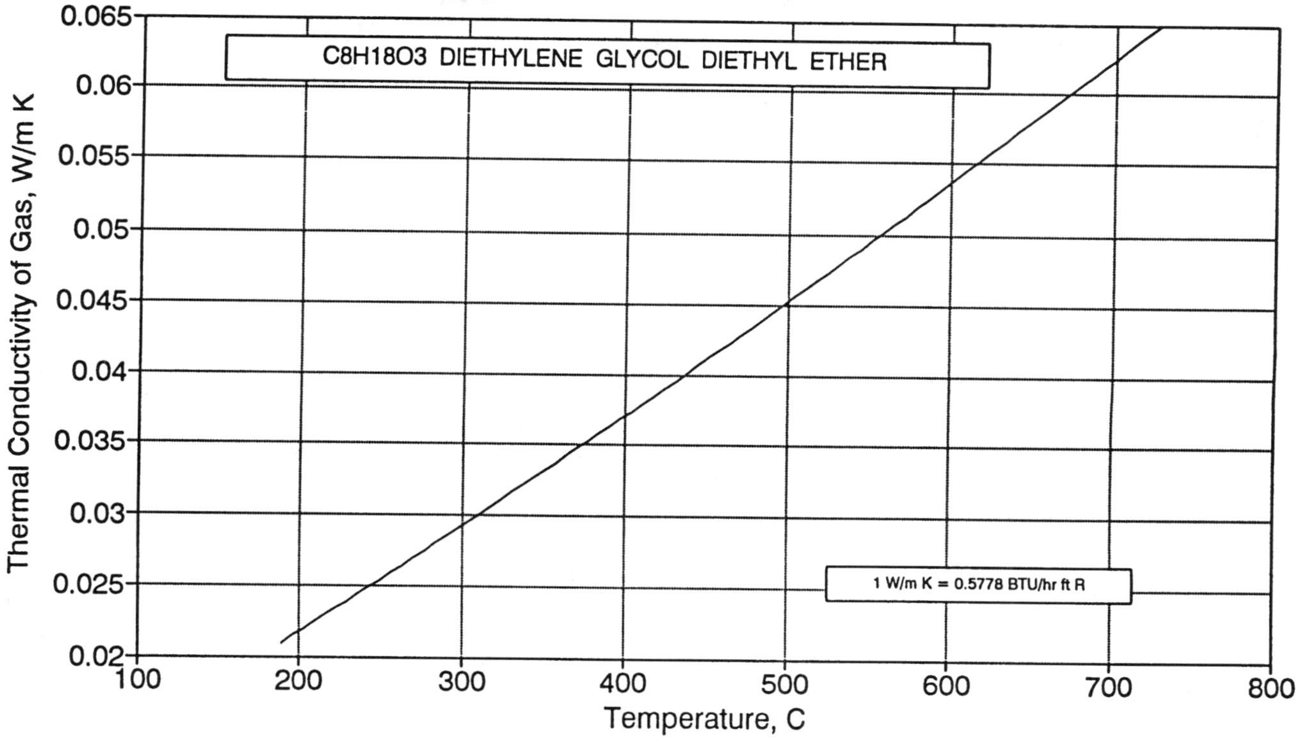
C8H18O3 DIETHYLENE GLYCOL DIETHYL ETHER
1 W/m K = 0.5778 BTU/hr ft R
Thermal Conductivity of Gas, W/m K
Temperature, C
0.065
0.06
0.055
0.05
0.045
0.04
0.035
0.03
0.025
0.02
100
200
300
400
500
600
700
800

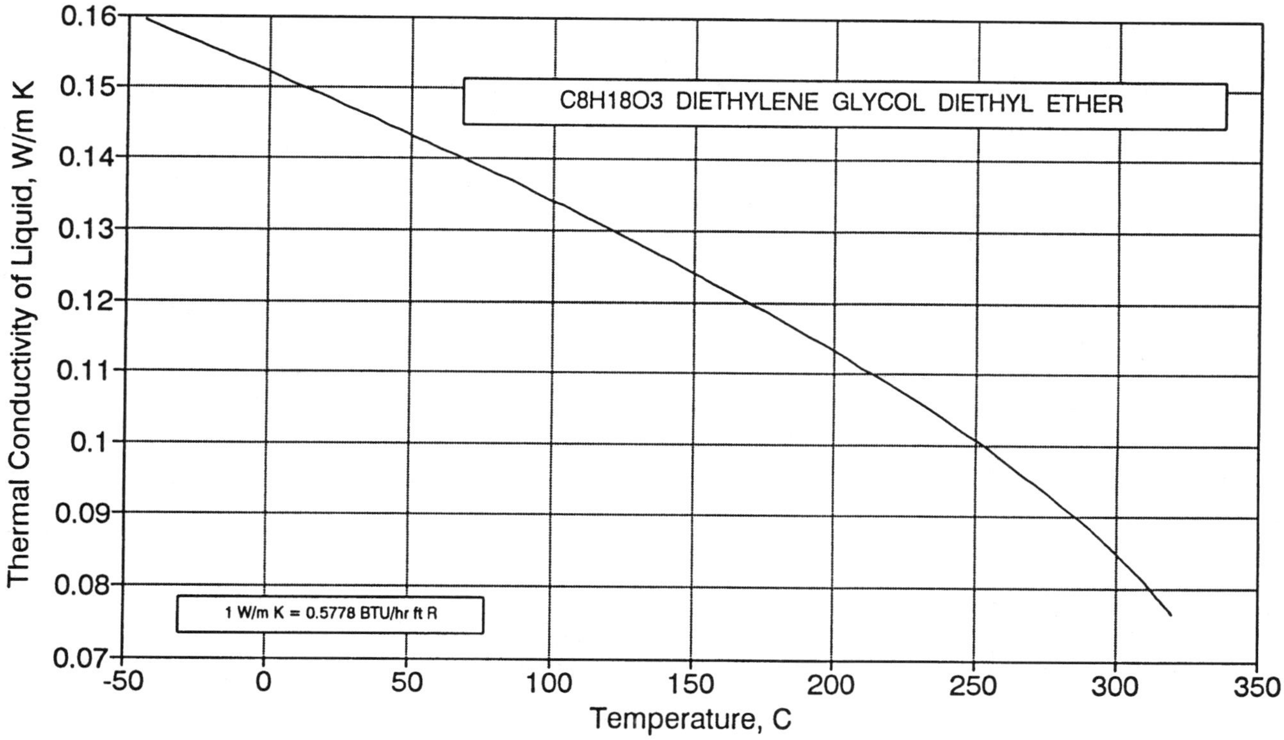
C8H18O3 DIETHYLENE GLYCOL DIETHYL ETHER
1 W/m K = 0.5778 BTU/hr ft R
Thermal Conductivity of Liquid, W/m K
Temperature, C
0.16
0.15
0.14
0.13
0.12
0.11
0.1
0.09
0.08
0.07
-50
0
50
100
150
200
250
300
350

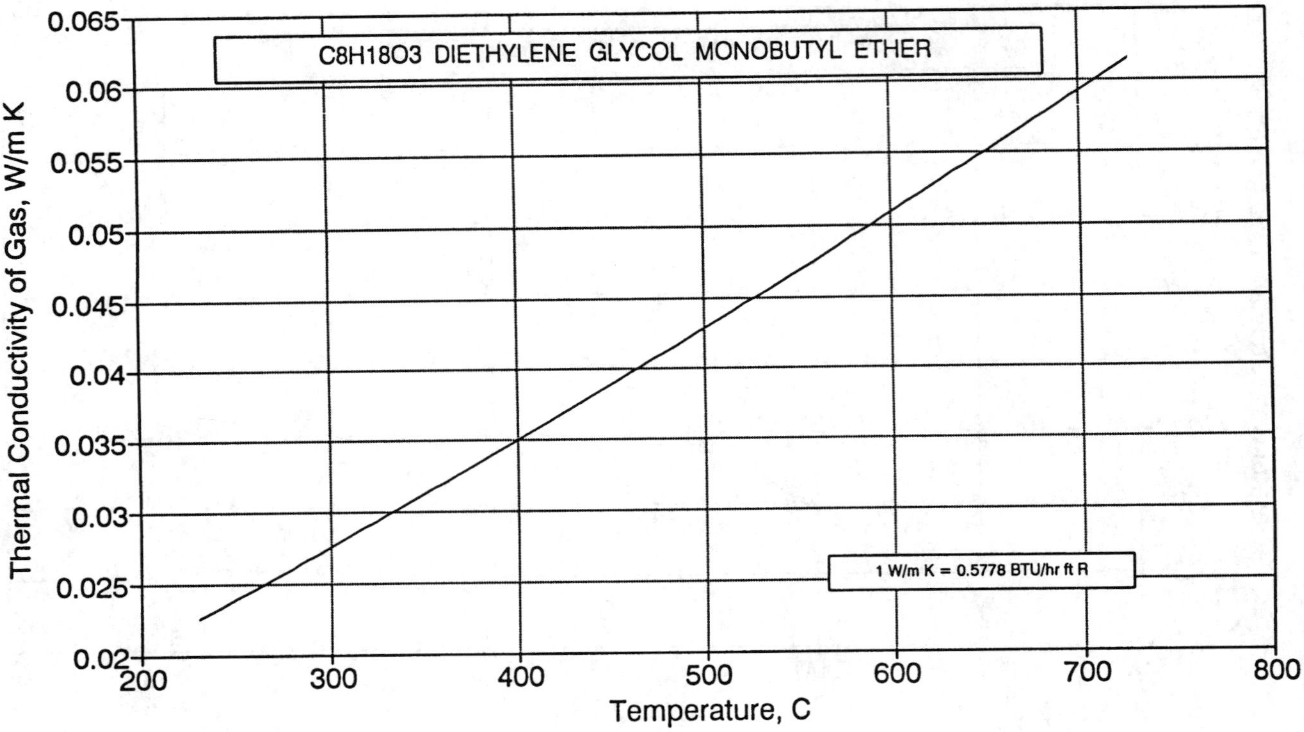
C8H18O3 DIETHYLENE GLYCOL MONOBUTYL ETHER
1 W/m K = 0.5778 BTU/hr ft R
Thermal Conductivity of Gas, W/m K
Temperature, C
0.065
0.06
0.055
0.05
0.045
0.04
0.035
0.03
0.025
0.02
200
300
400
500
600
700
800

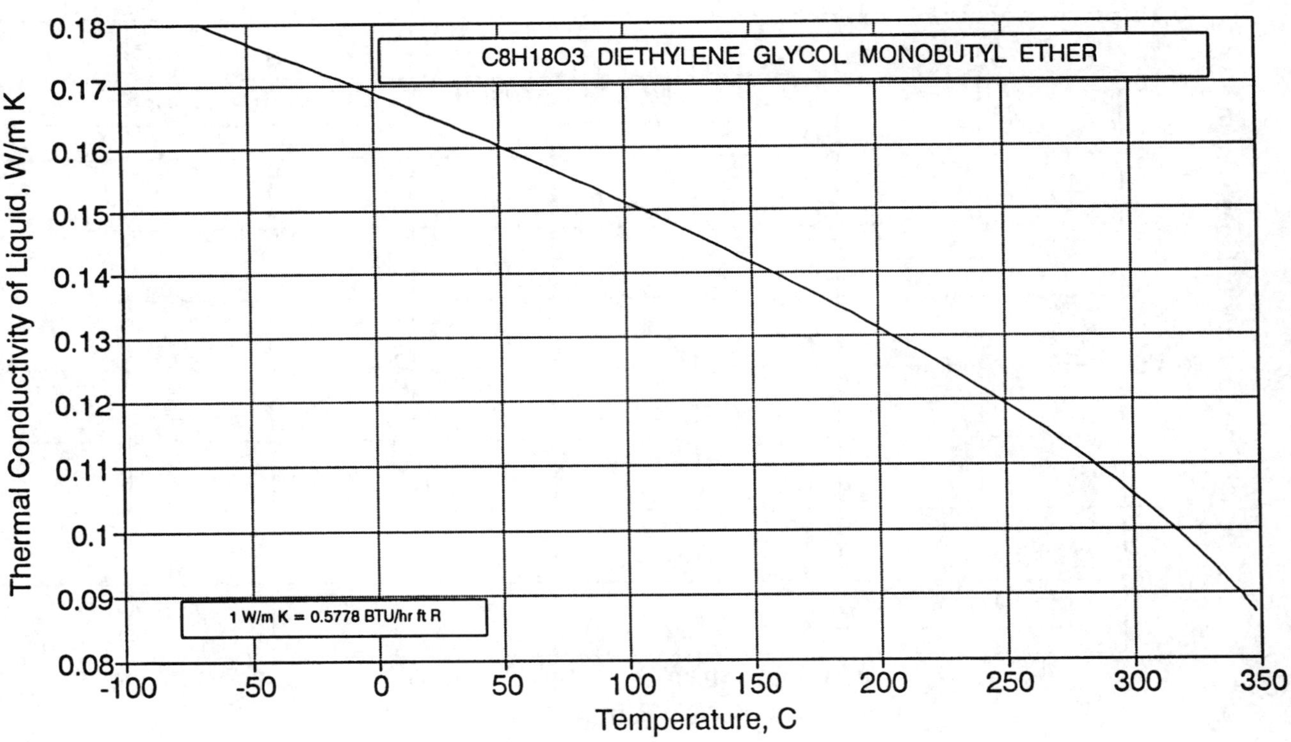
C8H18O3 DIETHYLENE GLYCOL MONOBUTYL ETHER
1 W/m K = 0.5778 BTU/hr ft R
Thermal Conductivity of Liquid, W/m K
Temperature, C
0.18
0.17
0.16
0.15
0.14
0.13
0.12
0.11
0.1
0.09
0.08
-100
-50
0
50
100
150
200
250
300
350

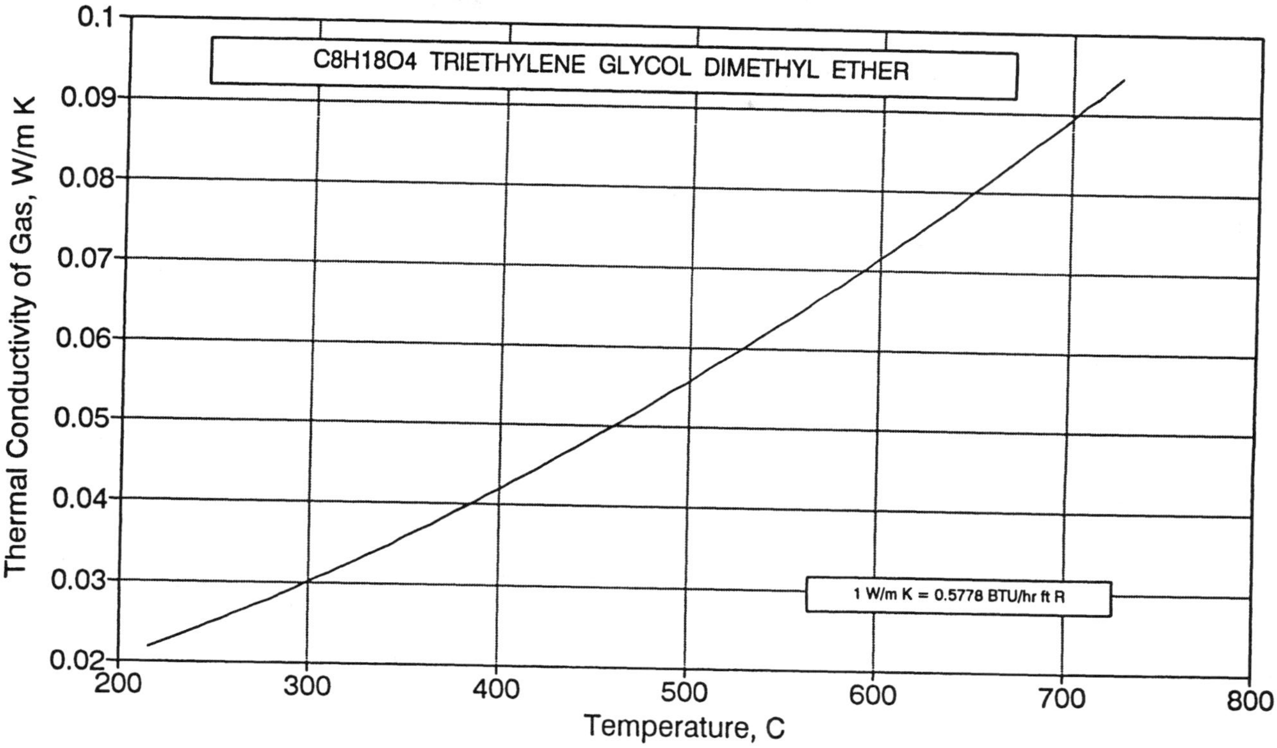
C8H18O4 TRIETHYLENE GLYCOL DIMETHYL ETHER
1 W/m K = 0.5778 BTU/hr ft R
Thermal Conductivity of Gas, W/m K
0.1
0.09
0.08
0.07
0.06
0.05
0.04
0.03
0.02
200
300
400
500
600
700
800
Temperature, C

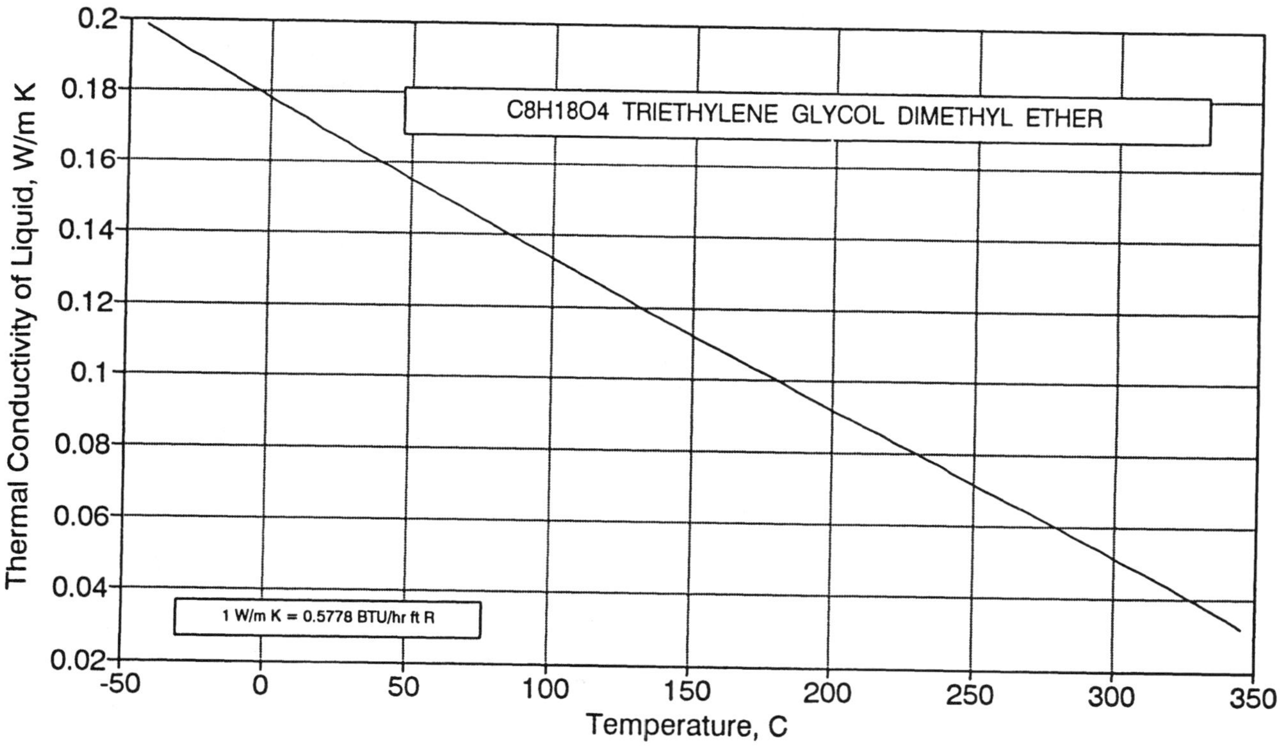
C8H18O4 TRIETHYLENE GLYCOL DIMETHYL ETHER
1 W/m K = 0.5778 BTU/hr ft R
Thermal Conductivity of Liquid, W/m K
0.2
0.18
0.16
0.14
0.12
0.1
0.08
0.06
0.04
0.02
-50
0
50
100
150
200
250
300
350
Temperature, C

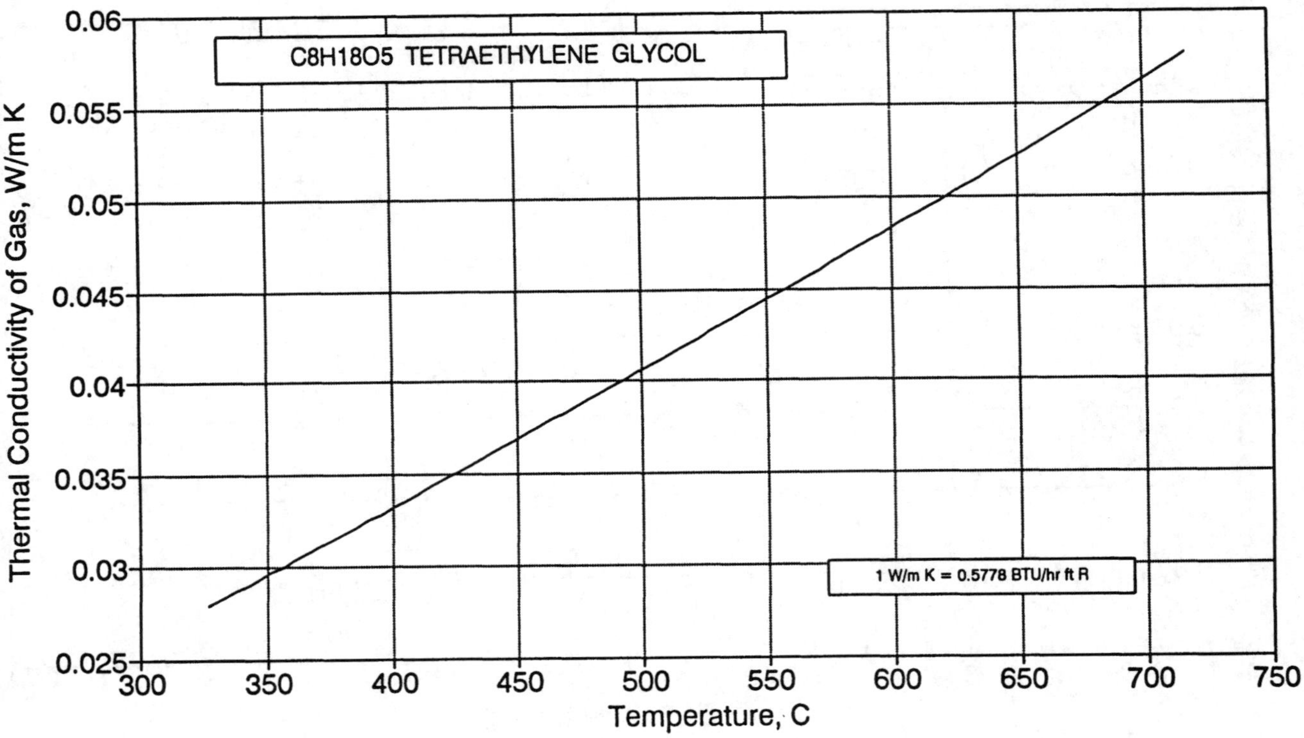
C8H18O5 TETRAETHYLENE GLYCOL
1 W/m K = 0.5778 BTU/hr ft R
Thermal Conductivity of Gas, W/m K
0.06
0.055
0.05
0.045
0.04
0.035
0.03
0.025
300
350
400
450
500
550
600
650
700
750
Temperature, C

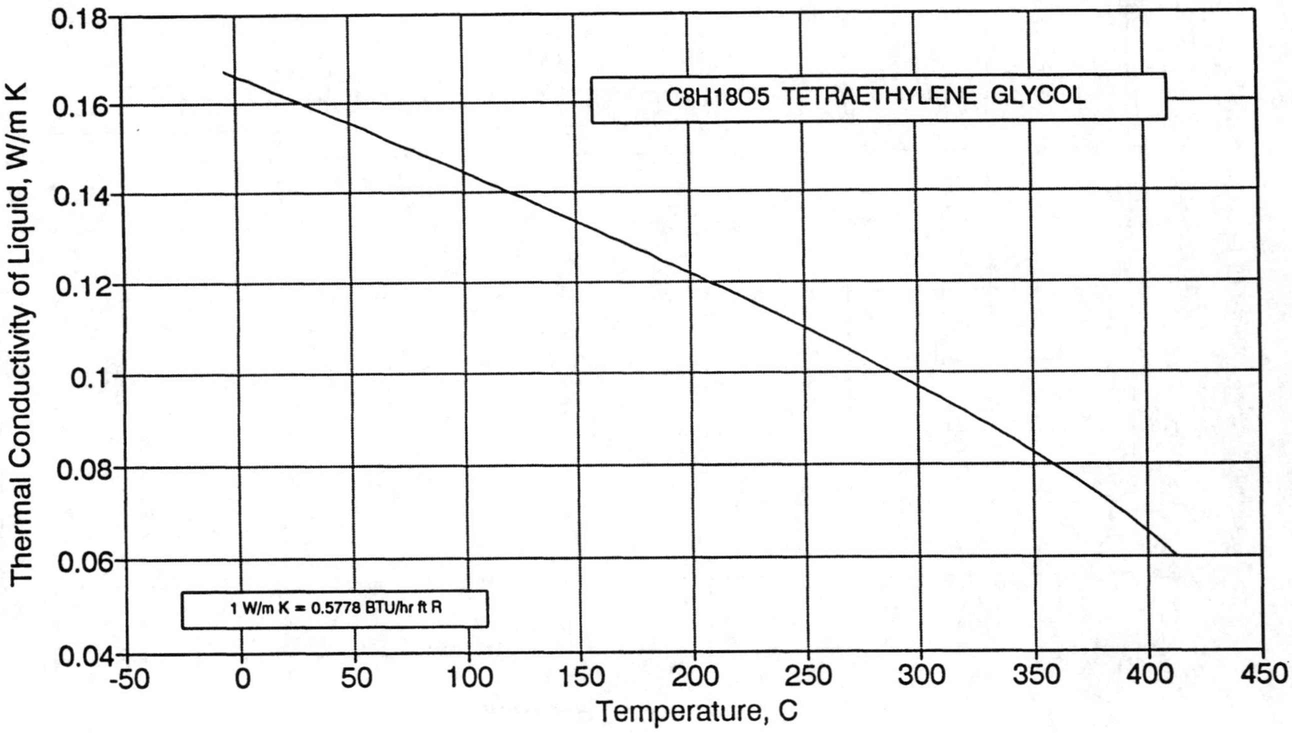
C8H18O5 TETRAETHYLENE GLYCOL
1 W/m K = 0.5778 BTU/hr ft R
Thermal Conductivity of Liquid, W/m K
0.18
0.16
0.14
0.12
0.1
0.08
0.06
0.04
-50
0
50
100
150
200
250
300
350
400
450
Temperature, C

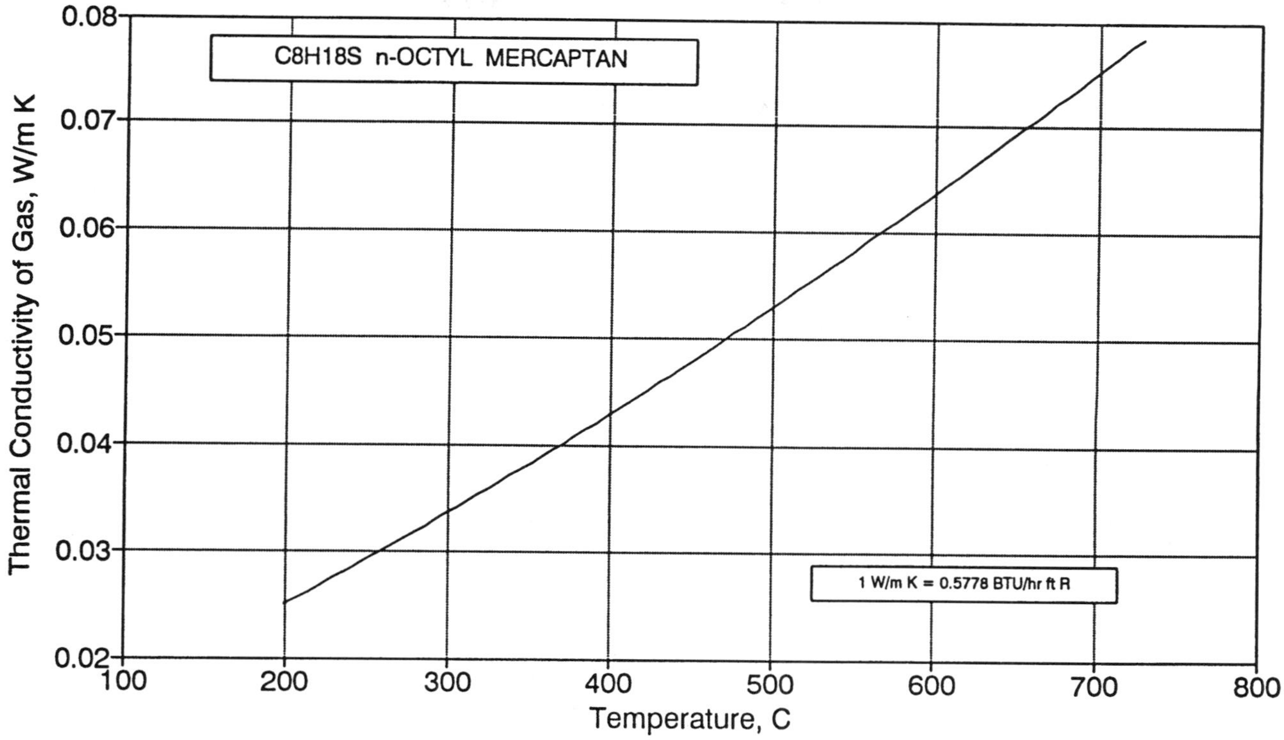
C8H18S n-OCTYL MERCAPTAN
1 W/m K = 0.5778 BTU/hr ft R
Thermal Conductivity of Gas, W/m K
0.08
0.07
0.06
0.05
0.04
0.03
0.02
100
200
300
400
500
600
700
800
Temperature, C

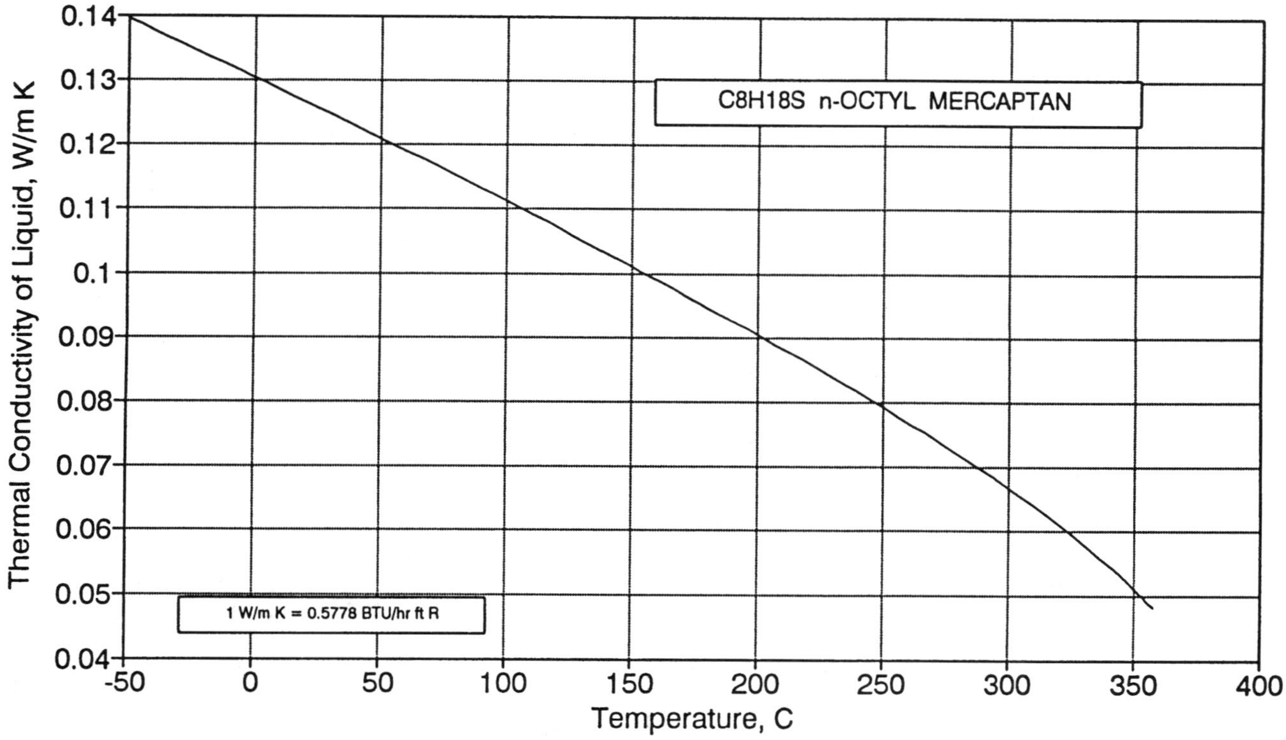
C8H18S n-OCTYL MERCAPTAN
1 W/m K = 0.5778 BTU/hr ft R
Thermal Conductivity of Liquid, W/m K
0.14
0.13
0.12
0.11
0.1
0.09
0.08
0.07
0.06
0.05
0.04
-50
0
50
100
150
200
250
300
350
400
Temperature, C

C8H18S tert-OCTYL MERCAPTAN

Gas thermal conductivity data are not available.

TF, K = 199.00

TB, K = 429.00

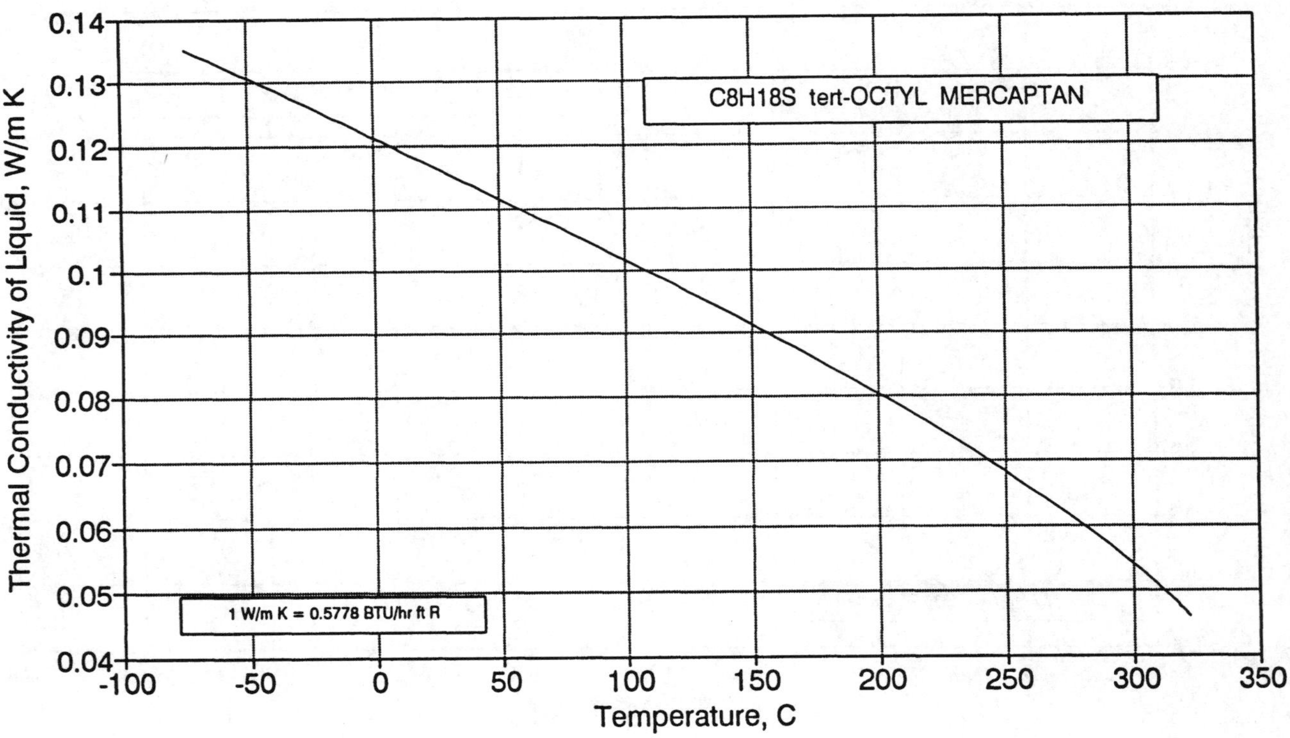

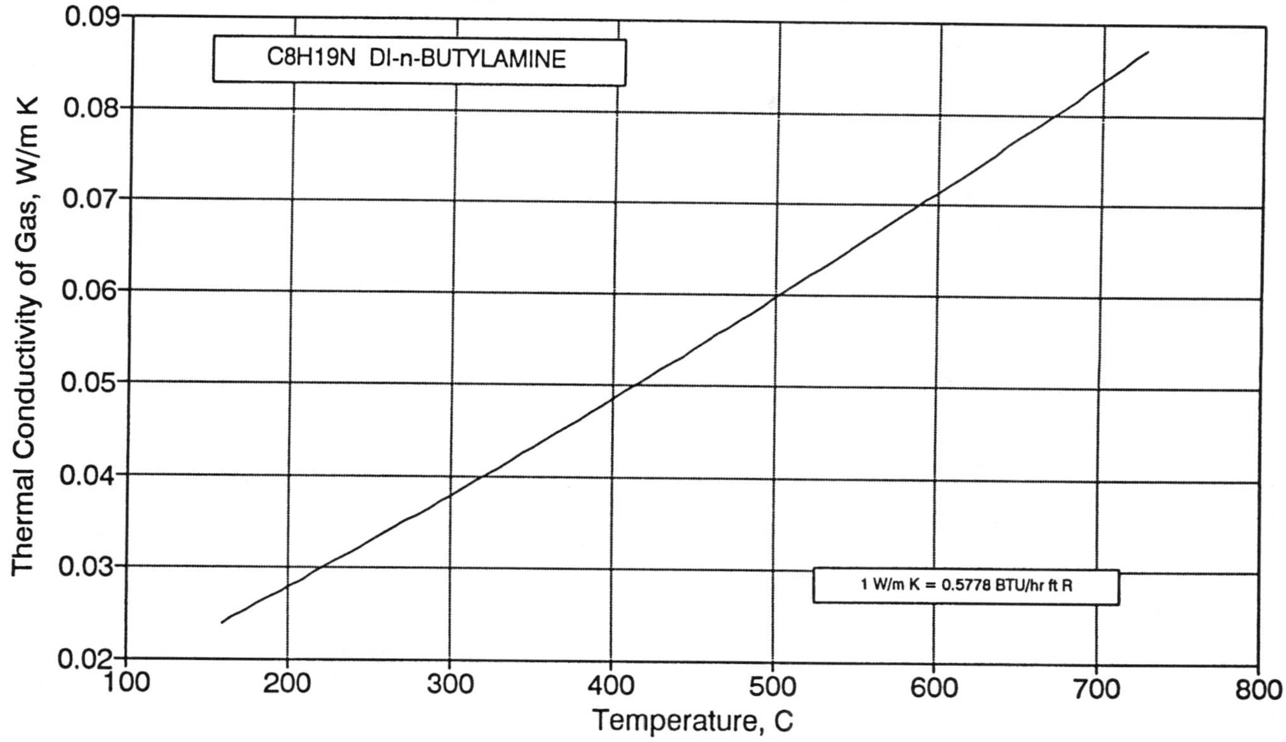
C8H19N DI-n-BUTYLAMINE
Thermal Conductivity of Gas, W/m K
0.09
0.08
0.07
0.06
0.05
0.04
0.03
0.02
1 W/m K = 0.5778 BTU/hr ft R
100
200
300
400
500
600
700
800
Temperature, C

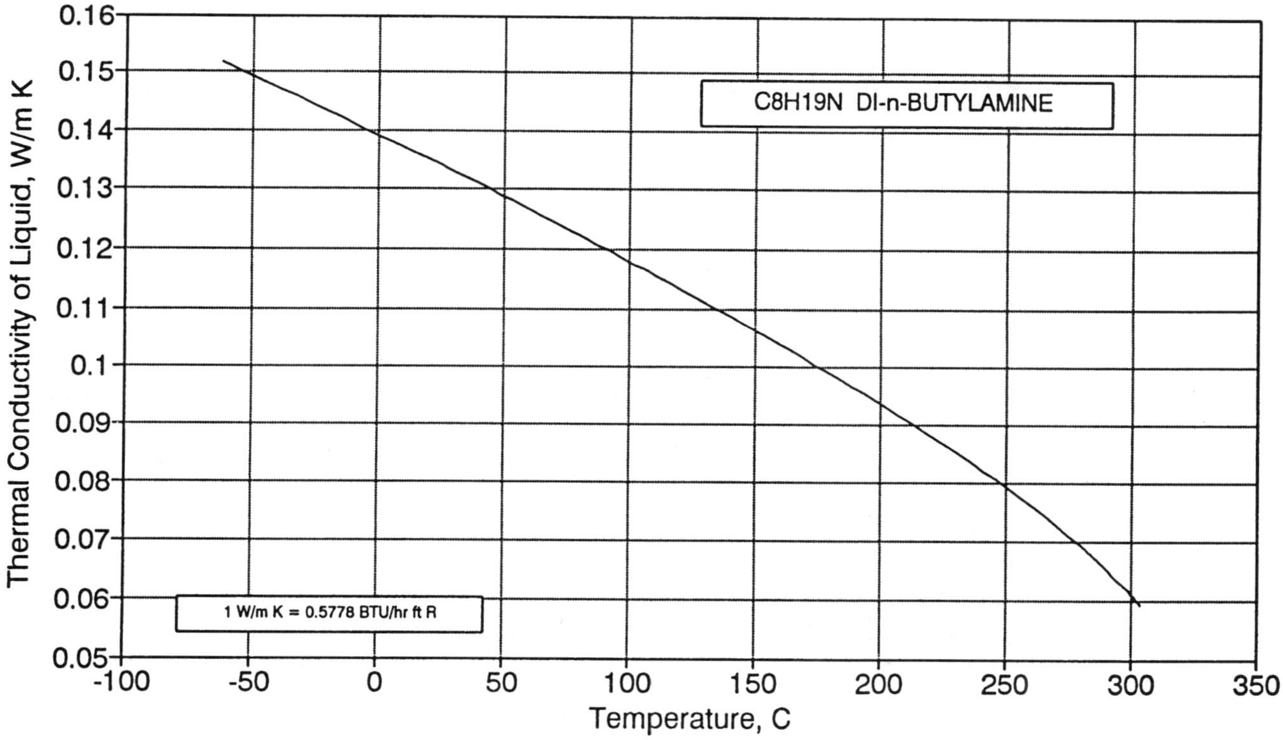
C8H19N DI-n-BUTYLAMINE
Thermal Conductivity of Liquid, W/m K
0.16
0.15
0.14
0.13
0.12
0.11
0.1
0.09
0.08
0.07
0.06
0.05
1 W/m K = 0.5778 BTU/hr ft R
-100
-50
0
50
100
150
200
250
300
350
Temperature, C

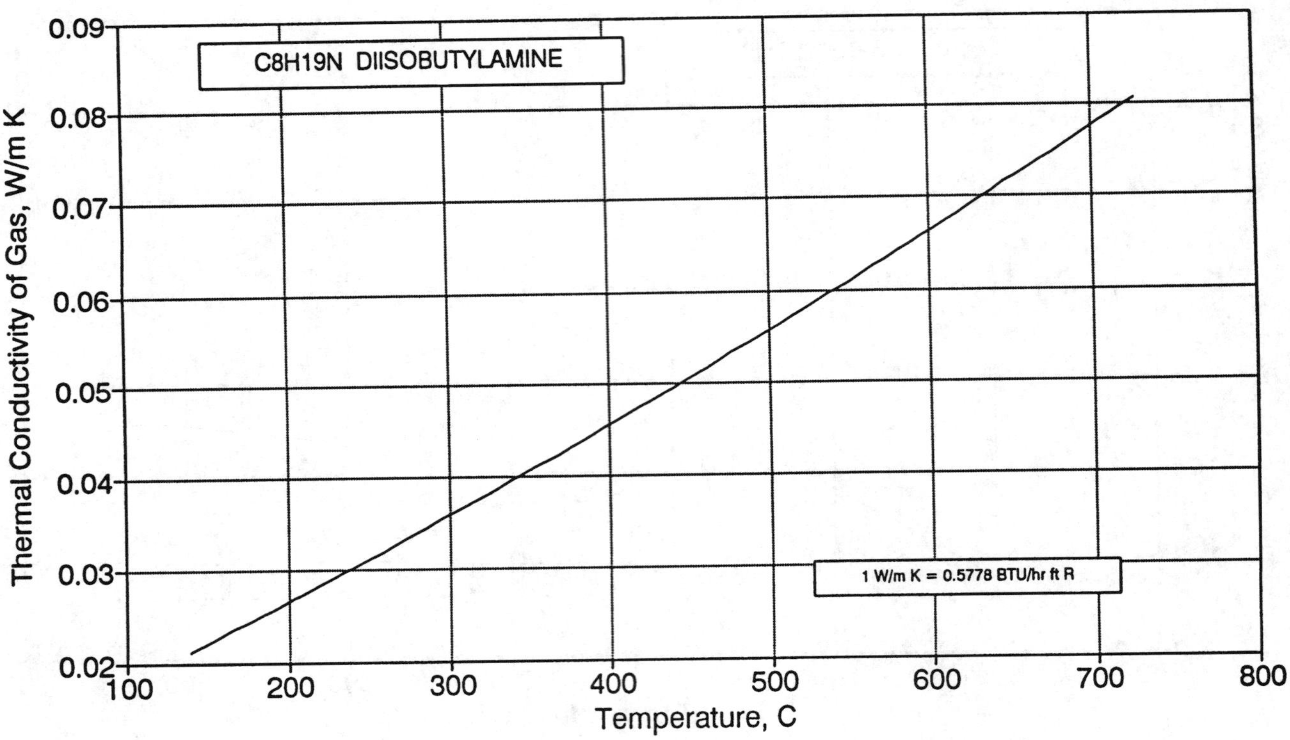
C8H19N DIISOBUTYLAMINE
1 W/m K = 0.5778 BTU/hr ft R
Thermal Conductivity of Gas, W/m K
0.09
0.08
0.07
0.06
0.05
0.04
0.03
0.02
100
200
300
400
500
600
700
800
Temperature, C

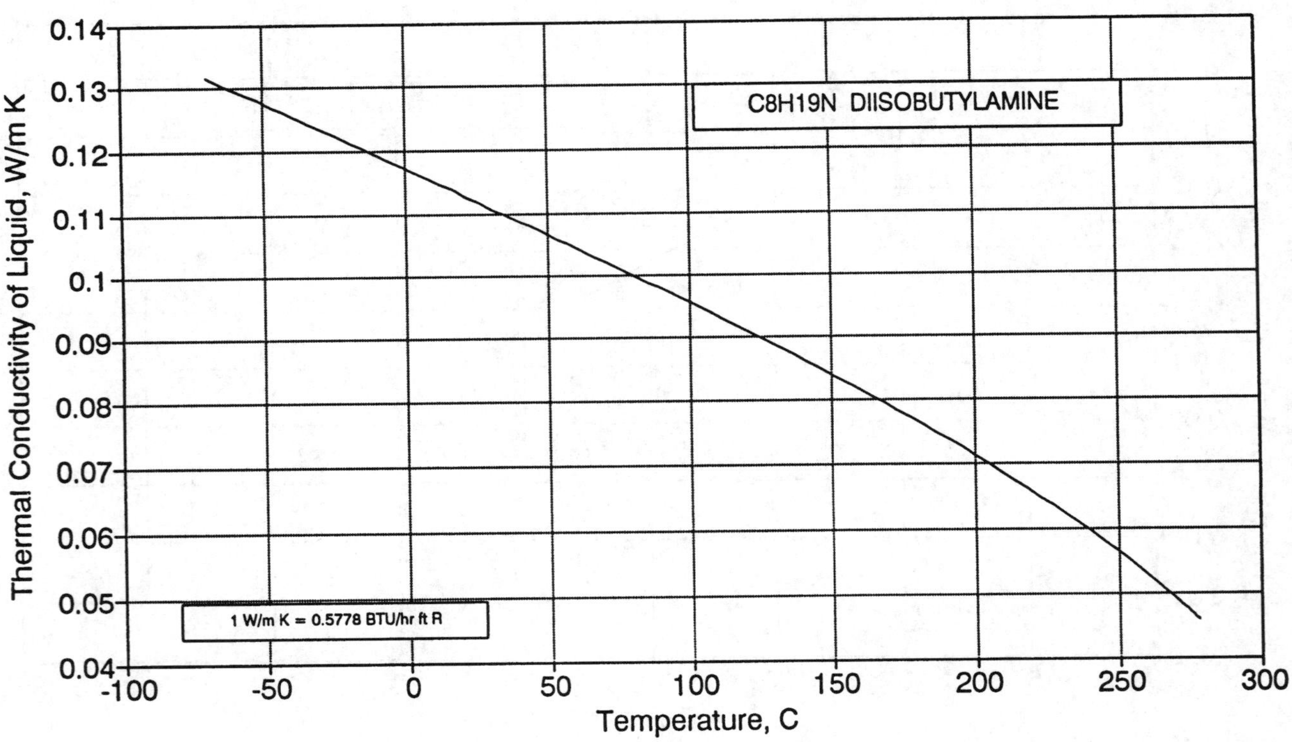
C8H19N DIISOBUTYLAMINE
1 W/m K = 0.5778 BTU/hr ft R
Thermal Conductivity of Liquid, W/m K
0.14
0.13
0.12
0.11
0.1
0.09
0.08
0.07
0.06
0.05
0.04
-100
-50
0
50
100
150
200
250
300
Temperature, C

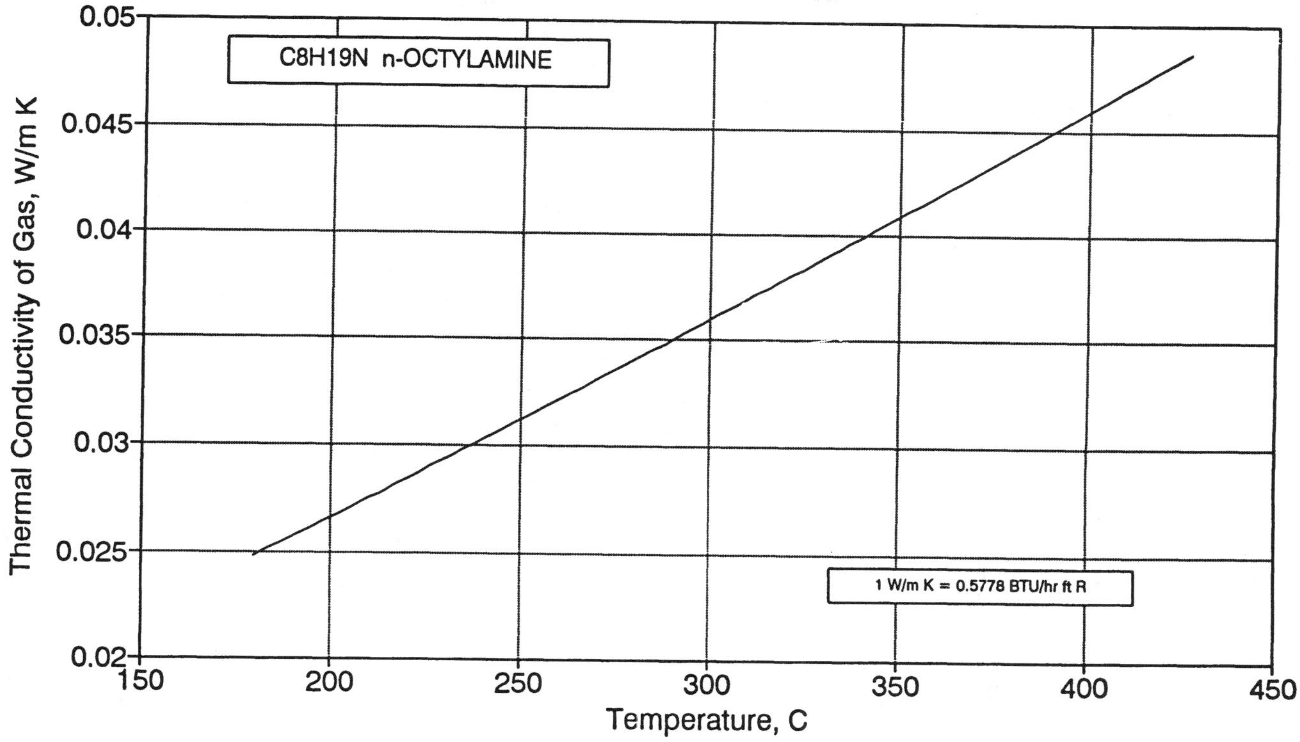
C8H19N n-OCTYLAMINE
1 W/m K = 0.5778 BTU/hr ft R
Thermal Conductivity of Gas, W/m K
0.05
0.045
0.04
0.035
0.03
0.025
0.02
150
200
250
300
350
400
450
Temperature, C

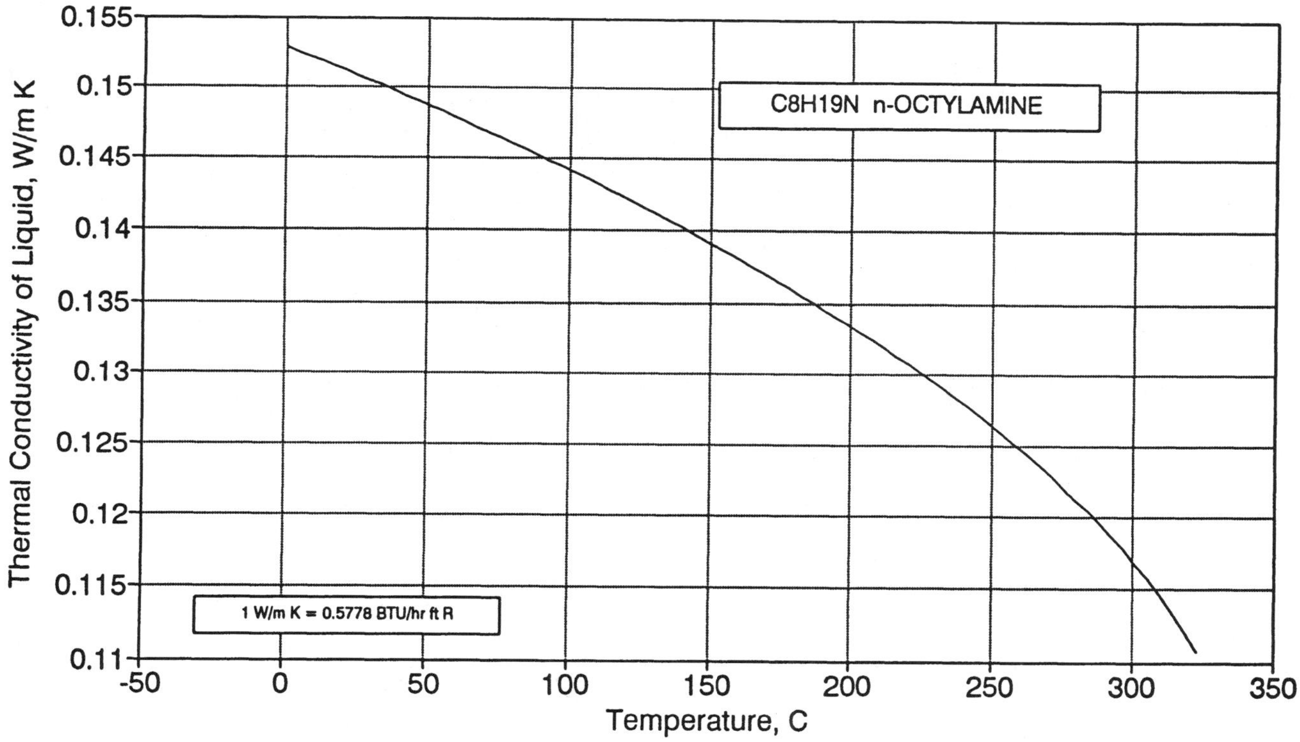
C8H19N n-OCTYLAMINE
1 W/m K = 0.5778 BTU/hr ft R
Thermal Conductivity of Liquid, W/m K
0.155
0.15
0.145
0.14
0.135
0.13
0.125
0.12
0.115
0.11
-50
0
50
100
150
200
250
300
350
Temperature, C

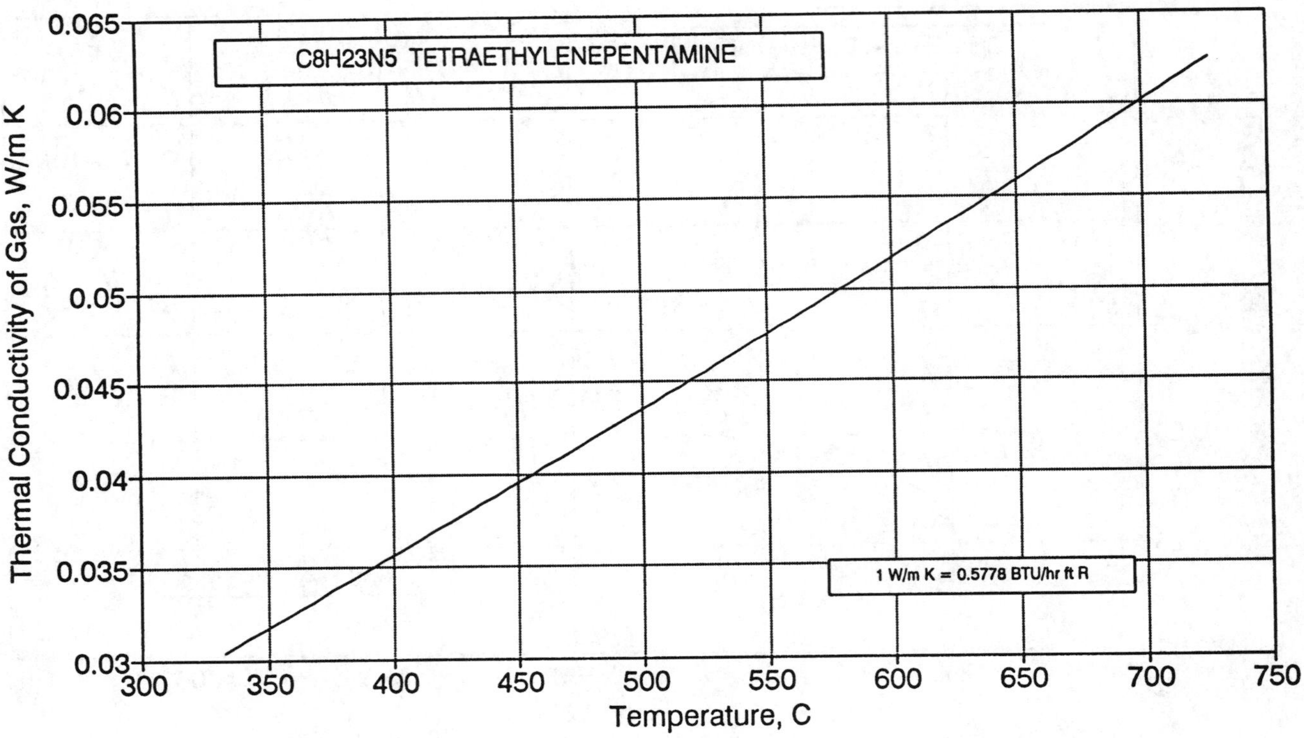
C8H23N5 TETRAETHYLENEPENTAMINE
1 W/m K = 0.5778 BTU/hr ft R
Thermal Conductivity of Gas, W/m K
0.065
0.06
0.055
0.05
0.045
0.04
0.035
0.03
300
350
400
450
500
550
600
650
700
750
Temperature, C

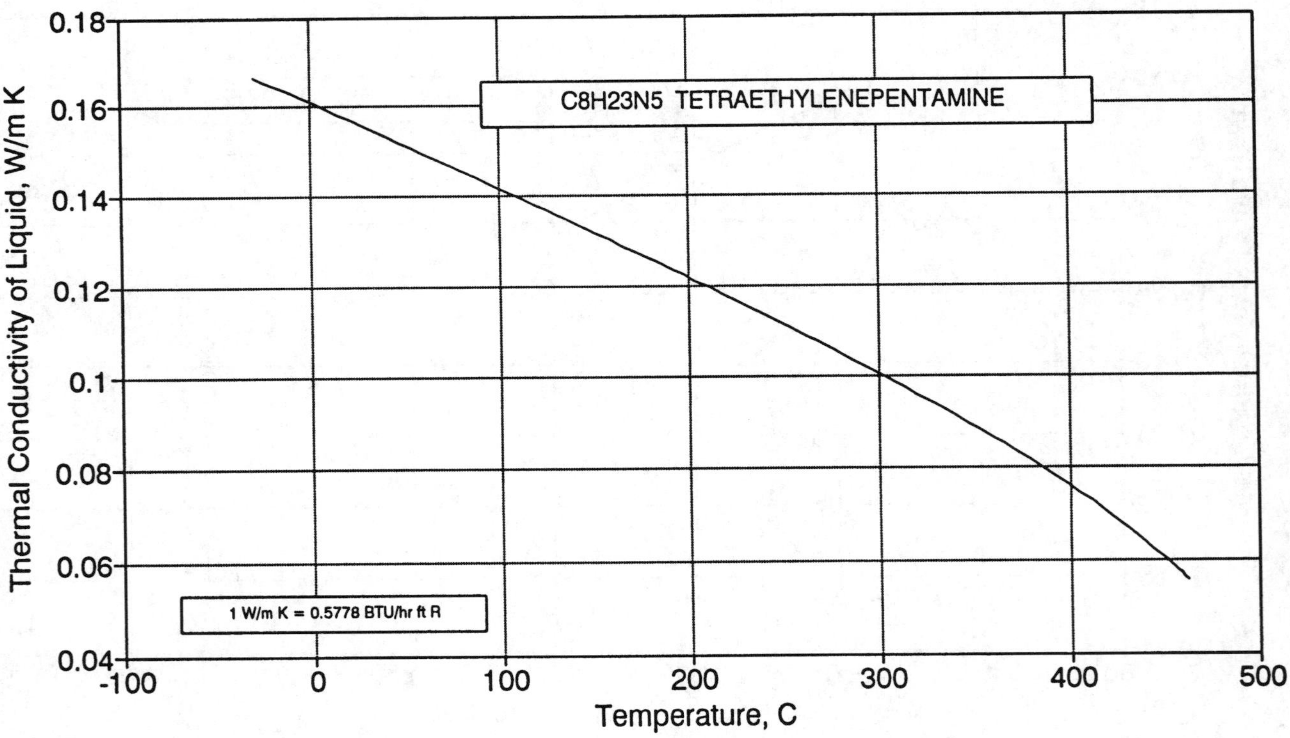
C8H23N5 TETRAETHYLENEPENTAMINE
1 W/m K = 0.5778 BTU/hr ft R
Thermal Conductivity of Liquid, W/m K
0.18
0.16
0.14
0.12
0.1
0.08
0.06
0.04
-100
0
100
200
300
400
500
Temperature, C

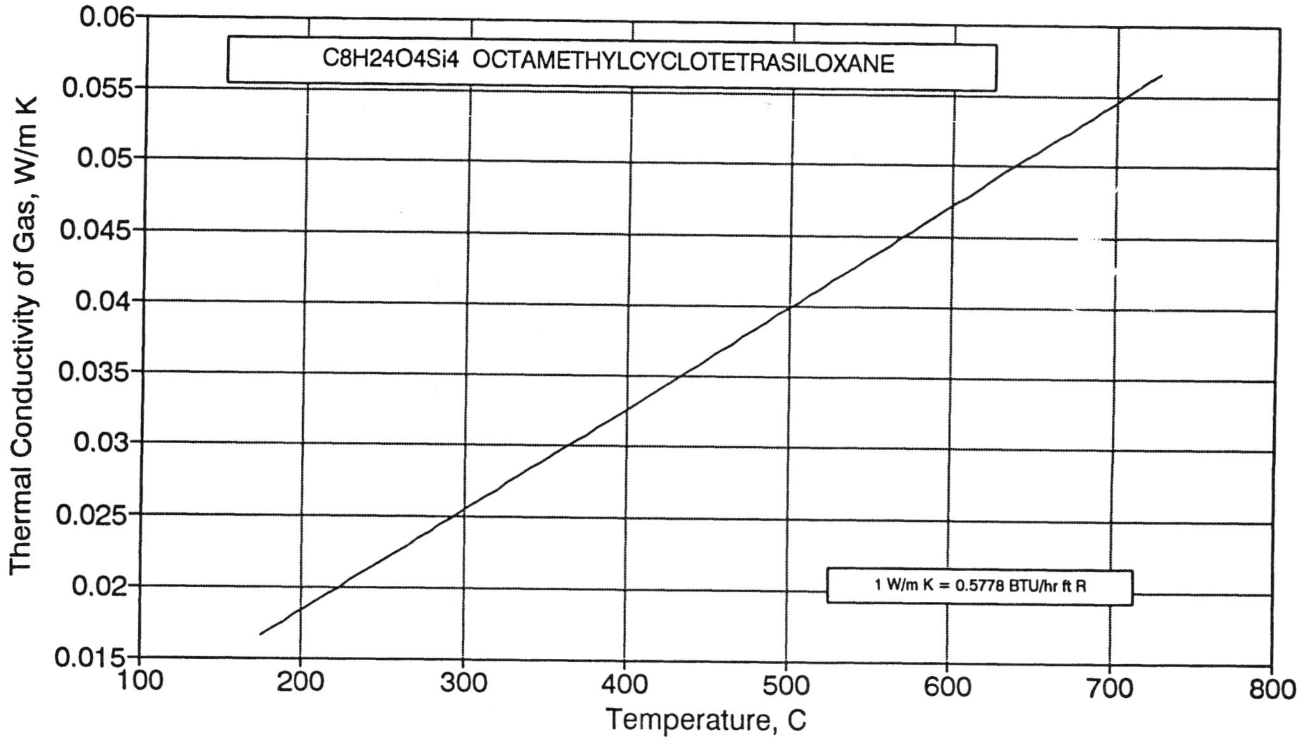
C8H24O4Si4 OCTAMETHYLCYCLOTETRASILOXANE
1 W/m K = 0.5778 BTU/hr ft R
Thermal Conductivity of Gas, W/m K
Temperature, C
0.06
0.055
0.05
0.045
0.04
0.035
0.03
0.025
0.02
0.015
100
200
300
400
500
600
700
800

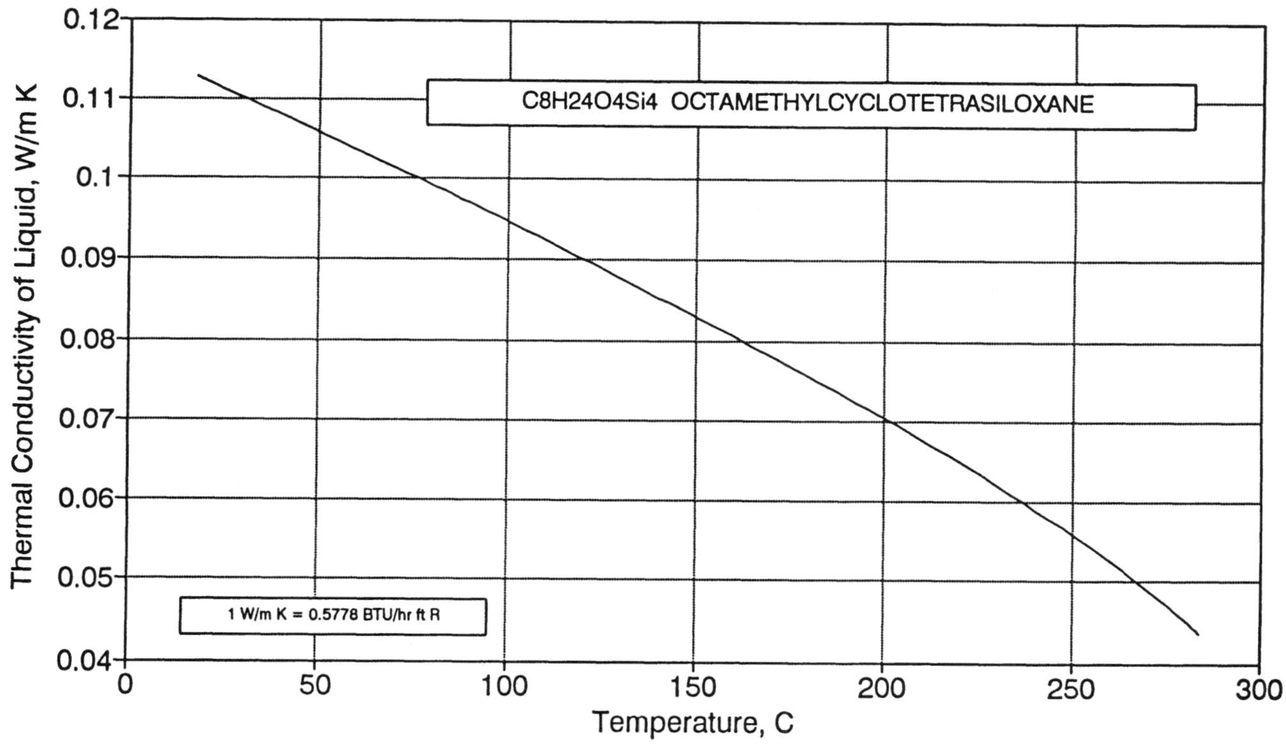
C8H24O4Si4 OCTAMETHYLCYCLOTETRASILOXANE
1 W/m K = 0.5778 BTU/hr ft R
Thermal Conductivity of Liquid, W/m K
Temperature, C
0.12
0.11
0.1
0.09
0.08
0.07
0.06
0.05
0.04
0
50
100
150
200
250
300

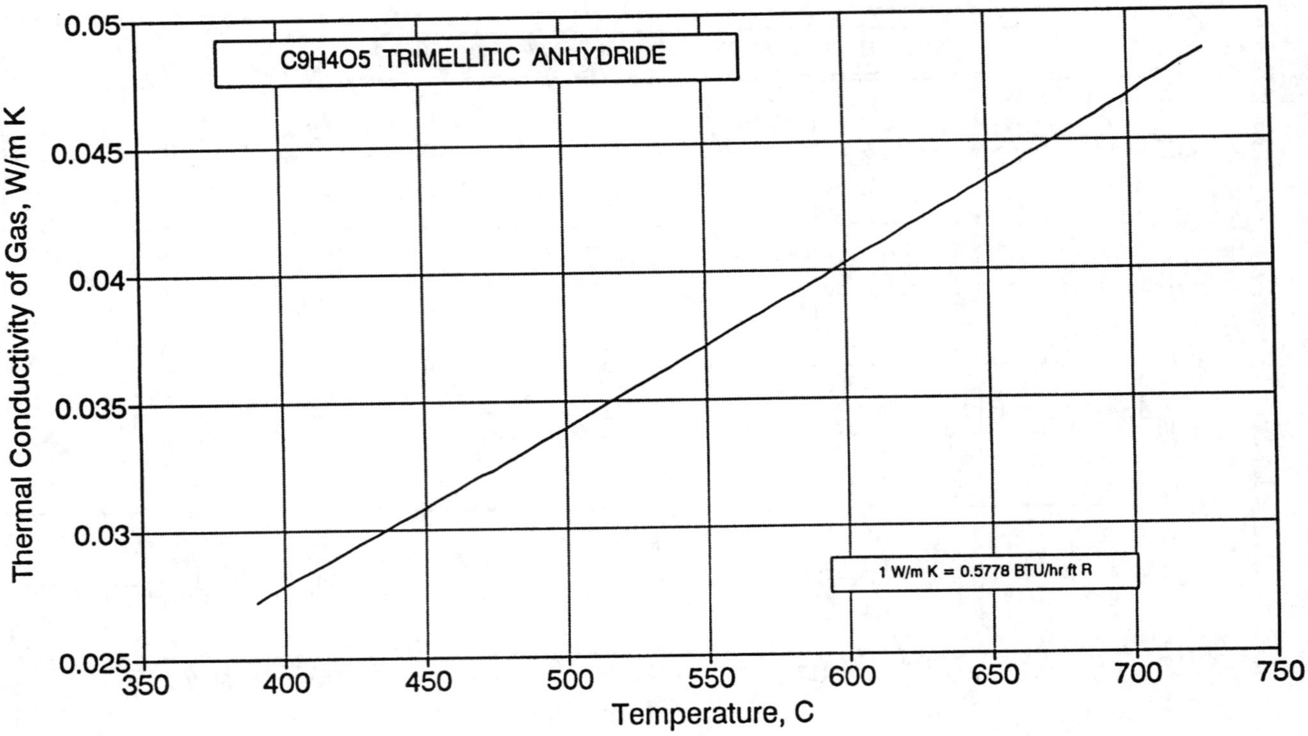
C9H4O5 TRIMELLITIC ANHYDRIDE
1 W/m K = 0.5778 BTU/hr ft R
Thermal Conductivity of Gas, W/m K
0.05
0.045
0.04
0.035
0.03
0.025
350
400
450
500
550
600
650
700
750
Temperature, C

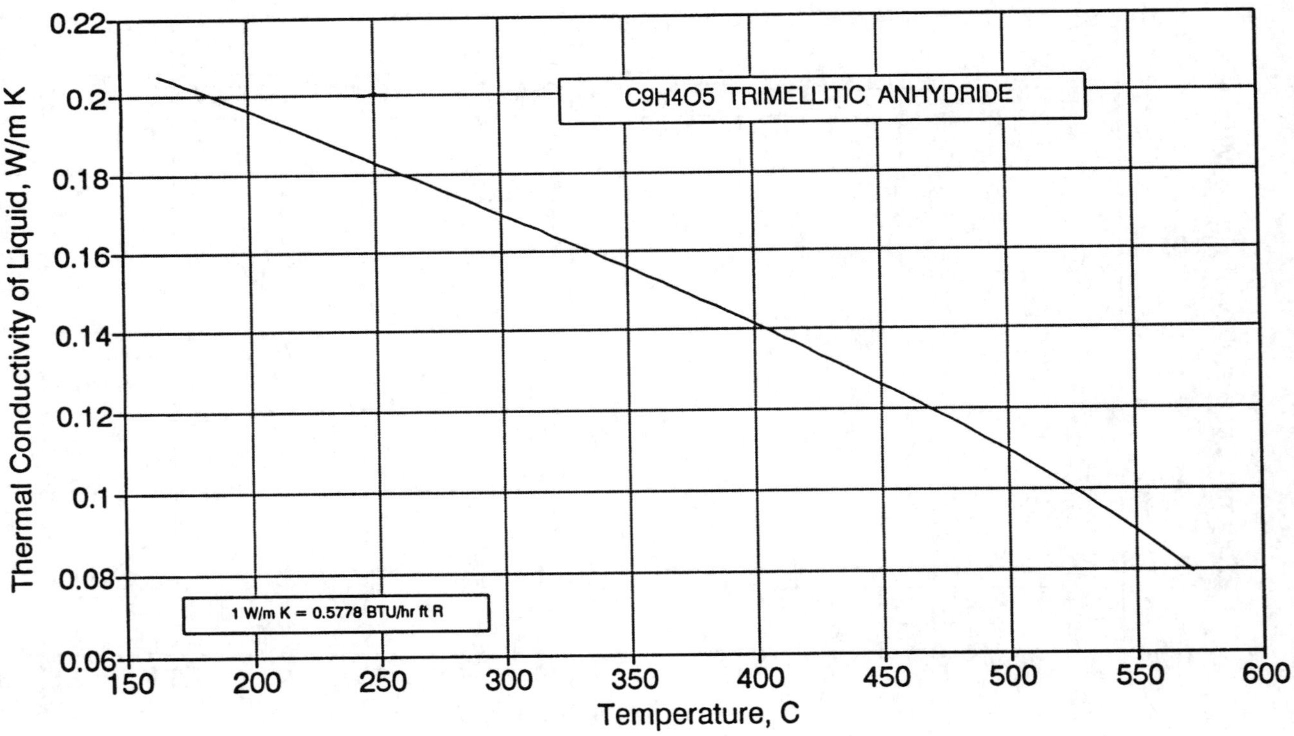
C9H4O5 TRIMELLITIC ANHYDRIDE
1 W/m K = 0.5778 BTU/hr ft R
Thermal Conductivity of Liquid, W/m K
0.22
0.2
0.18
0.16
0.14
0.12
0.1
0.08
0.06
150
200
250
300
350
400
450
500
550
600
Temperature, C

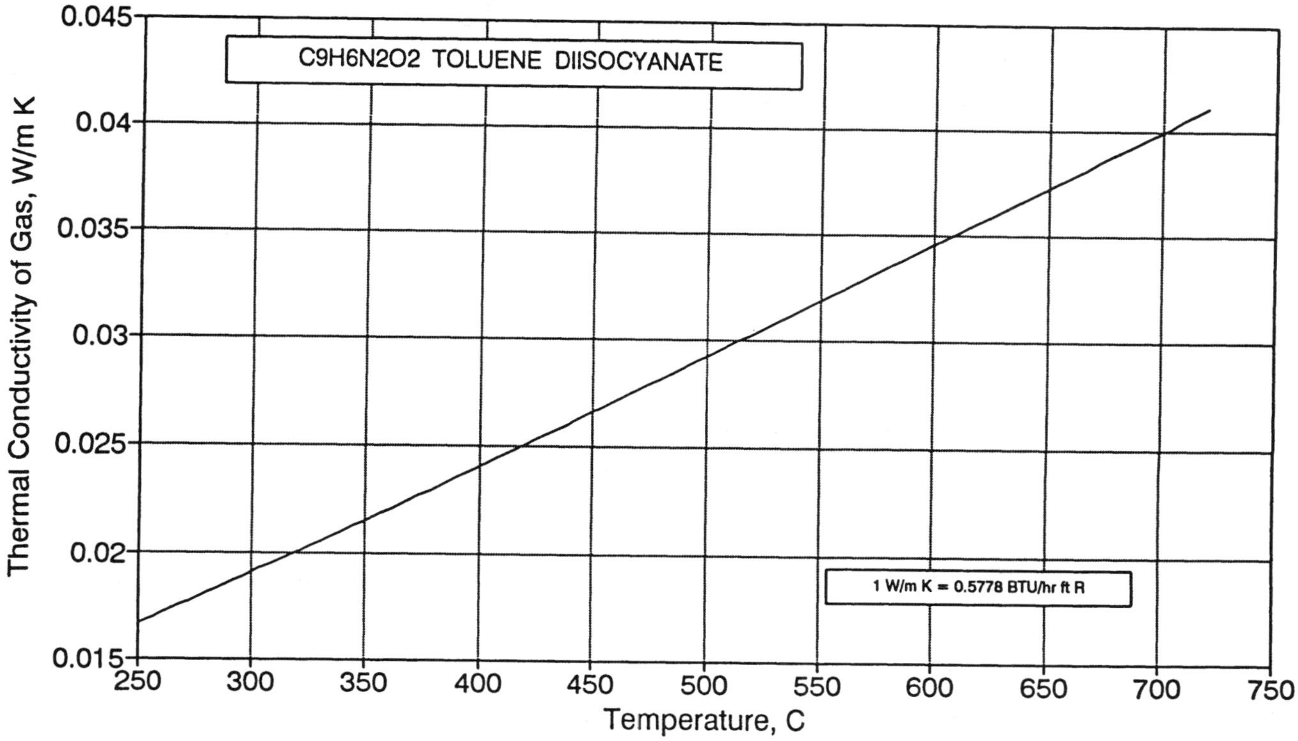

C9H6N2O2 TOLUENE DIISOCYANATE
1 W/m K = 0.5778 BTU/hr ft R
Thermal Conductivity of Gas, W/m K
0.045
0.04
0.035
0.03
0.025
0.02
0.015
250
300
350
400
450
500
550
600
650
700
750
Temperature, C

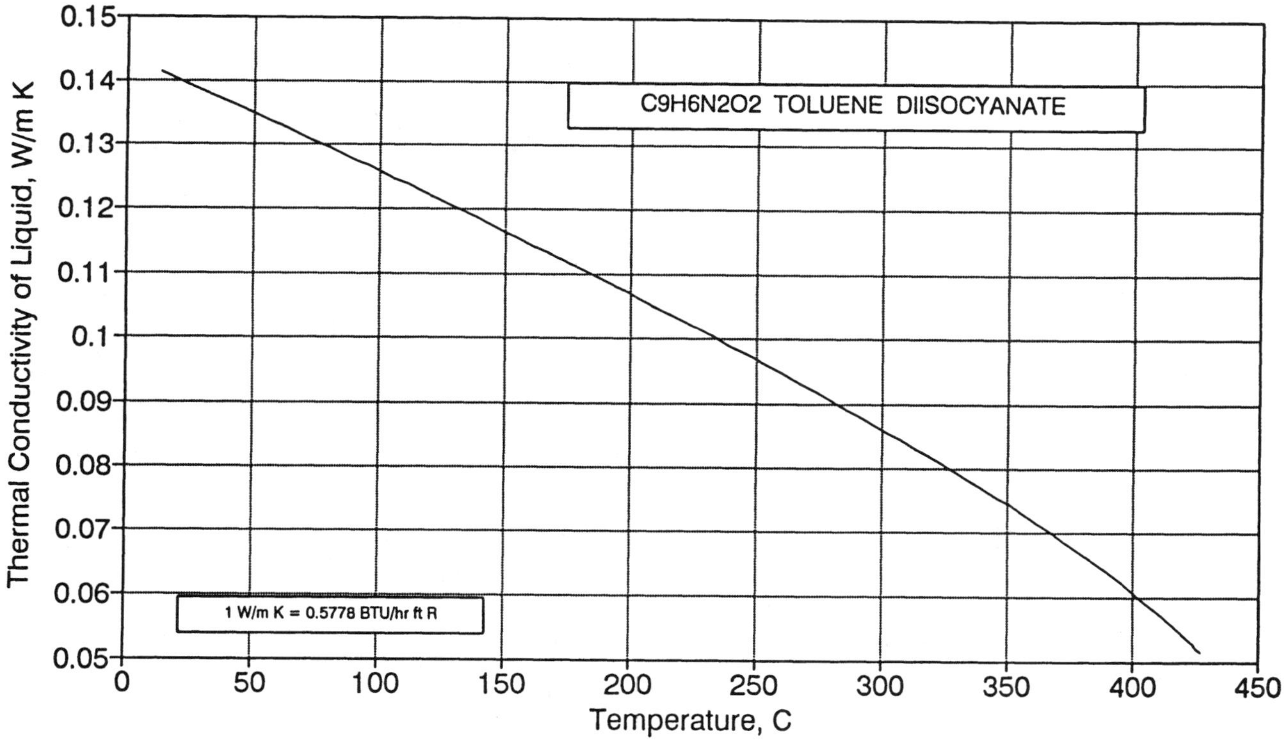

C9H6N2O2 TOLUENE DIISOCYANATE
1 W/m K = 0.5778 BTU/hr ft R
Thermal Conductivity of Liquid, W/m K
0.15
0.14
0.13
0.12
0.11
0.1
0.09
0.08
0.07
0.06
0.05
0
50
100
150
200
250
300
350
400
450
Temperature, C

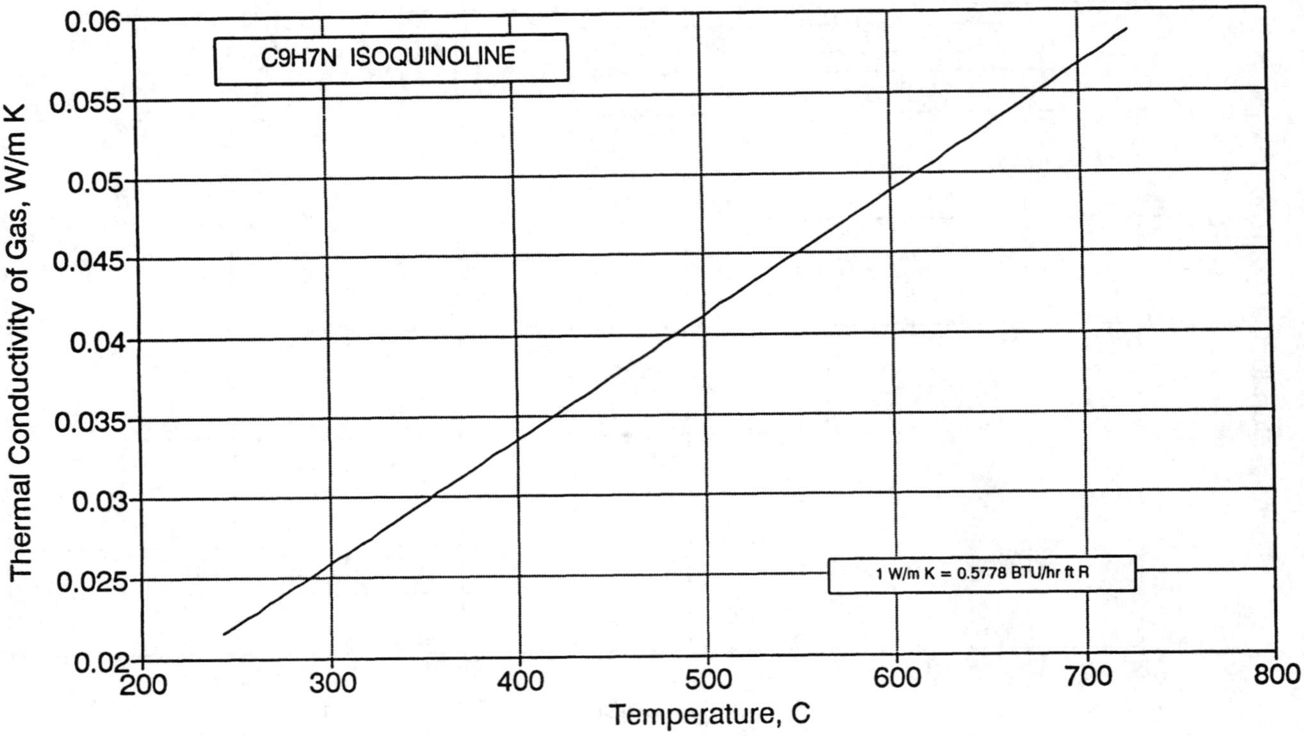
C9H7N ISOQUINOLINE
1 W/m K = 0.5778 BTU/hr ft R
Thermal Conductivity of Gas, W/m K
0.06
0.055
0.05
0.045
0.04
0.035
0.03
0.025
0.02
200
300
400
500
600
700
800
Temperature, C

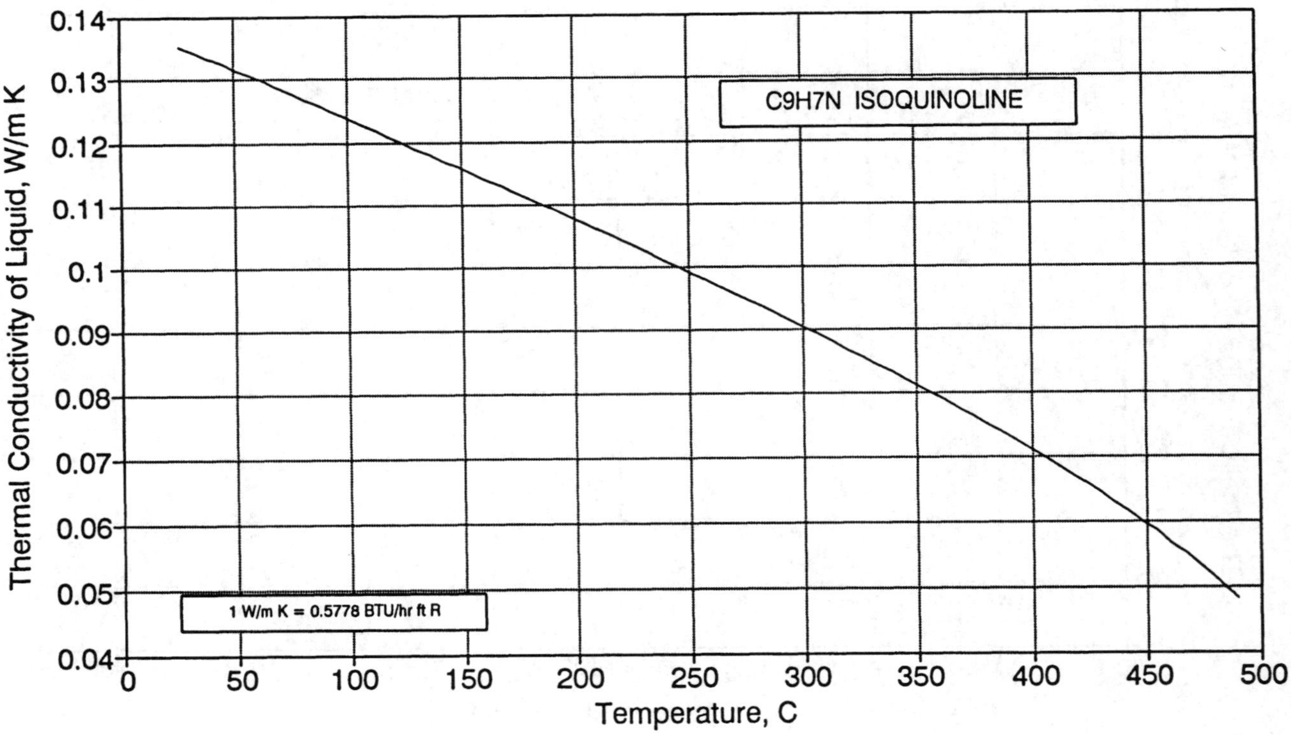
C9H7N ISOQUINOLINE
1 W/m K = 0.5778 BTU/hr ft R
Thermal Conductivity of Liquid, W/m K
0.14
0.13
0.12
0.11
0.1
0.09
0.08
0.07
0.06
0.05
0.04
0
50
100
150
200
250
300
350
400
450
500
Temperature, C

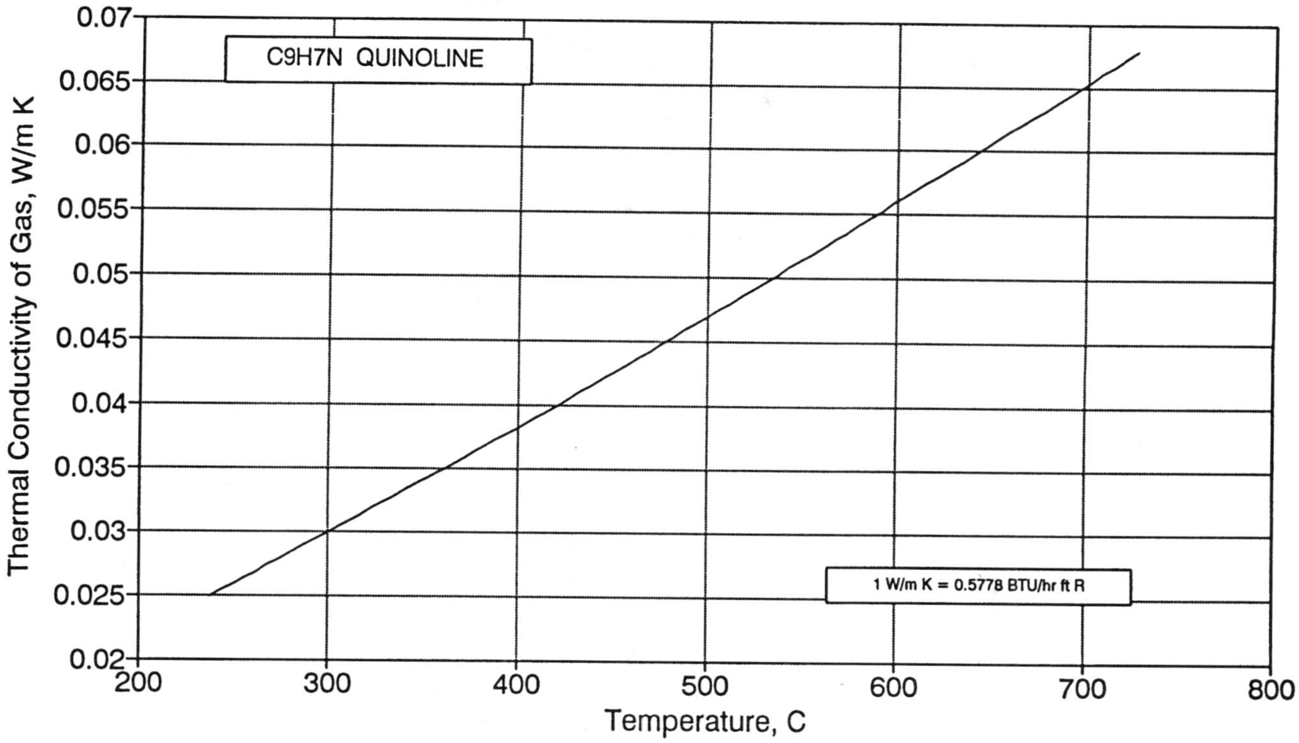
C9H7N QUINOLINE
1 W/m K = 0.5778 BTU/hr ft R
Thermal Conductivity of Gas, W/m K
Temperature, C
0.07
0.065
0.06
0.055
0.05
0.045
0.04
0.035
0.03
0.025
0.02
200
300
400
500
600
700
800

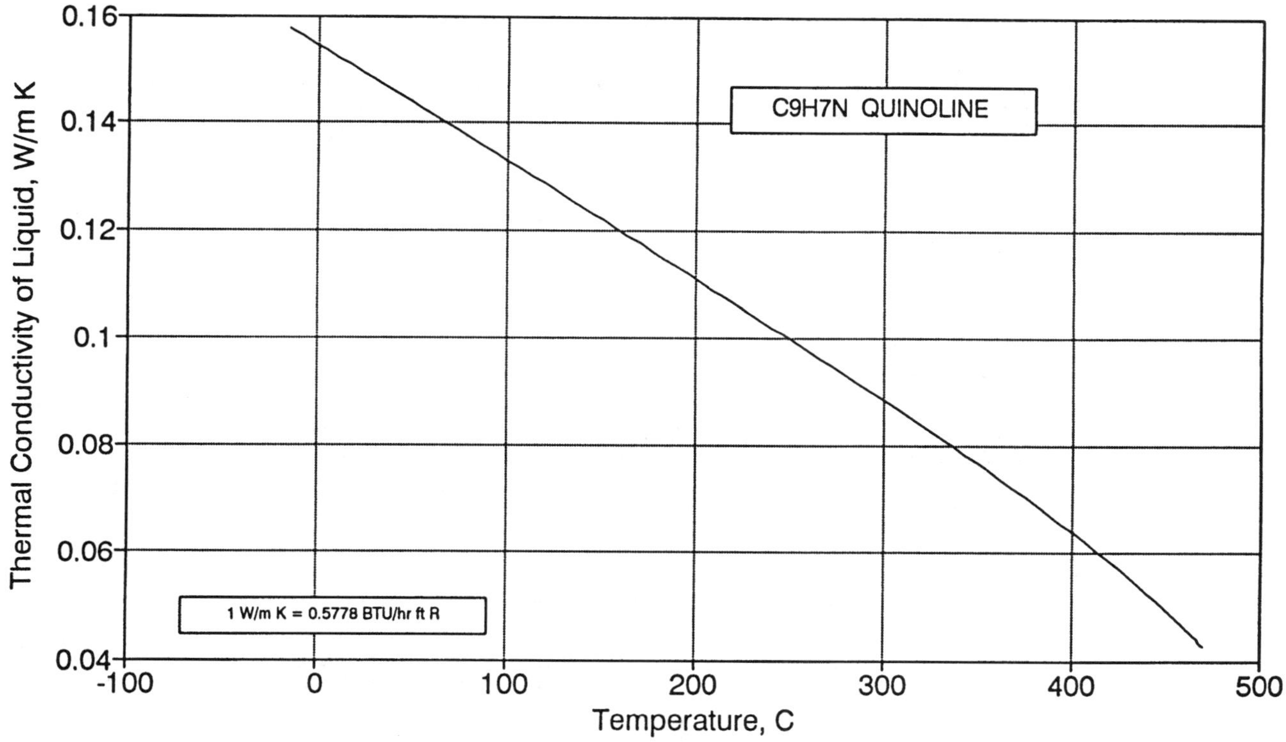
C9H7N QUINOLINE
1 W/m K = 0.5778 BTU/hr ft R
Thermal Conductivity of Liquid, W/m K
Temperature, C
0.16
0.14
0.12
0.1
0.08
0.06
0.04
-100
0
100
200
300
400
500

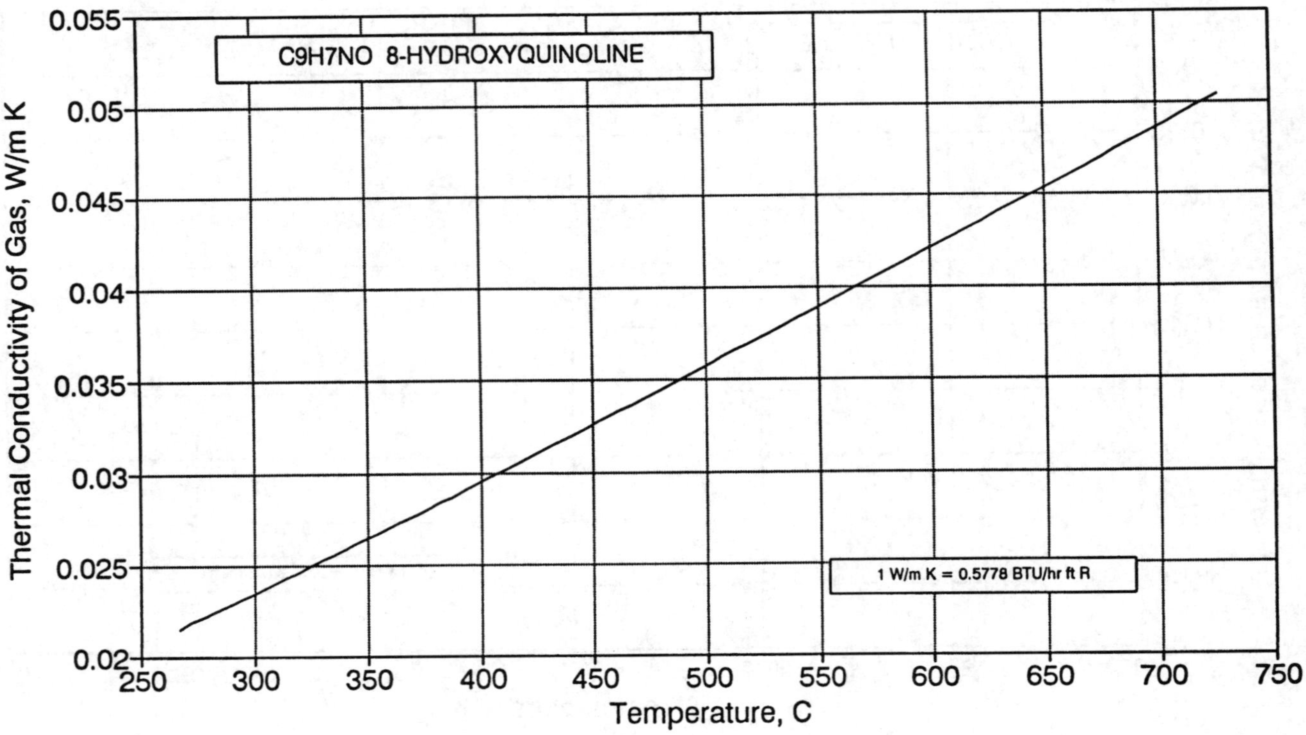
C9H7NO 8-HYDROXYQUINOLINE
Thermal Conductivity of Gas, W/m K
0.055
0.05
0.045
0.04
0.035
0.03
0.025
0.02
1 W/m K = 0.5778 BTU/hr ft R
250
300
350
400
450
500
550
600
650
700
750
Temperature, C

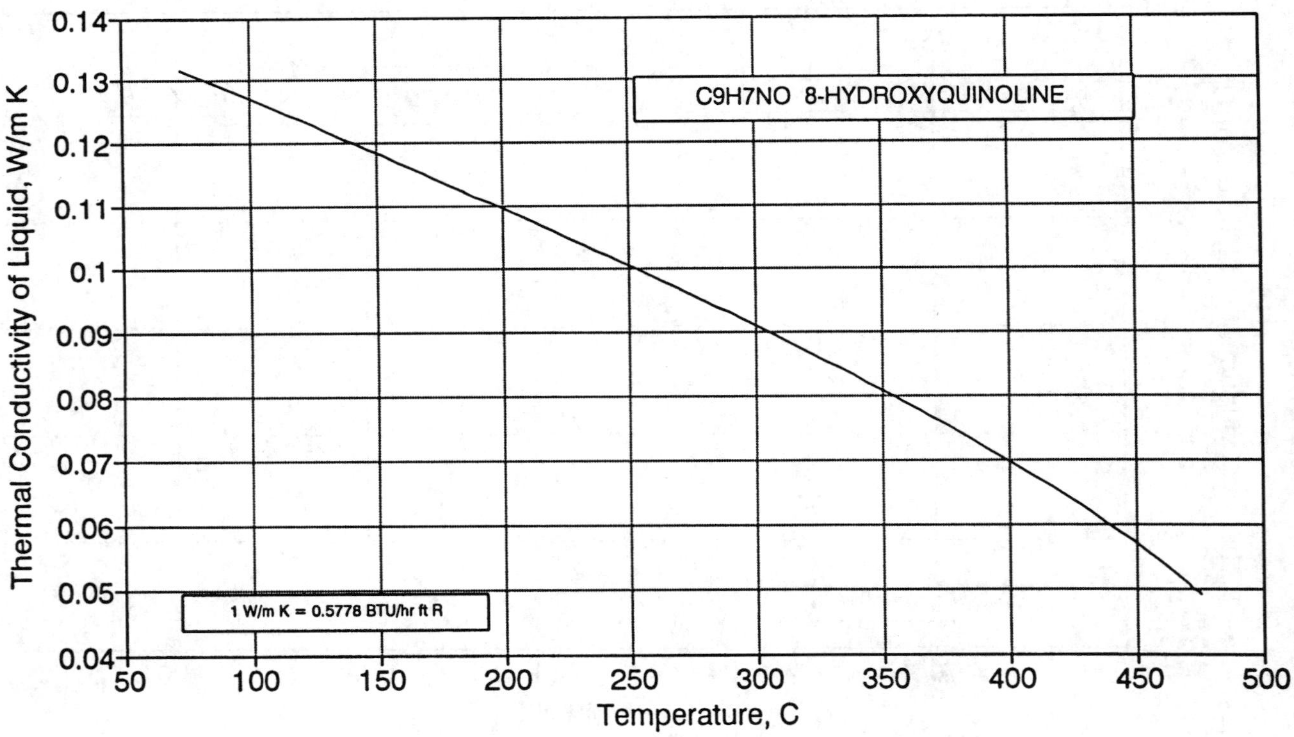
C9H7NO 8-HYDROXYQUINOLINE
Thermal Conductivity of Liquid, W/m K
0.14
0.13
0.12
0.11
0.1
0.09
0.08
0.07
0.06
0.05
0.04
1 W/m K = 0.5778 BTU/hr ft R
50
100
150
200
250
300
350
400
450
500
Temperature, C

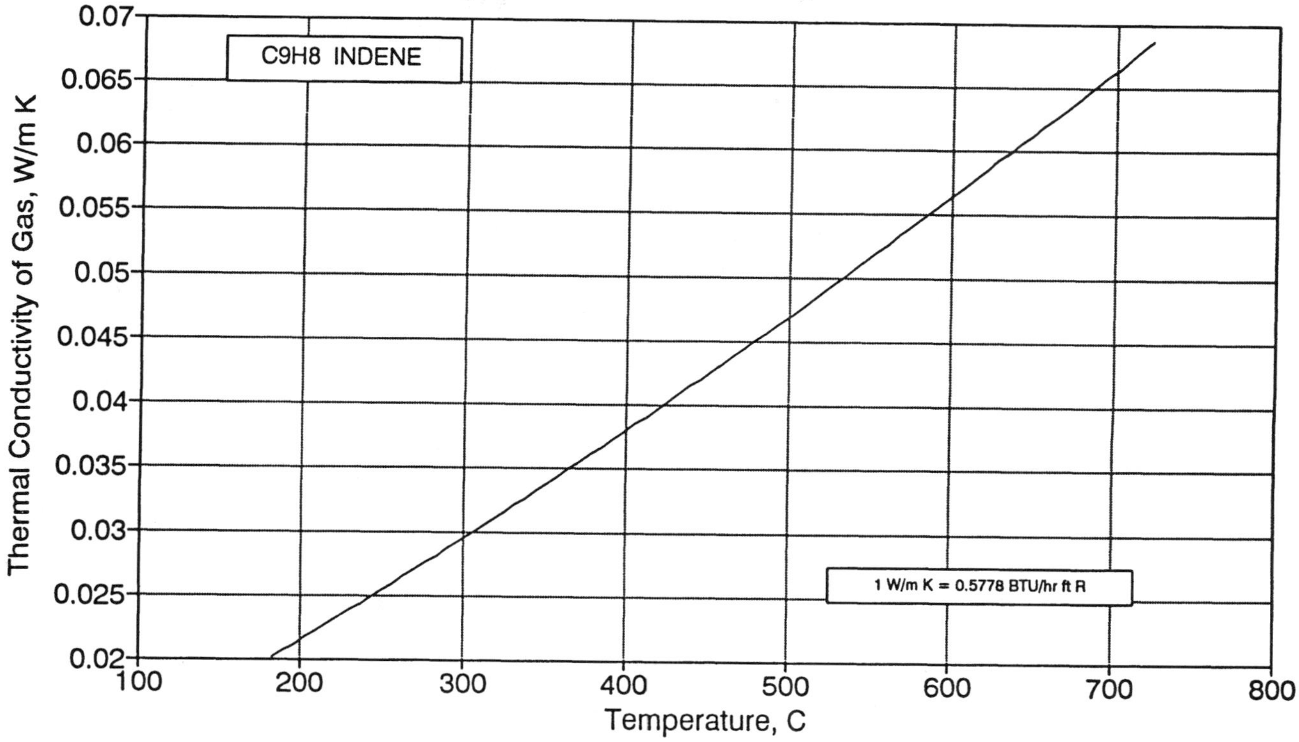
C9H8 INDENE
1 W/m K = 0.5778 BTU/hr ft R
Thermal Conductivity of Gas, W/m K
Temperature, C
0.07
0.065
0.06
0.055
0.05
0.045
0.04
0.035
0.03
0.025
0.02
100
200
300
400
500
600
700
800

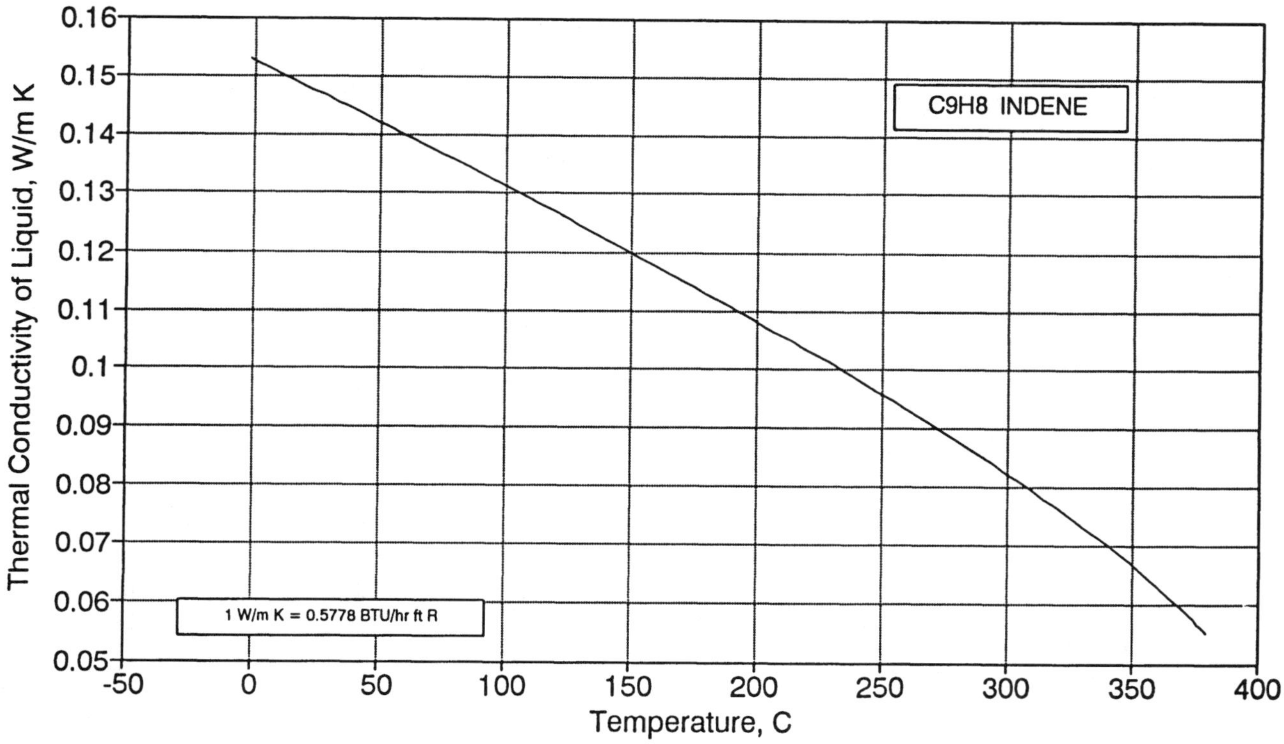
C9H8 INDENE
1 W/m K = 0.5778 BTU/hr ft R
Thermal Conductivity of Liquid, W/m K
Temperature, C
0.16
0.15
0.14
0.13
0.12
0.11
0.1
0.09
0.08
0.07
0.06
0.05
-50
0
50
100
150
200
250
300
350
400

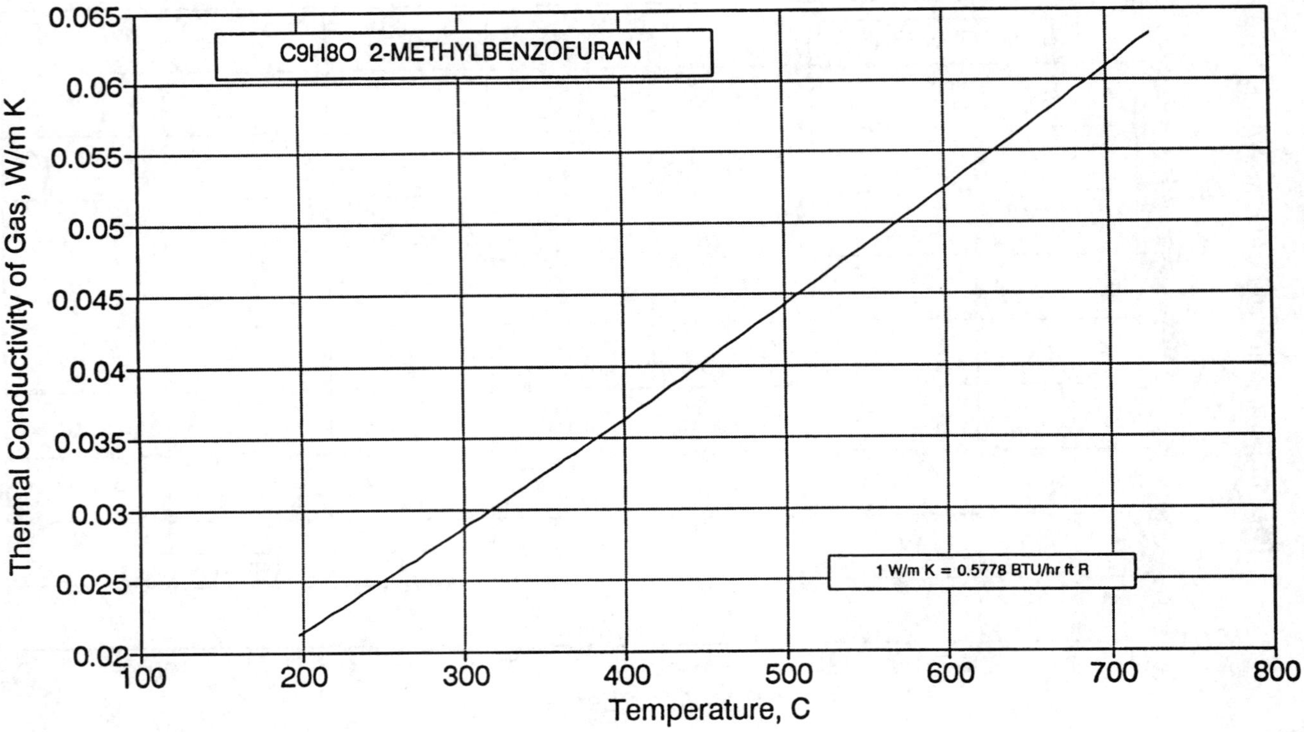
C9H8O 2-METHYLBENZOFURAN
1 W/m K = 0.5778 BTU/hr ft R
Thermal Conductivity of Gas, W/m K
0.065
0.06
0.055
0.05
0.045
0.04
0.035
0.03
0.025
0.02
100
200
300
400
500
600
700
800
Temperature, C

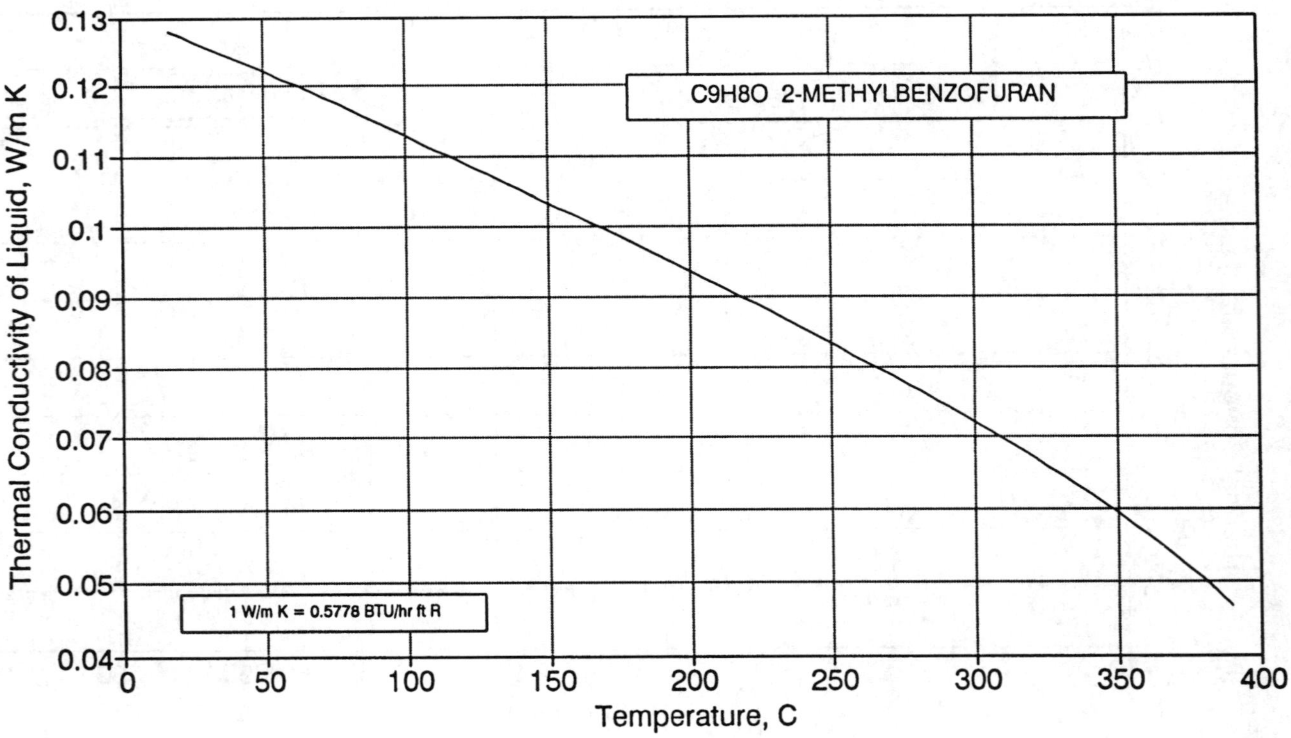
C9H8O 2-METHYLBENZOFURAN
1 W/m K = 0.5778 BTU/hr ft R
Thermal Conductivity of Liquid, W/m K
0.13
0.12
0.11
0.1
0.09
0.08
0.07
0.06
0.05
0.04
0
50
100
150
200
250
300
350
400
Temperature, C

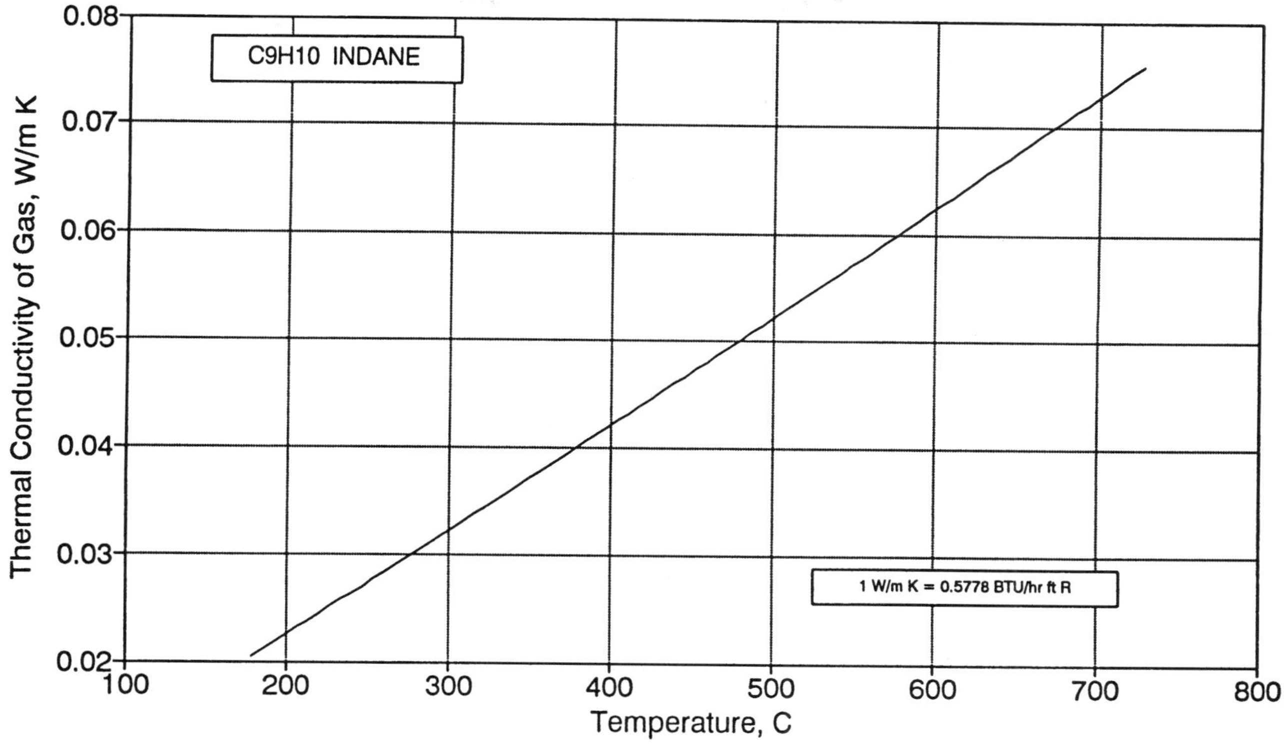
C9H10 INDANE
1 W/m K = 0.5778 BTU/hr ft R
Thermal Conductivity of Gas, W/m K
0.08
0.07
0.06
0.05
0.04
0.03
0.02
100
200
300
400
500
600
700
800
Temperature, C

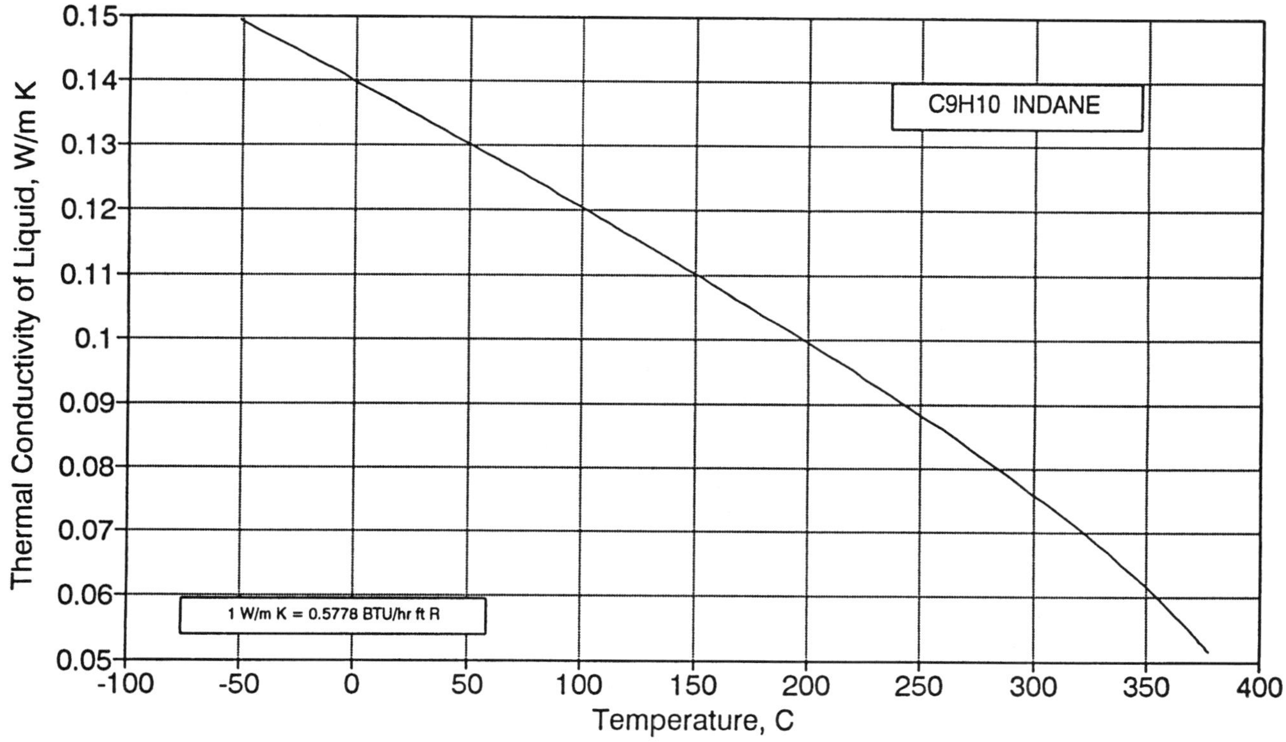
C9H10 INDANE
1 W/m K = 0.5778 BTU/hr ft R
Thermal Conductivity of Liquid, W/m K
0.15
0.14
0.13
0.12
0.11
0.1
0.09
0.08
0.07
0.06
0.05
-100
-50
0
50
100
150
200
250
300
350
400
Temperature, C

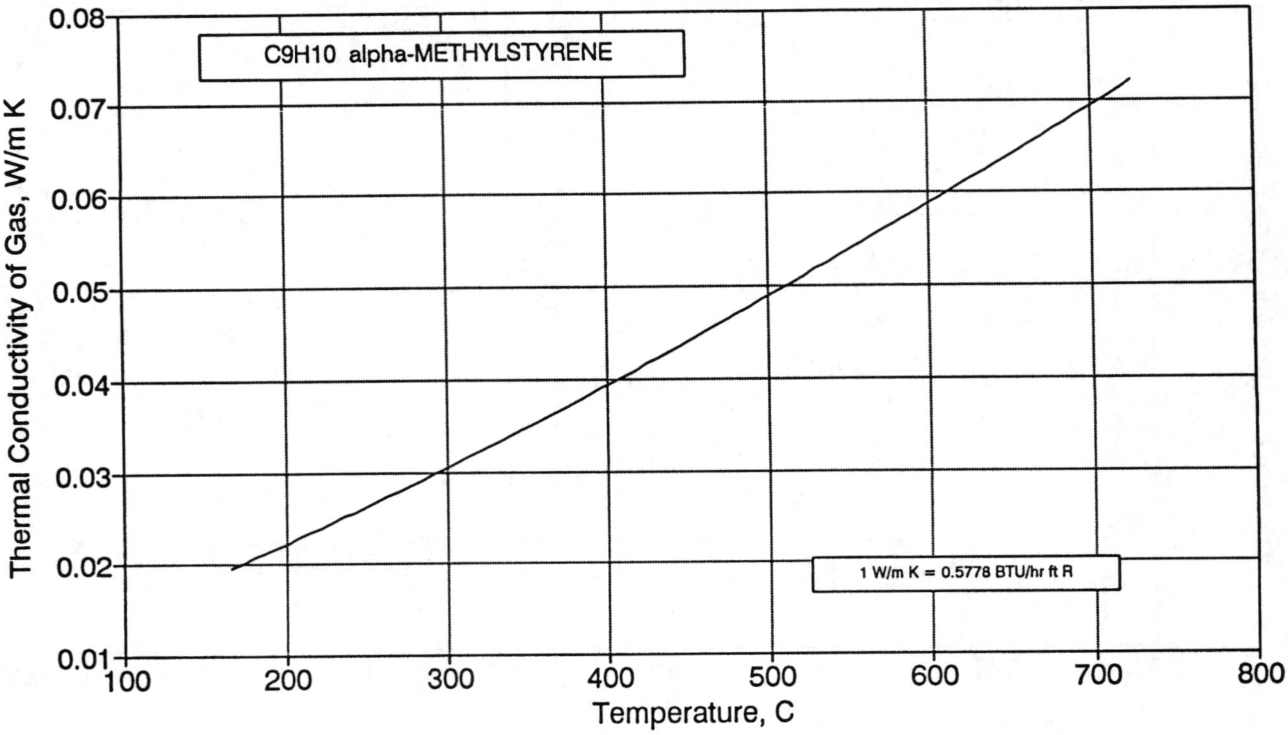
C9H10 alpha-METHYLSTYRENE
Thermal Conductivity of Gas, W/m K
0.08
0.07
0.06
0.05
0.04
0.03
0.02
0.01
1 W/m K = 0.5778 BTU/hr ft R
100
200
300
400
500
600
700
800
Temperature, C

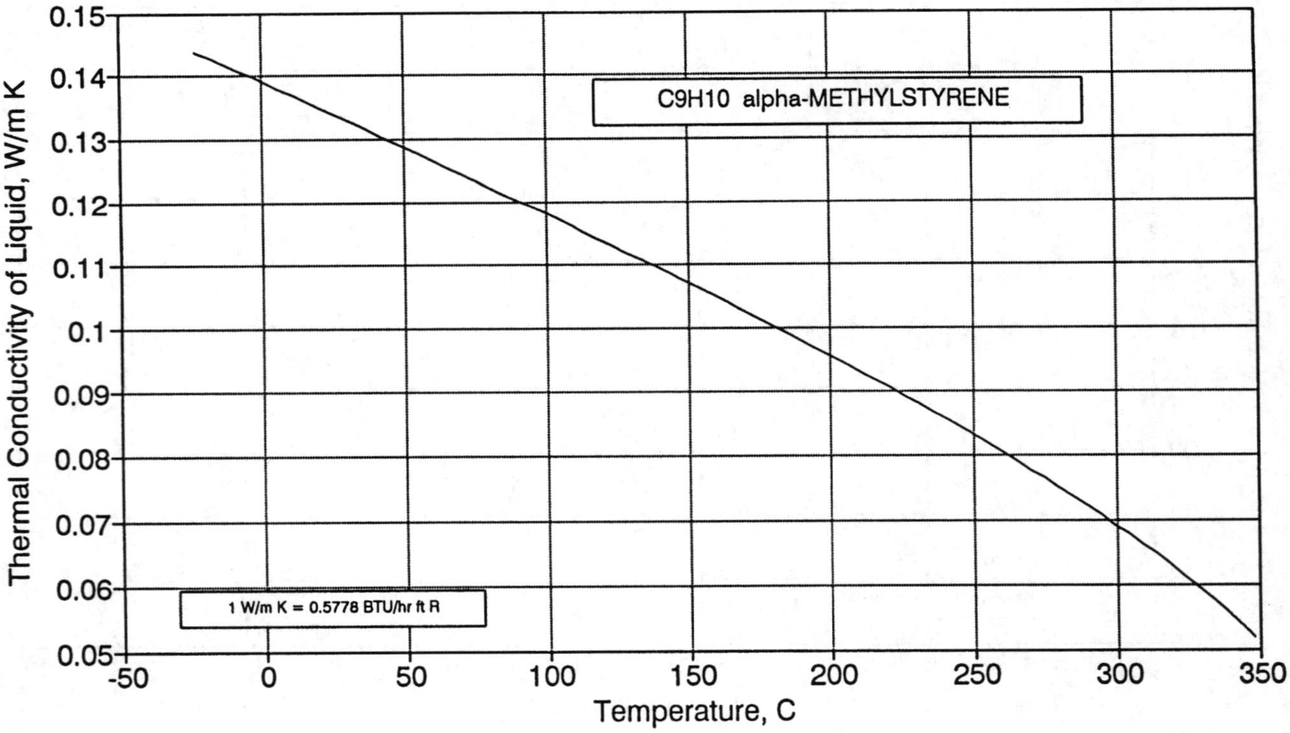
C9H10 alpha-METHYLSTYRENE
Thermal Conductivity of Liquid, W/m K
0.15
0.14
0.13
0.12
0.11
0.1
0.09
0.08
0.07
0.06
0.05
1 W/m K = 0.5778 BTU/hr ft R
-50
0
50
100
150
200
250
300
350
Temperature, C

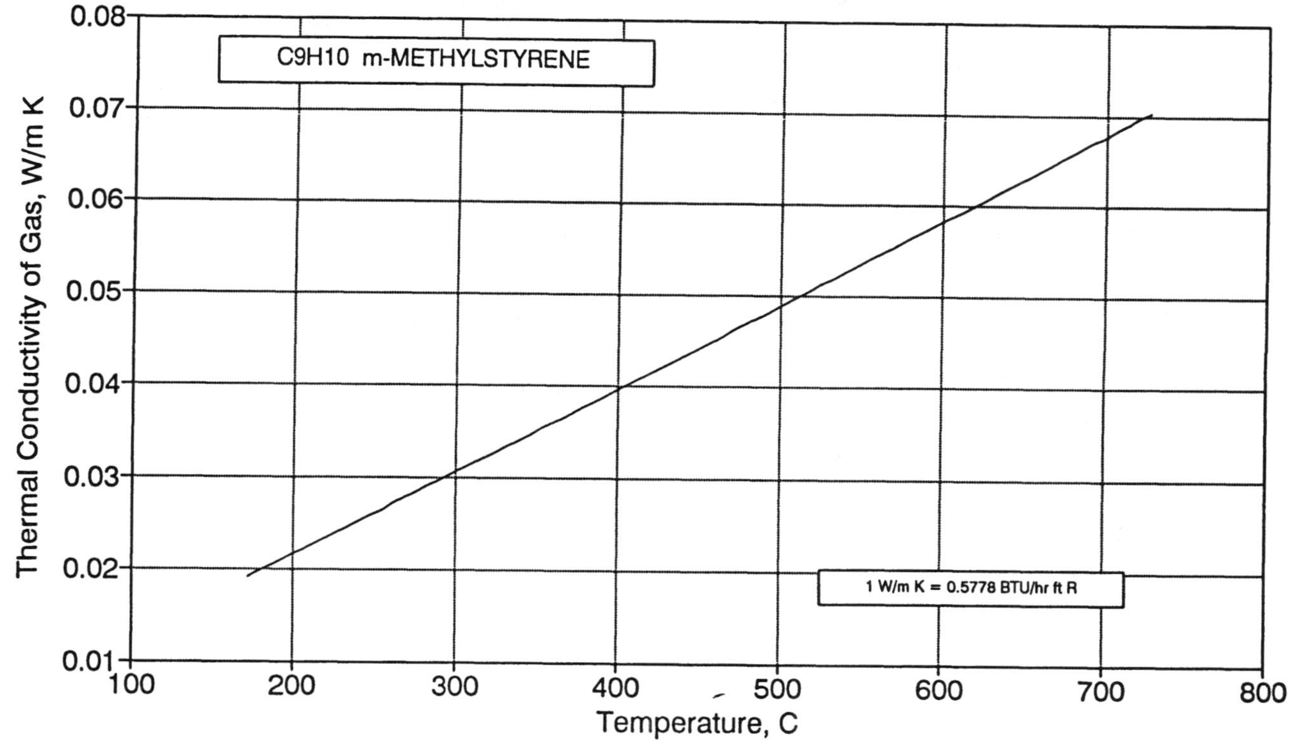
C9H10 m-METHYLSTYRENE
1 W/m K = 0.5778 BTU/hr ft R
Thermal Conductivity of Gas, W/m K
0.08
0.07
0.06
0.05
0.04
0.03
0.02
0.01
100
200
300
400
500
600
700
800
Temperature, C

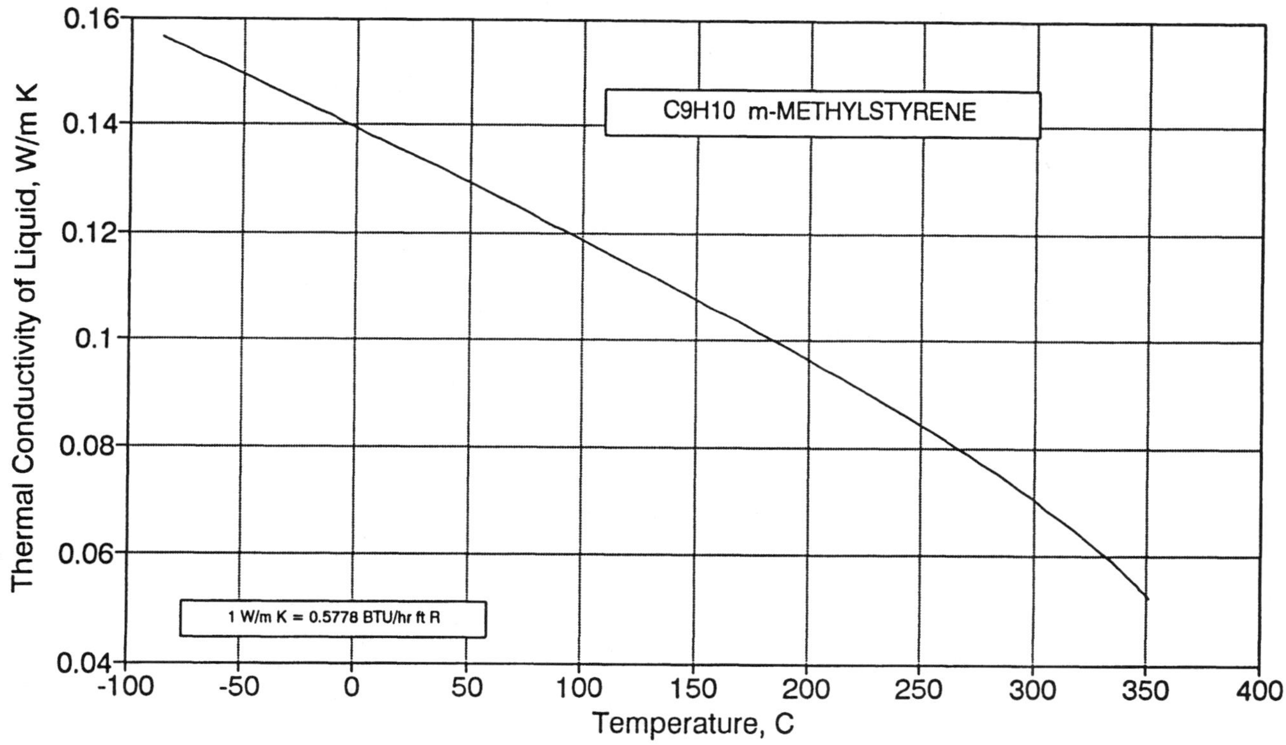
C9H10 m-METHYLSTYRENE
1 W/m K = 0.5778 BTU/hr ft R
Thermal Conductivity of Liquid, W/m K
0.16
0.14
0.12
0.1
0.08
0.06
0.04
-100
-50
0
50
100
150
200
250
300
350
400
Temperature, C

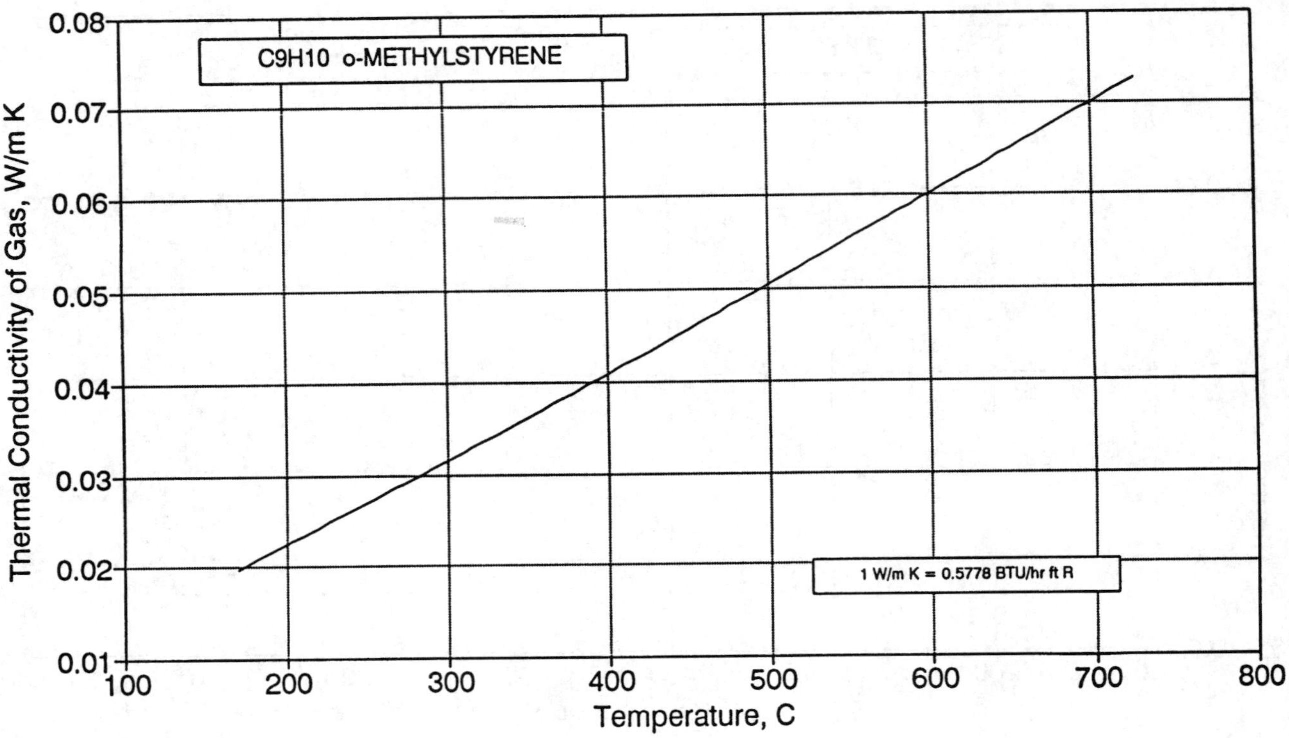
C9H10 o-METHYLSTYRENE
1 W/m K = 0.5778 BTU/hr ft R
Thermal Conductivity of Gas, W/m K
Temperature, C
0.08
0.07
0.06
0.05
0.04
0.03
0.02
0.01
100
200
300
400
500
600
700
800

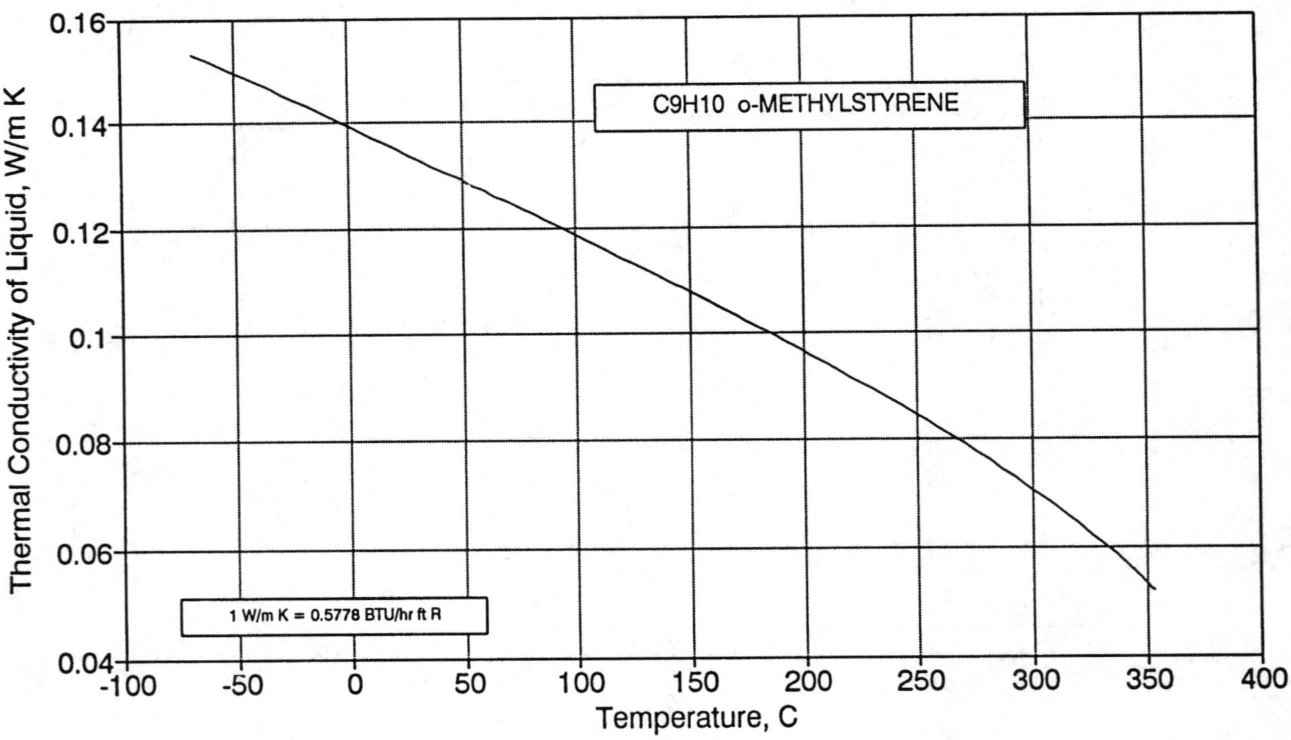
C9H10 o-METHYLSTYRENE
1 W/m K = 0.5778 BTU/hr ft R
Thermal Conductivity of Liquid, W/m K
Temperature, C
0.16
0.14
0.12
0.1
0.08
0.06
0.04
-100
-50
0
50
100
150
200
250
300
350
400

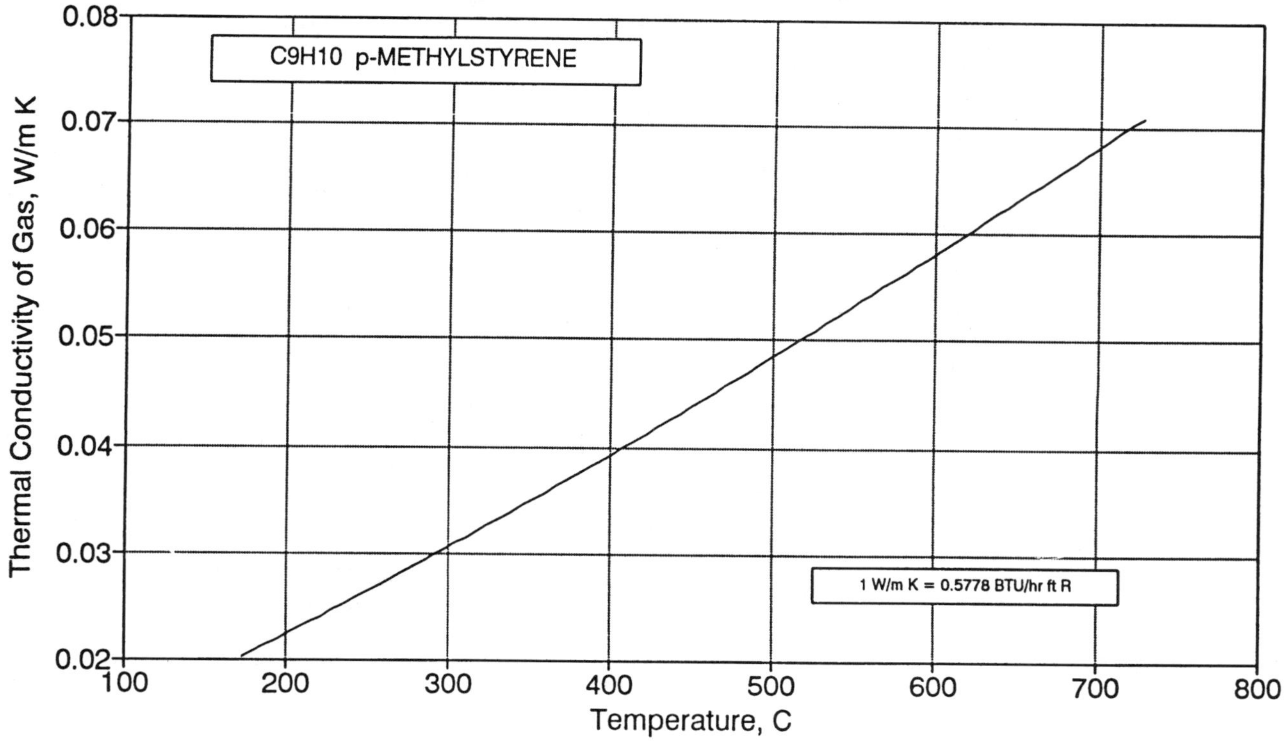
C9H10 p-METHYLSTYRENE
1 W/m K = 0.5778 BTU/hr ft R
Thermal Conductivity of Gas, W/m K
0.08
0.07
0.06
0.05
0.04
0.03
0.02
100
200
300
400
500
600
700
800
Temperature, C

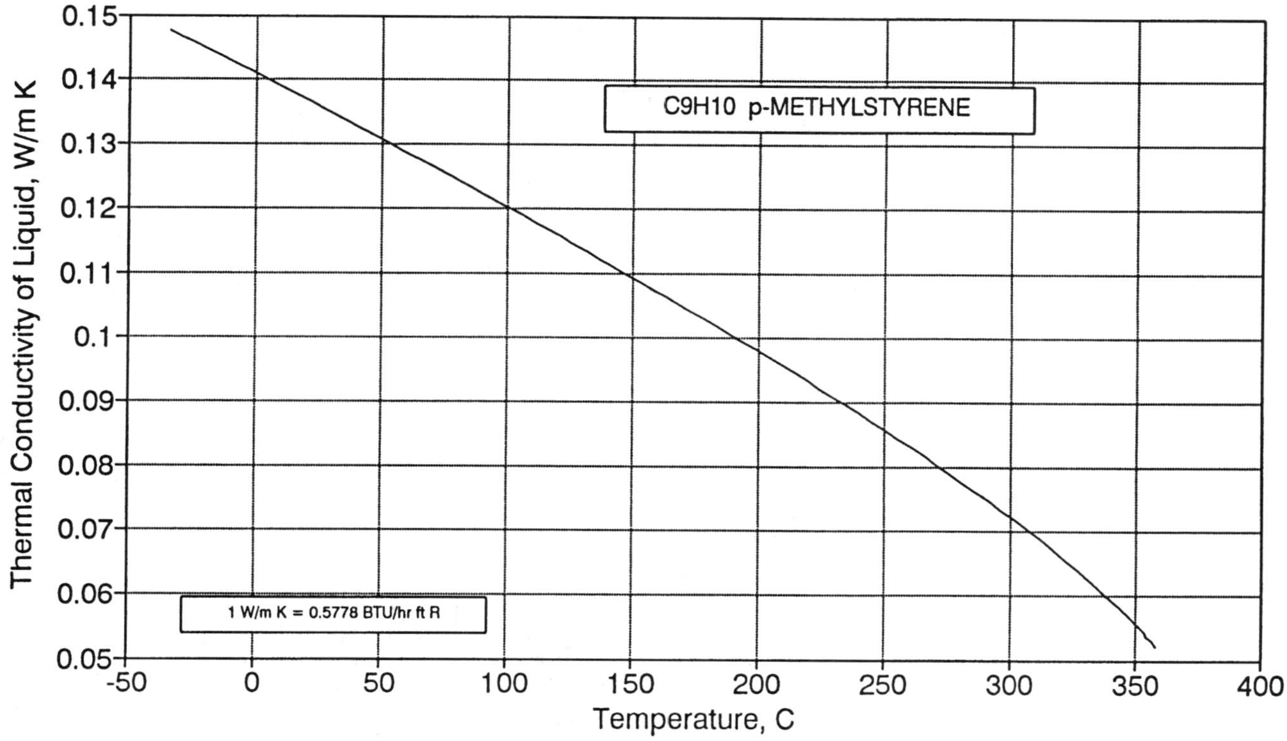
C9H10 p-METHYLSTYRENE
1 W/m K = 0.5778 BTU/hr ft R
Thermal Conductivity of Liquid, W/m K
0.15
0.14
0.13
0.12
0.11
0.1
0.09
0.08
0.07
0.06
0.05
-50
0
50
100
150
200
250
300
350
400
Temperature, C

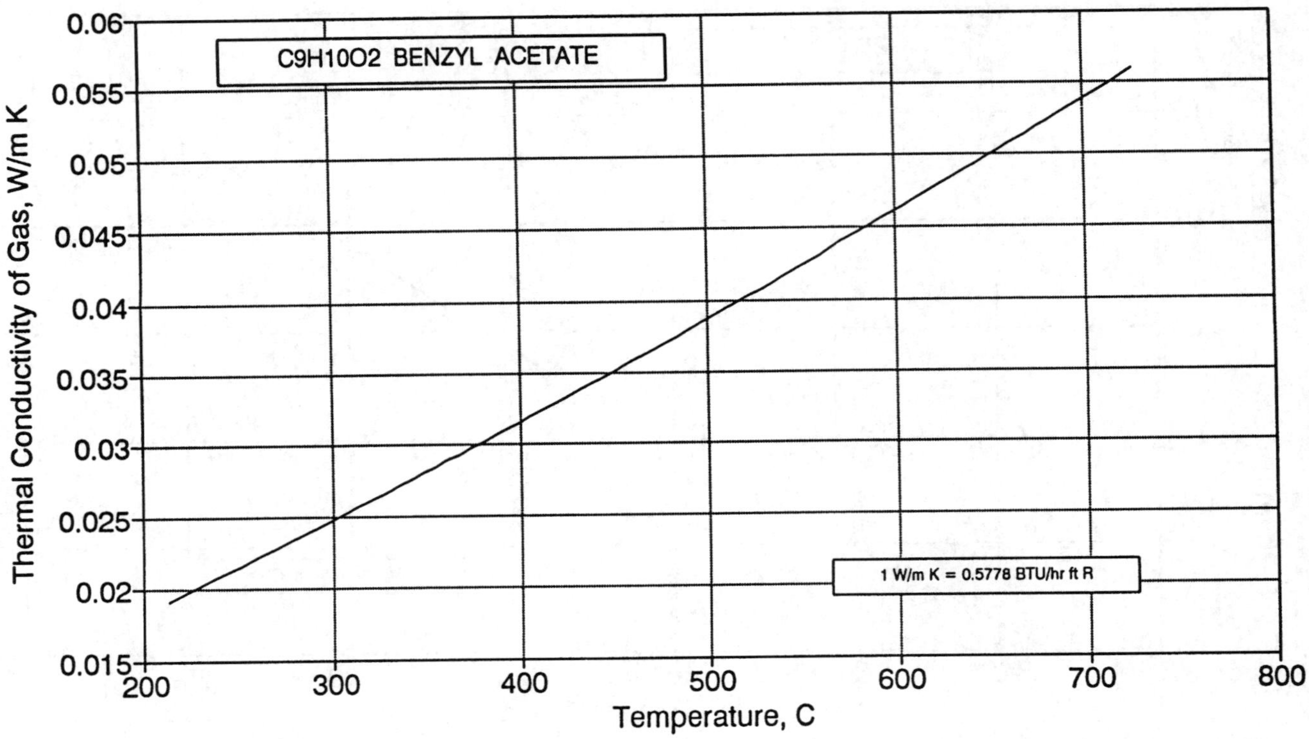
C9H10O2 BENZYL ACETATE
1 W/m K = 0.5778 BTU/hr ft R
Thermal Conductivity of Gas, W/m K
0.06
0.055
0.05
0.045
0.04
0.035
0.03
0.025
0.02
0.015
200
300
400
500
600
700
800
Temperature, C

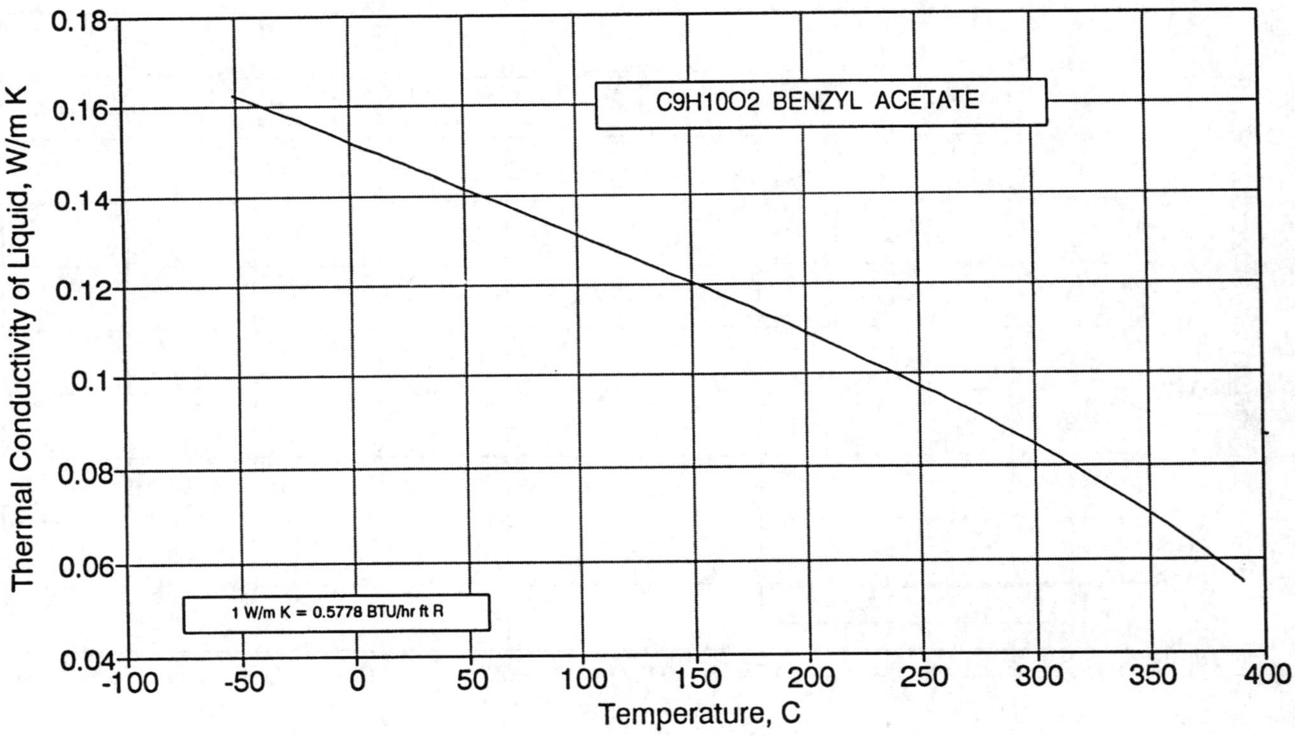
C9H10O2 BENZYL ACETATE
1 W/m K = 0.5778 BTU/hr ft R
Thermal Conductivity of Liquid, W/m K
0.18
0.16
0.14
0.12
0.1
0.08
0.06
0.04
-100
-50
0
50
100
150
200
250
300
350
400
Temperature, C

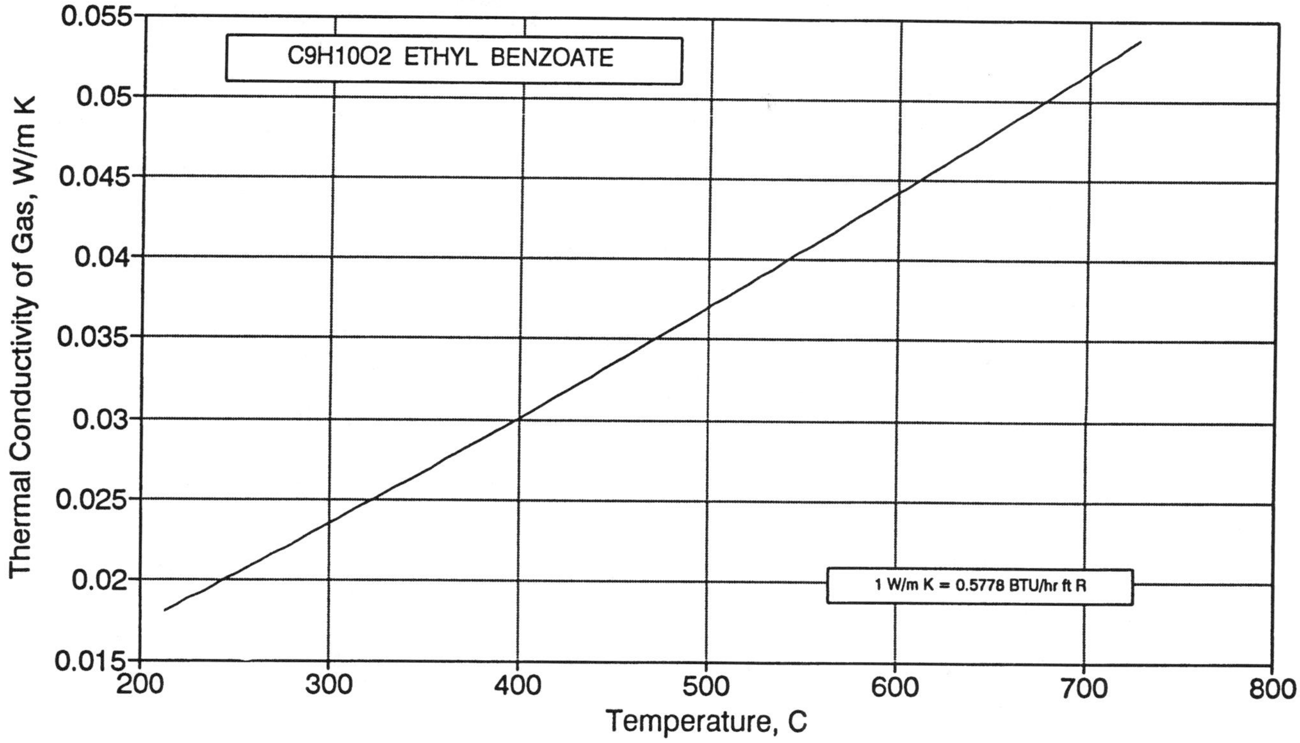
C9H10O2 ETHYL BENZOATE
1 W/m K = 0.5778 BTU/hr ft R
Thermal Conductivity of Gas, W/m K
0.055
0.05
0.045
0.04
0.035
0.03
0.025
0.02
0.015
200
300
400
500
600
700
800
Temperature, C

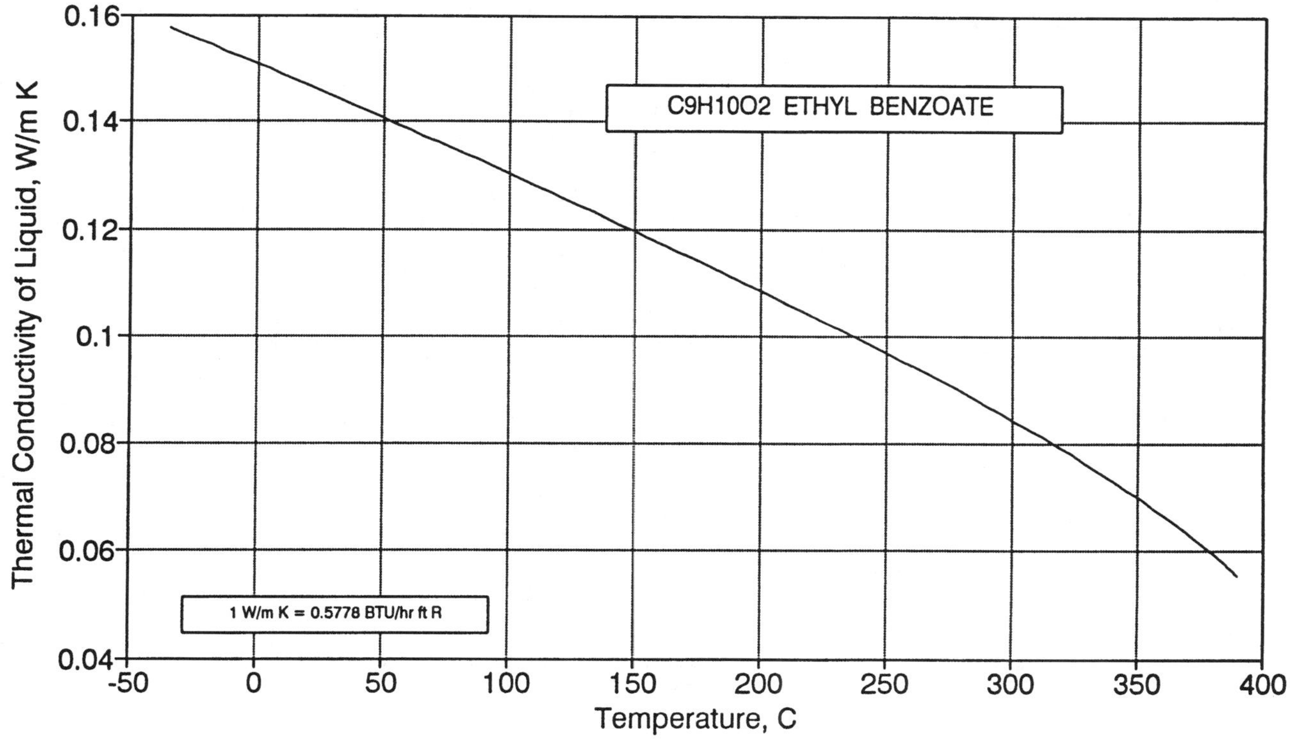
C9H10O2 ETHYL BENZOATE
1 W/m K = 0.5778 BTU/hr ft R
Thermal Conductivity of Liquid, W/m K
0.16
0.14
0.12
0.1
0.08
0.06
0.04
-50
0
50
100
150
200
250
300
350
400
Temperature, C

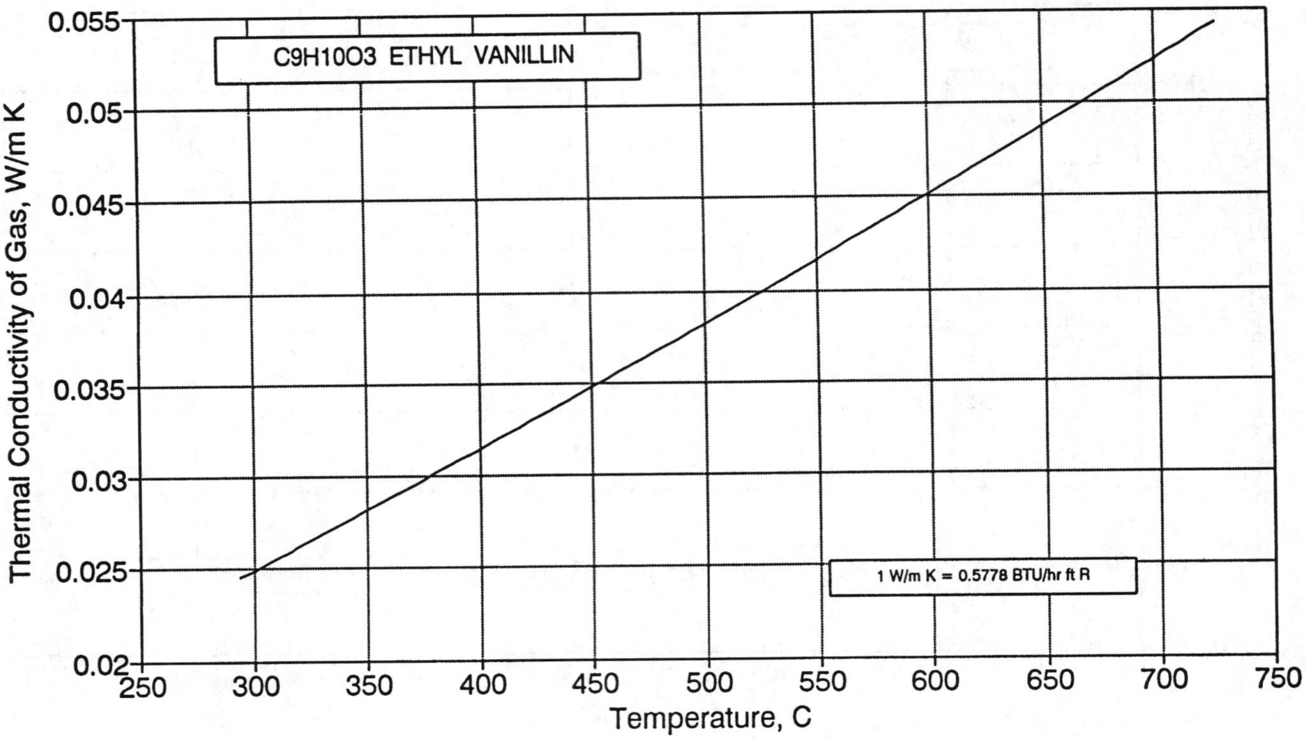
C9H10O3 ETHYL VANILLIN
1 W/m K = 0.5778 BTU/hr ft R
Thermal Conductivity of Gas, W/m K
0.055
0.05
0.045
0.04
0.035
0.03
0.025
0.02
250
300
350
400
450
500
550
600
650
700
750
Temperature, C

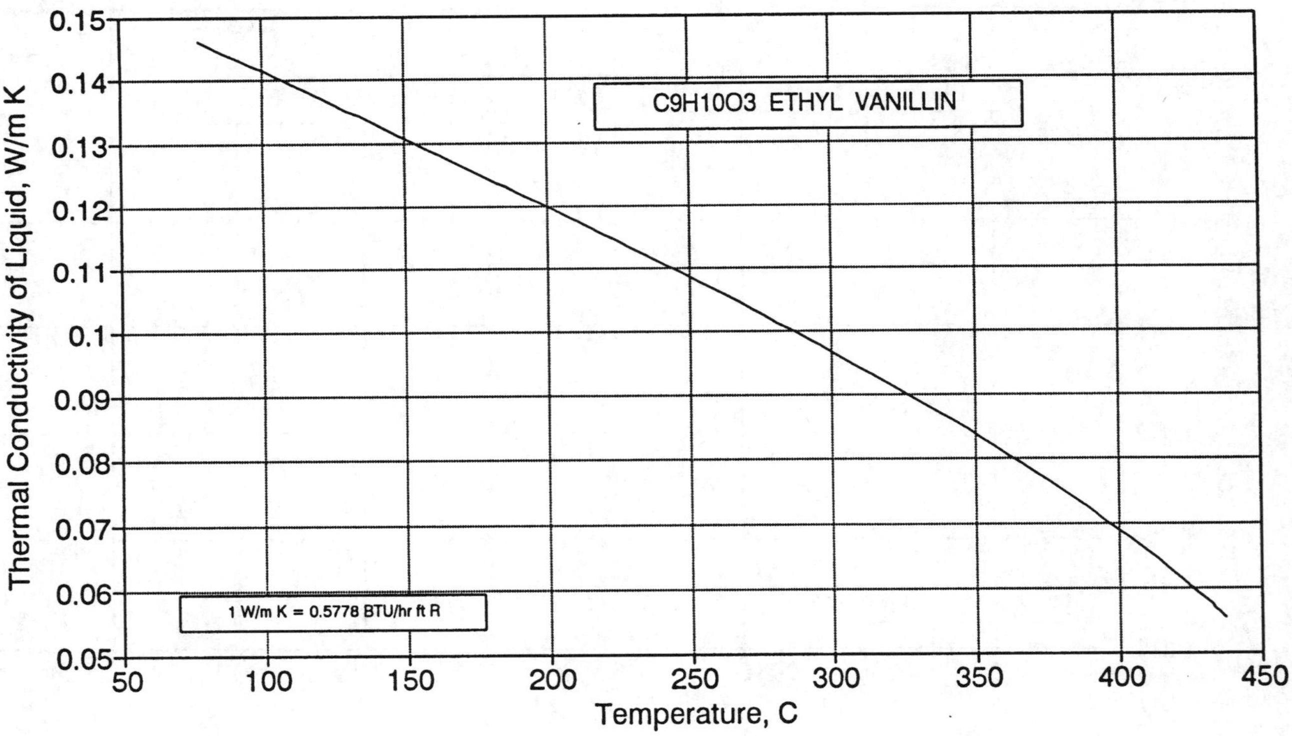
C9H10O3 ETHYL VANILLIN
1 W/m K = 0.5778 BTU/hr ft R
Thermal Conductivity of Liquid, W/m K
0.15
0.14
0.13
0.12
0.11
0.1
0.09
0.08
0.07
0.06
0.05
50
100
150
200
250
300
350
400
450
Temperature, C

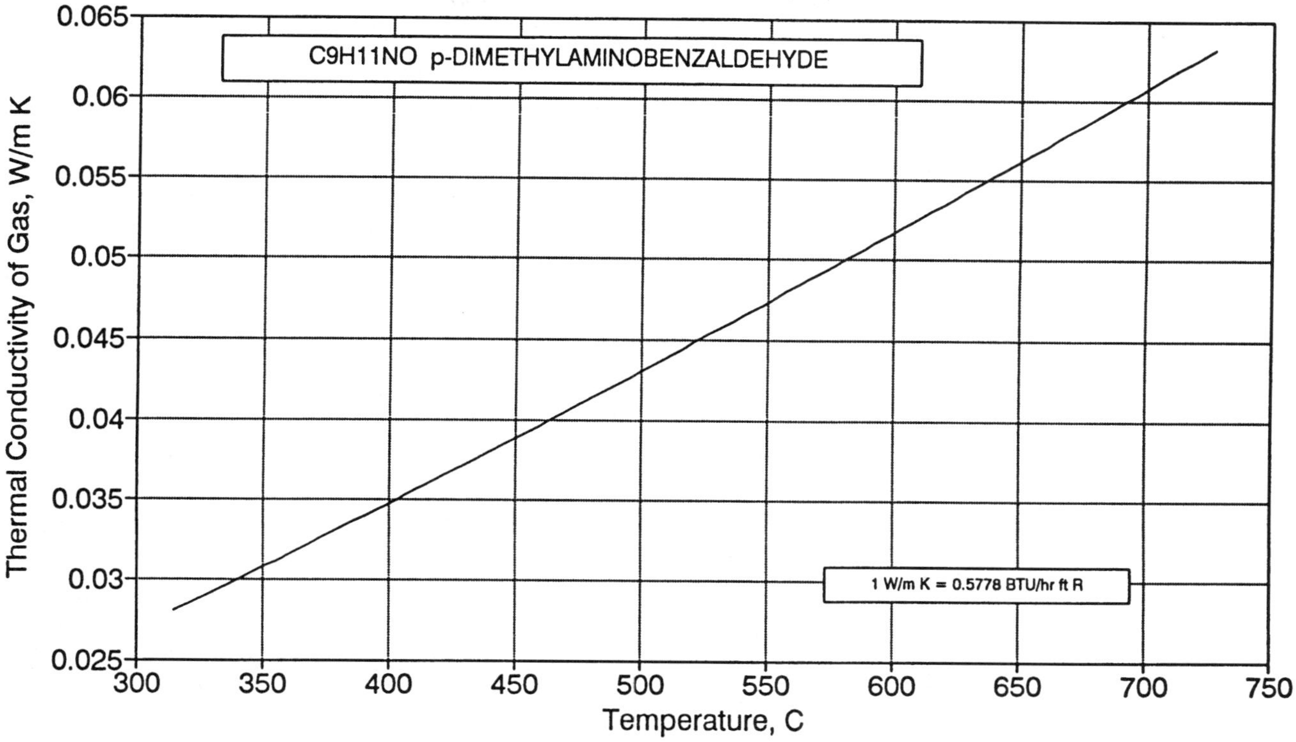
C9H11NO p-DIMETHYLAMINOBENZALDEHYDE
1 W/m K = 0.5778 BTU/hr ft R
Thermal Conductivity of Gas, W/m K
Temperature, C
0.065
0.06
0.055
0.05
0.045
0.04
0.035
0.03
0.025
300
350
400
450
500
550
600
650
700
750

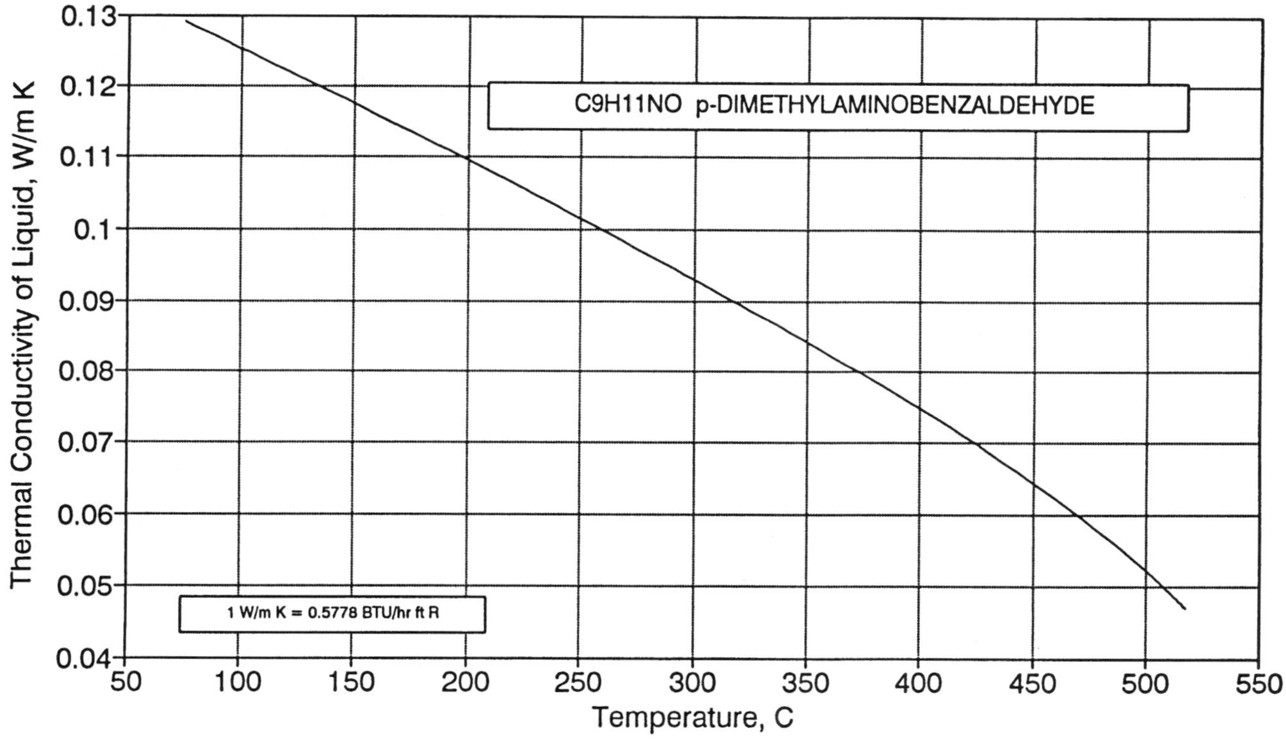
C9H11NO p-DIMETHYLAMINOBENZALDEHYDE
1 W/m K = 0.5778 BTU/hr ft R
Thermal Conductivity of Liquid, W/m K
Temperature, C
0.13
0.12
0.11
0.1
0.09
0.08
0.07
0.06
0.05
0.04
50
100
150
200
250
300
350
400
450
500
550

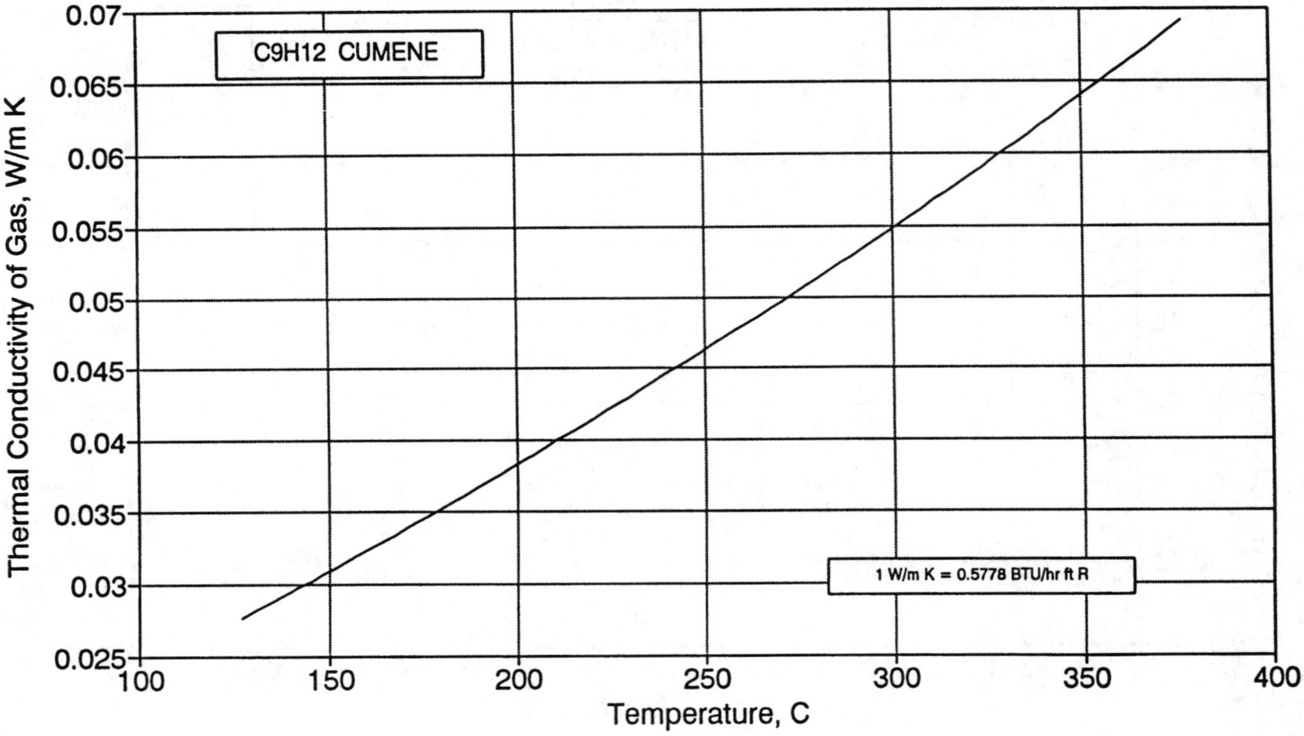
C9H12 CUMENE
1 W/m K = 0.5778 BTU/hr ft R
Thermal Conductivity of Gas, W/m K
Temperature, C
0.07
0.065
0.06
0.055
0.05
0.045
0.04
0.035
0.03
0.025
100
150
200
250
300
350
400

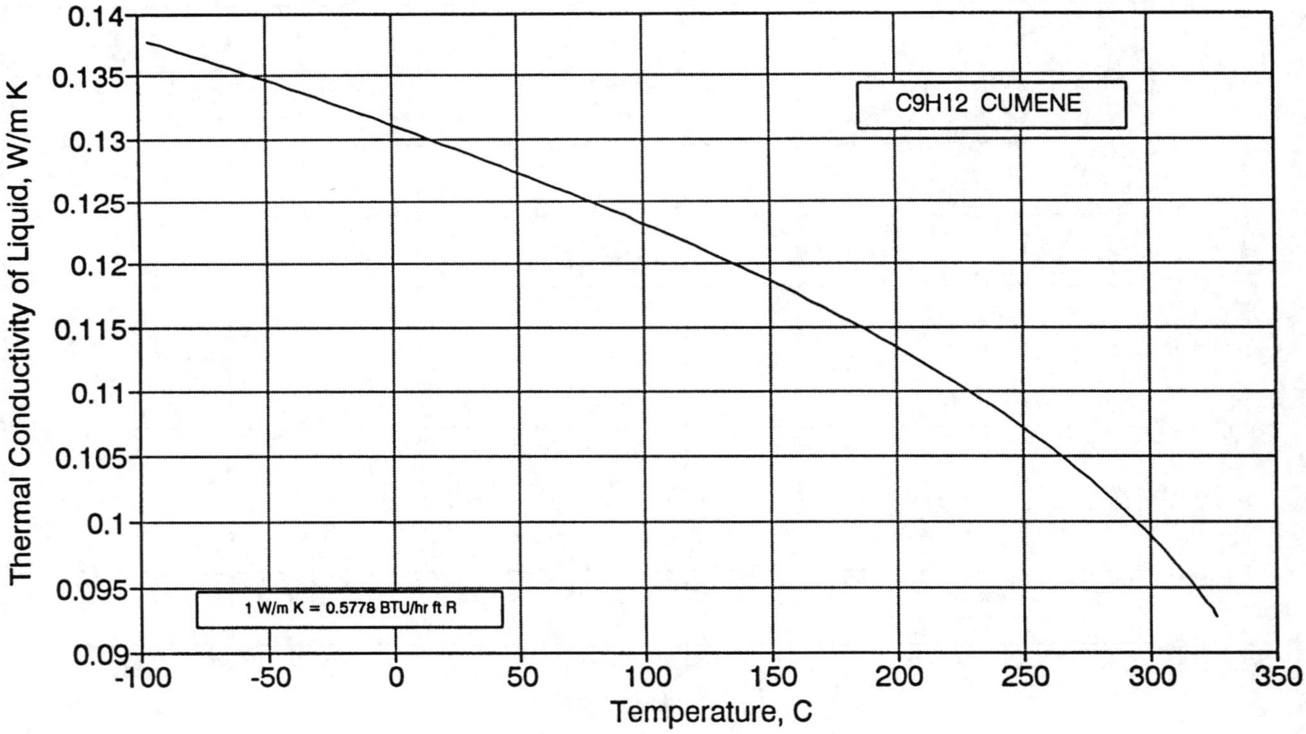
C9H12 CUMENE
1 W/m K = 0.5778 BTU/hr ft R
Thermal Conductivity of Liquid, W/m K
Temperature, C
0.14
0.135
0.13
0.125
0.12
0.115
0.11
0.105
0.1
0.095
0.09
-100
-50
0
50
100
150
200
250
300
350

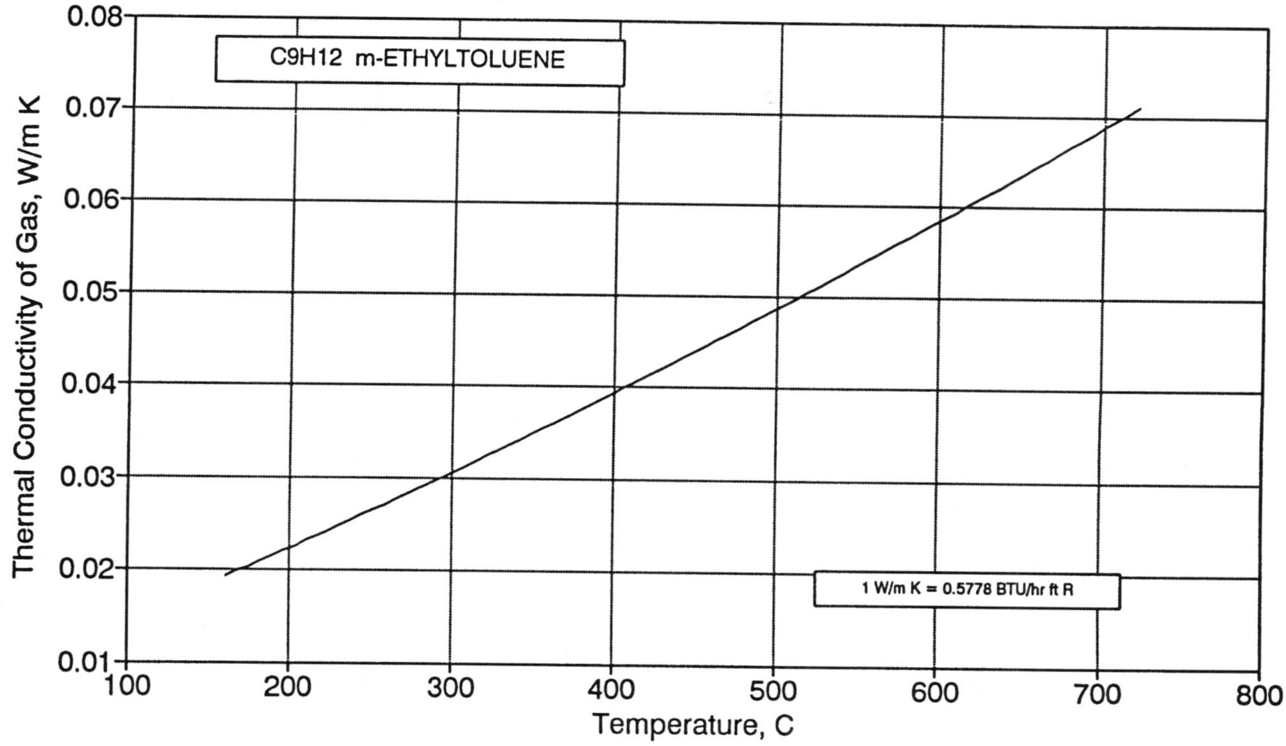
C9H12 m-ETHYLTOLUENE
1 W/m K = 0.5778 BTU/hr ft R
Thermal Conductivity of Gas, W/m K
0.08
0.07
0.06
0.05
0.04
0.03
0.02
0.01
100
200
300
400
500
600
700
800
Temperature, C

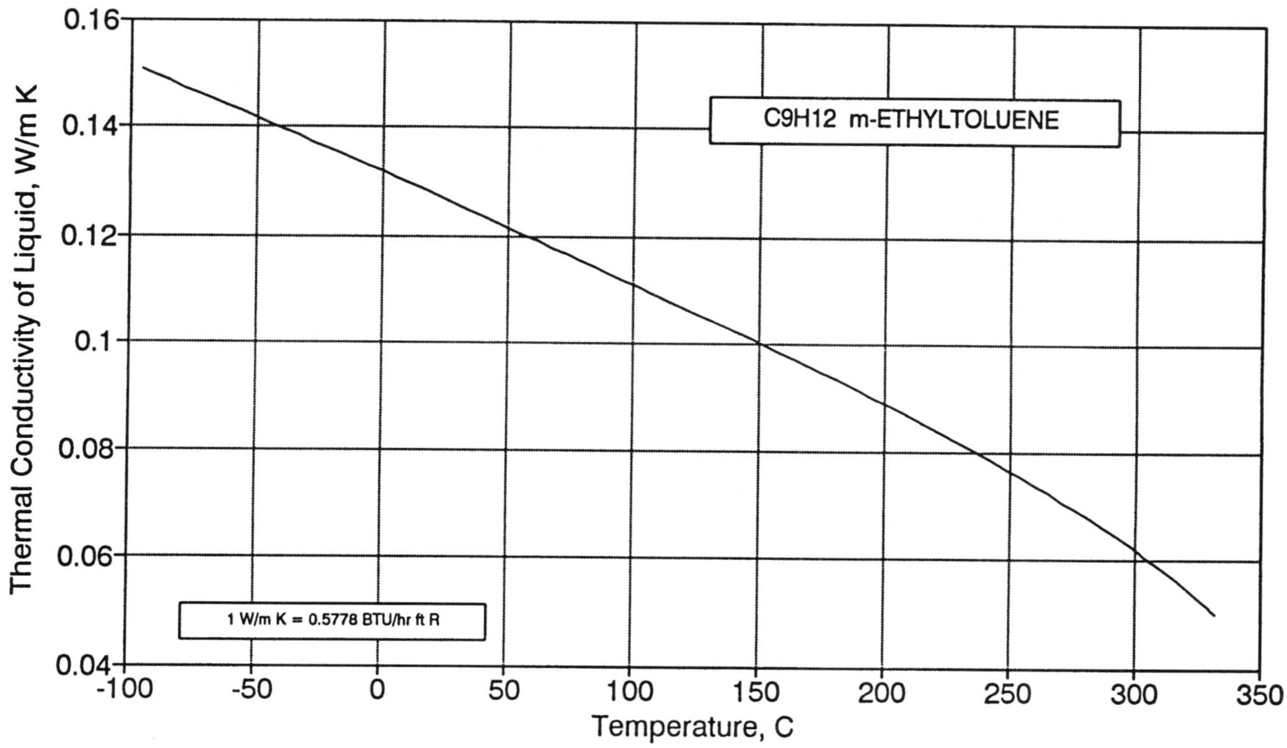
C9H12 m-ETHYLTOLUENE
1 W/m K = 0.5778 BTU/hr ft R
Thermal Conductivity of Liquid, W/m K
0.16
0.14
0.12
0.1
0.08
0.06
0.04
-100
-50
0
50
100
150
200
250
300
350
Temperature, C

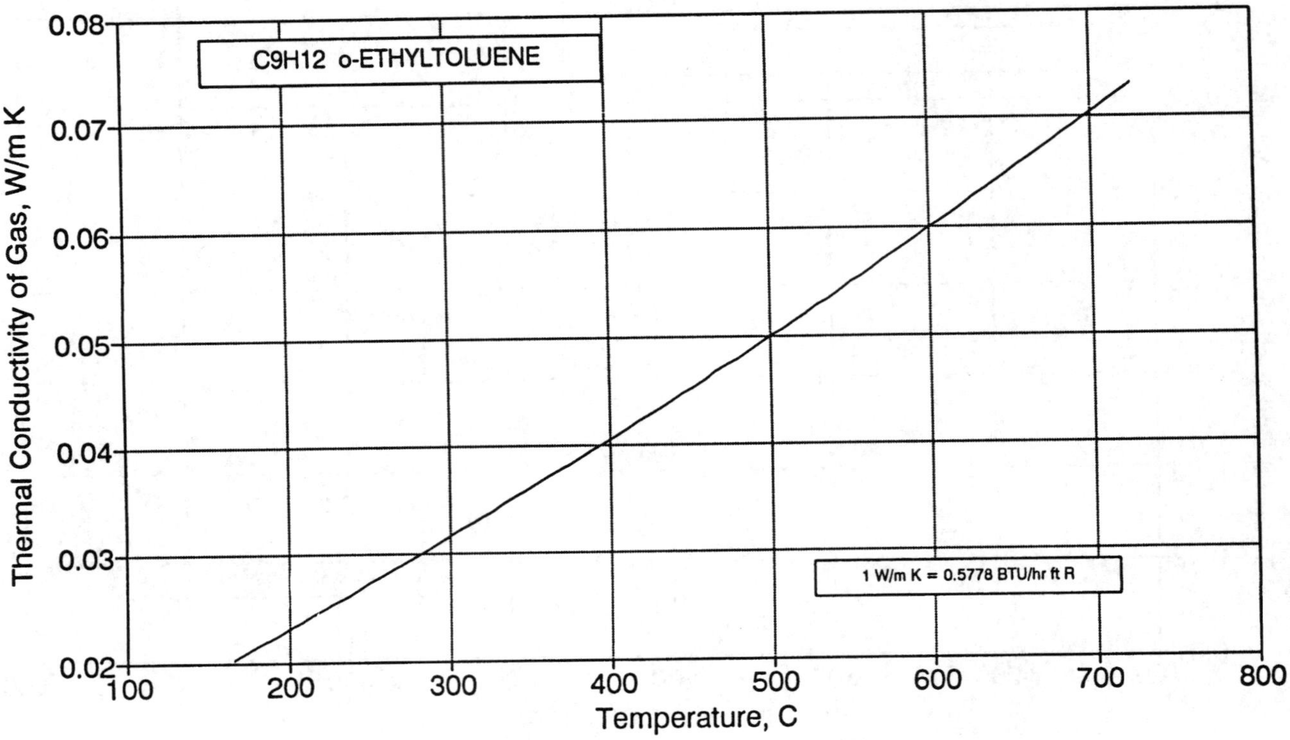

C9H12 o-ETHYLTOLUENE
1 W/m K = 0.5778 BTU/hr ft R
Thermal Conductivity of Gas, W/m K
0.08
0.07
0.06
0.05
0.04
0.03
0.02
100
200
300
400
500
600
700
800
Temperature, C

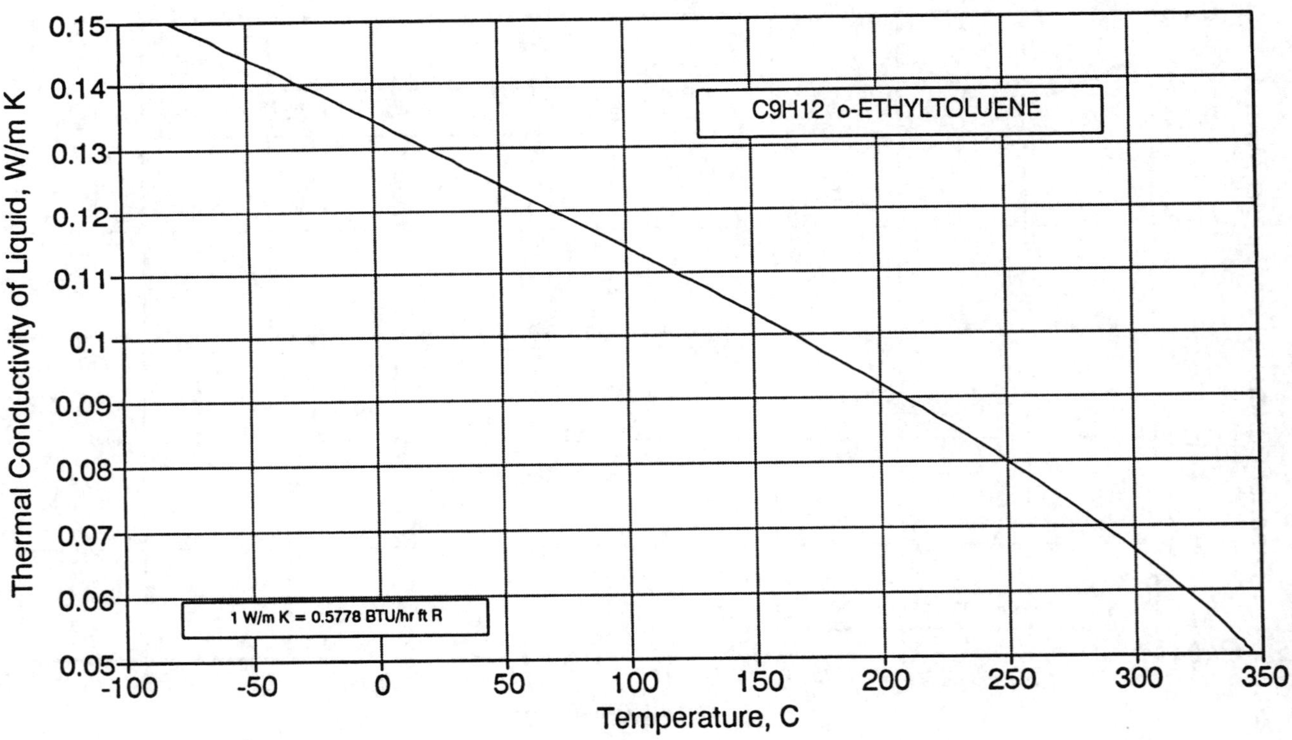

C9H12 o-ETHYLTOLUENE
1 W/m K = 0.5778 BTU/hr ft R
Thermal Conductivity of Liquid, W/m K
0.15
0.14
0.13
0.12
0.11
0.1
0.09
0.08
0.07
0.06
0.05
-100
-50
0
50
100
150
200
250
300
350
Temperature, C

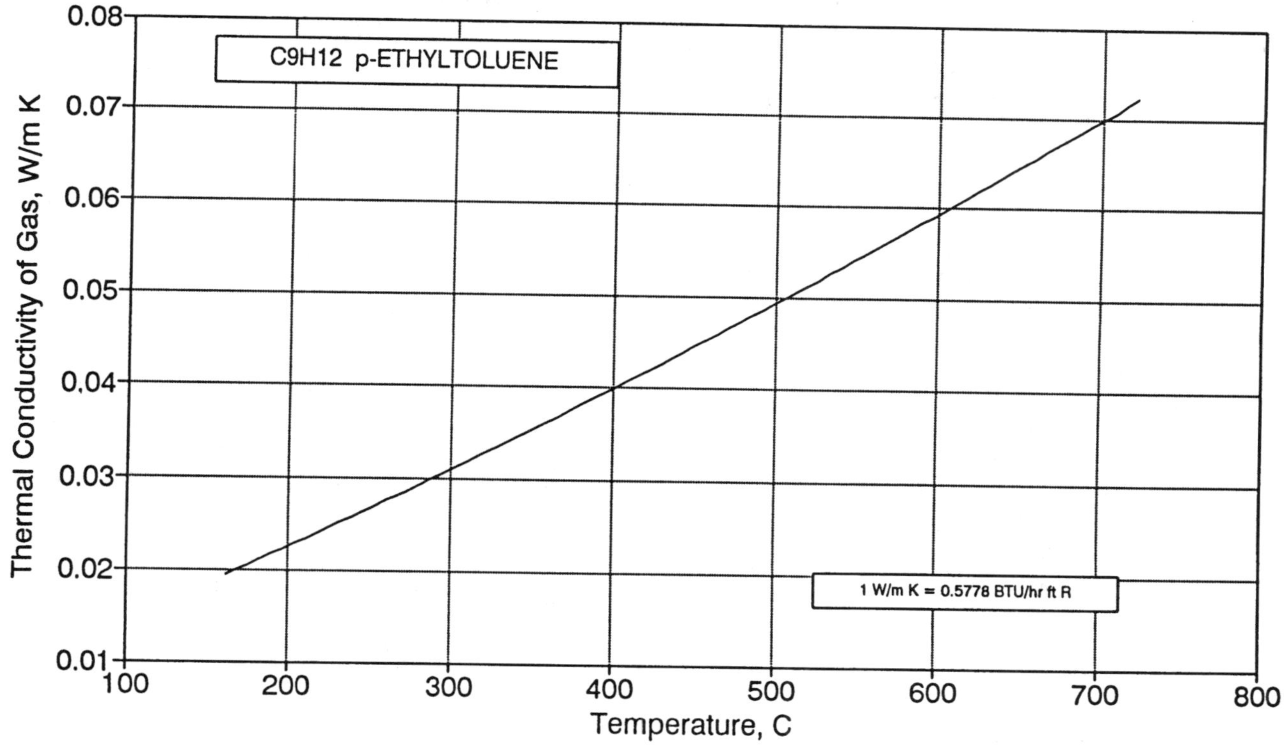
C9H12 p-ETHYLTOLUENE
1 W/m K = 0.5778 BTU/hr ft R
Thermal Conductivity of Gas, W/m K
0.08
0.07
0.06
0.05
0.04
0.03
0.02
0.01
100
200
300
400
500
600
700
800
Temperature, C

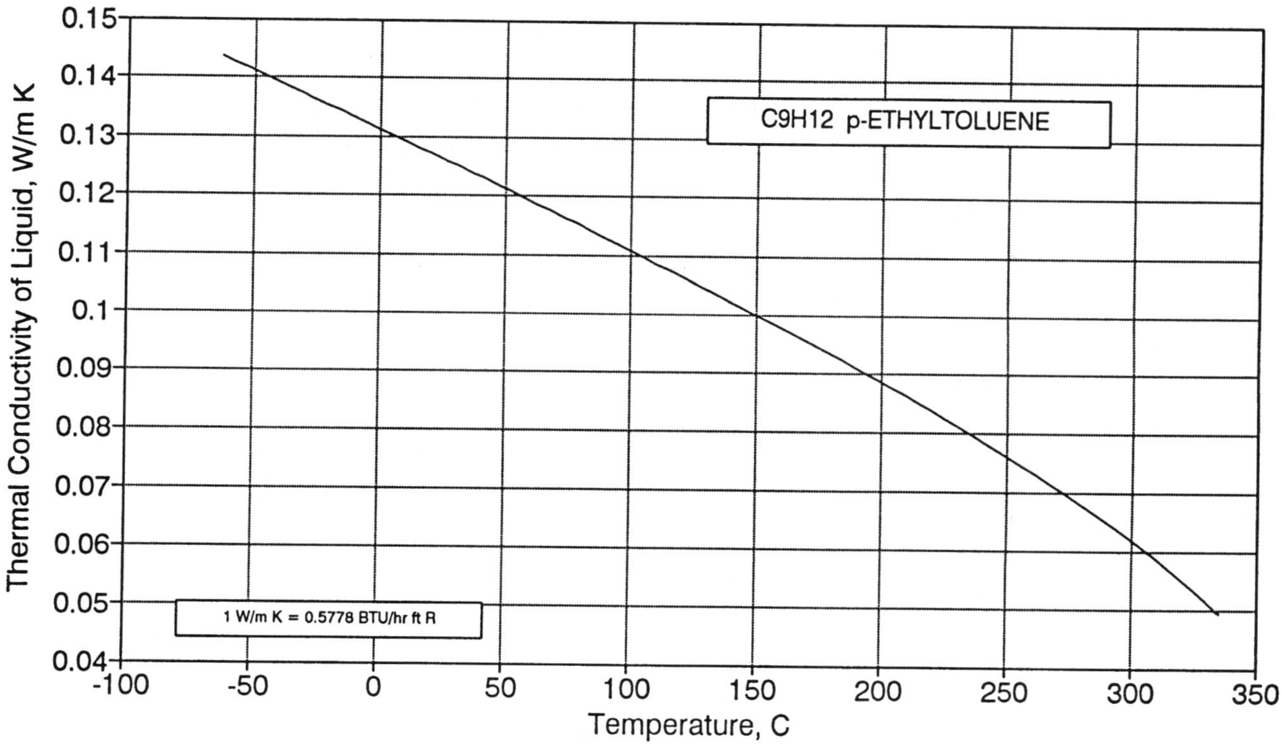
C9H12 p-ETHYLTOLUENE
1 W/m K = 0.5778 BTU/hr ft R
Thermal Conductivity of Liquid, W/m K
0.15
0.14
0.13
0.12
0.11
0.1
0.09
0.08
0.07
0.06
0.05
0.04
-100
-50
0
50
100
150
200
250
300
350
Temperature, C

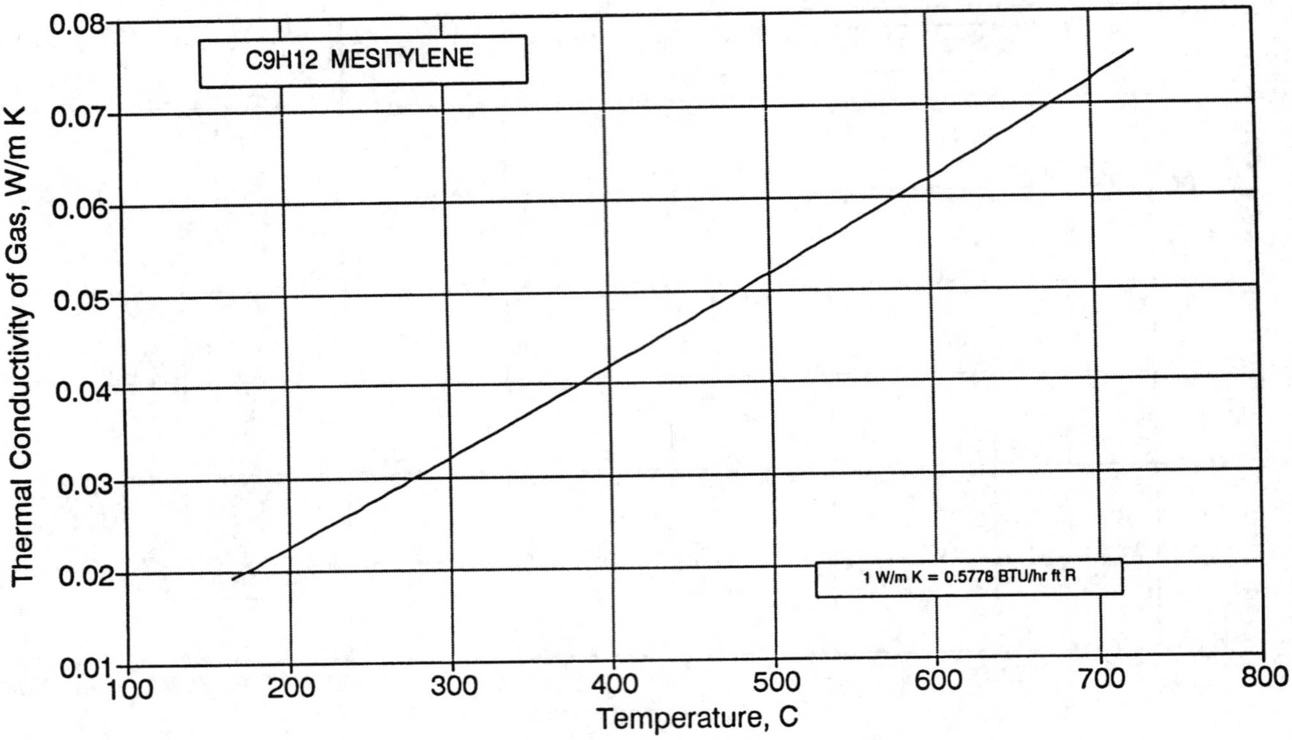

C9H12 MESITYLENE
1 W/m K = 0.5778 BTU/hr ft R
Thermal Conductivity of Gas, W/m K
0.08
0.07
0.06
0.05
0.04
0.03
0.02
0.01
100
200
300
400
500
600
700
800
Temperature, C

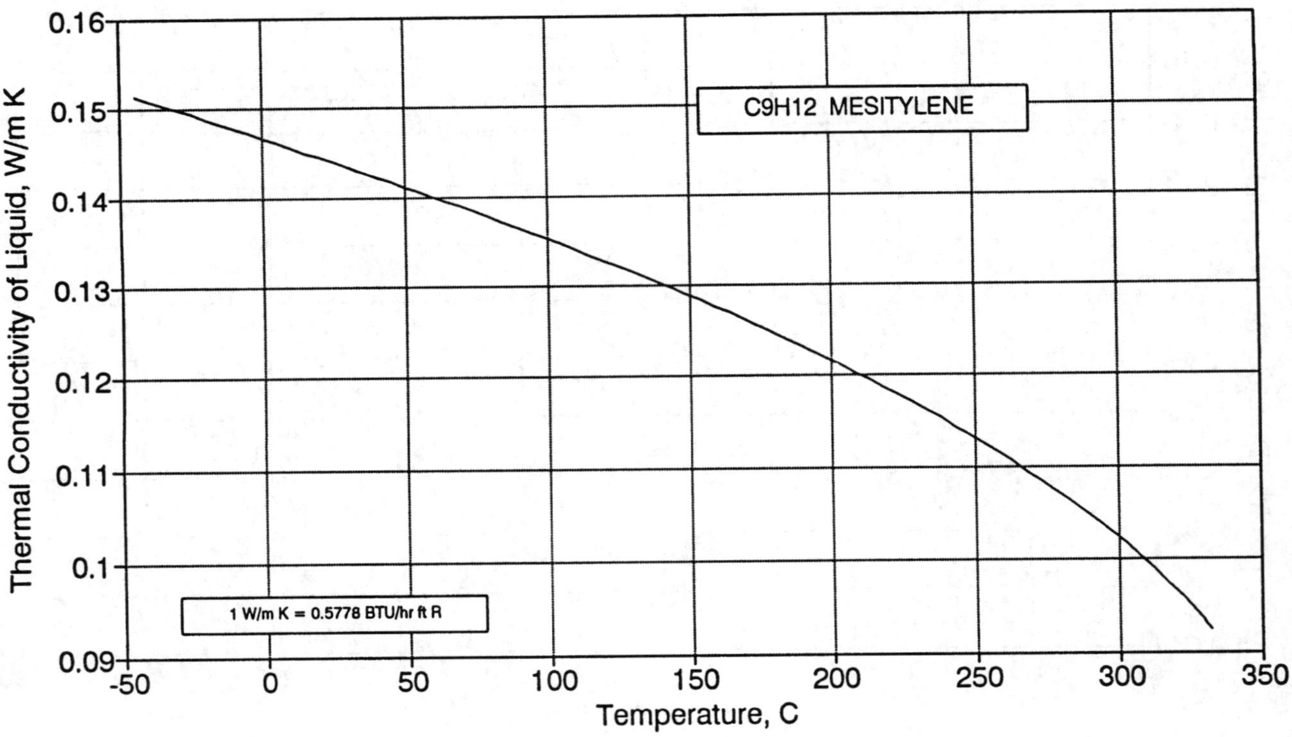

C9H12 MESITYLENE
1 W/m K = 0.5778 BTU/hr ft R
Thermal Conductivity of Liquid, W/m K
0.16
0.15
0.14
0.13
0.12
0.11
0.1
0.09
-50
0
50
100
150
200
250
300
350
Temperature, C

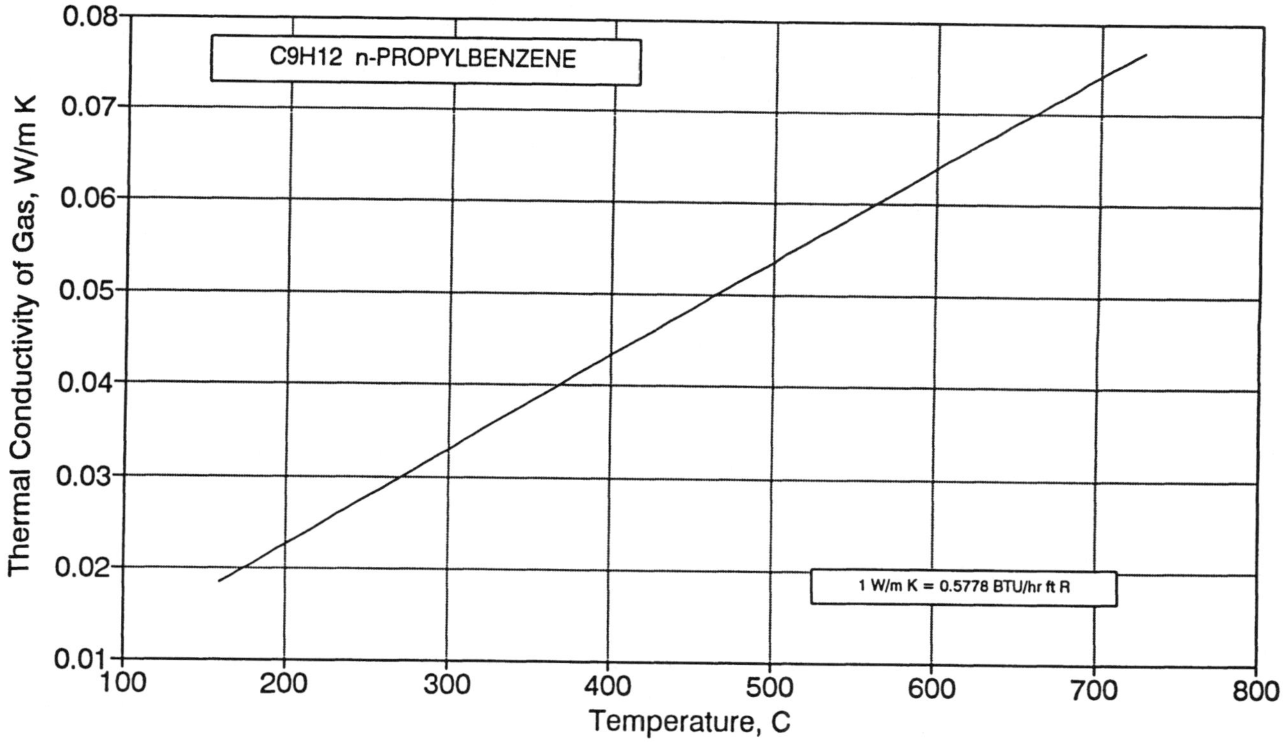
C9H12 n-PROPYLBENZENE
1 W/m K = 0.5778 BTU/hr ft R
Thermal Conductivity of Gas, W/m K
0.08
0.07
0.06
0.05
0.04
0.03
0.02
0.01
100
200
300
400
500
600
700
800
Temperature, C

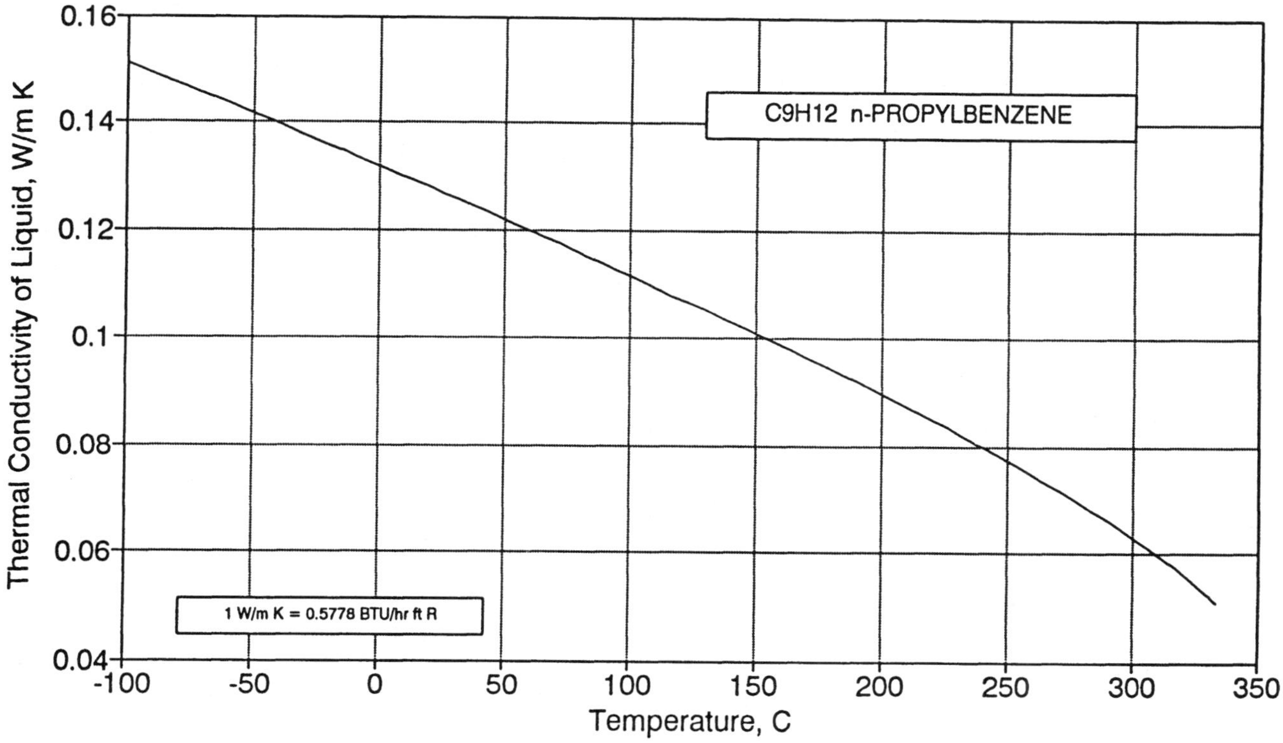
C9H12 n-PROPYLBENZENE
1 W/m K = 0.5778 BTU/hr ft R
Thermal Conductivity of Liquid, W/m K
0.16
0.14
0.12
0.1
0.08
0.06
0.04
-100
-50
0
50
100
150
200
250
300
350
Temperature, C

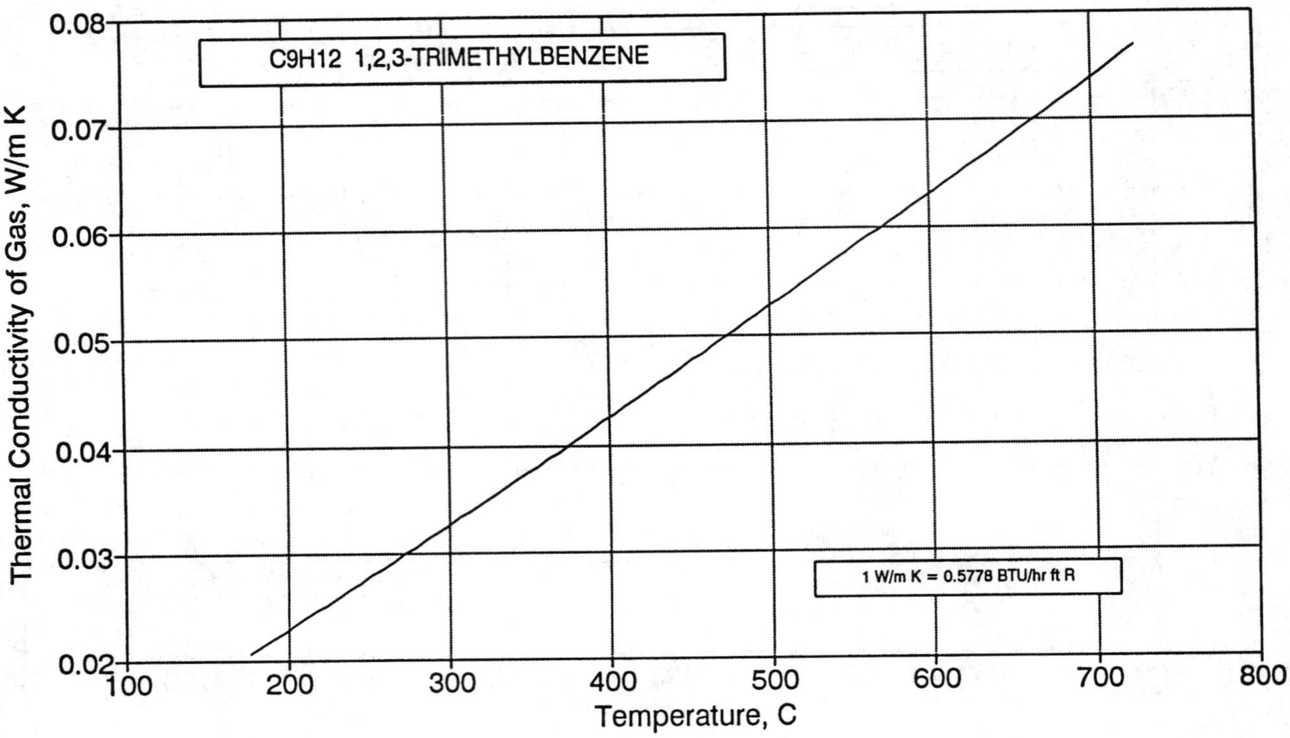
C9H12 1,2,3-TRIMETHYLBENZENE
1 W/m K = 0.5778 BTU/hr ft R
Thermal Conductivity of Gas, W/m K
0.08
0.07
0.06
0.05
0.04
0.03
0.02
100
200
300
400
500
600
700
800
Temperature, C

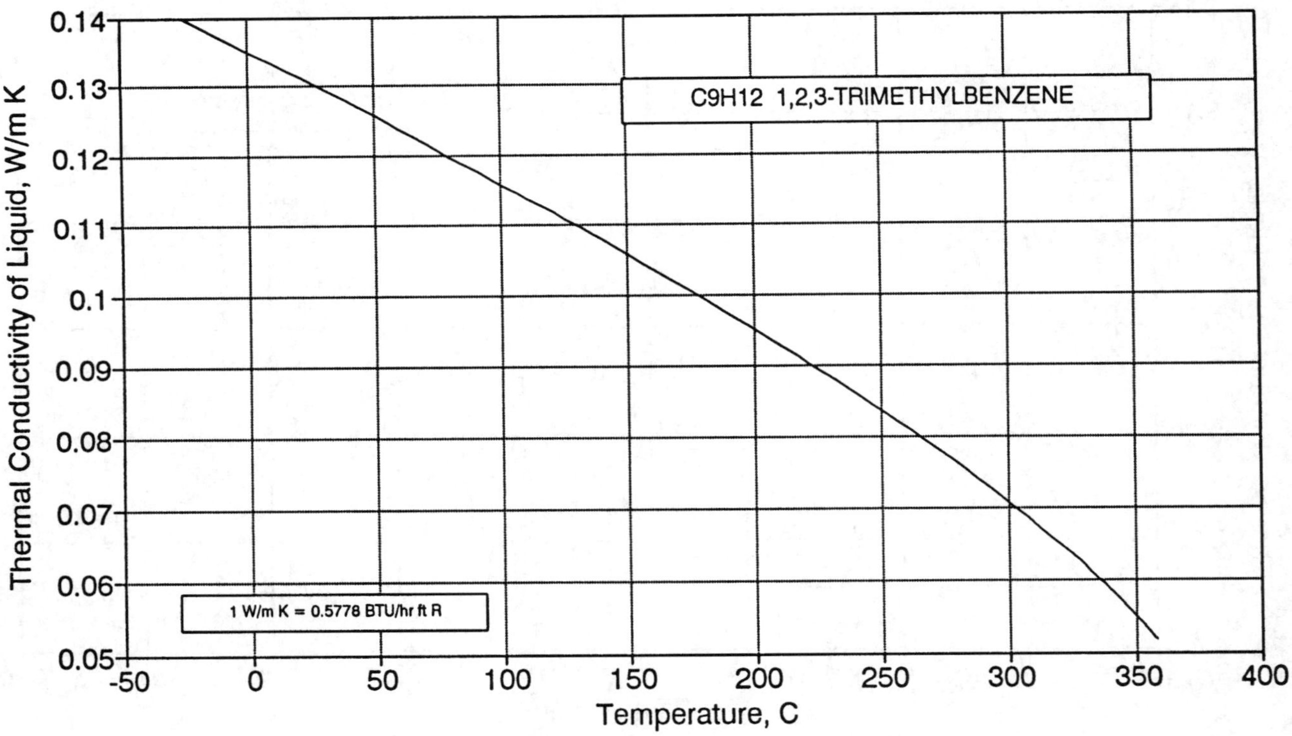
C9H12 1,2,3-TRIMETHYLBENZENE
1 W/m K = 0.5778 BTU/hr ft R
Thermal Conductivity of Liquid, W/m K
0.14
0.13
0.12
0.11
0.1
0.09
0.08
0.07
0.06
0.05
-50
0
50
100
150
200
250
300
350
400
Temperature, C

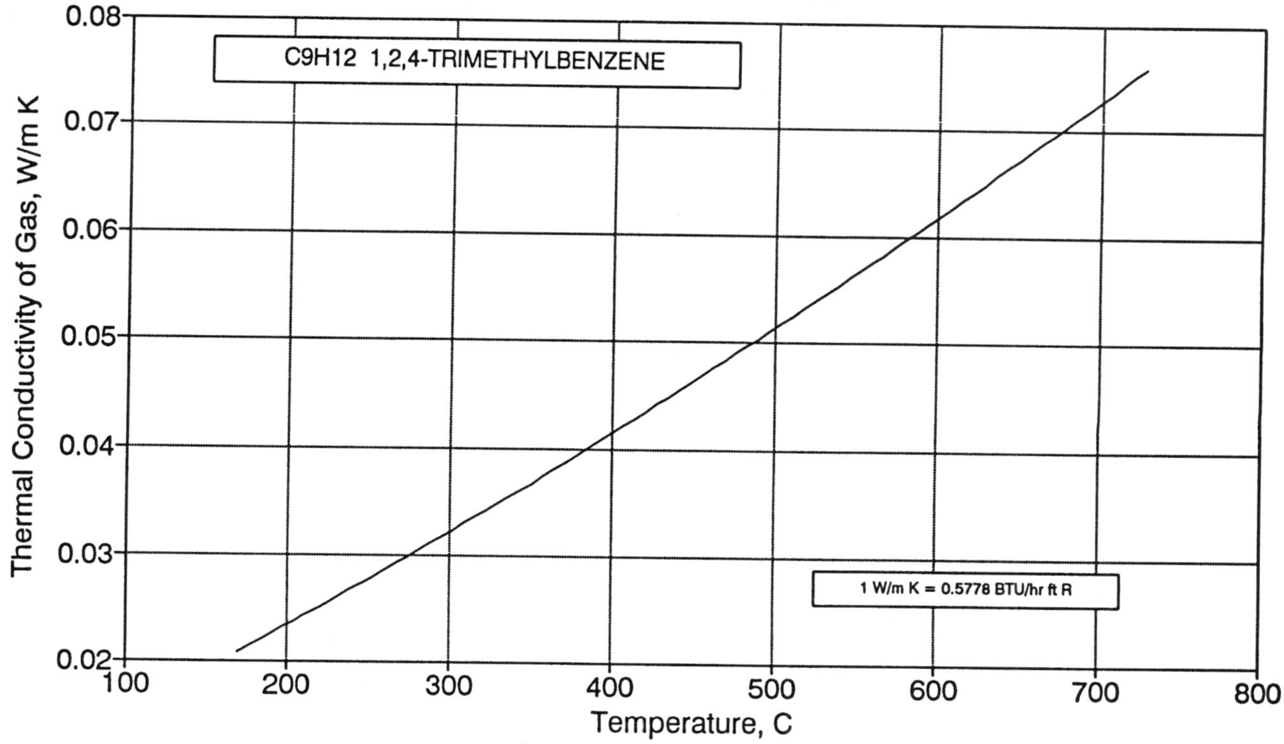
C9H12 1,2,4-TRIMETHYLBENZENE
1 W/m K = 0.5778 BTU/hr ft R
Thermal Conductivity of Gas, W/m K
0.08
0.07
0.06
0.05
0.04
0.03
0.02
100
200
300
400
500
600
700
800
Temperature, C

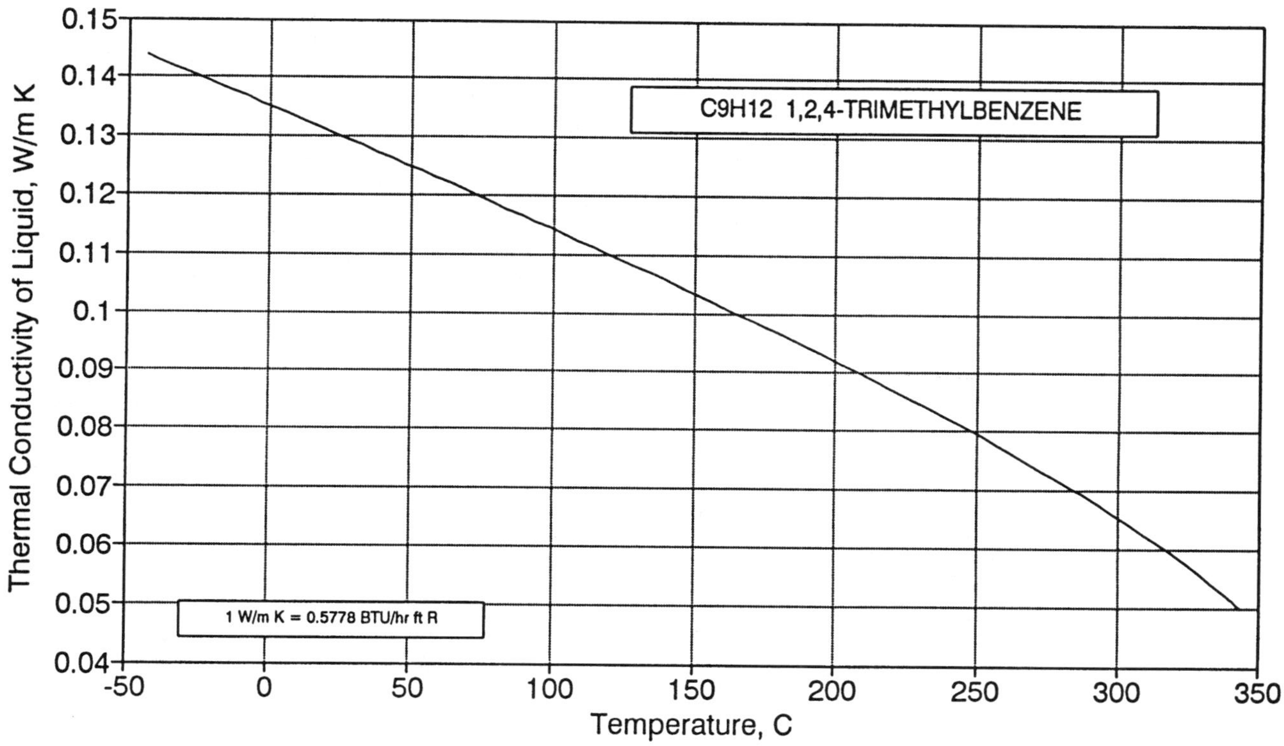
C9H12 1,2,4-TRIMETHYLBENZENE
1 W/m K = 0.5778 BTU/hr ft R
Thermal Conductivity of Liquid, W/m K
0.15
0.14
0.13
0.12
0.11
0.1
0.09
0.08
0.07
0.06
0.05
0.04
-50
0
50
100
150
200
250
300
350
Temperature, C

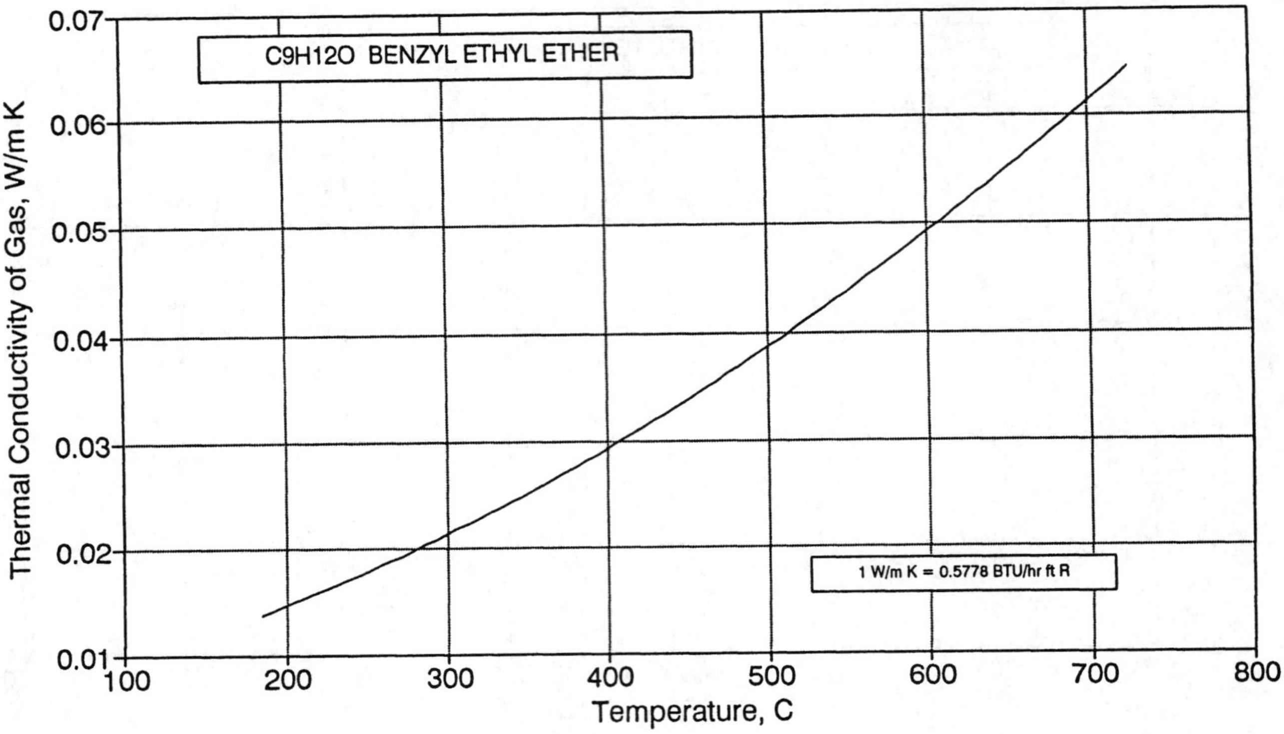
C9H12O BENZYL ETHYL ETHER
Thermal Conductivity of Gas, W/m K
0.07
0.06
0.05
0.04
0.03
0.02
0.01
100
200
300
400
500
600
700
800
Temperature, C
1 W/m K = 0.5778 BTU/hr ft R

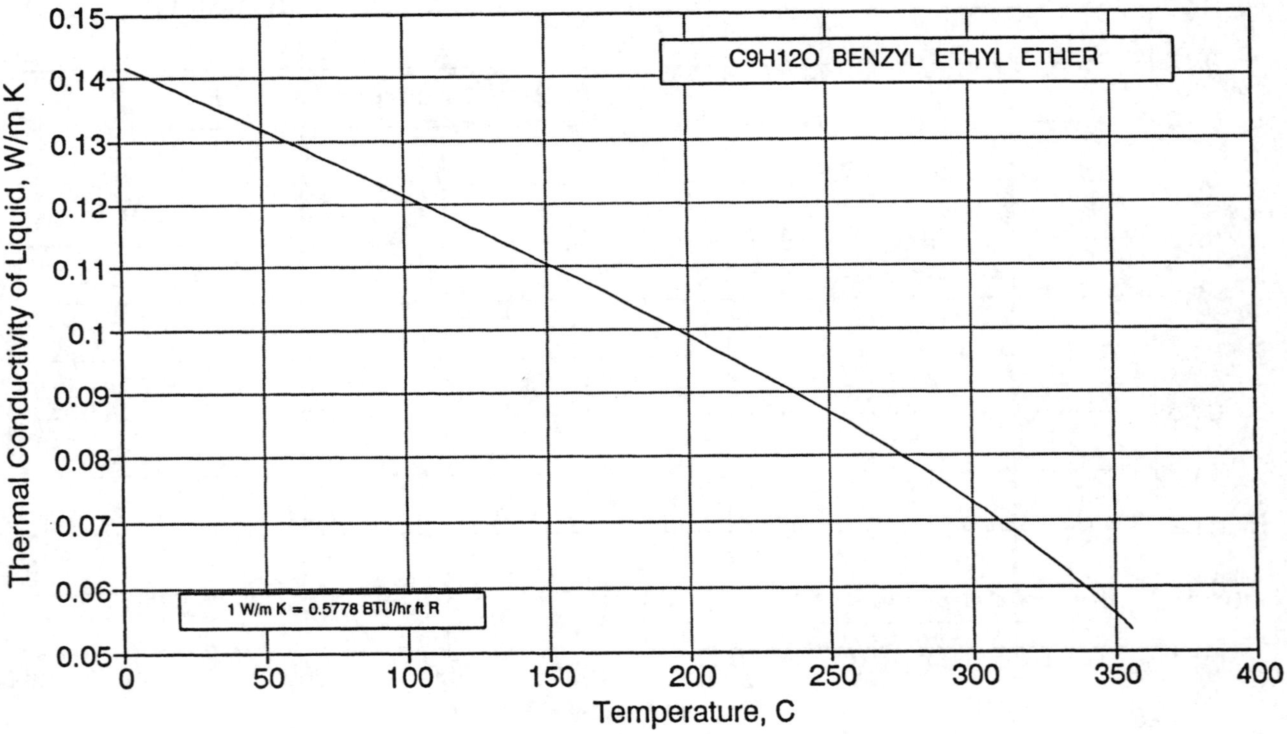
C9H12O BENZYL ETHYL ETHER
Thermal Conductivity of Liquid, W/m K
0.15
0.14
0.13
0.12
0.11
0.1
0.09
0.08
0.07
0.06
0.05
0
50
100
150
200
250
300
350
400
Temperature, C
1 W/m K = 0.5778 BTU/hr ft R

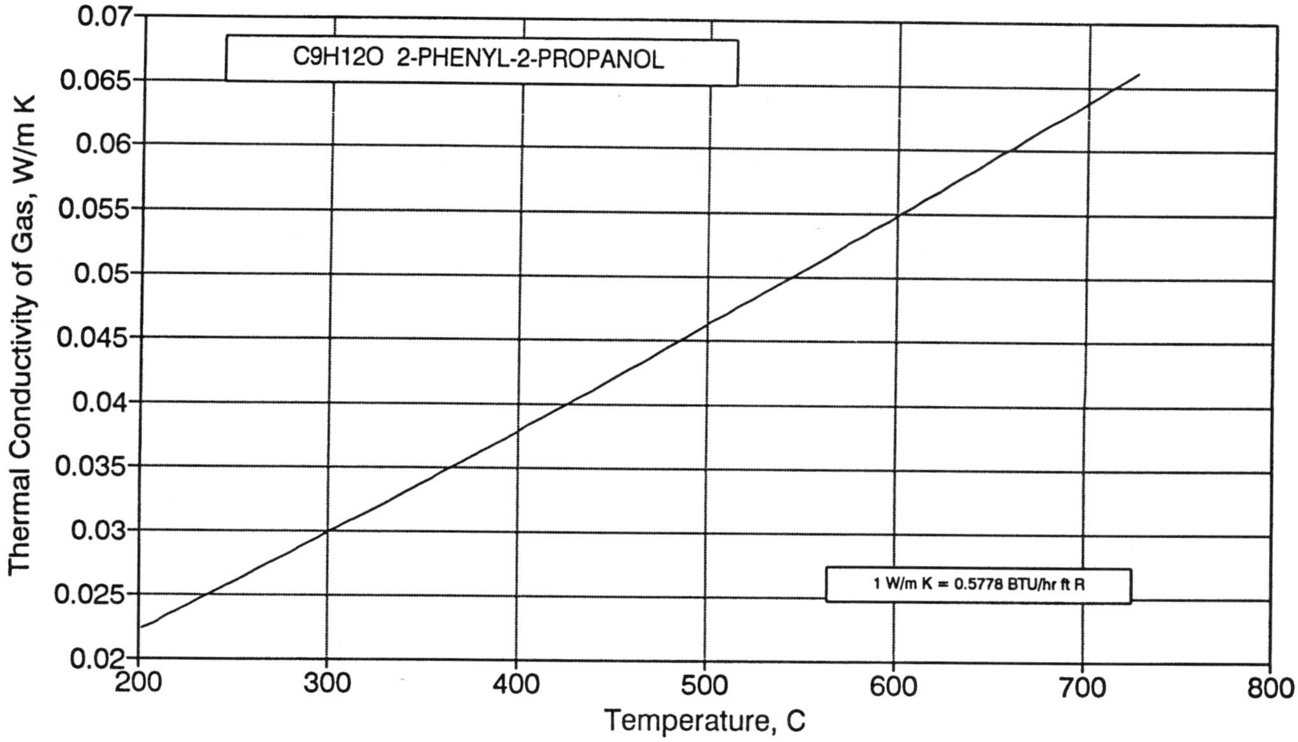
C9H12O 2-PHENYL-2-PROPANOL
1 W/m K = 0.5778 BTU/hr ft R
Thermal Conductivity of Gas, W/m K
Temperature, C
0.07
0.065
0.06
0.055
0.05
0.045
0.04
0.035
0.03
0.025
0.02
200
300
400
500
600
700
800

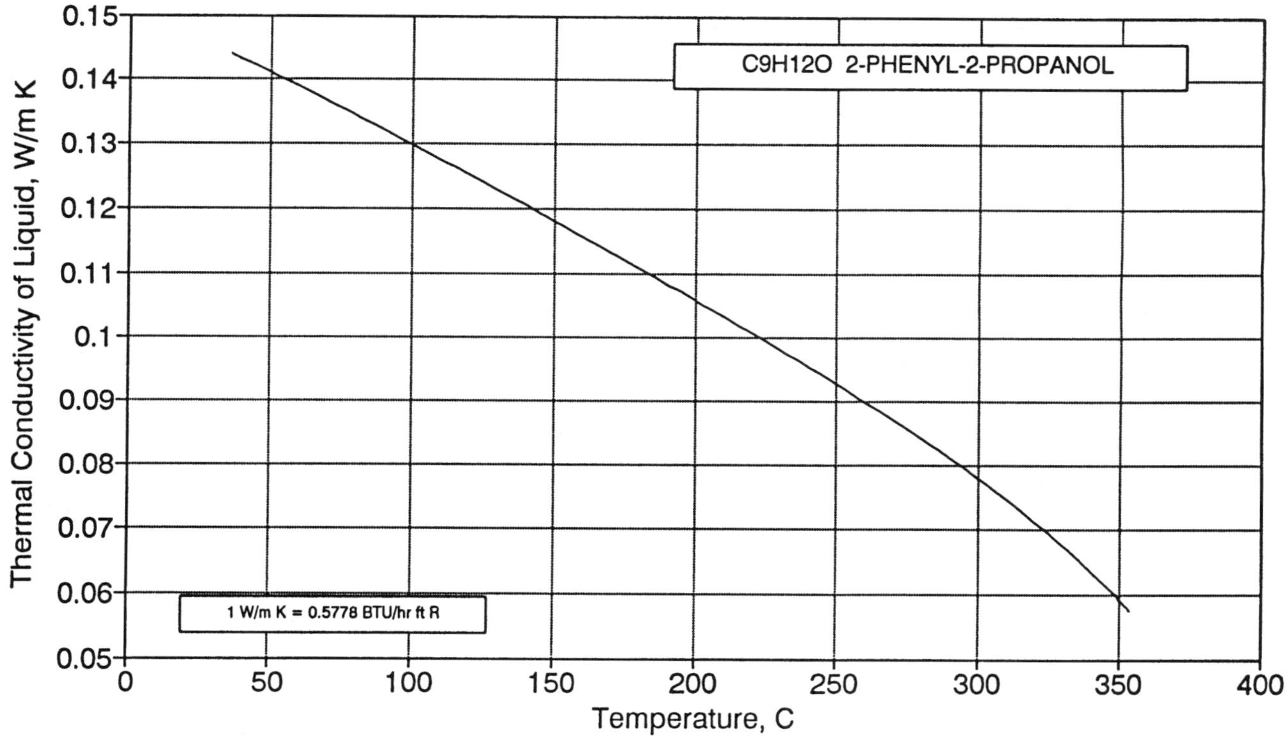
C9H12O 2-PHENYL-2-PROPANOL
1 W/m K = 0.5778 BTU/hr ft R
Thermal Conductivity of Liquid, W/m K
Temperature, C
0.15
0.14
0.13
0.12
0.11
0.1
0.09
0.08
0.07
0.06
0.05
0
50
100
150
200
250
300
350
400

C9H12O2 CUMENE HYDROPEROXIDE

Gas thermal conductivity data are not available.

TF, K = 264.26

TB, K = 442.70 (estimated)

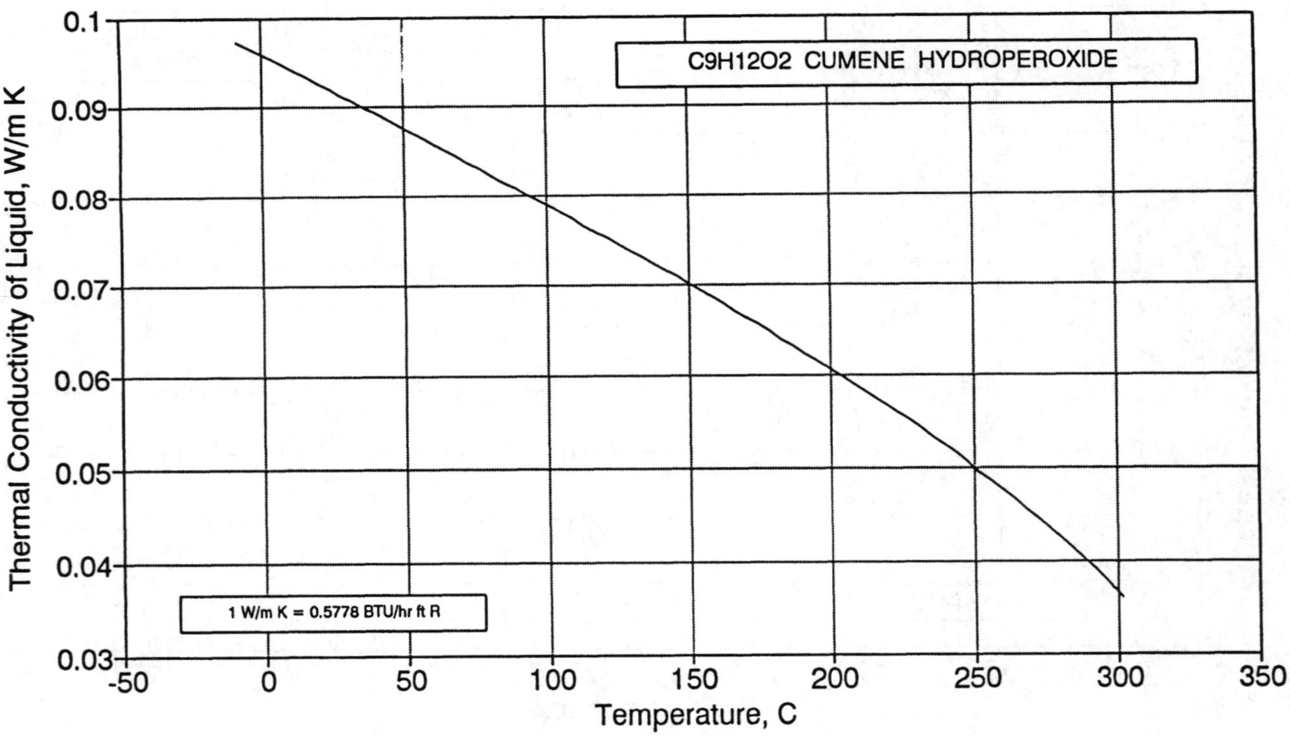

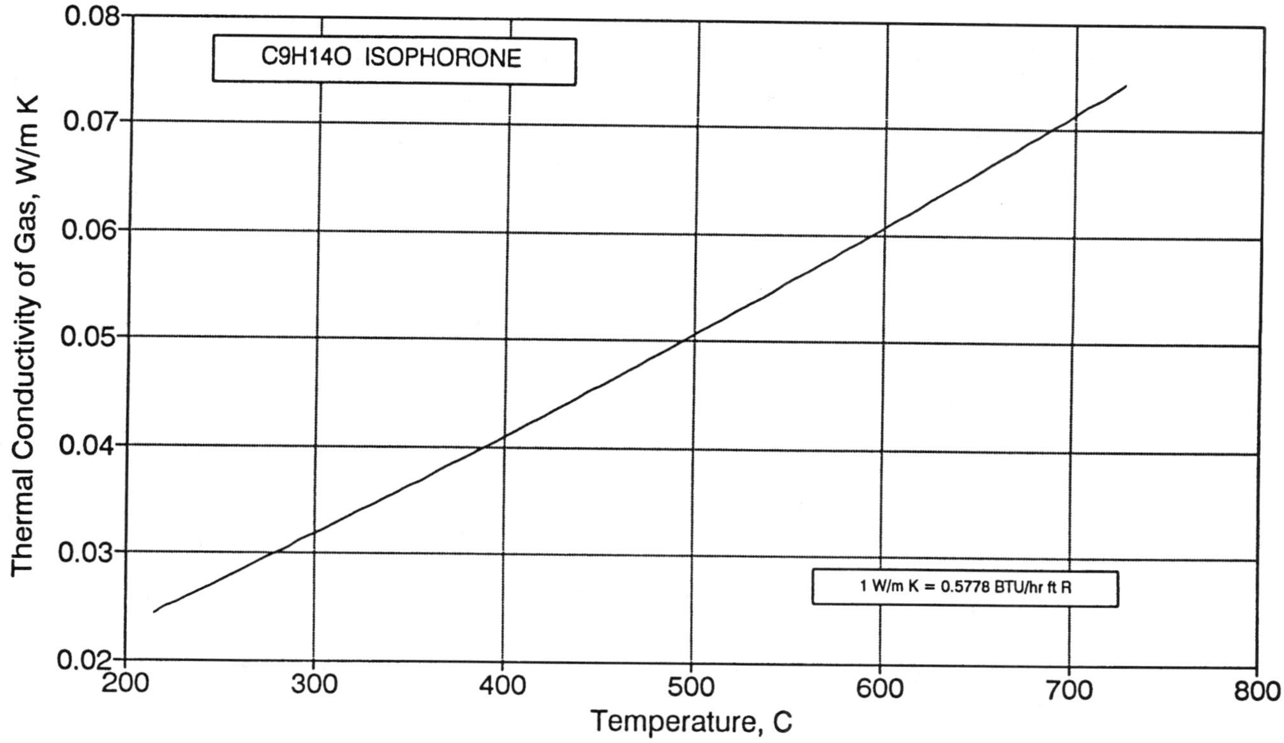
C9H14O ISOPHORONE
1 W/m K = 0.5778 BTU/hr ft R
Thermal Conductivity of Gas, W/m K
0.08
0.07
0.06
0.05
0.04
0.03
0.02
200
300
400
500
600
700
800
Temperature, C

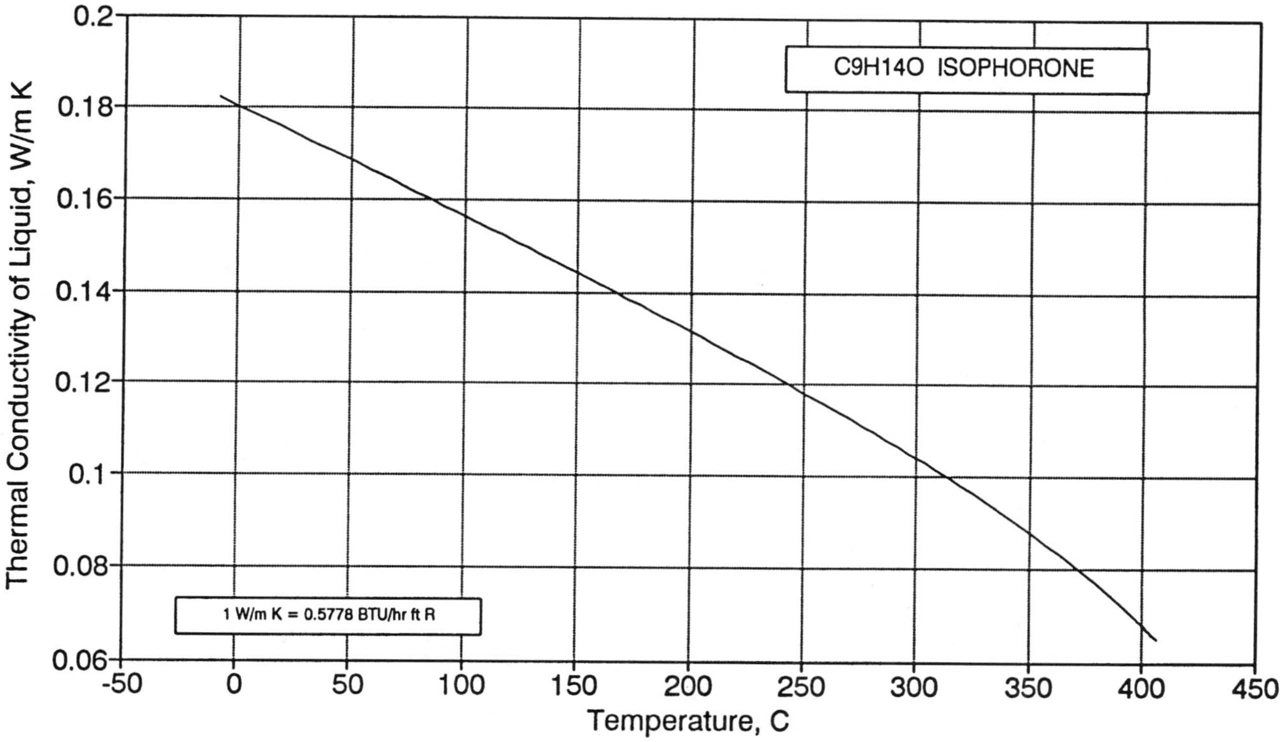
C9H14O ISOPHORONE
1 W/m K = 0.5778 BTU/hr ft R
Thermal Conductivity of Liquid, W/m K
0.2
0.18
0.16
0.14
0.12
0.1
0.08
0.06
-50
0
50
100
150
200
250
300
350
400
450
Temperature, C

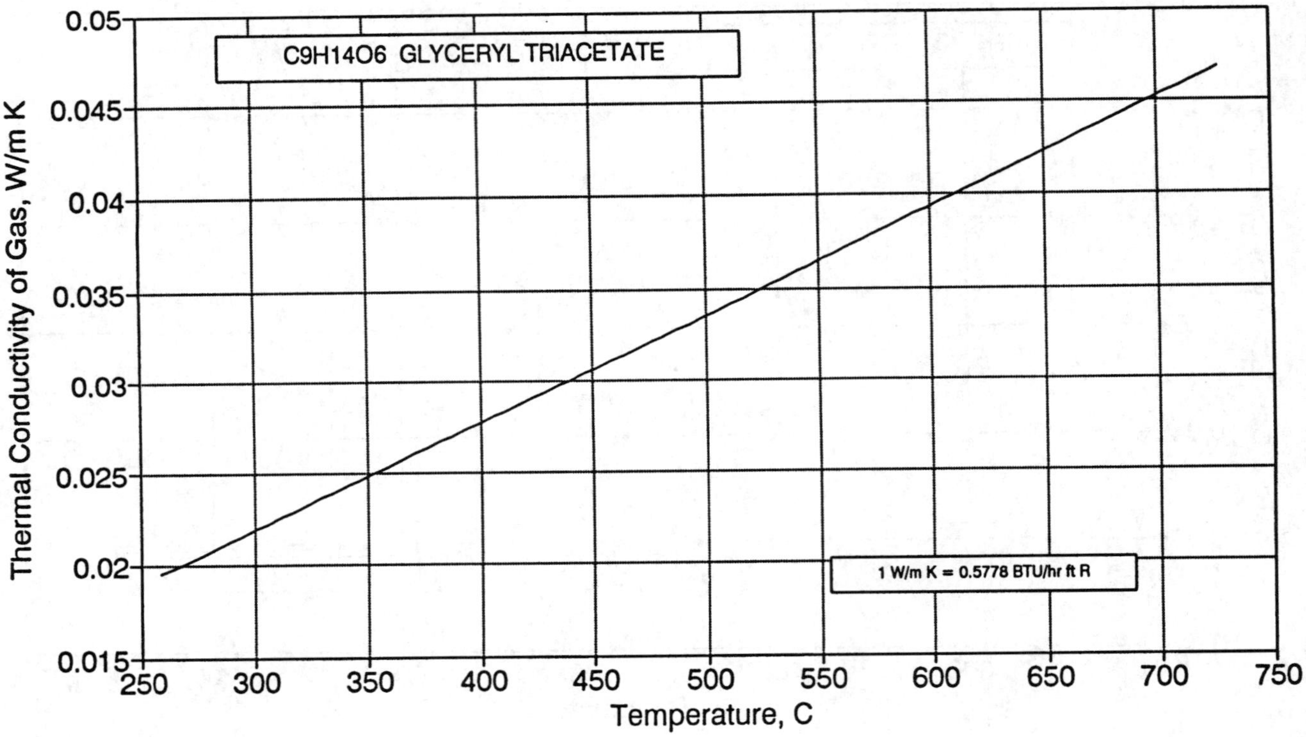
C9H14O6 GLYCERYL TRIACETATE
1 W/m K = 0.5778 BTU/hr ft R
Thermal Conductivity of Gas, W/m K
0.05
0.045
0.04
0.035
0.03
0.025
0.02
0.015
250
300
350
400
450
500
550
600
650
700
750
Temperature, C

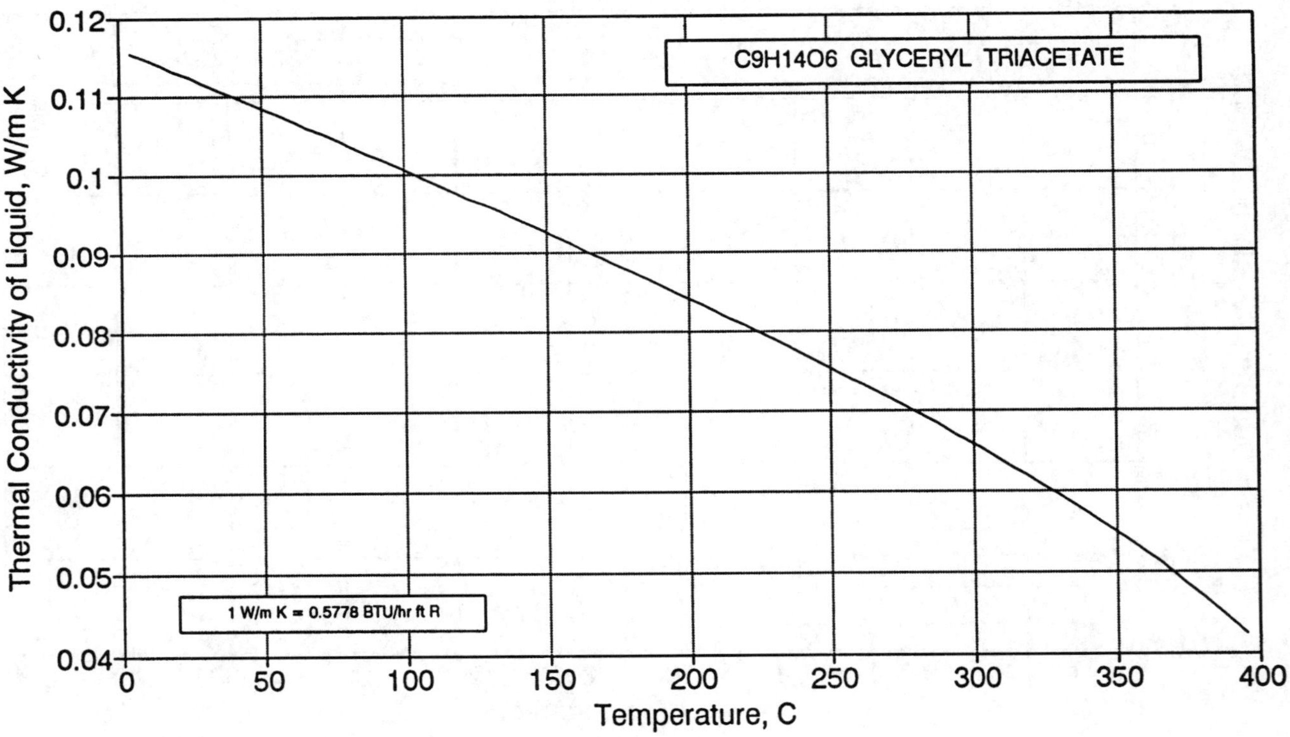
C9H14O6 GLYCERYL TRIACETATE
1 W/m K = 0.5778 BTU/hr ft R
Thermal Conductivity of Liquid, W/m K
0.12
0.11
0.1
0.09
0.08
0.07
0.06
0.05
0.04
0
50
100
150
200
250
300
350
400
Temperature, C

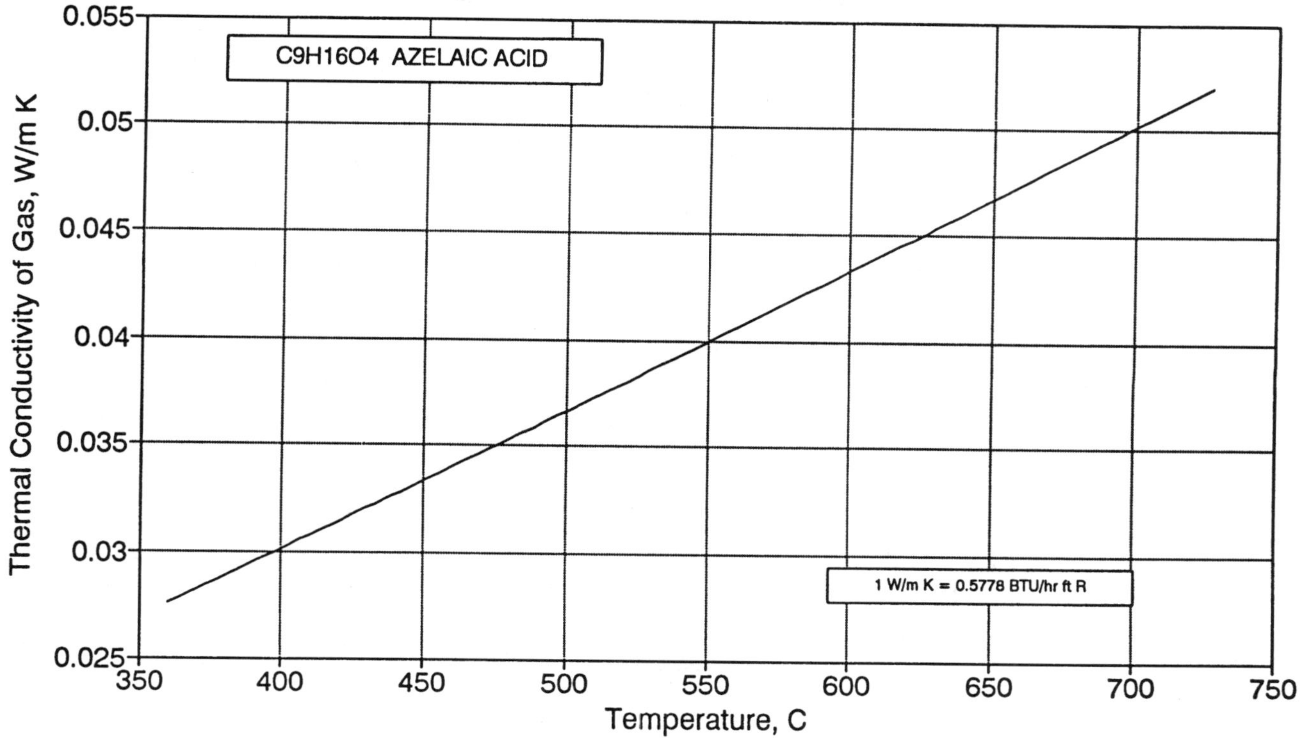
C9H16O4 AZELAIC ACID
1 W/m K = 0.5778 BTU/hr ft R
Thermal Conductivity of Gas, W/m K
Temperature, C
0.055
0.05
0.045
0.04
0.035
0.03
0.025
350
400
450
500
550
600
650
700
750

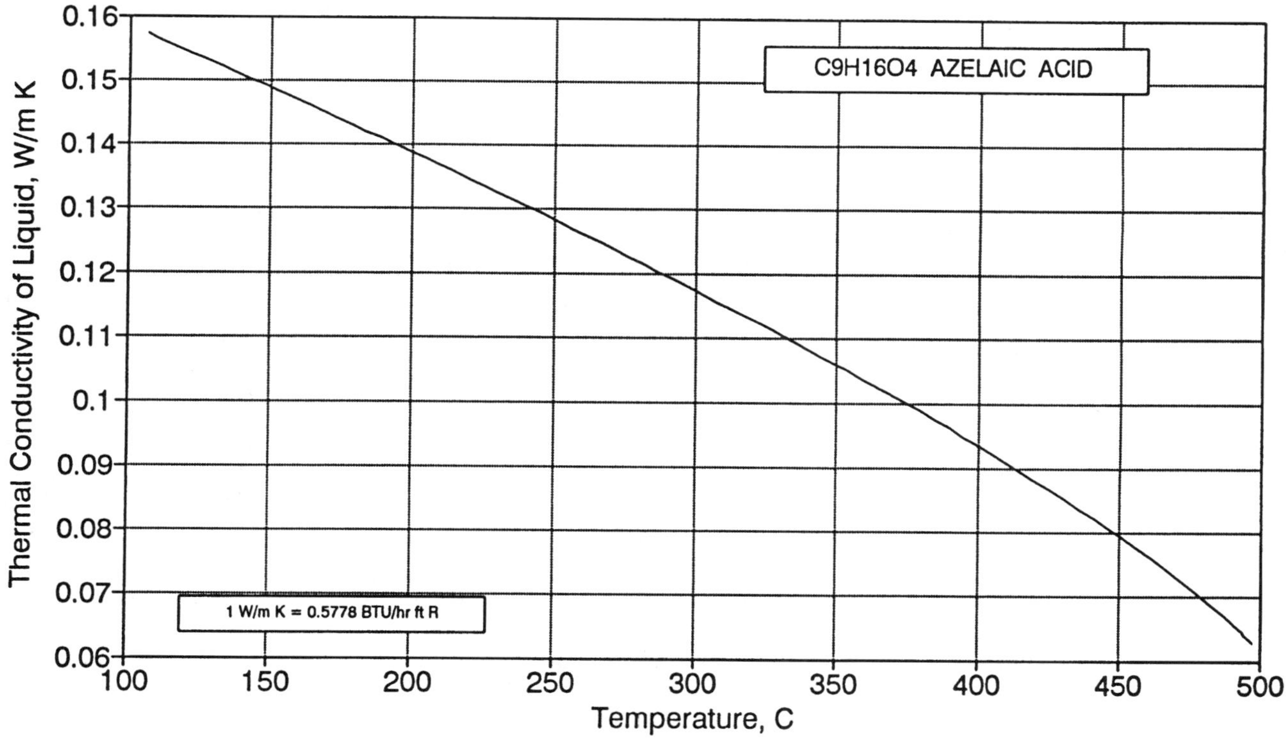
C9H16O4 AZELAIC ACID
1 W/m K = 0.5778 BTU/hr ft R
Thermal Conductivity of Liquid, W/m K
Temperature, C
0.16
0.15
0.14
0.13
0.12
0.11
0.1
0.09
0.08
0.07
0.06
100
150
200
250
300
350
400
450
500

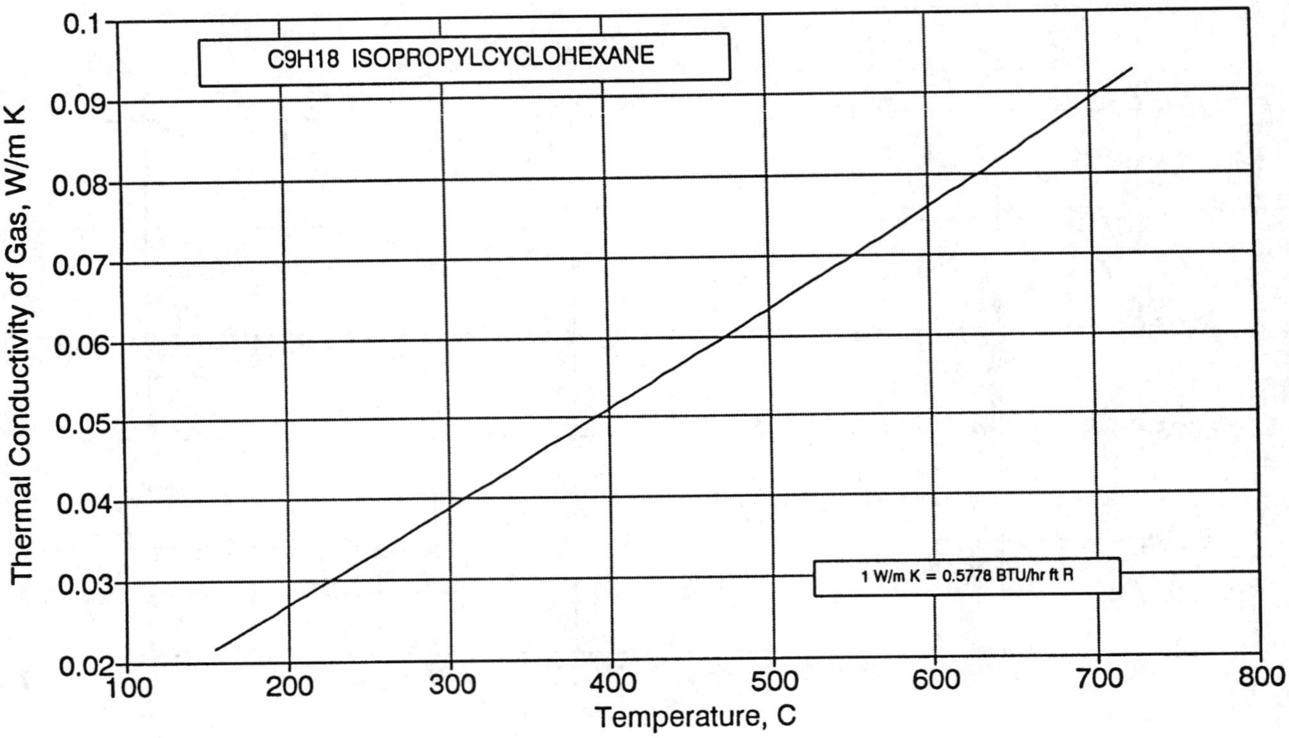
C9H18 ISOPROPYLCYCLOHEXANE
1 W/m K = 0.5778 BTU/hr ft R
Thermal Conductivity of Gas, W/m K
0.1
0.09
0.08
0.07
0.06
0.05
0.04
0.03
0.02
100
200
300
400
500
600
700
800
Temperature, C

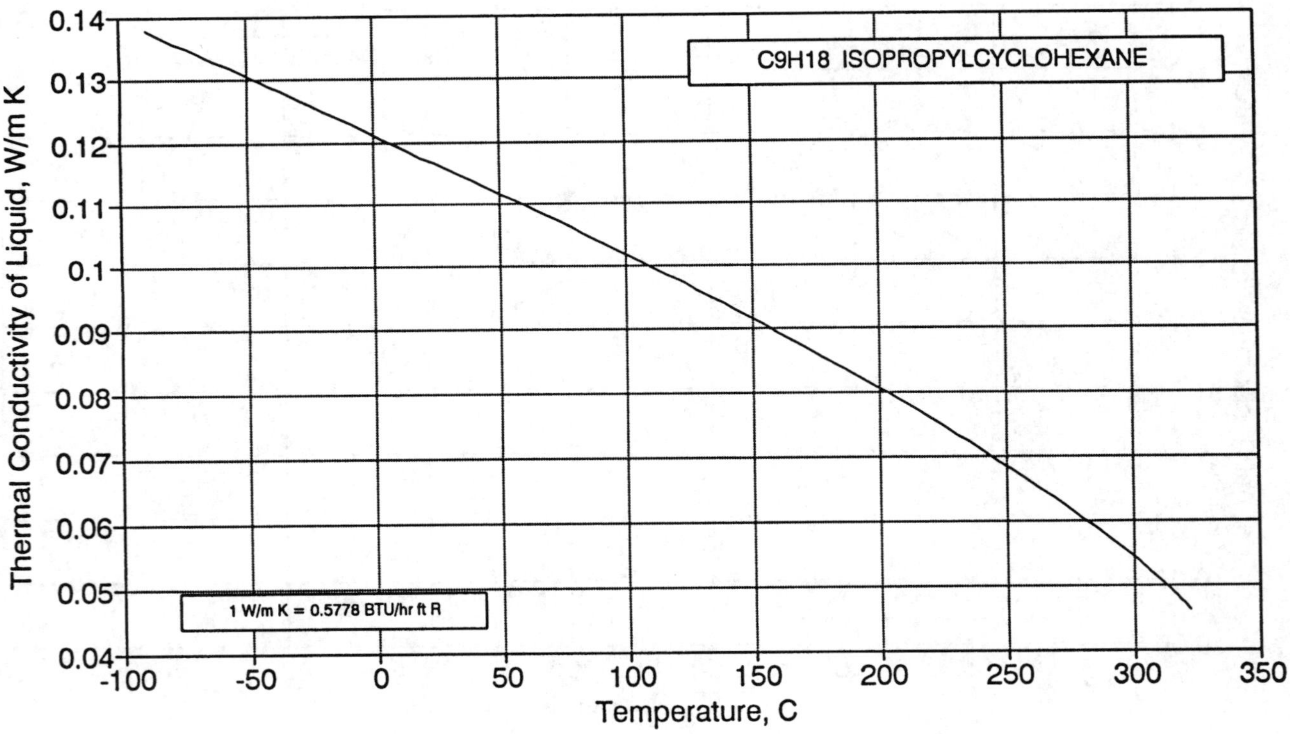
C9H18 ISOPROPYLCYCLOHEXANE
1 W/m K = 0.5778 BTU/hr ft R
Thermal Conductivity of Liquid, W/m K
0.14
0.13
0.12
0.11
0.1
0.09
0.08
0.07
0.06
0.05
0.04
-100
-50
0
50
100
150
200
250
300
350
Temperature, C

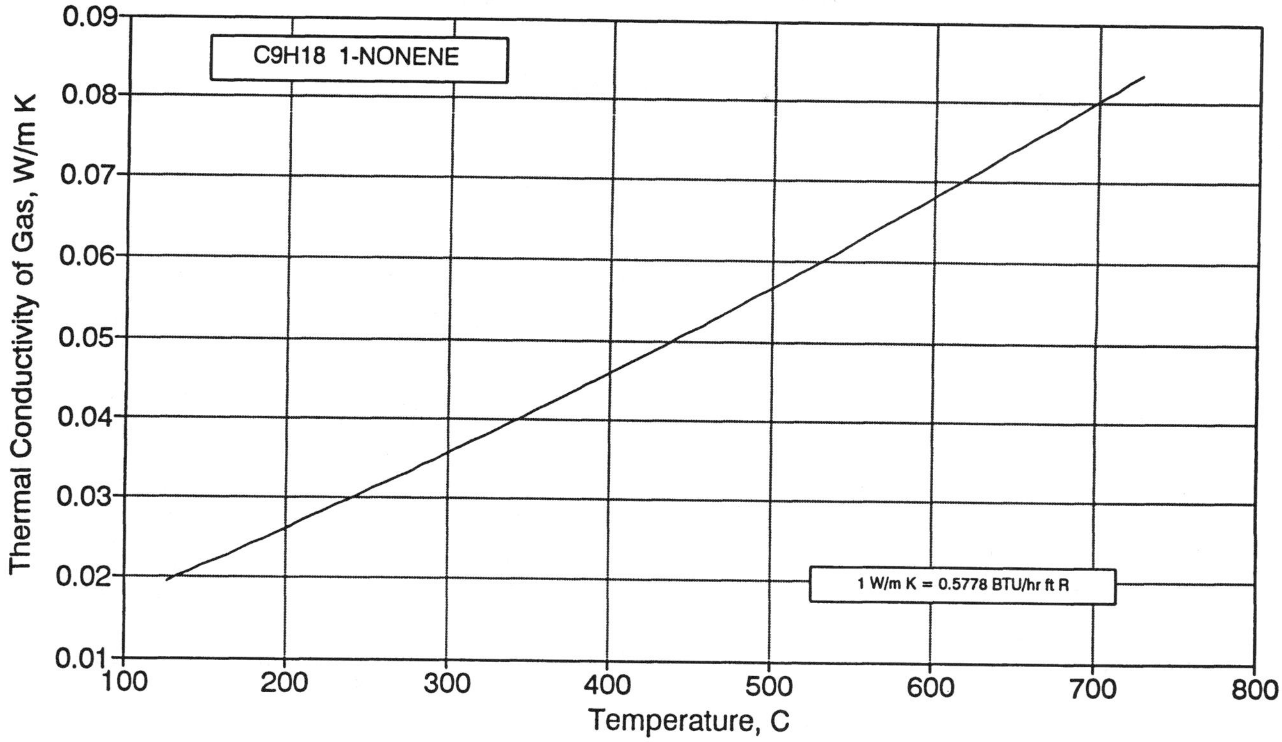
C9H18 1-NONENE
1 W/m K = 0.5778 BTU/hr ft R
Thermal Conductivity of Gas, W/m K
Temperature, C
0.09
0.08
0.07
0.06
0.05
0.04
0.03
0.02
0.01
100
200
300
400
500
600
700
800

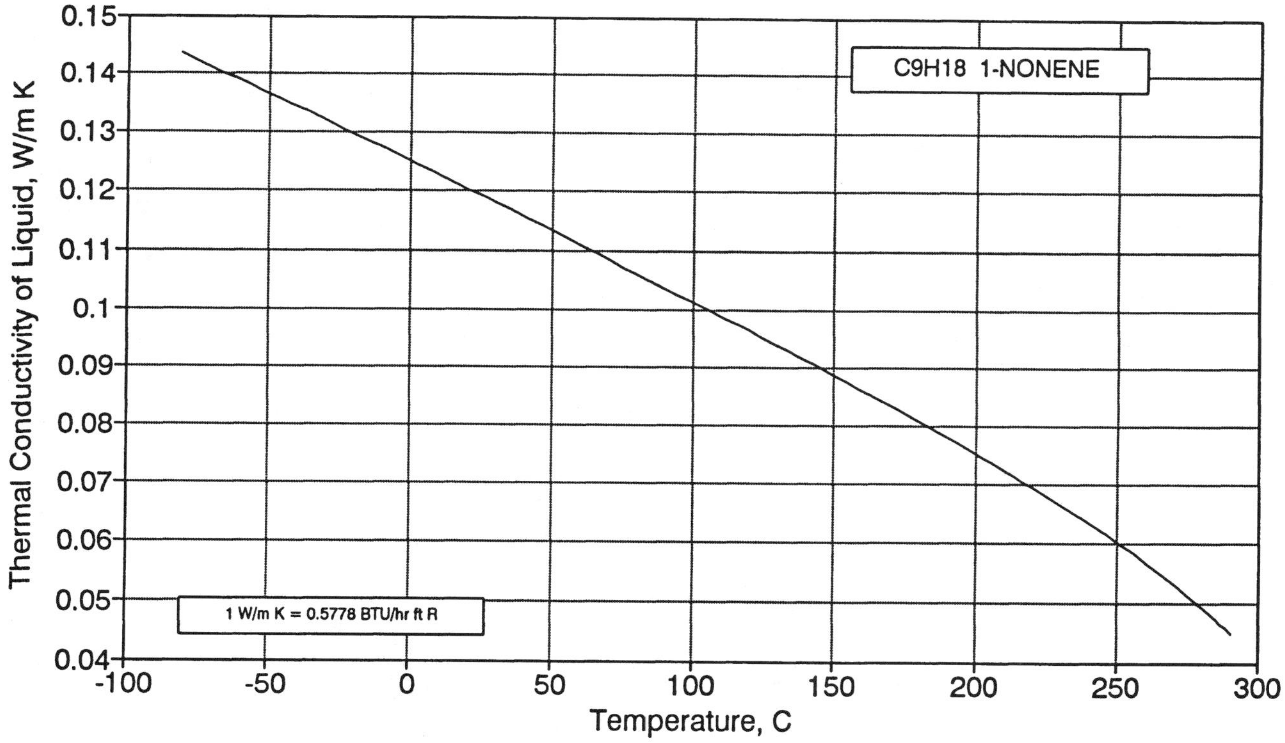
C9H18 1-NONENE
1 W/m K = 0.5778 BTU/hr ft R
Thermal Conductivity of Liquid, W/m K
Temperature, C
0.15
0.14
0.13
0.12
0.11
0.1
0.09
0.08
0.07
0.06
0.05
0.04
-100
-50
0
50
100
150
200
250
300

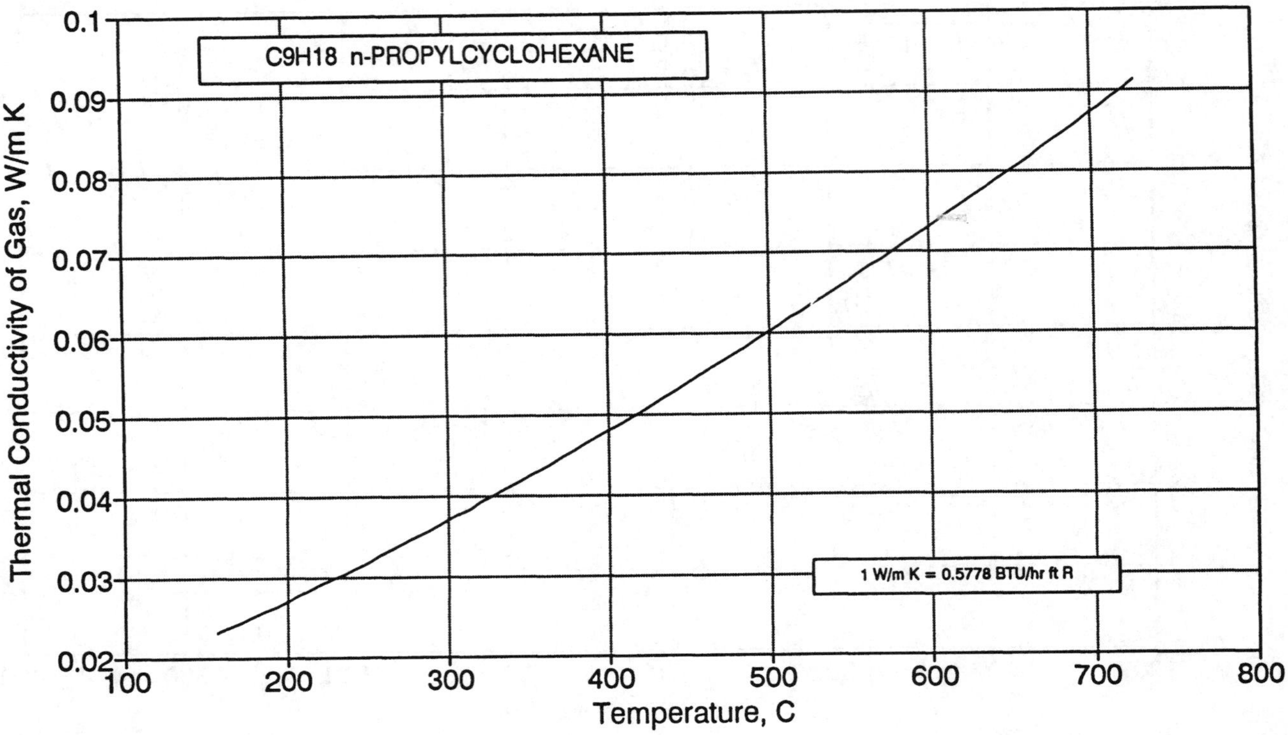
C9H18 n-PROPYLCYCLOHEXANE
Thermal Conductivity of Gas, W/m K
0.1
0.09
0.08
0.07
0.06
0.05
0.04
0.03
0.02
1 W/m K = 0.5778 BTU/hr ft R
100
200
300
400
500
600
700
800
Temperature, C

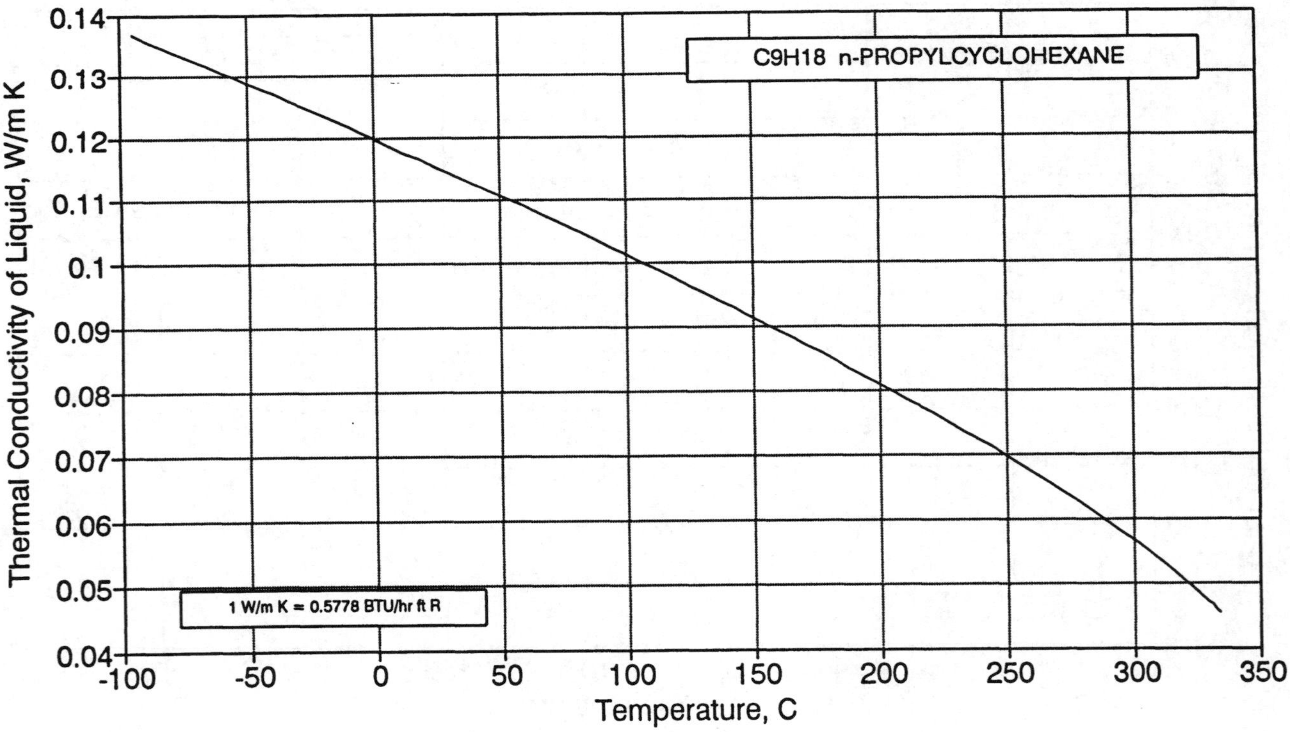
C9H18 n-PROPYLCYCLOHEXANE
Thermal Conductivity of Liquid, W/m K
0.14
0.13
0.12
0.11
0.1
0.09
0.08
0.07
0.06
0.05
0.04
1 W/m K = 0.5778 BTU/hr ft R
-100
-50
0
50
100
150
200
250
300
350
Temperature, C

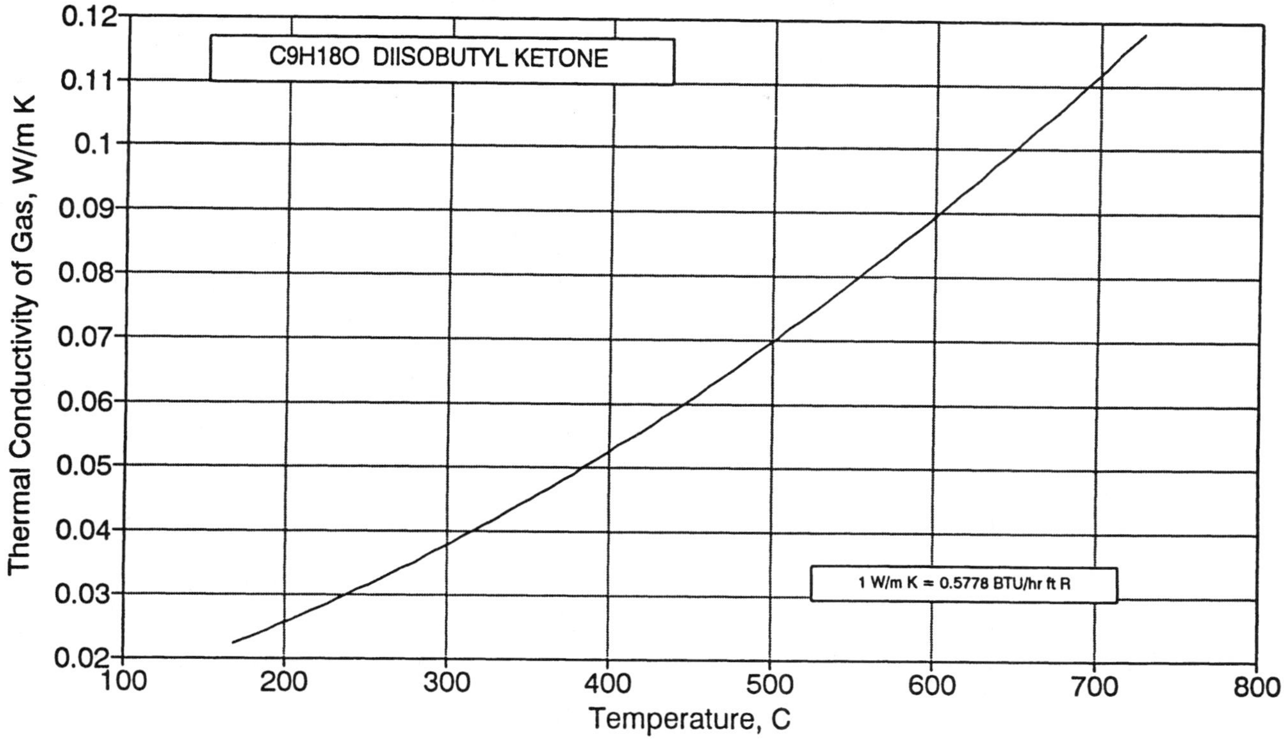
C9H18O DIISOBUTYL KETONE
1 W/m K = 0.5778 BTU/hr ft R
Thermal Conductivity of Gas, W/m K
0.12
0.11
0.1
0.09
0.08
0.07
0.06
0.05
0.04
0.03
0.02
100
200
300
400
500
600
700
800
Temperature, C

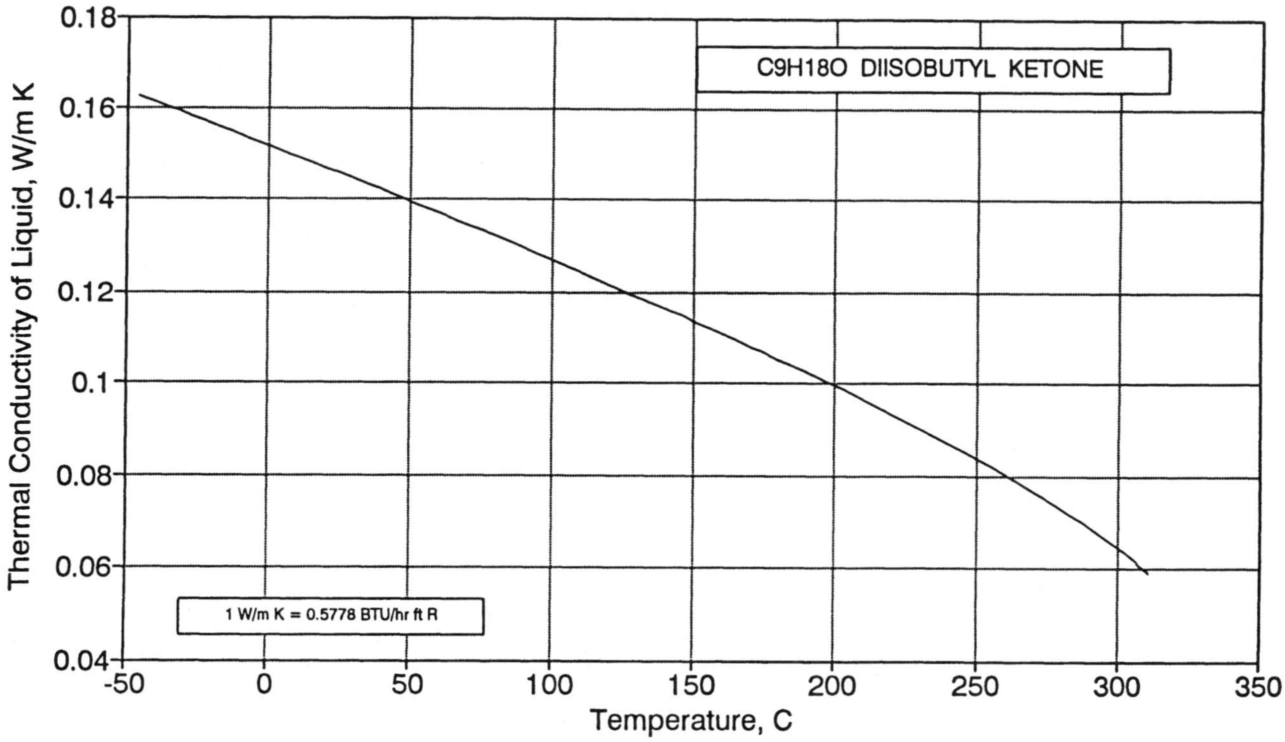
C9H18O DIISOBUTYL KETONE
1 W/m K = 0.5778 BTU/hr ft R
Thermal Conductivity of Liquid, W/m K
0.18
0.16
0.14
0.12
0.1
0.08
0.06
0.04
-50
0
50
100
150
200
250
300
350
Temperature, C

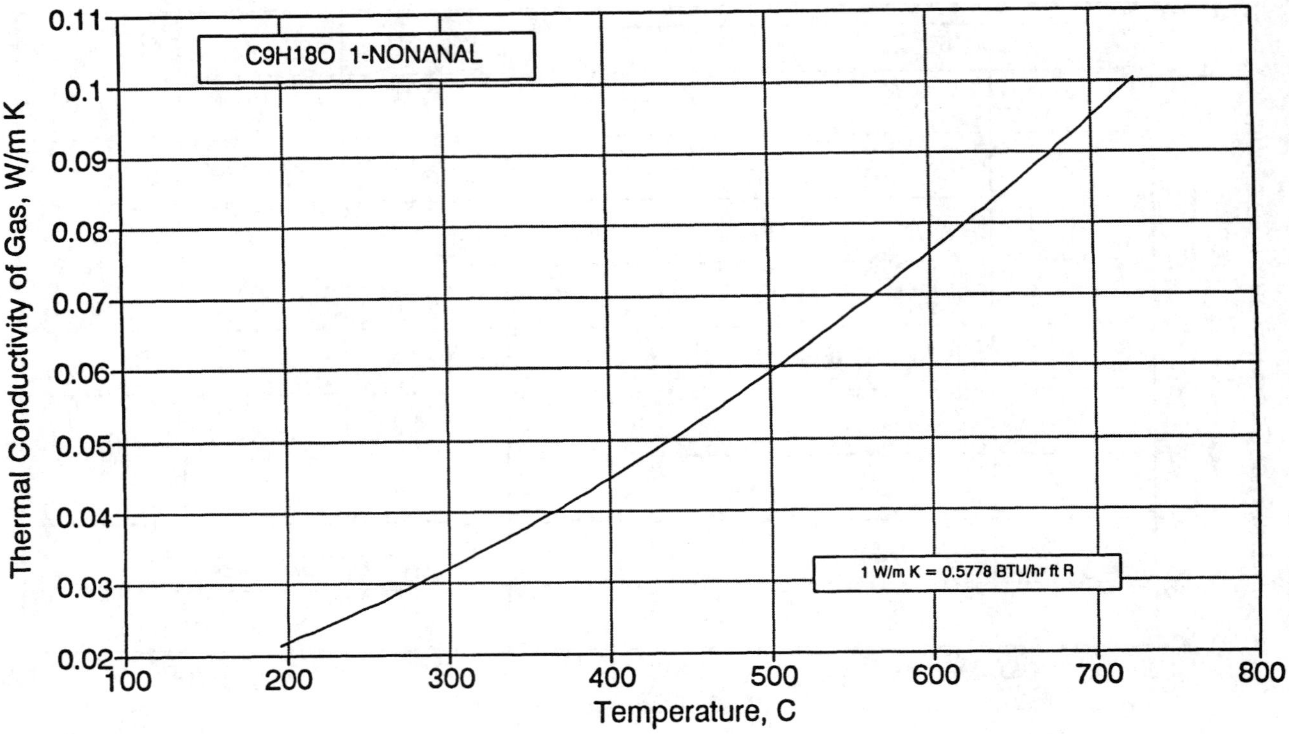
C9H18O 1-NONANAL
1 W/m K = 0.5778 BTU/hr ft R
Thermal Conductivity of Gas, W/m K
0.11
0.1
0.09
0.08
0.07
0.06
0.05
0.04
0.03
0.02
100
200
300
400
500
600
700
800
Temperature, C

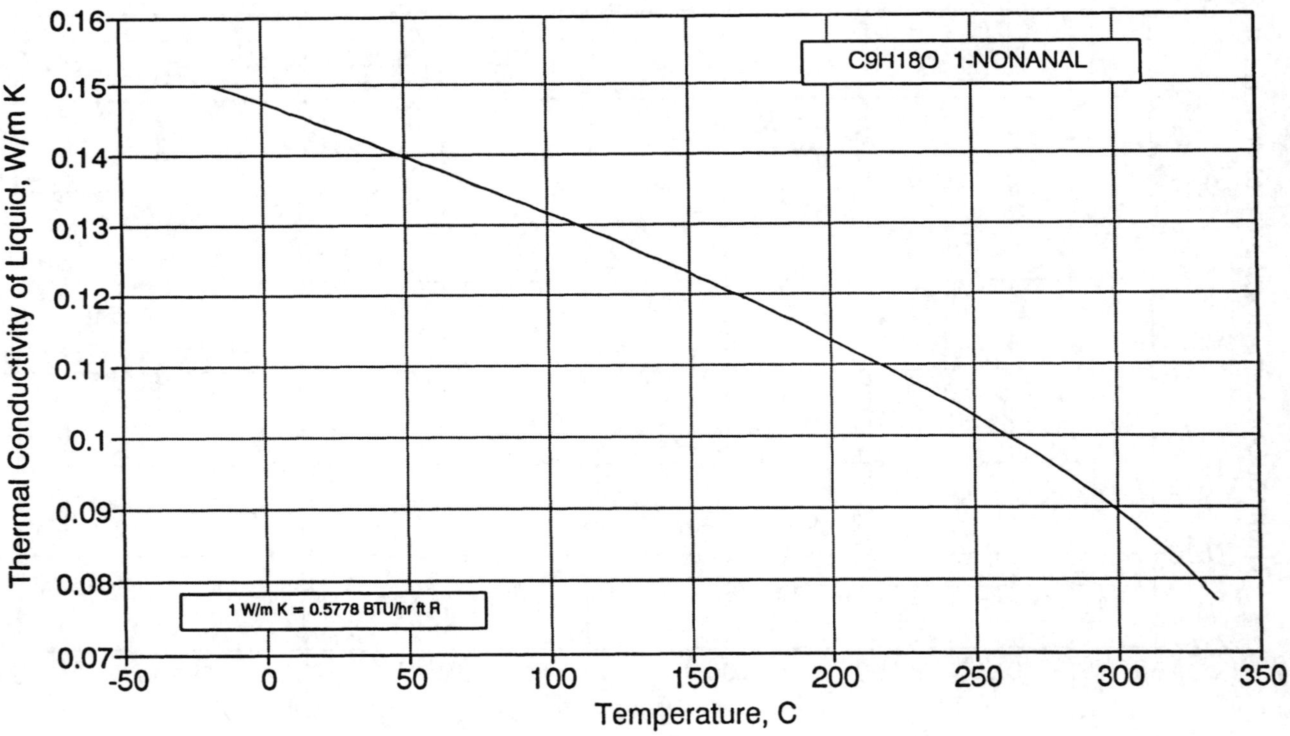
C9H18O 1-NONANAL
1 W/m K = 0.5778 BTU/hr ft R
Thermal Conductivity of Liquid, W/m K
0.16
0.15
0.14
0.13
0.12
0.11
0.1
0.09
0.08
0.07
-50
0
50
100
150
200
250
300
350
Temperature, C

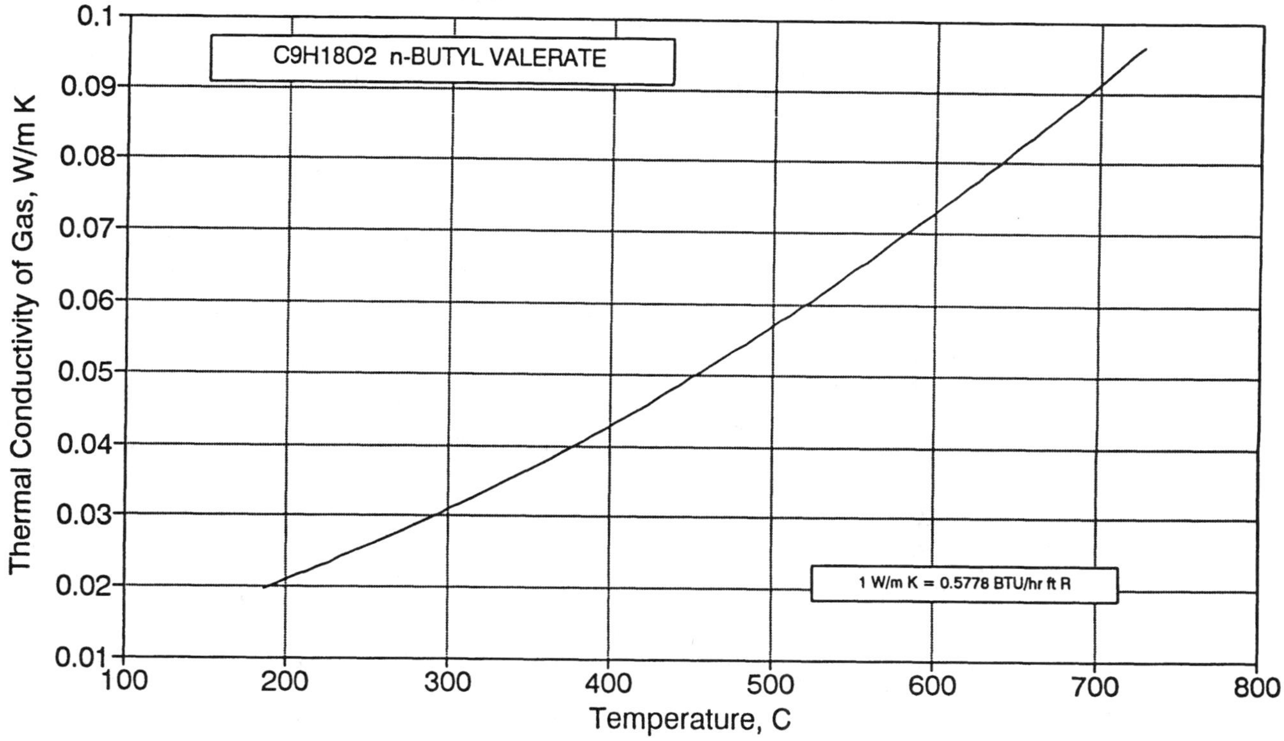
C9H18O2 n-BUTYL VALERATE
1 W/m K = 0.5778 BTU/hr ft R
Thermal Conductivity of Gas, W/m K
Temperature, C
0.1
0.09
0.08
0.07
0.06
0.05
0.04
0.03
0.02
0.01
100
200
300
400
500
600
700
800

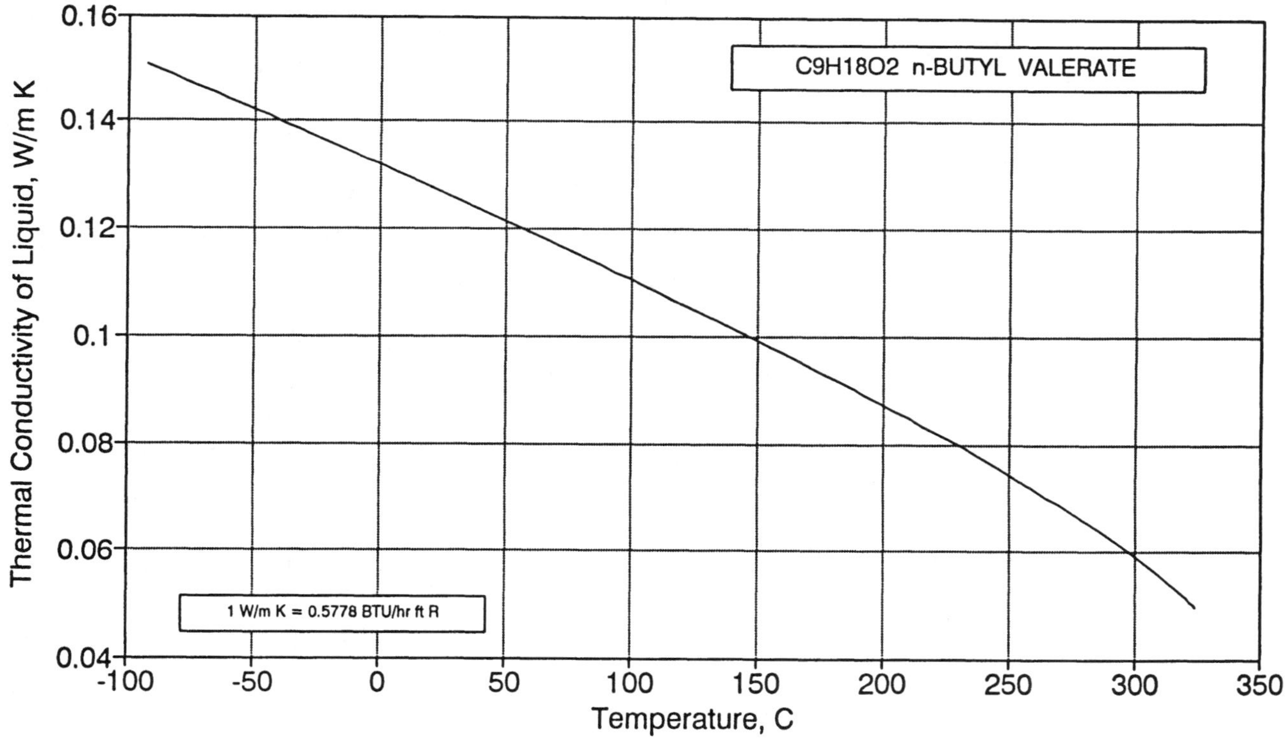
C9H18O2 n-BUTYL VALERATE
1 W/m K = 0.5778 BTU/hr ft R
Thermal Conductivity of Liquid, W/m K
Temperature, C
0.16
0.14
0.12
0.1
0.08
0.06
0.04
-100
-50
0
50
100
150
200
250
300
350

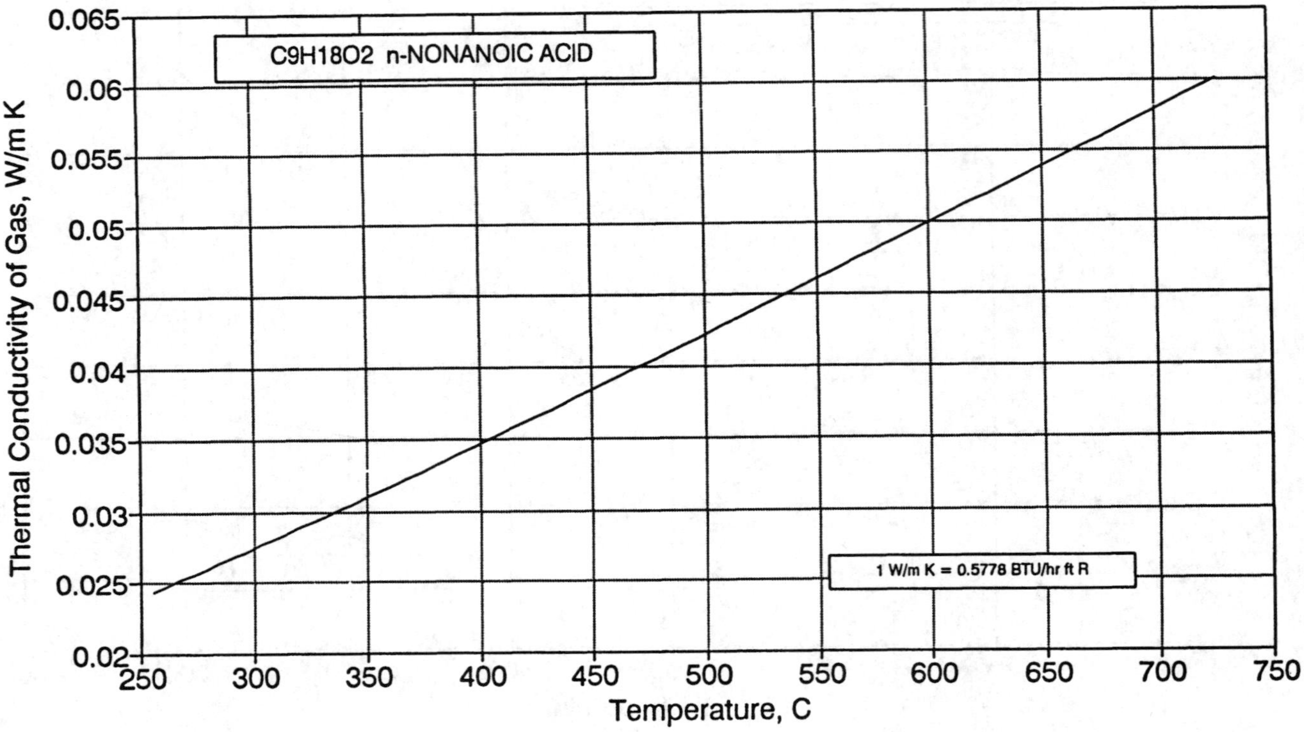
C9H18O2 n-NONANOIC ACID
1 W/m K = 0.5778 BTU/hr ft R
Thermal Conductivity of Gas, W/m K
0.065
0.06
0.055
0.05
0.045
0.04
0.035
0.03
0.025
0.02
250
300
350
400
450
500
550
600
650
700
750
Temperature, C

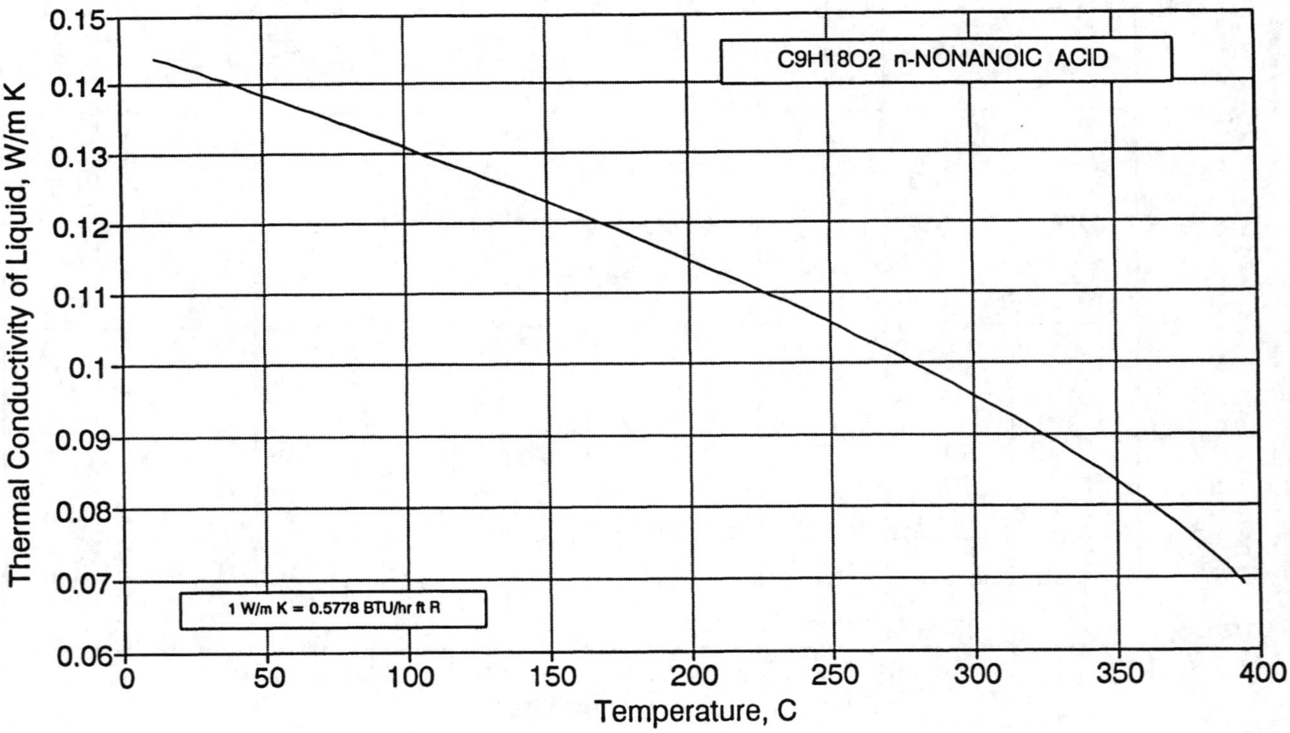
C9H18O2 n-NONANOIC ACID
1 W/m K = 0.5778 BTU/hr ft R
Thermal Conductivity of Liquid, W/m K
0.15
0.14
0.13
0.12
0.11
0.1
0.09
0.08
0.07
0.06
0
50
100
150
200
250
300
350
400
Temperature, C

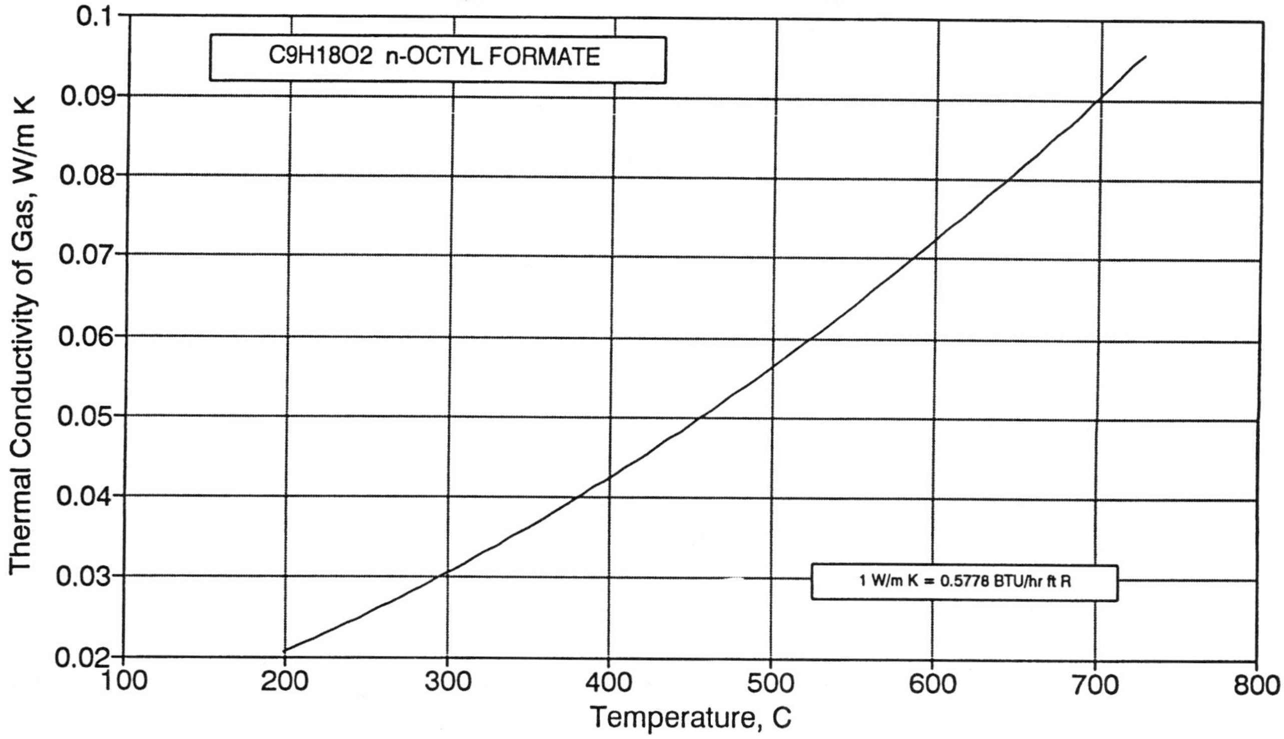
C9H18O2 n-OCTYL FORMATE
1 W/m K = 0.5778 BTU/hr ft R
Thermal Conductivity of Gas, W/m K
0.1
0.09
0.08
0.07
0.06
0.05
0.04
0.03
0.02
100
200
300
400
500
600
700
800
Temperature, C

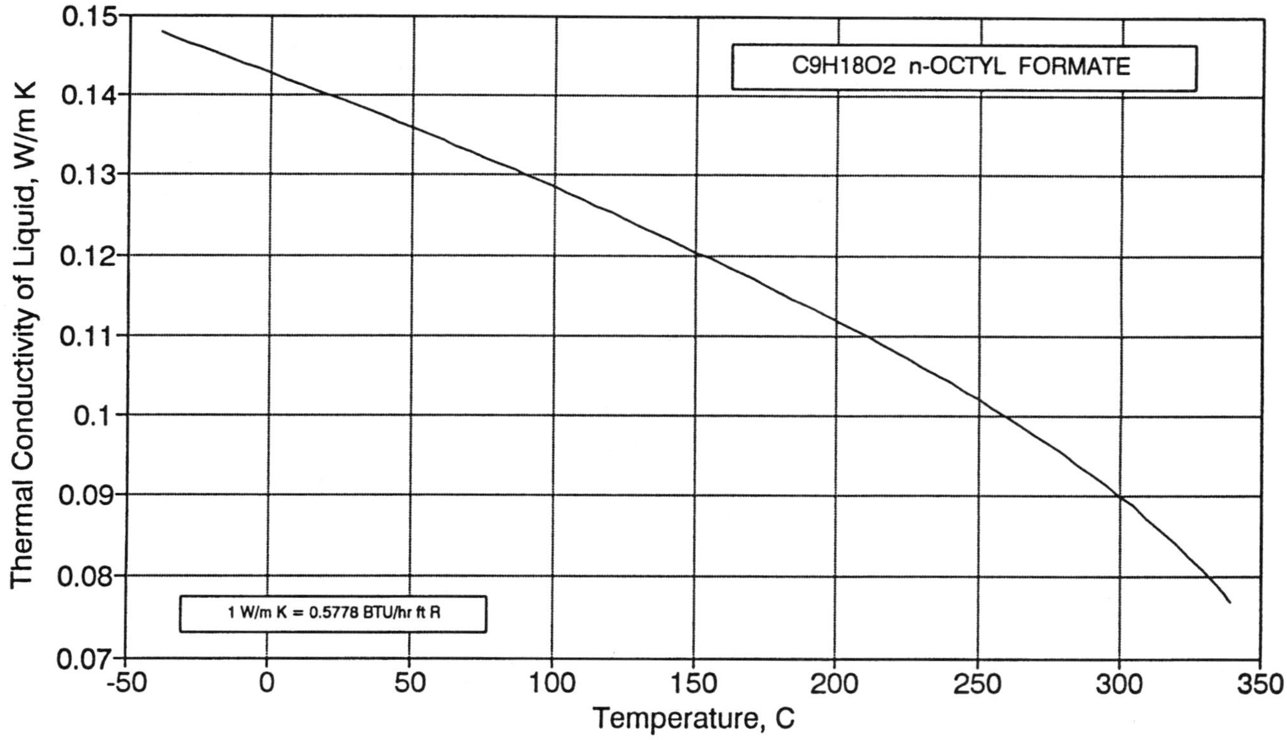
C9H18O2 n-OCTYL FORMATE
1 W/m K = 0.5778 BTU/hr ft R
Thermal Conductivity of Liquid, W/m K
0.15
0.14
0.13
0.12
0.11
0.1
0.09
0.08
0.07
-50
0
50
100
150
200
250
300
350
Temperature, C

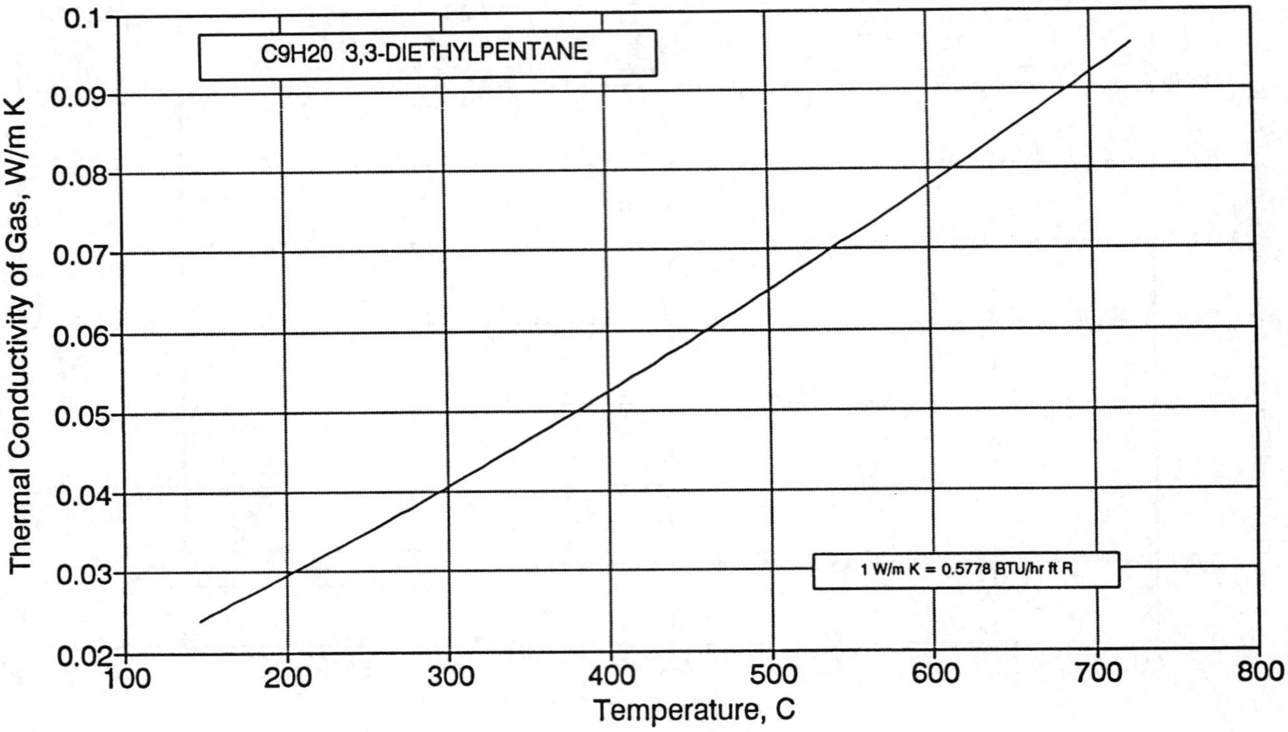
C9H20 3,3-DIETHYLPENTANE
1 W/m K = 0.5778 BTU/hr ft R
Thermal Conductivity of Gas, W/m K
0.1
0.09
0.08
0.07
0.06
0.05
0.04
0.03
0.02
100
200
300
400
500
600
700
800
Temperature, C

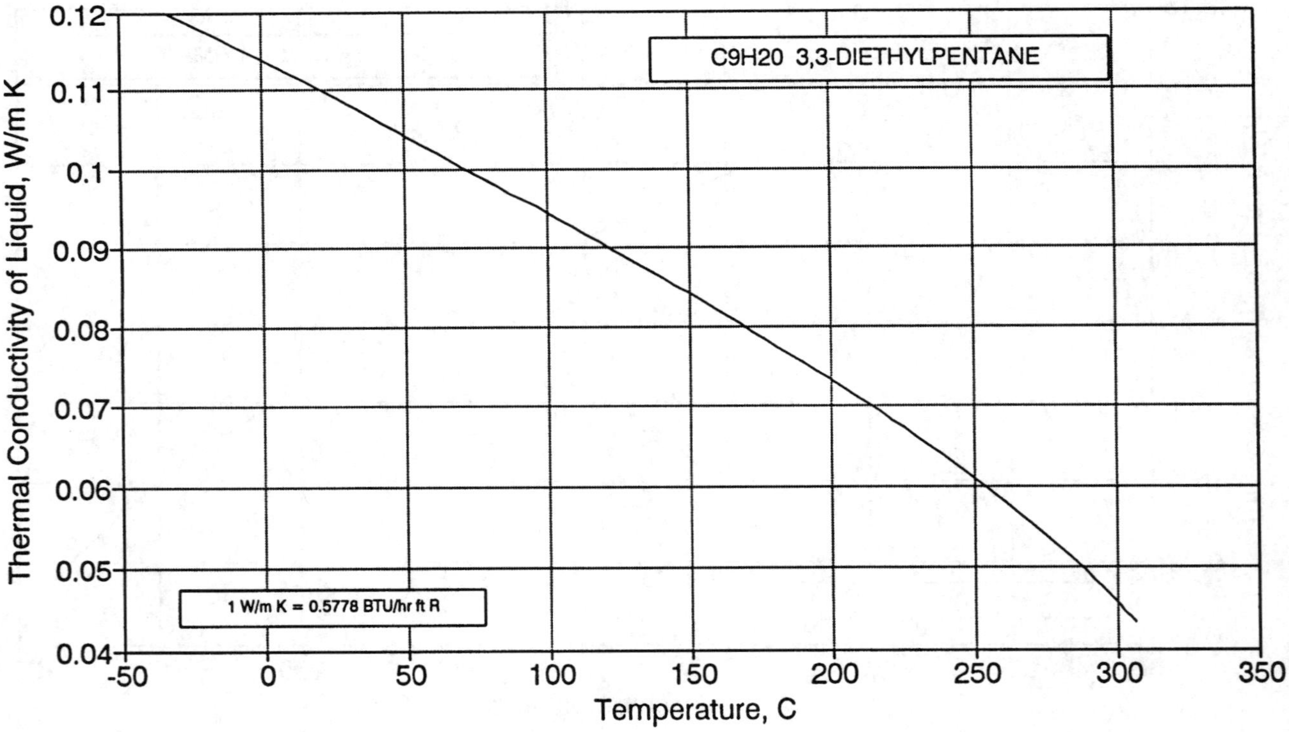
C9H20 3,3-DIETHYLPENTANE
1 W/m K = 0.5778 BTU/hr ft R
Thermal Conductivity of Liquid, W/m K
0.12
0.11
0.1
0.09
0.08
0.07
0.06
0.05
0.04
-50
0
50
100
150
200
250
300
350
Temperature, C

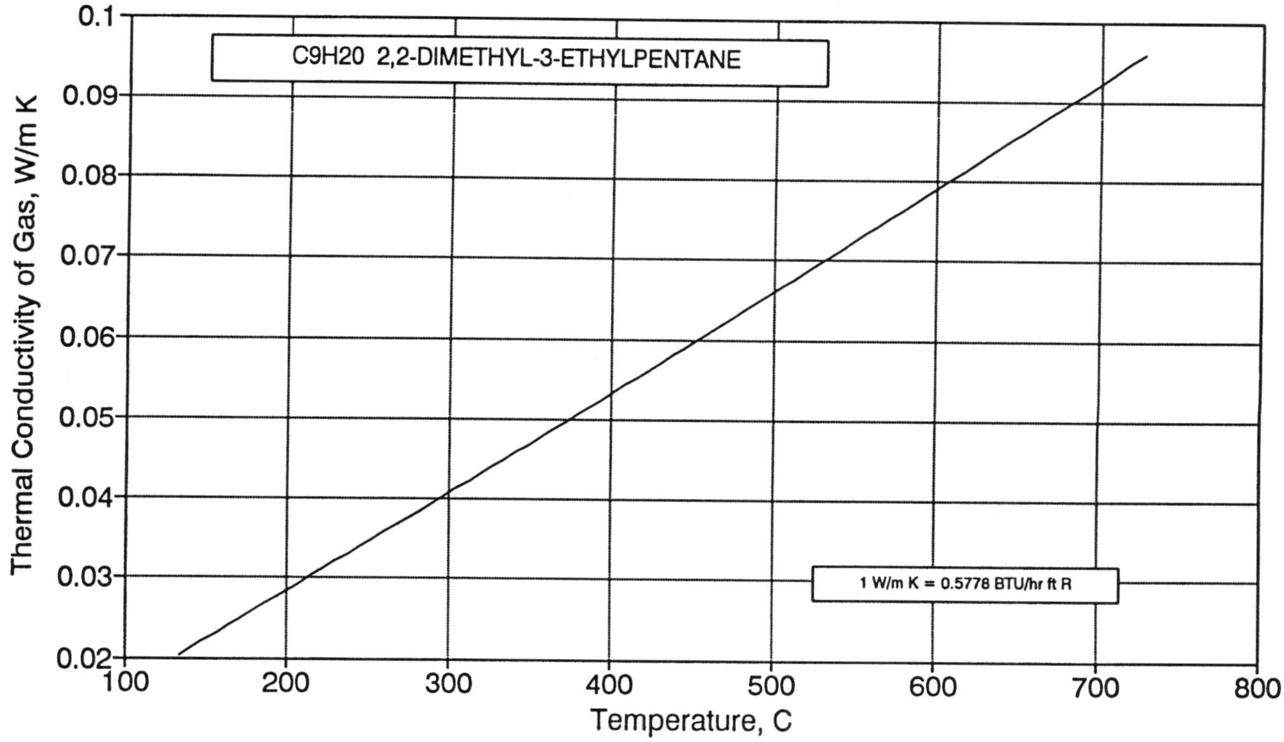
C9H20 2,2-DIMETHYL-3-ETHYLPENTANE
Thermal Conductivity of Gas, W/m K
0.1
0.09
0.08
0.07
0.06
0.05
0.04
0.03
0.02
1 W/m K = 0.5778 BTU/hr ft R
100
200
300
400
500
600
700
800
Temperature, C

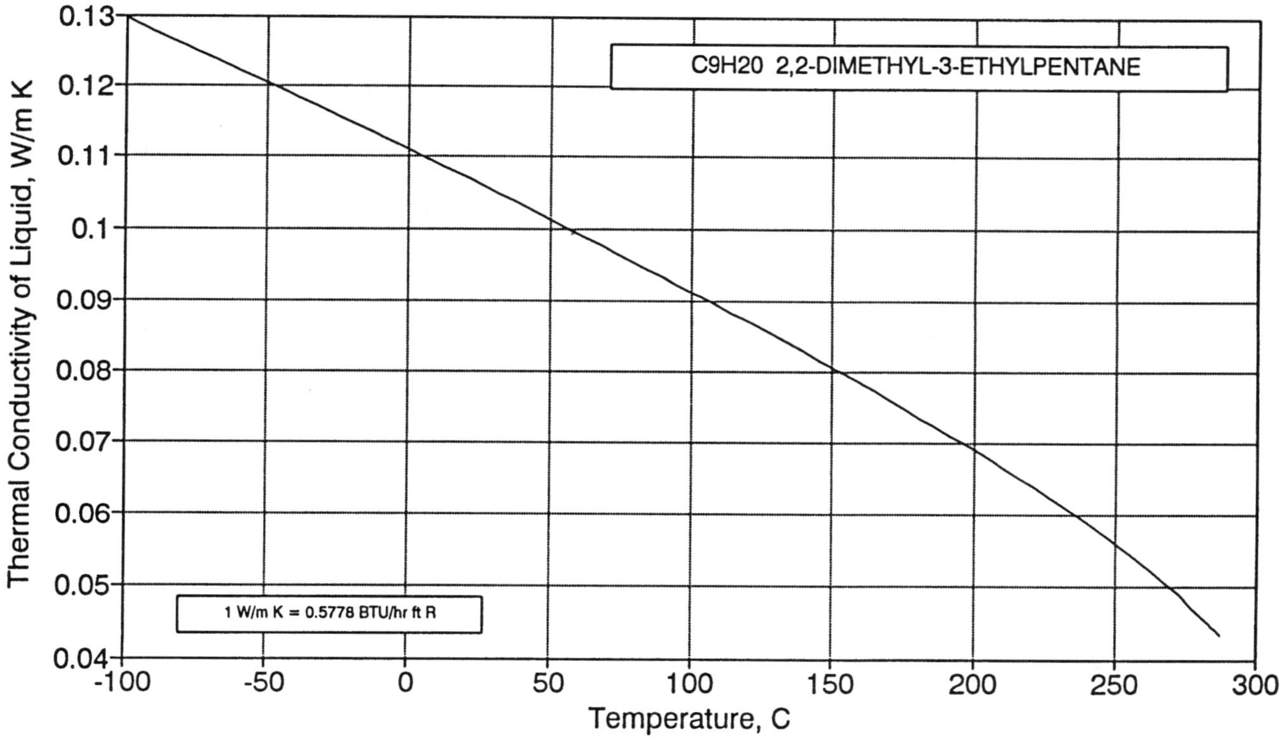
C9H20 2,2-DIMETHYL-3-ETHYLPENTANE
Thermal Conductivity of Liquid, W/m K
0.13
0.12
0.11
0.1
0.09
0.08
0.07
0.06
0.05
0.04
1 W/m K = 0.5778 BTU/hr ft R
-100
-50
0
50
100
150
200
250
300
Temperature, C

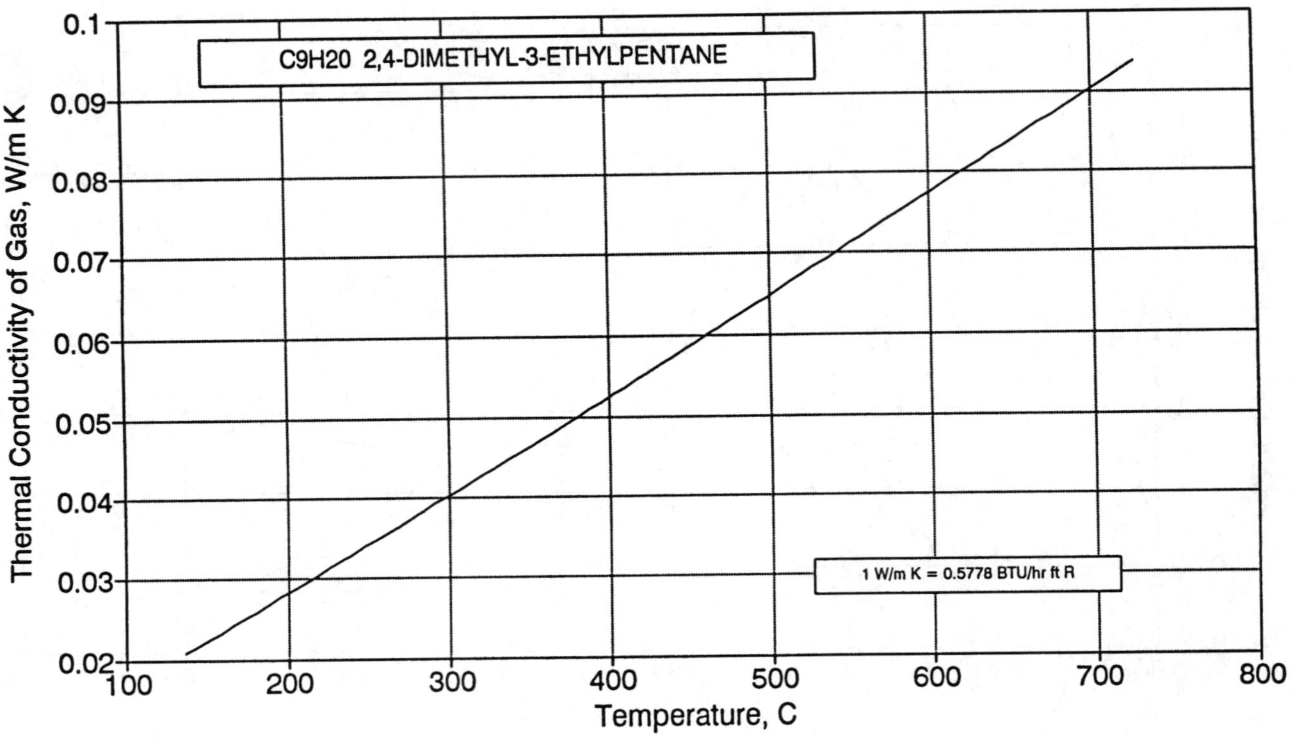
C9H20 2,4-DIMETHYL-3-ETHYLPENTANE
1 W/m K = 0.5778 BTU/hr ft R
Thermal Conductivity of Gas, W/m K
0.1
0.09
0.08
0.07
0.06
0.05
0.04
0.03
0.02
100
200
300
400
500
600
700
800
Temperature, C

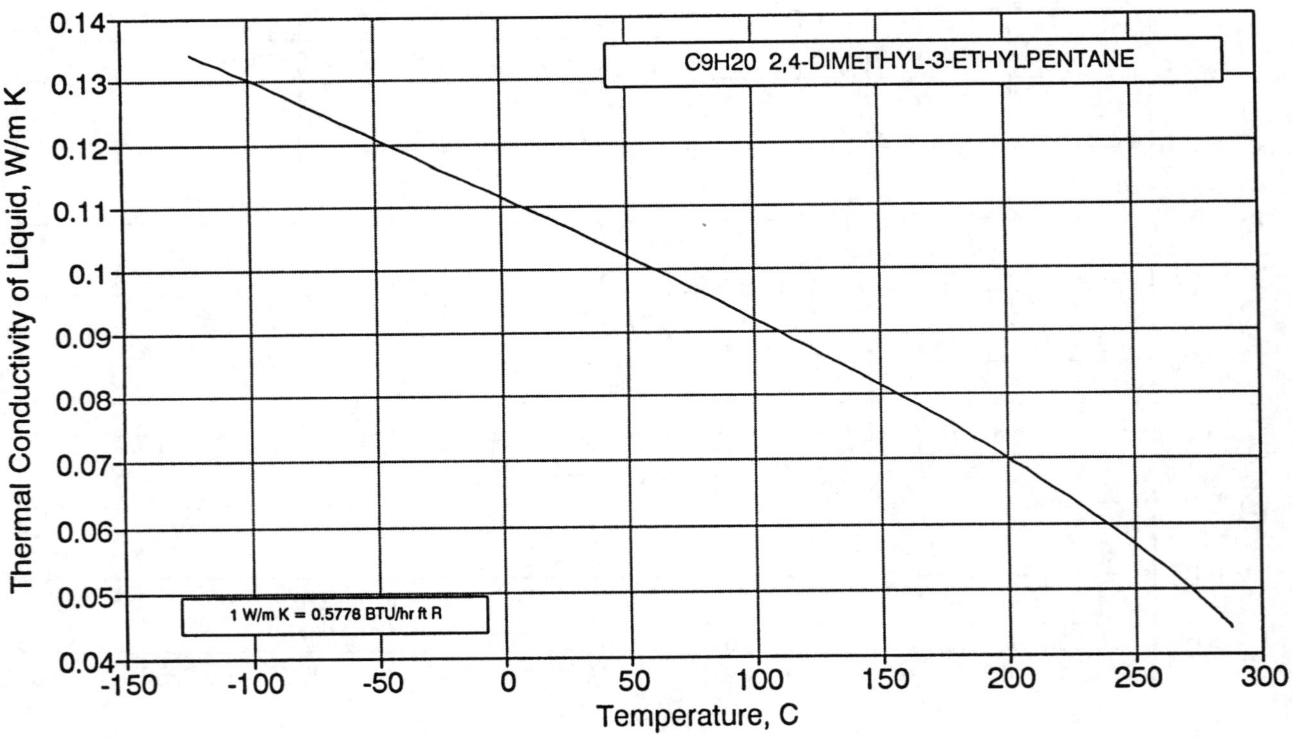
C9H20 2,4-DIMETHYL-3-ETHYLPENTANE
1 W/m K = 0.5778 BTU/hr ft R
Thermal Conductivity of Liquid, W/m K
0.14
0.13
0.12
0.11
0.1
0.09
0.08
0.07
0.06
0.05
0.04
-150
-100
-50
0
50
100
150
200
250
300
Temperature, C

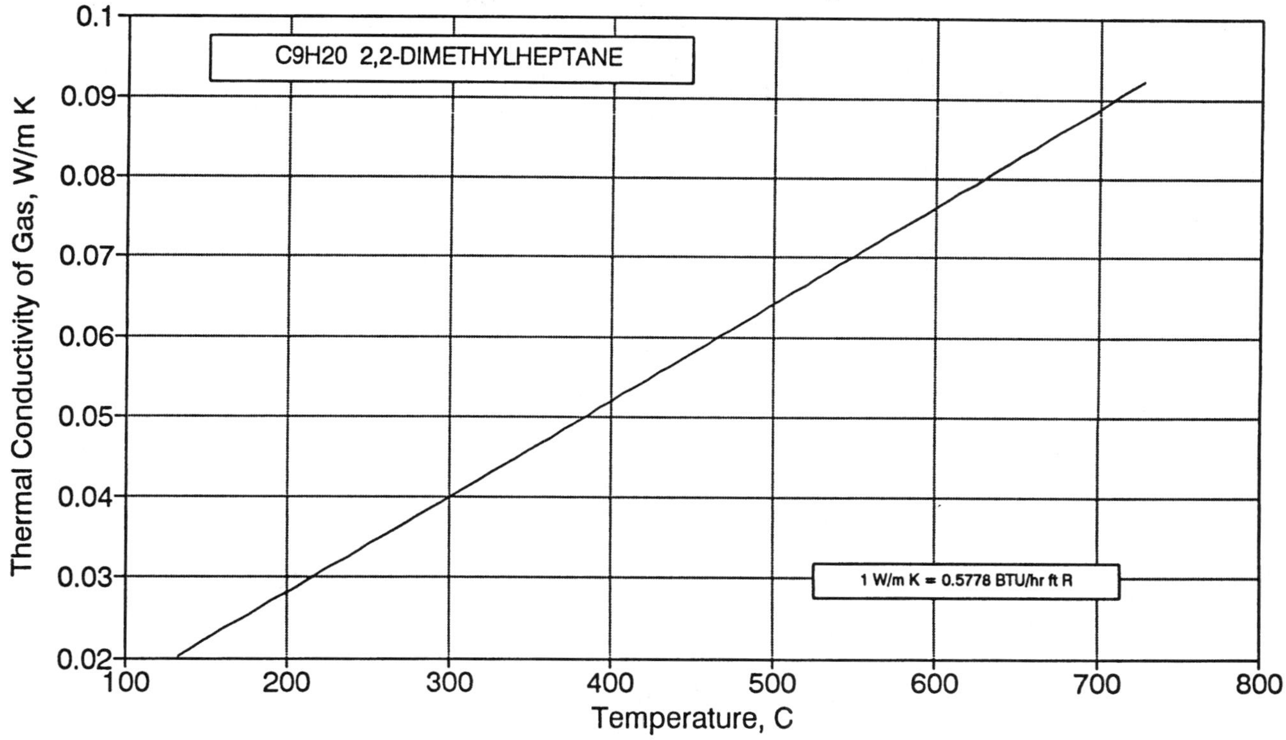
C9H20 2,2-DIMETHYLHEPTANE
1 W/m K = 0.5778 BTU/hr ft R
Thermal Conductivity of Gas, W/m K
Temperature, C
0.1
0.09
0.08
0.07
0.06
0.05
0.04
0.03
0.02
100
200
300
400
500
600
700
800

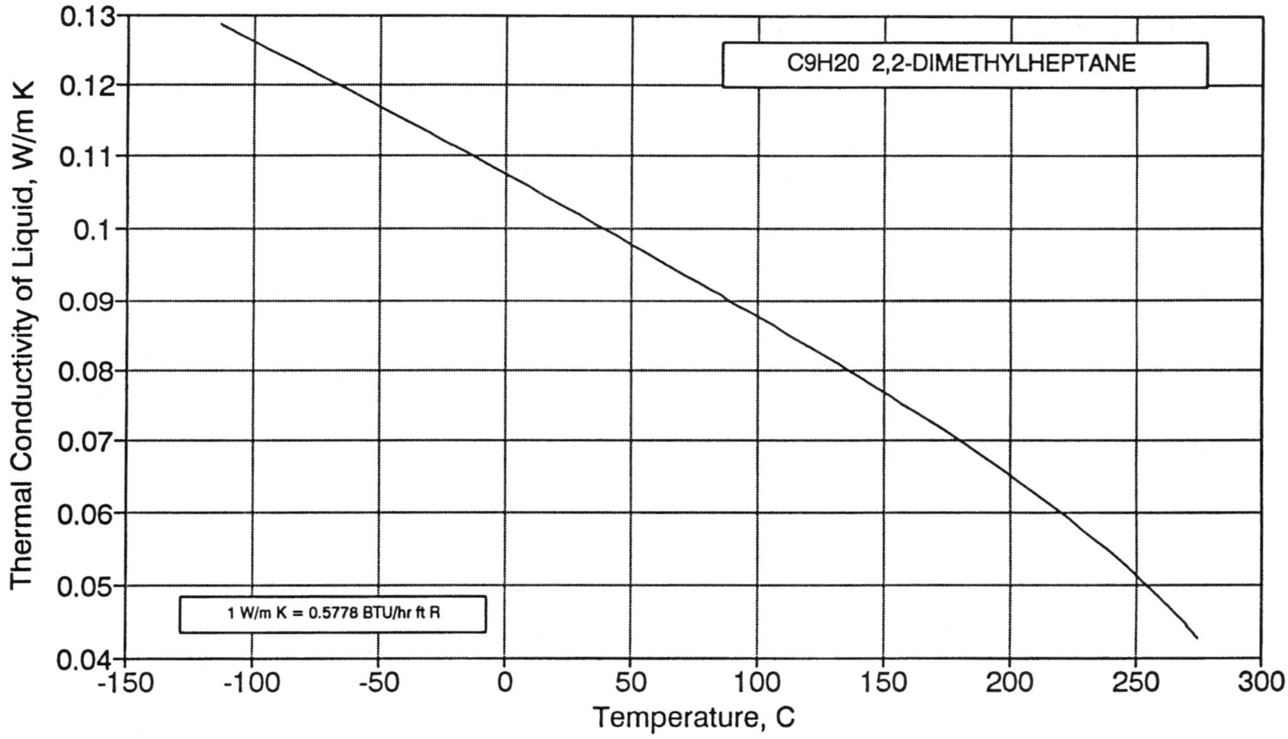
C9H20 2,2-DIMETHYLHEPTANE
1 W/m K = 0.5778 BTU/hr ft R
Thermal Conductivity of Liquid, W/m K
Temperature, C
0.13
0.12
0.11
0.1
0.09
0.08
0.07
0.06
0.05
0.04
-150
-100
-50
0
50
100
150
200
250
300

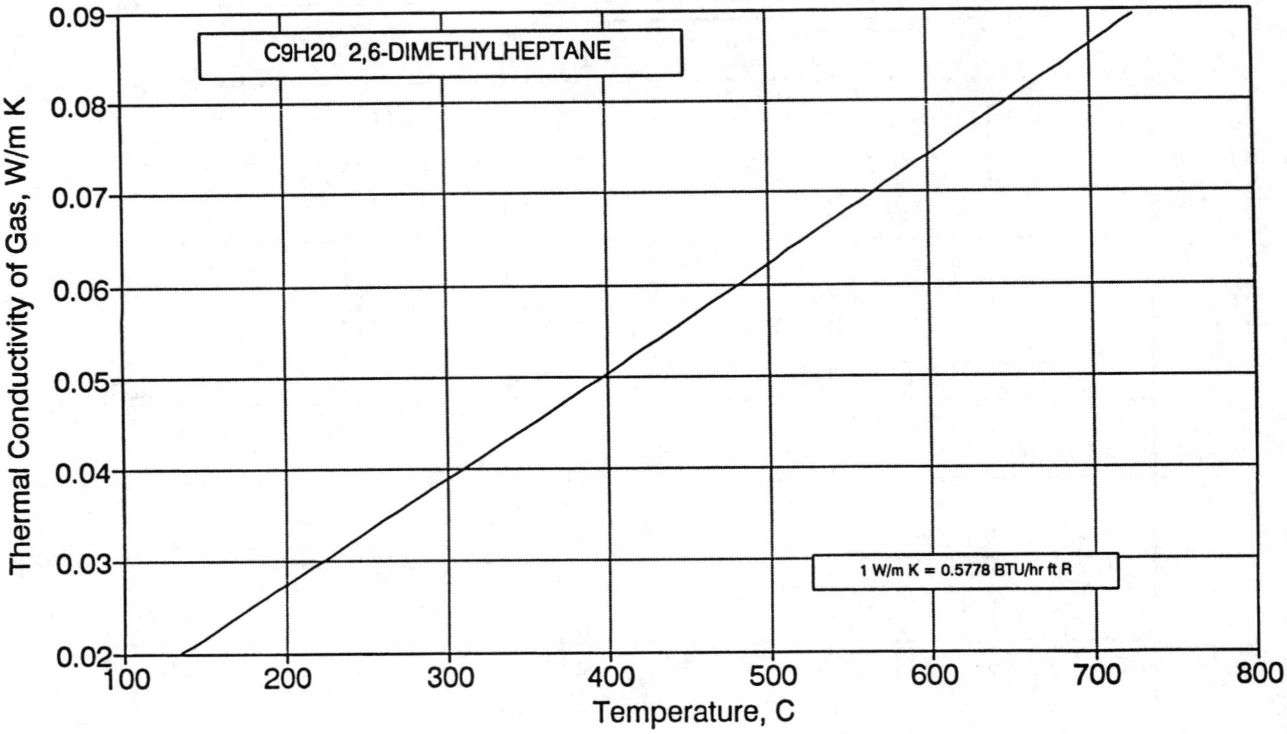
C9H20 2,6-DIMETHYLHEPTANE
1 W/m K = 0.5778 BTU/hr ft R
Thermal Conductivity of Gas, W/m K
0.09
0.08
0.07
0.06
0.05
0.04
0.03
0.02
100
200
300
400
500
600
700
800
Temperature, C

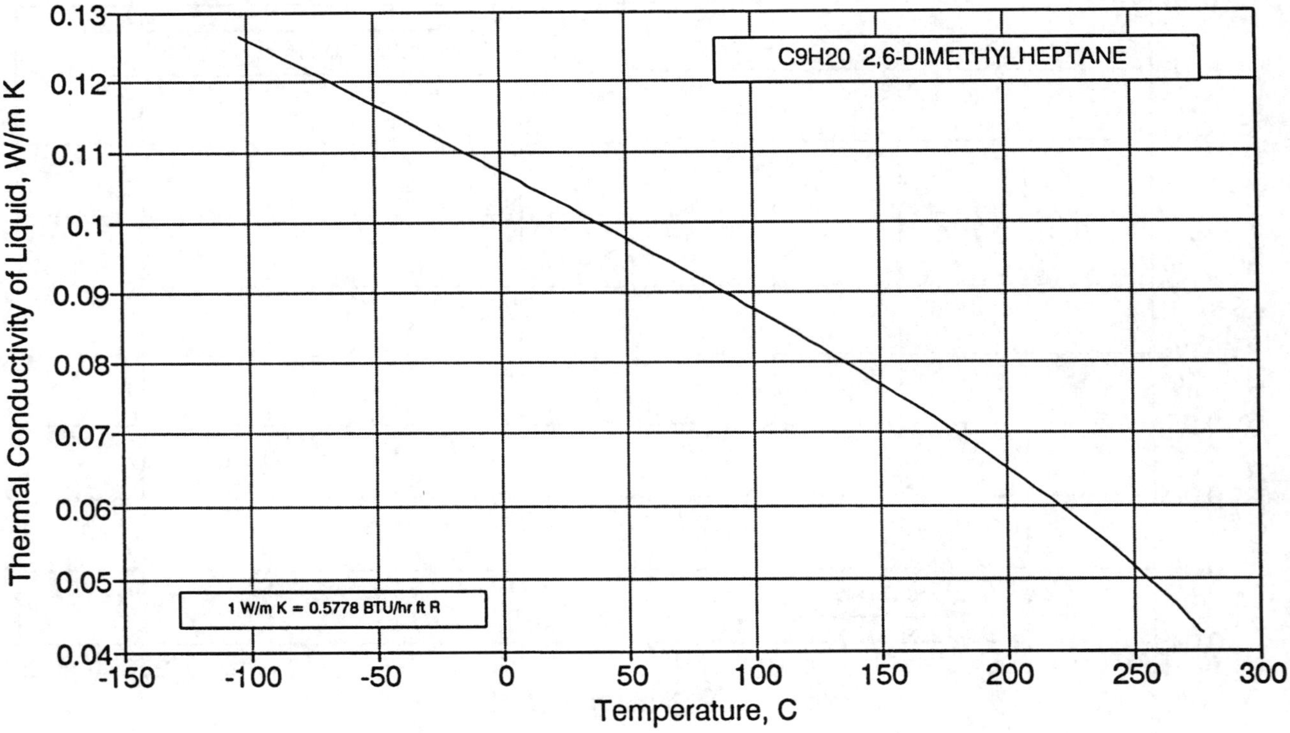
C9H20 2,6-DIMETHYLHEPTANE
1 W/m K = 0.5778 BTU/hr ft R
Thermal Conductivity of Liquid, W/m K
0.13
0.12
0.11
0.1
0.09
0.08
0.07
0.06
0.05
0.04
-150
-100
-50
0
50
100
150
200
250
300
Temperature, C

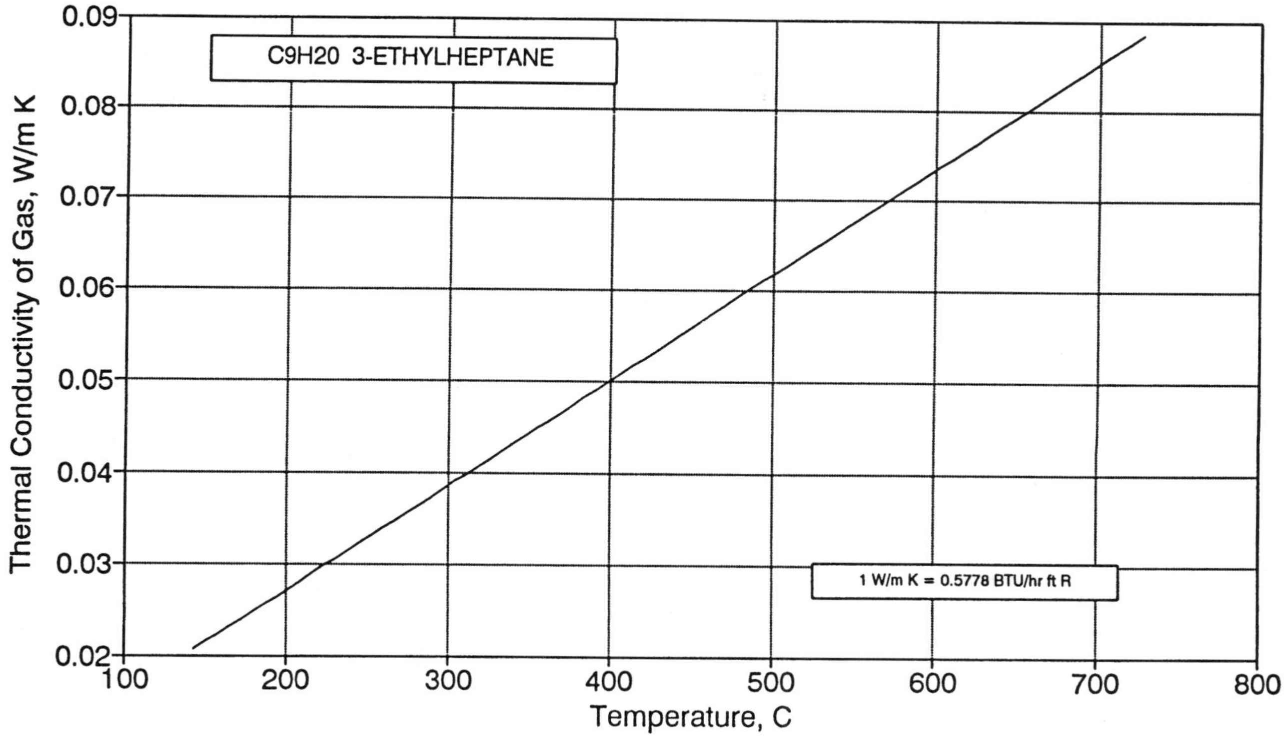

C9H20 3-ETHYLHEPTANE
1 W/m K = 0.5778 BTU/hr ft R
Thermal Conductivity of Gas, W/m K
0.09
0.08
0.07
0.06
0.05
0.04
0.03
0.02
100
200
300
400
500
600
700
800
Temperature, C

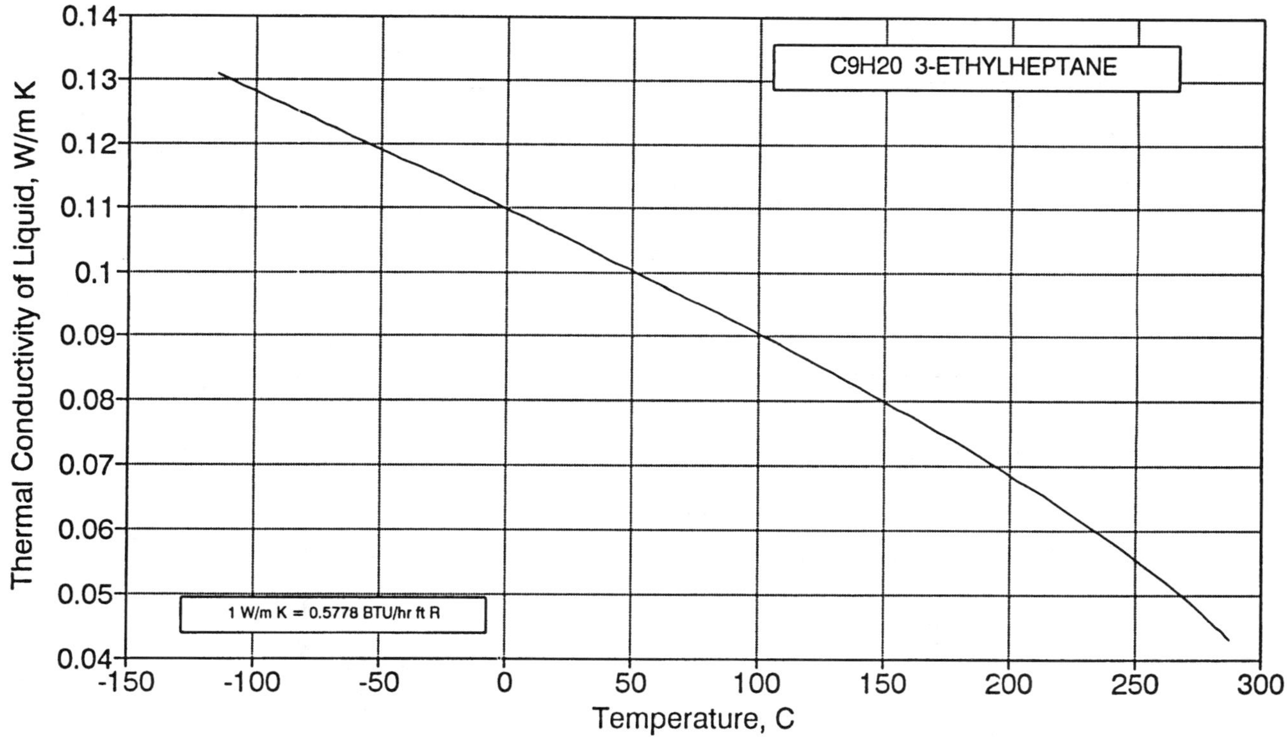

C9H20 3-ETHYLHEPTANE
1 W/m K = 0.5778 BTU/hr ft R
Thermal Conductivity of Liquid, W/m K
0.14
0.13
0.12
0.11
0.1
0.09
0.08
0.07
0.06
0.05
0.04
-150
-100
-50
0
50
100
150
200
250
300
Temperature, C

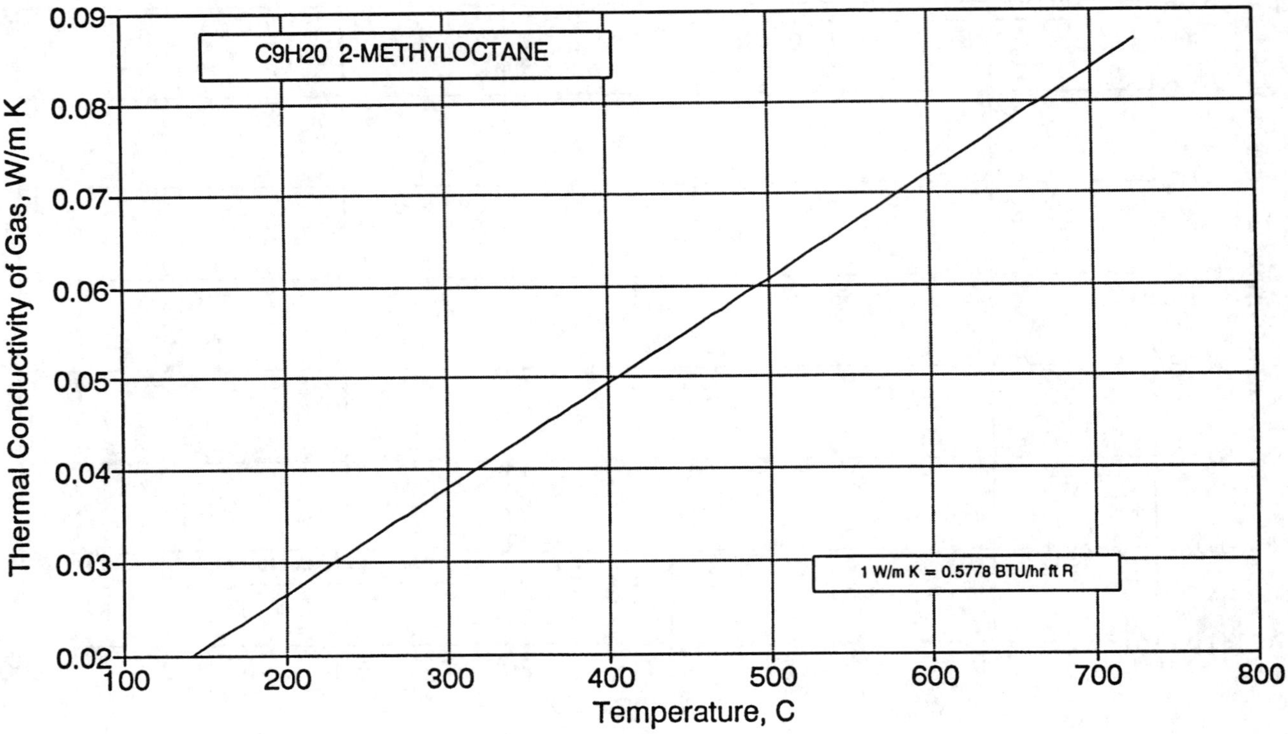
C9H20 2-METHYLOCTANE
1 W/m K = 0.5778 BTU/hr ft R
Thermal Conductivity of Gas, W/m K
0.09
0.08
0.07
0.06
0.05
0.04
0.03
0.02
100
200
300
400
500
600
700
800
Temperature, C

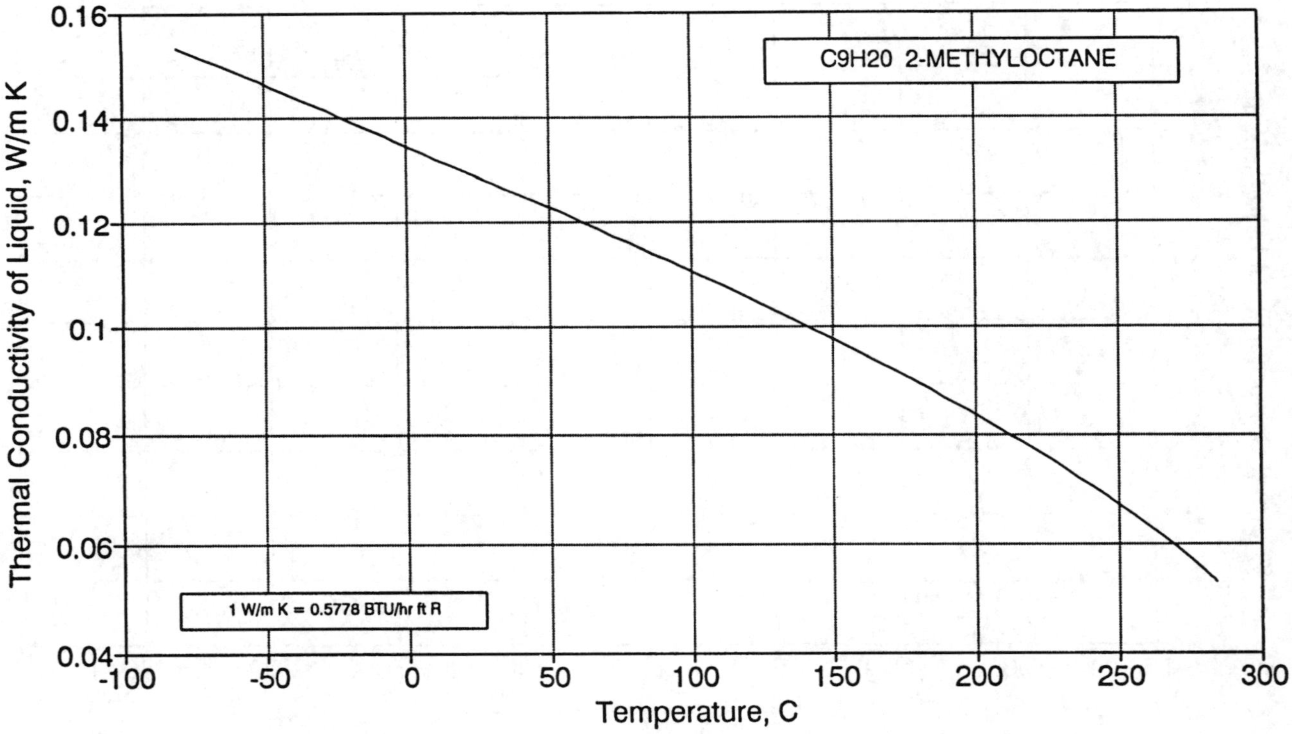
C9H20 2-METHYLOCTANE
1 W/m K = 0.5778 BTU/hr ft R
Thermal Conductivity of Liquid, W/m K
0.16
0.14
0.12
0.1
0.08
0.06
0.04
-100
-50
0
50
100
150
200
250
300
Temperature, C

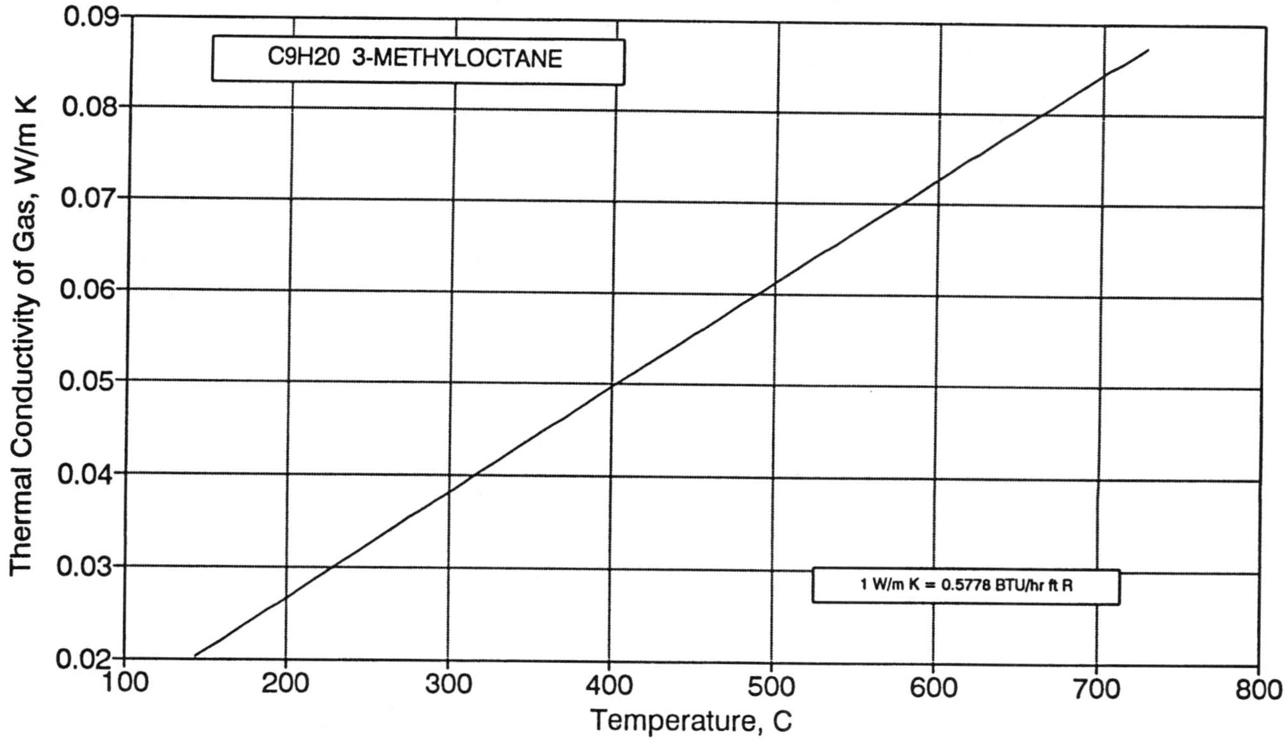
C9H20 3-METHYLOCTANE
1 W/m K = 0.5778 BTU/hr ft R
Thermal Conductivity of Gas, W/m K
0.09
0.08
0.07
0.06
0.05
0.04
0.03
0.02
100
200
300
400
500
600
700
800
Temperature, C

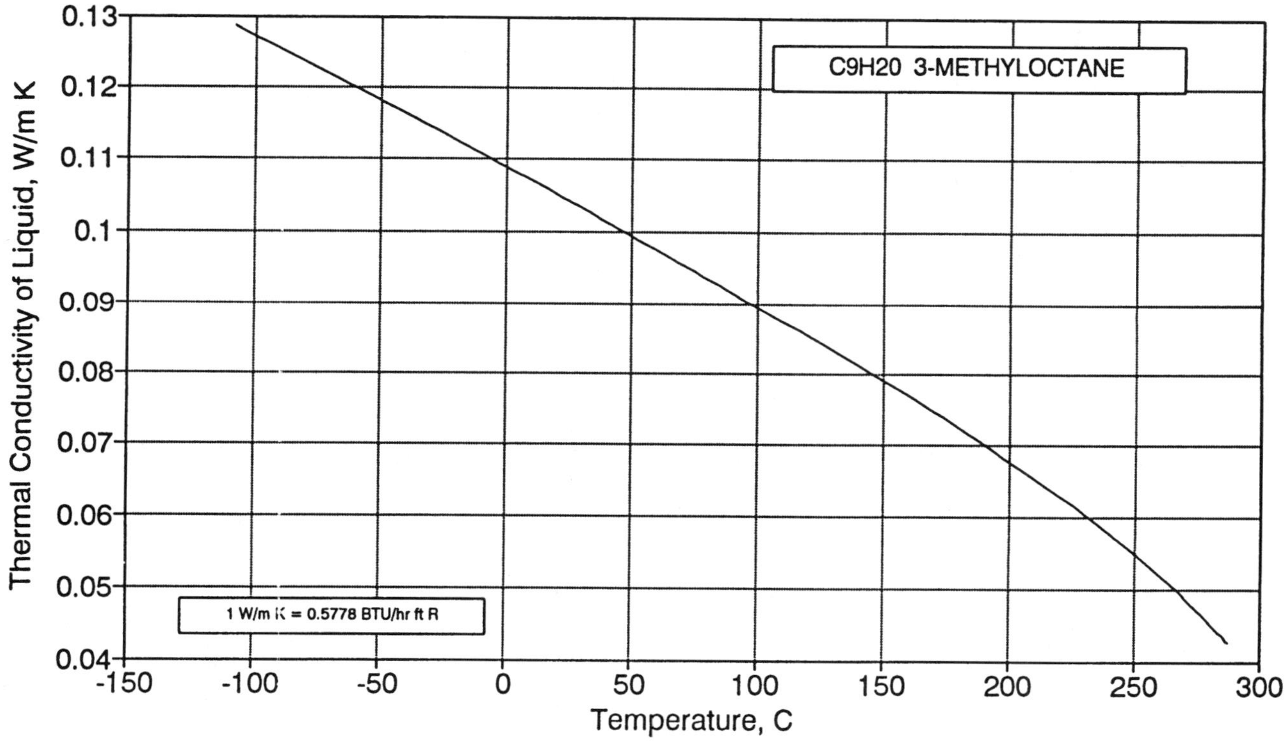
C9H20 3-METHYLOCTANE
1 W/m K = 0.5778 BTU/hr ft R
Thermal Conductivity of Liquid, W/m K
0.13
0.12
0.11
0.1
0.09
0.08
0.07
0.06
0.05
0.04
-150
-100
-50
0
50
100
150
200
250
300
Temperature, C

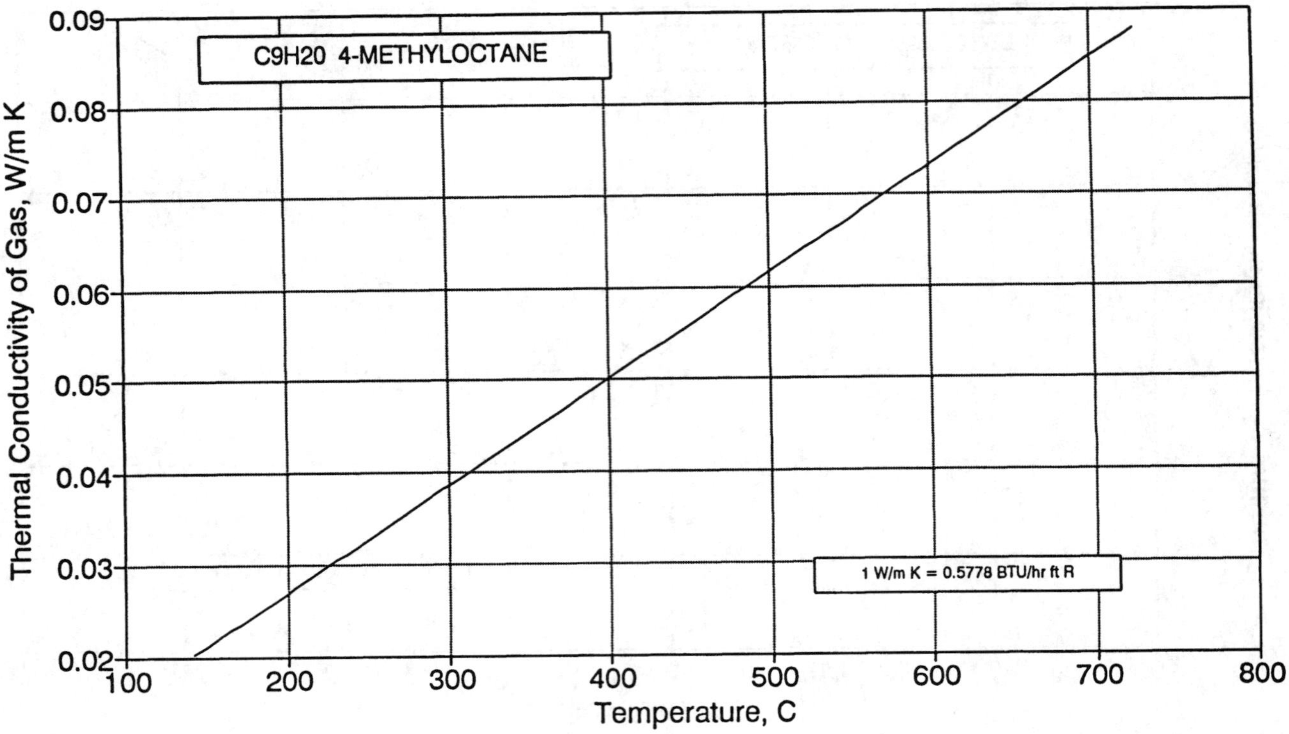
C9H20 4-METHYLOCTANE
1 W/m K = 0.5778 BTU/hr ft R
Thermal Conductivity of Gas, W/m K
0.09
0.08
0.07
0.06
0.05
0.04
0.03
0.02
100
200
300
400
500
600
700
800
Temperature, C

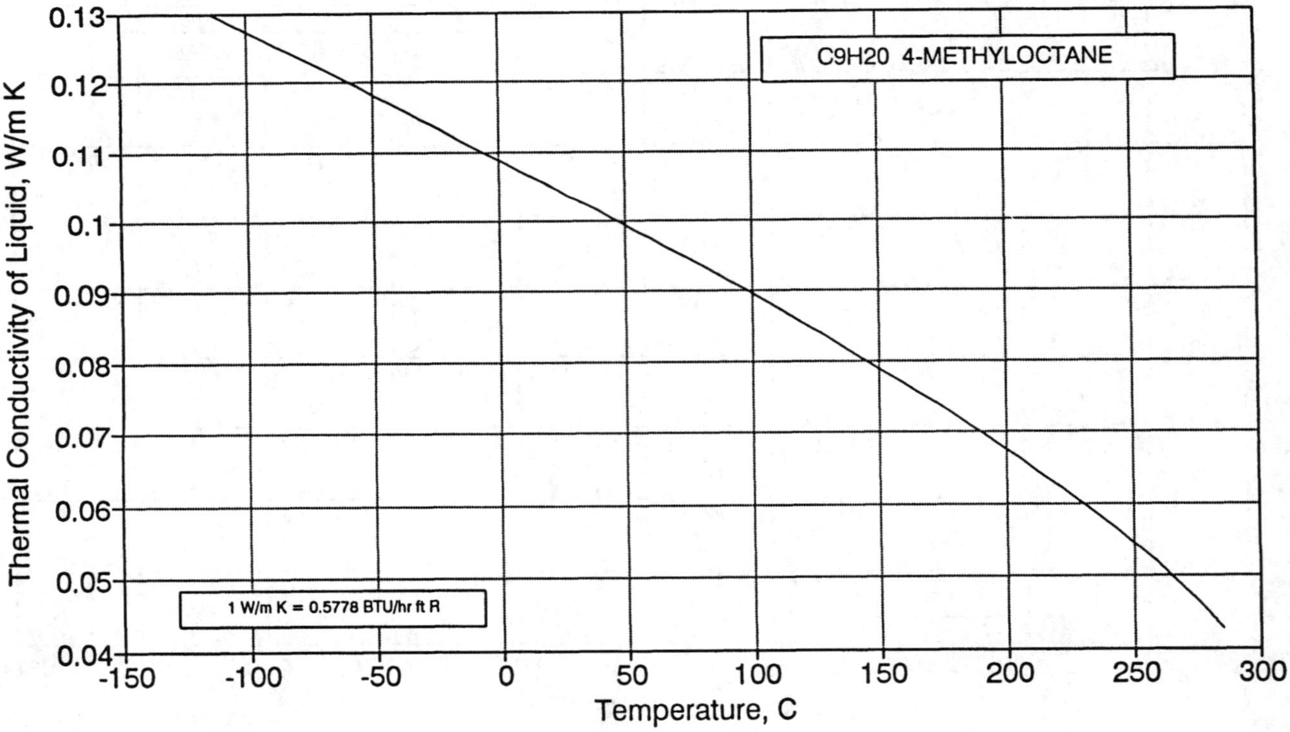
C9H20 4-METHYLOCTANE
1 W/m K = 0.5778 BTU/hr ft R
Thermal Conductivity of Liquid, W/m K
0.13
0.12
0.11
0.1
0.09
0.08
0.07
0.06
0.05
0.04
-150
-100
-50
0
50
100
150
200
250
300
Temperature, C

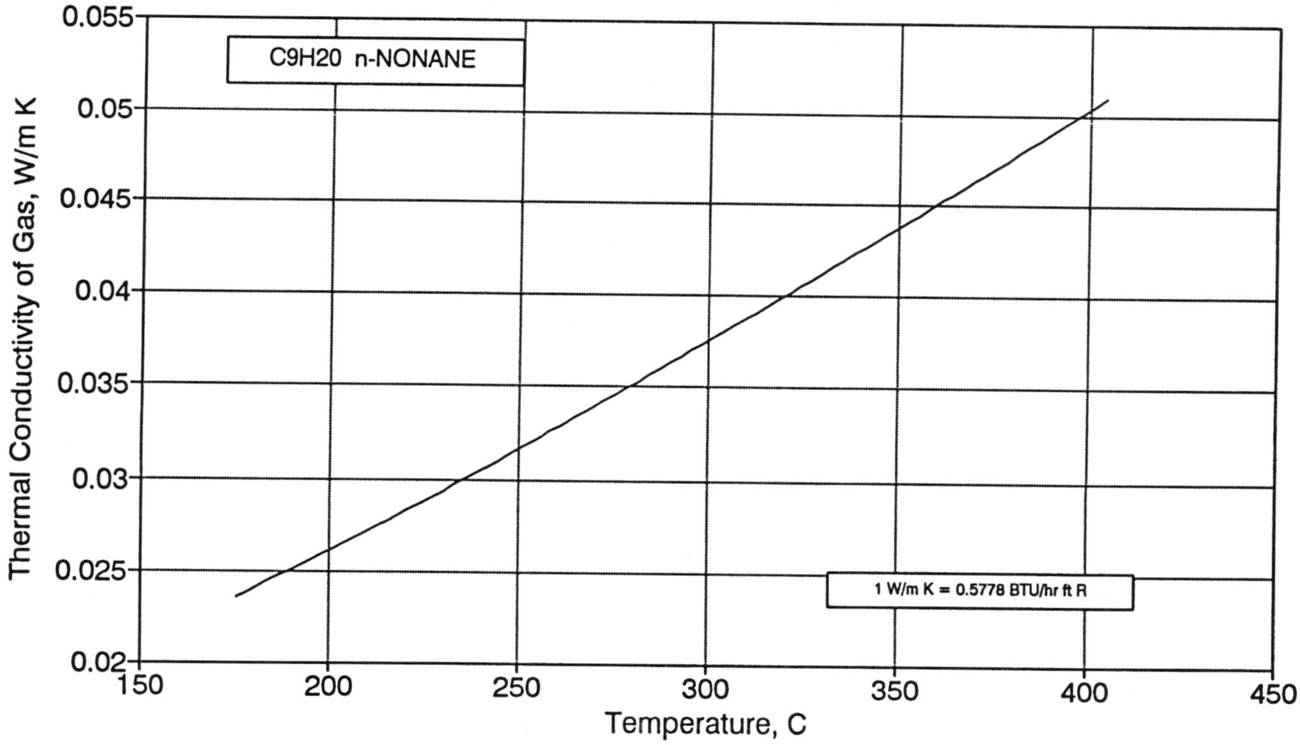
C9H20 n-NONANE
1 W/m K = 0.5778 BTU/hr ft R
Thermal Conductivity of Gas, W/m K
0.055
0.05
0.045
0.04
0.035
0.03
0.025
0.02
150
200
250
300
350
400
450
Temperature, C

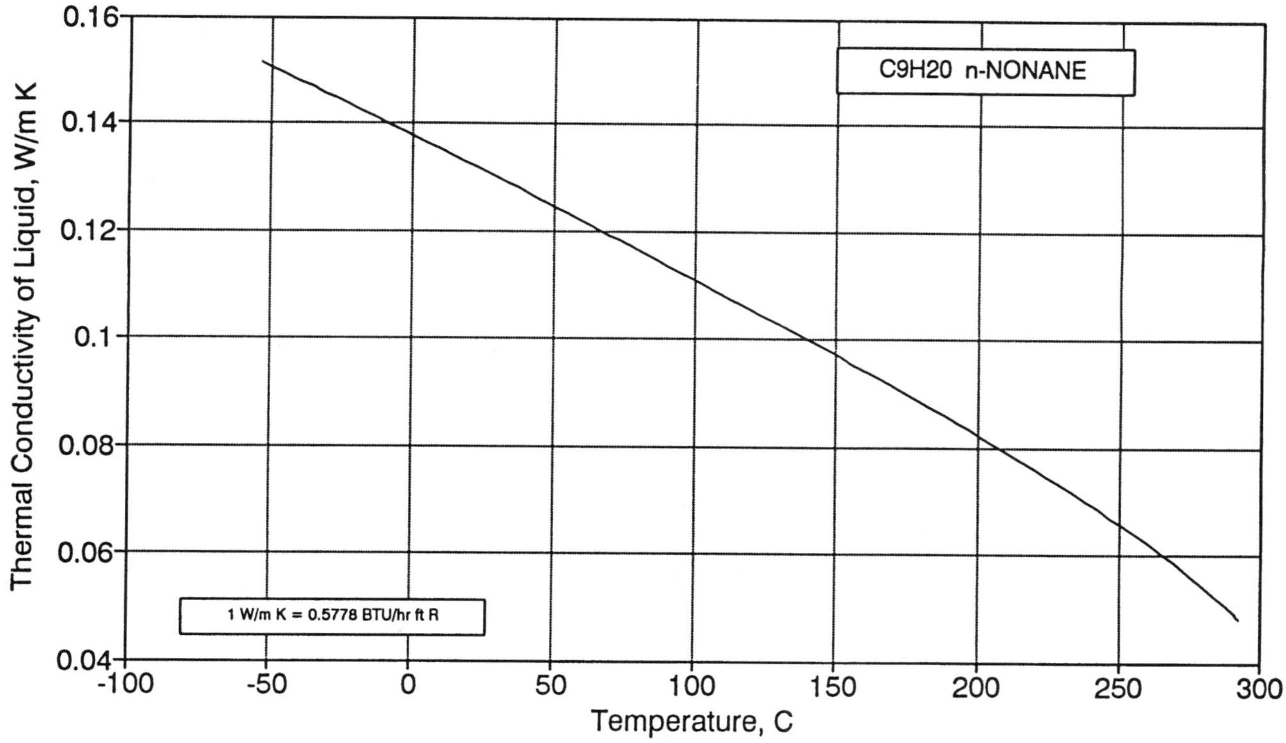
C9H20 n-NONANE
1 W/m K = 0.5778 BTU/hr ft R
Thermal Conductivity of Liquid, W/m K
0.16
0.14
0.12
0.1
0.08
0.06
0.04
-100
-50
0
50
100
150
200
250
300
Temperature, C

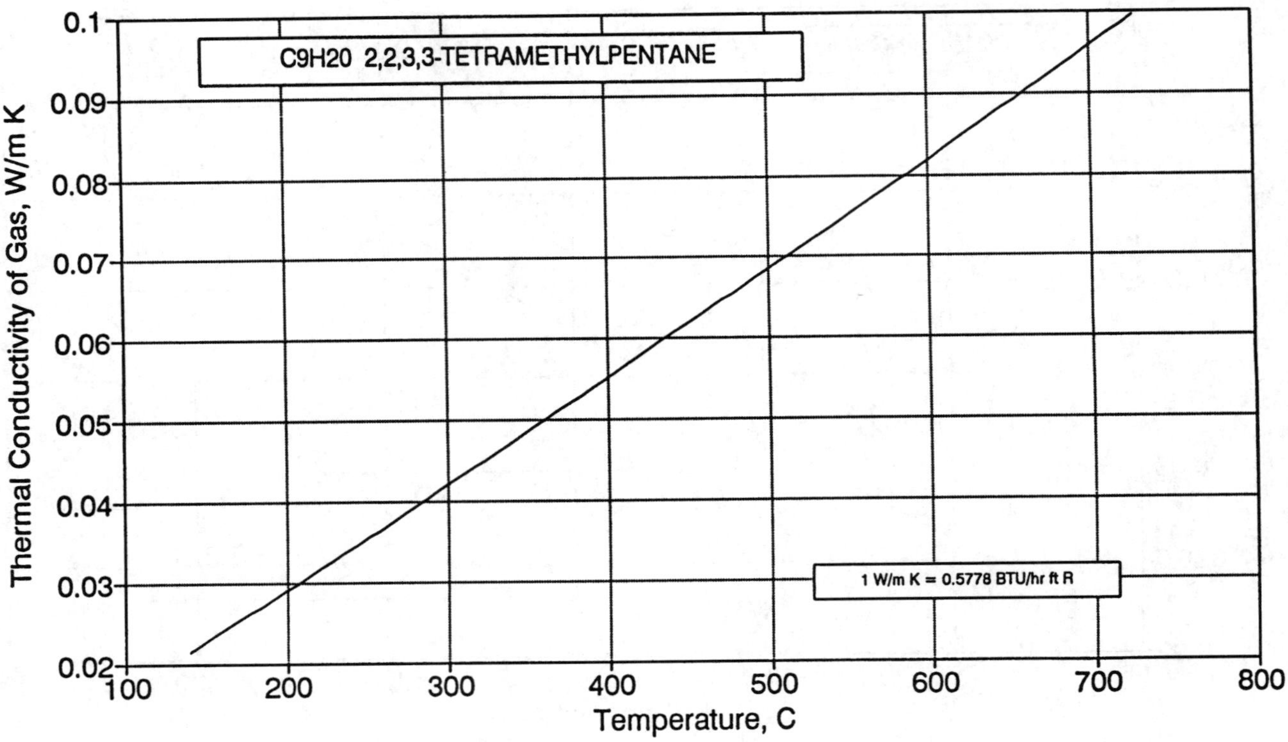
C9H20 2,2,3,3-TETRAMETHYLPENTANE
1 W/m K = 0.5778 BTU/hr ft R
Thermal Conductivity of Gas, W/m K
0.1
0.09
0.08
0.07
0.06
0.05
0.04
0.03
0.02
100
200
300
400
500
600
700
800
Temperature, C

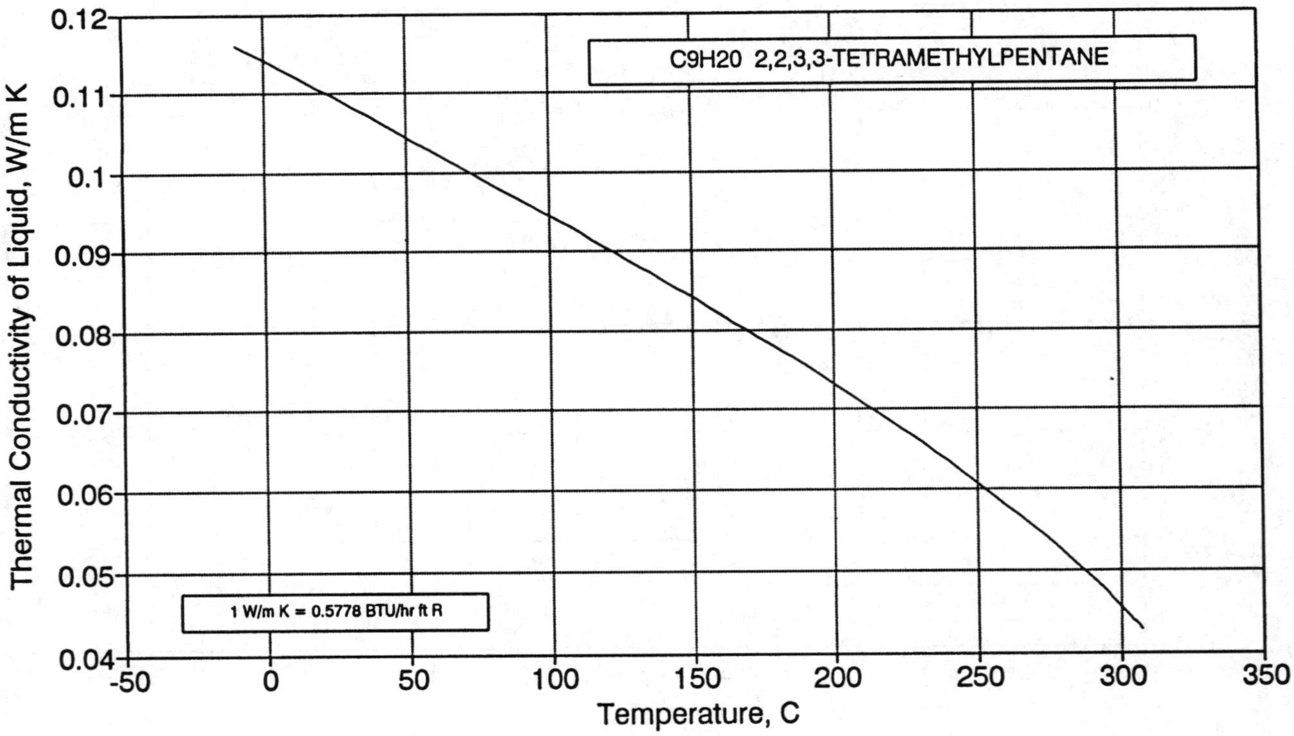
C9H20 2,2,3,3-TETRAMETHYLPENTANE
1 W/m K = 0.5778 BTU/hr ft R
Thermal Conductivity of Liquid, W/m K
0.12
0.11
0.1
0.09
0.08
0.07
0.06
0.05
0.04
-50
0
50
100
150
200
250
300
350
Temperature, C

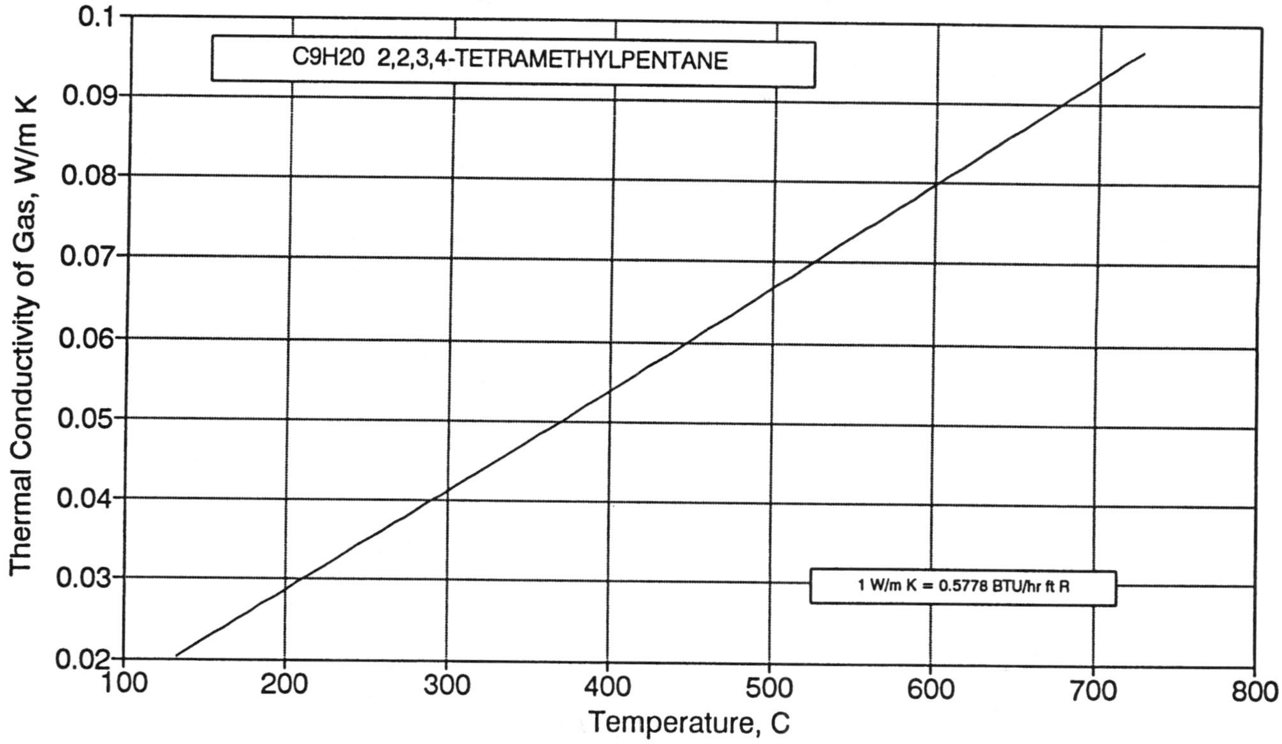
C9H20 2,2,3,4-TETRAMETHYLPENTANE
1 W/m K = 0.5778 BTU/hr ft R
Thermal Conductivity of Gas, W/m K
0.1
0.09
0.08
0.07
0.06
0.05
0.04
0.03
0.02
100
200
300
400
500
600
700
800
Temperature, C

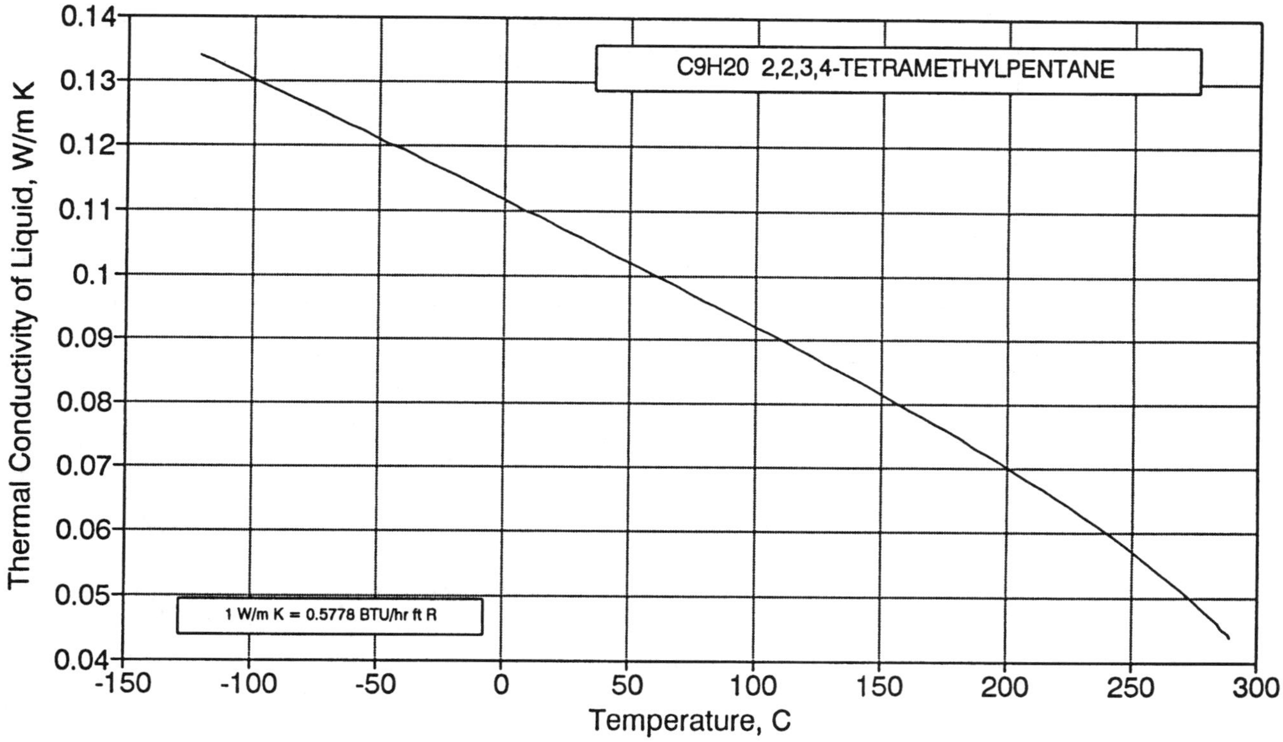
C9H20 2,2,3,4-TETRAMETHYLPENTANE
1 W/m K = 0.5778 BTU/hr ft R
Thermal Conductivity of Liquid, W/m K
0.14
0.13
0.12
0.11
0.1
0.09
0.08
0.07
0.06
0.05
0.04
-150
-100
-50
0
50
100
150
200
250
300
Temperature, C

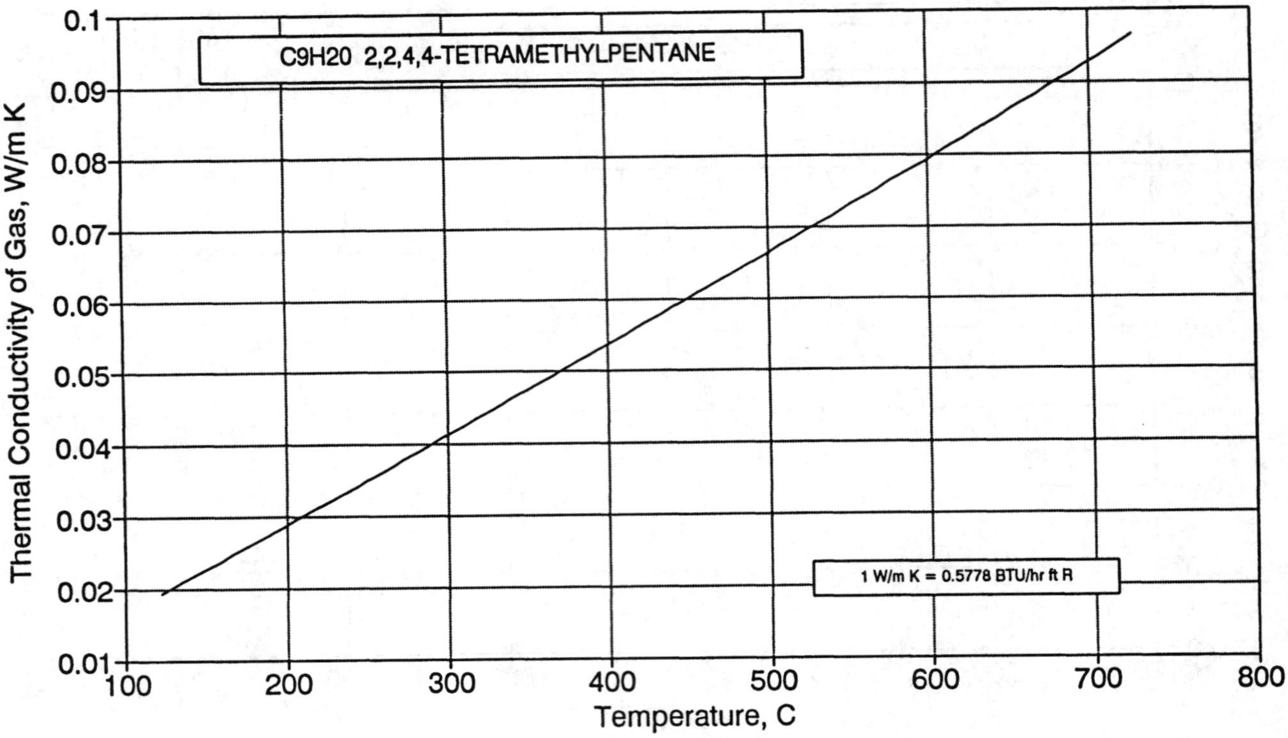
C9H20 2,2,4,4-TETRAMETHYLPENTANE
Thermal Conductivity of Gas, W/m K
Temperature, C
1 W/m K = 0.5778 BTU/hr ft R
0.1
0.09
0.08
0.07
0.06
0.05
0.04
0.03
0.02
0.01
100
200
300
400
500
600
700
800

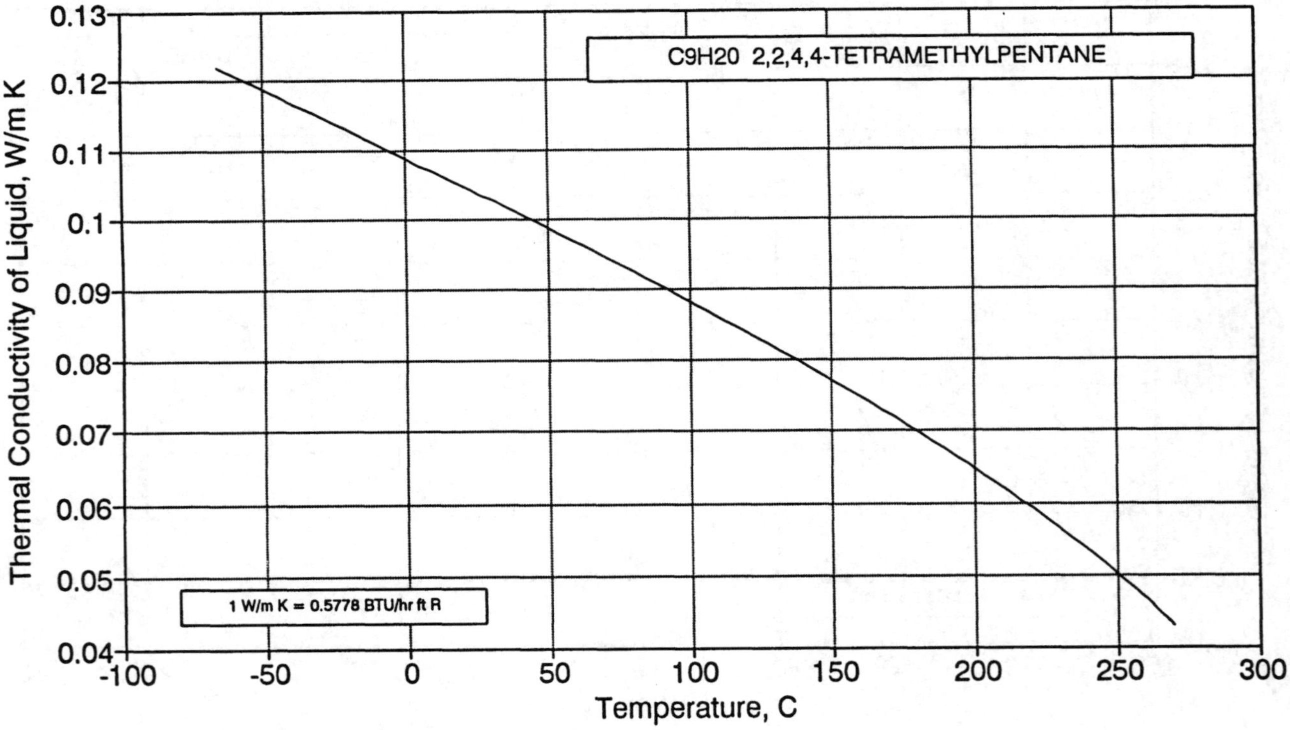
C9H20 2,2,4,4-TETRAMETHYLPENTANE
Thermal Conductivity of Liquid, W/m K
Temperature, C
1 W/m K = 0.5778 BTU/hr ft R
0.13
0.12
0.11
0.1
0.09
0.08
0.07
0.06
0.05
0.04
-100
-50
0
50
100
150
200
250
300

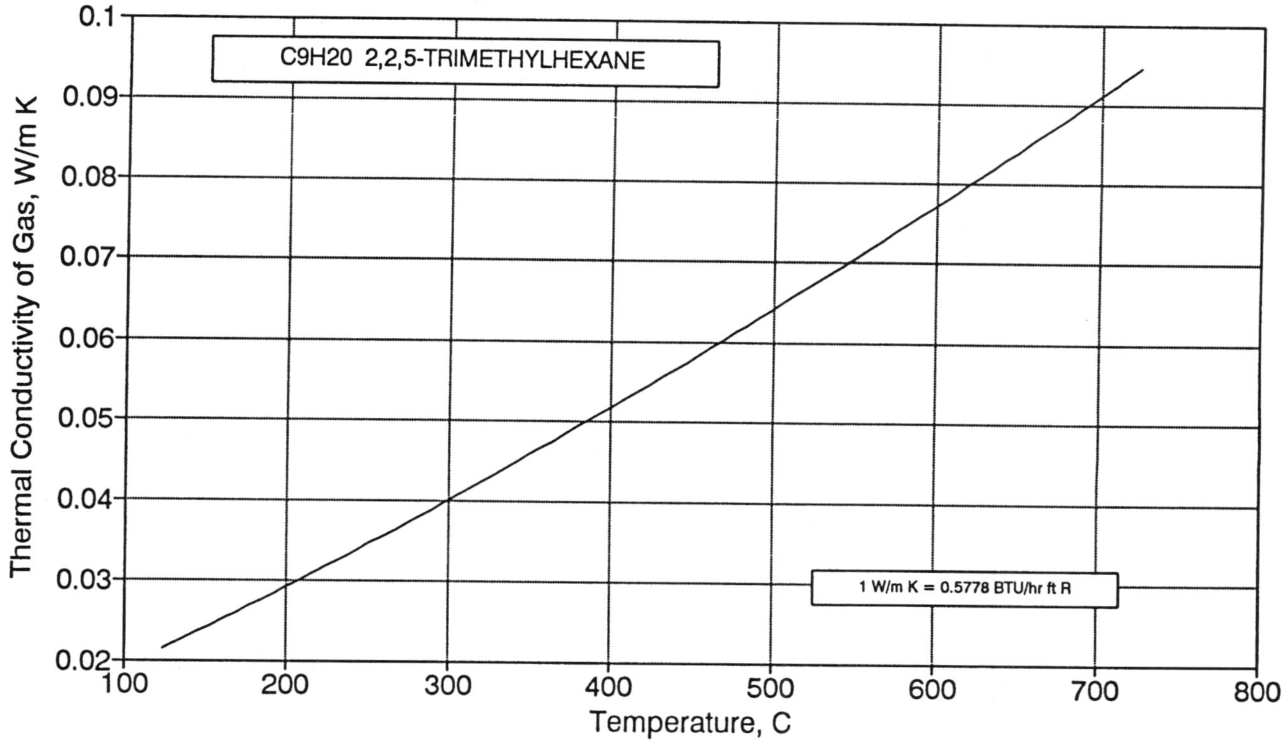
C9H20 2,2,5-TRIMETHYLHEXANE
1 W/m K = 0.5778 BTU/hr ft R
Thermal Conductivity of Gas, W/m K
0.1
0.09
0.08
0.07
0.06
0.05
0.04
0.03
0.02
100
200
300
400
500
600
700
800
Temperature, C

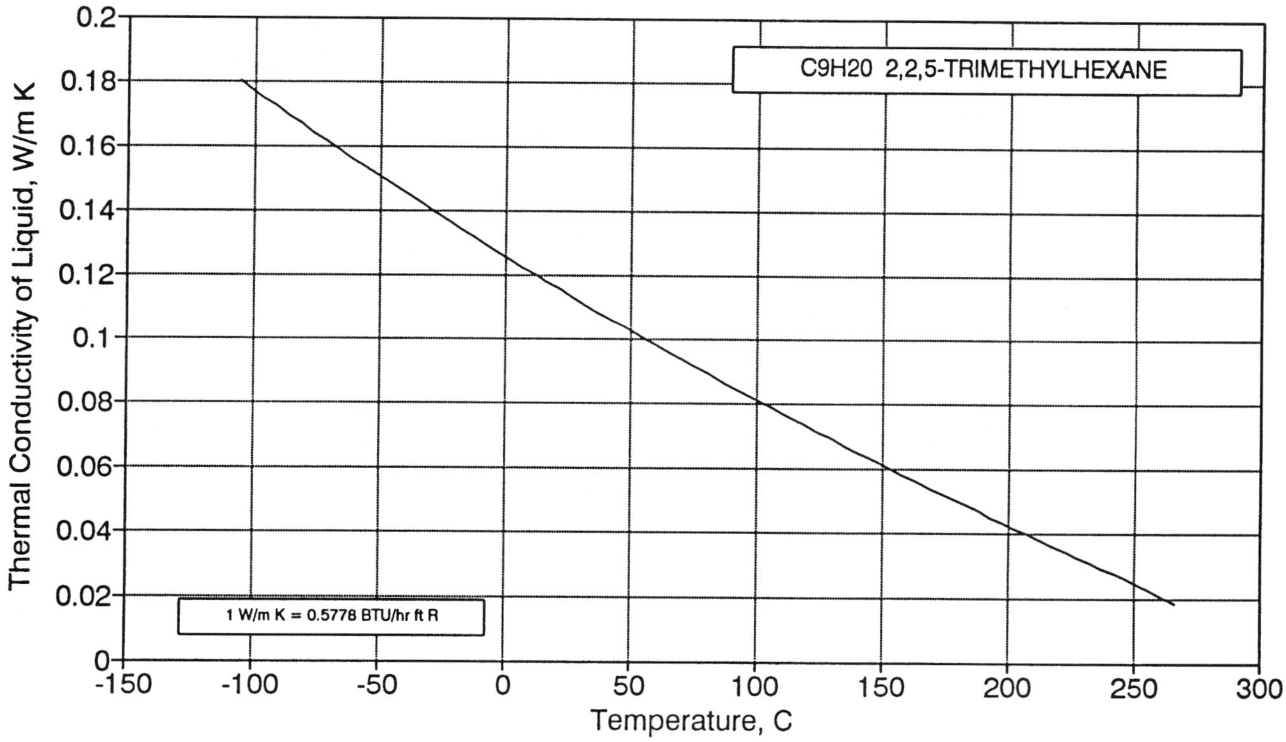
C9H20 2,2,5-TRIMETHYLHEXANE
1 W/m K = 0.5778 BTU/hr ft R
Thermal Conductivity of Liquid, W/m K
0.2
0.18
0.16
0.14
0.12
0.1
0.08
0.06
0.04
0.02
0
-150
-100
-50
0
50
100
150
200
250
300
Temperature, C

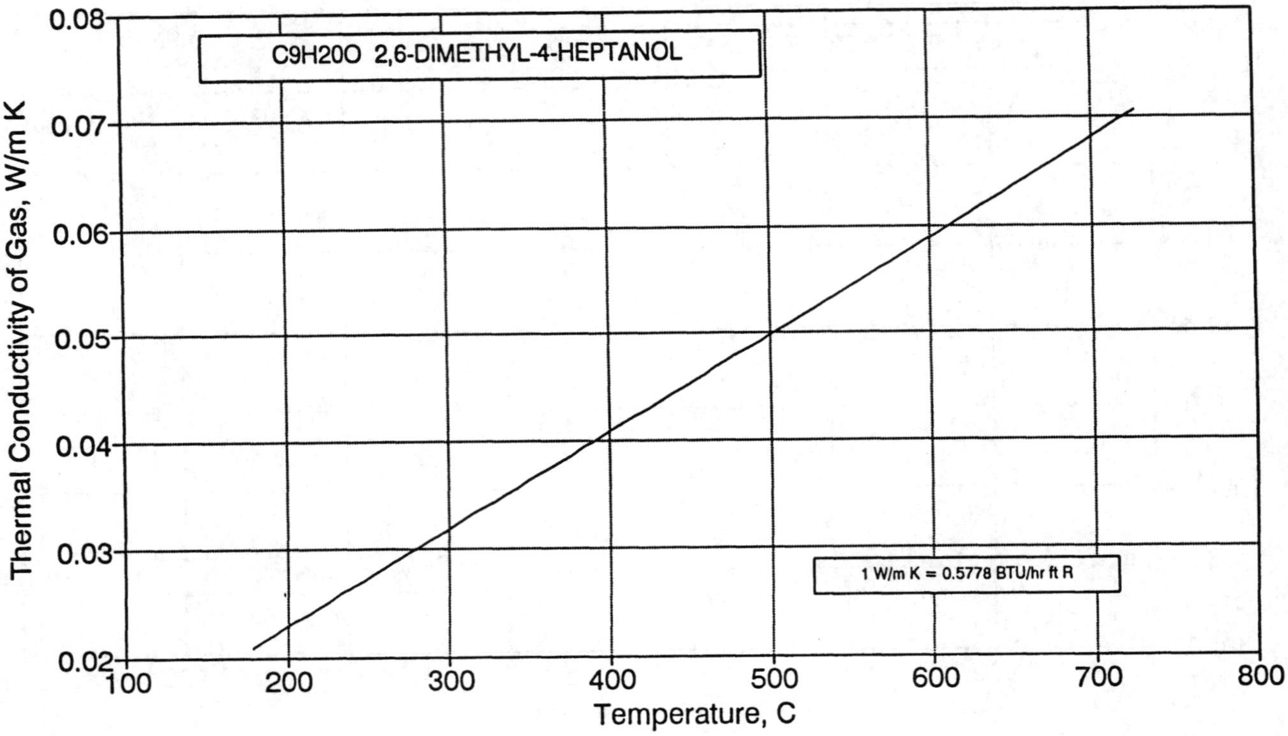
C9H20O 2,6-DIMETHYL-4-HEPTANOL
Thermal Conductivity of Gas, W/m K
0.08
0.07
0.06
0.05
0.04
0.03
0.02
100
200
300
400
500
600
700
800
Temperature, C
1 W/m K = 0.5778 BTU/hr ft R

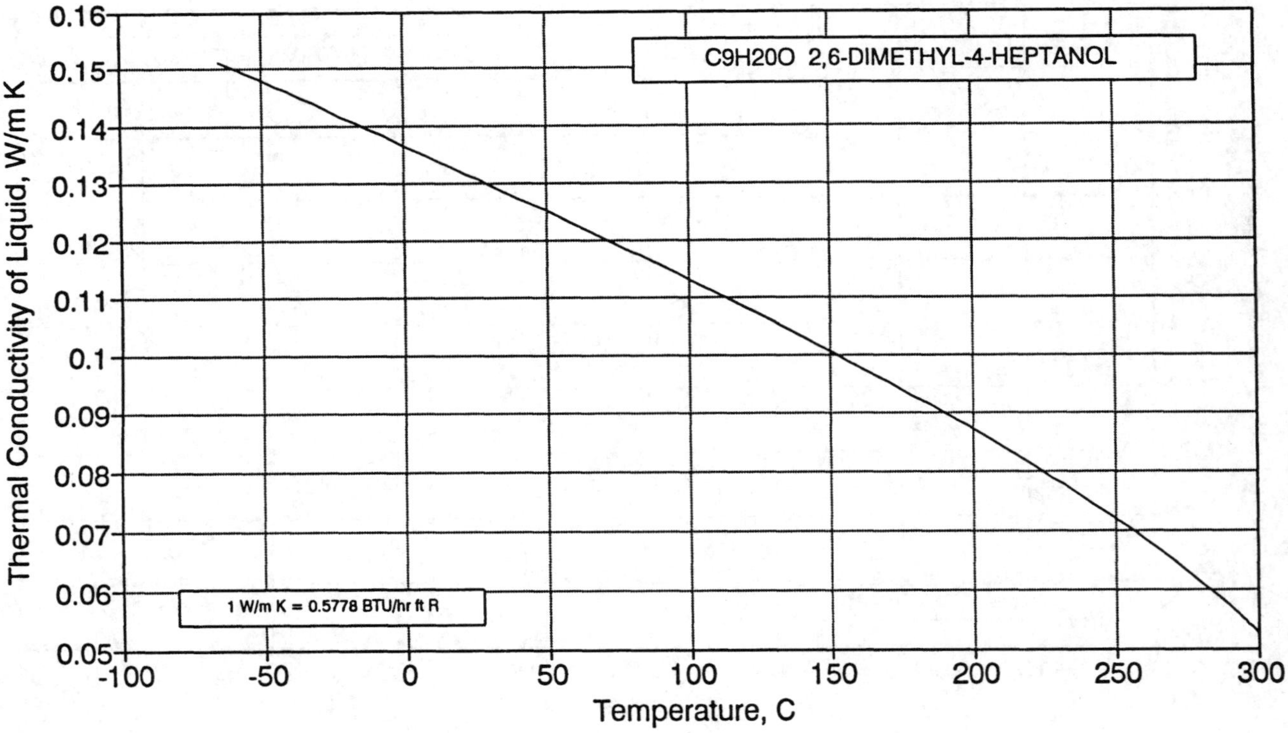
C9H20O 2,6-DIMETHYL-4-HEPTANOL
Thermal Conductivity of Liquid, W/m K
0.16
0.15
0.14
0.13
0.12
0.11
0.1
0.09
0.08
0.07
0.06
0.05
-100
-50
0
50
100
150
200
250
300
Temperature, C
1 W/m K = 0.5778 BTU/hr ft R

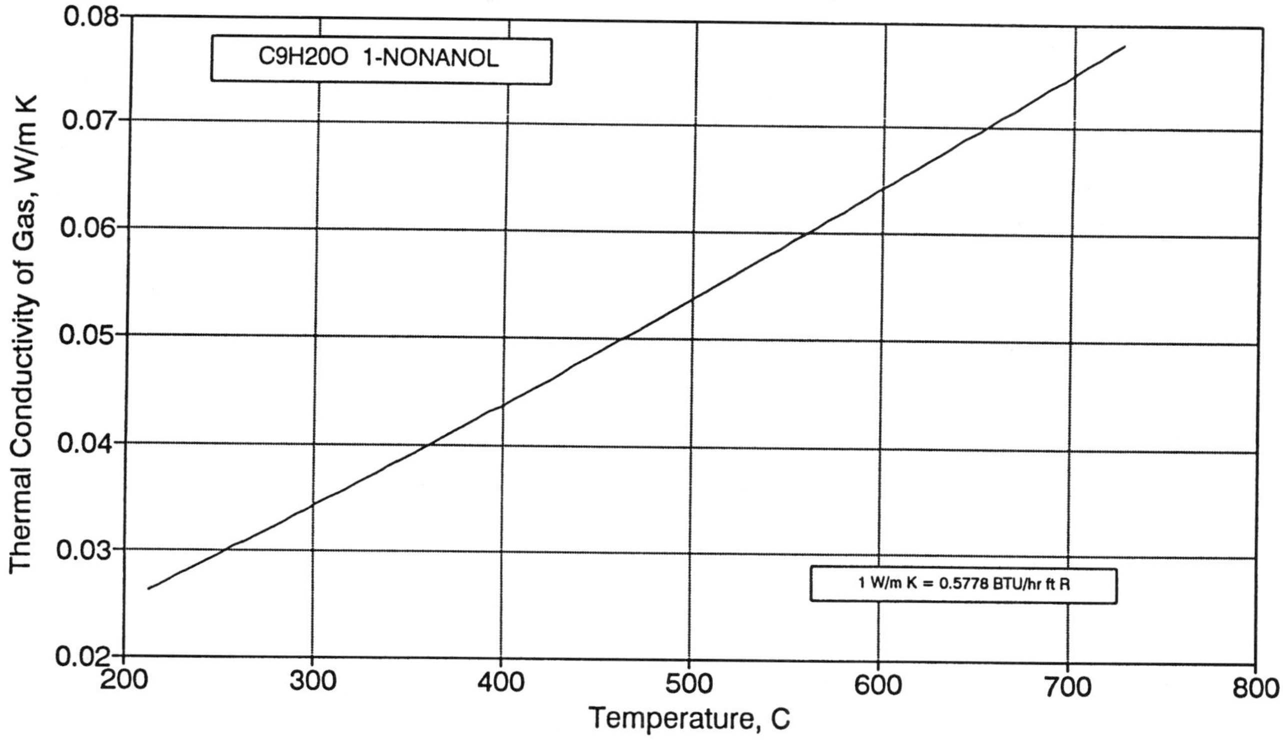
C9H20O 1-NONANOL
Thermal Conductivity of Gas, W/m K
0.08
0.07
0.06
0.05
0.04
0.03
0.02
1 W/m K = 0.5778 BTU/hr ft R
200
300
400
500
600
700
800
Temperature, C

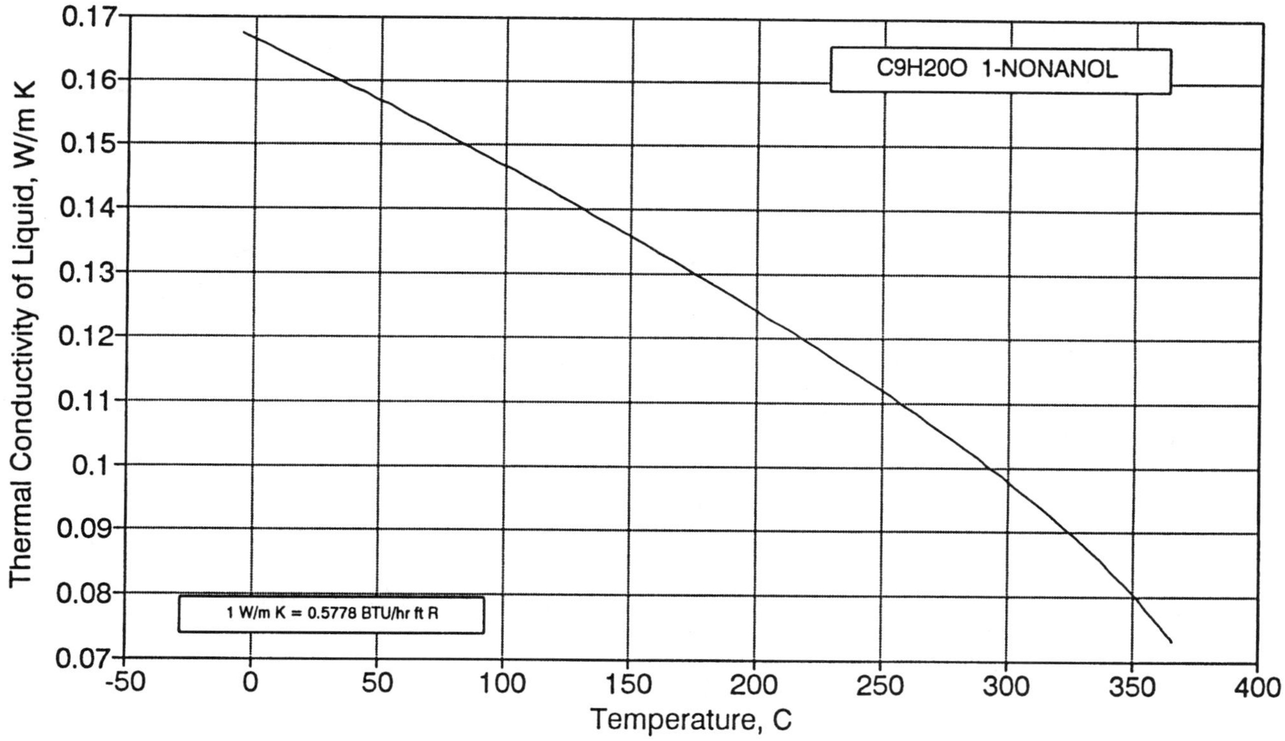
C9H20O 1-NONANOL
Thermal Conductivity of Liquid, W/m K
0.17
0.16
0.15
0.14
0.13
0.12
0.11
0.1
0.09
0.08
0.07
1 W/m K = 0.5778 BTU/hr ft R
-50
0
50
100
150
200
250
300
350
400
Temperature, C

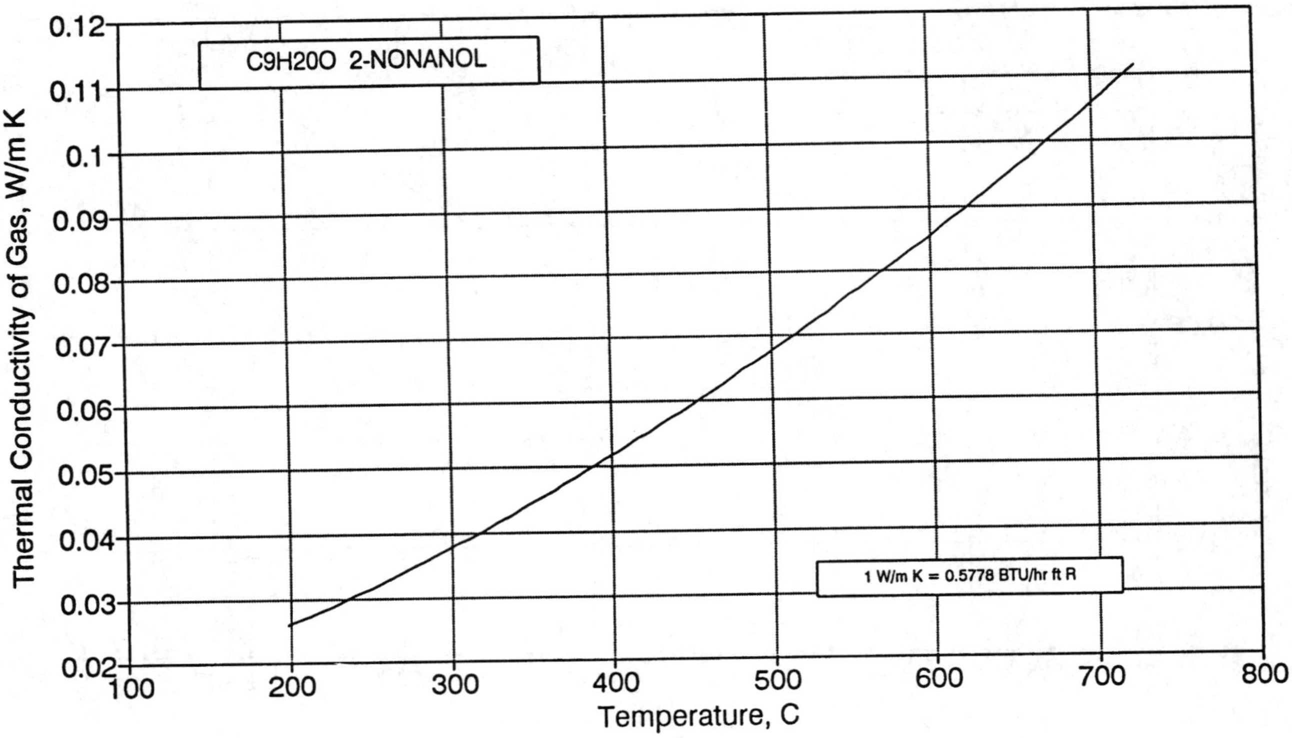
C9H20O 2-NONANOL
1 W/m K = 0.5778 BTU/hr ft R
Thermal Conductivity of Gas, W/m K
0.12
0.11
0.1
0.09
0.08
0.07
0.06
0.05
0.04
0.03
0.02
100
200
300
400
500
600
700
800
Temperature, C

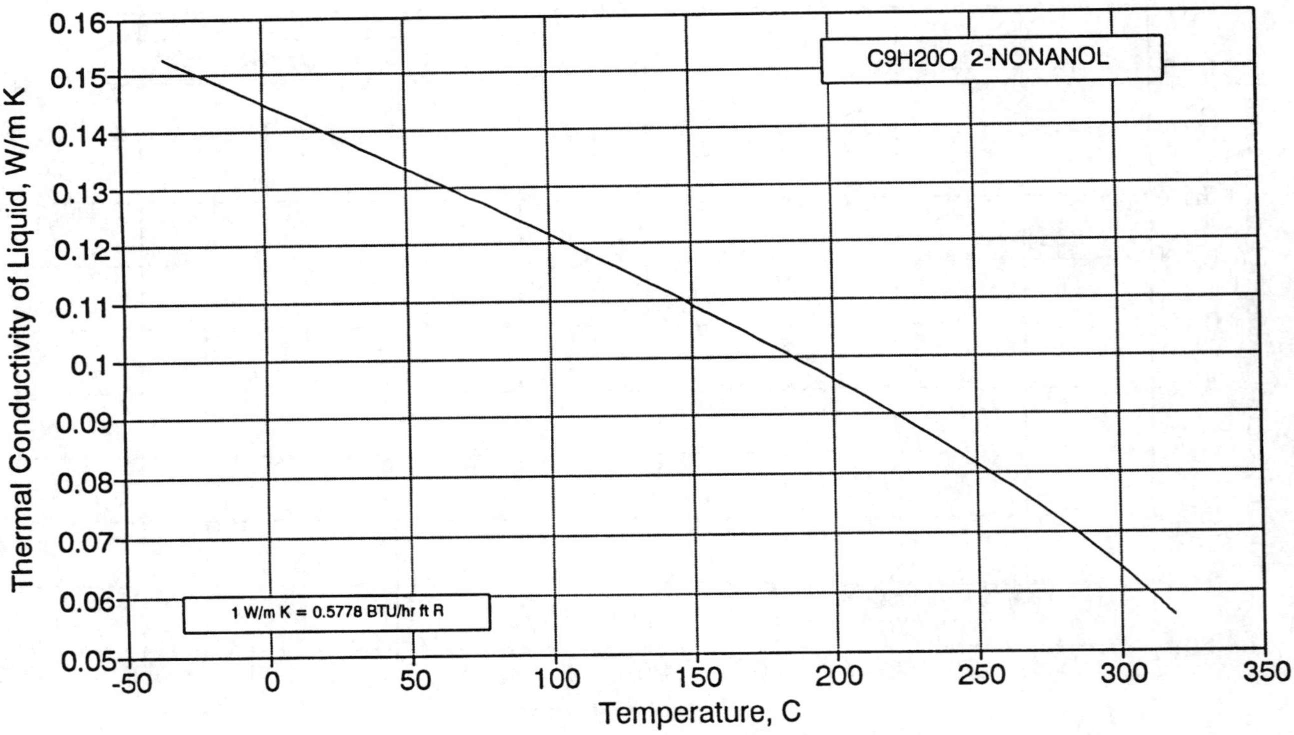
C9H20O 2-NONANOL
1 W/m K = 0.5778 BTU/hr ft R
Thermal Conductivity of Liquid, W/m K
0.16
0.15
0.14
0.13
0.12
0.11
0.1
0.09
0.08
0.07
0.06
0.05
-50
0
50
100
150
200
250
300
350
Temperature, C

C9H20S n-NONYL MERCAPTAN

Gas thermal conductivity data are not available.

TF, K = 253.05

TB, K = 492.95 (experimental)

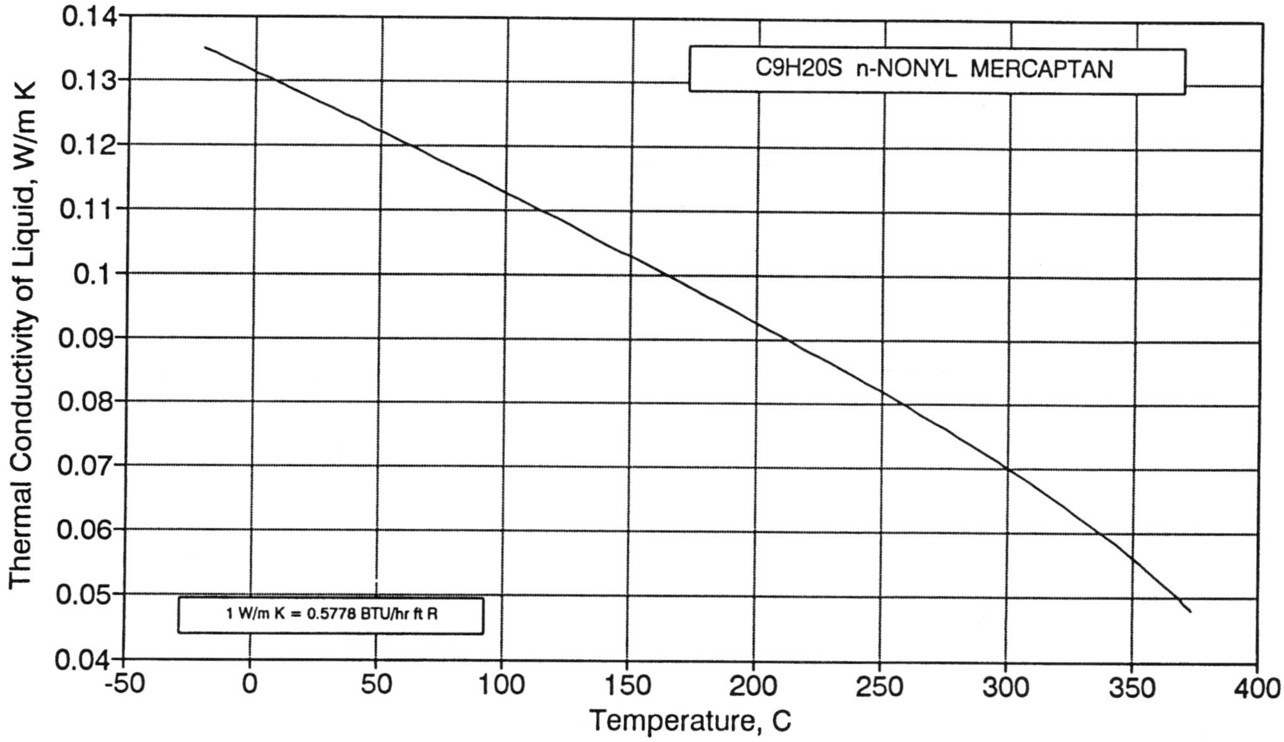

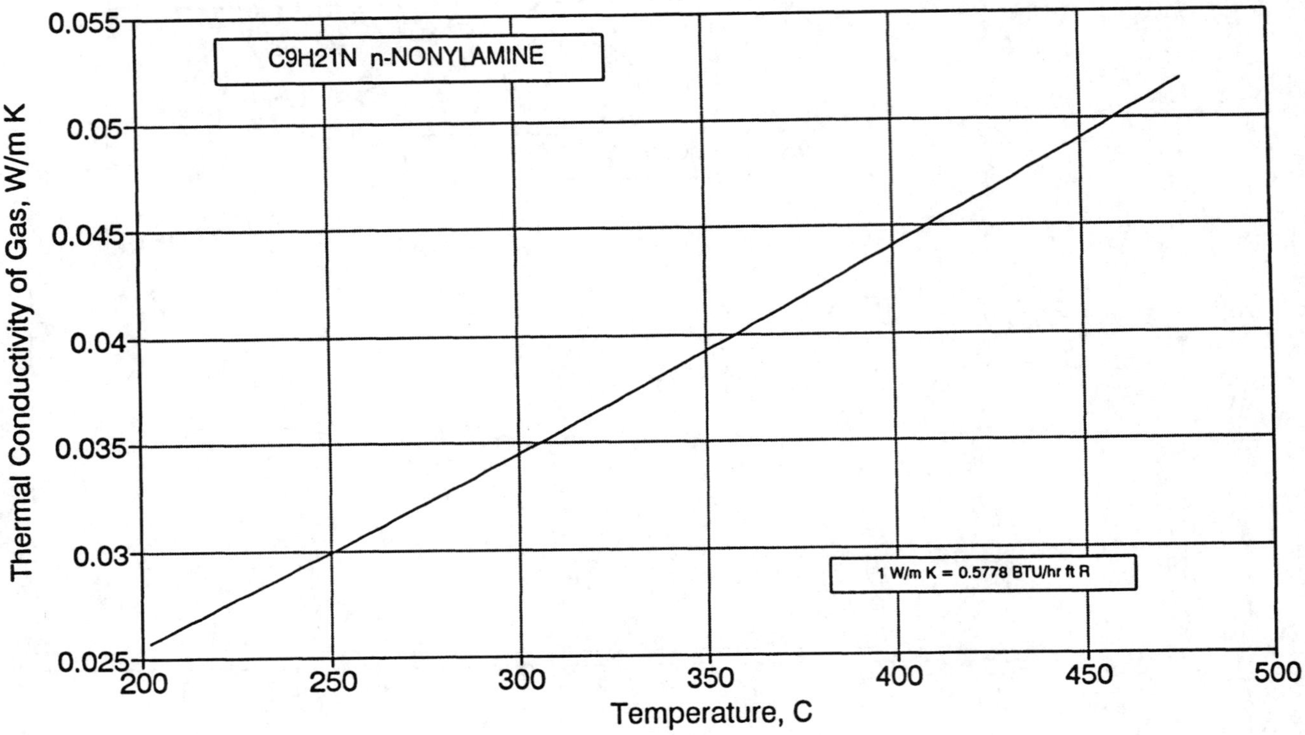
C9H21N n-NONYLAMINE
1 W/m K = 0.5778 BTU/hr ft R
Thermal Conductivity of Gas, W/m K
0.055
0.05
0.045
0.04
0.035
0.03
0.025
200
250
300
350
400
450
500
Temperature, C

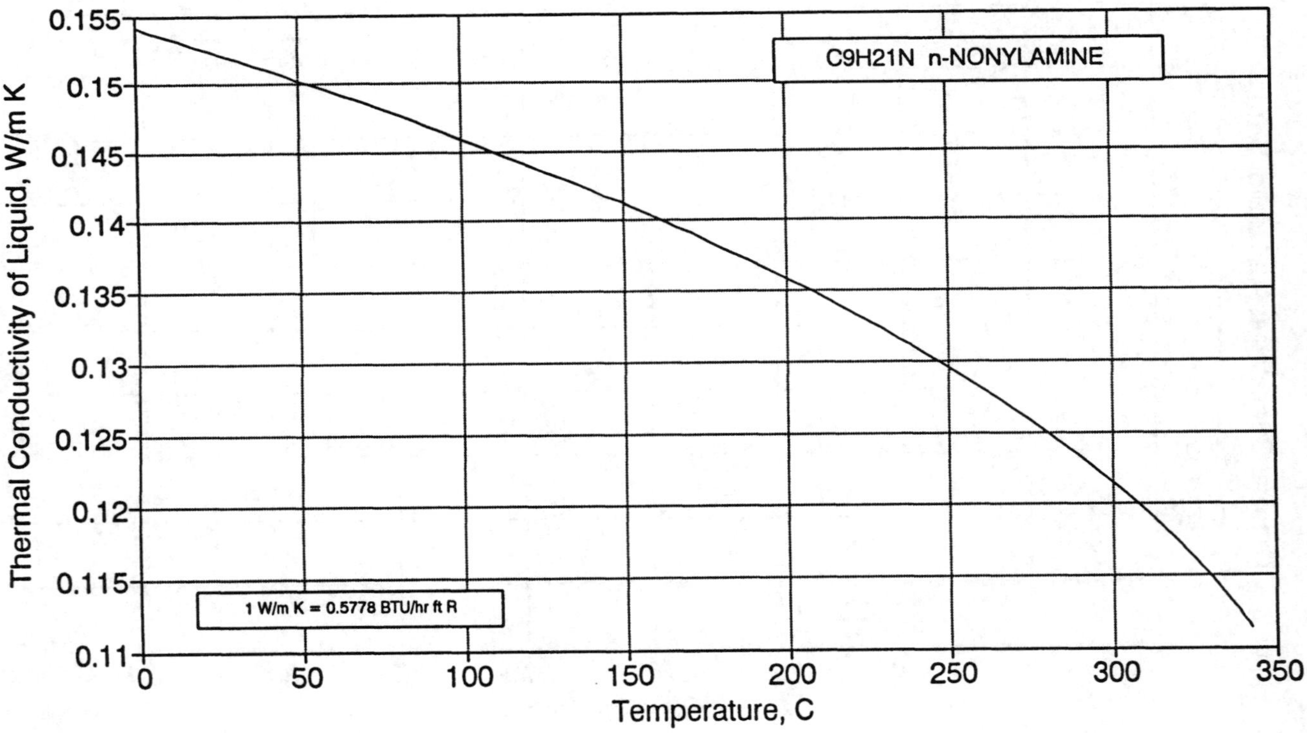
C9H21N n-NONYLAMINE
1 W/m K = 0.5778 BTU/hr ft R
Thermal Conductivity of Liquid, W/m K
0.155
0.15
0.145
0.14
0.135
0.13
0.125
0.12
0.115
0.11
0
50
100
150
200
250
300
350
Temperature, C

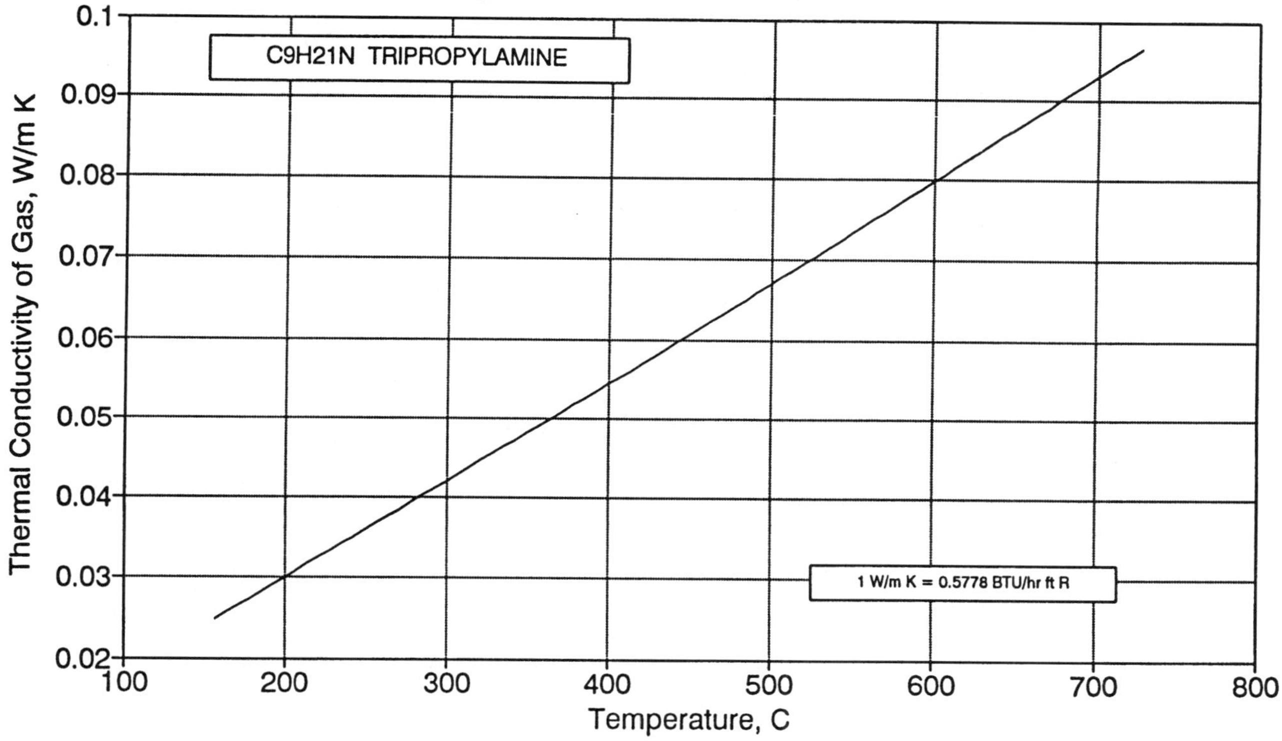
C9H21N TRIPROPYLAMINE
1 W/m K = 0.5778 BTU/hr ft R
Thermal Conductivity of Gas, W/m K
0.1
0.09
0.08
0.07
0.06
0.05
0.04
0.03
0.02
100
200
300
400
500
600
700
800
Temperature, C

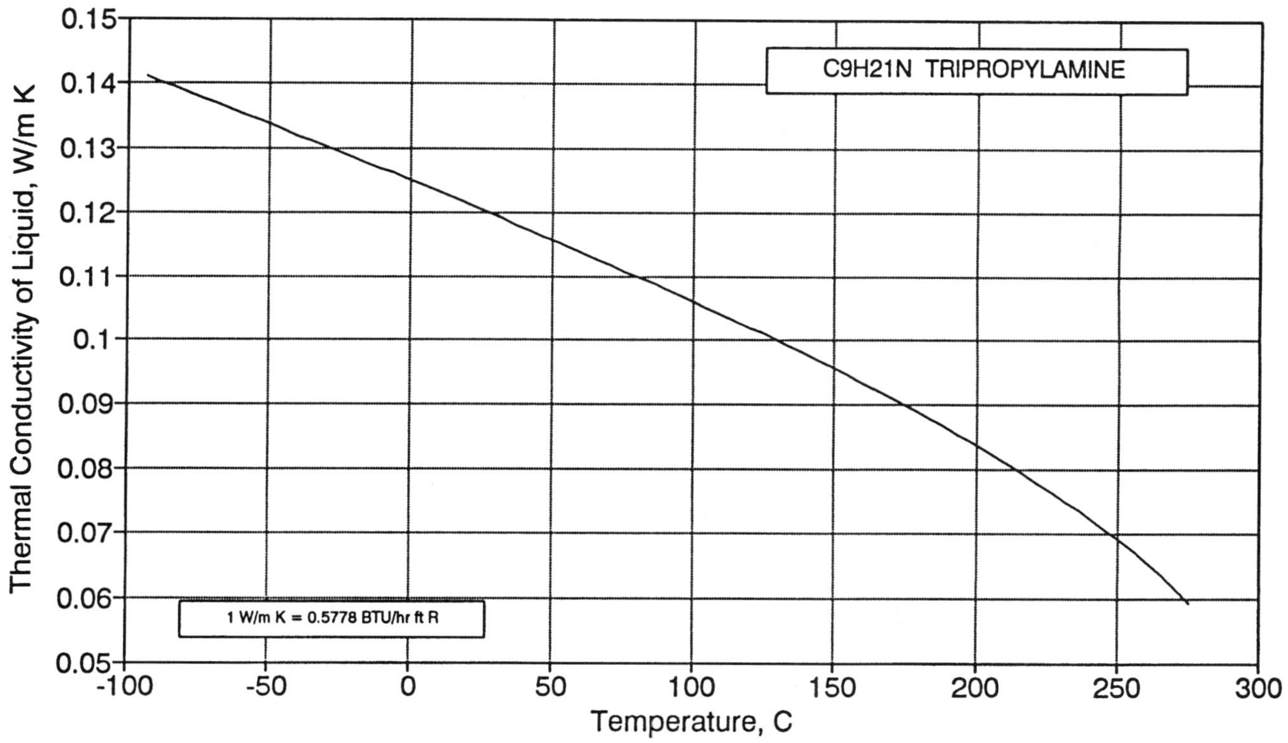
C9H21N TRIPROPYLAMINE
1 W/m K = 0.5778 BTU/hr ft R
Thermal Conductivity of Liquid, W/m K
0.15
0.14
0.13
0.12
0.11
0.1
0.09
0.08
0.07
0.06
0.05
-100
-50
0
50
100
150
200
250
300
Temperature, C

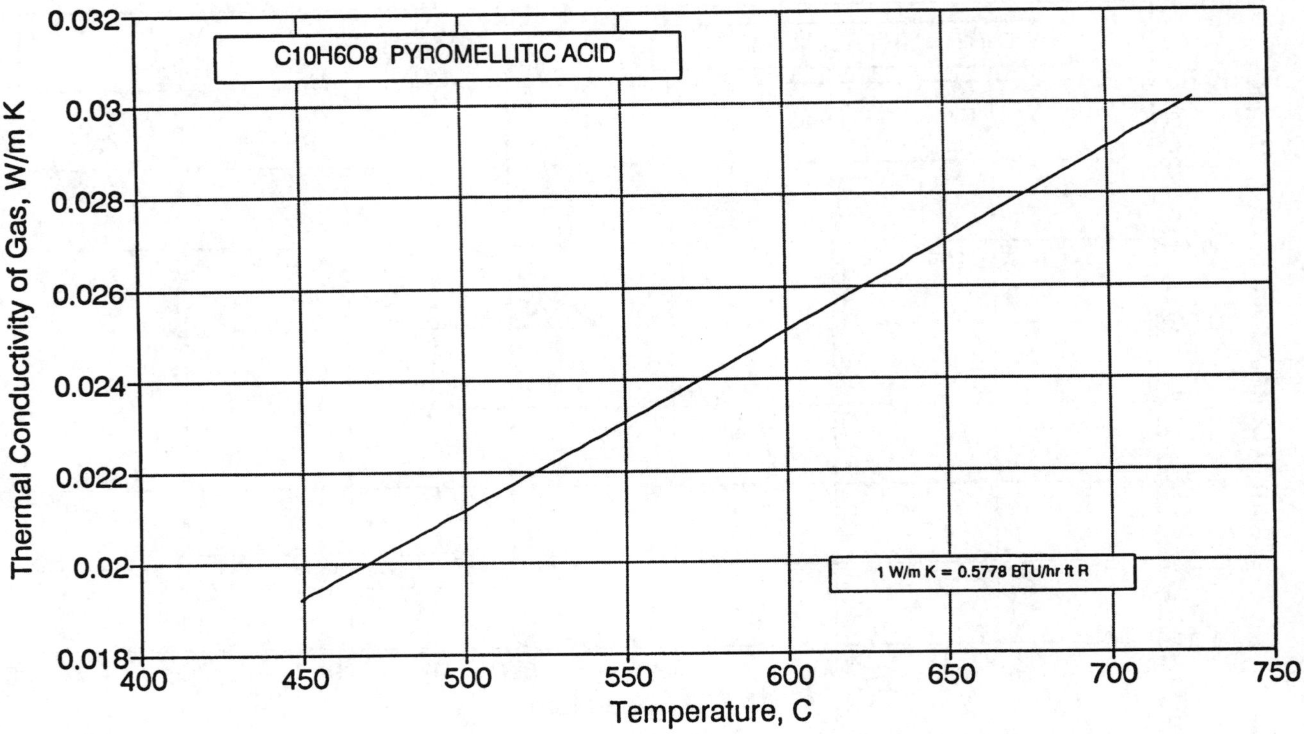
C10H6O8 PYROMELLITIC ACID
1 W/m K = 0.5778 BTU/hr ft R
Thermal Conductivity of Gas, W/m K
0.032
0.03
0.028
0.026
0.024
0.022
0.02
0.018
400
450
500
550
600
650
700
750
Temperature, C

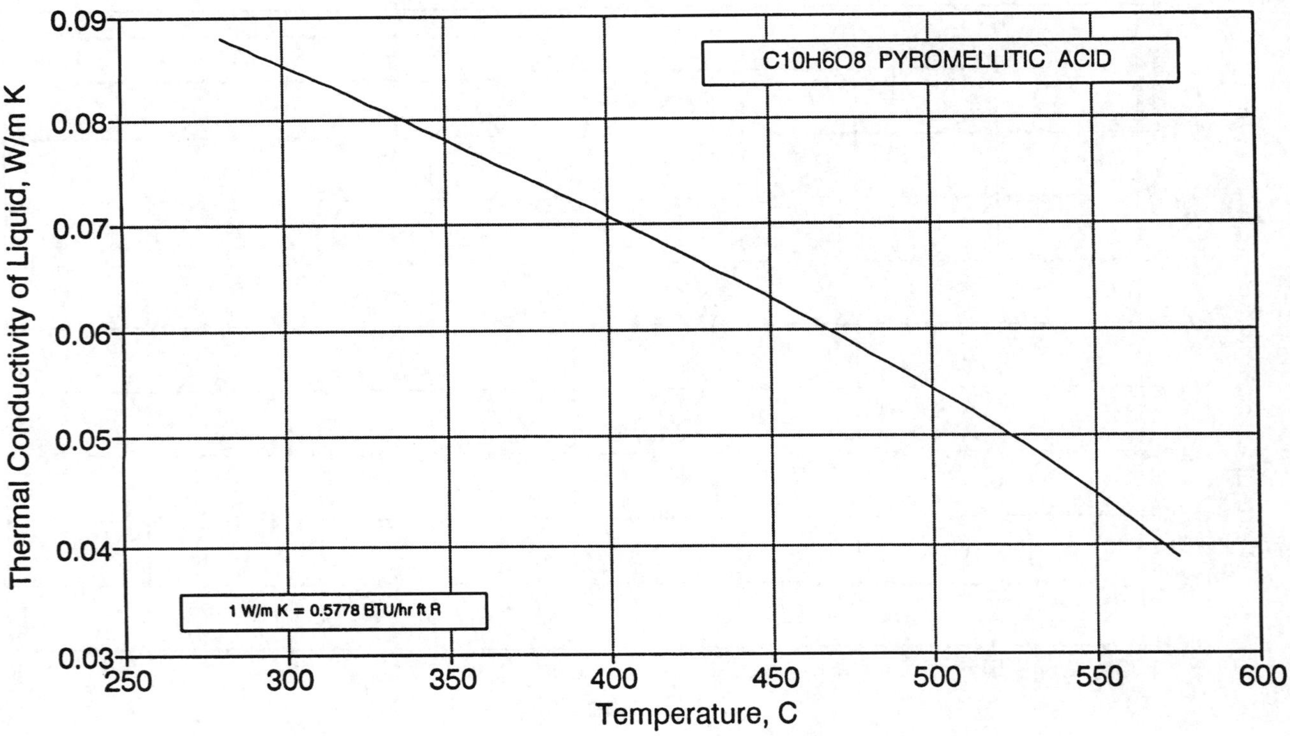
C10H6O8 PYROMELLITIC ACID
1 W/m K = 0.5778 BTU/hr ft R
Thermal Conductivity of Liquid, W/m K
0.09
0.08
0.07
0.06
0.05
0.04
0.03
250
300
350
400
450
500
550
600
Temperature, C

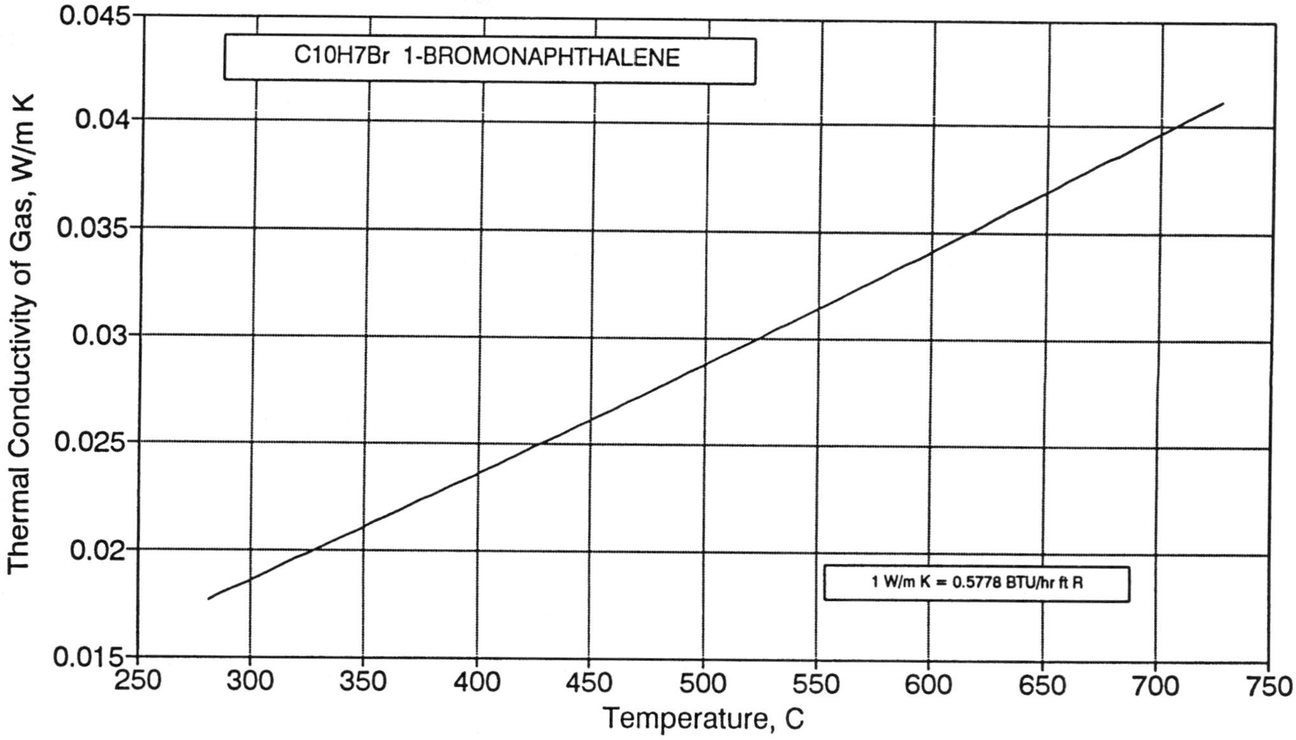
C10H7Br 1-BROMONAPHTHALENE
1 W/m K = 0.5778 BTU/hr ft R
Thermal Conductivity of Gas, W/m K
0.045
0.04
0.035
0.03
0.025
0.02
0.015
250
300
350
400
450
500
550
600
650
700
750
Temperature, C

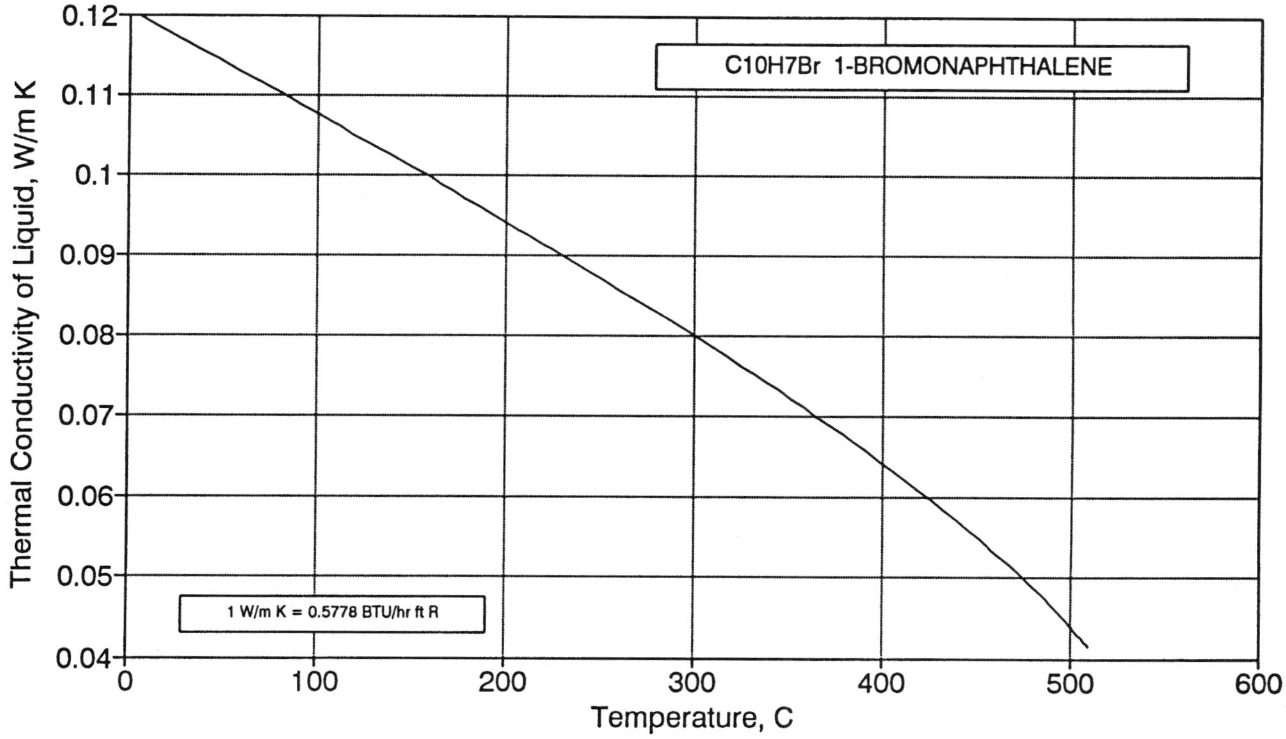
C10H7Br 1-BROMONAPHTHALENE
1 W/m K = 0.5778 BTU/hr ft R
Thermal Conductivity of Liquid, W/m K
0.12
0.11
0.1
0.09
0.08
0.07
0.06
0.05
0.04
0
100
200
300
400
500
600
Temperature, C

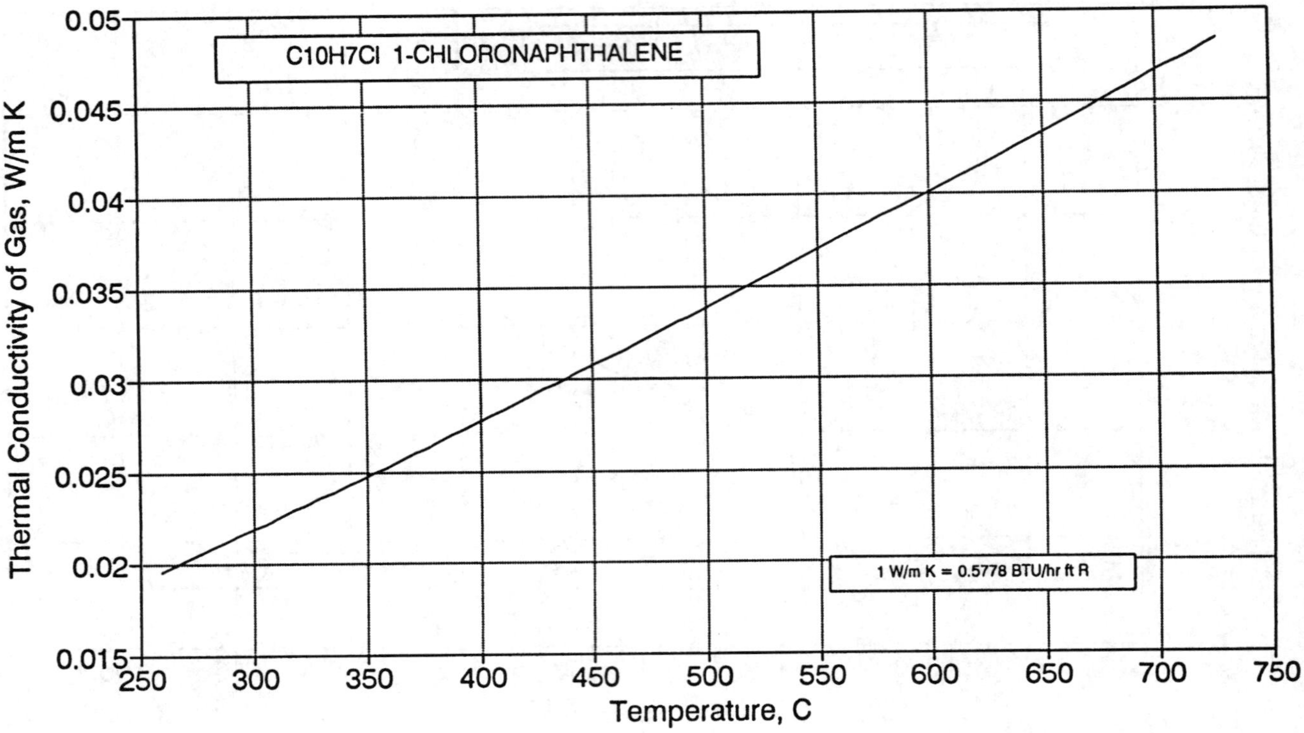
C10H7Cl 1-CHLORONAPHTHALENE
Thermal Conductivity of Gas, W/m K
0.05
0.045
0.04
0.035
0.03
0.025
0.02
0.015
250
300
350
400
450
500
550
600
650
700
750
Temperature, C
1 W/m K = 0.5778 BTU/hr ft R

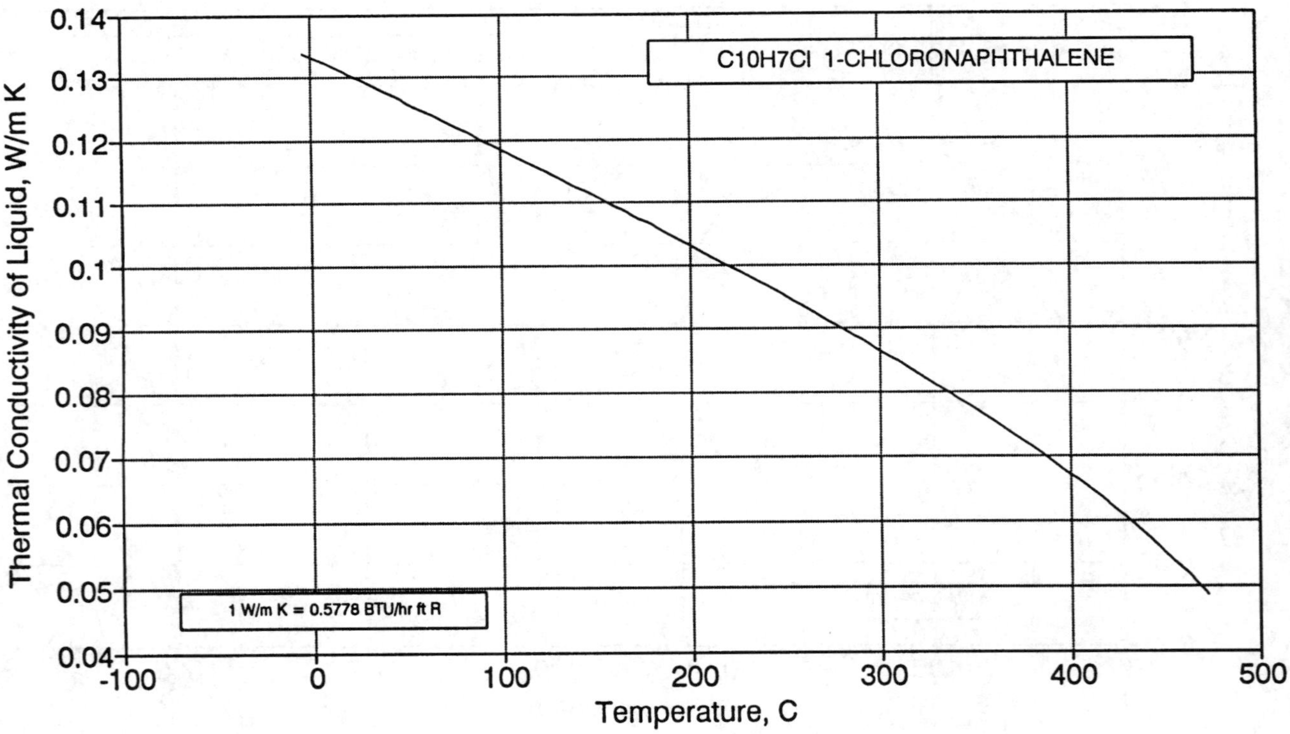
C10H7Cl 1-CHLORONAPHTHALENE
Thermal Conductivity of Liquid, W/m K
0.14
0.13
0.12
0.11
0.1
0.09
0.08
0.07
0.06
0.05
0.04
-100
0
100
200
300
400
500
Temperature, C
1 W/m K = 0.5778 BTU/hr ft R

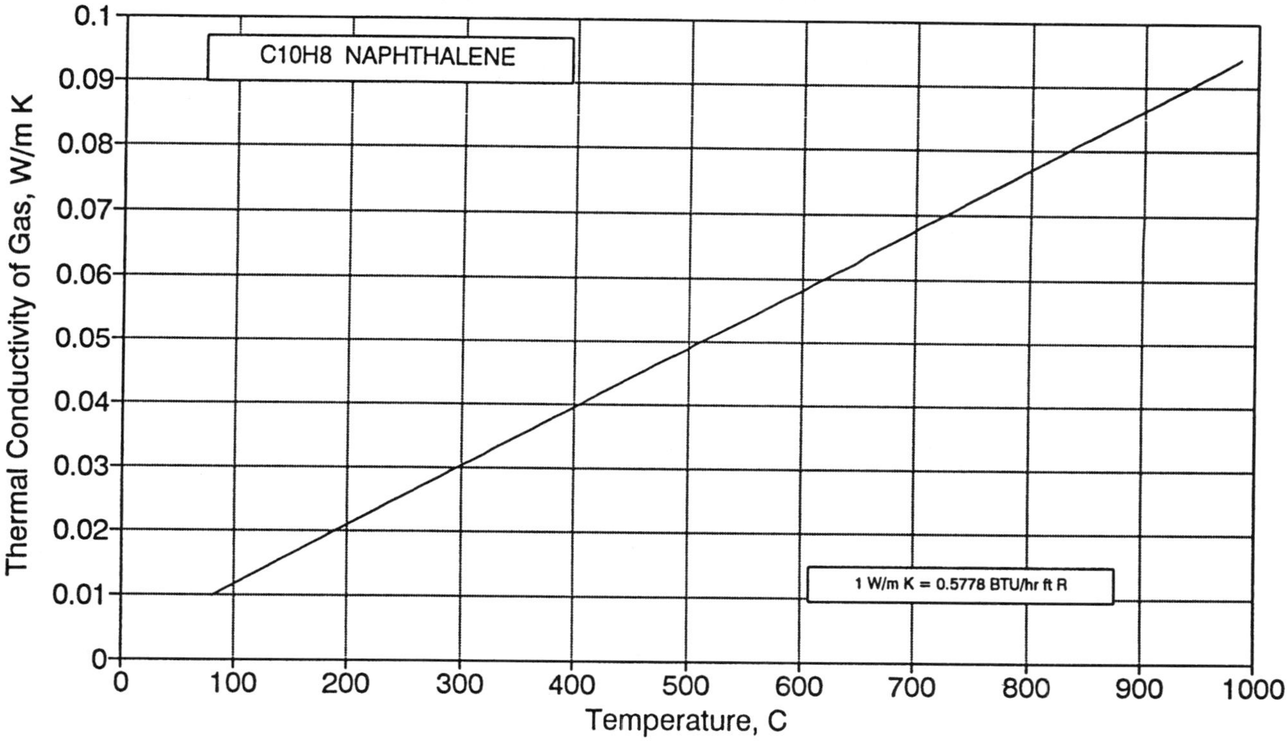
C10H8 NAPHTHALENE
1 W/m K = 0.5778 BTU/hr ft R
Thermal Conductivity of Gas, W/m K
0.1
0.09
0.08
0.07
0.06
0.05
0.04
0.03
0.02
0.01
0
0
100
200
300
400
500
600
700
800
900
1000
Temperature, C

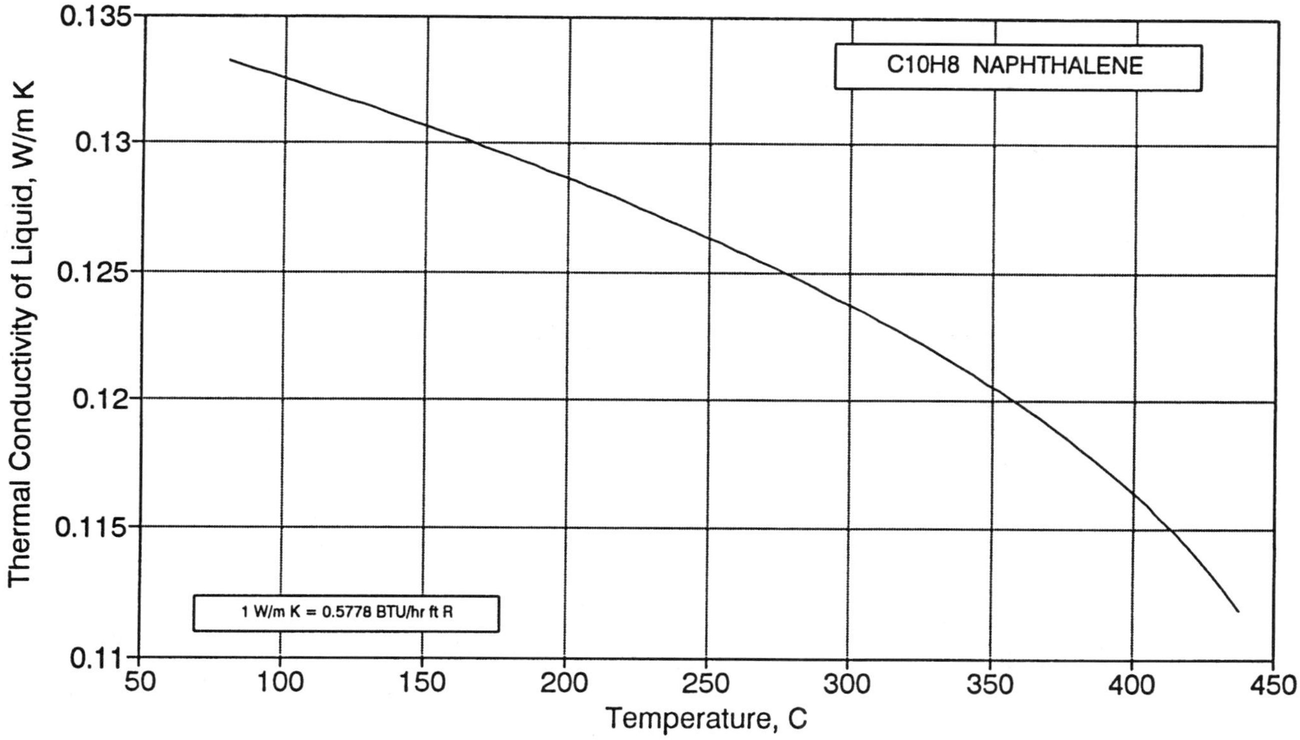
C10H8 NAPHTHALENE
1 W/m K = 0.5778 BTU/hr ft R
Thermal Conductivity of Liquid, W/m K
0.135
0.13
0.125
0.12
0.115
0.11
50
100
150
200
250
300
350
400
450
Temperature, C

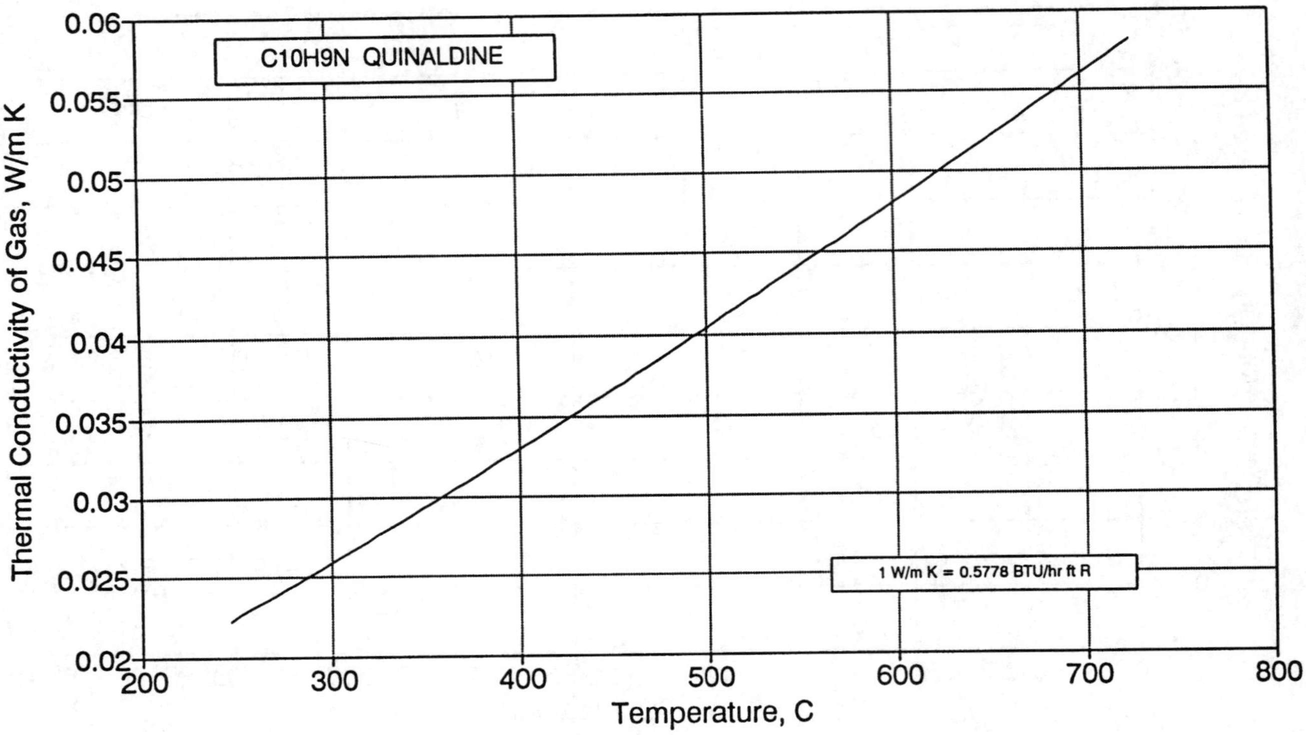
C10H9N QUINALDINE
1 W/m K = 0.5778 BTU/hr ft R
Thermal Conductivity of Gas, W/m K
0.06
0.055
0.05
0.045
0.04
0.035
0.03
0.025
0.02
200
300
400
500
600
700
800
Temperature, C

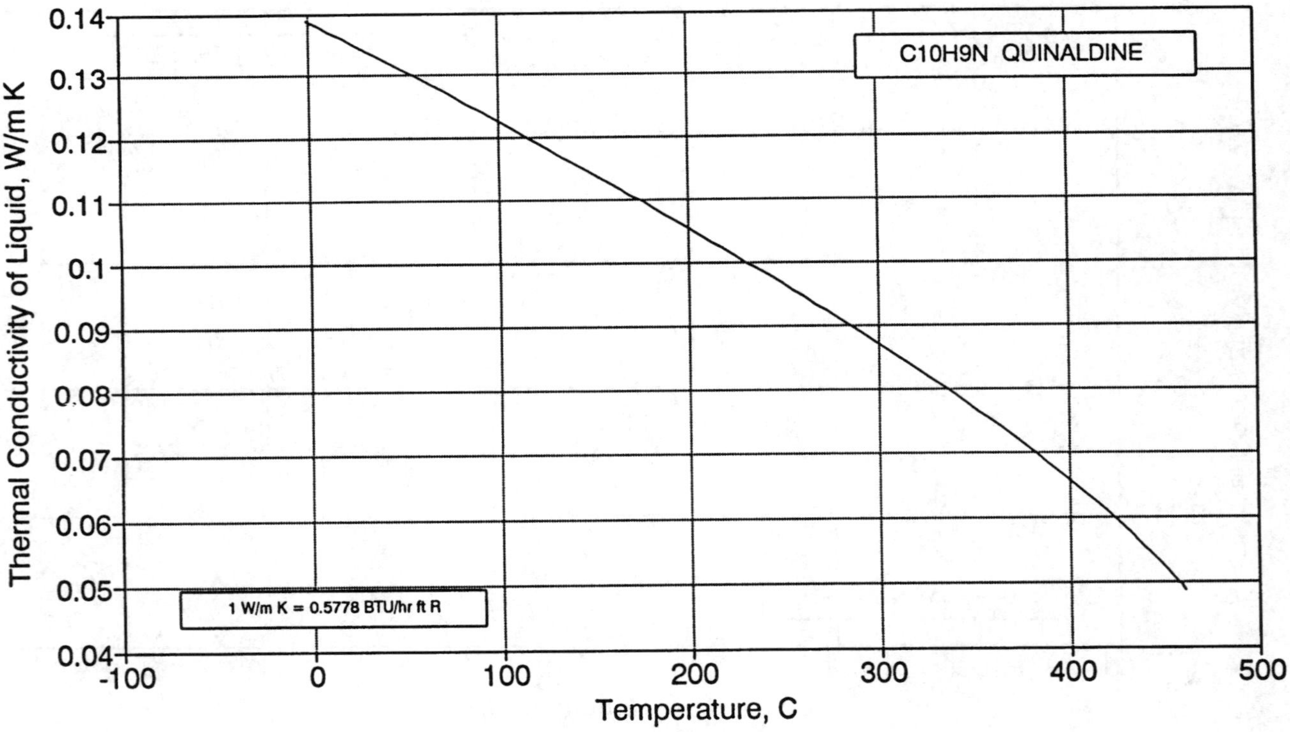
C10H9N QUINALDINE
1 W/m K = 0.5778 BTU/hr ft R
Thermal Conductivity of Liquid, W/m K
0.14
0.13
0.12
0.11
0.1
0.09
0.08
0.07
0.06
0.05
0.04
-100
0
100
200
300
400
500
Temperature, C

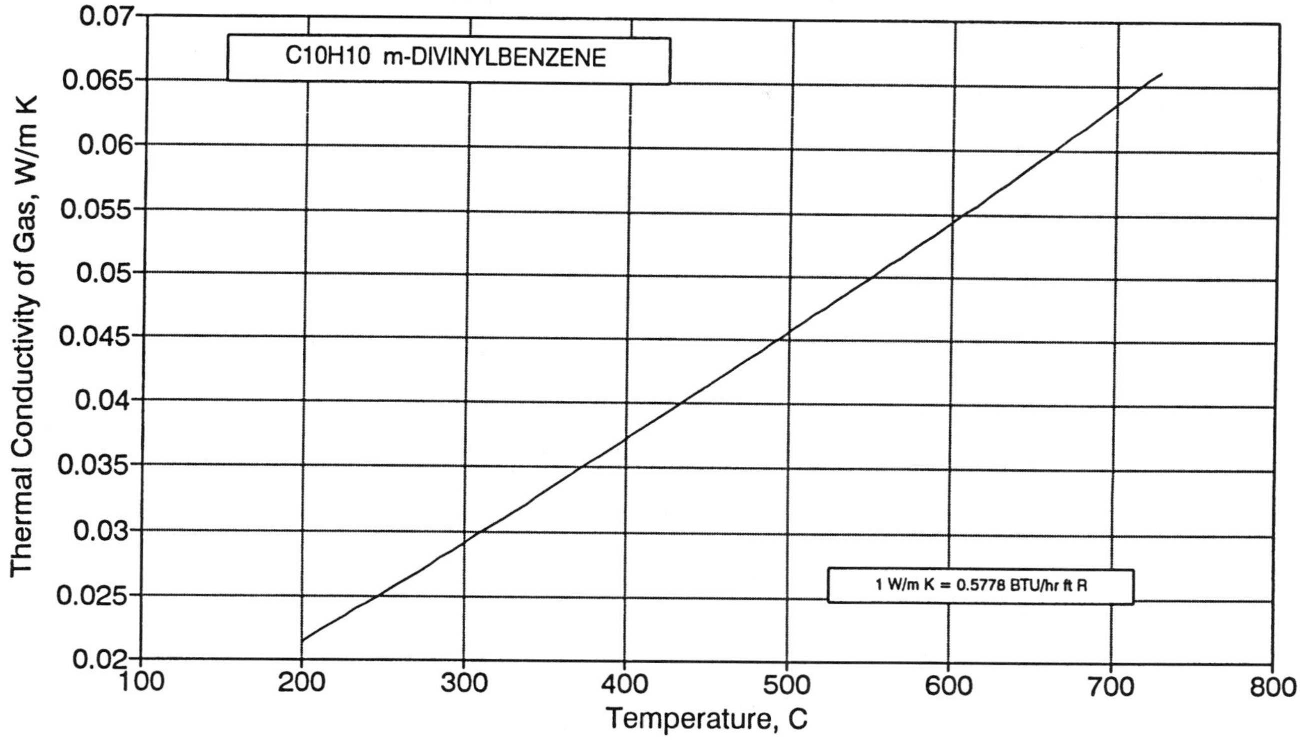
C10H10 m-DIVINYLBENZENE
1 W/m K = 0.5778 BTU/hr ft R
Thermal Conductivity of Gas, W/m K
0.07
0.065
0.06
0.055
0.05
0.045
0.04
0.035
0.03
0.025
0.02
100
200
300
400
500
600
700
800
Temperature, C

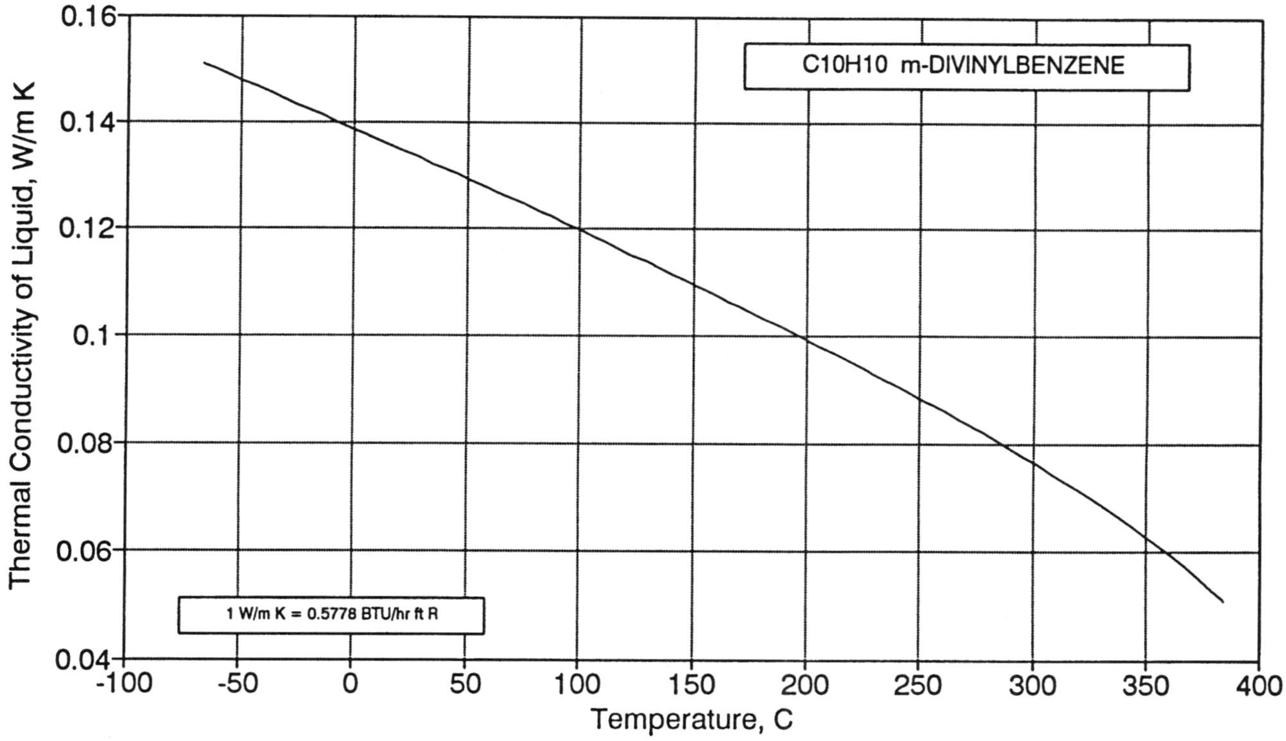
C10H10 m-DIVINYLBENZENE
1 W/m K = 0.5778 BTU/hr ft R
Thermal Conductivity of Liquid, W/m K
0.16
0.14
0.12
0.1
0.08
0.06
0.04
-100
-50
0
50
100
150
200
250
300
350
400
Temperature, C

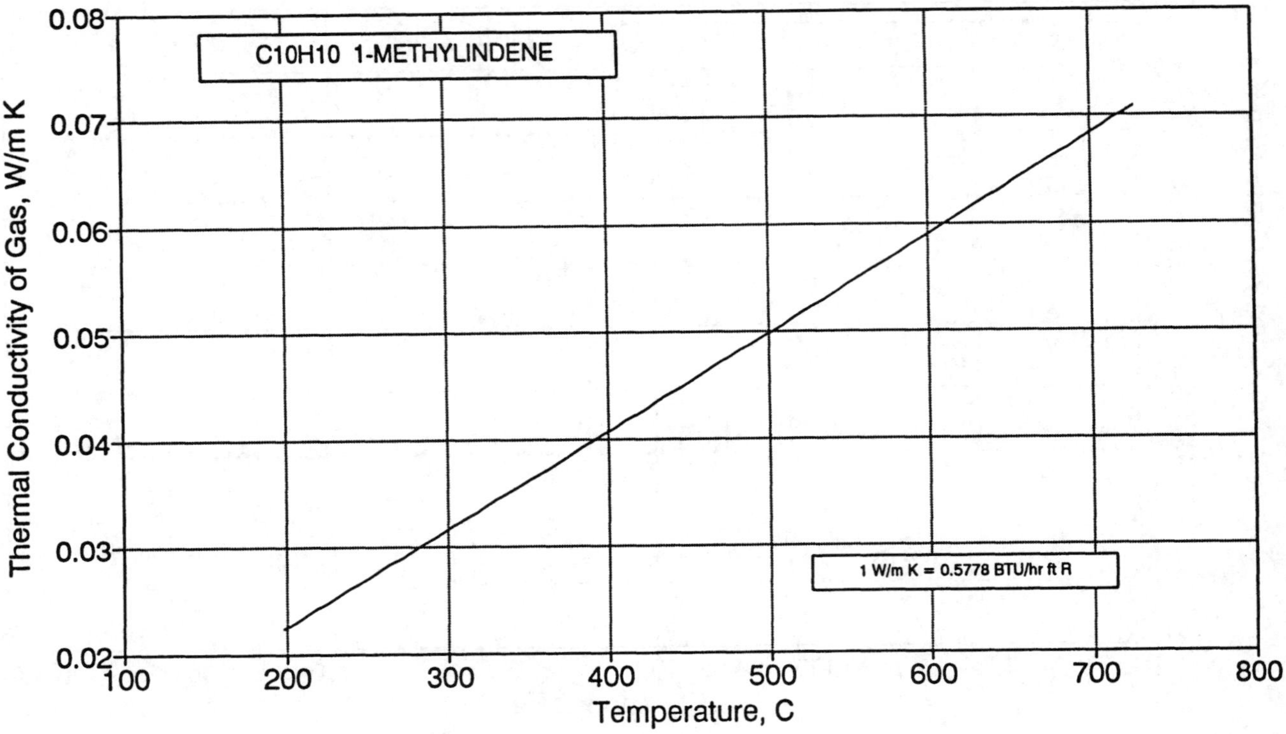
C10H10 1-METHYLINDENE
Thermal Conductivity of Gas, W/m K
0.08
0.07
0.06
0.05
0.04
0.03
0.02
100
200
300
400
500
600
700
800
Temperature, C
1 W/m K = 0.5778 BTU/hr ft R

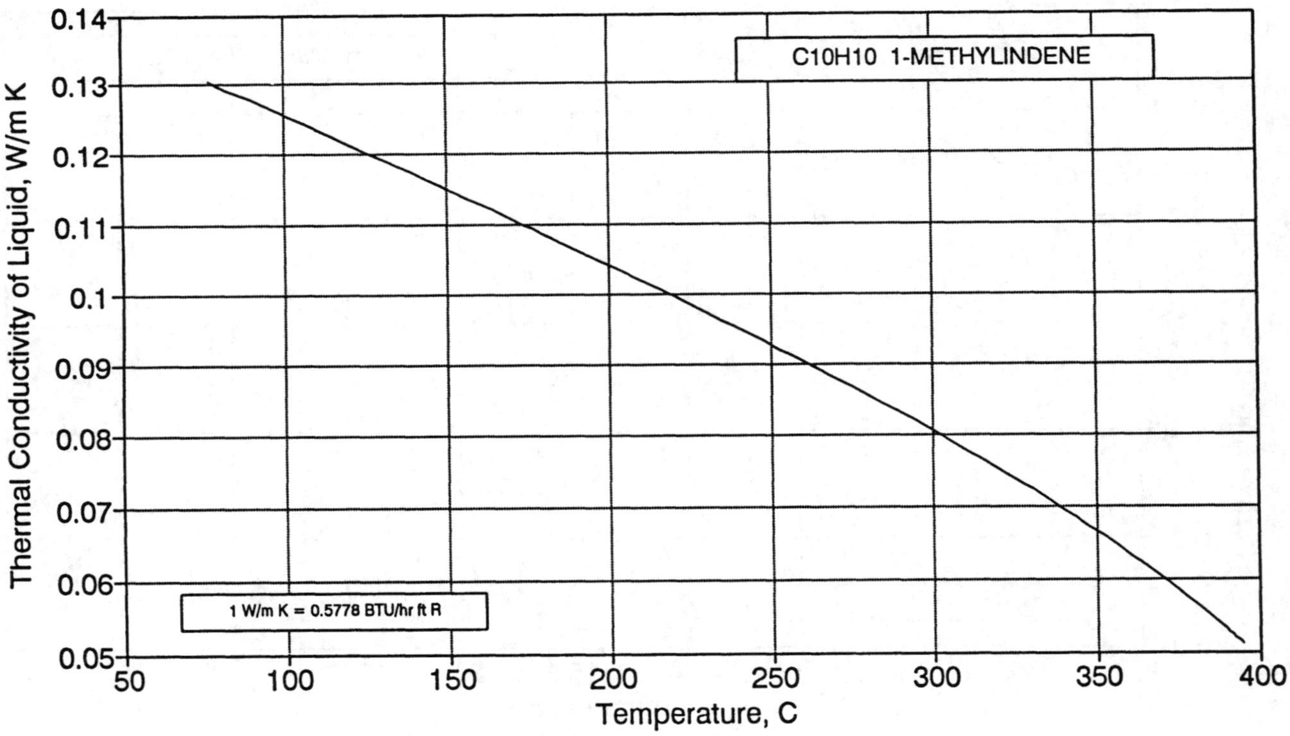
C10H10 1-METHYLINDENE
Thermal Conductivity of Liquid, W/m K
0.14
0.13
0.12
0.11
0.1
0.09
0.08
0.07
0.06
0.05
50
100
150
200
250
300
350
400
Temperature, C
1 W/m K = 0.5778 BTU/hr ft R

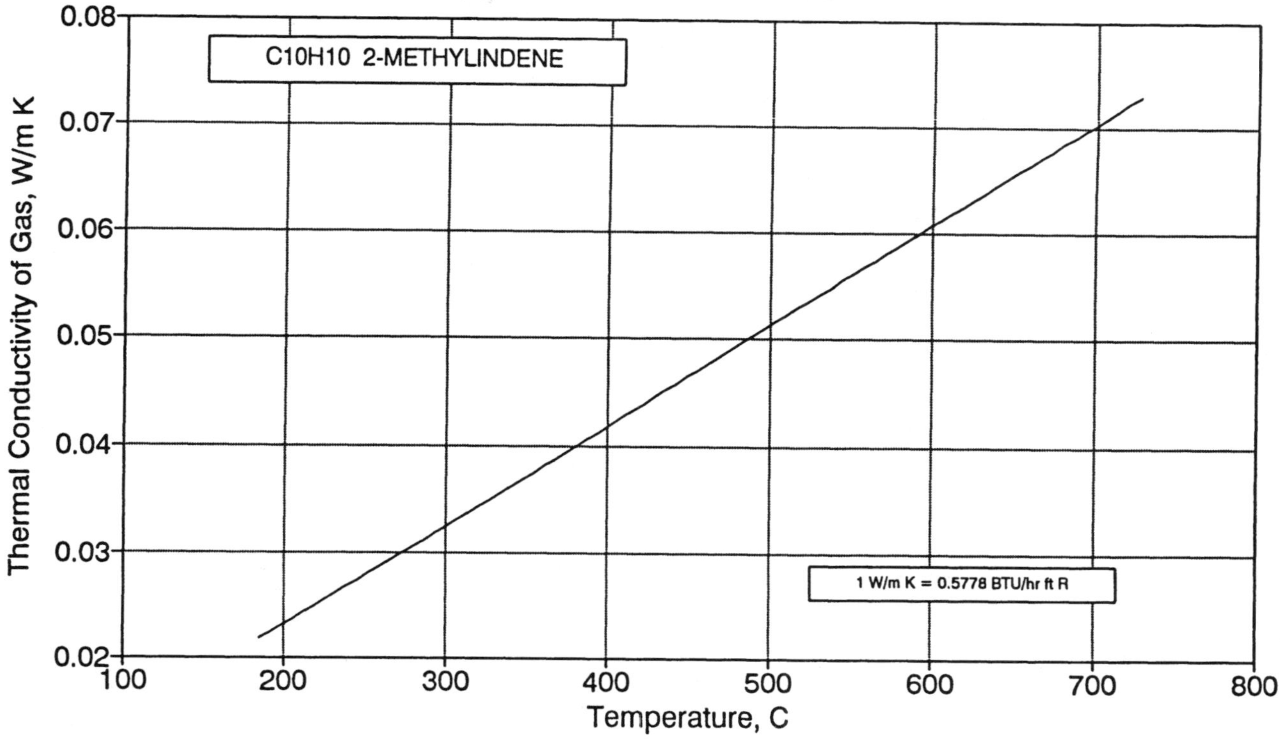
C10H10 2-METHYLINDENE
1 W/m K = 0.5778 BTU/hr ft R
Thermal Conductivity of Gas, W/m K
0.08
0.07
0.06
0.05
0.04
0.03
0.02
100
200
300
400
500
600
700
800
Temperature, C

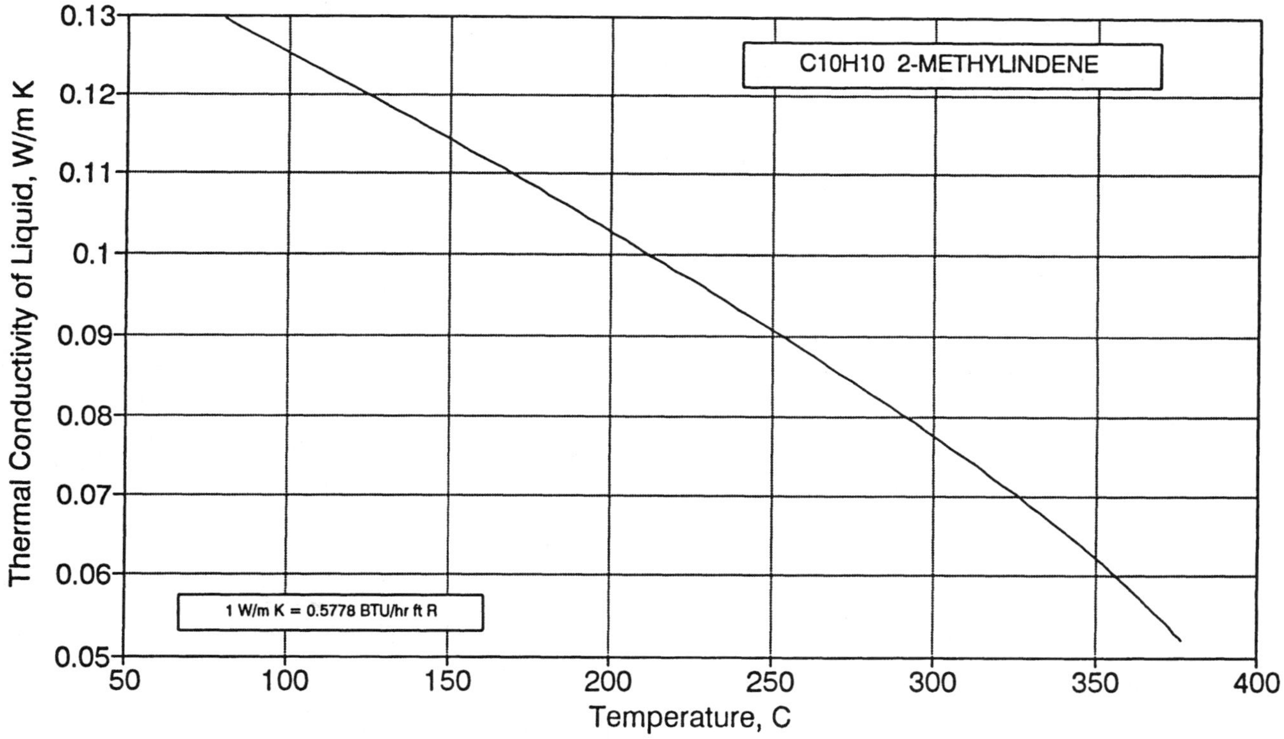
C10H10 2-METHYLINDENE
1 W/m K = 0.5778 BTU/hr ft R
Thermal Conductivity of Liquid, W/m K
0.13
0.12
0.11
0.1
0.09
0.08
0.07
0.06
0.05
50
100
150
200
250
300
350
400
Temperature, C

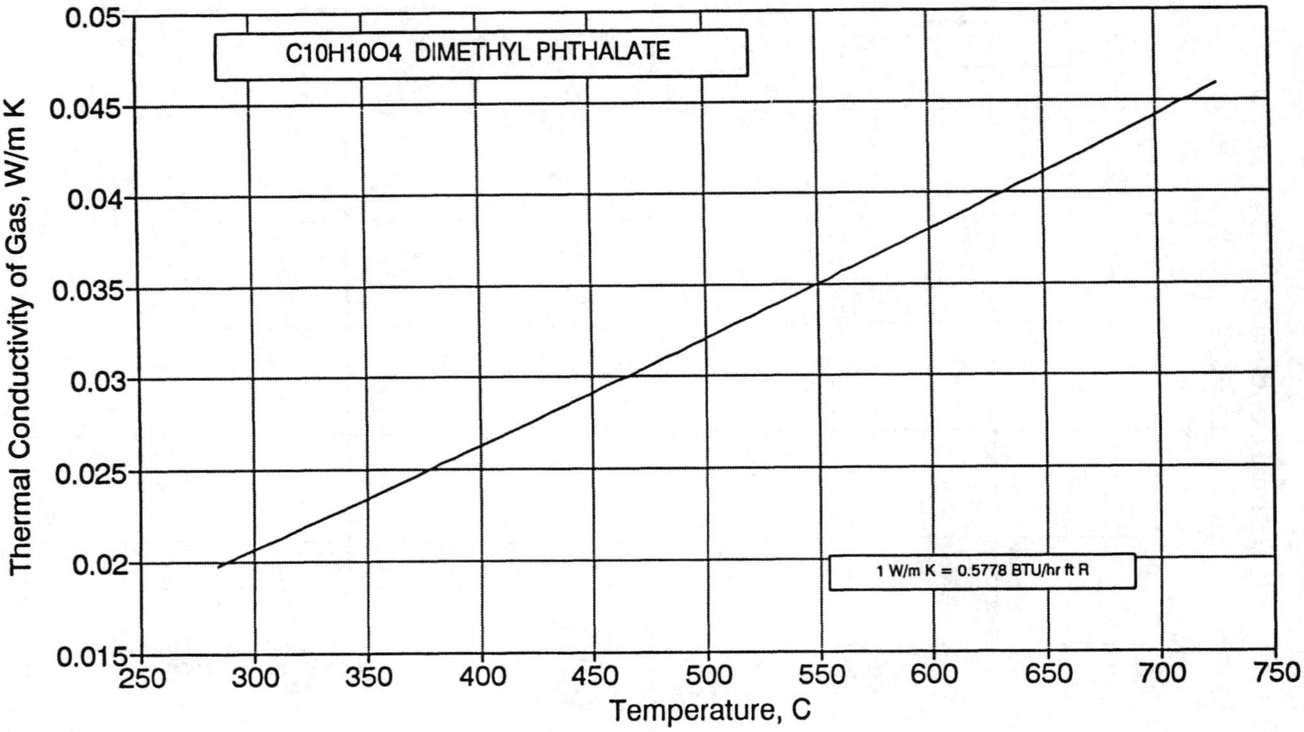

C10H10O4 DIMETHYL PHTHALATE
1 W/m K = 0.5778 BTU/hr ft R
Thermal Conductivity of Gas, W/m K
0.05
0.045
0.04
0.035
0.03
0.025
0.02
0.015
250
300
350
400
450
500
550
600
650
700
750
Temperature, C

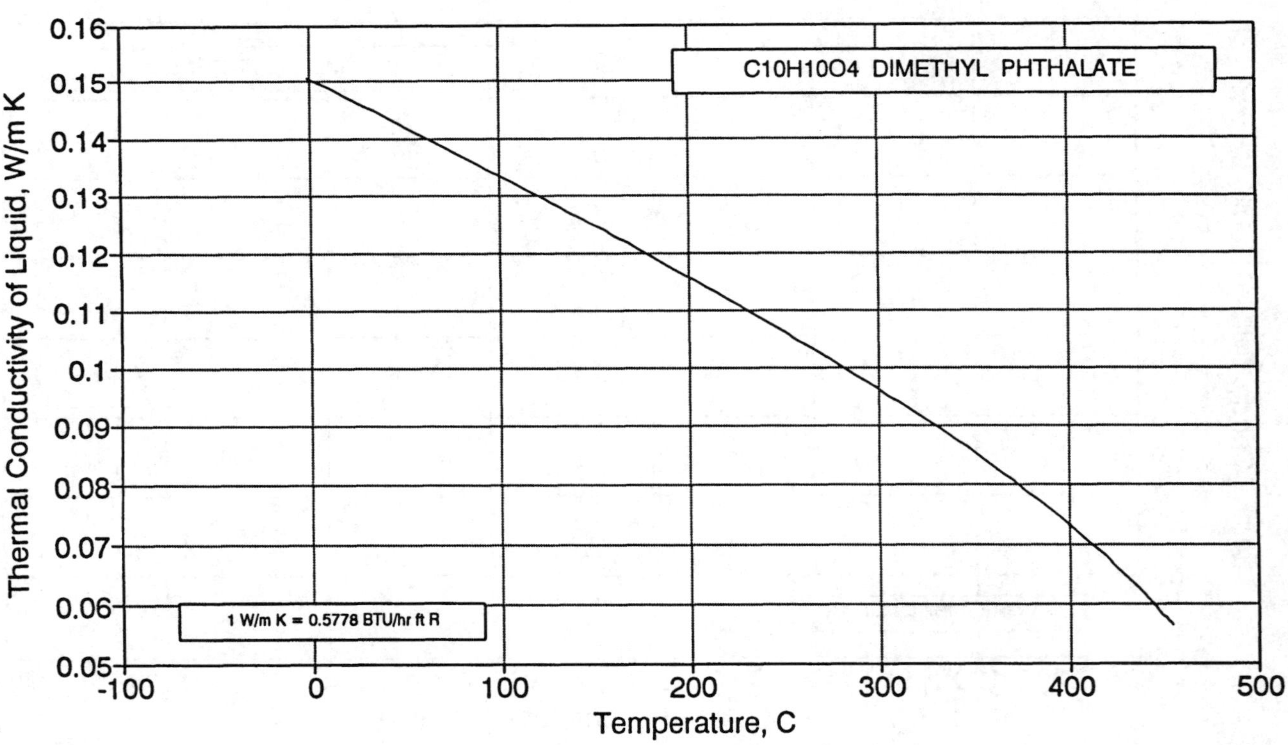

C10H10O4 DIMETHYL PHTHALATE
1 W/m K = 0.5778 BTU/hr ft R
Thermal Conductivity of Liquid, W/m K
0.16
0.15
0.14
0.13
0.12
0.11
0.1
0.09
0.08
0.07
0.06
0.05
-100
0
100
200
300
400
500
Temperature, C

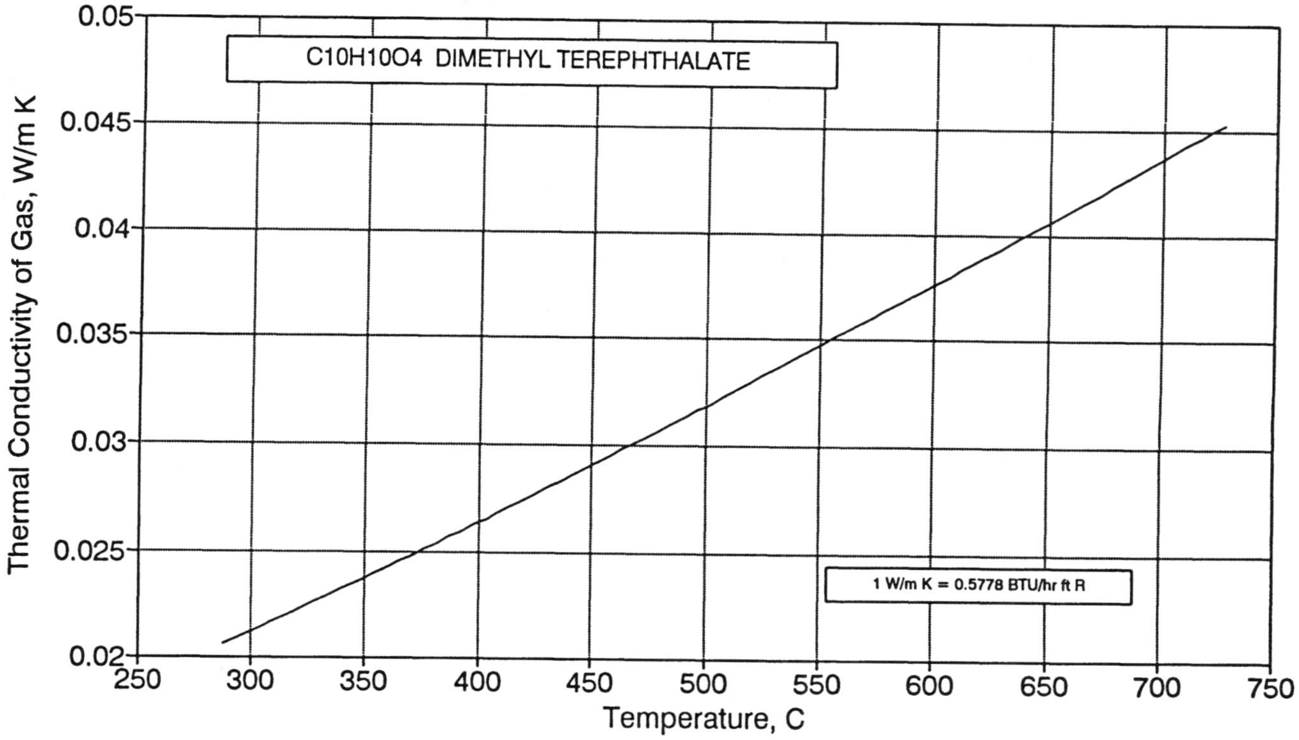
C10H10O4 DIMETHYL TEREPHTHALATE
1 W/m K = 0.5778 BTU/hr ft R
Thermal Conductivity of Gas, W/m K
0.05
0.045
0.04
0.035
0.03
0.025
0.02
250
300
350
400
450
500
550
600
650
700
750
Temperature, C

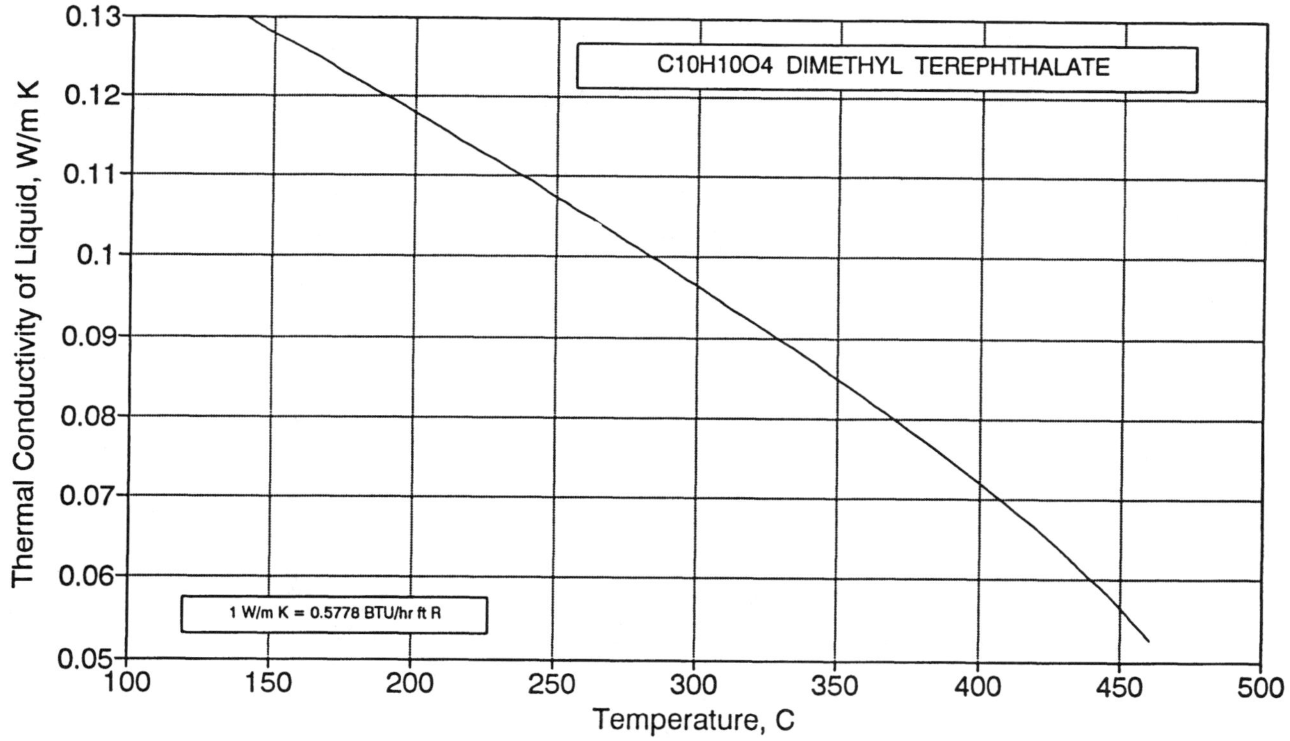
C10H10O4 DIMETHYL TEREPHTHALATE
1 W/m K = 0.5778 BTU/hr ft R
Thermal Conductivity of Liquid, W/m K
0.13
0.12
0.11
0.1
0.09
0.08
0.07
0.06
0.05
100
150
200
250
300
350
400
450
500
Temperature, C

C10H12 DICYCLOPENTADIENE

Gas thermal conductivity data are not available.

TF, K = 307.00

TB, K = 443.00 (experimental)

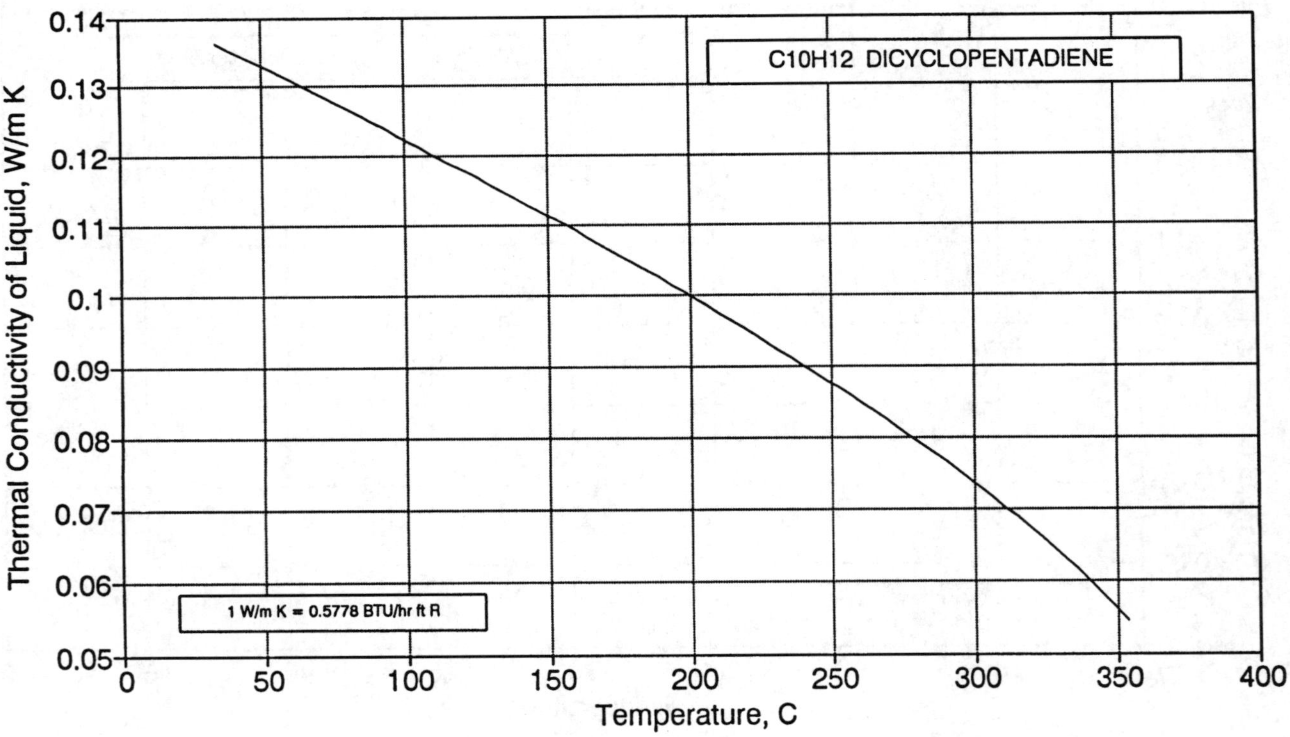

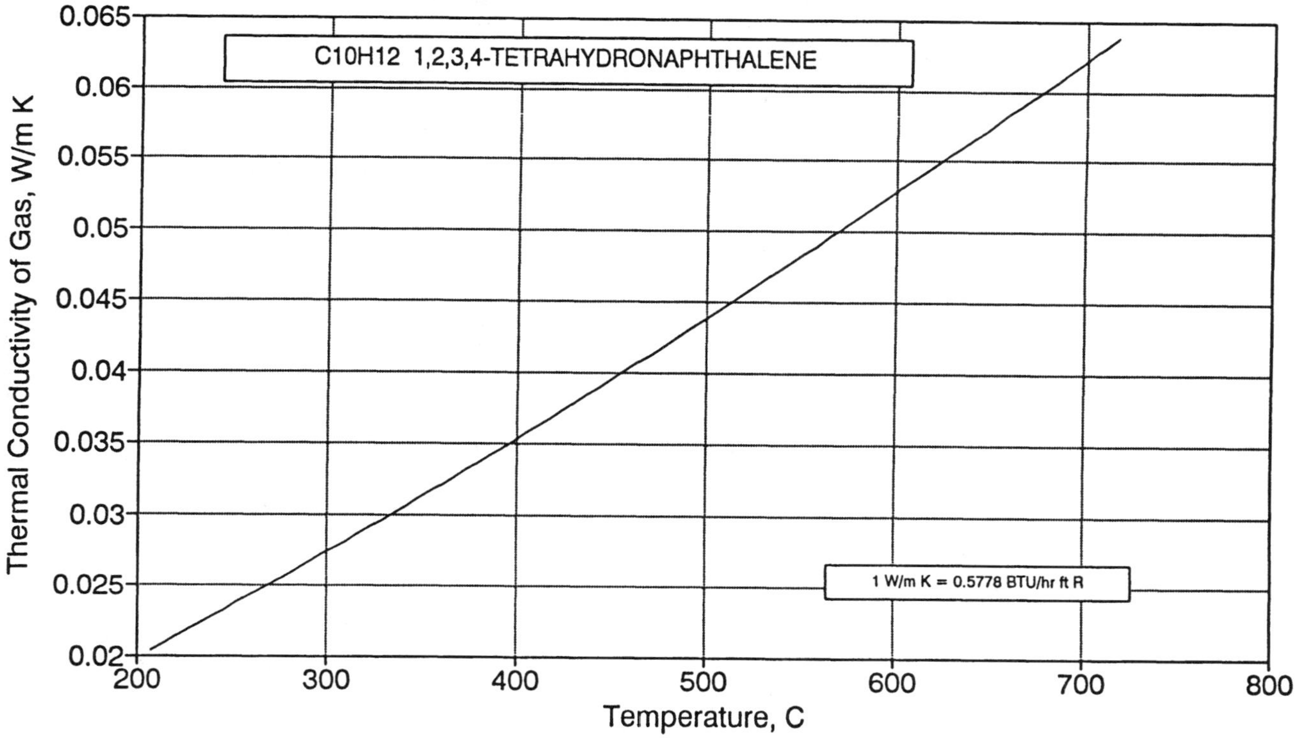
C10H12 1,2,3,4-TETRAHYDRONAPHTHALENE
1 W/m K = 0.5778 BTU/hr ft R
Thermal Conductivity of Gas, W/m K
Temperature, C
0.065
0.06
0.055
0.05
0.045
0.04
0.035
0.03
0.025
0.02
200
300
400
500
600
700
800

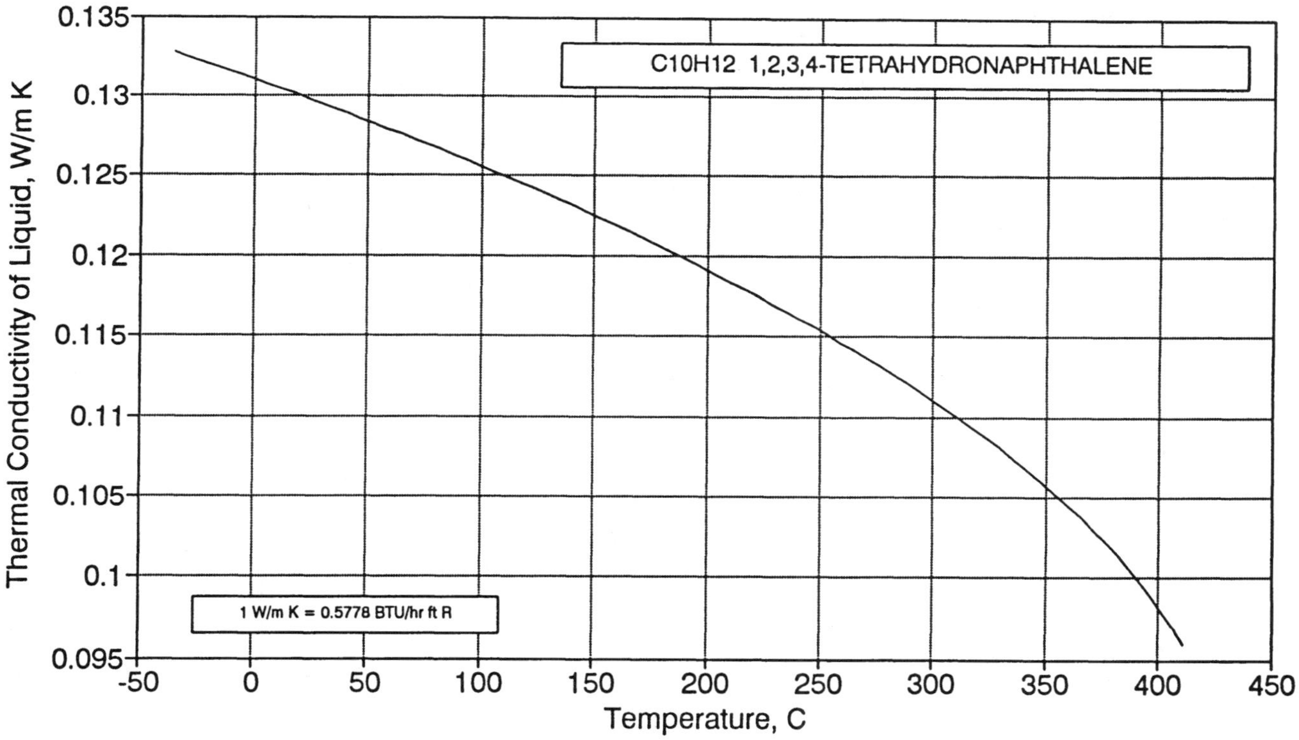
C10H12 1,2,3,4-TETRAHYDRONAPHTHALENE
1 W/m K = 0.5778 BTU/hr ft R
Thermal Conductivity of Liquid, W/m K
Temperature, C
0.135
0.13
0.125
0.12
0.115
0.11
0.105
0.1
0.095
-50
0
50
100
150
200
250
300
350
400
450

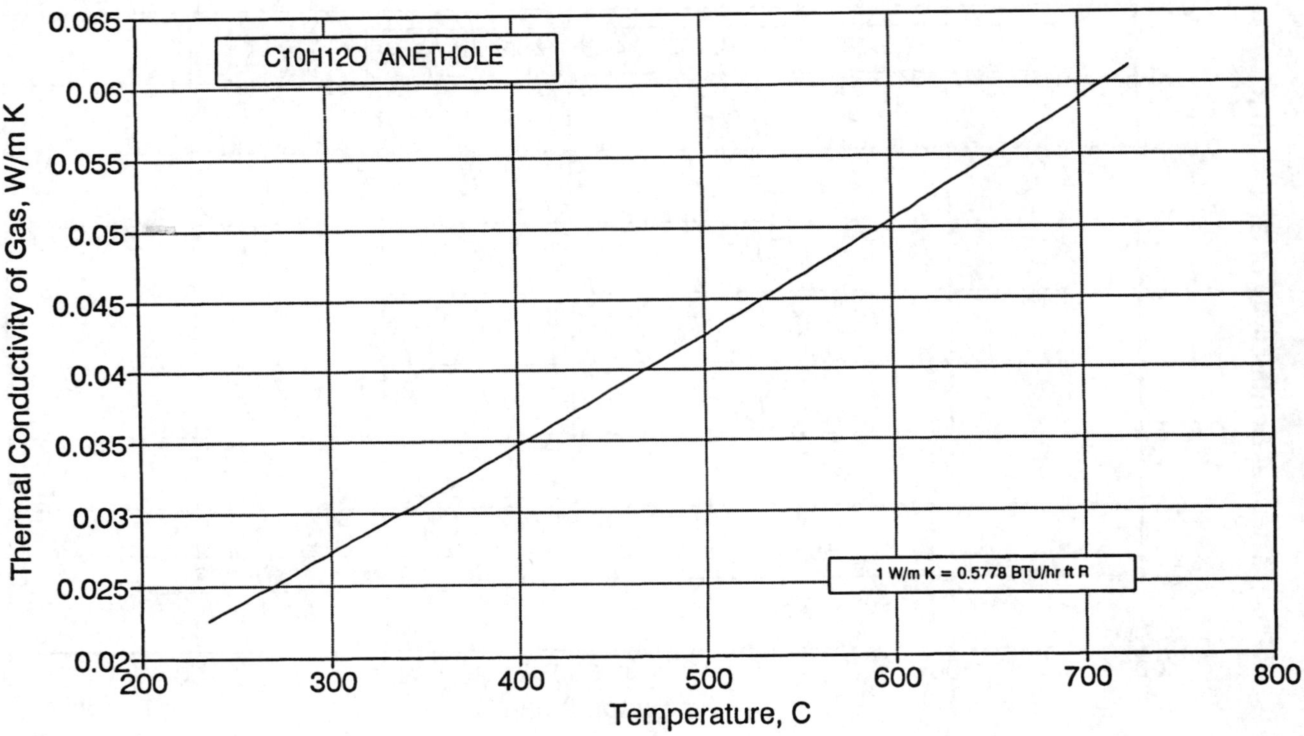

C10H12O ANETHOLE

Liquid thermal conductivity data are not available.

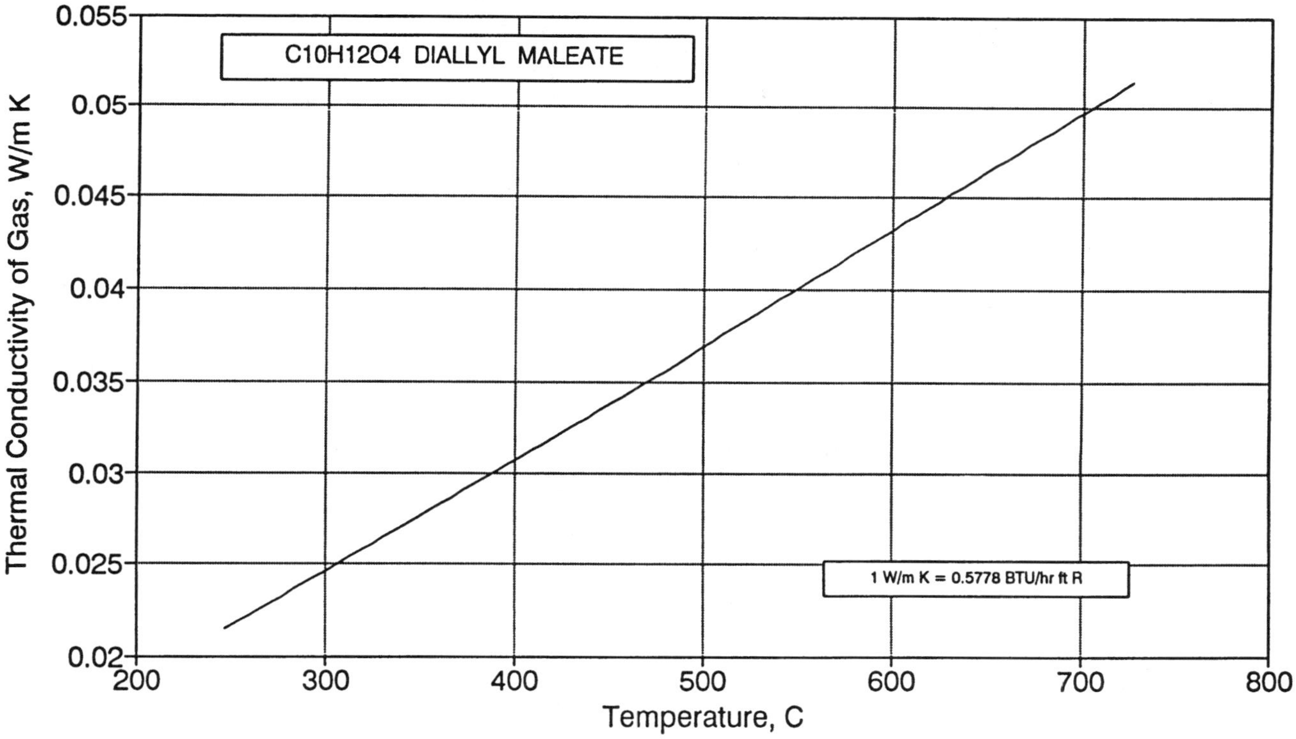
C10H12O4 DIALLYL MALEATE
1 W/m K = 0.5778 BTU/hr ft R
Thermal Conductivity of Gas, W/m K
0.055
0.05
0.045
0.04
0.035
0.03
0.025
0.02
200
300
400
500
600
700
800
Temperature, C

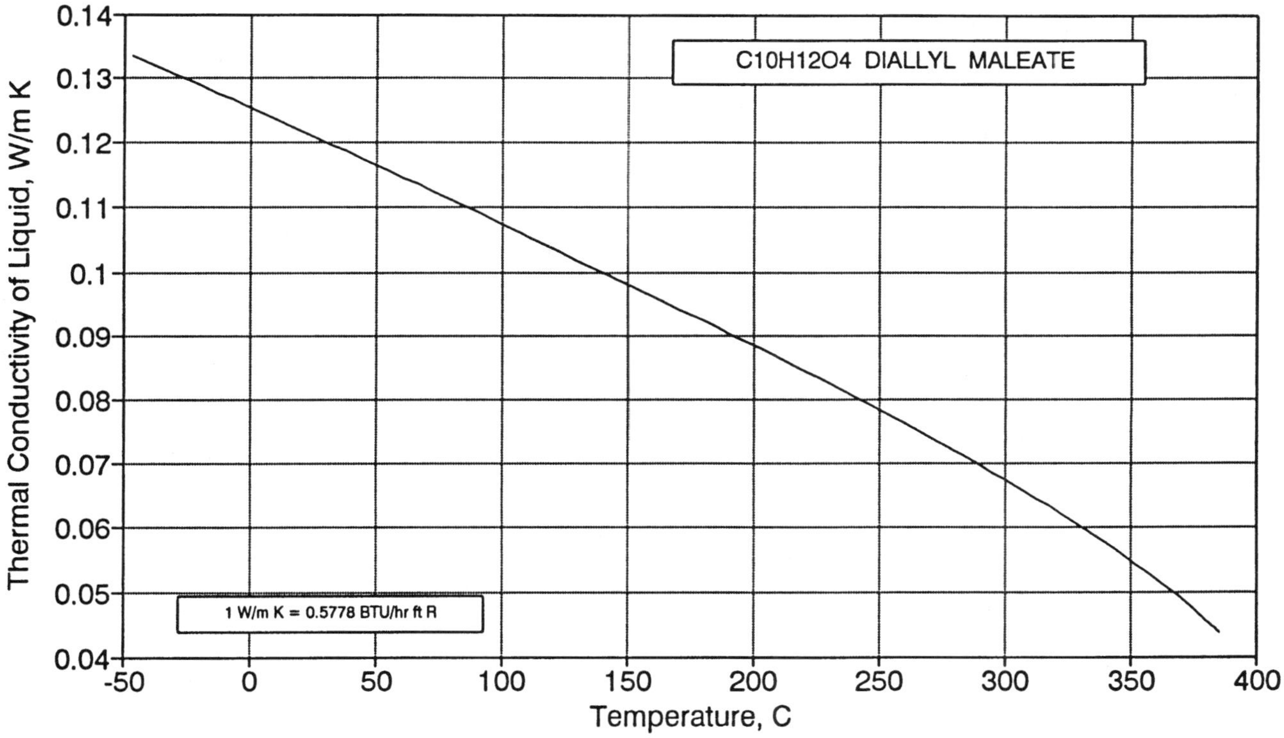
C10H12O4 DIALLYL MALEATE
1 W/m K = 0.5778 BTU/hr ft R
Thermal Conductivity of Liquid, W/m K
0.14
0.13
0.12
0.11
0.1
0.09
0.08
0.07
0.06
0.05
0.04
-50
0
50
100
150
200
250
300
350
400
Temperature, C

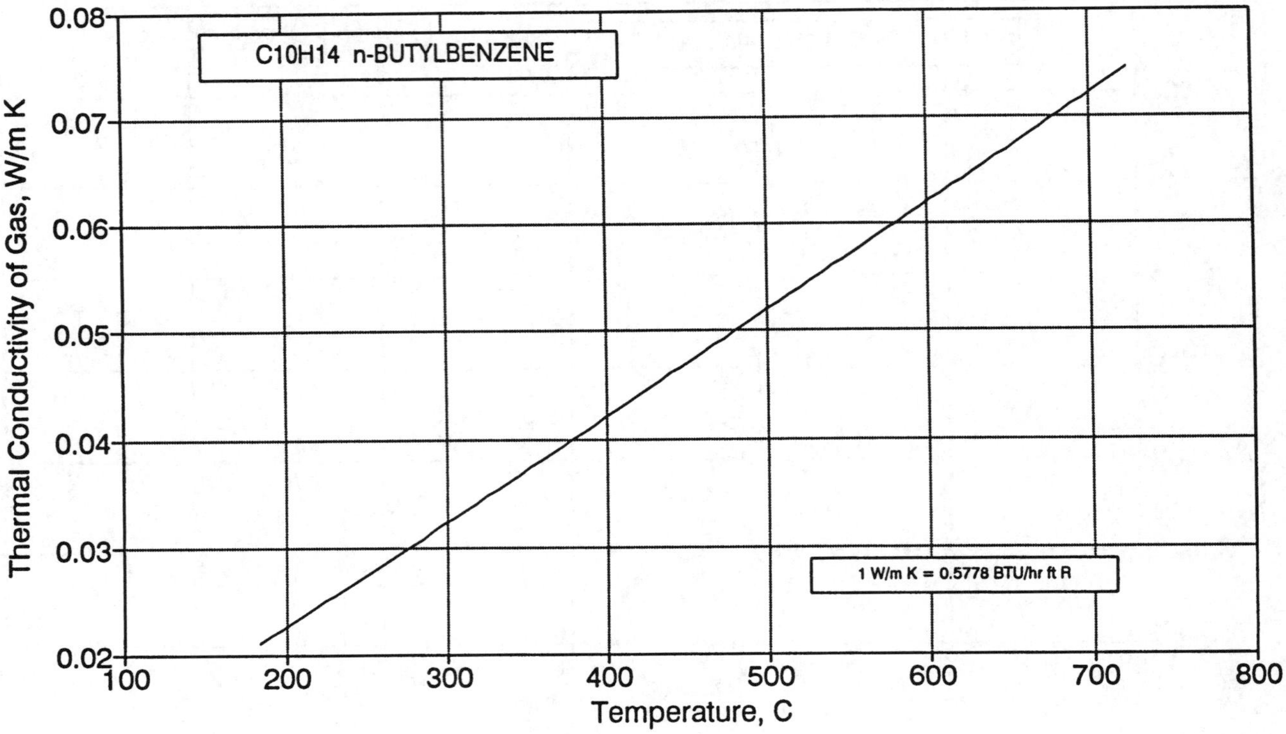
C10H14 n-BUTYLBENZENE
1 W/m K = 0.5778 BTU/hr ft R
Thermal Conductivity of Gas, W/m K
0.08
0.07
0.06
0.05
0.04
0.03
0.02
100
200
300
400
500
600
700
800
Temperature, C

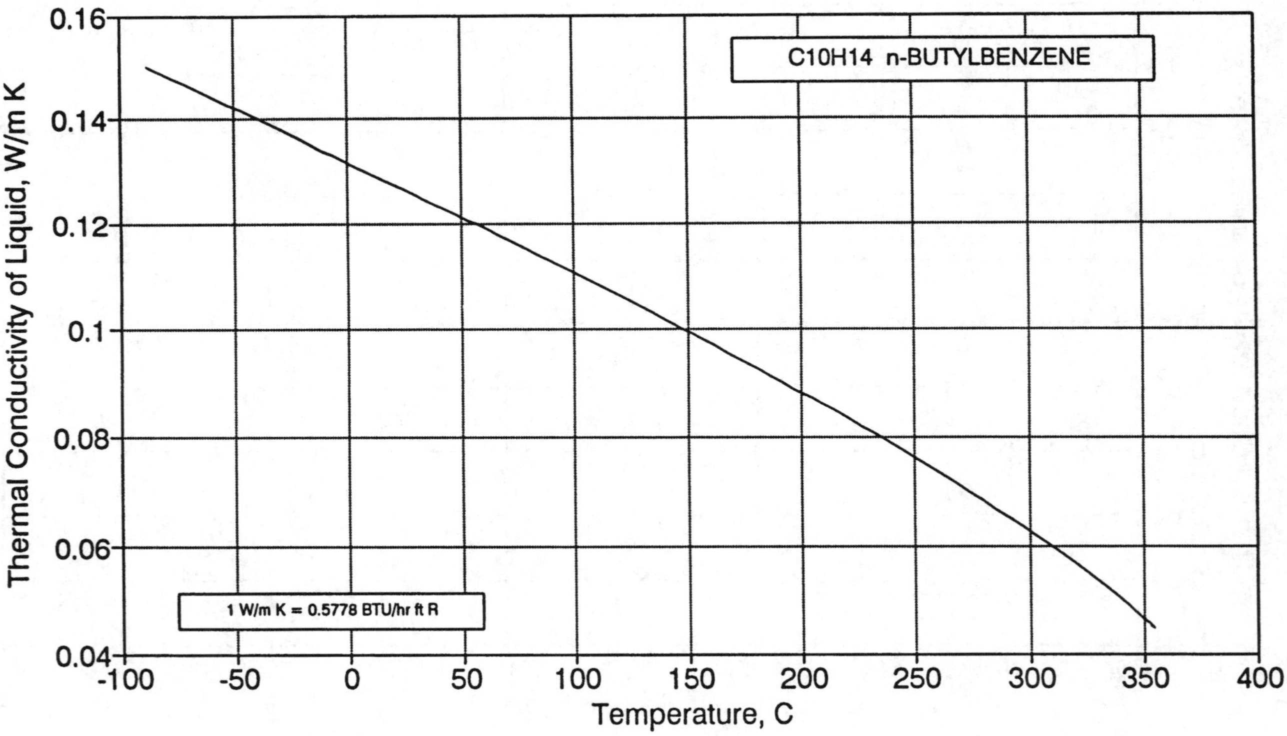
C10H14 n-BUTYLBENZENE
1 W/m K = 0.5778 BTU/hr ft R
Thermal Conductivity of Liquid, W/m K
0.16
0.14
0.12
0.1
0.08
0.06
0.04
-100
-50
0
50
100
150
200
250
300
350
400
Temperature, C

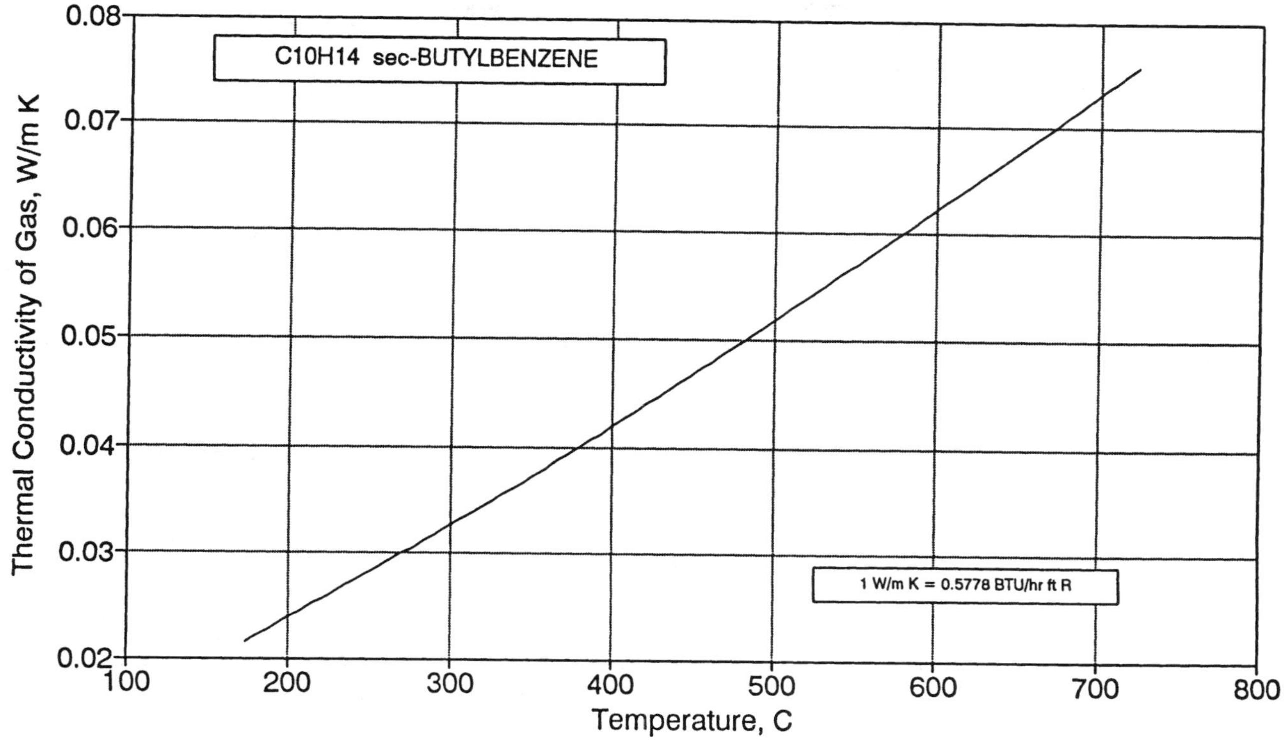
C10H14 sec-BUTYLBENZENE
1 W/m K = 0.5778 BTU/hr ft R
Thermal Conductivity of Gas, W/m K
0.08
0.07
0.06
0.05
0.04
0.03
0.02
100
200
300
400
500
600
700
800
Temperature, C

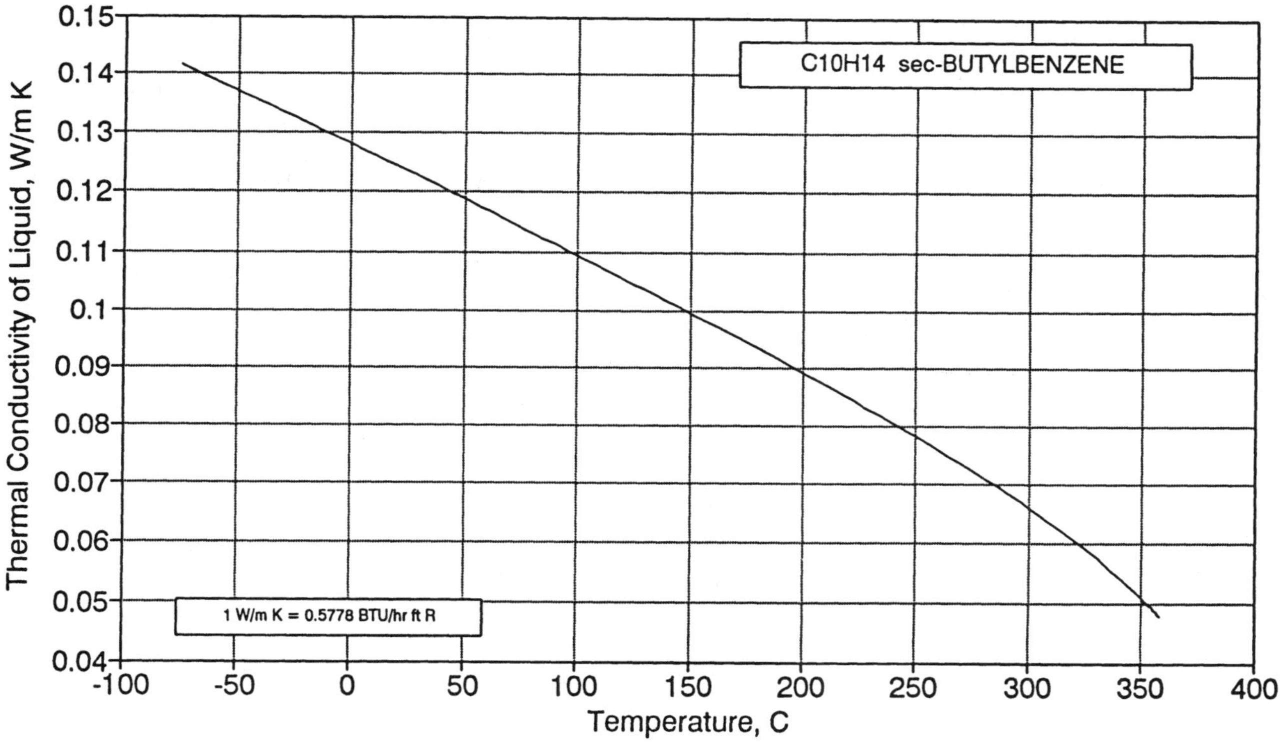
C10H14 sec-BUTYLBENZENE
1 W/m K = 0.5778 BTU/hr ft R
Thermal Conductivity of Liquid, W/m K
0.15
0.14
0.13
0.12
0.11
0.1
0.09
0.08
0.07
0.06
0.05
0.04
-100
-50
0
50
100
150
200
250
300
350
400
Temperature, C

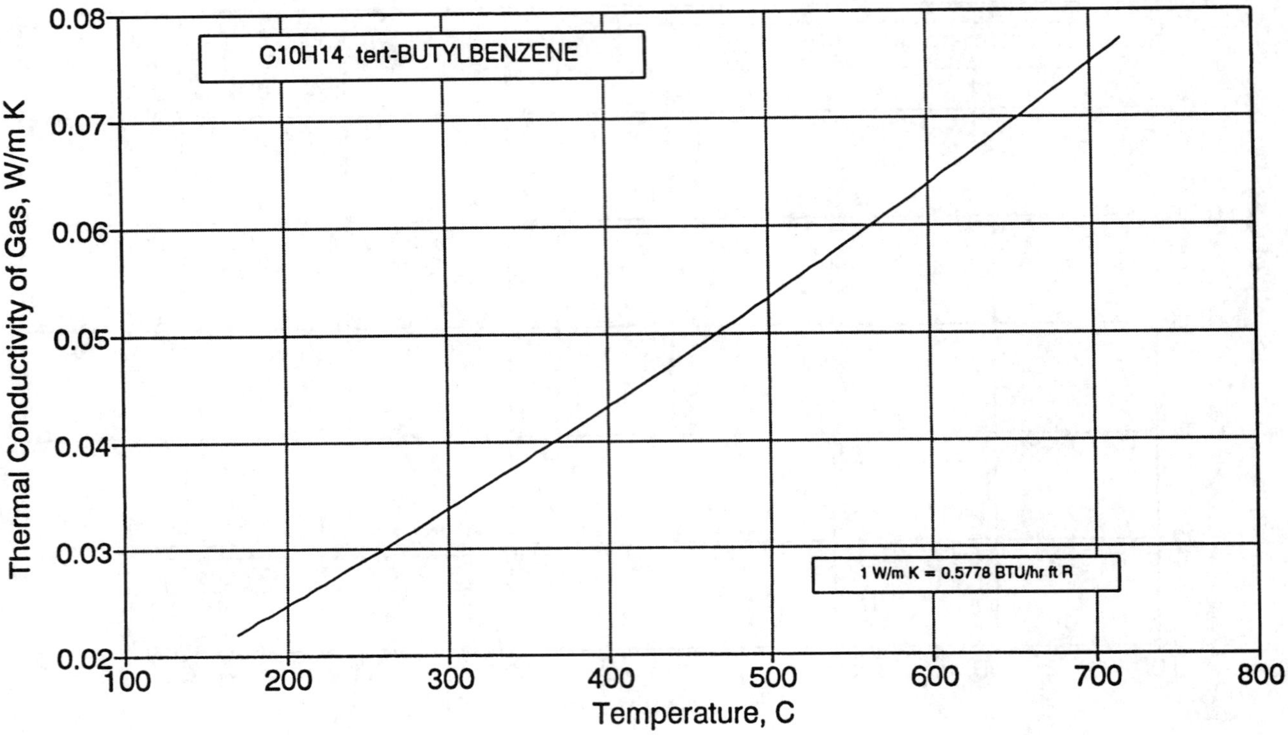
C10H14 tert-BUTYLBENZENE
1 W/m K = 0.5778 BTU/hr ft R
Thermal Conductivity of Gas, W/m K
0.08
0.07
0.06
0.05
0.04
0.03
0.02
100
200
300
400
500
600
700
800
Temperature, C

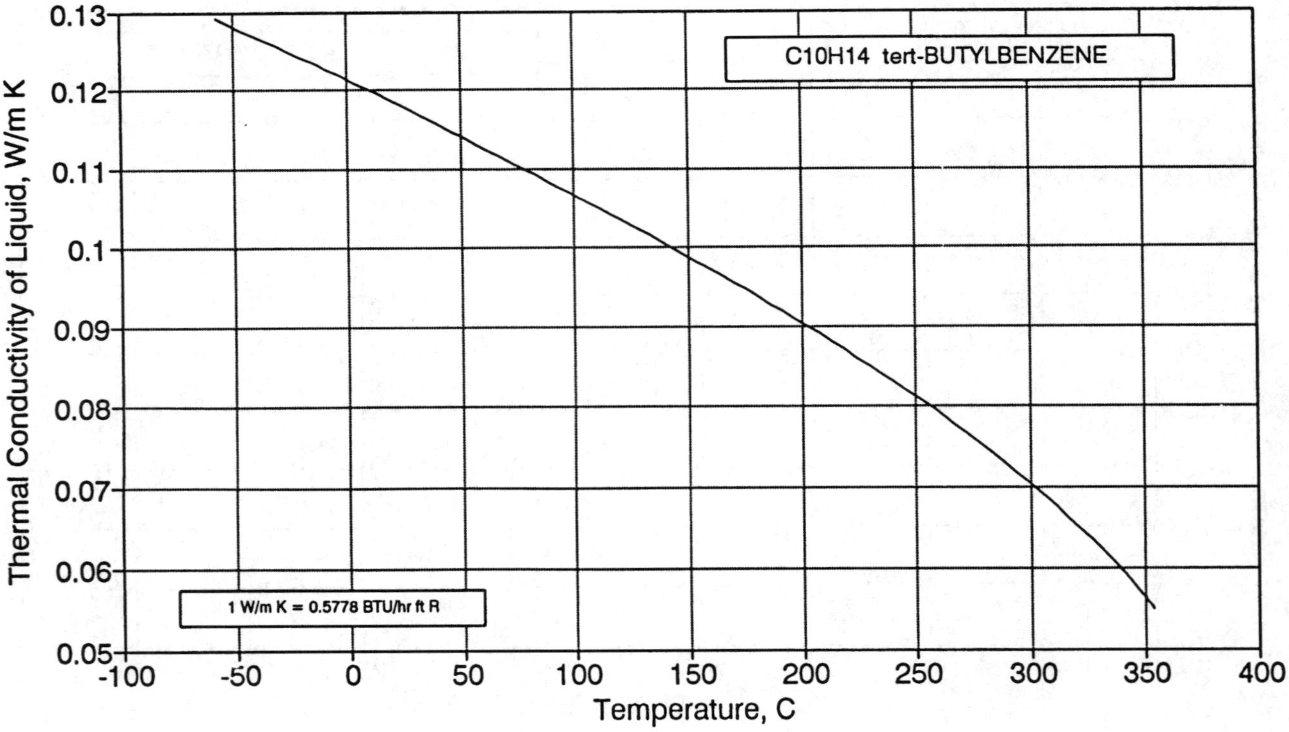
C10H14 tert-BUTYLBENZENE
1 W/m K = 0.5778 BTU/hr ft R
Thermal Conductivity of Liquid, W/m K
0.13
0.12
0.11
0.1
0.09
0.08
0.07
0.06
0.05
-100
-50
0
50
100
150
200
250
300
350
400
Temperature, C

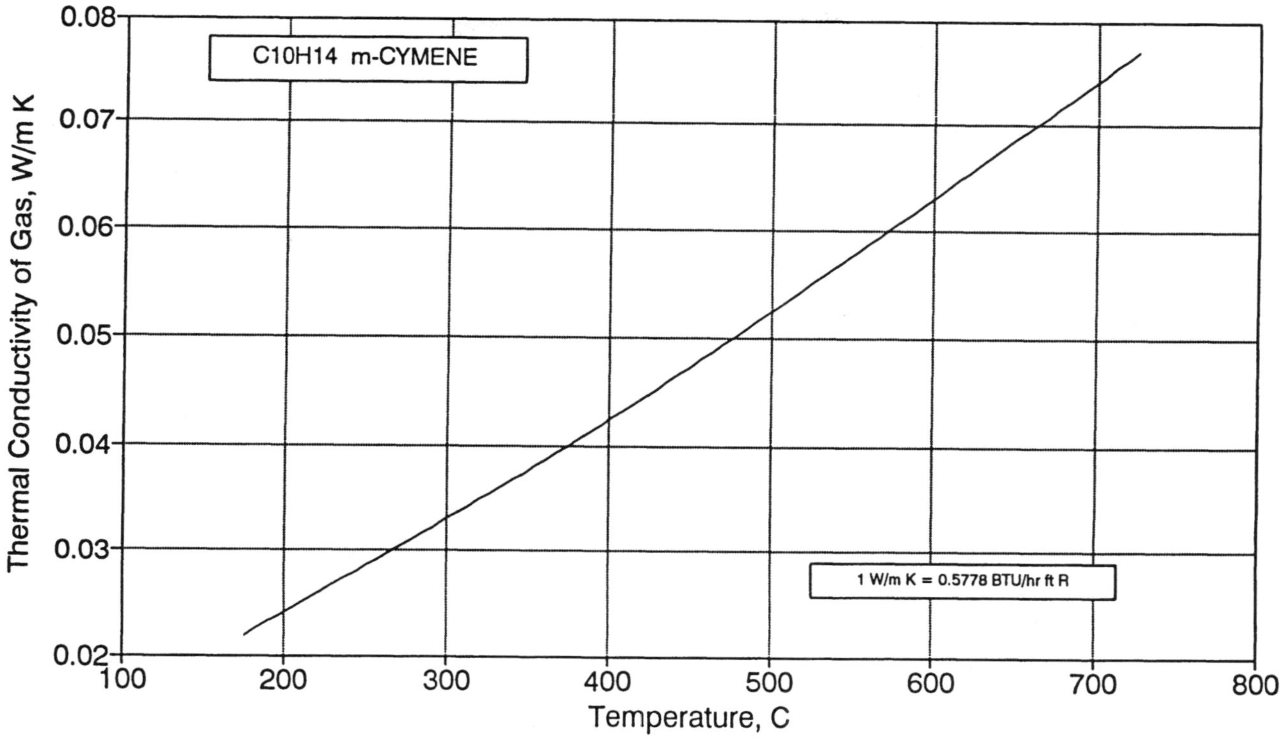
C10H14 m-CYMENE
Thermal Conductivity of Gas, W/m K
0.08
0.07
0.06
0.05
0.04
0.03
0.02
1 W/m K = 0.5778 BTU/hr ft R
100
200
300
400
500
600
700
800
Temperature, C

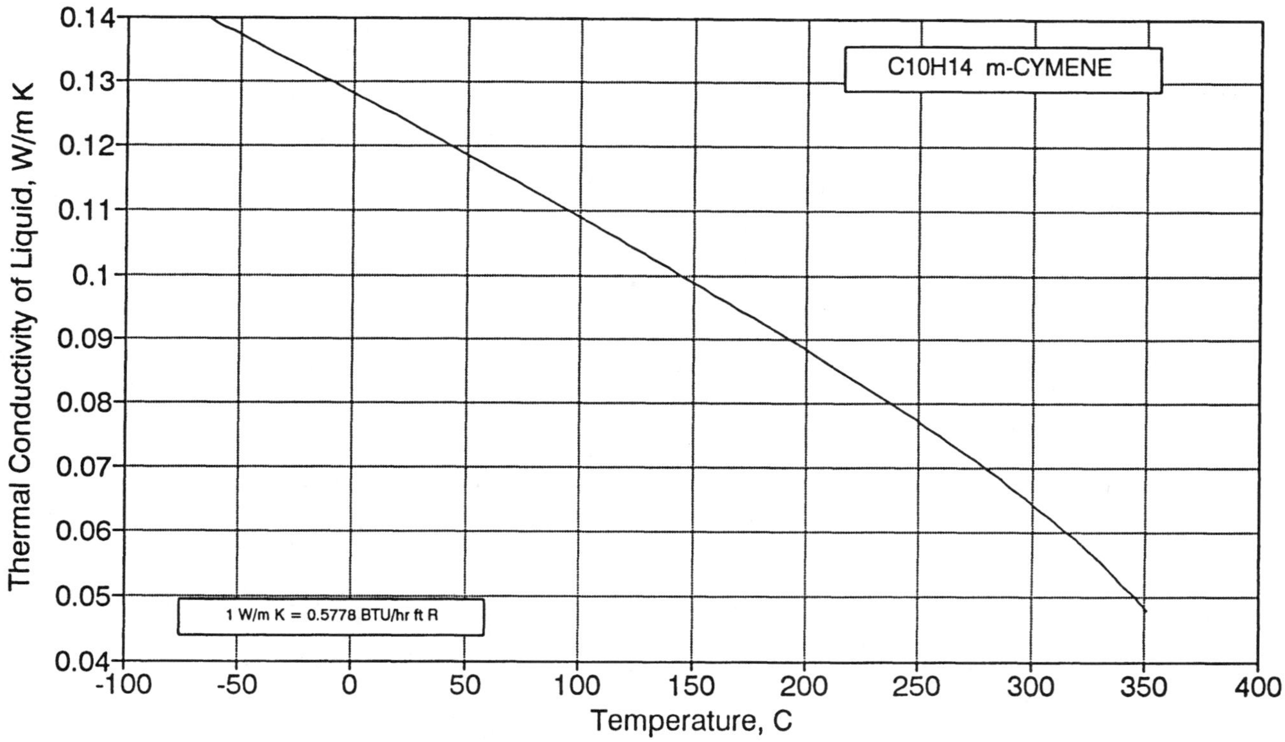
C10H14 m-CYMENE
Thermal Conductivity of Liquid, W/m K
0.14
0.13
0.12
0.11
0.1
0.09
0.08
0.07
0.06
0.05
0.04
1 W/m K = 0.5778 BTU/hr ft R
-100
-50
0
50
100
150
200
250
300
350
400
Temperature, C

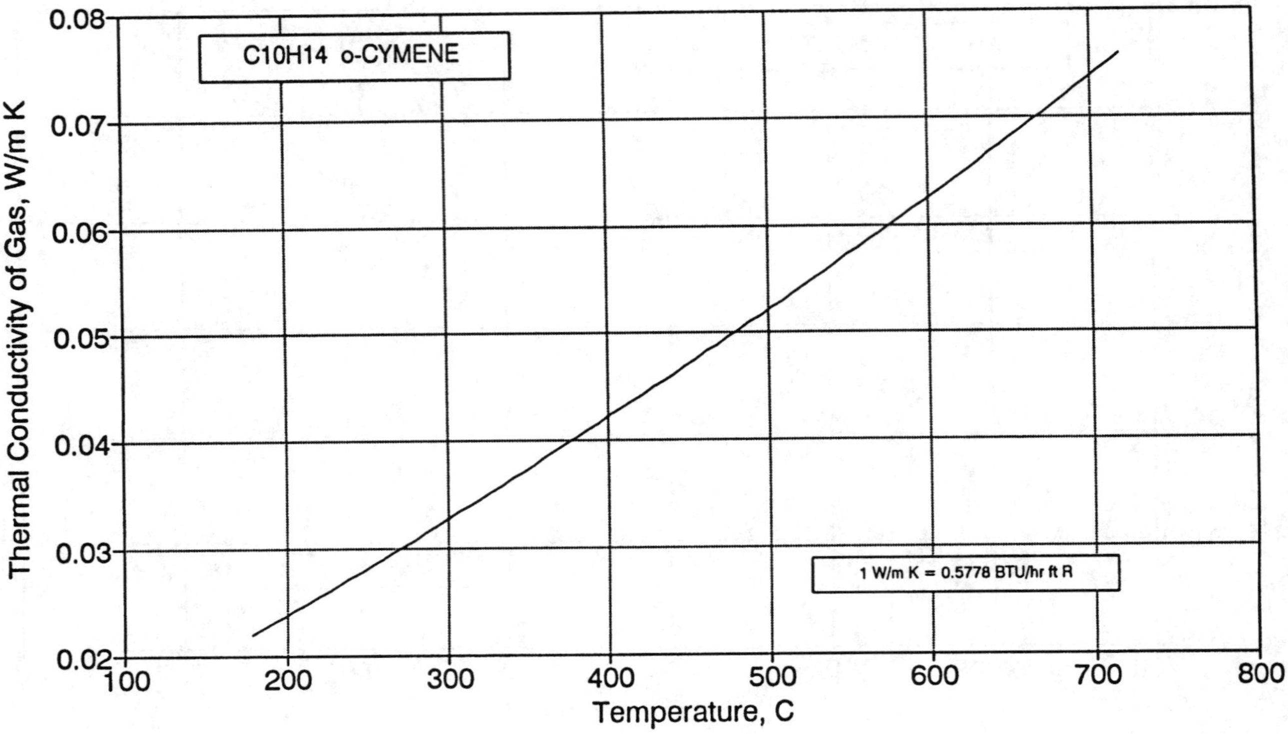
C10H14 o-CYMENE
1 W/m K = 0.5778 BTU/hr ft R
Thermal Conductivity of Gas, W/m K
Temperature, C
0.08
0.07
0.06
0.05
0.04
0.03
0.02
100
200
300
400
500
600
700
800

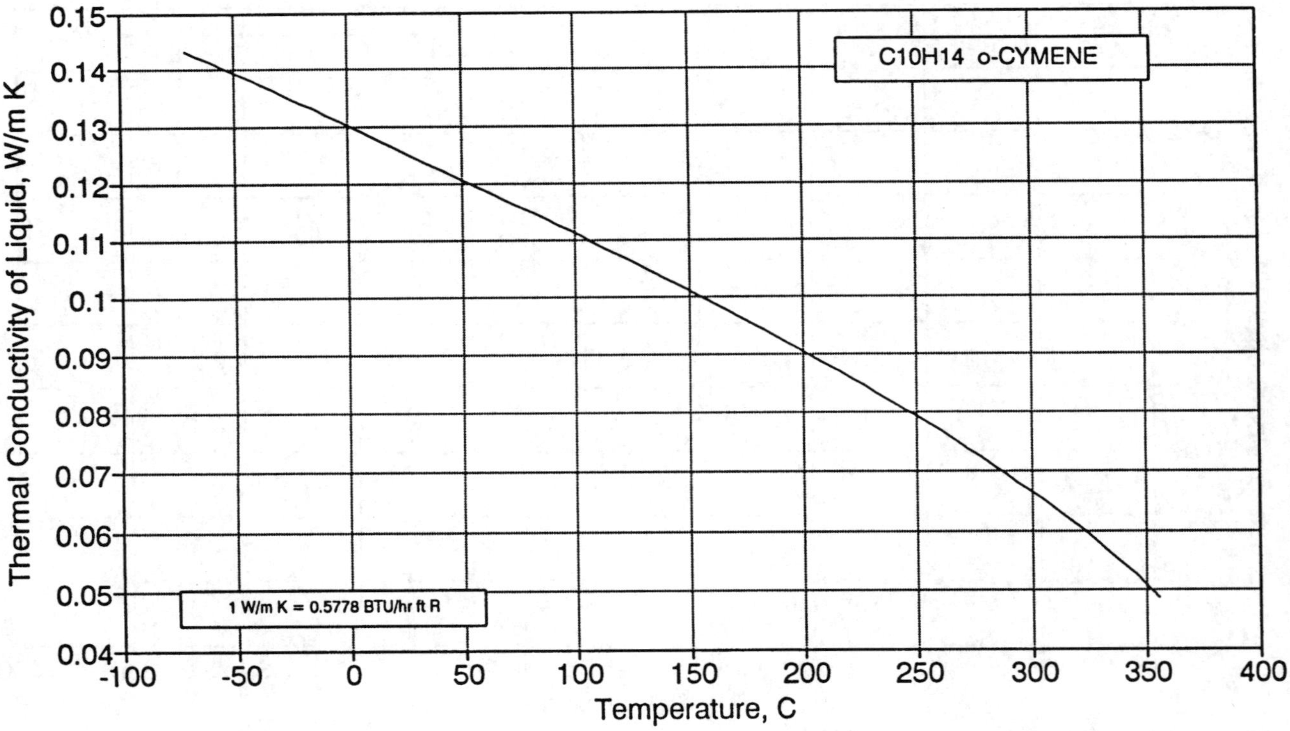
C10H14 o-CYMENE
1 W/m K = 0.5778 BTU/hr ft R
Thermal Conductivity of Liquid, W/m K
Temperature, C
0.15
0.14
0.13
0.12
0.11
0.1
0.09
0.08
0.07
0.06
0.05
0.04
-100
-50
0
50
100
150
200
250
300
350
400

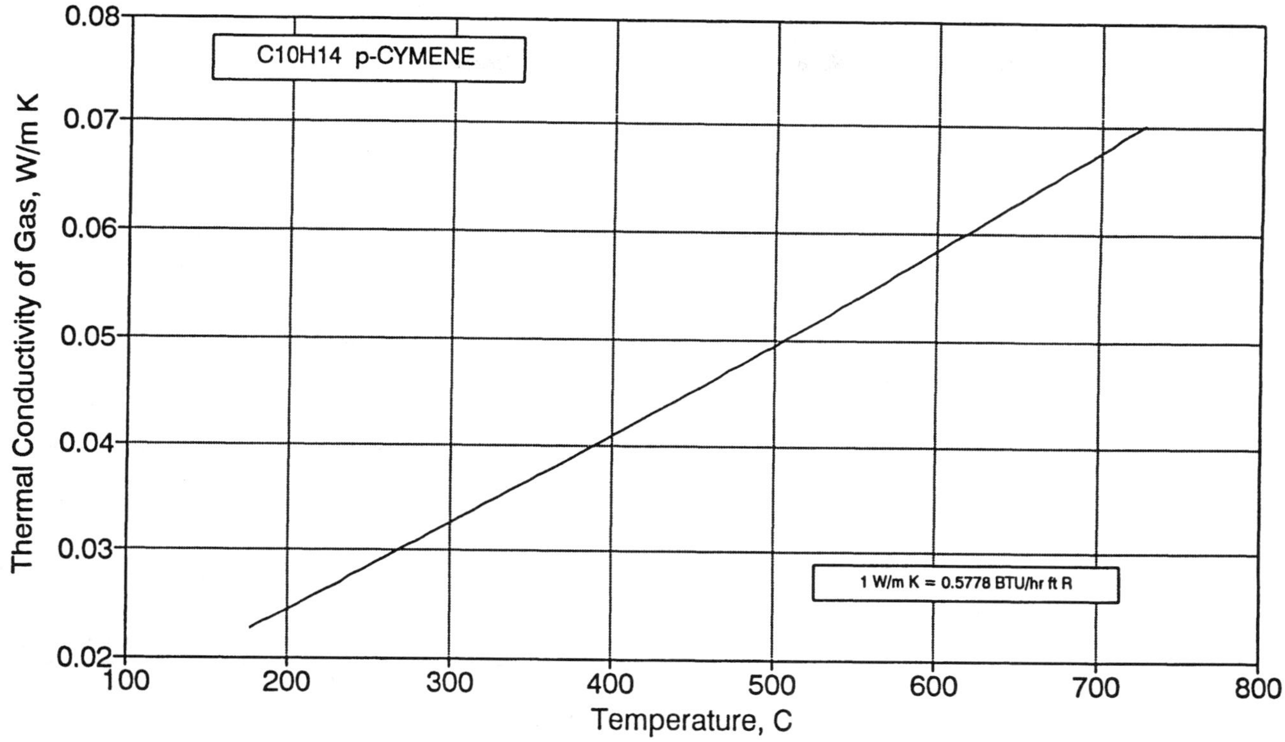
C10H14 p-CYMENE
1 W/m K = 0.5778 BTU/hr ft R
Thermal Conductivity of Gas, W/m K
0.08
0.07
0.06
0.05
0.04
0.03
0.02
100
200
300
400
500
600
700
800
Temperature, C

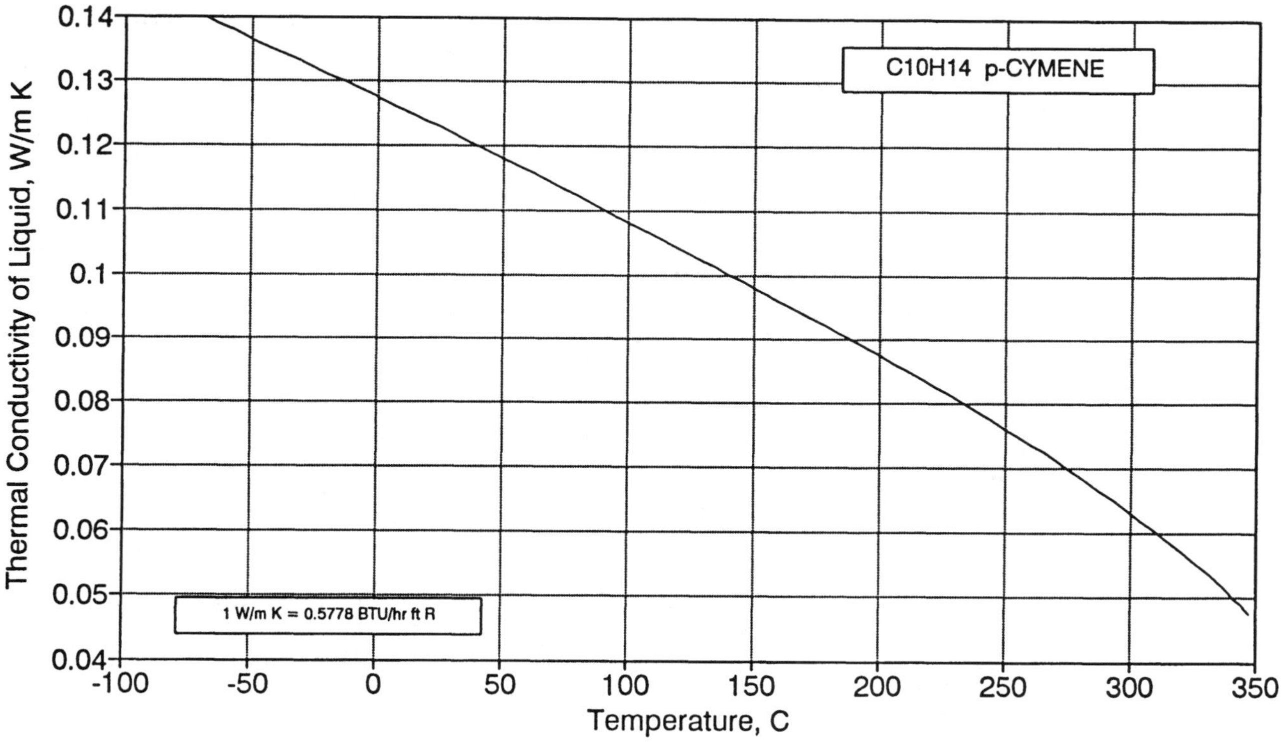
C10H14 p-CYMENE
1 W/m K = 0.5778 BTU/hr ft R
Thermal Conductivity of Liquid, W/m K
0.14
0.13
0.12
0.11
0.1
0.09
0.08
0.07
0.06
0.05
0.04
-100
-50
0
50
100
150
200
250
300
350
Temperature, C

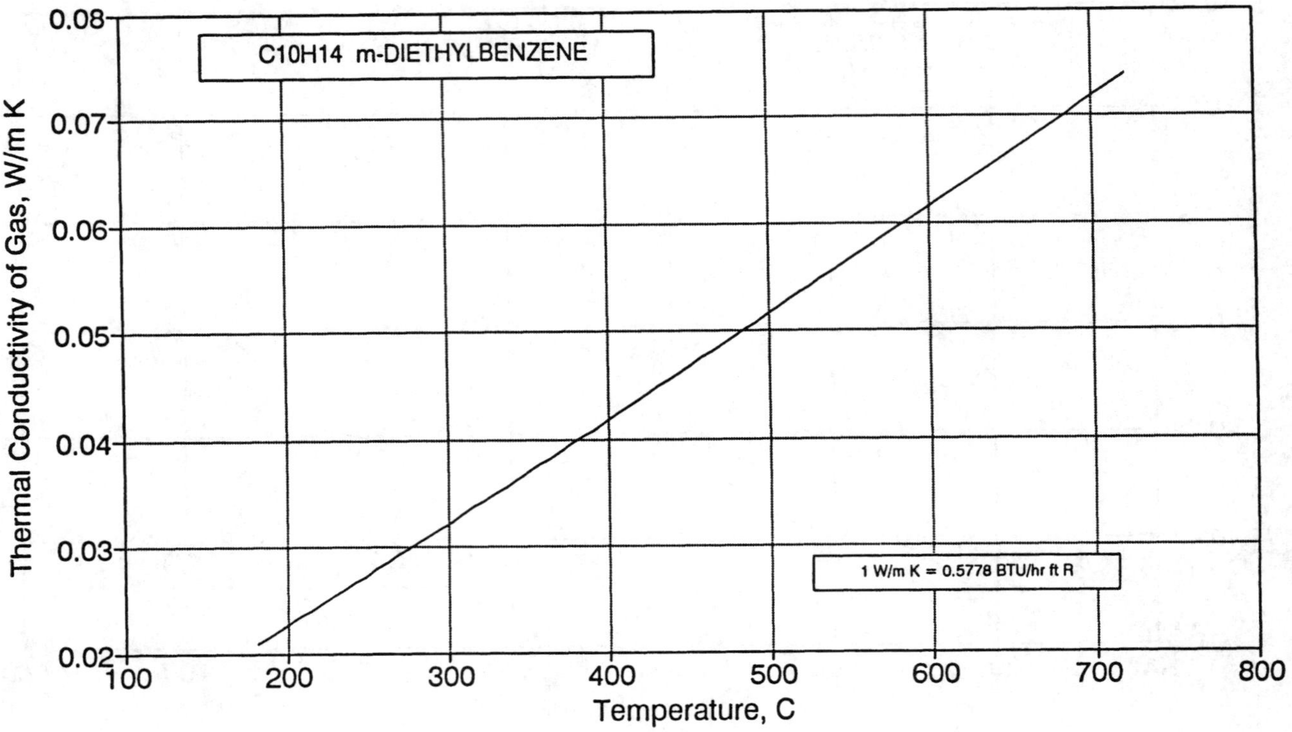
C10H14 m-DIETHYLBENZENE
1 W/m K = 0.5778 BTU/hr ft R
Thermal Conductivity of Gas, W/m K
0.08
0.07
0.06
0.05
0.04
0.03
0.02
100
200
300
400
500
600
700
800
Temperature, C

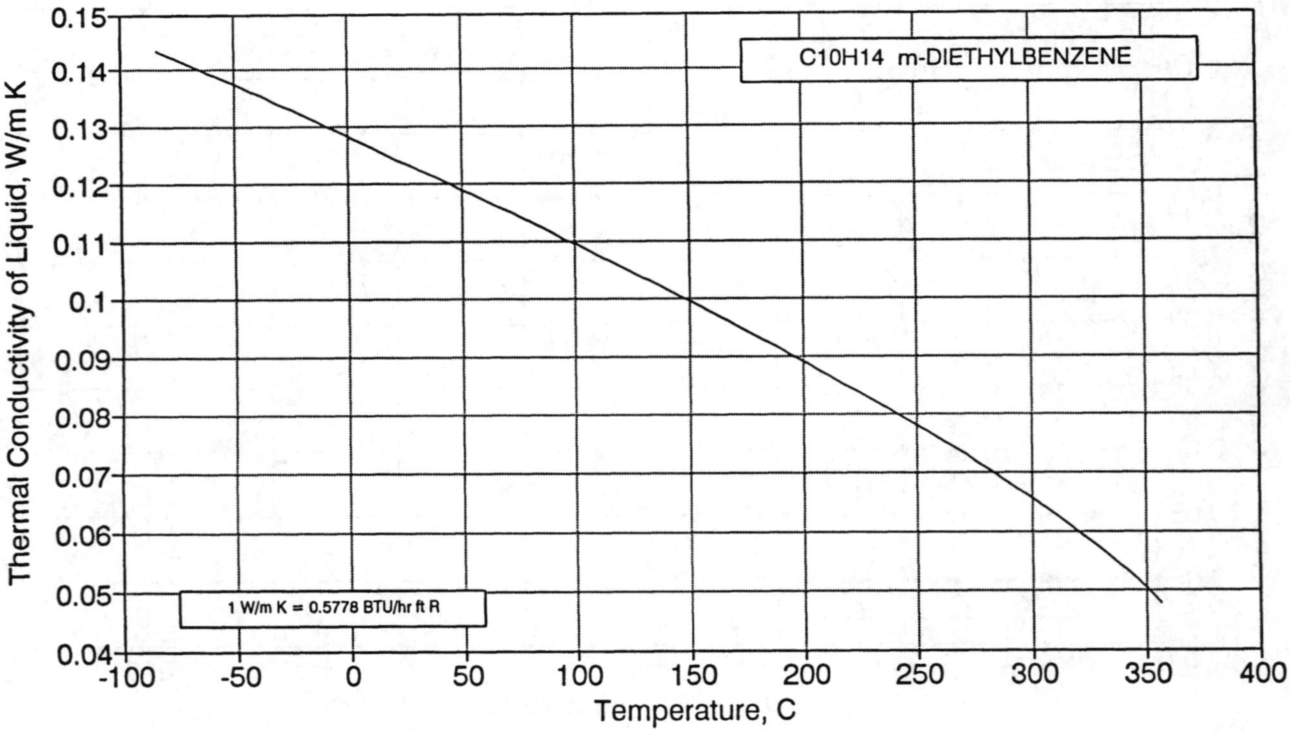
C10H14 m-DIETHYLBENZENE
1 W/m K = 0.5778 BTU/hr ft R
Thermal Conductivity of Liquid, W/m K
0.15
0.14
0.13
0.12
0.11
0.1
0.09
0.08
0.07
0.06
0.05
0.04
-100
-50
0
50
100
150
200
250
300
350
400
Temperature, C

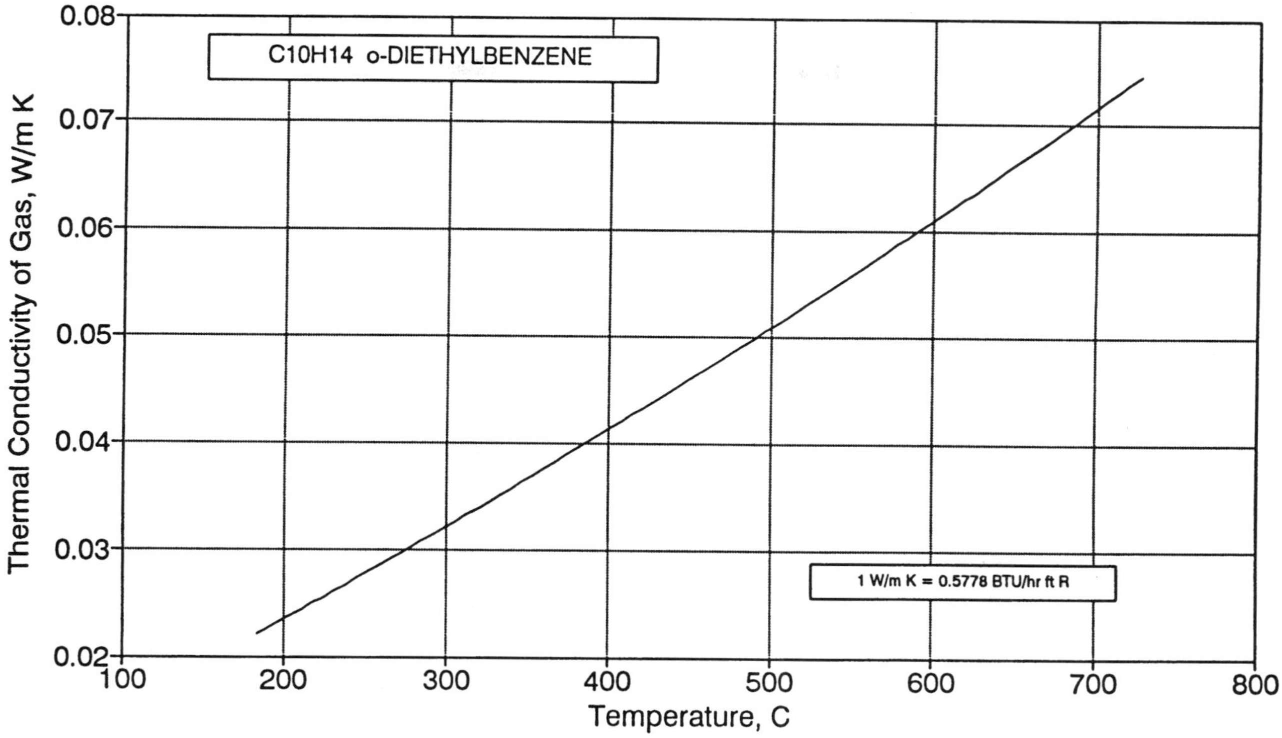
C10H14 o-DIETHYLBENZENE
1 W/m K = 0.5778 BTU/hr ft R
Thermal Conductivity of Gas, W/m K
0.08
0.07
0.06
0.05
0.04
0.03
0.02
100
200
300
400
500
600
700
800
Temperature, C

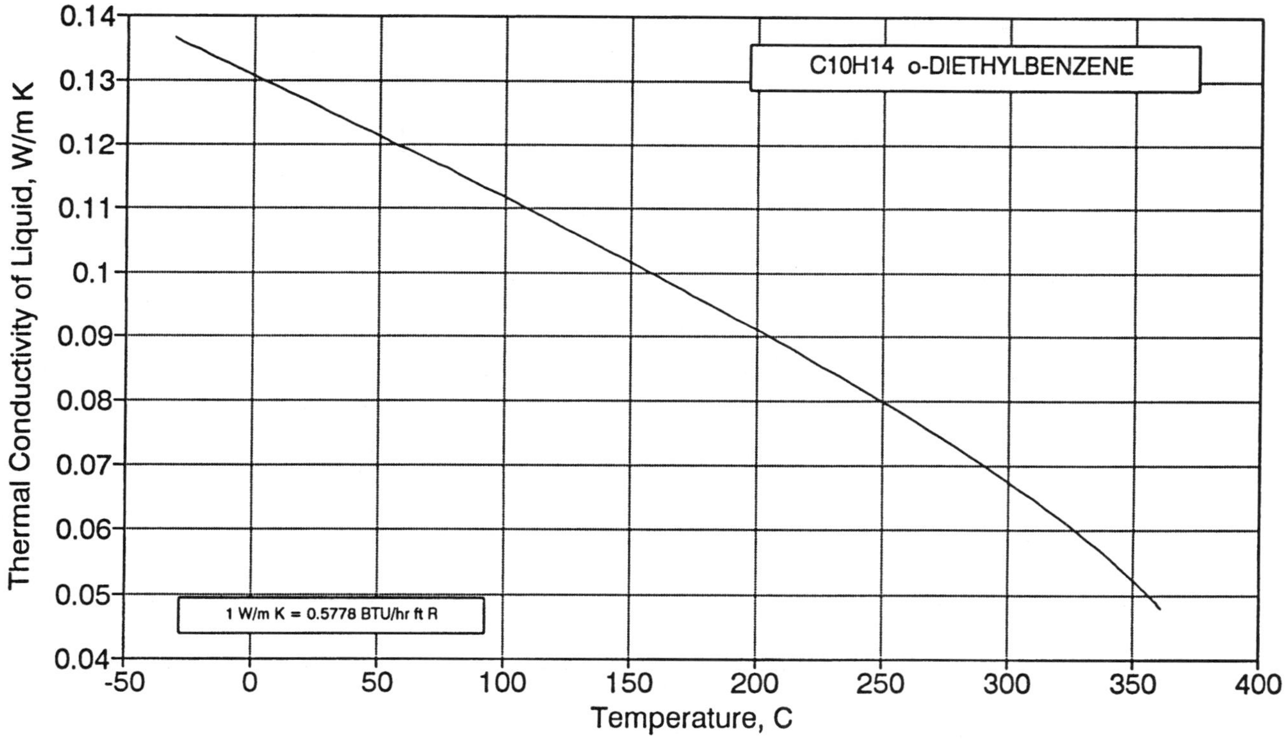
C10H14 o-DIETHYLBENZENE
1 W/m K = 0.5778 BTU/hr ft R
Thermal Conductivity of Liquid, W/m K
0.14
0.13
0.12
0.11
0.1
0.09
0.08
0.07
0.06
0.05
0.04
-50
0
50
100
150
200
250
300
350
400
Temperature, C

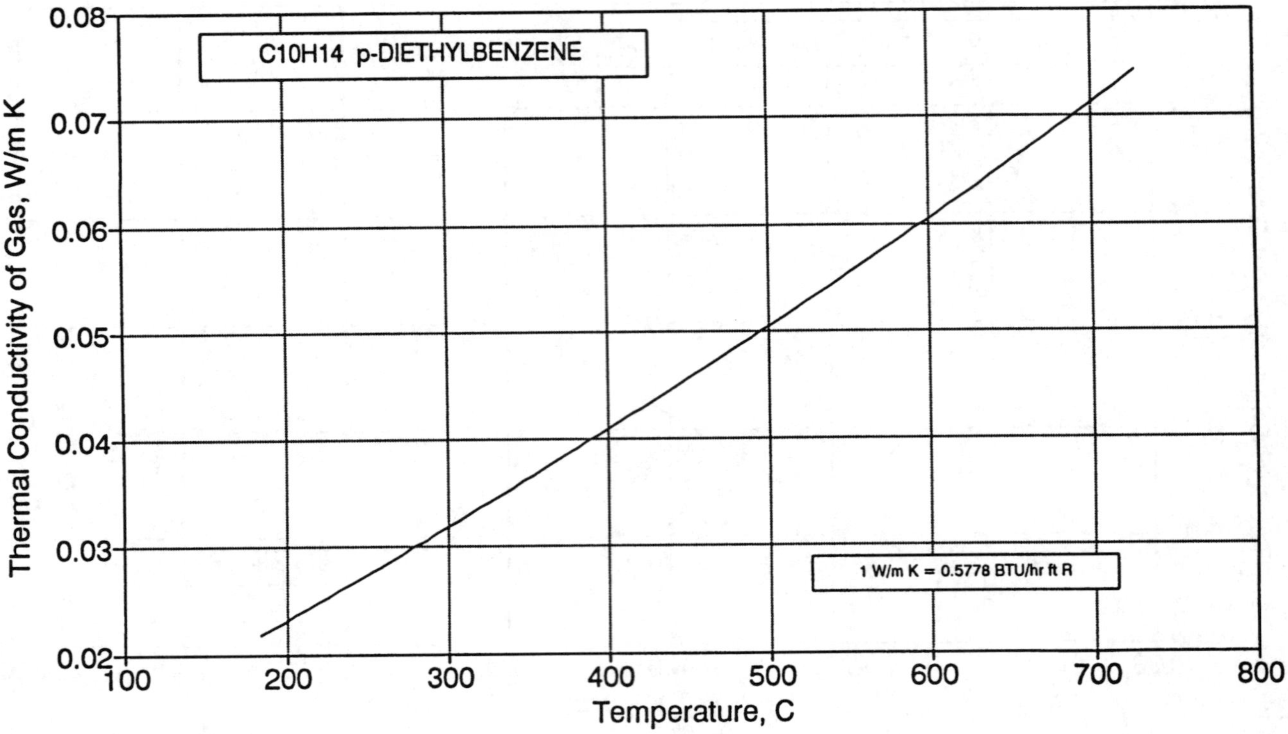
C10H14 p-DIETHYLBENZENE
Thermal Conductivity of Gas, W/m K
0.08
0.07
0.06
0.05
0.04
0.03
0.02
1 W/m K = 0.5778 BTU/hr ft R
100
200
300
400
500
600
700
800
Temperature, C

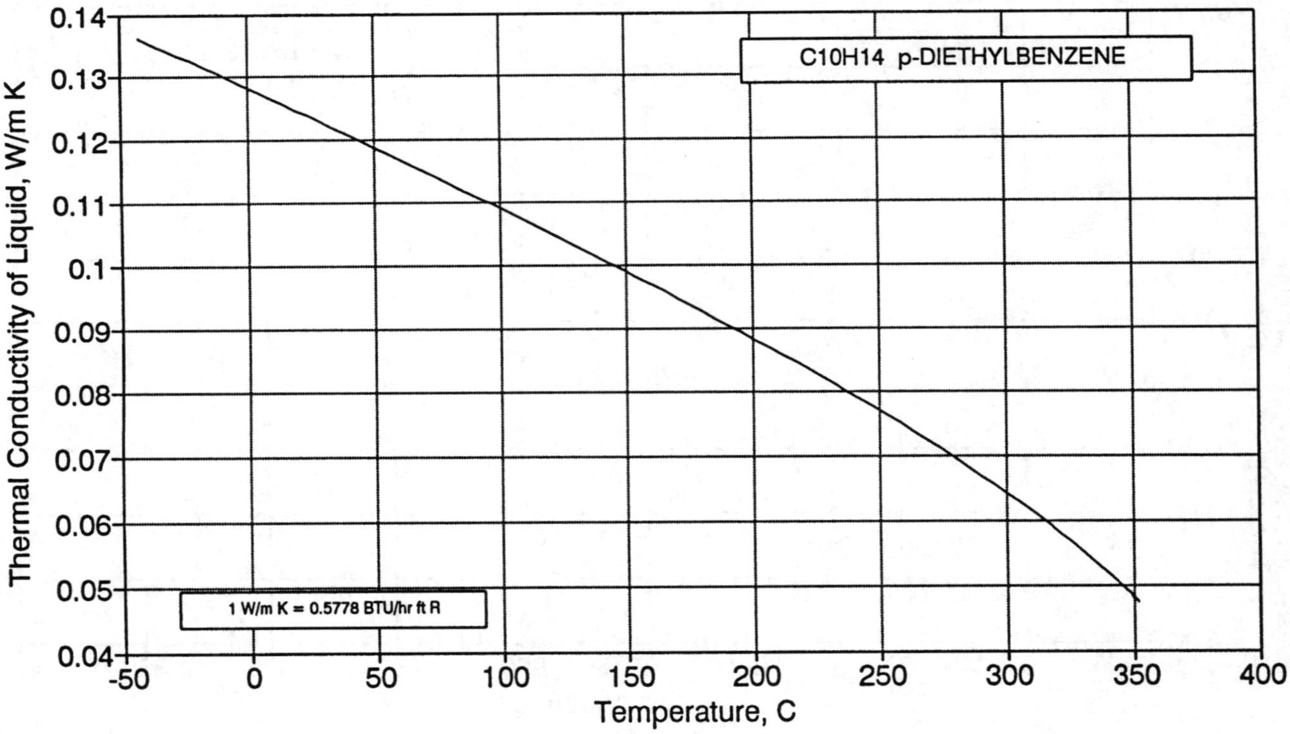
C10H14 p-DIETHYLBENZENE
Thermal Conductivity of Liquid, W/m K
0.14
0.13
0.12
0.11
0.1
0.09
0.08
0.07
0.06
0.05
0.04
1 W/m K = 0.5778 BTU/hr ft R
-50
0
50
100
150
200
250
300
350
400
Temperature, C

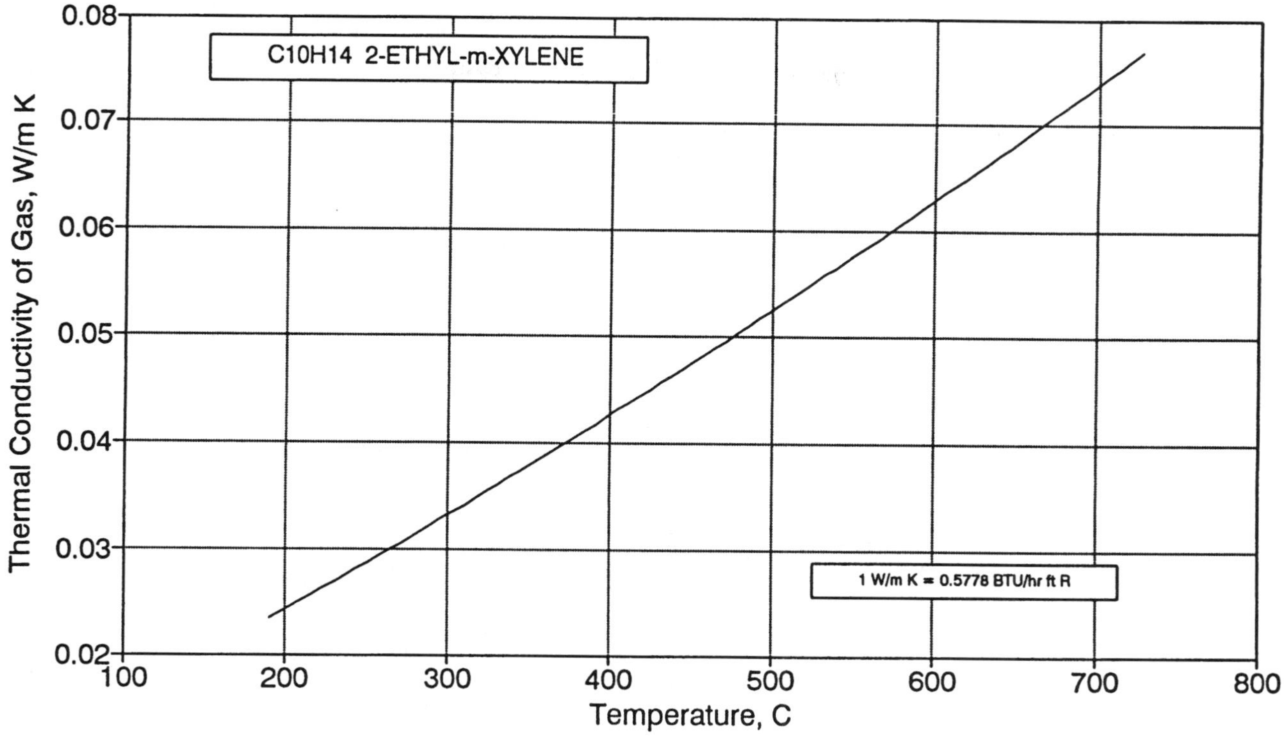
C10H14 2-ETHYL-m-XYLENE
1 W/m K = 0.5778 BTU/hr ft R
Thermal Conductivity of Gas, W/m K
0.08
0.07
0.06
0.05
0.04
0.03
0.02
100
200
300
400
500
600
700
800
Temperature, C

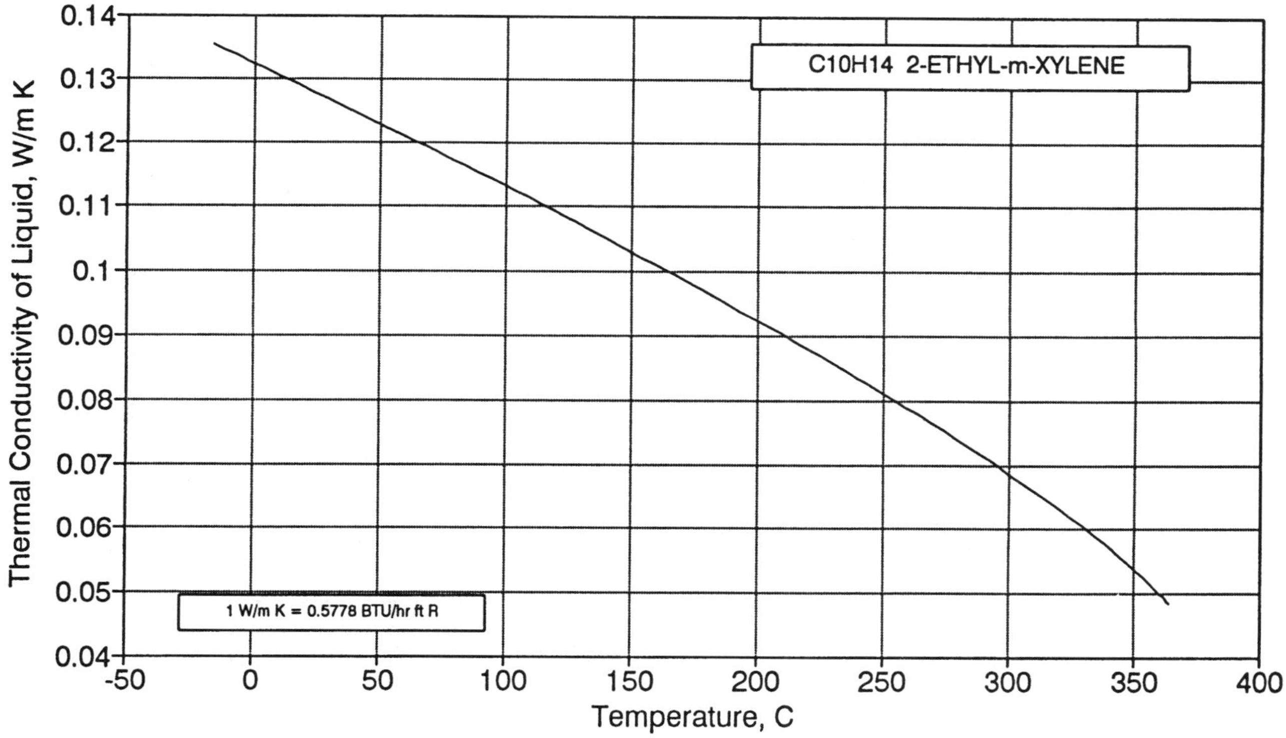
C10H14 2-ETHYL-m-XYLENE
1 W/m K = 0.5778 BTU/hr ft R
Thermal Conductivity of Liquid, W/m K
0.14
0.13
0.12
0.11
0.1
0.09
0.08
0.07
0.06
0.05
0.04
-50
0
50
100
150
200
250
300
350
400
Temperature, C

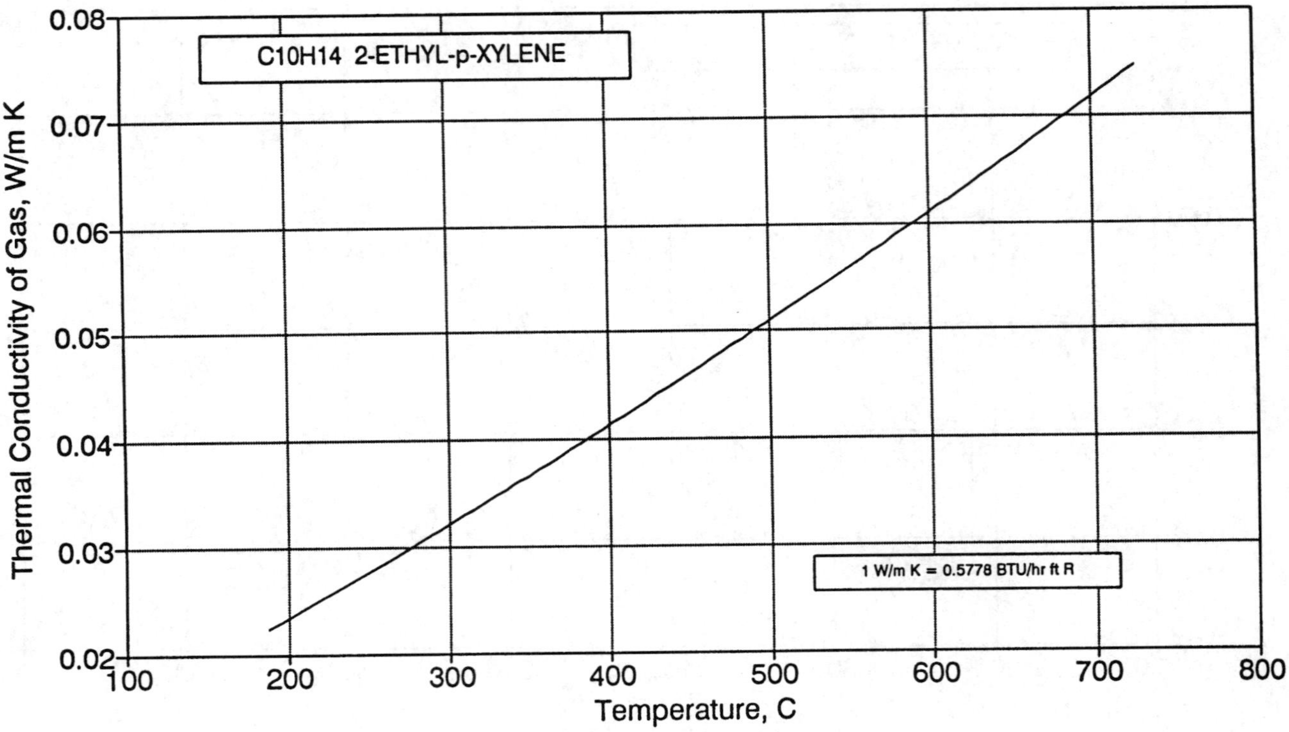
C10H14 2-ETHYL-p-XYLENE
1 W/m K = 0.5778 BTU/hr ft R
Thermal Conductivity of Gas, W/m K
0.08
0.07
0.06
0.05
0.04
0.03
0.02
100
200
300
400
500
600
700
800
Temperature, C

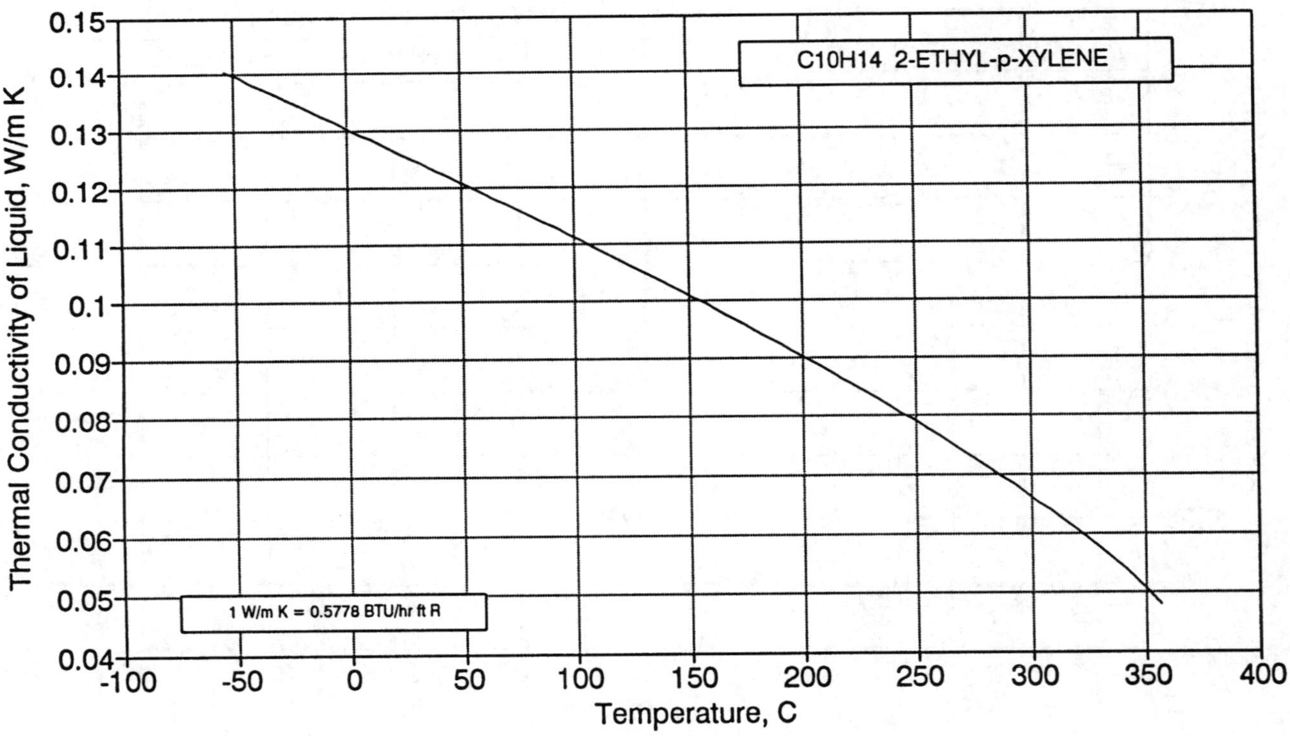
C10H14 2-ETHYL-p-XYLENE
1 W/m K = 0.5778 BTU/hr ft R
Thermal Conductivity of Liquid, W/m K
0.15
0.14
0.13
0.12
0.11
0.1
0.09
0.08
0.07
0.06
0.05
0.04
-100
-50
0
50
100
150
200
250
300
350
400
Temperature, C

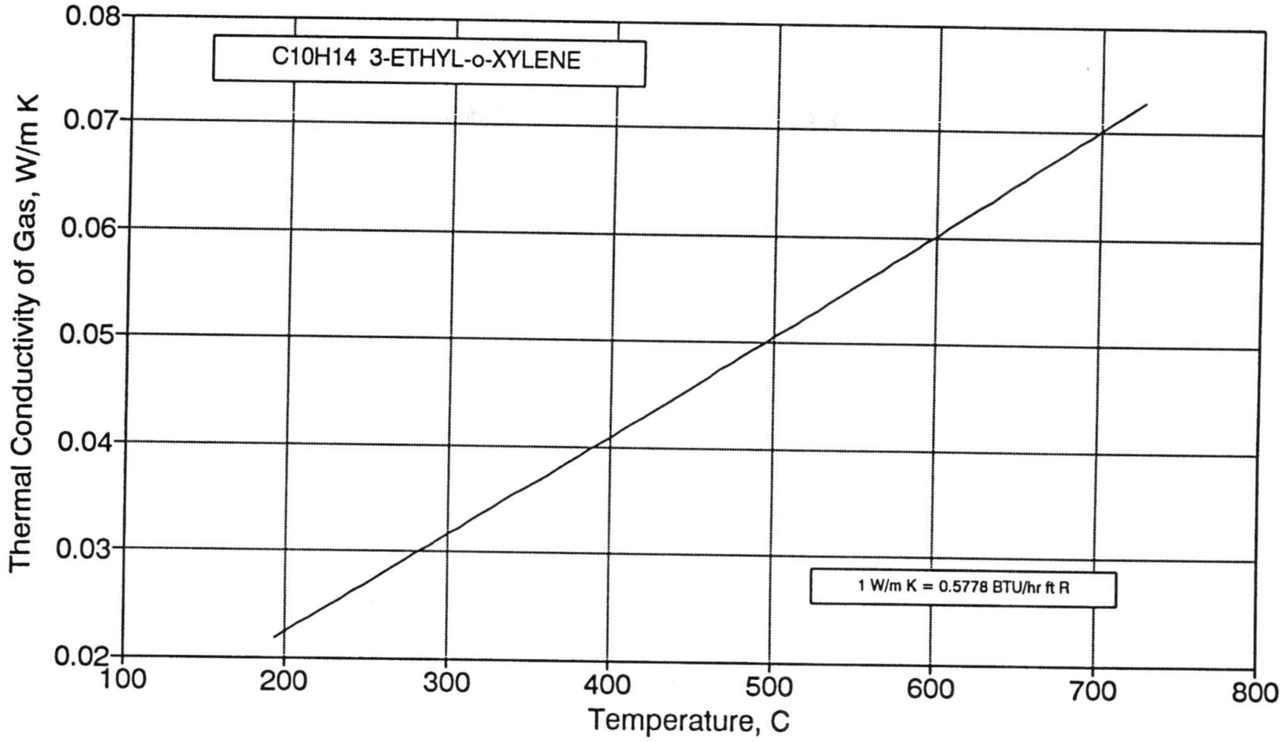
C10H14 3-ETHYL-o-XYLENE
Thermal Conductivity of Gas, W/m K
0.08
0.07
0.06
0.05
0.04
0.03
0.02
1 W/m K = 0.5778 BTU/hr ft R
100
200
300
400
500
600
700
800
Temperature, C

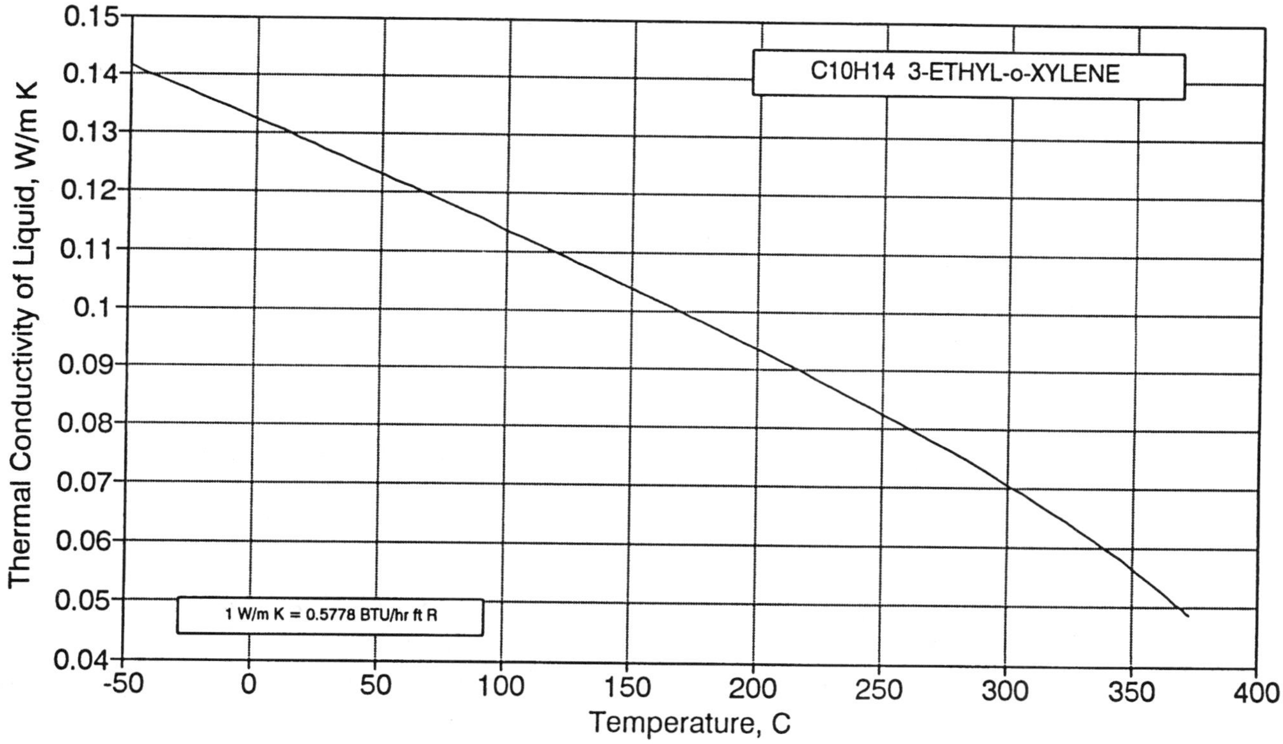
C10H14 3-ETHYL-o-XYLENE
Thermal Conductivity of Liquid, W/m K
0.15
0.14
0.13
0.12
0.11
0.1
0.09
0.08
0.07
0.06
0.05
0.04
1 W/m K = 0.5778 BTU/hr ft R
-50
0
50
100
150
200
250
300
350
400
Temperature, C

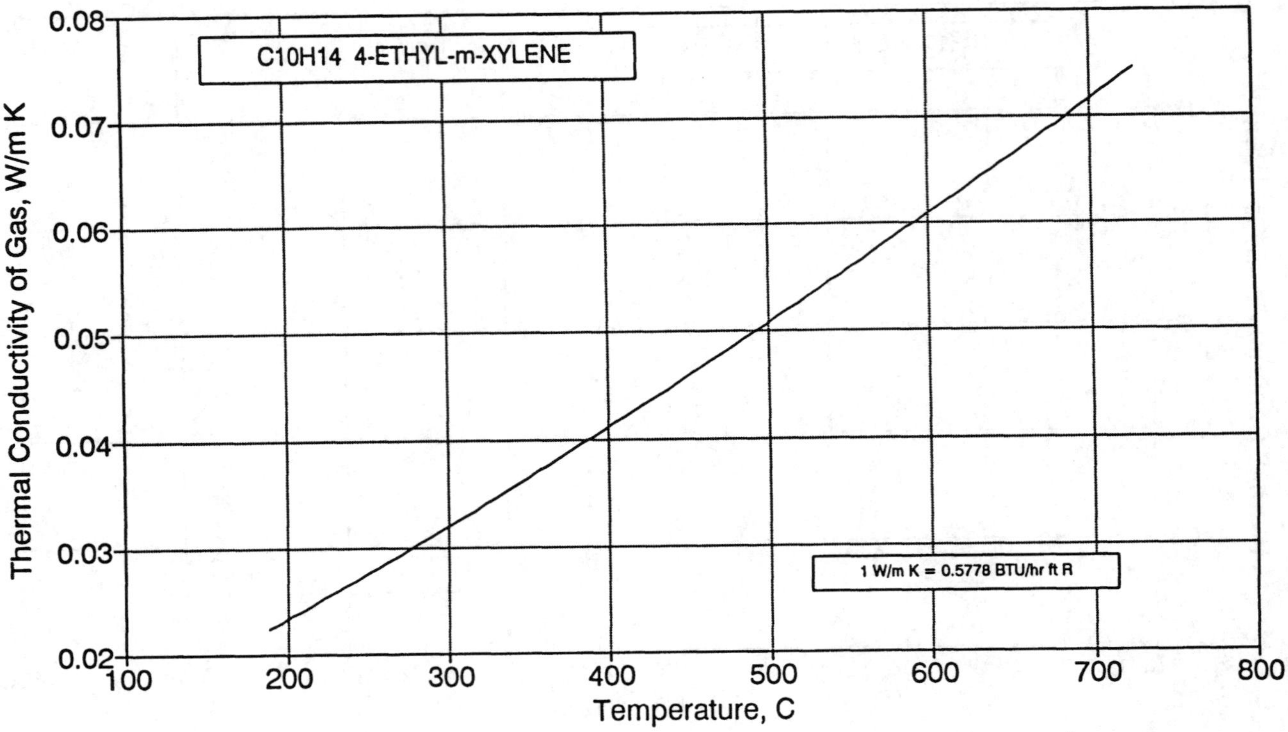
C10H14 4-ETHYL-m-XYLENE
1 W/m K = 0.5778 BTU/hr ft R
Thermal Conductivity of Gas, W/m K
Temperature, C
0.08
0.07
0.06
0.05
0.04
0.03
0.02
100
200
300
400
500
600
700
800

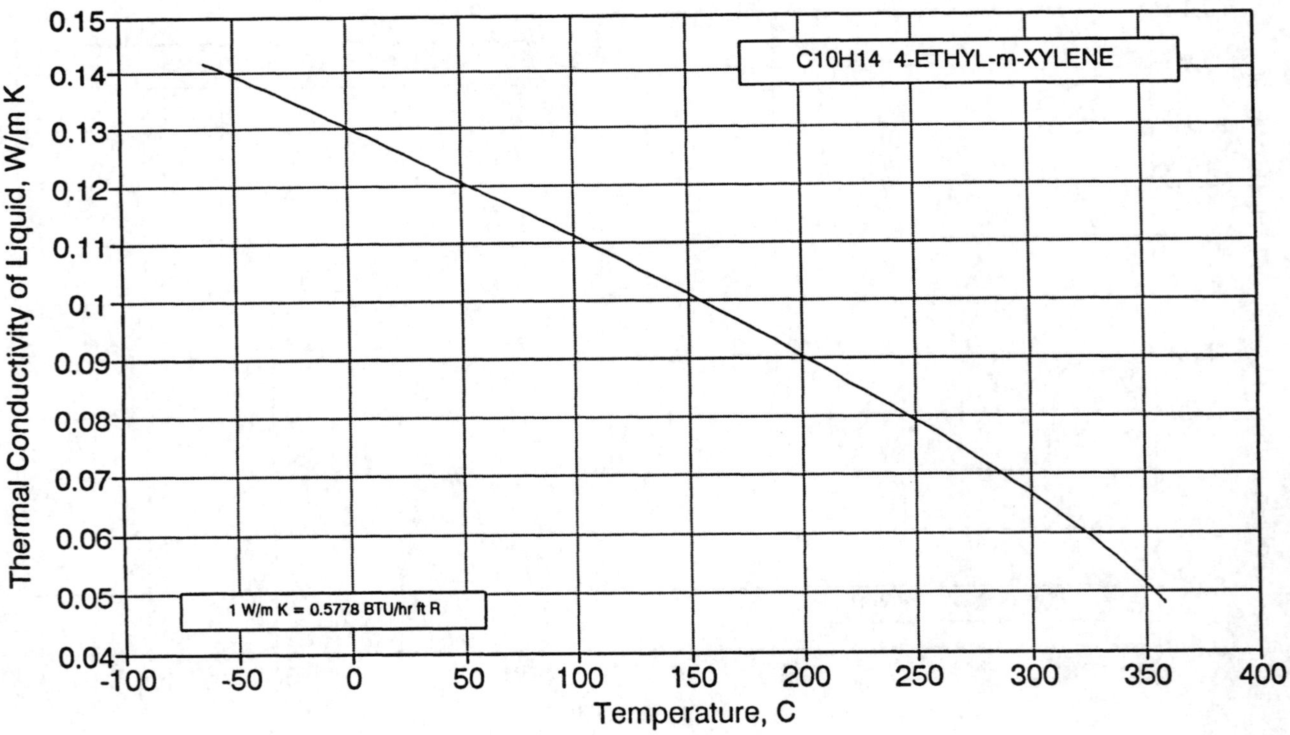
C10H14 4-ETHYL-m-XYLENE
1 W/m K = 0.5778 BTU/hr ft R
Thermal Conductivity of Liquid, W/m K
Temperature, C
0.15
0.14
0.13
0.12
0.11
0.1
0.09
0.08
0.07
0.06
0.05
0.04
-100
-50
0
50
100
150
200
250
300
350
400

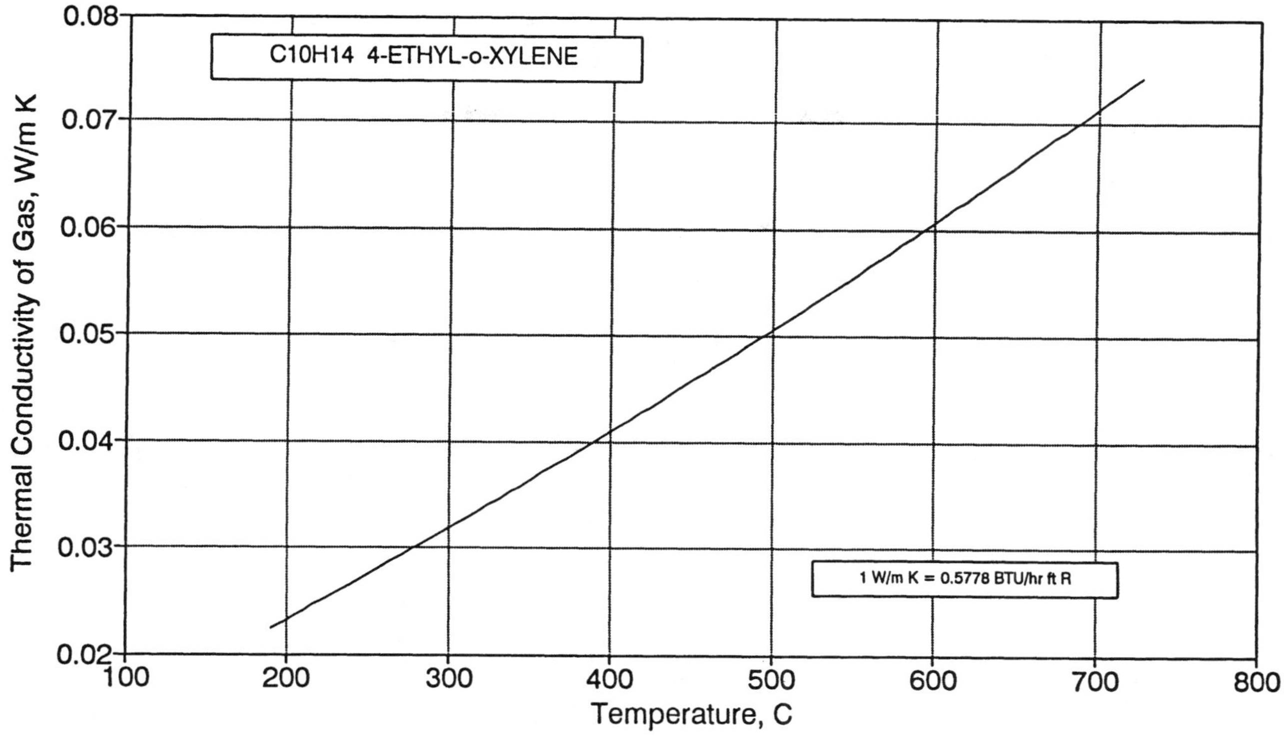
C10H14 4-ETHYL-o-XYLENE
1 W/m K = 0.5778 BTU/hr ft R
Thermal Conductivity of Gas, W/m K
0.08
0.07
0.06
0.05
0.04
0.03
0.02
100
200
300
400
500
600
700
800
Temperature, C

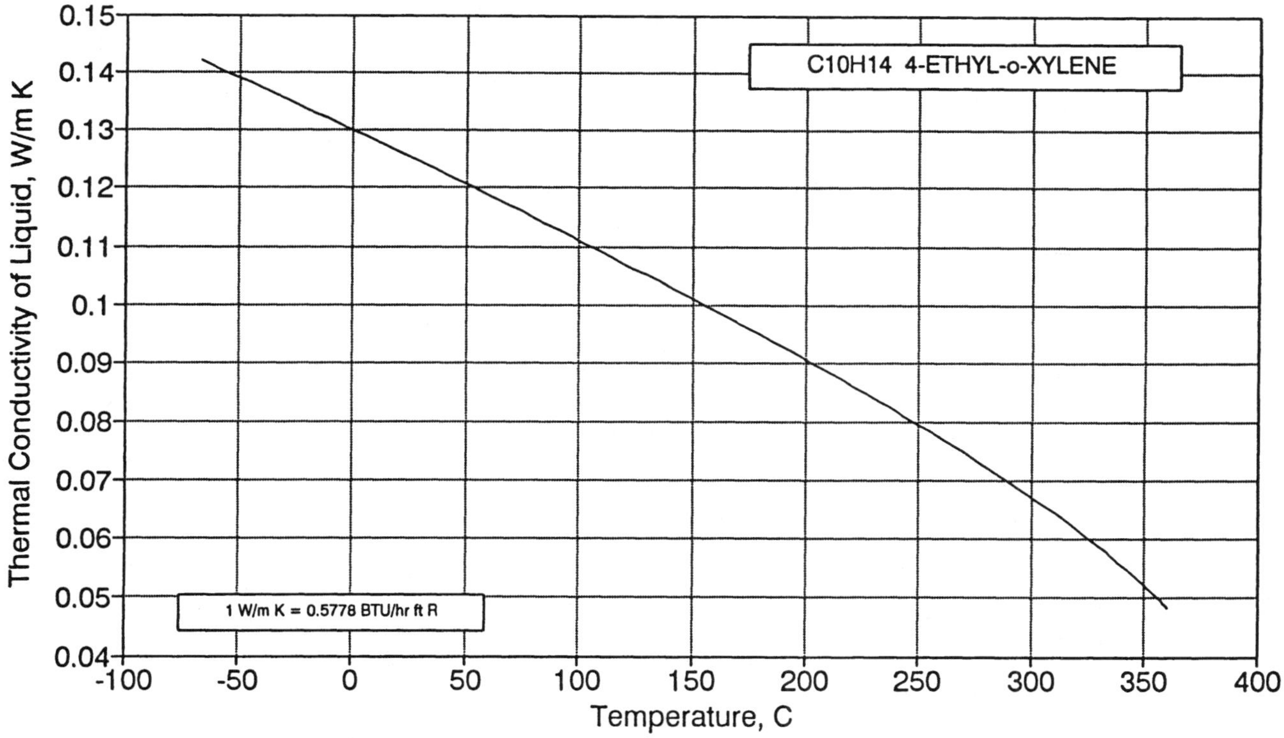
C10H14 4-ETHYL-o-XYLENE
1 W/m K = 0.5778 BTU/hr ft R
Thermal Conductivity of Liquid, W/m K
0.15
0.14
0.13
0.12
0.11
0.1
0.09
0.08
0.07
0.06
0.05
0.04
-100
-50
0
50
100
150
200
250
300
350
400
Temperature, C

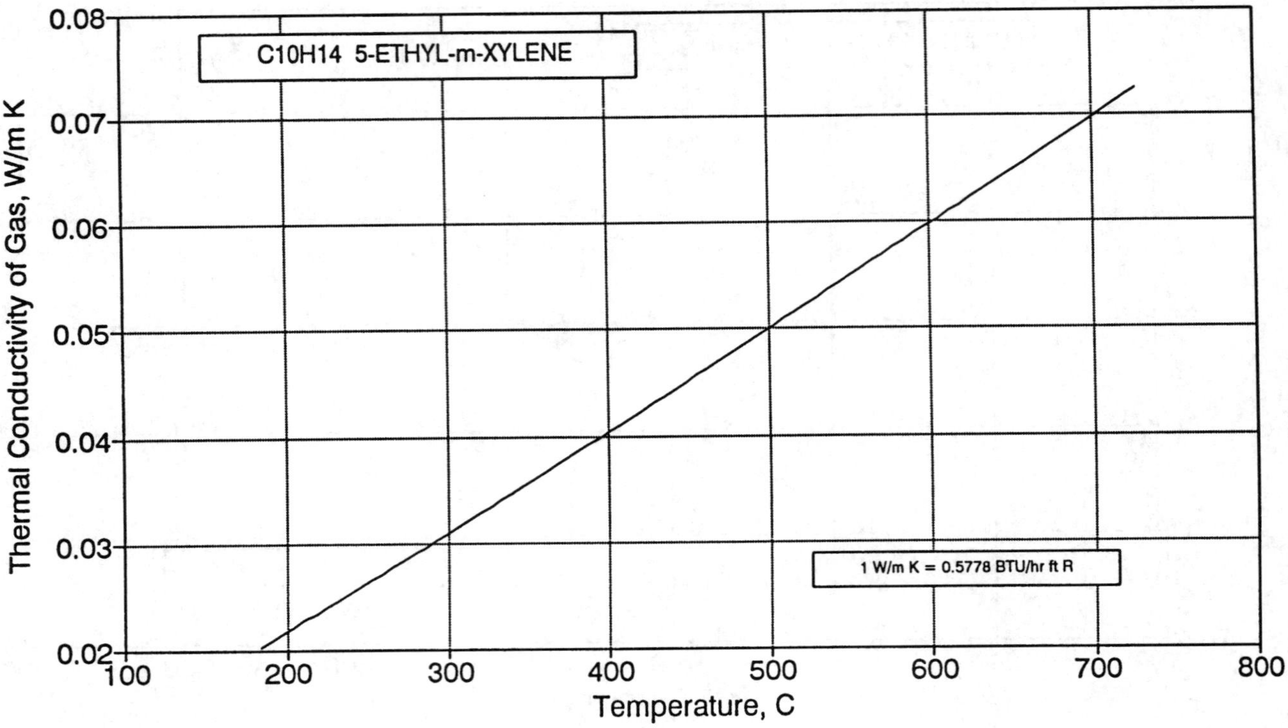
C10H14 5-ETHYL-m-XYLENE
1 W/m K = 0.5778 BTU/hr ft R
Thermal Conductivity of Gas, W/m K
0.08
0.07
0.06
0.05
0.04
0.03
0.02
100
200
300
400
500
600
700
800
Temperature, C

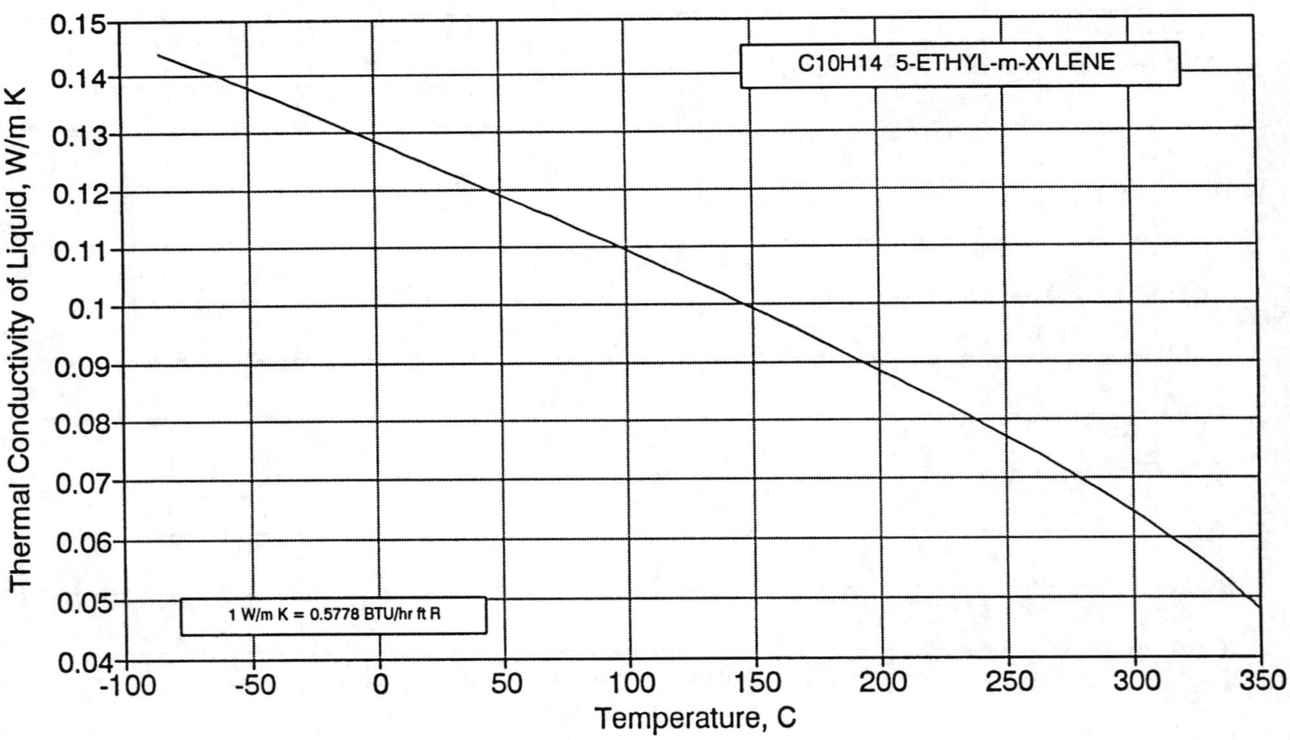
C10H14 5-ETHYL-m-XYLENE
1 W/m K = 0.5778 BTU/hr ft R
Thermal Conductivity of Liquid, W/m K
0.15
0.14
0.13
0.12
0.11
0.1
0.09
0.08
0.07
0.06
0.05
0.04
-100
-50
0
50
100
150
200
250
300
350
Temperature, C

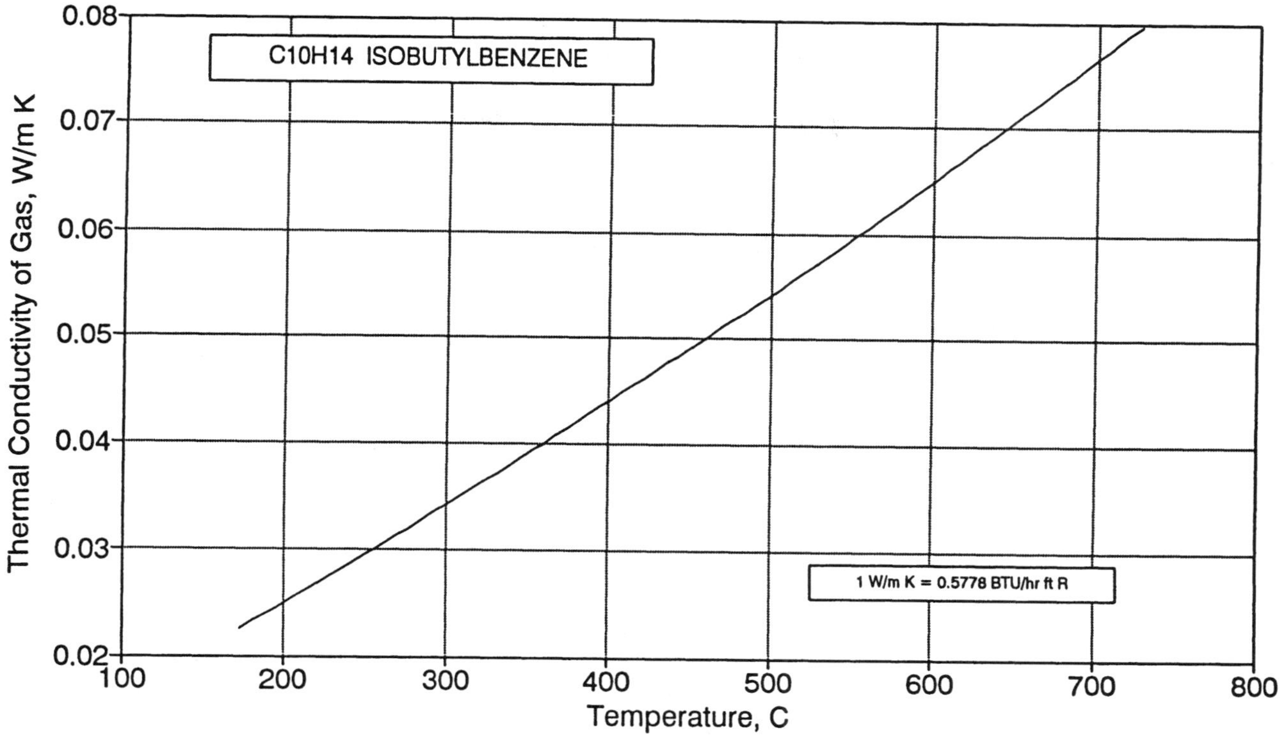
C10H14 ISOBUTYLBENZENE
1 W/m K = 0.5778 BTU/hr ft R
Thermal Conductivity of Gas, W/m K
0.08
0.07
0.06
0.05
0.04
0.03
0.02
100
200
300
400
500
600
700
800
Temperature, C

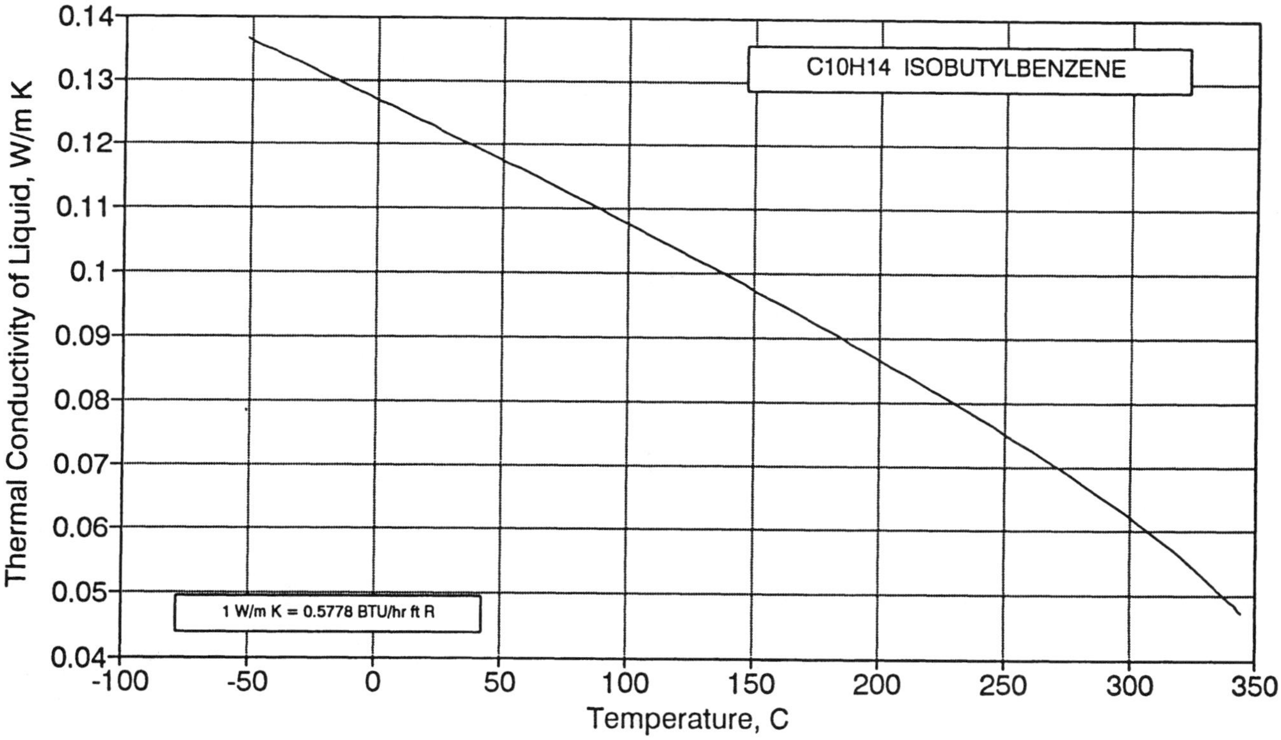
C10H14 ISOBUTYLBENZENE
1 W/m K = 0.5778 BTU/hr ft R
Thermal Conductivity of Liquid, W/m K
0.14
0.13
0.12
0.11
0.1
0.09
0.08
0.07
0.06
0.05
0.04
-100
-50
0
50
100
150
200
250
300
350
Temperature, C

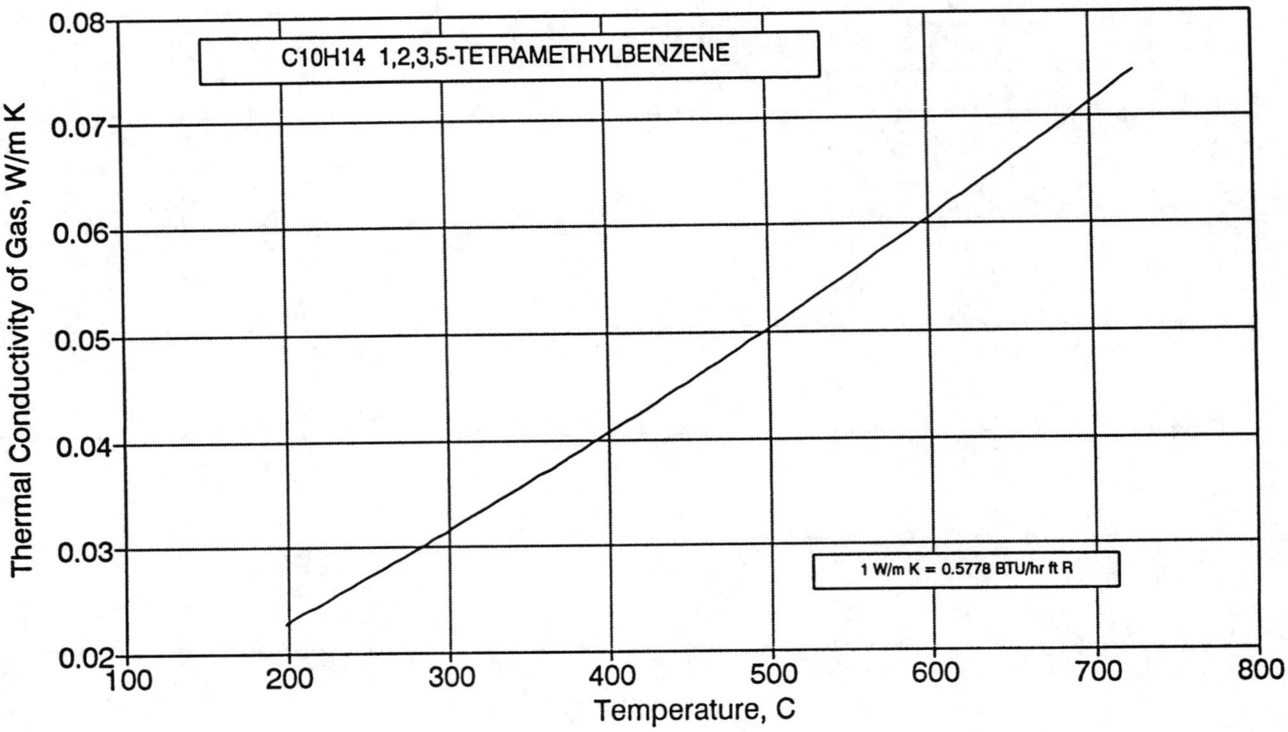
C10H14 1,2,3,5-TETRAMETHYLBENZENE
Thermal Conductivity of Gas, W/m K
0.08
0.07
0.06
0.05
0.04
0.03
0.02
100
200
300
400
500
600
700
800
Temperature, C
1 W/m K = 0.5778 BTU/hr ft R

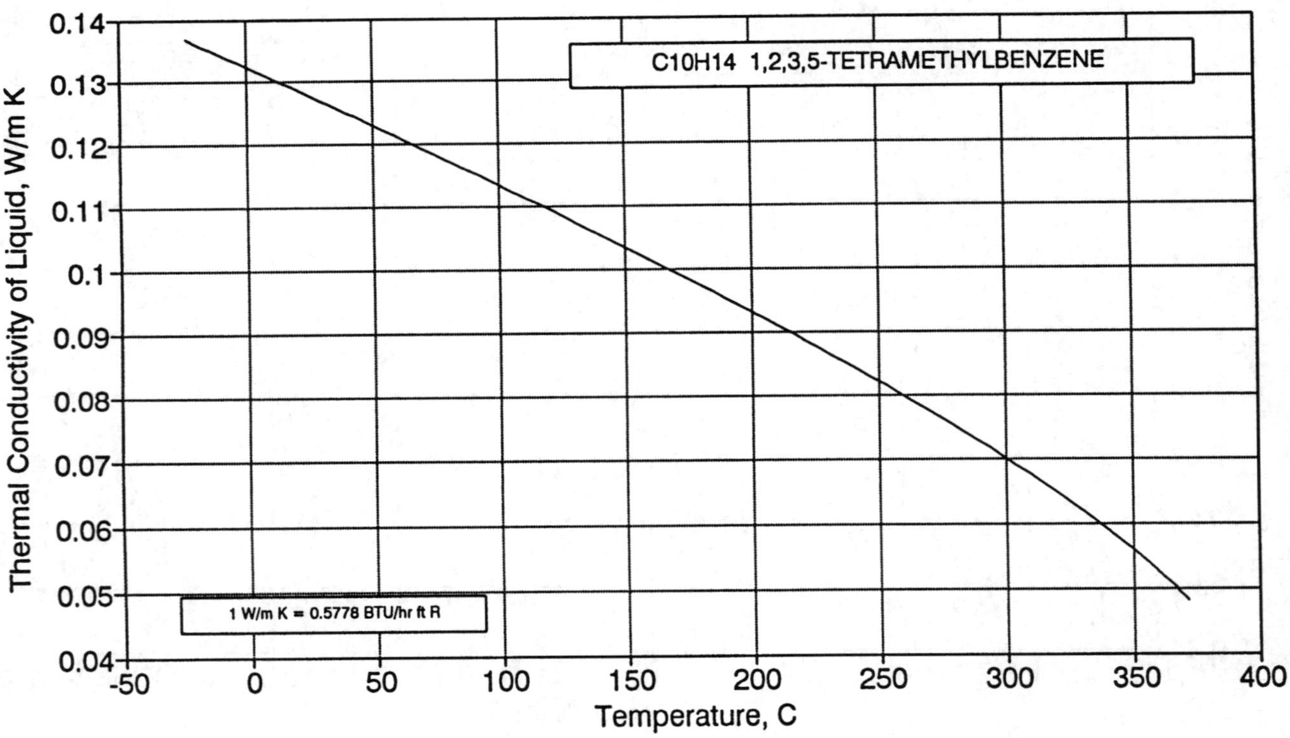
C10H14 1,2,3,5-TETRAMETHYLBENZENE
Thermal Conductivity of Liquid, W/m K
0.14
0.13
0.12
0.11
0.1
0.09
0.08
0.07
0.06
0.05
0.04
-50
0
50
100
150
200
250
300
350
400
Temperature, C
1 W/m K = 0.5778 BTU/hr ft R

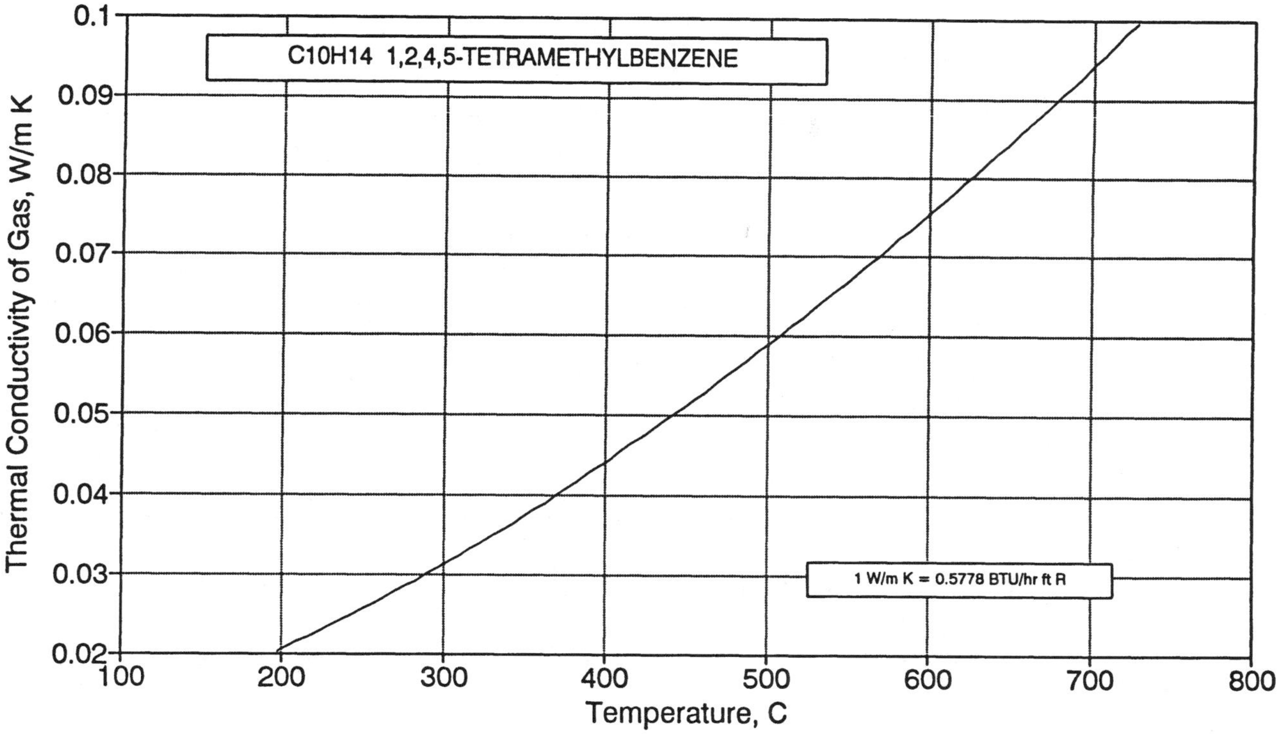
C10H14 1,2,4,5-TETRAMETHYLBENZENE
1 W/m K = 0.5778 BTU/hr ft R
Thermal Conductivity of Gas, W/m K
Temperature, C
0.1
0.09
0.08
0.07
0.06
0.05
0.04
0.03
0.02
100
200
300
400
500
600
700
800

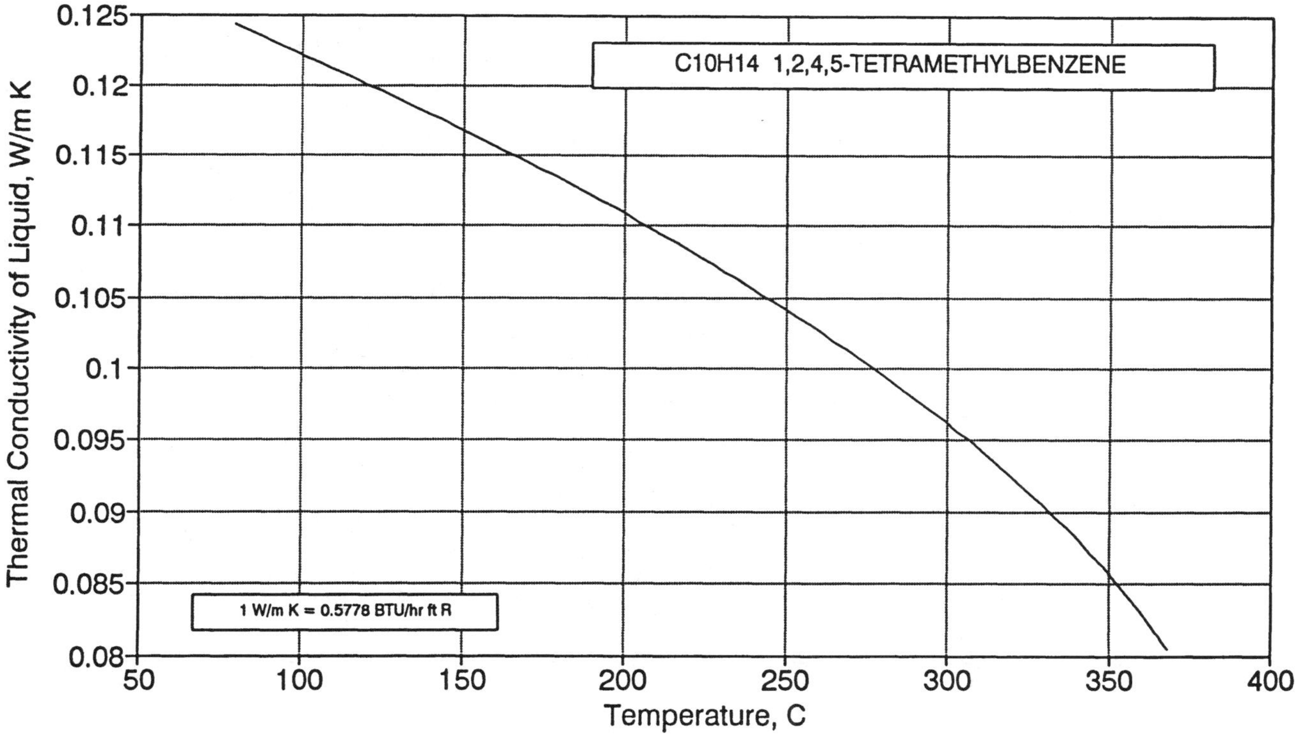
C10H14 1,2,4,5-TETRAMETHYLBENZENE
1 W/m K = 0.5778 BTU/hr ft R
Thermal Conductivity of Liquid, W/m K
Temperature, C
0.125
0.12
0.115
0.11
0.105
0.1
0.095
0.09
0.085
0.08
50
100
150
200
250
300
350
400

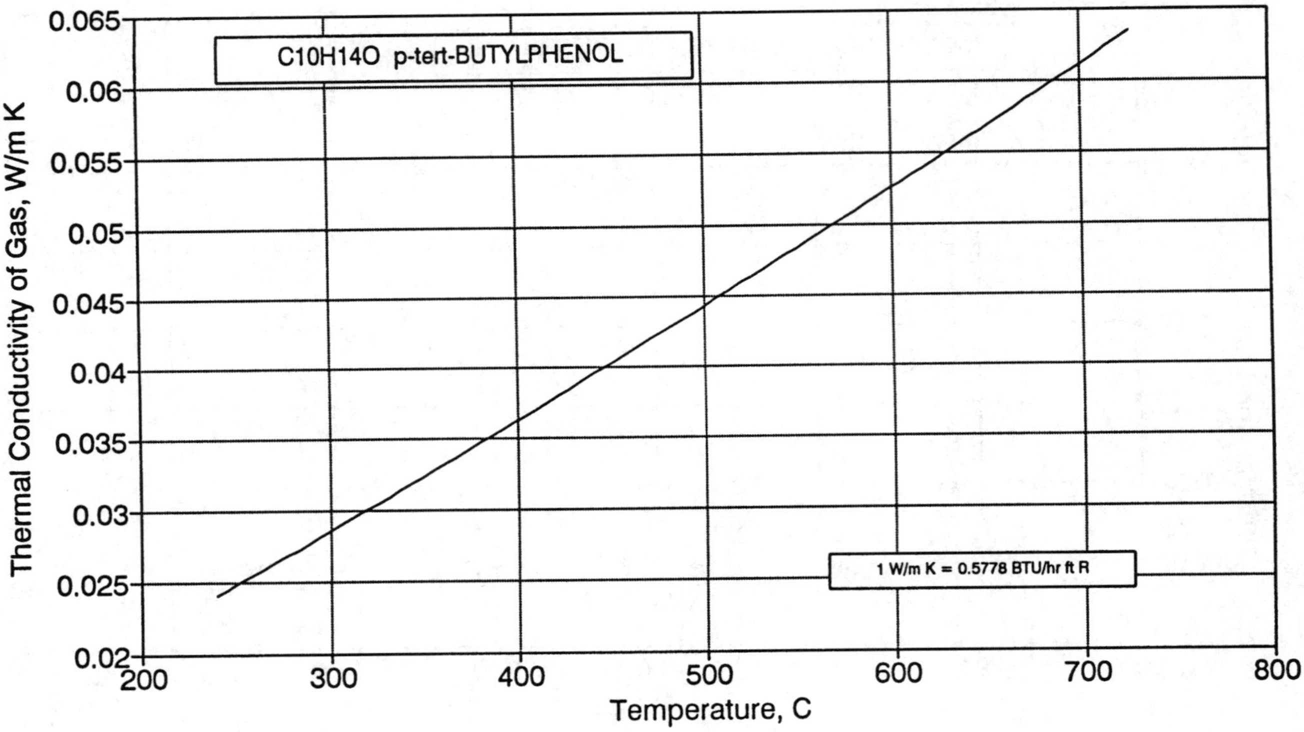
C10H14O p-tert-BUTYLPHENOL
1 W/m K = 0.5778 BTU/hr ft R
Thermal Conductivity of Gas, W/m K
0.065
0.06
0.055
0.05
0.045
0.04
0.035
0.03
0.025
0.02
200
300
400
500
600
700
800
Temperature, C

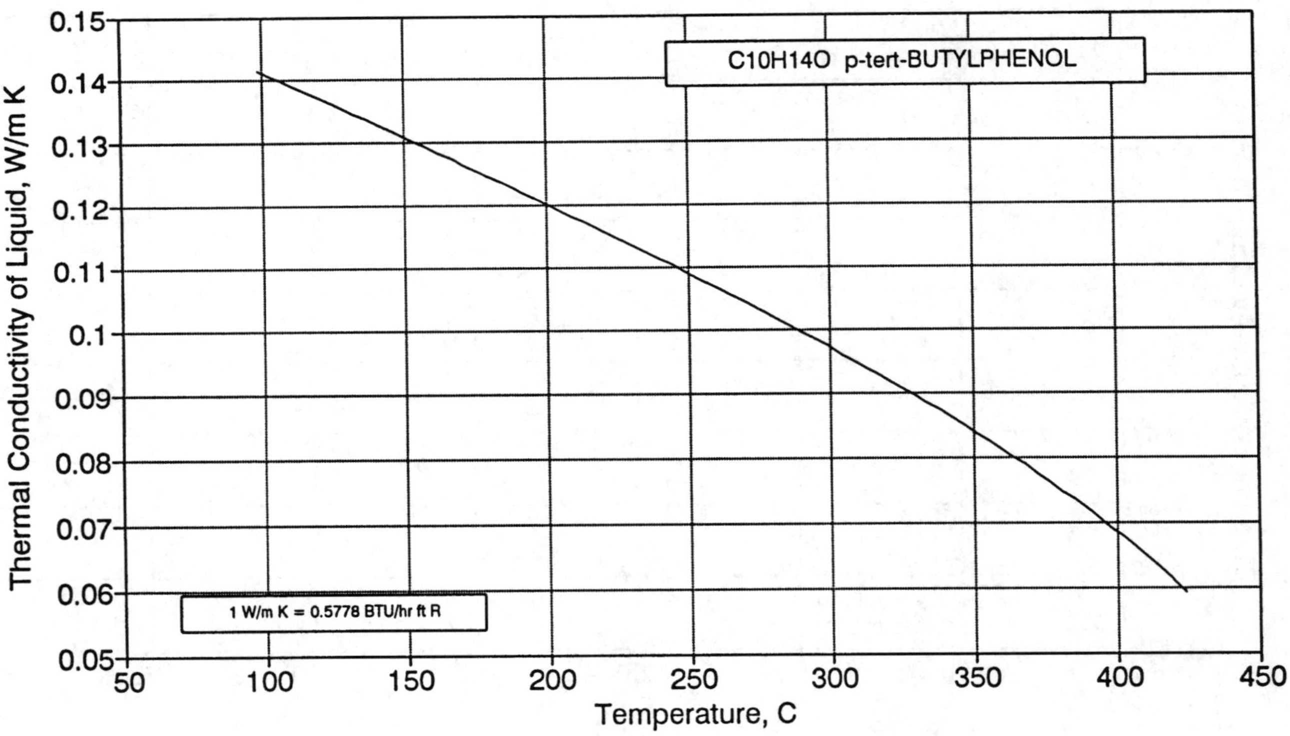
C10H14O p-tert-BUTYLPHENOL
1 W/m K = 0.5778 BTU/hr ft R
Thermal Conductivity of Liquid, W/m K
0.15
0.14
0.13
0.12
0.11
0.1
0.09
0.08
0.07
0.06
0.05
50
100
150
200
250
300
350
400
450
Temperature, C

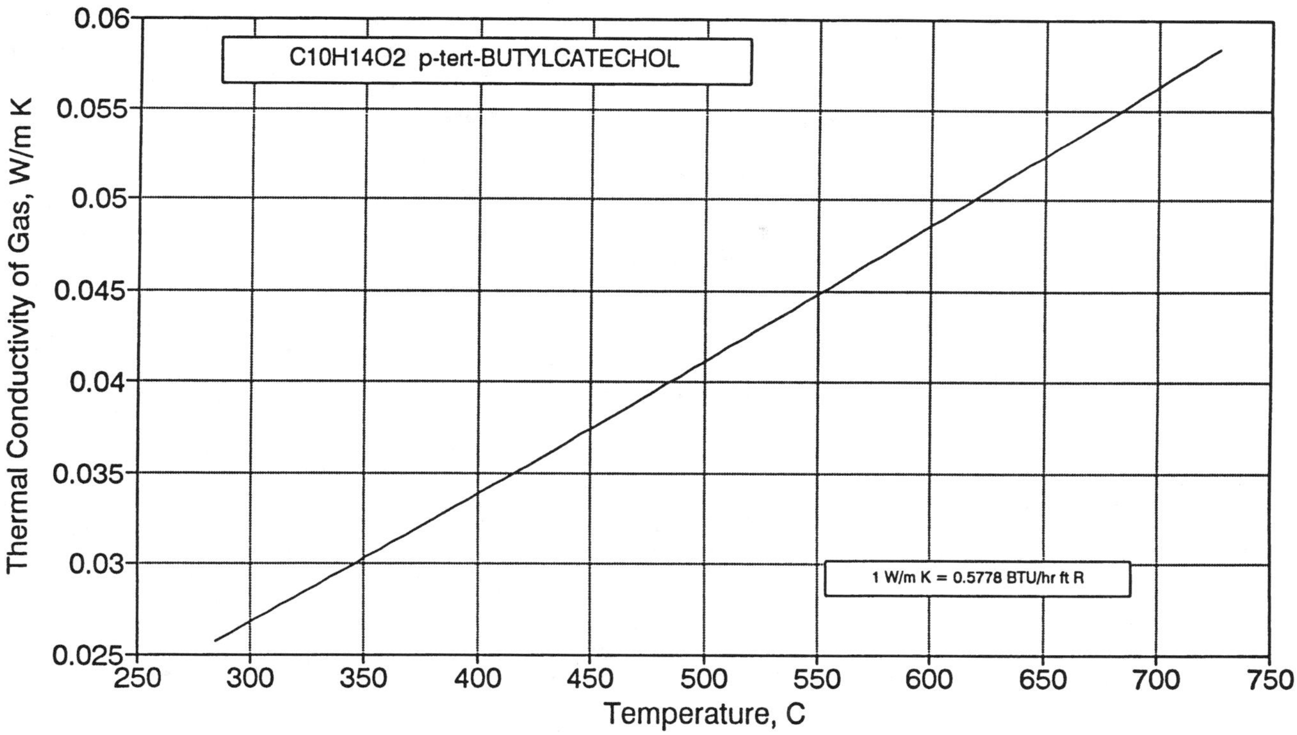
C10H14O2 p-tert-BUTYLCATECHOL
1 W/m K = 0.5778 BTU/hr ft R
Thermal Conductivity of Gas, W/m K
Temperature, C
0.06
0.055
0.05
0.045
0.04
0.035
0.03
0.025
250
300
350
400
450
500
550
600
650
700
750

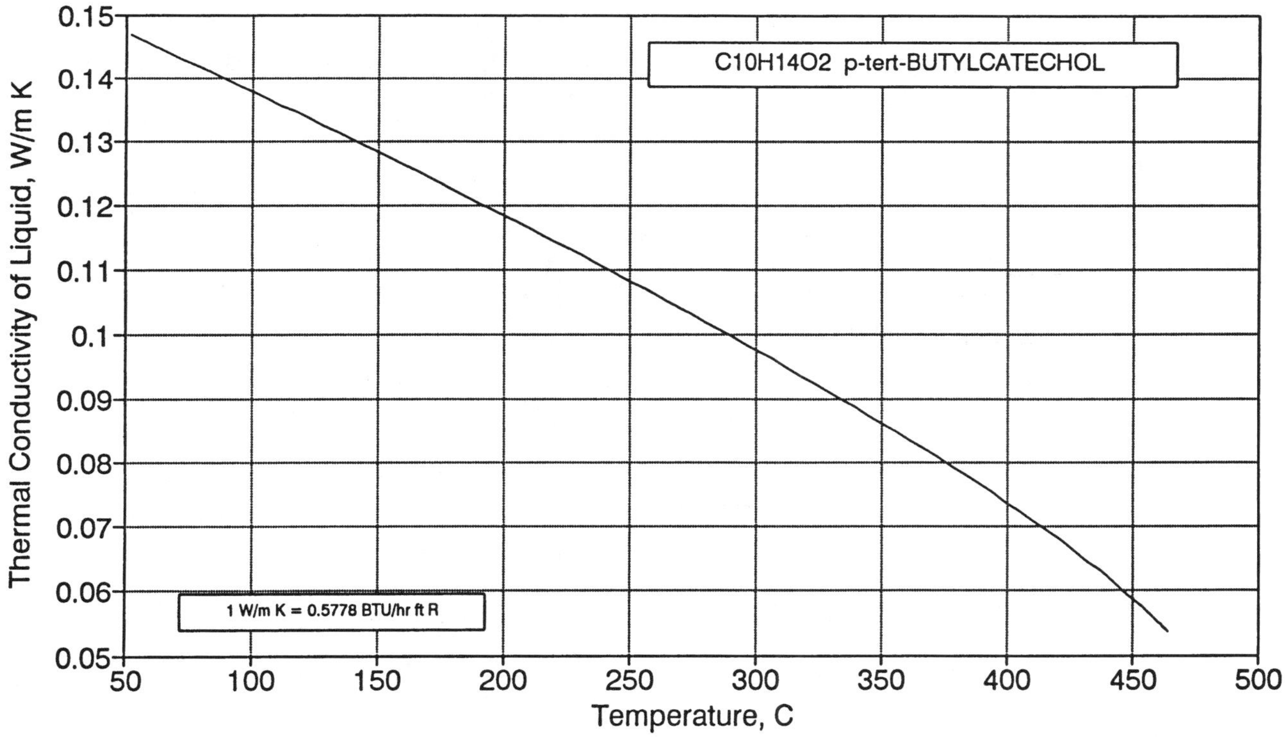
C10H14O2 p-tert-BUTYLCATECHOL
1 W/m K = 0.5778 BTU/hr ft R
Thermal Conductivity of Liquid, W/m K
Temperature, C
0.15
0.14
0.13
0.12
0.11
0.1
0.09
0.08
0.07
0.06
0.05
50
100
150
200
250
300
350
400
450
500

C10H15N N,N-DIETHYLANILINE

Gas thermal conductivity data are not available .

TF , K = 235.15

TB , K = 489.42 (experimental)

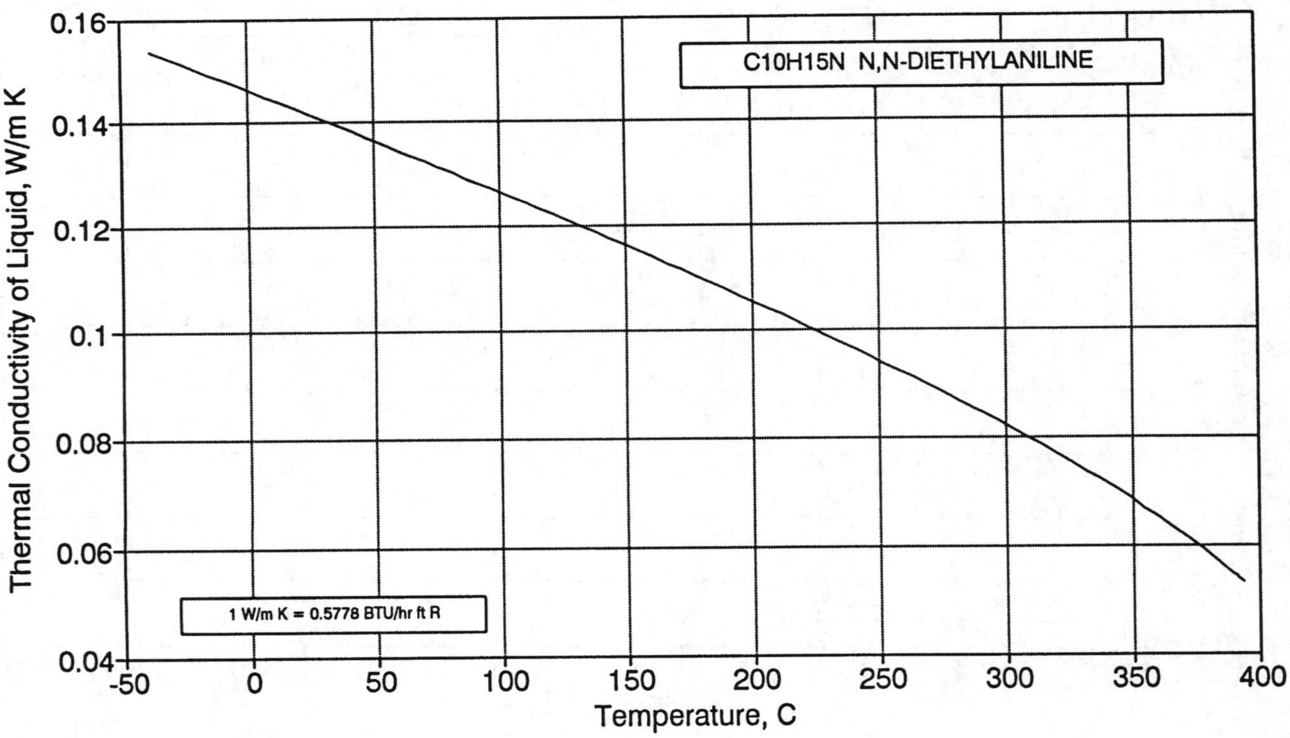

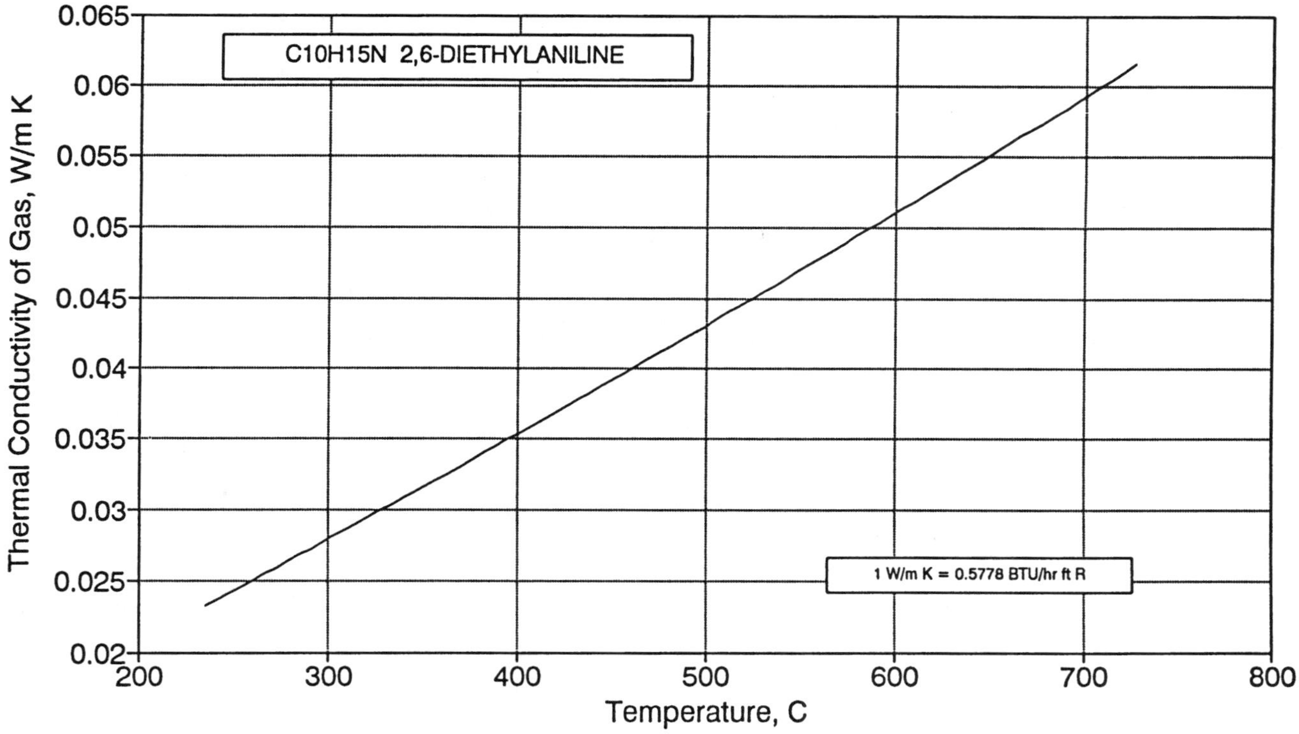
C10H15N 2,6-DIETHYLANILINE
1 W/m K = 0.5778 BTU/hr ft R
Thermal Conductivity of Gas, W/m K
Temperature, C
0.065
0.06
0.055
0.05
0.045
0.04
0.035
0.03
0.025
0.02
200
300
400
500
600
700
800

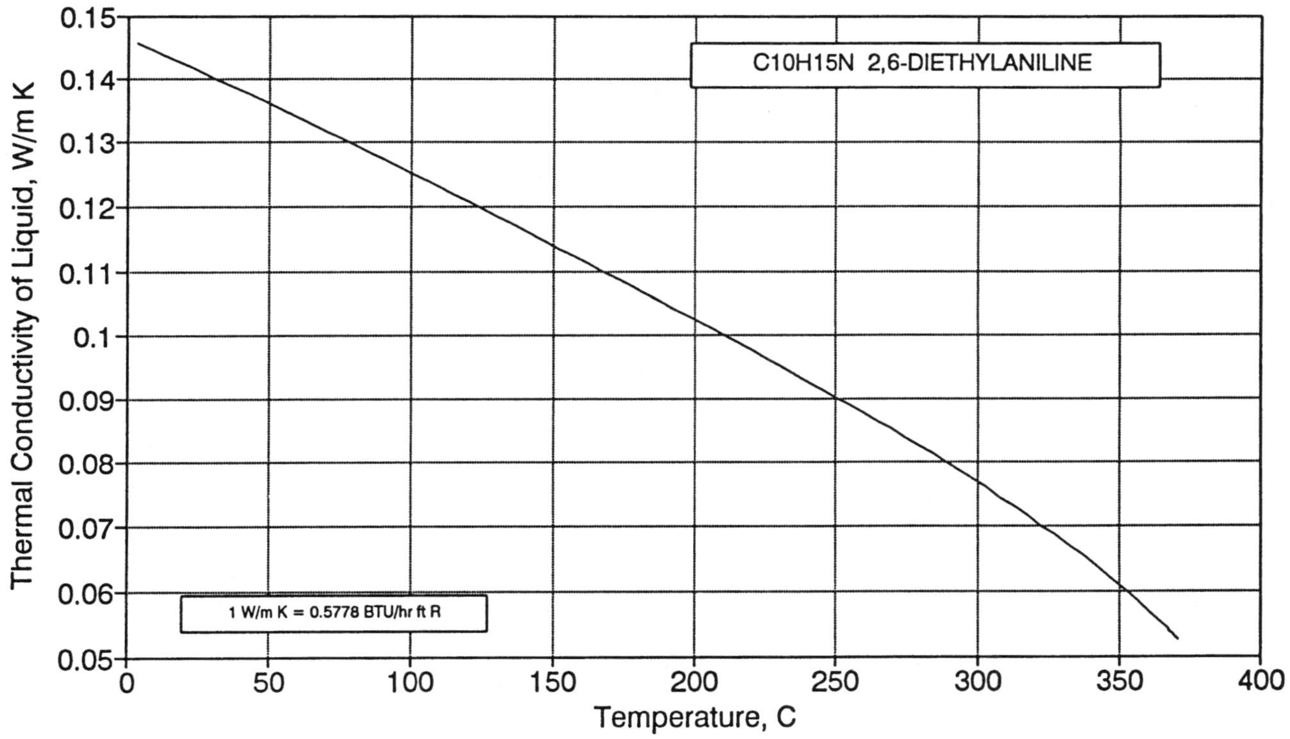
C10H15N 2,6-DIETHYLANILINE
1 W/m K = 0.5778 BTU/hr ft R
Thermal Conductivity of Liquid, W/m K
Temperature, C
0.15
0.14
0.13
0.12
0.11
0.1
0.09
0.08
0.07
0.06
0.05
0
50
100
150
200
250
300
350
400

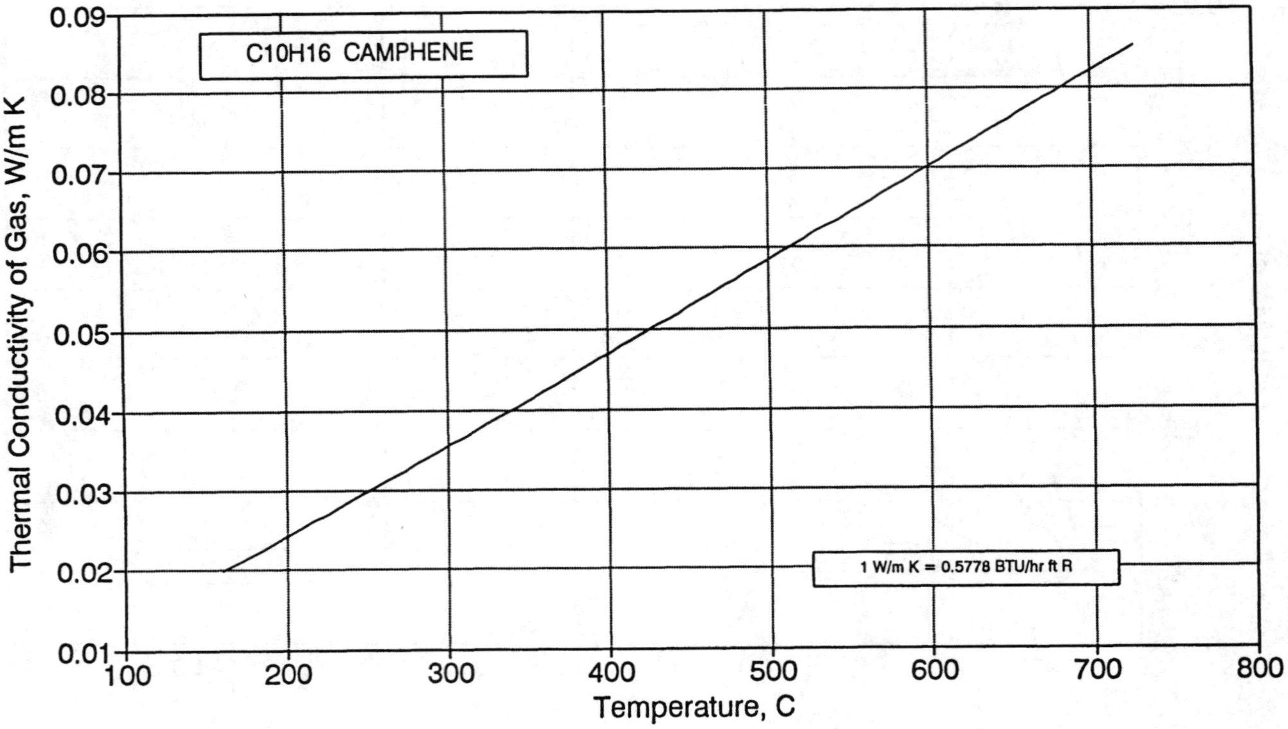

C10H16 CAMPHENE
1 W/m K = 0.5778 BTU/hr ft R
Thermal Conductivity of Gas, W/m K
0.09
0.08
0.07
0.06
0.05
0.04
0.03
0.02
0.01
100
200
300
400
500
600
700
800
Temperature, C

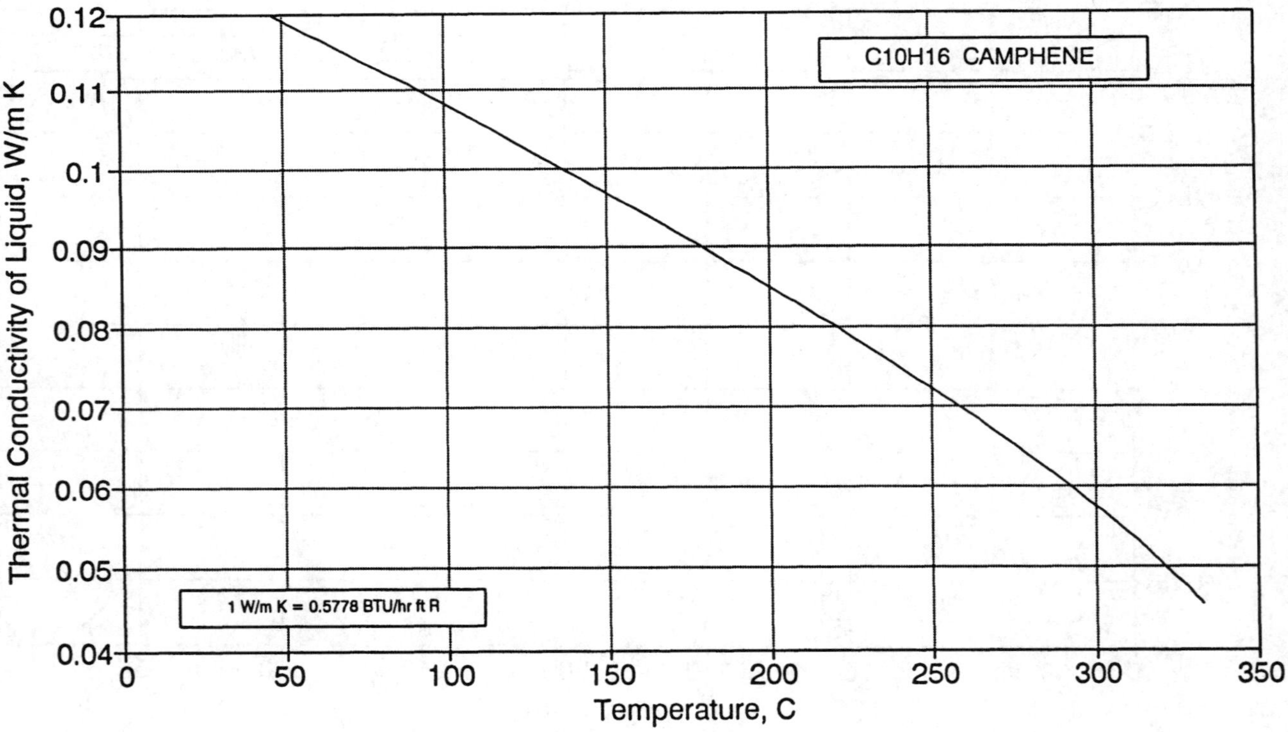

C10H16 CAMPHENE
1 W/m K = 0.5778 BTU/hr ft R
Thermal Conductivity of Liquid, W/m K
0.12
0.11
0.1
0.09
0.08
0.07
0.06
0.05
0.04
0
50
100
150
200
250
300
350
Temperature, C

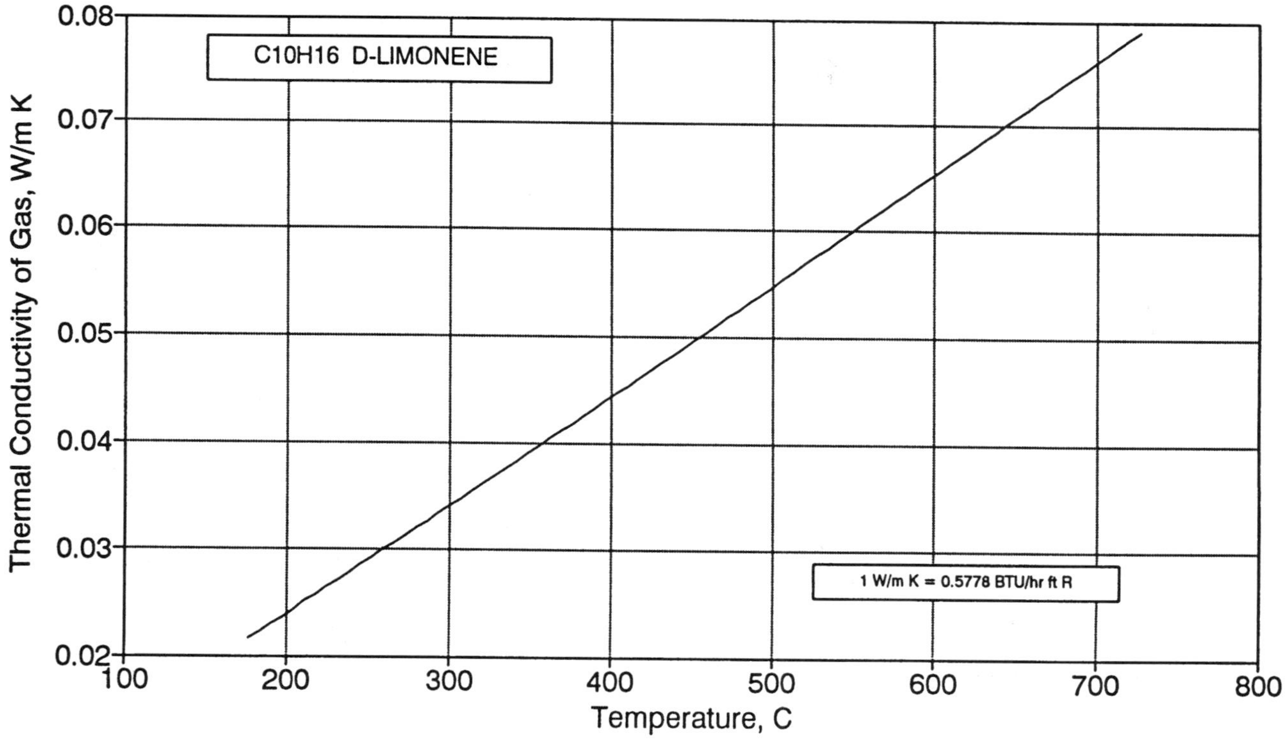
C10H16 D-LIMONENE
1 W/m K = 0.5778 BTU/hr ft R
Thermal Conductivity of Gas, W/m K
Temperature, C
0.08
0.07
0.06
0.05
0.04
0.03
0.02
100
200
300
400
500
600
700
800

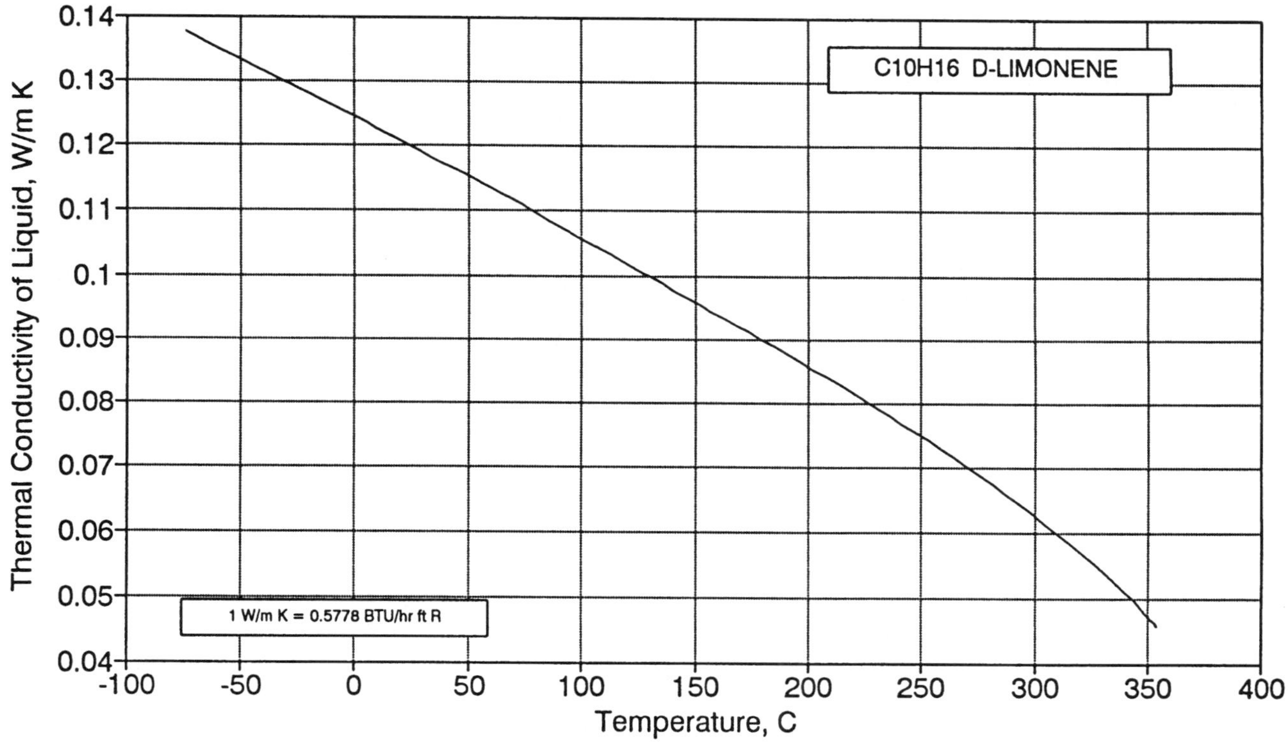
C10H16 D-LIMONENE
1 W/m K = 0.5778 BTU/hr ft R
Thermal Conductivity of Liquid, W/m K
Temperature, C
0.14
0.13
0.12
0.11
0.1
0.09
0.08
0.07
0.06
0.05
0.04
-100
-50
0
50
100
150
200
250
300
350
400

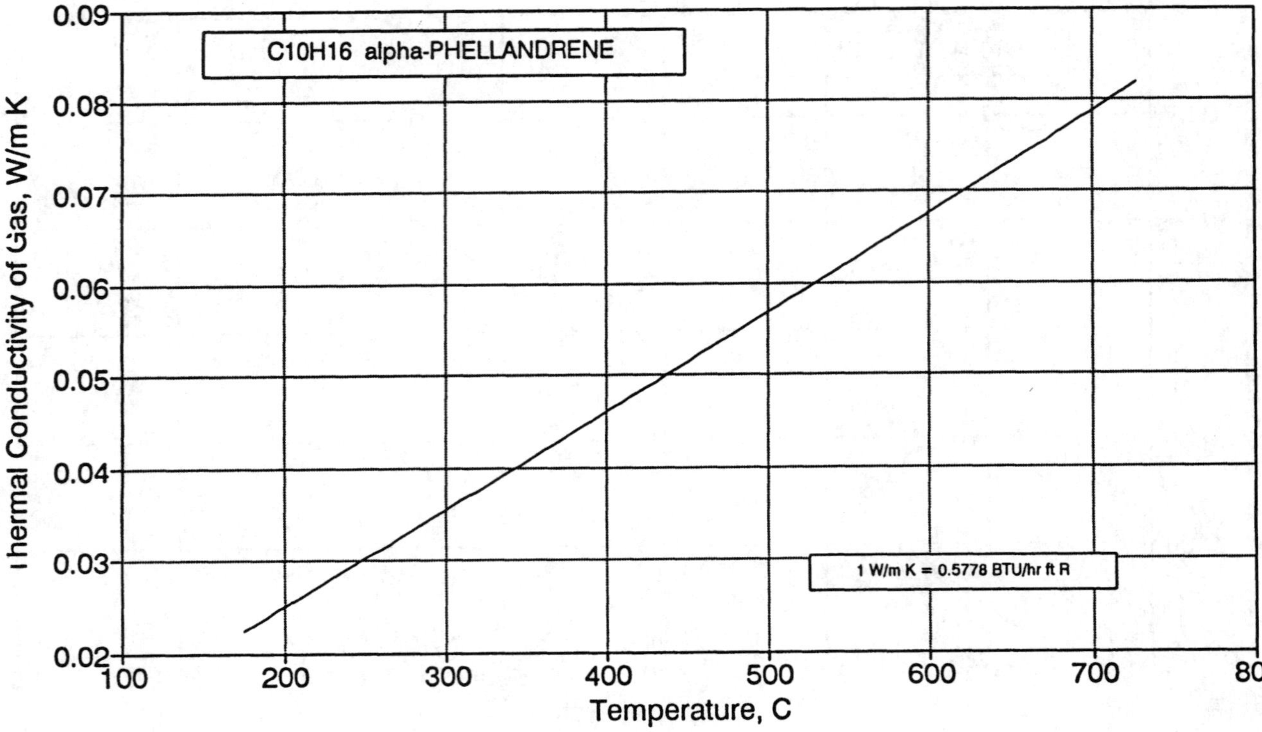
C10H16 alpha-PHELLANDRENE
1 W/m K = 0.5778 BTU/hr ft R
Thermal Conductivity of Gas, W/m K
Temperature, C
0.09
0.08
0.07
0.06
0.05
0.04
0.03
0.02
100
200
300
400
500
600
700
800

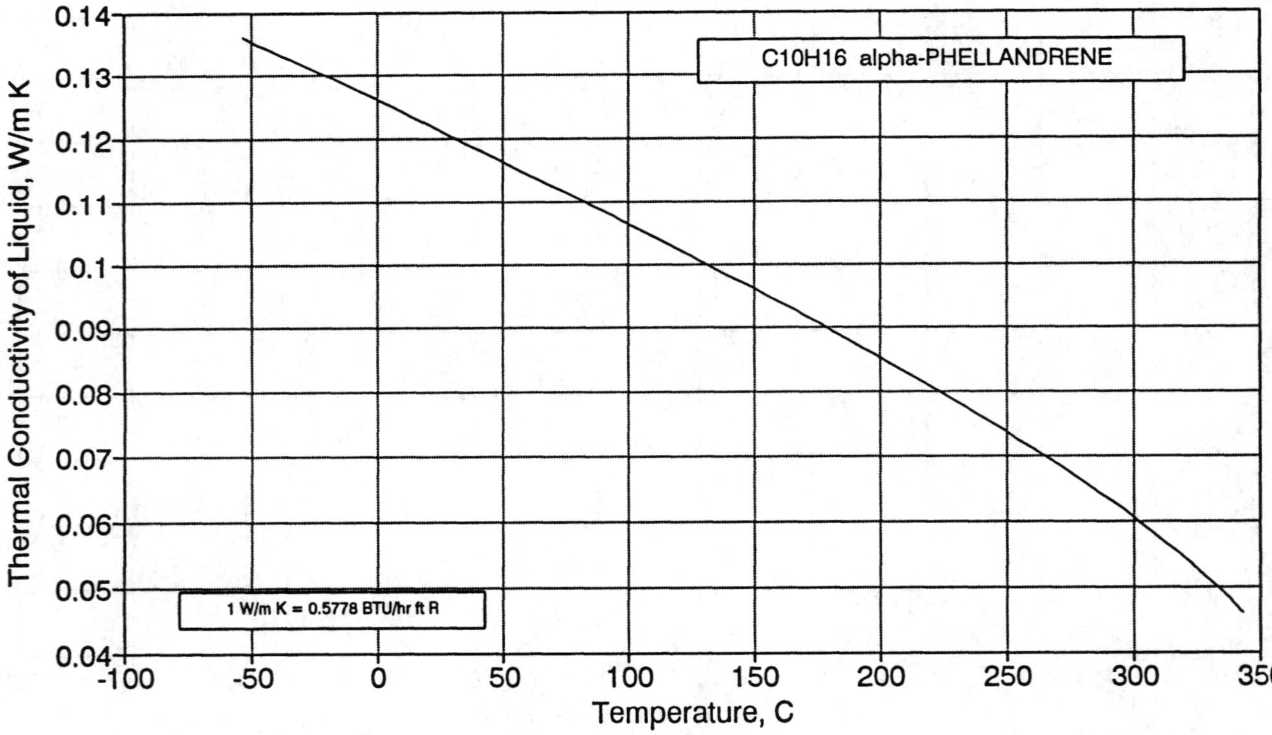
C10H16 alpha-PHELLANDRENE
1 W/m K = 0.5778 BTU/hr ft R
Thermal Conductivity of Liquid, W/m K
Temperature, C
0.14
0.13
0.12
0.11
0.1
0.09
0.08
0.07
0.06
0.05
0.04
-100
-50
0
50
100
150
200
250
300
350

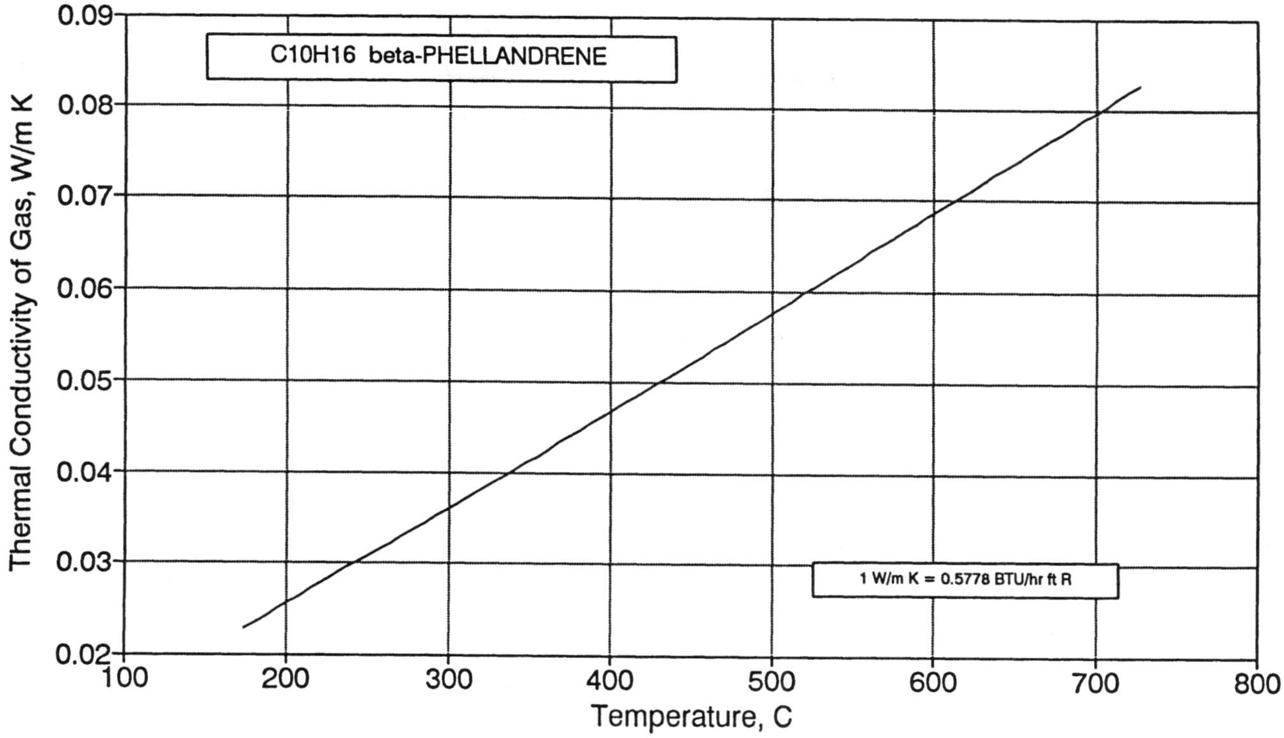
C10H16 beta-PHELLANDRENE
1 W/m K = 0.5778 BTU/hr ft R
Thermal Conductivity of Gas, W/m K
Temperature, C
0.09
0.08
0.07
0.06
0.05
0.04
0.03
0.02
100
200
300
400
500
600
700
800

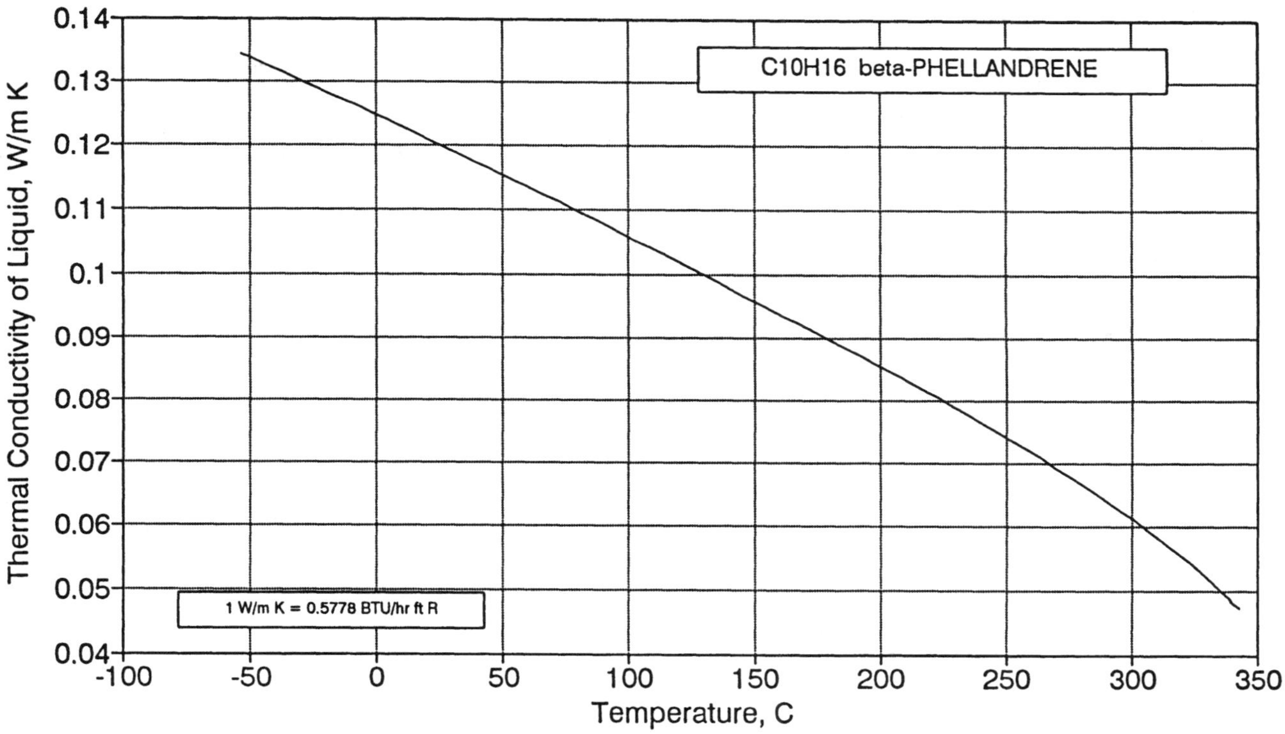
C10H16 beta-PHELLANDRENE
1 W/m K = 0.5778 BTU/hr ft R
Thermal Conductivity of Liquid, W/m K
Temperature, C
0.14
0.13
0.12
0.11
0.1
0.09
0.08
0.07
0.06
0.05
0.04
-100
-50
0
50
100
150
200
250
300
350

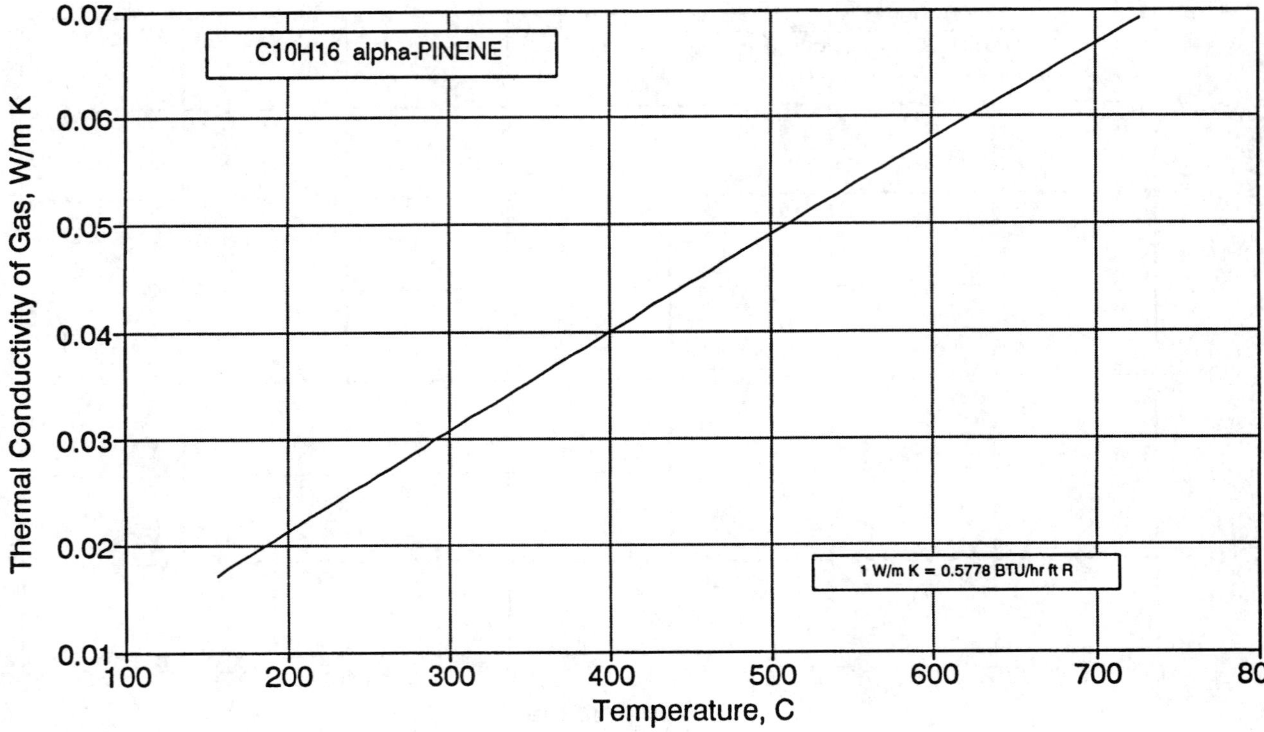
C10H16 alpha-PINENE
1 W/m K = 0.5778 BTU/hr ft R
Thermal Conductivity of Gas, W/m K
0.07
0.06
0.05
0.04
0.03
0.02
0.01
100
200
300
400
500
600
700
800
Temperature, C

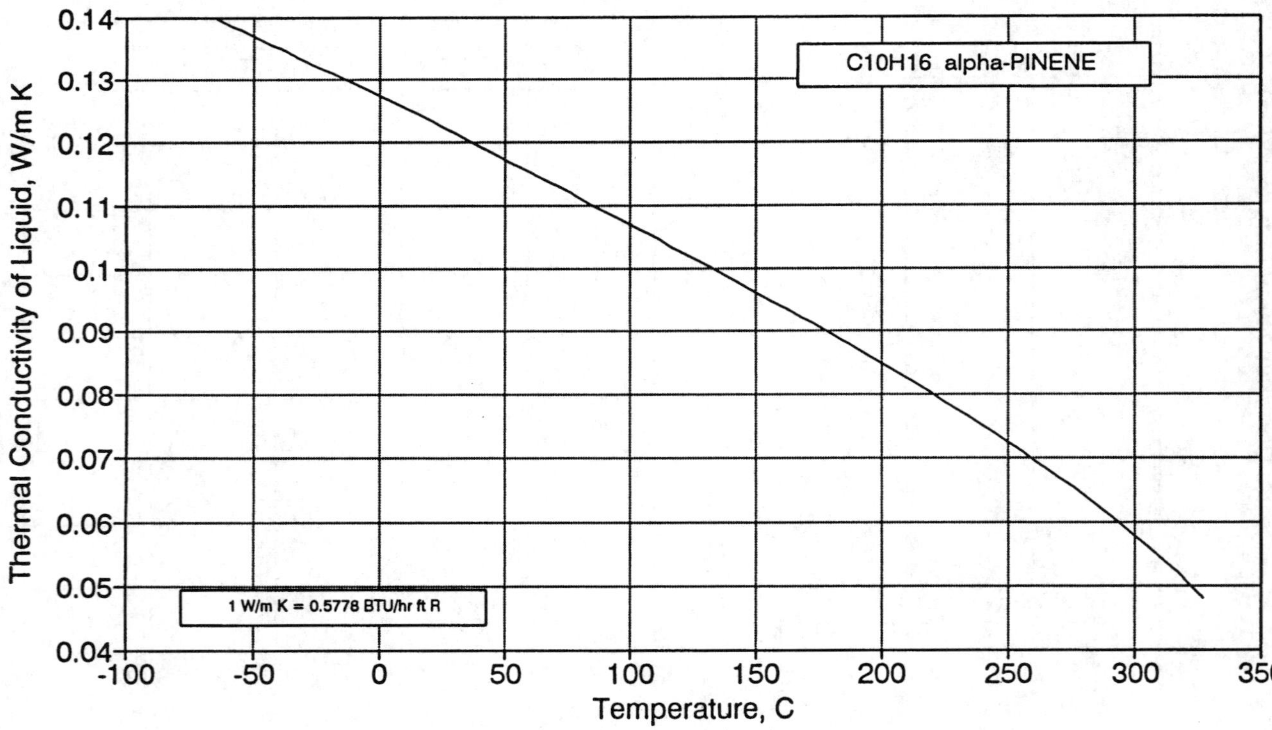
C10H16 alpha-PINENE
1 W/m K = 0.5778 BTU/hr ft R
Thermal Conductivity of Liquid, W/m K
0.14
0.13
0.12
0.11
0.1
0.09
0.08
0.07
0.06
0.05
0.04
-100
-50
0
50
100
150
200
250
300
350
Temperature, C

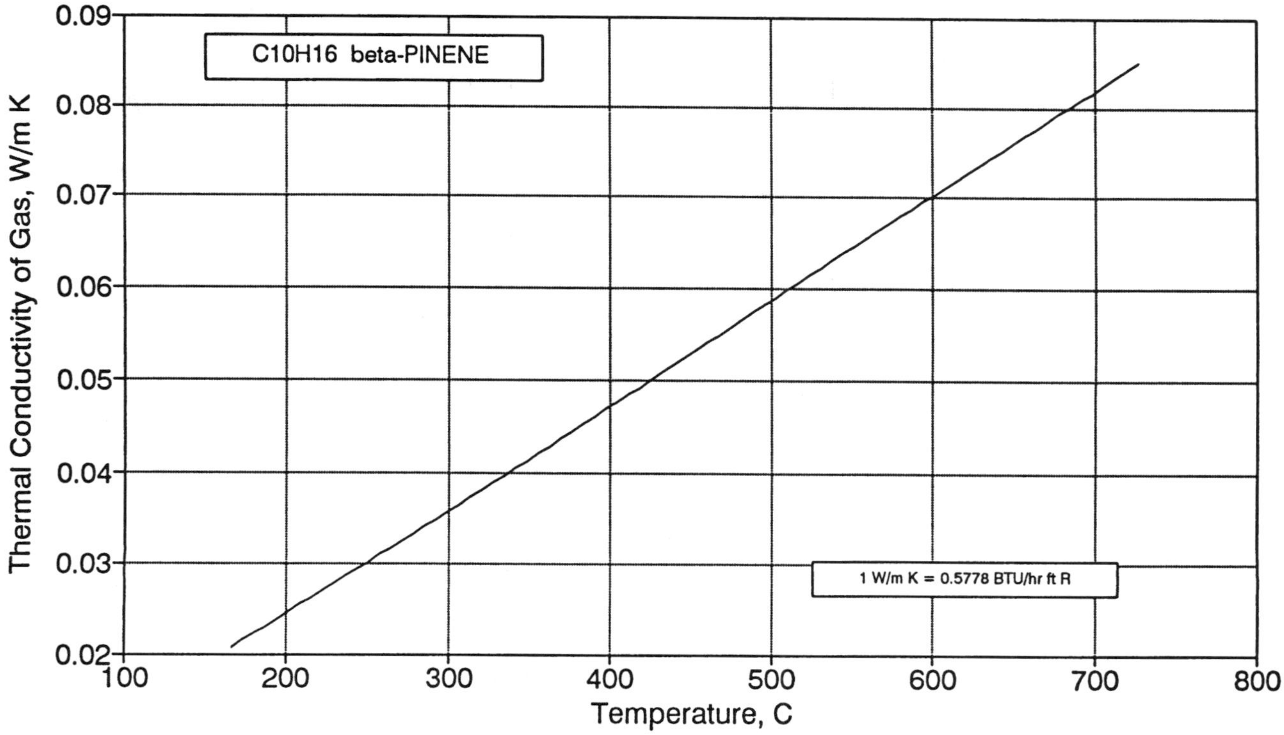

C10H16 beta-PINENE
1 W/m K = 0.5778 BTU/hr ft R
Thermal Conductivity of Gas, W/m K
0.09
0.08
0.07
0.06
0.05
0.04
0.03
0.02
100
200
300
400
500
600
700
800
Temperature, C

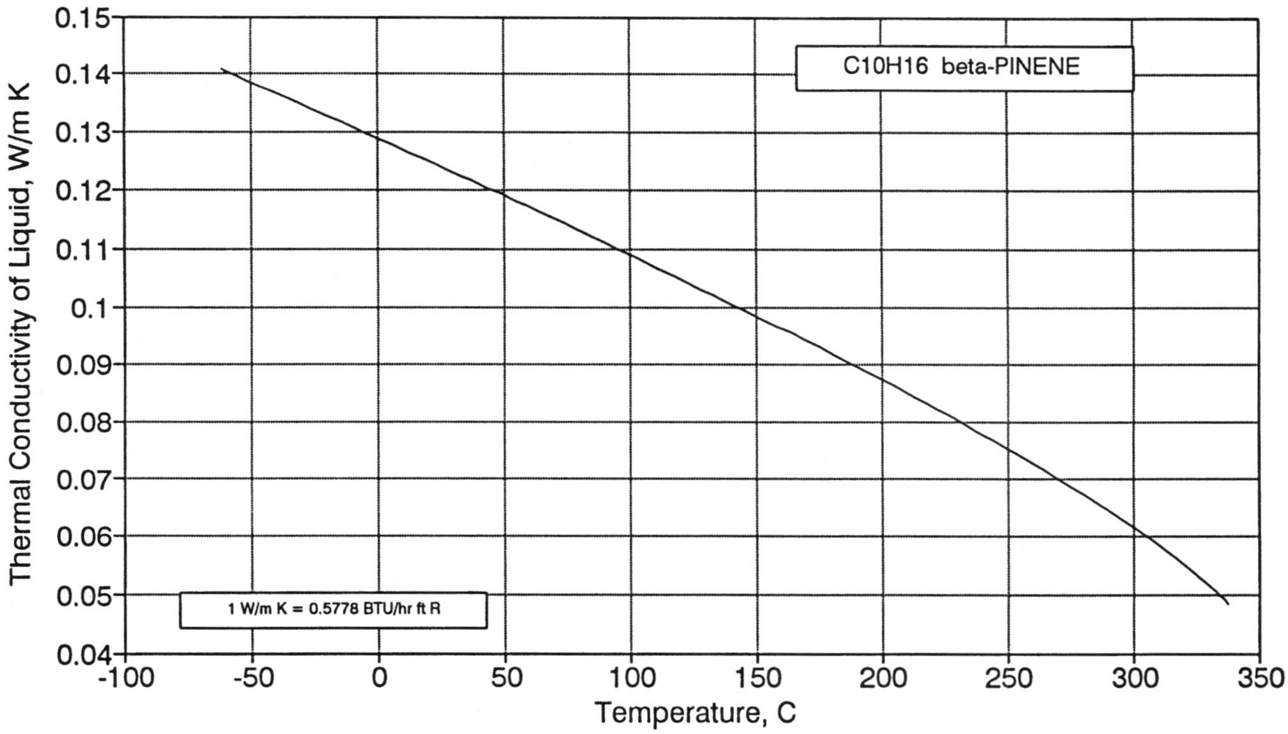

C10H16 beta-PINENE
1 W/m K = 0.5778 BTU/hr ft R
Thermal Conductivity of Liquid, W/m K
0.15
0.14
0.13
0.12
0.11
0.1
0.09
0.08
0.07
0.06
0.05
0.04
-100
-50
0
50
100
150
200
250
300
350
Temperature, C

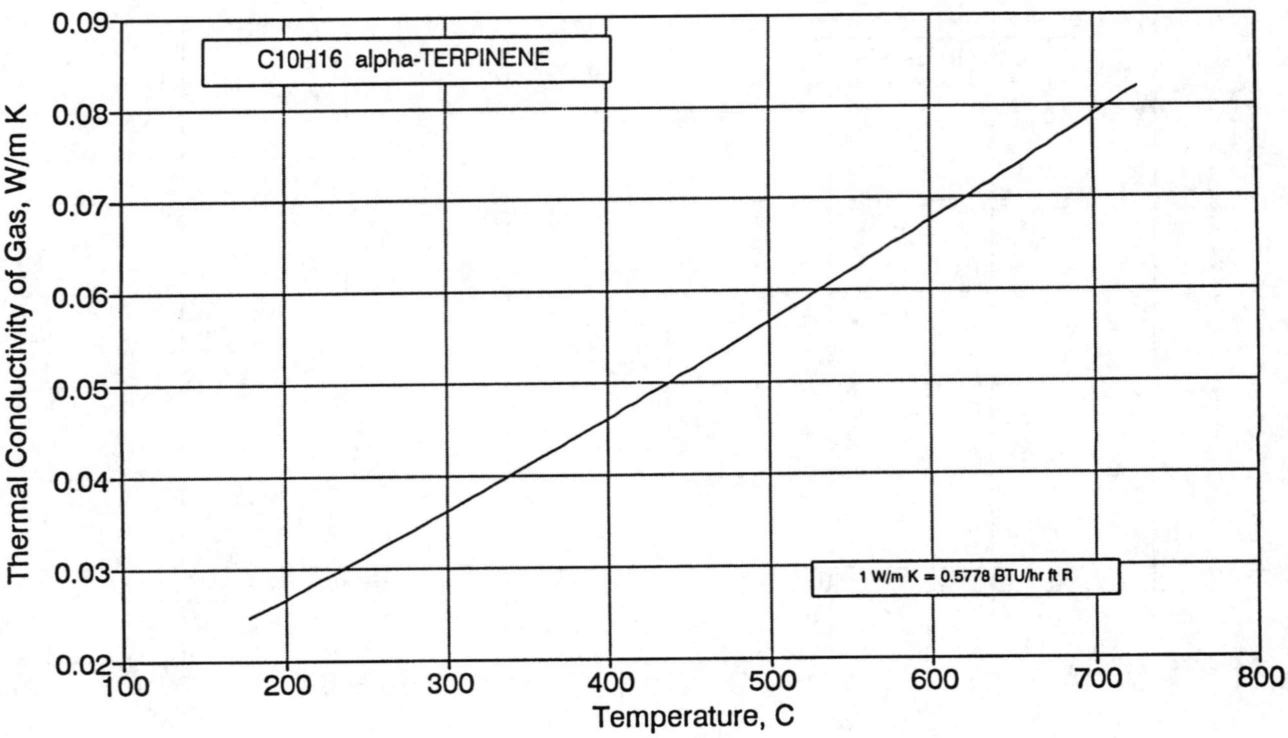

C10H16 alpha-TERPINENE
1 W/m K = 0.5778 BTU/hr ft R
Thermal Conductivity of Gas, W/m K
Temperature, C
0.09
0.08
0.07
0.06
0.05
0.04
0.03
0.02
100
200
300
400
500
600
700
800

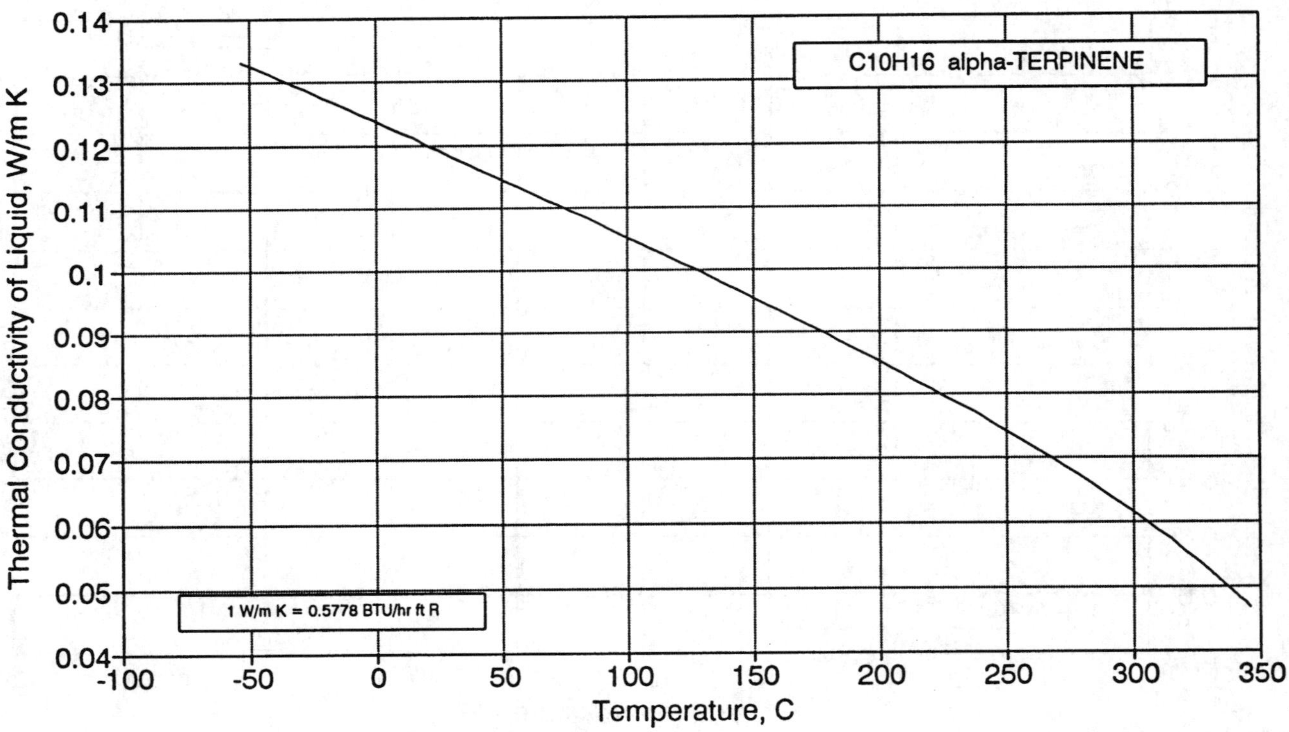

C10H16 alpha-TERPINENE
1 W/m K = 0.5778 BTU/hr ft R
Thermal Conductivity of Liquid, W/m K
Temperature, C
0.14
0.13
0.12
0.11
0.1
0.09
0.08
0.07
0.06
0.05
0.04
-100
-50
0
50
100
150
200
250
300
350

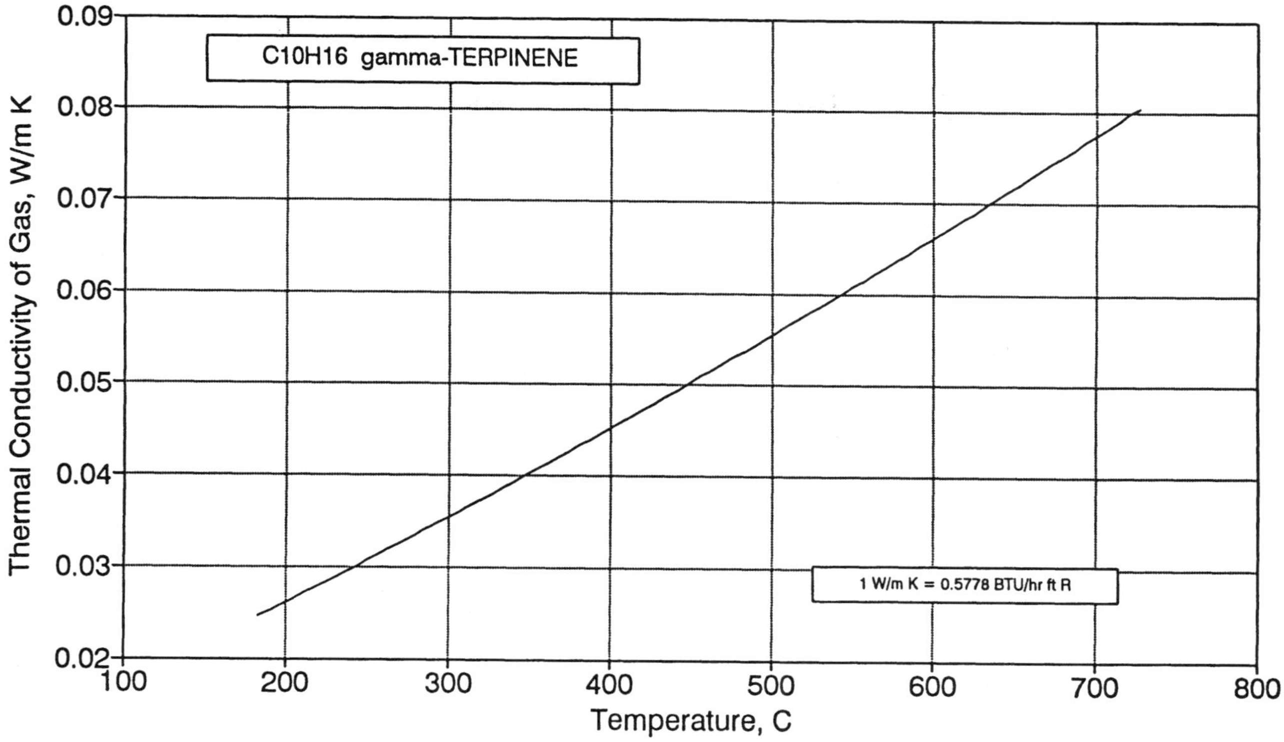

C10H16 gamma-TERPINENE
1 W/m K = 0.5778 BTU/hr ft R
Thermal Conductivity of Gas, W/m K
0.09
0.08
0.07
0.06
0.05
0.04
0.03
0.02
100
200
300
400
500
600
700
800
Temperature, C

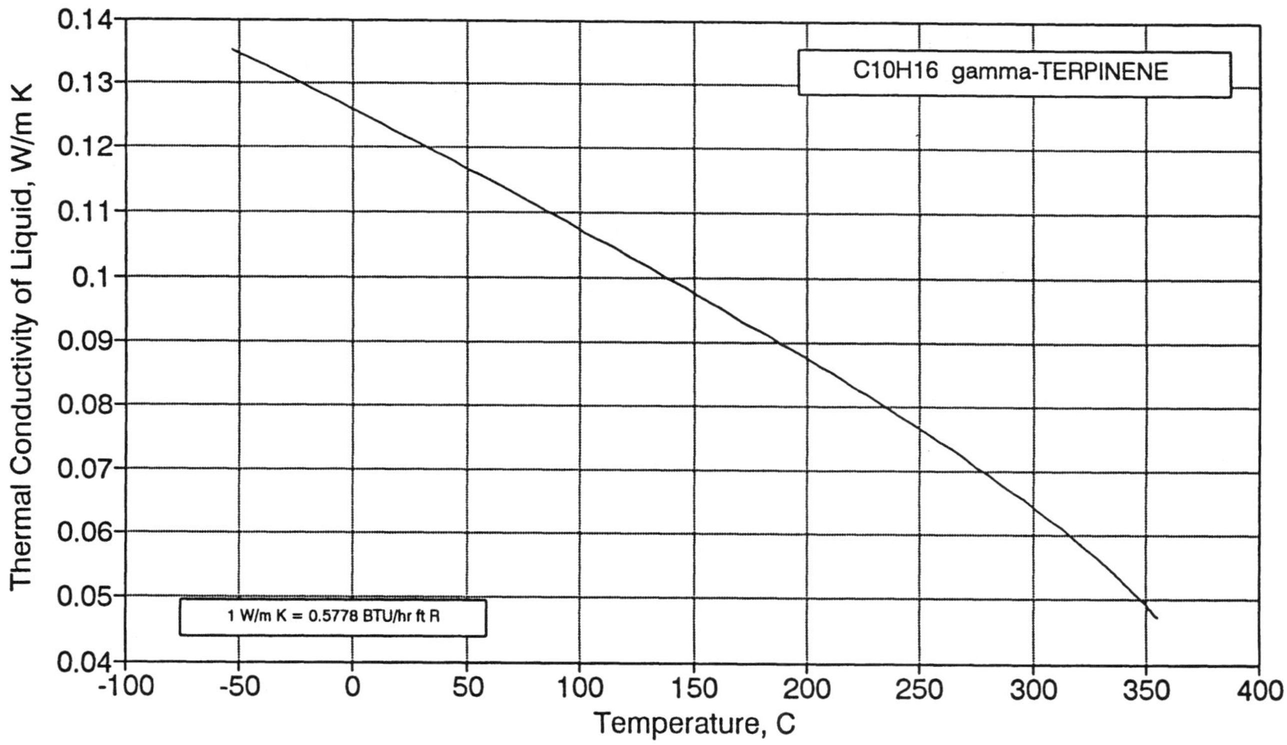

C10H16 gamma-TERPINENE
1 W/m K = 0.5778 BTU/hr ft R
Thermal Conductivity of Liquid, W/m K
0.14
0.13
0.12
0.11
0.1
0.09
0.08
0.07
0.06
0.05
0.04
-100
-50
0
50
100
150
200
250
300
350
400
Temperature, C

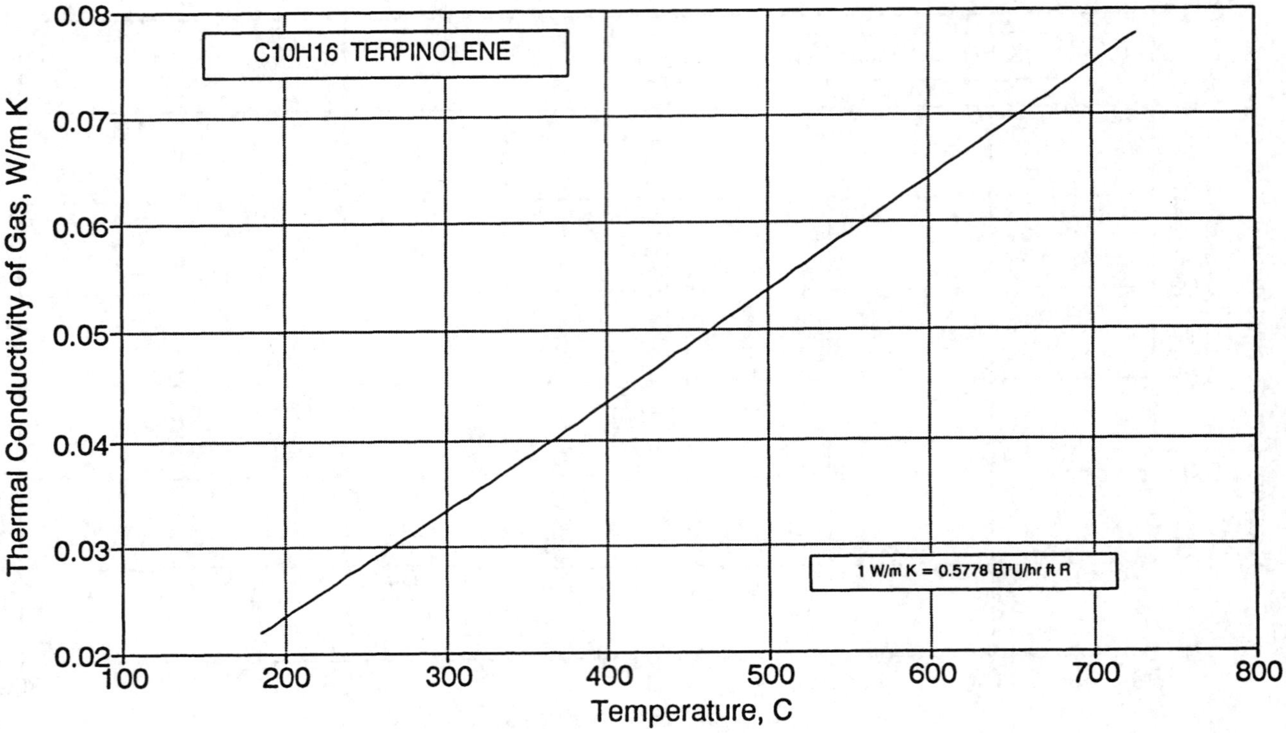

C10H16 TERPINOLENE
1 W/m K = 0.5778 BTU/hr ft R
Thermal Conductivity of Gas, W/m K
0.08
0.07
0.06
0.05
0.04
0.03
0.02
100
200
300
400
500
600
700
800
Temperature, C

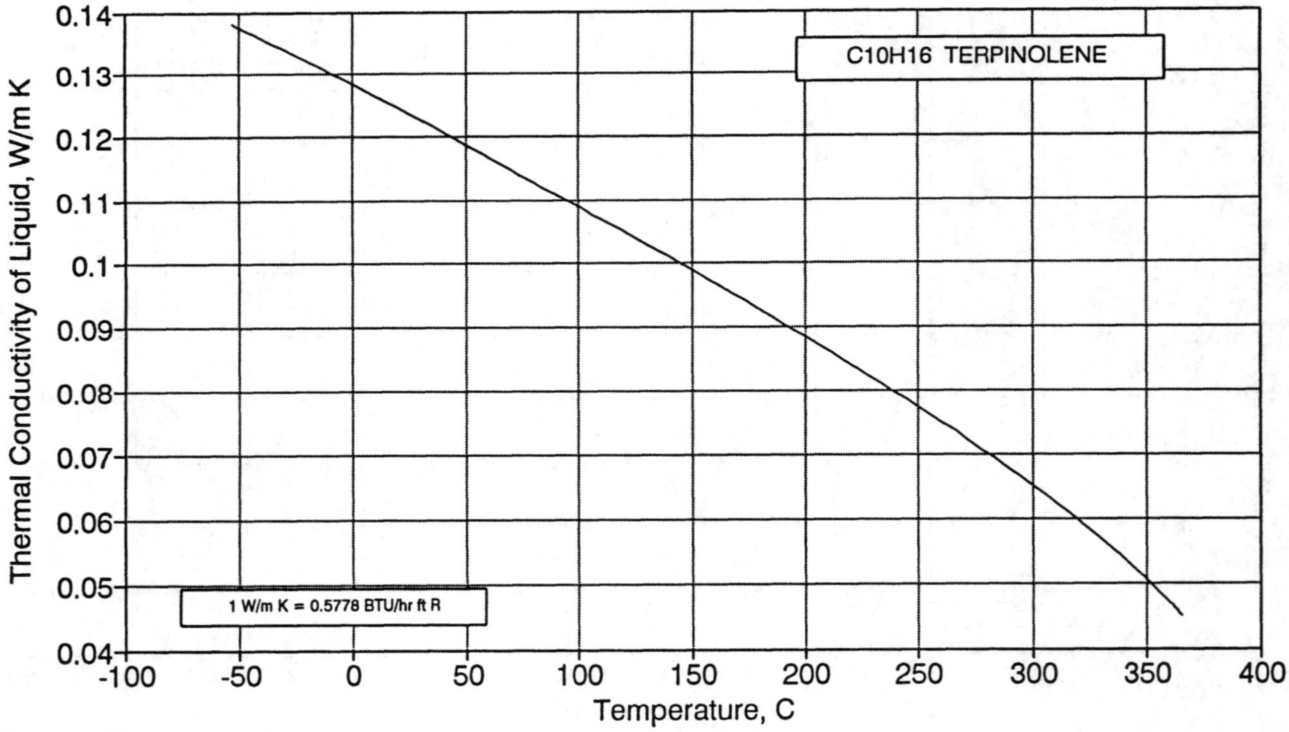

C10H16 TERPINOLENE
1 W/m K = 0.5778 BTU/hr ft R
Thermal Conductivity of Liquid, W/m K
0.14
0.13
0.12
0.11
0.1
0.09
0.08
0.07
0.06
0.05
0.04
-100
-50
0
50
100
150
200
250
300
350
400
Temperature, C

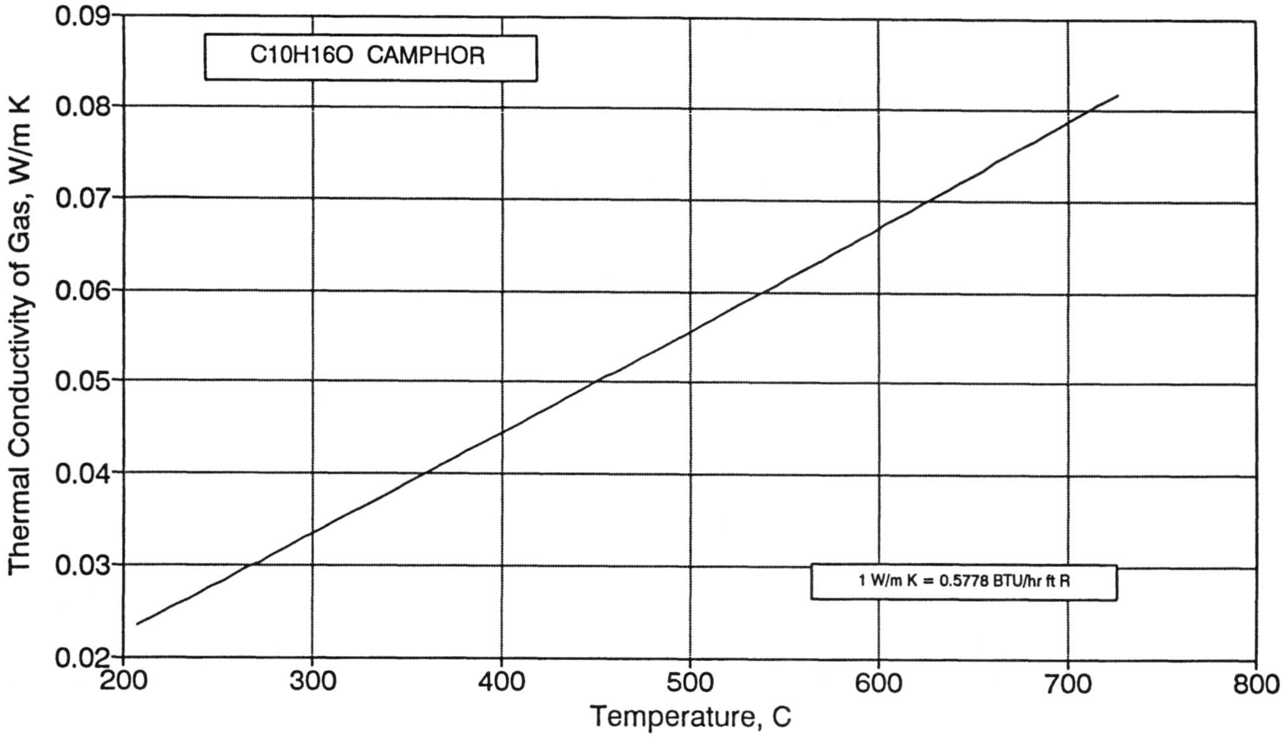
C10H16O CAMPHOR
Thermal Conductivity of Gas, W/m K
0.09
0.08
0.07
0.06
0.05
0.04
0.03
0.02
1 W/m K = 0.5778 BTU/hr ft R
200
300
400
500
600
700
800
Temperature, C

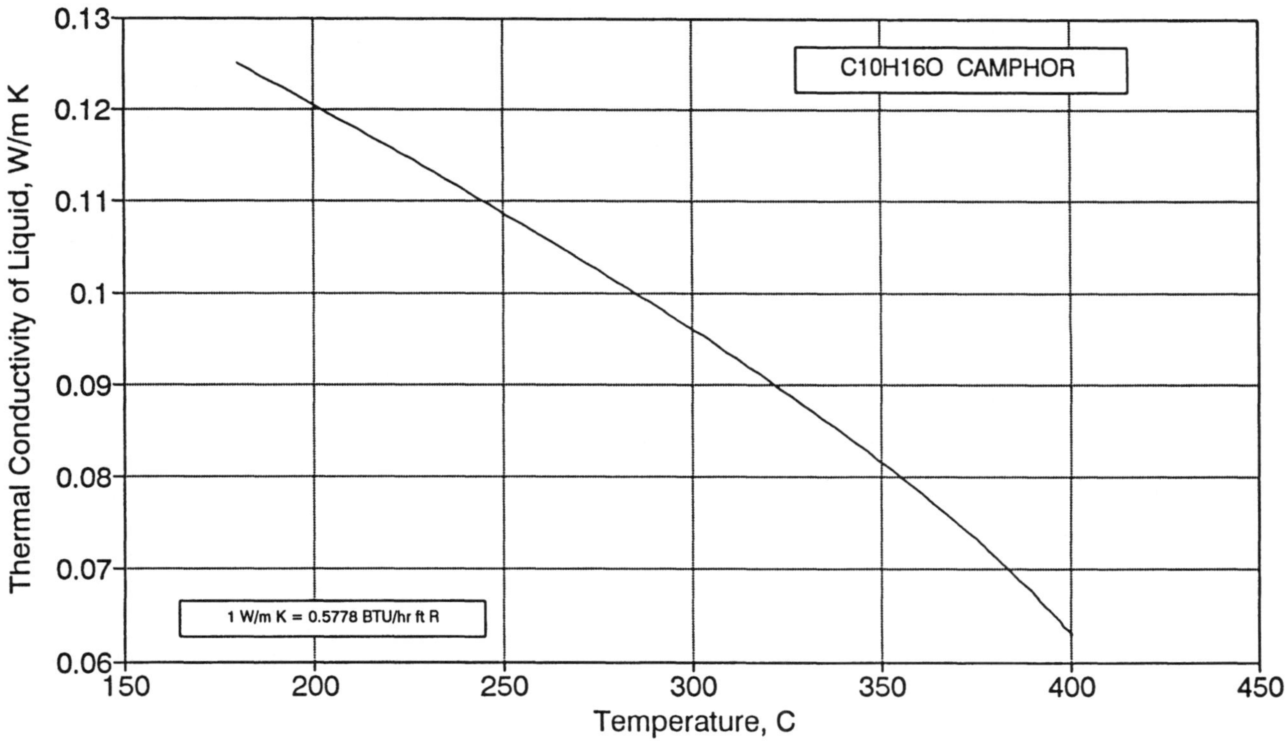
C10H16O CAMPHOR
Thermal Conductivity of Liquid, W/m K
0.13
0.12
0.11
0.1
0.09
0.08
0.07
0.06
1 W/m K = 0.5778 BTU/hr ft R
150
200
250
300
350
400
450
Temperature, C

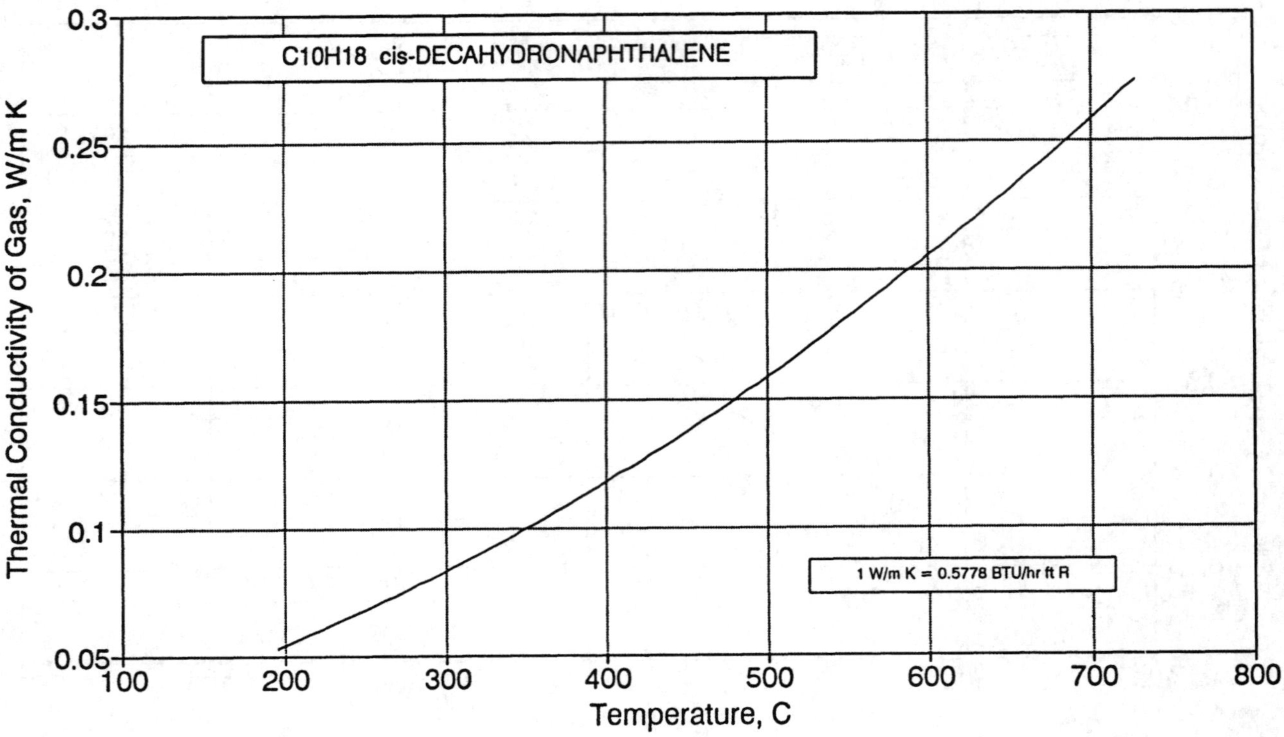
C10H18 cis-DECAHYDRONAPHTHALENE
1 W/m K = 0.5778 BTU/hr ft R
Thermal Conductivity of Gas, W/m K
0.3
0.25
0.2
0.15
0.1
0.05
100
200
300
400
500
600
700
800
Temperature, C

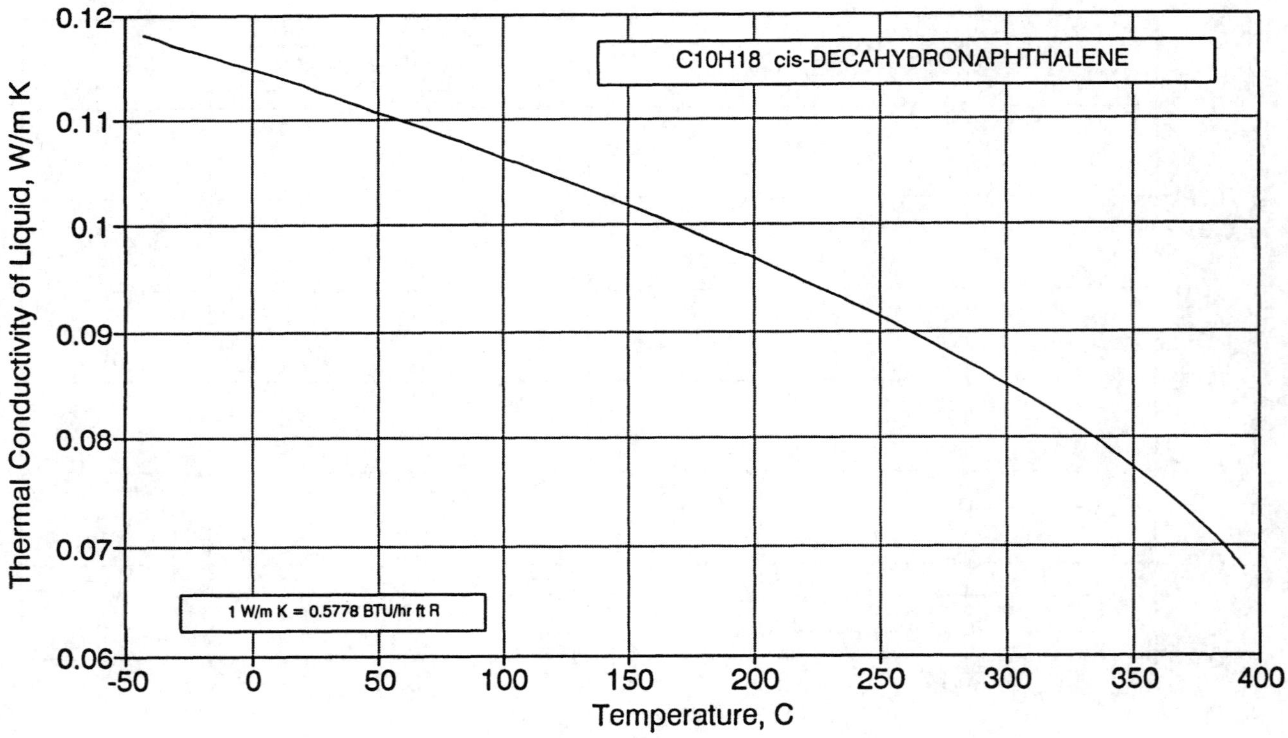
C10H18 cis-DECAHYDRONAPHTHALENE
1 W/m K = 0.5778 BTU/hr ft R
Thermal Conductivity of Liquid, W/m K
0.12
0.11
0.1
0.09
0.08
0.07
0.06
-50
0
50
100
150
200
250
300
350
400
Temperature, C

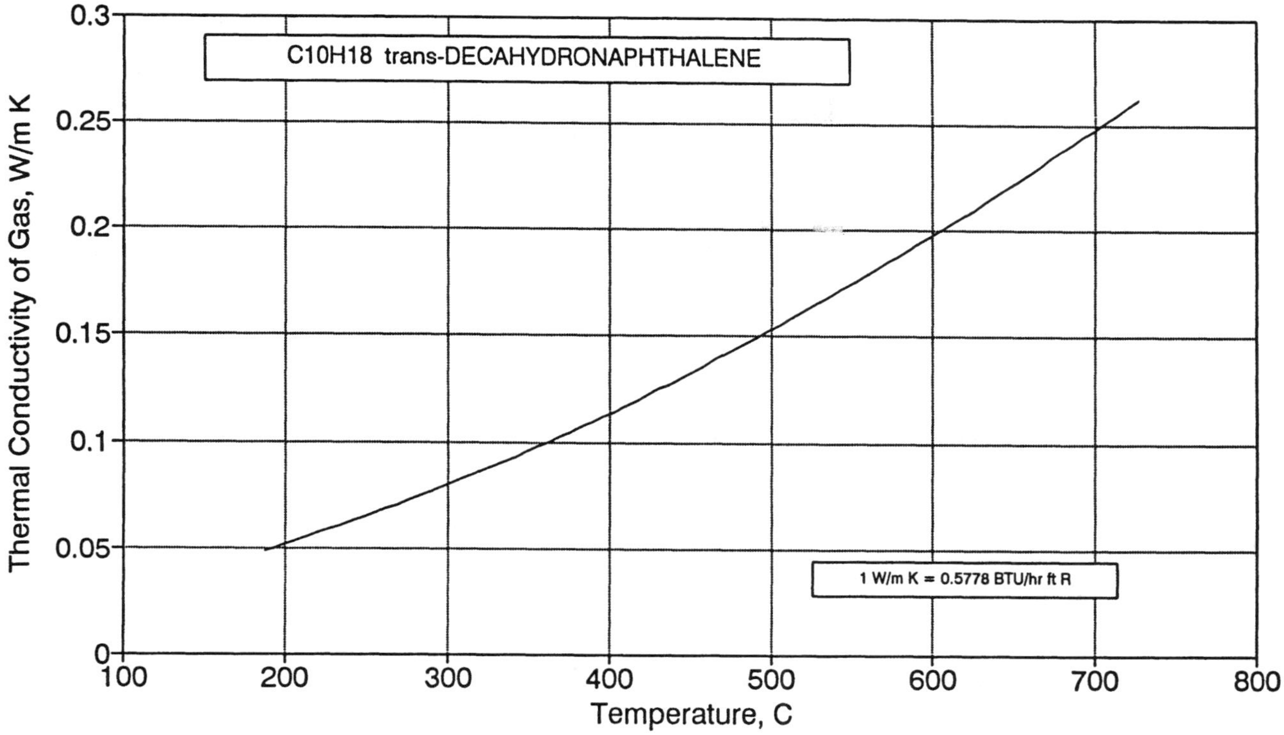
C10H18 trans-DECAHYDRONAPHTHALENE
1 W/m K = 0.5778 BTU/hr ft R
Thermal Conductivity of Gas, W/m K
Temperature, C
0.3
0.25
0.2
0.15
0.1
0.05
0
100
200
300
400
500
600
700
800

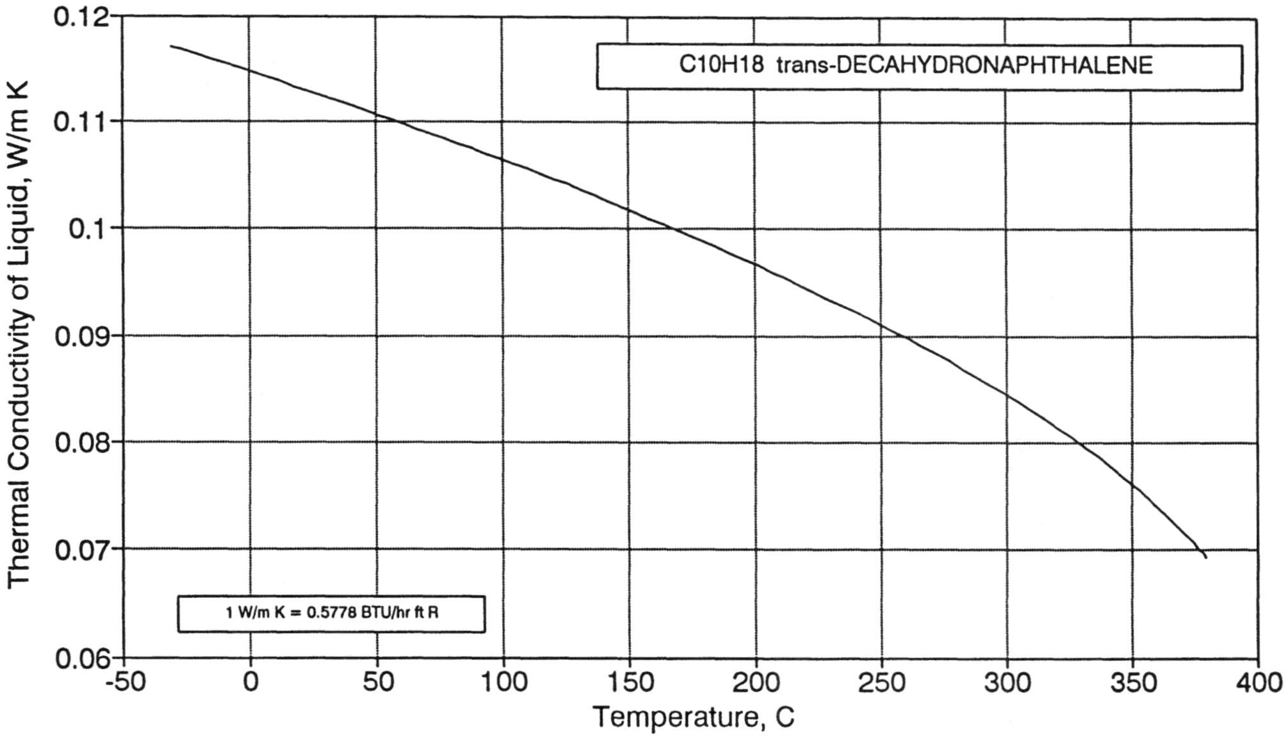
C10H18 trans-DECAHYDRONAPHTHALENE
1 W/m K = 0.5778 BTU/hr ft R
Thermal Conductivity of Liquid, W/m K
Temperature, C
0.12
0.11
0.1
0.09
0.08
0.07
0.06
-50
0
50
100
150
200
250
300
350
400

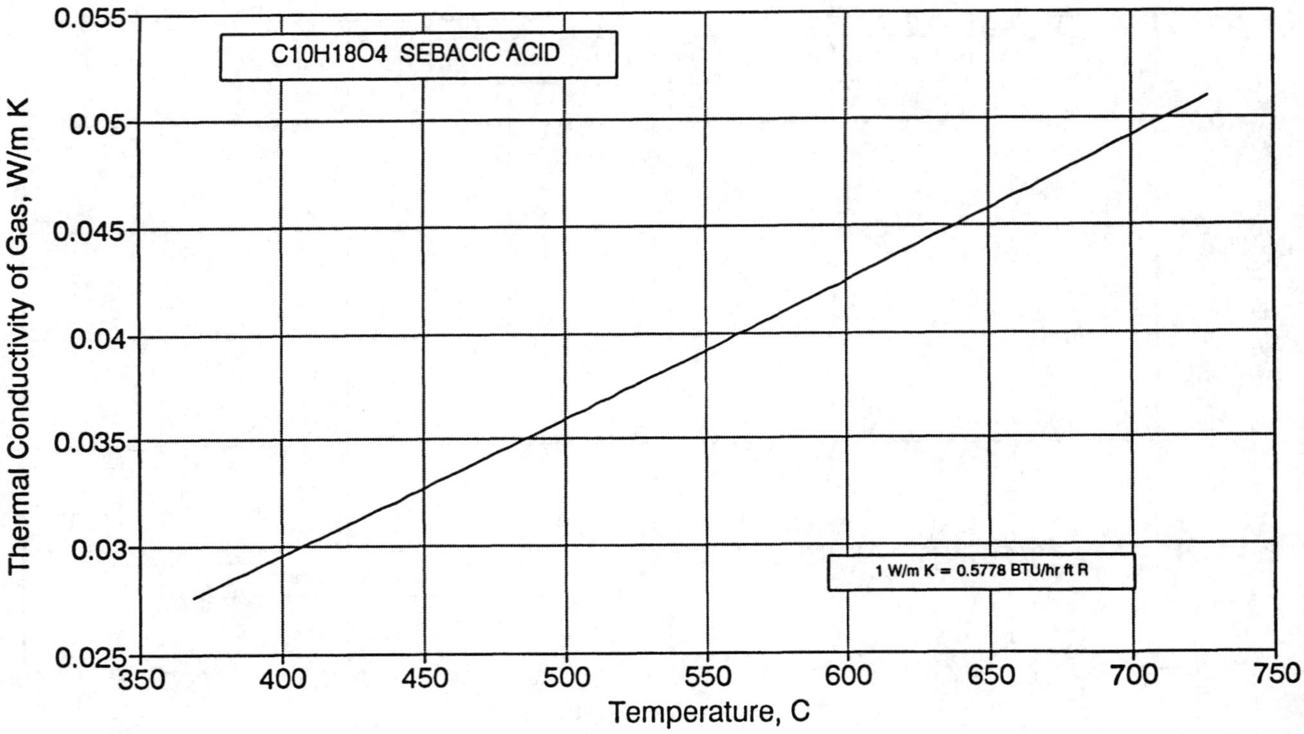
C10H18O4 SEBACIC ACID
1 W/m K = 0.5778 BTU/hr ft R
Thermal Conductivity of Gas, W/m K
0.055
0.05
0.045
0.04
0.035
0.03
0.025
350
400
450
500
550
600
650
700
750
Temperature, C

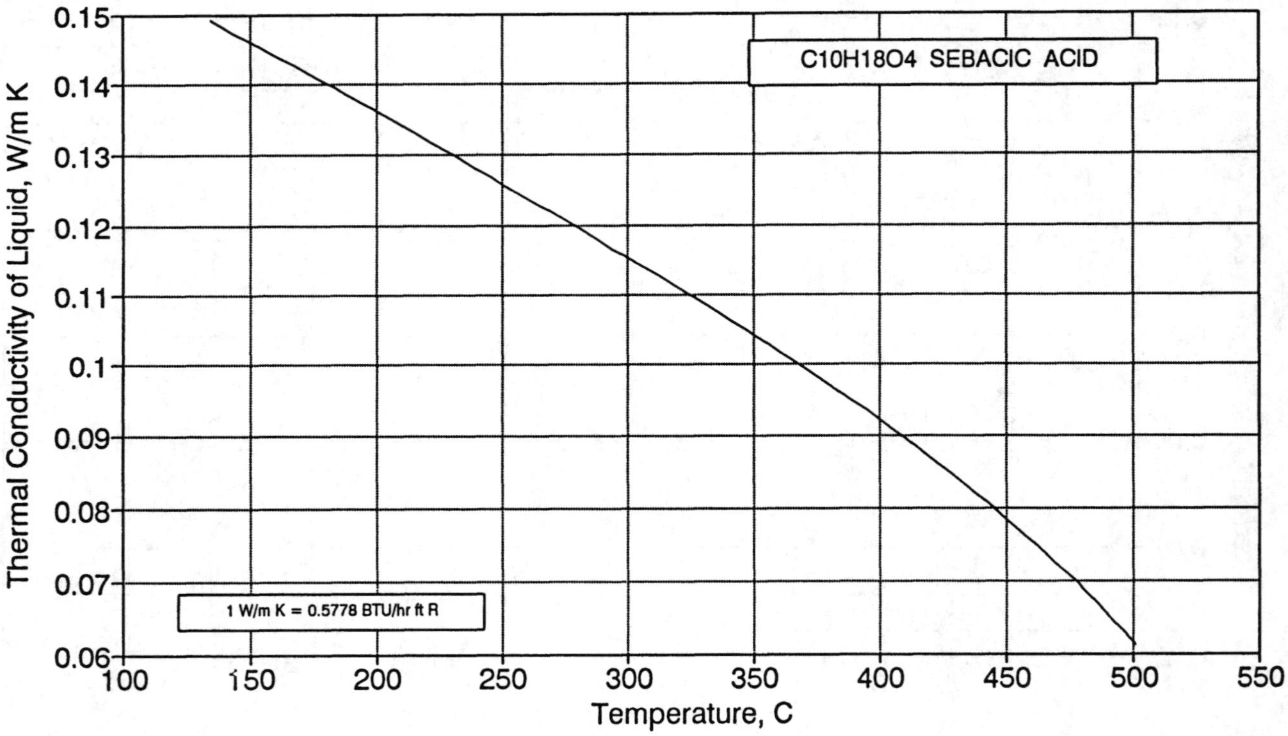
C10H18O4 SEBACIC ACID
1 W/m K = 0.5778 BTU/hr ft R
Thermal Conductivity of Liquid, W/m K
0.15
0.14
0.13
0.12
0.11
0.1
0.09
0.08
0.07
0.06
100
150
200
250
300
350
400
450
500
550
Temperature, C

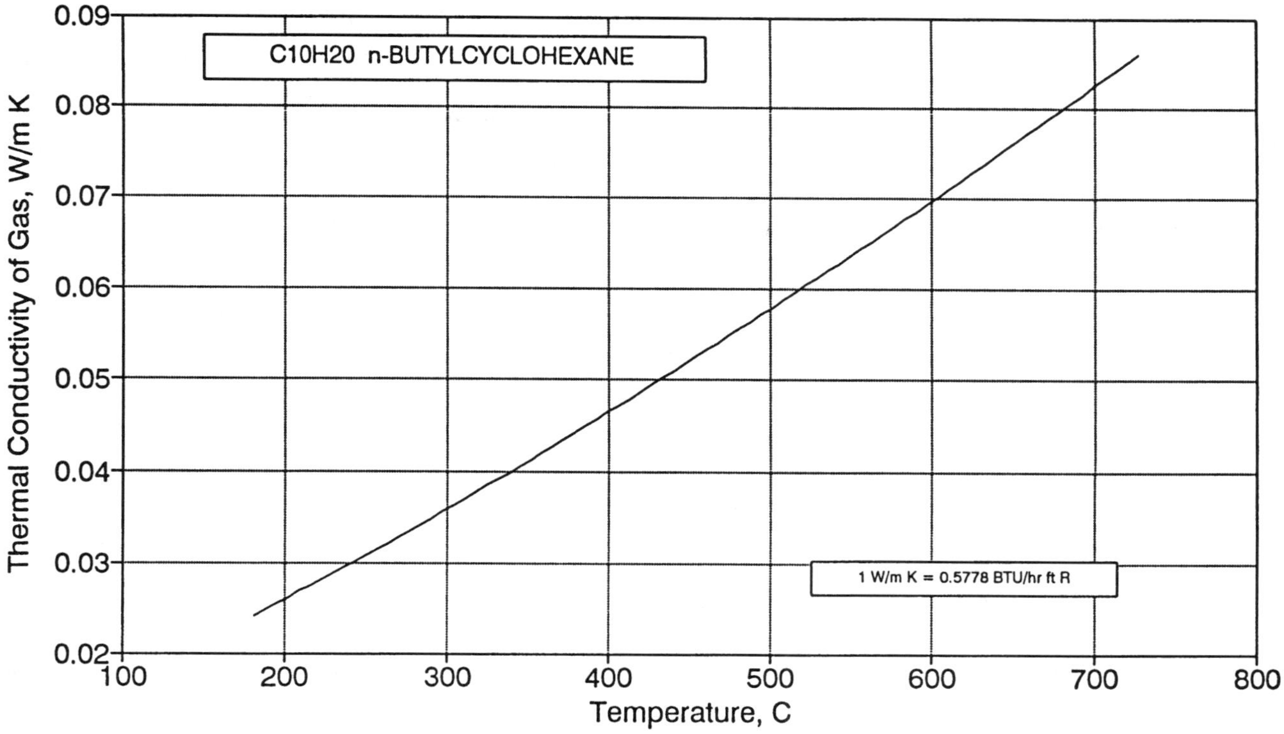
C10H20 n-BUTYLCYCLOHEXANE
1 W/m K = 0.5778 BTU/hr ft R
Thermal Conductivity of Gas, W/m K
0.09
0.08
0.07
0.06
0.05
0.04
0.03
0.02
100
200
300
400
500
600
700
800
Temperature, C

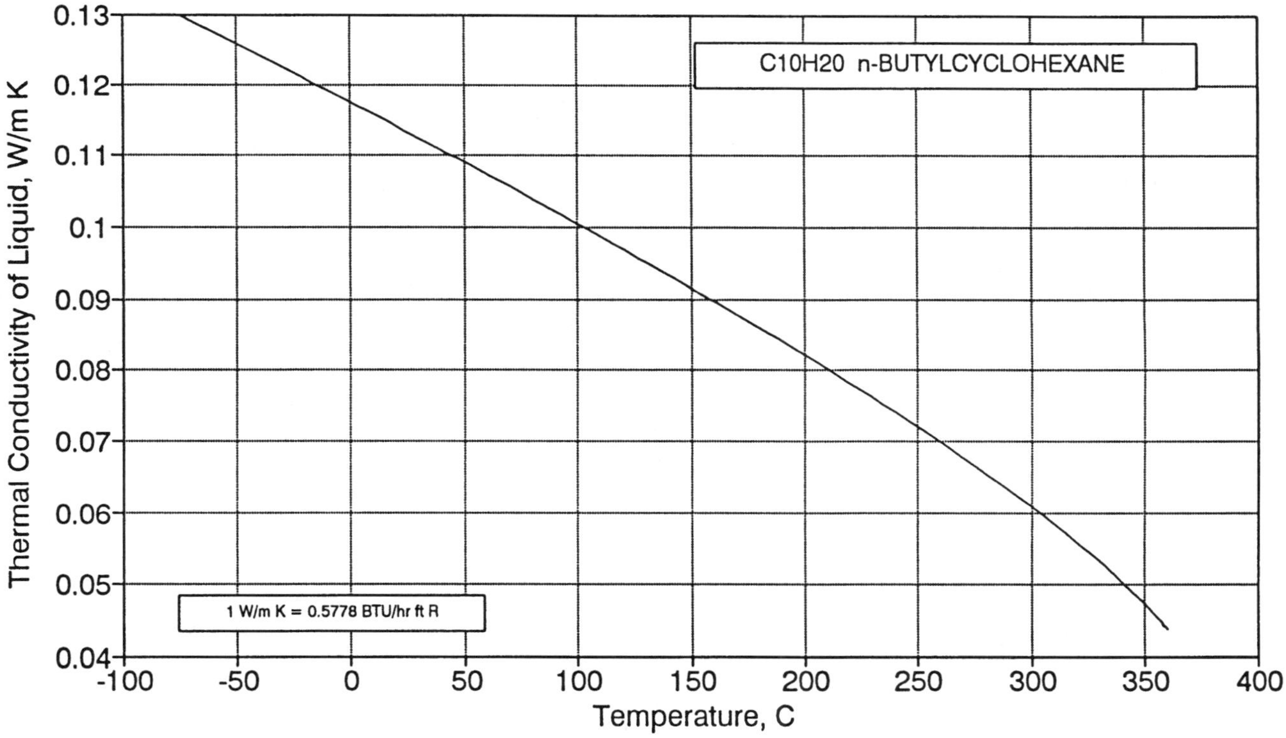
C10H20 n-BUTYLCYCLOHEXANE
1 W/m K = 0.5778 BTU/hr ft R
Thermal Conductivity of Liquid, W/m K
0.13
0.12
0.11
0.1
0.09
0.08
0.07
0.06
0.05
0.04
-100
-50
0
50
100
150
200
250
300
350
400
Temperature, C

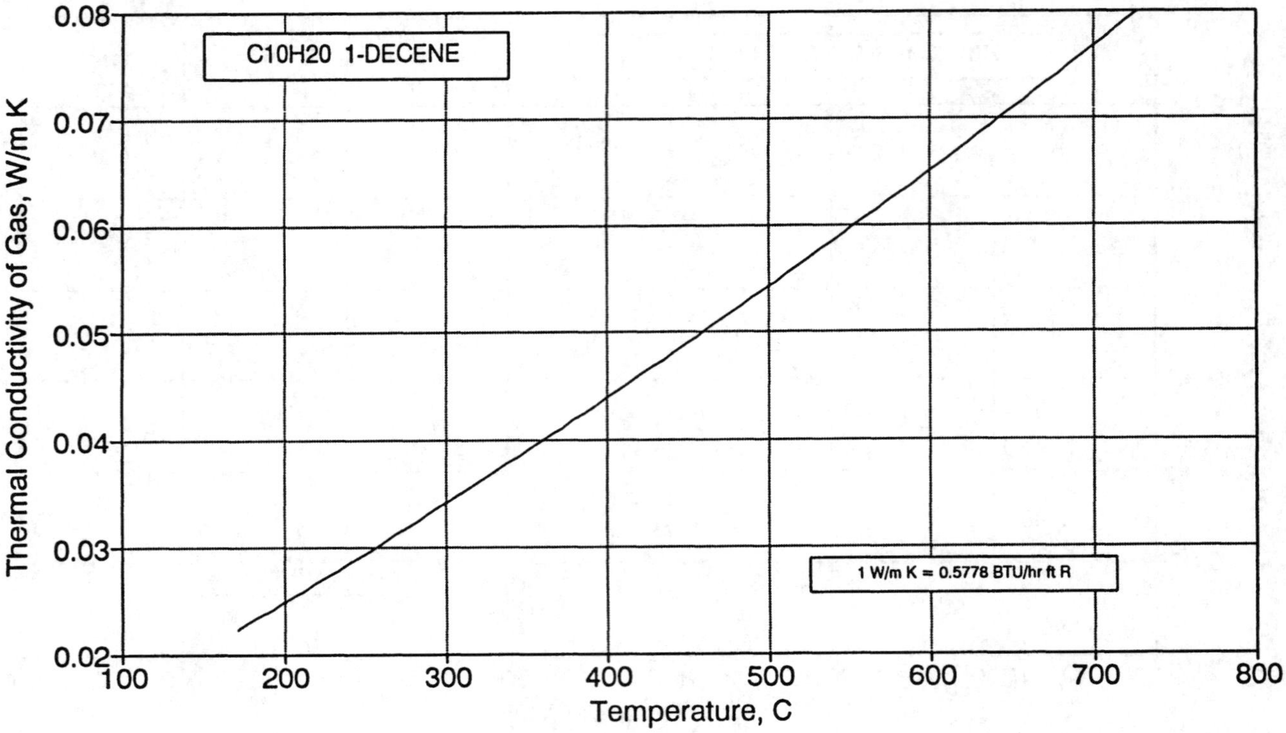
C10H20 1-DECENE
1 W/m K = 0.5778 BTU/hr ft R
Thermal Conductivity of Gas, W/m K
Temperature, C
0.08
0.07
0.06
0.05
0.04
0.03
0.02
100
200
300
400
500
600
700
800

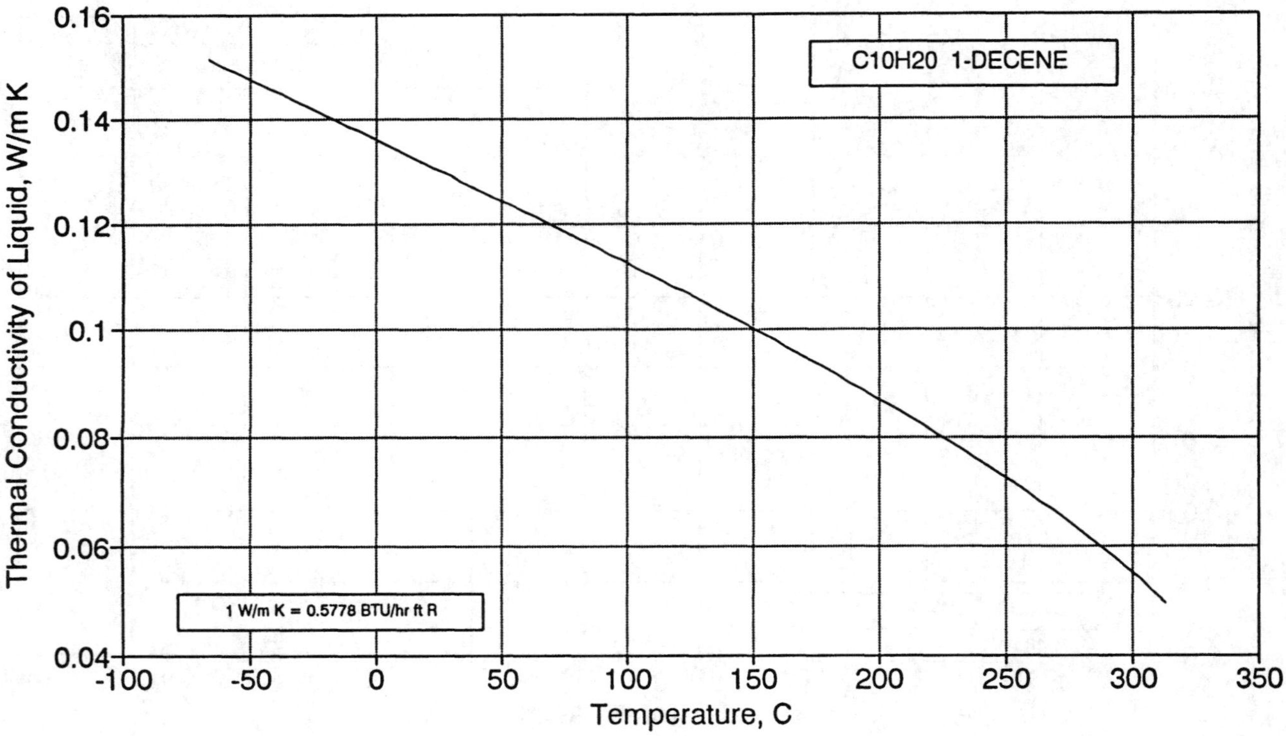
C10H20 1-DECENE
1 W/m K = 0.5778 BTU/hr ft R
Thermal Conductivity of Liquid, W/m K
Temperature, C
0.16
0.14
0.12
0.1
0.08
0.06
0.04
-100
-50
0
50
100
150
200
250
300
350

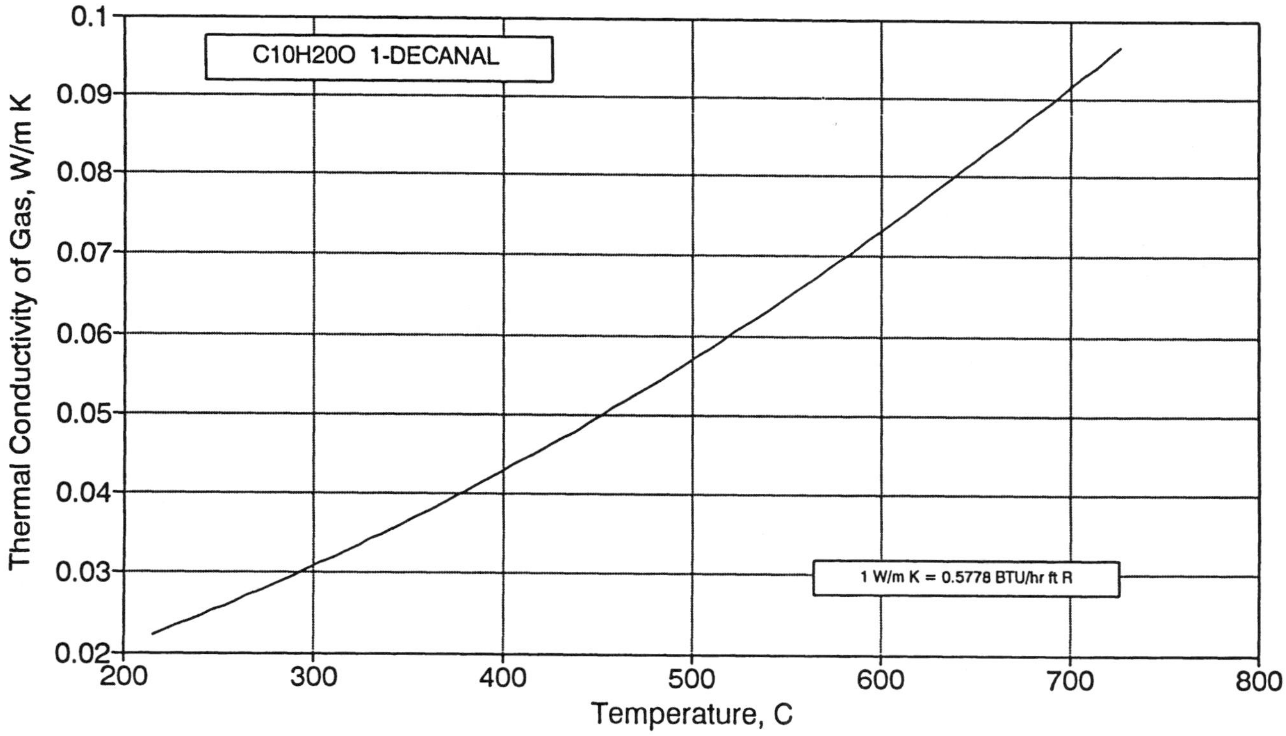
C10H20O 1-DECANAL
1 W/m K = 0.5778 BTU/hr ft R
Thermal Conductivity of Gas, W/m K
Temperature, C
0.1
0.09
0.08
0.07
0.06
0.05
0.04
0.03
0.02
200
300
400
500
600
700
800

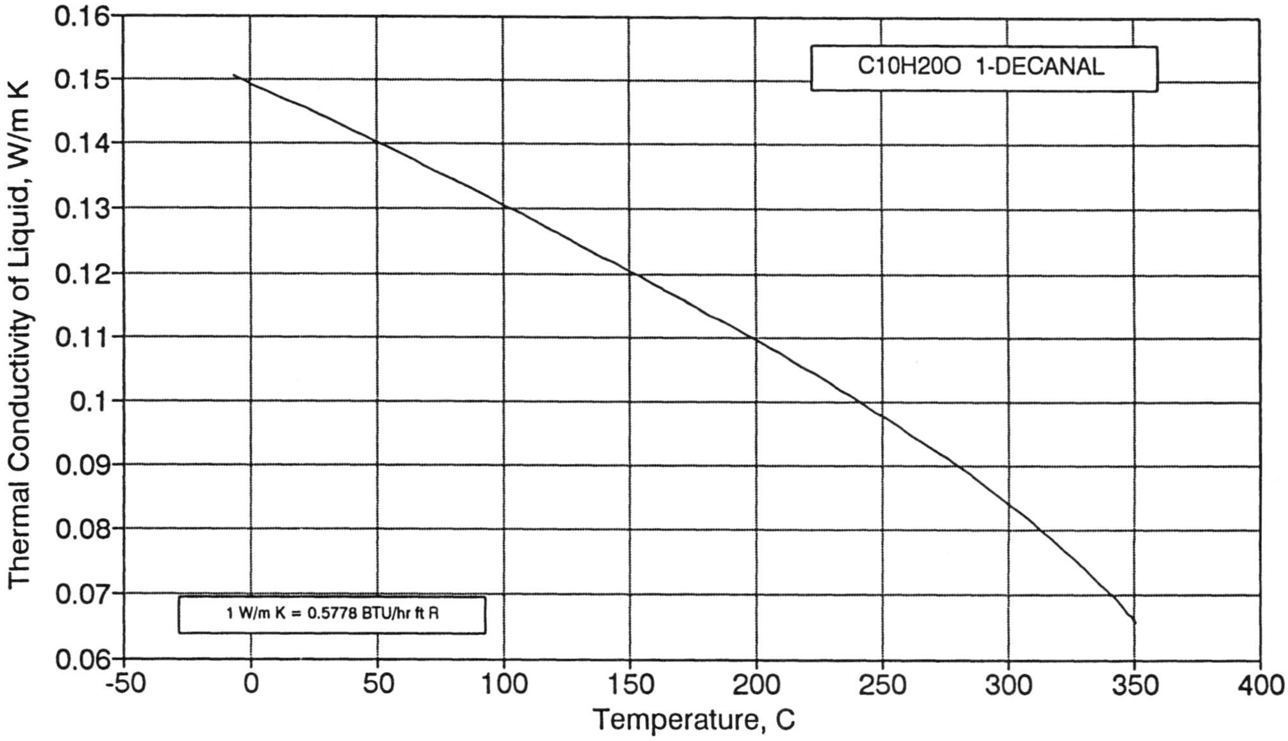
C10H20O 1-DECANAL
1 W/m K = 0.5778 BTU/hr ft R
Thermal Conductivity of Liquid, W/m K
Temperature, C
0.16
0.15
0.14
0.13
0.12
0.11
0.1
0.09
0.08
0.07
0.06
-50
0
50
100
150
200
250
300
350
400

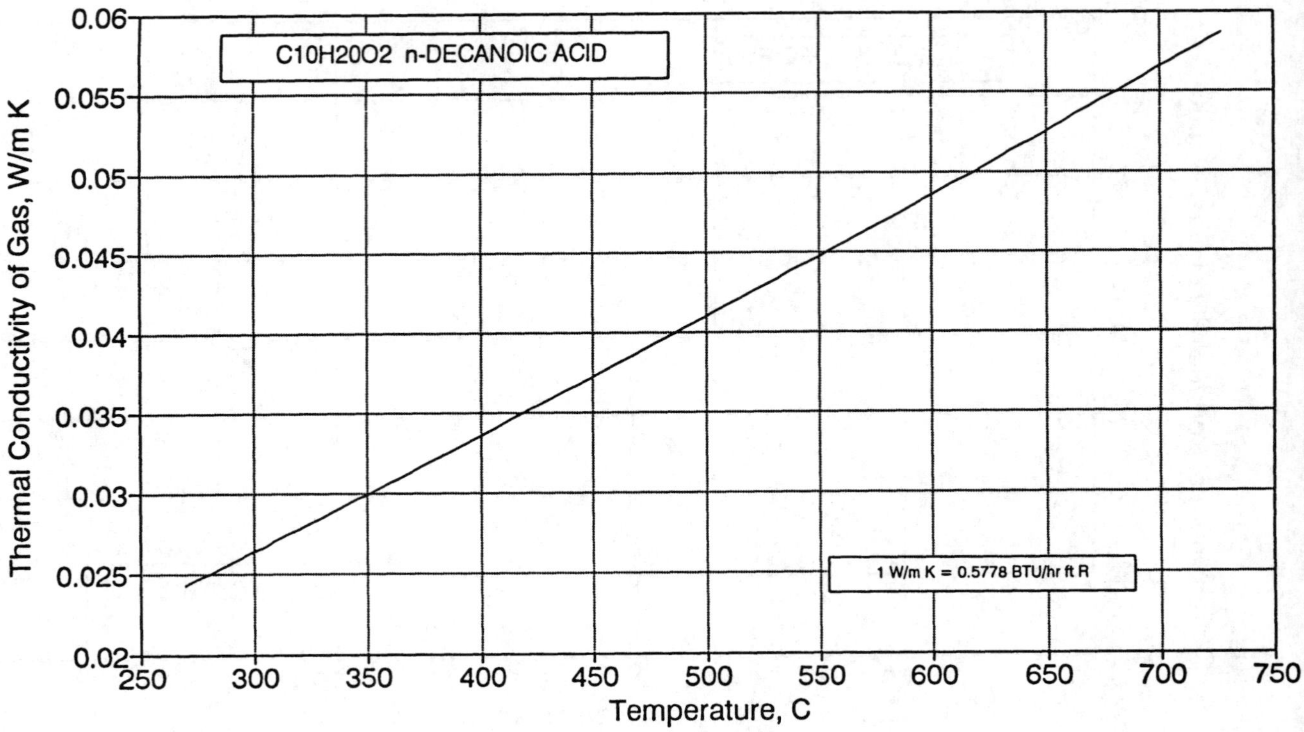
C10H20O2 n-DECANOIC ACID
Thermal Conductivity of Gas, W/m K
0.06
0.055
0.05
0.045
0.04
0.035
0.03
0.025
0.02
250
300
350
400
450
500
550
600
650
700
750
Temperature, C
1 W/m K = 0.5778 BTU/hr ft R

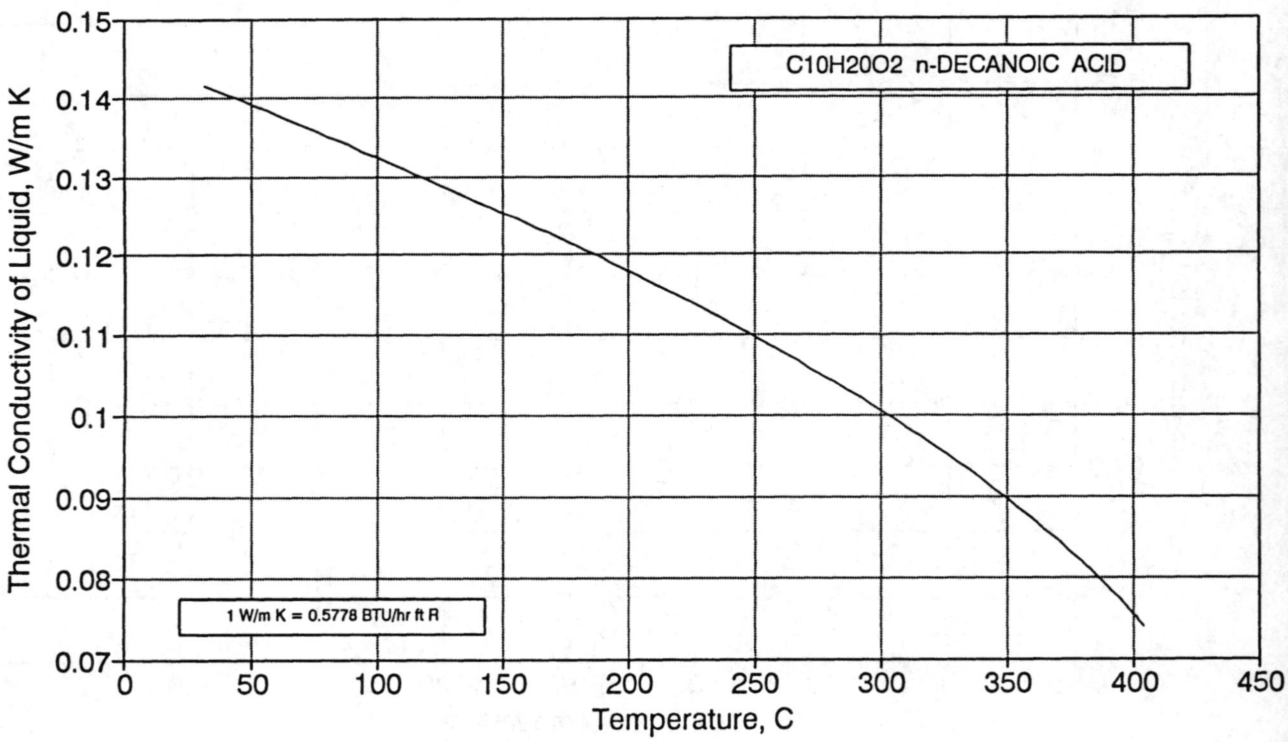
C10H20O2 n-DECANOIC ACID
Thermal Conductivity of Liquid, W/m K
0.15
0.14
0.13
0.12
0.11
0.1
0.09
0.08
0.07
0
50
100
150
200
250
300
350
400
450
Temperature, C
1 W/m K = 0.5778 BTU/hr ft R

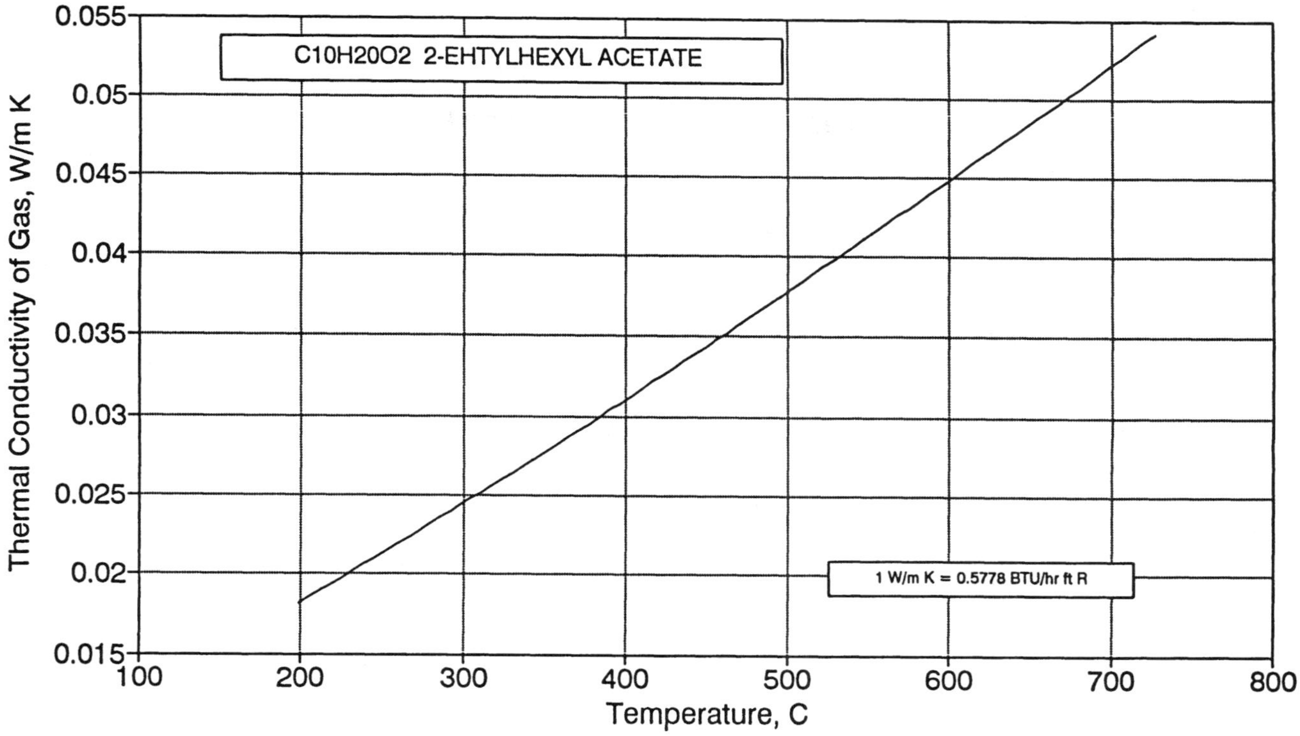
C10H20O2 2-EHTYLHEXYL ACETATE
1 W/m K = 0.5778 BTU/hr ft R
Thermal Conductivity of Gas, W/m K
Temperature, C
0.055
0.05
0.045
0.04
0.035
0.03
0.025
0.02
0.015
100
200
300
400
500
600
700
800

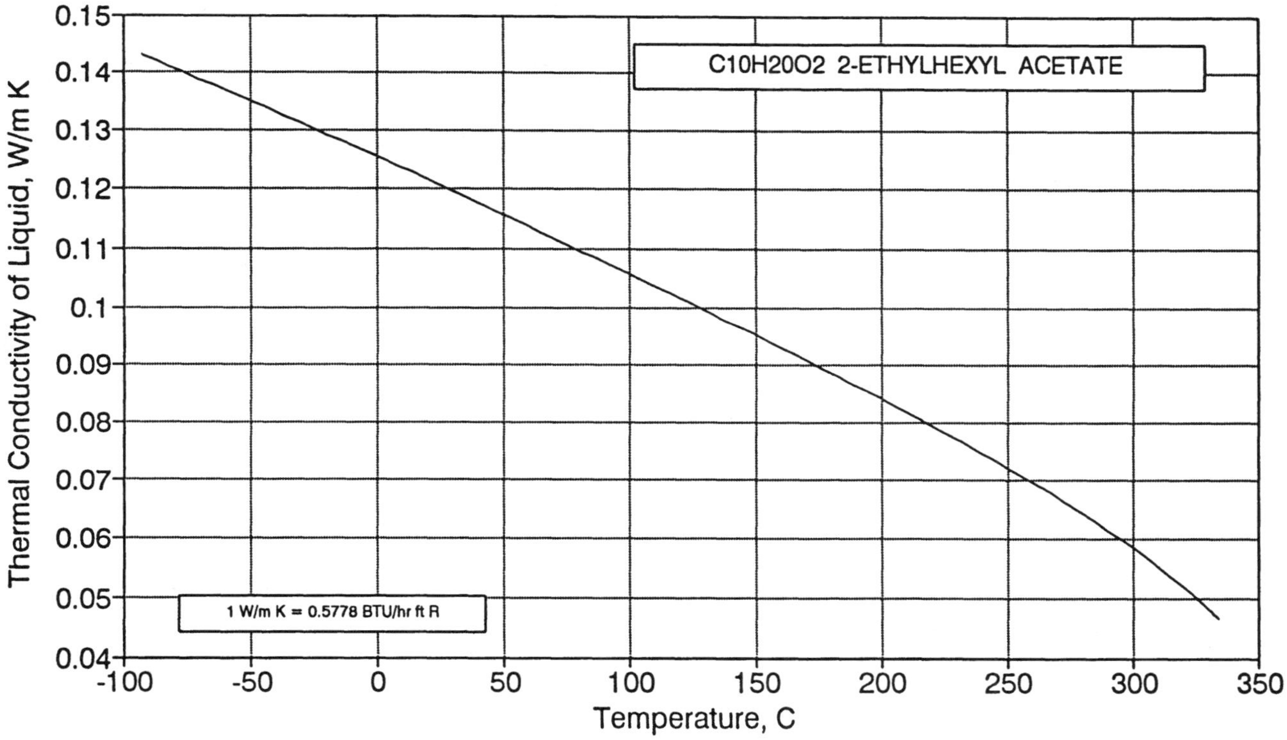
C10H20O2 2-ETHYLHEXYL ACETATE
1 W/m K = 0.5778 BTU/hr ft R
Thermal Conductivity of Liquid, W/m K
Temperature, C
0.15
0.14
0.13
0.12
0.11
0.1
0.09
0.08
0.07
0.06
0.05
0.04
-100
-50
0
50
100
150
200
250
300
350

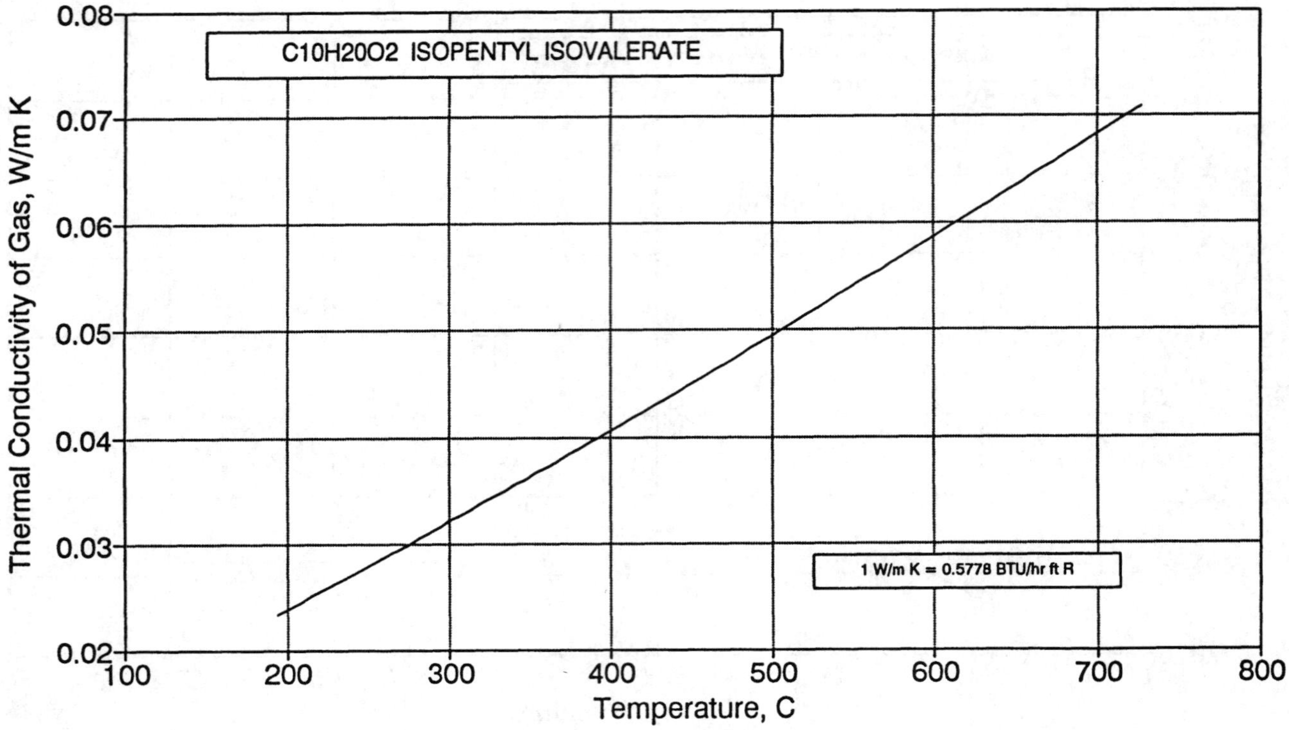
C10H20O2 ISOPENTYL ISOVALERATE
Thermal Conductivity of Gas, W/m K
0.08
0.07
0.06
0.05
0.04
0.03
0.02
1 W/m K = 0.5778 BTU/hr ft R
100
200
300
400
500
600
700
800
Temperature, C

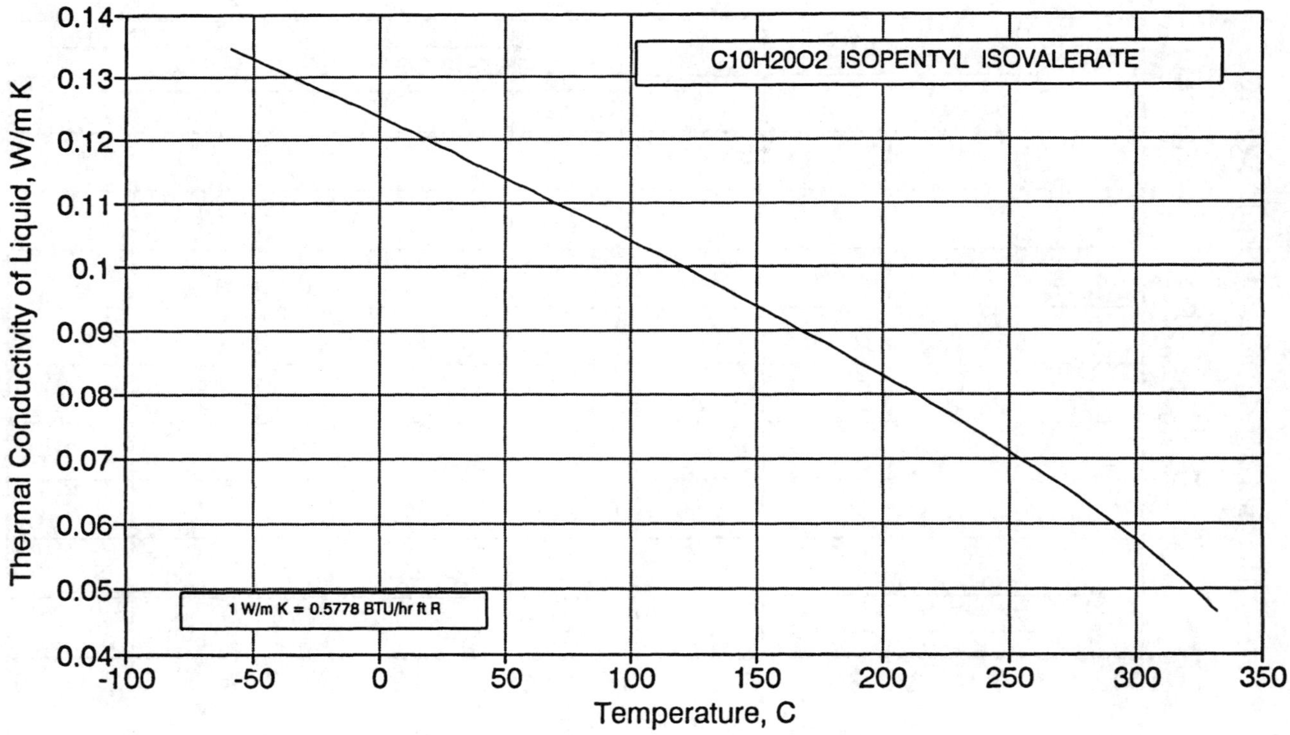
C10H20O2 ISOPENTYL ISOVALERATE
Thermal Conductivity of Liquid, W/m K
0.14
0.13
0.12
0.11
0.1
0.09
0.08
0.07
0.06
0.05
0.04
1 W/m K = 0.5778 BTU/hr ft R
-100
-50
0
50
100
150
200
250
300
350
Temperature, C

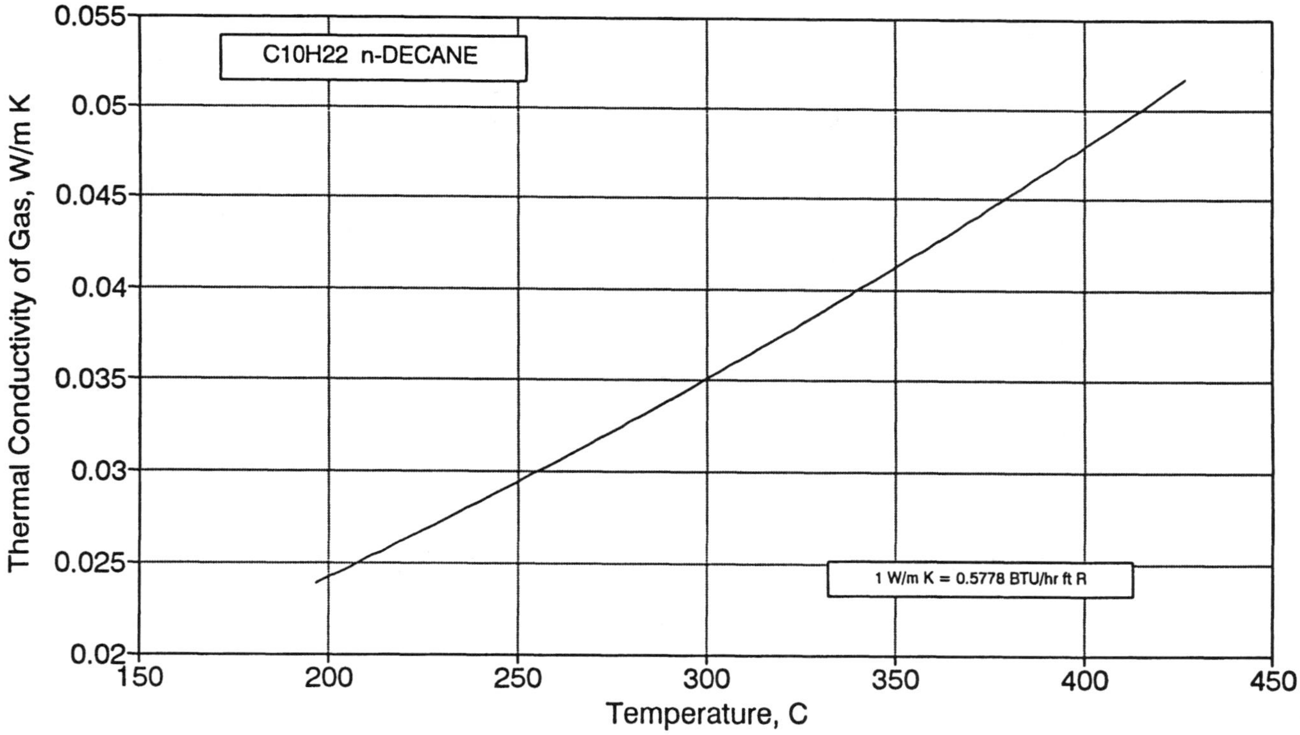

C10H22 n-DECANE
1 W/m K = 0.5778 BTU/hr ft R
Thermal Conductivity of Gas, W/m K
0.055
0.05
0.045
0.04
0.035
0.03
0.025
0.02
150
200
250
300
350
400
450
Temperature, C

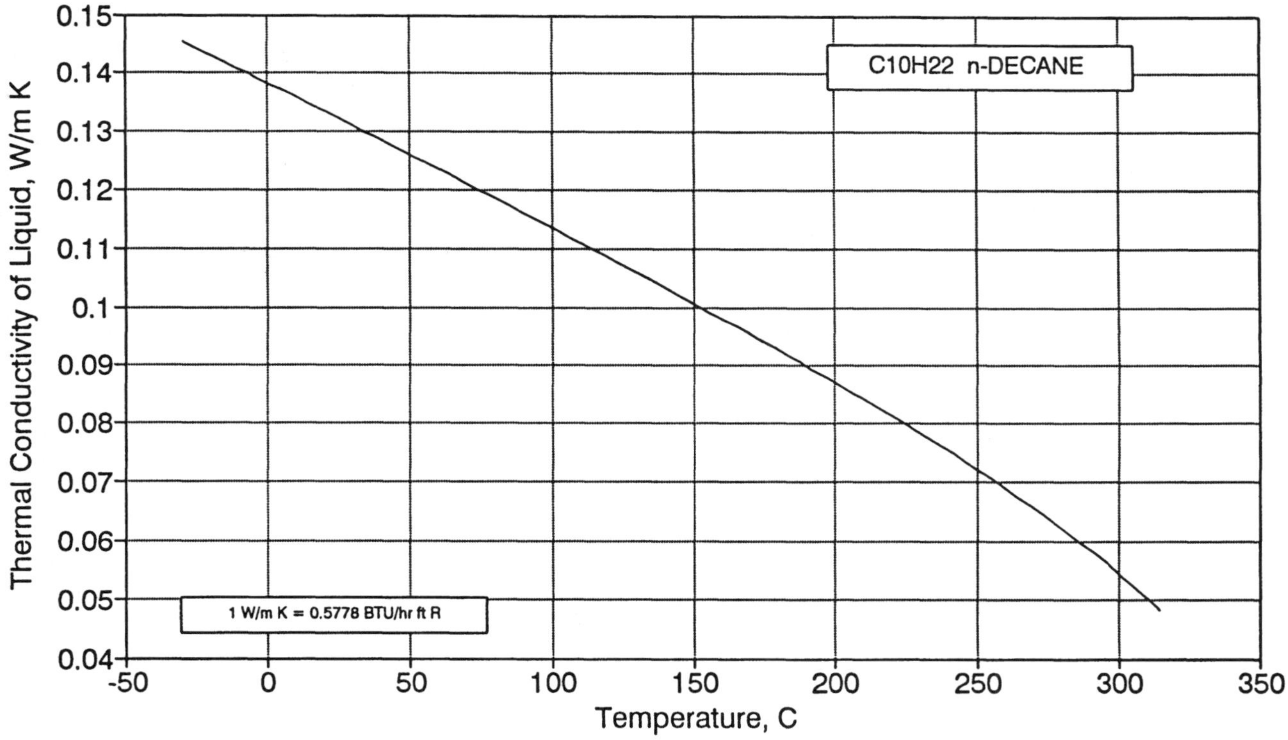

C10H22 n-DECANE
1 W/m K = 0.5778 BTU/hr ft R
Thermal Conductivity of Liquid, W/m K
0.15
0.14
0.13
0.12
0.11
0.1
0.09
0.08
0.07
0.06
0.05
0.04
-50
0
50
100
150
200
250
300
350
Temperature, C

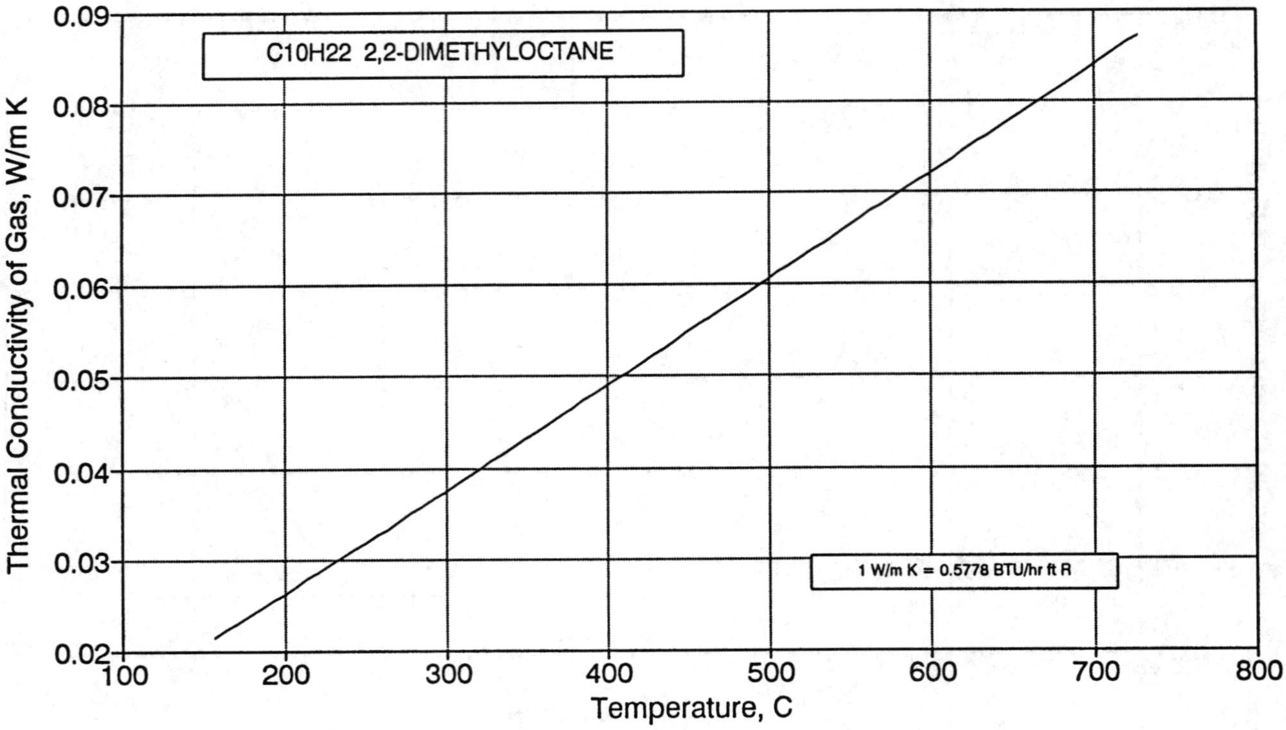
C10H22 2,2-DIMETHYLOCTANE
1 W/m K = 0.5778 BTU/hr ft R
Thermal Conductivity of Gas, W/m K
0.09
0.08
0.07
0.06
0.05
0.04
0.03
0.02
100
200
300
400
500
600
700
800
Temperature, C

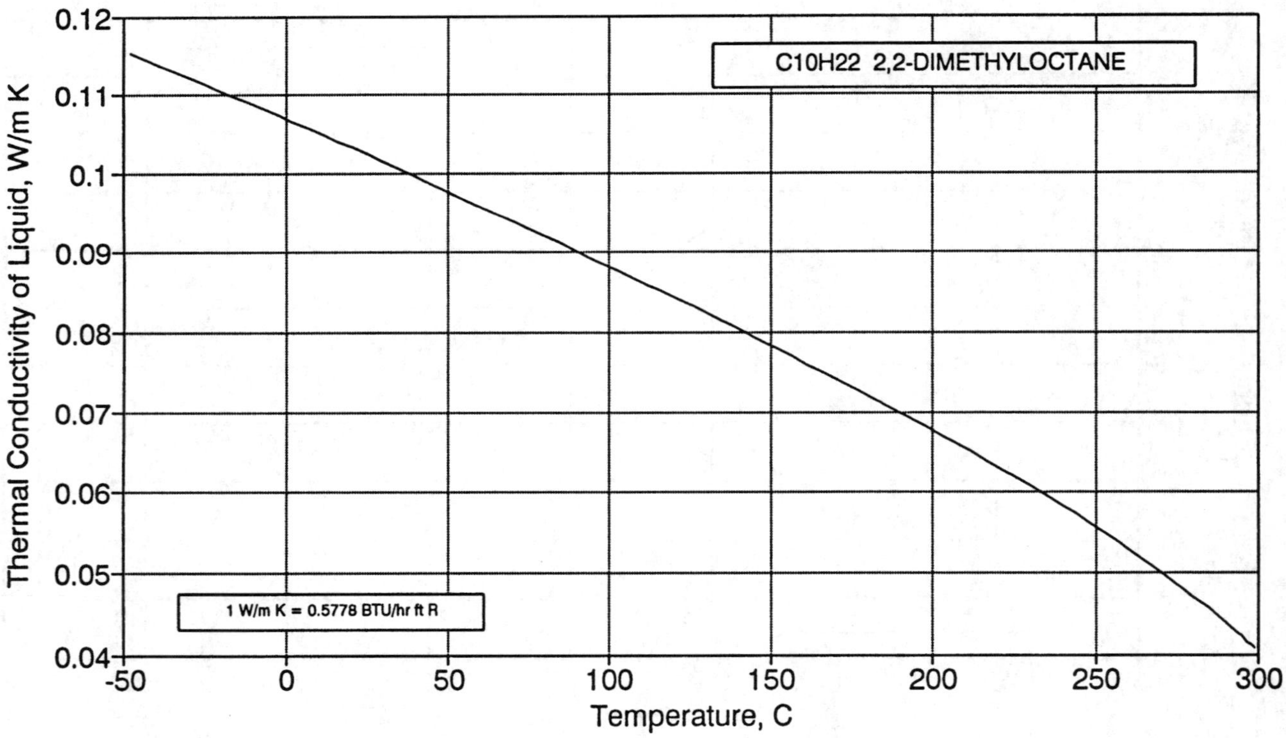
C10H22 2,2-DIMETHYLOCTANE
1 W/m K = 0.5778 BTU/hr ft R
Thermal Conductivity of Liquid, W/m K
0.12
0.11
0.1
0.09
0.08
0.07
0.06
0.05
0.04
-50
0
50
100
150
200
250
300
Temperature, C

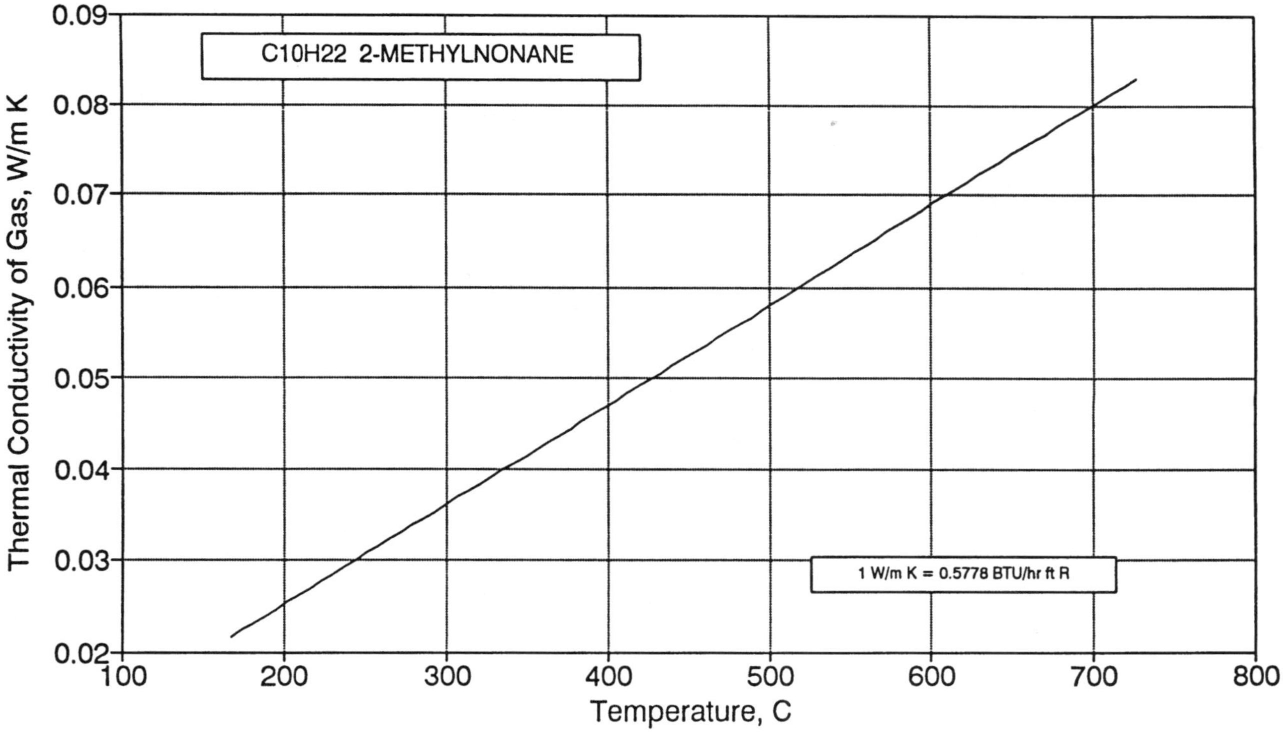
C10H22 2-METHYLNONANE
1 W/m K = 0.5778 BTU/hr ft R
Thermal Conductivity of Gas, W/m K
Temperature, C
0.09
0.08
0.07
0.06
0.05
0.04
0.03
0.02
100
200
300
400
500
600
700
800

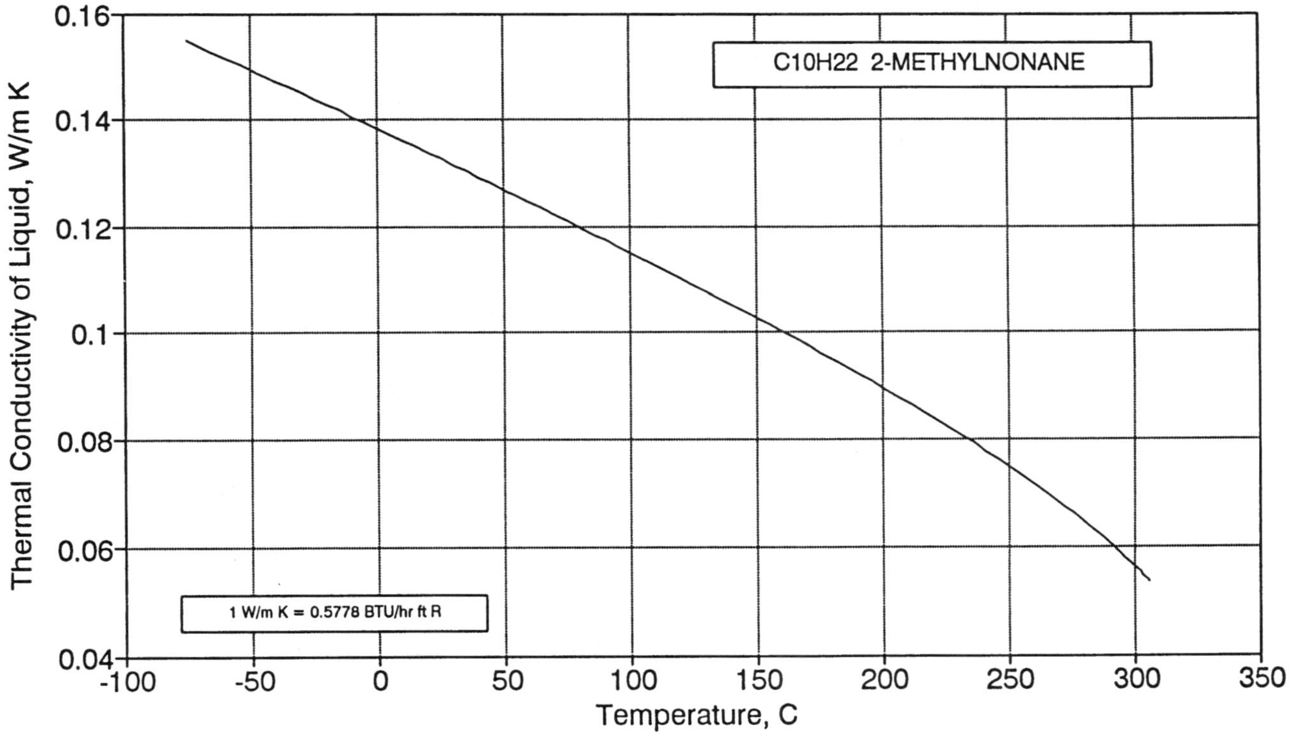
C10H22 2-METHYLNONANE
1 W/m K = 0.5778 BTU/hr ft R
Thermal Conductivity of Liquid, W/m K
Temperature, C
0.16
0.14
0.12
0.1
0.08
0.06
0.04
-100
-50
0
50
100
150
200
250
300
350

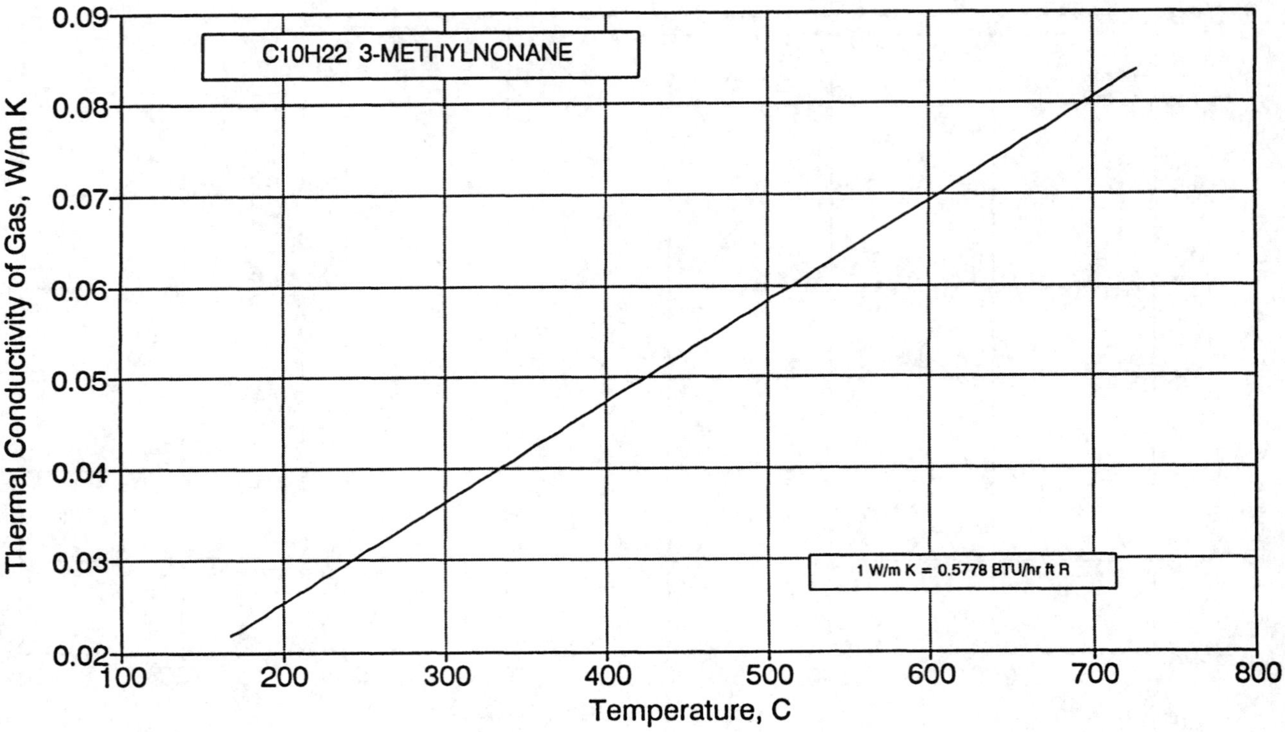
C10H22 3-METHYLNONANE
1 W/m K = 0.5778 BTU/hr ft R
Thermal Conductivity of Gas, W/m K
0.09
0.08
0.07
0.06
0.05
0.04
0.03
0.02
100
200
300
400
500
600
700
800
Temperature, C

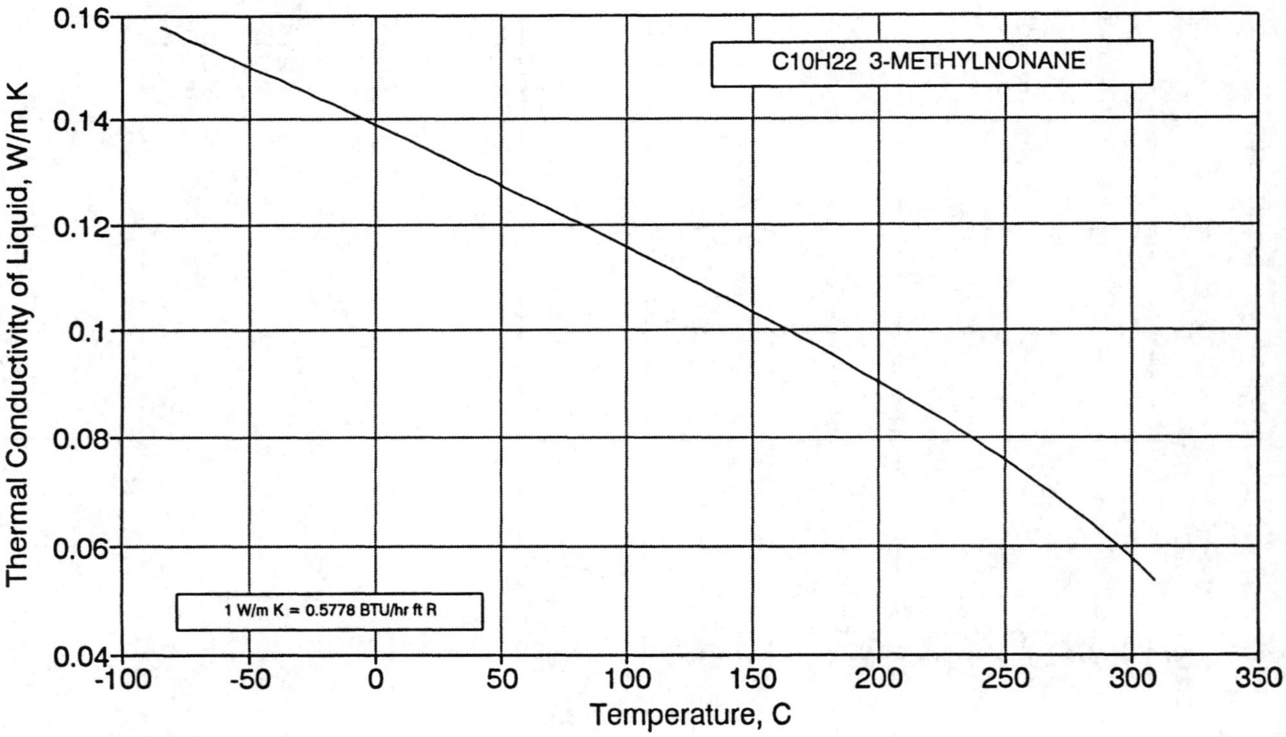
C10H22 3-METHYLNONANE
1 W/m K = 0.5778 BTU/hr ft R
Thermal Conductivity of Liquid, W/m K
0.16
0.14
0.12
0.1
0.08
0.06
0.04
-100
-50
0
50
100
150
200
250
300
350
Temperature, C

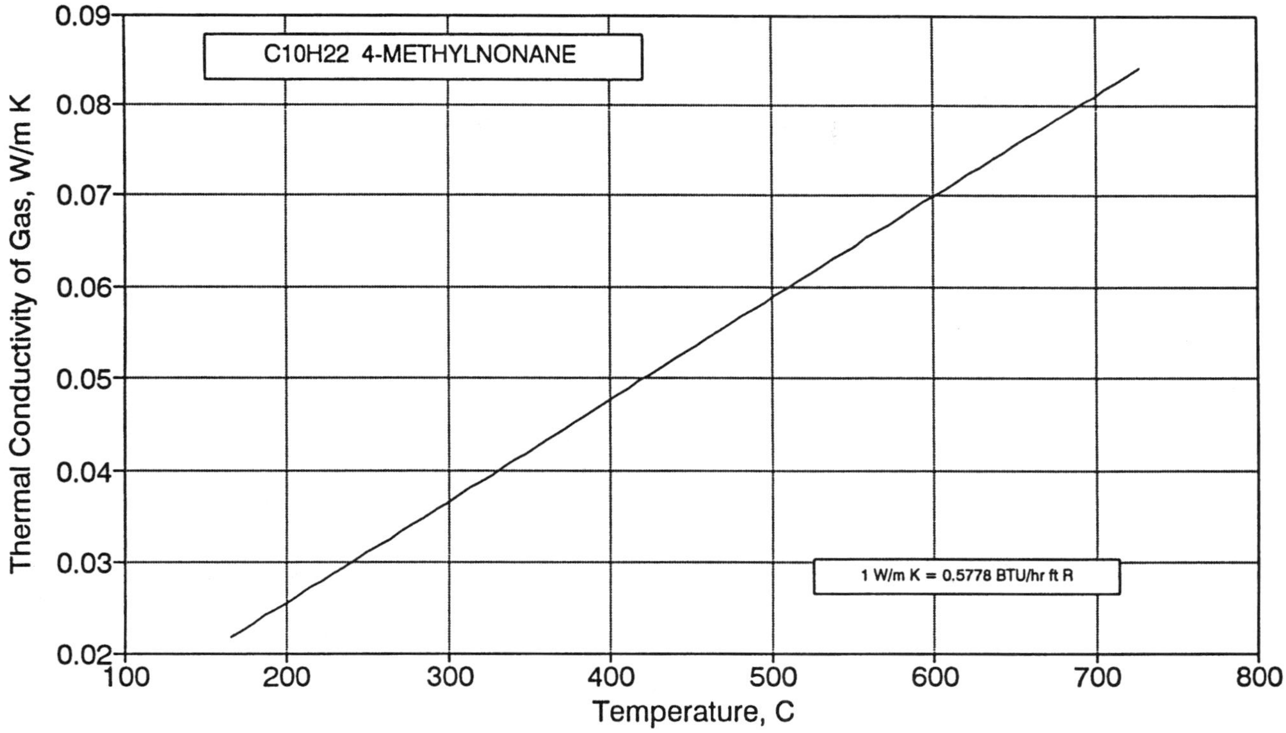
C10H22 4-METHYLNONANE
1 W/m K = 0.5778 BTU/hr ft R
Thermal Conductivity of Gas, W/m K
0.09
0.08
0.07
0.06
0.05
0.04
0.03
0.02
100
200
300
400
500
600
700
800
Temperature, C

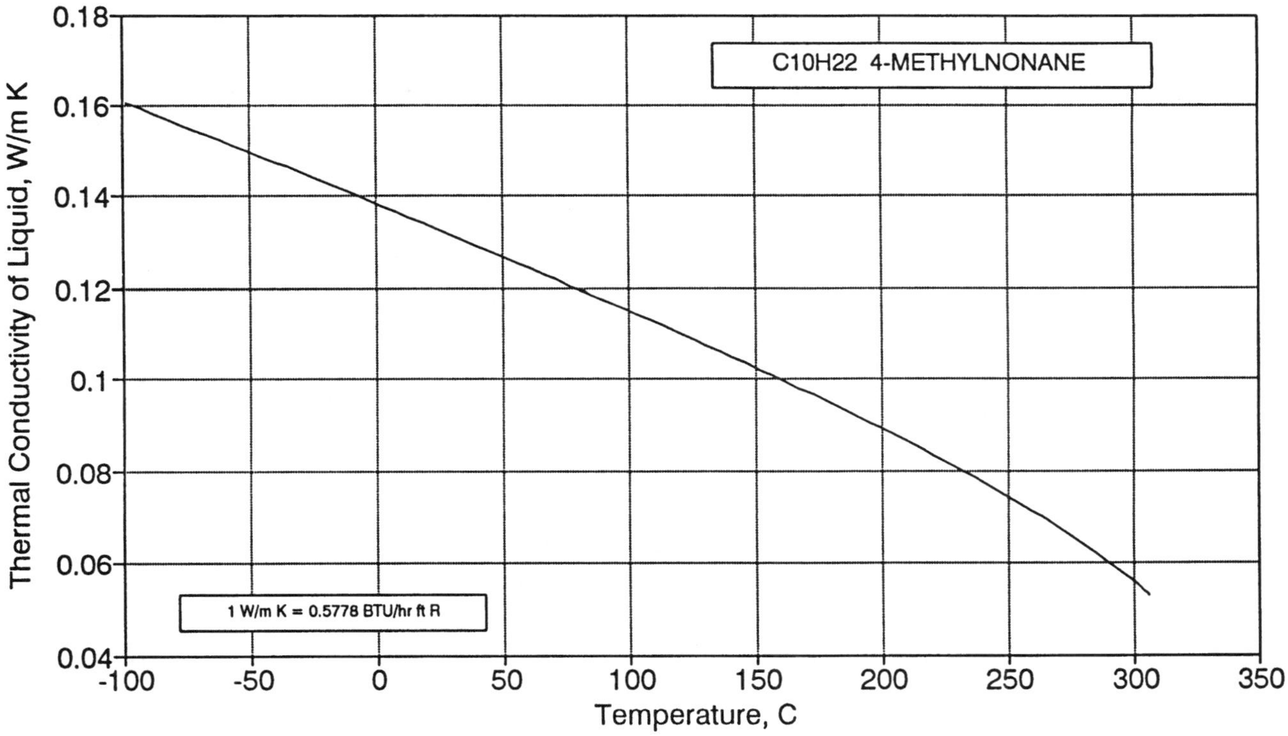
C10H22 4-METHYLNONANE
1 W/m K = 0.5778 BTU/hr ft R
Thermal Conductivity of Liquid, W/m K
0.18
0.16
0.14
0.12
0.1
0.08
0.06
0.04
-100
-50
0
50
100
150
200
250
300
350
Temperature, C

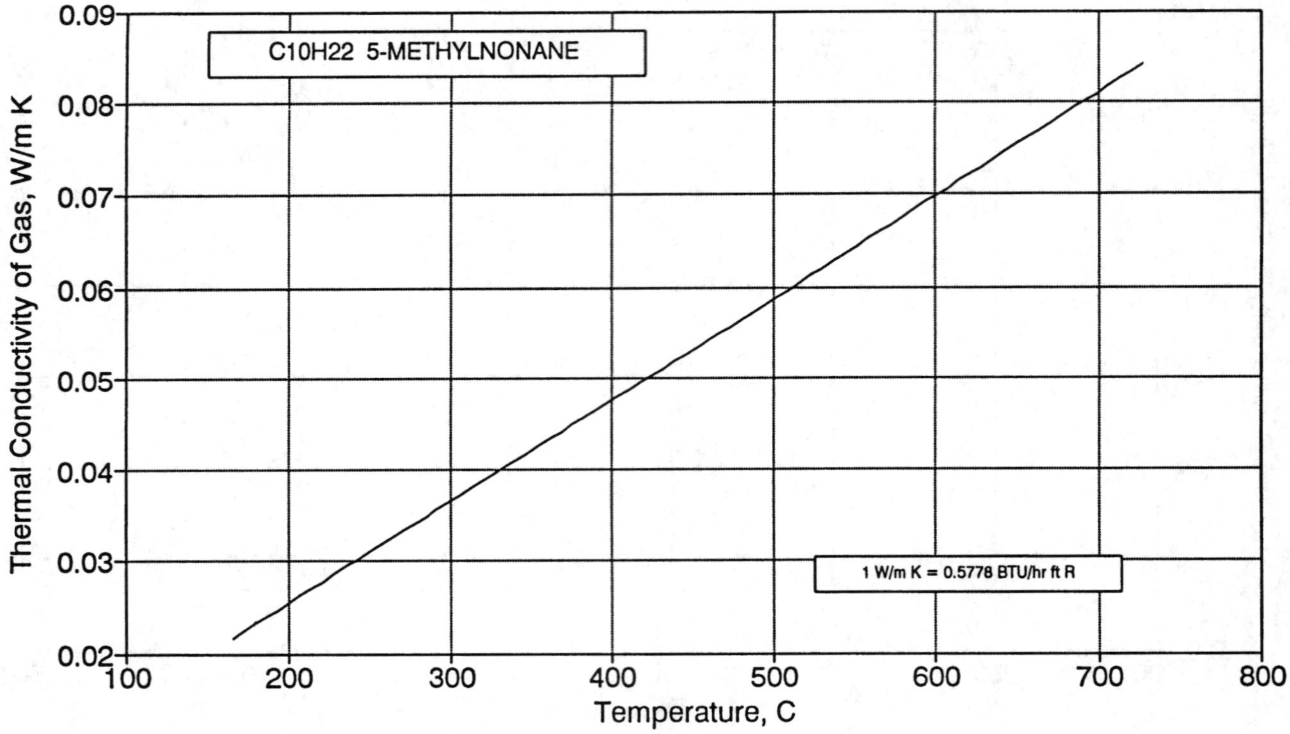
C10H22 5-METHYLNONANE
1 W/m K = 0.5778 BTU/hr ft R
Thermal Conductivity of Gas, W/m K
Temperature, C
0.09
0.08
0.07
0.06
0.05
0.04
0.03
0.02
100
200
300
400
500
600
700
800

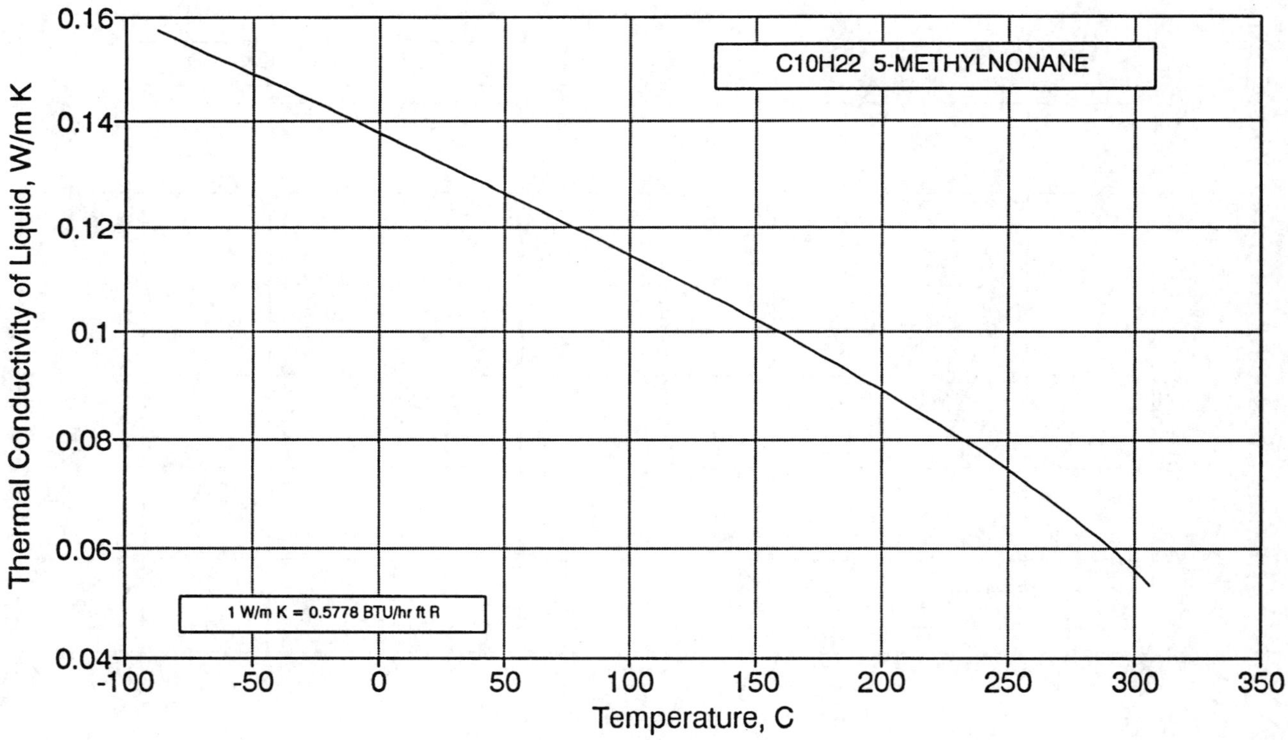
C10H22 5-METHYLNONANE
1 W/m K = 0.5778 BTU/hr ft R
Thermal Conductivity of Liquid, W/m K
Temperature, C
0.16
0.14
0.12
0.1
0.08
0.06
0.04
-100
-50
0
50
100
150
200
250
300
350

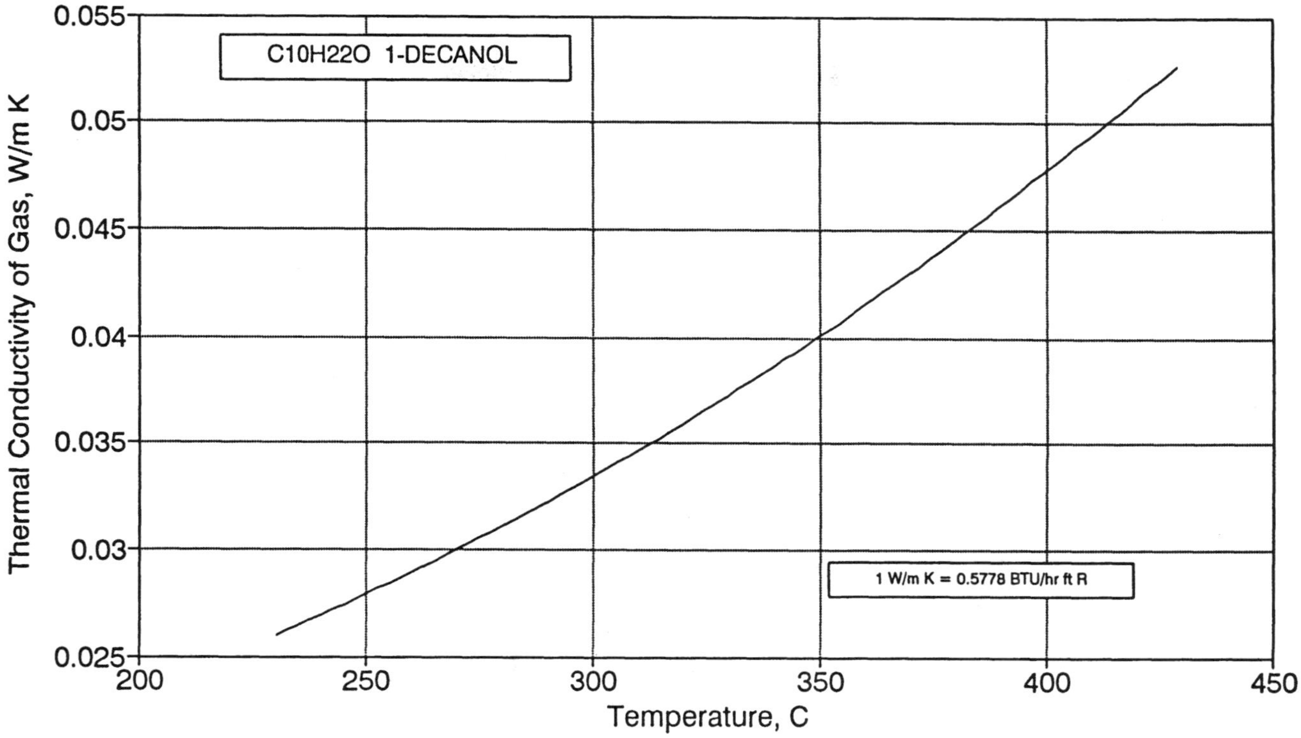
C10H22O 1-DECANOL
1 W/m K = 0.5778 BTU/hr ft R
Thermal Conductivity of Gas, W/m K
0.055
0.05
0.045
0.04
0.035
0.03
0.025
200
250
300
350
400
450
Temperature, C

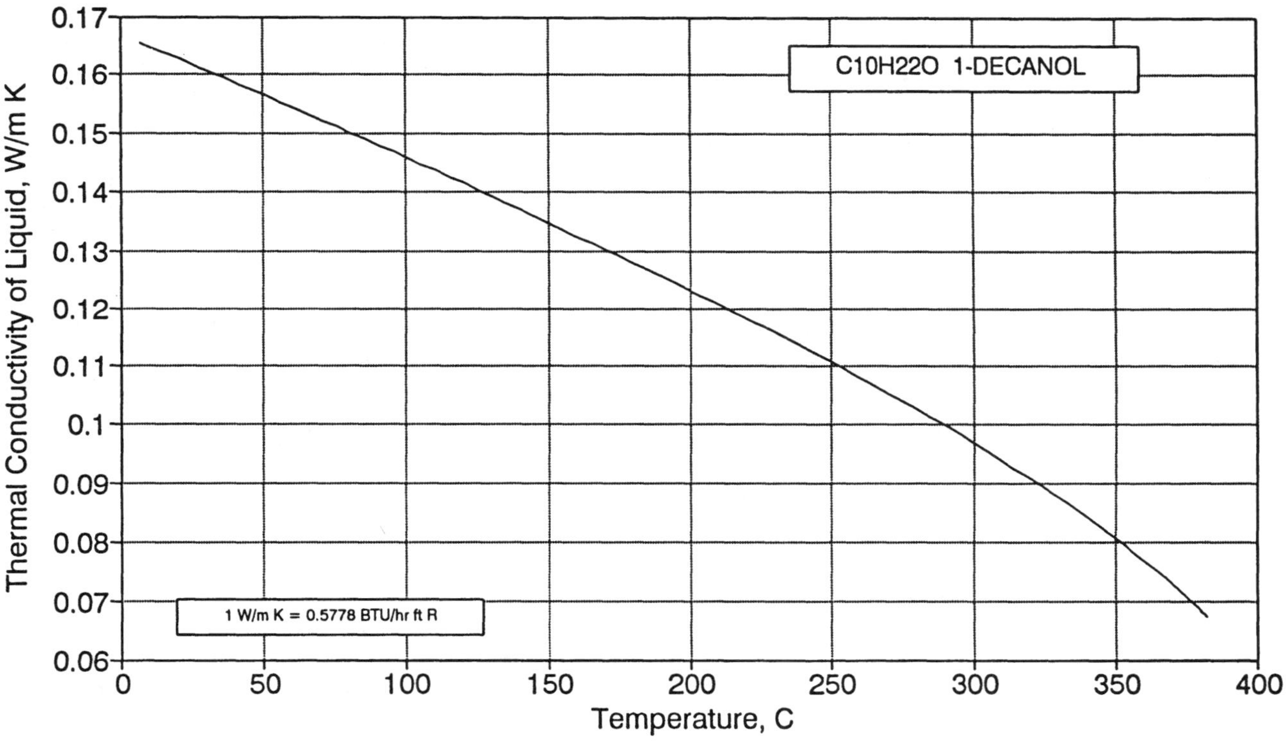
C10H22O 1-DECANOL
1 W/m K = 0.5778 BTU/hr ft R
Thermal Conductivity of Liquid, W/m K
0.17
0.16
0.15
0.14
0.13
0.12
0.11
0.1
0.09
0.08
0.07
0.06
0
50
100
150
200
250
300
350
400
Temperature, C

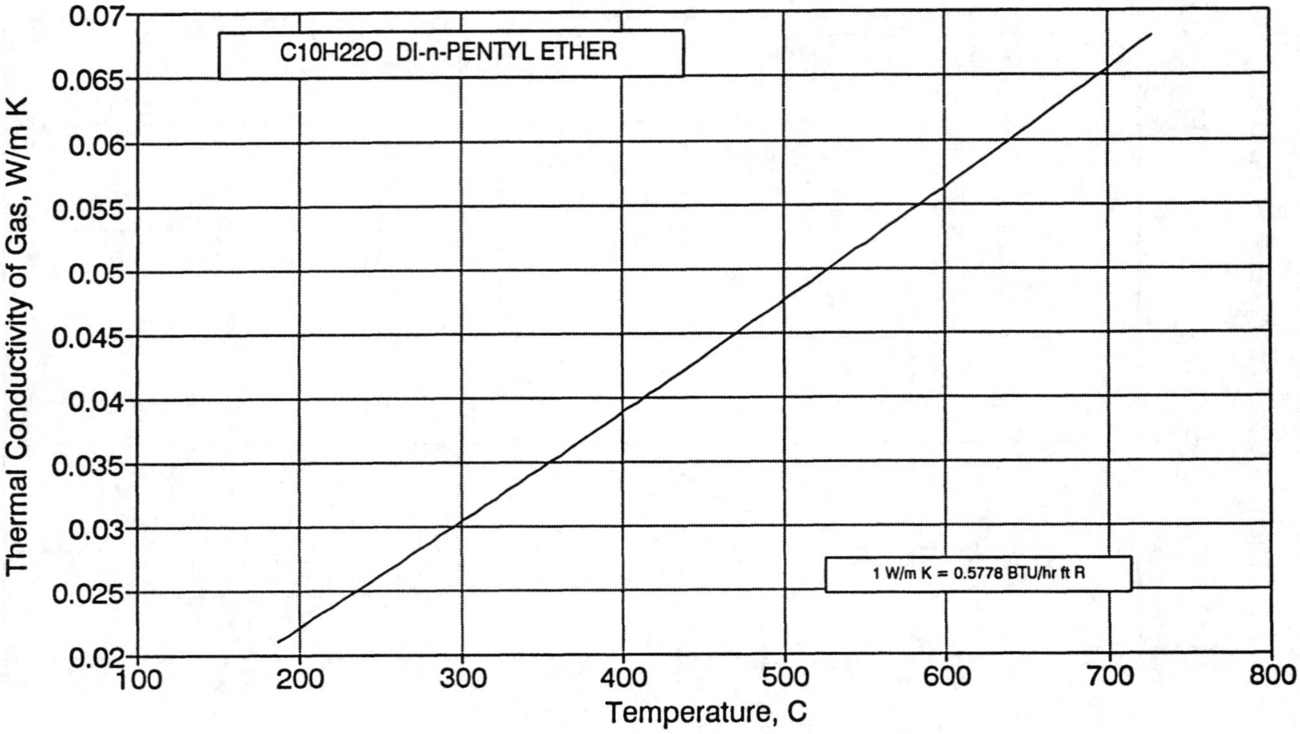
C10H22O DI-n-PENTYL ETHER
1 W/m K = 0.5778 BTU/hr ft R
Thermal Conductivity of Gas, W/m K
0.07
0.065
0.06
0.055
0.05
0.045
0.04
0.035
0.03
0.025
0.02
100
200
300
400
500
600
700
800
Temperature, C

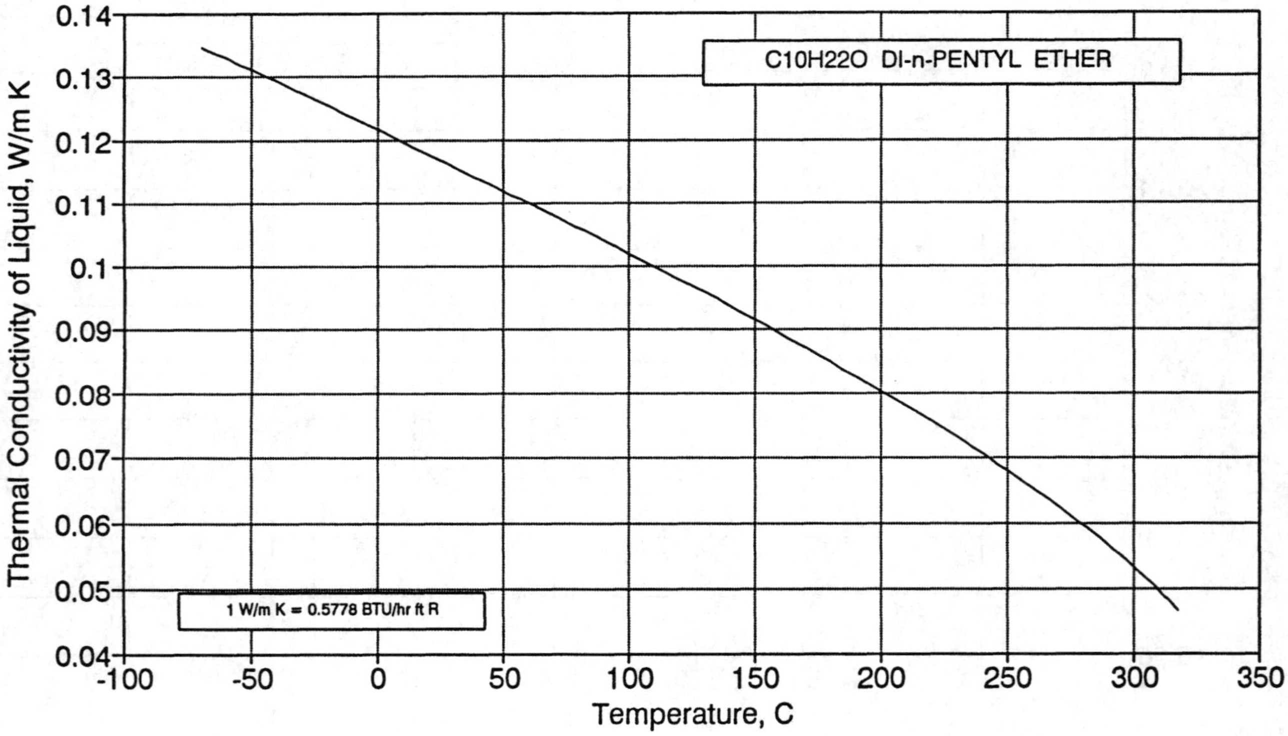
C10H22O DI-n-PENTYL ETHER
1 W/m K = 0.5778 BTU/hr ft R
Thermal Conductivity of Liquid, W/m K
0.14
0.13
0.12
0.11
0.1
0.09
0.08
0.07
0.06
0.05
0.04
-100
-50
0
50
100
150
200
250
300
350
Temperature, C

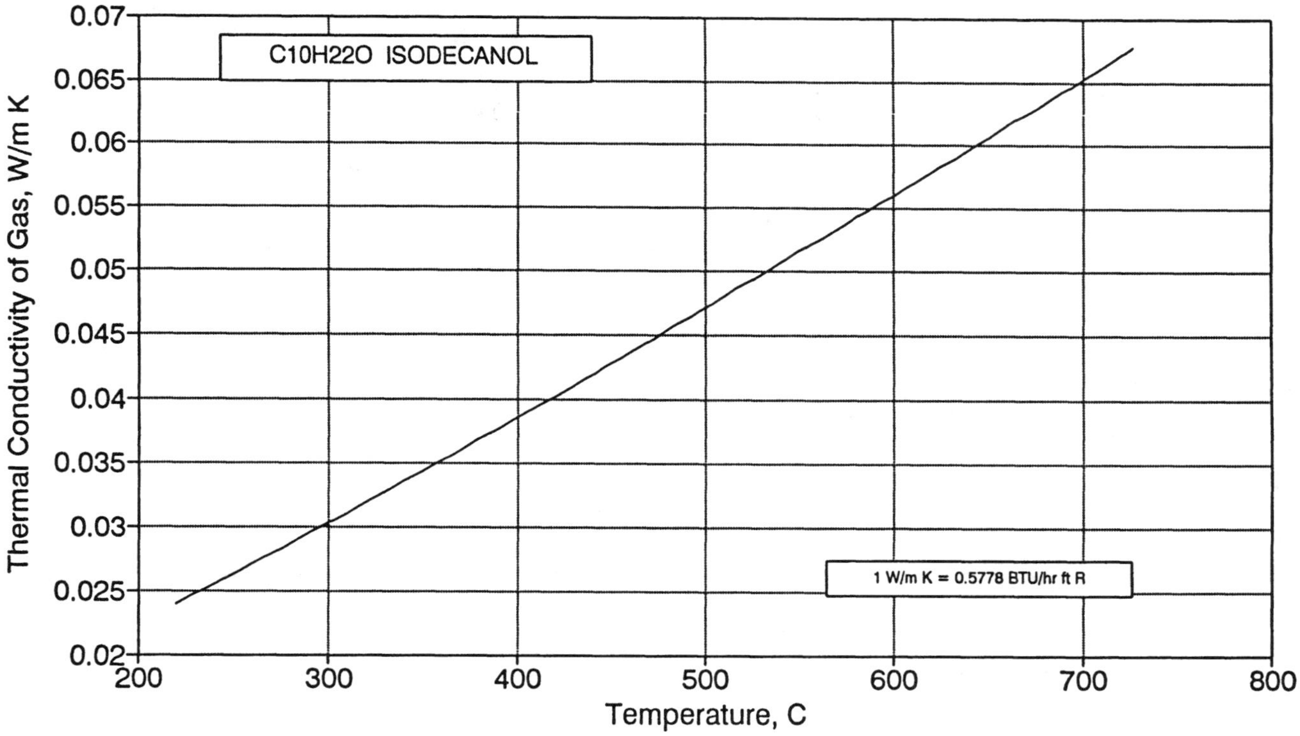
C10H22O ISODECANOL
1 W/m K = 0.5778 BTU/hr ft R
Thermal Conductivity of Gas, W/m K
0.07
0.065
0.06
0.055
0.05
0.045
0.04
0.035
0.03
0.025
0.02
200
300
400
500
600
700
800
Temperature, C

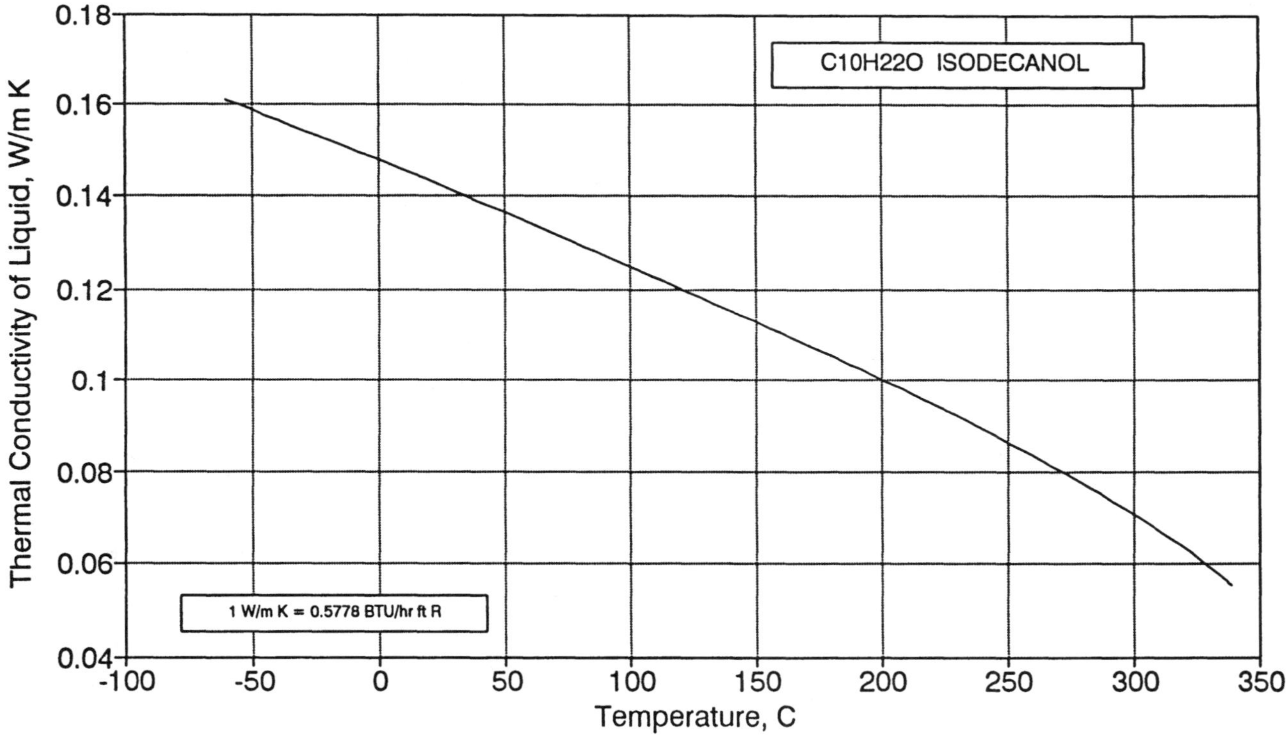
C10H22O ISODECANOL
1 W/m K = 0.5778 BTU/hr ft R
Thermal Conductivity of Liquid, W/m K
0.18
0.16
0.14
0.12
0.1
0.08
0.06
0.04
-100
-50
0
50
100
150
200
250
300
350
Temperature, C

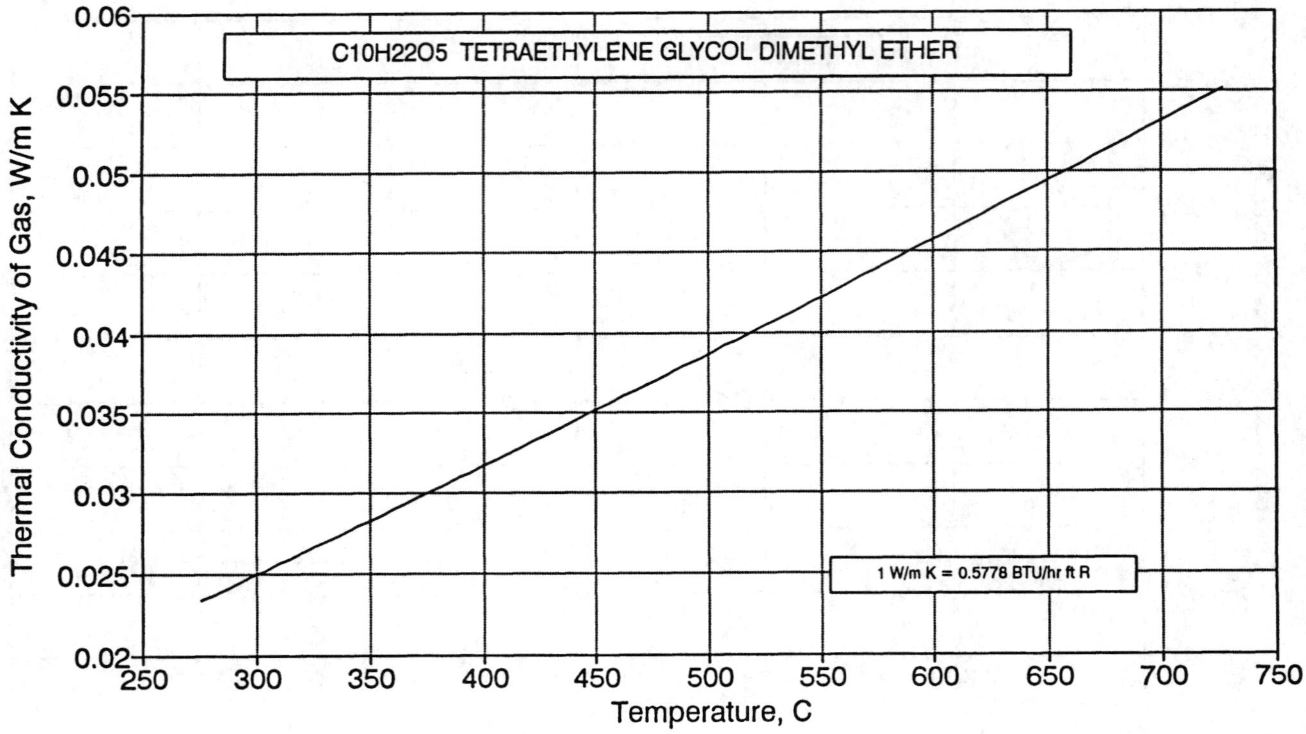
C10H22O5 TETRAETHYLENE GLYCOL DIMETHYL ETHER
Thermal Conductivity of Gas, W/m K
0.06
0.055
0.05
0.045
0.04
0.035
0.03
0.025
0.02
250
300
350
400
450
500
550
600
650
700
750
Temperature, C
1 W/m K = 0.5778 BTU/hr ft R

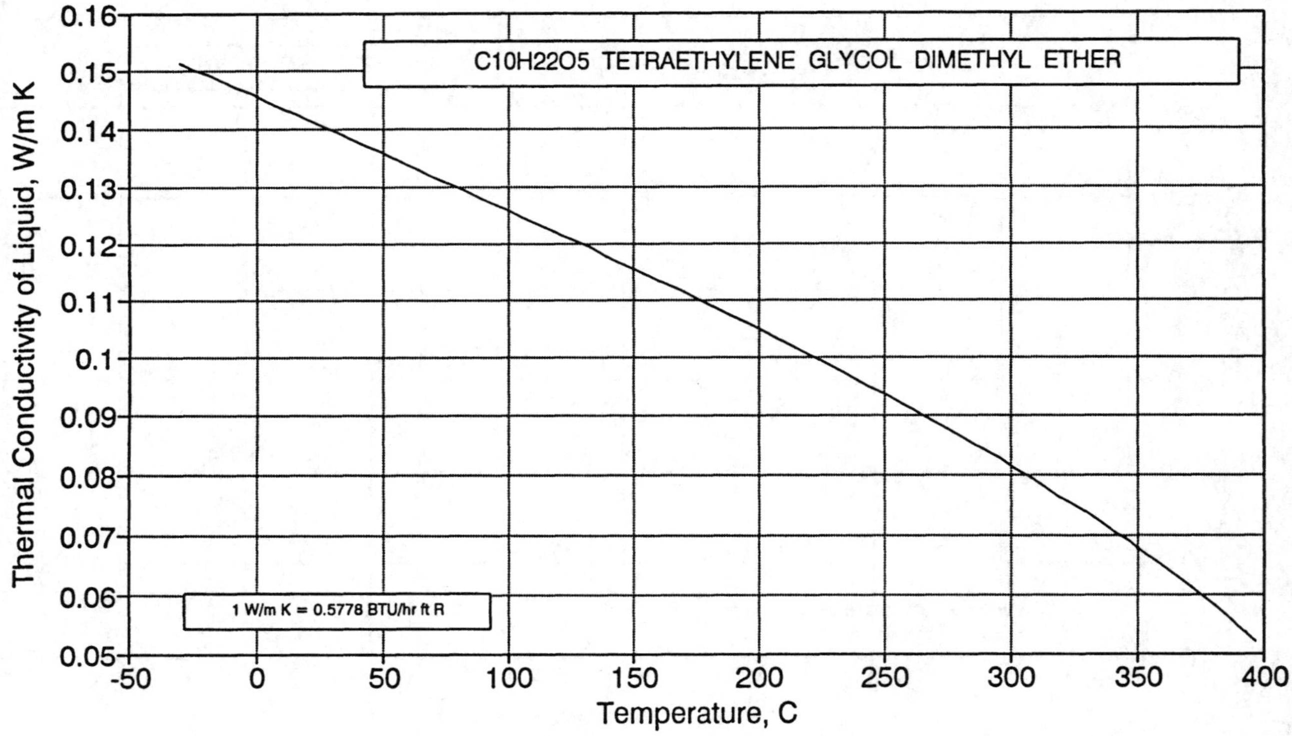
C10H22O5 TETRAETHYLENE GLYCOL DIMETHYL ETHER
Thermal Conductivity of Liquid, W/m K
0.16
0.15
0.14
0.13
0.12
0.11
0.1
0.09
0.08
0.07
0.06
0.05
-50
0
50
100
150
200
250
300
350
400
Temperature, C
1 W/m K = 0.5778 BTU/hr ft R

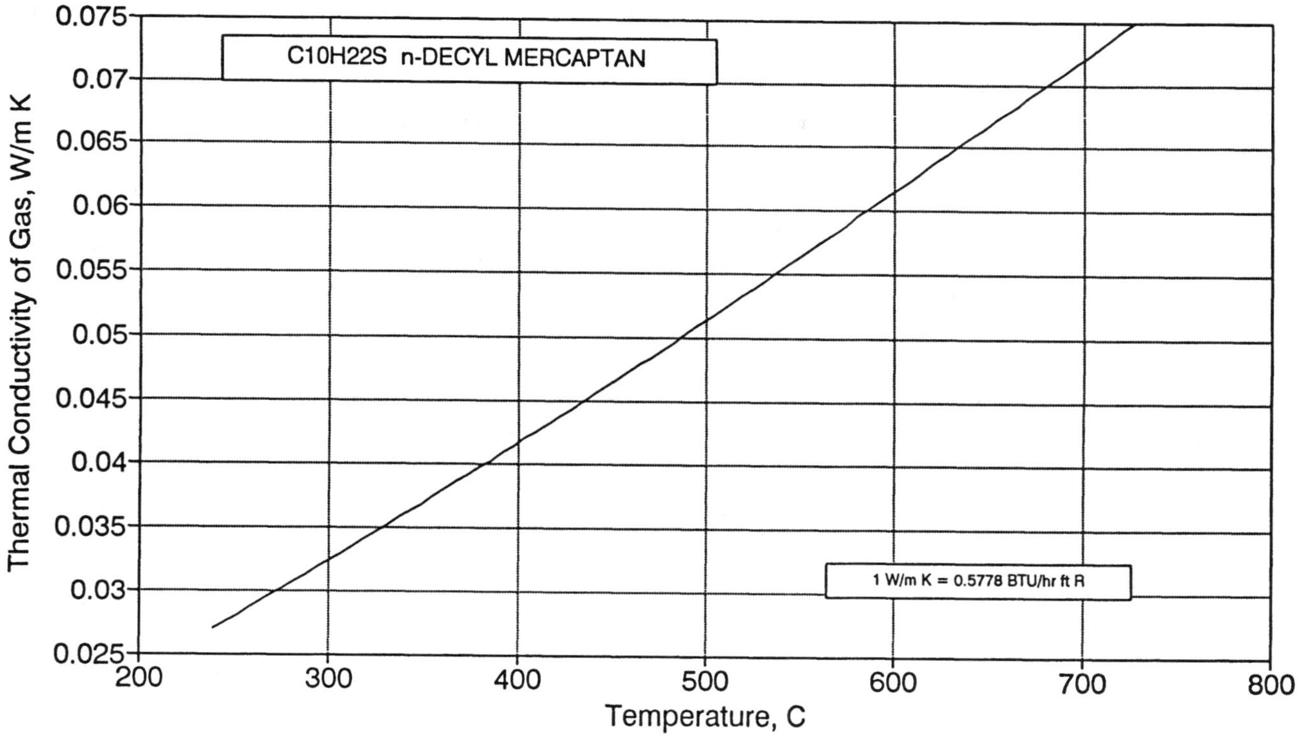
C10H22S n-DECYL MERCAPTAN
1 W/m K = 0.5778 BTU/hr ft R
Thermal Conductivity of Gas, W/m K
Temperature, C
0.075
0.07
0.065
0.06
0.055
0.05
0.045
0.04
0.035
0.03
0.025
200
300
400
500
600
700
800

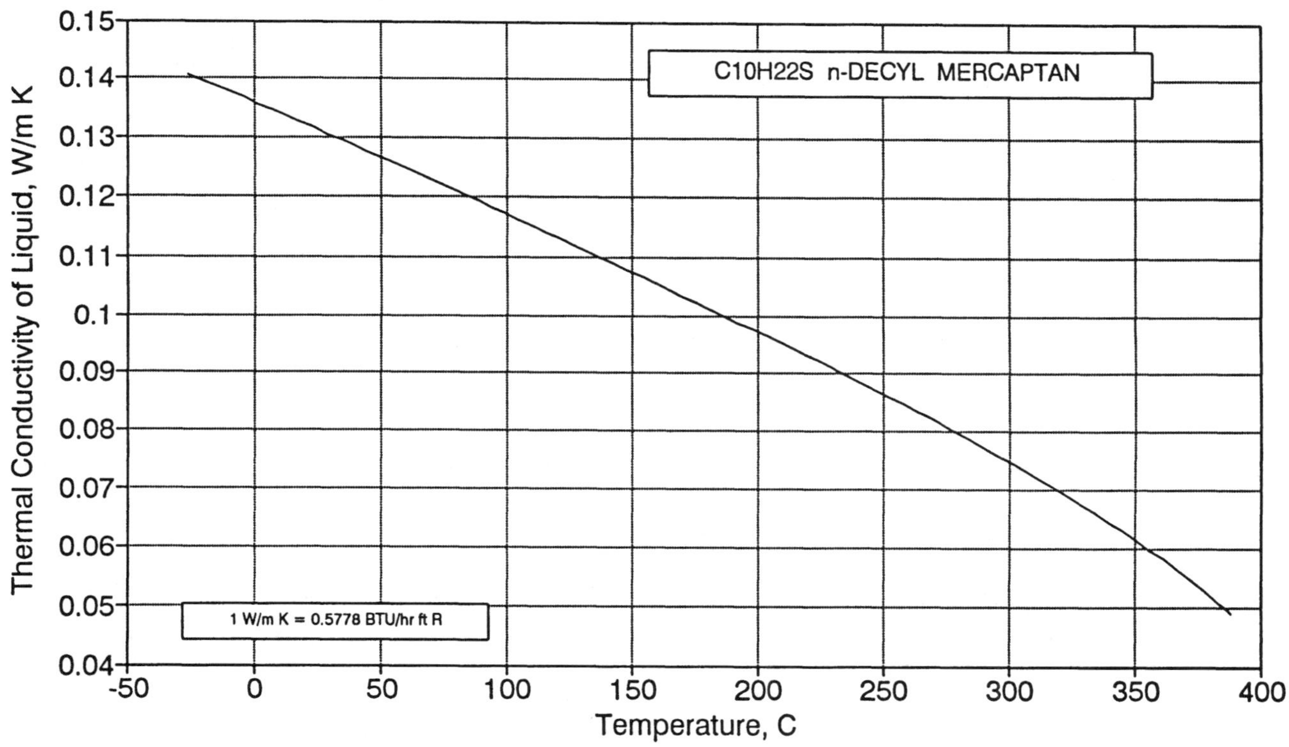
C10H22S n-DECYL MERCAPTAN
1 W/m K = 0.5778 BTU/hr ft R
Thermal Conductivity of Liquid, W/m K
Temperature, C
0.15
0.14
0.13
0.12
0.11
0.1
0.09
0.08
0.07
0.06
0.05
0.04
-50
0
50
100
150
200
250
300
350
400

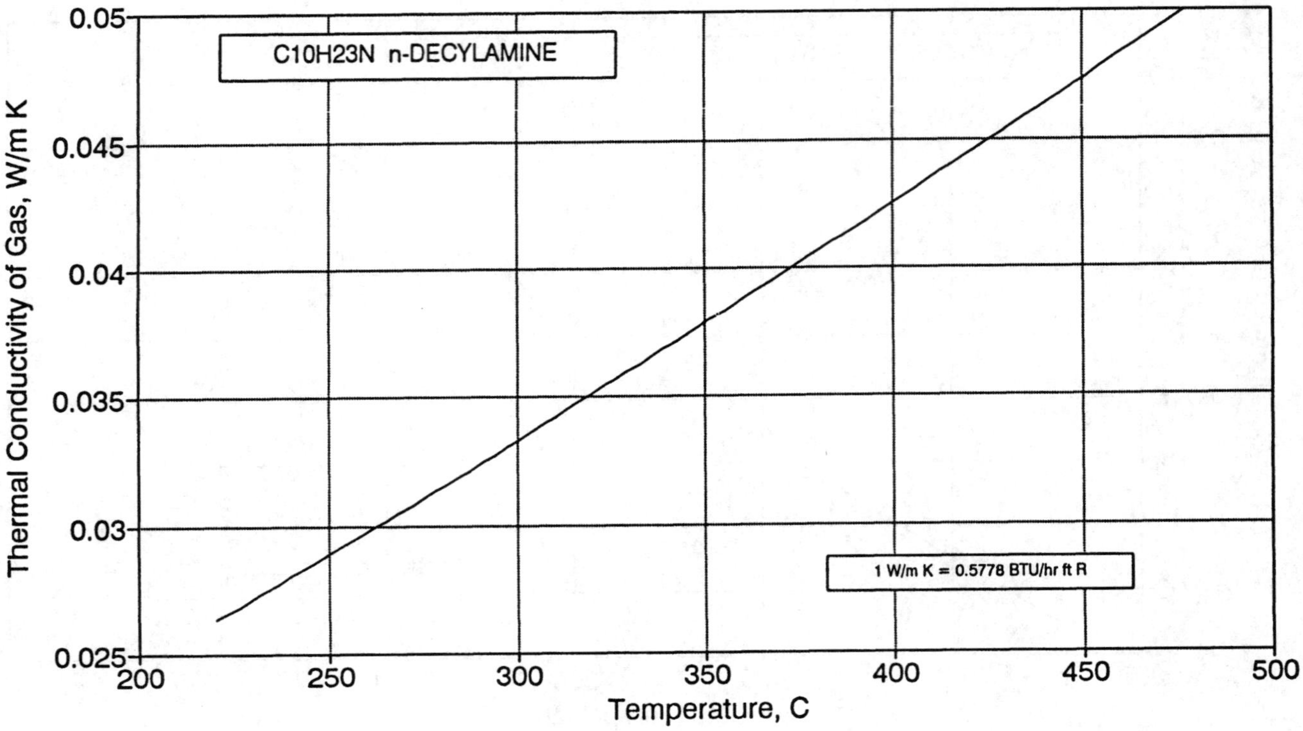
C10H23N n-DECYLAMINE
Thermal Conductivity of Gas, W/m K
0.05
0.045
0.04
0.035
0.03
0.025
200
250
300
350
400
450
500
Temperature, C
1 W/m K = 0.5778 BTU/hr ft R

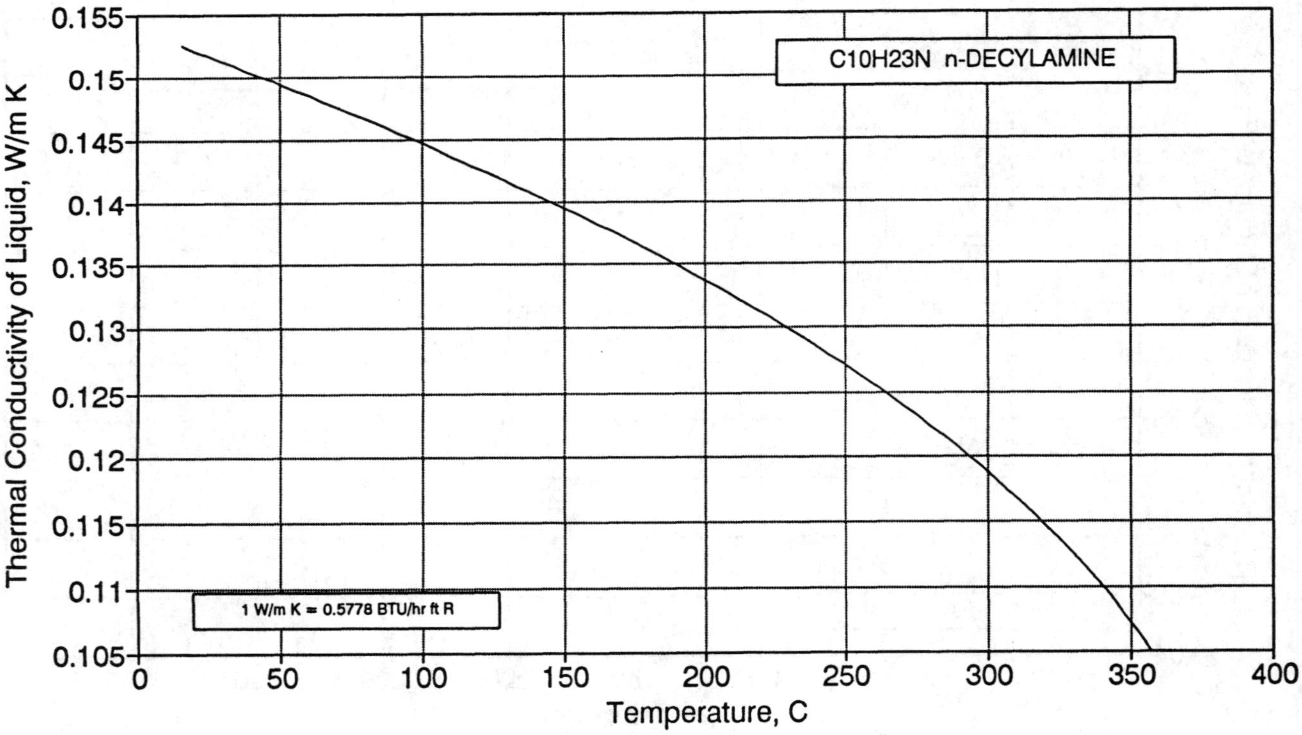
C10H23N n-DECYLAMINE
Thermal Conductivity of Liquid, W/m K
0.155
0.15
0.145
0.14
0.135
0.13
0.125
0.12
0.115
0.11
0.105
0
50
100
150
200
250
300
350
400
Temperature, C
1 W/m K = 0.5778 BTU/hr ft R

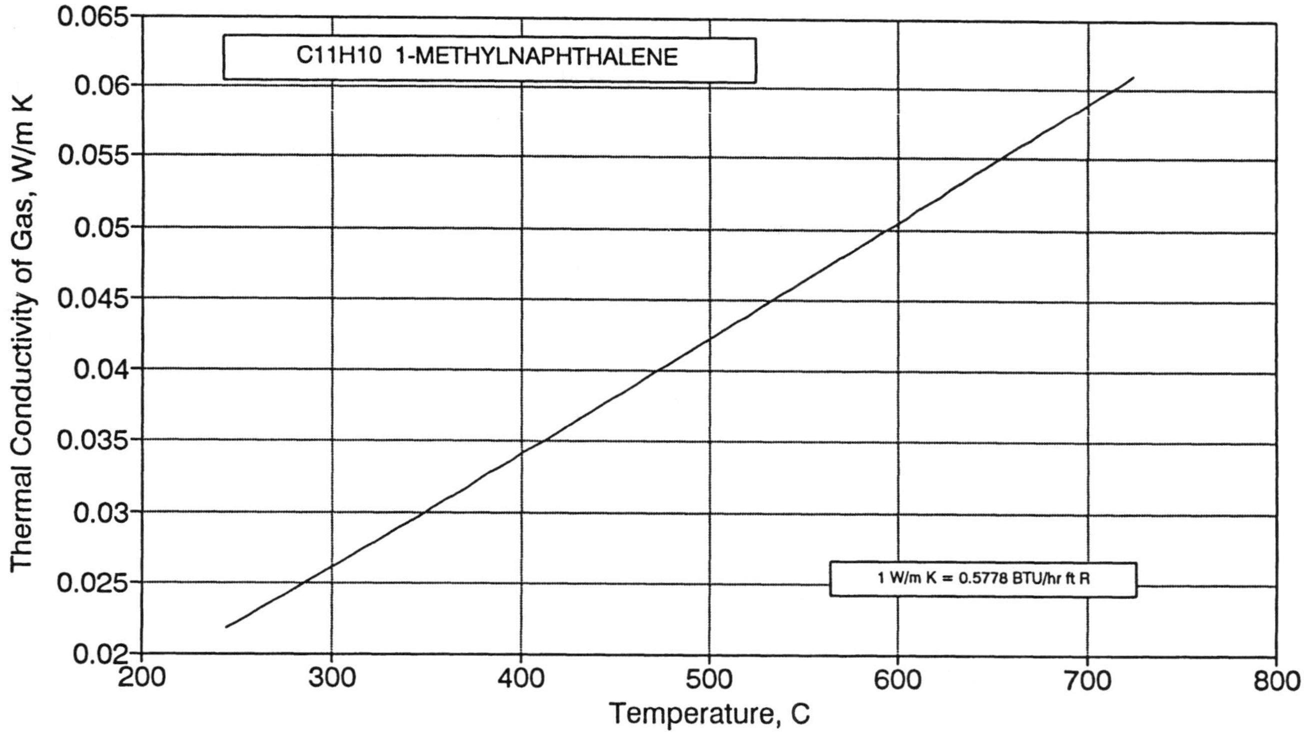
C11H10 1-METHYLNAPHTHALENE
1 W/m K = 0.5778 BTU/hr ft R
Thermal Conductivity of Gas, W/m K
0.065
0.06
0.055
0.05
0.045
0.04
0.035
0.03
0.025
0.02
200
300
400
500
600
700
800
Temperature, C

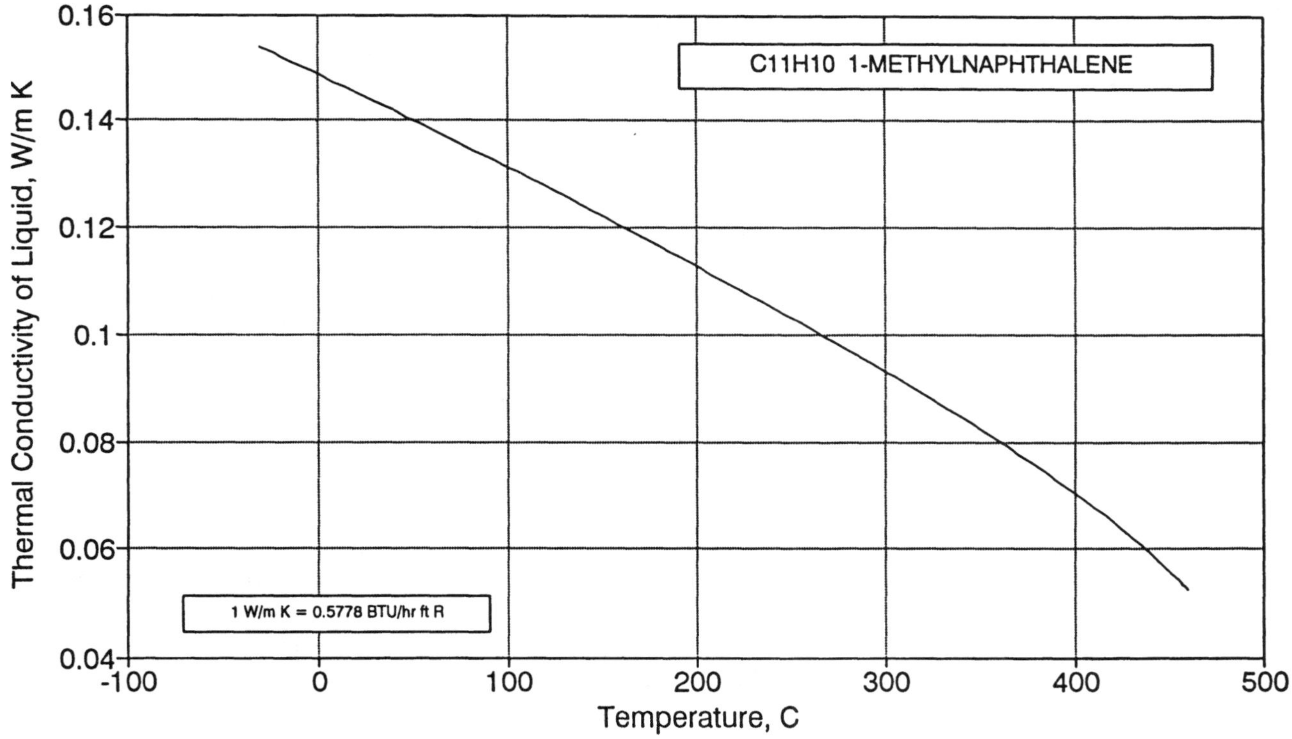
C11H10 1-METHYLNAPHTHALENE
1 W/m K = 0.5778 BTU/hr ft R
Thermal Conductivity of Liquid, W/m K
0.16
0.14
0.12
0.1
0.08
0.06
0.04
-100
0
100
200
300
400
500
Temperature, C

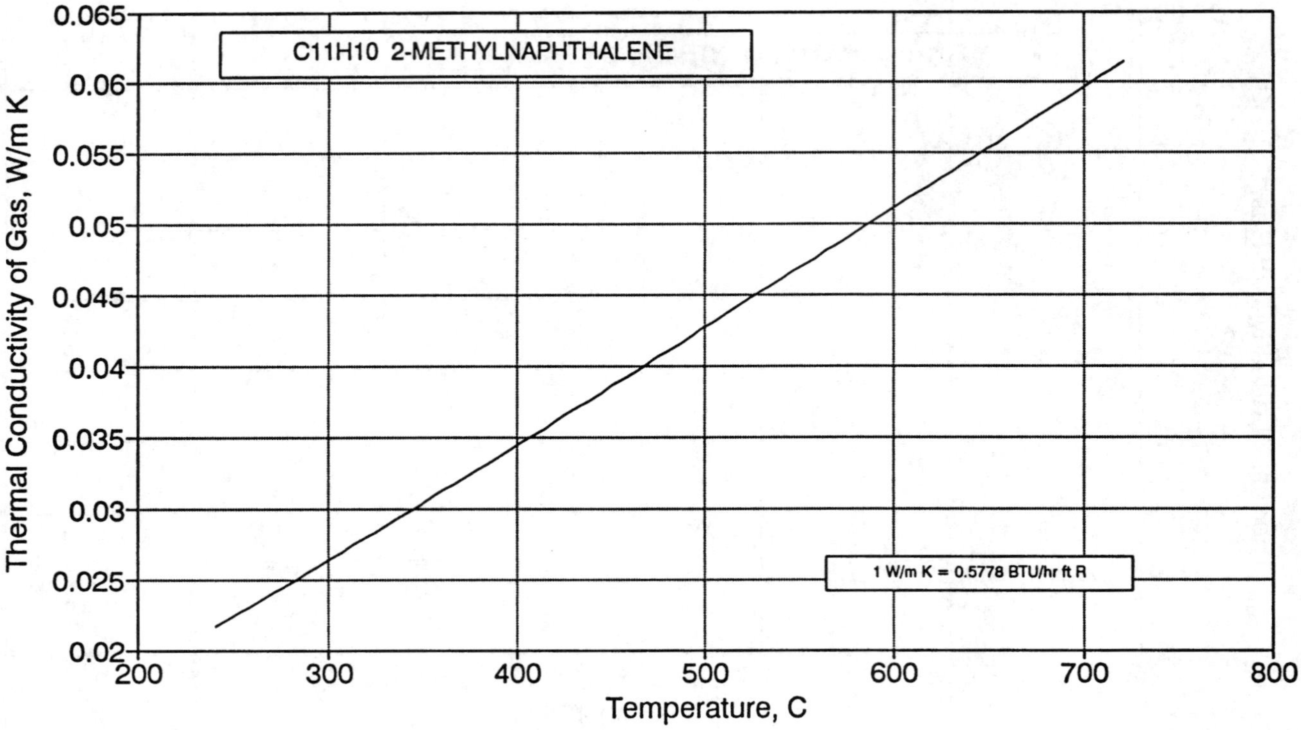
C11H10 2-METHYLNAPHTHALENE
1 W/m K = 0.5778 BTU/hr ft R
Thermal Conductivity of Gas, W/m K
Temperature, C
0.065
0.06
0.055
0.05
0.045
0.04
0.035
0.03
0.025
0.02
200
300
400
500
600
700
800

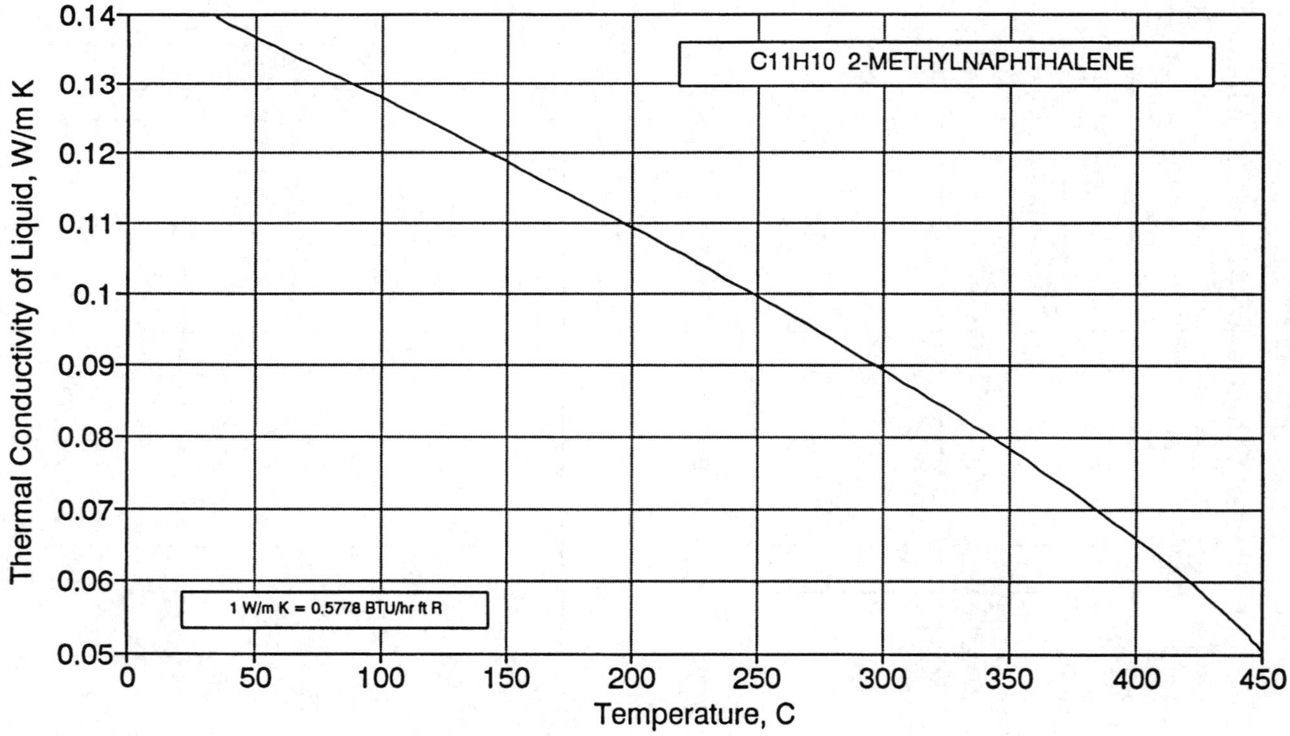
C11H10 2-METHYLNAPHTHALENE
1 W/m K = 0.5778 BTU/hr ft R
Thermal Conductivity of Liquid, W/m K
Temperature, C
0.14
0.13
0.12
0.11
0.1
0.09
0.08
0.07
0.06
0.05
0
50
100
150
200
250
300
350
400
450

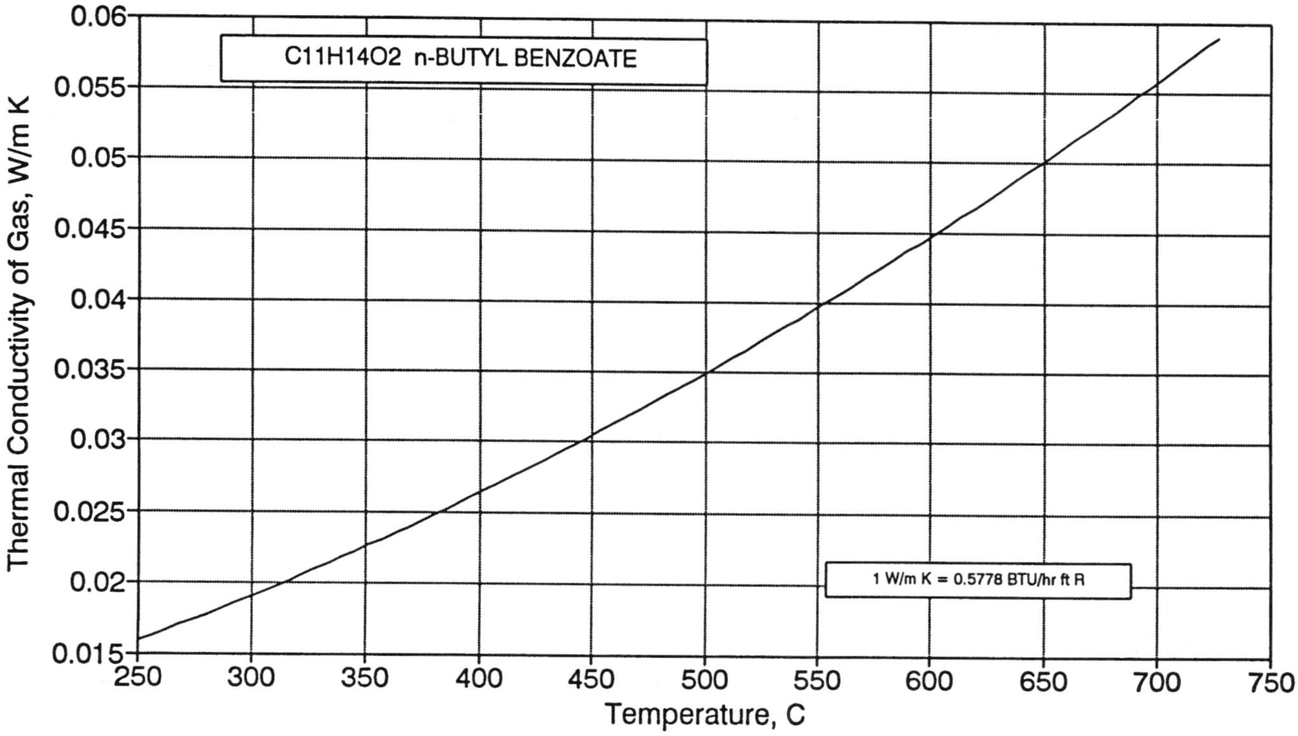

C11H14O2 n-BUTYL BENZOATE
1 W/m K = 0.5778 BTU/hr ft R
Thermal Conductivity of Gas, W/m K
0.06
0.055
0.05
0.045
0.04
0.035
0.03
0.025
0.02
0.015
250
300
350
400
450
500
550
600
650
700
750
Temperature, C

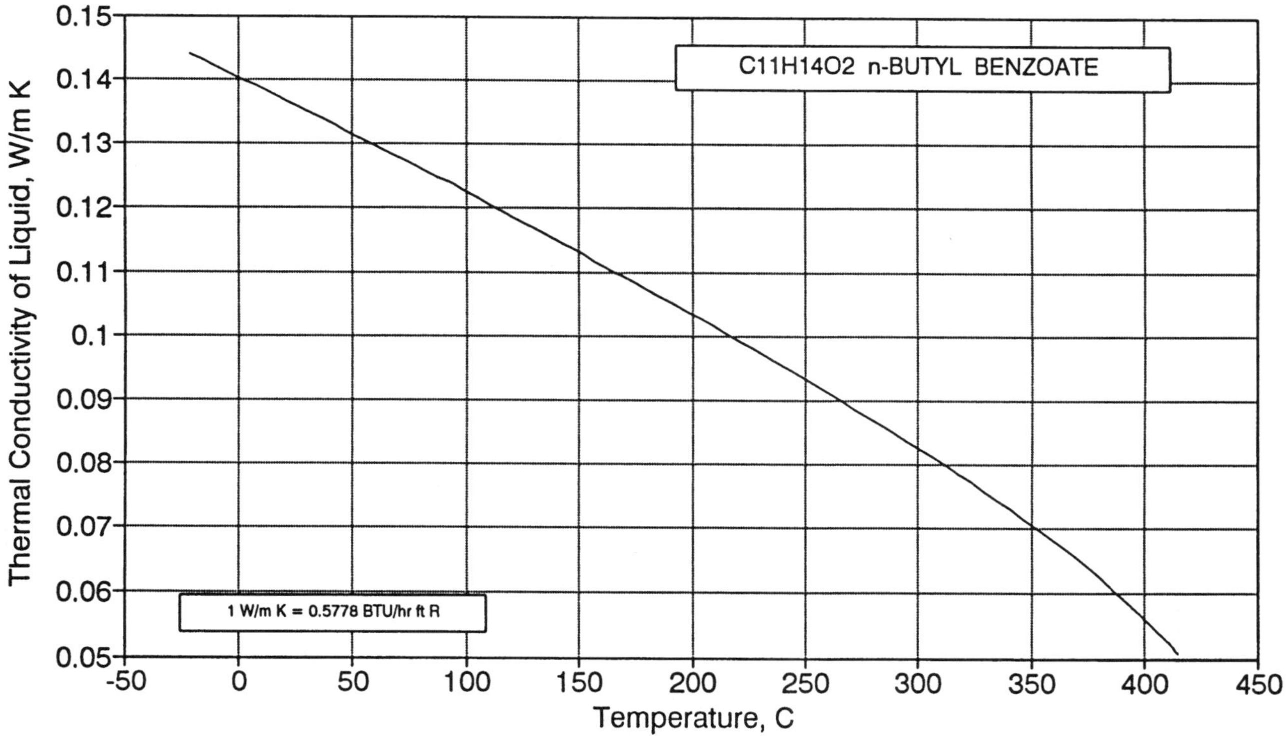

C11H14O2 n-BUTYL BENZOATE
1 W/m K = 0.5778 BTU/hr ft R
Thermal Conductivity of Liquid, W/m K
0.15
0.14
0.13
0.12
0.11
0.1
0.09
0.08
0.07
0.06
0.05
-50
0
50
100
150
200
250
300
350
400
450
Temperature, C

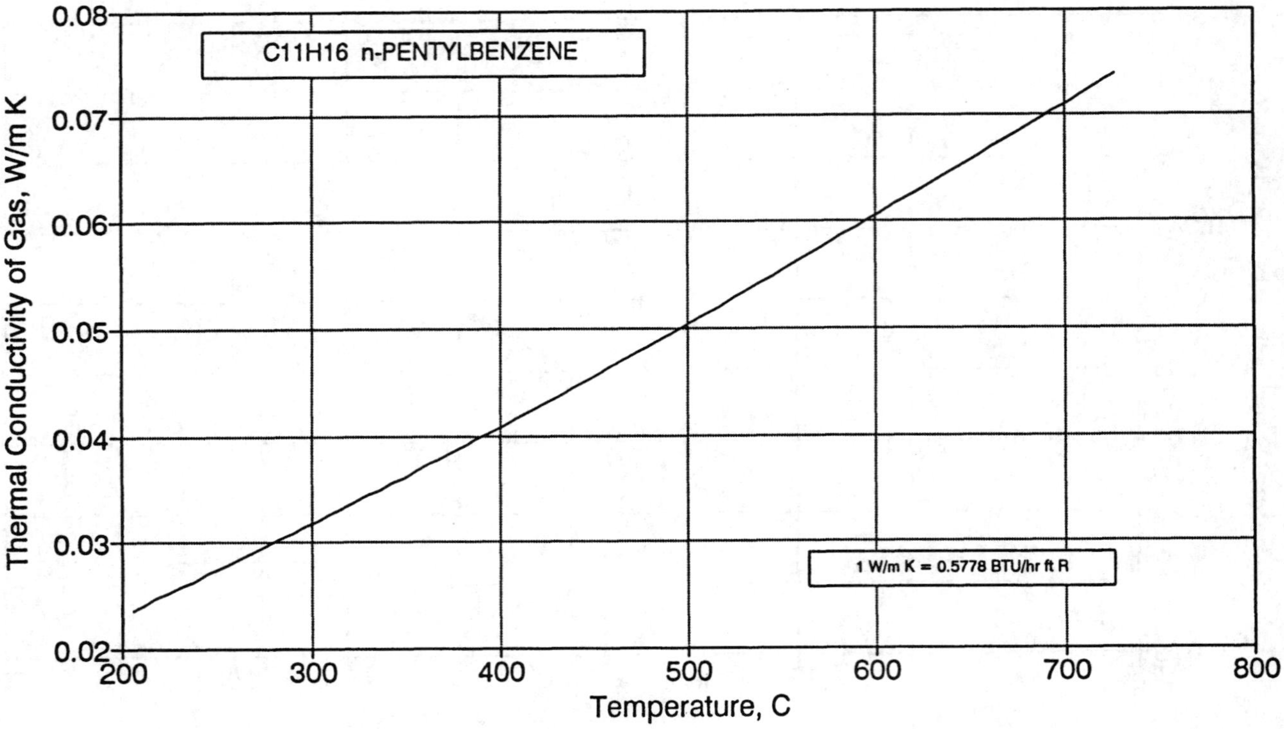
C11H16 n-PENTYLBENZENE
1 W/m K = 0.5778 BTU/hr ft R
Thermal Conductivity of Gas, W/m K
0.08
0.07
0.06
0.05
0.04
0.03
0.02
200
300
400
500
600
700
800
Temperature, C

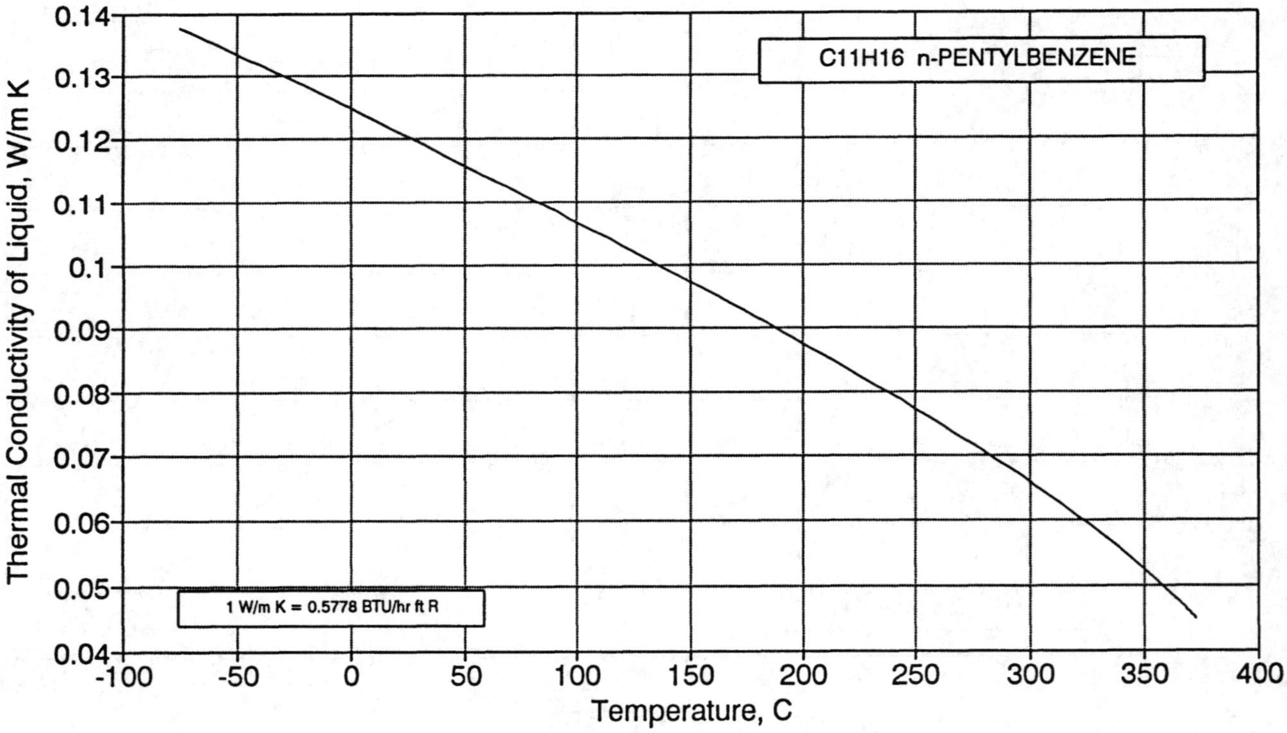
C11H16 n-PENTYLBENZENE
1 W/m K = 0.5778 BTU/hr ft R
Thermal Conductivity of Liquid, W/m K
0.14
0.13
0.12
0.11
0.1
0.09
0.08
0.07
0.06
0.05
0.04
-100
-50
0
50
100
150
200
250
300
350
400
Temperature, C

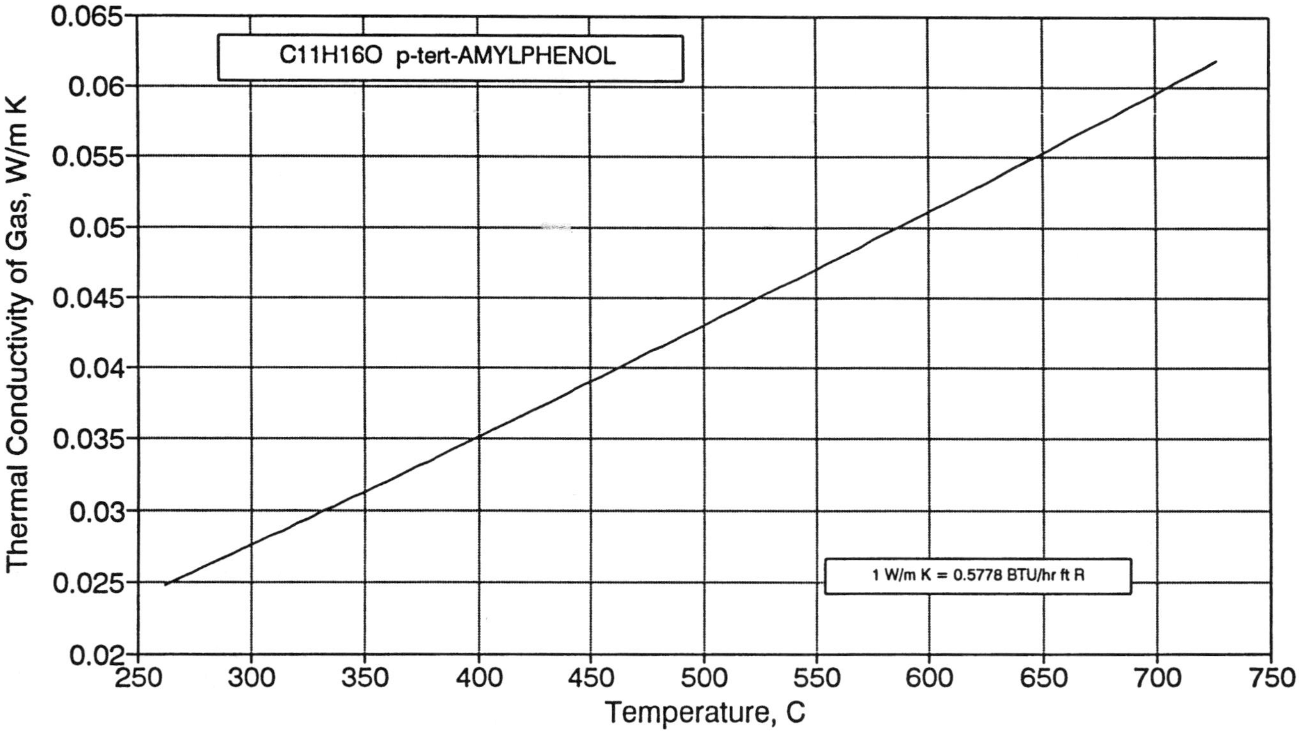
C11H16O p-tert-AMYLPHENOL
1 W/m K = 0.5778 BTU/hr ft R
Thermal Conductivity of Gas, W/m K
0.065
0.06
0.055
0.05
0.045
0.04
0.035
0.03
0.025
0.02
250
300
350
400
450
500
550
600
650
700
750
Temperature, C

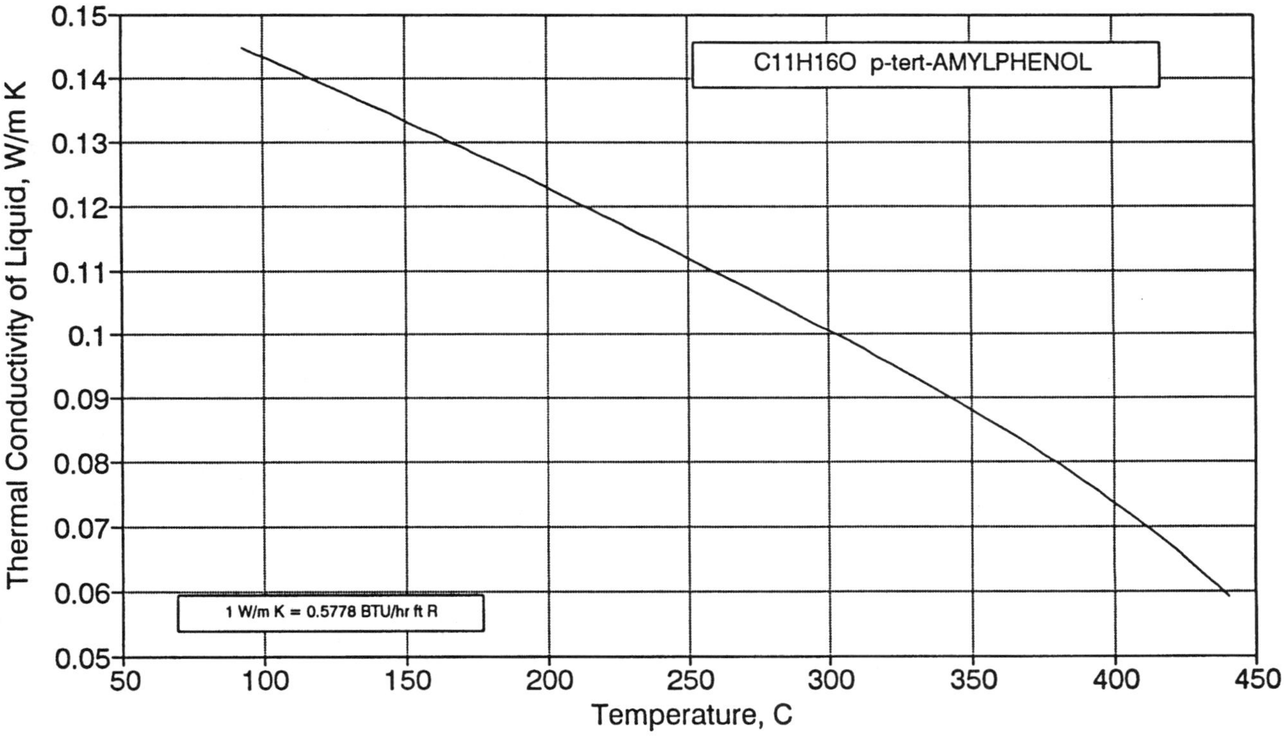
C11H16O p-tert-AMYLPHENOL
1 W/m K = 0.5778 BTU/hr ft R
Thermal Conductivity of Liquid, W/m K
0.15
0.14
0.13
0.12
0.11
0.1
0.09
0.08
0.07
0.06
0.05
50
100
150
200
250
300
350
400
450
Temperature, C

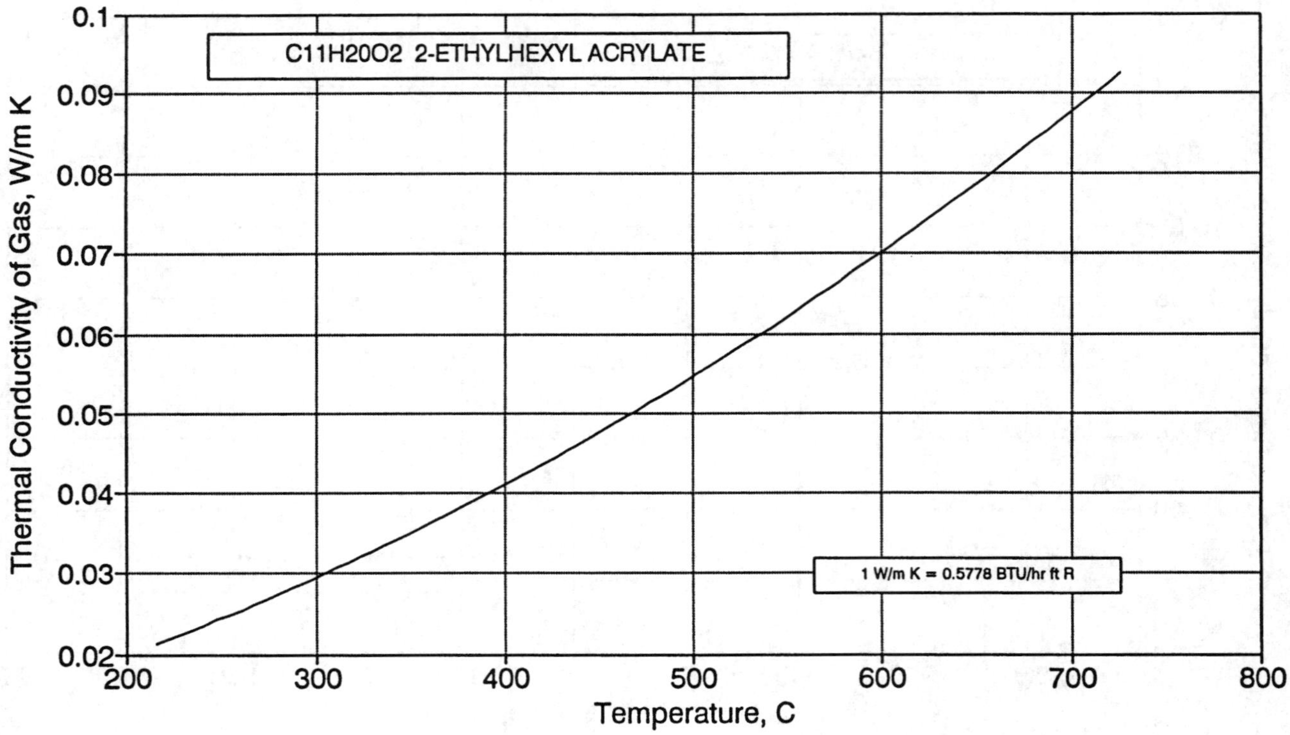
C11H20O2 2-ETHYLHEXYL ACRYLATE
Thermal Conductivity of Gas, W/m K
Temperature, C
1 W/m K = 0.5778 BTU/hr ft R
0.1
0.09
0.08
0.07
0.06
0.05
0.04
0.03
0.02
200
300
400
500
600
700
800

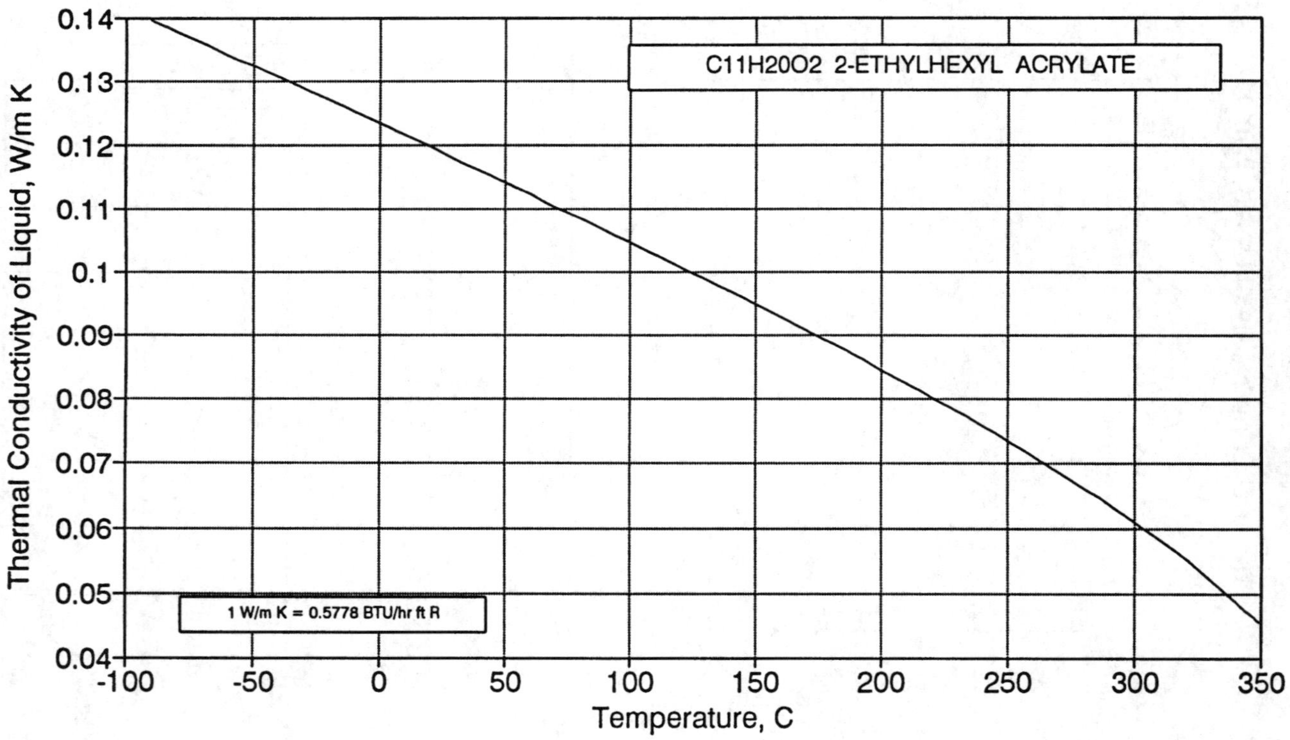
C11H20O2 2-ETHYLHEXYL ACRYLATE
Thermal Conductivity of Liquid, W/m K
Temperature, C
1 W/m K = 0.5778 BTU/hr ft R
0.14
0.13
0.12
0.11
0.1
0.09
0.08
0.07
0.06
0.05
0.04
-100
-50
0
50
100
150
200
250
300
350

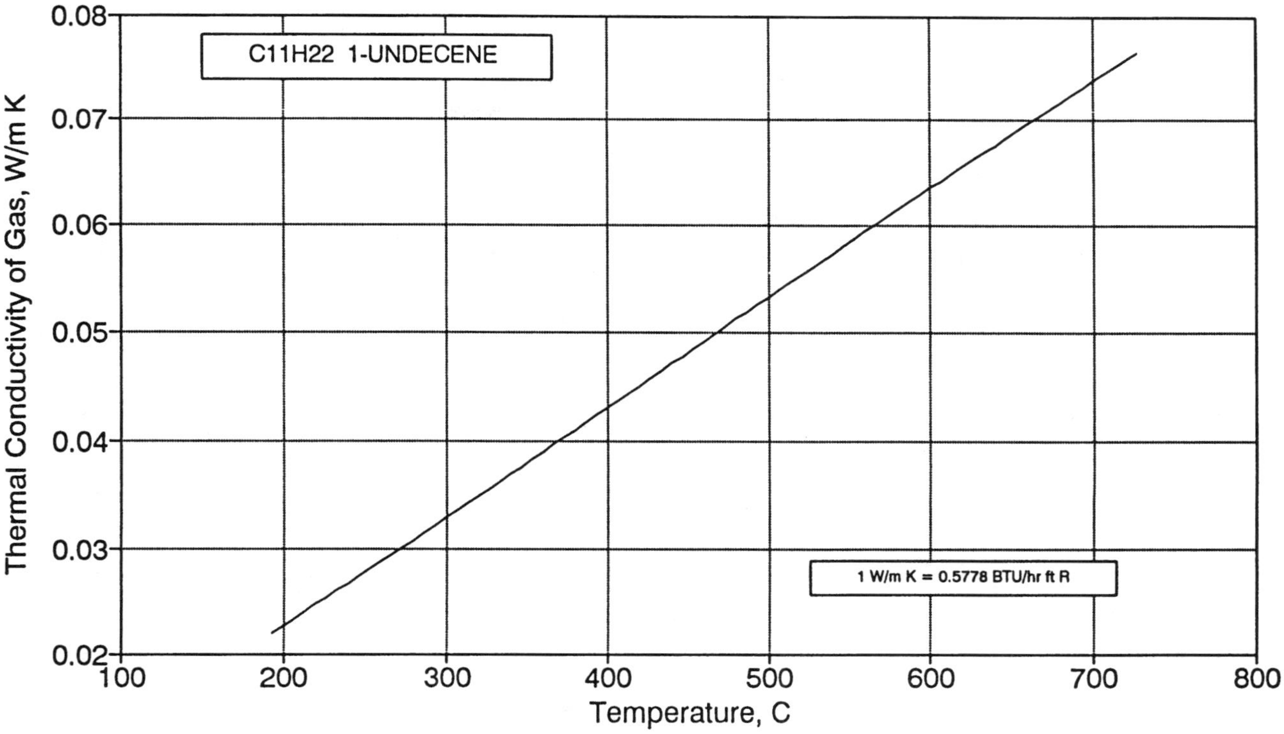

C11H22 1-UNDECENE
1 W/m K = 0.5778 BTU/hr ft R
Thermal Conductivity of Gas, W/m K
0.08
0.07
0.06
0.05
0.04
0.03
0.02
100
200
300
400
500
600
700
800
Temperature, C

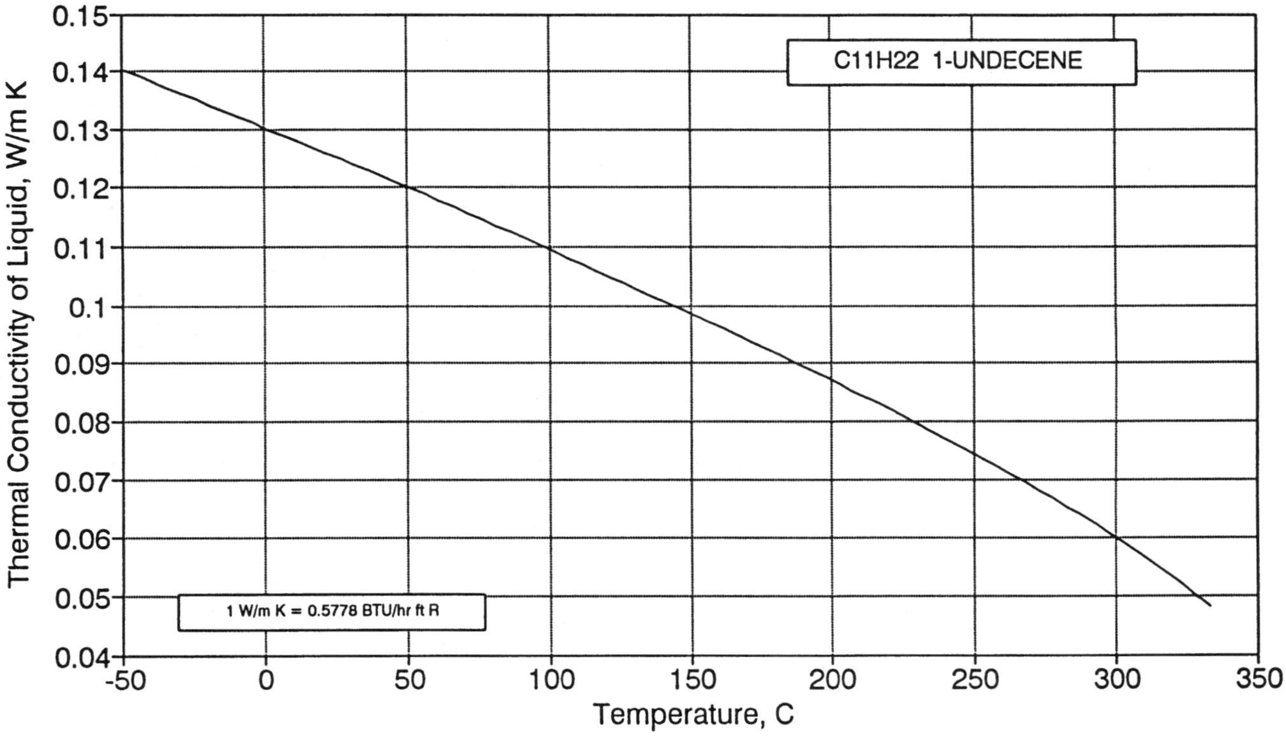

C11H22 1-UNDECENE
1 W/m K = 0.5778 BTU/hr ft R
Thermal Conductivity of Liquid, W/m K
0.15
0.14
0.13
0.12
0.11
0.1
0.09
0.08
0.07
0.06
0.05
0.04
-50
0
50
100
150
200
250
300
350
Temperature, C

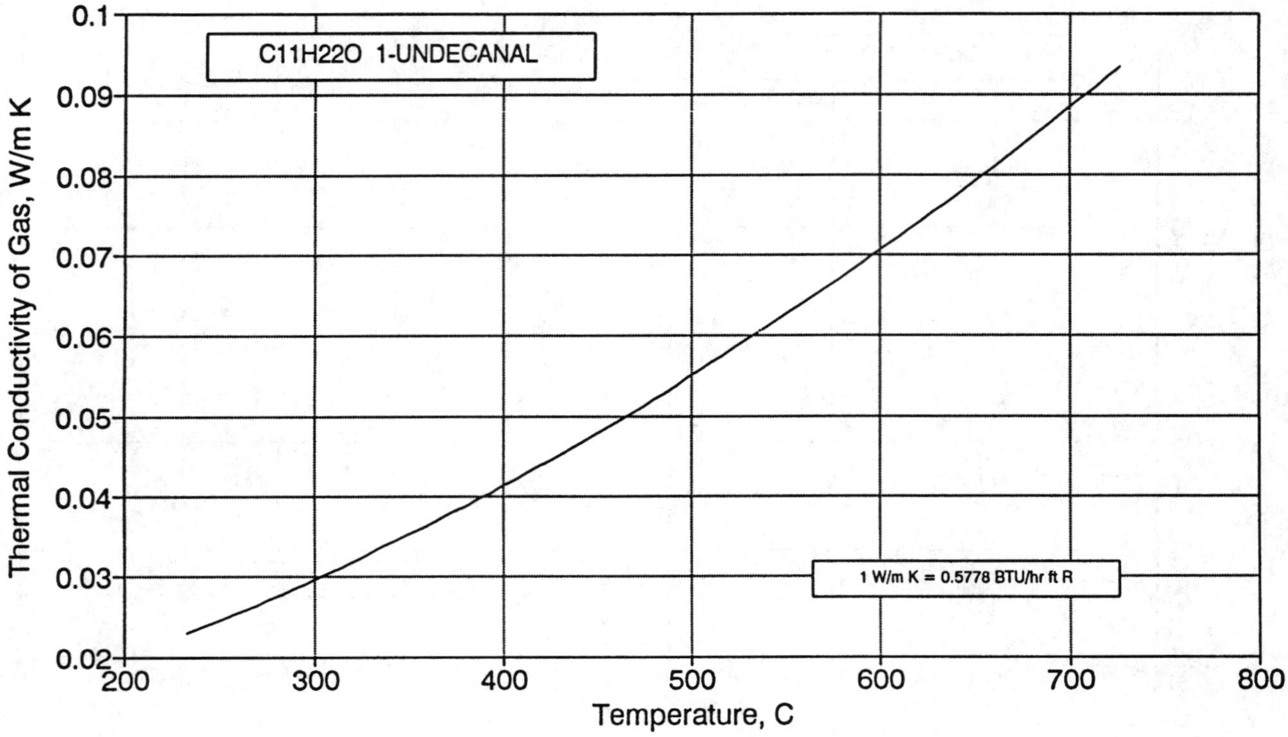
C11H22O 1-UNDECANAL
1 W/m K = 0.5778 BTU/hr ft R
Thermal Conductivity of Gas, W/m K
0.1
0.09
0.08
0.07
0.06
0.05
0.04
0.03
0.02
200
300
400
500
600
700
800
Temperature, C

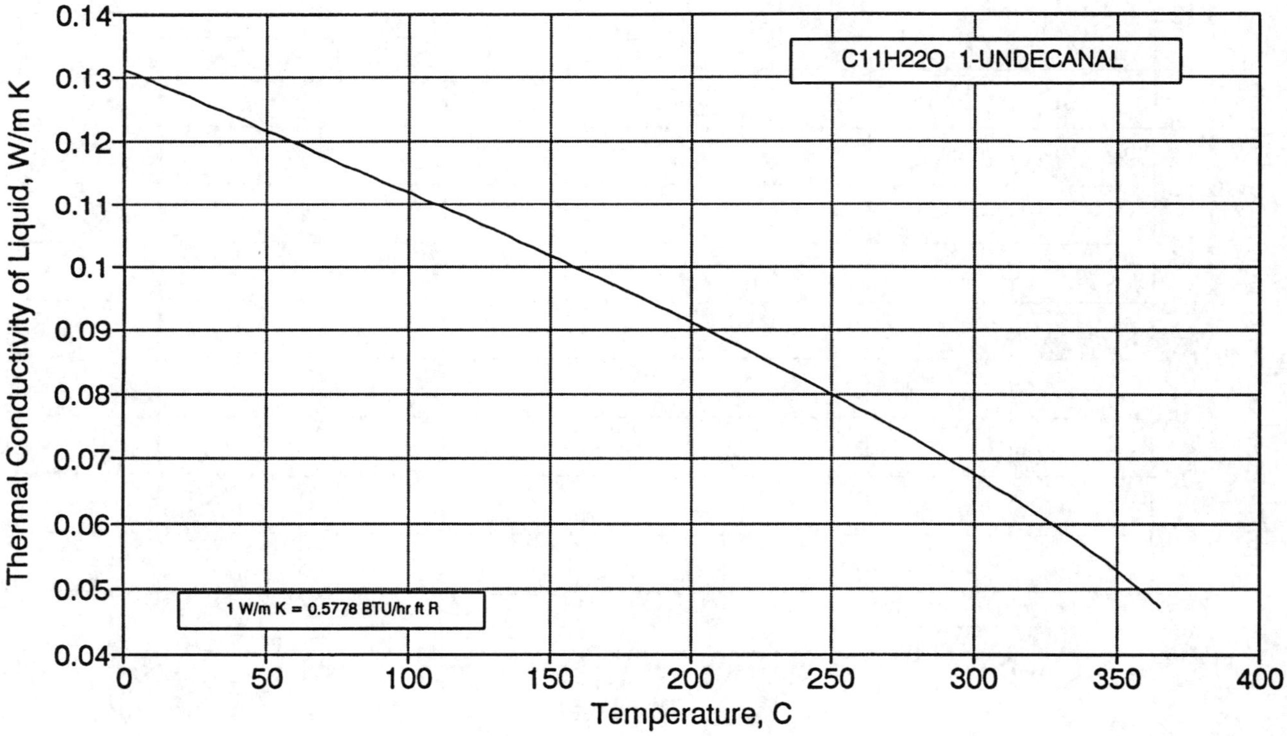
C11H22O 1-UNDECANAL
1 W/m K = 0.5778 BTU/hr ft R
Thermal Conductivity of Liquid, W/m K
0.14
0.13
0.12
0.11
0.1
0.09
0.08
0.07
0.06
0.05
0.04
0
50
100
150
200
250
300
350
400
Temperature, C

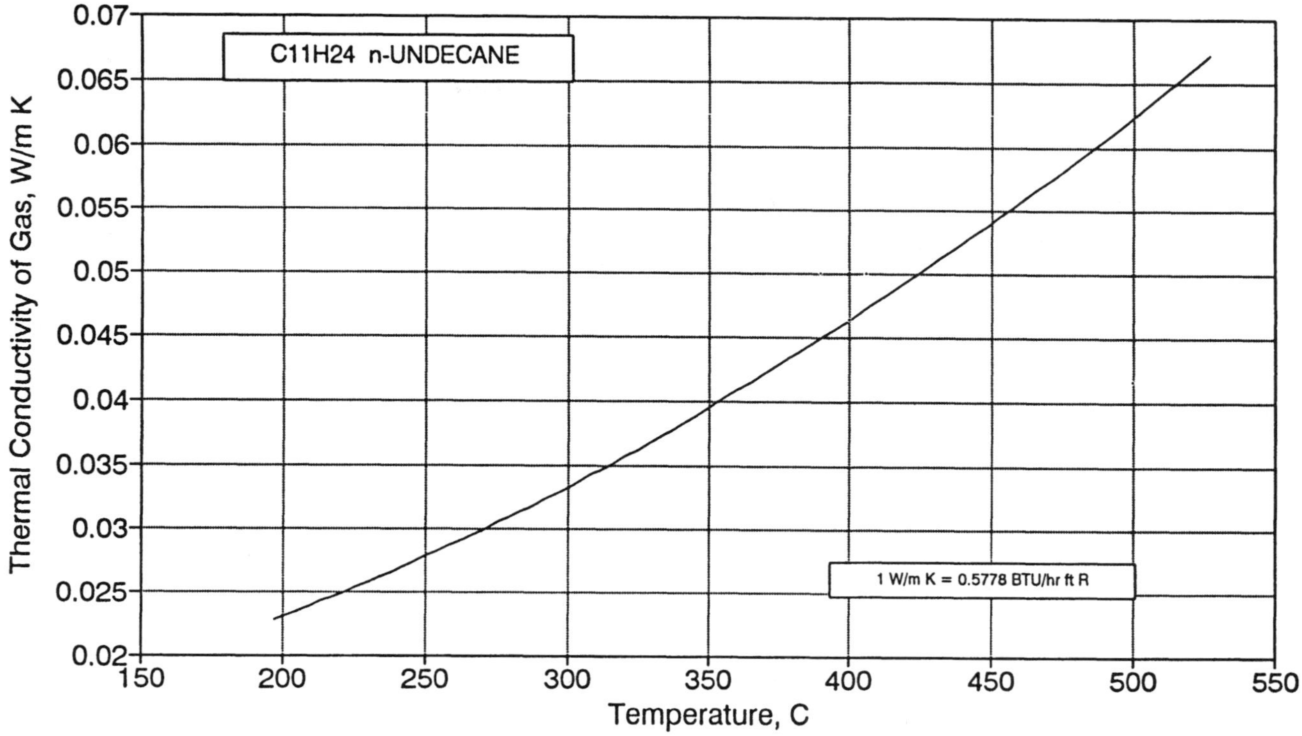
C11H24 n-UNDECANE
1 W/m K = 0.5778 BTU/hr ft R
Thermal Conductivity of Gas, W/m K
Temperature, C
0.07
0.065
0.06
0.055
0.05
0.045
0.04
0.035
0.03
0.025
0.02
150
200
250
300
350
400
450
500
550

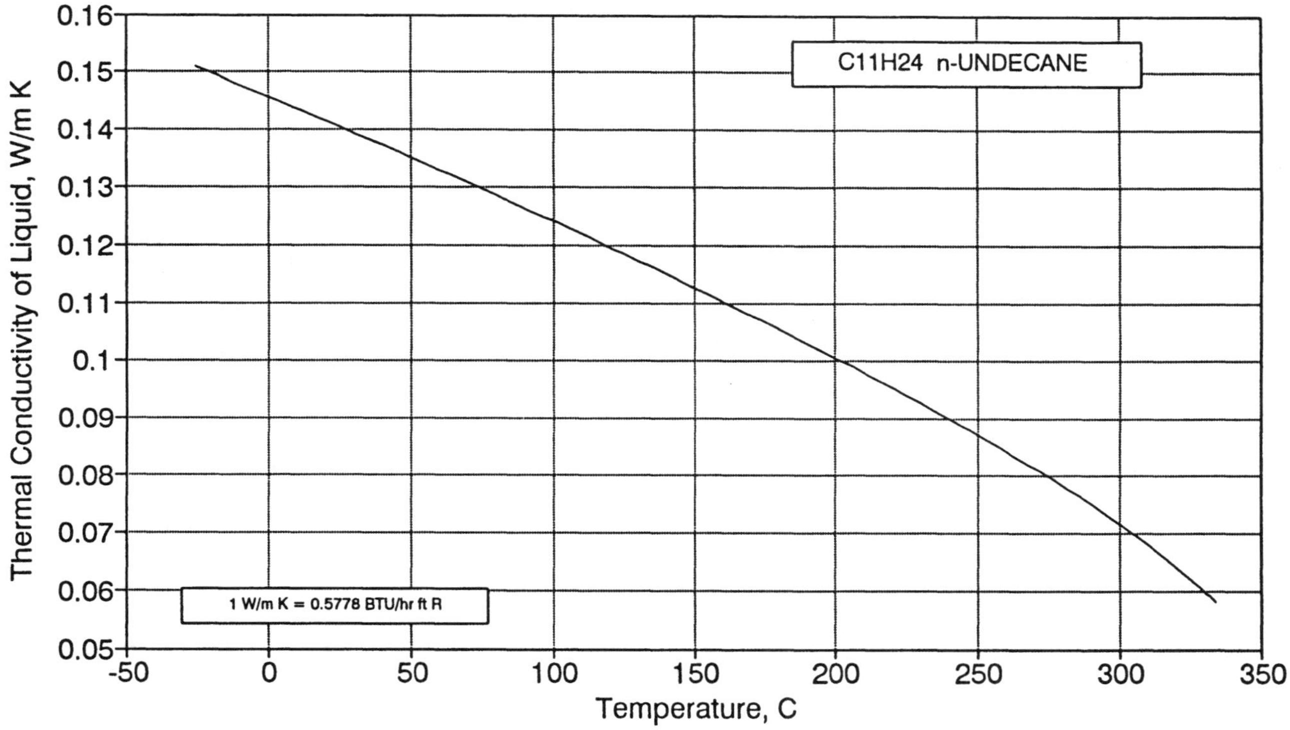
C11H24 n-UNDECANE
1 W/m K = 0.5778 BTU/hr ft R
Thermal Conductivity of Liquid, W/m K
Temperature, C
0.16
0.15
0.14
0.13
0.12
0.11
0.1
0.09
0.08
0.07
0.06
0.05
-50
0
50
100
150
200
250
300
350

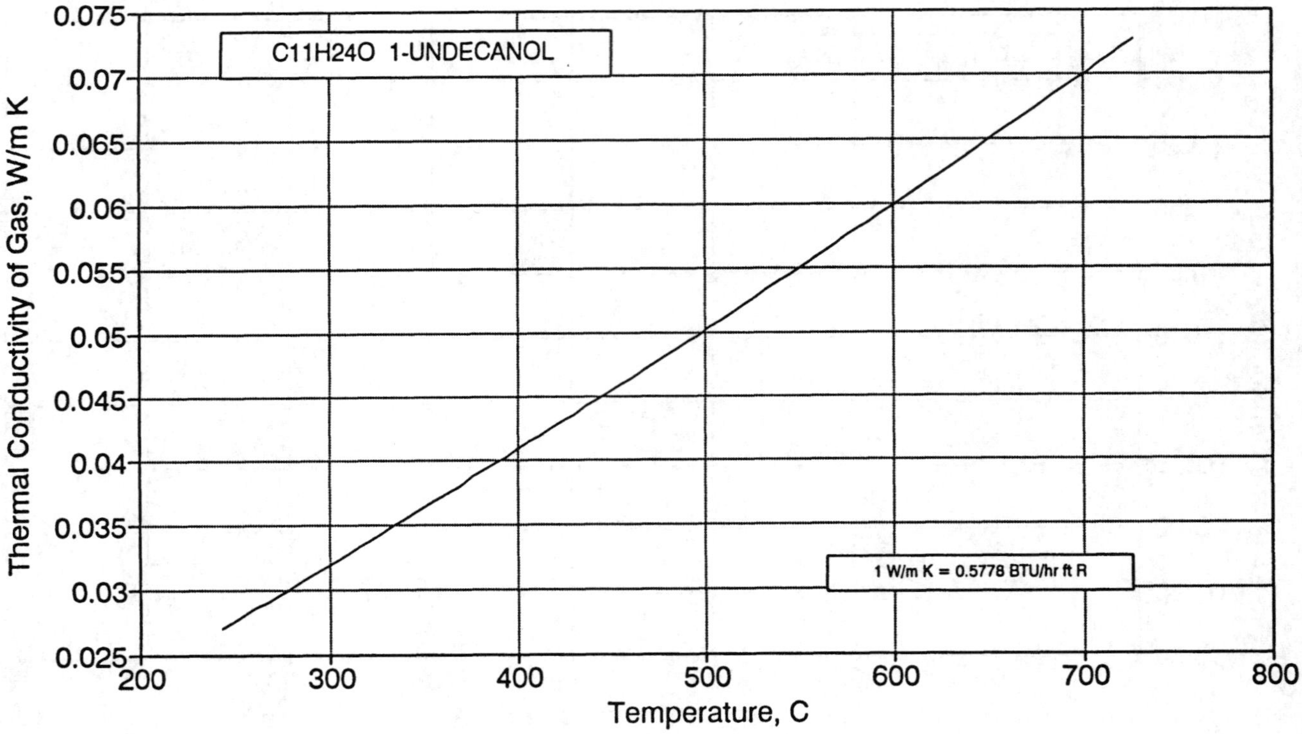
C11H24O 1-UNDECANOL
1 W/m K = 0.5778 BTU/hr ft R
Thermal Conductivity of Gas, W/m K
0.075
0.07
0.065
0.06
0.055
0.05
0.045
0.04
0.035
0.03
0.025
200
300
400
500
600
700
800
Temperature, C

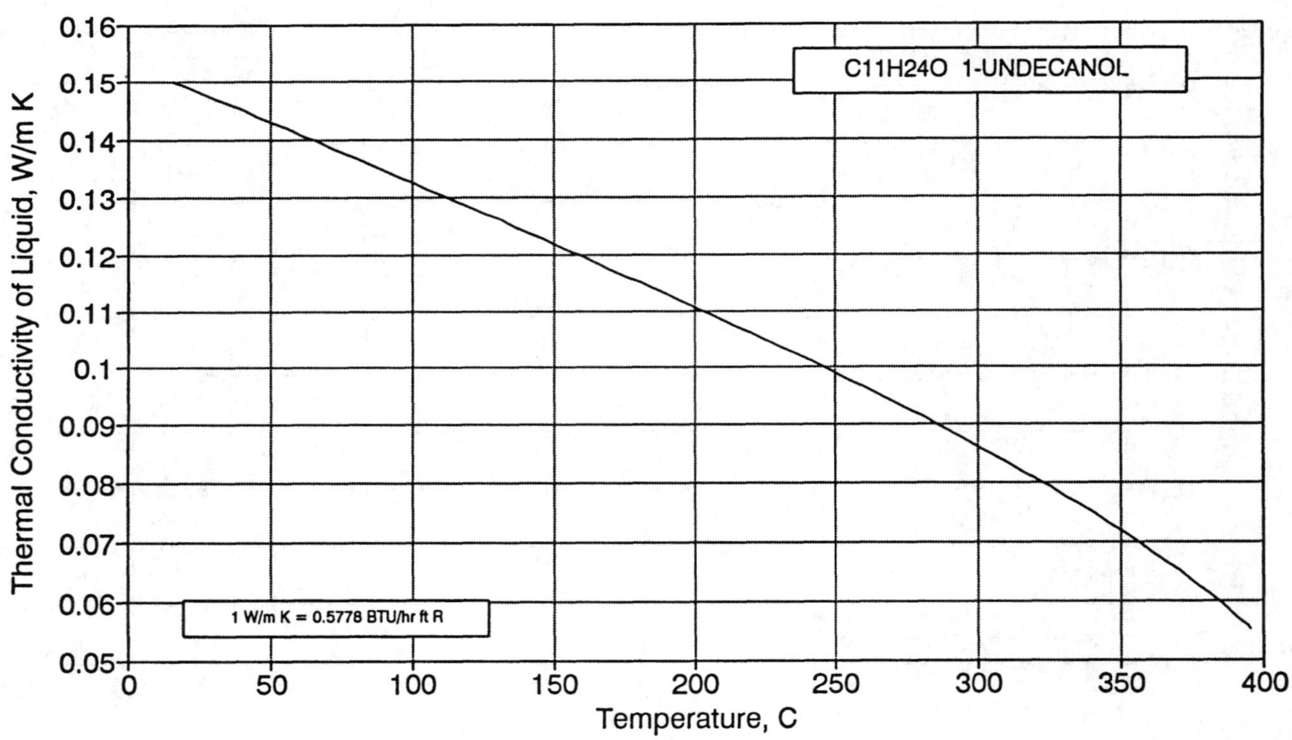
C11H24O 1-UNDECANOL
1 W/m K = 0.5778 BTU/hr ft R
Thermal Conductivity of Liquid, W/m K
0.16
0.15
0.14
0.13
0.12
0.11
0.1
0.09
0.08
0.07
0.06
0.05
0
50
100
150
200
250
300
350
400
Temperature, C

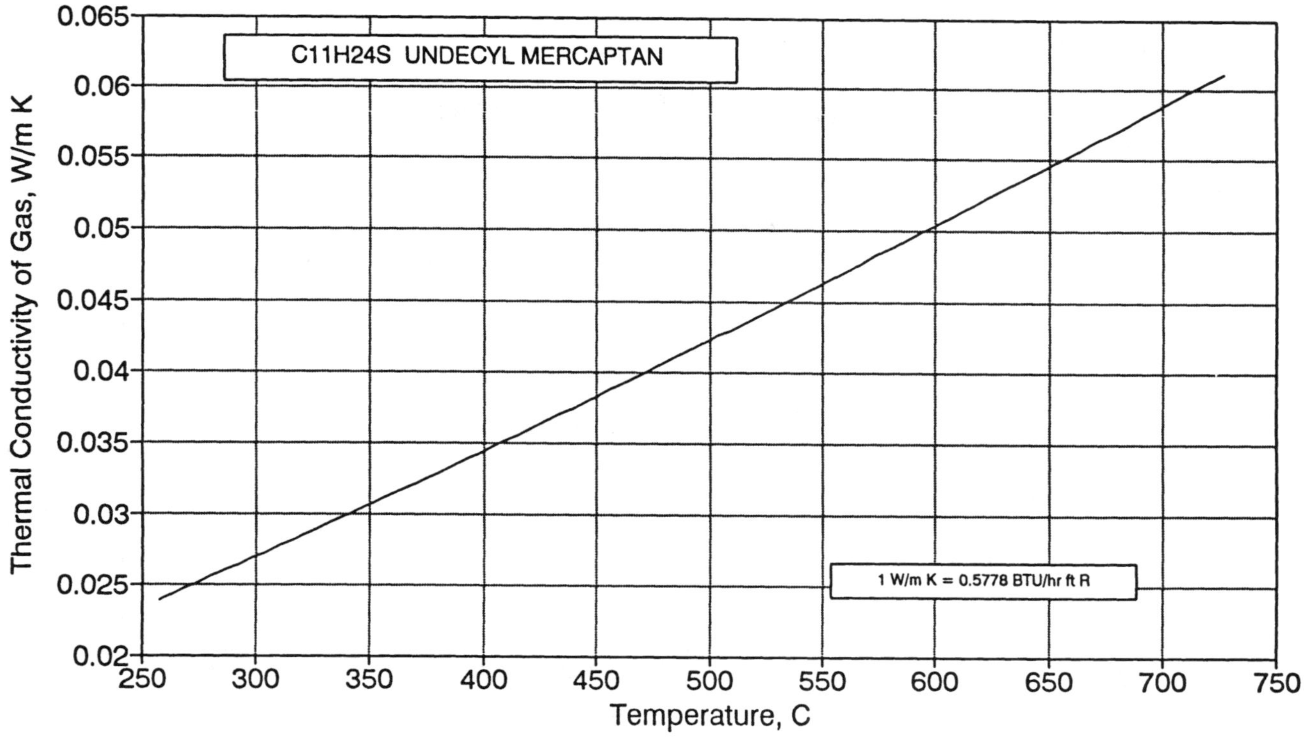
C11H24S UNDECYL MERCAPTAN
Thermal Conductivity of Gas, W/m K
0.065
0.06
0.055
0.05
0.045
0.04
0.035
0.03
0.025
0.02
250
300
350
400
450
500
550
600
650
700
750
Temperature, C
1 W/m K = 0.5778 BTU/hr ft R

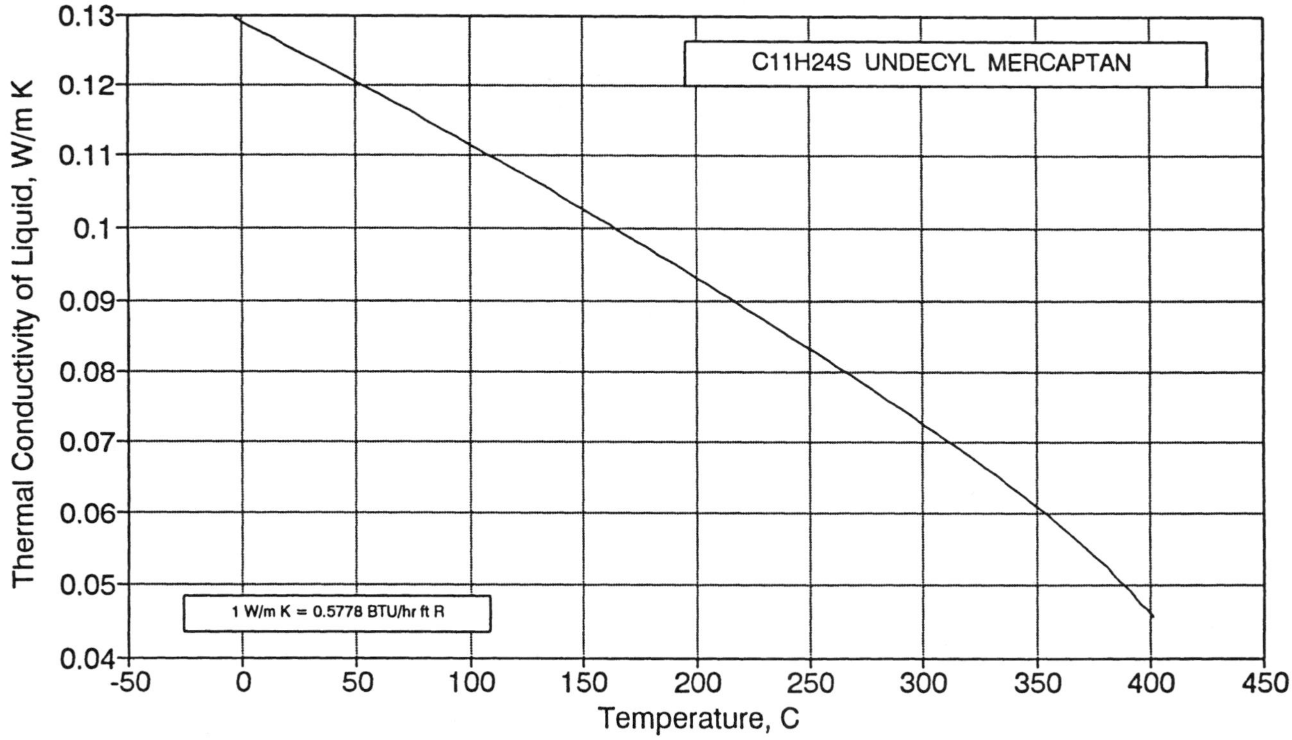
C11H24S UNDECYL MERCAPTAN
Thermal Conductivity of Liquid, W/m K
0.13
0.12
0.11
0.1
0.09
0.08
0.07
0.06
0.05
0.04
-50
0
50
100
150
200
250
300
350
400
450
Temperature, C
1 W/m K = 0.5778 BTU/hr ft R

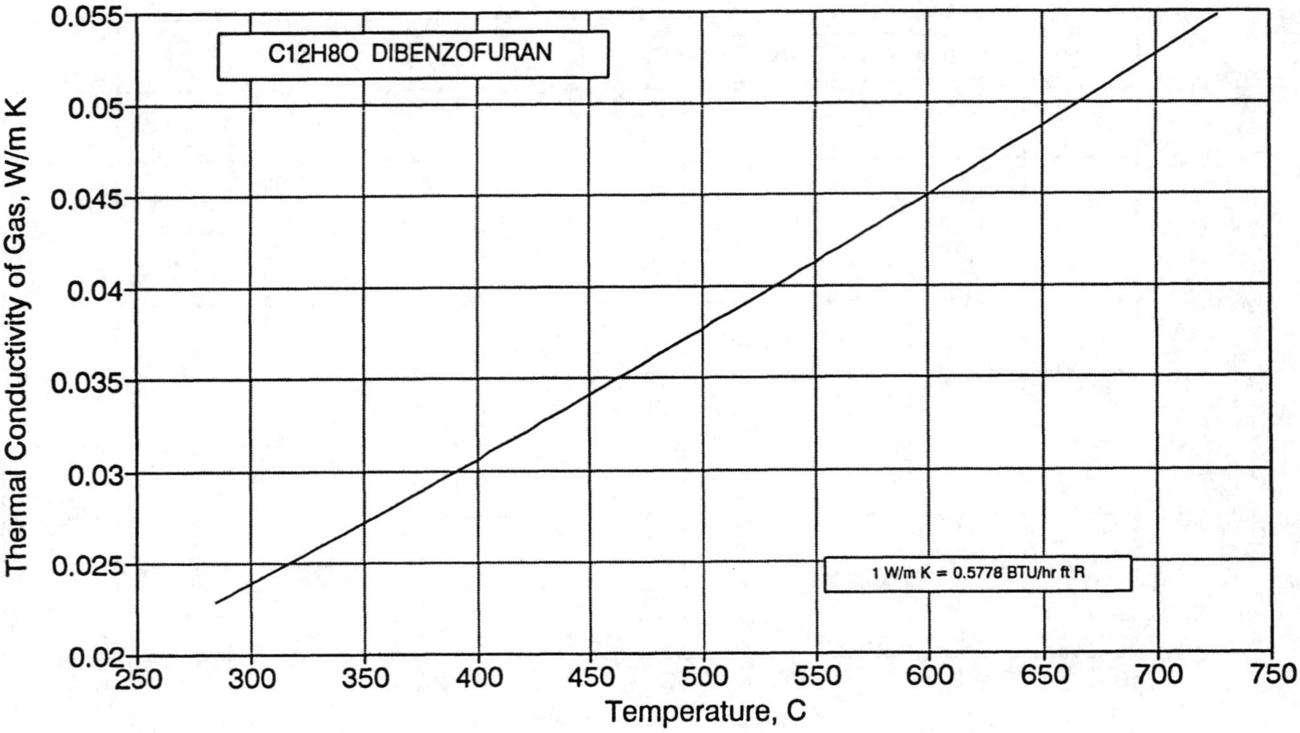
C12H8O DIBENZOFURAN
1 W/m K = 0.5778 BTU/hr ft R
Thermal Conductivity of Gas, W/m K
Temperature, C
0.055
0.05
0.045
0.04
0.035
0.03
0.025
0.02
250
300
350
400
450
500
550
600
650
700
750

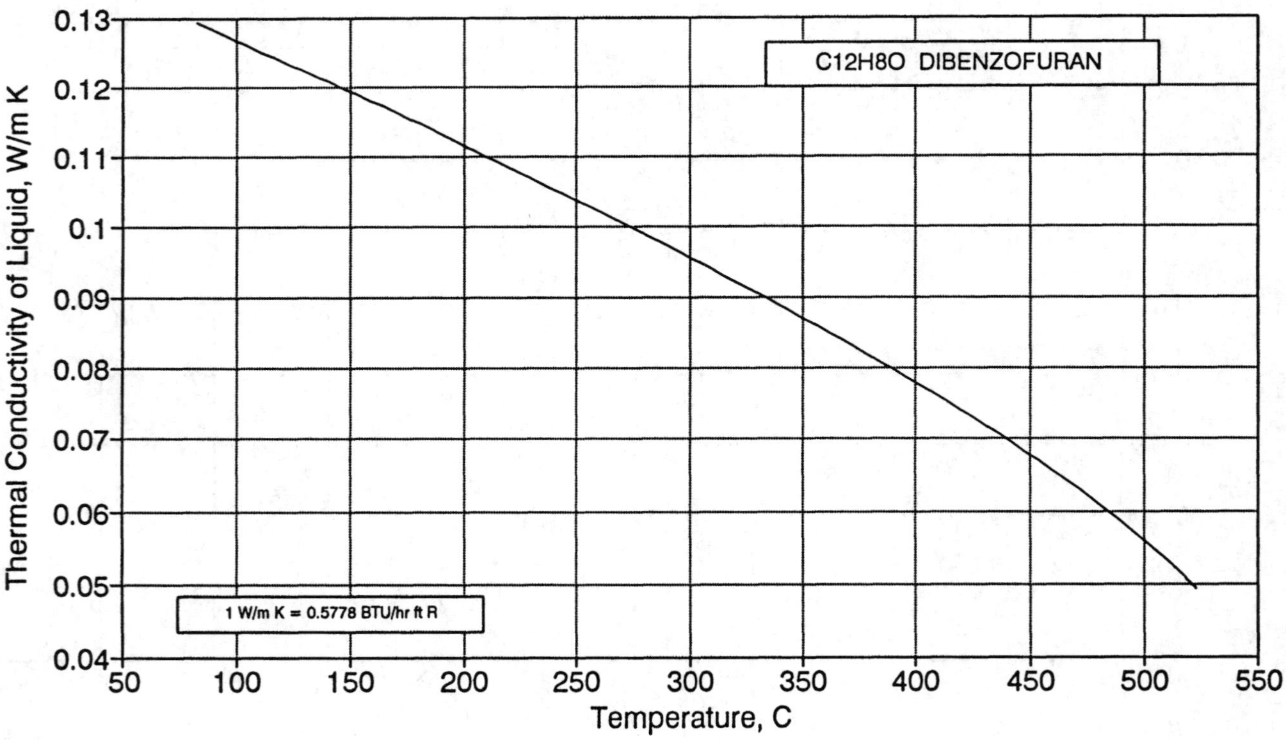
C12H8O DIBENZOFURAN
1 W/m K = 0.5778 BTU/hr ft R
Thermal Conductivity of Liquid, W/m K
Temperature, C
0.13
0.12
0.11
0.1
0.09
0.08
0.07
0.06
0.05
0.04
50
100
150
200
250
300
350
400
450
500
550

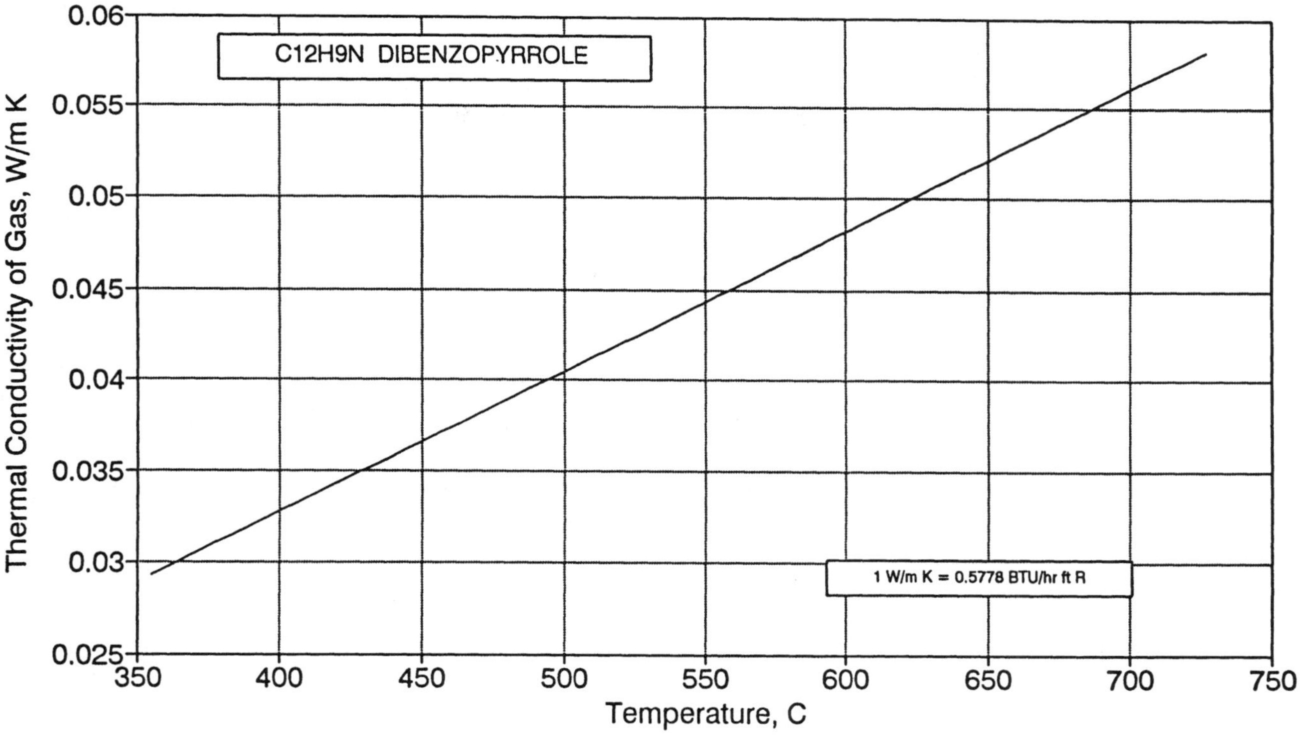
C12H9N DIBENZOPYRROLE
1 W/m K = 0.5778 BTU/hr ft R
Thermal Conductivity of Gas, W/m K
0.06
0.055
0.05
0.045
0.04
0.035
0.03
0.025
350
400
450
500
550
600
650
700
750
Temperature, C

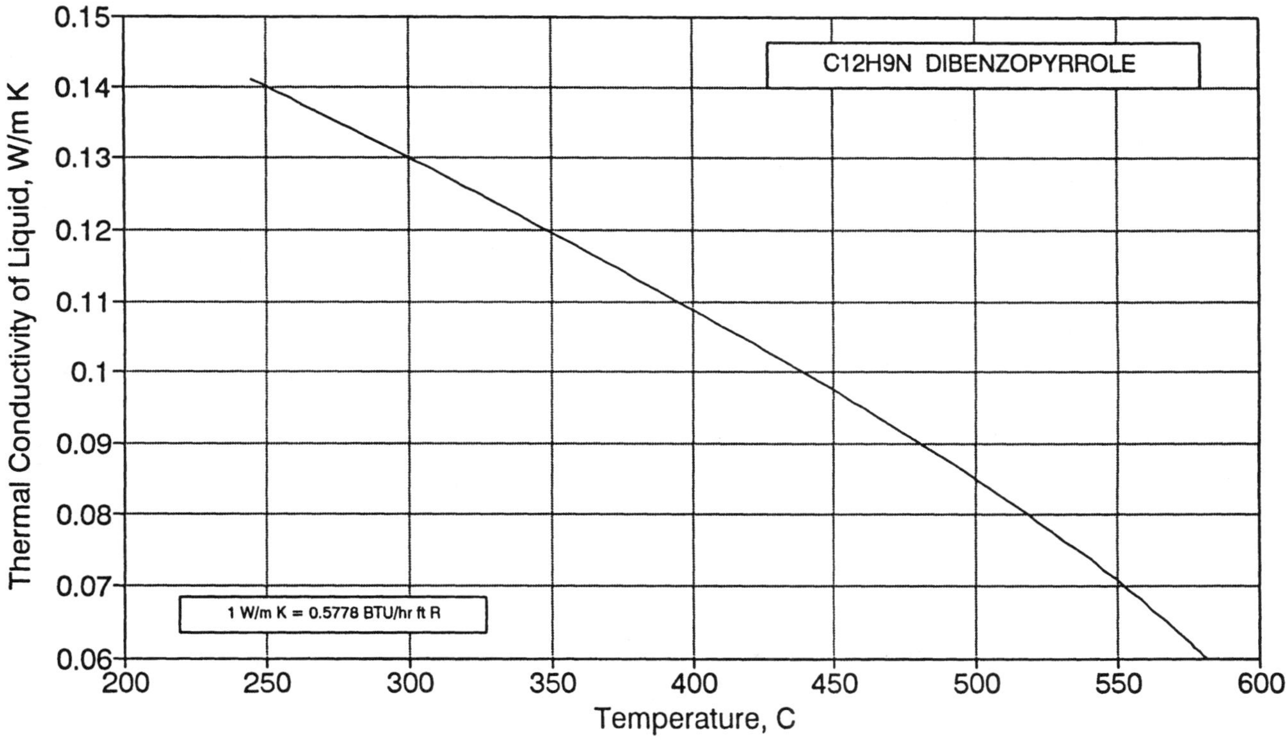
C12H9N DIBENZOPYRROLE
1 W/m K = 0.5778 BTU/hr ft R
Thermal Conductivity of Liquid, W/m K
0.15
0.14
0.13
0.12
0.11
0.1
0.09
0.08
0.07
0.06
200
250
300
350
400
450
500
550
600
Temperature, C

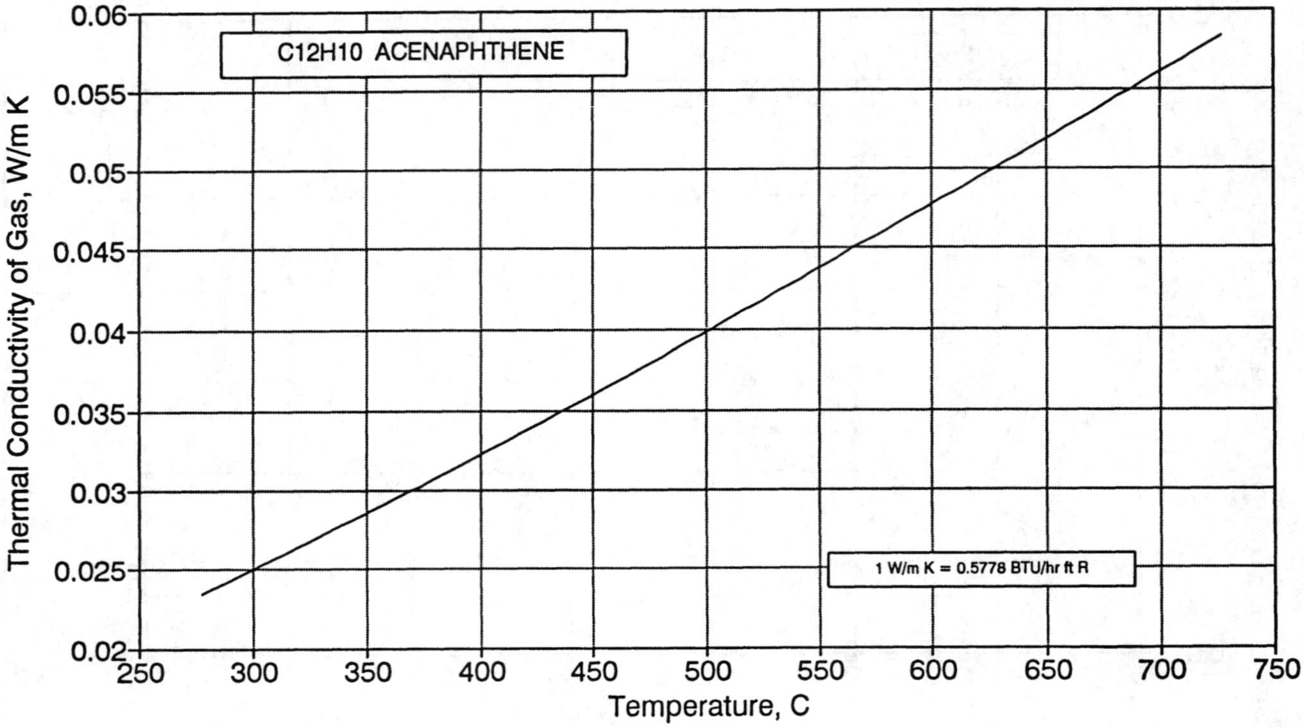
C12H10 ACENAPHTHENE
1 W/m K = 0.5778 BTU/hr ft R
Thermal Conductivity of Gas, W/m K
0.06
0.055
0.05
0.045
0.04
0.035
0.03
0.025
0.02
250
300
350
400
450
500
550
600
650
700
750
Temperature, C

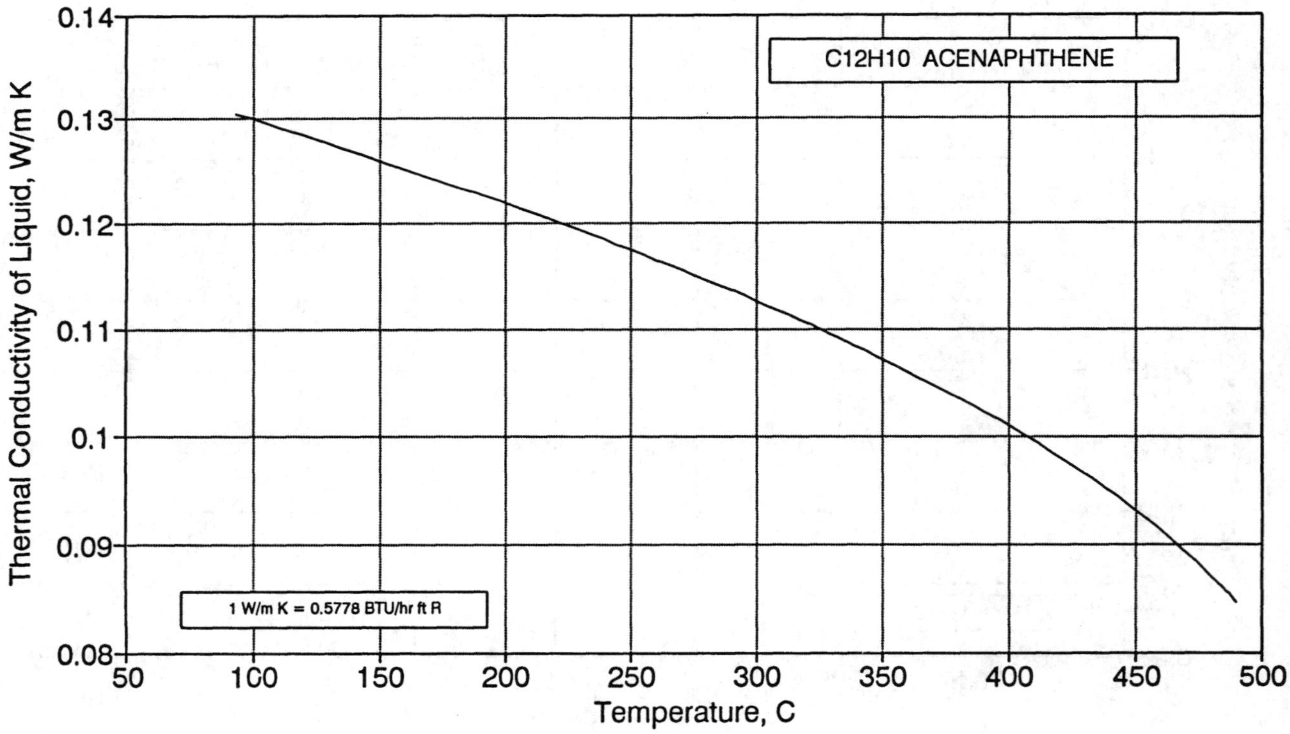
C12H10 ACENAPHTHENE
1 W/m K = 0.5778 BTU/hr ft R
Thermal Conductivity of Liquid, W/m K
0.14
0.13
0.12
0.11
0.1
0.09
0.08
50
100
150
200
250
300
350
400
450
500
Temperature, C

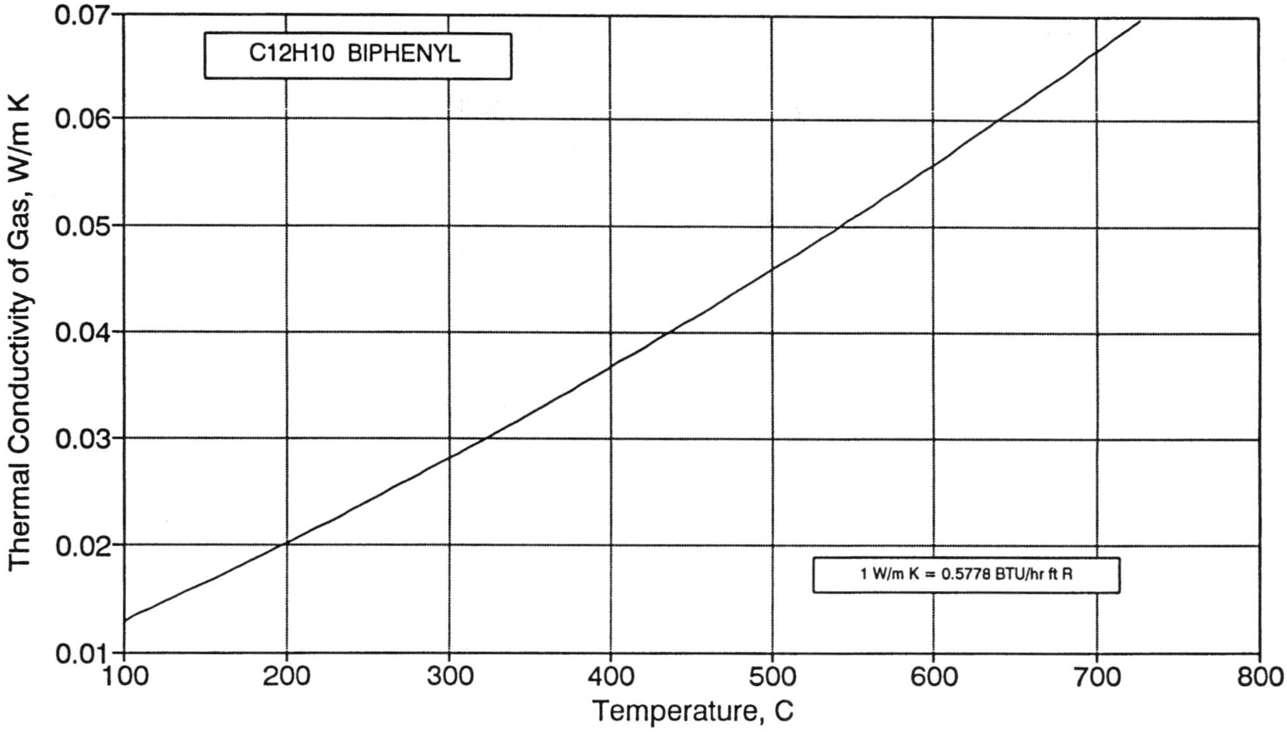
C12H10 BIPHENYL
1 W/m K = 0.5778 BTU/hr ft R
Thermal Conductivity of Gas, W/m K
Temperature, C
0.07
0.06
0.05
0.04
0.03
0.02
0.01
100
200
300
400
500
600
700
800

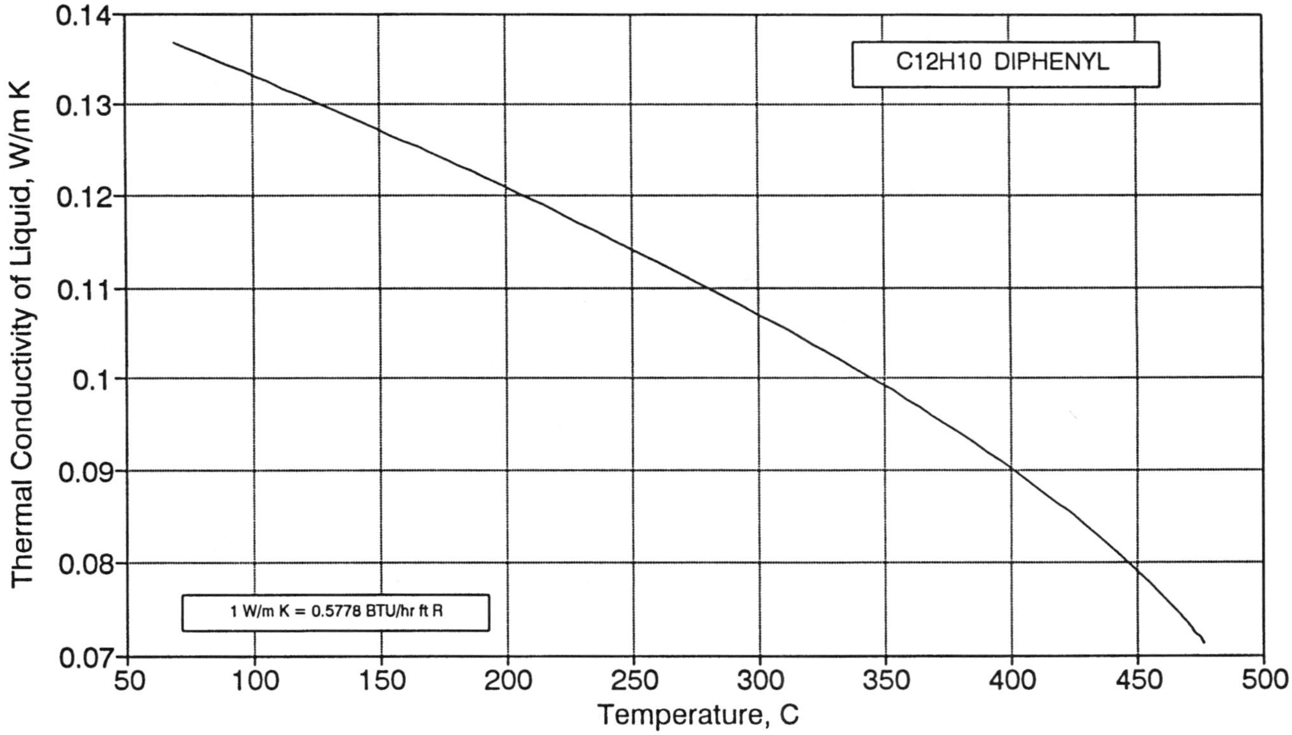
C12H10 DIPHENYL
1 W/m K = 0.5778 BTU/hr ft R
Thermal Conductivity of Liquid, W/m K
Temperature, C
0.14
0.13
0.12
0.11
0.1
0.09
0.08
0.07
50
100
150
200
250
300
350
400
450
500

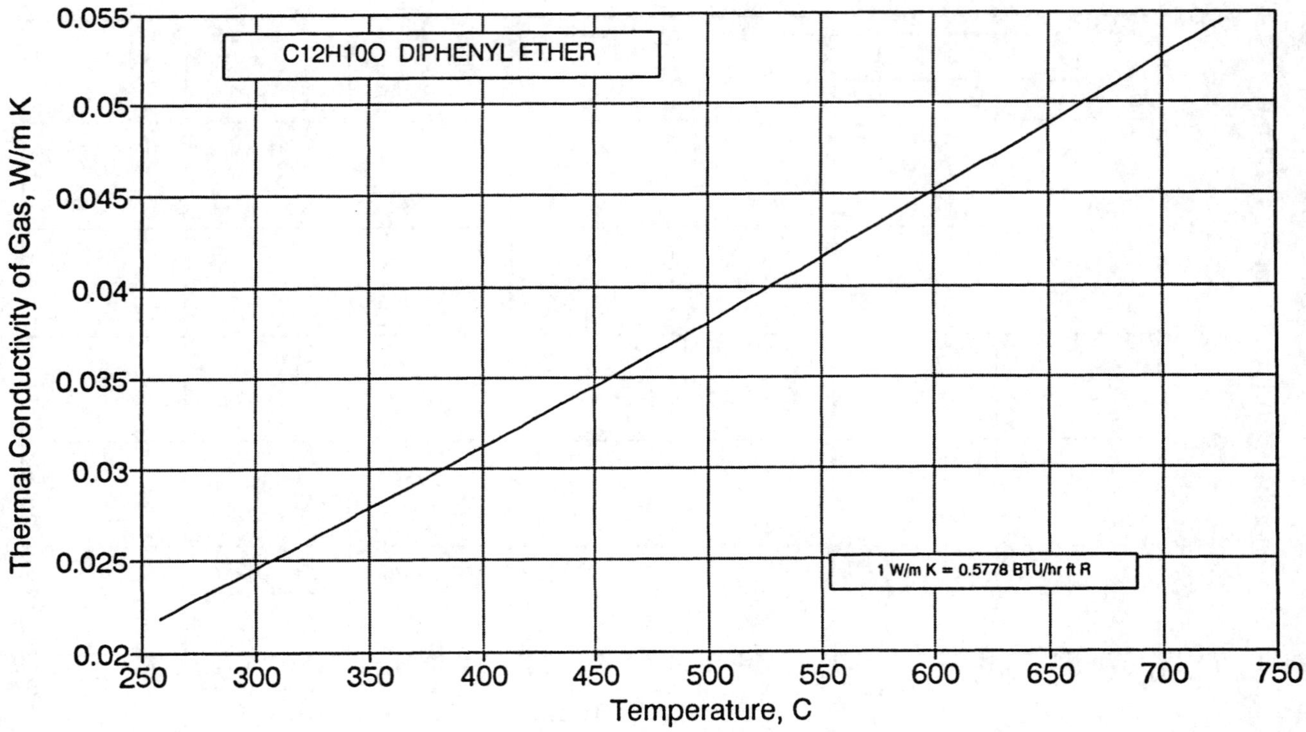
C12H10O DIPHENYL ETHER
1 W/m K = 0.5778 BTU/hr ft R
Thermal Conductivity of Gas, W/m K
Temperature, C
0.055
0.05
0.045
0.04
0.035
0.03
0.025
0.02
250
300
350
400
450
500
550
600
650
700
750

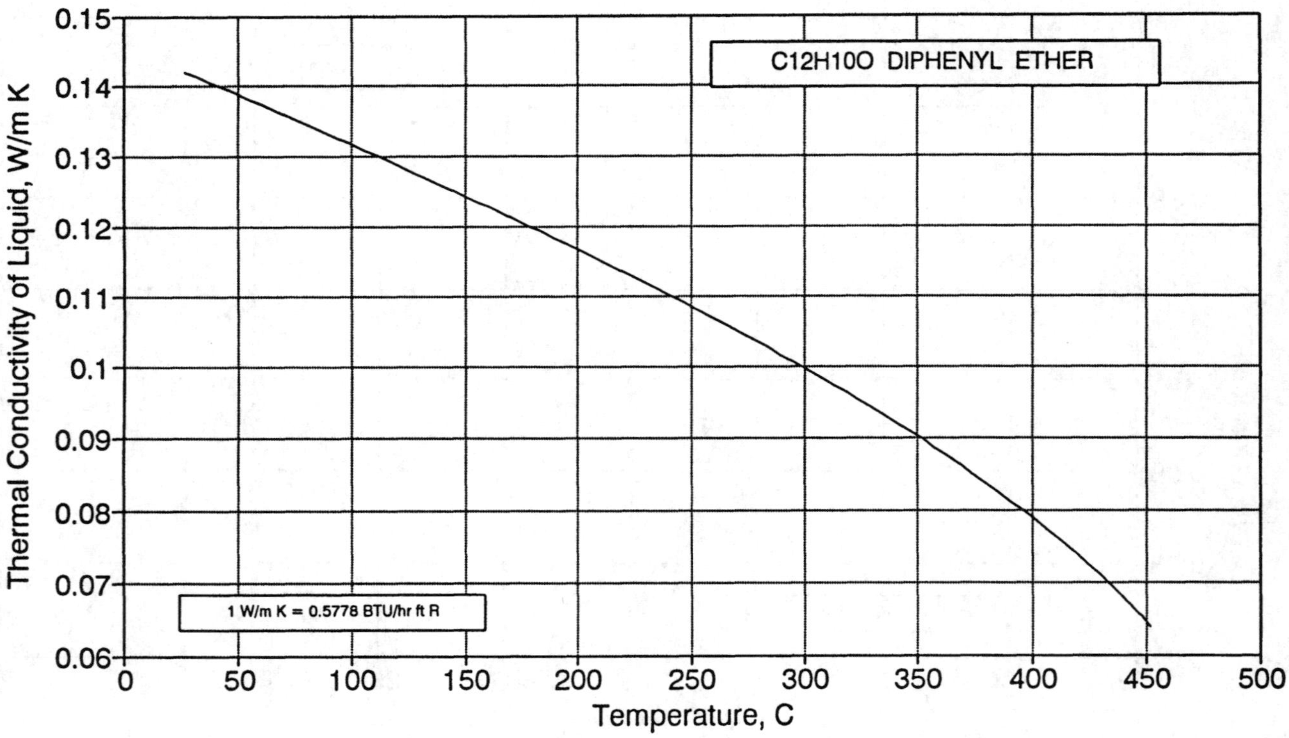
C12H10O DIPHENYL ETHER
1 W/m K = 0.5778 BTU/hr ft R
Thermal Conductivity of Liquid, W/m K
Temperature, C
0.15
0.14
0.13
0.12
0.11
0.1
0.09
0.08
0.07
0.06
0
50
100
150
200
250
300
350
400
450
500

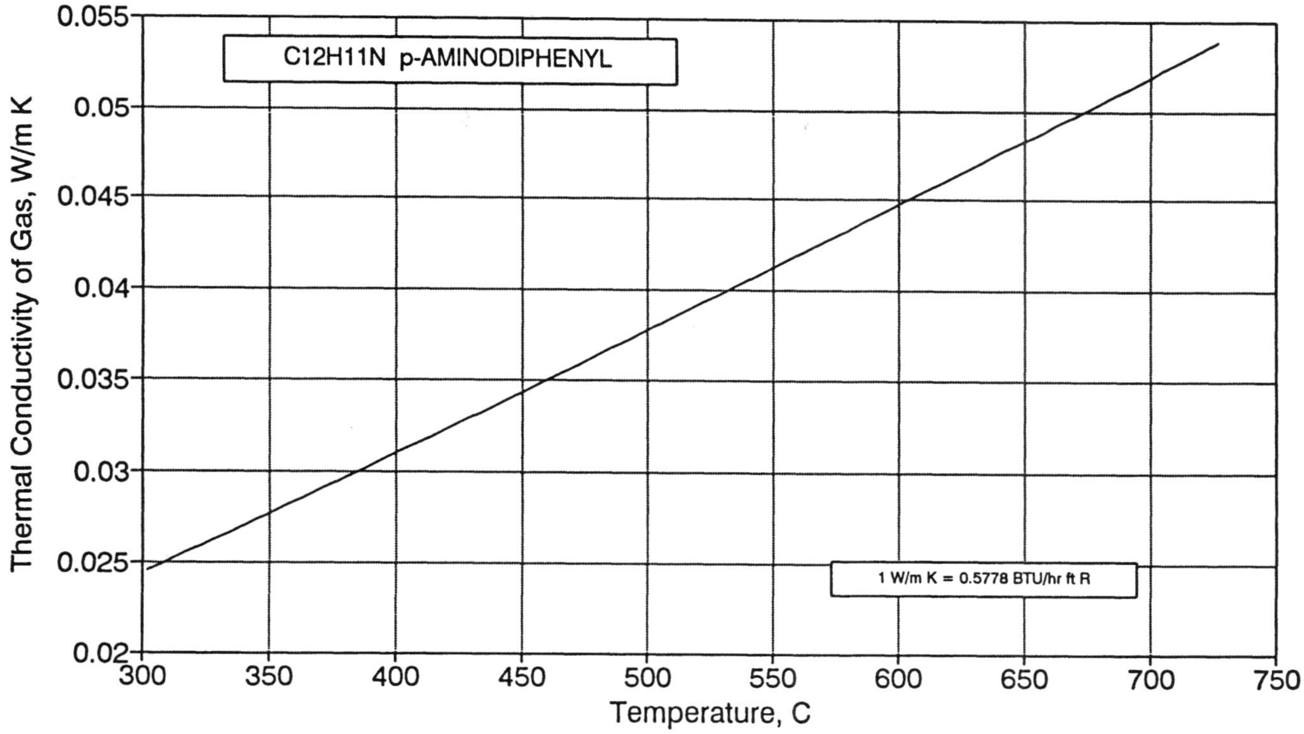
C12H11N p-AMINODIPHENYL
1 W/m K = 0.5778 BTU/hr ft R
Thermal Conductivity of Gas, W/m K
0.055
0.05
0.045
0.04
0.035
0.03
0.025
0.02
300
350
400
450
500
550
600
650
700
750
Temperature, C

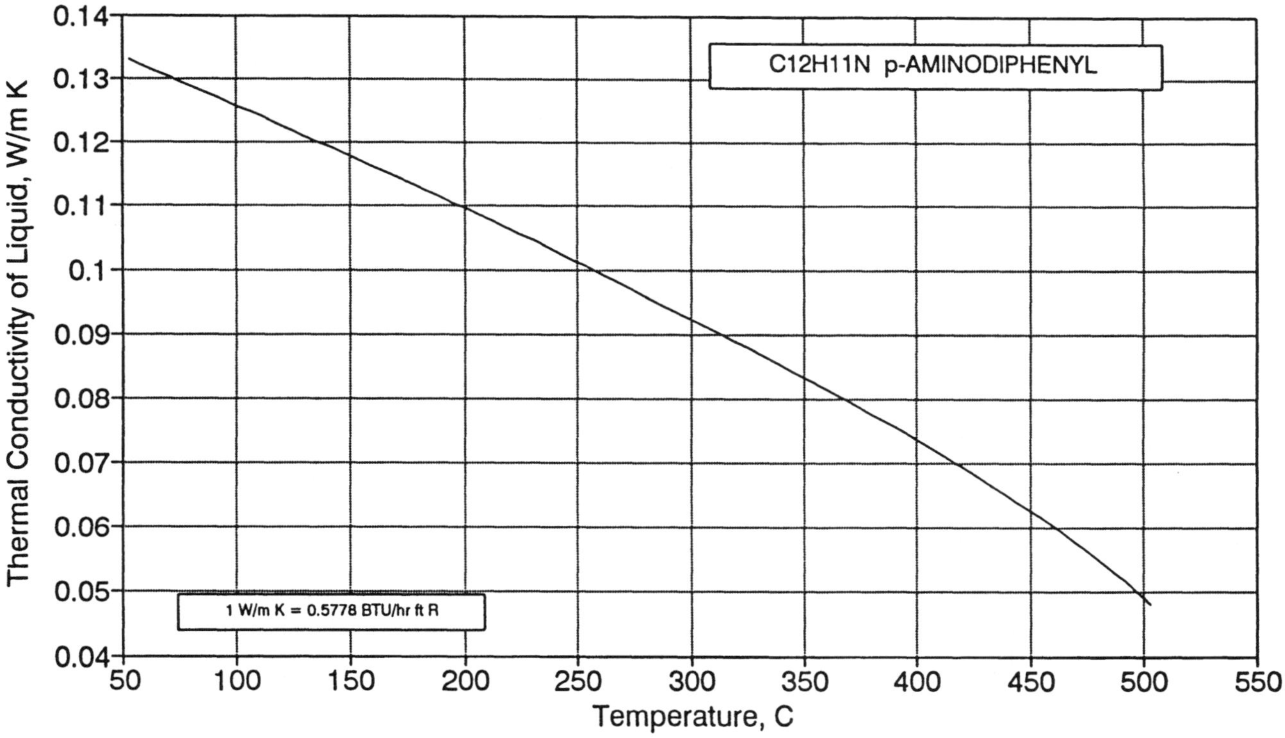
C12H11N p-AMINODIPHENYL
1 W/m K = 0.5778 BTU/hr ft R
Thermal Conductivity of Liquid, W/m K
0.14
0.13
0.12
0.11
0.1
0.09
0.08
0.07
0.06
0.05
0.04
50
100
150
200
250
300
350
400
450
500
550
Temperature, C

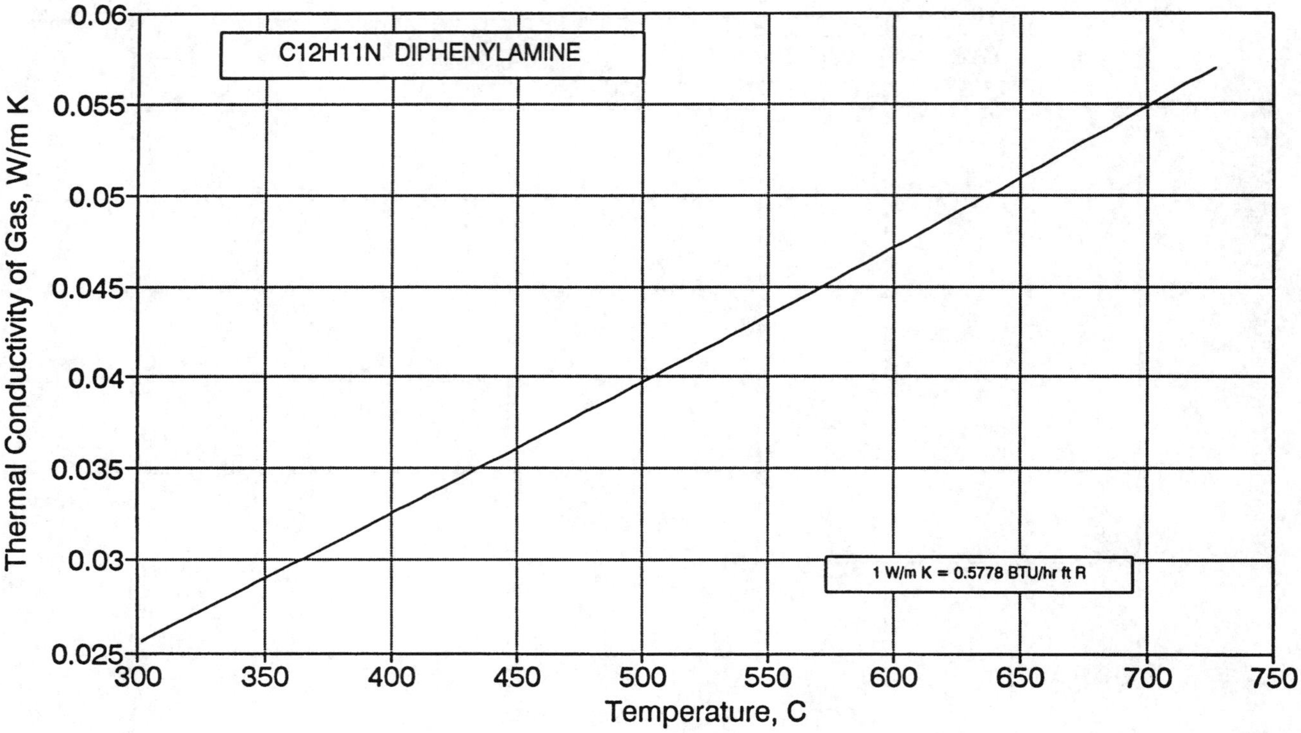
C12H11N DIPHENYLAMINE
1 W/m K = 0.5778 BTU/hr ft R
Thermal Conductivity of Gas, W/m K
0.06
0.055
0.05
0.045
0.04
0.035
0.03
0.025
300
350
400
450
500
550
600
650
700
750
Temperature, C

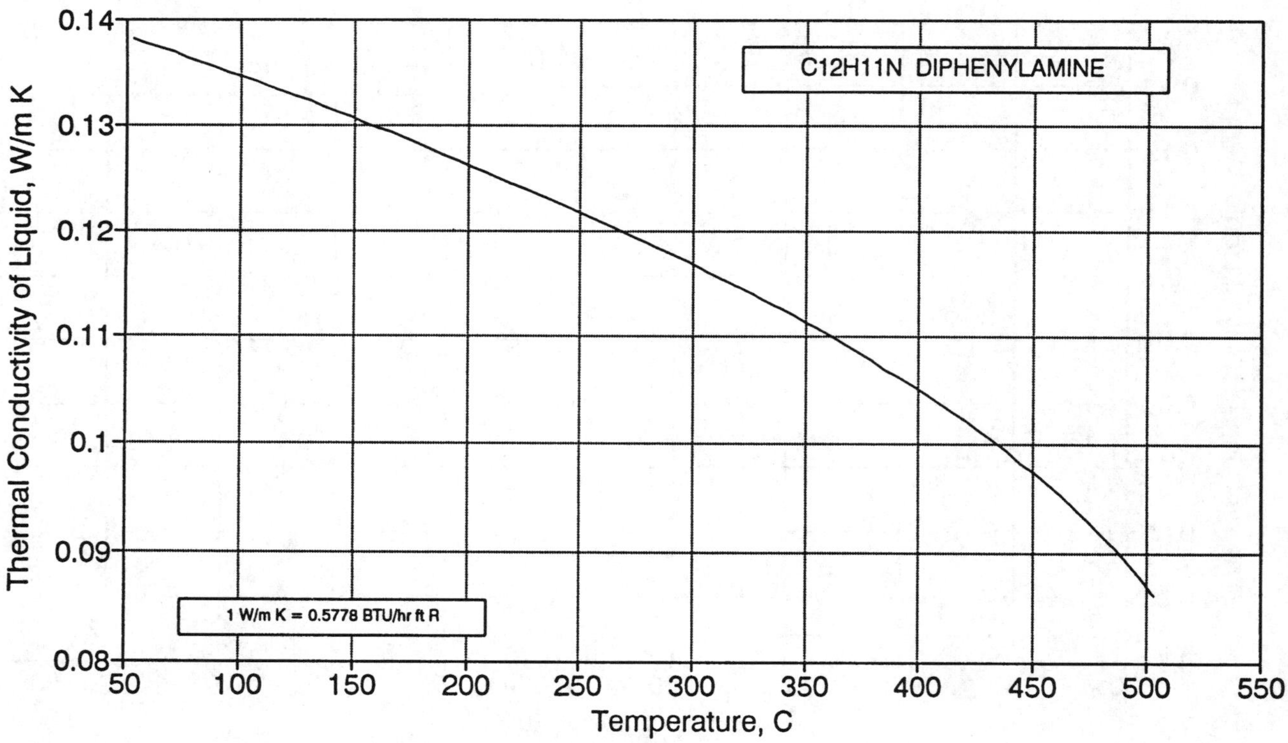
C12H11N DIPHENYLAMINE
1 W/m K = 0.5778 BTU/hr ft R
Thermal Conductivity of Liquid, W/m K
0.14
0.13
0.12
0.11
0.1
0.09
0.08
50
100
150
200
250
300
350
400
450
500
550
Temperature, C

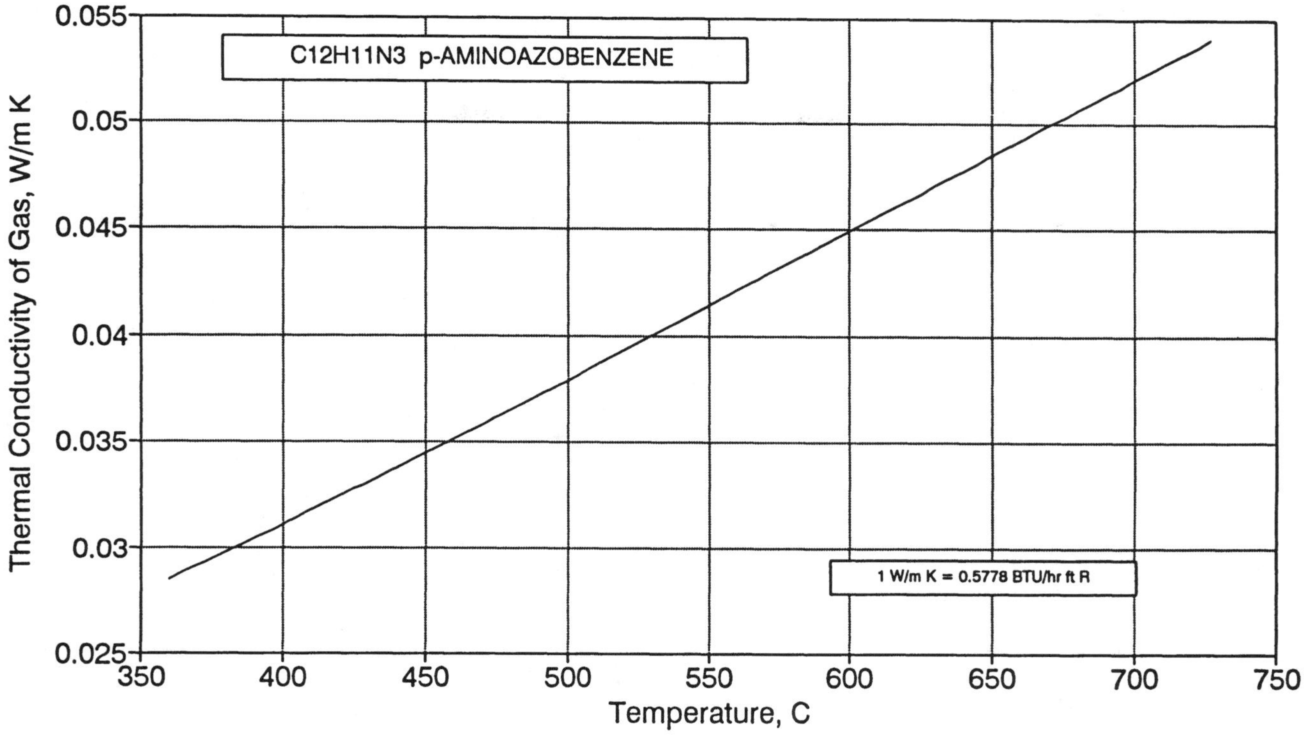
C12H11N3 p-AMINOAZOBENZENE
1 W/m K = 0.5778 BTU/hr ft R
Thermal Conductivity of Gas, W/m K
0.055
0.05
0.045
0.04
0.035
0.03
0.025
350
400
450
500
550
600
650
700
750
Temperature, C

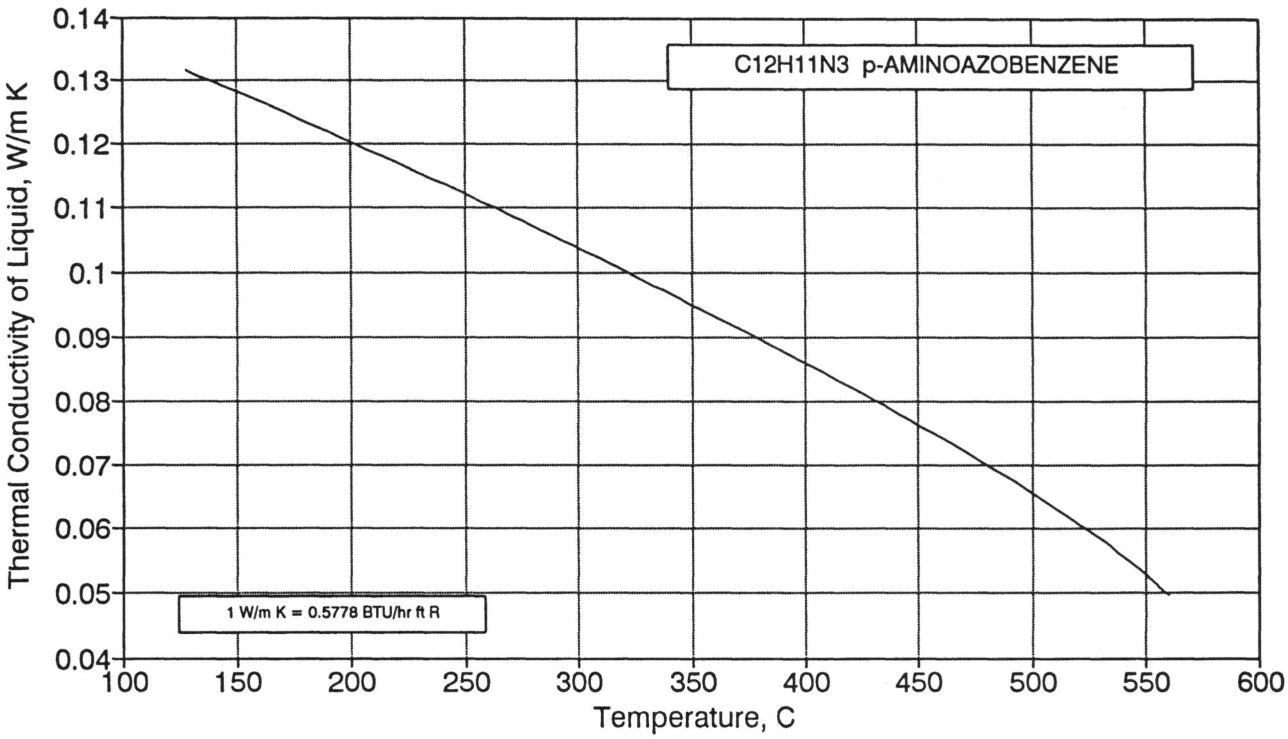
C12H11N3 p-AMINOAZOBENZENE
1 W/m K = 0.5778 BTU/hr ft R
Thermal Conductivity of Liquid, W/m K
0.14
0.13
0.12
0.11
0.1
0.09
0.08
0.07
0.06
0.05
0.04
100
150
200
250
300
350
400
450
500
550
600
Temperature, C

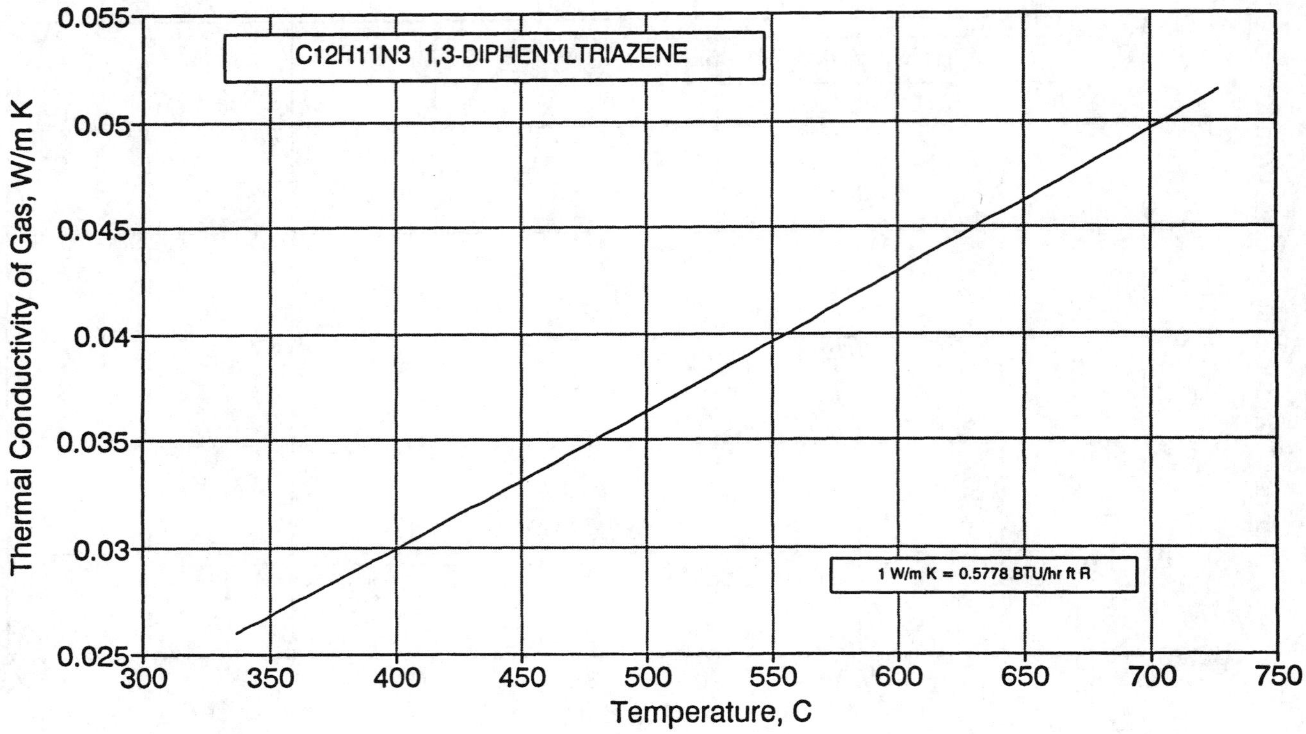
C12H11N3 1,3-DIPHENYLTRIAZENE
1 W/m K = 0.5778 BTU/hr ft R
Thermal Conductivity of Gas, W/m K
0.055
0.05
0.045
0.04
0.035
0.03
0.025
300
350
400
450
500
550
600
650
700
750
Temperature, C

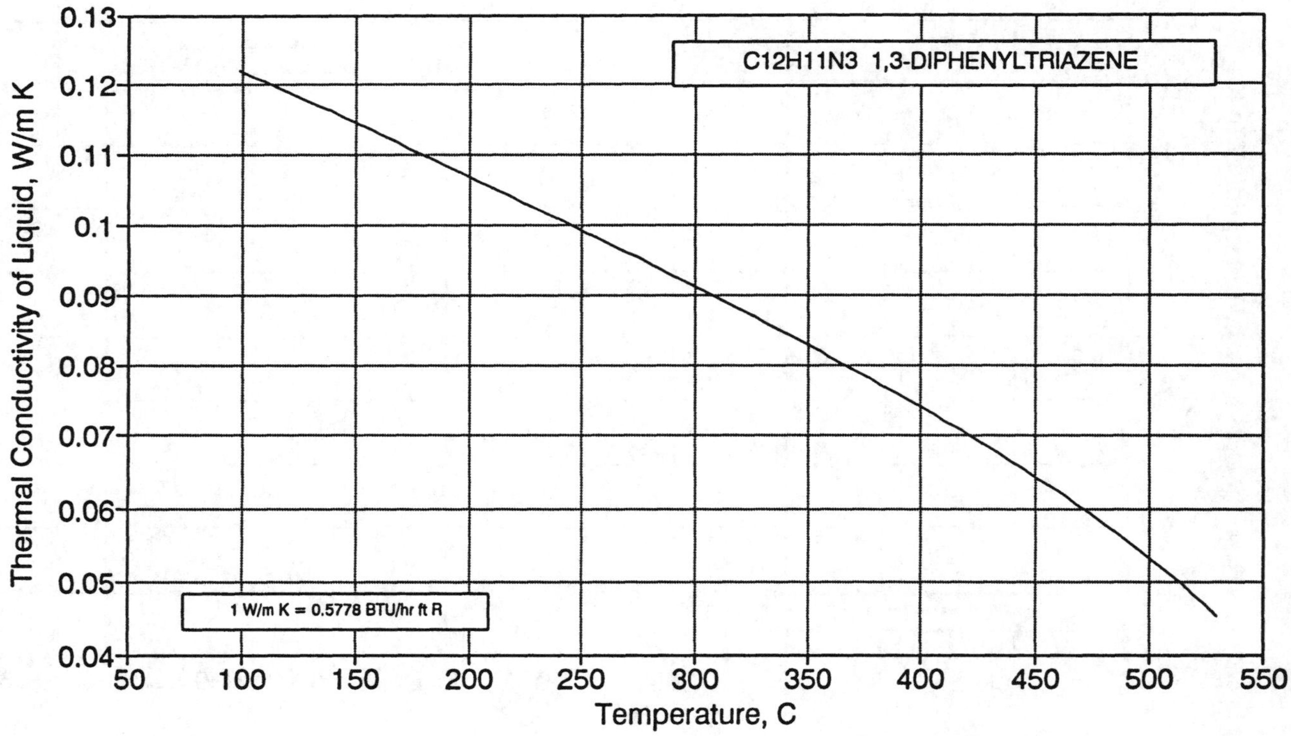
C12H11N3 1,3-DIPHENYLTRIAZENE
1 W/m K = 0.5778 BTU/hr ft R
Thermal Conductivity of Liquid, W/m K
0.13
0.12
0.11
0.1
0.09
0.08
0.07
0.06
0.05
0.04
50
100
150
200
250
300
350
400
450
500
550
Temperature, C

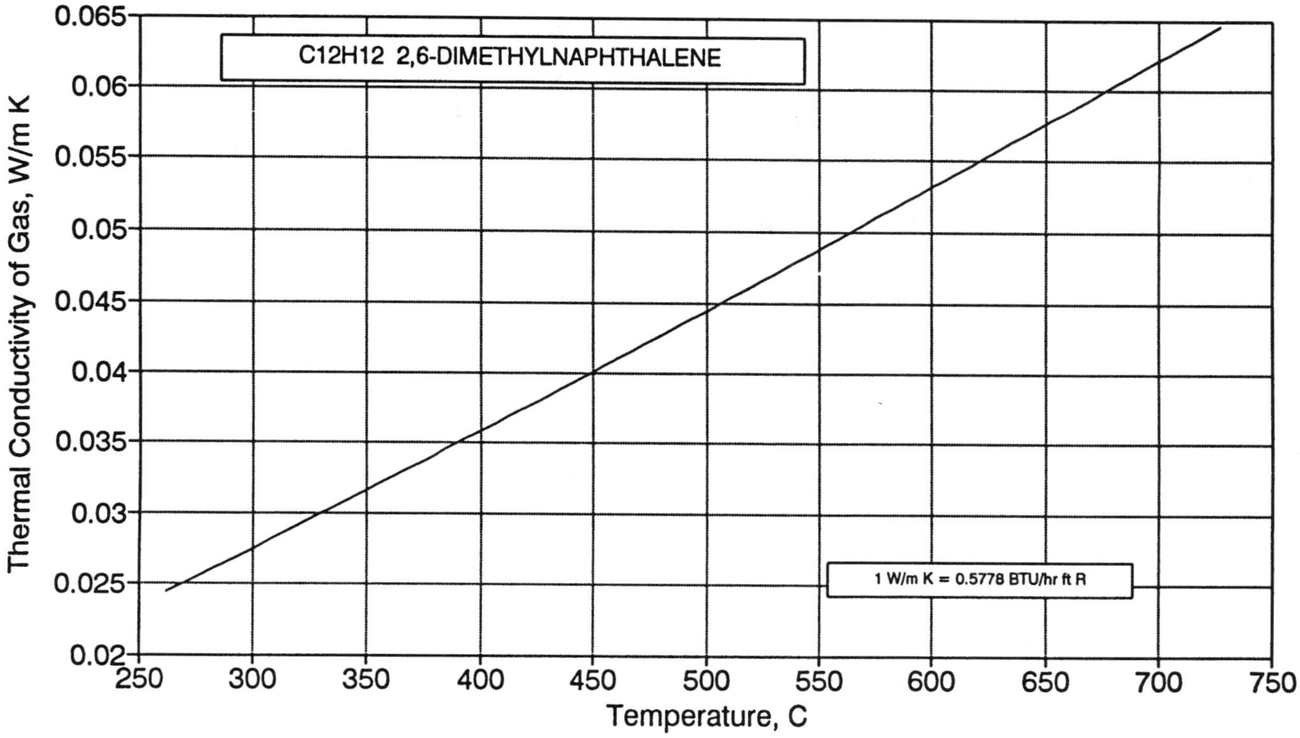
C12H12 2,6-DIMETHYLNAPHTHALENE
1 W/m K = 0.5778 BTU/hr ft R
Thermal Conductivity of Gas, W/m K
0.065
0.06
0.055
0.05
0.045
0.04
0.035
0.03
0.025
0.02
250
300
350
400
450
500
550
600
650
700
750
Temperature, C

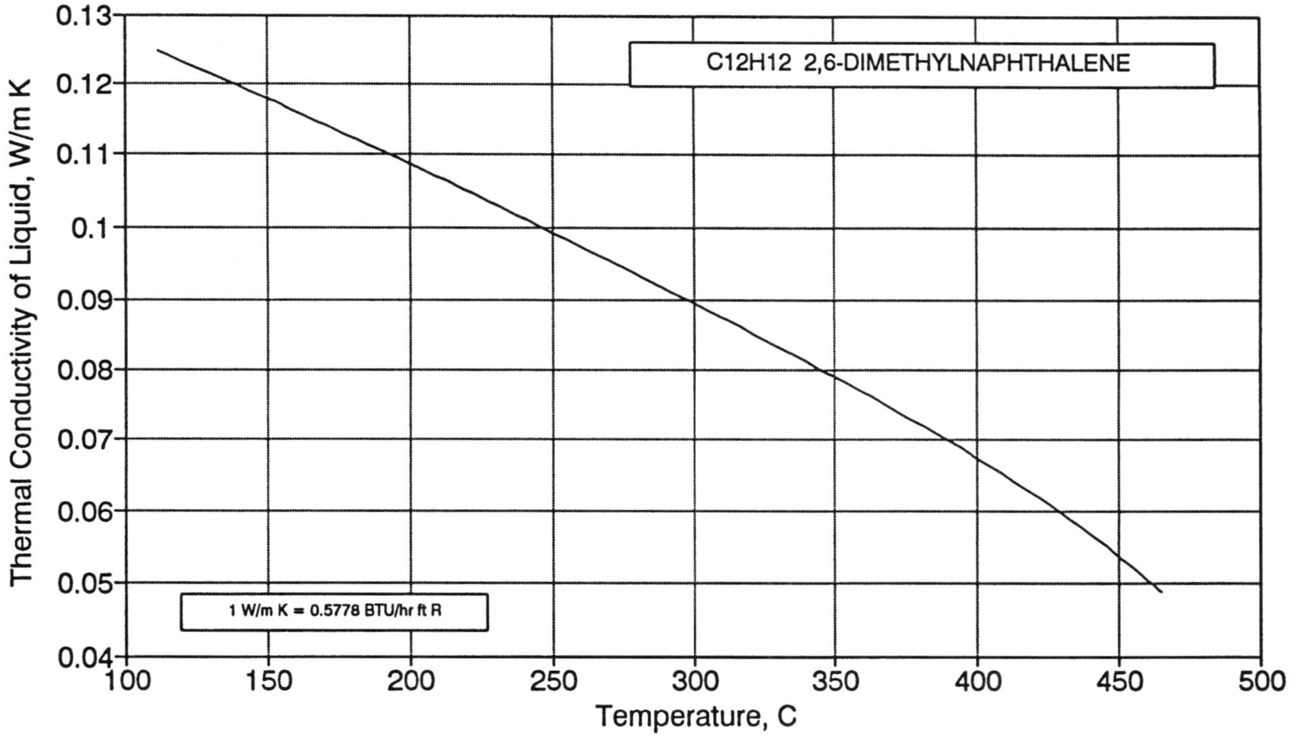
C12H12 2,6-DIMETHYLNAPHTHALENE
1 W/m K = 0.5778 BTU/hr ft R
Thermal Conductivity of Liquid, W/m K
0.13
0.12
0.11
0.1
0.09
0.08
0.07
0.06
0.05
0.04
100
150
200
250
300
350
400
450
500
Temperature, C

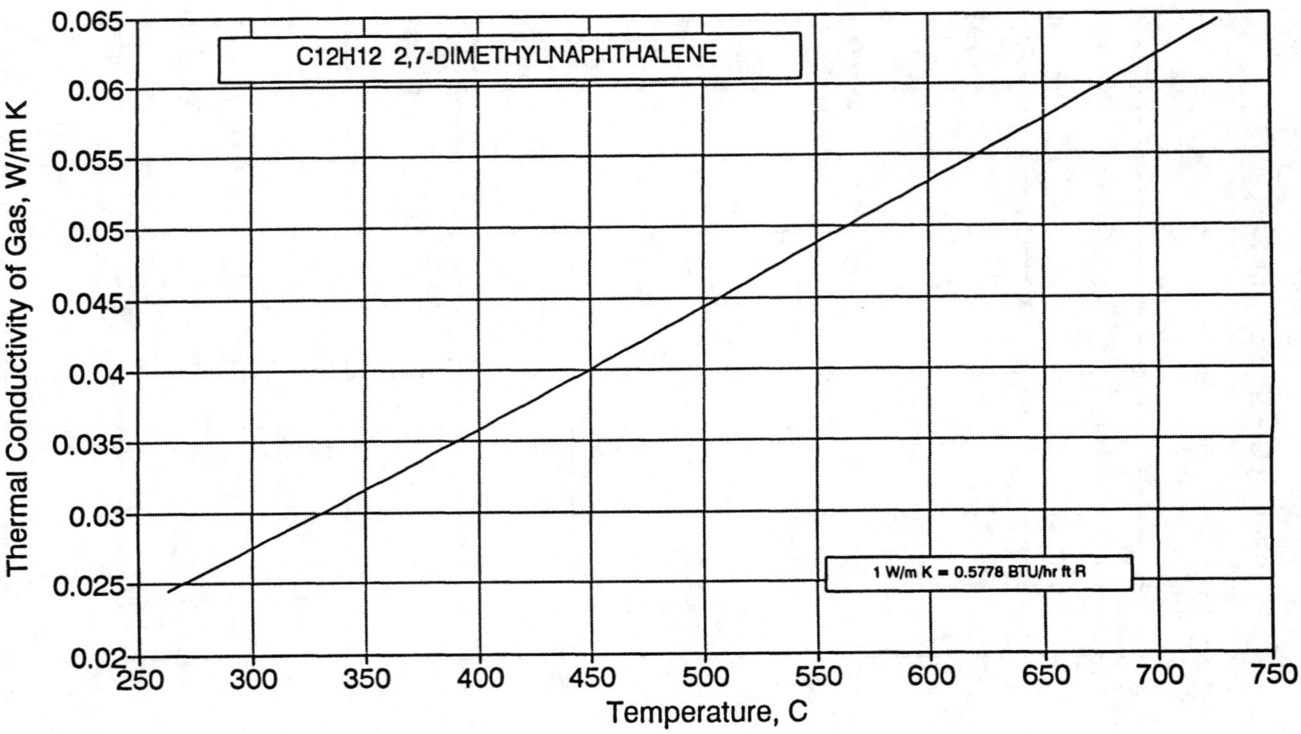
C12H12 2,7-DIMETHYLNAPHTHALENE
1 W/m K = 0.5778 BTU/hr ft R
Thermal Conductivity of Gas, W/m K
Temperature, C
0.065
0.06
0.055
0.05
0.045
0.04
0.035
0.03
0.025
0.02
250
300
350
400
450
500
550
600
650
700
750

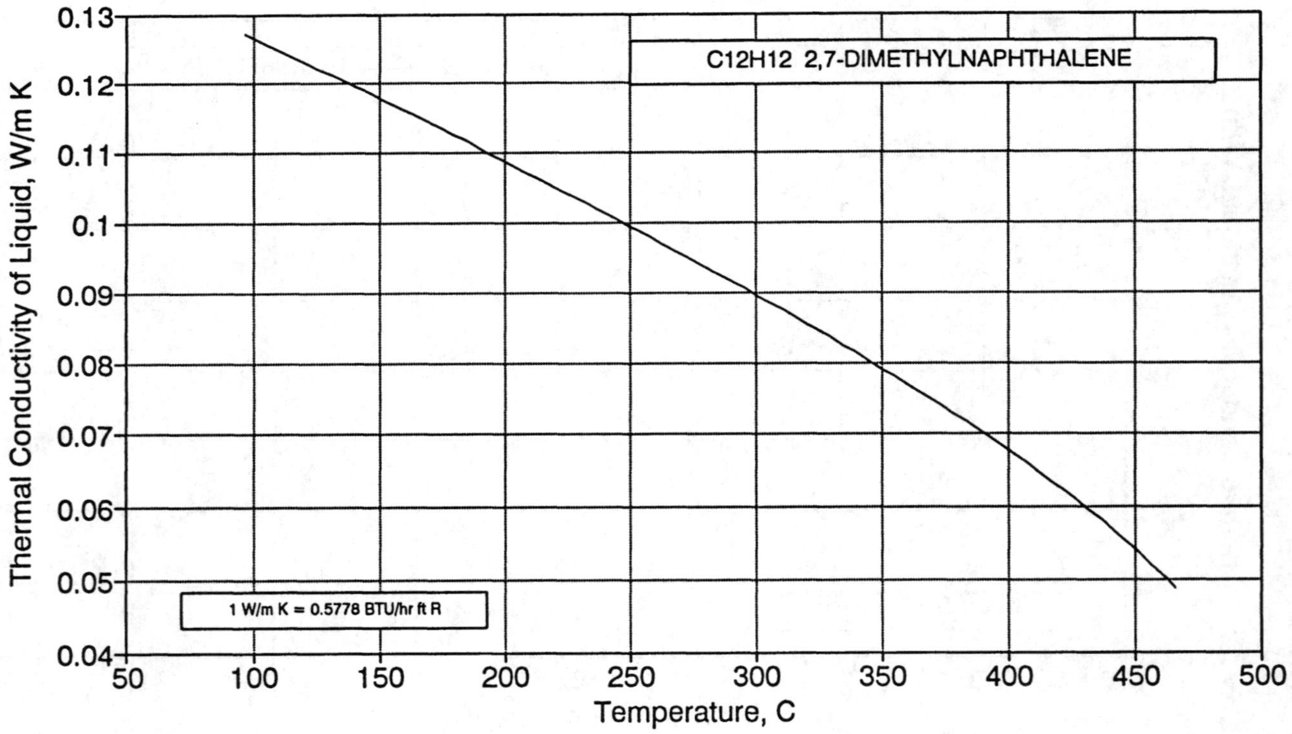
C12H12 2,7-DIMETHYLNAPHTHALENE
1 W/m K = 0.5778 BTU/hr ft R
Thermal Conductivity of Liquid, W/m K
Temperature, C
0.13
0.12
0.11
0.1
0.09
0.08
0.07
0.06
0.05
0.04
50
100
150
200
250
300
350
400
450
500

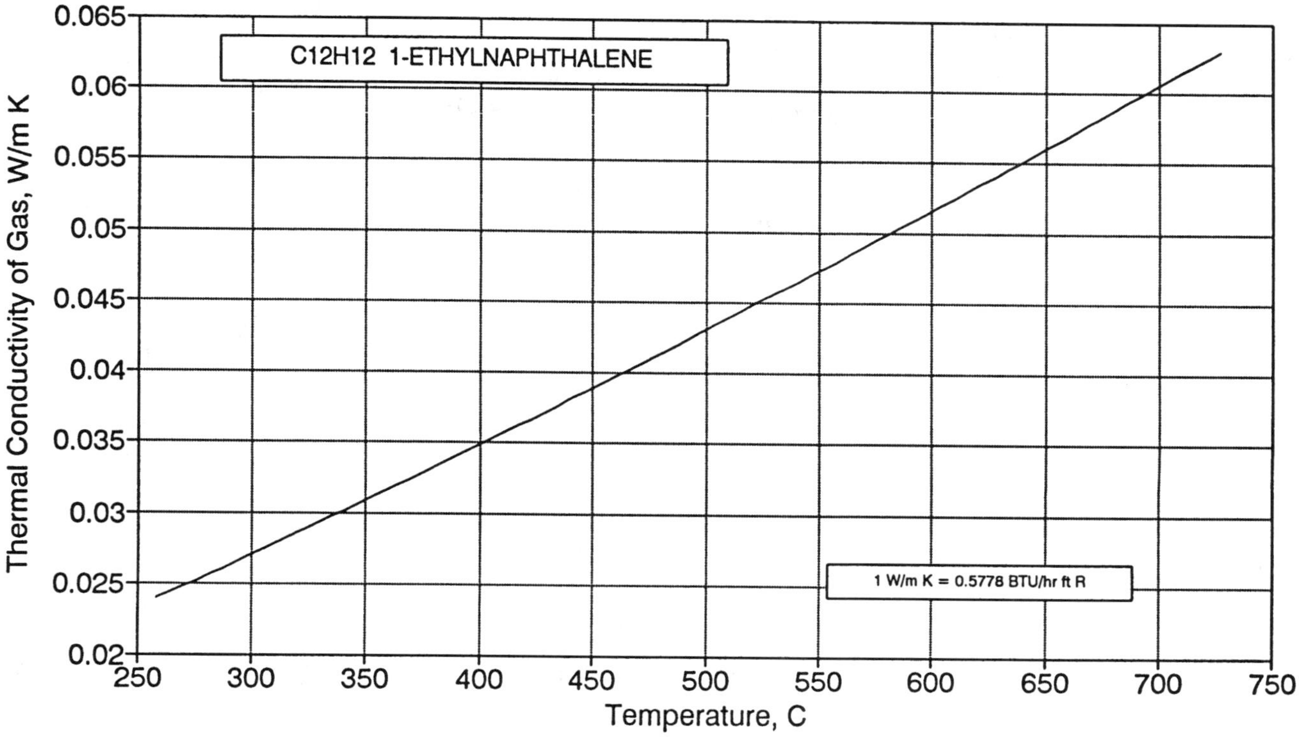
C12H12 1-ETHYLNAPHTHALENE
1 W/m K = 0.5778 BTU/hr ft R
Thermal Conductivity of Gas, W/m K
Temperature, C
0.065
0.06
0.055
0.05
0.045
0.04
0.035
0.03
0.025
0.02
250
300
350
400
450
500
550
600
650
700
750

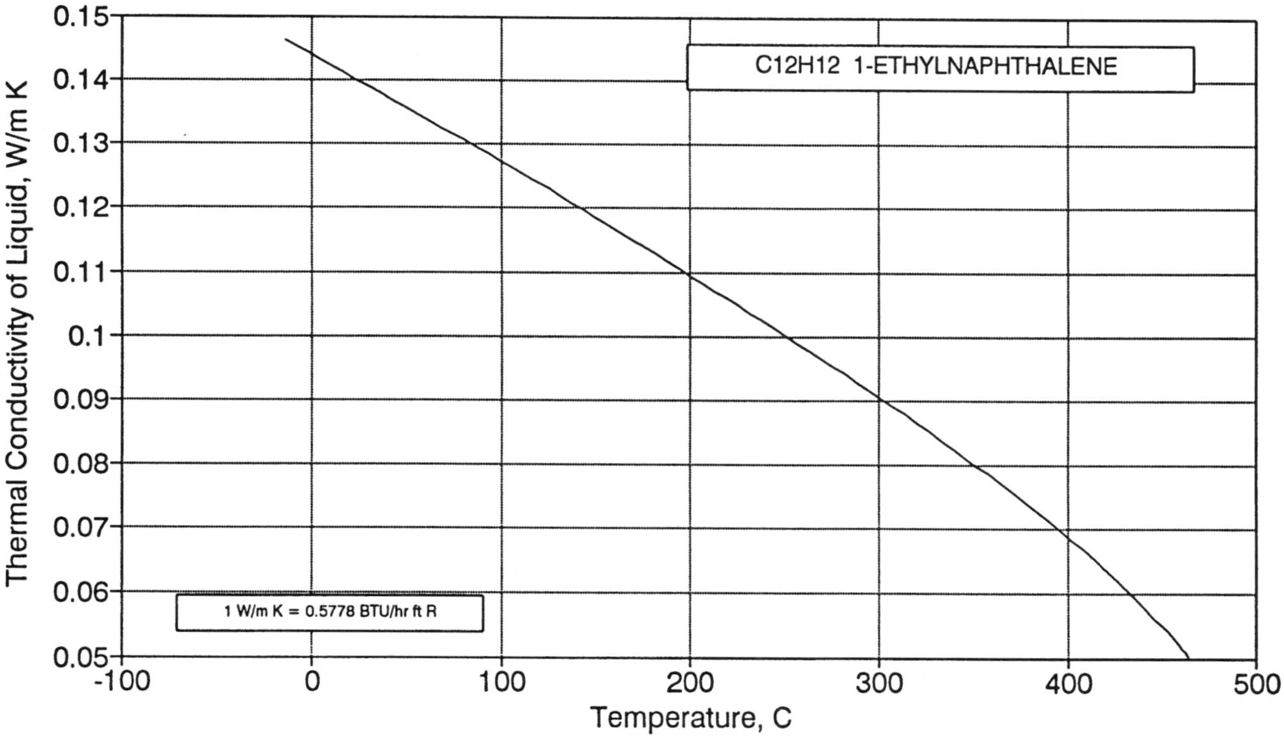
C12H12 1-ETHYLNAPHTHALENE
1 W/m K = 0.5778 BTU/hr ft R
Thermal Conductivity of Liquid, W/m K
Temperature, C
0.15
0.14
0.13
0.12
0.11
0.1
0.09
0.08
0.07
0.06
0.05
-100
0
100
200
300
400
500

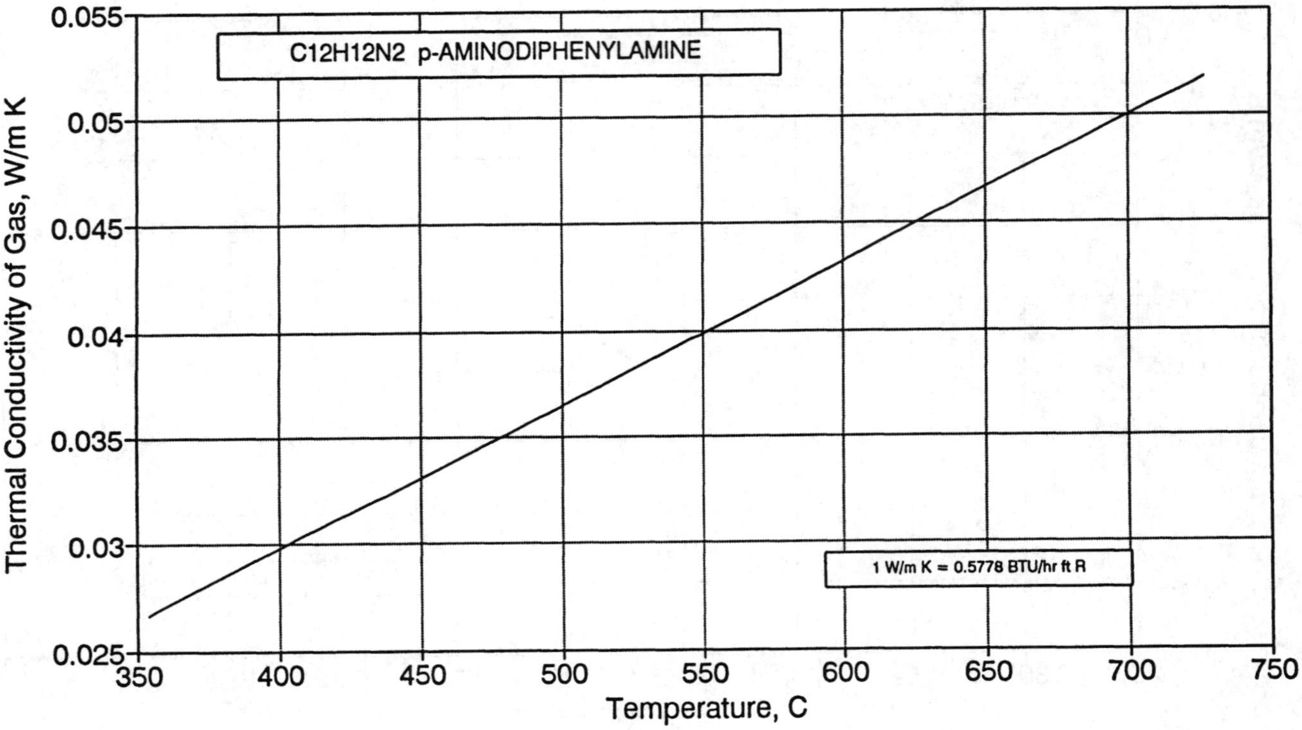
C12H12N2 p-AMINODIPHENYLAMINE
1 W/m K = 0.5778 BTU/hr ft R
Thermal Conductivity of Gas, W/m K
Temperature, C
0.025
0.03
0.035
0.04
0.045
0.05
0.055
350
400
450
500
550
600
650
700
750

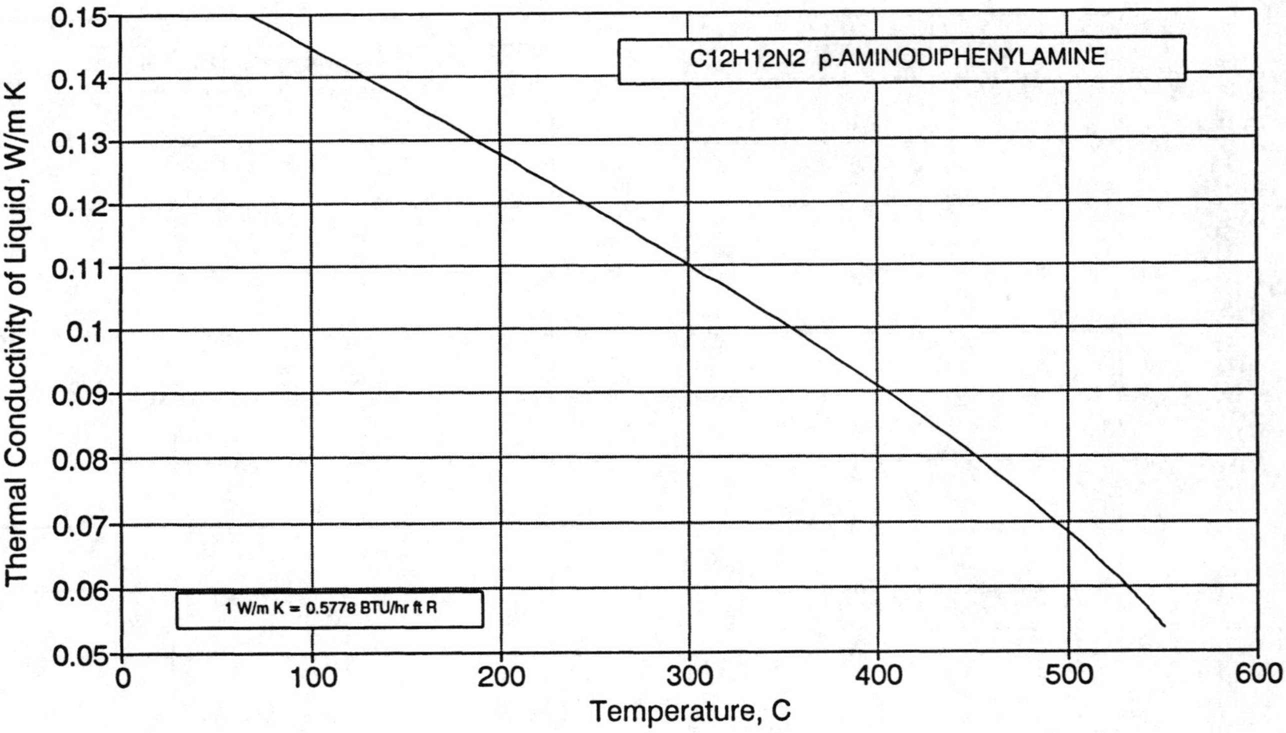
C12H12N2 p-AMINODIPHENYLAMINE
1 W/m K = 0.5778 BTU/hr ft R
Thermal Conductivity of Liquid, W/m K
Temperature, C
0.05
0.06
0.07
0.08
0.09
0.1
0.11
0.12
0.13
0.14
0.15
0
100
200
300
400
500
600

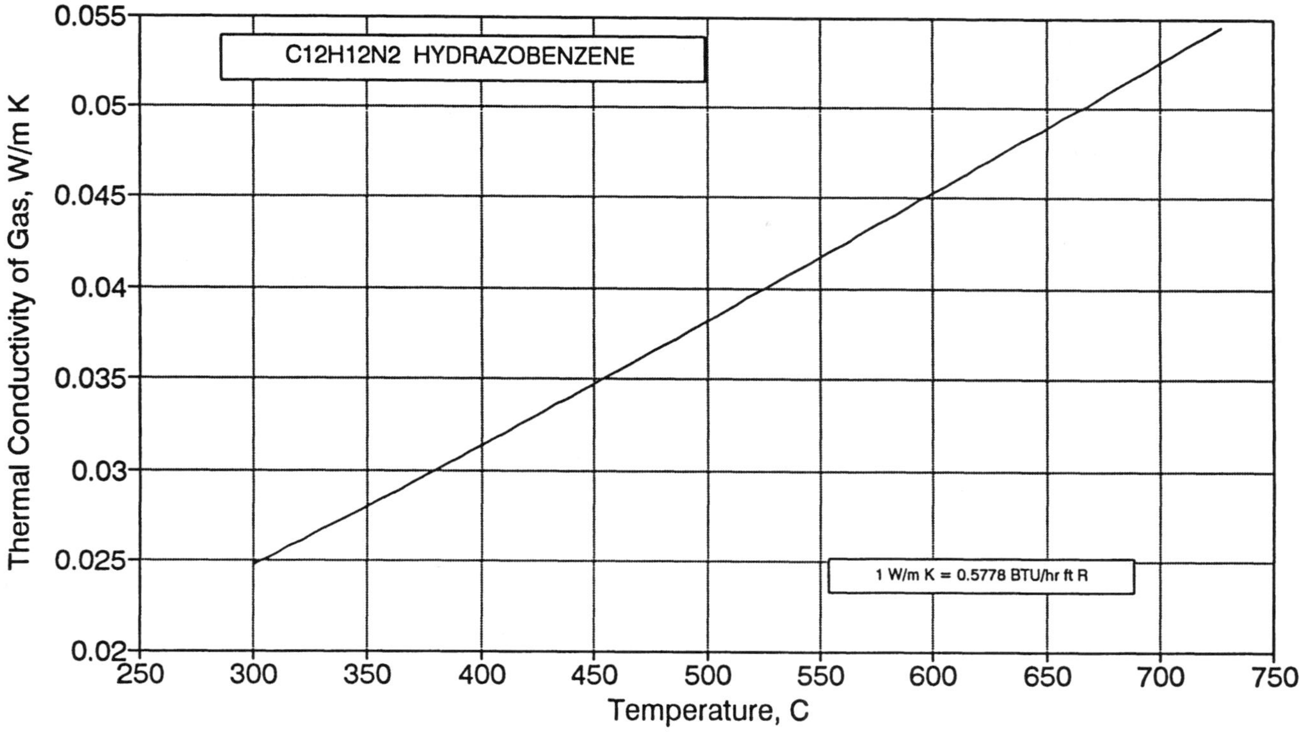

C12H12N2 HYDRAZOBENZENE

Liquid thermal conductivity data are not available .

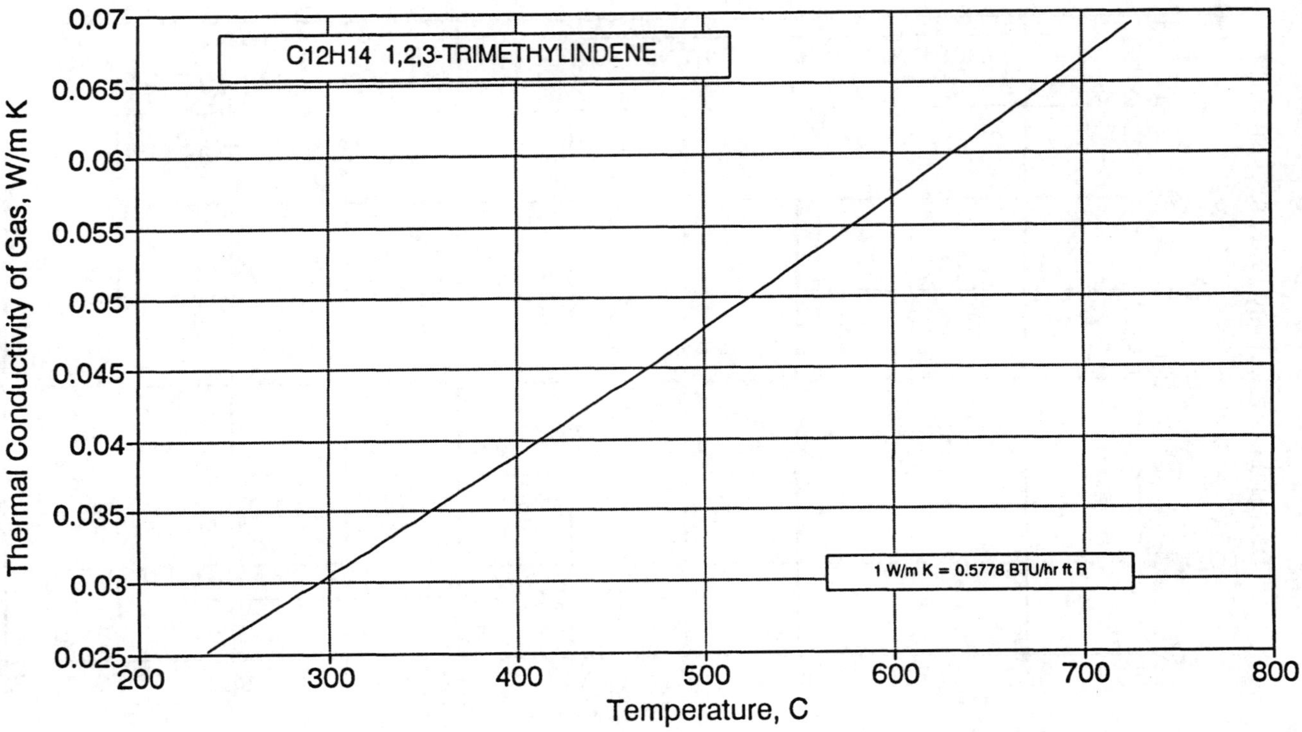
C12H14 1,2,3-TRIMETHYLINDENE
1 W/m K = 0.5778 BTU/hr ft R
Thermal Conductivity of Gas, W/m K
0.07
0.065
0.06
0.055
0.05
0.045
0.04
0.035
0.03
0.025
200
300
400
500
600
700
800
Temperature, C

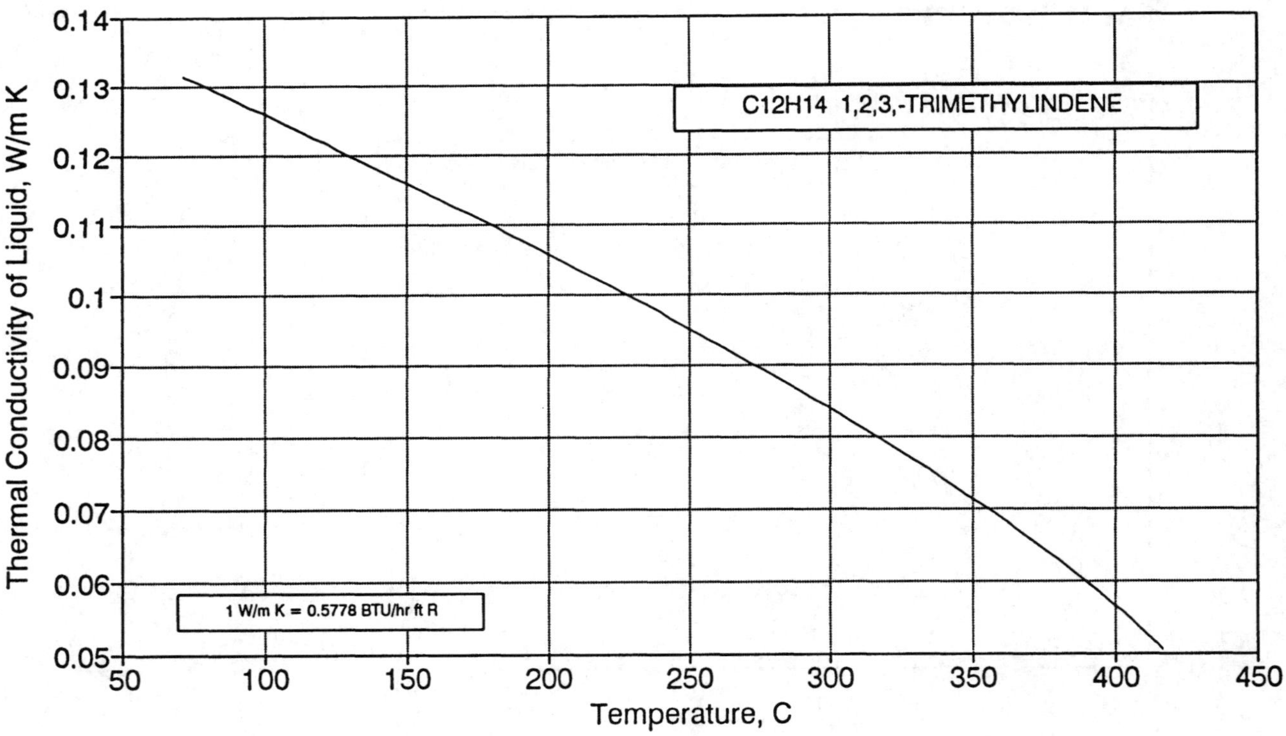
C12H14 1,2,3,-TRIMETHYLINDENE
1 W/m K = 0.5778 BTU/hr ft R
Thermal Conductivity of Liquid, W/m K
0.14
0.13
0.12
0.11
0.1
0.09
0.08
0.07
0.06
0.05
50
100
150
200
250
300
350
400
450
Temperature, C

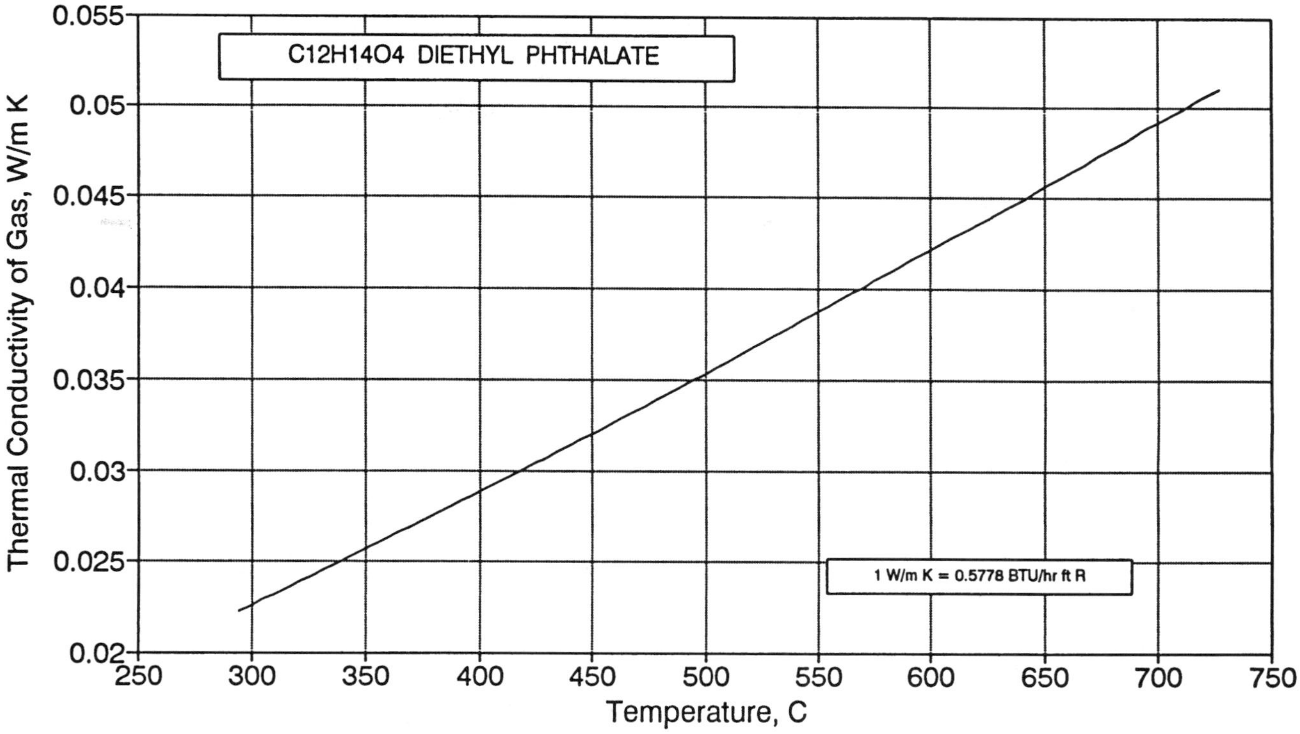
C12H14O4 DIETHYL PHTHALATE
1 W/m K = 0.5778 BTU/hr ft R
Thermal Conductivity of Gas, W/m K
Temperature, C
0.055
0.05
0.045
0.04
0.035
0.03
0.025
0.02
250
300
350
400
450
500
550
600
650
700
750

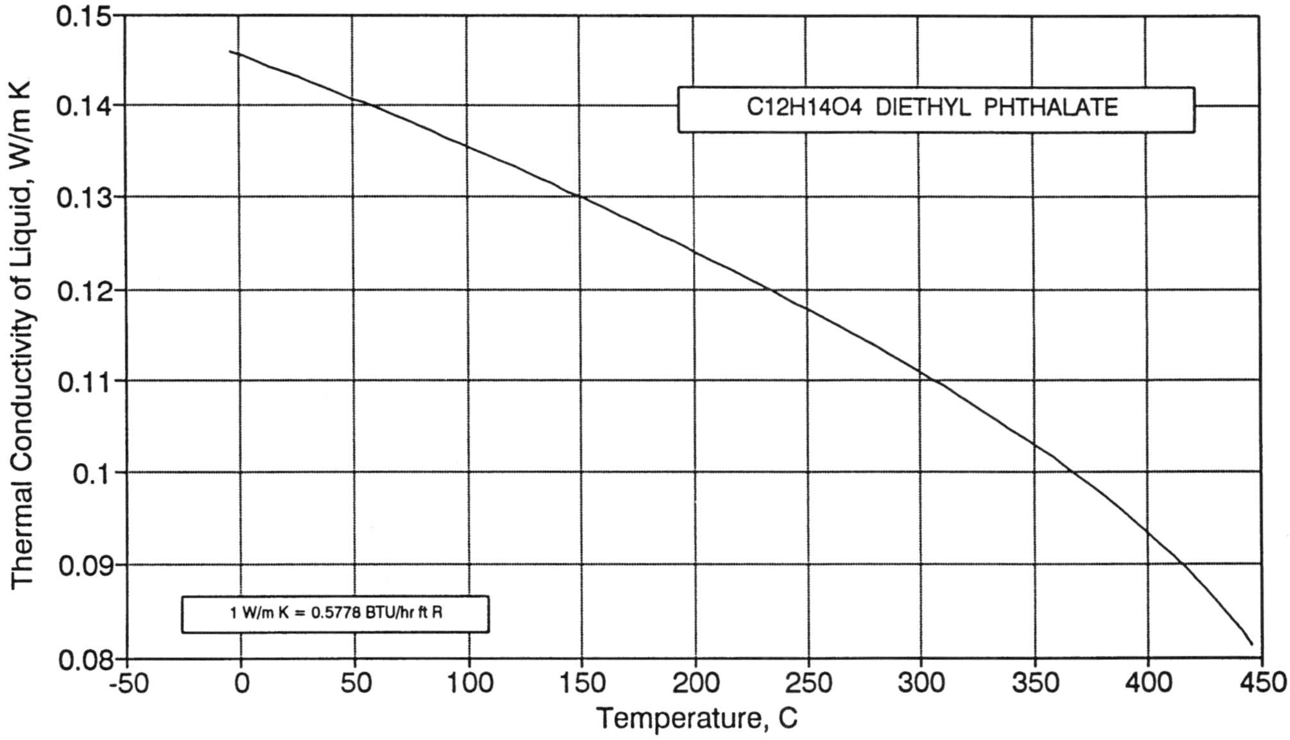
C12H14O4 DIETHYL PHTHALATE
1 W/m K = 0.5778 BTU/hr ft R
Thermal Conductivity of Liquid, W/m K
Temperature, C
0.15
0.14
0.13
0.12
0.11
0.1
0.09
0.08
-50
0
50
100
150
200
250
300
350
400
450

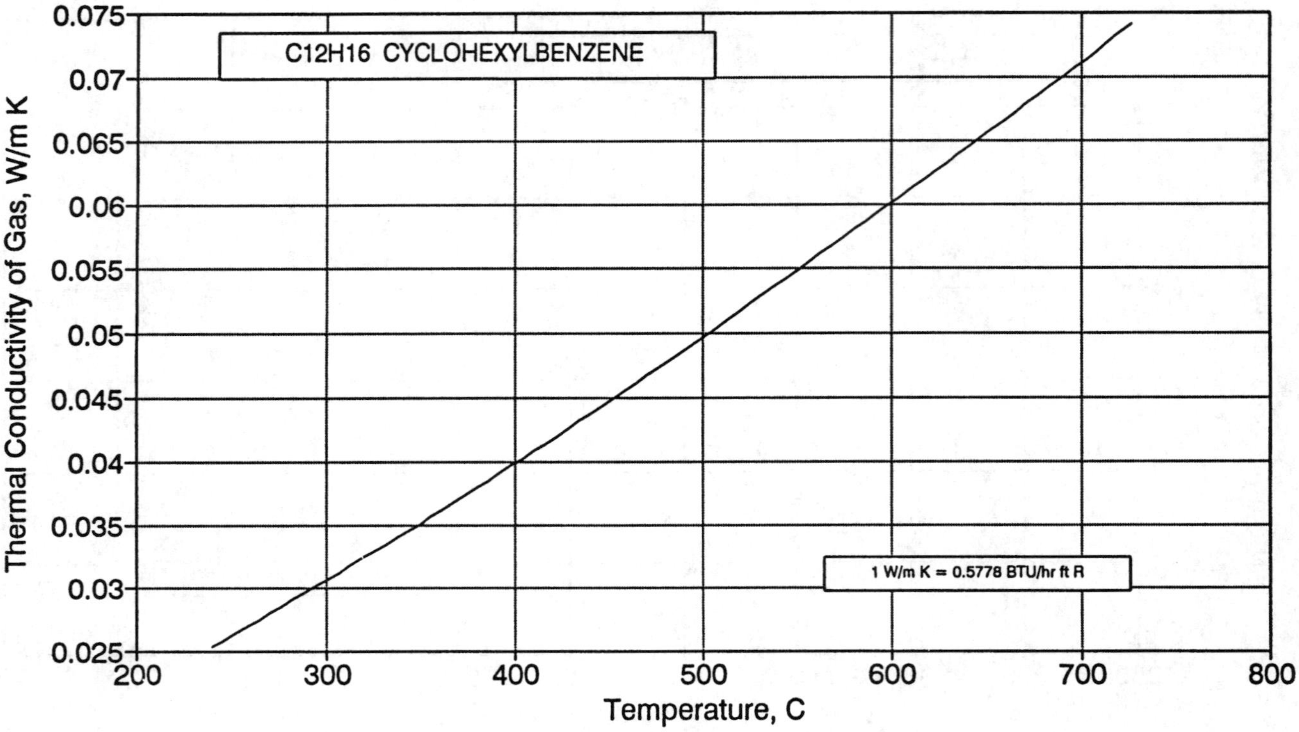
C12H16 CYCLOHEXYLBENZENE
1 W/m K = 0.5778 BTU/hr ft R
Thermal Conductivity of Gas, W/m K
0.075
0.07
0.065
0.06
0.055
0.05
0.045
0.04
0.035
0.03
0.025
200
300
400
500
600
700
800
Temperature, C

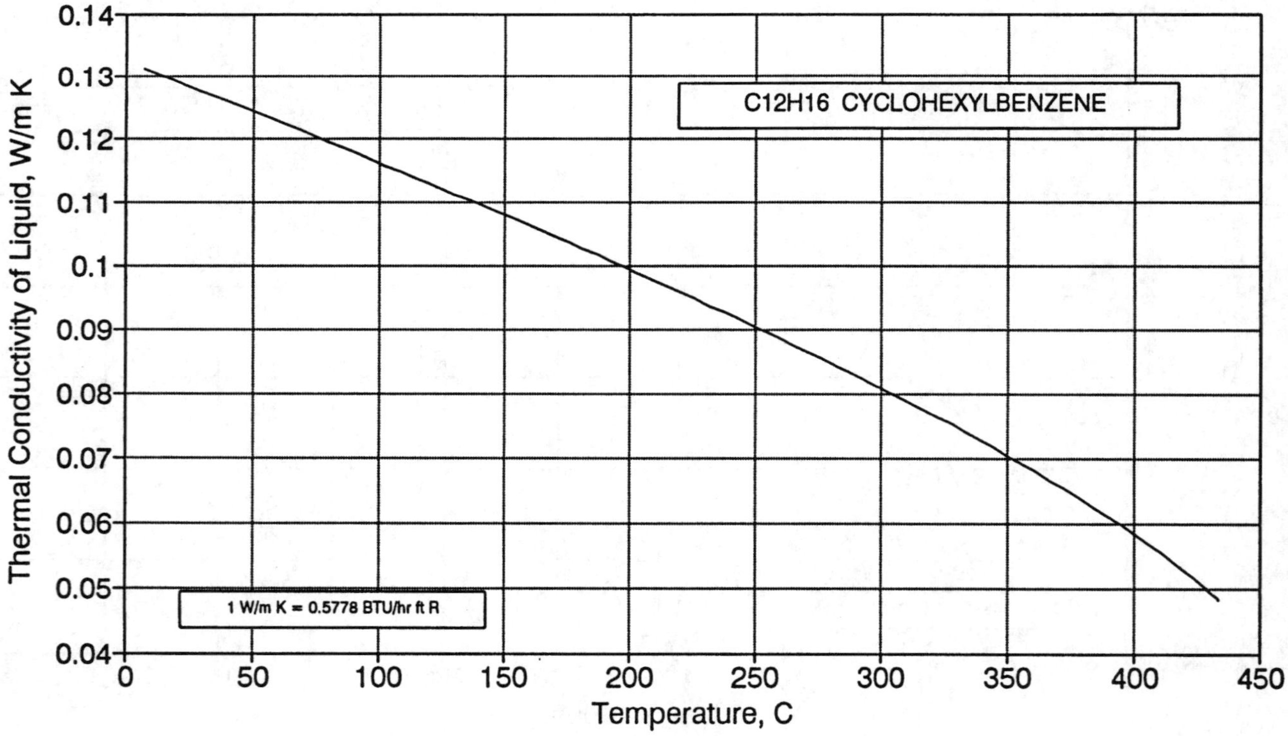
C12H16 CYCLOHEXYLBENZENE
1 W/m K = 0.5778 BTU/hr ft R
Thermal Conductivity of Liquid, W/m K
0.14
0.13
0.12
0.11
0.1
0.09
0.08
0.07
0.06
0.05
0.04
0
50
100
150
200
250
300
350
400
450
Temperature, C

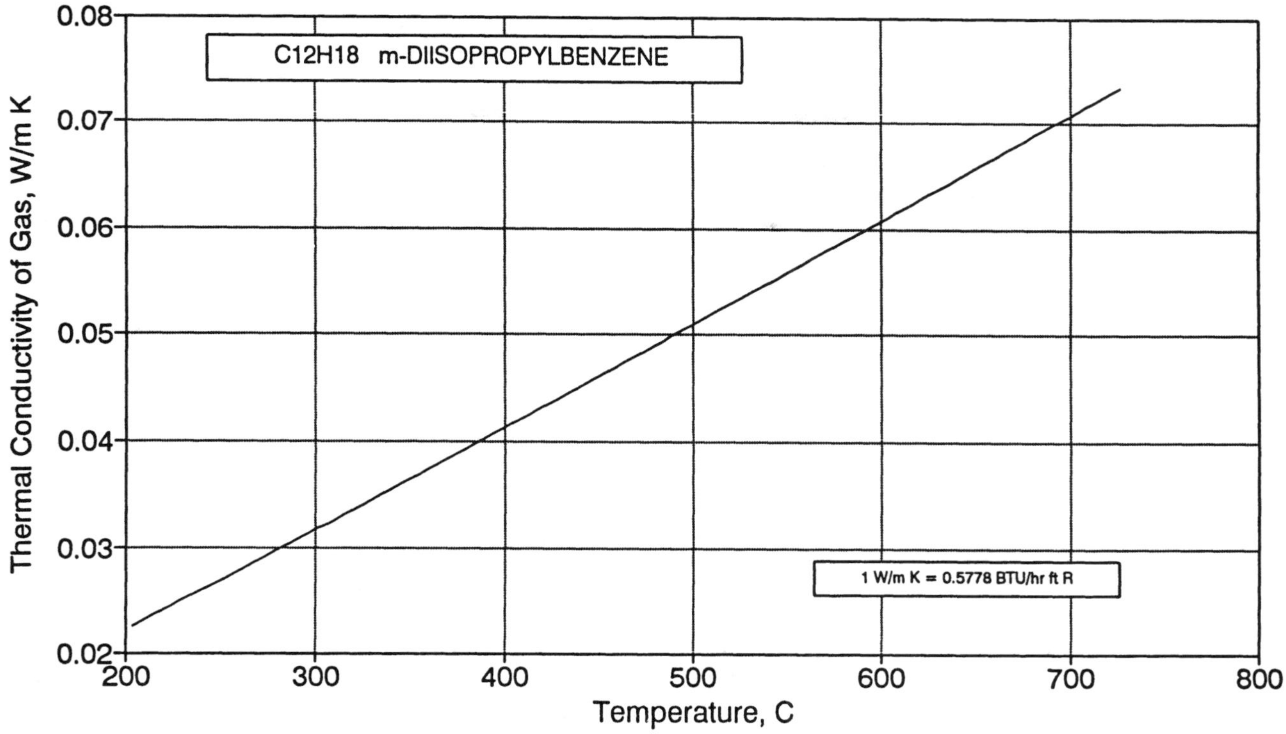
C12H18 m-DIISOPROPYLBENZENE
1 W/m K = 0.5778 BTU/hr ft R
Thermal Conductivity of Gas, W/m K
0.08
0.07
0.06
0.05
0.04
0.03
0.02
200
300
400
500
600
700
800
Temperature, C

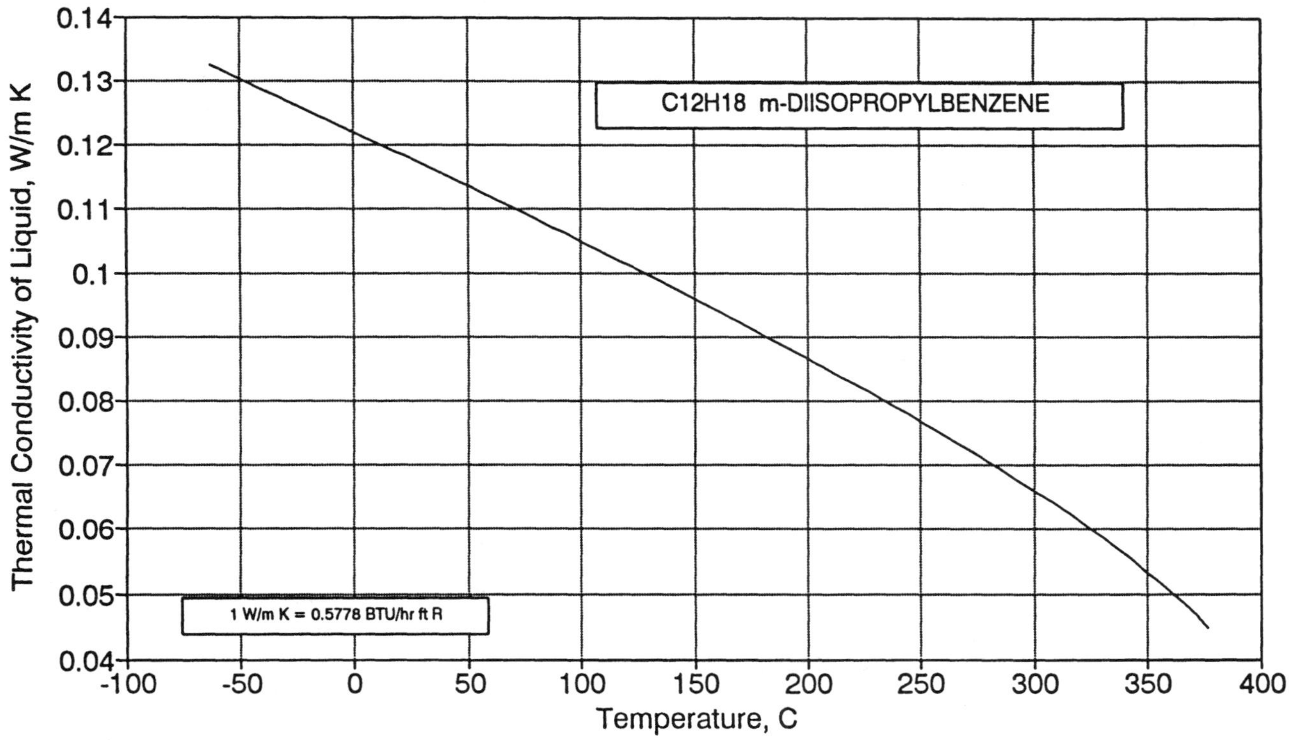
C12H18 m-DIISOPROPYLBENZENE
1 W/m K = 0.5778 BTU/hr ft R
Thermal Conductivity of Liquid, W/m K
0.14
0.13
0.12
0.11
0.1
0.09
0.08
0.07
0.06
0.05
0.04
-100
-50
0
50
100
150
200
250
300
350
400
Temperature, C

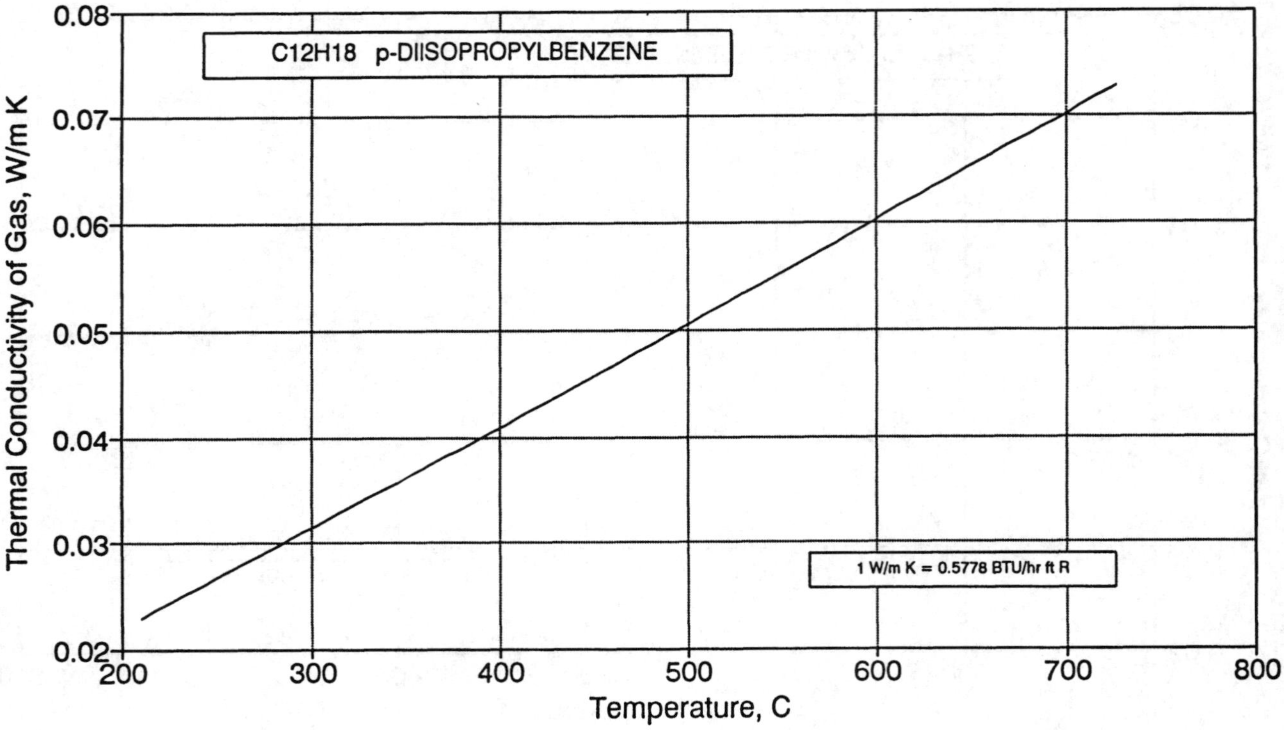
C12H18 p-DIISOPROPYLBENZENE
1 W/m K = 0.5778 BTU/hr ft R
Thermal Conductivity of Gas, W/m K
Temperature, C
0.08
0.07
0.06
0.05
0.04
0.03
0.02
200
300
400
500
600
700
800

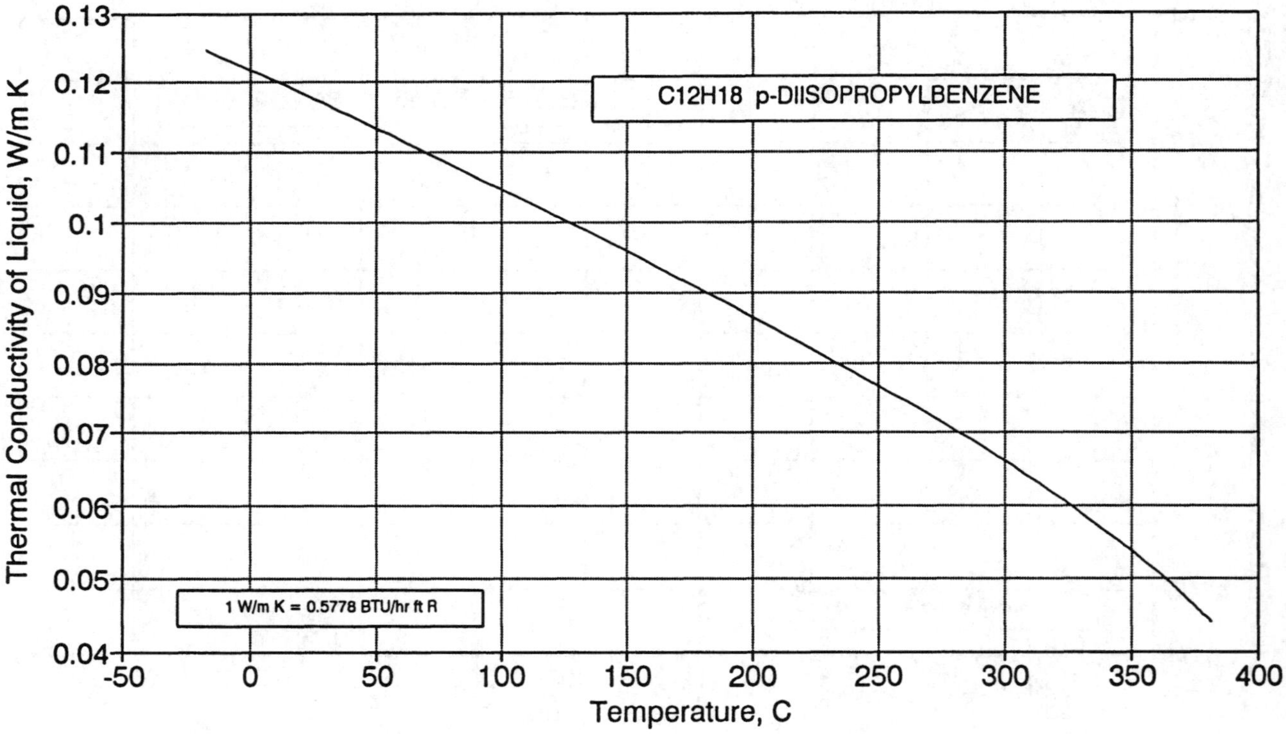
C12H18 p-DIISOPROPYLBENZENE
1 W/m K = 0.5778 BTU/hr ft R
Thermal Conductivity of Liquid, W/m K
Temperature, C
0.13
0.12
0.11
0.1
0.09
0.08
0.07
0.06
0.05
0.04
-50
0
50
100
150
200
250
300
350
400

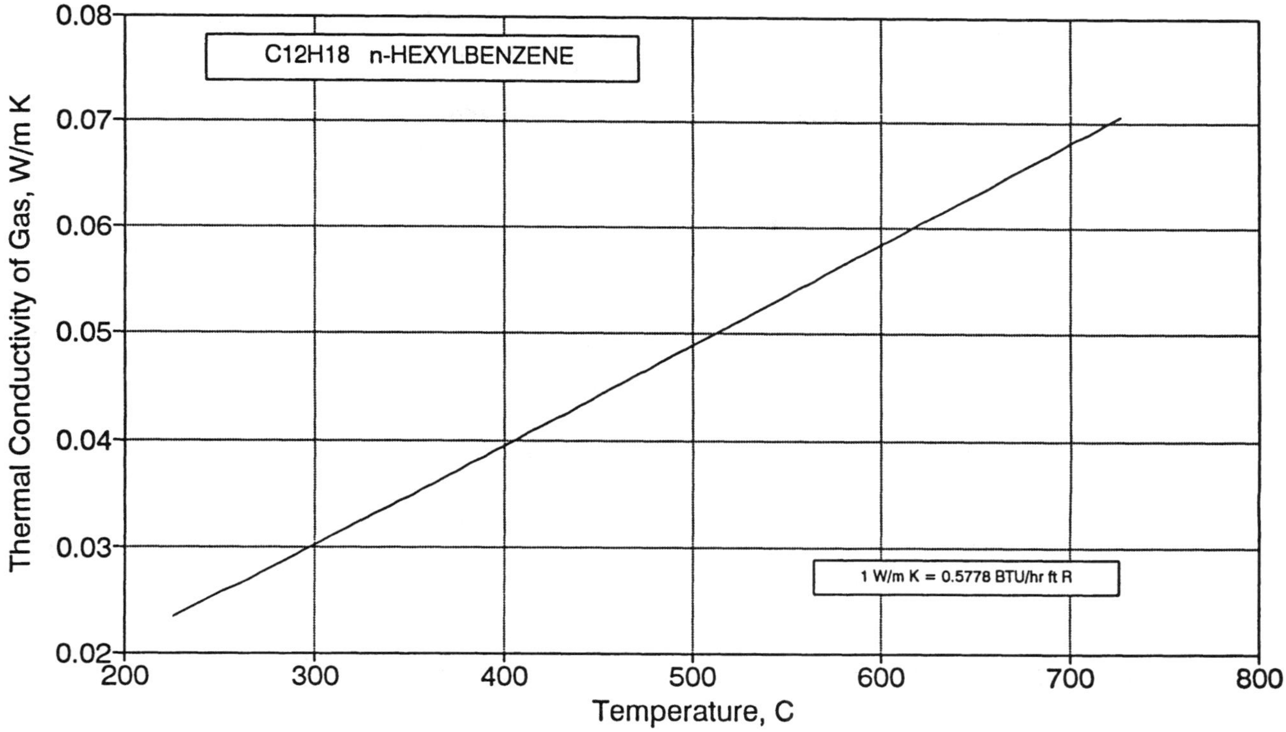
C12H18 n-HEXYLBENZENE
1 W/m K = 0.5778 BTU/hr ft R
Thermal Conductivity of Gas, W/m K
Temperature, C
0.08
0.07
0.06
0.05
0.04
0.03
0.02
200
300
400
500
600
700
800

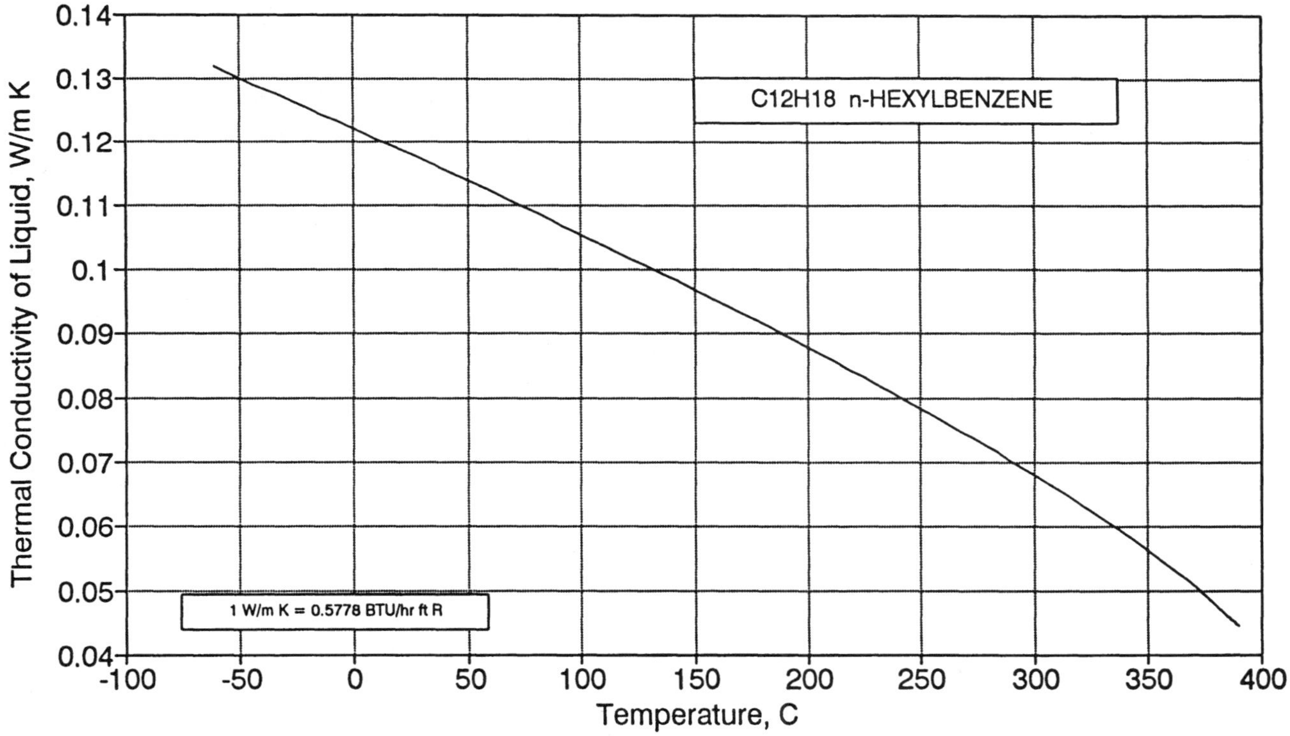
C12H18 n-HEXYLBENZENE
1 W/m K = 0.5778 BTU/hr ft R
Thermal Conductivity of Liquid, W/m K
Temperature, C
0.14
0.13
0.12
0.11
0.1
0.09
0.08
0.07
0.06
0.05
0.04
-100
-50
0
50
100
150
200
250
300
350
400

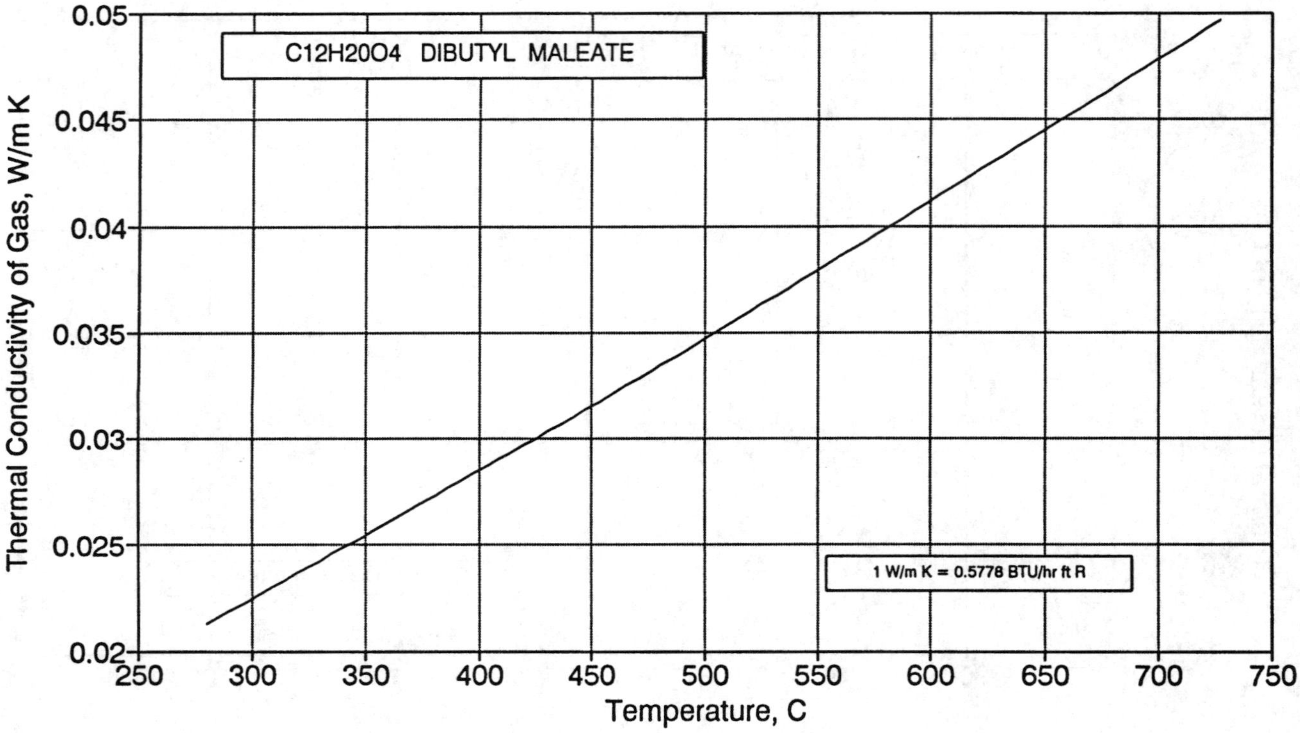
C12H20O4 DIBUTYL MALEATE
1 W/m K = 0.5778 BTU/hr ft R
Thermal Conductivity of Gas, W/m K
0.05
0.045
0.04
0.035
0.03
0.025
0.02
250
300
350
400
450
500
550
600
650
700
750
Temperature, C

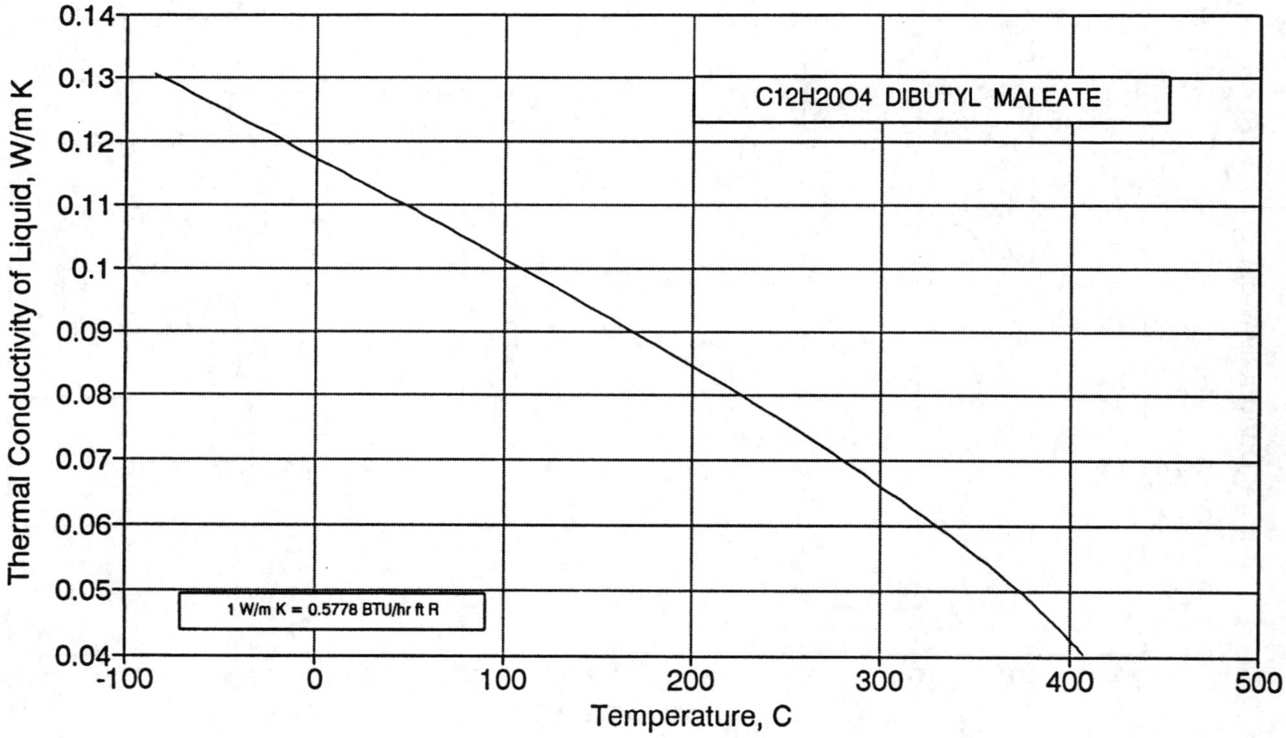
C12H20O4 DIBUTYL MALEATE
1 W/m K = 0.5778 BTU/hr ft R
Thermal Conductivity of Liquid, W/m K
0.14
0.13
0.12
0.11
0.1
0.09
0.08
0.07
0.06
0.05
0.04
-100
0
100
200
300
400
500
Temperature, C

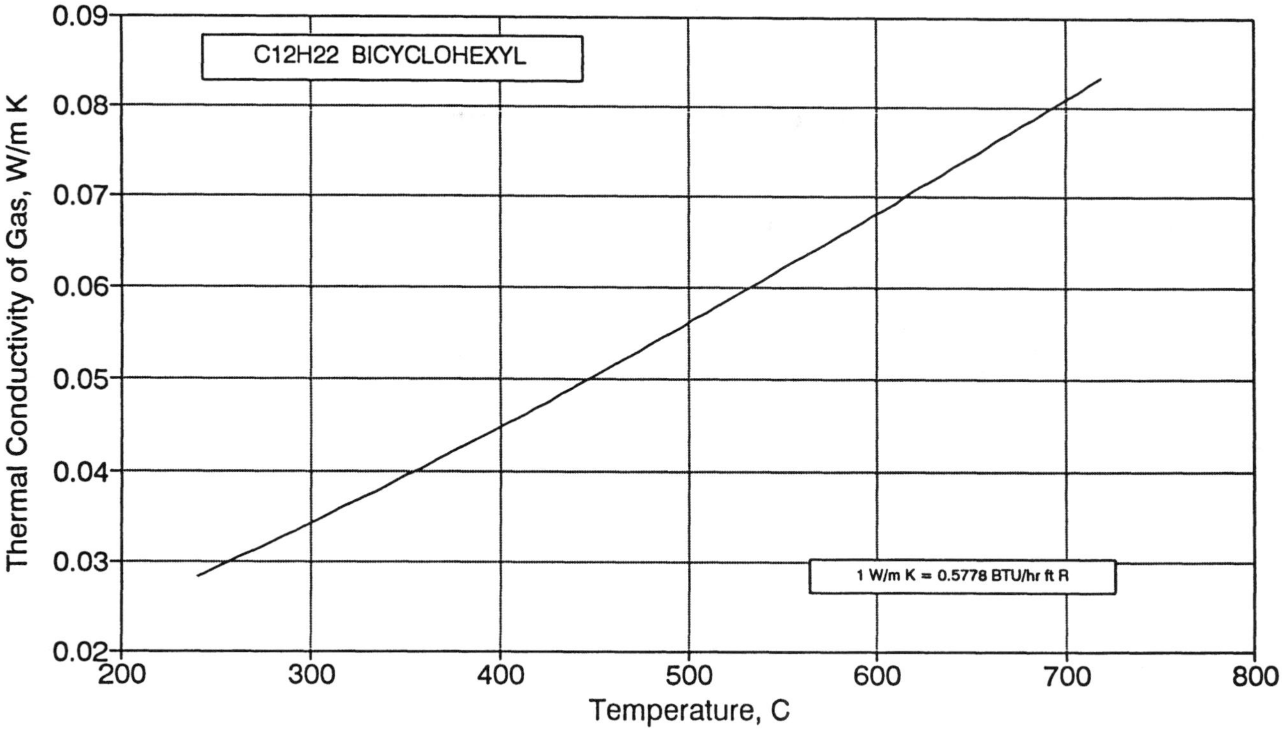
C12H22 BICYCLOHEXYL
1 W/m K = 0.5778 BTU/hr ft R
Thermal Conductivity of Gas, W/m K
0.09
0.08
0.07
0.06
0.05
0.04
0.03
0.02
200
300
400
500
600
700
800
Temperature, C

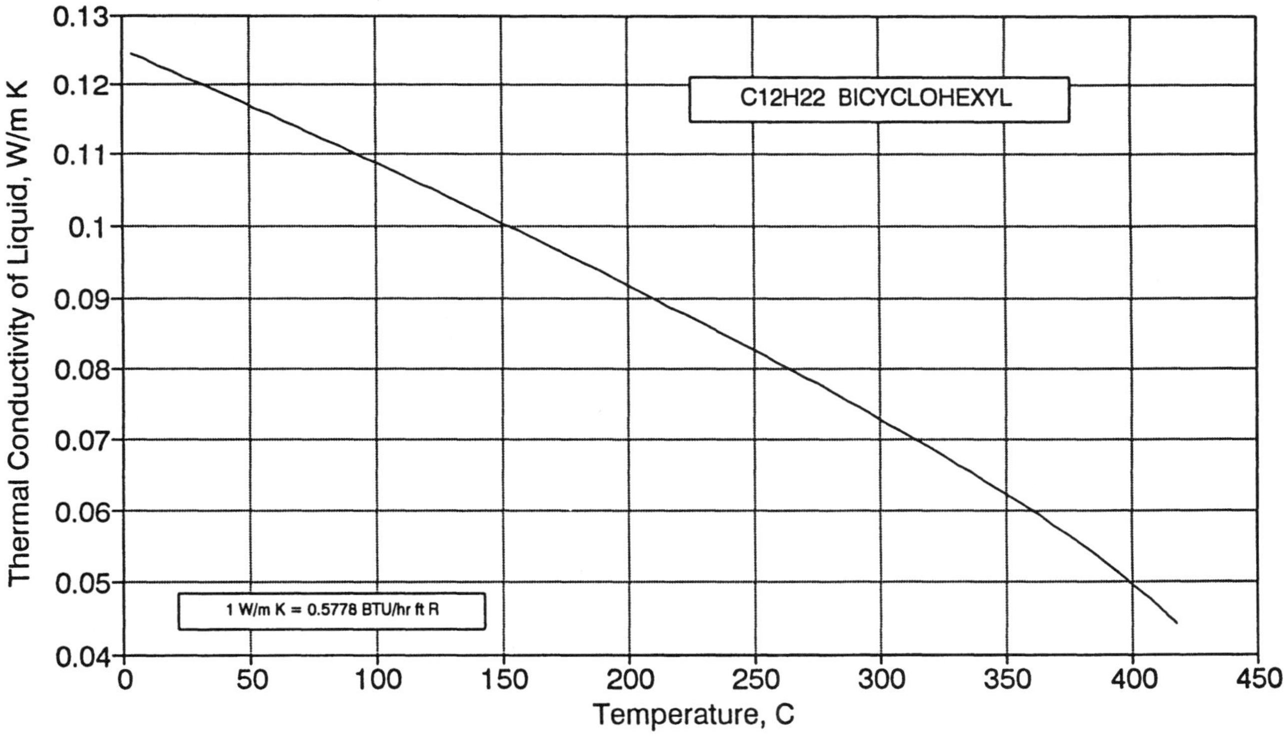
C12H22 BICYCLOHEXYL
1 W/m K = 0.5778 BTU/hr ft R
Thermal Conductivity of Liquid, W/m K
0.13
0.12
0.11
0.1
0.09
0.08
0.07
0.06
0.05
0.04
0
50
100
150
200
250
300
350
400
450
Temperature, C

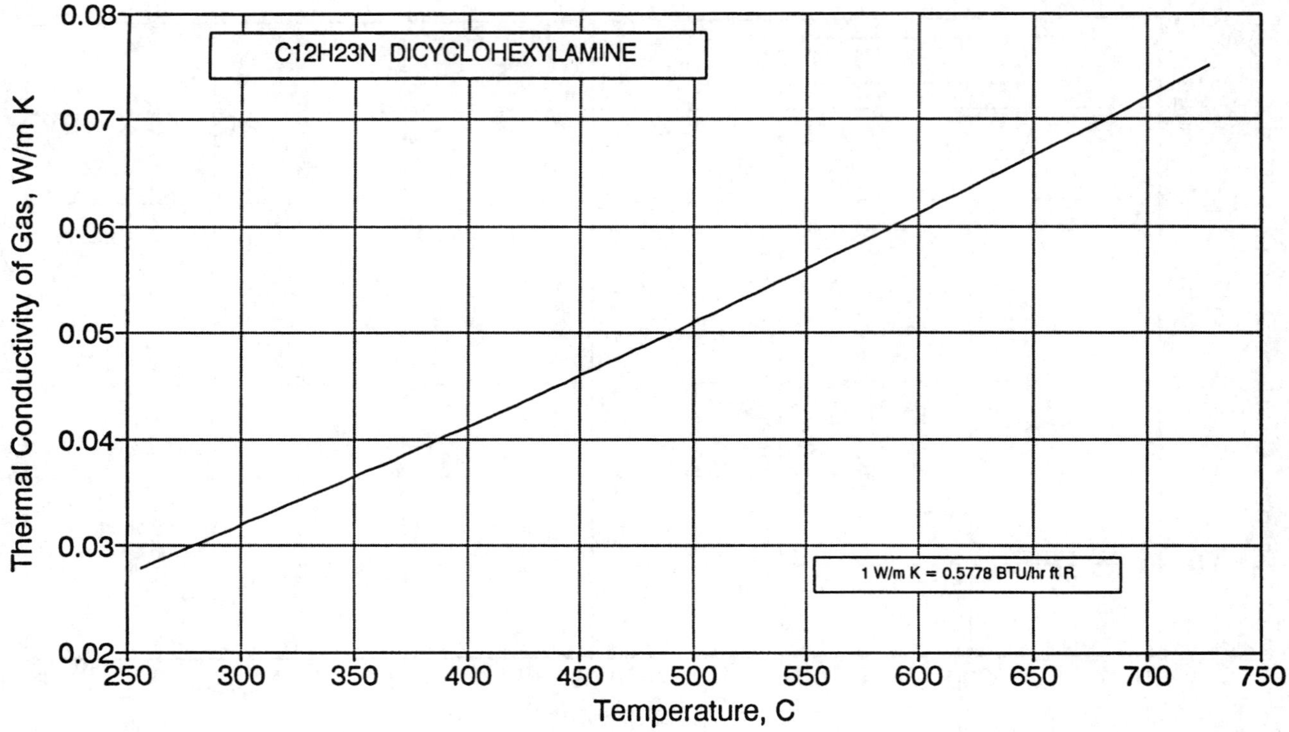
C12H23N DICYCLOHEXYLAMINE
1 W/m K = 0.5778 BTU/hr ft R
Thermal Conductivity of Gas, W/m K
0.08
0.07
0.06
0.05
0.04
0.03
0.02
250
300
350
400
450
500
550
600
650
700
750
Temperature, C

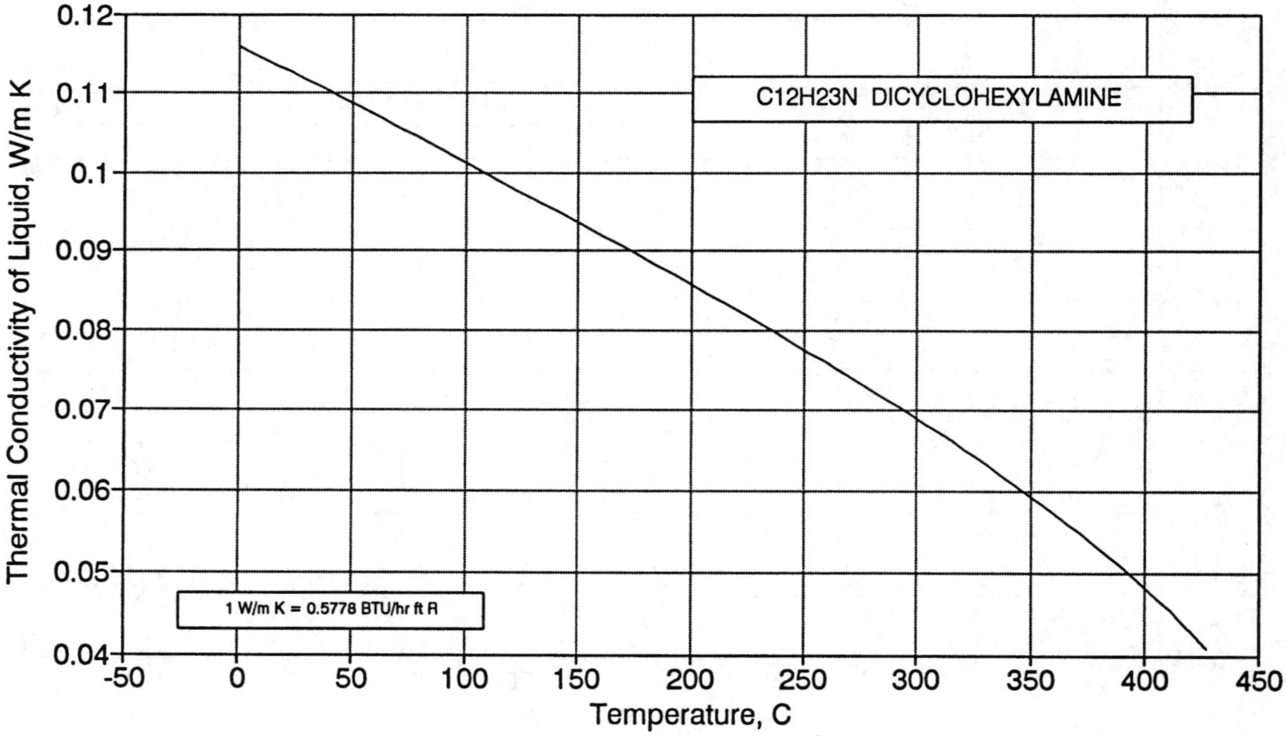
C12H23N DICYCLOHEXYLAMINE
1 W/m K = 0.5778 BTU/hr ft R
Thermal Conductivity of Liquid, W/m K
0.12
0.11
0.1
0.09
0.08
0.07
0.06
0.05
0.04
-50
0
50
100
150
200
250
300
350
400
450
Temperature, C

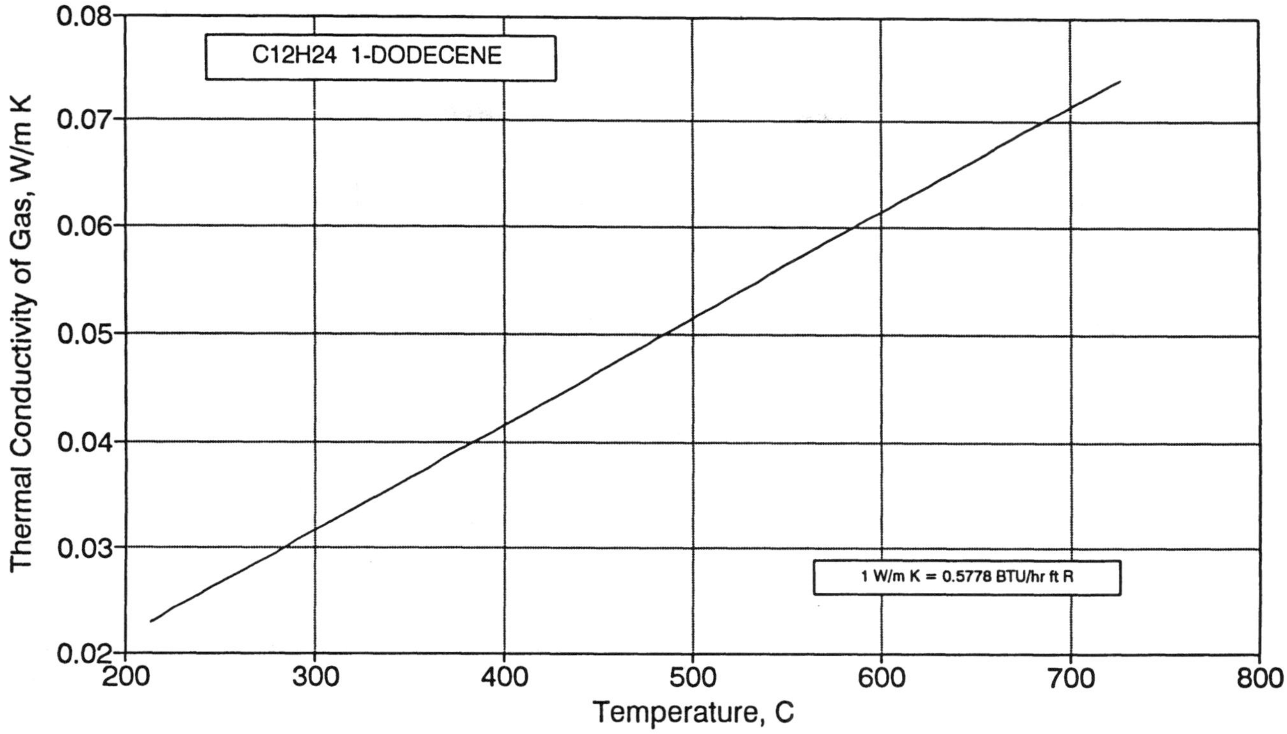
C12H24 1-DODECENE
1 W/m K = 0.5778 BTU/hr ft R
Thermal Conductivity of Gas, W/m K
Temperature, C
0.08
0.07
0.06
0.05
0.04
0.03
0.02
200
300
400
500
600
700
800

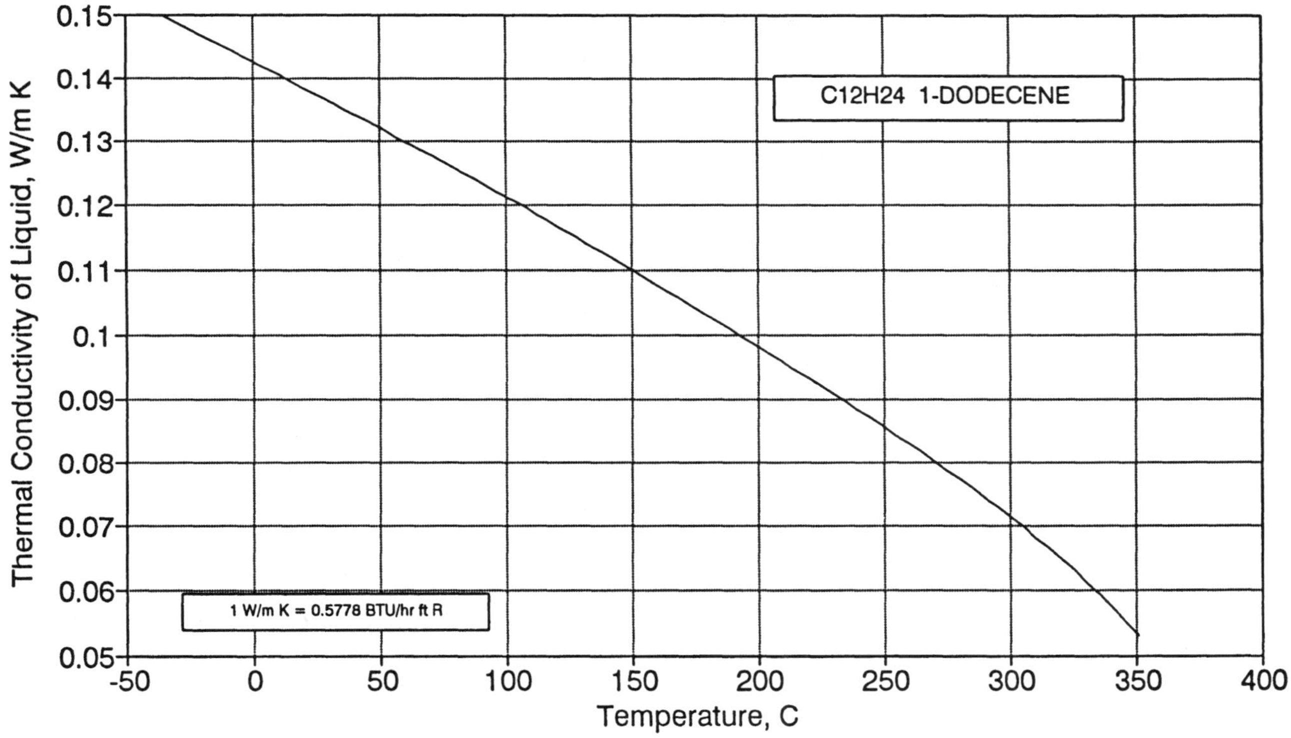
C12H24 1-DODECENE
1 W/m K = 0.5778 BTU/hr ft R
Thermal Conductivity of Liquid, W/m K
Temperature, C
0.15
0.14
0.13
0.12
0.11
0.1
0.09
0.08
0.07
0.06
0.05
-50
0
50
100
150
200
250
300
350
400

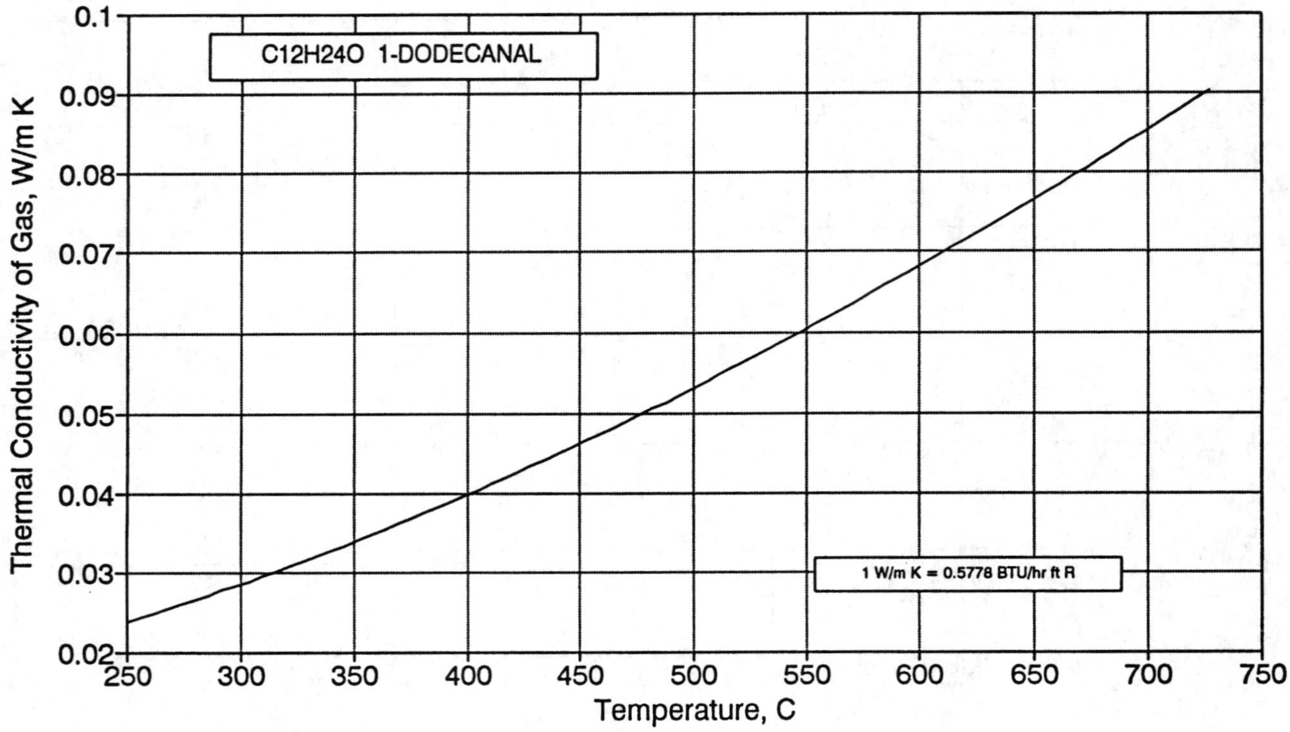
C12H24O 1-DODECANAL
1 W/m K = 0.5778 BTU/hr ft R
Thermal Conductivity of Gas, W/m K
Temperature, C
0.1
0.09
0.08
0.07
0.06
0.05
0.04
0.03
0.02
250
300
350
400
450
500
550
600
650
700
750

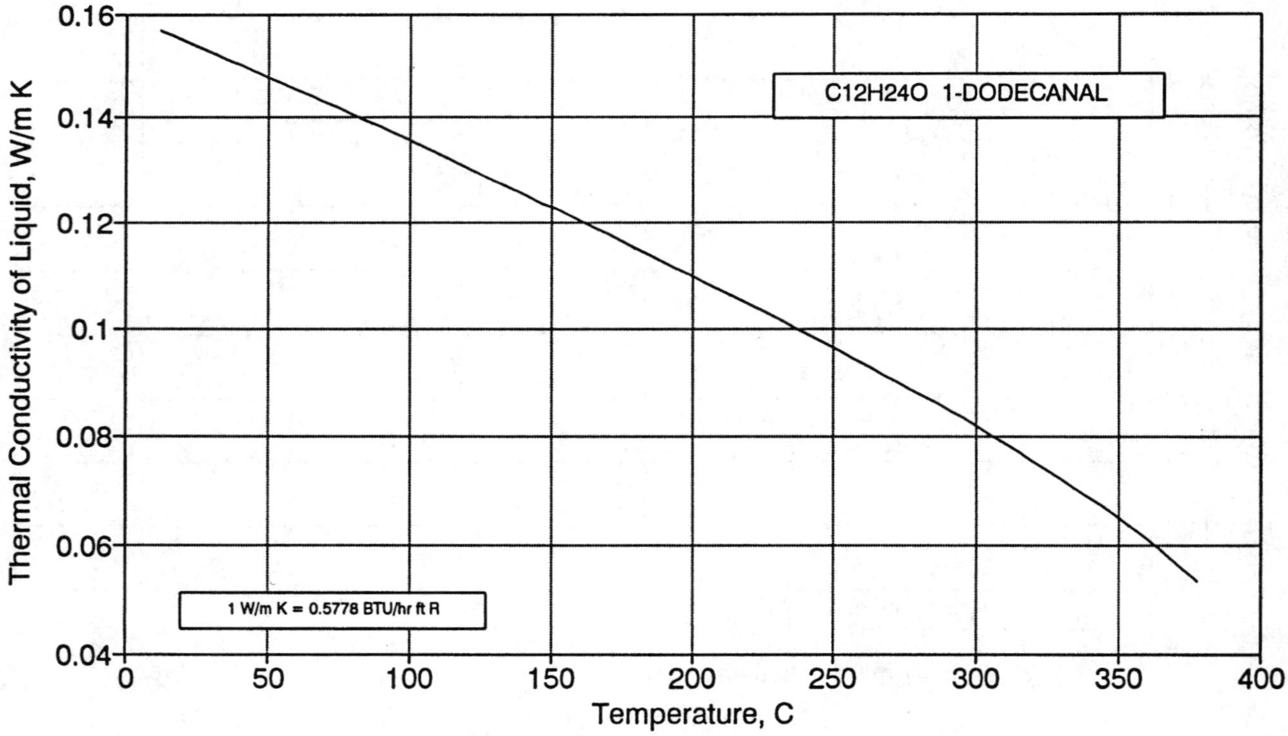
C12H24O 1-DODECANAL
1 W/m K = 0.5778 BTU/hr ft R
Thermal Conductivity of Liquid, W/m K
Temperature, C
0.16
0.14
0.12
0.1
0.08
0.06
0.04
0
50
100
150
200
250
300
350
400

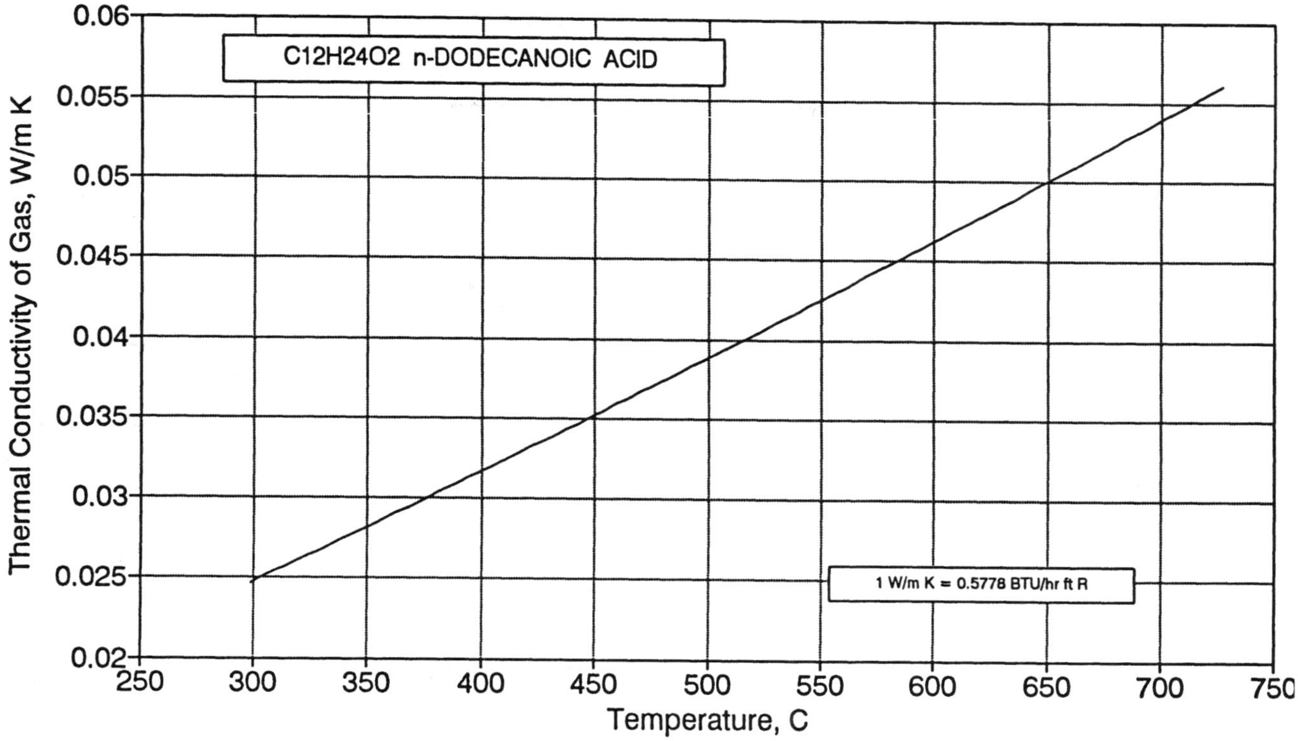
C12H24O2 n-DODECANOIC ACID
1 W/m K = 0.5778 BTU/hr ft R
Thermal Conductivity of Gas, W/m K
0.06
0.055
0.05
0.045
0.04
0.035
0.03
0.025
0.02
250
300
350
400
450
500
550
600
650
700
750
Temperature, C

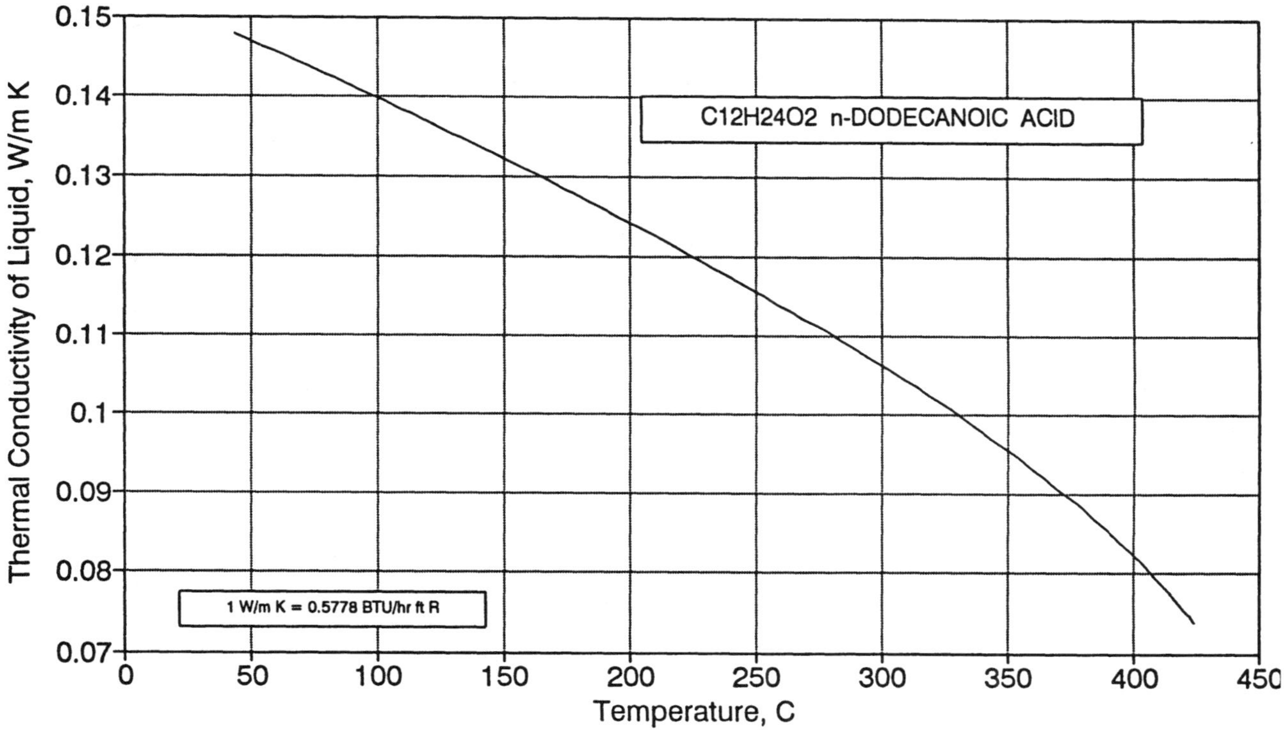
C12H24O2 n-DODECANOIC ACID
1 W/m K = 0.5778 BTU/hr ft R
Thermal Conductivity of Liquid, W/m K
0.15
0.14
0.13
0.12
0.11
0.1
0.09
0.08
0.07
0
50
100
150
200
250
300
350
400
450
Temperature, C

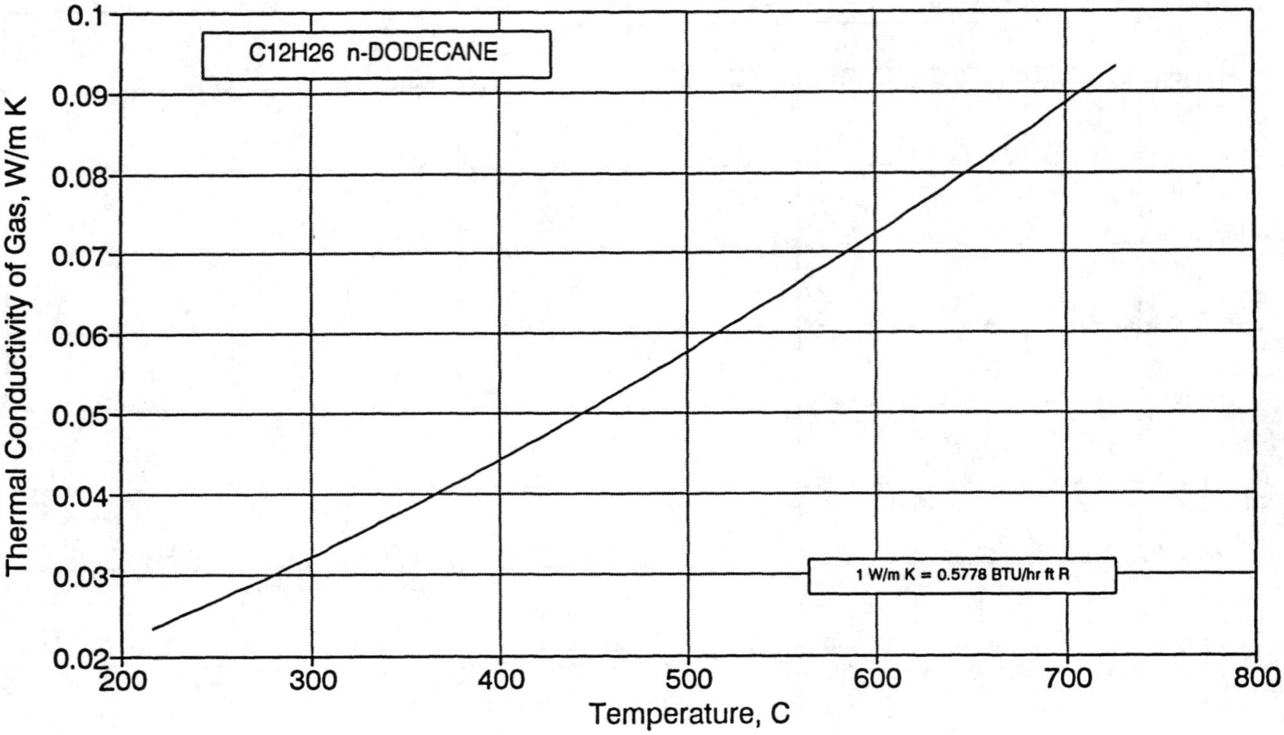
C12H26 n-DODECANE
Thermal Conductivity of Gas, W/m K
0.1
0.09
0.08
0.07
0.06
0.05
0.04
0.03
0.02
200
300
400
500
600
700
800
Temperature, C
1 W/m K = 0.5778 BTU/hr ft R

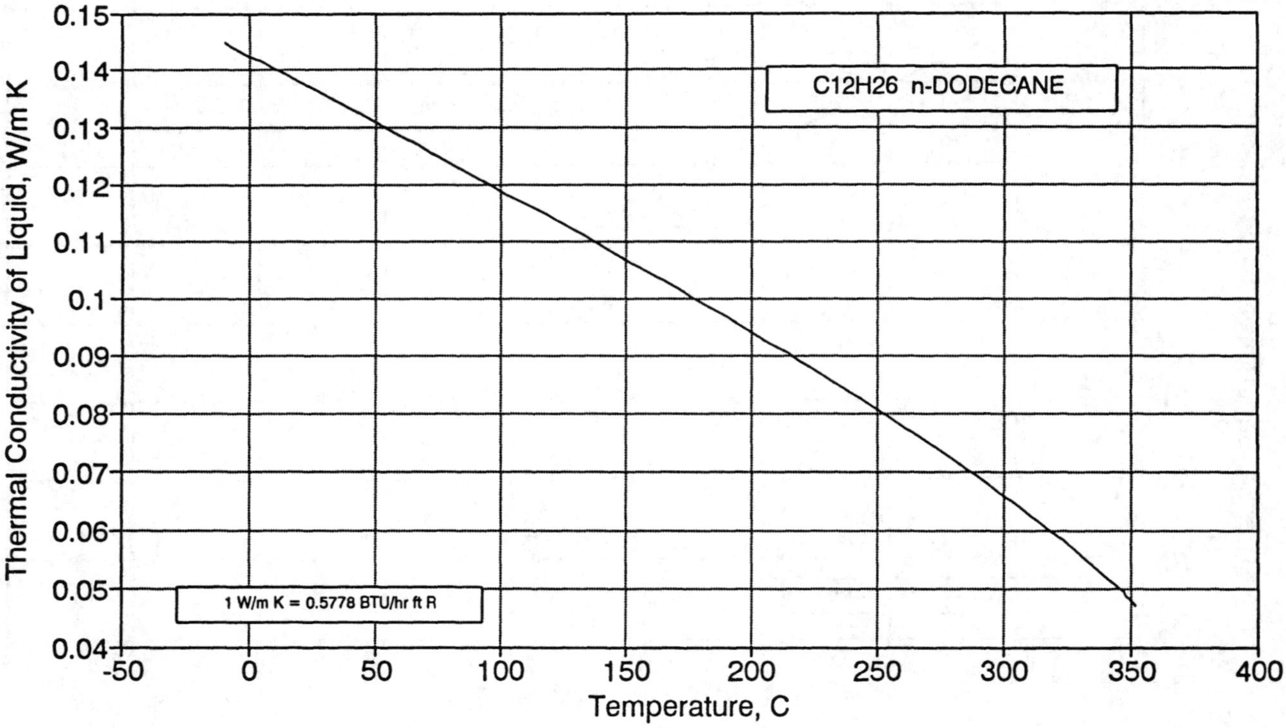
C12H26 n-DODECANE
Thermal Conductivity of Liquid, W/m K
0.15
0.14
0.13
0.12
0.11
0.1
0.09
0.08
0.07
0.06
0.05
0.04
-50
0
50
100
150
200
250
300
350
400
Temperature, C
1 W/m K = 0.5778 BTU/hr ft R

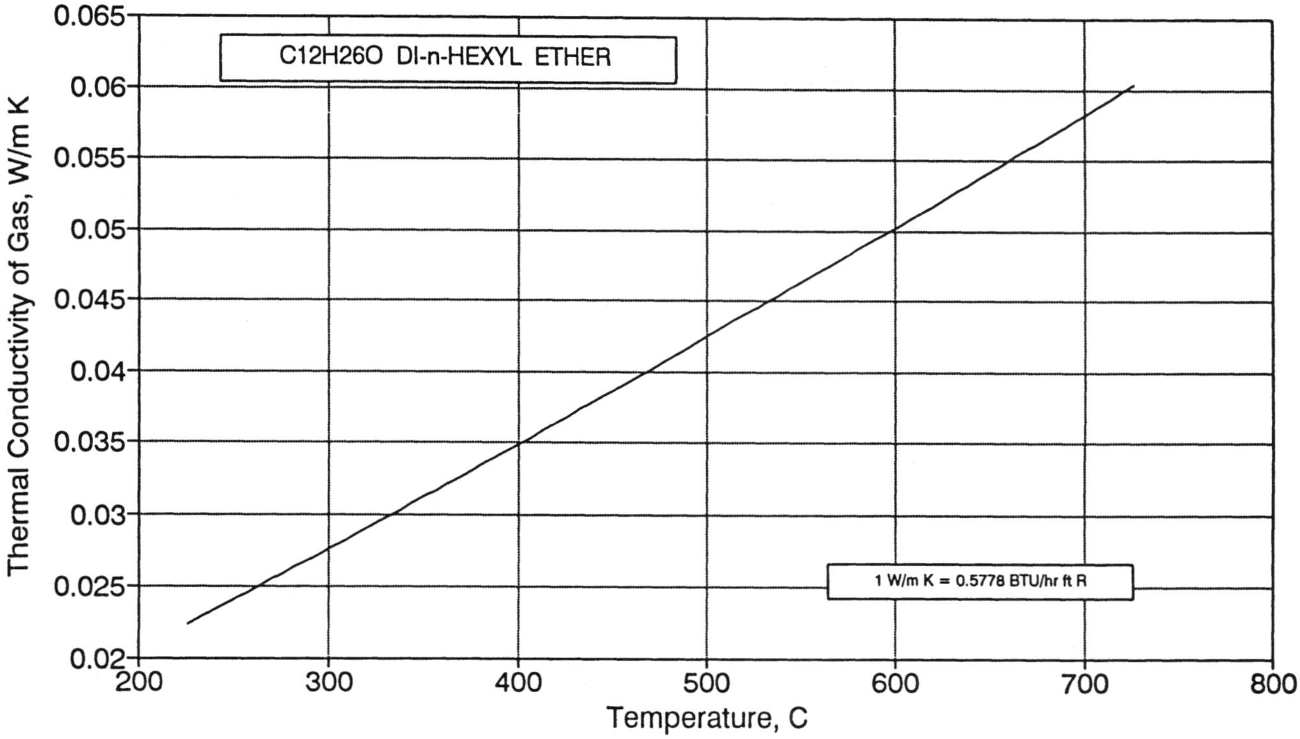
C12H26O DI-n-HEXYL ETHER
1 W/m K = 0.5778 BTU/hr ft R
Thermal Conductivity of Gas, W/m K
0.065
0.06
0.055
0.05
0.045
0.04
0.035
0.03
0.025
0.02
200
300
400
500
600
700
800
Temperature, C

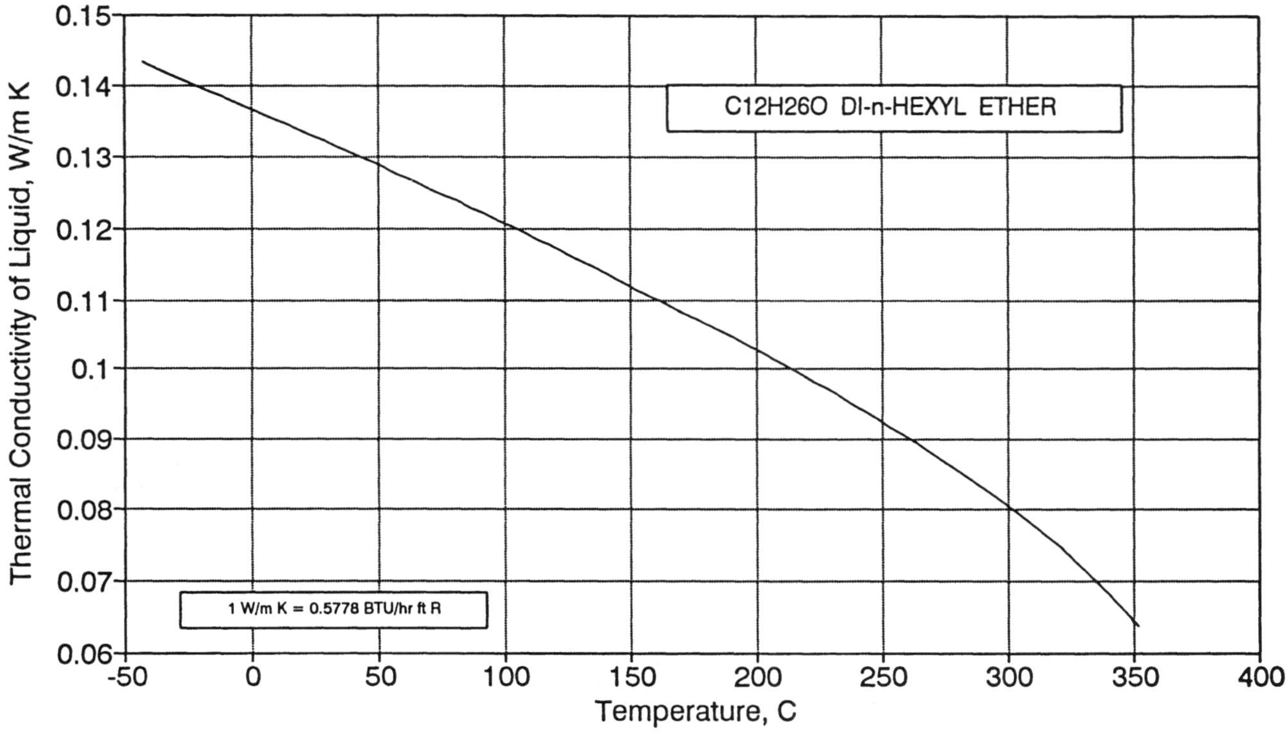
C12H26O DI-n-HEXYL ETHER
1 W/m K = 0.5778 BTU/hr ft R
Thermal Conductivity of Liquid, W/m K
0.15
0.14
0.13
0.12
0.11
0.1
0.09
0.08
0.07
0.06
-50
0
50
100
150
200
250
300
350
400
Temperature, C

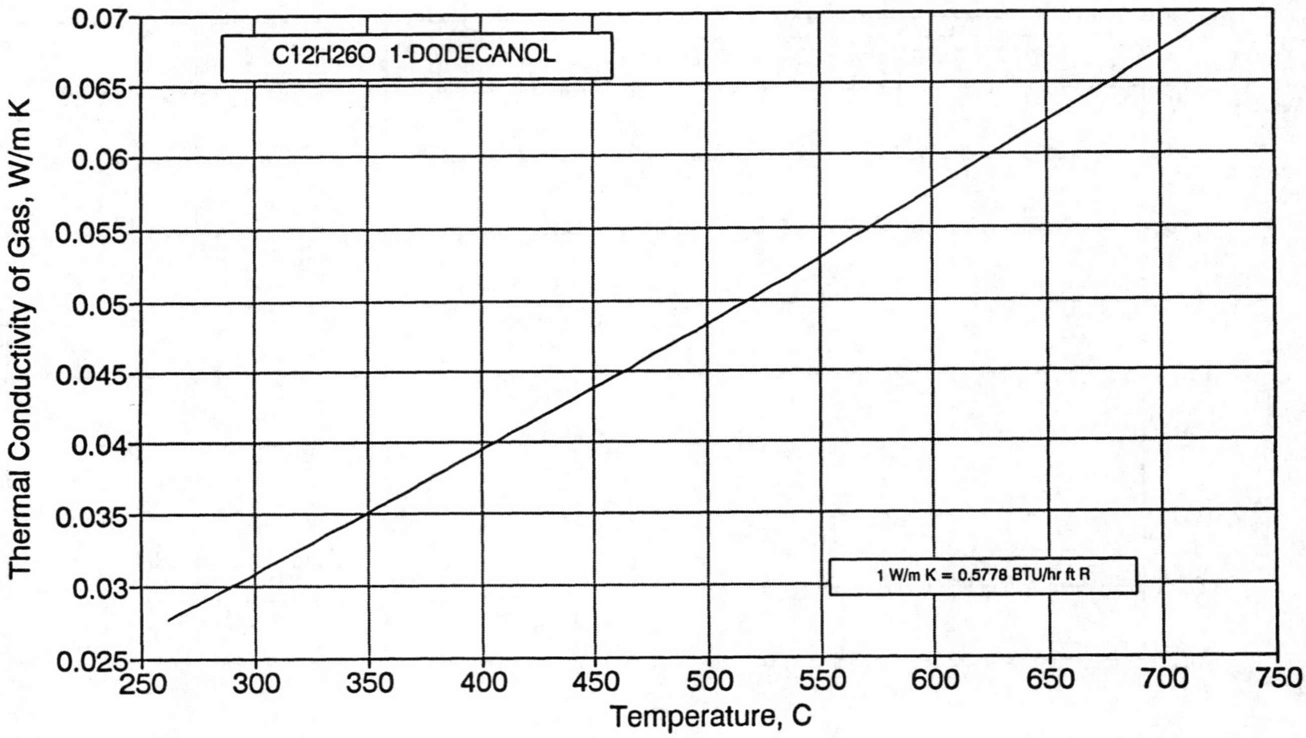
C12H26O 1-DODECANOL
1 W/m K = 0.5778 BTU/hr ft R
Thermal Conductivity of Gas, W/m K
0.07
0.065
0.06
0.055
0.05
0.045
0.04
0.035
0.03
0.025
250
300
350
400
450
500
550
600
650
700
750
Temperature, C

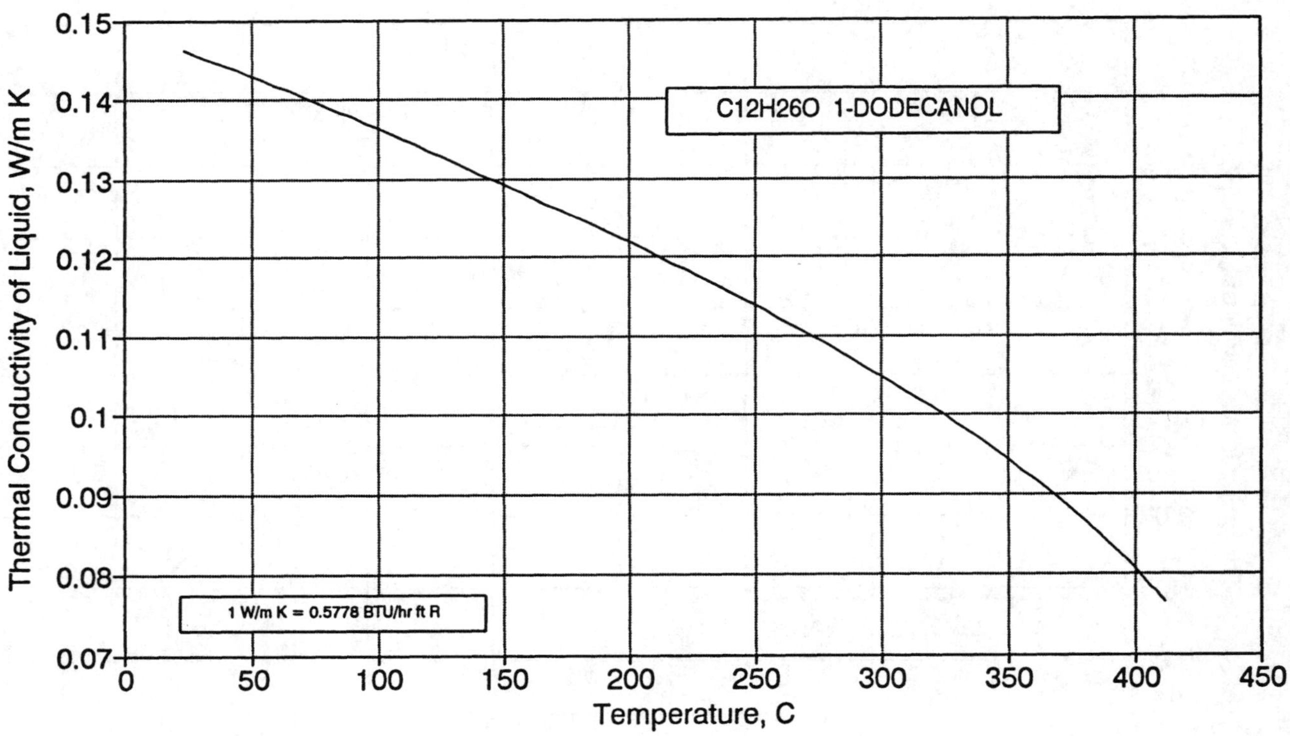
C12H26O 1-DODECANOL
1 W/m K = 0.5778 BTU/hr ft R
Thermal Conductivity of Liquid, W/m K
0.15
0.14
0.13
0.12
0.11
0.1
0.09
0.08
0.07
0
50
100
150
200
250
300
350
400
450
Temperature, C

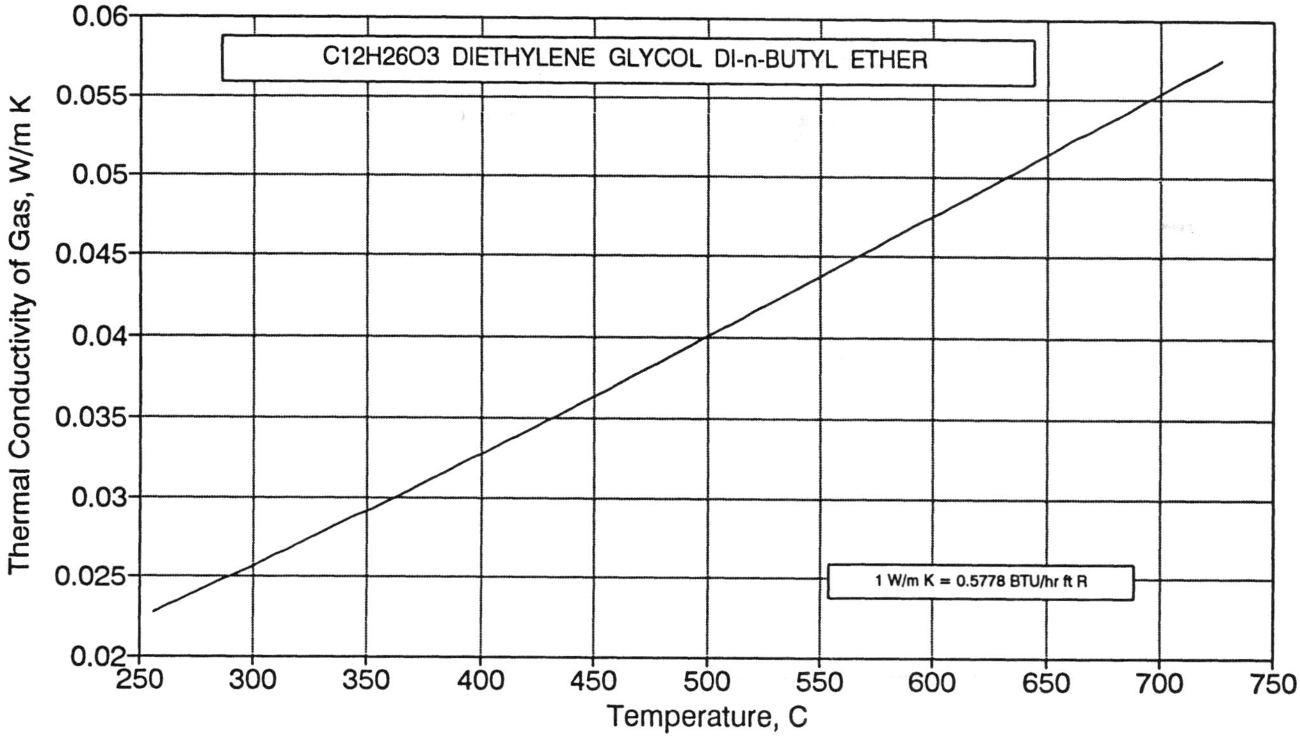
C12H26O3 DIETHYLENE GLYCOL DI-n-BUTYL ETHER
Thermal Conductivity of Gas, W/m K
0.06
0.055
0.05
0.045
0.04
0.035
0.03
0.025
0.02
250
300
350
400
450
500
550
600
650
700
750
Temperature, C
1 W/m K = 0.5778 BTU/hr ft R

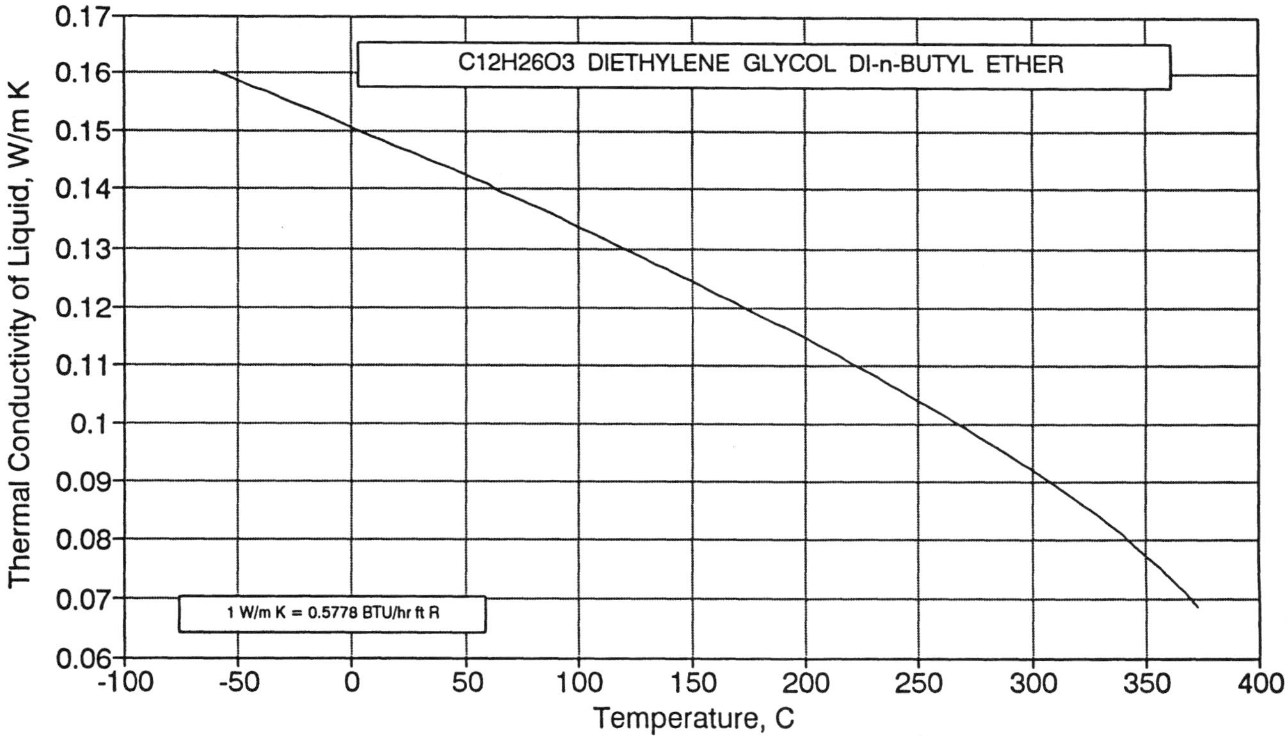
C12H26O3 DIETHYLENE GLYCOL DI-n-BUTYL ETHER
Thermal Conductivity of Liquid, W/m K
0.17
0.16
0.15
0.14
0.13
0.12
0.11
0.1
0.09
0.08
0.07
0.06
-100
-50
0
50
100
150
200
250
300
350
400
Temperature, C
1 W/m K = 0.5778 BTU/hr ft R

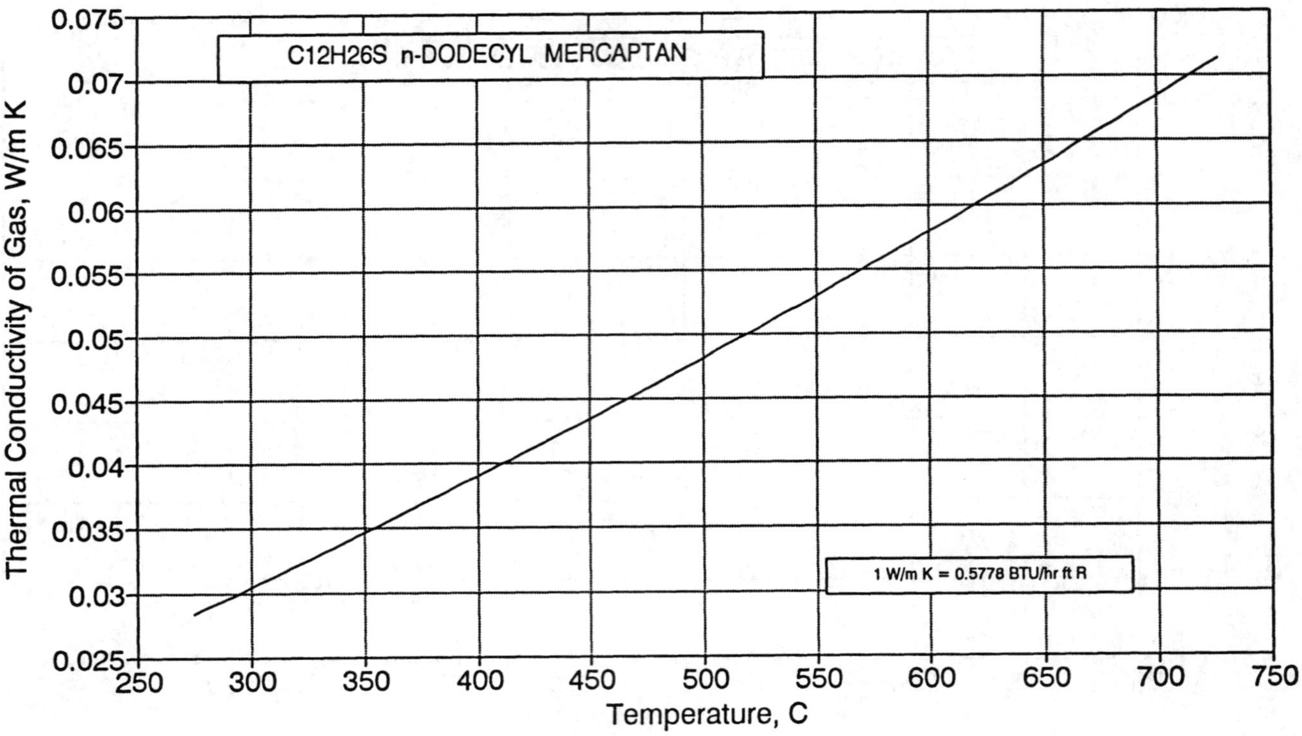
C12H26S n-DODECYL MERCAPTAN
1 W/m K = 0.5778 BTU/hr ft R
Thermal Conductivity of Gas, W/m K
Temperature, C
0.075
0.07
0.065
0.06
0.055
0.05
0.045
0.04
0.035
0.03
0.025
250
300
350
400
450
500
550
600
650
700
750

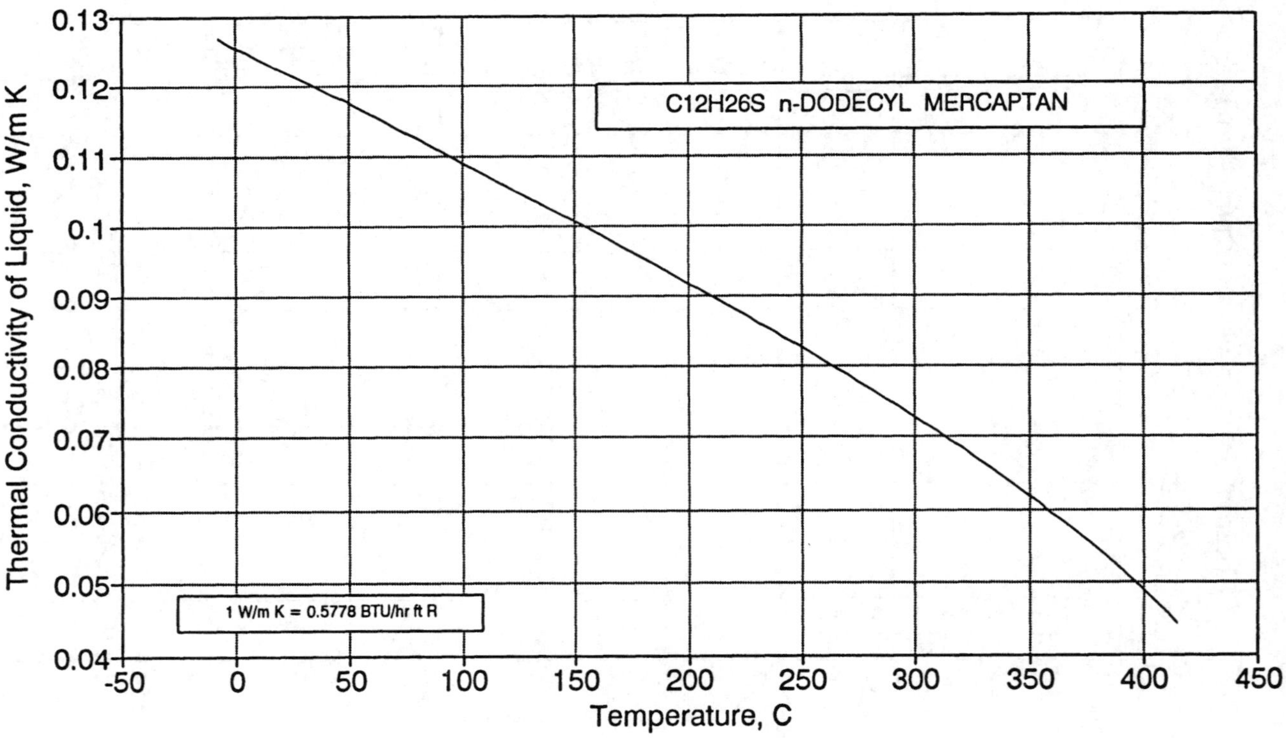
C12H26S n-DODECYL MERCAPTAN
1 W/m K = 0.5778 BTU/hr ft R
Thermal Conductivity of Liquid, W/m K
Temperature, C
0.13
0.12
0.11
0.1
0.09
0.08
0.07
0.06
0.05
0.04
-50
0
50
100
150
200
250
300
350
400
450

$C_{12}H_{27}BO_3$ TRI-n-BUTYL BORATE

Gas thermal conductivity data are not available .

TF , K = 203.15

TB , K = 506.65

$C_{12}H_{27}BO_3$ TRI-n-BUTYL BORATE

Liquid thermal conductivity data are not available .

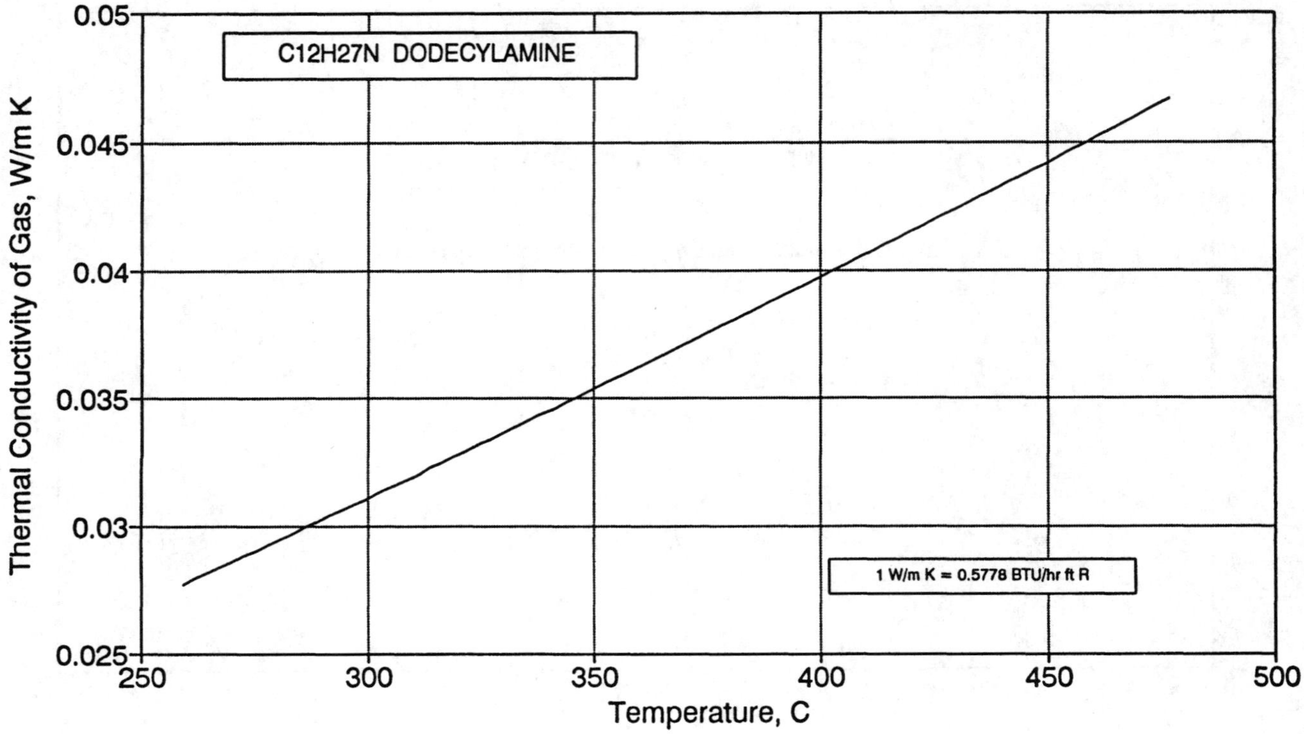
C12H27N DODECYLAMINE
1 W/m K = 0.5778 BTU/hr ft R
Thermal Conductivity of Gas, W/m K
Temperature, C
0.05
0.045
0.04
0.035
0.03
0.025
250
300
350
400
450
500

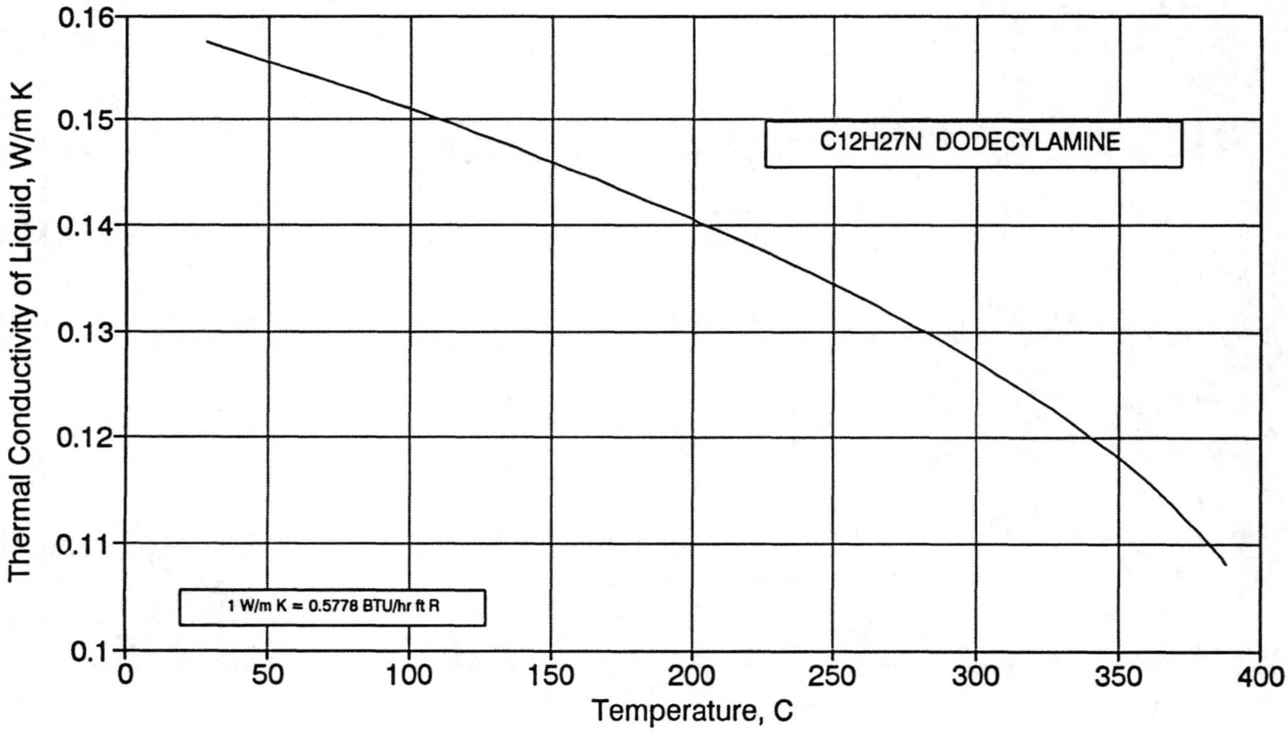
C12H27N DODECYLAMINE
1 W/m K = 0.5778 BTU/hr ft R
Thermal Conductivity of Liquid, W/m K
Temperature, C
0.16
0.15
0.14
0.13
0.12
0.11
0.1
0
50
100
150
200
250
300
350
400

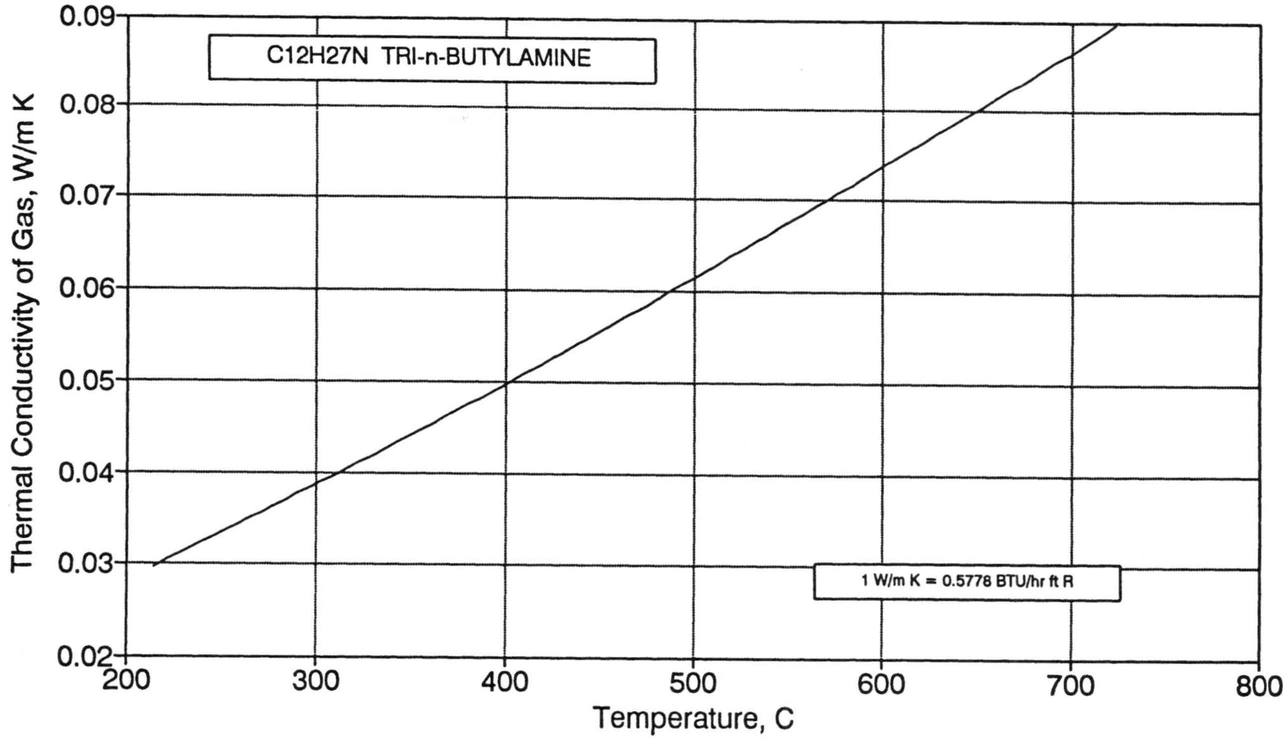
C12H27N TRI-n-BUTYLAMINE
1 W/m K = 0.5778 BTU/hr ft R
Thermal Conductivity of Gas, W/m K
0.09
0.08
0.07
0.06
0.05
0.04
0.03
0.02
200
300
400
500
600
700
800
Temperature, C

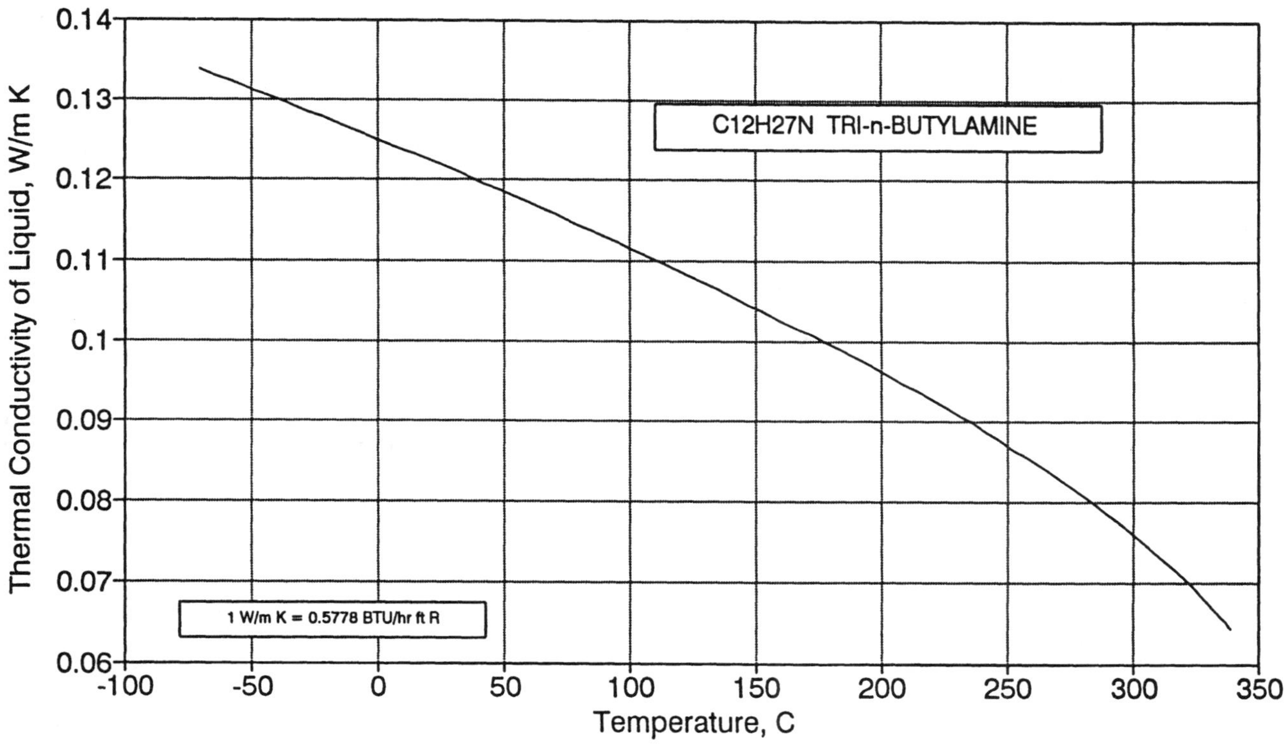
C12H27N TRI-n-BUTYLAMINE
1 W/m K = 0.5778 BTU/hr ft R
Thermal Conductivity of Liquid, W/m K
0.14
0.13
0.12
0.11
0.1
0.09
0.08
0.07
0.06
-100
-50
0
50
100
150
200
250
300
350
Temperature, C

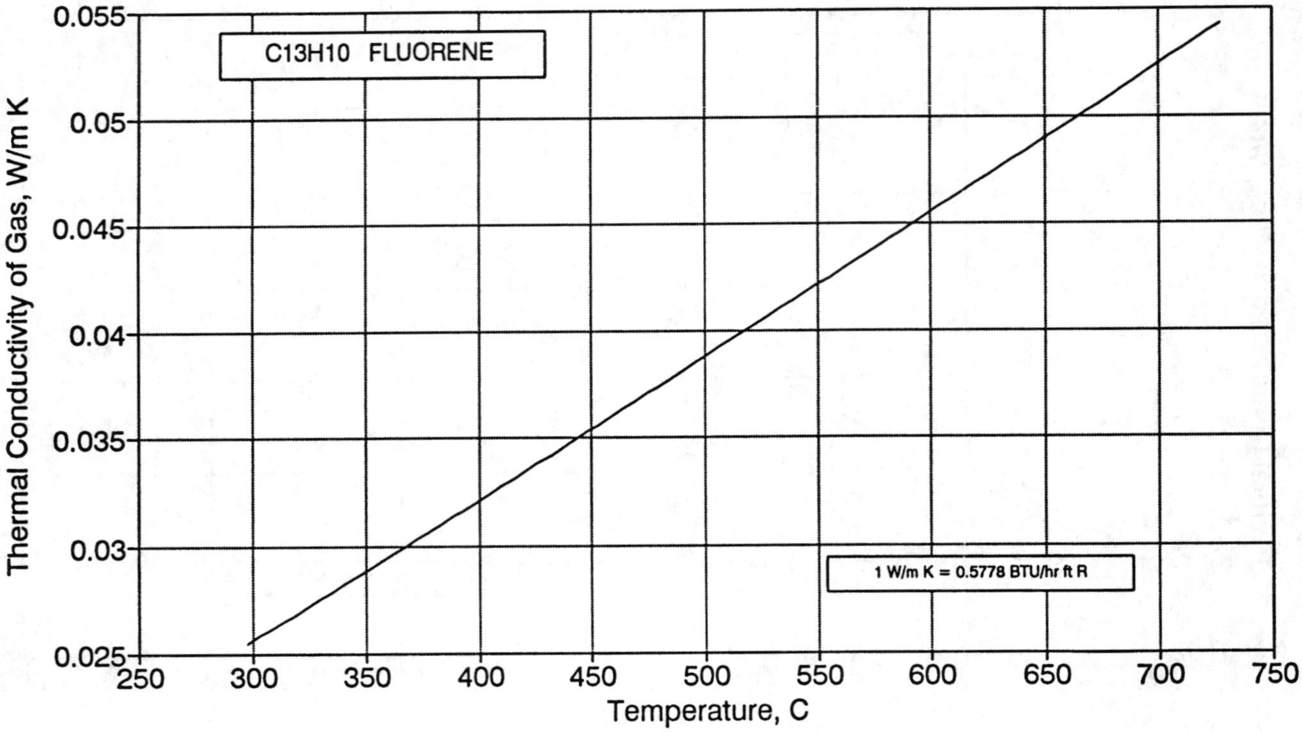
C13H10 FLUORENE
1 W/m K = 0.5778 BTU/hr ft R
Thermal Conductivity of Gas, W/m K
0.055
0.05
0.045
0.04
0.035
0.03
0.025
250
300
350
400
450
500
550
600
650
700
750
Temperature, C

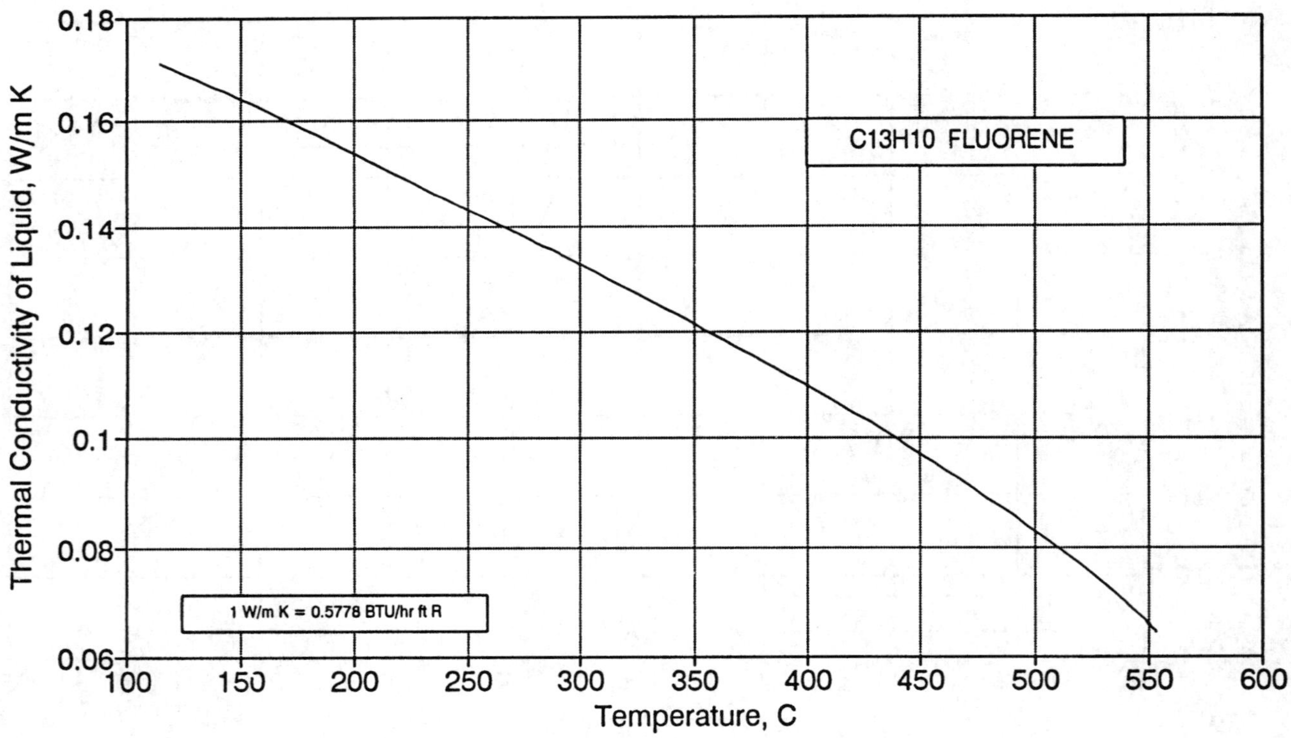
C13H10 FLUORENE
1 W/m K = 0.5778 BTU/hr ft R
Thermal Conductivity of Liquid, W/m K
0.18
0.16
0.14
0.12
0.1
0.08
0.06
100
150
200
250
300
350
400
450
500
550
600
Temperature, C

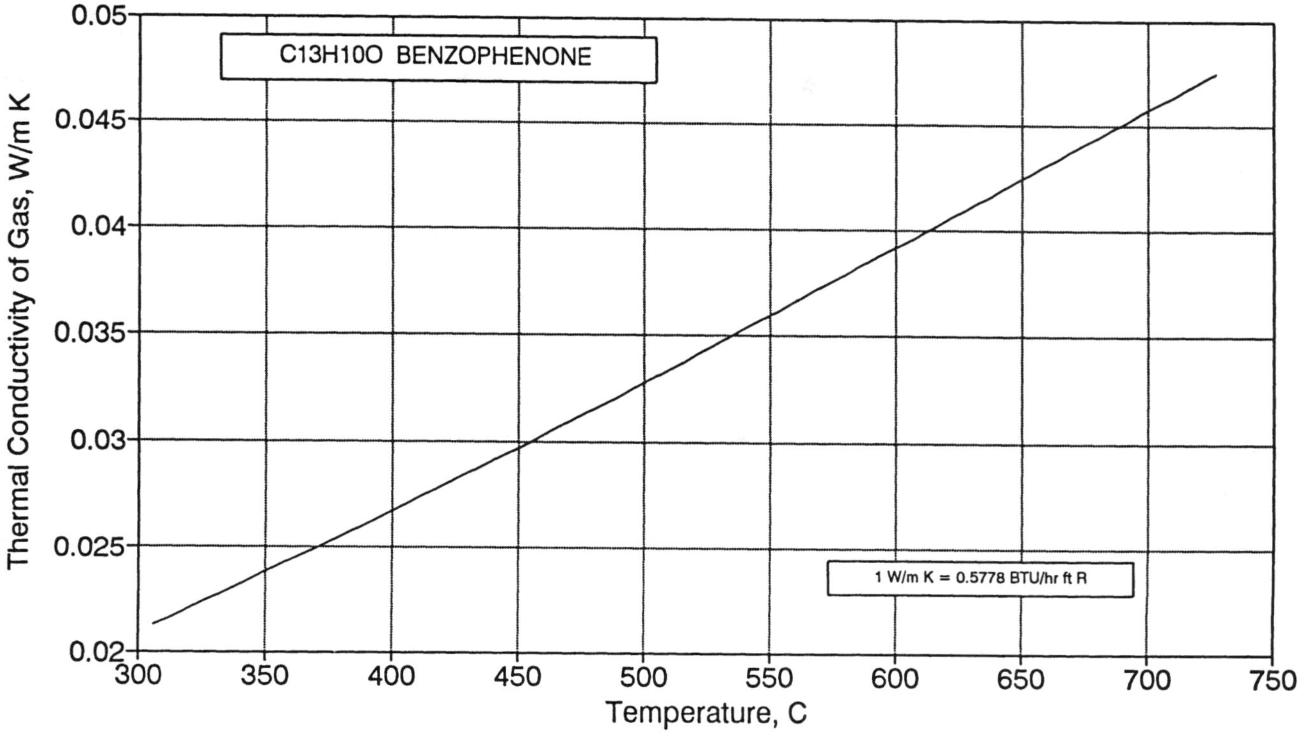
C13H10O BENZOPHENONE
1 W/m K = 0.5778 BTU/hr ft R
Thermal Conductivity of Gas, W/m K
0.05
0.045
0.04
0.035
0.03
0.025
0.02
300
350
400
450
500
550
600
650
700
750
Temperature, C

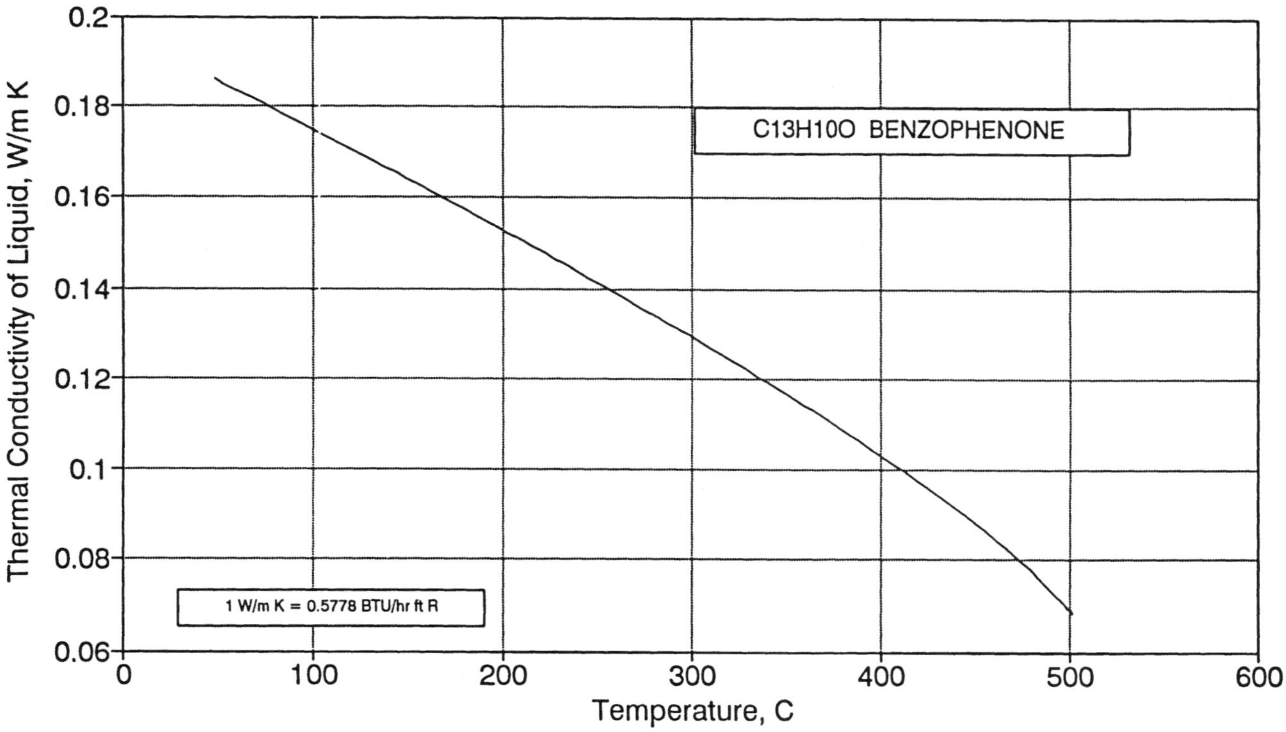
C13H10O BENZOPHENONE
1 W/m K = 0.5778 BTU/hr ft R
Thermal Conductivity of Liquid, W/m K
0.2
0.18
0.16
0.14
0.12
0.1
0.08
0.06
0
100
200
300
400
500
600
Temperature, C

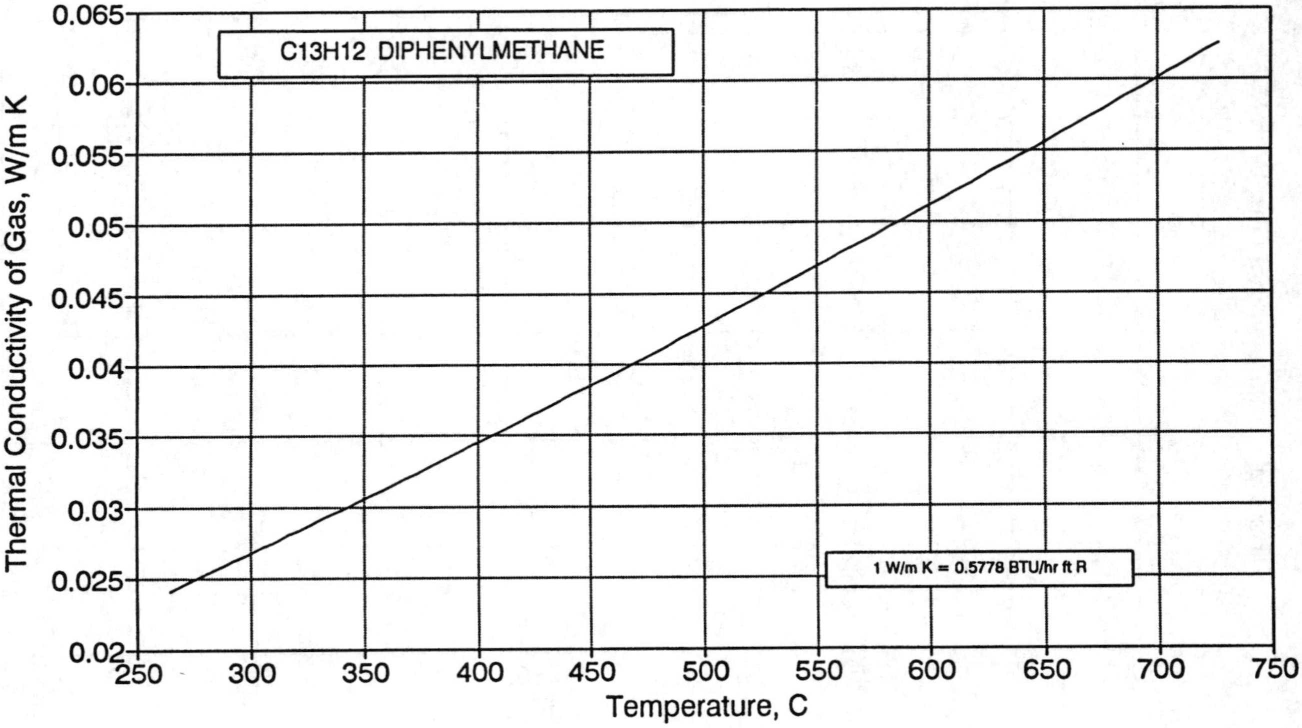
C13H12 DIPHENYLMETHANE
1 W/m K = 0.5778 BTU/hr ft R
Thermal Conductivity of Gas, W/m K
Temperature, C
0.065
0.06
0.055
0.05
0.045
0.04
0.035
0.03
0.025
0.02
250
300
350
400
450
500
550
600
650
700
750

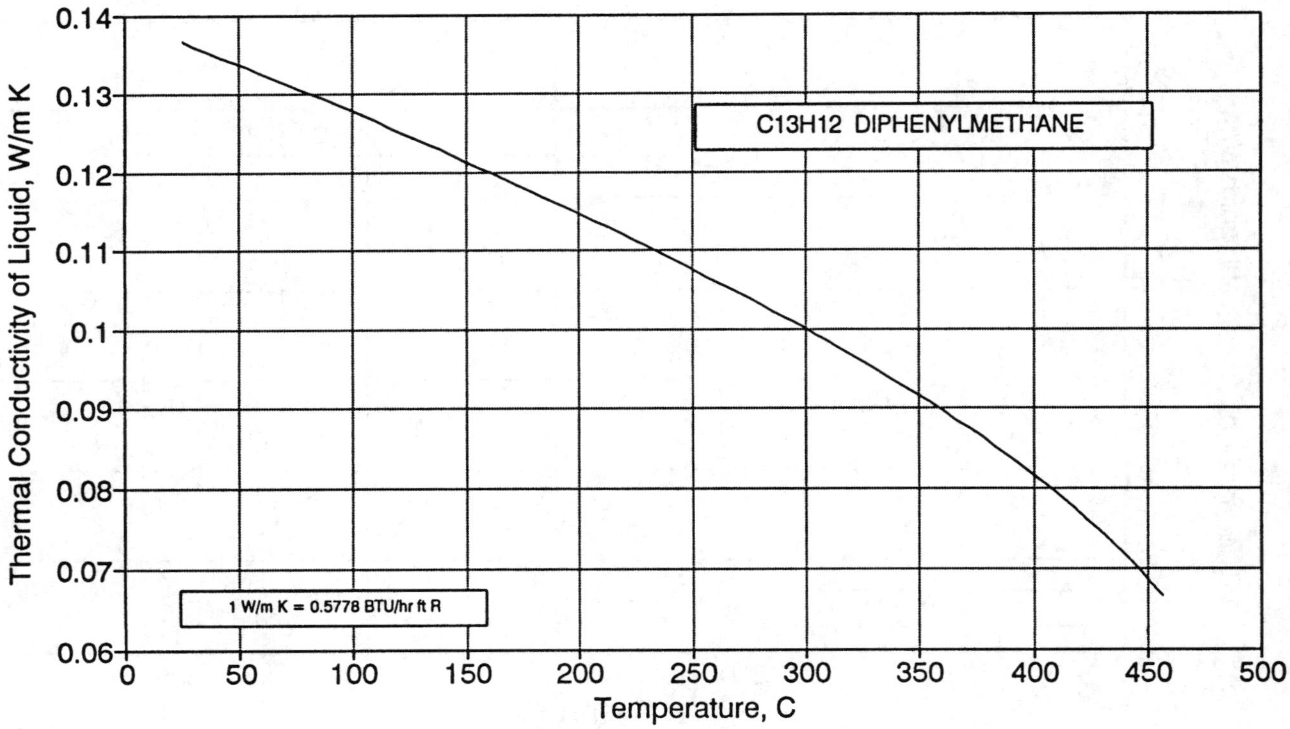
C13H12 DIPHENYLMETHANE
1 W/m K = 0.5778 BTU/hr ft R
Thermal Conductivity of Liquid, W/m K
Temperature, C
0.14
0.13
0.12
0.11
0.1
0.09
0.08
0.07
0.06
0
50
100
150
200
250
300
350
400
450
500

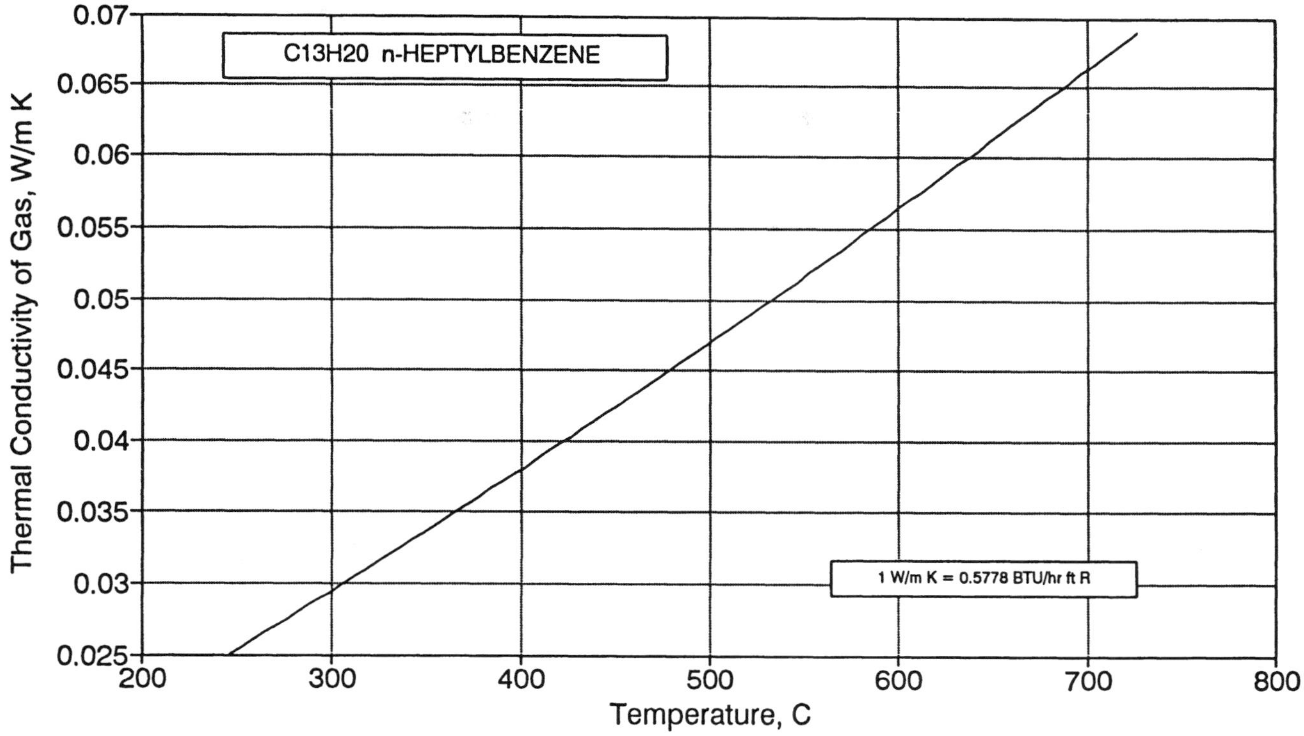
C13H20 n-HEPTYLBENZENE
1 W/m K = 0.5778 BTU/hr ft R
Thermal Conductivity of Gas, W/m K
0.07
0.065
0.06
0.055
0.05
0.045
0.04
0.035
0.03
0.025
200
300
400
500
600
700
800
Temperature, C

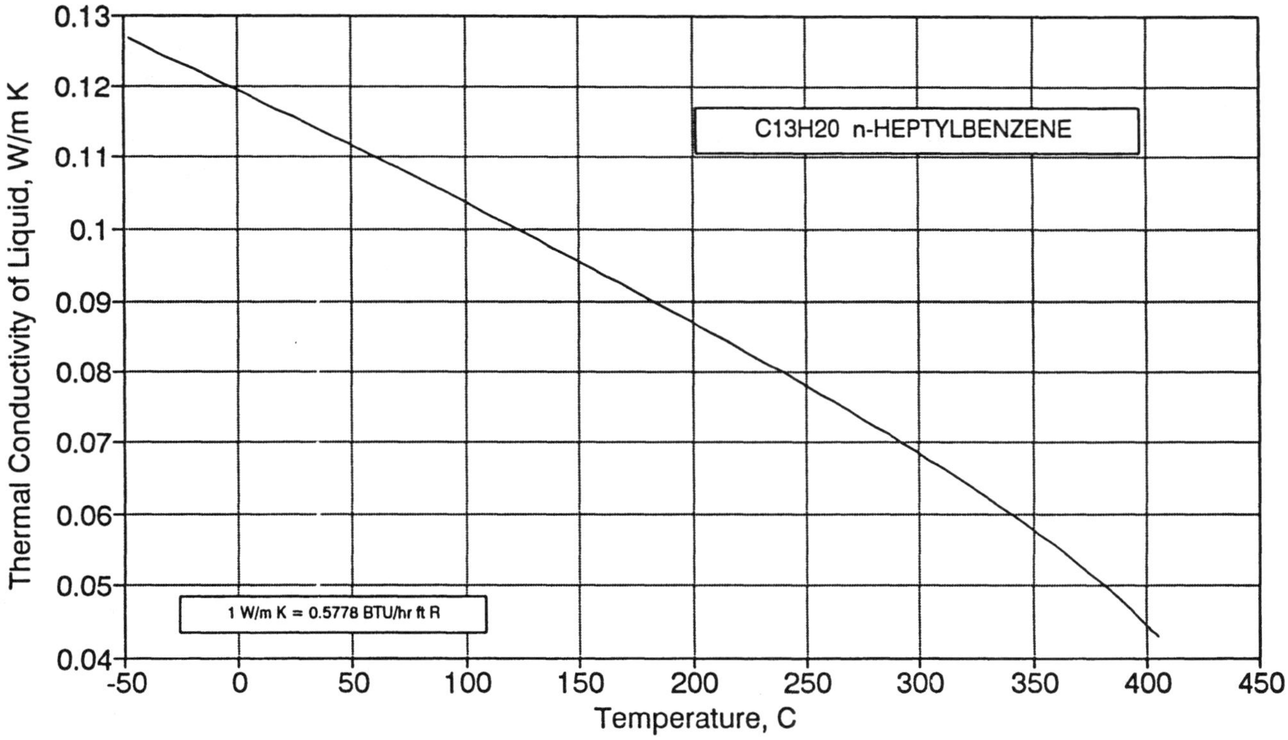
C13H20 n-HEPTYLBENZENE
1 W/m K = 0.5778 BTU/hr ft R
Thermal Conductivity of Liquid, W/m K
0.13
0.12
0.11
0.1
0.09
0.08
0.07
0.06
0.05
0.04
-50
0
50
100
150
200
250
300
350
400
450
Temperature, C

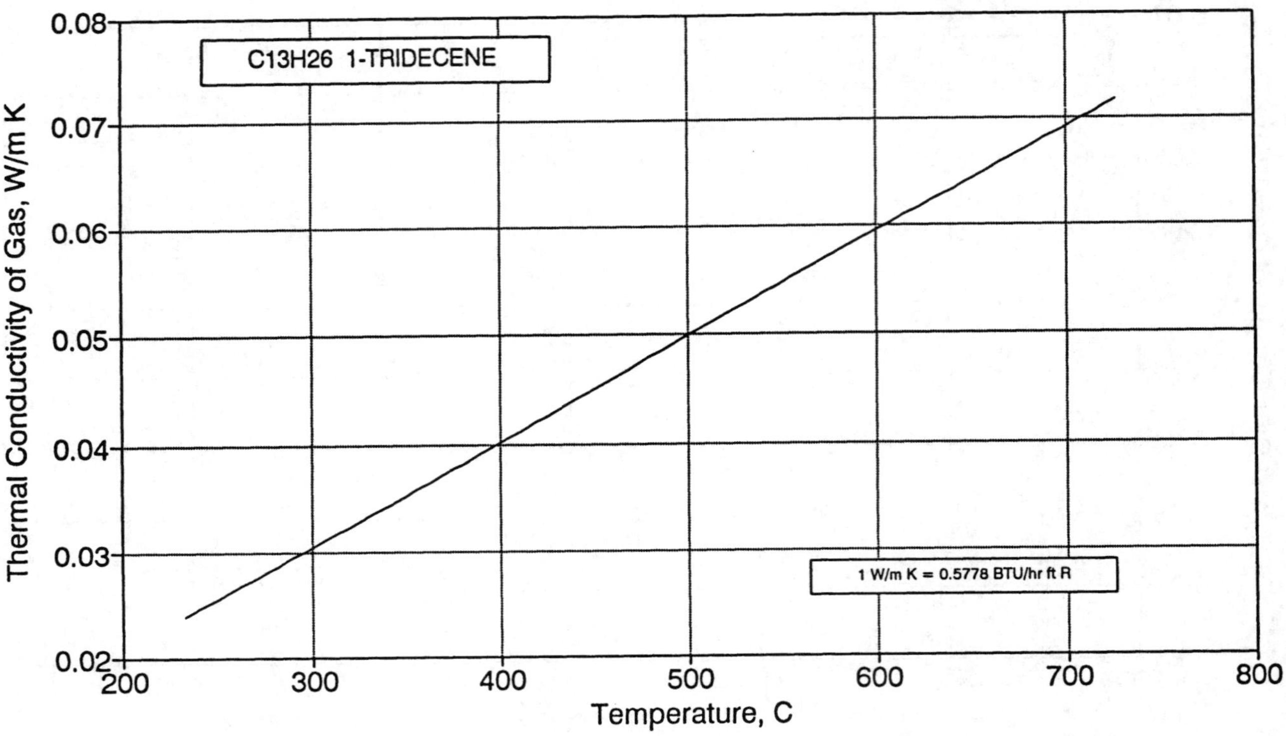
C13H26 1-TRIDECENE
1 W/m K = 0.5778 BTU/hr ft R
Thermal Conductivity of Gas, W/m K
0.08
0.07
0.06
0.05
0.04
0.03
0.02
200
300
400
500
600
700
800
Temperature, C

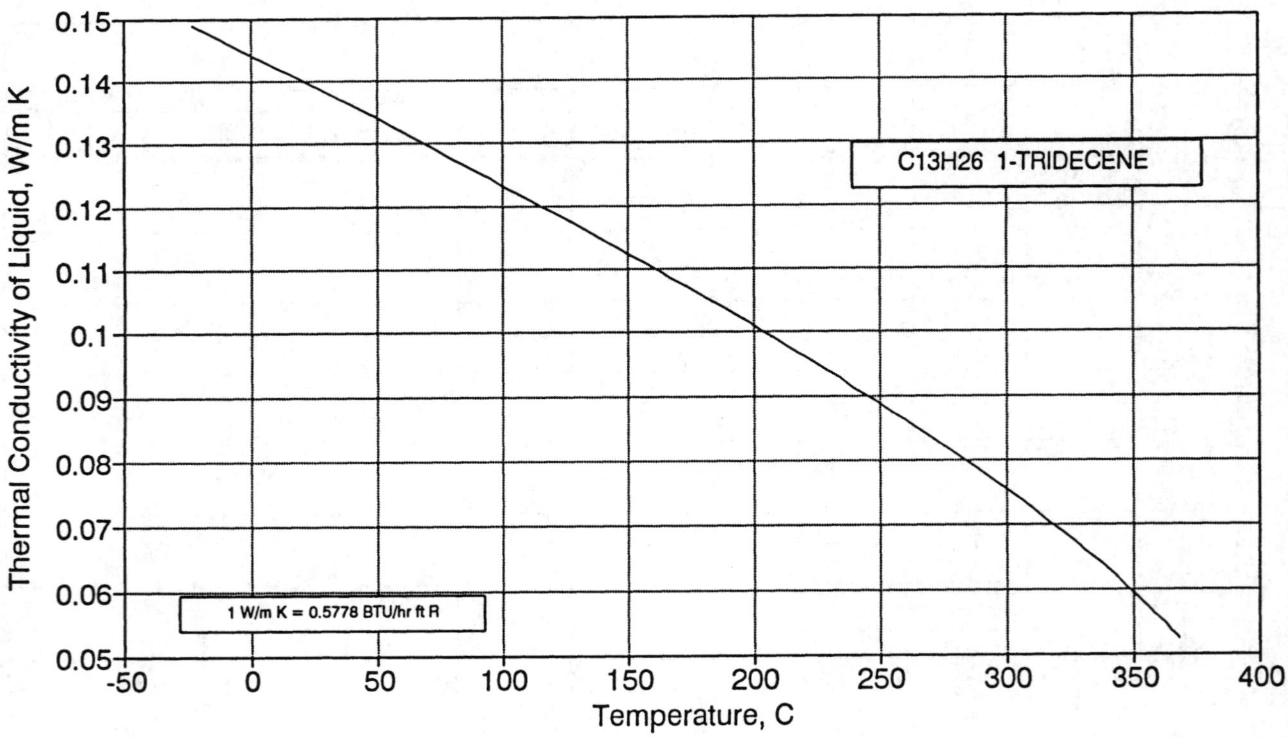
C13H26 1-TRIDECENE
1 W/m K = 0.5778 BTU/hr ft R
Thermal Conductivity of Liquid, W/m K
0.15
0.14
0.13
0.12
0.11
0.1
0.09
0.08
0.07
0.06
0.05
-50
0
50
100
150
200
250
300
350
400
Temperature, C

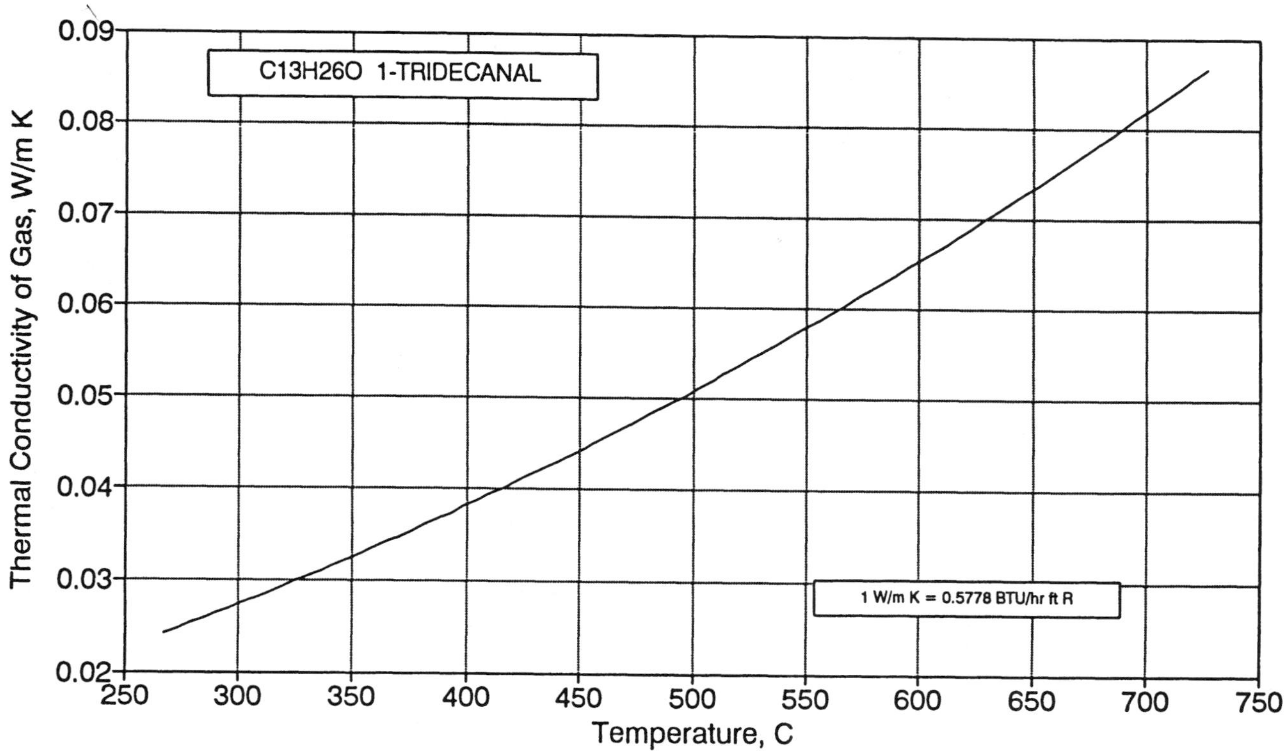
C13H26O 1-TRIDECANAL
1 W/m K = 0.5778 BTU/hr ft R
Thermal Conductivity of Gas, W/m K
0.09
0.08
0.07
0.06
0.05
0.04
0.03
0.02
250
300
350
400
450
500
550
600
650
700
750
Temperature, C

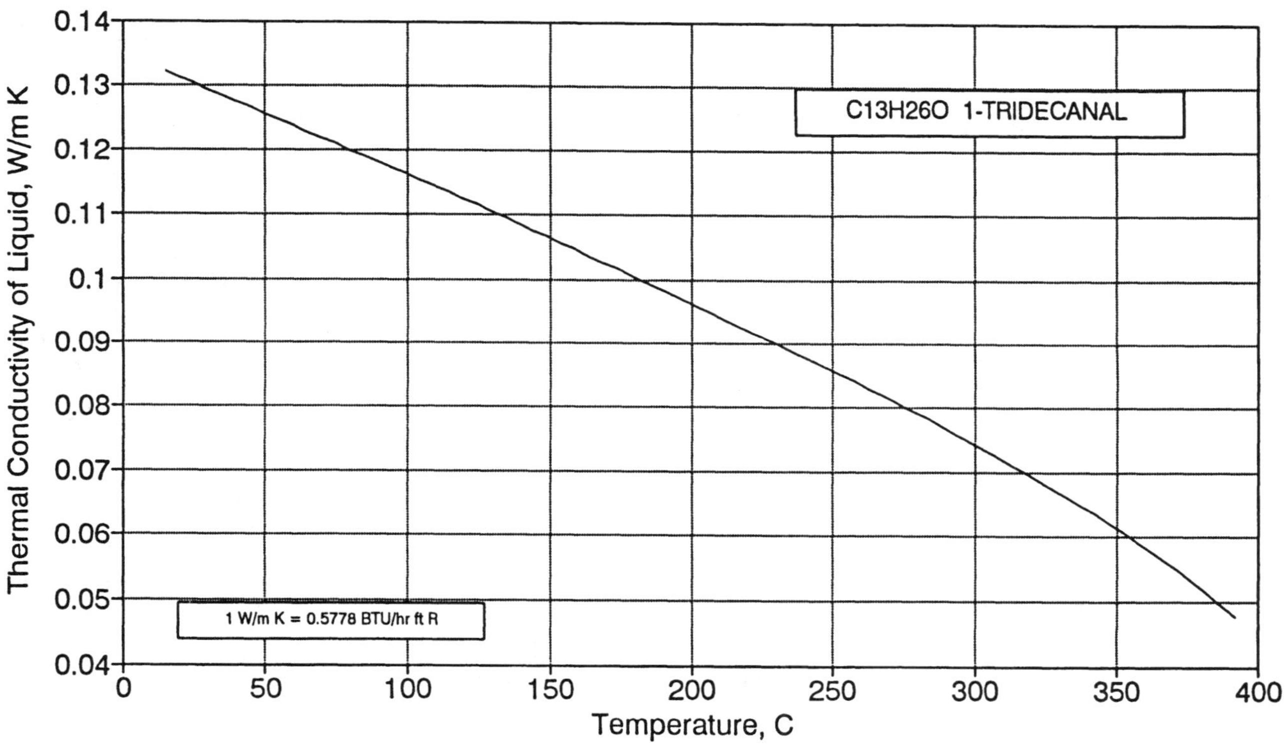
C13H26O 1-TRIDECANAL
1 W/m K = 0.5778 BTU/hr ft R
Thermal Conductivity of Liquid, W/m K
0.14
0.13
0.12
0.11
0.1
0.09
0.08
0.07
0.06
0.05
0.04
0
50
100
150
200
250
300
350
400
Temperature, C

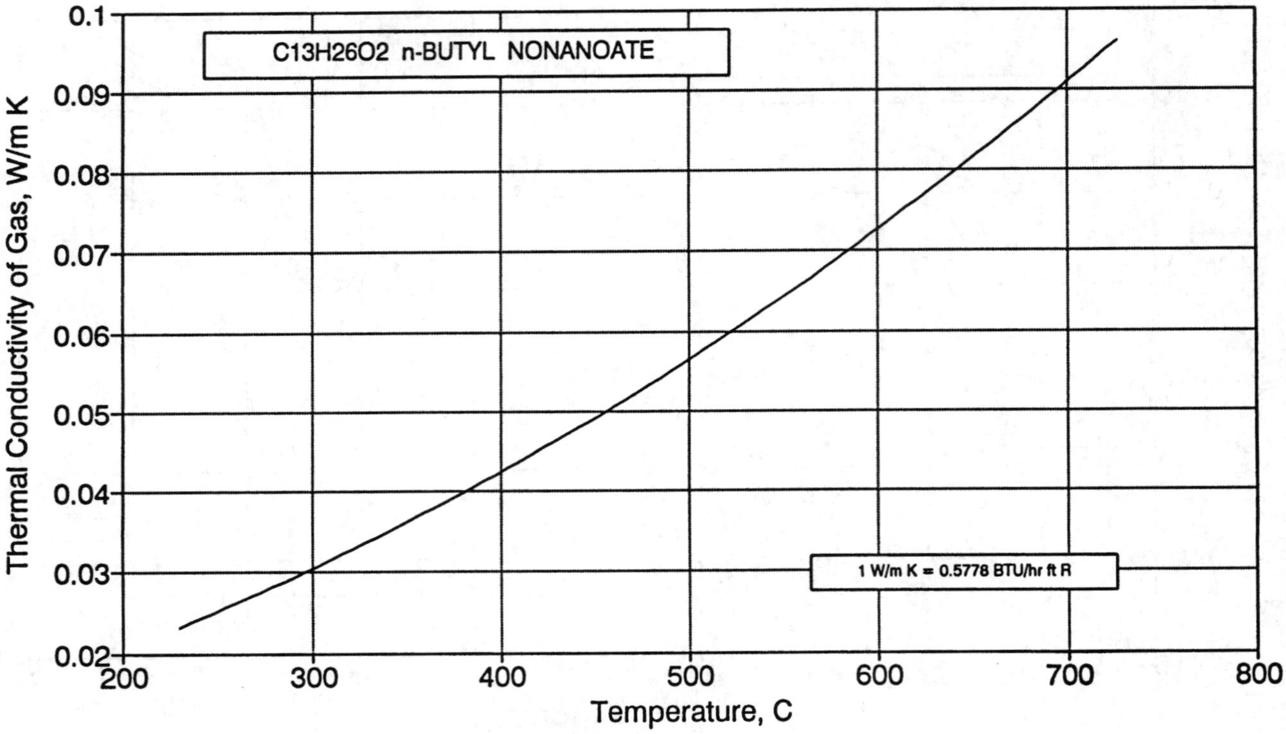
C13H26O2 n-BUTYL NONANOATE
1 W/m K = 0.5778 BTU/hr ft R
Thermal Conductivity of Gas, W/m K
0.1
0.09
0.08
0.07
0.06
0.05
0.04
0.03
0.02
200
300
400
500
600
700
800
Temperature, C

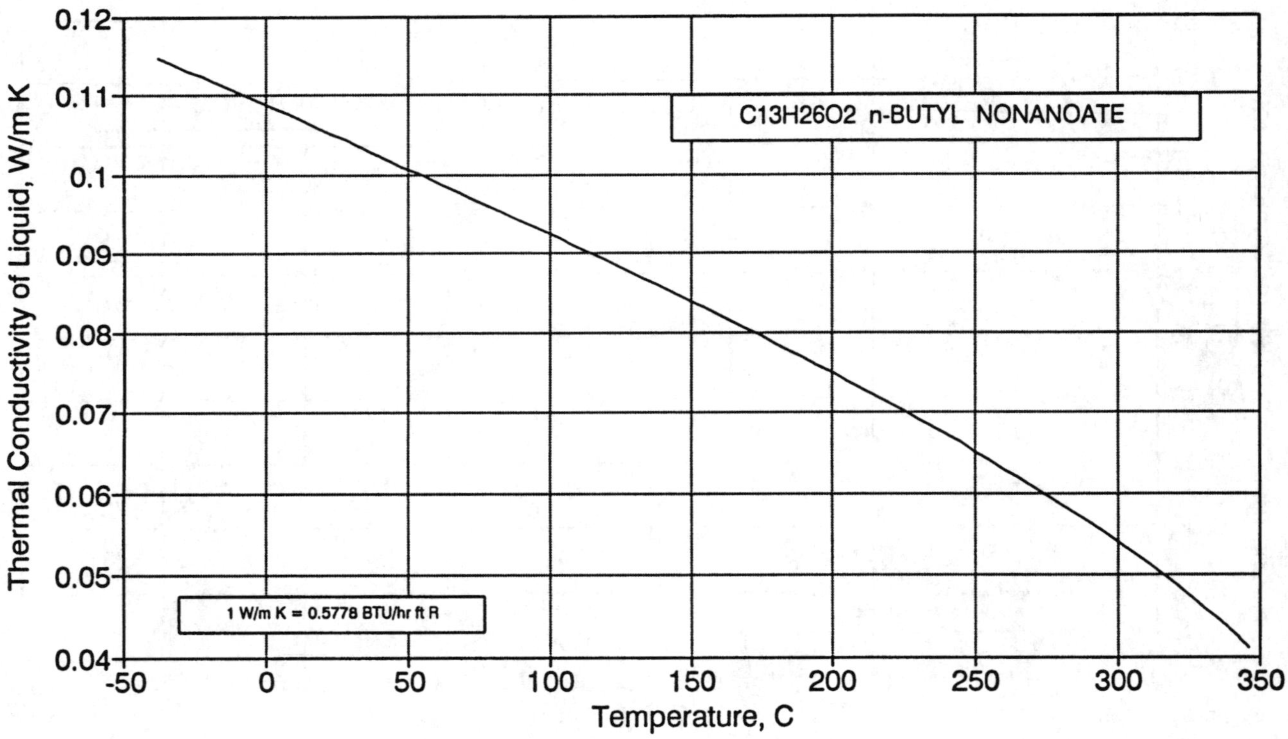
C13H26O2 n-BUTYL NONANOATE
1 W/m K = 0.5778 BTU/hr ft R
Thermal Conductivity of Liquid, W/m K
0.12
0.11
0.1
0.09
0.08
0.07
0.06
0.05
0.04
-50
0
50
100
150
200
250
300
350
Temperature, C

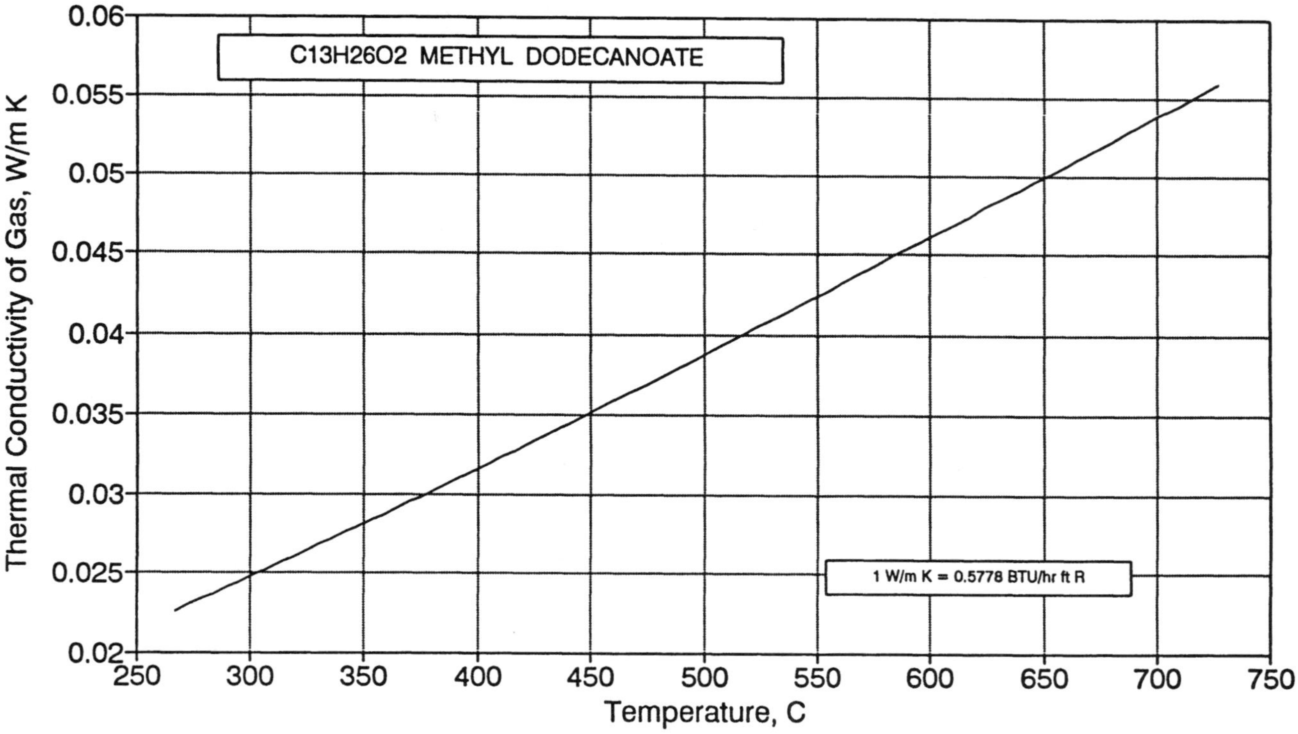
C13H26O2 METHYL DODECANOATE
1 W/m K = 0.5778 BTU/hr ft R
Thermal Conductivity of Gas, W/m K
0.06
0.055
0.05
0.045
0.04
0.035
0.03
0.025
0.02
250
300
350
400
450
500
550
600
650
700
750
Temperature, C

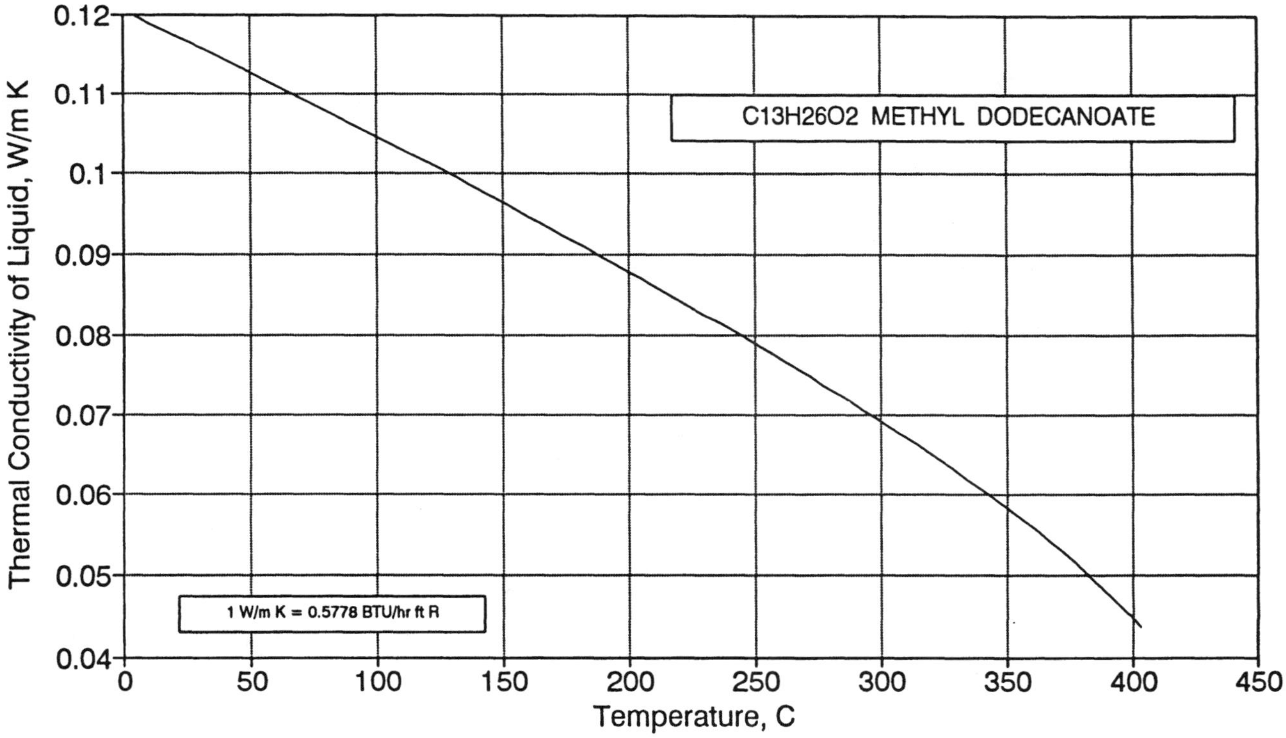
C13H26O2 METHYL DODECANOATE
1 W/m K = 0.5778 BTU/hr ft R
Thermal Conductivity of Liquid, W/m K
0.12
0.11
0.1
0.09
0.08
0.07
0.06
0.05
0.04
0
50
100
150
200
250
300
350
400
450
Temperature, C

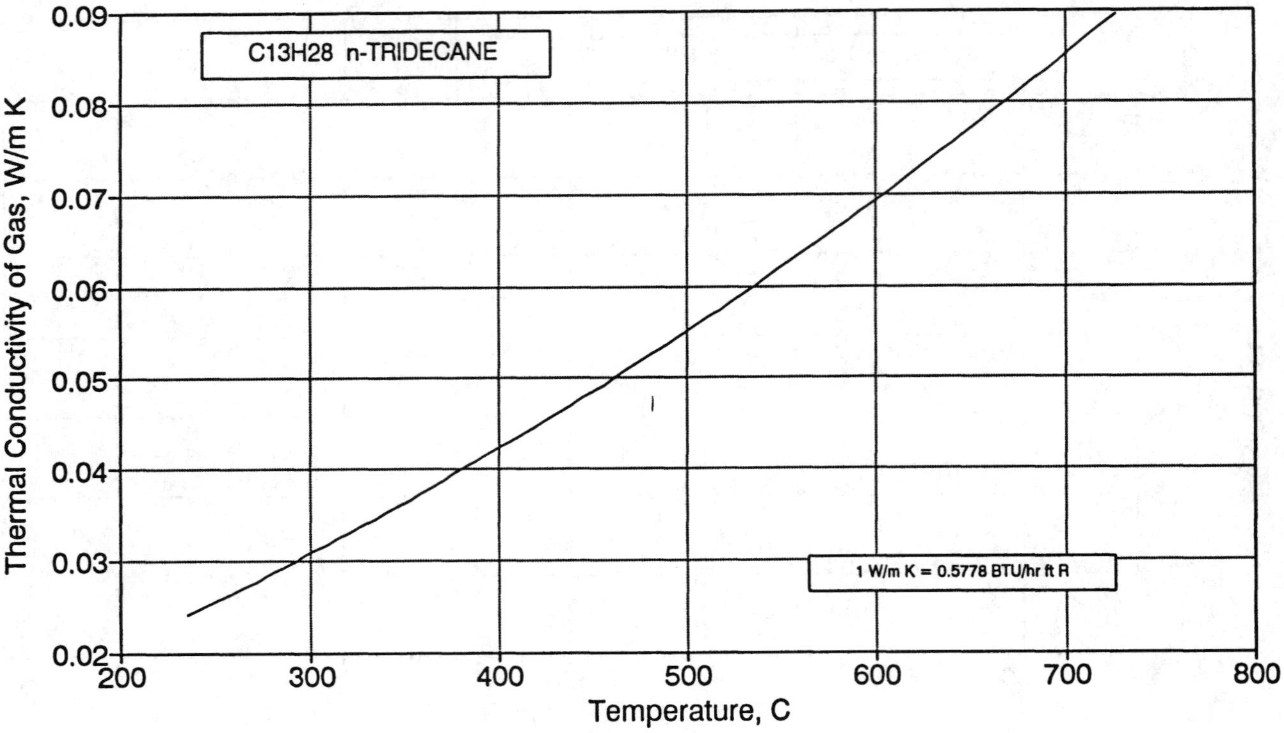
C13H28 n-TRIDECANE
1 W/m K = 0.5778 BTU/hr ft R
Thermal Conductivity of Gas, W/m K
0.09
0.08
0.07
0.06
0.05
0.04
0.03
0.02
200
300
400
500
600
700
800
Temperature, C

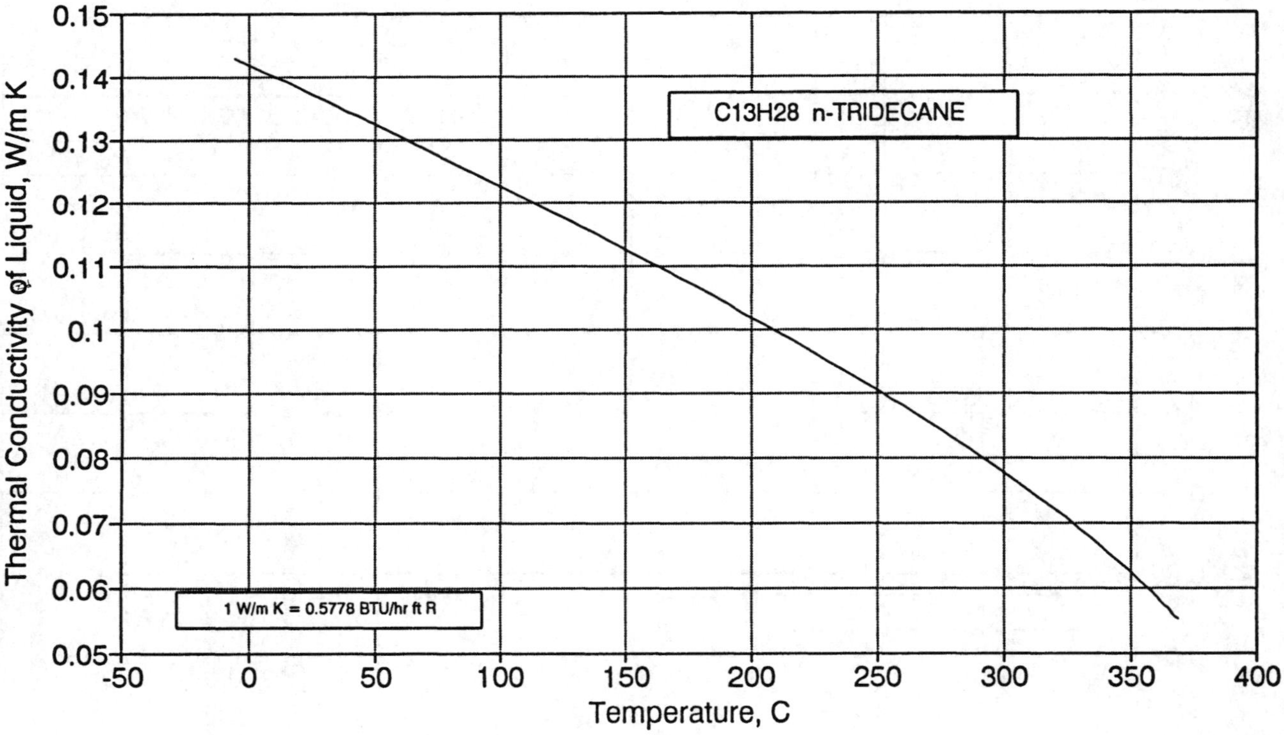
C13H28 n-TRIDECANE
1 W/m K = 0.5778 BTU/hr ft R
Thermal Conductivity of Liquid, W/m K
0.15
0.14
0.13
0.12
0.11
0.1
0.09
0.08
0.07
0.06
0.05
-50
0
50
100
150
200
250
300
350
400
Temperature, C

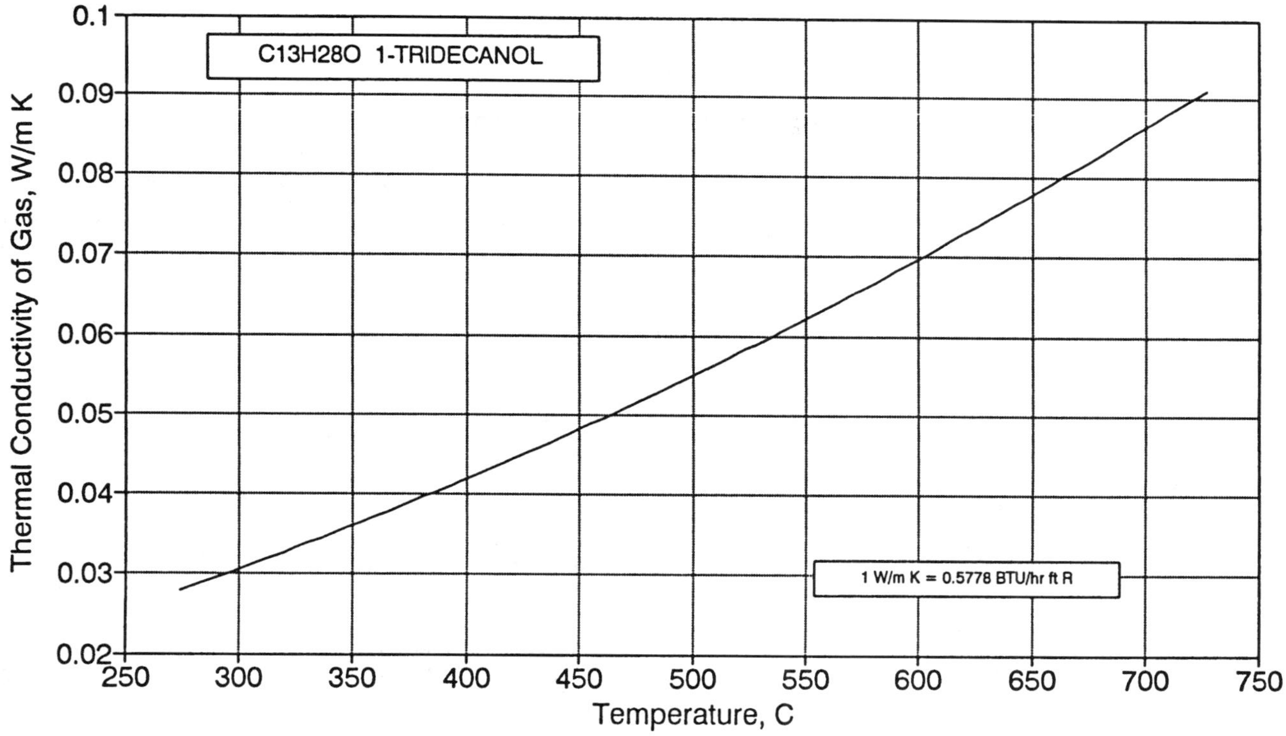
C13H28O 1-TRIDECANOL
1 W/m K = 0.5778 BTU/hr ft R
Thermal Conductivity of Gas, W/m K
0.1
0.09
0.08
0.07
0.06
0.05
0.04
0.03
0.02
250
300
350
400
450
500
550
600
650
700
750
Temperature, C

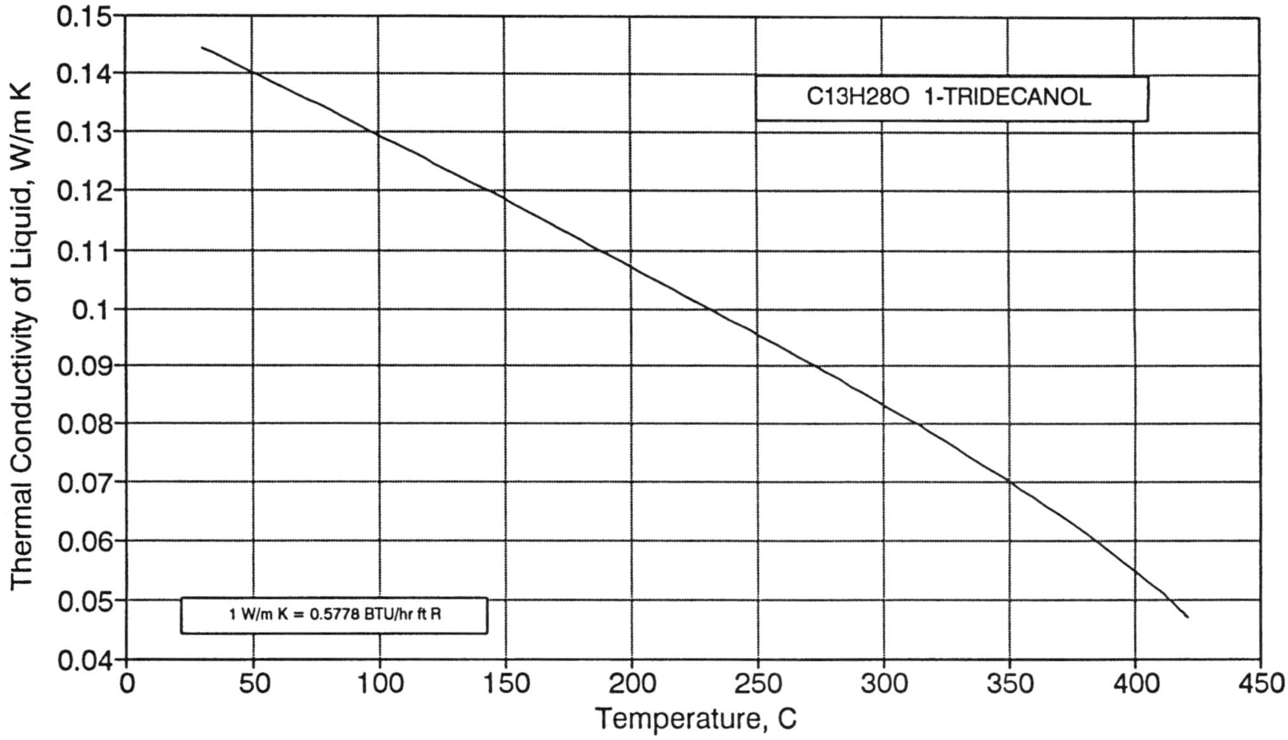
C13H28O 1-TRIDECANOL
1 W/m K = 0.5778 BTU/hr ft R
Thermal Conductivity of Liquid, W/m K
0.15
0.14
0.13
0.12
0.11
0.1
0.09
0.08
0.07
0.06
0.05
0.04
0
50
100
150
200
250
300
350
400
450
Temperature, C

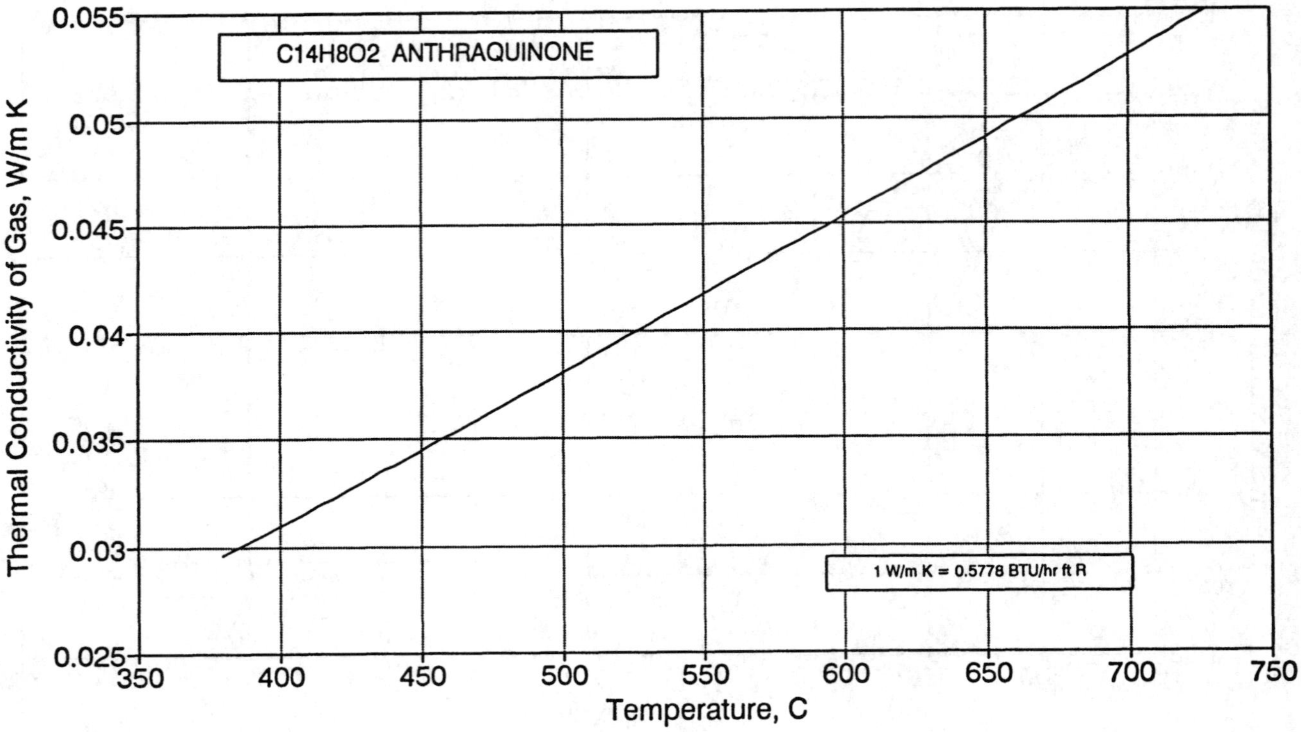
C14H8O2 ANTHRAQUINONE
1 W/m K = 0.5778 BTU/hr ft R
Thermal Conductivity of Gas, W/m K
0.055
0.05
0.045
0.04
0.035
0.03
0.025
350
400
450
500
550
600
650
700
750
Temperature, C

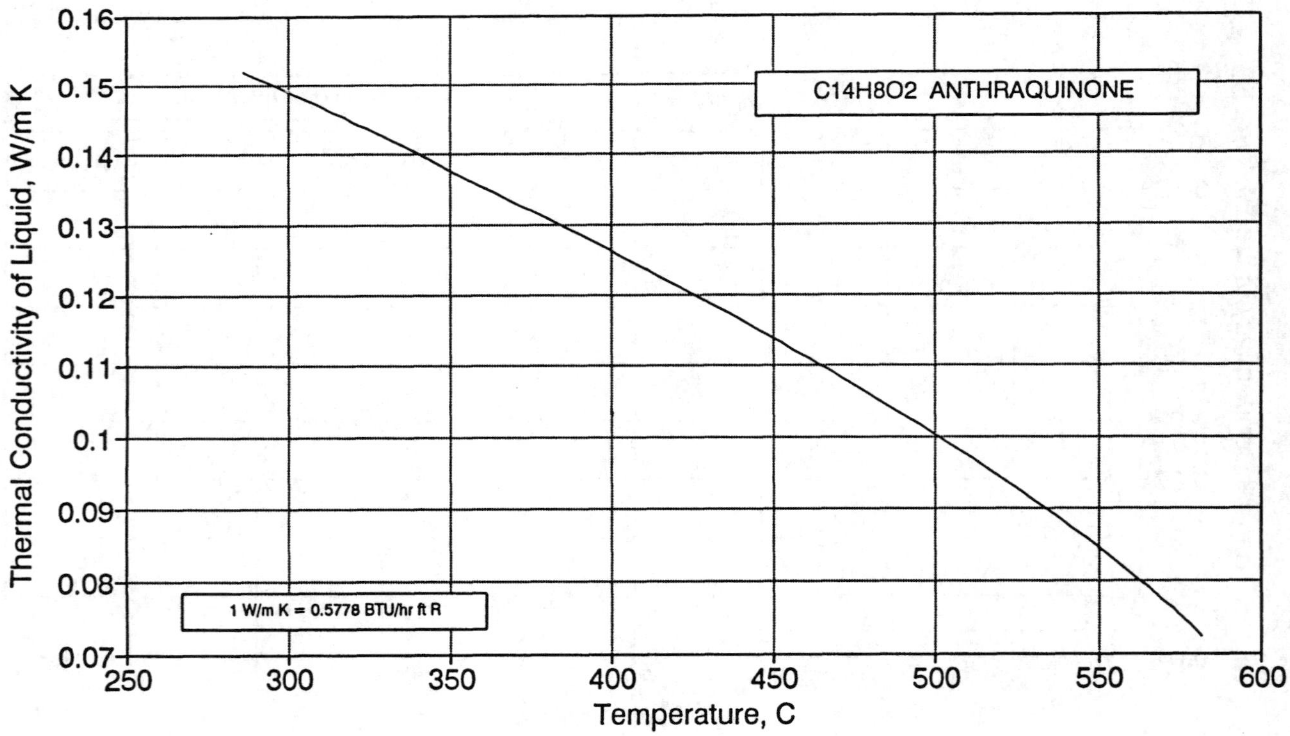
C14H8O2 ANTHRAQUINONE
1 W/m K = 0.5778 BTU/hr ft R
Thermal Conductivity of Liquid, W/m K
0.16
0.15
0.14
0.13
0.12
0.11
0.1
0.09
0.08
0.07
250
300
350
400
450
500
550
600
Temperature, C

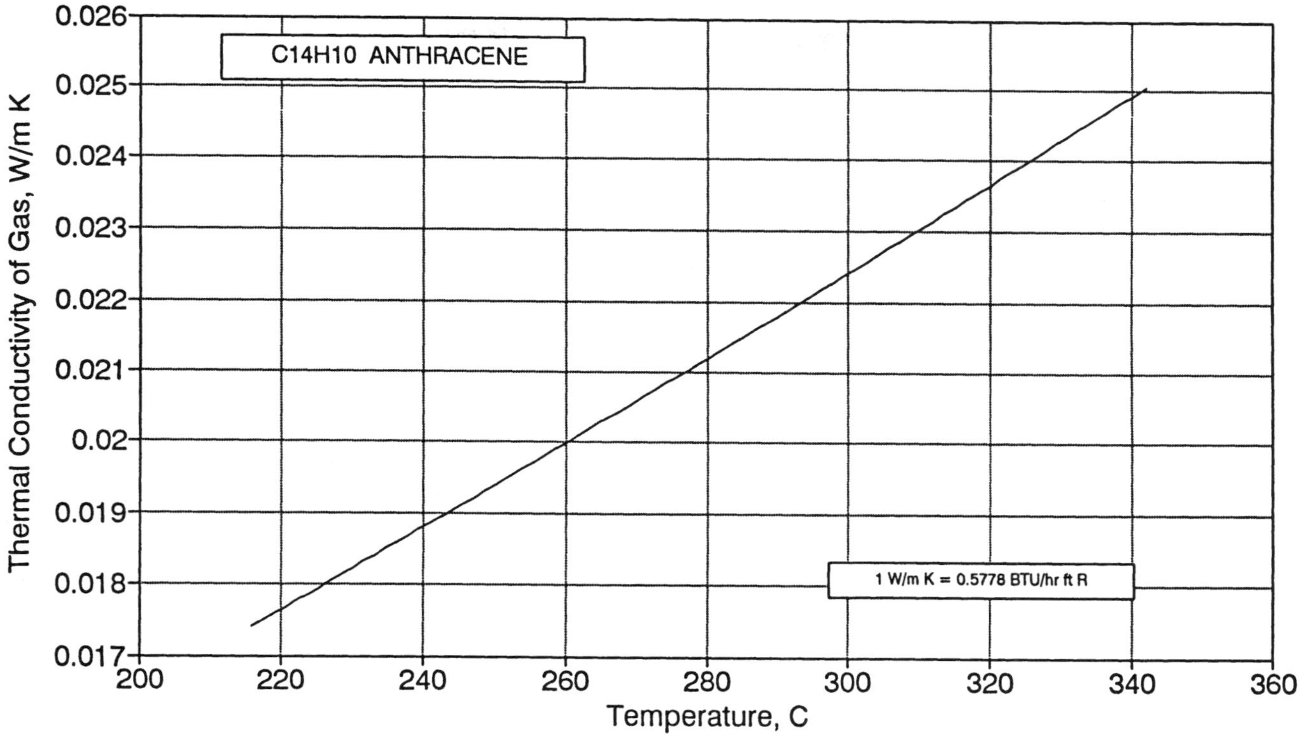
C14H10 ANTHRACENE
1 W/m K = 0.5778 BTU/hr ft R
Thermal Conductivity of Gas, W/m K
0.026
0.025
0.024
0.023
0.022
0.021
0.02
0.019
0.018
0.017
200
220
240
260
280
300
320
340
360
Temperature, C

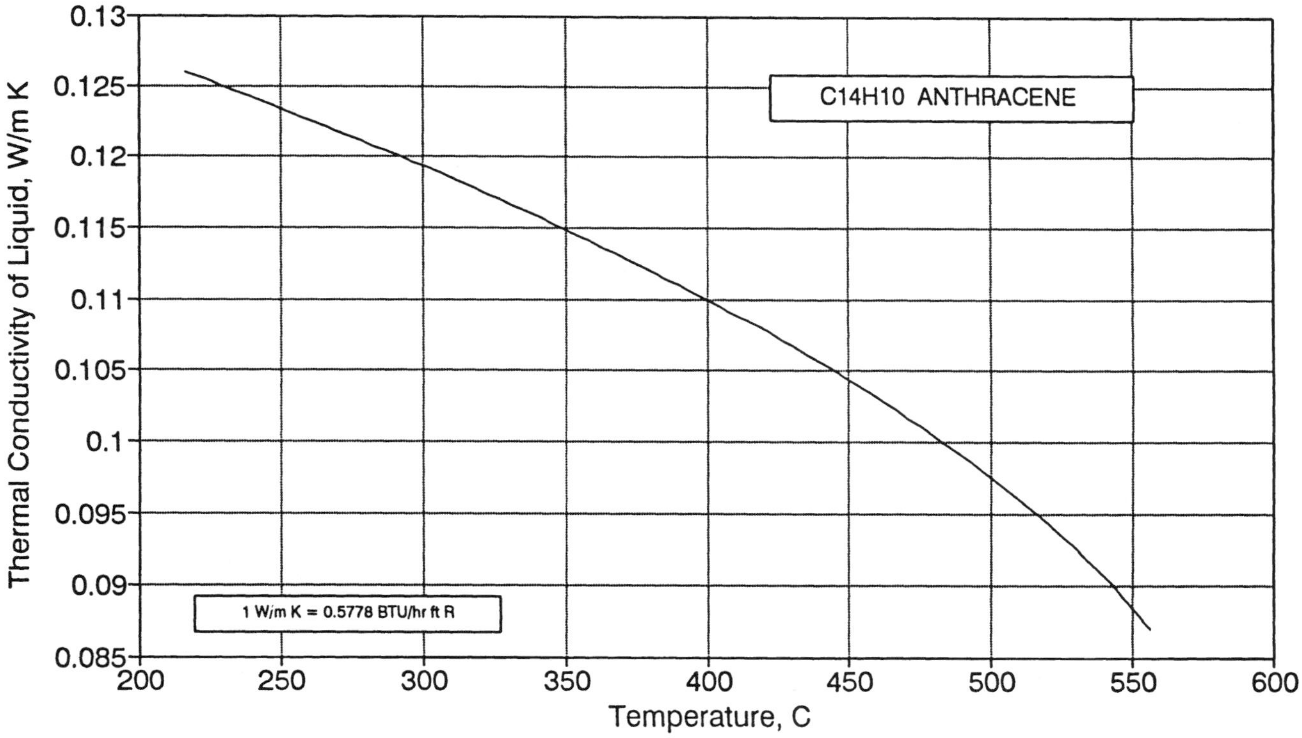
C14H10 ANTHRACENE
1 W/m K = 0.5778 BTU/hr ft R
Thermal Conductivity of Liquid, W/m K
0.13
0.125
0.12
0.115
0.11
0.105
0.1
0.095
0.09
0.085
200
250
300
350
400
450
500
550
600
Temperature, C

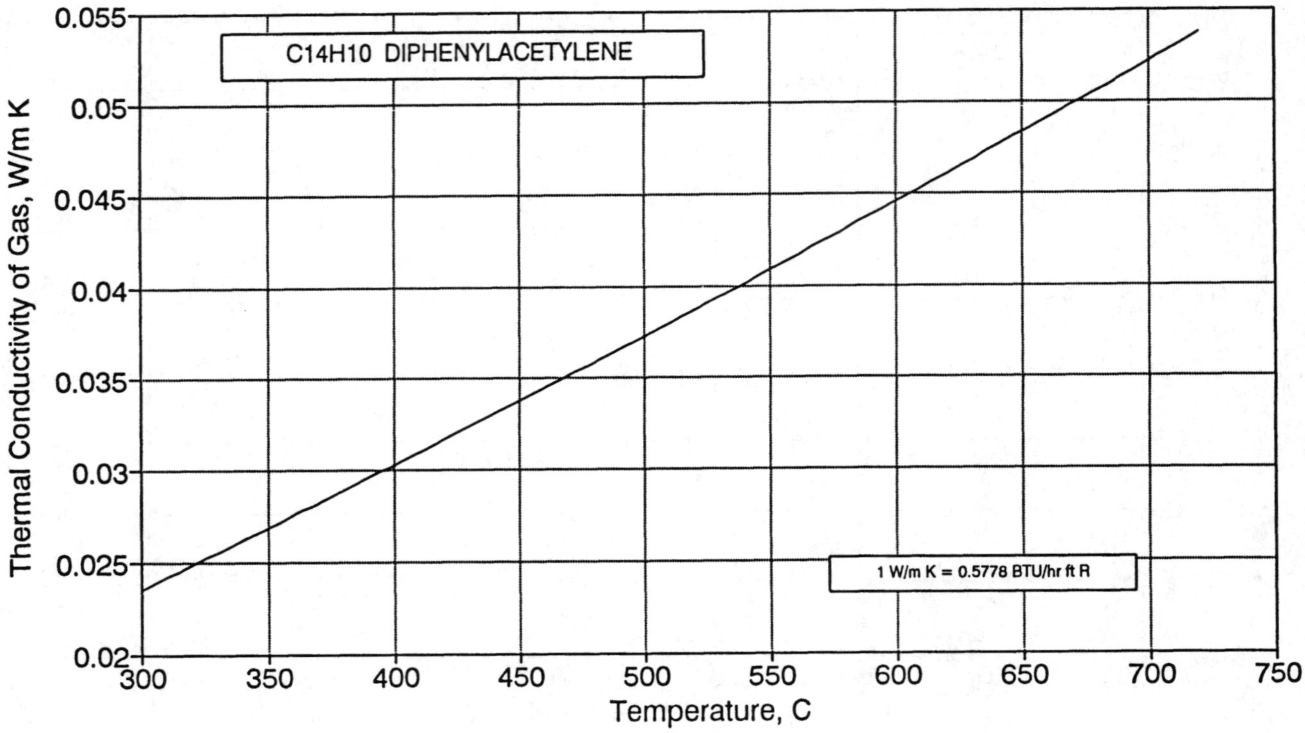
C14H10 DIPHENYLACETYLENE
1 W/m K = 0.5778 BTU/hr ft R
Thermal Conductivity of Gas, W/m K
0.055
0.05
0.045
0.04
0.035
0.03
0.025
0.02
300
350
400
450
500
550
600
650
700
750
Temperature, C

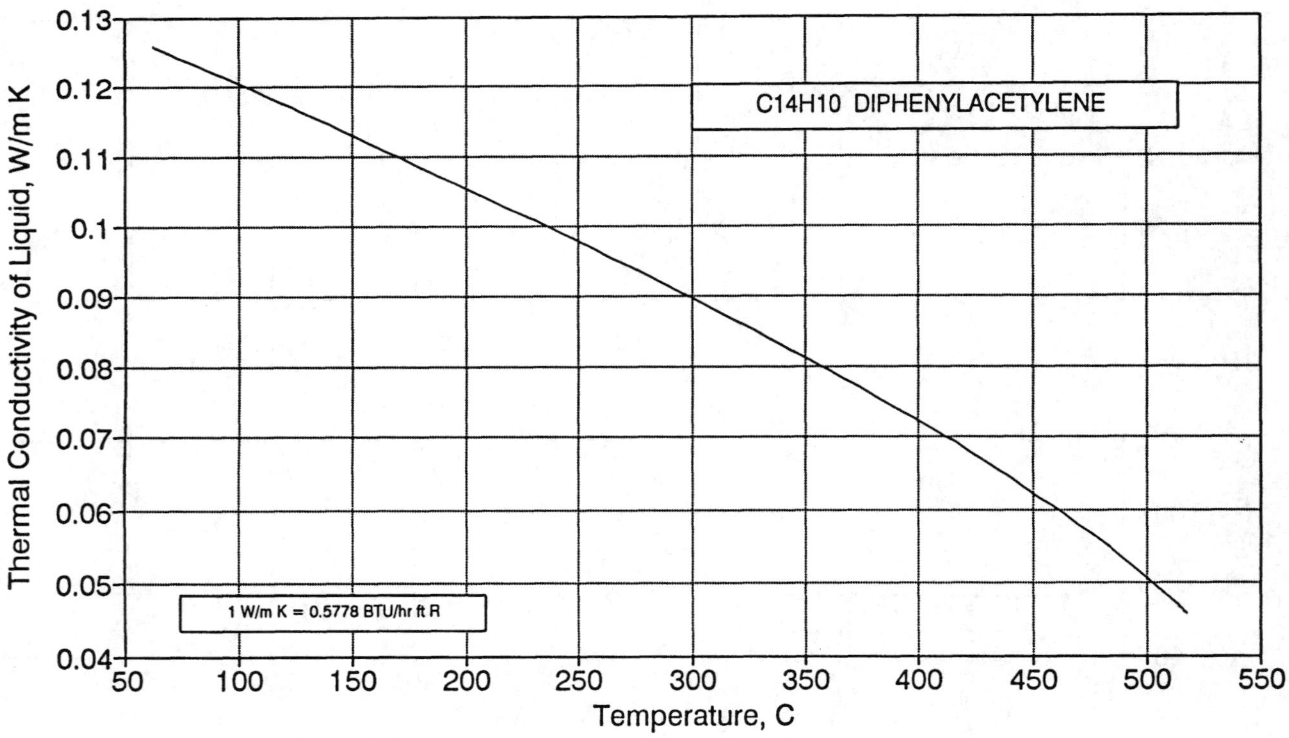
C14H10 DIPHENYLACETYLENE
1 W/m K = 0.5778 BTU/hr ft R
Thermal Conductivity of Liquid, W/m K
0.13
0.12
0.11
0.1
0.09
0.08
0.07
0.06
0.05
0.04
50
100
150
200
250
300
350
400
450
500
550
Temperature, C

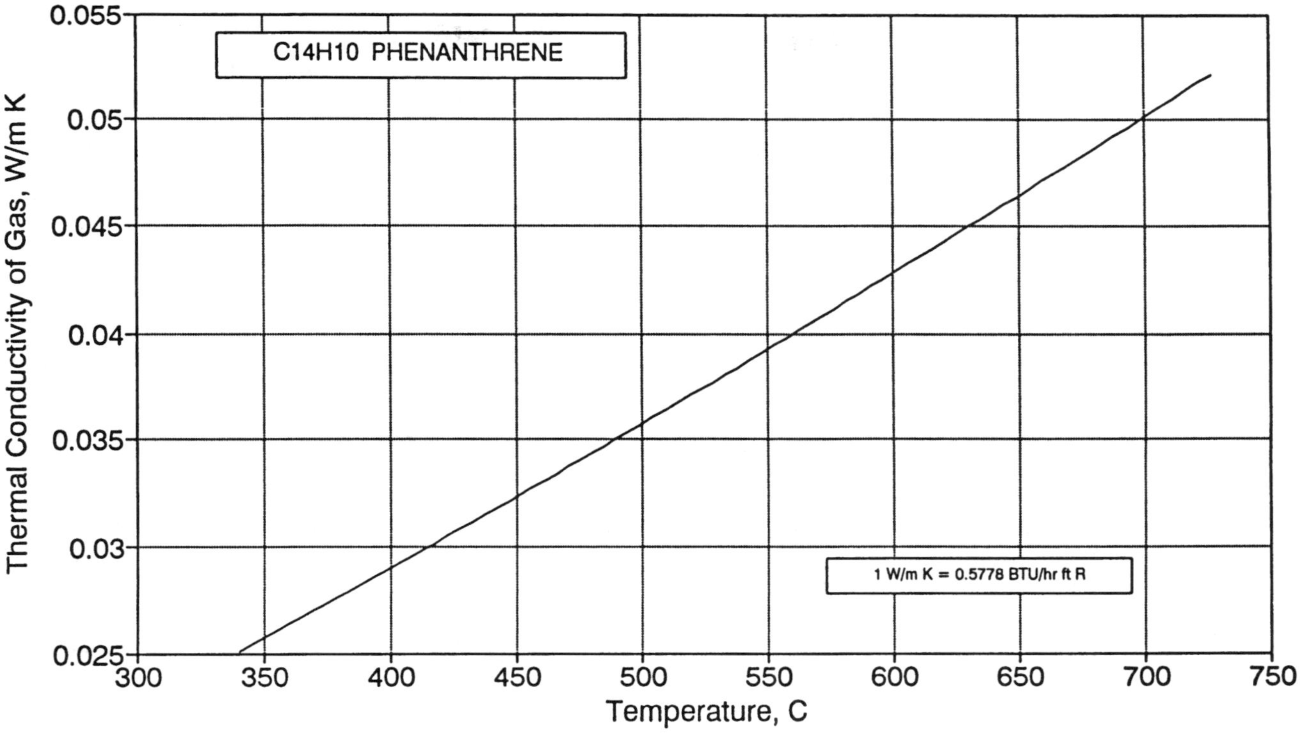
C14H10 PHENANTHRENE
1 W/m K = 0.5778 BTU/hr ft R
Thermal Conductivity of Gas, W/m K
0.055
0.05
0.045
0.04
0.035
0.03
0.025
300
350
400
450
500
550
600
650
700
750
Temperature, C

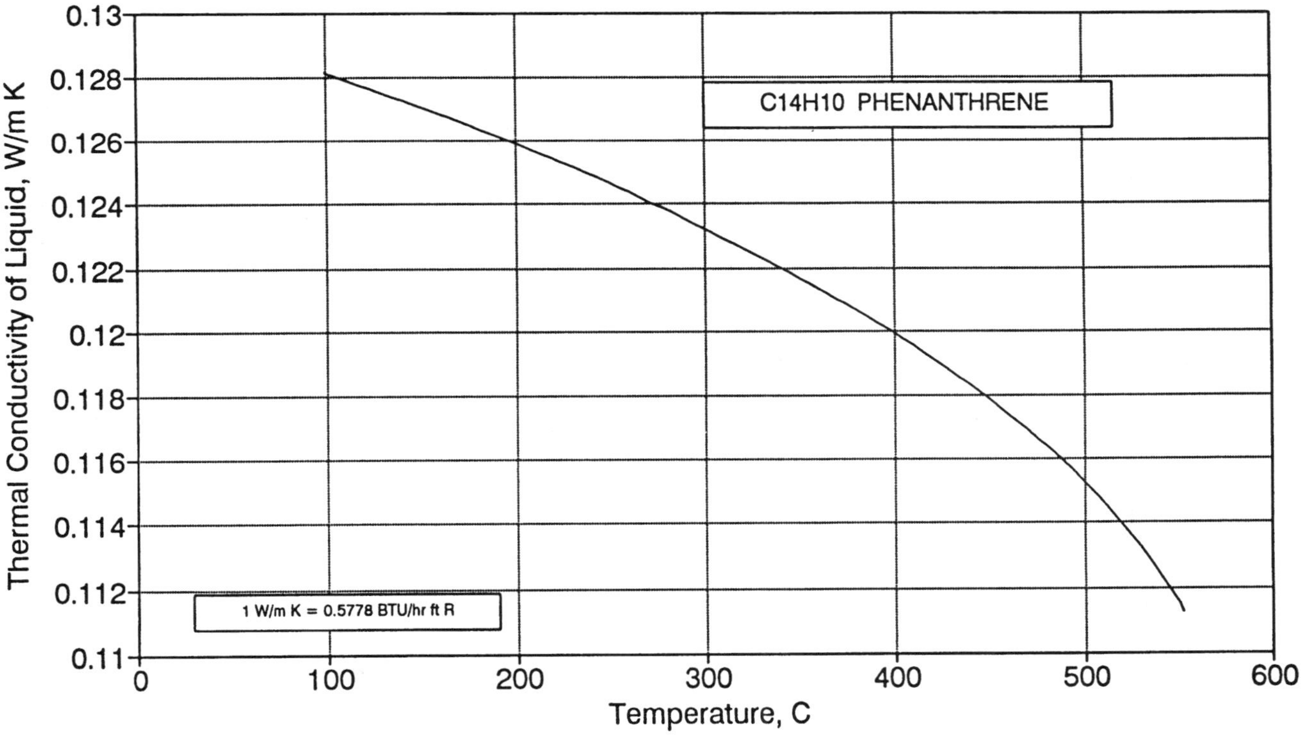
C14H10 PHENANTHRENE
1 W/m K = 0.5778 BTU/hr ft R
Thermal Conductivity of Liquid, W/m K
0.13
0.128
0.126
0.124
0.122
0.12
0.118
0.116
0.114
0.112
0.11
0
100
200
300
400
500
600
Temperature, C

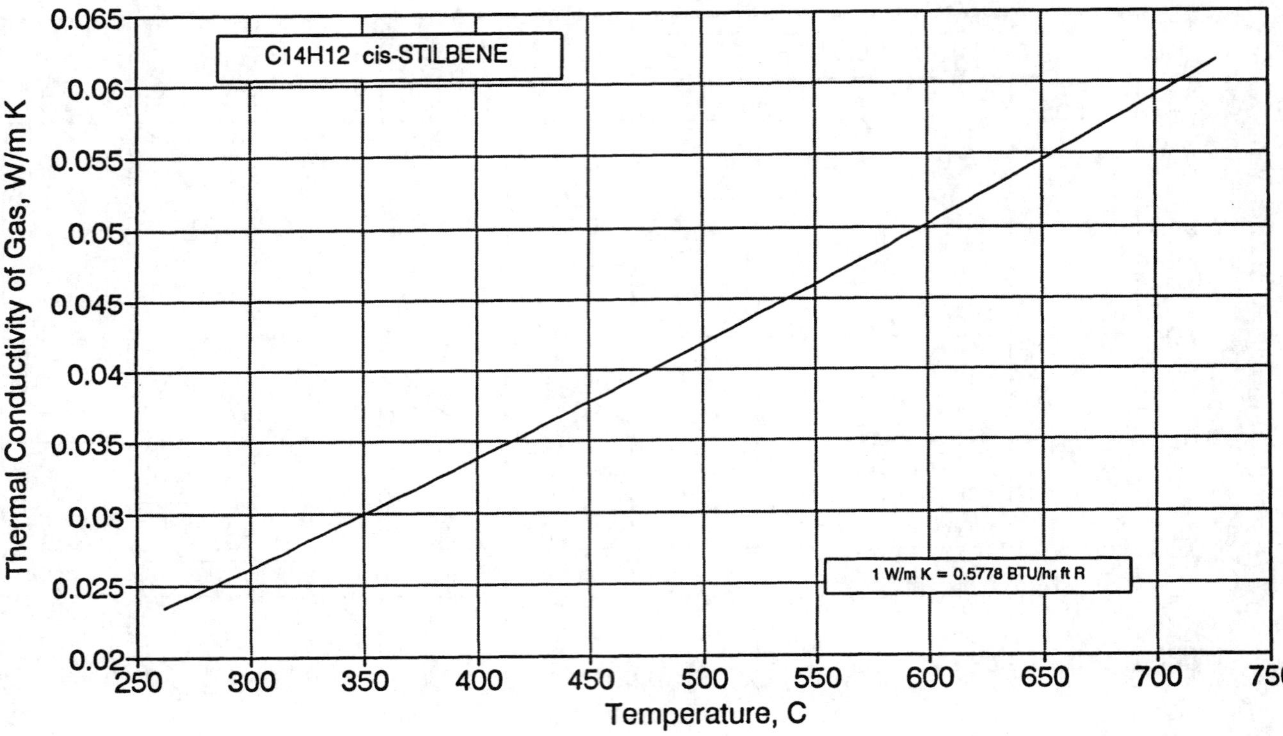
C14H12 cis-STILBENE
1 W/m K = 0.5778 BTU/hr ft R
Thermal Conductivity of Gas, W/m K
0.065
0.06
0.055
0.05
0.045
0.04
0.035
0.03
0.025
0.02
250
300
350
400
450
500
550
600
650
700
750
Temperature, C

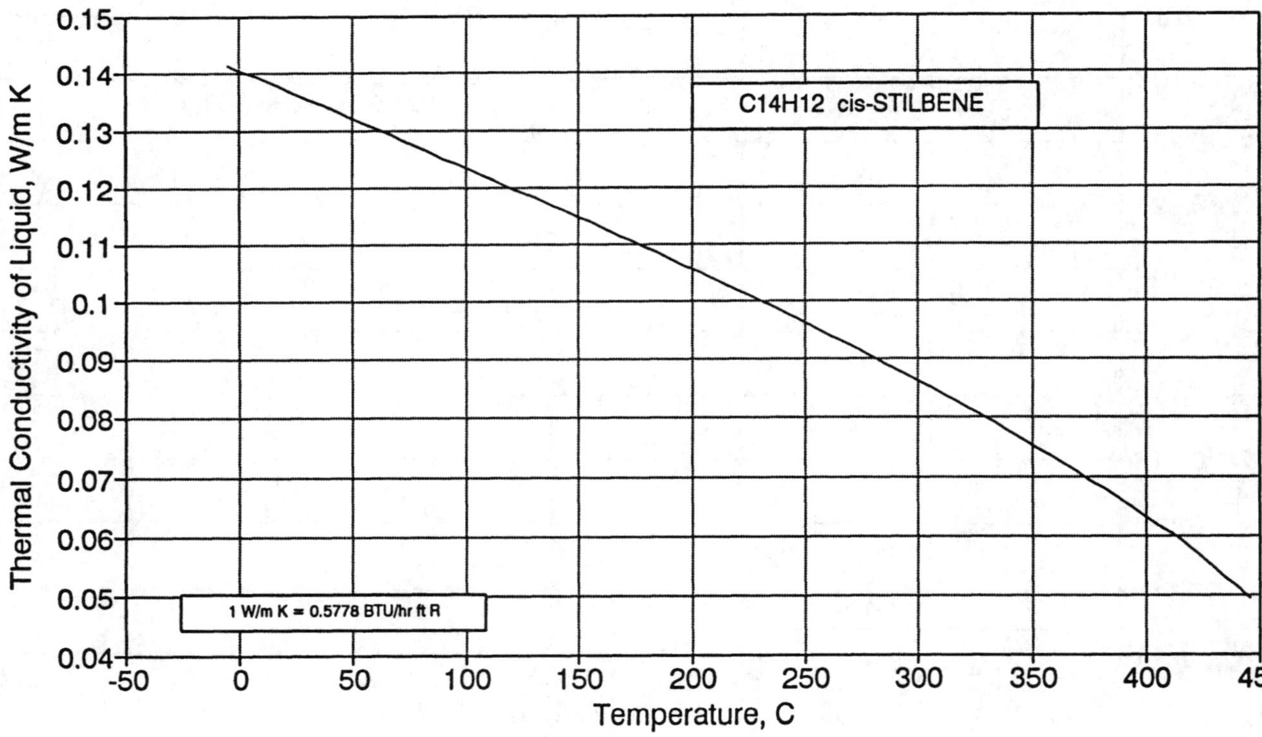
C14H12 cis-STILBENE
1 W/m K = 0.5778 BTU/hr ft R
Thermal Conductivity of Liquid, W/m K
0.15
0.14
0.13
0.12
0.11
0.1
0.09
0.08
0.07
0.06
0.05
0.04
-50
0
50
100
150
200
250
300
350
400
450
Temperature, C

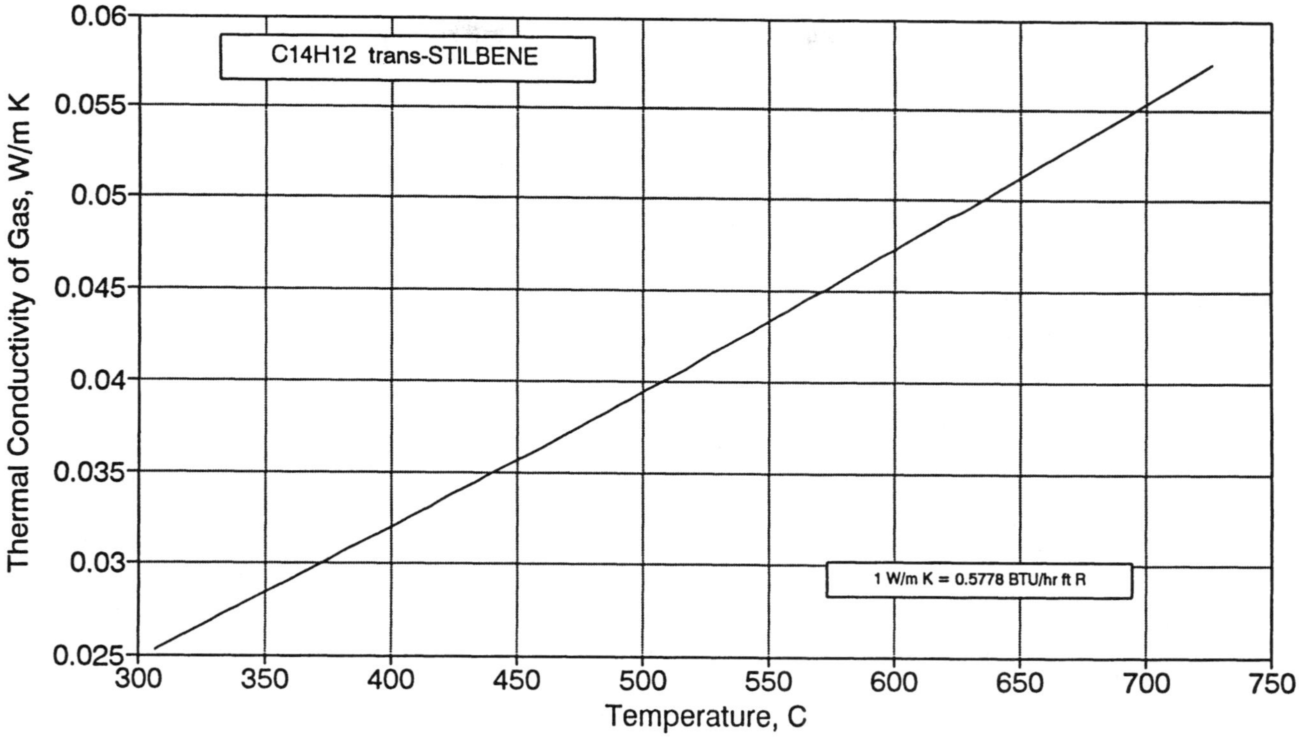
C14H12 trans-STILBENE
1 W/m K = 0.5778 BTU/hr ft R
Thermal Conductivity of Gas, W/m K
0.06
0.055
0.05
0.045
0.04
0.035
0.03
0.025
300
350
400
450
500
550
600
650
700
750
Temperature, C

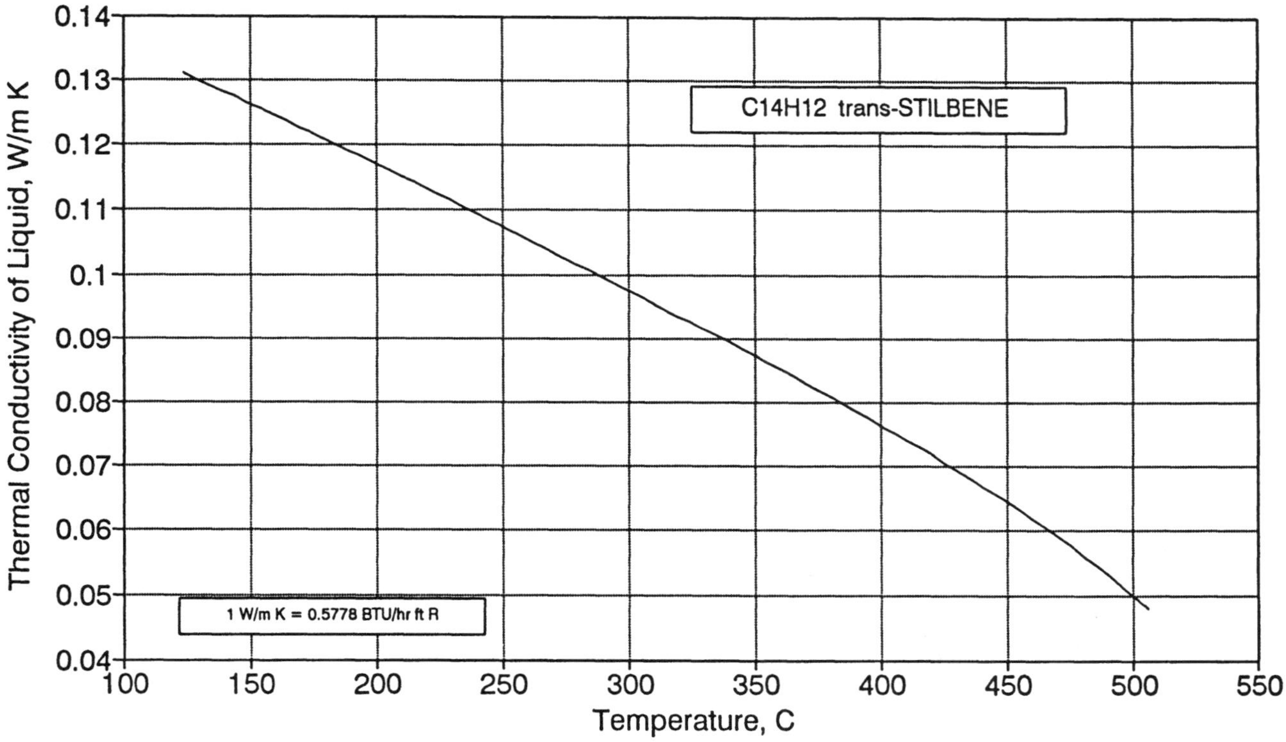
C14H12 trans-STILBENE
1 W/m K = 0.5778 BTU/hr ft R
Thermal Conductivity of Liquid, W/m K
0.14
0.13
0.12
0.11
0.1
0.09
0.08
0.07
0.06
0.05
0.04
100
150
200
250
300
350
400
450
500
550
Temperature, C

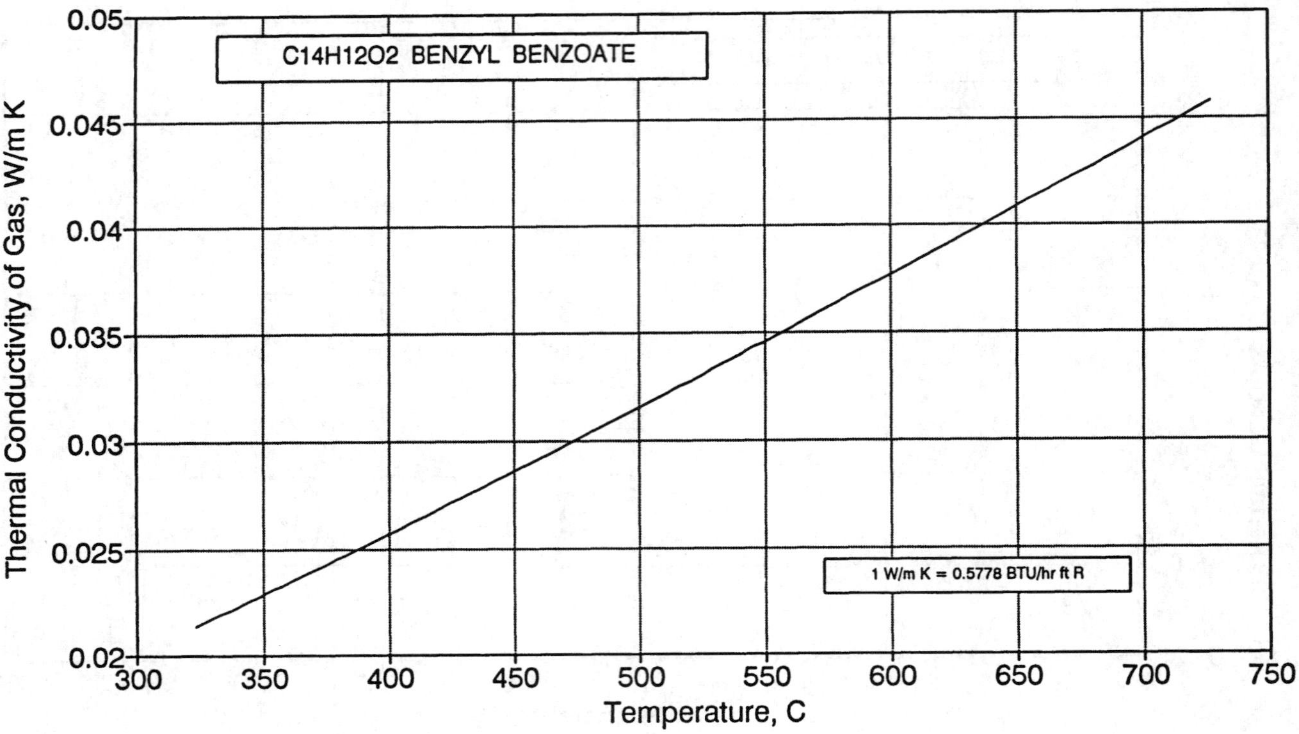
C14H12O2 BENZYL BENZOATE
1 W/m K = 0.5778 BTU/hr ft R
Thermal Conductivity of Gas, W/m K
Temperature, C

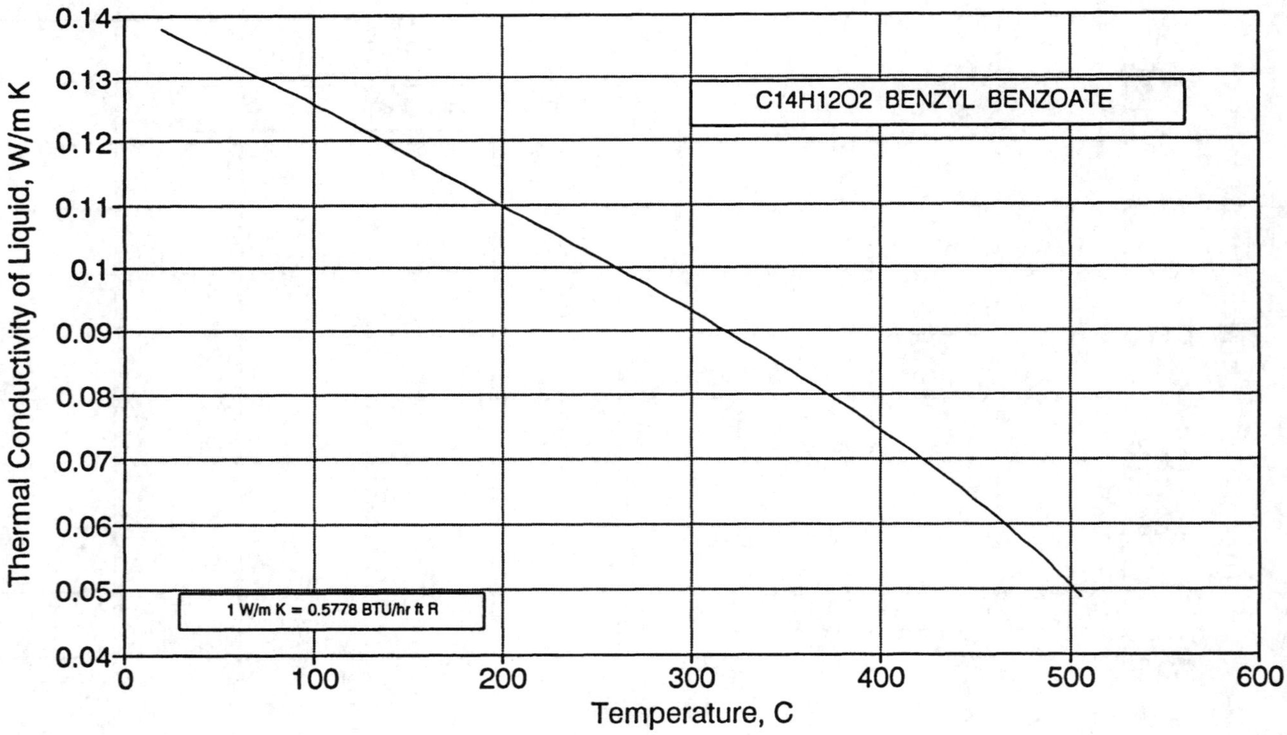
C14H12O2 BENZYL BENZOATE
1 W/m K = 0.5778 BTU/hr ft R
Thermal Conductivity of Liquid, W/m K
Temperature, C

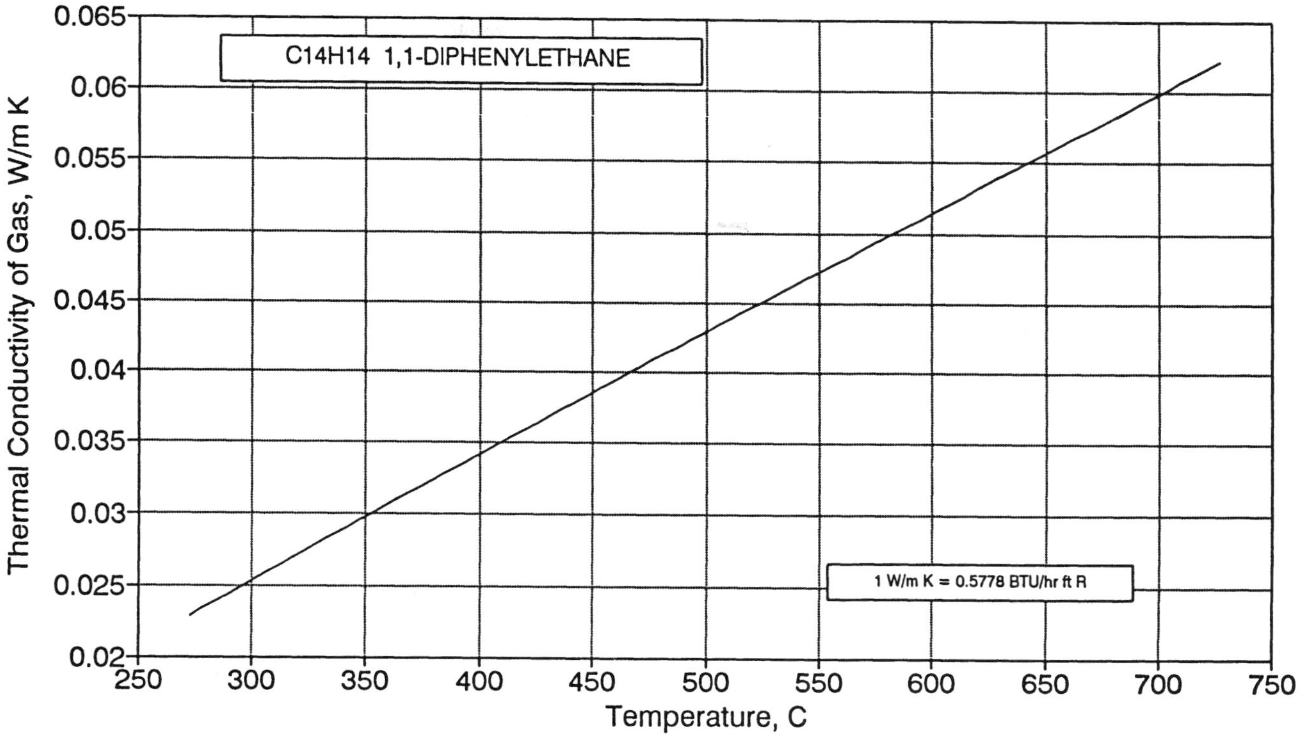
C14H14 1,1-DIPHENYLETHANE
1 W/m K = 0.5778 BTU/hr ft R
Thermal Conductivity of Gas, W/m K
0.065
0.06
0.055
0.05
0.045
0.04
0.035
0.03
0.025
0.02
250
300
350
400
450
500
550
600
650
700
750
Temperature, C

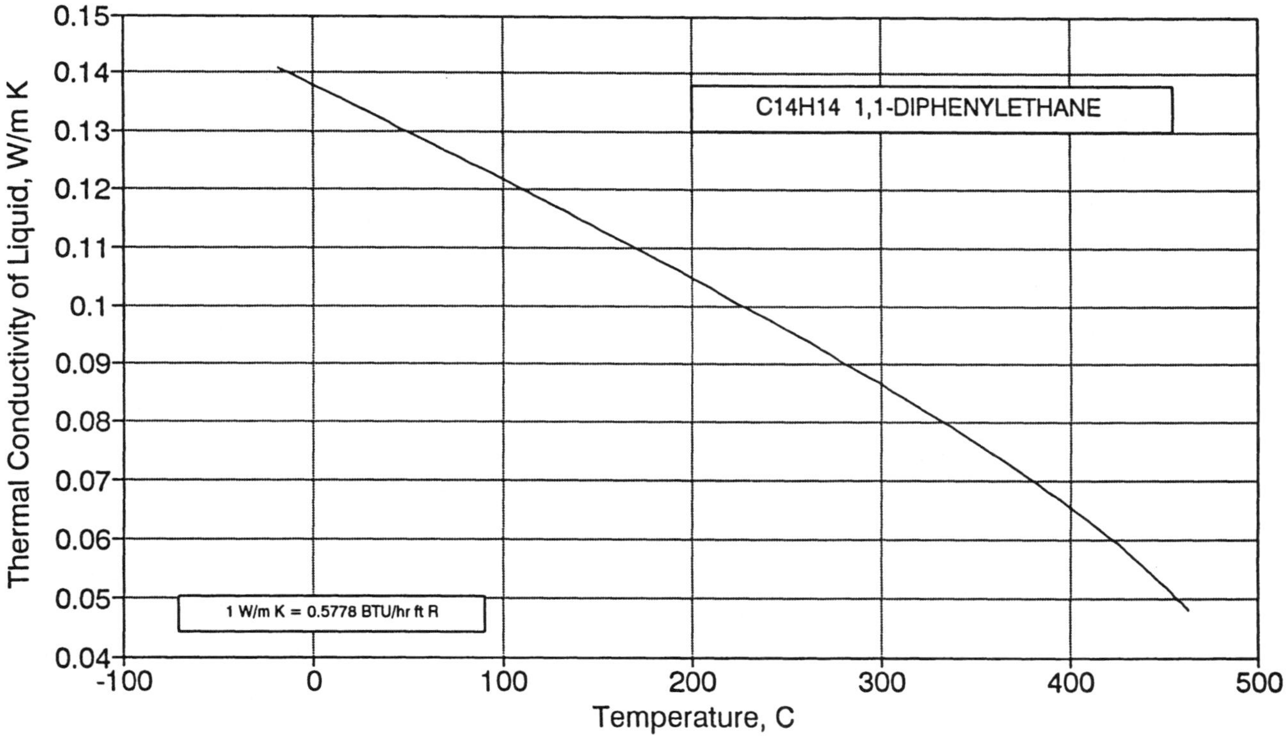
C14H14 1,1-DIPHENYLETHANE
1 W/m K = 0.5778 BTU/hr ft R
Thermal Conductivity of Liquid, W/m K
0.15
0.14
0.13
0.12
0.11
0.1
0.09
0.08
0.07
0.06
0.05
0.04
-100
0
100
200
300
400
500
Temperature, C

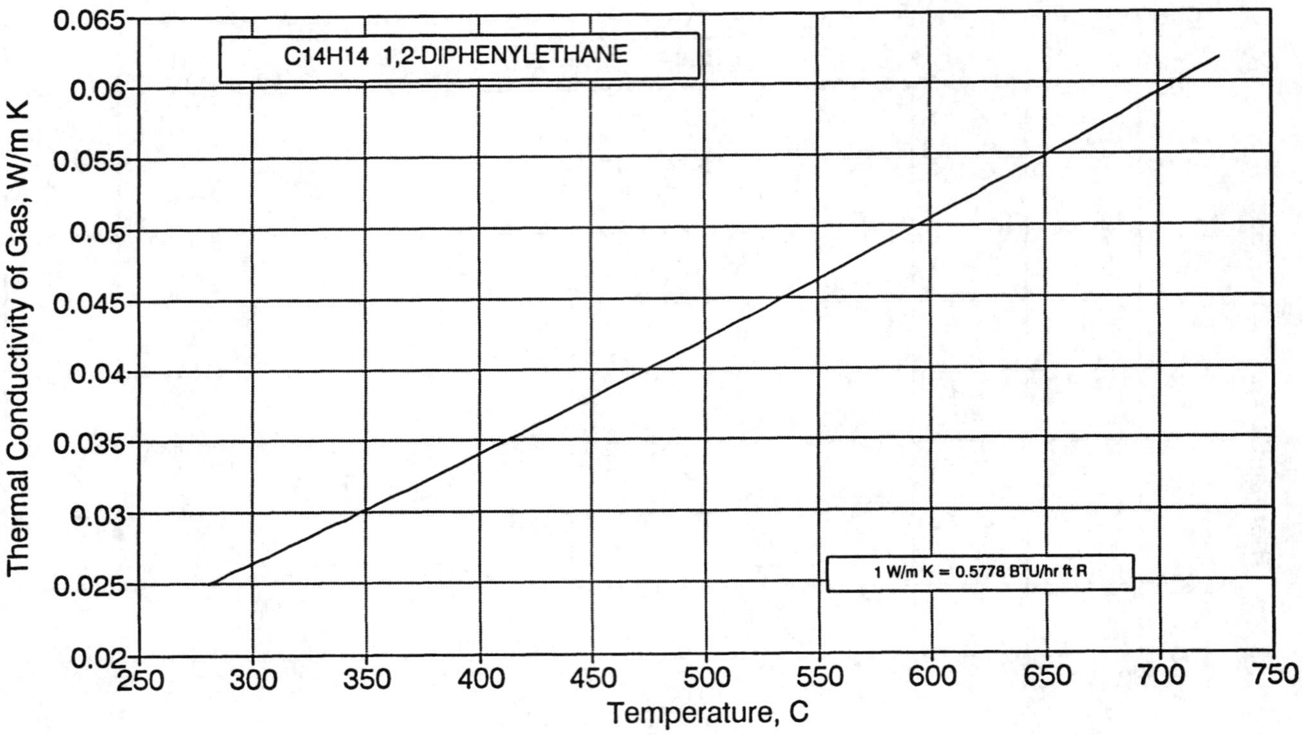
C14H14 1,2-DIPHENYLETHANE
1 W/m K = 0.5778 BTU/hr ft R
Thermal Conductivity of Gas, W/m K
0.065
0.06
0.055
0.05
0.045
0.04
0.035
0.03
0.025
0.02
250
300
350
400
450
500
550
600
650
700
750
Temperature, C

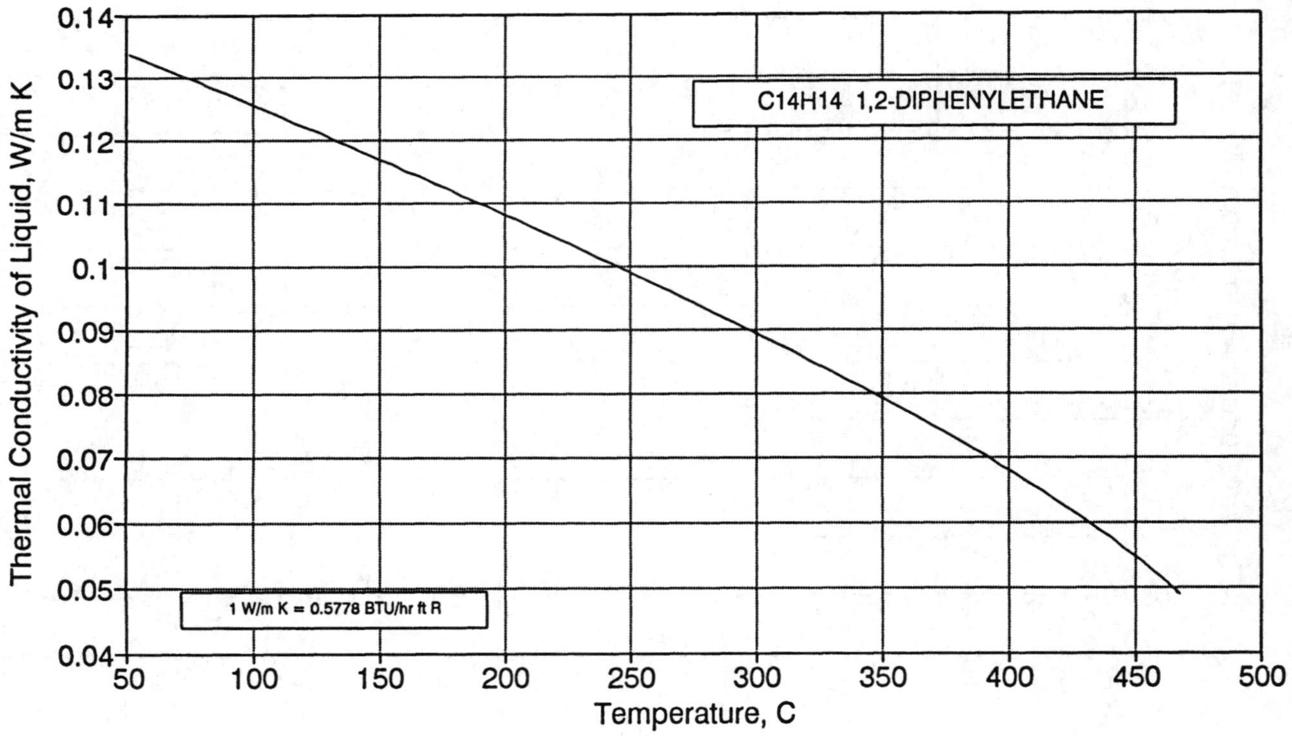
C14H14 1,2-DIPHENYLETHANE
1 W/m K = 0.5778 BTU/hr ft R
Thermal Conductivity of Liquid, W/m K
0.14
0.13
0.12
0.11
0.1
0.09
0.08
0.07
0.06
0.05
0.04
50
100
150
200
250
300
350
400
450
500
Temperature, C

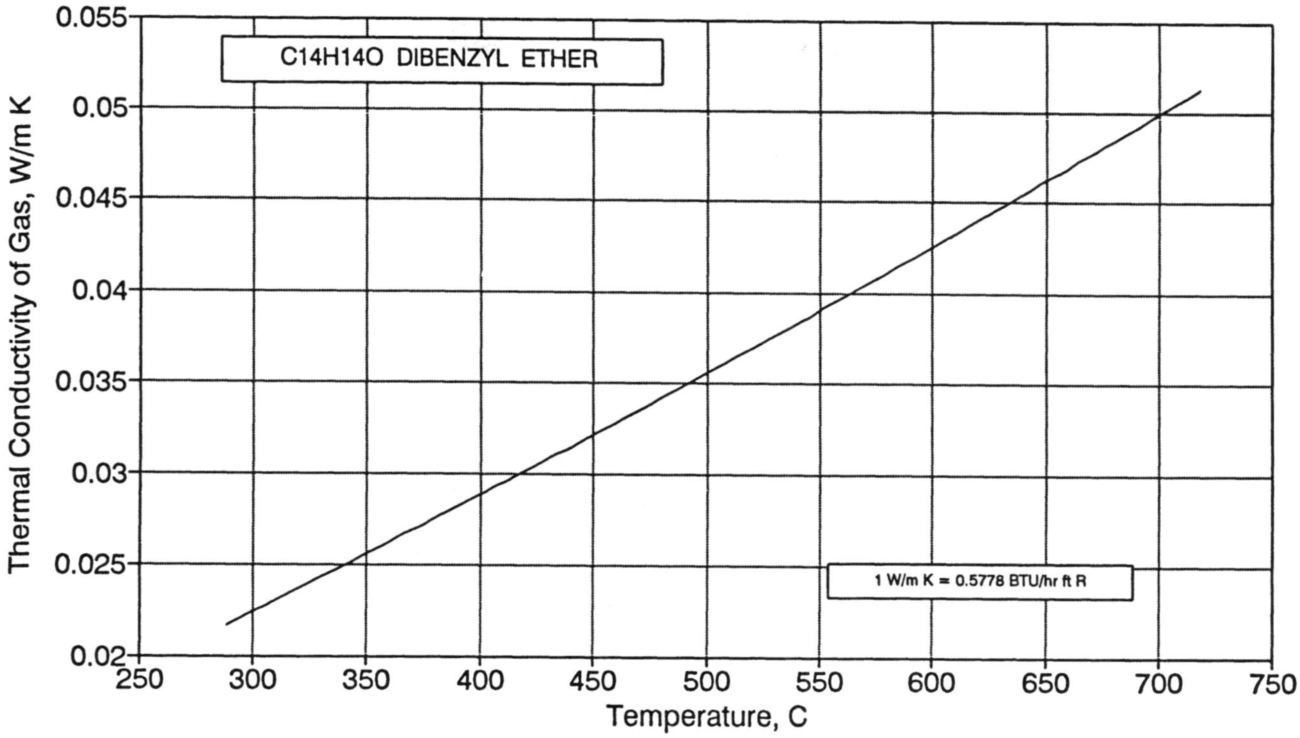
C14H14O DIBENZYL ETHER
1 W/m K = 0.5778 BTU/hr ft R
Thermal Conductivity of Gas, W/m K
0.055
0.05
0.045
0.04
0.035
0.03
0.025
0.02
250
300
350
400
450
500
550
600
650
700
750
Temperature, C

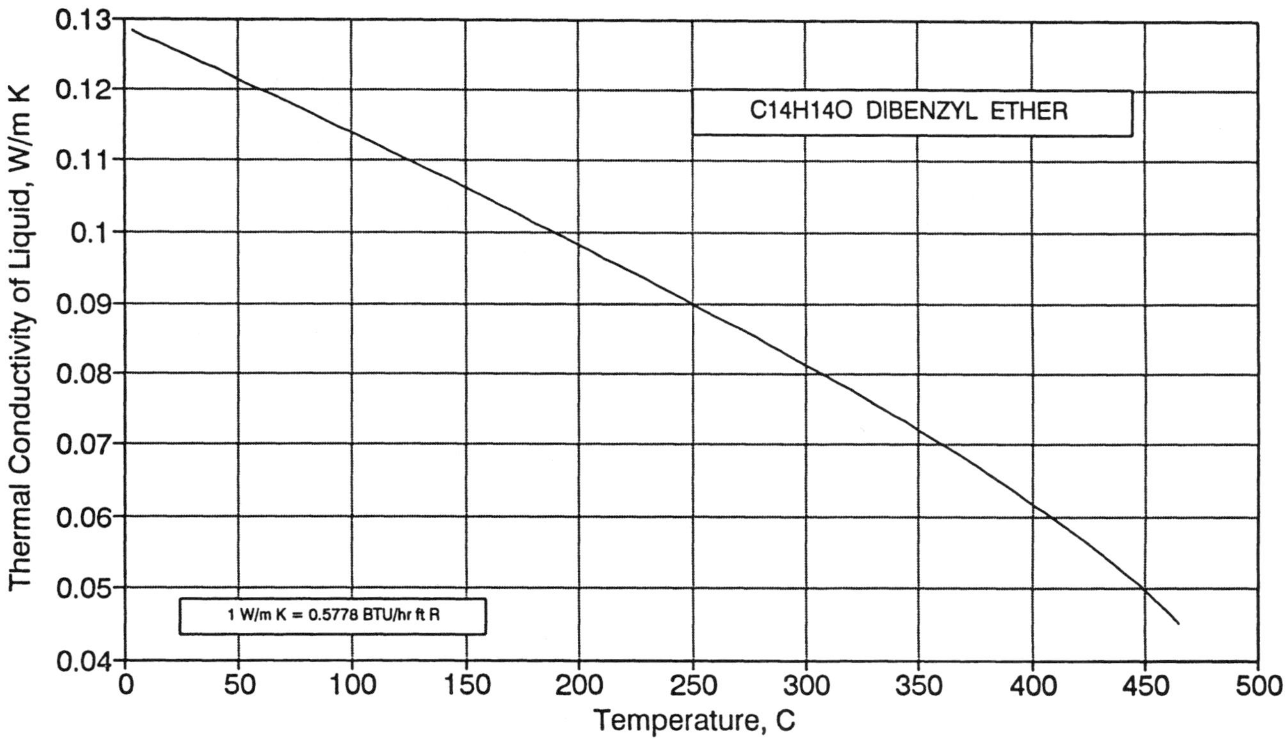
C14H14O DIBENZYL ETHER
1 W/m K = 0.5778 BTU/hr ft R
Thermal Conductivity of Liquid, W/m K
0.13
0.12
0.11
0.1
0.09
0.08
0.07
0.06
0.05
0.04
0
50
100
150
200
250
300
350
400
450
500
Temperature, C

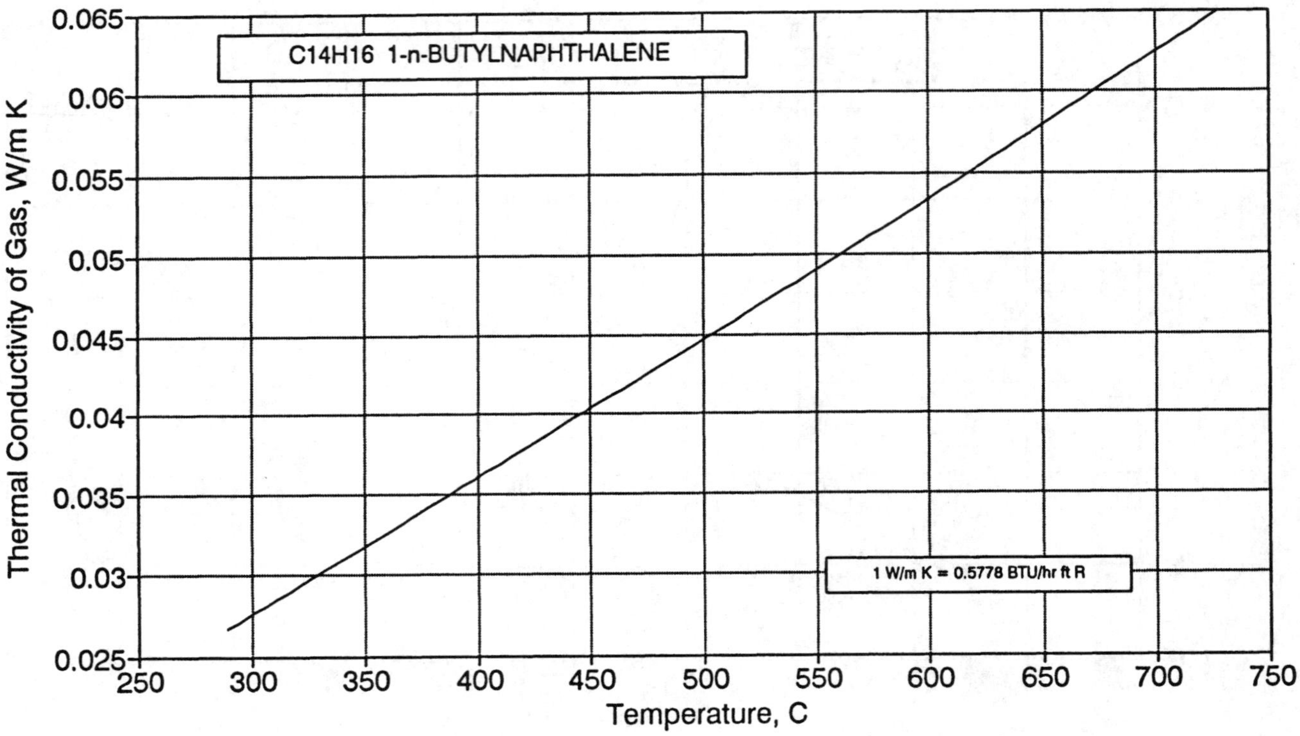
C14H16 1-n-BUTYLNAPHTHALENE
Thermal Conductivity of Gas, W/m K
0.065
0.06
0.055
0.05
0.045
0.04
0.035
0.03
0.025
1 W/m K = 0.5778 BTU/hr ft R
250
300
350
400
450
500
550
600
650
700
750
Temperature, C

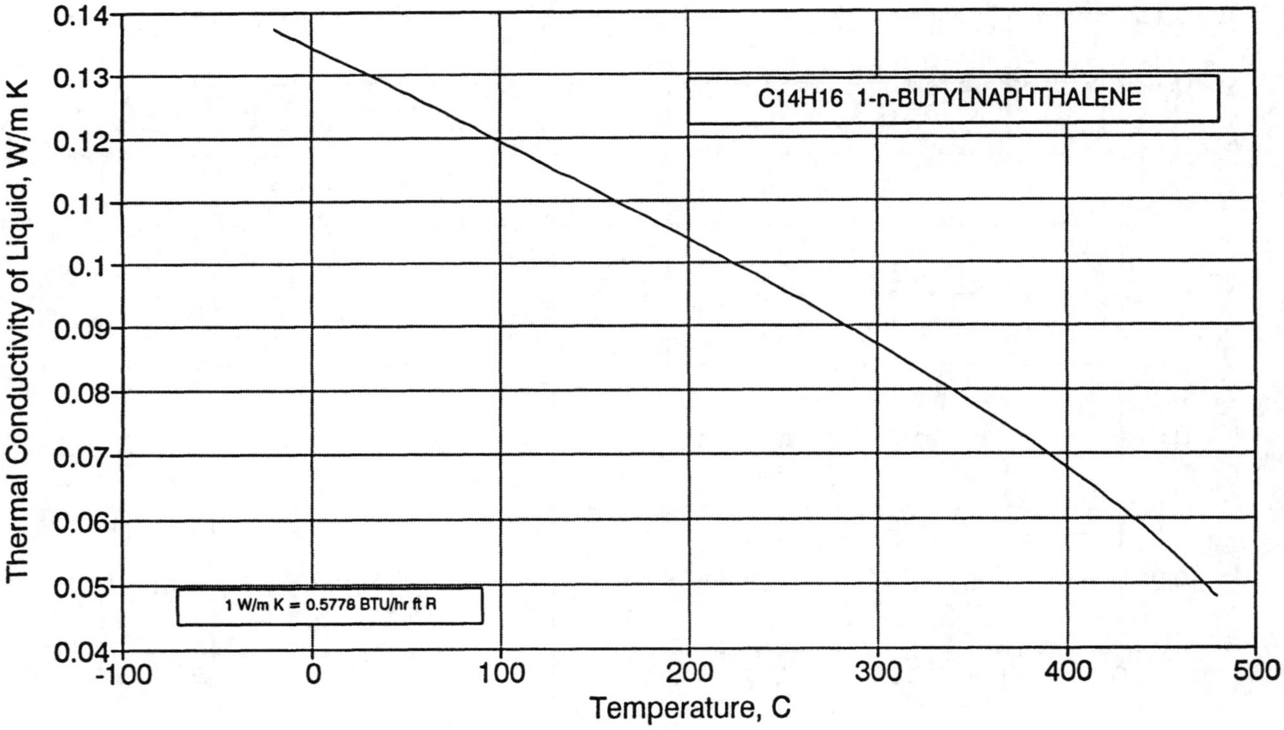
C14H16 1-n-BUTYLNAPHTHALENE
Thermal Conductivity of Liquid, W/m K
0.14
0.13
0.12
0.11
0.1
0.09
0.08
0.07
0.06
0.05
0.04
1 W/m K = 0.5778 BTU/hr ft R
-100
0
100
200
300
400
500
Temperature, C

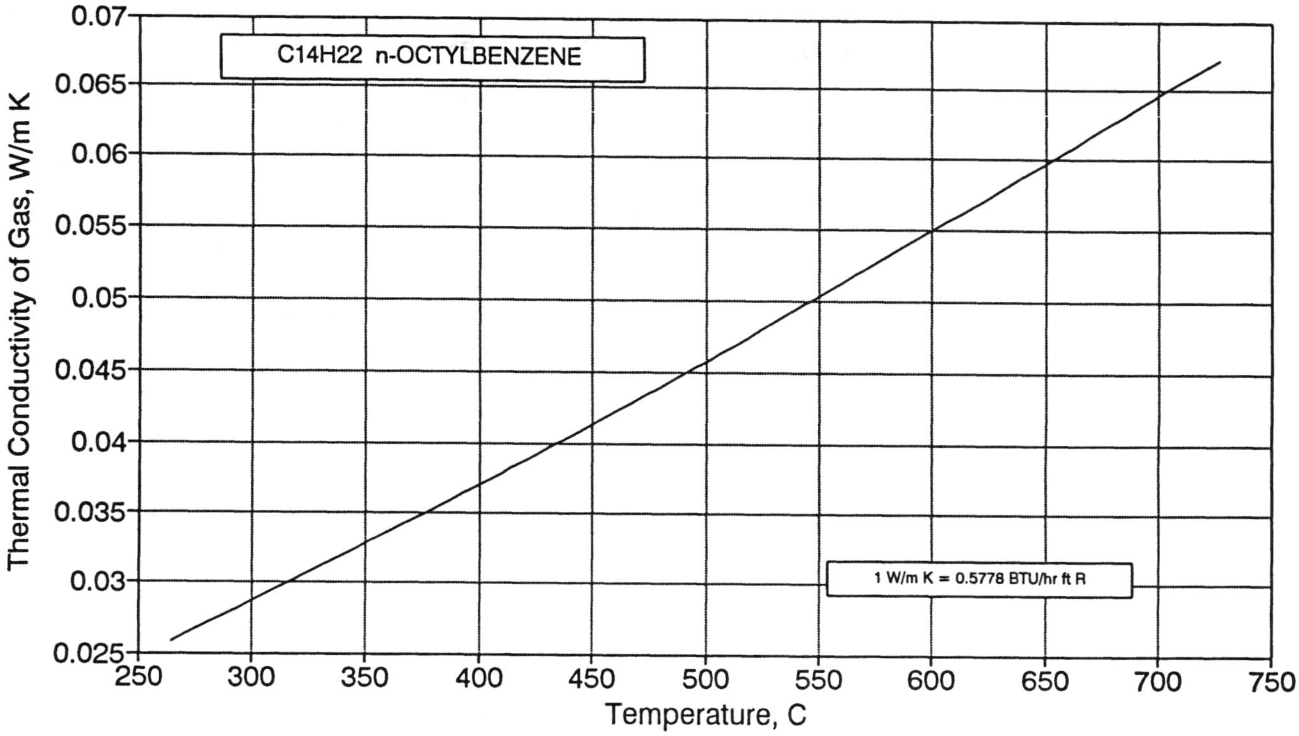
C14H22 n-OCTYLBENZENE
1 W/m K = 0.5778 BTU/hr ft R
Thermal Conductivity of Gas, W/m K
0.07
0.065
0.06
0.055
0.05
0.045
0.04
0.035
0.03
0.025
250
300
350
400
450
500
550
600
650
700
750
Temperature, C

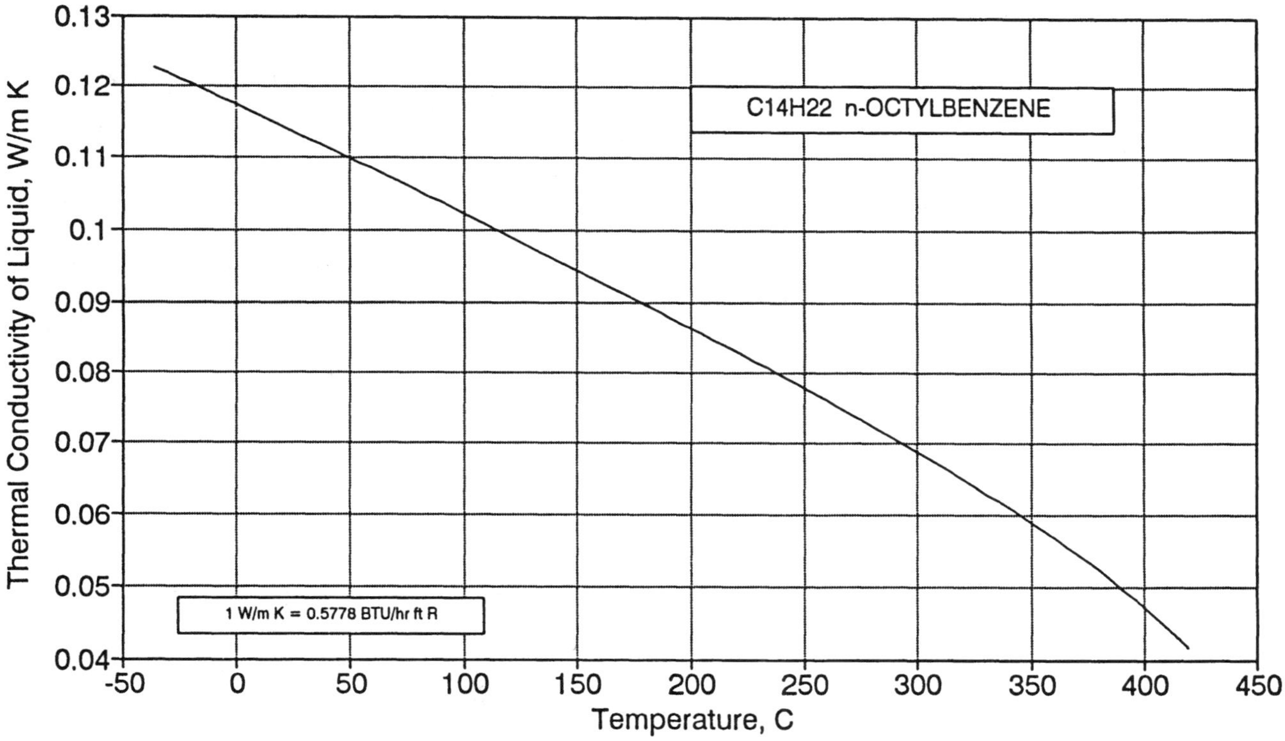
C14H22 n-OCTYLBENZENE
1 W/m K = 0.5778 BTU/hr ft R
Thermal Conductivity of Liquid, W/m K
0.13
0.12
0.11
0.1
0.09
0.08
0.07
0.06
0.05
0.04
-50
0
50
100
150
200
250
300
350
400
450
Temperature, C

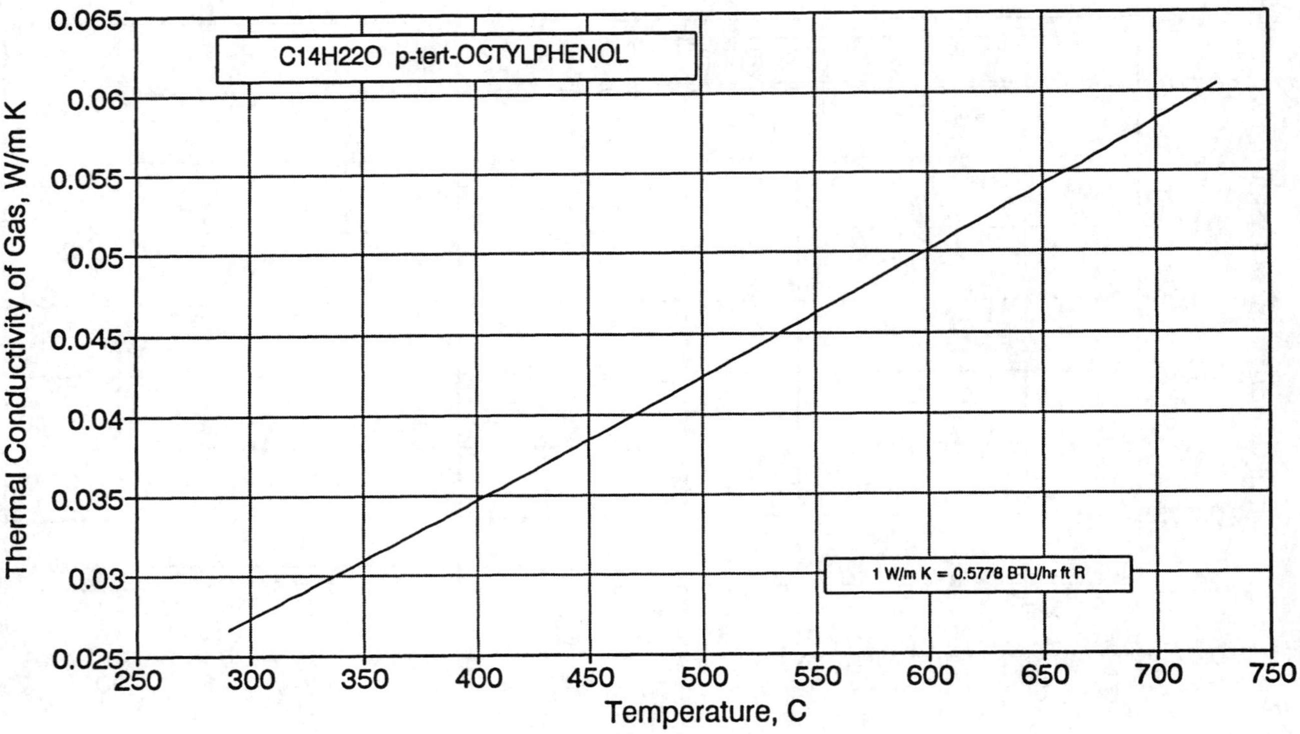
C14H22O p-tert-OCTYLPHENOL
1 W/m K = 0.5778 BTU/hr ft R
Thermal Conductivity of Gas, W/m K
0.065
0.06
0.055
0.05
0.045
0.04
0.035
0.03
0.025
250
300
350
400
450
500
550
600
650
700
750
Temperature, C

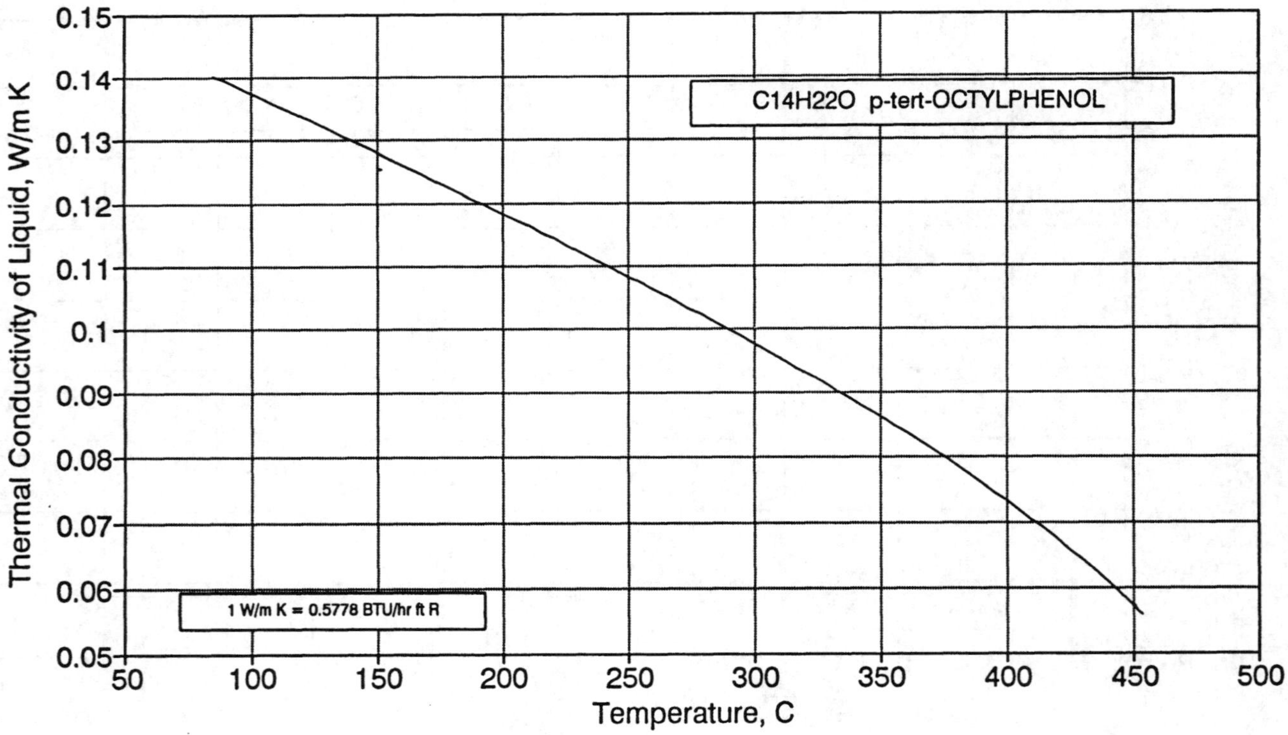
C14H22O p-tert-OCTYLPHENOL
1 W/m K = 0.5778 BTU/hr ft R
Thermal Conductivity of Liquid, W/m K
0.15
0.14
0.13
0.12
0.11
0.1
0.09
0.08
0.07
0.06
0.05
50
100
150
200
250
300
350
400
450
500
Temperature, C

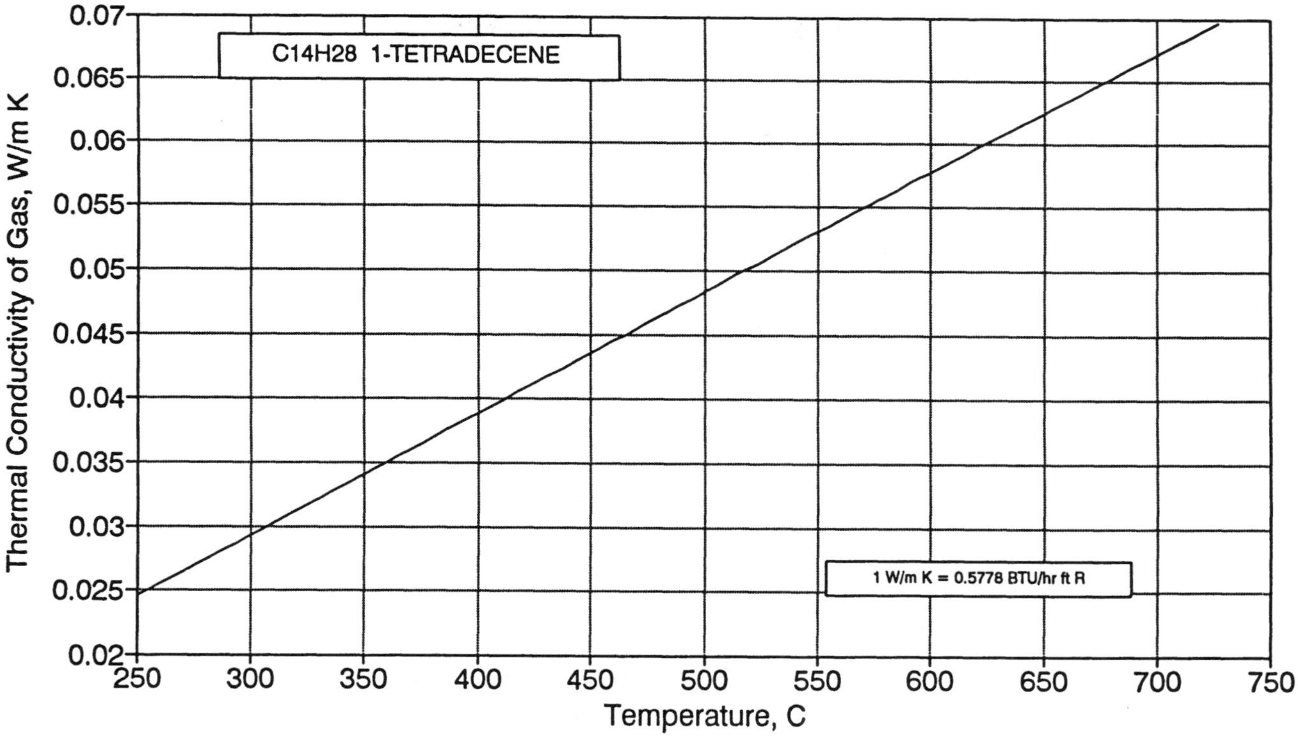
C14H28 1-TETRADECENE
1 W/m K = 0.5778 BTU/hr ft R
Thermal Conductivity of Gas, W/m K
0.07
0.065
0.06
0.055
0.05
0.045
0.04
0.035
0.03
0.025
0.02
250
300
350
400
450
500
550
600
650
700
750
Temperature, C

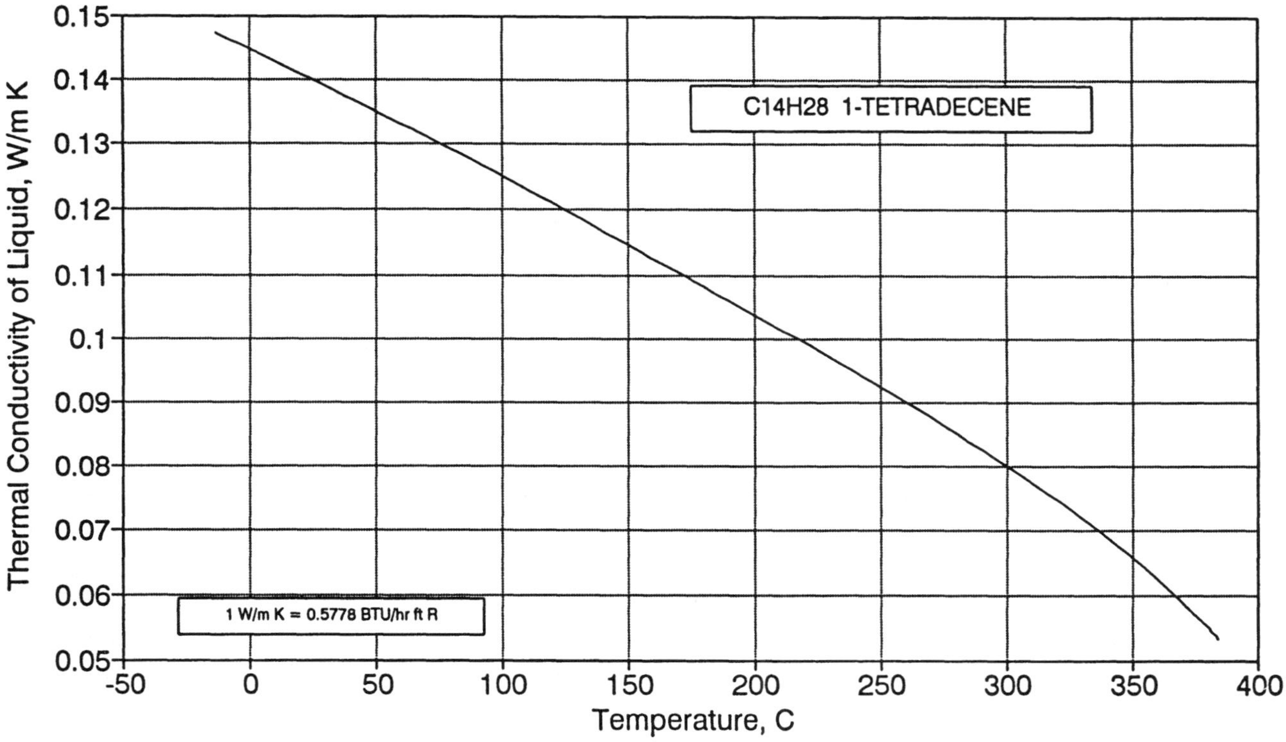
C14H28 1-TETRADECENE
1 W/m K = 0.5778 BTU/hr ft R
Thermal Conductivity of Liquid, W/m K
0.15
0.14
0.13
0.12
0.11
0.1
0.09
0.08
0.07
0.06
0.05
-50
0
50
100
150
200
250
300
350
400
Temperature, C

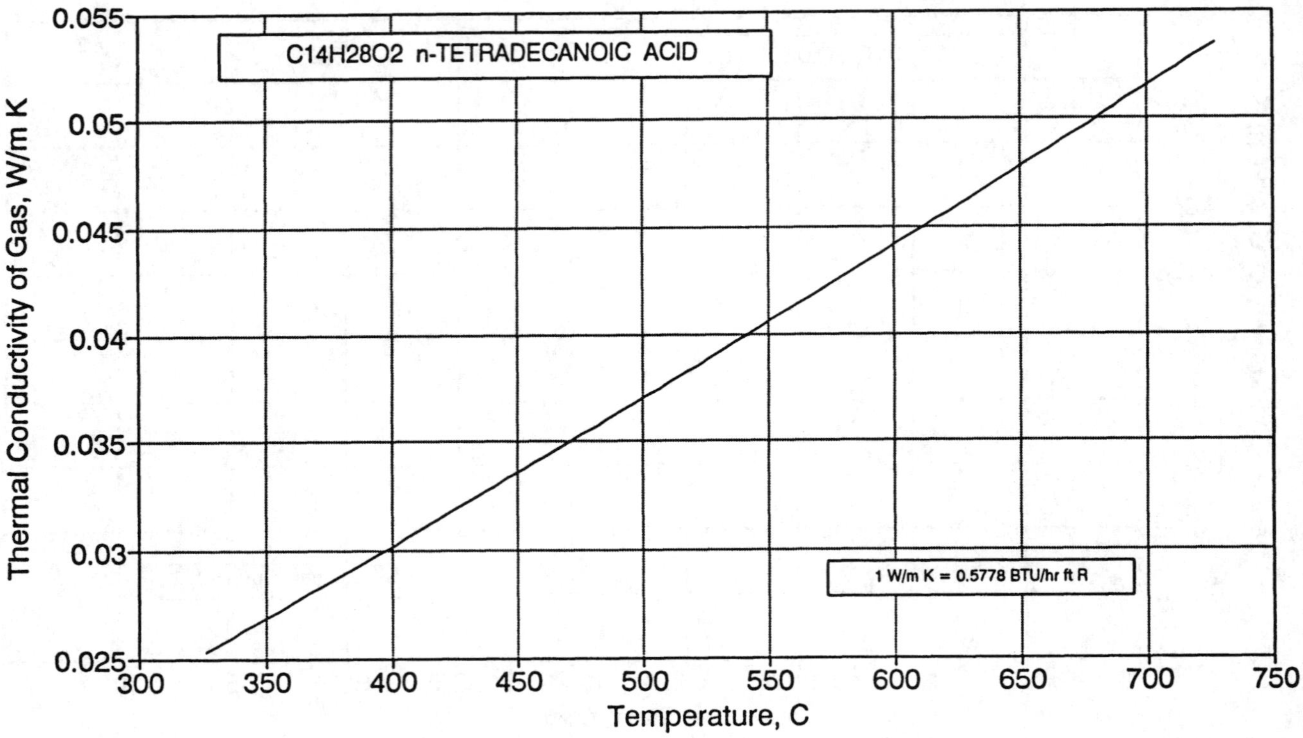

C14H28O2 n-TETRADECANOIC ACID
1 W/m K = 0.5778 BTU/hr ft R
Thermal Conductivity of Gas, W/m K
0.055
0.05
0.045
0.04
0.035
0.03
0.025
300
350
400
450
500
550
600
650
700
750
Temperature, C

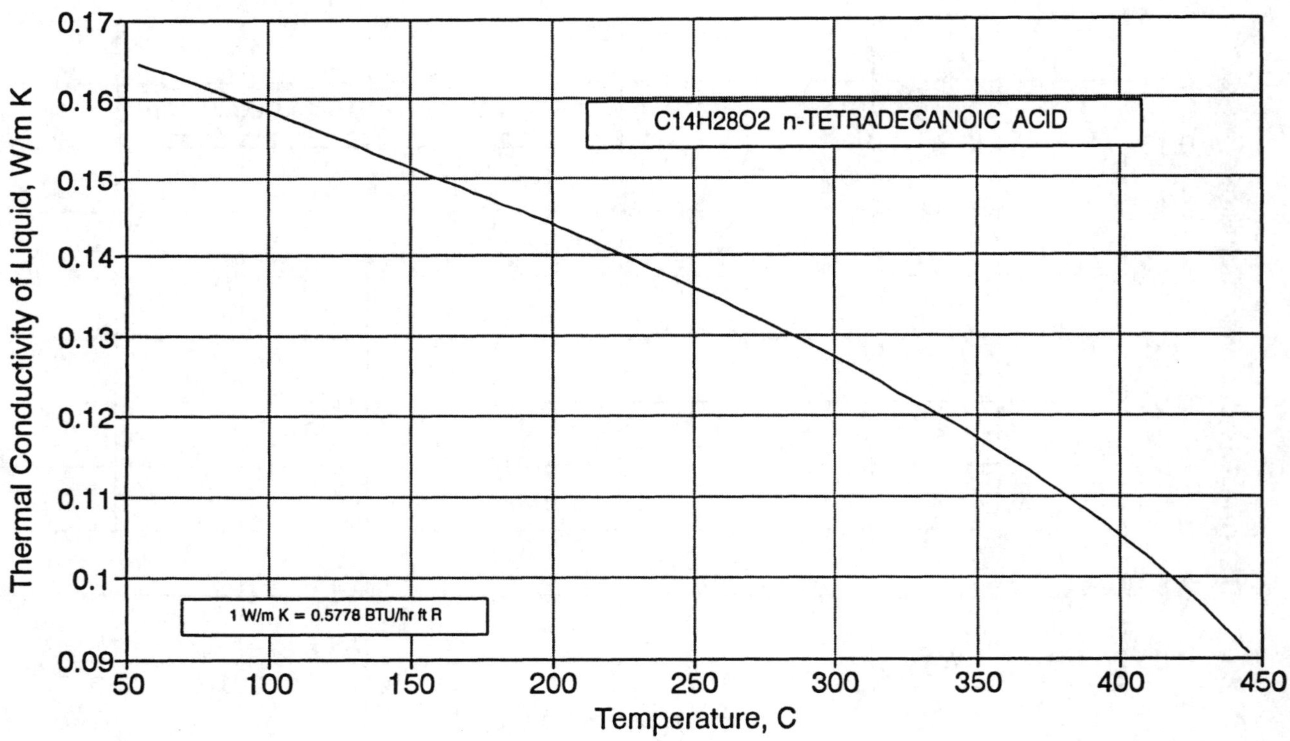

C14H28O2 n-TETRADECANOIC ACID
1 W/m K = 0.5778 BTU/hr ft R
Thermal Conductivity of Liquid, W/m K
0.17
0.16
0.15
0.14
0.13
0.12
0.11
0.1
0.09
50
100
150
200
250
300
350
400
450
Temperature, C

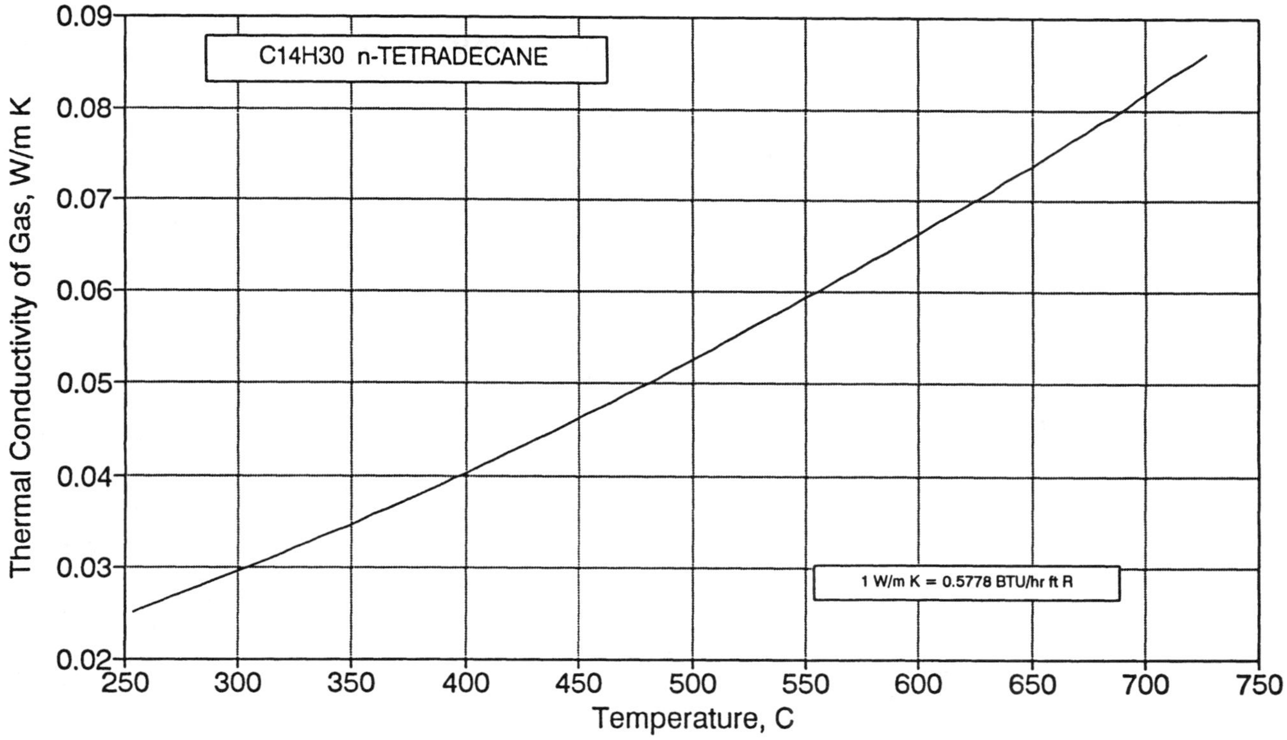
C14H30 n-TETRADECANE
1 W/m K = 0.5778 BTU/hr ft R
Thermal Conductivity of Gas, W/m K
Temperature, C
0.09
0.08
0.07
0.06
0.05
0.04
0.03
0.02
250
300
350
400
450
500
550
600
650
700
750

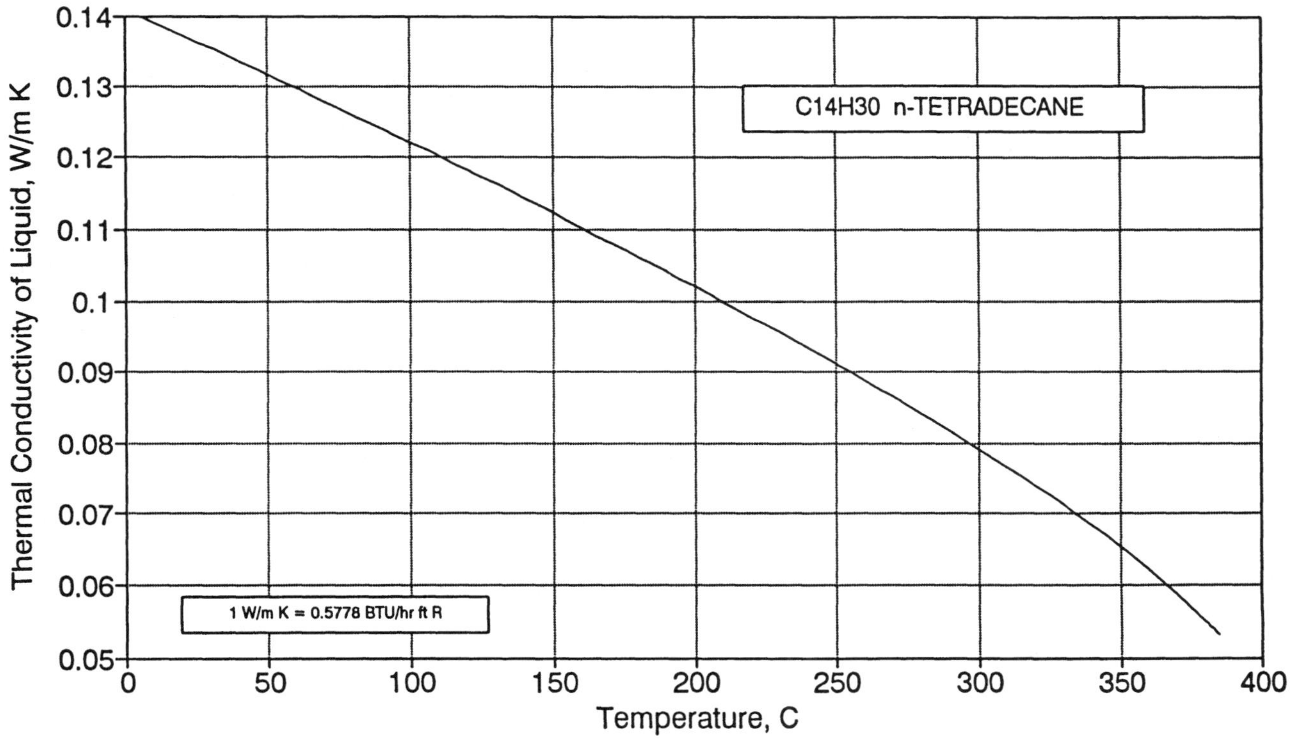
C14H30 n-TETRADECANE
1 W/m K = 0.5778 BTU/hr ft R
Thermal Conductivity of Liquid, W/m K
Temperature, C
0.14
0.13
0.12
0.11
0.1
0.09
0.08
0.07
0.06
0.05
0
50
100
150
200
250
300
350
400

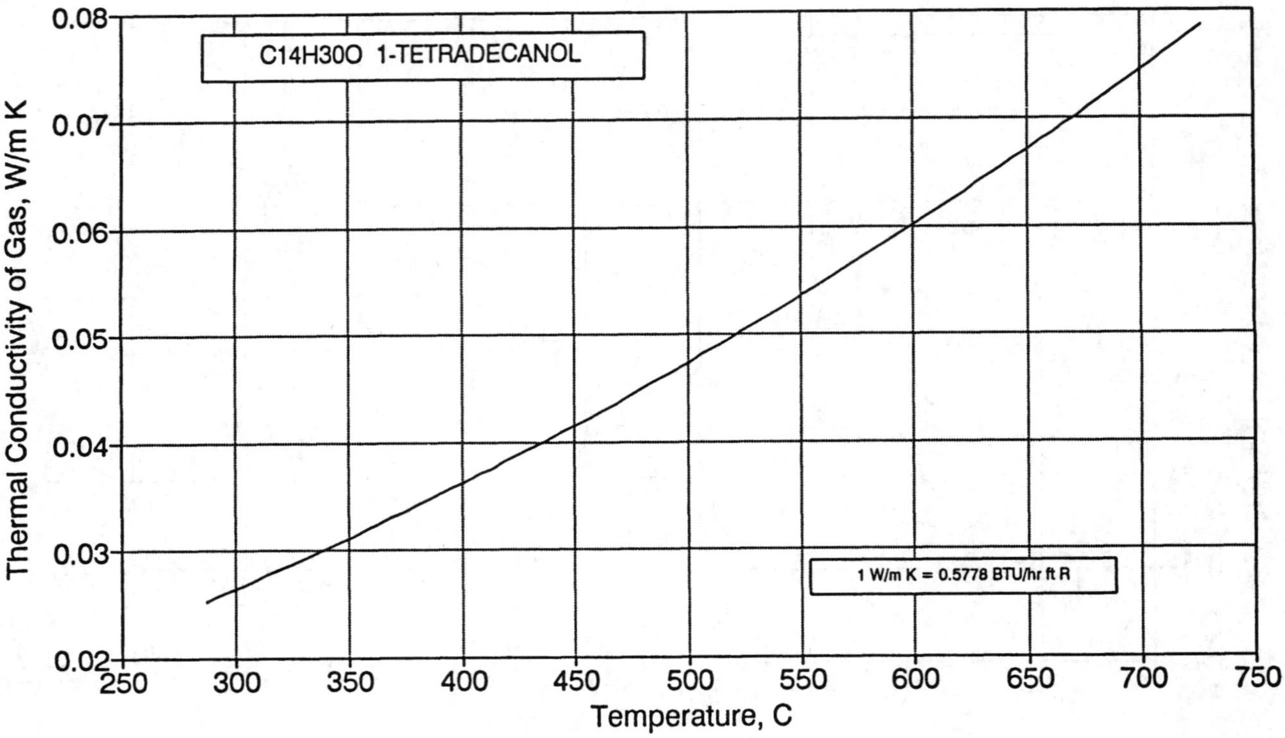
C14H30O 1-TETRADECANOL
1 W/m K = 0.5778 BTU/hr ft R
Thermal Conductivity of Gas, W/m K
0.08
0.07
0.06
0.05
0.04
0.03
0.02
250
300
350
400
450
500
550
600
650
700
750
Temperature, C

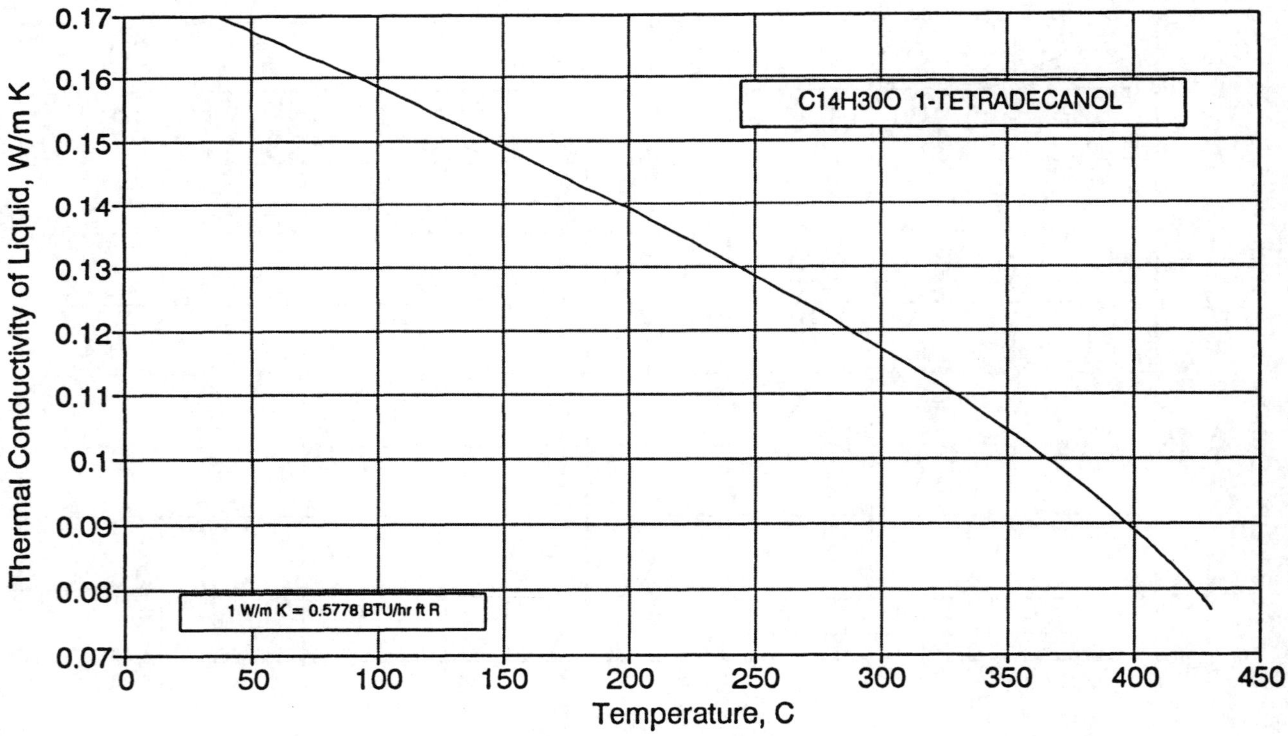
C14H30O 1-TETRADECANOL
1 W/m K = 0.5778 BTU/hr ft R
Thermal Conductivity of Liquid, W/m K
0.17
0.16
0.15
0.14
0.13
0.12
0.11
0.1
0.09
0.08
0.07
0
50
100
150
200
250
300
350
400
450
Temperature, C

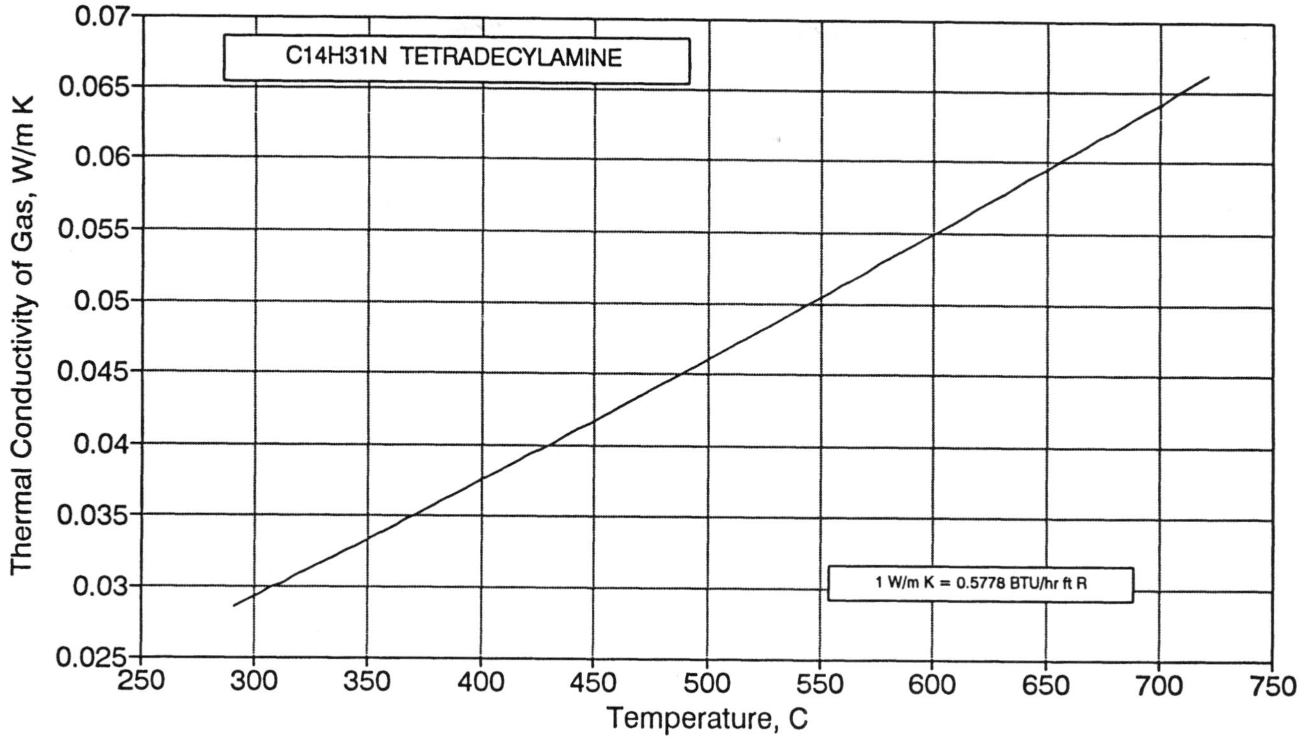
C14H31N TETRADECYLAMINE
1 W/m K = 0.5778 BTU/hr ft R
Thermal Conductivity of Gas, W/m K
Temperature, C
0.07
0.065
0.06
0.055
0.05
0.045
0.04
0.035
0.03
0.025
250
300
350
400
450
500
550
600
650
700
750

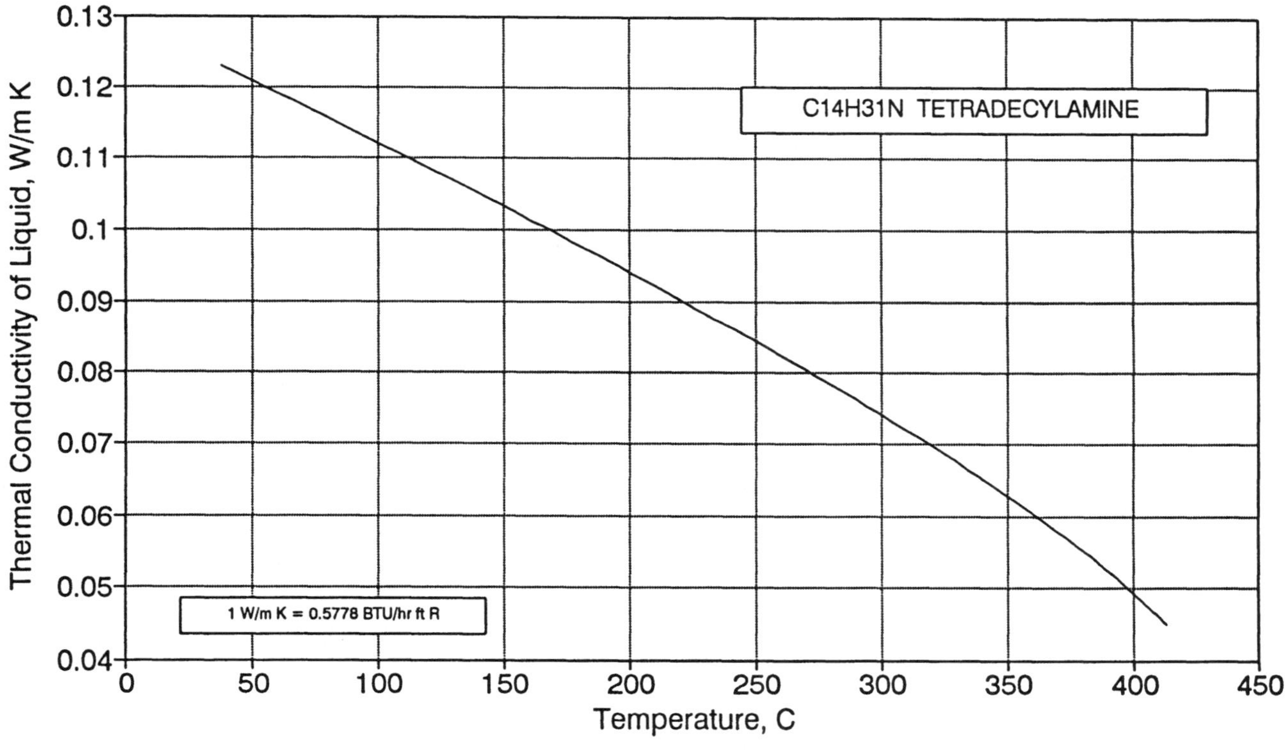
C14H31N TETRADECYLAMINE
1 W/m K = 0.5778 BTU/hr ft R
Thermal Conductivity of Liquid, W/m K
Temperature, C
0.13
0.12
0.11
0.1
0.09
0.08
0.07
0.06
0.05
0.04
0
50
100
150
200
250
300
350
400
450

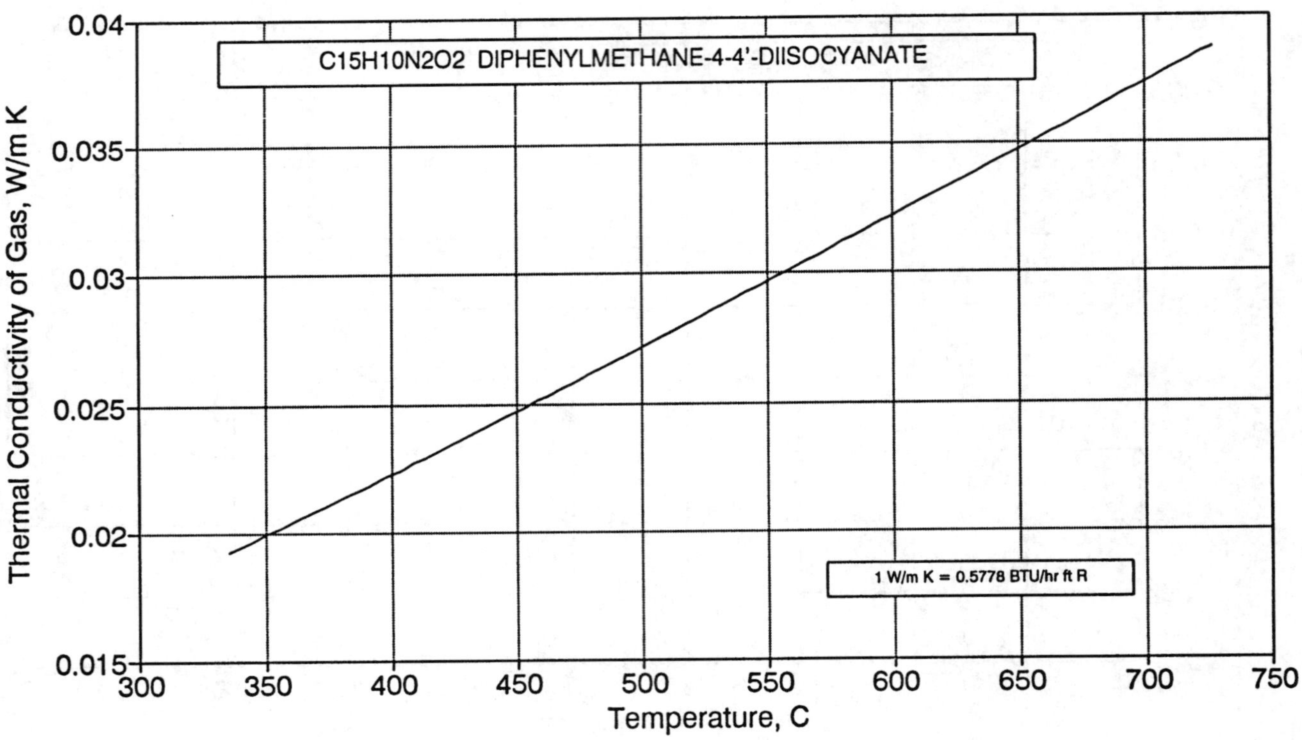
C15H10N2O2 DIPHENYLMETHANE-4-4'-DIISOCYANATE
1 W/m K = 0.5778 BTU/hr ft R
Thermal Conductivity of Gas, W/m K
0.04
0.035
0.03
0.025
0.02
0.015
300
350
400
450
500
550
600
650
700
750
Temperature, C

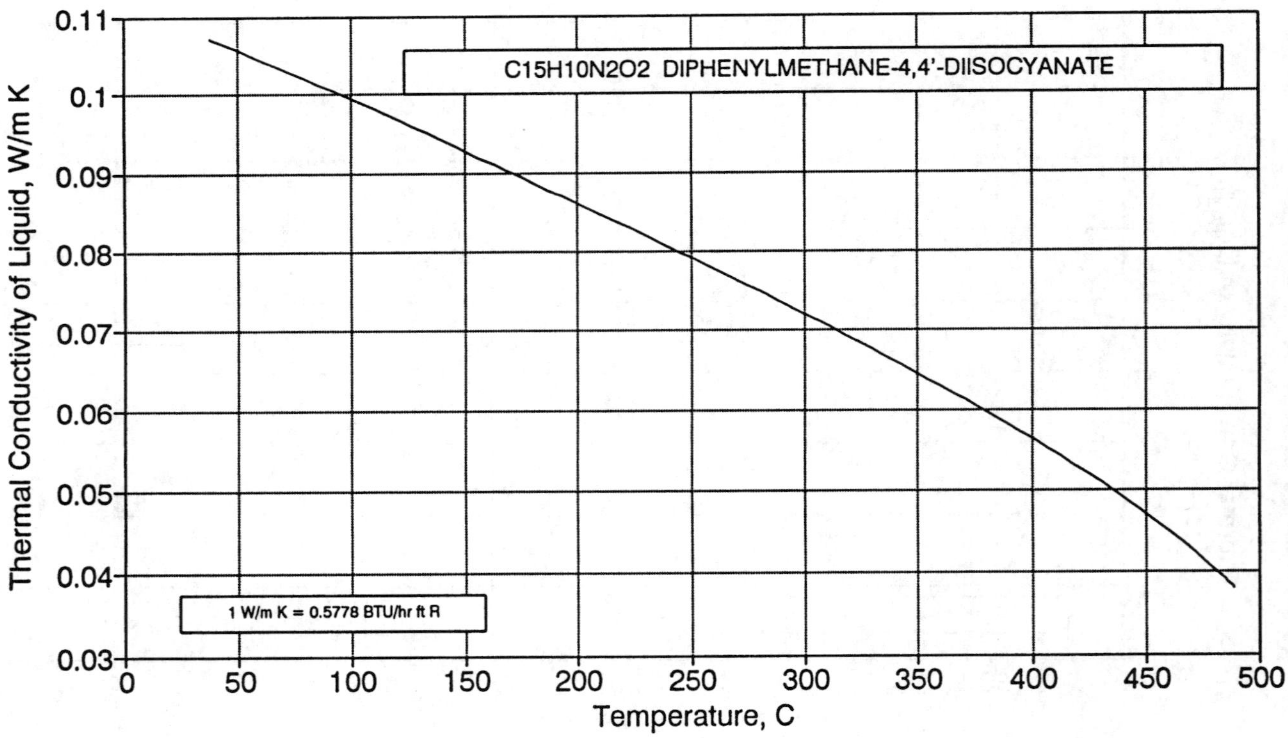
C15H10N2O2 DIPHENYLMETHANE-4,4'-DIISOCYANATE
1 W/m K = 0.5778 BTU/hr ft R
Thermal Conductivity of Liquid, W/m K
0.11
0.1
0.09
0.08
0.07
0.06
0.05
0.04
0.03
0
50
100
150
200
250
300
350
400
450
500
Temperature, C

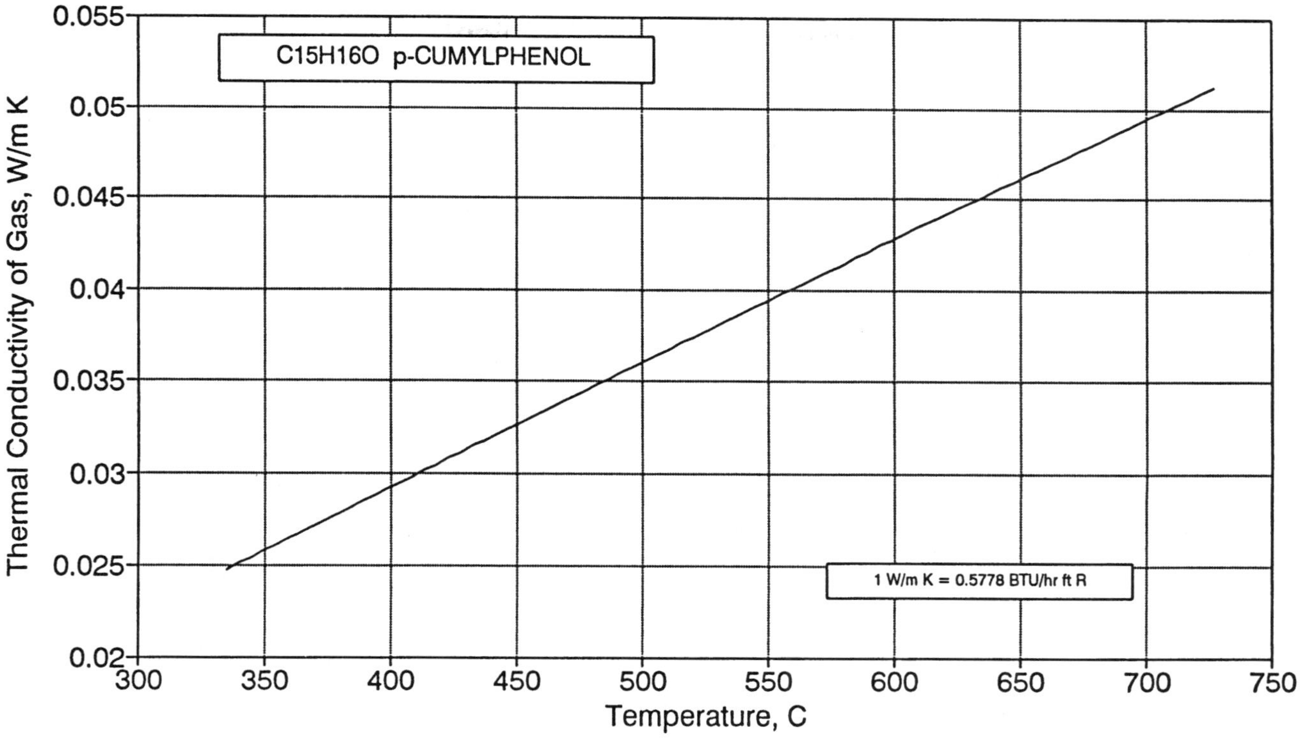
C15H16O p-CUMYLPHENOL
1 W/m K = 0.5778 BTU/hr ft R
Thermal Conductivity of Gas, W/m K
Temperature, C
0.055
0.05
0.045
0.04
0.035
0.03
0.025
0.02
300
350
400
450
500
550
600
650
700
750

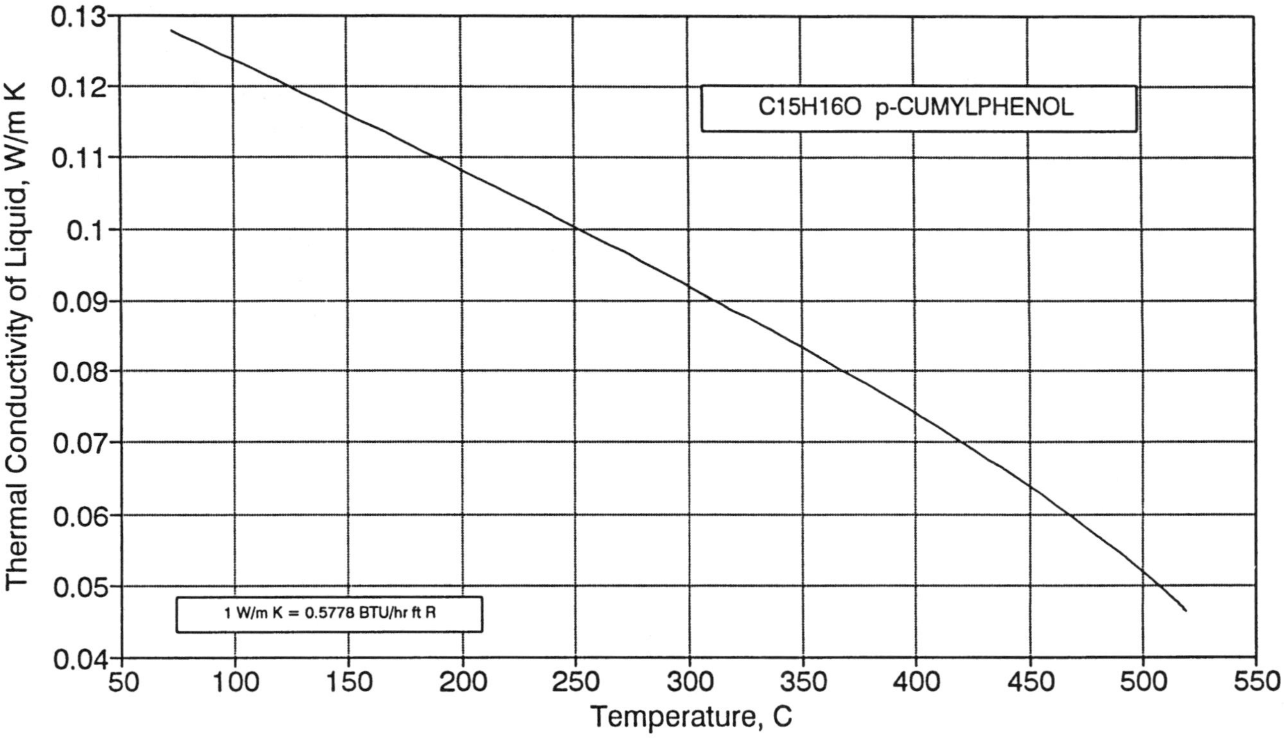
C15H16O p-CUMYLPHENOL
1 W/m K = 0.5778 BTU/hr ft R
Thermal Conductivity of Liquid, W/m K
Temperature, C
0.13
0.12
0.11
0.1
0.09
0.08
0.07
0.06
0.05
0.04
50
100
150
200
250
300
350
400
450
500
550

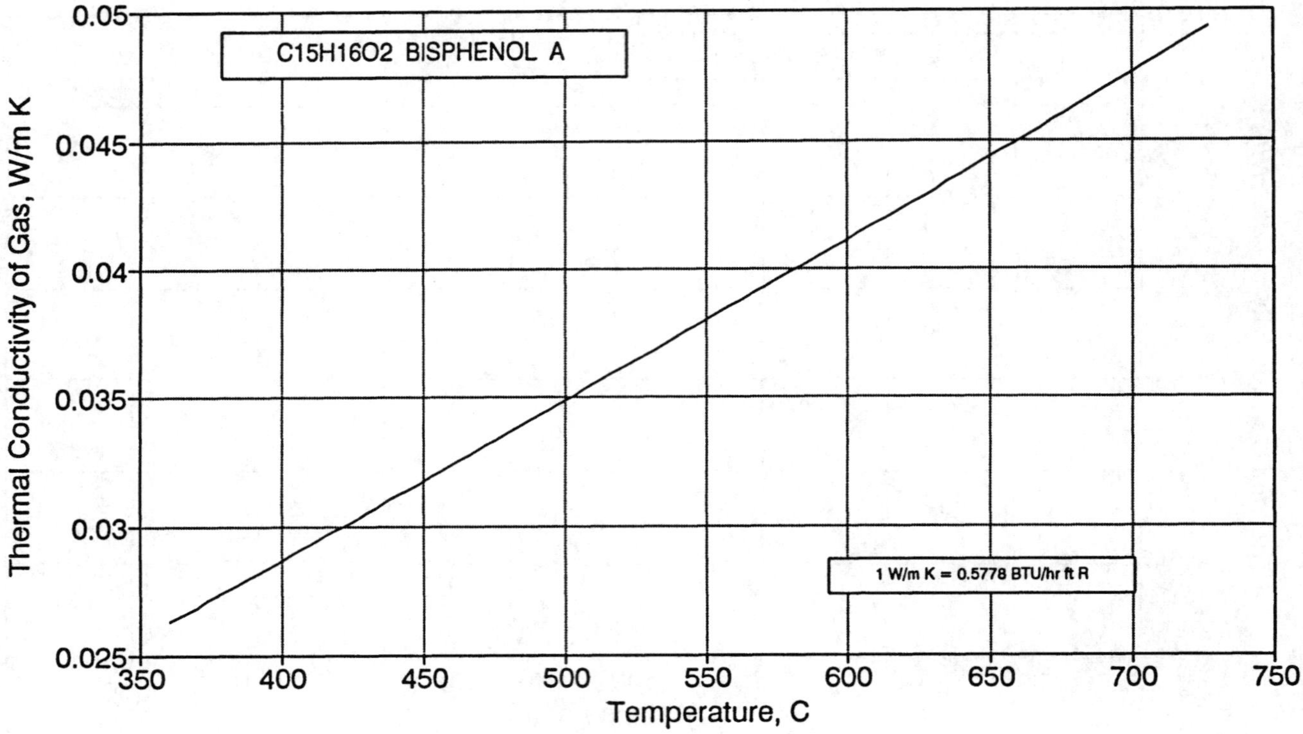
C15H16O2 BISPHENOL A
1 W/m K = 0.5778 BTU/hr ft R
Thermal Conductivity of Gas, W/m K
0.05
0.045
0.04
0.035
0.03
0.025
350
400
450
500
550
600
650
700
750
Temperature, C

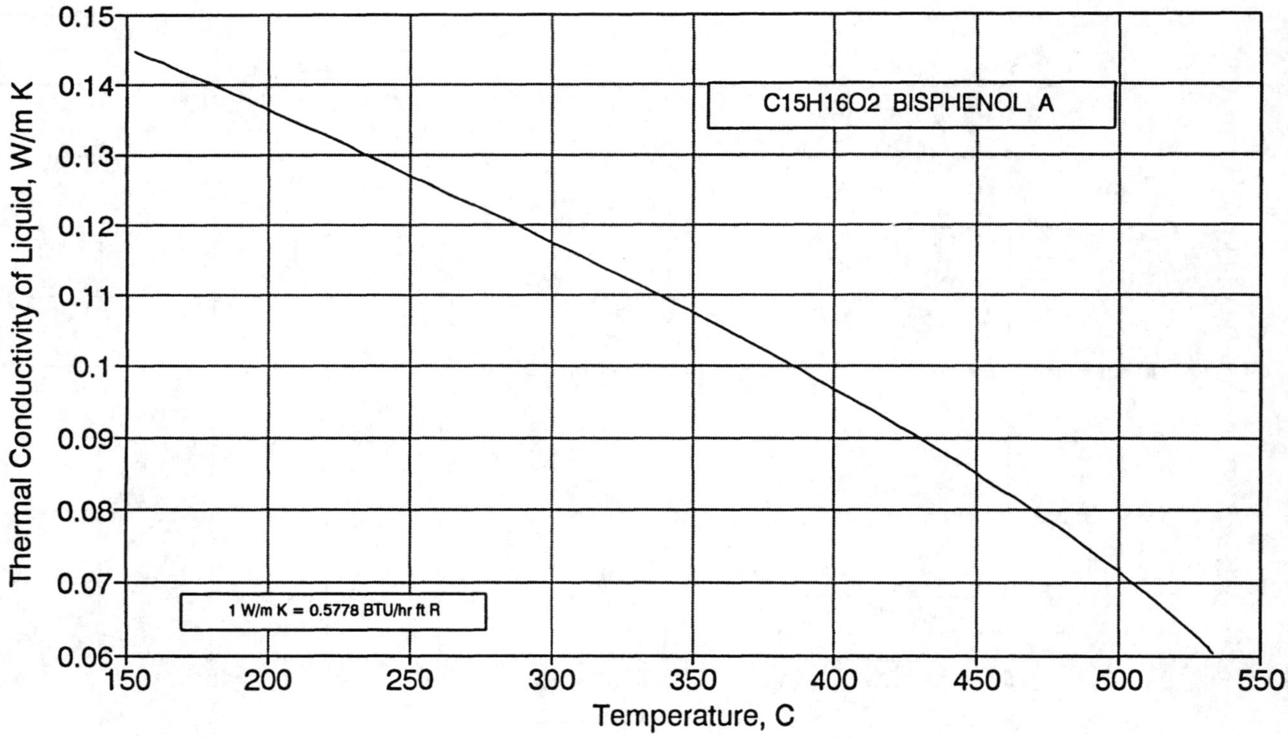
C15H16O2 BISPHENOL A
1 W/m K = 0.5778 BTU/hr ft R
Thermal Conductivity of Liquid, W/m K
0.15
0.14
0.13
0.12
0.11
0.1
0.09
0.08
0.07
0.06
150
200
250
300
350
400
450
500
550
Temperature, C

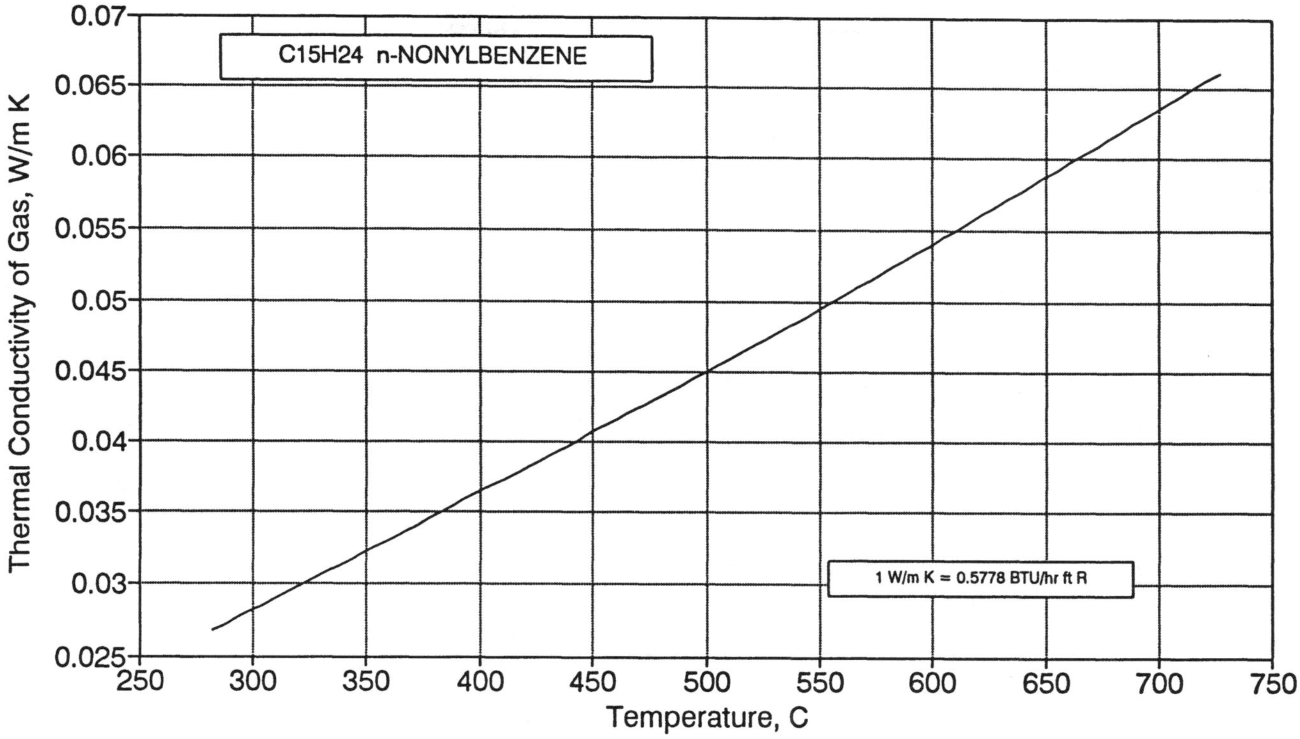
C15H24 n-NONYLBENZENE
1 W/m K = 0.5778 BTU/hr ft R
Thermal Conductivity of Gas, W/m K
0.07
0.065
0.06
0.055
0.05
0.045
0.04
0.035
0.03
0.025
250
300
350
400
450
500
550
600
650
700
750
Temperature, C

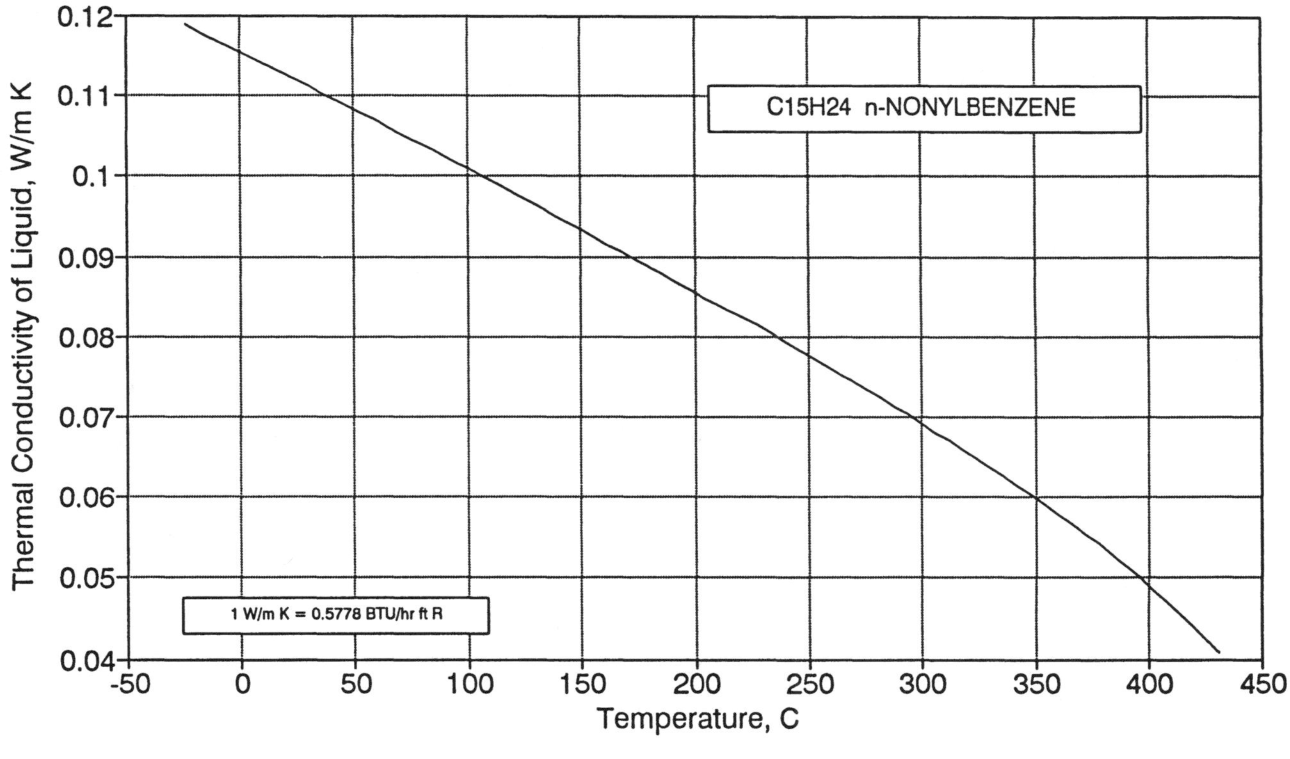
C15H24 n-NONYLBENZENE
1 W/m K = 0.5778 BTU/hr ft R
Thermal Conductivity of Liquid, W/m K
0.12
0.11
0.1
0.09
0.08
0.07
0.06
0.05
0.04
-50
0
50
100
150
200
250
300
350
400
450
Temperature, C

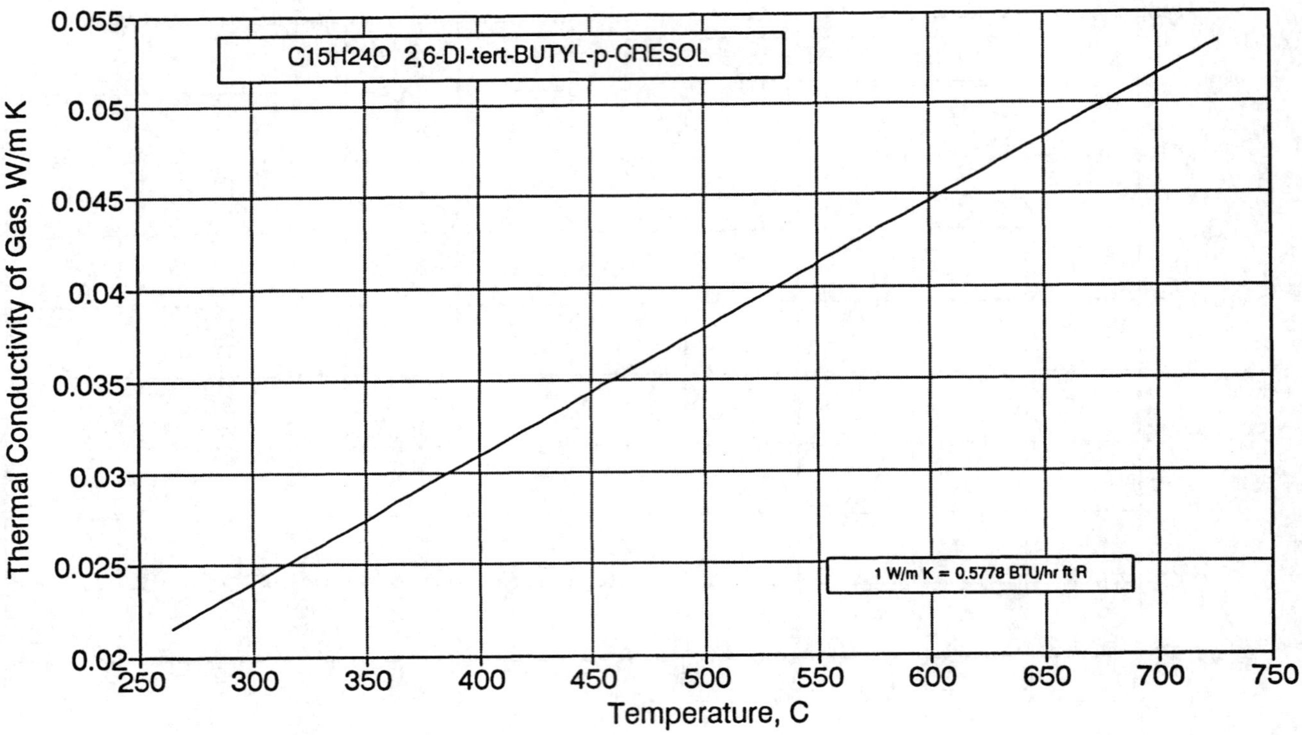
C15H24O 2,6-DI-tert-BUTYL-p-CRESOL
1 W/m K = 0.5778 BTU/hr ft R
Thermal Conductivity of Gas, W/m K
0.055
0.05
0.045
0.04
0.035
0.03
0.025
0.02
250
300
350
400
450
500
550
600
650
700
750
Temperature, C

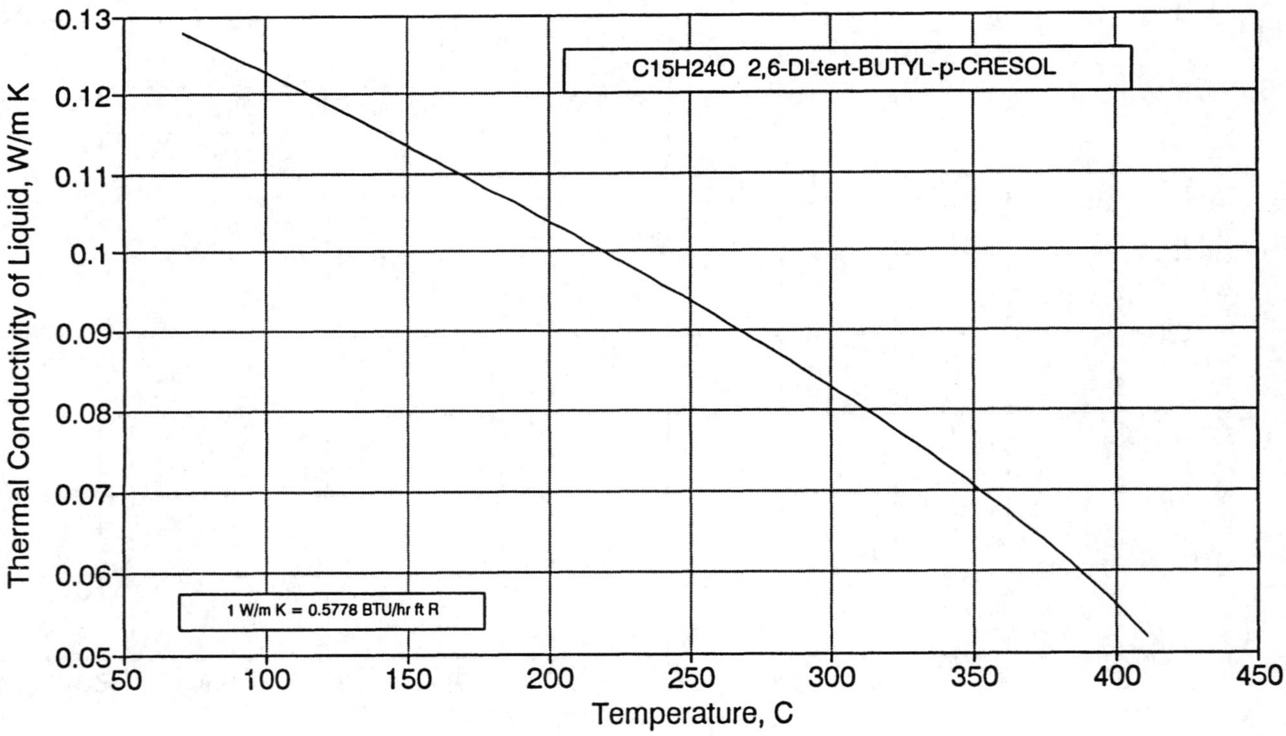
C15H24O 2,6-DI-tert-BUTYL-p-CRESOL
1 W/m K = 0.5778 BTU/hr ft R
Thermal Conductivity of Liquid, W/m K
0.13
0.12
0.11
0.1
0.09
0.08
0.07
0.06
0.05
50
100
150
200
250
300
350
400
450
Temperature, C

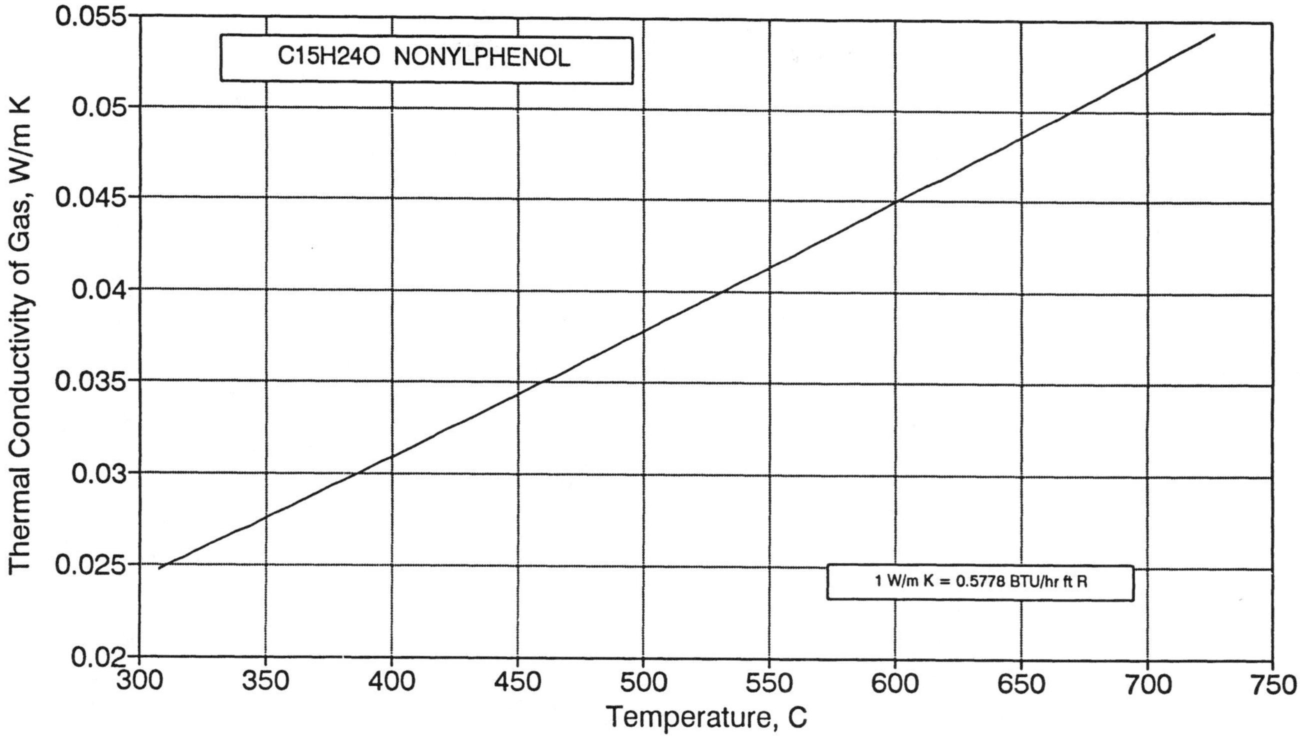
C15H24O NONYLPHENOL
1 W/m K = 0.5778 BTU/hr ft R
Thermal Conductivity of Gas, W/m K
0.055
0.05
0.045
0.04
0.035
0.03
0.025
0.02
300
350
400
450
500
550
600
650
700
750
Temperature, C

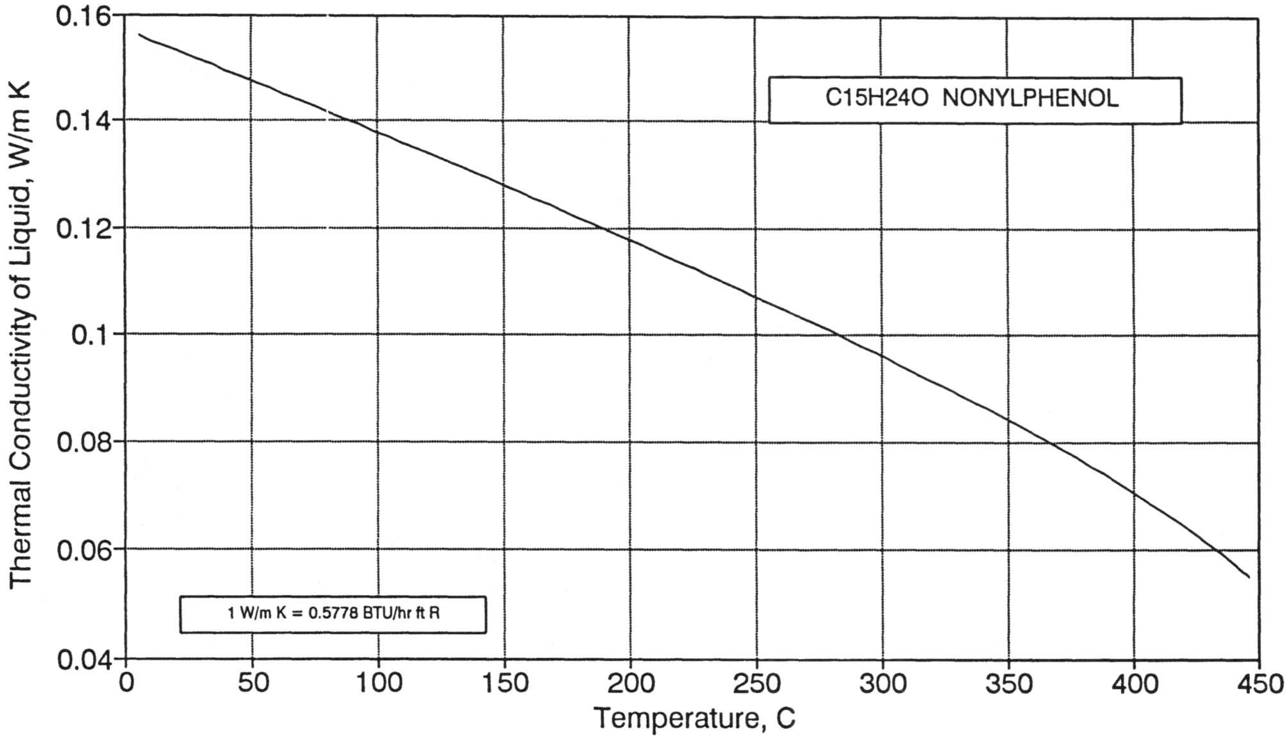
C15H24O NONYLPHENOL
1 W/m K = 0.5778 BTU/hr ft R
Thermal Conductivity of Liquid, W/m K
0.16
0.14
0.12
0.1
0.08
0.06
0.04
0
50
100
150
200
250
300
350
400
450
Temperature, C

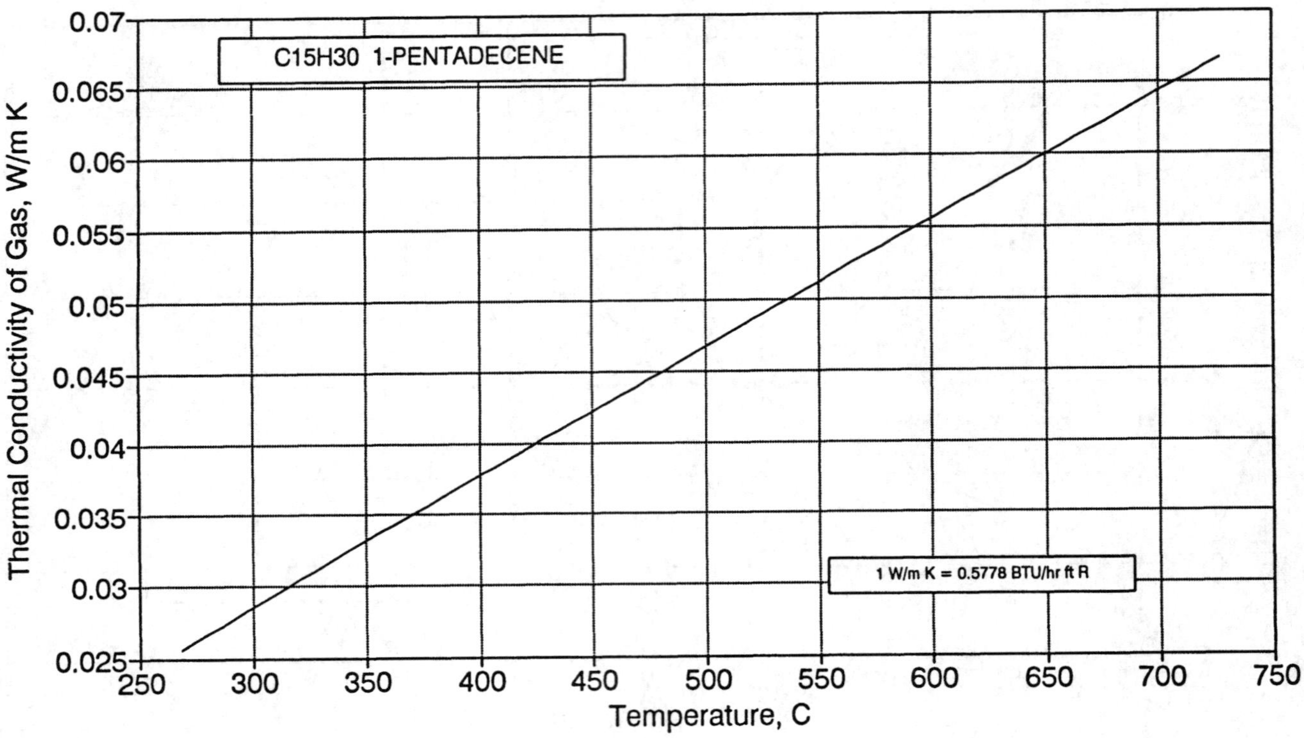
C15H30 1-PENTADECENE
1 W/m K = 0.5778 BTU/hr ft R
Thermal Conductivity of Gas, W/m K
Temperature, C
0.07
0.065
0.06
0.055
0.05
0.045
0.04
0.035
0.03
0.025
250
300
350
400
450
500
550
600
650
700
750

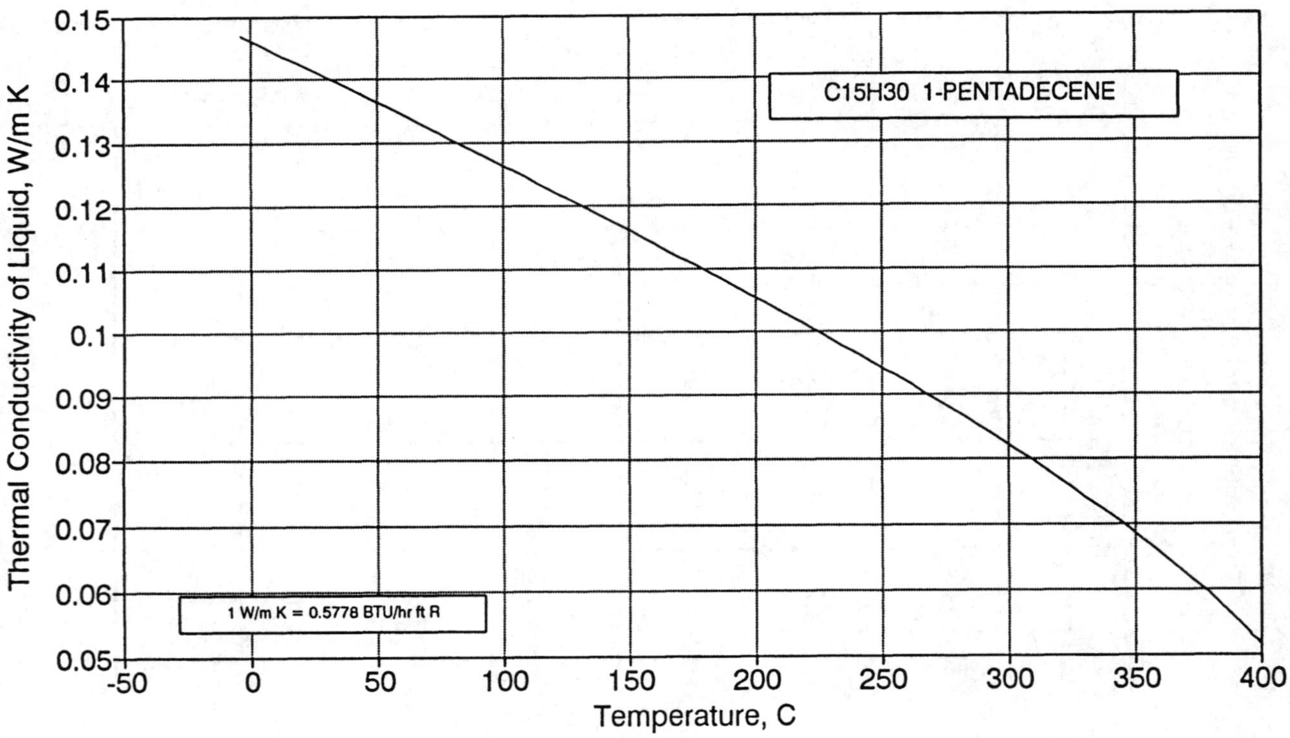
C15H30 1-PENTADECENE
1 W/m K = 0.5778 BTU/hr ft R
Thermal Conductivity of Liquid, W/m K
Temperature, C
0.15
0.14
0.13
0.12
0.11
0.1
0.09
0.08
0.07
0.06
0.05
-50
0
50
100
150
200
250
300
350
400

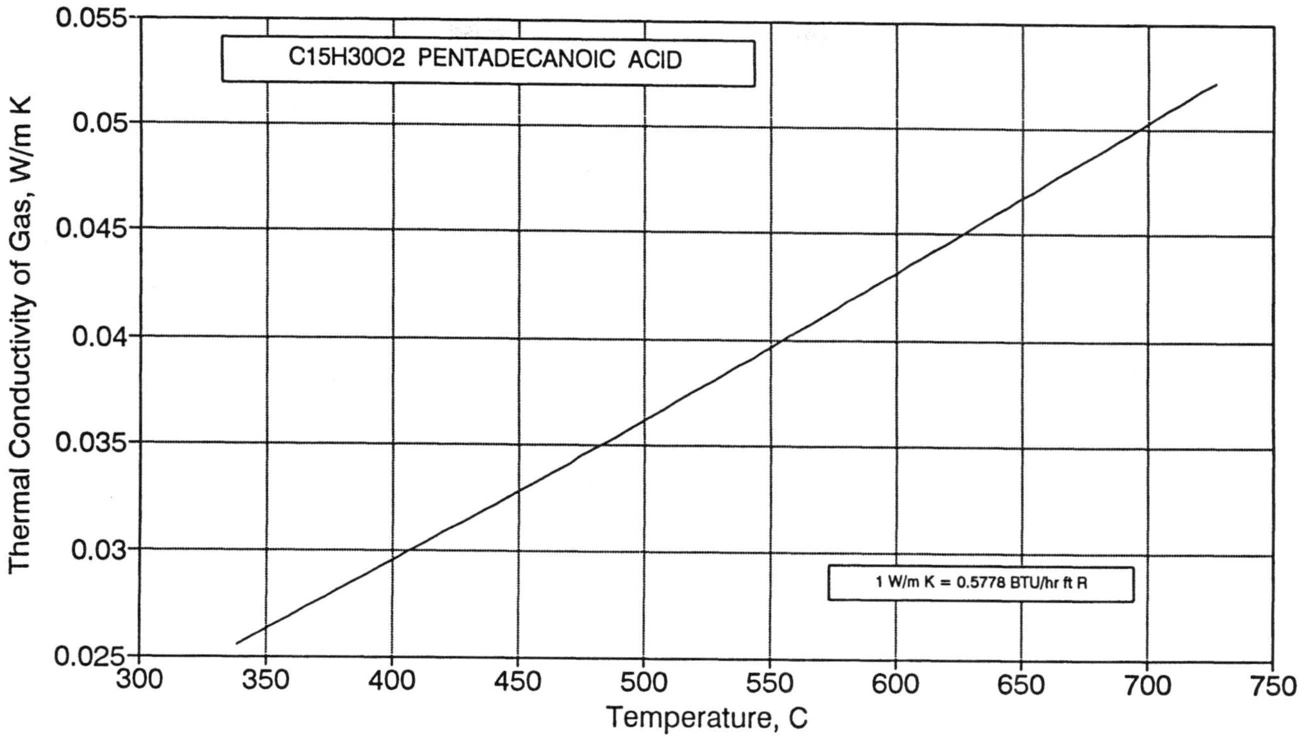
C15H30O2 PENTADECANOIC ACID
1 W/m K = 0.5778 BTU/hr ft R
Thermal Conductivity of Gas, W/m K
0.055
0.05
0.045
0.04
0.035
0.03
0.025
300
350
400
450
500
550
600
650
700
750
Temperature, C

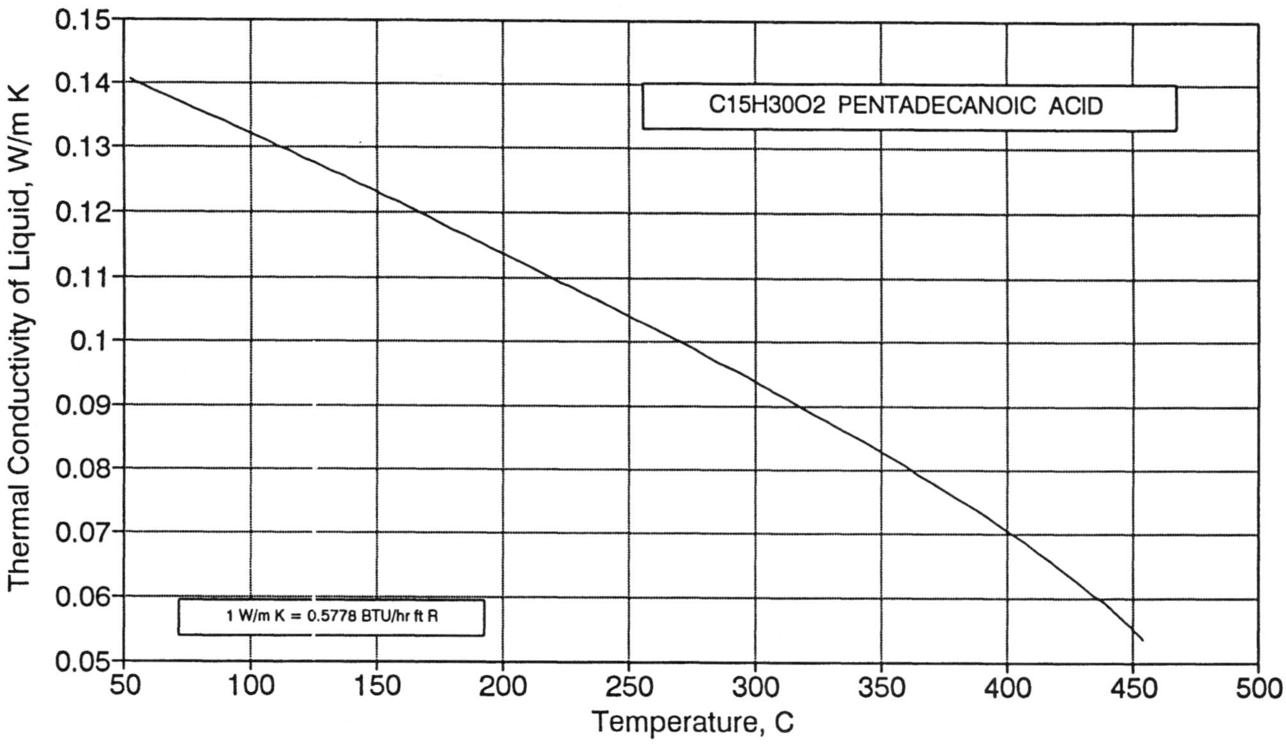
C15H30O2 PENTADECANOIC ACID
1 W/m K = 0.5778 BTU/hr ft R
Thermal Conductivity of Liquid, W/m K
0.15
0.14
0.13
0.12
0.11
0.1
0.09
0.08
0.07
0.06
0.05
50
100
150
200
250
300
350
400
450
500
Temperature, C

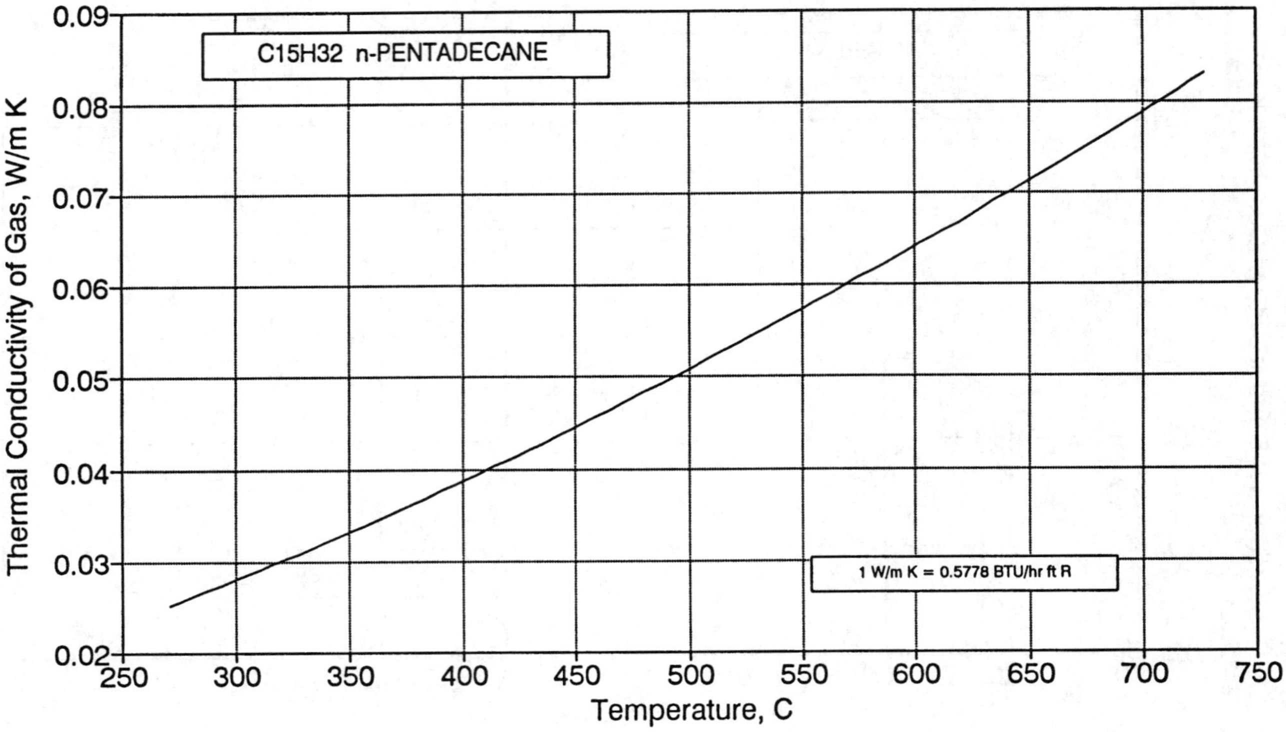
C15H32 n-PENTADECANE
1 W/m K = 0.5778 BTU/hr ft R
Thermal Conductivity of Gas, W/m K
0.09
0.08
0.07
0.06
0.05
0.04
0.03
0.02
250
300
350
400
450
500
550
600
650
700
750
Temperature, C

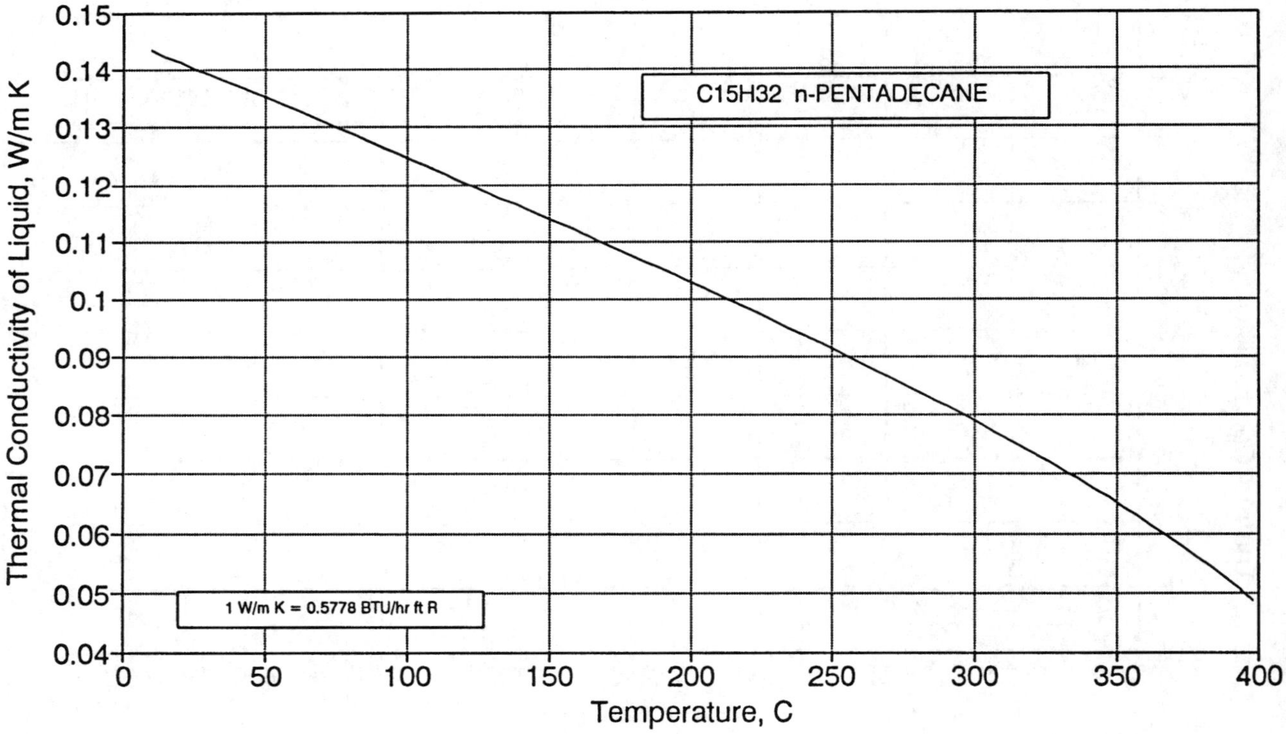
C15H32 n-PENTADECANE
1 W/m K = 0.5778 BTU/hr ft R
Thermal Conductivity of Liquid, W/m K
0.15
0.14
0.13
0.12
0.11
0.1
0.09
0.08
0.07
0.06
0.05
0.04
0
50
100
150
200
250
300
350
400
Temperature, C

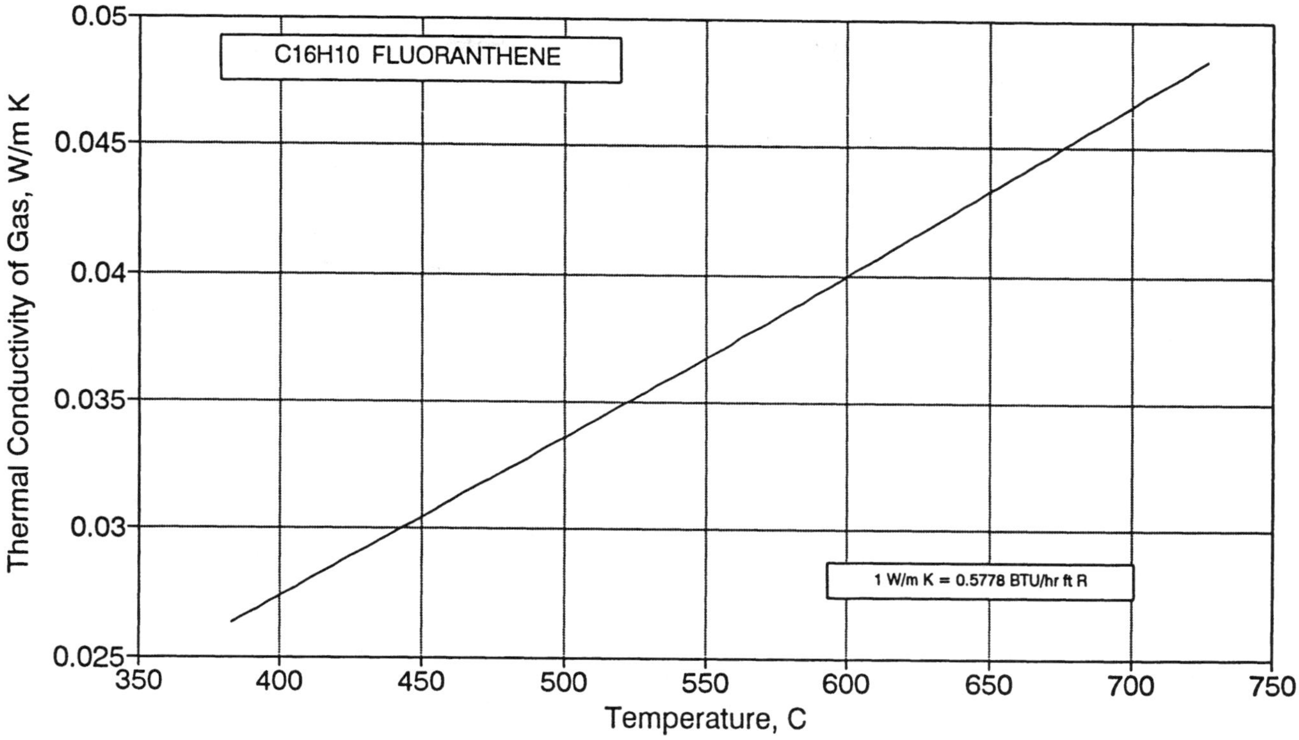
C16H10 FLUORANTHENE
1 W/m K = 0.5778 BTU/hr ft R
Thermal Conductivity of Gas, W/m K
0.05
0.045
0.04
0.035
0.03
0.025
350
400
450
500
550
600
650
700
750
Temperature, C

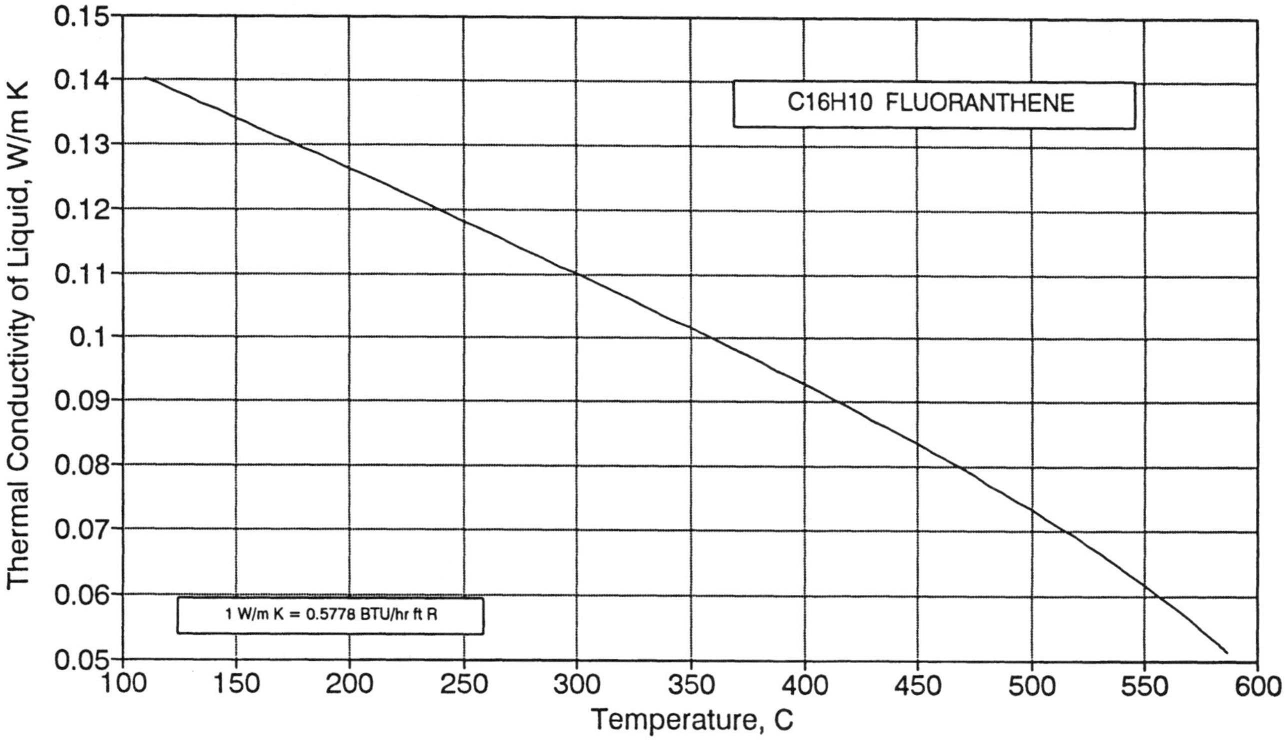
C16H10 FLUORANTHENE
1 W/m K = 0.5778 BTU/hr ft R
Thermal Conductivity of Liquid, W/m K
0.15
0.14
0.13
0.12
0.11
0.1
0.09
0.08
0.07
0.06
0.05
100
150
200
250
300
350
400
450
500
550
600
Temperature, C

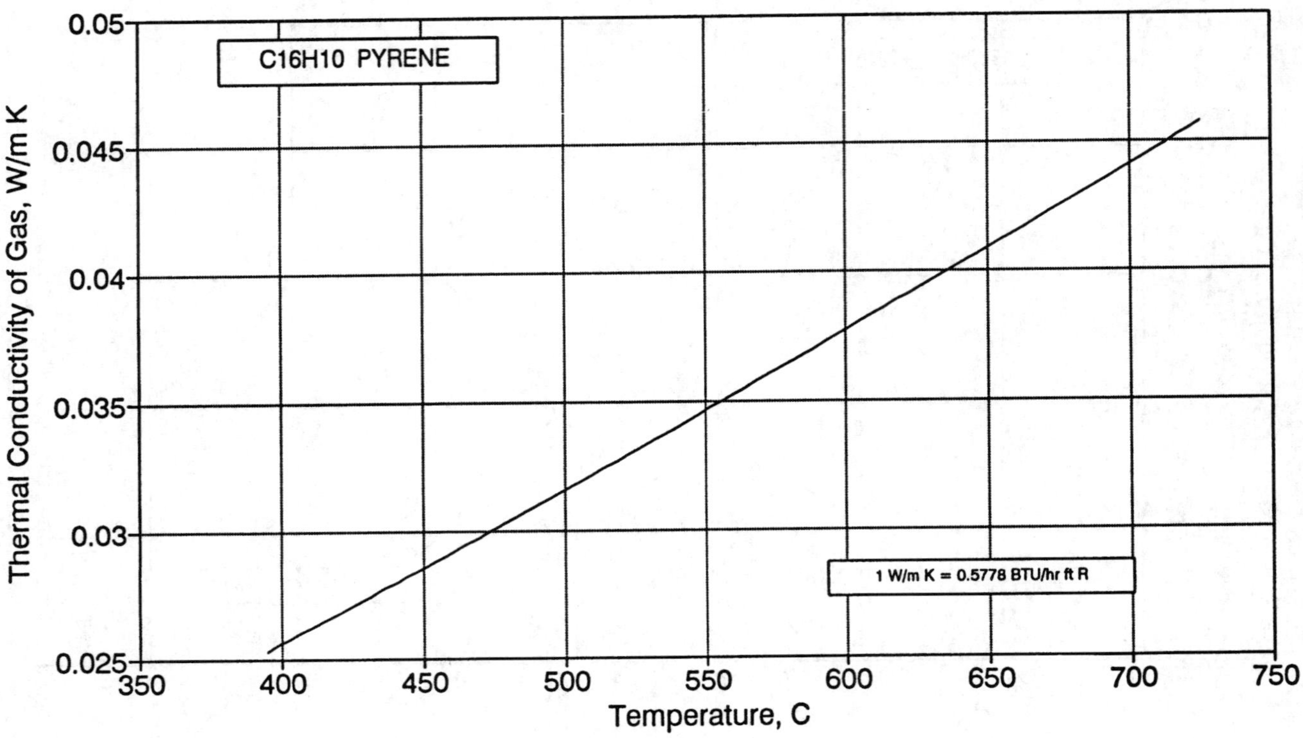
C16H10 PYRENE
1 W/m K = 0.5778 BTU/hr ft R
Thermal Conductivity of Gas, W/m K
Temperature, C
0.05
0.045
0.04
0.035
0.03
0.025
350
400
450
500
550
600
650
700
750

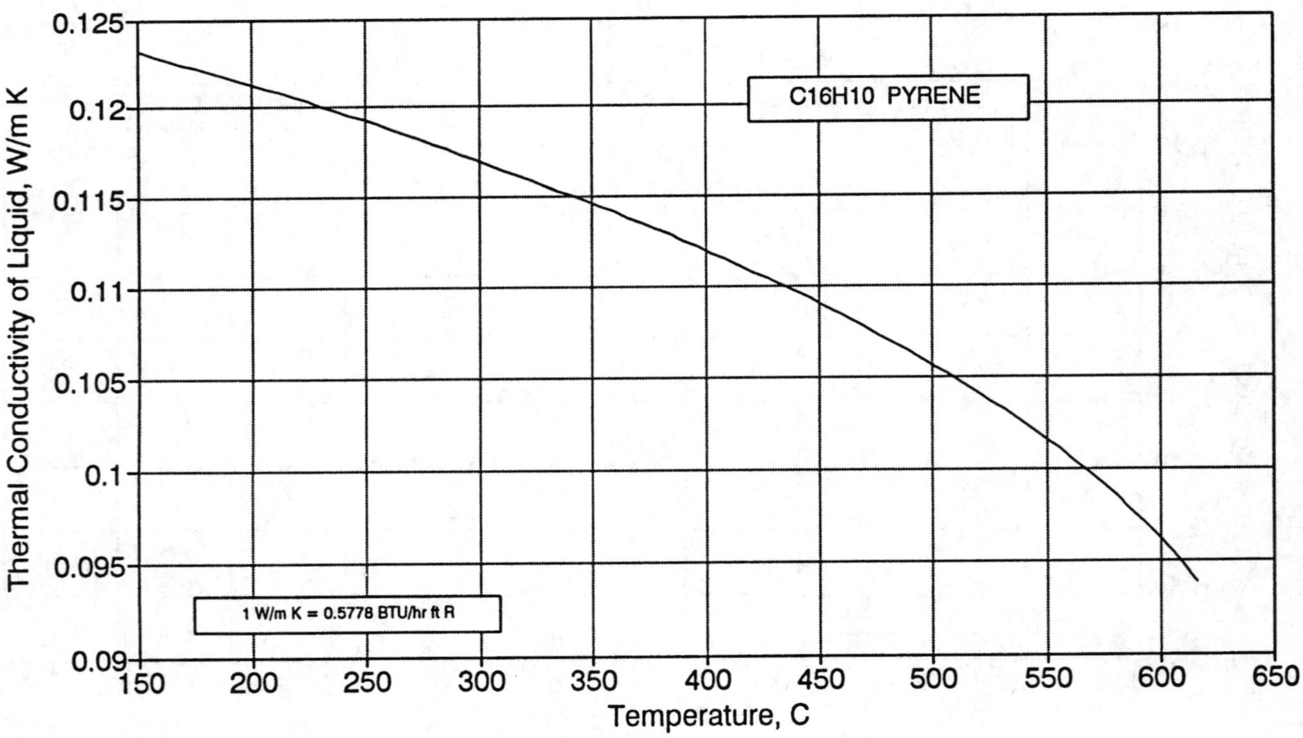
C16H10 PYRENE
1 W/m K = 0.5778 BTU/hr ft R
Thermal Conductivity of Liquid, W/m K
Temperature, C
0.125
0.12
0.115
0.11
0.105
0.1
0.095
0.09
150
200
250
300
350
400
450
500
550
600
650

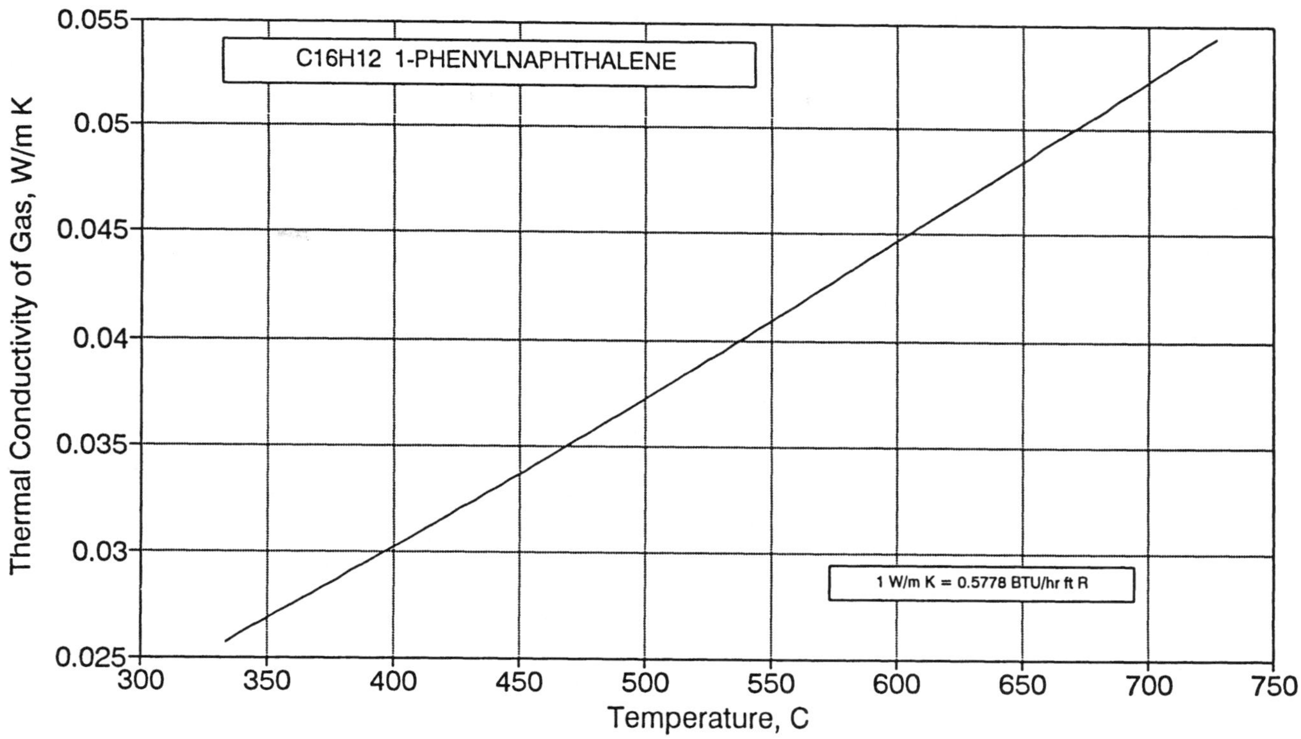
C16H12 1-PHENYLNAPHTHALENE
1 W/m K = 0.5778 BTU/hr ft R
Thermal Conductivity of Gas, W/m K
0.055
0.05
0.045
0.04
0.035
0.03
0.025
300
350
400
450
500
550
600
650
700
750
Temperature, C

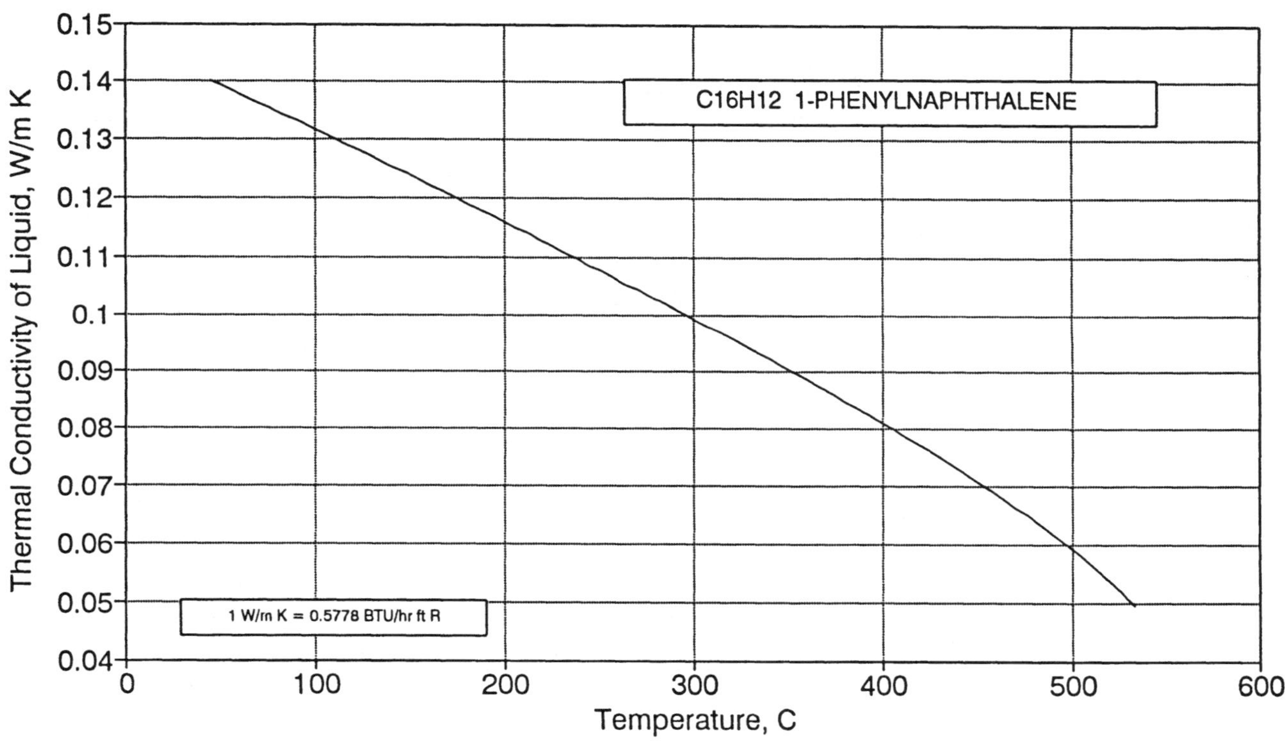
C16H12 1-PHENYLNAPHTHALENE
1 W/m K = 0.5778 BTU/hr ft R
Thermal Conductivity of Liquid, W/m K
0.15
0.14
0.13
0.12
0.11
0.1
0.09
0.08
0.07
0.06
0.05
0.04
0
100
200
300
400
500
600
Temperature, C

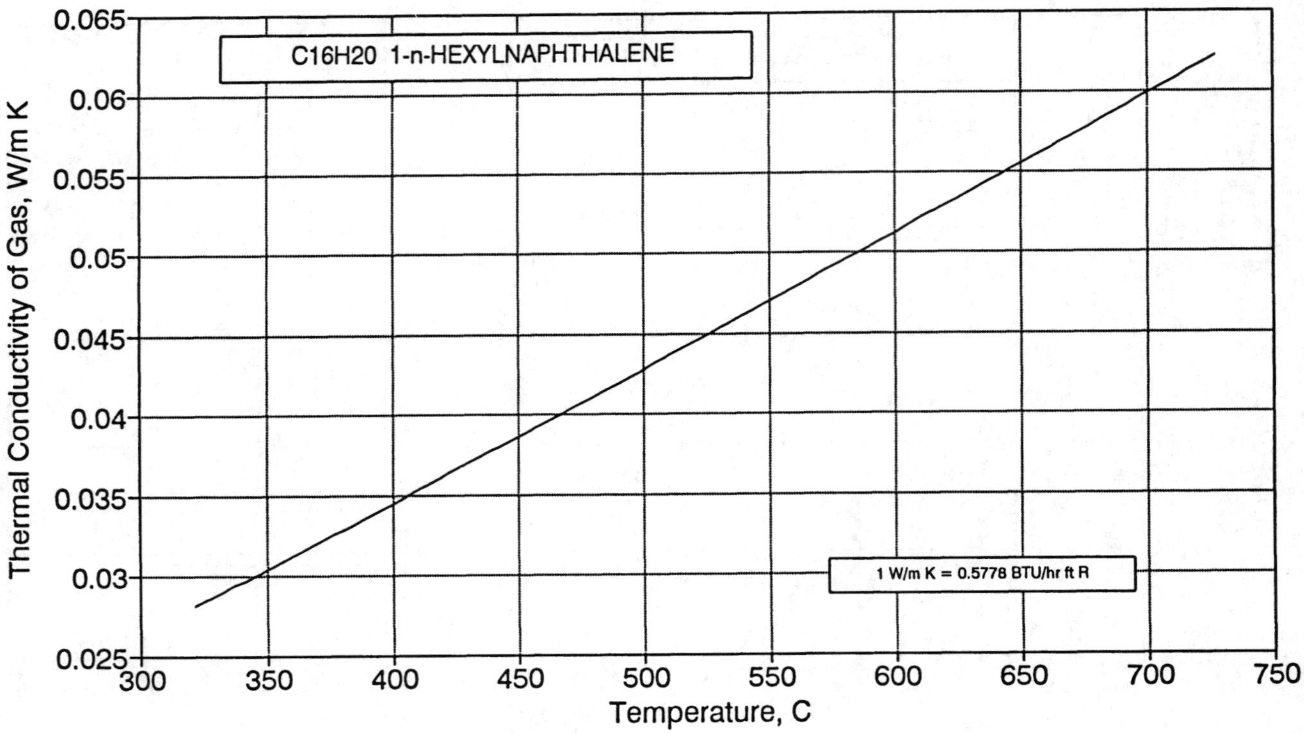
C16H20 1-n-HEXYLNAPHTHALENE
Thermal Conductivity of Gas, W/m K
0.065
0.06
0.055
0.05
0.045
0.04
0.035
0.03
0.025
1 W/m K = 0.5778 BTU/hr ft R
300
350
400
450
500
550
600
650
700
750
Temperature, C

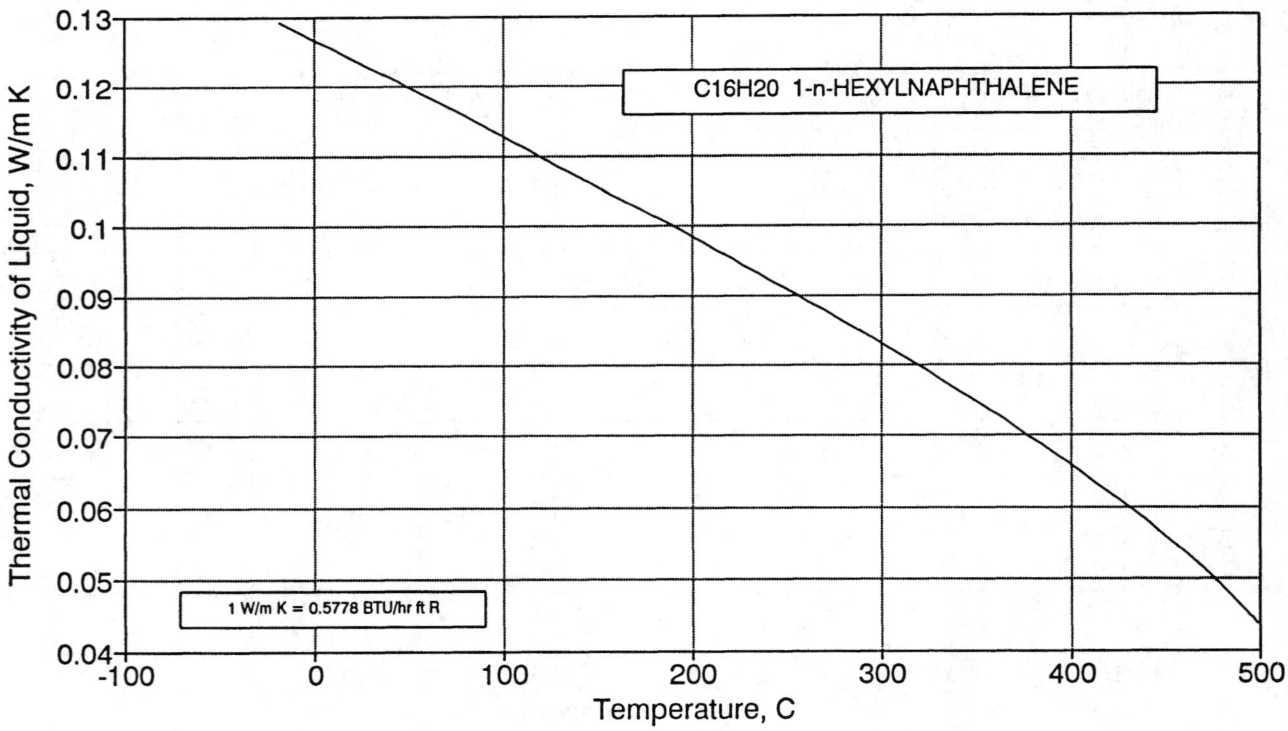
C16H20 1-n-HEXYLNAPHTHALENE
Thermal Conductivity of Liquid, W/m K
0.13
0.12
0.11
0.1
0.09
0.08
0.07
0.06
0.05
0.04
1 W/m K = 0.5778 BTU/hr ft R
-100
0
100
200
300
400
500
Temperature, C

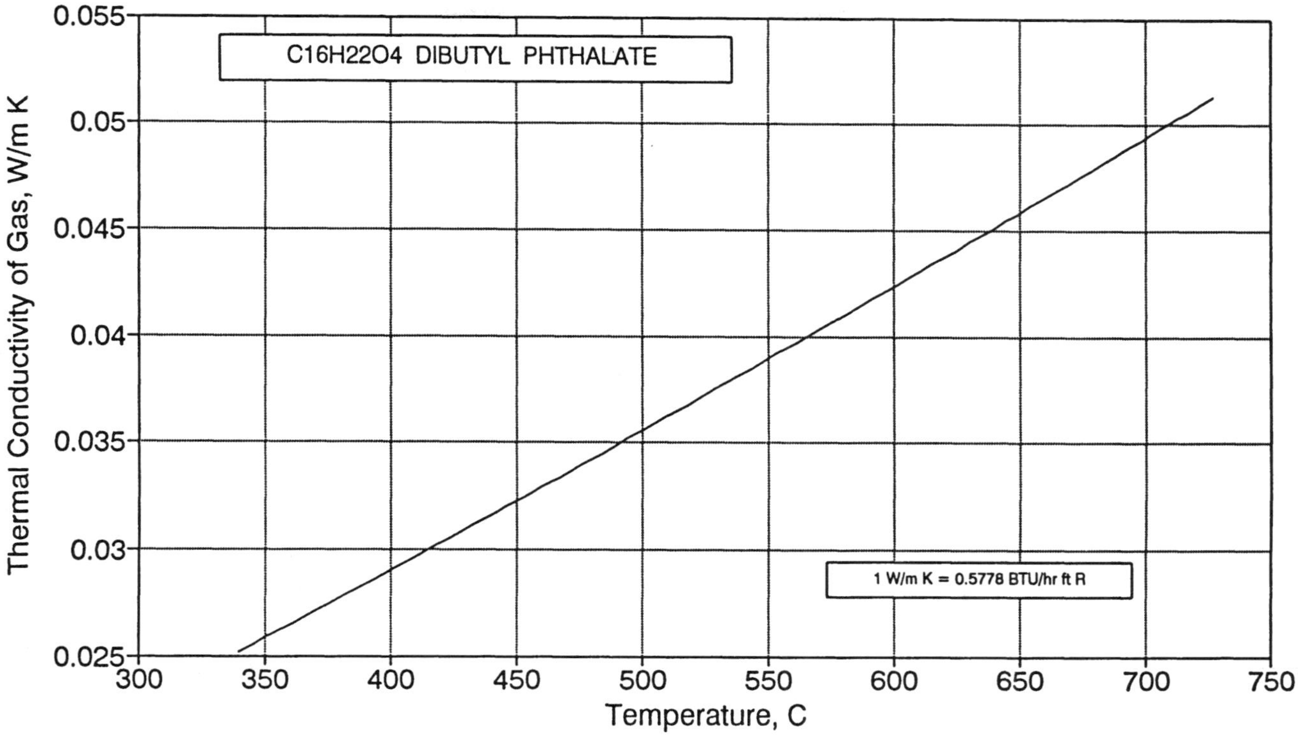
C16H22O4 DIBUTYL PHTHALATE
1 W/m K = 0.5778 BTU/hr ft R
Thermal Conductivity of Gas, W/m K
0.055
0.05
0.045
0.04
0.035
0.03
0.025
300
350
400
450
500
550
600
650
700
750
Temperature, C

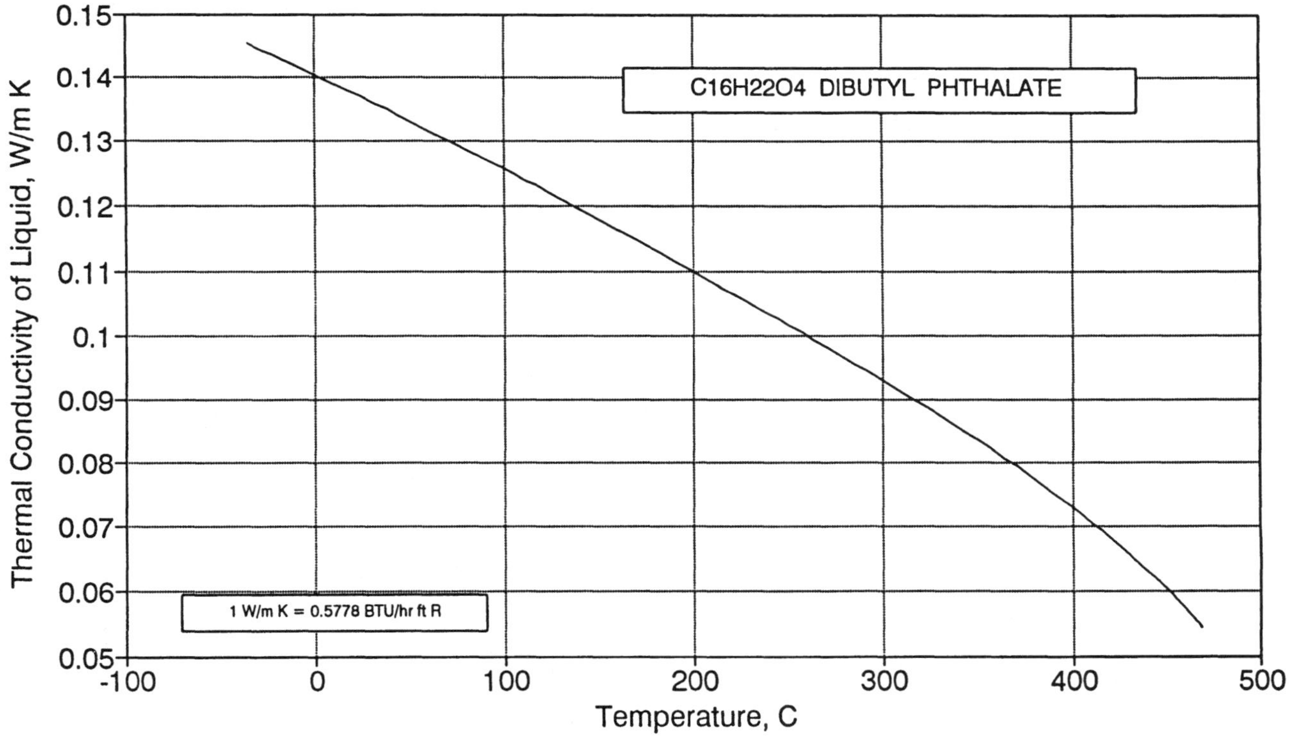
C16H22O4 DIBUTYL PHTHALATE
1 W/m K = 0.5778 BTU/hr ft R
Thermal Conductivity of Liquid, W/m K
0.15
0.14
0.13
0.12
0.11
0.1
0.09
0.08
0.07
0.06
0.05
-100
0
100
200
300
400
500
Temperature, C

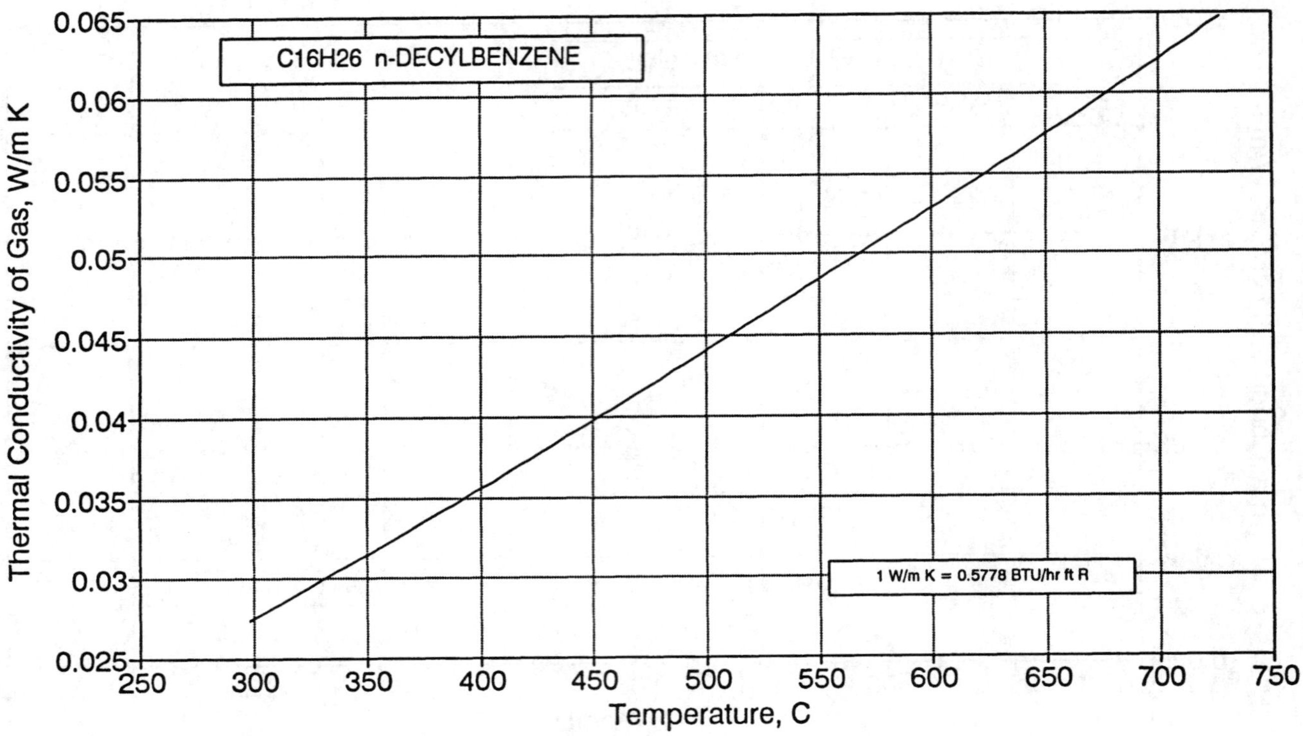
C16H26 n-DECYLBENZENE
1 W/m K = 0.5778 BTU/hr ft R
Thermal Conductivity of Gas, W/m K
0.065
0.06
0.055
0.05
0.045
0.04
0.035
0.03
0.025
250
300
350
400
450
500
550
600
650
700
750
Temperature, C

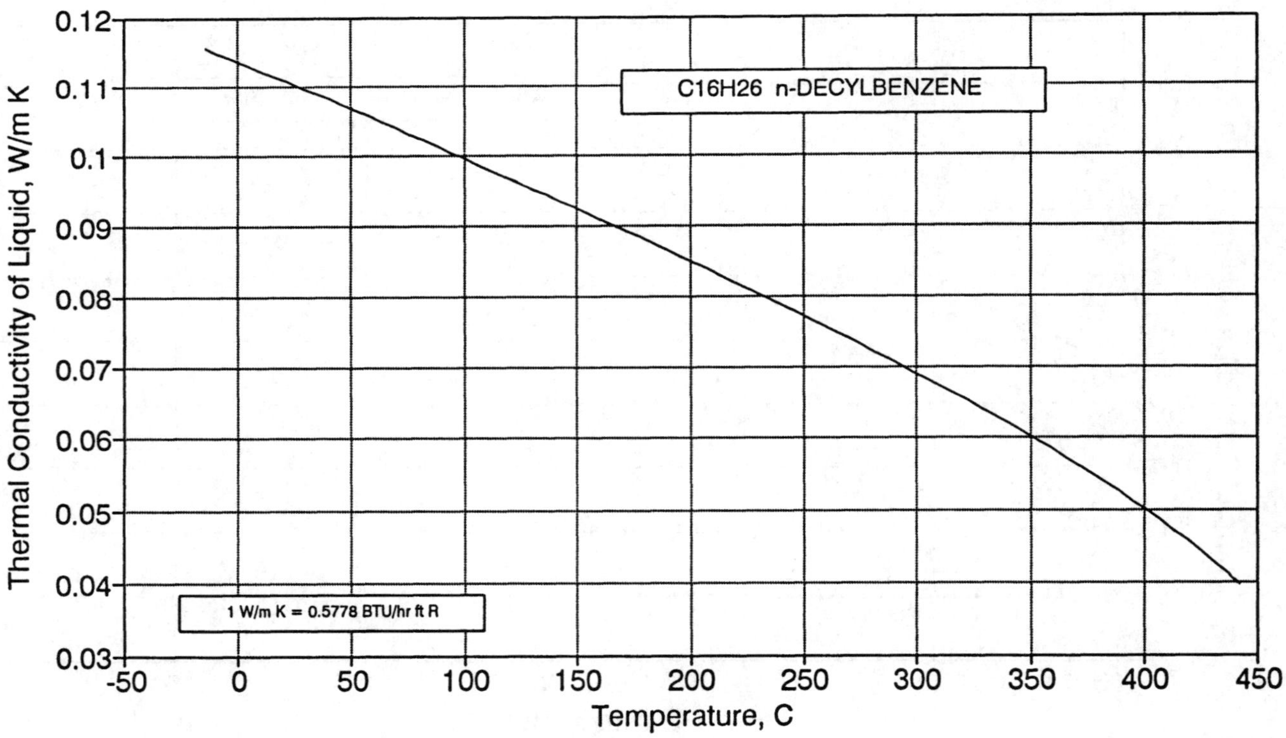
C16H26 n-DECYLBENZENE
1 W/m K = 0.5778 BTU/hr ft R
Thermal Conductivity of Liquid, W/m K
0.12
0.11
0.1
0.09
0.08
0.07
0.06
0.05
0.04
0.03
-50
0
50
100
150
200
250
300
350
400
450
Temperature, C

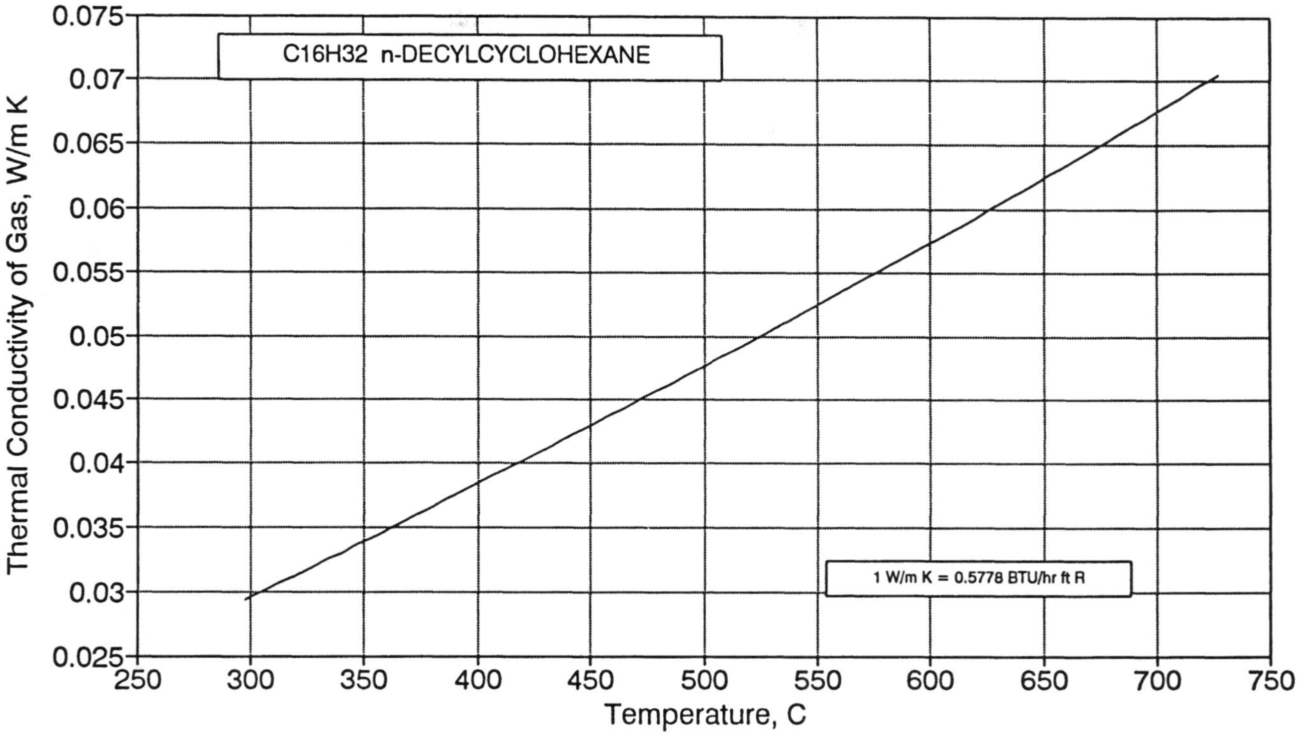
C16H32 n-DECYLCYCLOHEXANE
1 W/m K = 0.5778 BTU/hr ft R
Thermal Conductivity of Gas, W/m K
0.075
0.07
0.065
0.06
0.055
0.05
0.045
0.04
0.035
0.03
0.025
250
300
350
400
450
500
550
600
650
700
750
Temperature, C

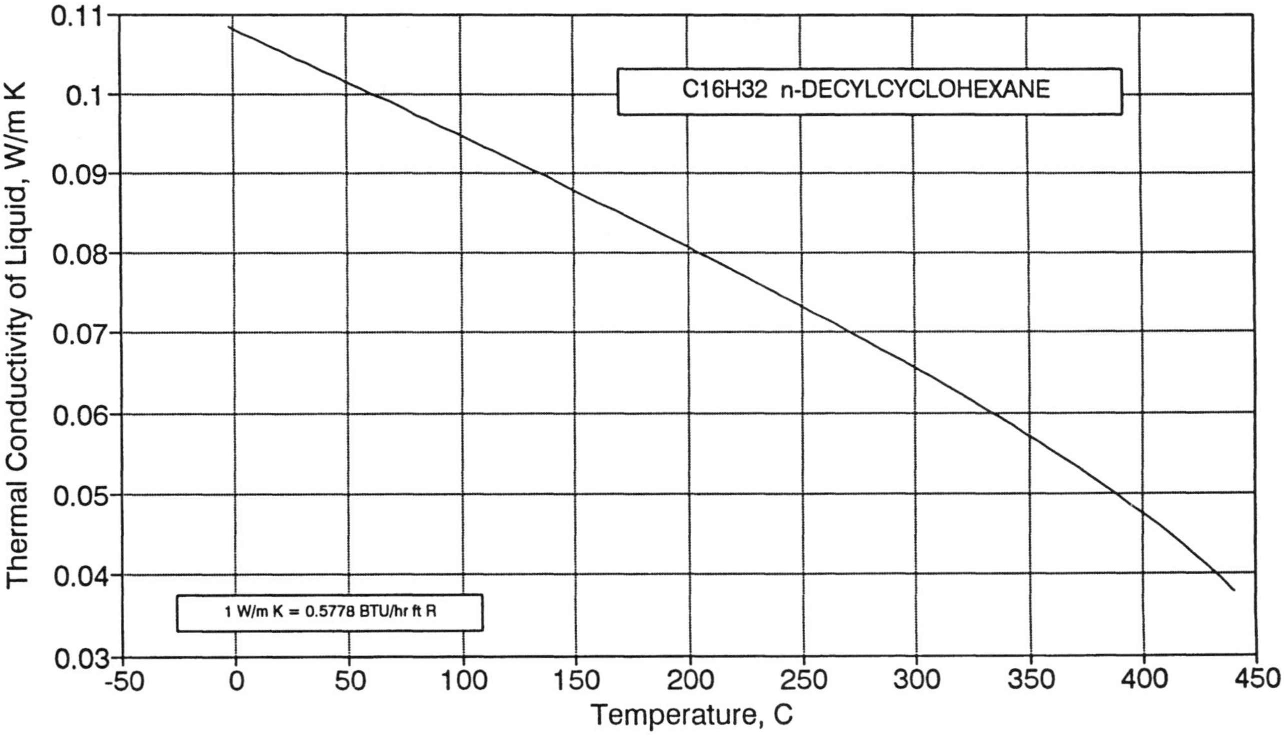
C16H32 n-DECYLCYCLOHEXANE
1 W/m K = 0.5778 BTU/hr ft R
Thermal Conductivity of Liquid, W/m K
0.11
0.1
0.09
0.08
0.07
0.06
0.05
0.04
0.03
-50
0
50
100
150
200
250
300
350
400
450
Temperature, C

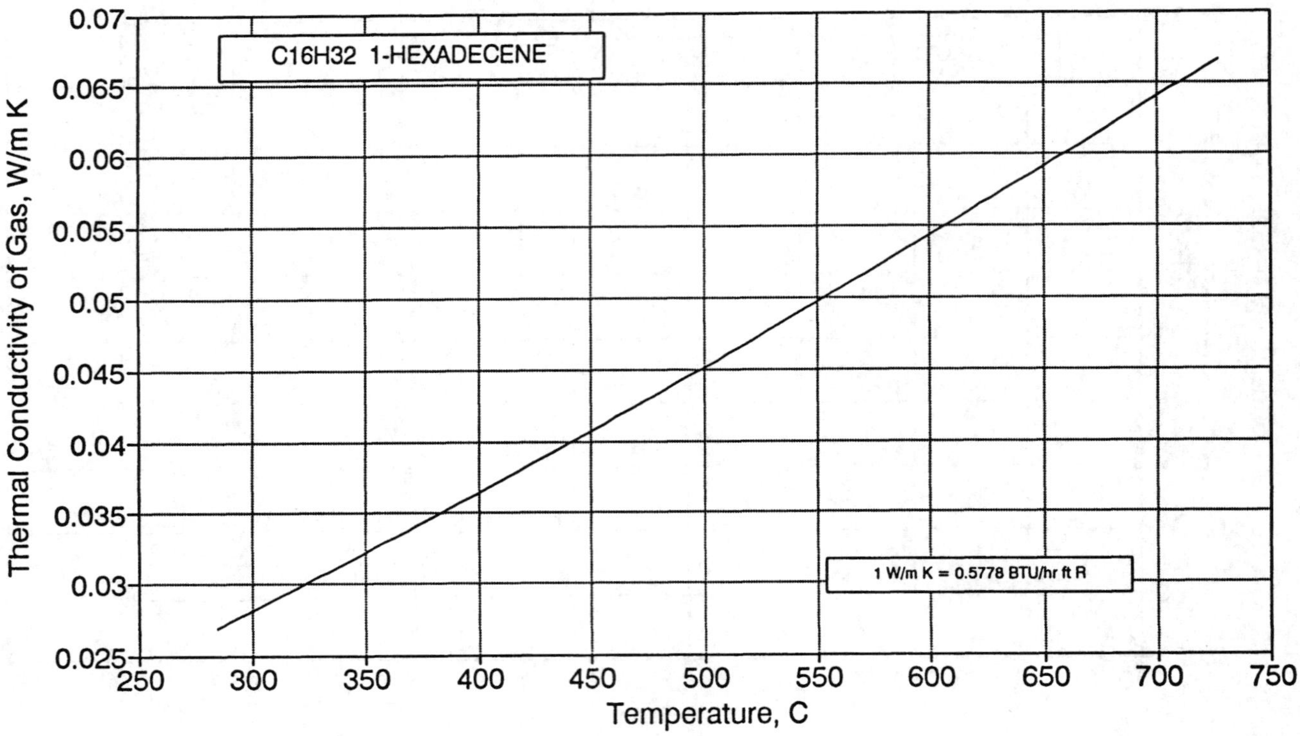
C16H32 1-HEXADECENE
1 W/m K = 0.5778 BTU/hr ft R
Thermal Conductivity of Gas, W/m K
0.07
0.065
0.06
0.055
0.05
0.045
0.04
0.035
0.03
0.025
250
300
350
400
450
500
550
600
650
700
750
Temperature, C

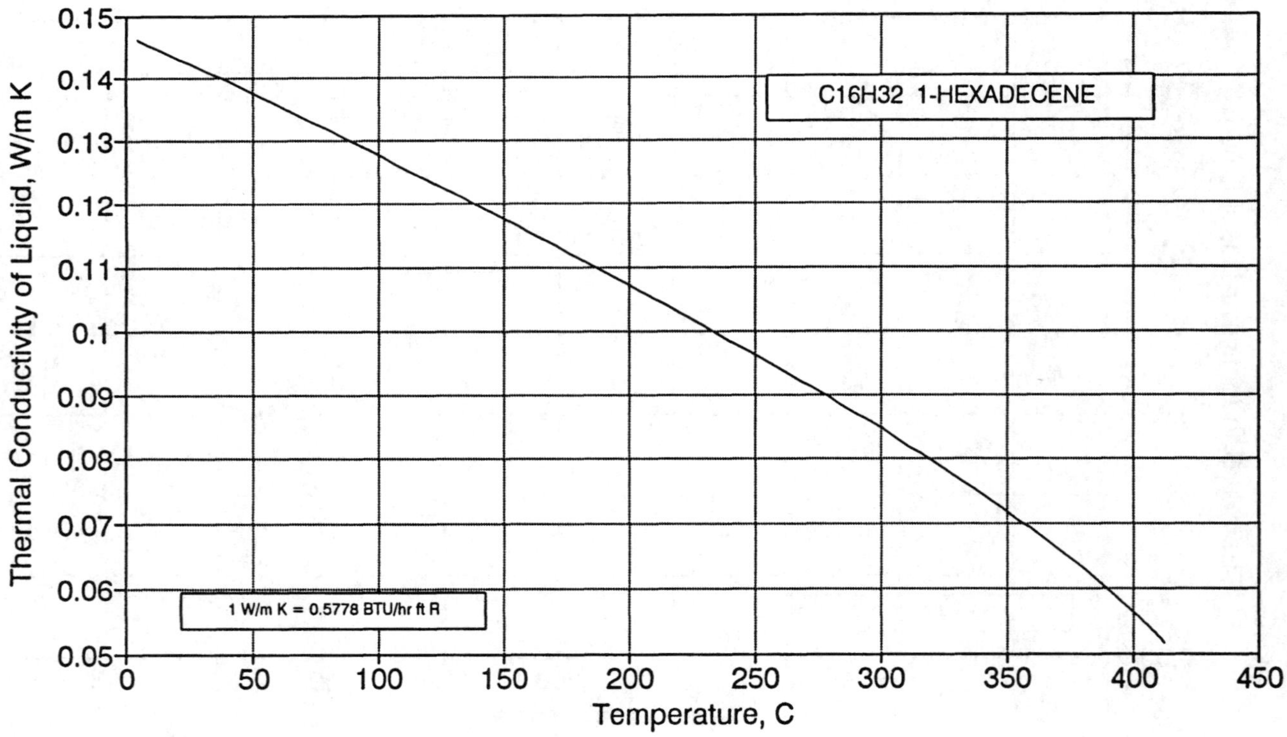
C16H32 1-HEXADECENE
1 W/m K = 0.5778 BTU/hr ft R
Thermal Conductivity of Liquid, W/m K
0.15
0.14
0.13
0.12
0.11
0.1
0.09
0.08
0.07
0.06
0.05
0
50
100
150
200
250
300
350
400
450
Temperature, C

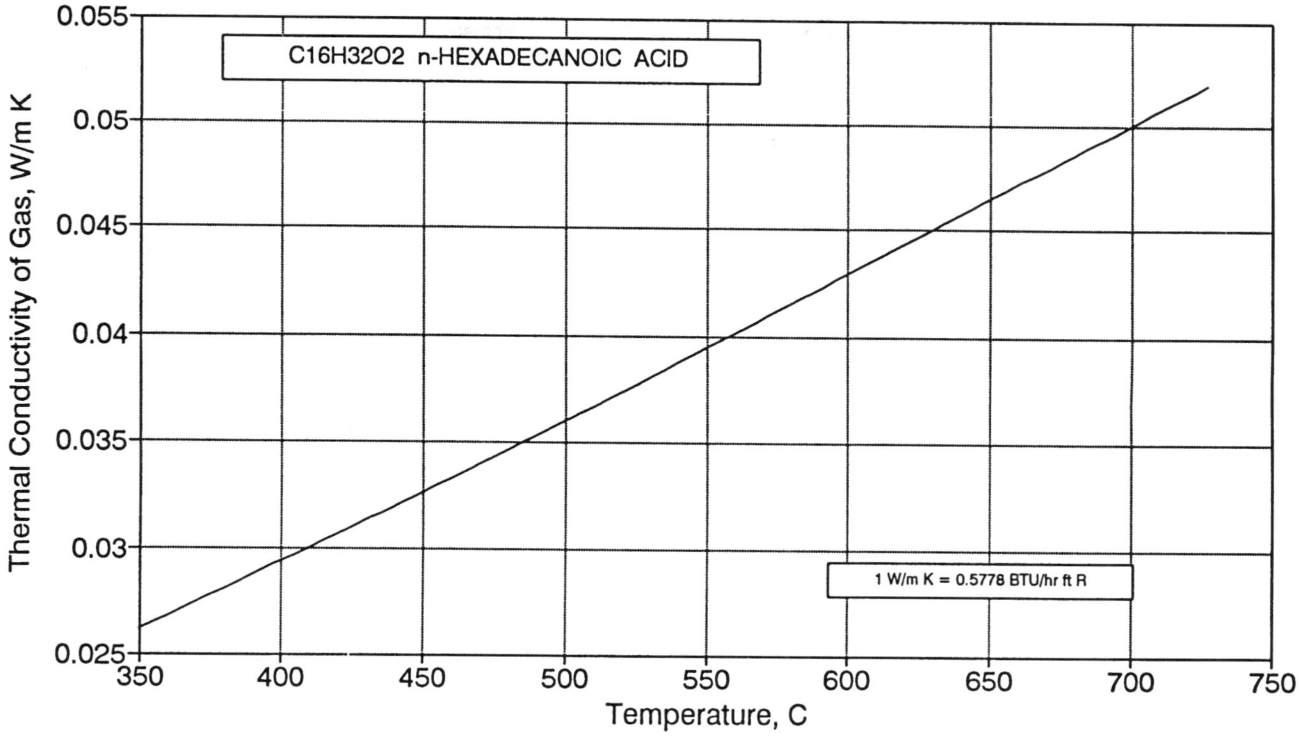
C16H32O2 n-HEXADECANOIC ACID
Thermal Conductivity of Gas, W/m K
Temperature, C
1 W/m K = 0.5778 BTU/hr ft R
0.055
0.05
0.045
0.04
0.035
0.03
0.025
350
400
450
500
550
600
650
700
750

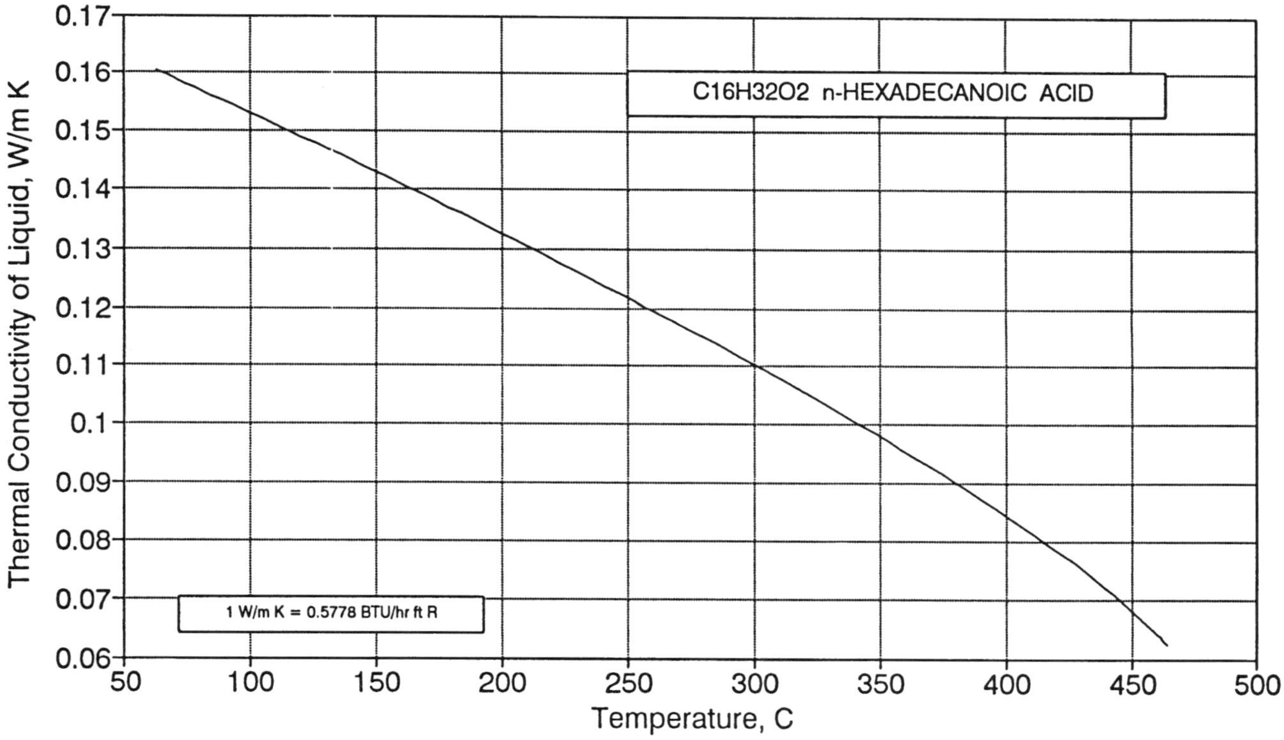
C16H32O2 n-HEXADECANOIC ACID
Thermal Conductivity of Liquid, W/m K
Temperature, C
1 W/m K = 0.5778 BTU/hr ft R
0.17
0.16
0.15
0.14
0.13
0.12
0.11
0.1
0.09
0.08
0.07
0.06
50
100
150
200
250
300
350
400
450
500

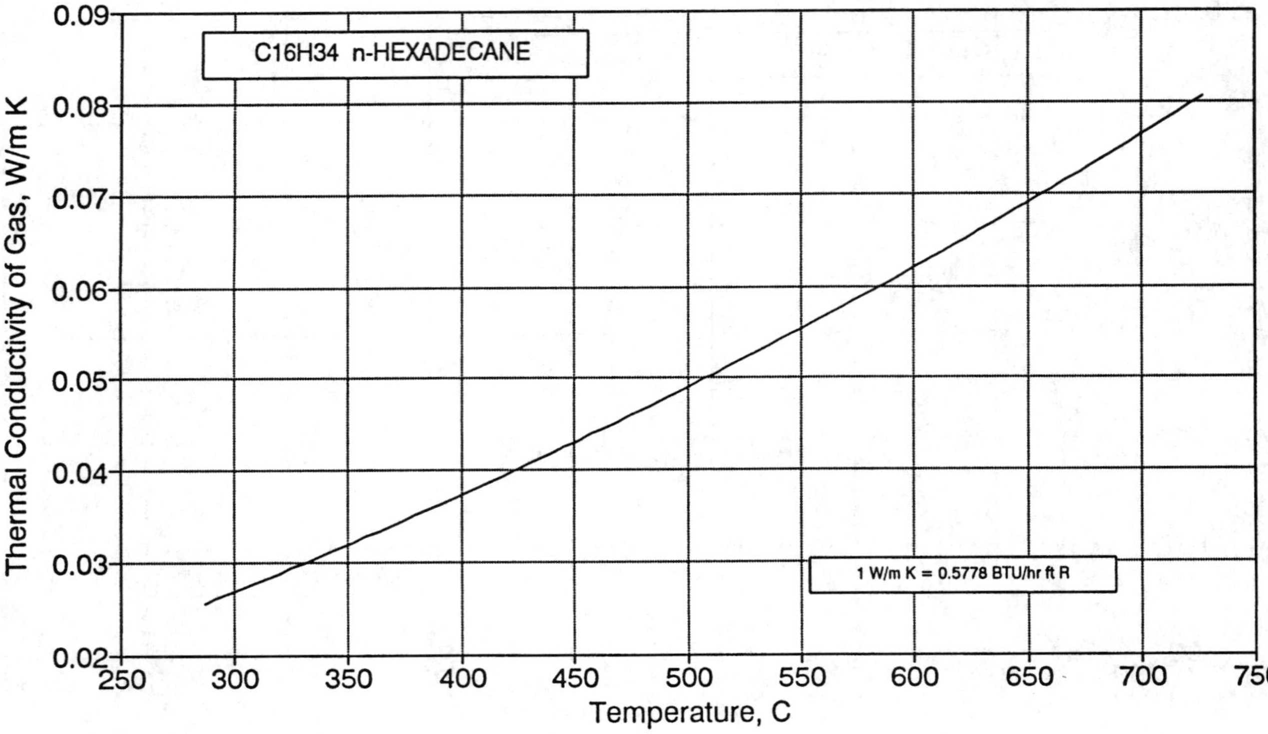
C16H34 n-HEXADECANE
1 W/m K = 0.5778 BTU/hr ft R
Thermal Conductivity of Gas, W/m K
0.09
0.08
0.07
0.06
0.05
0.04
0.03
0.02
250
300
350
400
450
500
550
600
650
700
750
Temperature, C

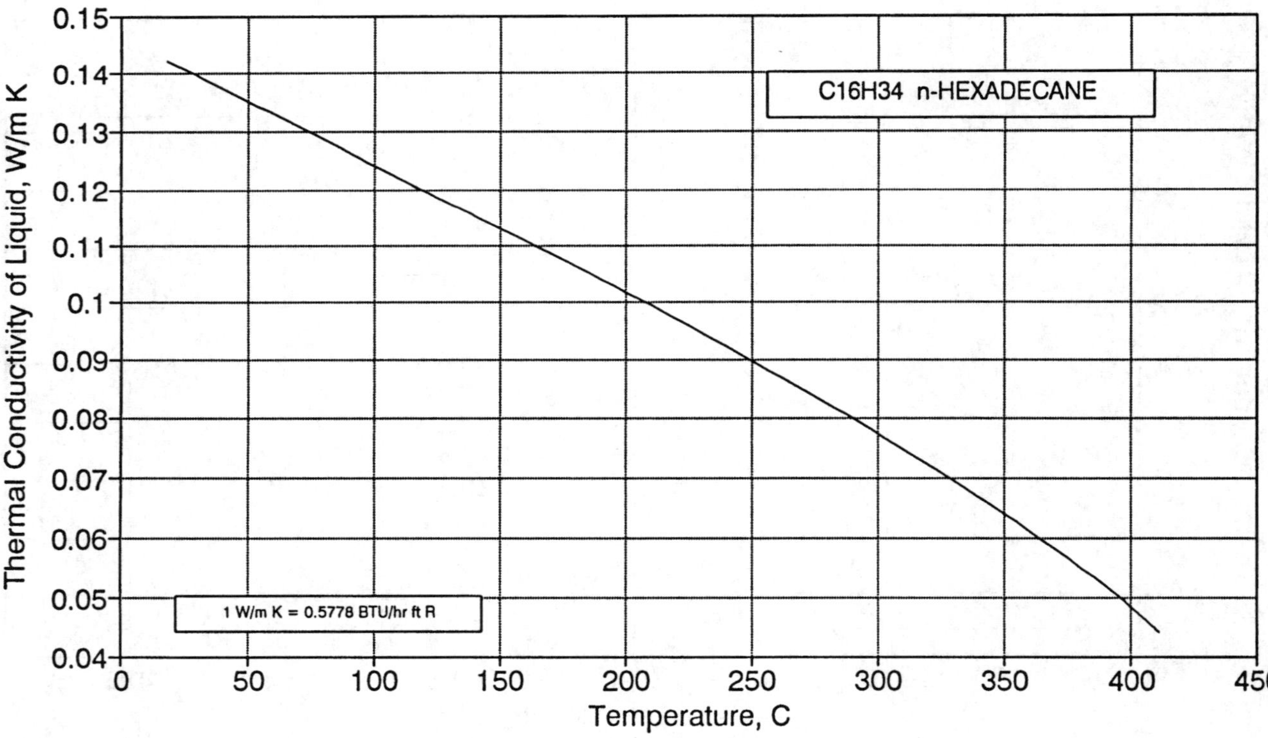
C16H34 n-HEXADECANE
1 W/m K = 0.5778 BTU/hr ft R
Thermal Conductivity of Liquid, W/m K
0.15
0.14
0.13
0.12
0.11
0.1
0.09
0.08
0.07
0.06
0.05
0.04
0
50
100
150
200
250
300
350
400
450
Temperature, C

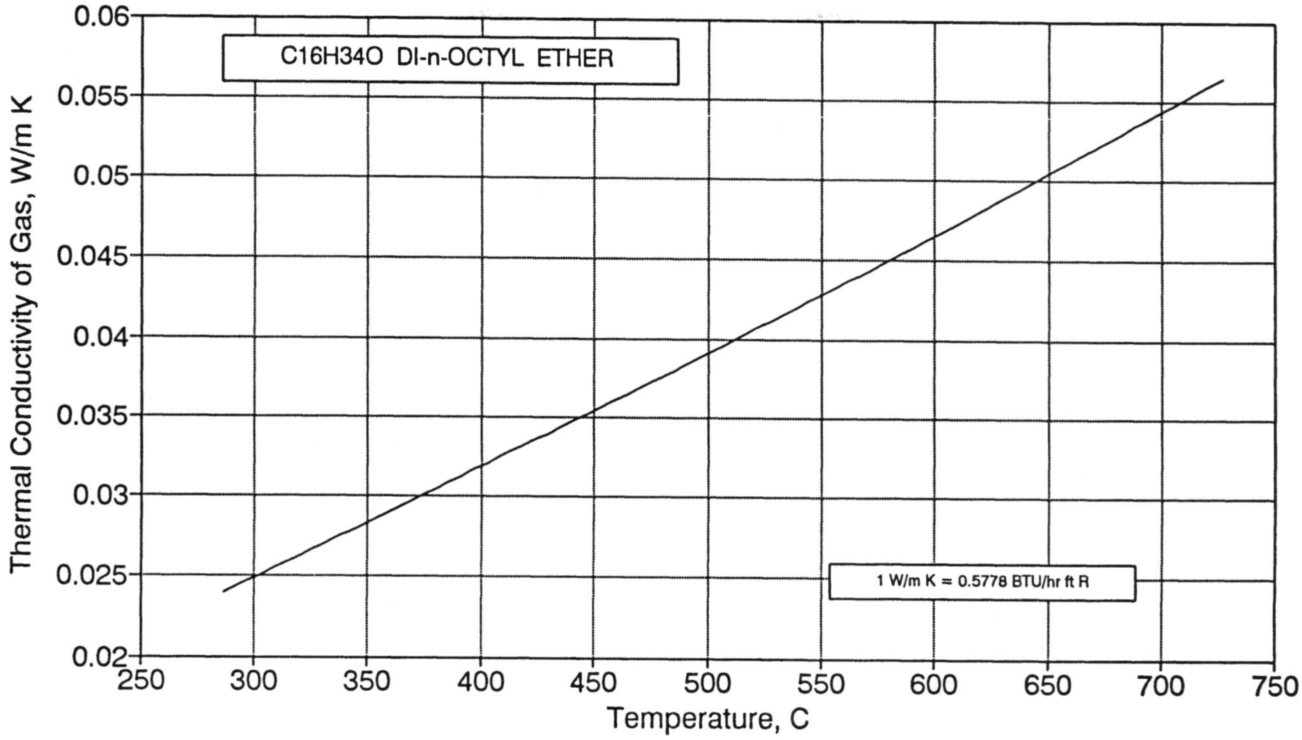
C16H34O DI-n-OCTYL ETHER
1 W/m K = 0.5778 BTU/hr ft R
Thermal Conductivity of Gas, W/m K
0.06
0.055
0.05
0.045
0.04
0.035
0.03
0.025
0.02
250
300
350
400
450
500
550
600
650
700
750
Temperature, C

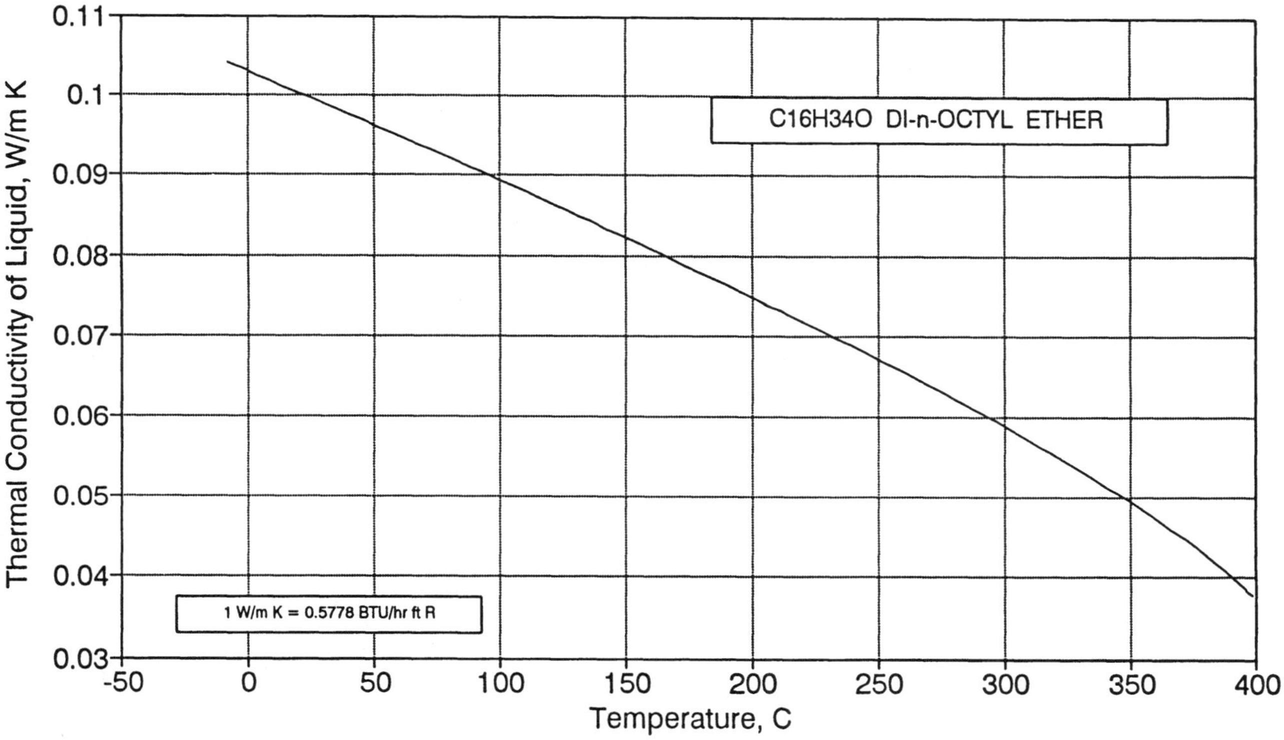
C16H34O DI-n-OCTYL ETHER
1 W/m K = 0.5778 BTU/hr ft R
Thermal Conductivity of Liquid, W/m K
0.11
0.1
0.09
0.08
0.07
0.06
0.05
0.04
0.03
-50
0
50
100
150
200
250
300
350
400
Temperature, C

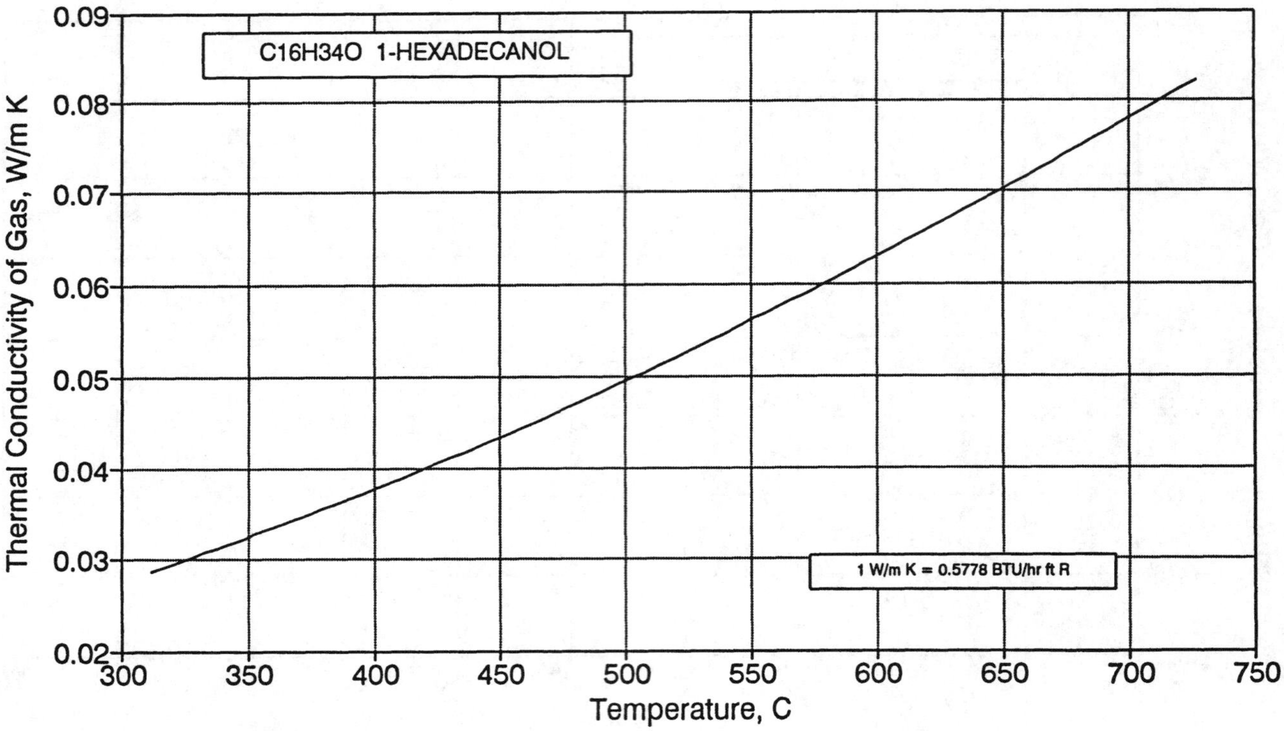
C16H34O 1-HEXADECANOL
1 W/m K = 0.5778 BTU/hr ft R
Thermal Conductivity of Gas, W/m K
0.09
0.08
0.07
0.06
0.05
0.04
0.03
0.02
300
350
400
450
500
550
600
650
700
750
Temperature, C

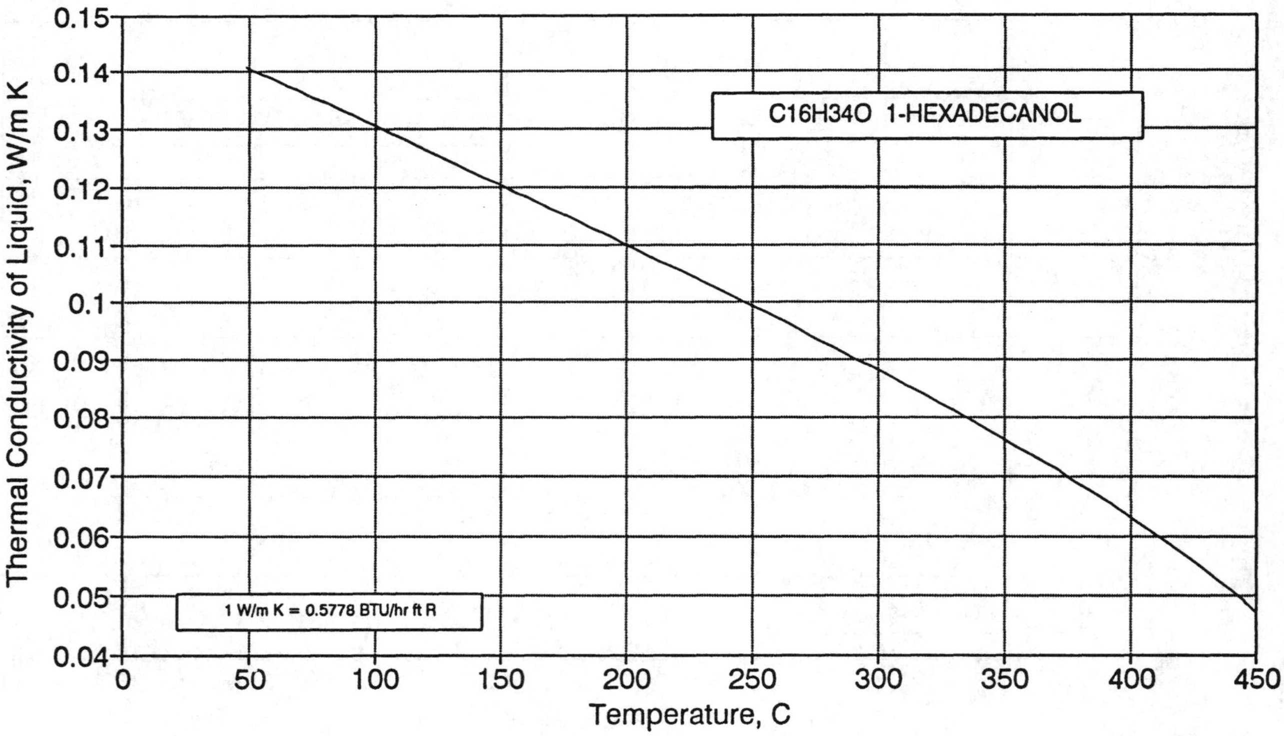
C16H34O 1-HEXADECANOL
1 W/m K = 0.5778 BTU/hr ft R
Thermal Conductivity of Liquid, W/m K
0.15
0.14
0.13
0.12
0.11
0.1
0.09
0.08
0.07
0.06
0.05
0.04
0
50
100
150
200
250
300
350
400
450
Temperature, C

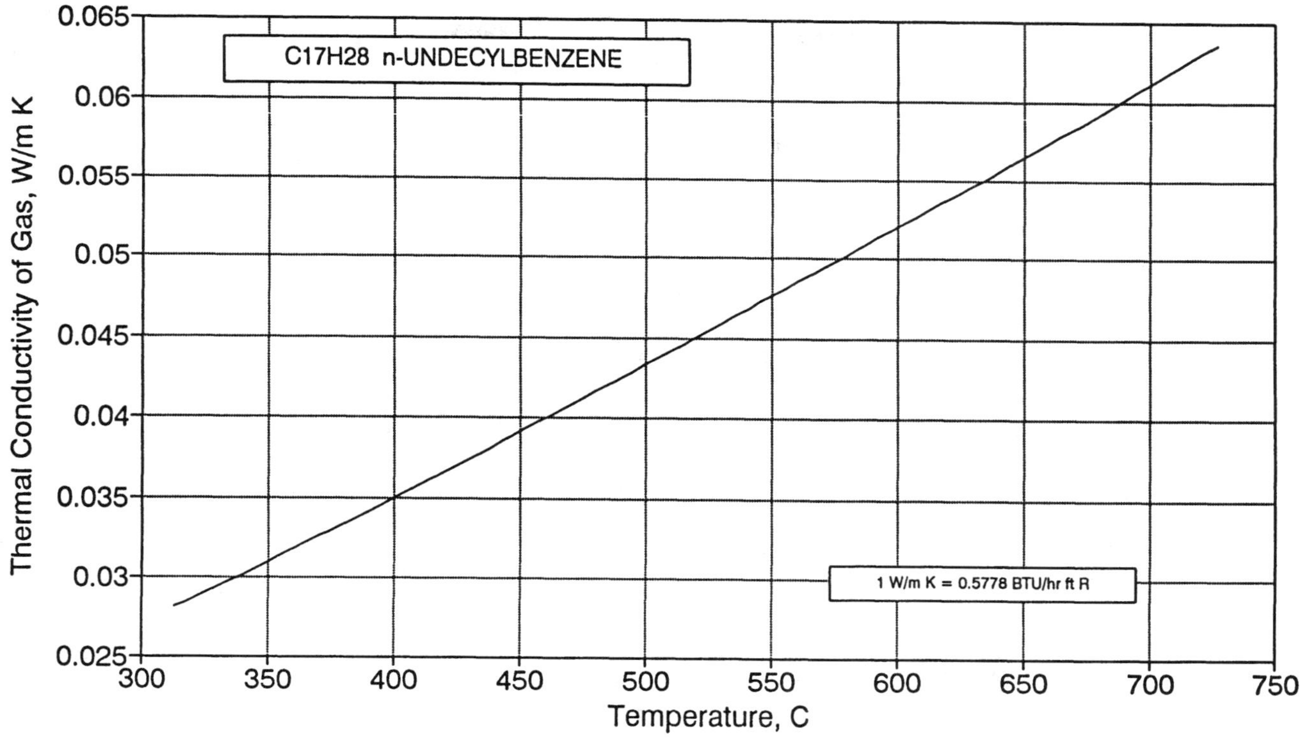
C17H28 n-UNDECYLBENZENE
1 W/m K = 0.5778 BTU/hr ft R
Thermal Conductivity of Gas, W/m K
Temperature, C
0.065
0.06
0.055
0.05
0.045
0.04
0.035
0.03
0.025
300
350
400
450
500
550
600
650
700
750

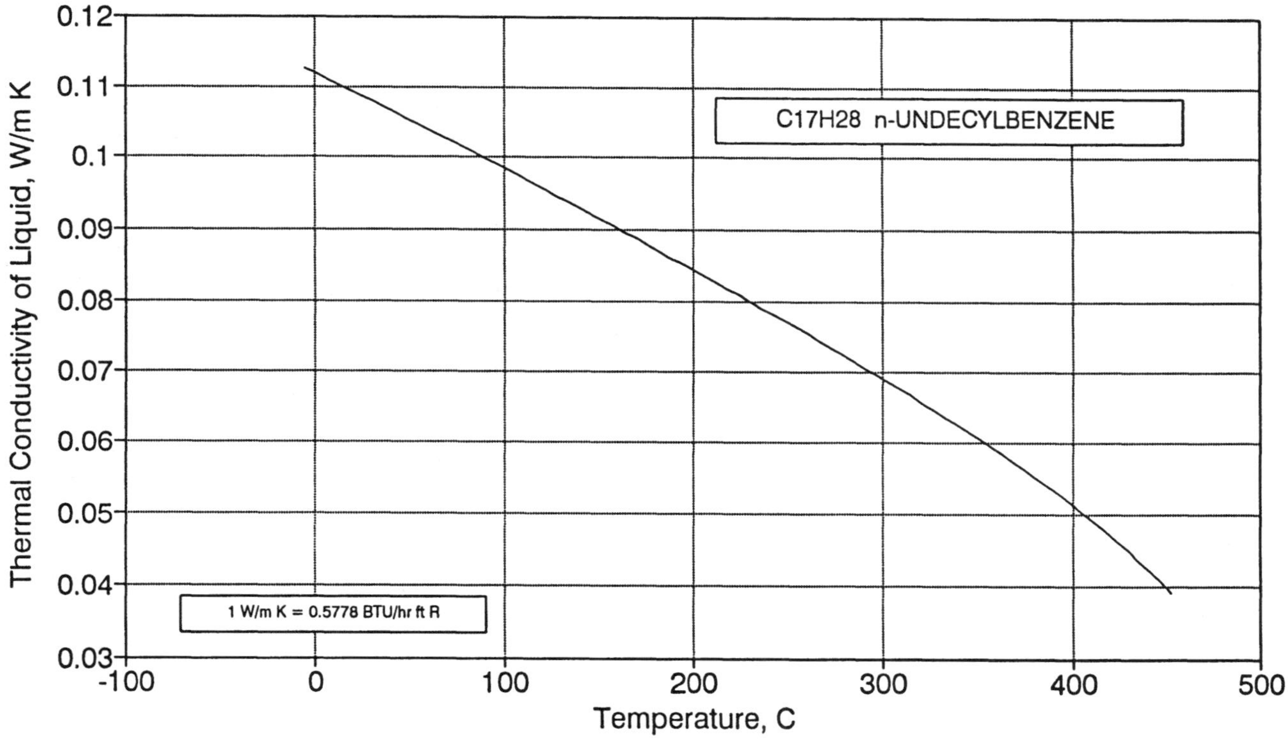
C17H28 n-UNDECYLBENZENE
1 W/m K = 0.5778 BTU/hr ft R
Thermal Conductivity of Liquid, W/m K
Temperature, C
0.12
0.11
0.1
0.09
0.08
0.07
0.06
0.05
0.04
0.03
-100
0
100
200
300
400
500

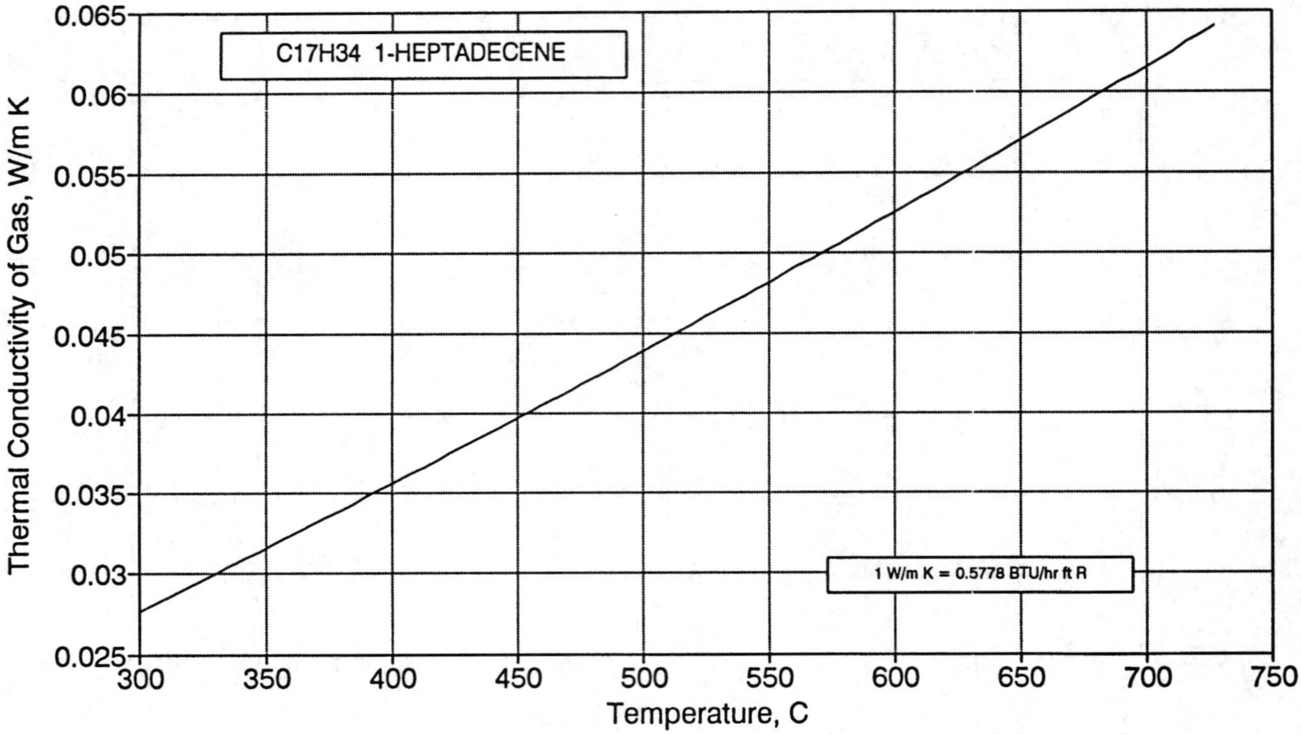
C17H34 1-HEPTADECENE
1 W/m K = 0.5778 BTU/hr ft R
Thermal Conductivity of Gas, W/m K
Temperature, C
0.065
0.06
0.055
0.05
0.045
0.04
0.035
0.03
0.025
300
350
400
450
500
550
600
650
700
750

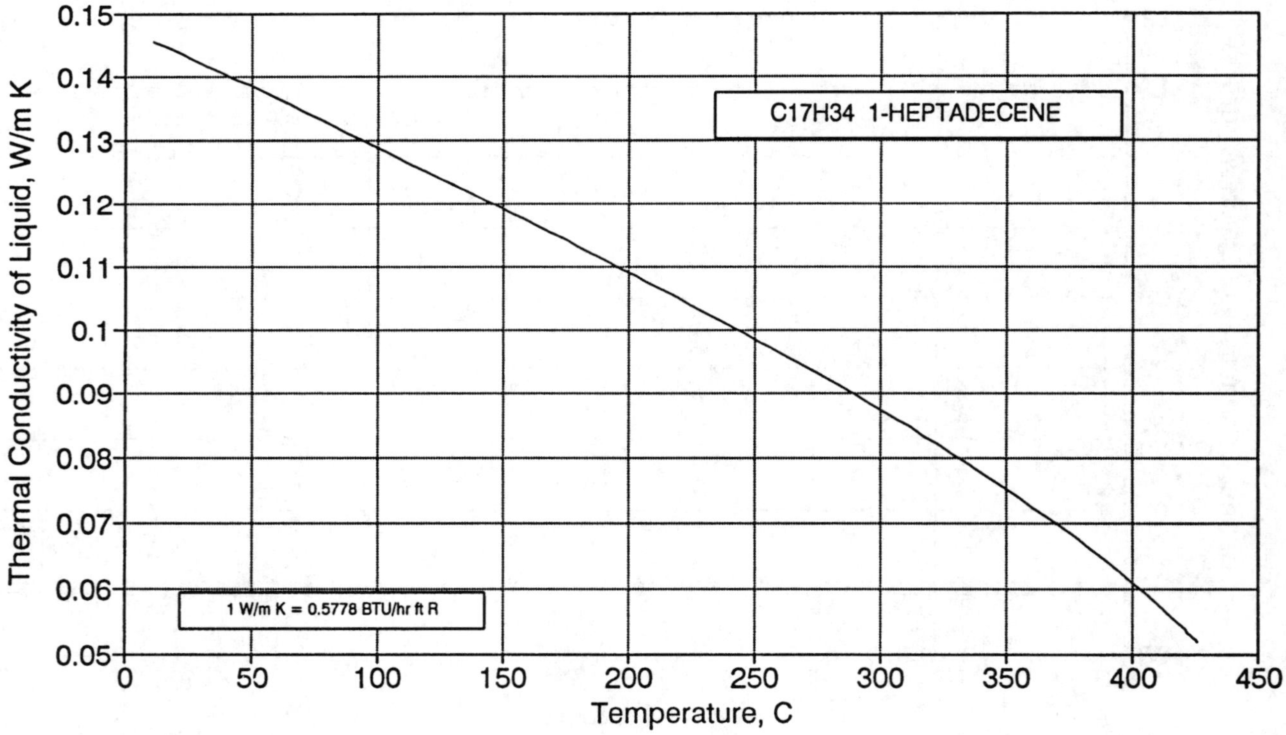
C17H34 1-HEPTADECENE
1 W/m K = 0.5778 BTU/hr ft R
Thermal Conductivity of Liquid, W/m K
Temperature, C
0.15
0.14
0.13
0.12
0.11
0.1
0.09
0.08
0.07
0.06
0.05
0
50
100
150
200
250
300
350
400
450

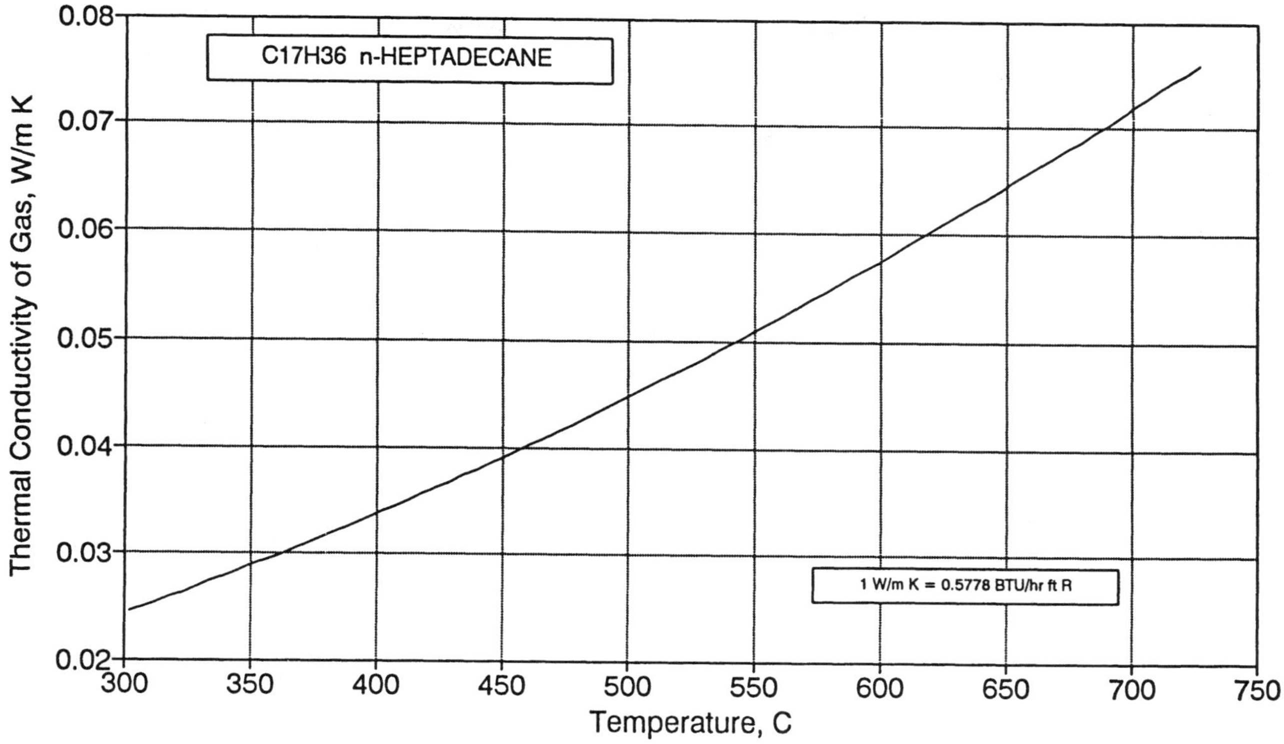
C17H36 n-HEPTADECANE
1 W/m K = 0.5778 BTU/hr ft R
Thermal Conductivity of Gas, W/m K
0.08
0.07
0.06
0.05
0.04
0.03
0.02
300
350
400
450
500
550
600
650
700
750
Temperature, C

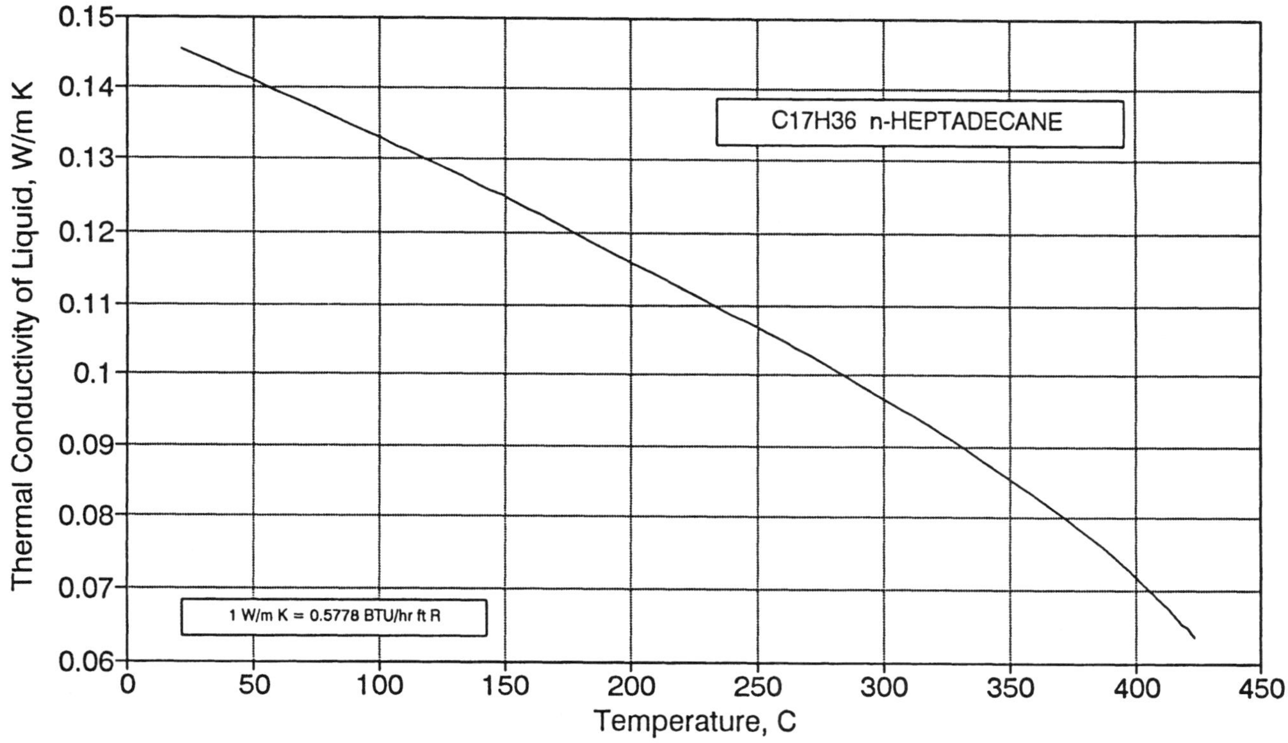
C17H36 n-HEPTADECANE
1 W/m K = 0.5778 BTU/hr ft R
Thermal Conductivity of Liquid, W/m K
0.15
0.14
0.13
0.12
0.11
0.1
0.09
0.08
0.07
0.06
0
50
100
150
200
250
300
350
400
450
Temperature, C

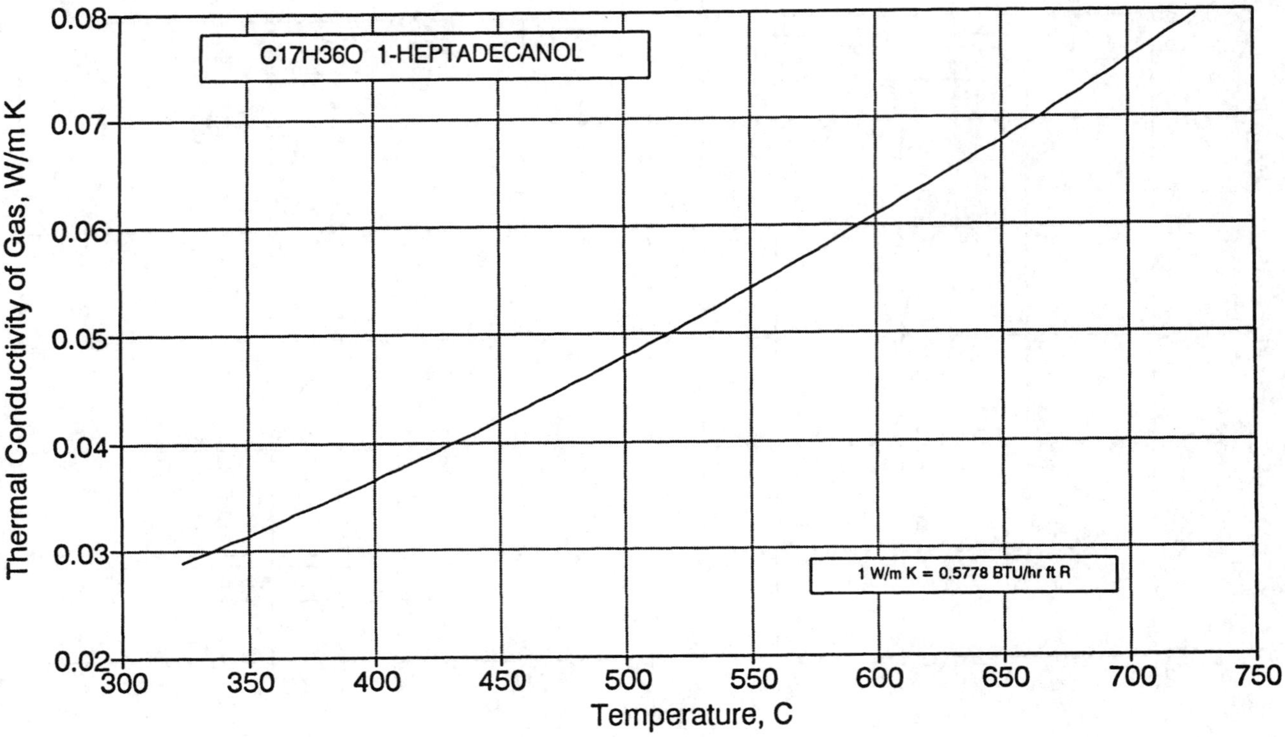
C17H36O 1-HEPTADECANOL
Thermal Conductivity of Gas, W/m K
0.08
0.07
0.06
0.05
0.04
0.03
0.02
1 W/m K = 0.5778 BTU/hr ft R
300
350
400
450
500
550
600
650
700
750
Temperature, C

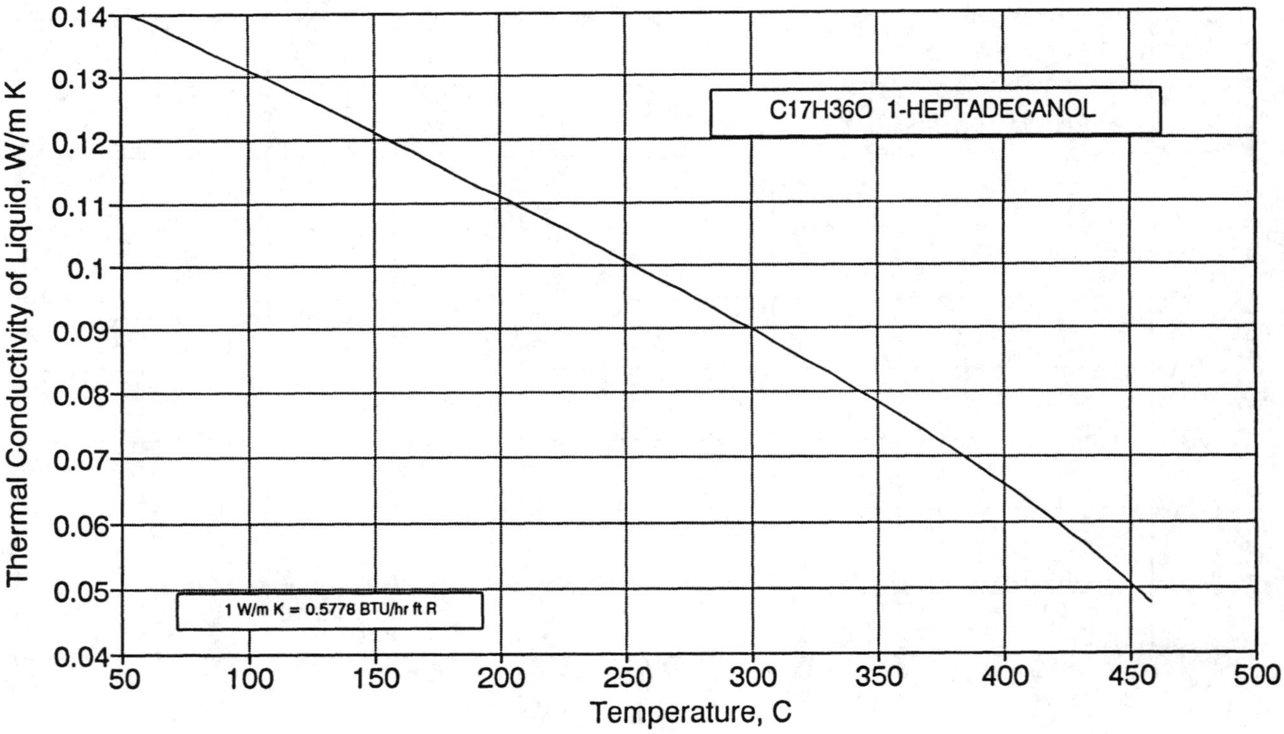
C17H36O 1-HEPTADECANOL
Thermal Conductivity of Liquid, W/m K
0.14
0.13
0.12
0.11
0.1
0.09
0.08
0.07
0.06
0.05
0.04
1 W/m K = 0.5778 BTU/hr ft R
50
100
150
200
250
300
350
400
450
500
Temperature, C

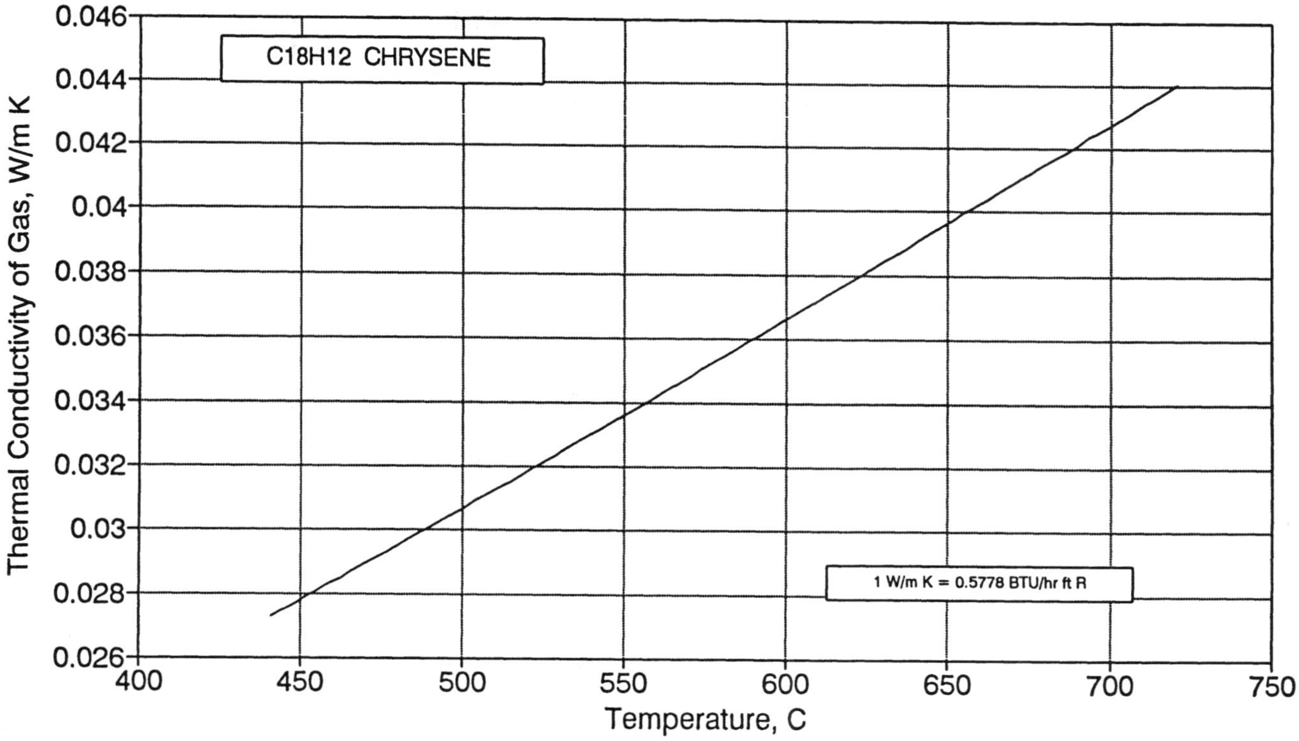
C18H12 CHRYSENE
1 W/m K = 0.5778 BTU/hr ft R
Thermal Conductivity of Gas, W/m K
0.046
0.044
0.042
0.04
0.038
0.036
0.034
0.032
0.03
0.028
0.026
400
450
500
550
600
650
700
750
Temperature, C

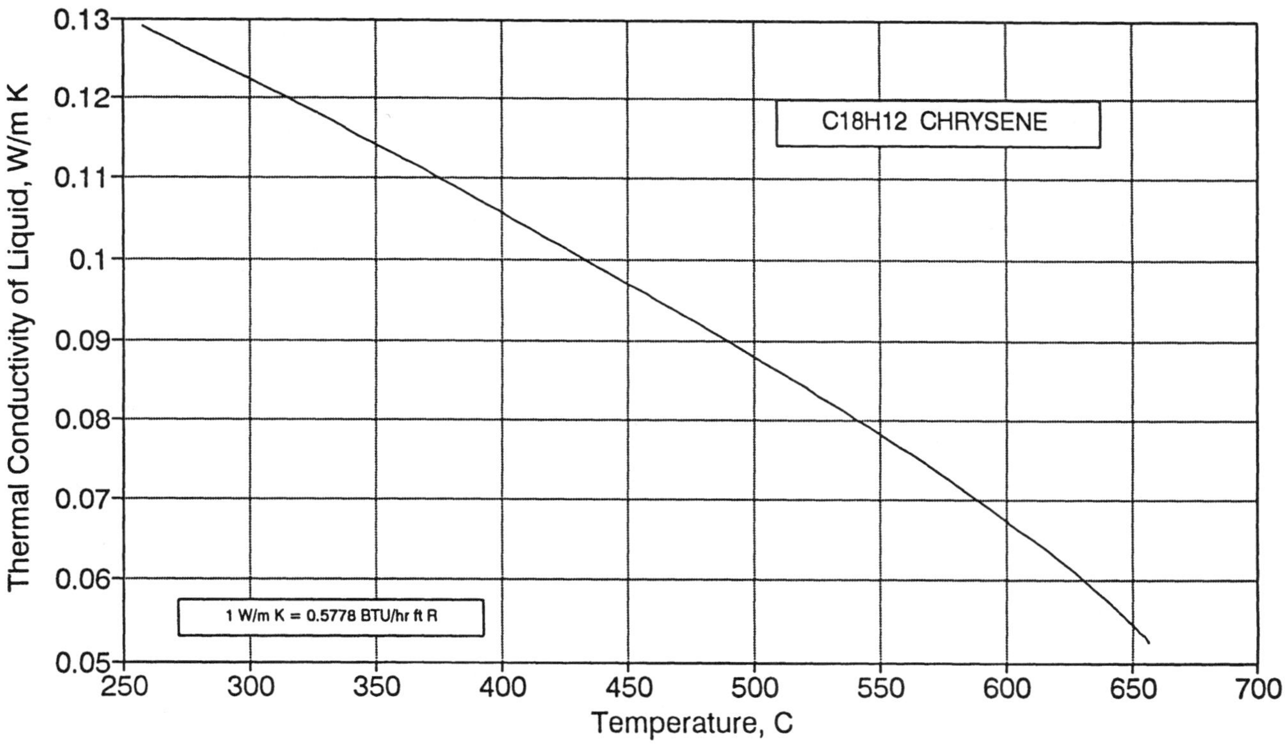
C18H12 CHRYSENE
1 W/m K = 0.5778 BTU/hr ft R
Thermal Conductivity of Liquid, W/m K
0.13
0.12
0.11
0.1
0.09
0.08
0.07
0.06
0.05
250
300
350
400
450
500
550
600
650
700
Temperature, C

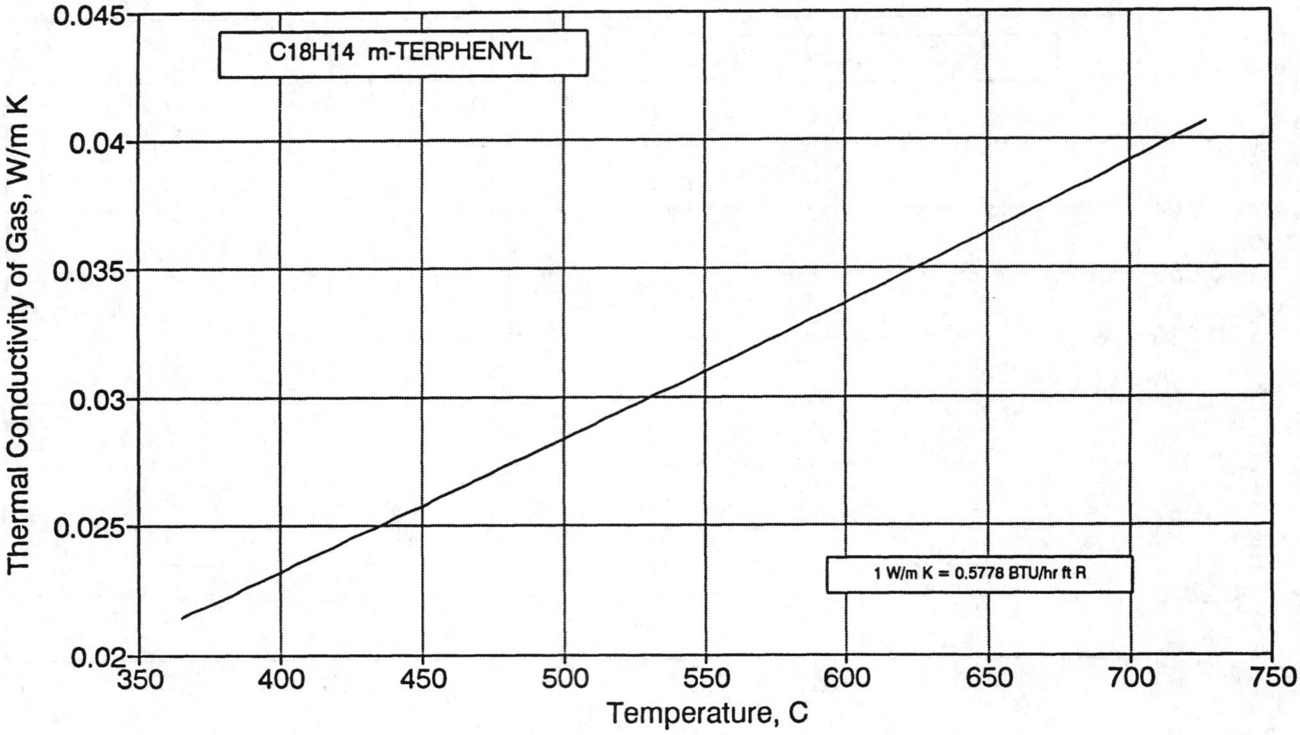
C18H14 m-TERPHENYL
1 W/m K = 0.5778 BTU/hr ft R
Thermal Conductivity of Gas, W/m K
0.045
0.04
0.035
0.03
0.025
0.02
350
400
450
500
550
600
650
700
750
Temperature, C

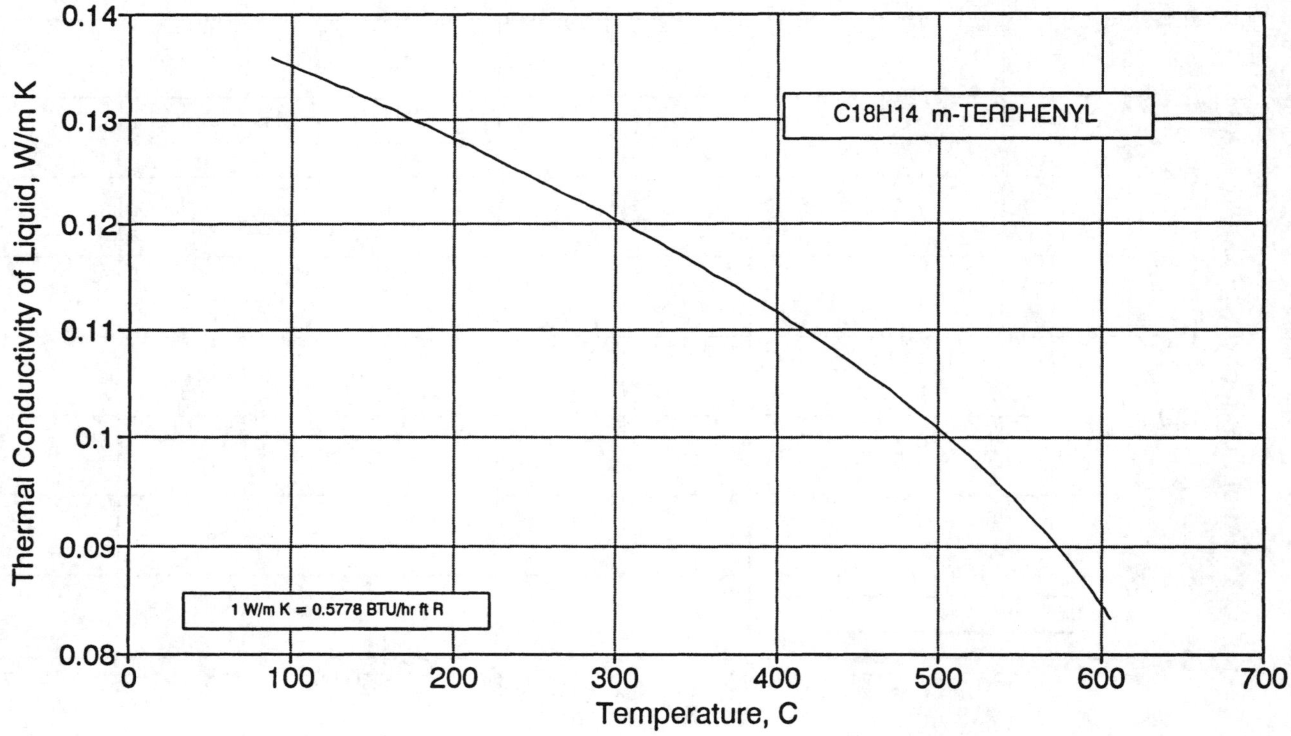
C18H14 m-TERPHENYL
1 W/m K = 0.5778 BTU/hr ft R
Thermal Conductivity of Liquid, W/m K
0.14
0.13
0.12
0.11
0.1
0.09
0.08
0
100
200
300
400
500
600
700
Temperature, C

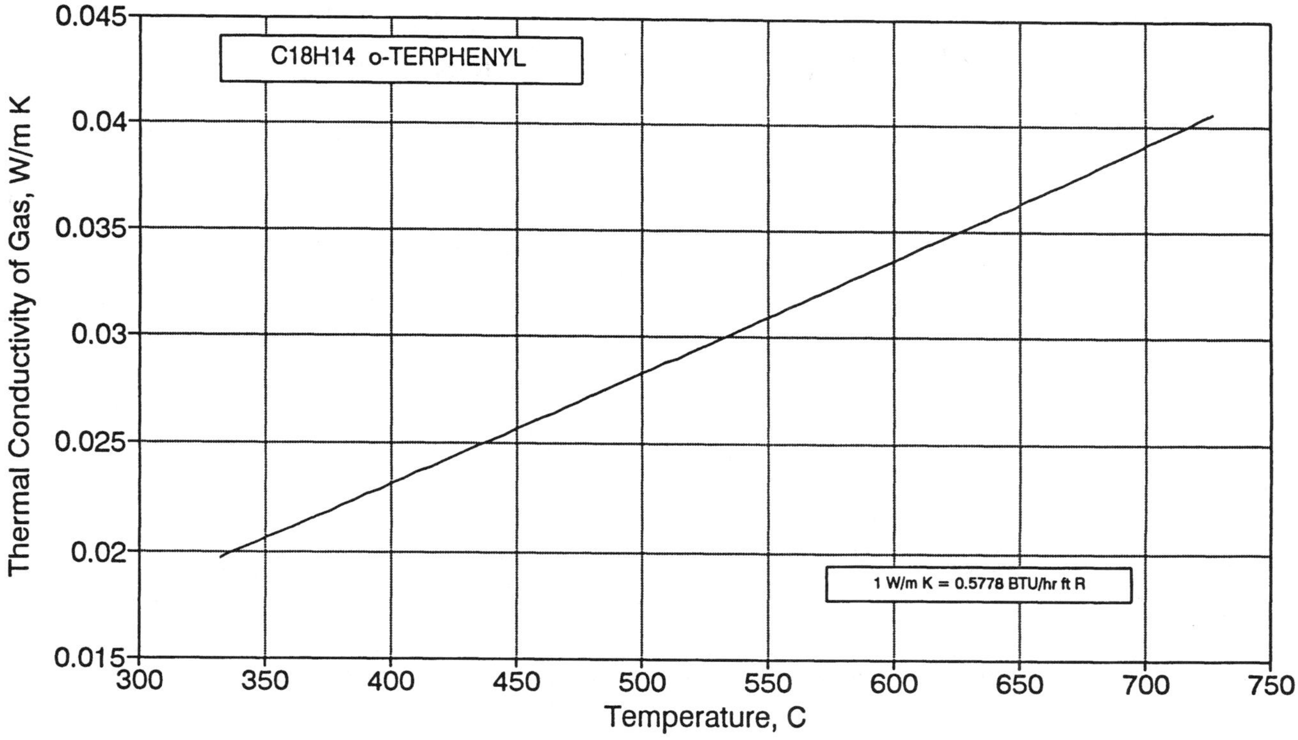
C18H14 o-TERPHENYL
1 W/m K = 0.5778 BTU/hr ft R
Thermal Conductivity of Gas, W/m K
0.045
0.04
0.035
0.03
0.025
0.02
0.015
300
350
400
450
500
550
600
650
700
750
Temperature, C

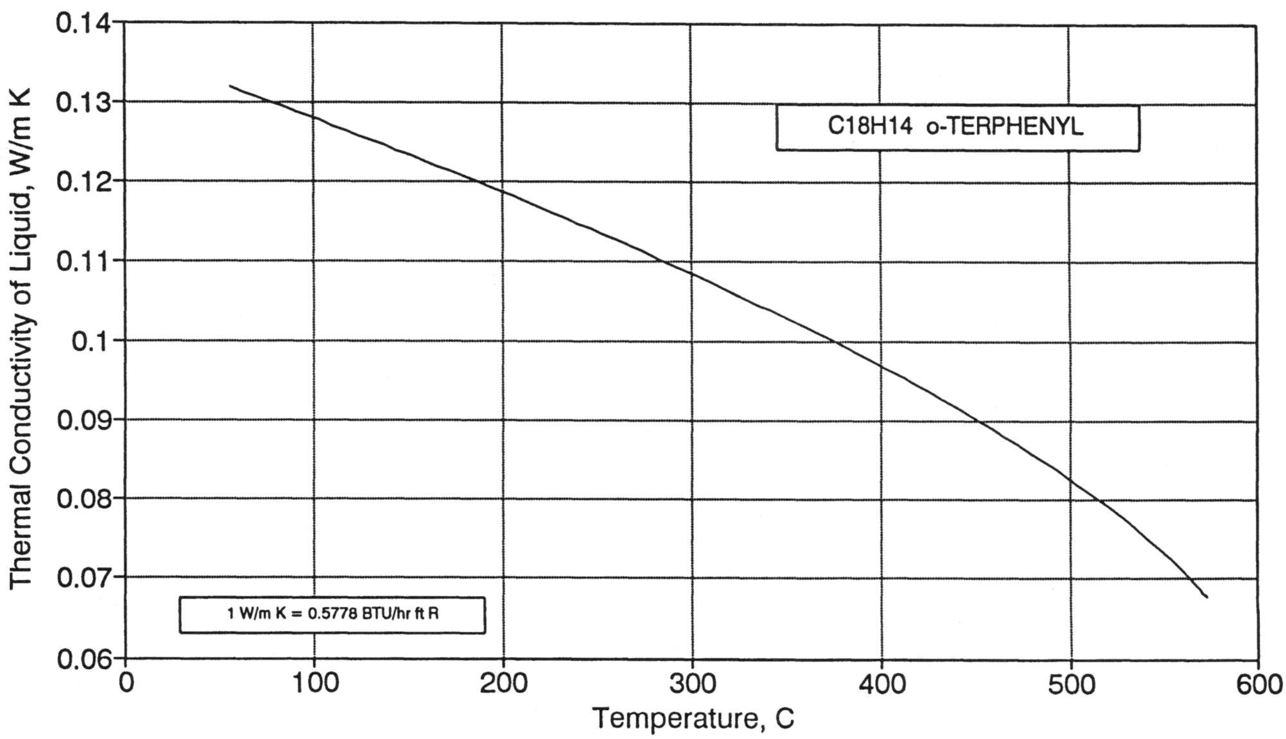
C18H14 o-TERPHENYL
1 W/m K = 0.5778 BTU/hr ft R
Thermal Conductivity of Liquid, W/m K
0.14
0.13
0.12
0.11
0.1
0.09
0.08
0.07
0.06
0
100
200
300
400
500
600
Temperature, C

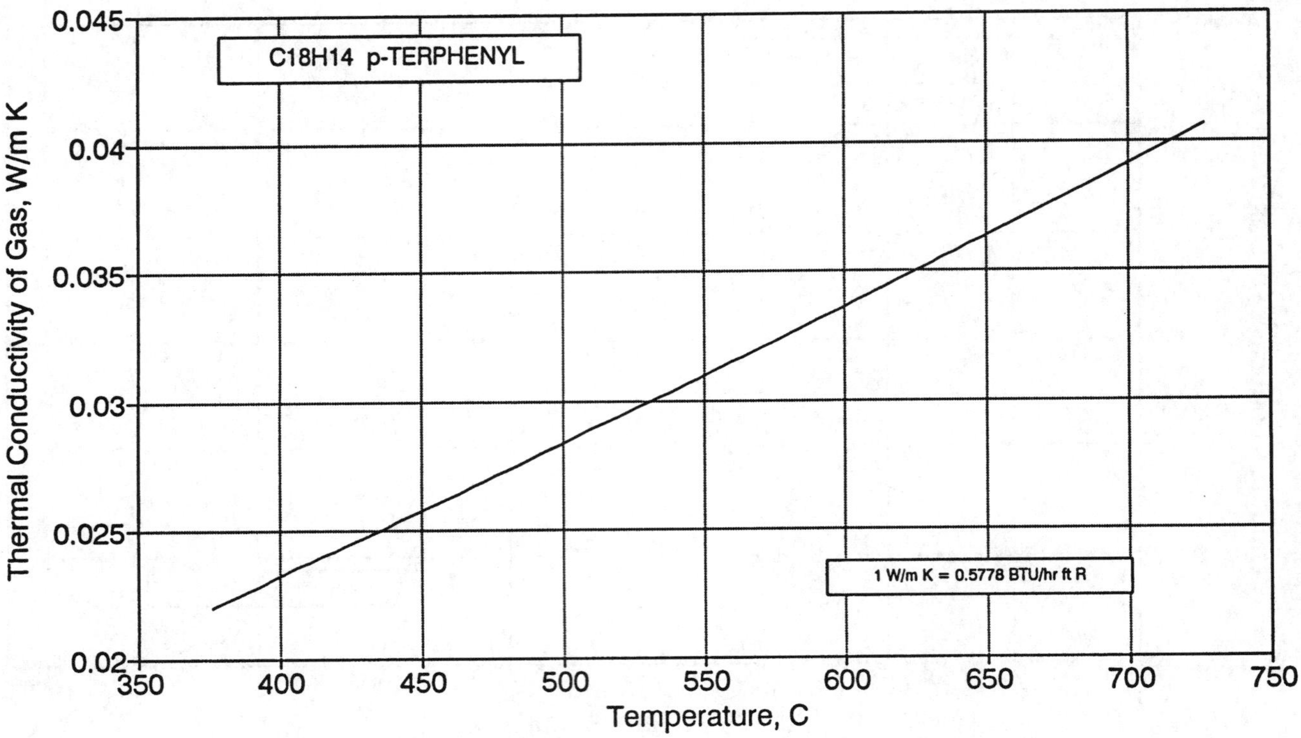

C18H14 p-TERPHENYL
1 W/m K = 0.5778 BTU/hr ft R
Thermal Conductivity of Gas, W/m K
0.045
0.04
0.035
0.03
0.025
0.02
350
400
450
500
550
600
650
700
750
Temperature, C

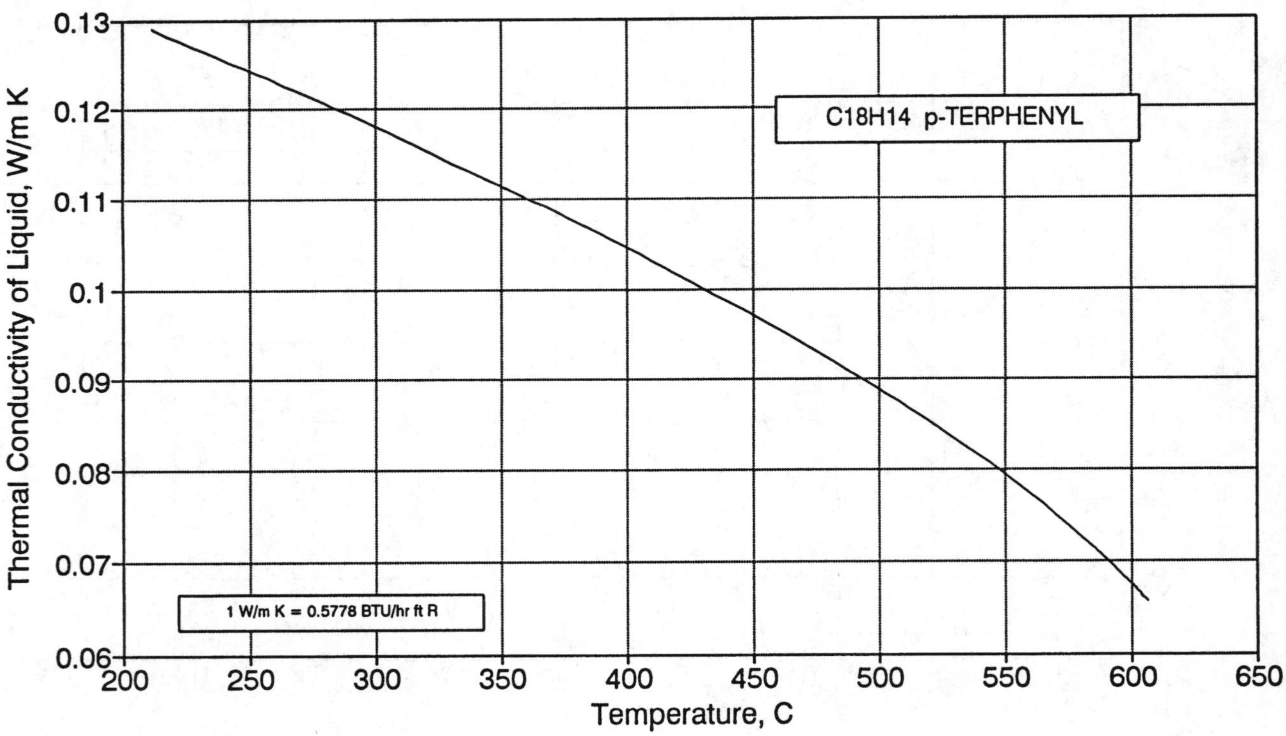

C18H14 p-TERPHENYL
1 W/m K = 0.5778 BTU/hr ft R
Thermal Conductivity of Liquid, W/m K
0.13
0.12
0.11
0.1
0.09
0.08
0.07
0.06
200
250
300
350
400
450
500
550
600
650
Temperature, C

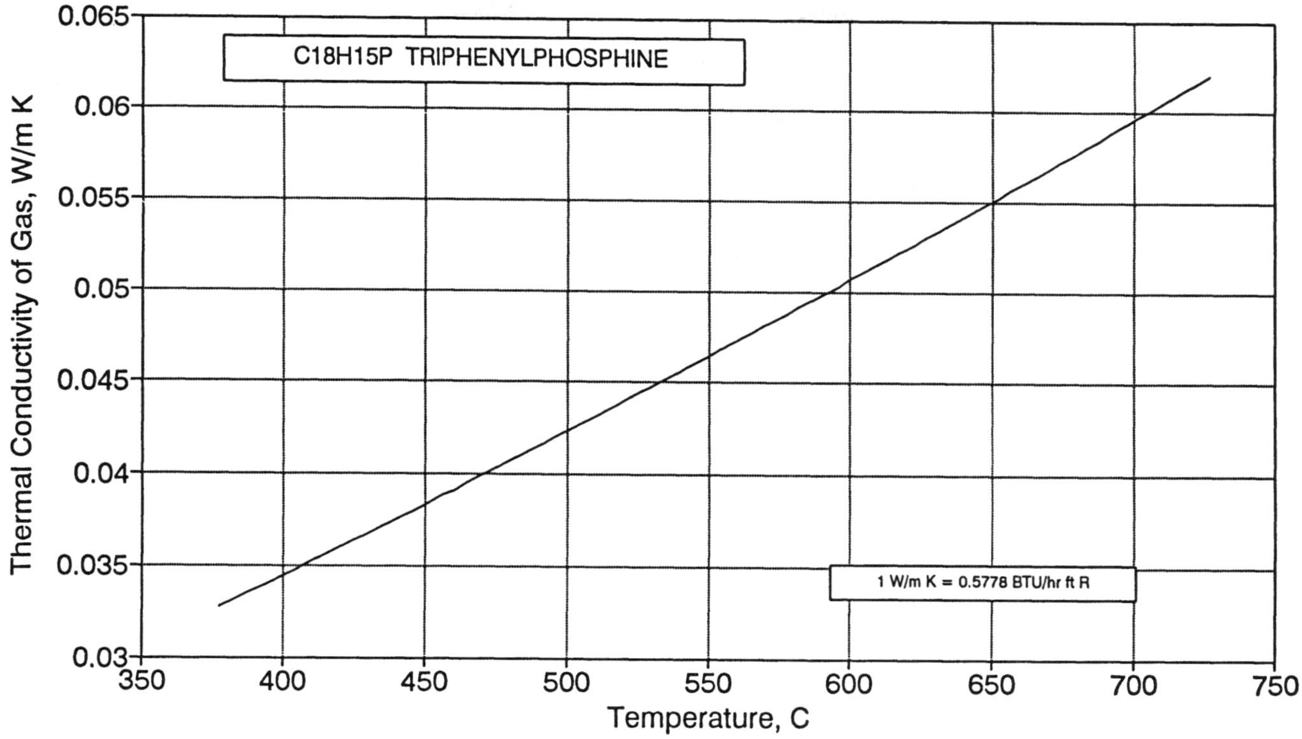
C18H15P TRIPHENYLPHOSPHINE
1 W/m K = 0.5778 BTU/hr ft R
Thermal Conductivity of Gas, W/m K
Temperature, C
0.065
0.06
0.055
0.05
0.045
0.04
0.035
0.03
350
400
450
500
550
600
650
700
750

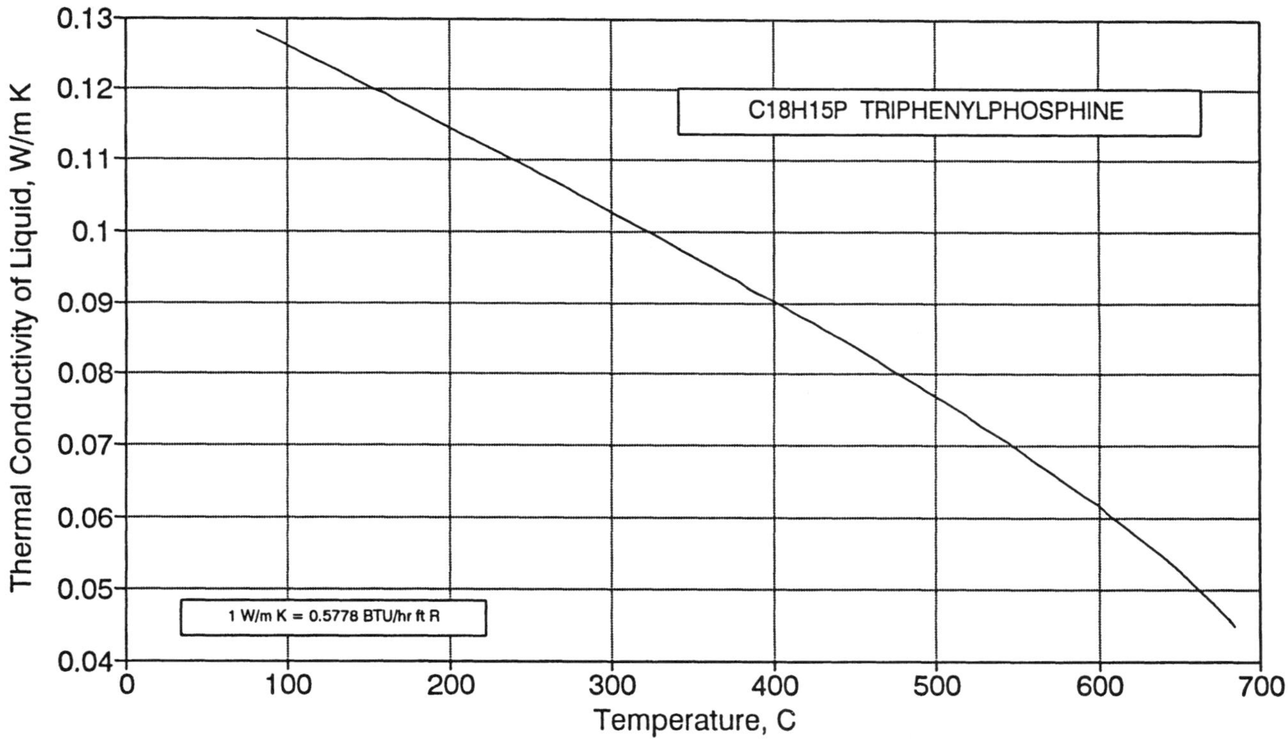
C18H15P TRIPHENYLPHOSPHINE
1 W/m K = 0.5778 BTU/hr ft R
Thermal Conductivity of Liquid, W/m K
Temperature, C
0.13
0.12
0.11
0.1
0.09
0.08
0.07
0.06
0.05
0.04
0
100
200
300
400
500
600
700

C18H15O4P TRIPHENYL PHOSPHATE

Gas thermal conductivity data are not available.

TF, K = 323.15

TB, K = 686.65

C18H15O4P TRIPHENYL PHOSPHATE

Liquid thermal conductivity data are not available.

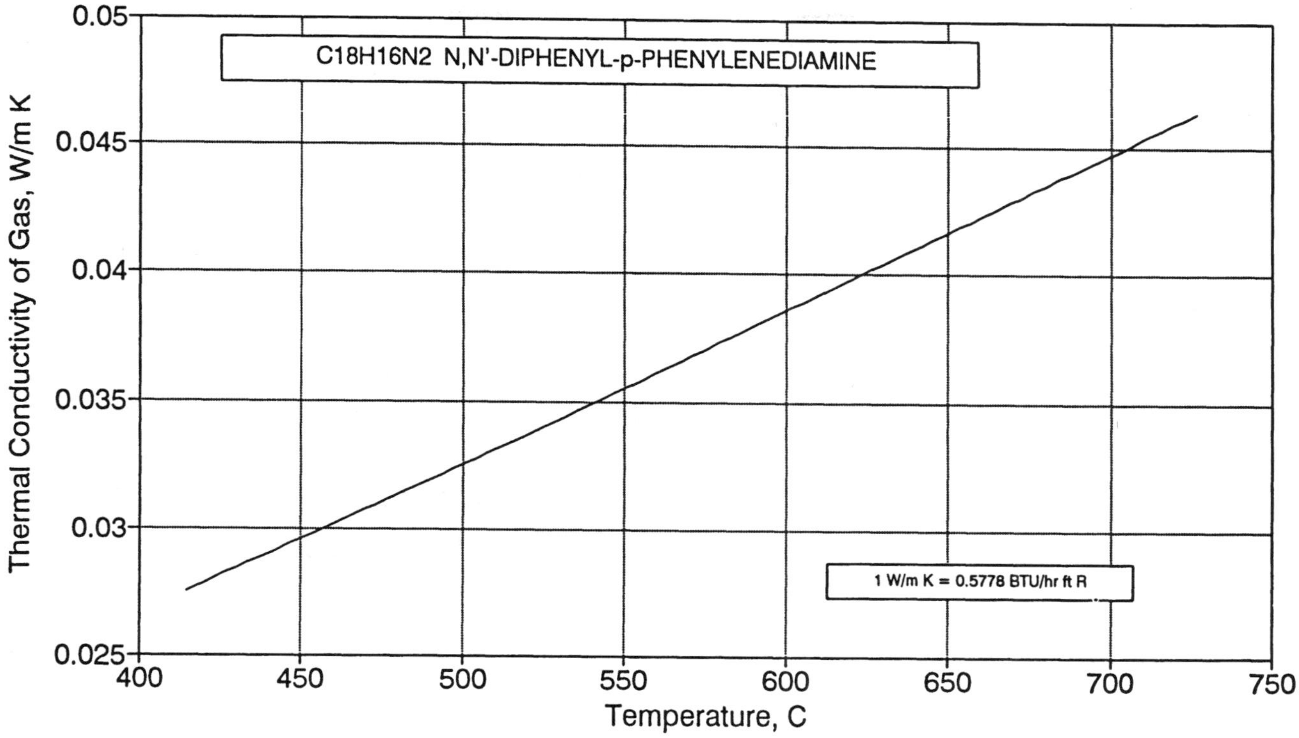
C18H16N2 N,N'-DIPHENYL-p-PHENYLENEDIAMINE
Thermal Conductivity of Gas, W/m K
0.05
0.045
0.04
0.035
0.03
0.025
400
450
500
550
600
650
700
750
Temperature, C
1 W/m K = 0.5778 BTU/hr ft R

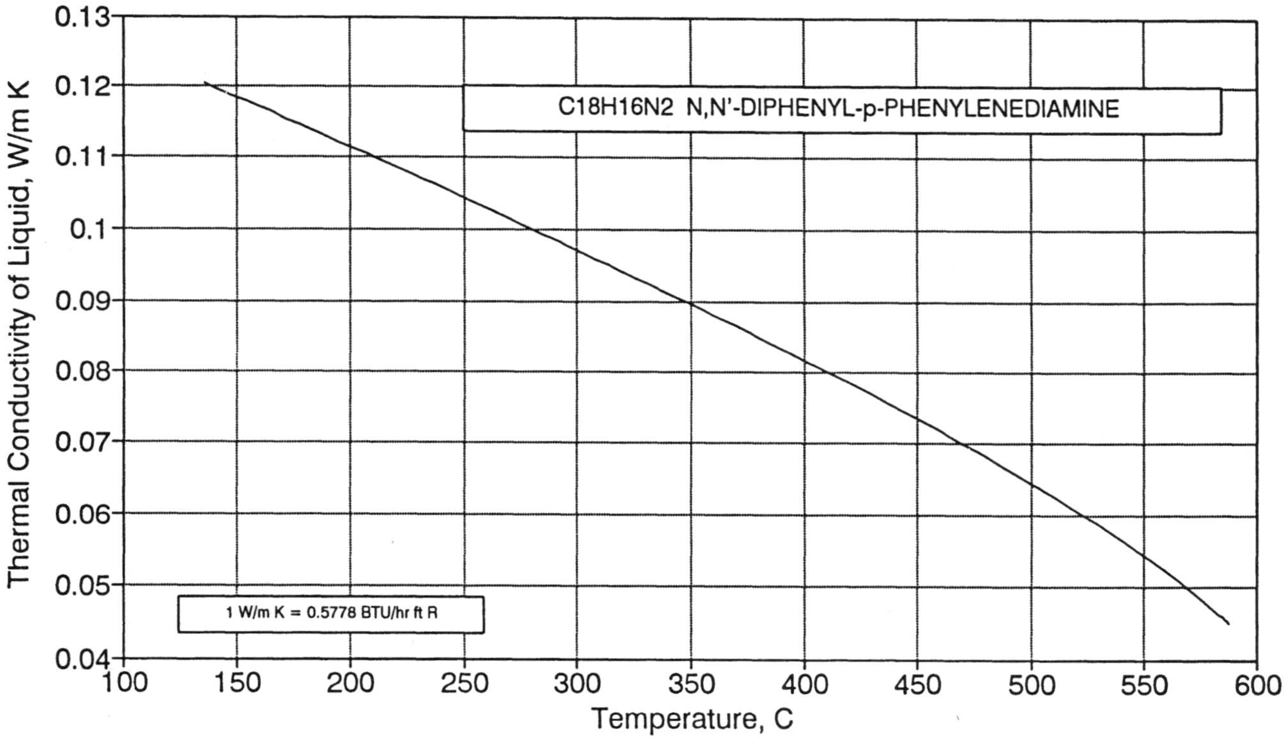
C18H16N2 N,N'-DIPHENYL-p-PHENYLENEDIAMINE
Thermal Conductivity of Liquid, W/m K
0.13
0.12
0.11
0.1
0.09
0.08
0.07
0.06
0.05
0.04
100
150
200
250
300
350
400
450
500
550
600
Temperature, C
1 W/m K = 0.5778 BTU/hr ft R

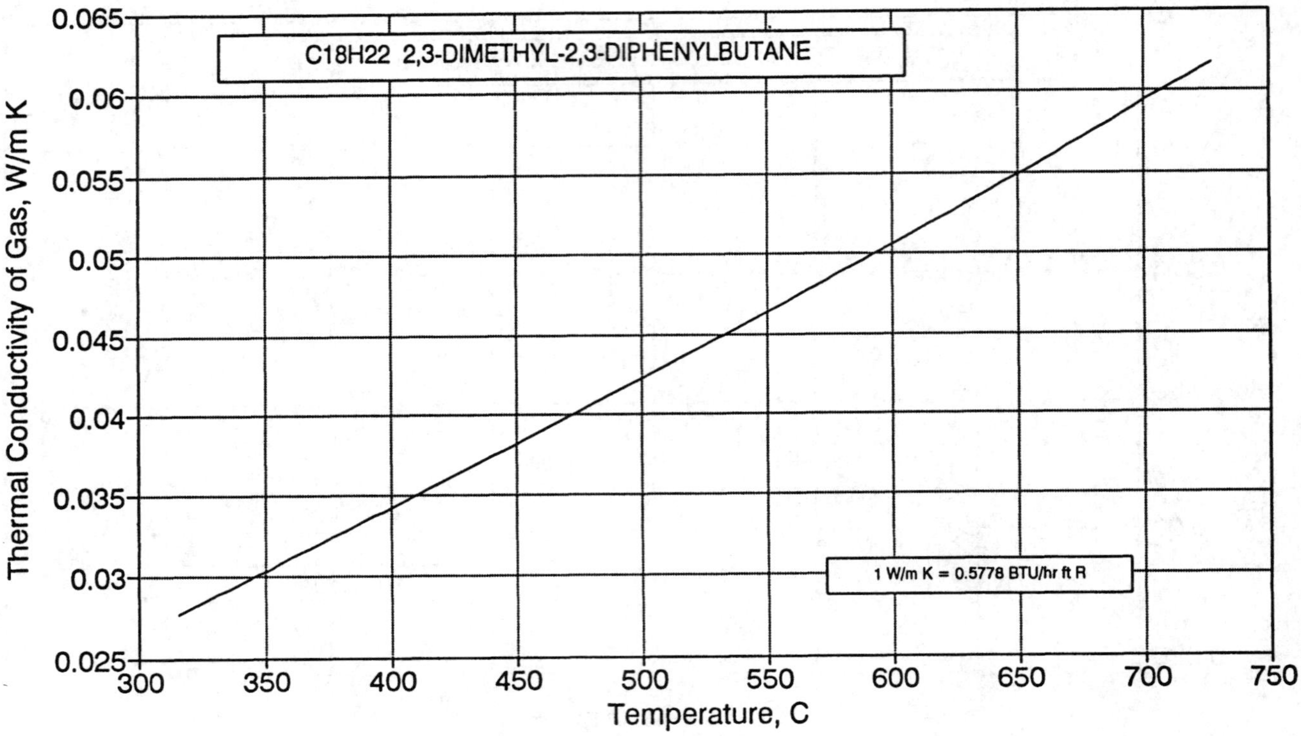
C18H22 2,3-DIMETHYL-2,3-DIPHENYLBUTANE
1 W/m K = 0.5778 BTU/hr ft R
Thermal Conductivity of Gas, W/m K
0.065
0.06
0.055
0.05
0.045
0.04
0.035
0.03
0.025
300
350
400
450
500
550
600
650
700
750
Temperature, C

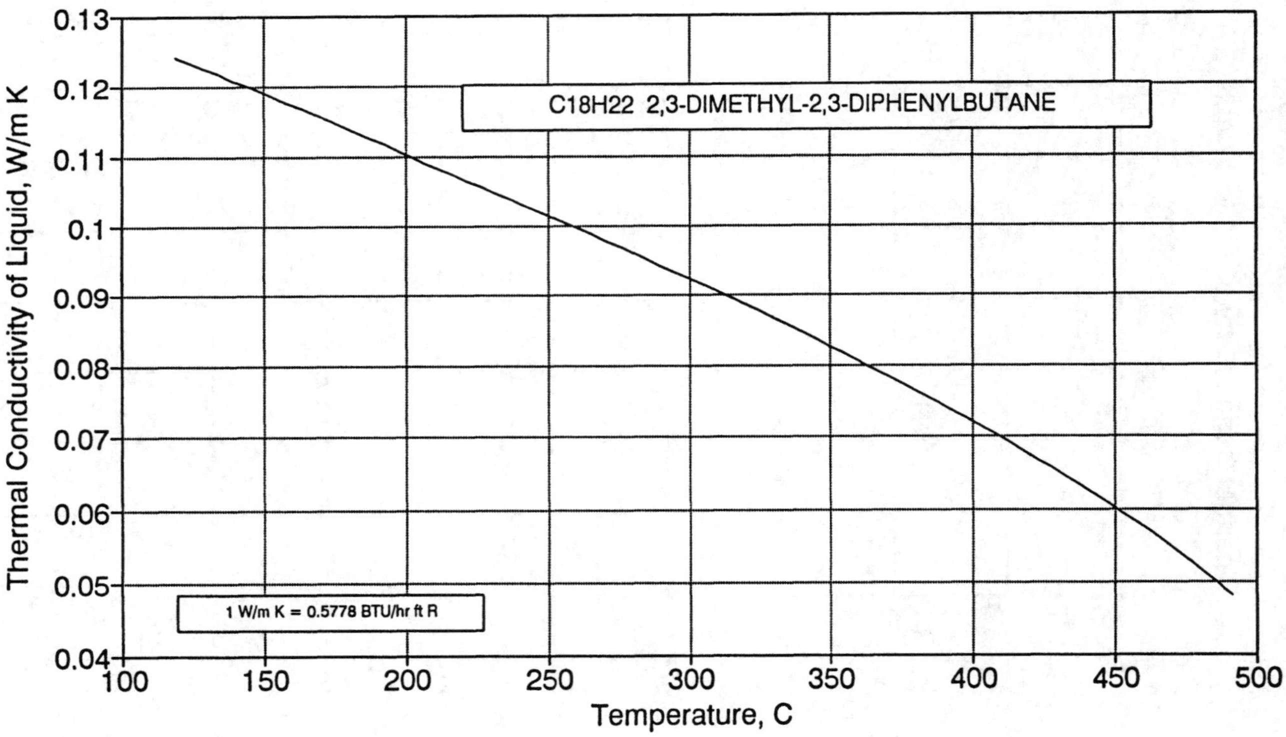
C18H22 2,3-DIMETHYL-2,3-DIPHENYLBUTANE
1 W/m K = 0.5778 BTU/hr ft R
Thermal Conductivity of Liquid, W/m K
0.13
0.12
0.11
0.1
0.09
0.08
0.07
0.06
0.05
0.04
100
150
200
250
300
350
400
450
500
Temperature, C

C18H22O2 DICUMYL PEROXIDE

Gas thermal conductivity data are not available.

TF, K = 311.15

TB, K = 669.00

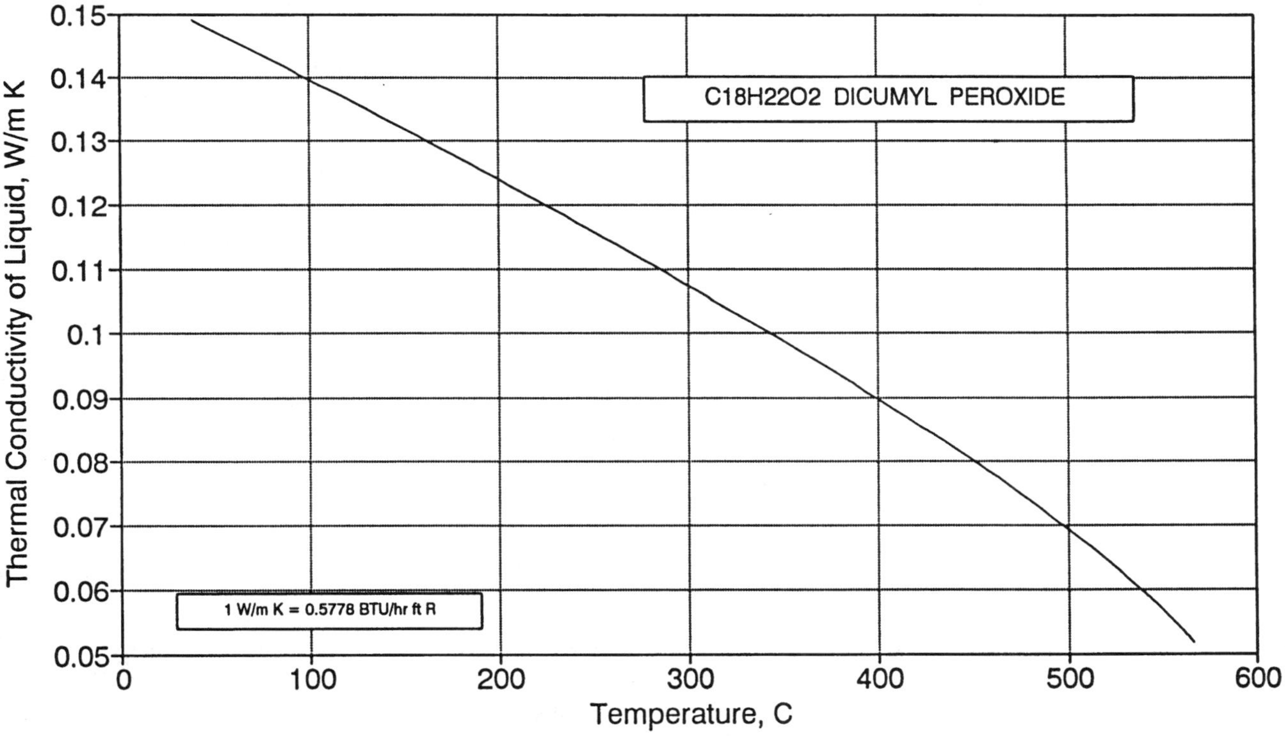

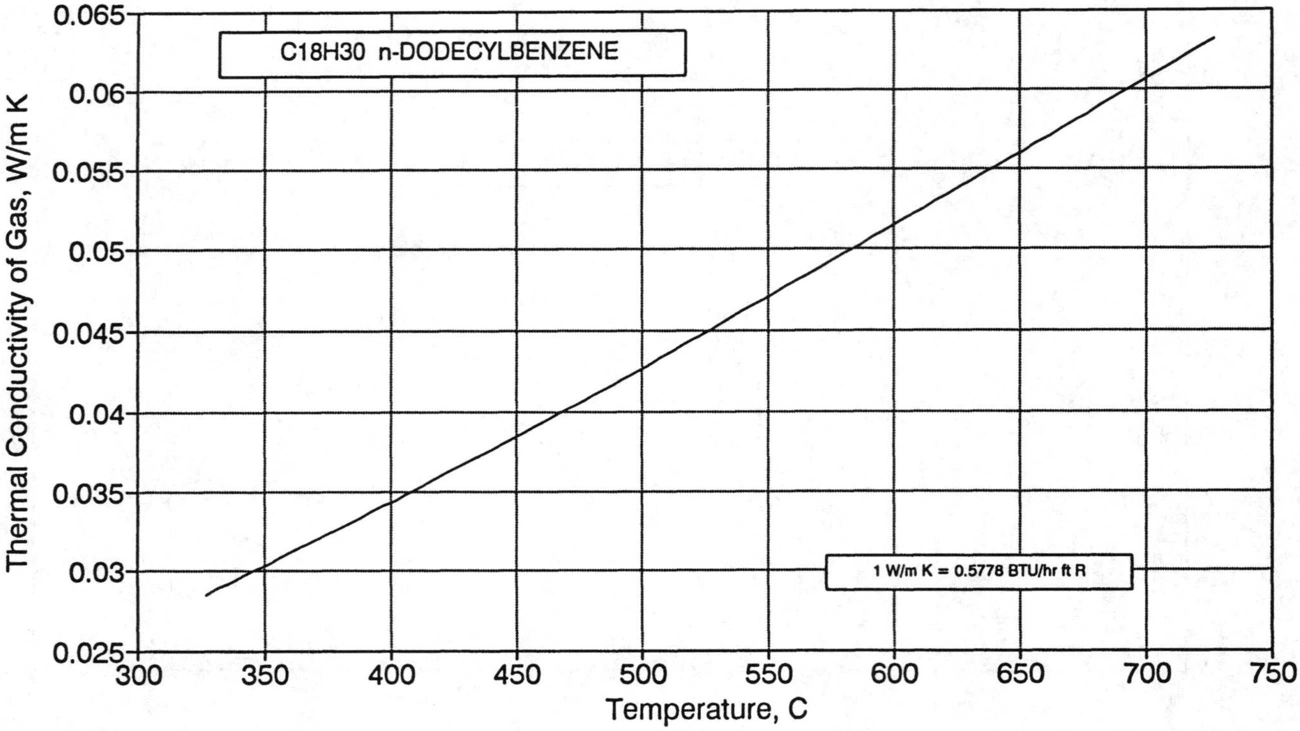
C18H30 n-DODECYLBENZENE
1 W/m K = 0.5778 BTU/hr ft R
Thermal Conductivity of Gas, W/m K
0.065
0.06
0.055
0.05
0.045
0.04
0.035
0.03
0.025
300
350
400
450
500
550
600
650
700
750
Temperature, C

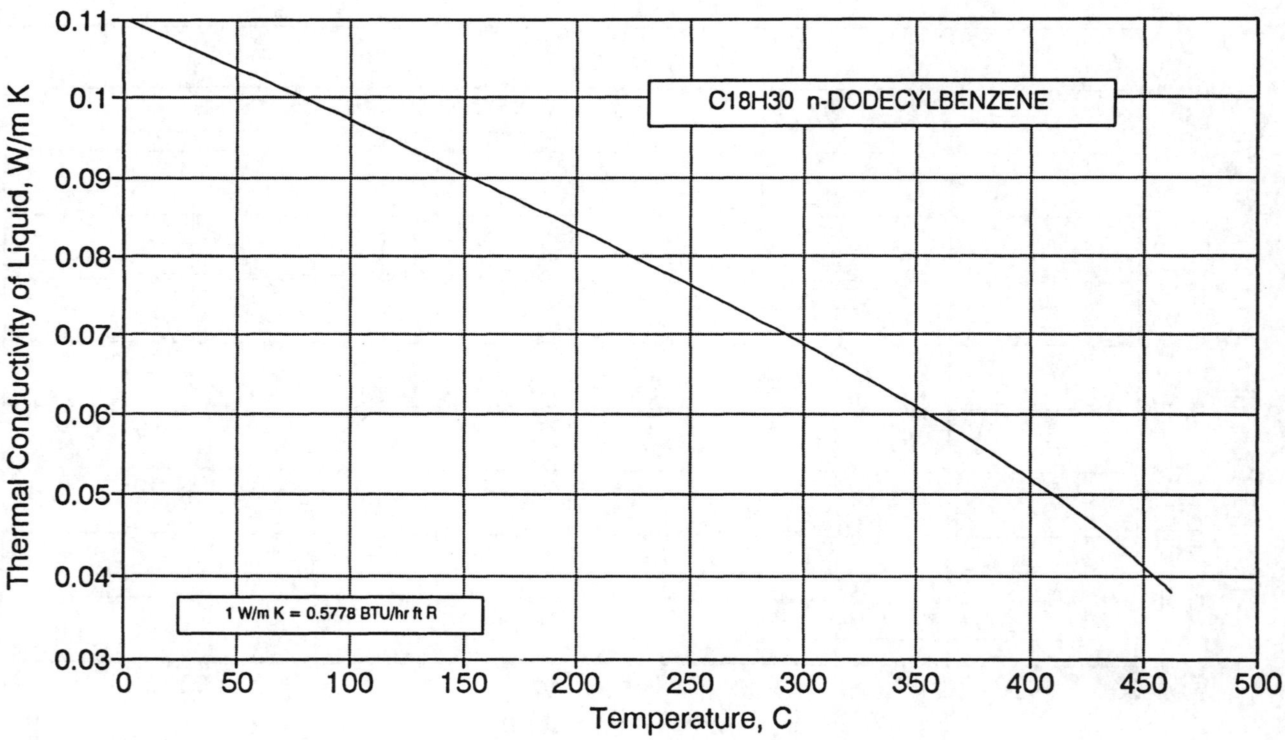
C18H30 n-DODECYLBENZENE
1 W/m K = 0.5778 BTU/hr ft R
Thermal Conductivity of Liquid, W/m K
0.11
0.1
0.09
0.08
0.07
0.06
0.05
0.04
0.03
0
50
100
150
200
250
300
350
400
450
500
Temperature, C

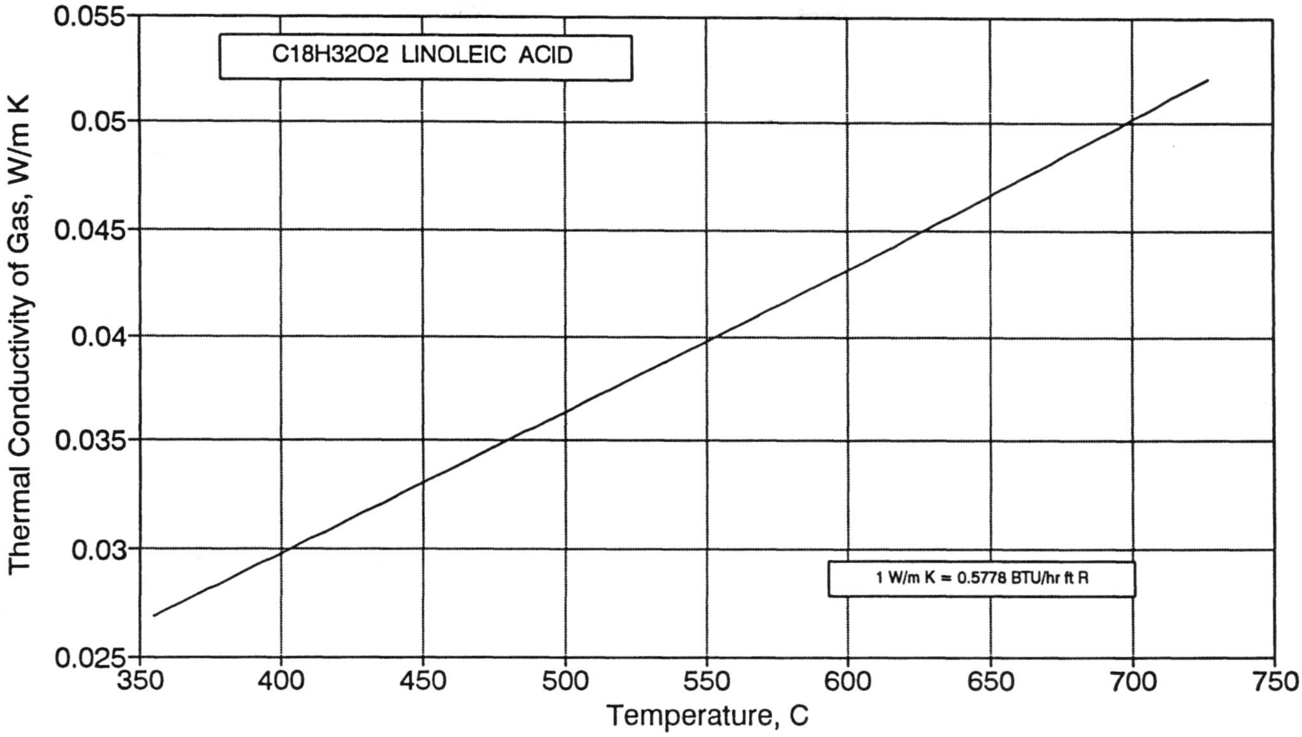

C18H32O2 LINOLEIC ACID
1 W/m K = 0.5778 BTU/hr ft R
Thermal Conductivity of Gas, W/m K
0.055
0.05
0.045
0.04
0.035
0.03
0.025
350
400
450
500
550
600
650
700
750
Temperature, C

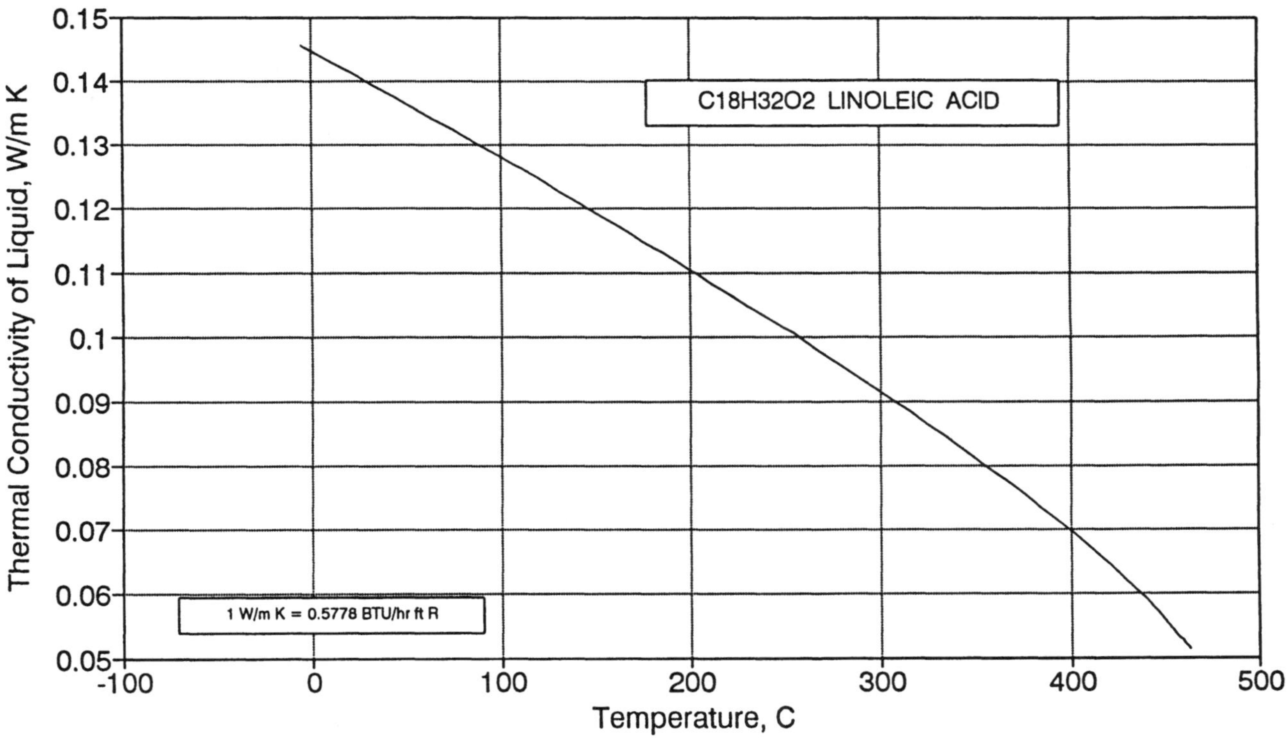

C18H32O2 LINOLEIC ACID
1 W/m K = 0.5778 BTU/hr ft R
Thermal Conductivity of Liquid, W/m K
0.15
0.14
0.13
0.12
0.11
0.1
0.09
0.08
0.07
0.06
0.05
-100
0
100
200
300
400
500
Temperature, C

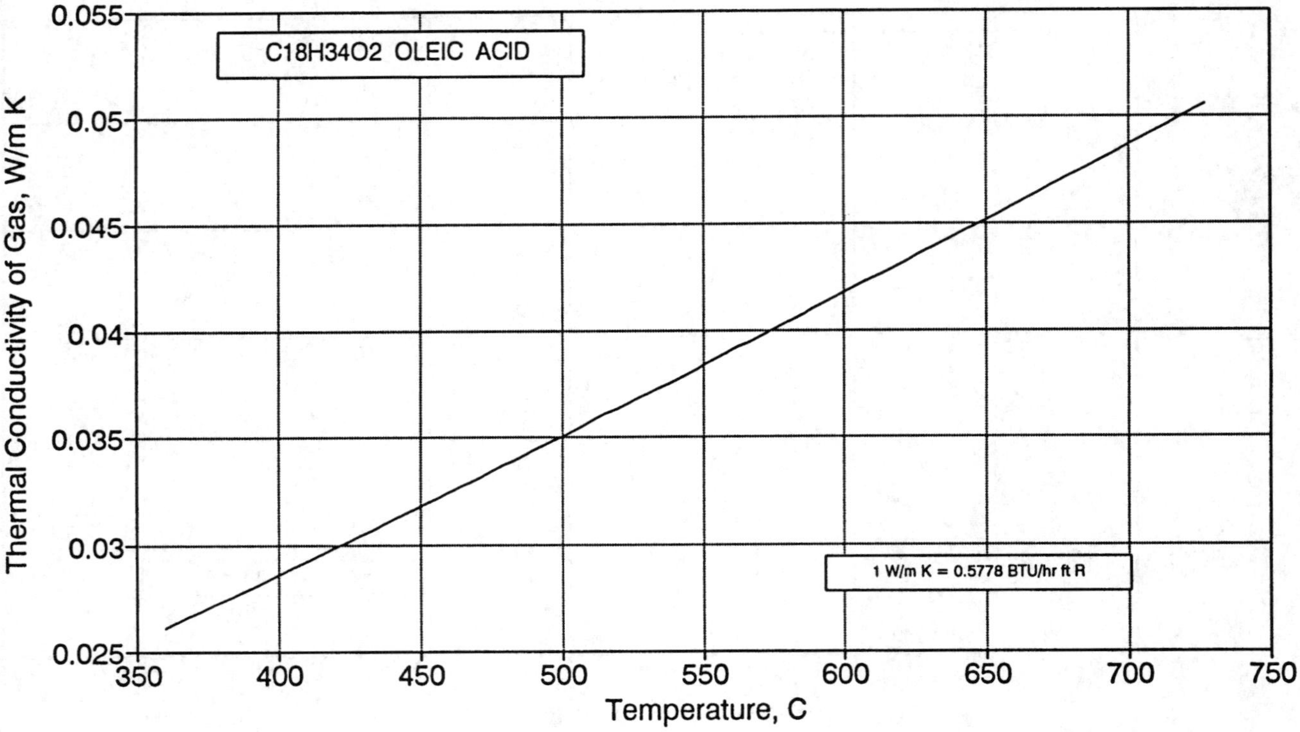
C18H34O2 OLEIC ACID
Thermal Conductivity of Gas, W/m K
0.055
0.05
0.045
0.04
0.035
0.03
0.025
350
400
450
500
550
600
650
700
750
Temperature, C
1 W/m K = 0.5778 BTU/hr ft R

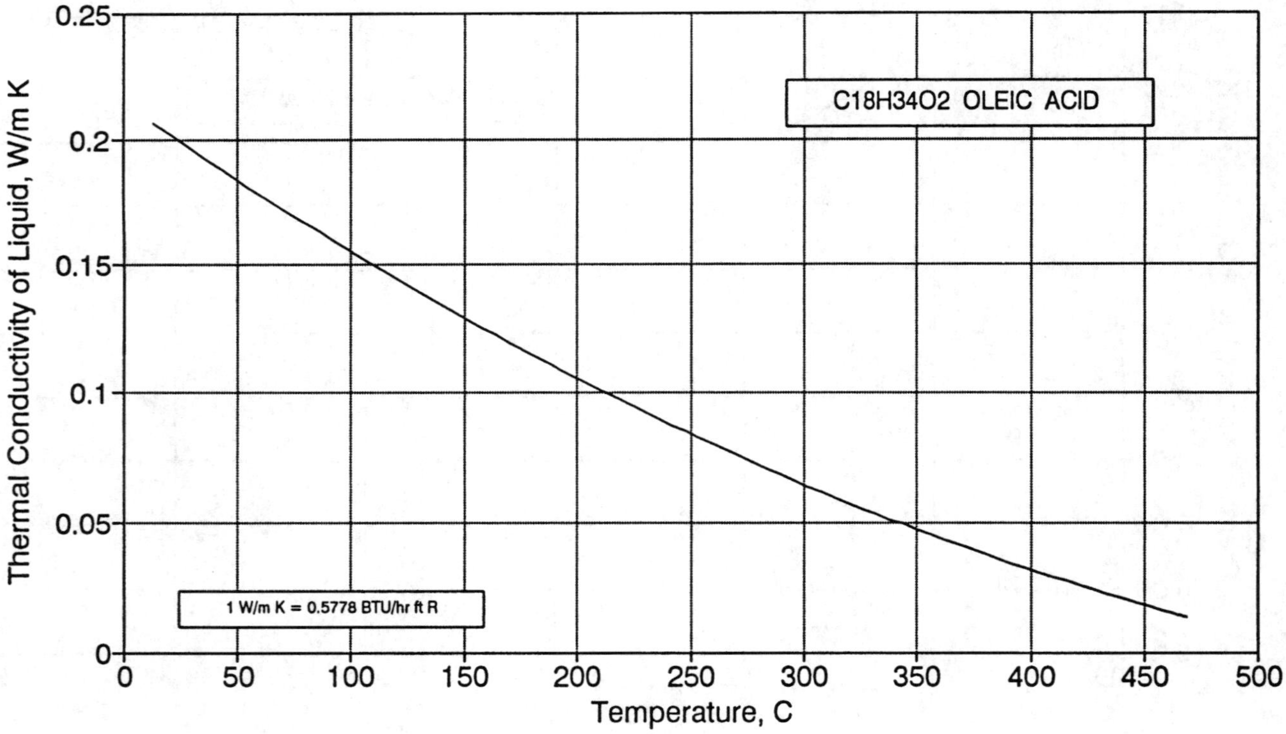
C18H34O2 OLEIC ACID
Thermal Conductivity of Liquid, W/m K
0.25
0.2
0.15
0.1
0.05
0
0
50
100
150
200
250
300
350
400
450
500
Temperature, C
1 W/m K = 0.5778 BTU/hr ft R

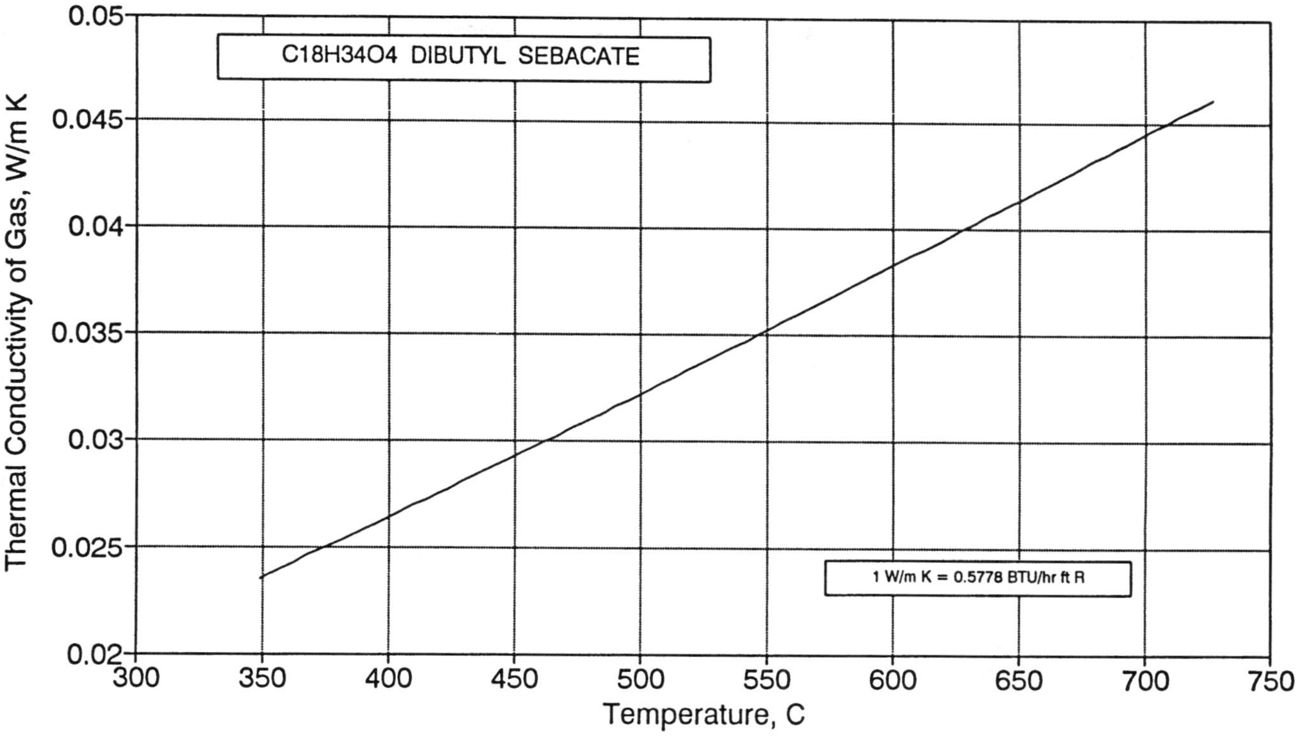
C18H34O4 DIBUTYL SEBACATE
1 W/m K = 0.5778 BTU/hr ft R
Thermal Conductivity of Gas, W/m K
0.05
0.045
0.04
0.035
0.03
0.025
0.02
300
350
400
450
500
550
600
650
700
750
Temperature, C

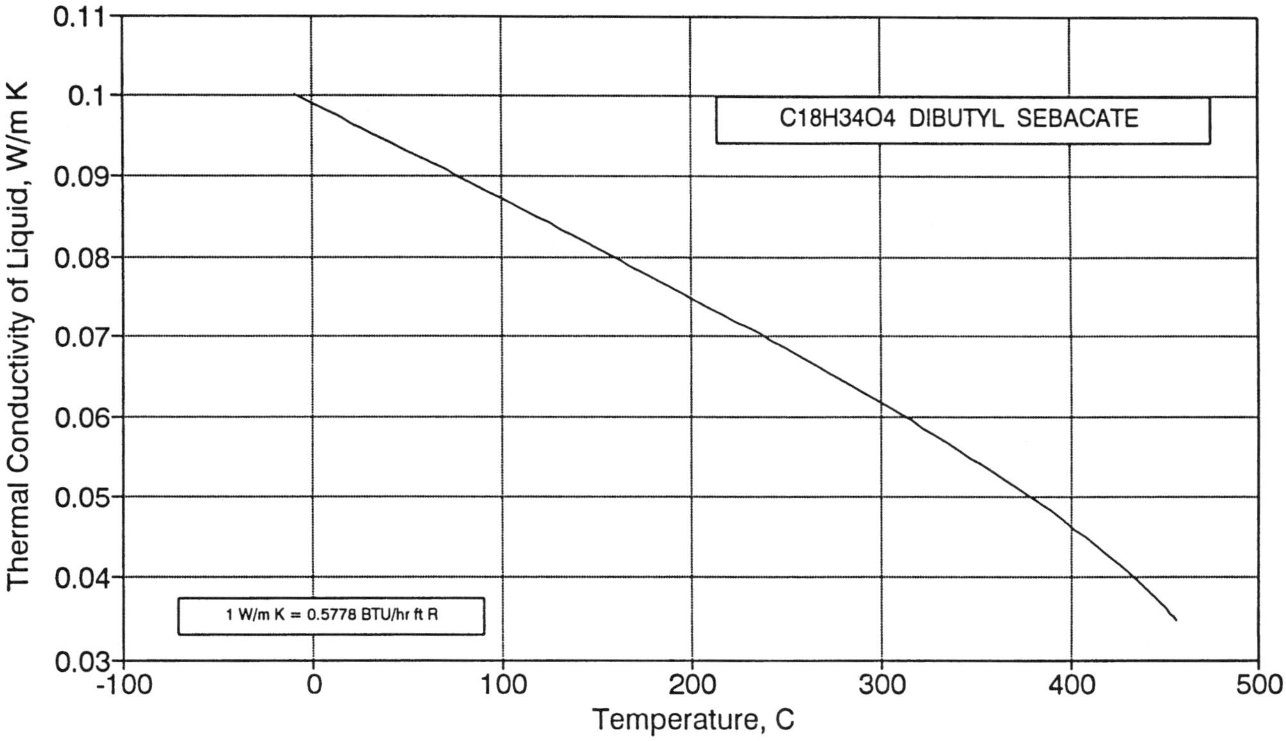
C18H34O4 DIBUTYL SEBACATE
1 W/m K = 0.5778 BTU/hr ft R
Thermal Conductivity of Liquid, W/m K
0.11
0.1
0.09
0.08
0.07
0.06
0.05
0.04
0.03
-100
0
100
200
300
400
500
Temperature, C

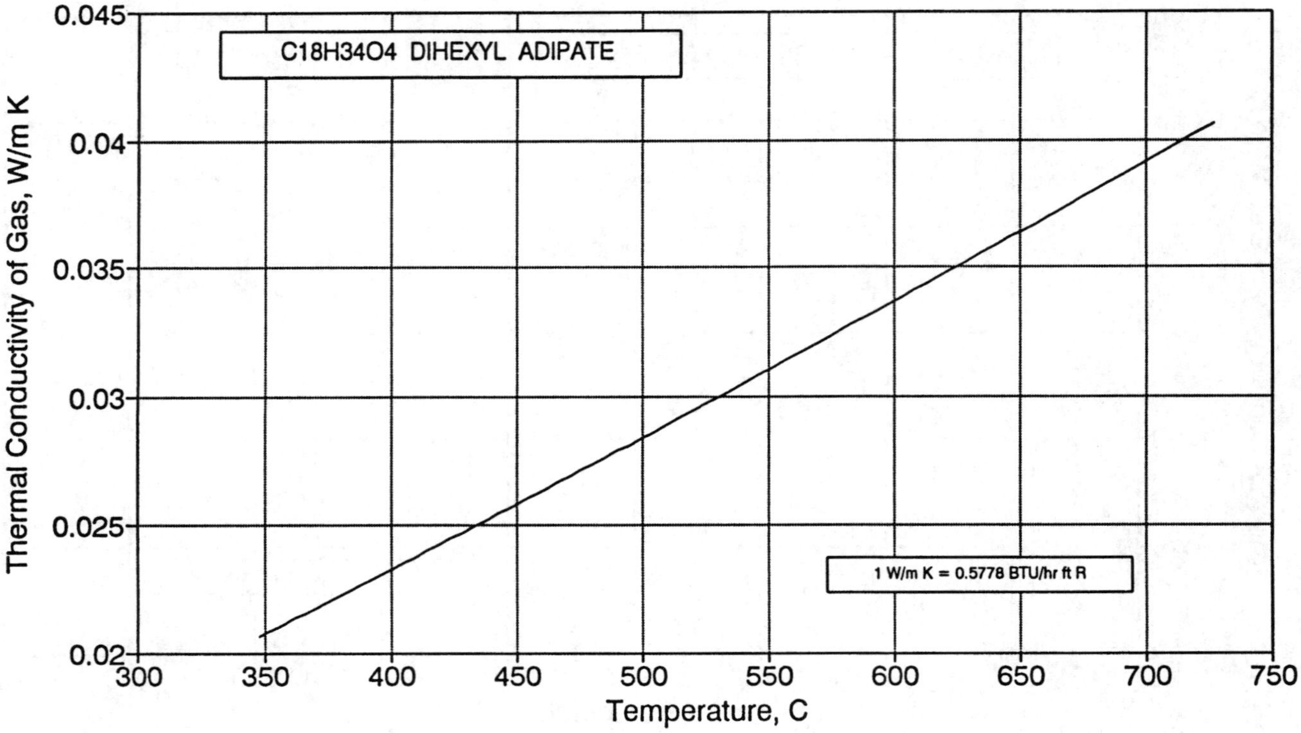
C18H34O4 DIHEXYL ADIPATE
1 W/m K = 0.5778 BTU/hr ft R
Thermal Conductivity of Gas, W/m K
0.045
0.04
0.035
0.03
0.025
0.02
300
350
400
450
500
550
600
650
700
750
Temperature, C

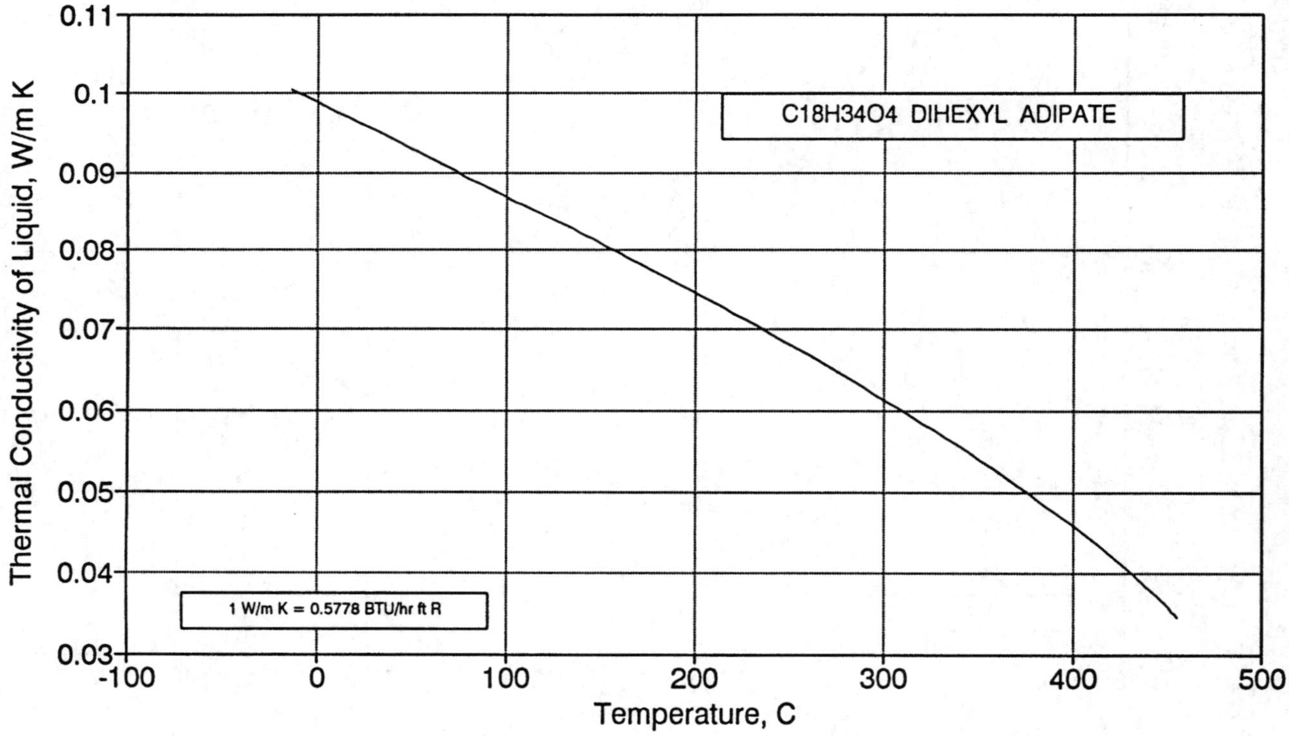
C18H34O4 DIHEXYL ADIPATE
1 W/m K = 0.5778 BTU/hr ft R
Thermal Conductivity of Liquid, W/m K
0.11
0.1
0.09
0.08
0.07
0.06
0.05
0.04
0.03
-100
0
100
200
300
400
500
Temperature, C

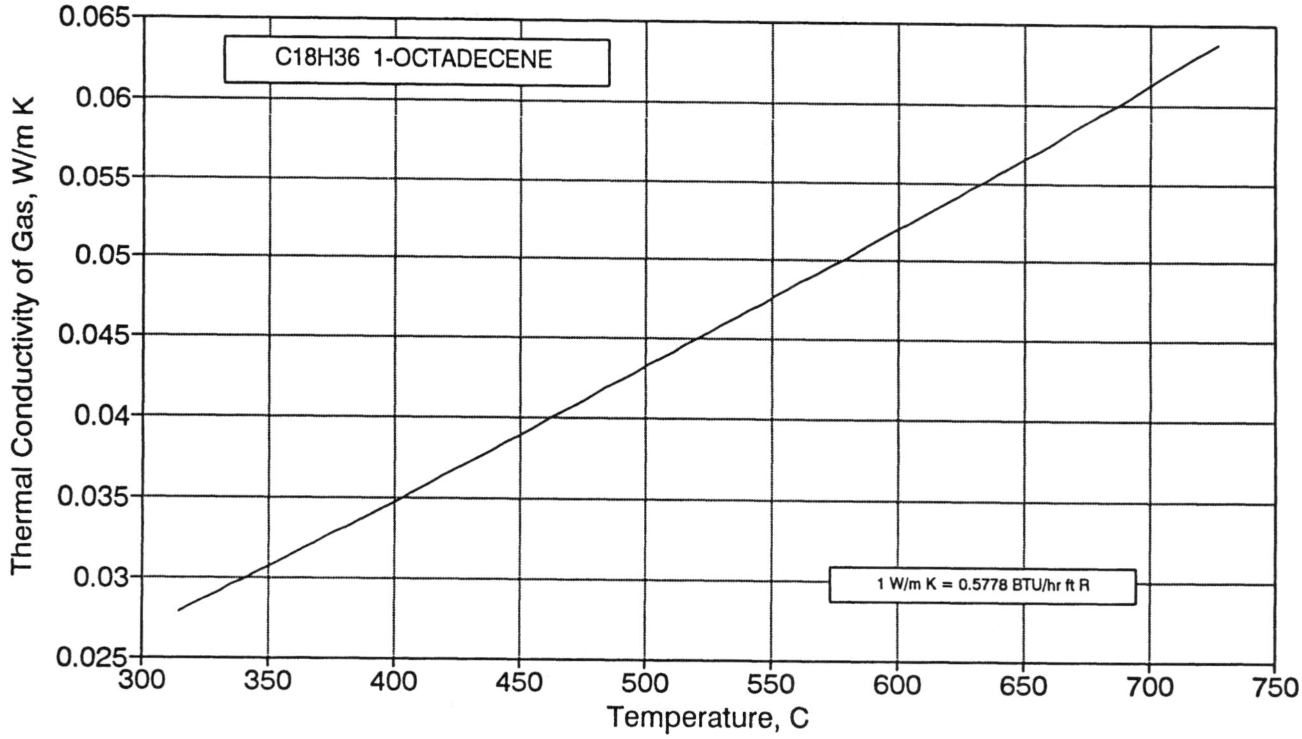
C18H36 1-OCTADECENE
1 W/m K = 0.5778 BTU/hr ft R
Thermal Conductivity of Gas, W/m K
Temperature, C
0.065
0.06
0.055
0.05
0.045
0.04
0.035
0.03
0.025
300
350
400
450
500
550
600
650
700
750

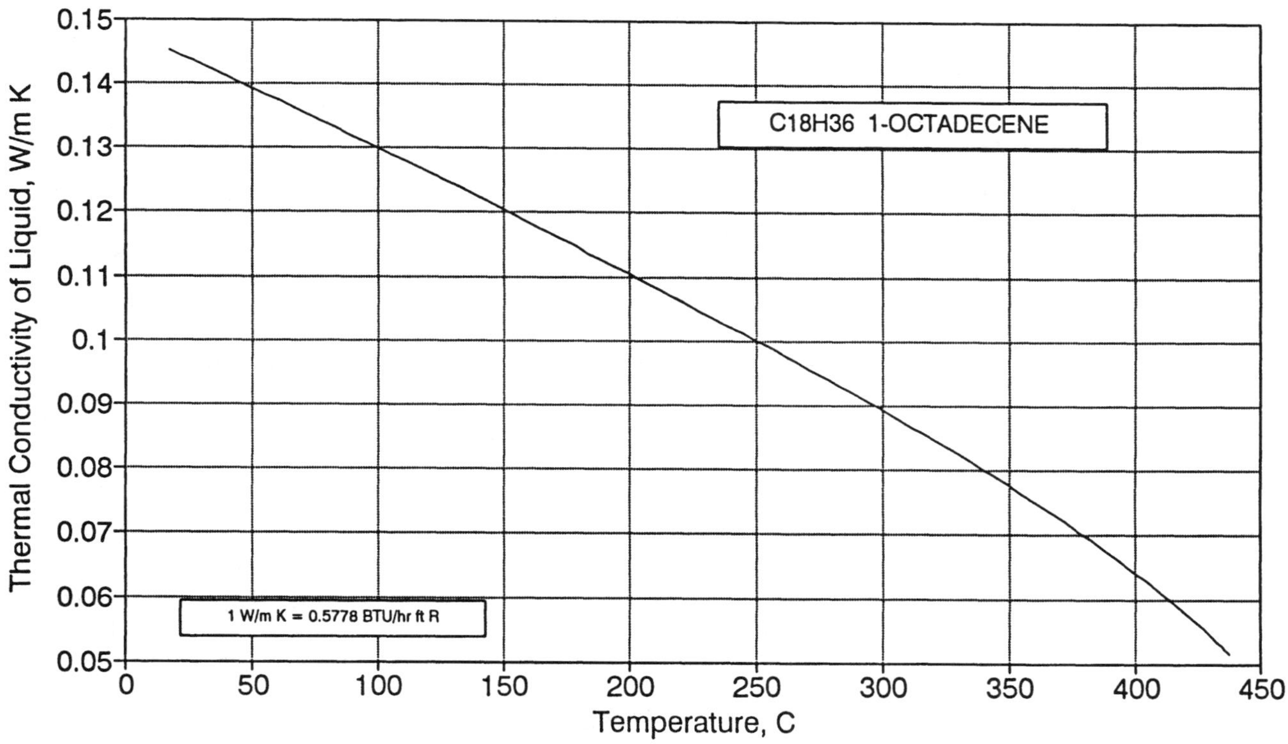
C18H36 1-OCTADECENE
1 W/m K = 0.5778 BTU/hr ft R
Thermal Conductivity of Liquid, W/m K
Temperature, C
0.15
0.14
0.13
0.12
0.11
0.1
0.09
0.08
0.07
0.06
0.05
0
50
100
150
200
250
300
350
400
450

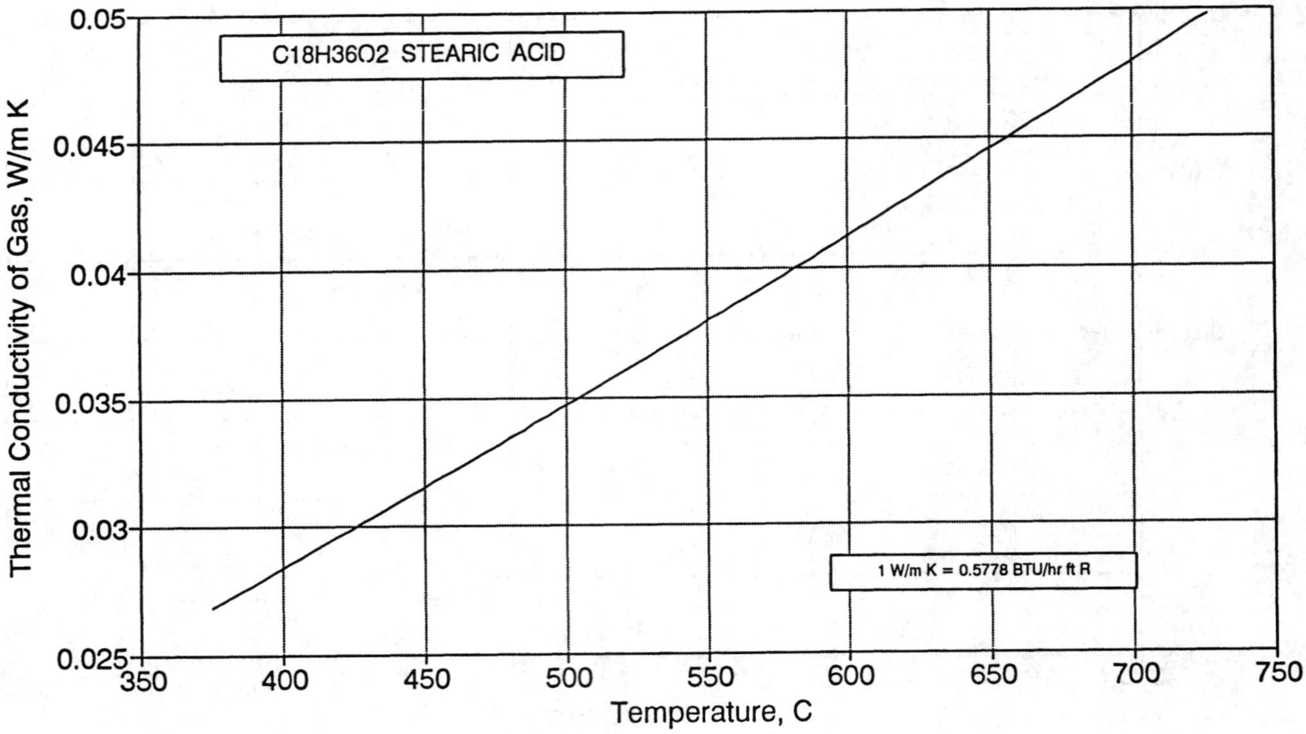
C18H36O2 STEARIC ACID
1 W/m K = 0.5778 BTU/hr ft R
Thermal Conductivity of Gas, W/m K
0.05
0.045
0.04
0.035
0.03
0.025
350
400
450
500
550
600
650
700
750
Temperature, C

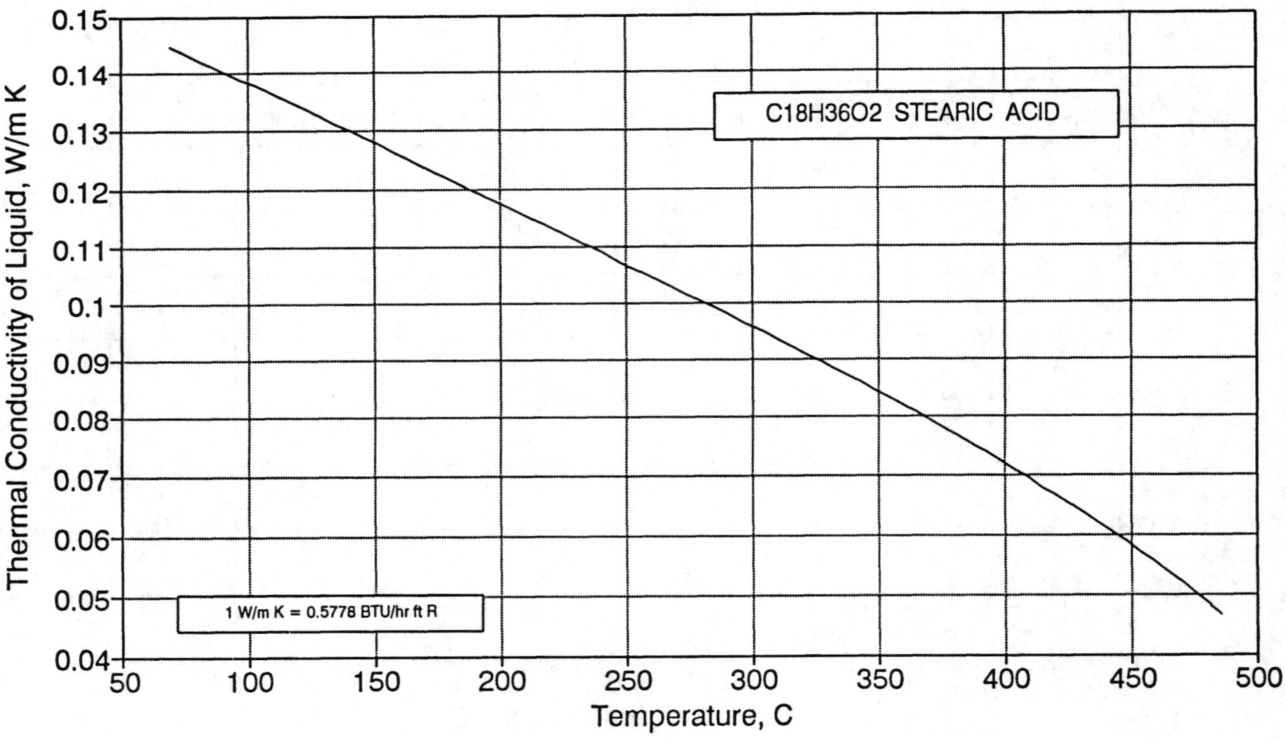
C18H36O2 STEARIC ACID
1 W/m K = 0.5778 BTU/hr ft R
Thermal Conductivity of Liquid, W/m K
0.15
0.14
0.13
0.12
0.11
0.1
0.09
0.08
0.07
0.06
0.05
0.04
50
100
150
200
250
300
350
400
450
500
Temperature, C

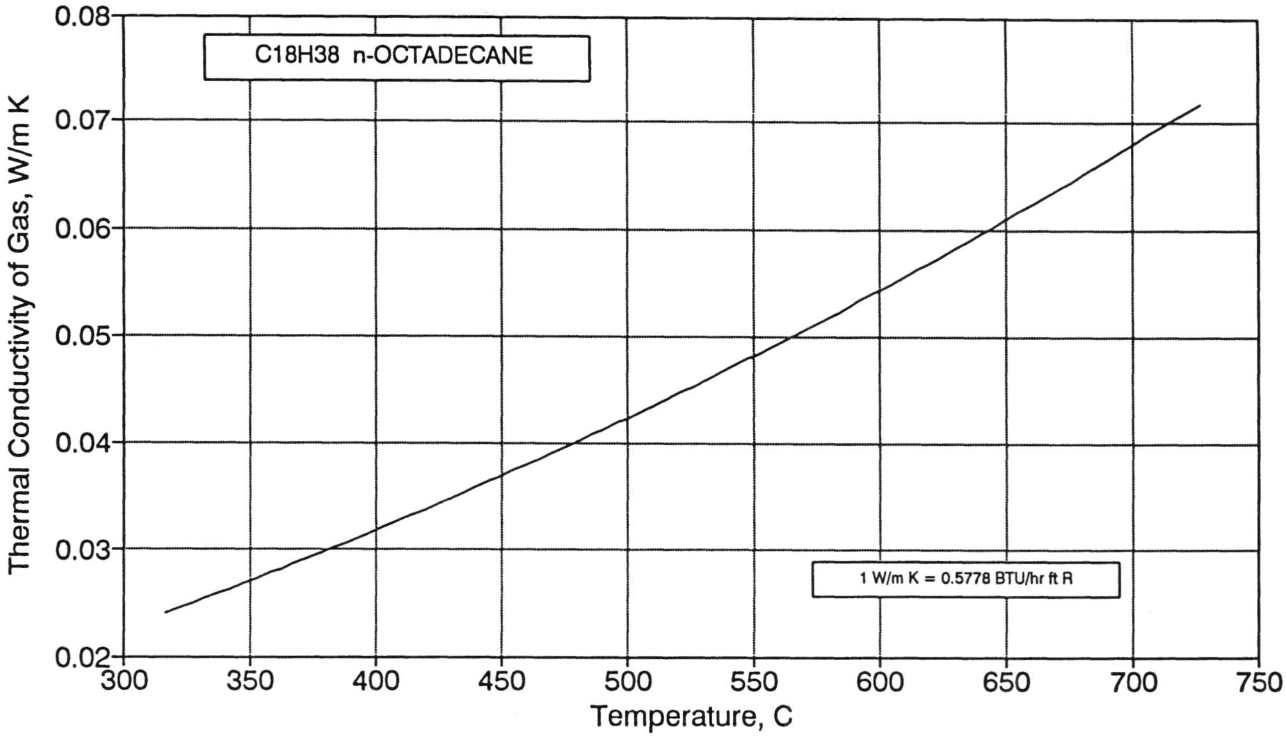
C18H38 n-OCTADECANE
1 W/m K = 0.5778 BTU/hr ft R
Thermal Conductivity of Gas, W/m K
0.08
0.07
0.06
0.05
0.04
0.03
0.02
300
350
400
450
500
550
600
650
700
750
Temperature, C

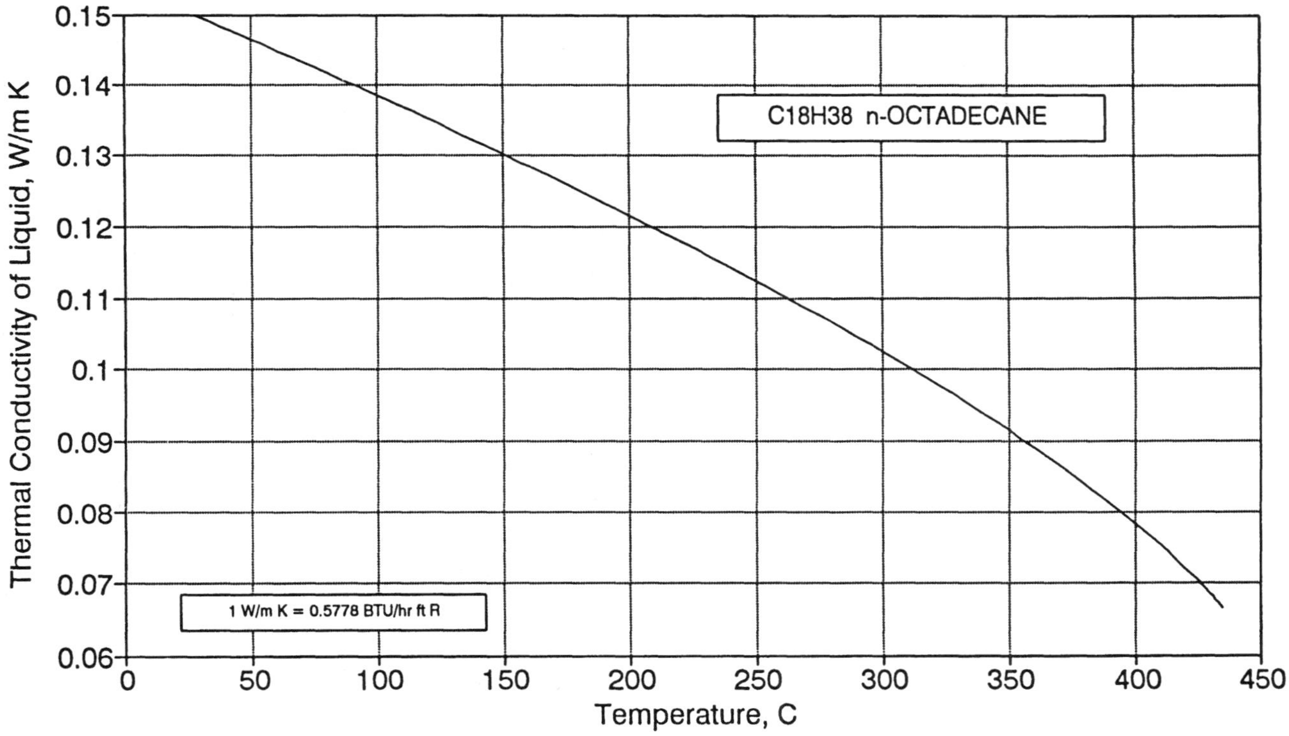
C18H38 n-OCTADECANE
1 W/m K = 0.5778 BTU/hr ft R
Thermal Conductivity of Liquid, W/m K
0.15
0.14
0.13
0.12
0.11
0.1
0.09
0.08
0.07
0.06
0
50
100
150
200
250
300
350
400
450
Temperature, C

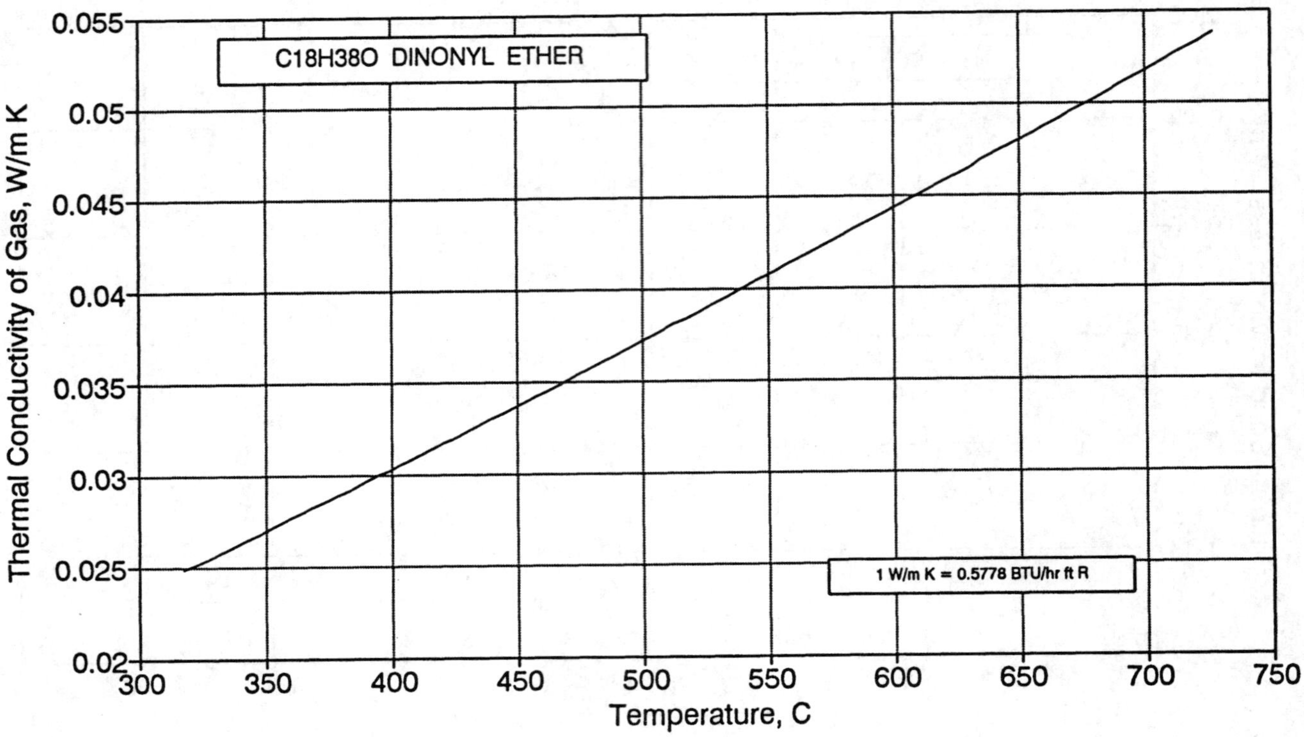
C18H38O DINONYL ETHER
1 W/m K = 0.5778 BTU/hr ft R
Thermal Conductivity of Gas, W/m K
0.055
0.05
0.045
0.04
0.035
0.03
0.025
0.02
300
350
400
450
500
550
600
650
700
750
Temperature, C

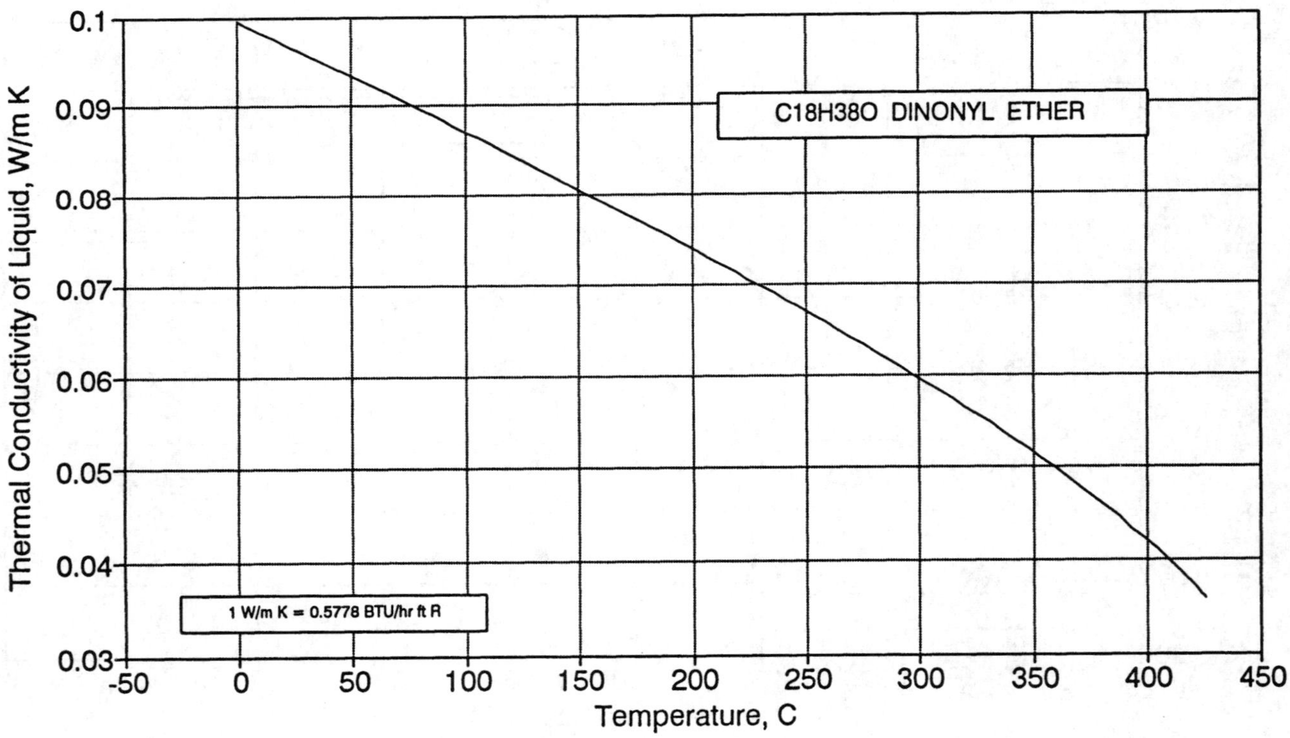
C18H38O DINONYL ETHER
1 W/m K = 0.5778 BTU/hr ft R
Thermal Conductivity of Liquid, W/m K
0.1
0.09
0.08
0.07
0.06
0.05
0.04
0.03
-50
0
50
100
150
200
250
300
350
400
450
Temperature, C

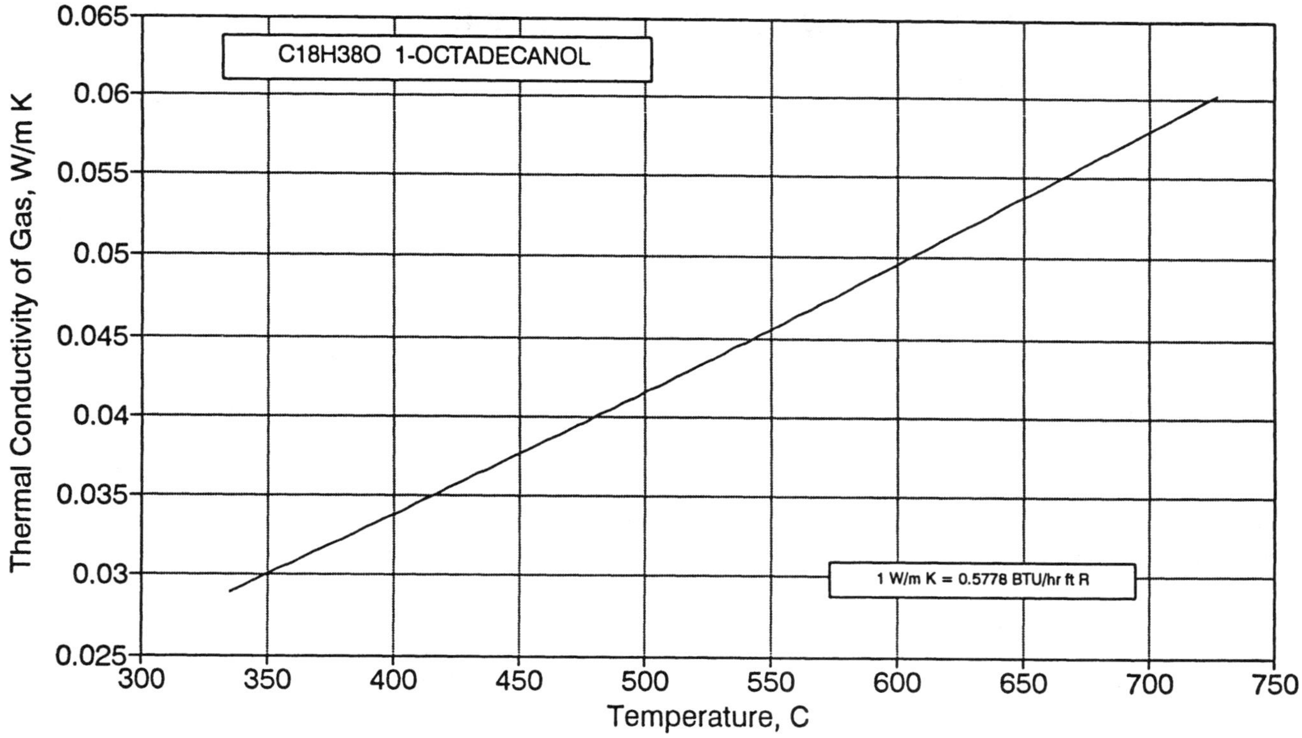
C18H38O 1-OCTADECANOL
1 W/m K = 0.5778 BTU/hr ft R
Thermal Conductivity of Gas, W/m K
0.065
0.06
0.055
0.05
0.045
0.04
0.035
0.03
0.025
300
350
400
450
500
550
600
650
700
750
Temperature, C

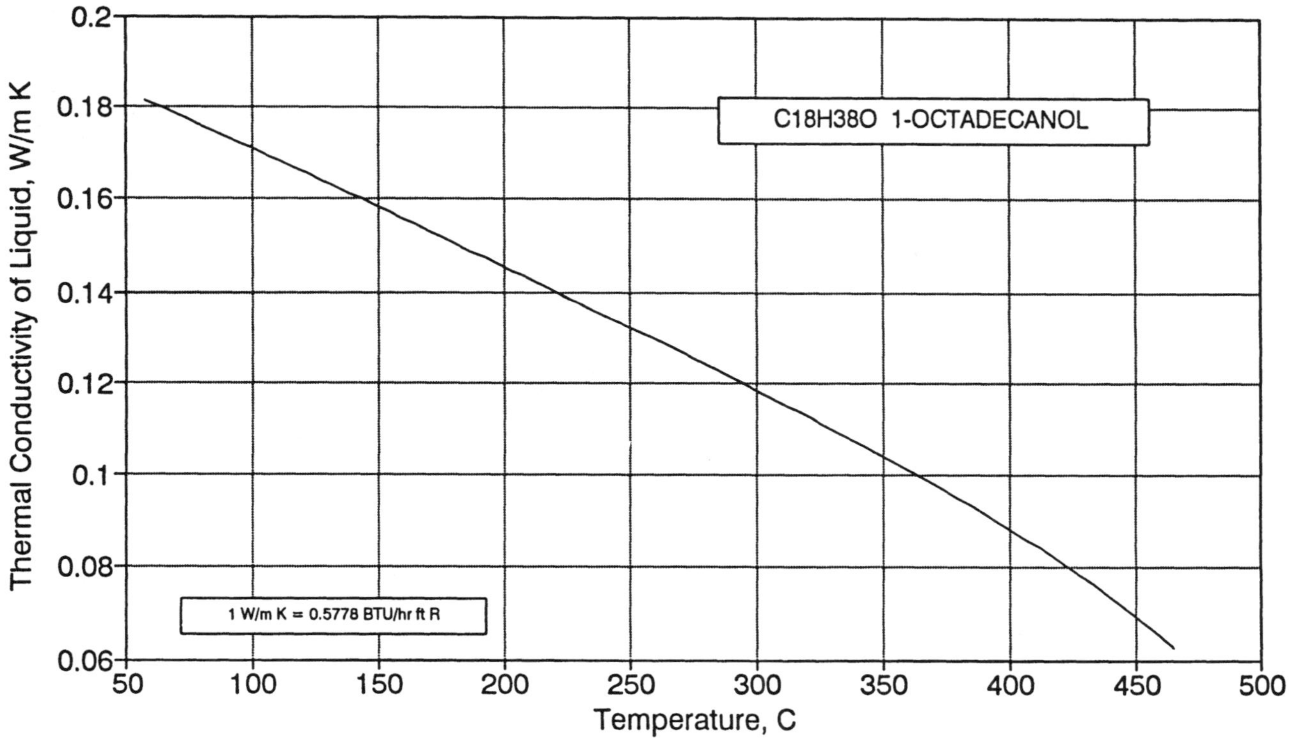
C18H38O 1-OCTADECANOL
1 W/m K = 0.5778 BTU/hr ft R
Thermal Conductivity of Liquid, W/m K
0.2
0.18
0.16
0.14
0.12
0.1
0.08
0.06
50
100
150
200
250
300
350
400
450
500
Temperature, C

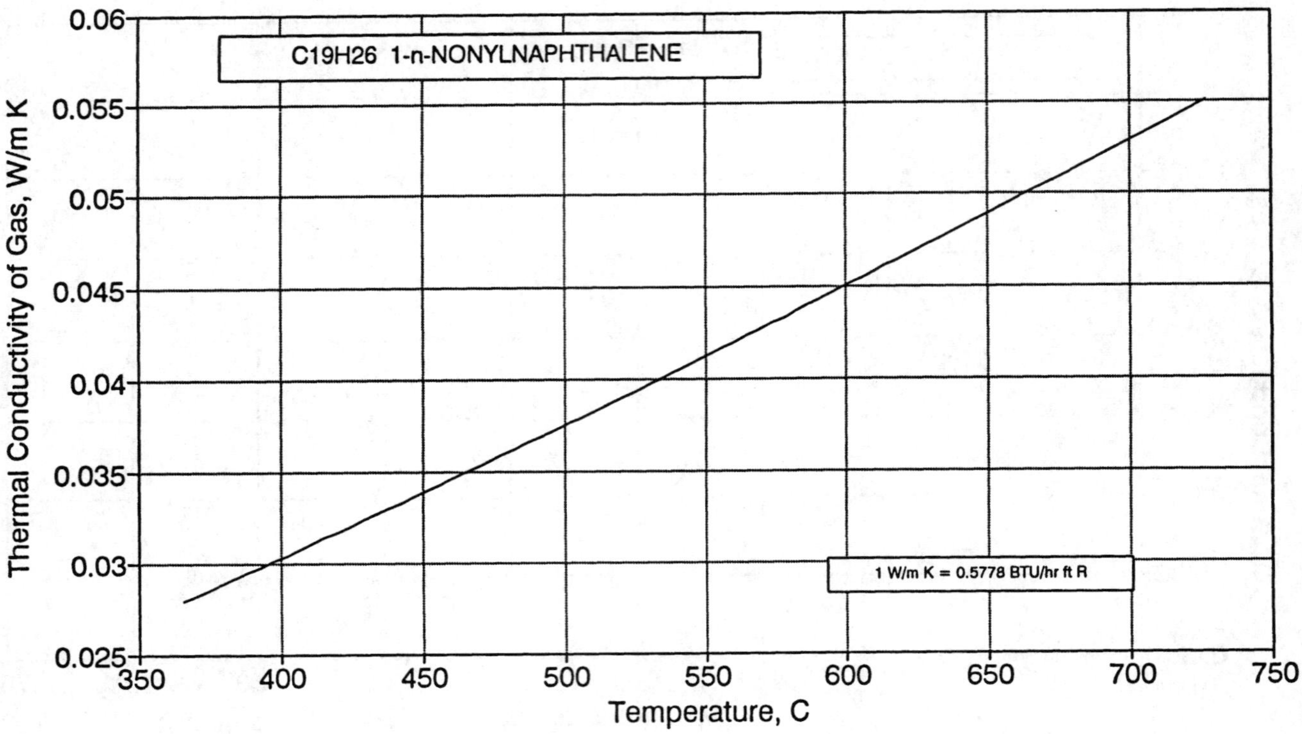
C19H26 1-n-NONYLNAPHTHALENE
1 W/m K = 0.5778 BTU/hr ft R
Thermal Conductivity of Gas, W/m K
0.06
0.055
0.05
0.045
0.04
0.035
0.03
0.025
350
400
450
500
550
600
650
700
750
Temperature, C

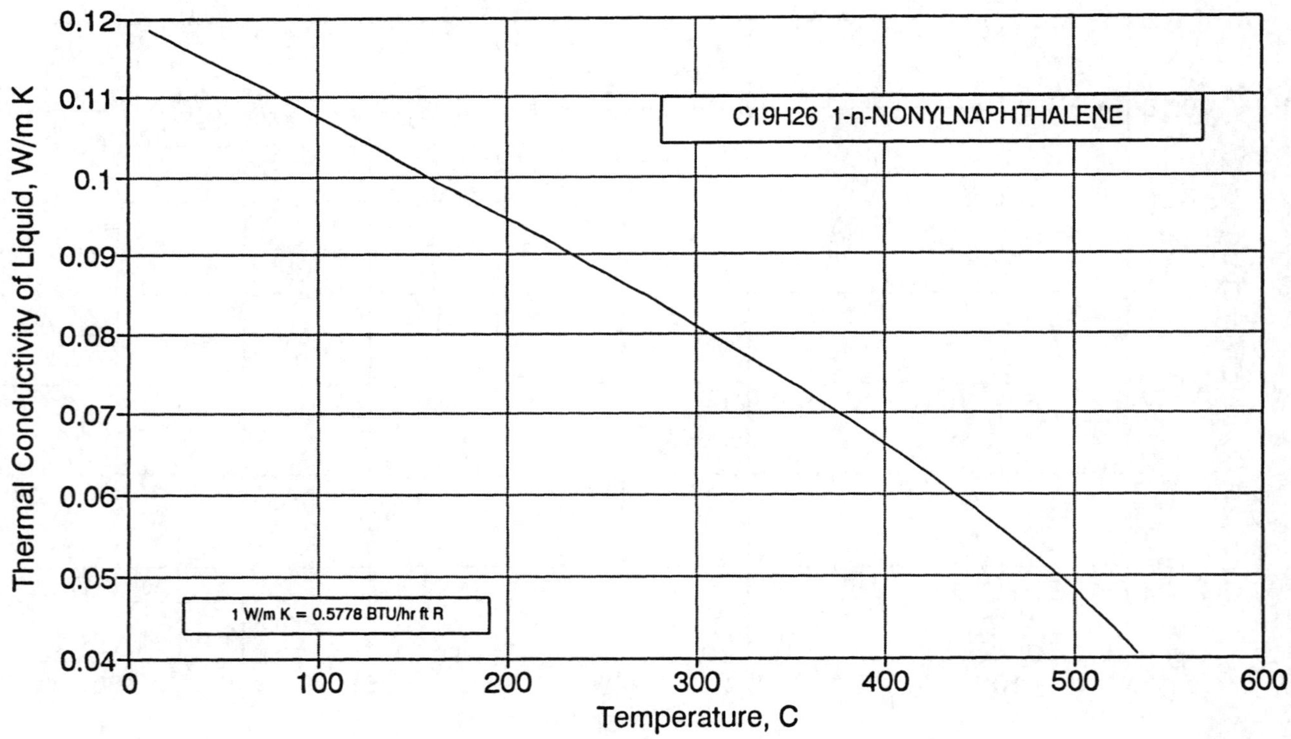
C19H26 1-n-NONYLNAPHTHALENE
1 W/m K = 0.5778 BTU/hr ft R
Thermal Conductivity of Liquid, W/m K
0.12
0.11
0.1
0.09
0.08
0.07
0.06
0.05
0.04
0
100
200
300
400
500
600
Temperature, C

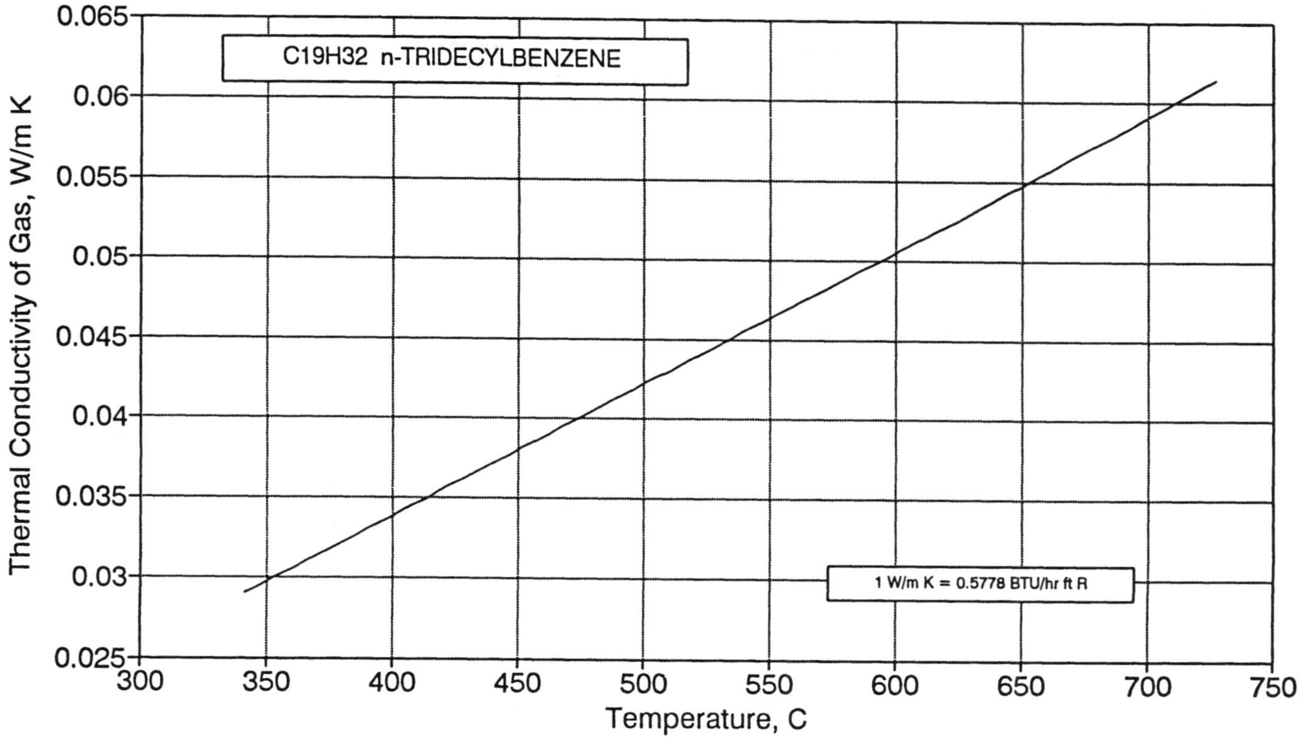
C19H32 n-TRIDECYLBENZENE
1 W/m K = 0.5778 BTU/hr ft R
Thermal Conductivity of Gas, W/m K
0.065
0.06
0.055
0.05
0.045
0.04
0.035
0.03
0.025
300
350
400
450
500
550
600
650
700
750
Temperature, C

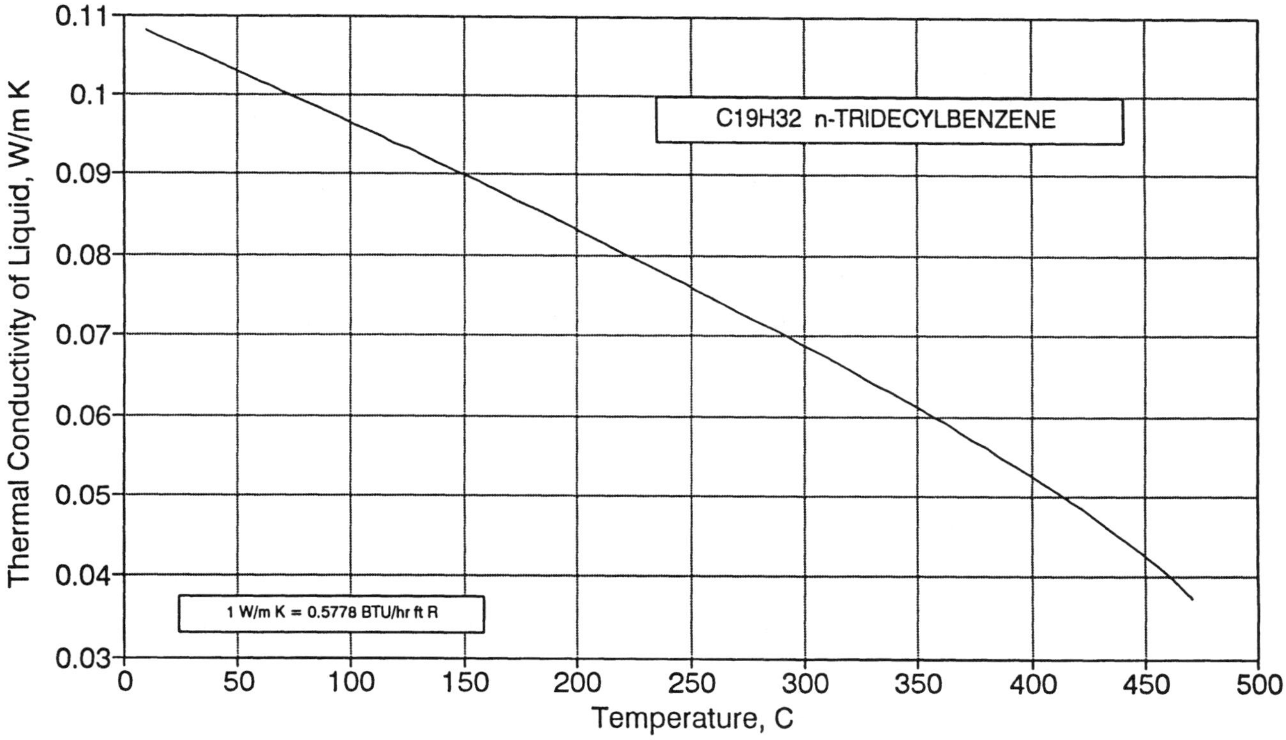
C19H32 n-TRIDECYLBENZENE
1 W/m K = 0.5778 BTU/hr ft R
Thermal Conductivity of Liquid, W/m K
0.11
0.1
0.09
0.08
0.07
0.06
0.05
0.04
0.03
0
50
100
150
200
250
300
350
400
450
500
Temperature, C

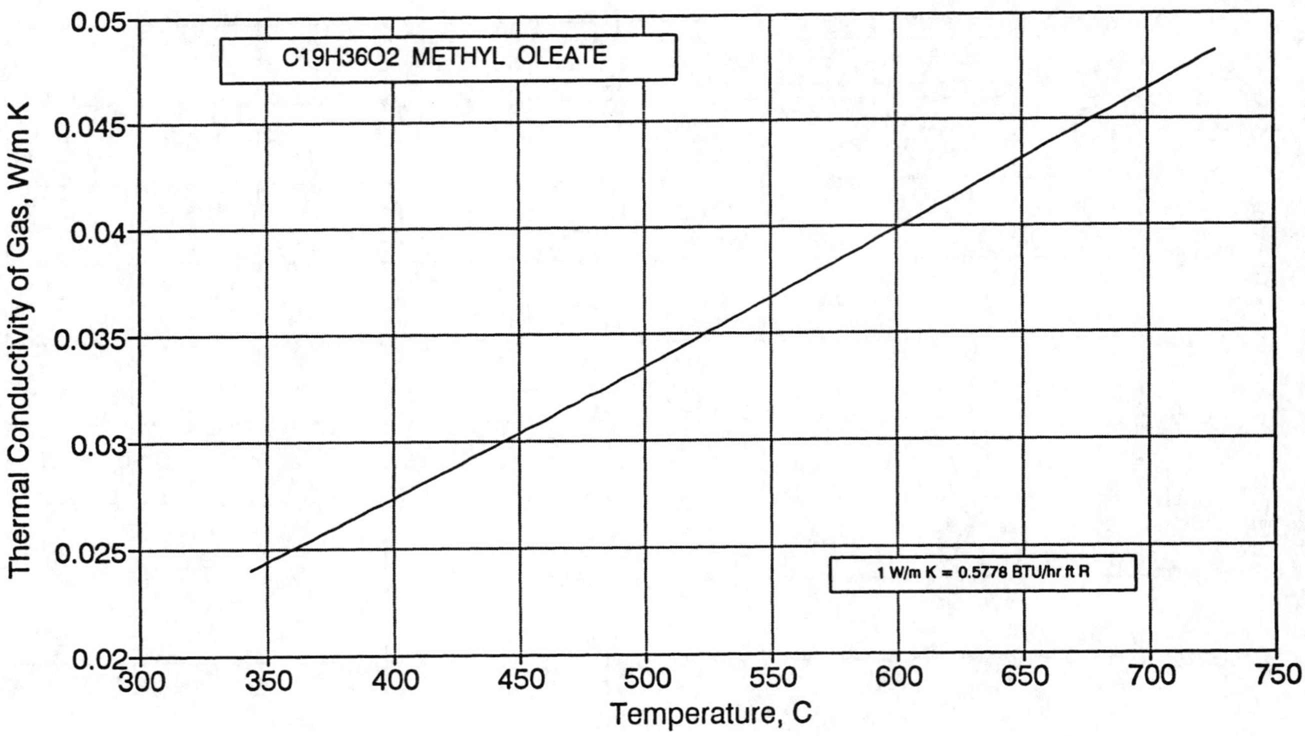
C19H36O2 METHYL OLEATE
1 W/m K = 0.5778 BTU/hr ft R
Thermal Conductivity of Gas, W/m K
0.05
0.045
0.04
0.035
0.03
0.025
0.02
300
350
400
450
500
550
600
650
700
750
Temperature, C

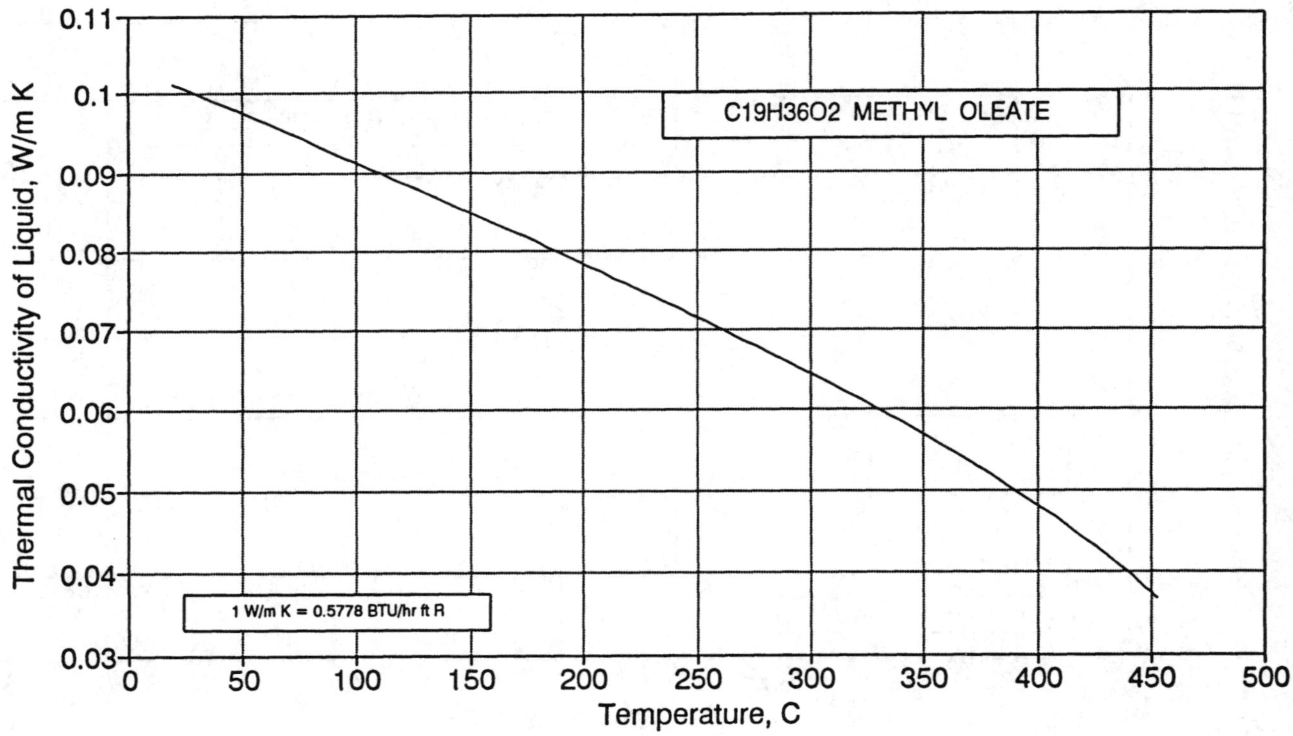
C19H36O2 METHYL OLEATE
1 W/m K = 0.5778 BTU/hr ft R
Thermal Conductivity of Liquid, W/m K
0.11
0.1
0.09
0.08
0.07
0.06
0.05
0.04
0.03
0
50
100
150
200
250
300
350
400
450
500
Temperature, C

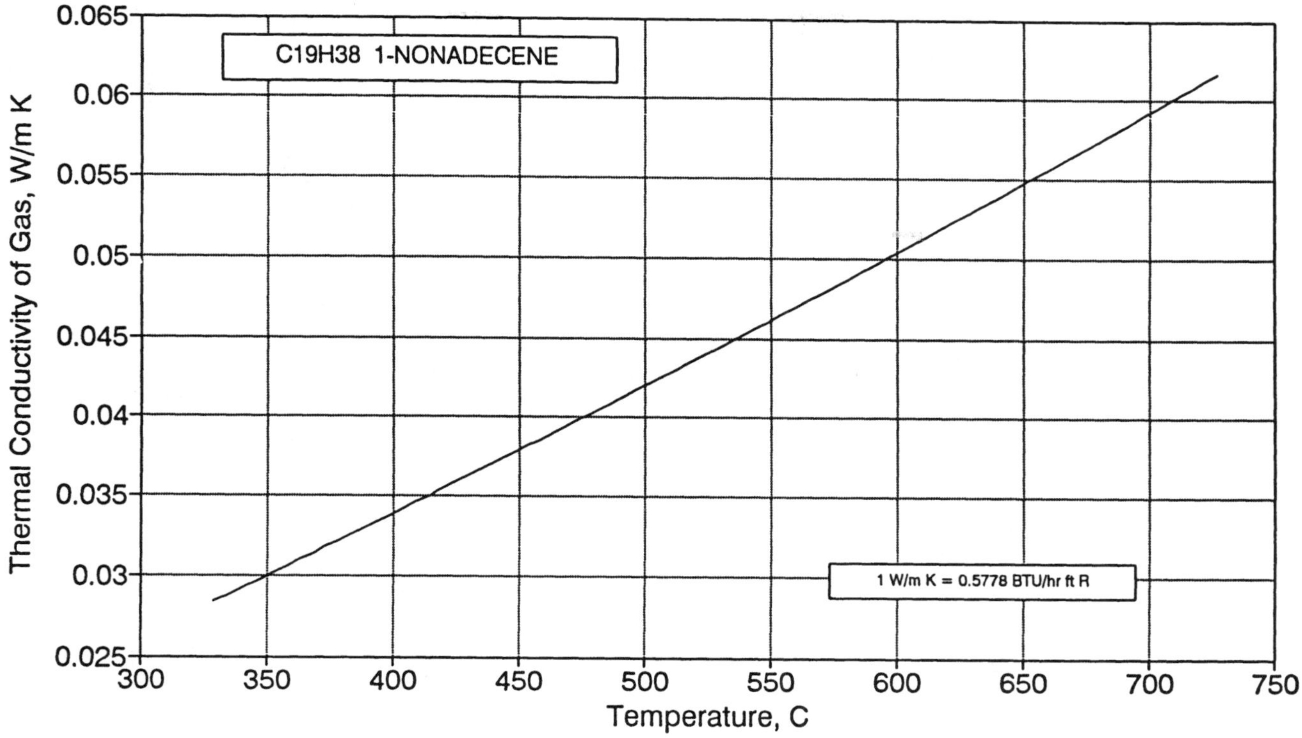
C19H38 1-NONADECENE
Thermal Conductivity of Gas, W/m K
0.065
0.06
0.055
0.05
0.045
0.04
0.035
0.03
0.025
1 W/m K = 0.5778 BTU/hr ft R
300
350
400
450
500
550
600
650
700
750
Temperature, C

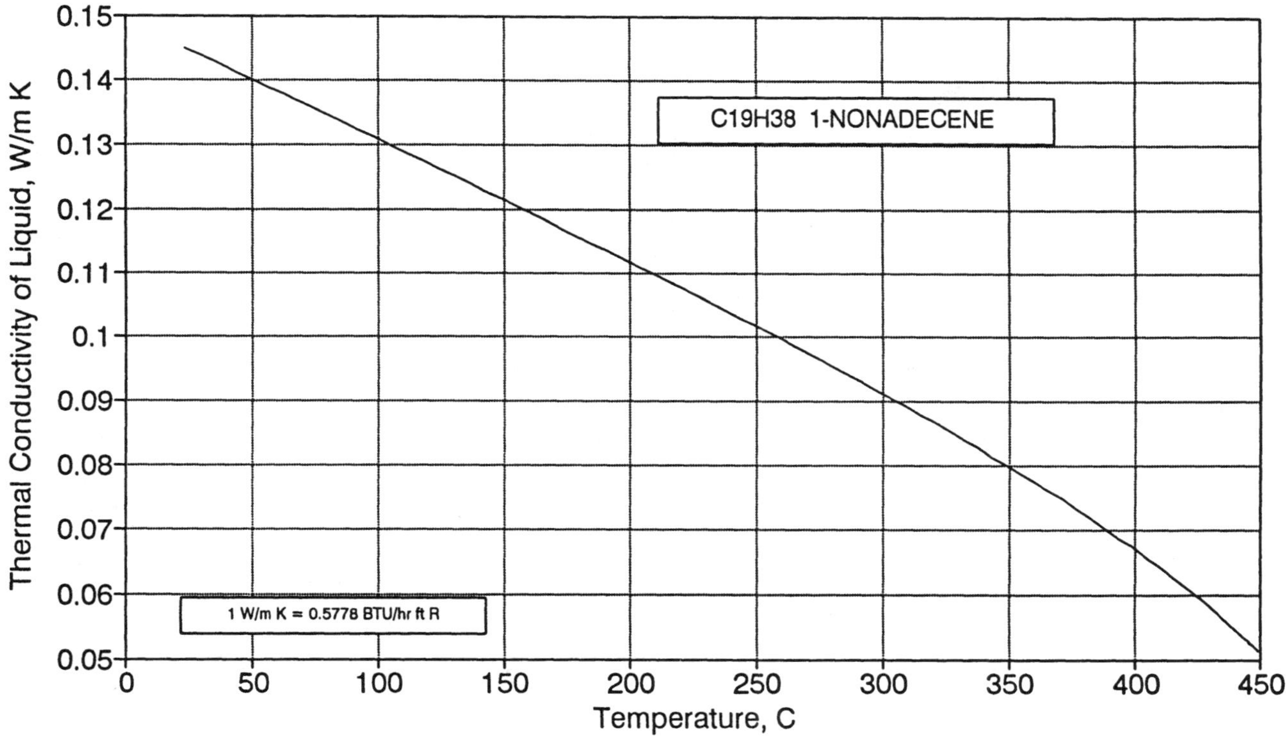
C19H38 1-NONADECENE
Thermal Conductivity of Liquid, W/m K
0.15
0.14
0.13
0.12
0.11
0.1
0.09
0.08
0.07
0.06
0.05
1 W/m K = 0.5778 BTU/hr ft R
0
50
100
150
200
250
300
350
400
450
Temperature, C

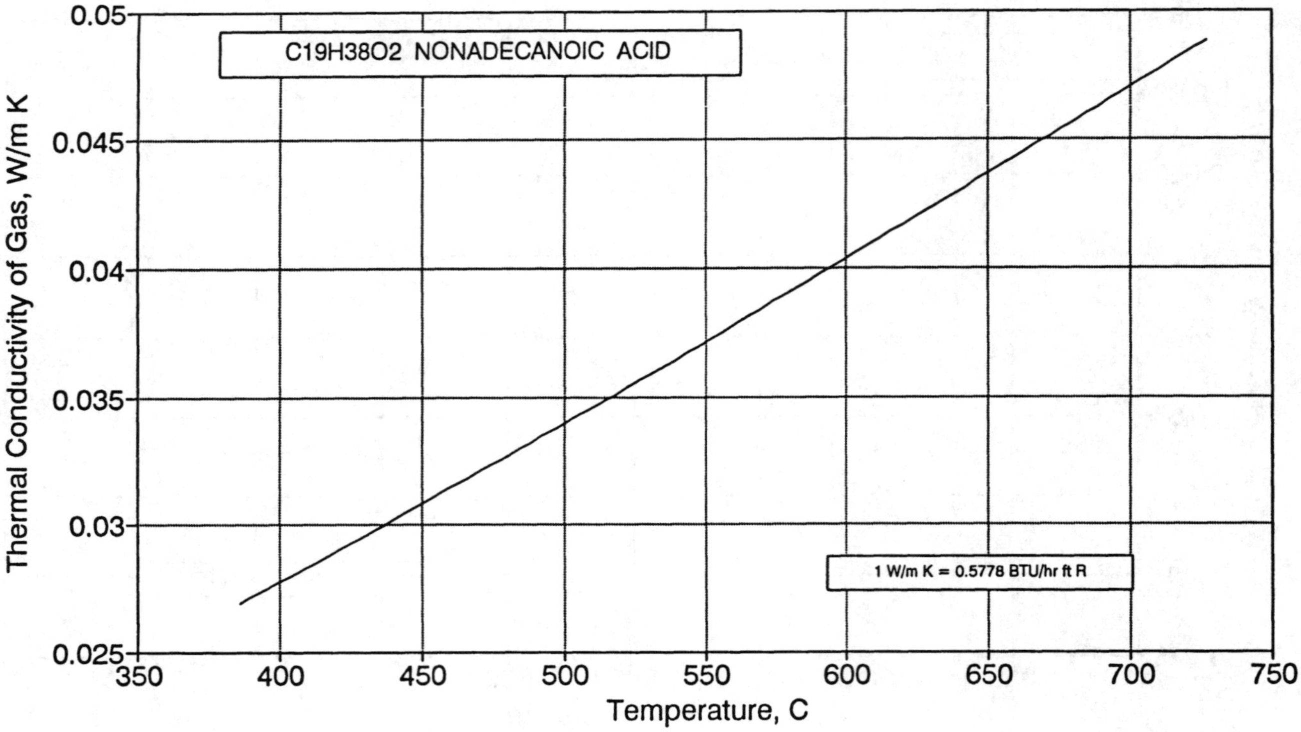
C19H38O2 NONADECANOIC ACID
1 W/m K = 0.5778 BTU/hr ft R
Thermal Conductivity of Gas, W/m K
Temperature, C
0.05
0.045
0.04
0.035
0.03
0.025
350
400
450
500
550
600
650
700
750

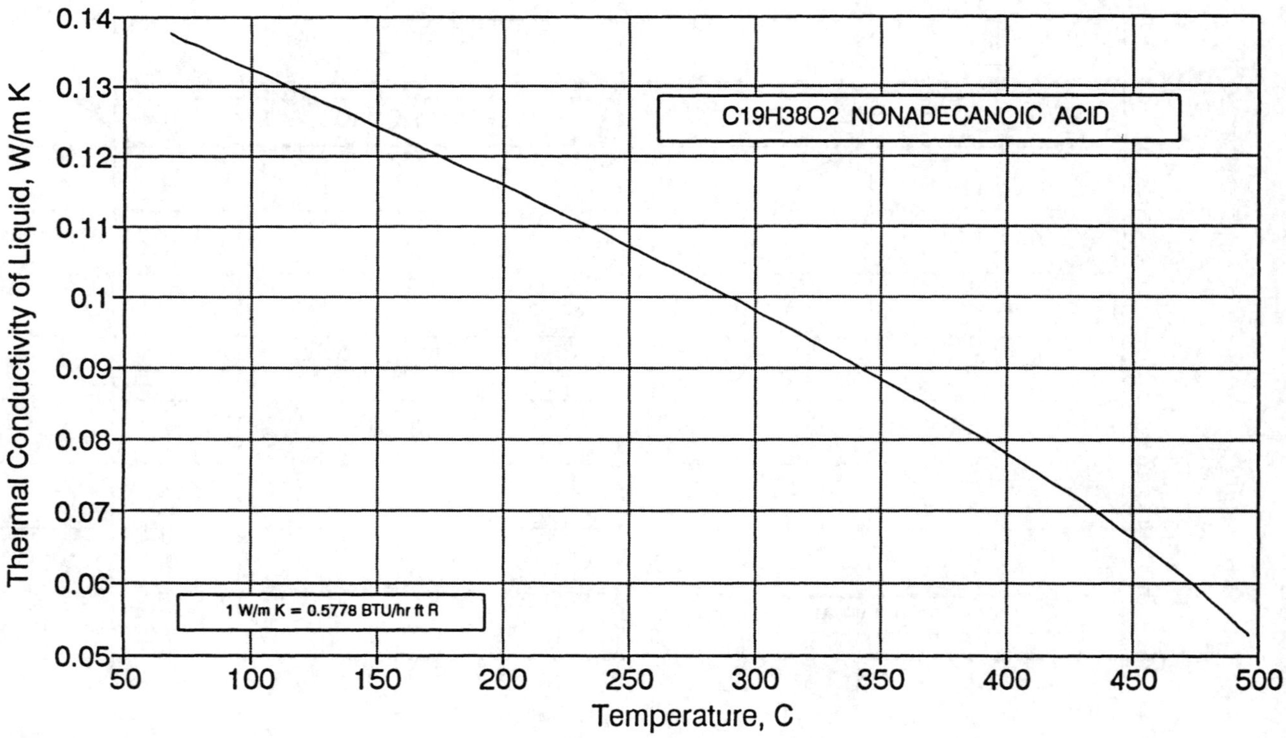
C19H38O2 NONADECANOIC ACID
1 W/m K = 0.5778 BTU/hr ft R
Thermal Conductivity of Liquid, W/m K
Temperature, C
0.14
0.13
0.12
0.11
0.1
0.09
0.08
0.07
0.06
0.05
50
100
150
200
250
300
350
400
450
500

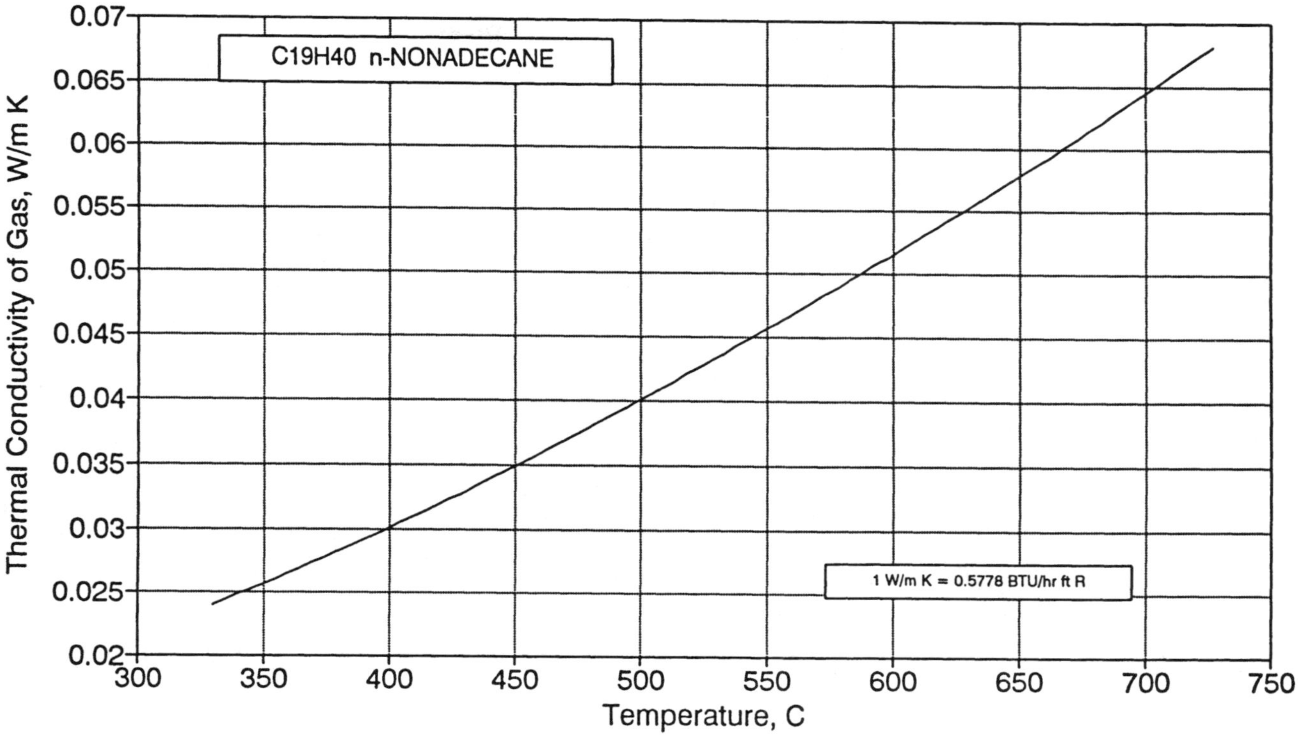
C19H40 n-NONADECANE
1 W/m K = 0.5778 BTU/hr ft R
Thermal Conductivity of Gas, W/m K
0.07
0.065
0.06
0.055
0.05
0.045
0.04
0.035
0.03
0.025
0.02
300
350
400
450
500
550
600
650
700
750
Temperature, C

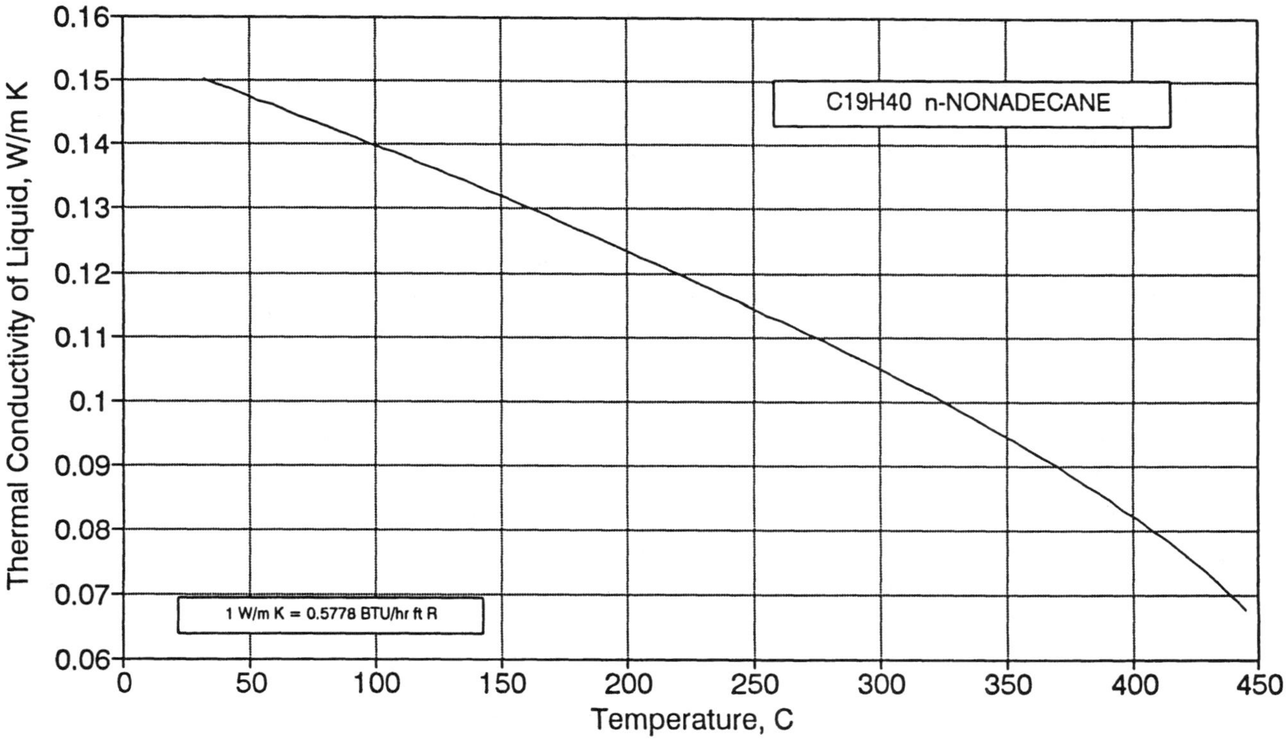
C19H40 n-NONADECANE
1 W/m K = 0.5778 BTU/hr ft R
Thermal Conductivity of Liquid, W/m K
0.16
0.15
0.14
0.13
0.12
0.11
0.1
0.09
0.08
0.07
0.06
0
50
100
150
200
250
300
350
400
450
Temperature, C

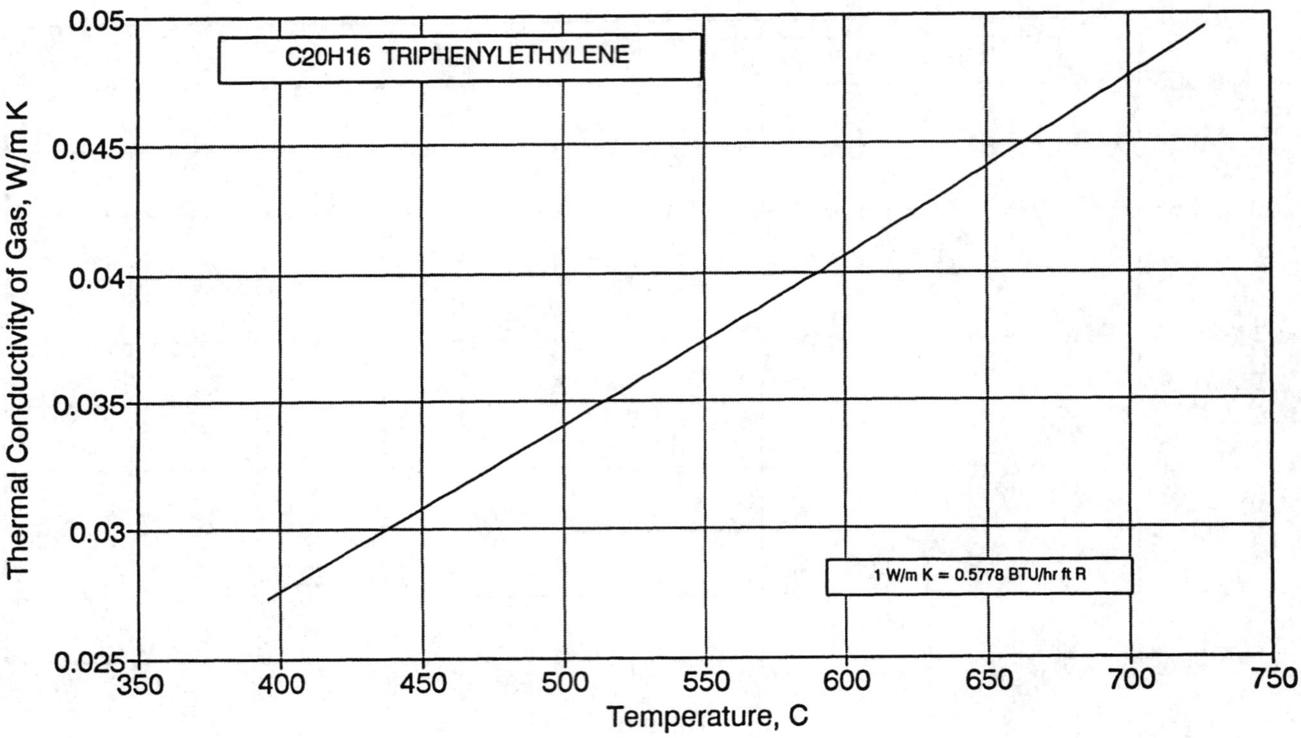
C20H16 TRIPHENYLETHYLENE
1 W/m K = 0.5778 BTU/hr ft R
Thermal Conductivity of Gas, W/m K
0.05
0.045
0.04
0.035
0.03
0.025
350
400
450
500
550
600
650
700
750
Temperature, C

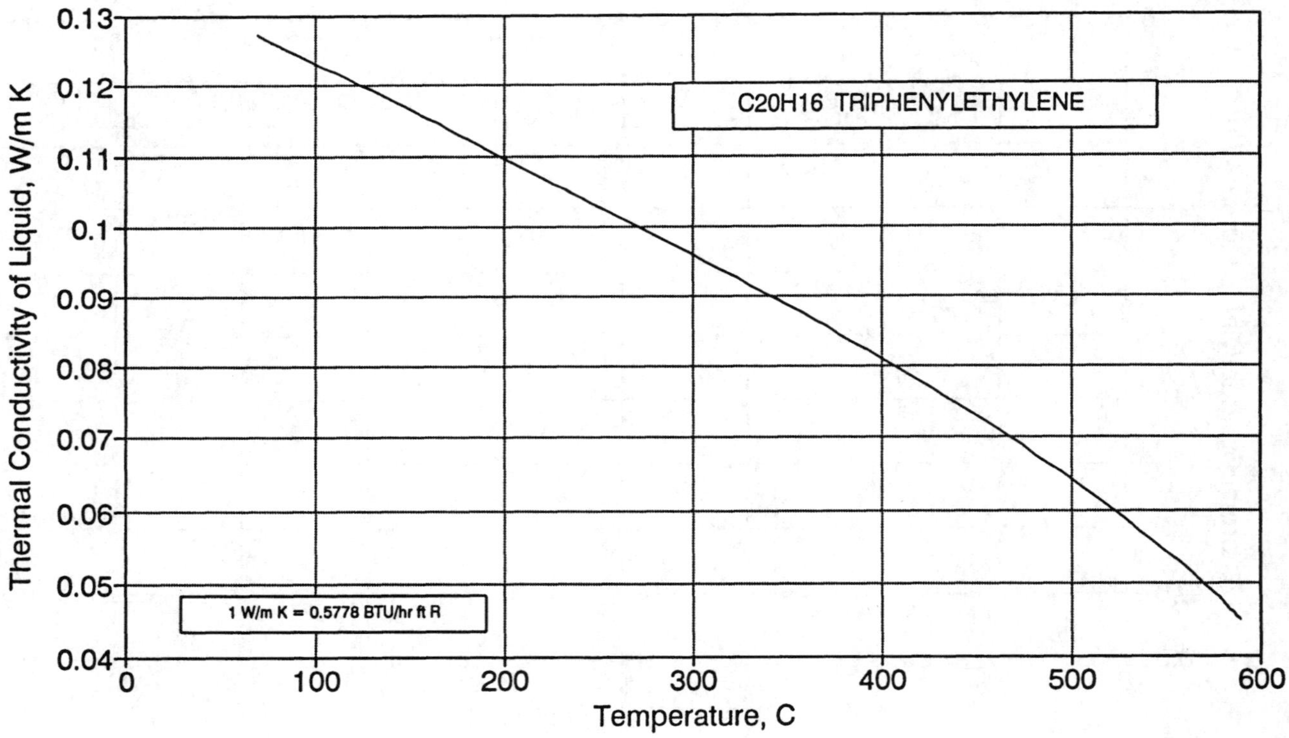
C20H16 TRIPHENYLETHYLENE
1 W/m K = 0.5778 BTU/hr ft R
Thermal Conductivity of Liquid, W/m K
0.13
0.12
0.11
0.1
0.09
0.08
0.07
0.06
0.05
0.04
0
100
200
300
400
500
600
Temperature, C

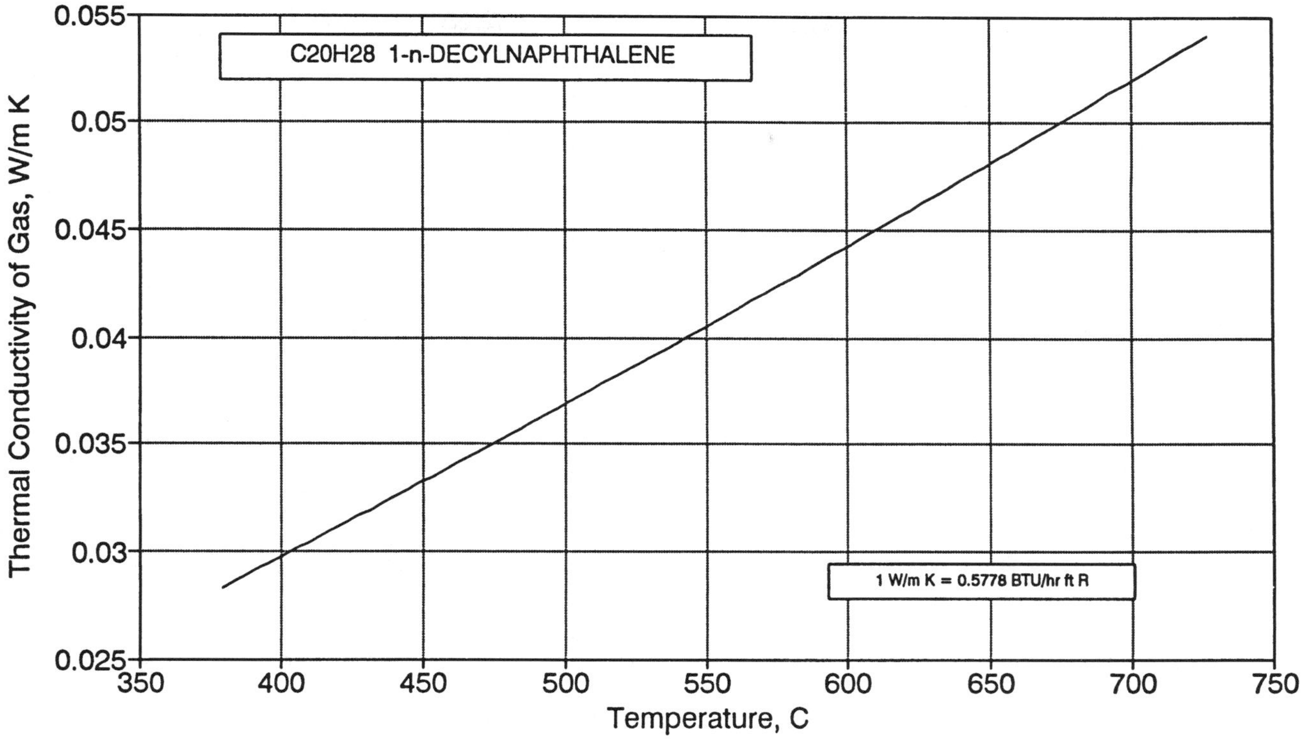
C20H28 1-n-DECYLNAPHTHALENE
1 W/m K = 0.5778 BTU/hr ft R
Thermal Conductivity of Gas, W/m K
0.055
0.05
0.045
0.04
0.035
0.03
0.025
350
400
450
500
550
600
650
700
750
Temperature, C

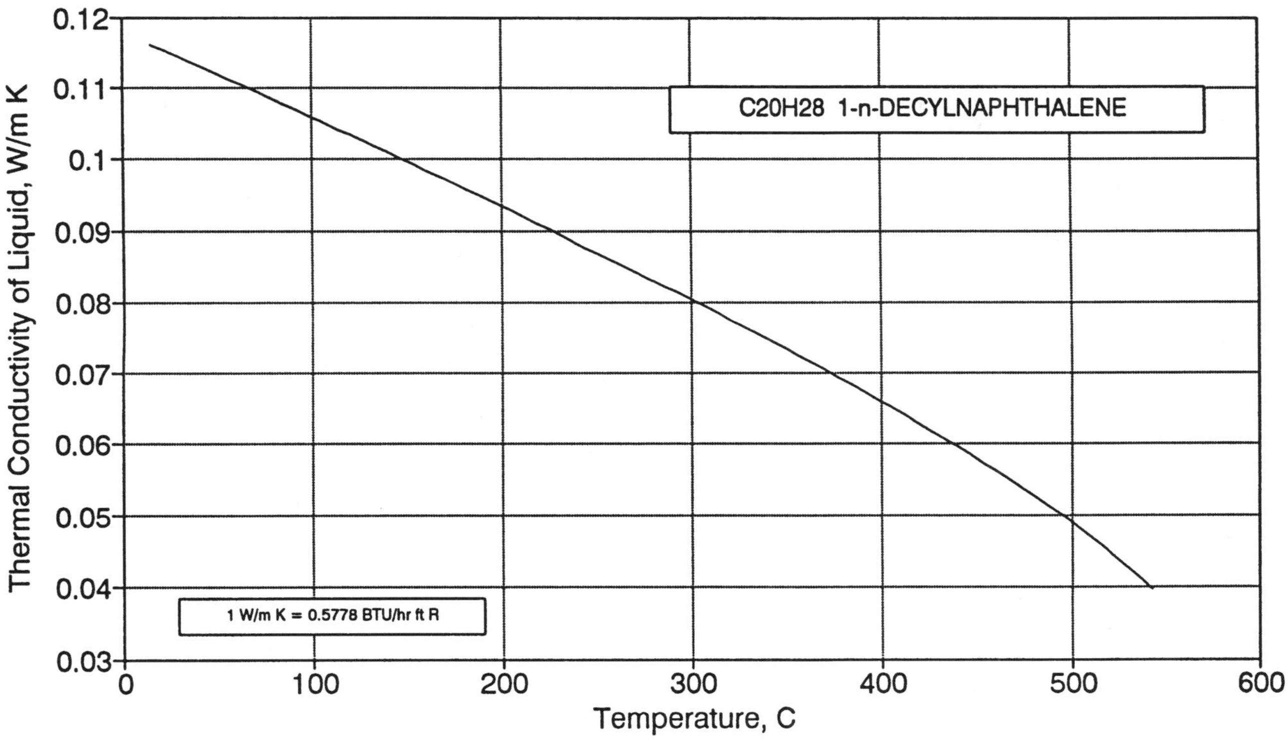
C20H28 1-n-DECYLNAPHTHALENE
1 W/m K = 0.5778 BTU/hr ft R
Thermal Conductivity of Liquid, W/m K
0.12
0.11
0.1
0.09
0.08
0.07
0.06
0.05
0.04
0.03
0
100
200
300
400
500
600
Temperature, C

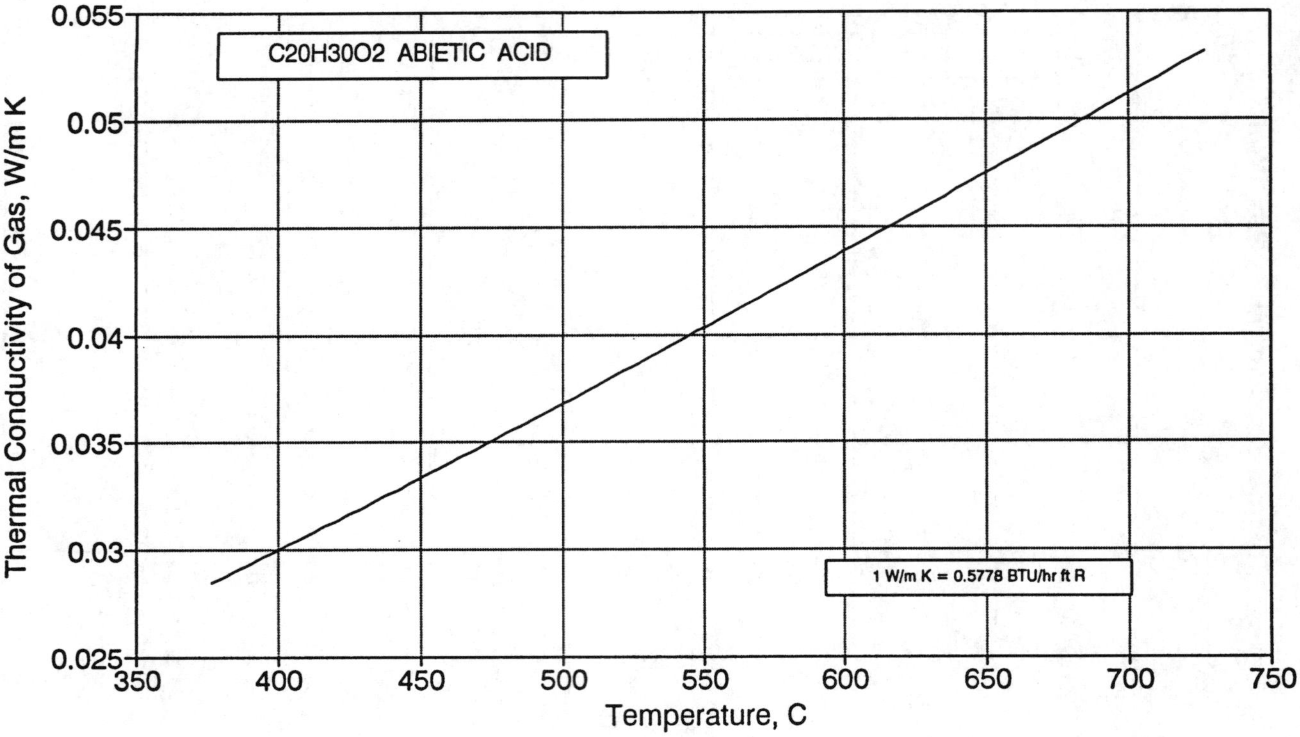
C20H30O2 ABIETIC ACID
1 W/m K = 0.5778 BTU/hr ft R
Thermal Conductivity of Gas, W/m K
0.055
0.05
0.045
0.04
0.035
0.03
0.025
350
400
450
500
550
600
650
700
750
Temperature, C

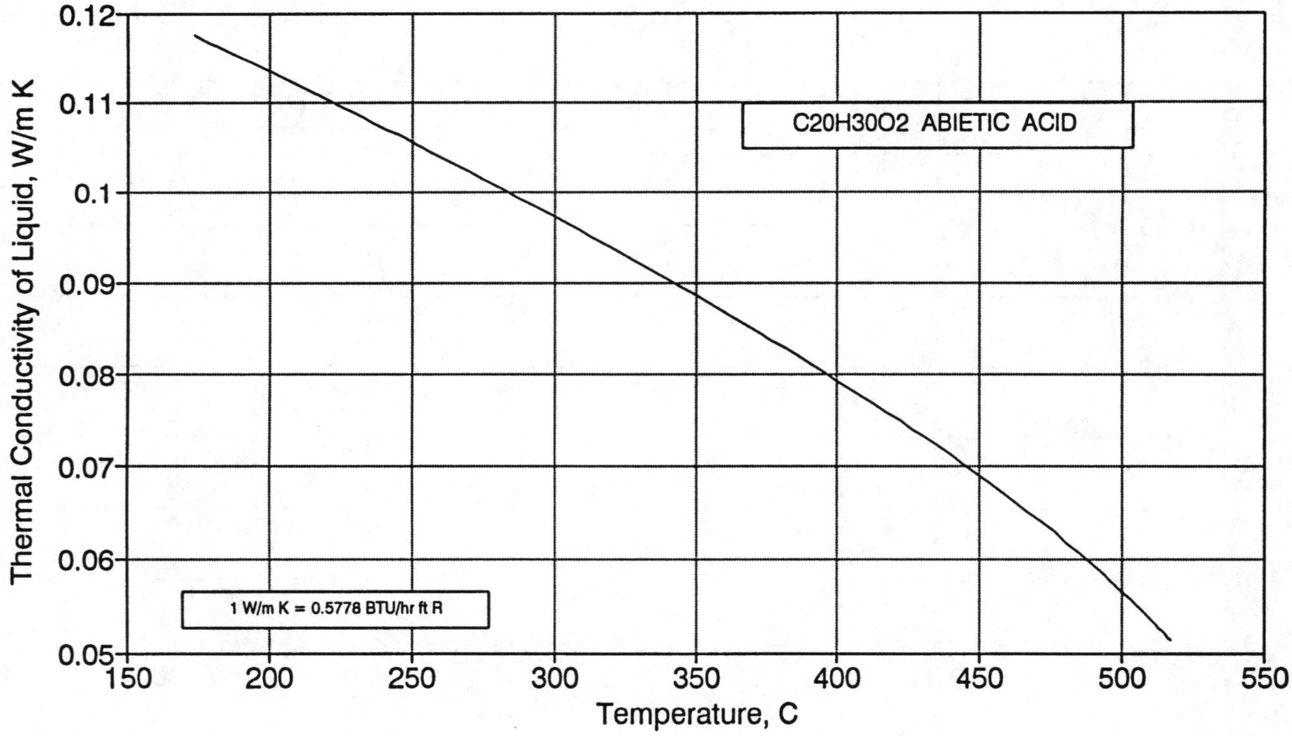
C20H30O2 ABIETIC ACID
1 W/m K = 0.5778 BTU/hr ft R
Thermal Conductivity of Liquid, W/m K
0.12
0.11
0.1
0.09
0.08
0.07
0.06
0.05
150
200
250
300
350
400
450
500
550
Temperature, C

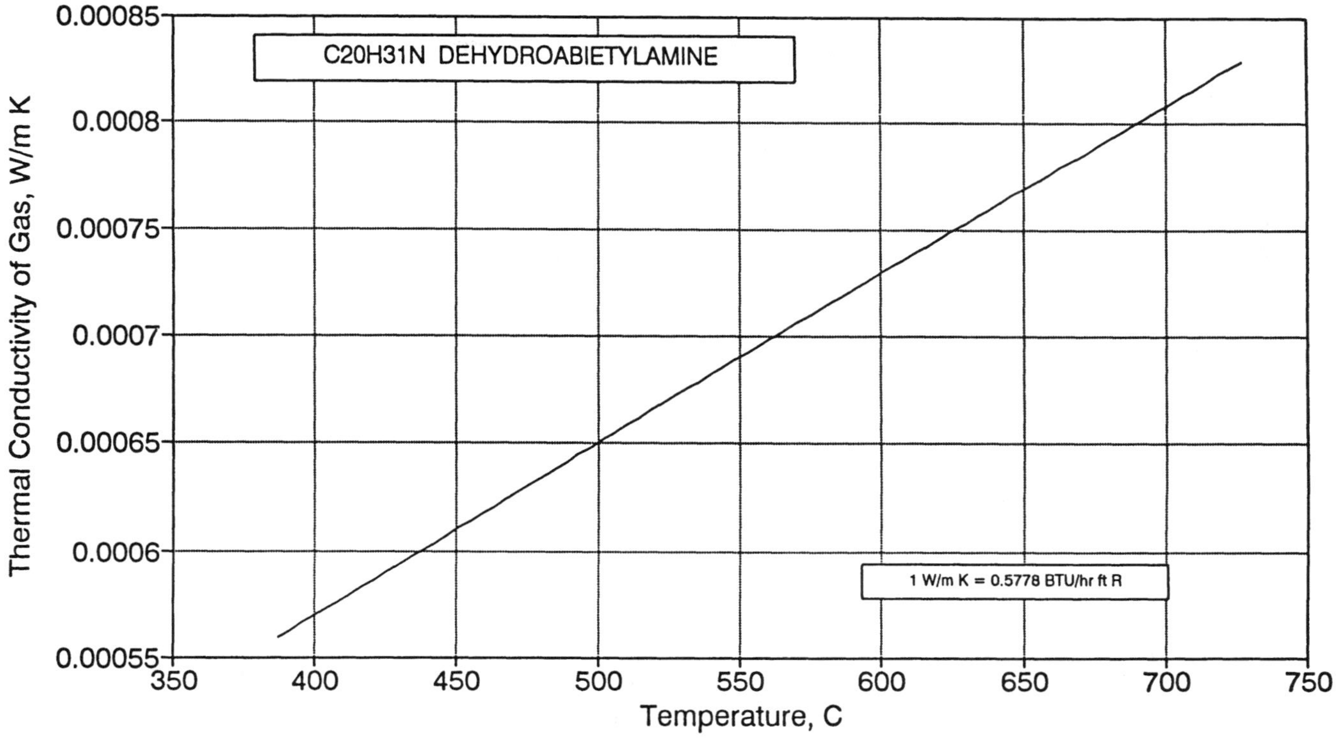
C20H31N DEHYDROABIETYLAMINE
1 W/m K = 0.5778 BTU/hr ft R
Thermal Conductivity of Gas, W/m K
0.00085
0.0008
0.00075
0.0007
0.00065
0.0006
0.00055
350
400
450
500
550
600
650
700
750
Temperature, C

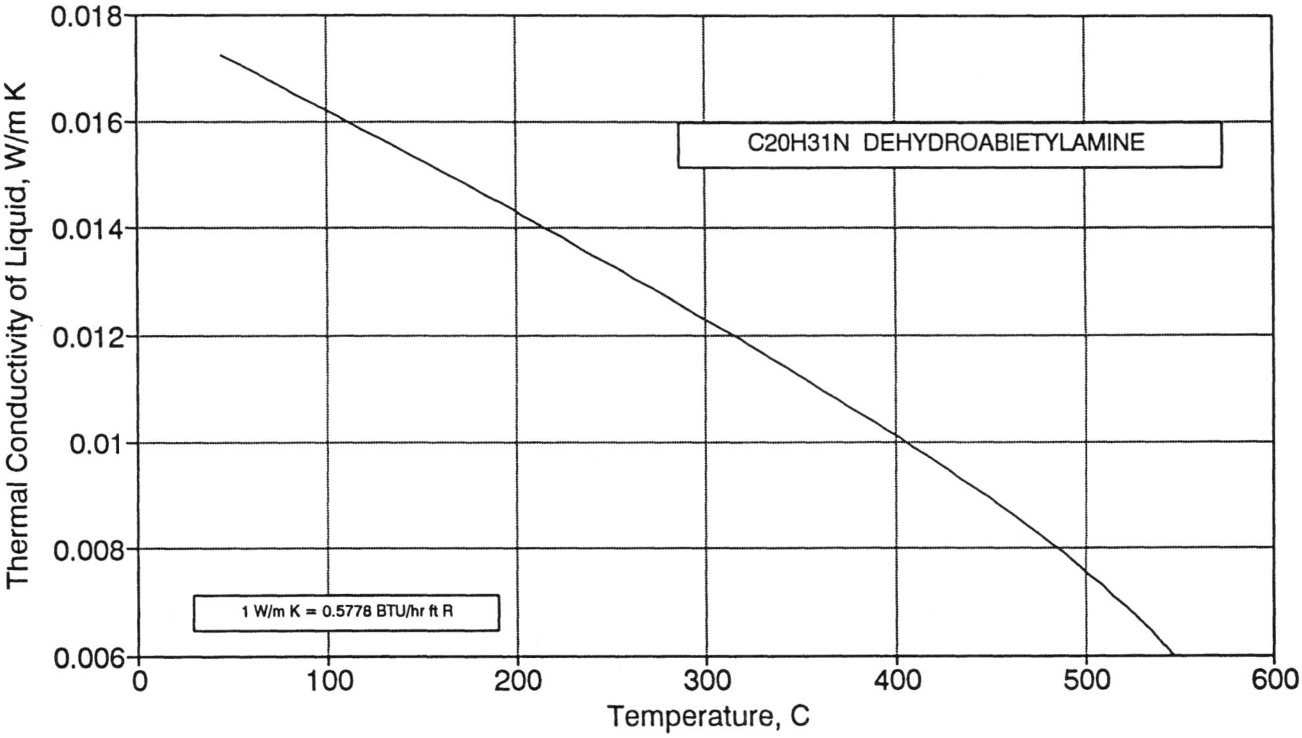
C20H31N DEHYDROABIETYLAMINE
1 W/m K = 0.5778 BTU/hr ft R
Thermal Conductivity of Liquid, W/m K
0.018
0.016
0.014
0.012
0.01
0.008
0.006
0
100
200
300
400
500
600
Temperature, C

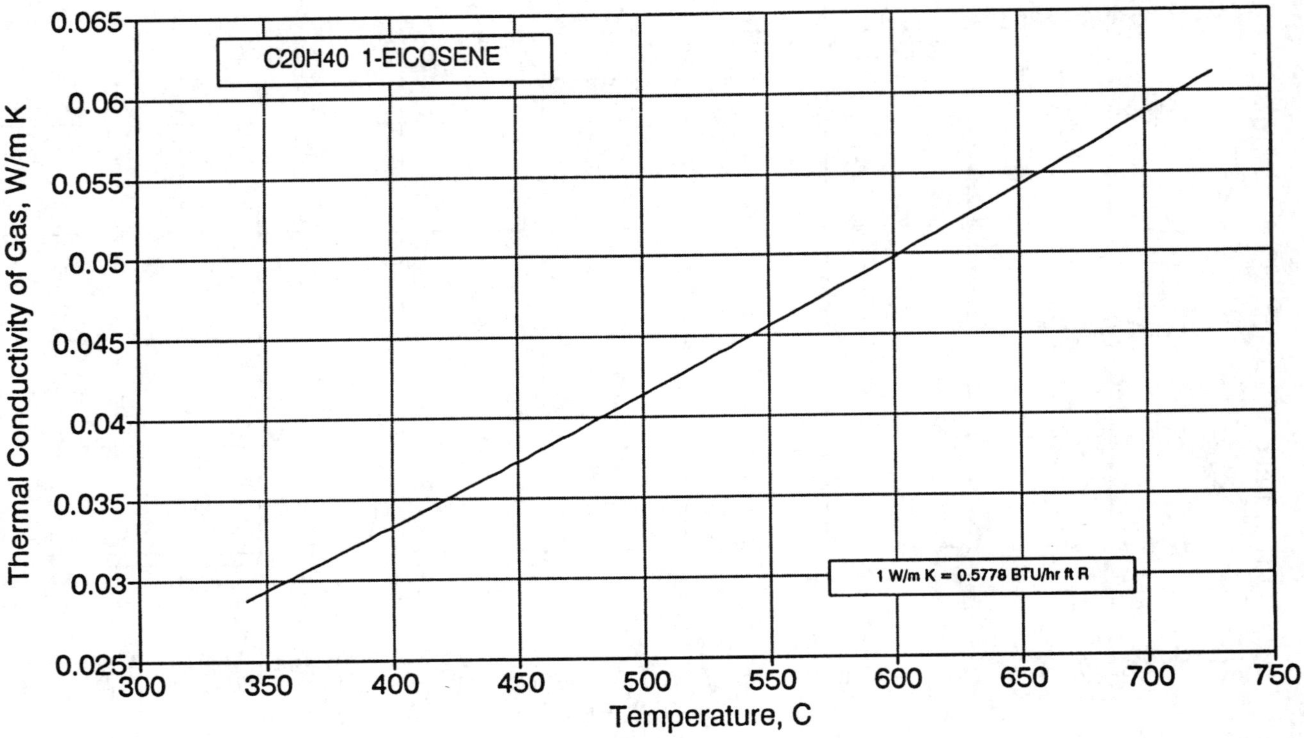
C20H40 1-EICOSENE
1 W/m K = 0.5778 BTU/hr ft R
Thermal Conductivity of Gas, W/m K
0.065
0.06
0.055
0.05
0.045
0.04
0.035
0.03
0.025
300
350
400
450
500
550
600
650
700
750
Temperature, C

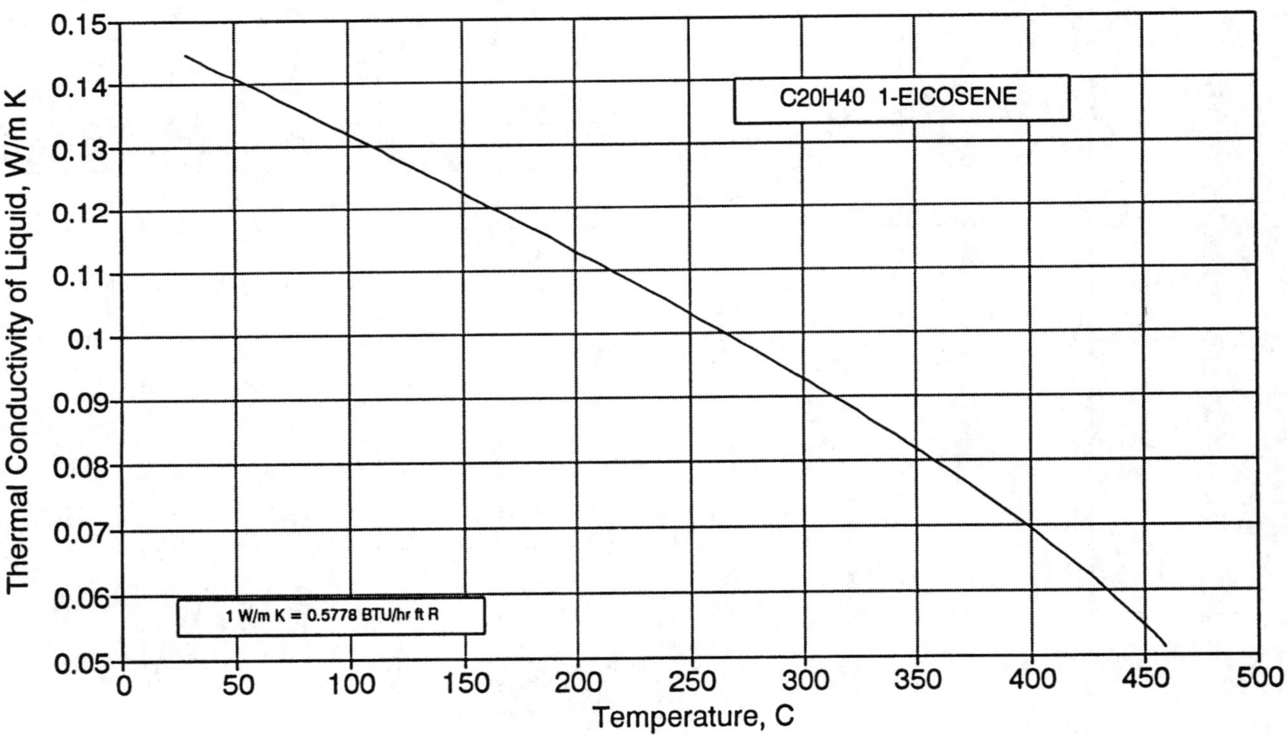
C20H40 1-EICOSENE
1 W/m K = 0.5778 BTU/hr ft R
Thermal Conductivity of Liquid, W/m K
0.15
0.14
0.13
0.12
0.11
0.1
0.09
0.08
0.07
0.06
0.05
0
50
100
150
200
250
300
350
400
450
500
Temperature, C

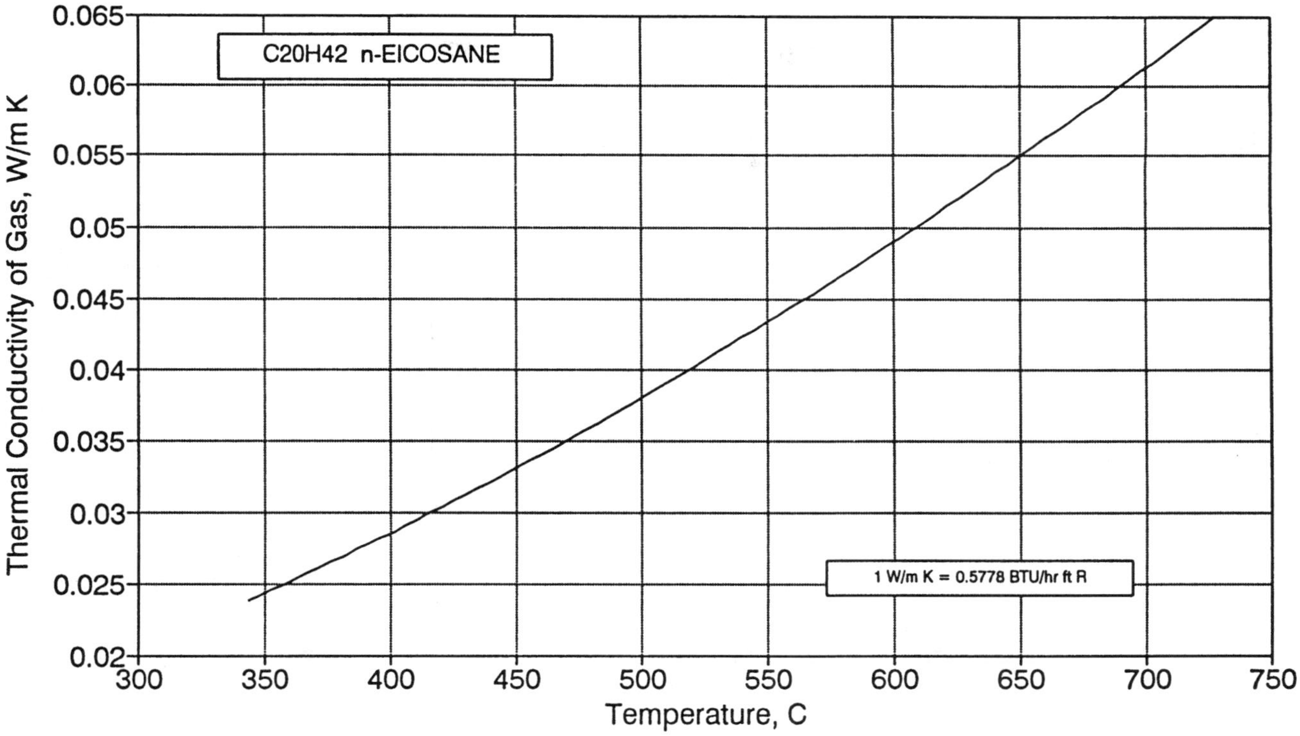
C20H42 n-EICOSANE
1 W/m K = 0.5778 BTU/hr ft R
Thermal Conductivity of Gas, W/m K
Temperature, C
0.065
0.06
0.055
0.05
0.045
0.04
0.035
0.03
0.025
0.02
300
350
400
450
500
550
600
650
700
750

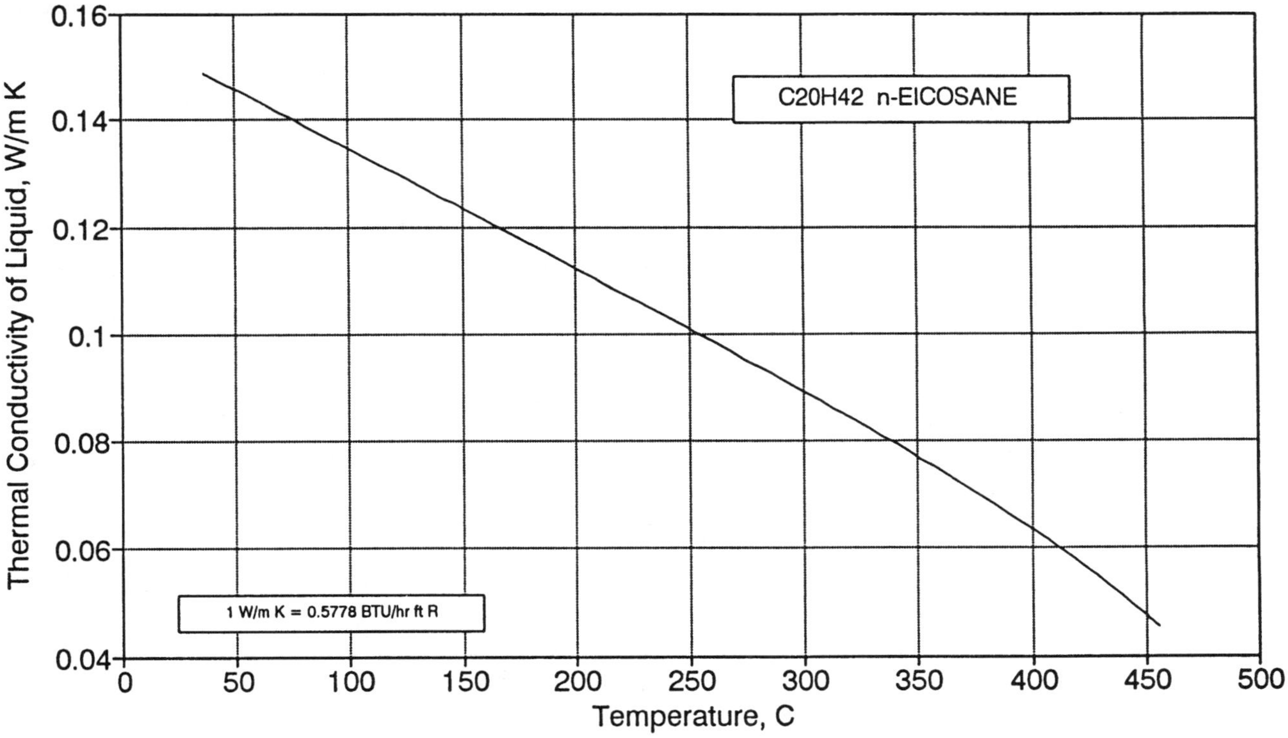
C20H42 n-EICOSANE
1 W/m K = 0.5778 BTU/hr ft R
Thermal Conductivity of Liquid, W/m K
Temperature, C
0.16
0.14
0.12
0.1
0.08
0.06
0.04
0
50
100
150
200
250
300
350
400
450
500

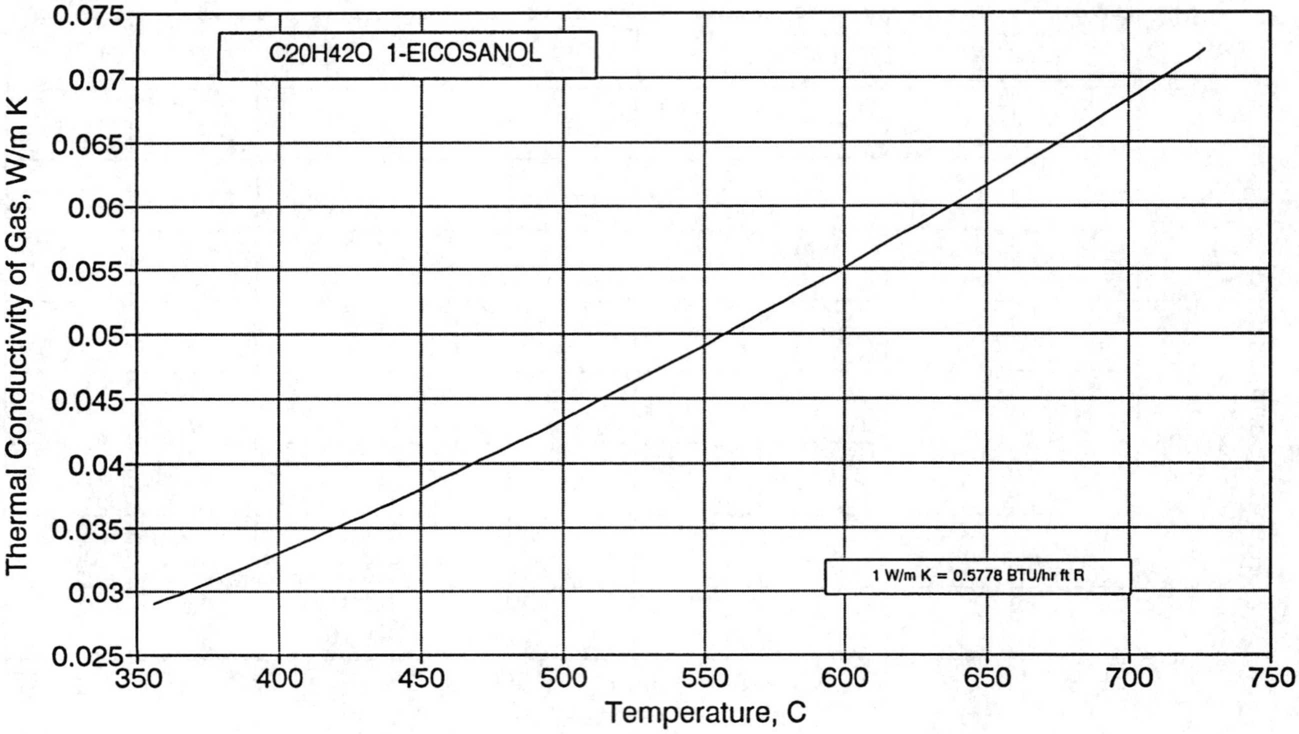
C20H42O 1-EICOSANOL
1 W/m K = 0.5778 BTU/hr ft R
Thermal Conductivity of Gas, W/m K
0.075
0.07
0.065
0.06
0.055
0.05
0.045
0.04
0.035
0.03
0.025
350
400
450
500
550
600
650
700
750
Temperature, C

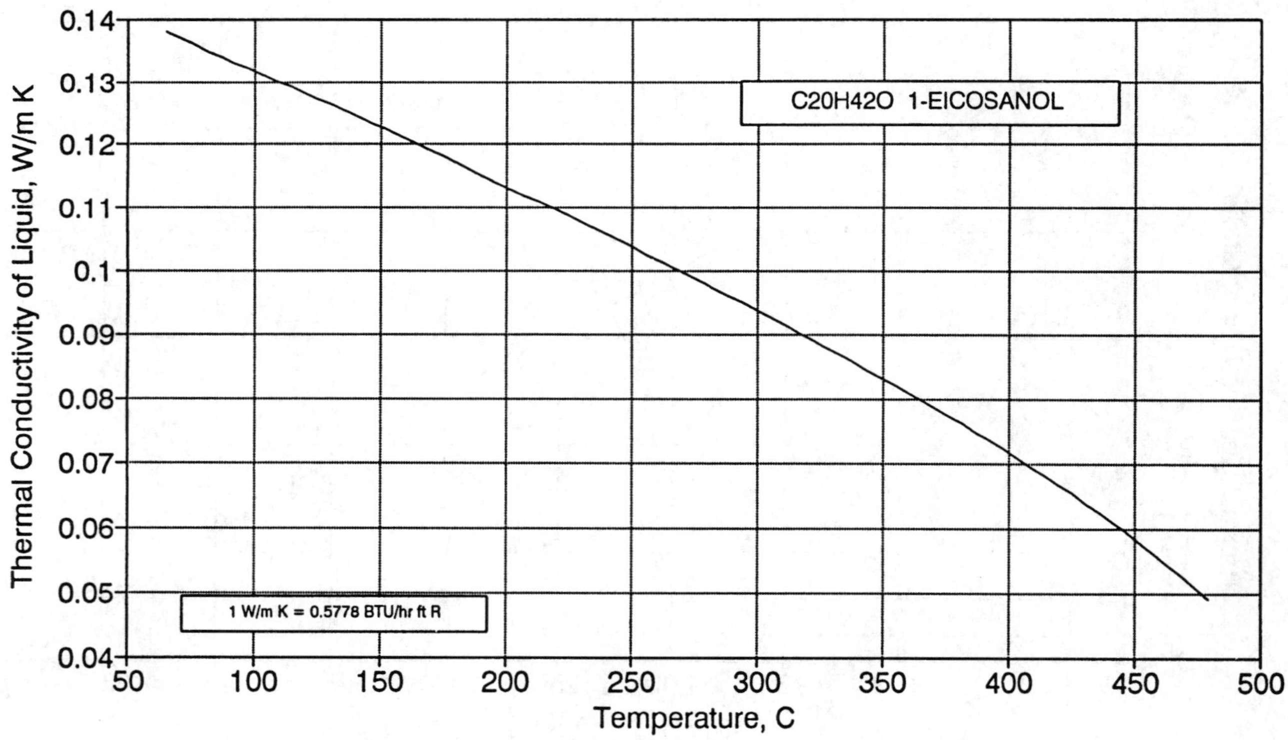
C20H42O 1-EICOSANOL
1 W/m K = 0.5778 BTU/hr ft R
Thermal Conductivity of Liquid, W/m K
0.14
0.13
0.12
0.11
0.1
0.09
0.08
0.07
0.06
0.05
0.04
50
100
150
200
250
300
350
400
450
500
Temperature, C

C21H21O4P TRI-o-CRESYL PHOSPHATE

Gas thermal conductivity data are not available.

TF, K = 240.15

TB, K = not acailable

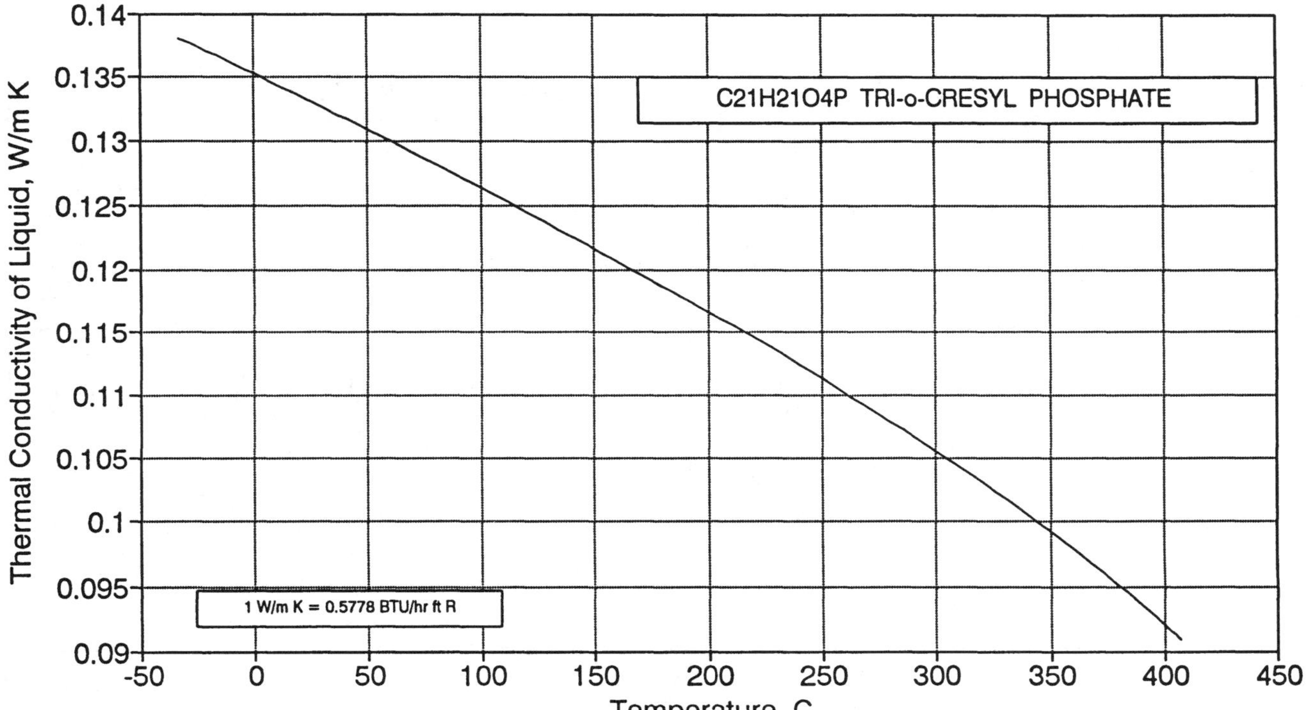

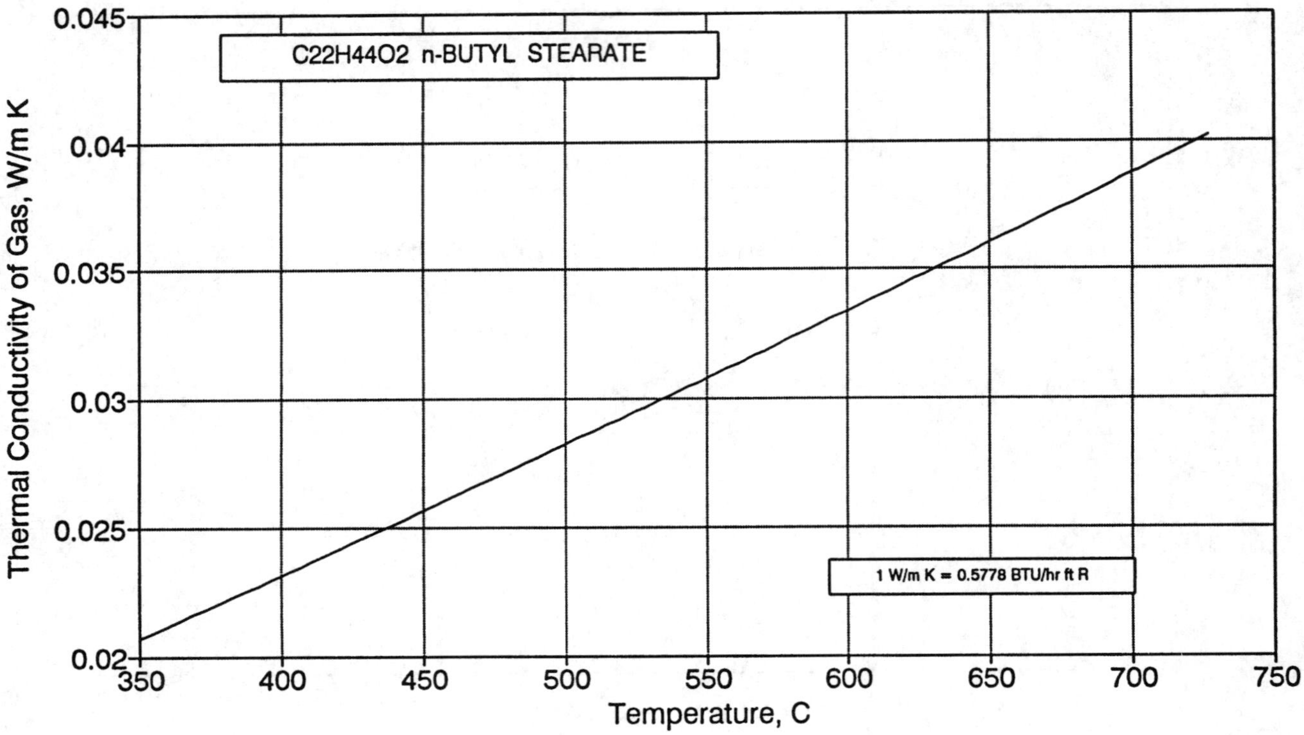
C22H44O2 n-BUTYL STEARATE
1 W/m K = 0.5778 BTU/hr ft R
Thermal Conductivity of Gas, W/m K
Temperature, C
0.045
0.04
0.035
0.03
0.025
0.02
350
400
450
500
550
600
650
700
750

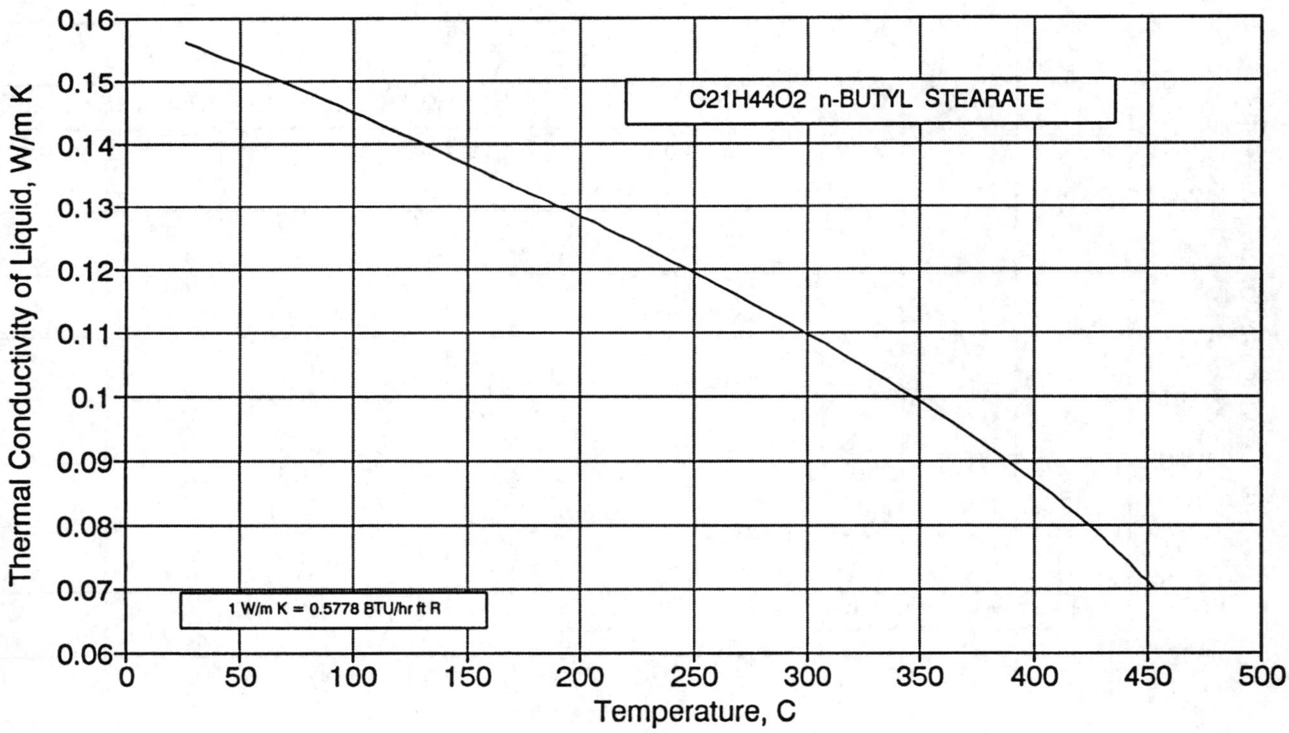
C21H44O2 n-BUTYL STEARATE
1 W/m K = 0.5778 BTU/hr ft R
Thermal Conductivity of Liquid, W/m K
Temperature, C
0.16
0.15
0.14
0.13
0.12
0.11
0.1
0.09
0.08
0.07
0.06
0
50
100
150
200
250
300
350
400
450
500

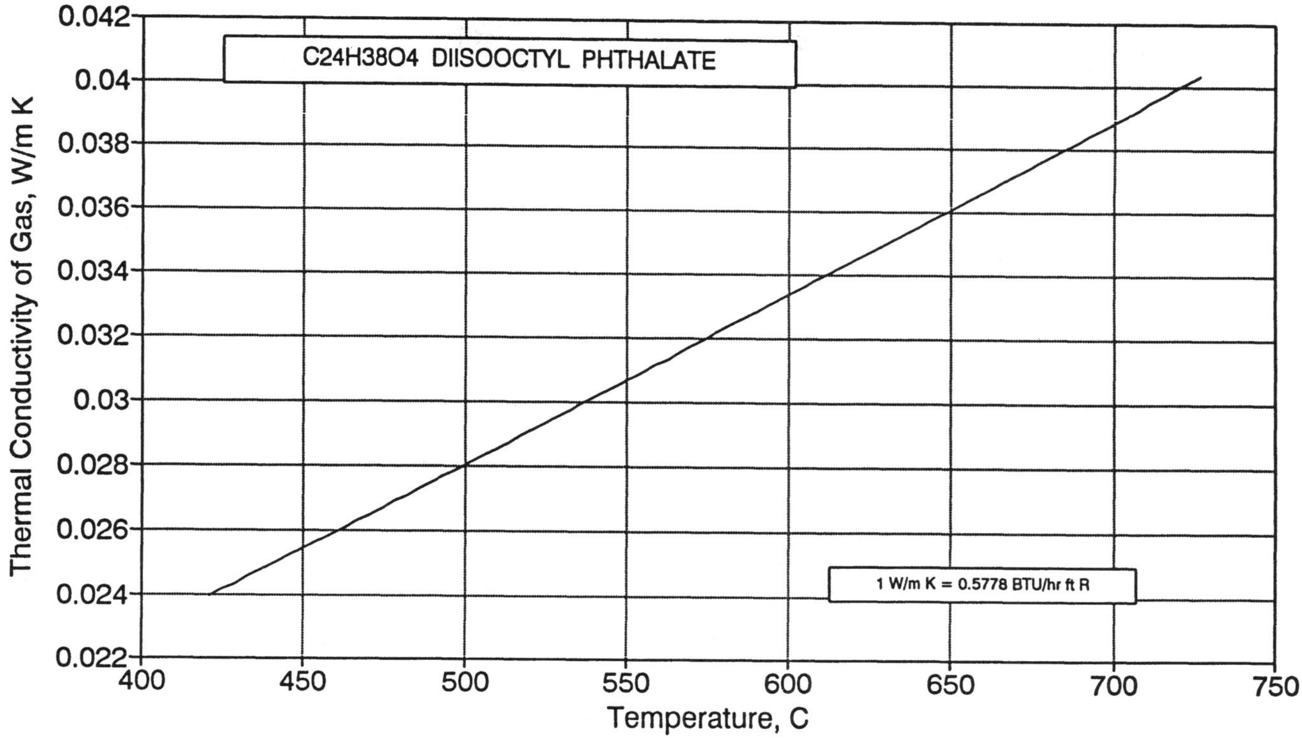
C24H38O4 DIISOOCTYL PHTHALATE
1 W/m K = 0.5778 BTU/hr ft R
Thermal Conductivity of Gas, W/m K
0.042
0.04
0.038
0.036
0.034
0.032
0.03
0.028
0.026
0.024
0.022
400
450
500
550
600
650
700
750
Temperature, C

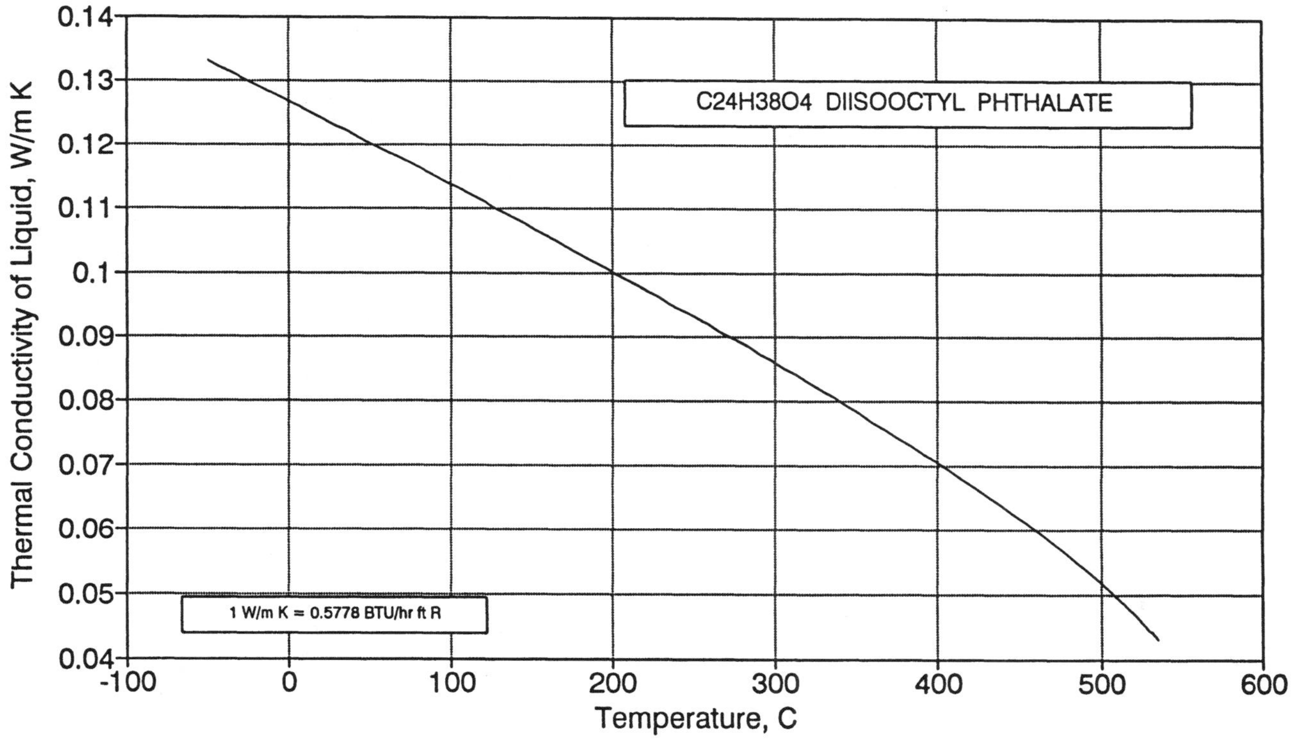
C24H38O4 DIISOOCTYL PHTHALATE
1 W/m K = 0.5778 BTU/hr ft R
Thermal Conductivity of Liquid, W/m K
0.14
0.13
0.12
0.11
0.1
0.09
0.08
0.07
0.06
0.05
0.04
-100
0
100
200
300
400
500
600
Temperature, C

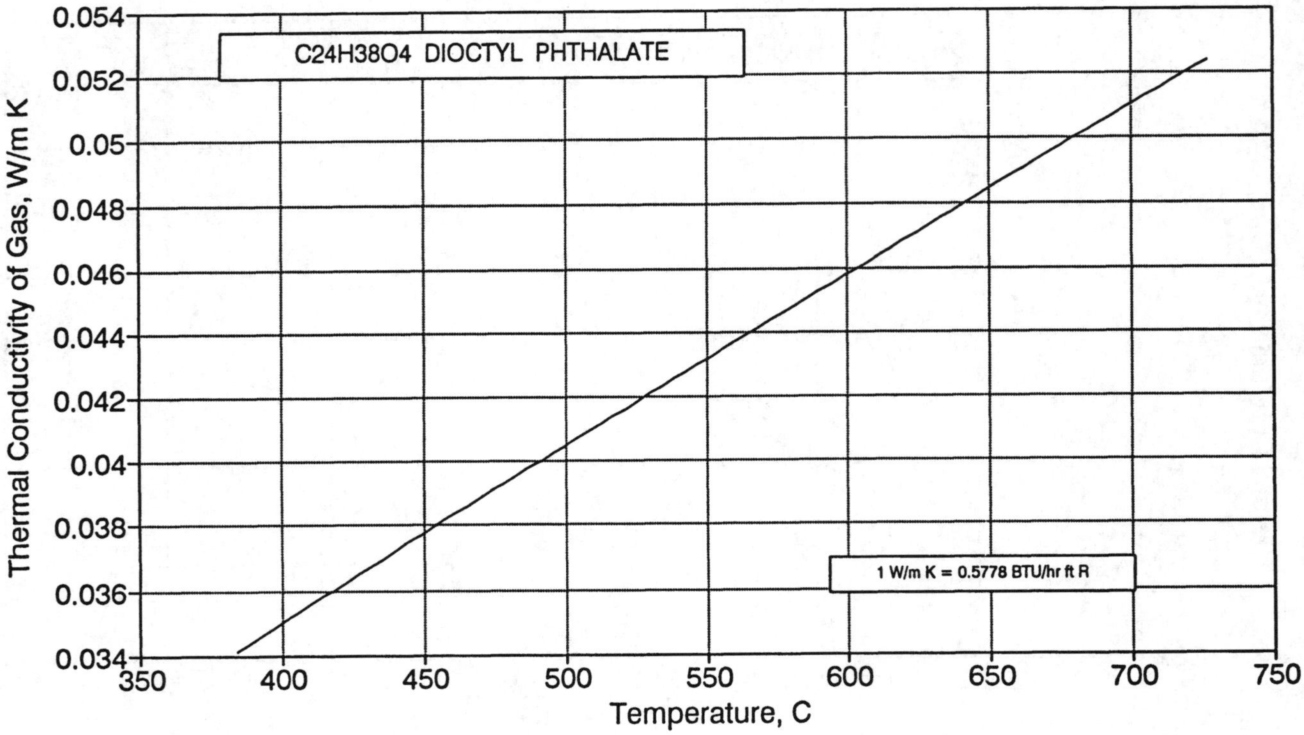
C24H38O4 DIOCTYL PHTHALATE
1 W/m K = 0.5778 BTU/hr ft R
Thermal Conductivity of Gas, W/m K
0.054
0.052
0.05
0.048
0.046
0.044
0.042
0.04
0.038
0.036
0.034
350
400
450
500
550
600
650
700
750
Temperature, C

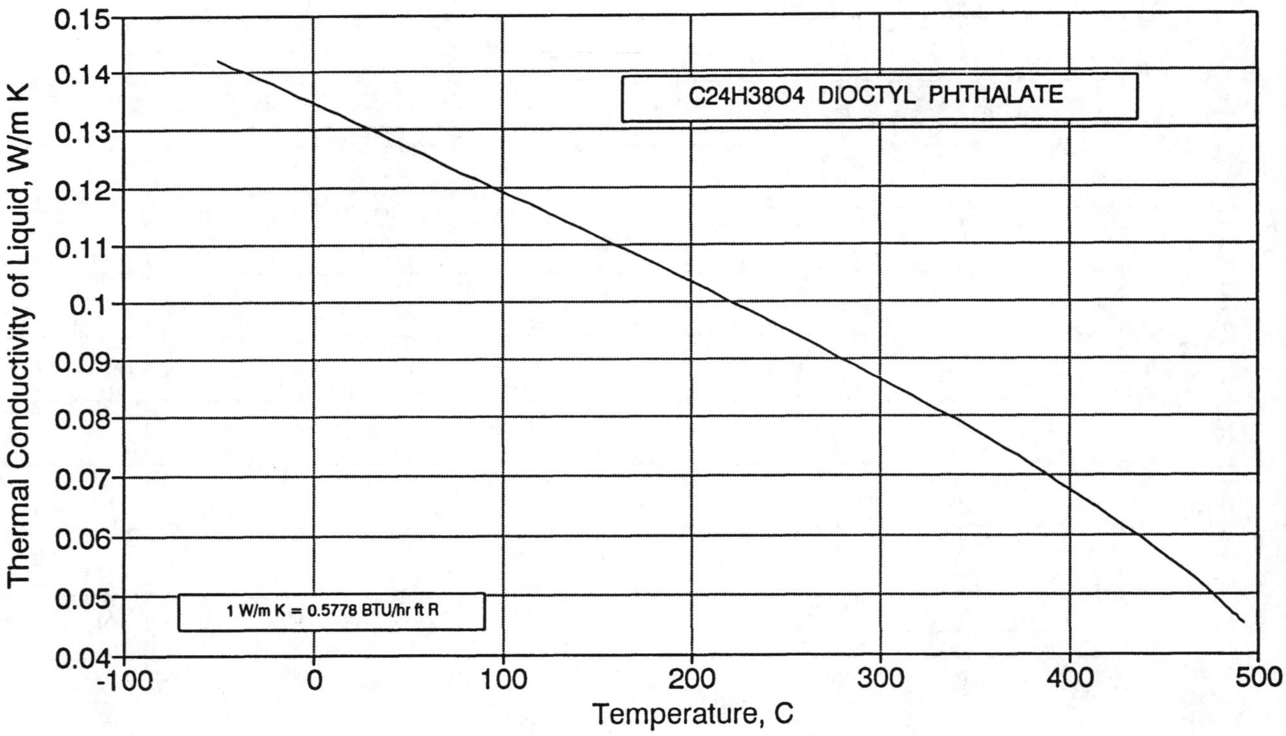
C24H38O4 DIOCTYL PHTHALATE
1 W/m K = 0.5778 BTU/hr ft R
Thermal Conductivity of Liquid, W/m K
0.15
0.14
0.13
0.12
0.11
0.1
0.09
0.08
0.07
0.06
0.05
0.04
-100
0
100
200
300
400
500
Temperature, C

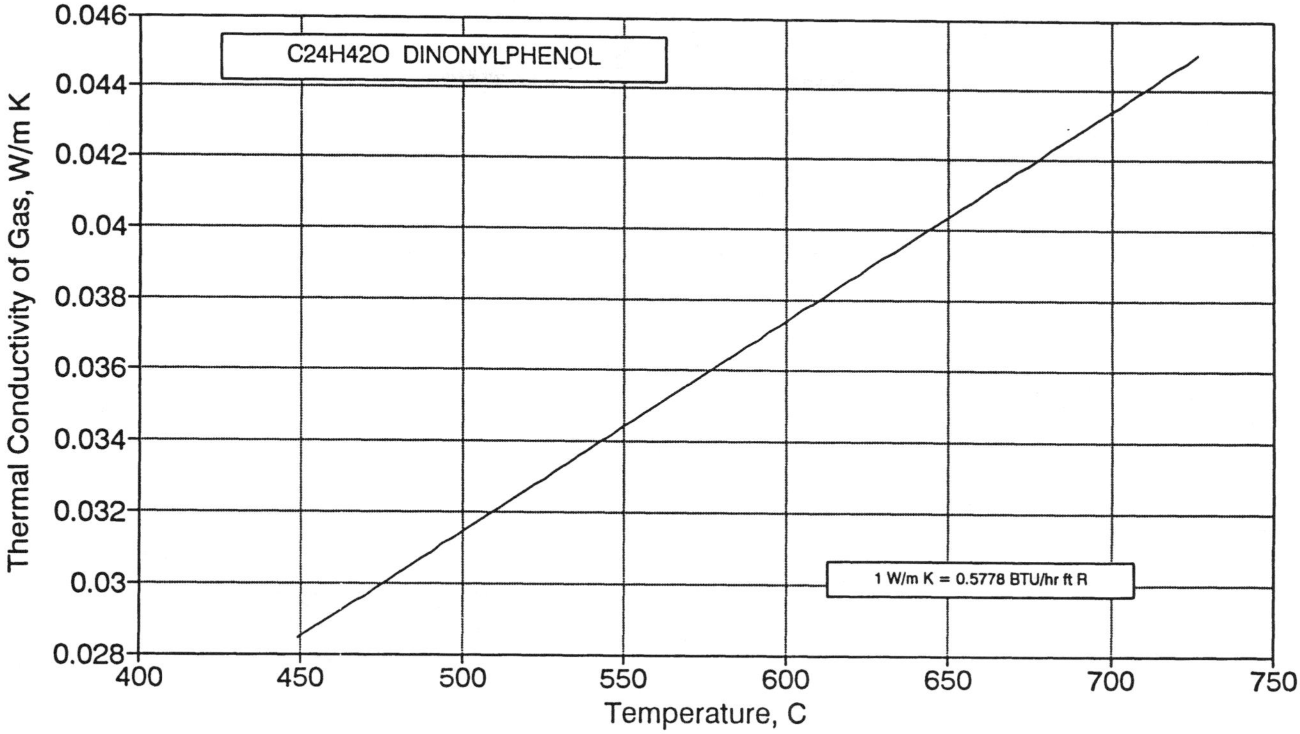
C24H42O DINONYLPHENOL
1 W/m K = 0.5778 BTU/hr ft R
Thermal Conductivity of Gas, W/m K
Temperature, C
0.046
0.044
0.042
0.04
0.038
0.036
0.034
0.032
0.03
0.028
400
450
500
550
600
650
700
750

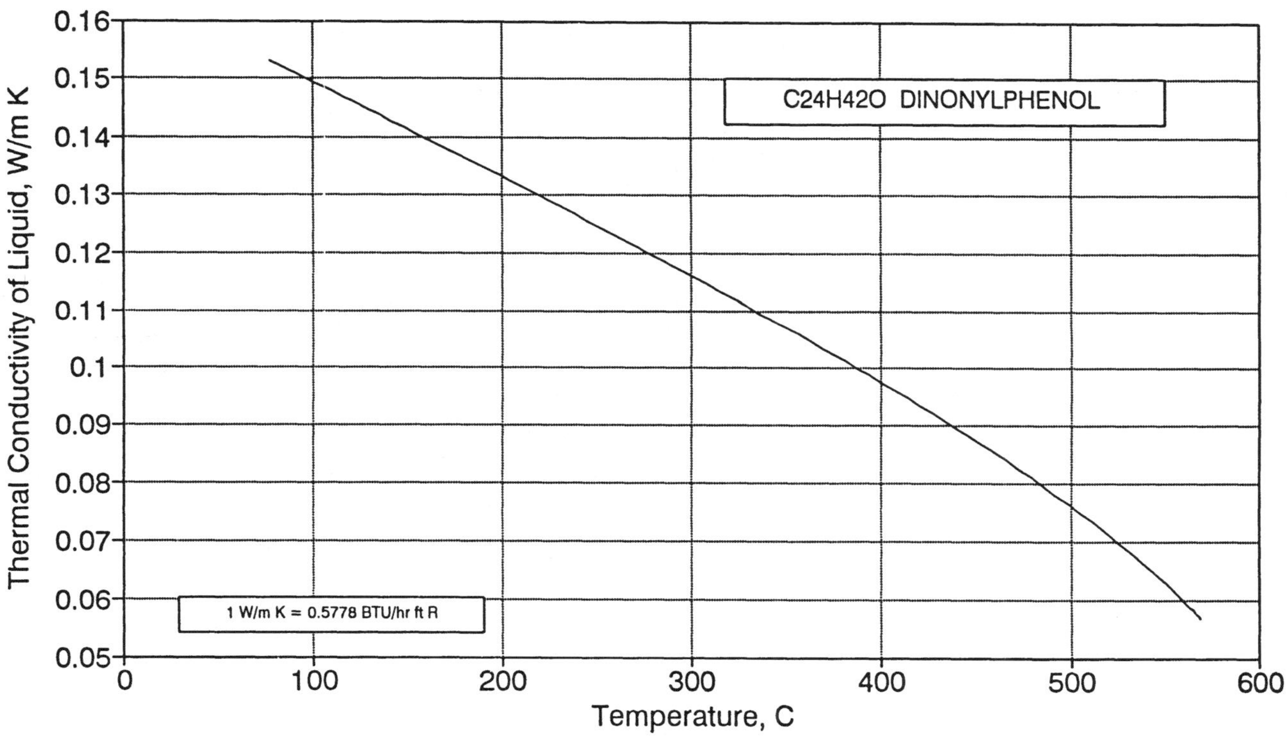
C24H42O DINONYLPHENOL
1 W/m K = 0.5778 BTU/hr ft R
Thermal Conductivity of Liquid, W/m K
Temperature, C
0.16
0.15
0.14
0.13
0.12
0.11
0.1
0.09
0.08
0.07
0.06
0.05
0
100
200
300
400
500
600

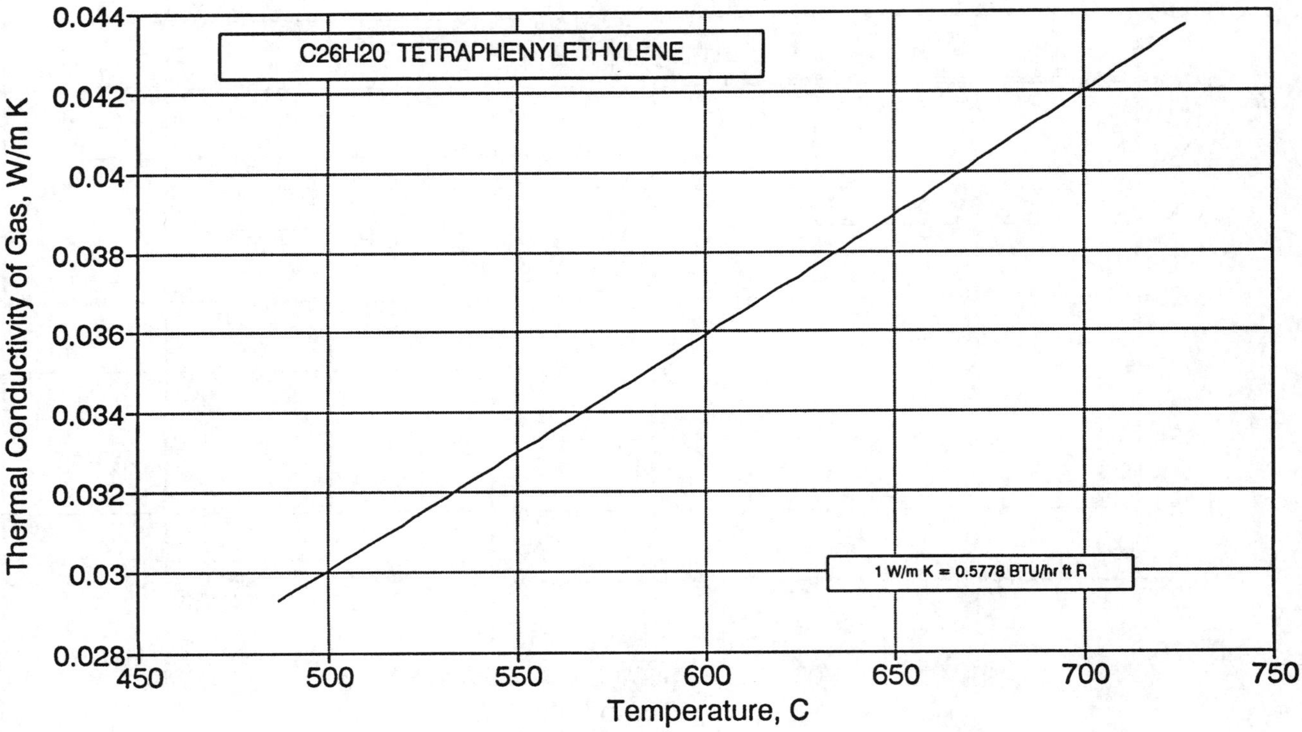
C26H20 TETRAPHENYLETHYLENE
1 W/m K = 0.5778 BTU/hr ft R
Thermal Conductivity of Gas, W/m K
Temperature, C
0.044
0.042
0.04
0.038
0.036
0.034
0.032
0.03
0.028
450
500
550
600
650
700
750

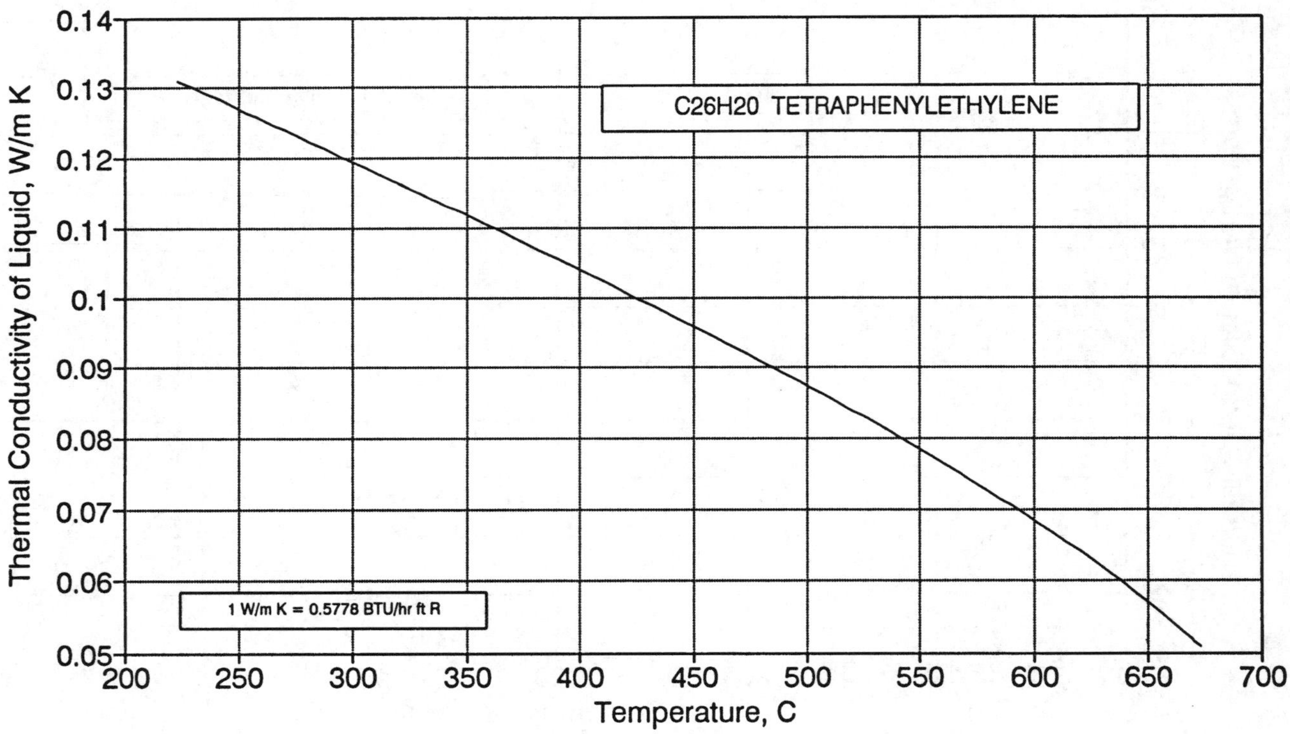
C26H20 TETRAPHENYLETHYLENE
1 W/m K = 0.5778 BTU/hr ft R
Thermal Conductivity of Liquid, W/m K
Temperature, C
0.14
0.13
0.12
0.11
0.1
0.09
0.08
0.07
0.06
0.05
200
250
300
350
400
450
500
550
600
650
700

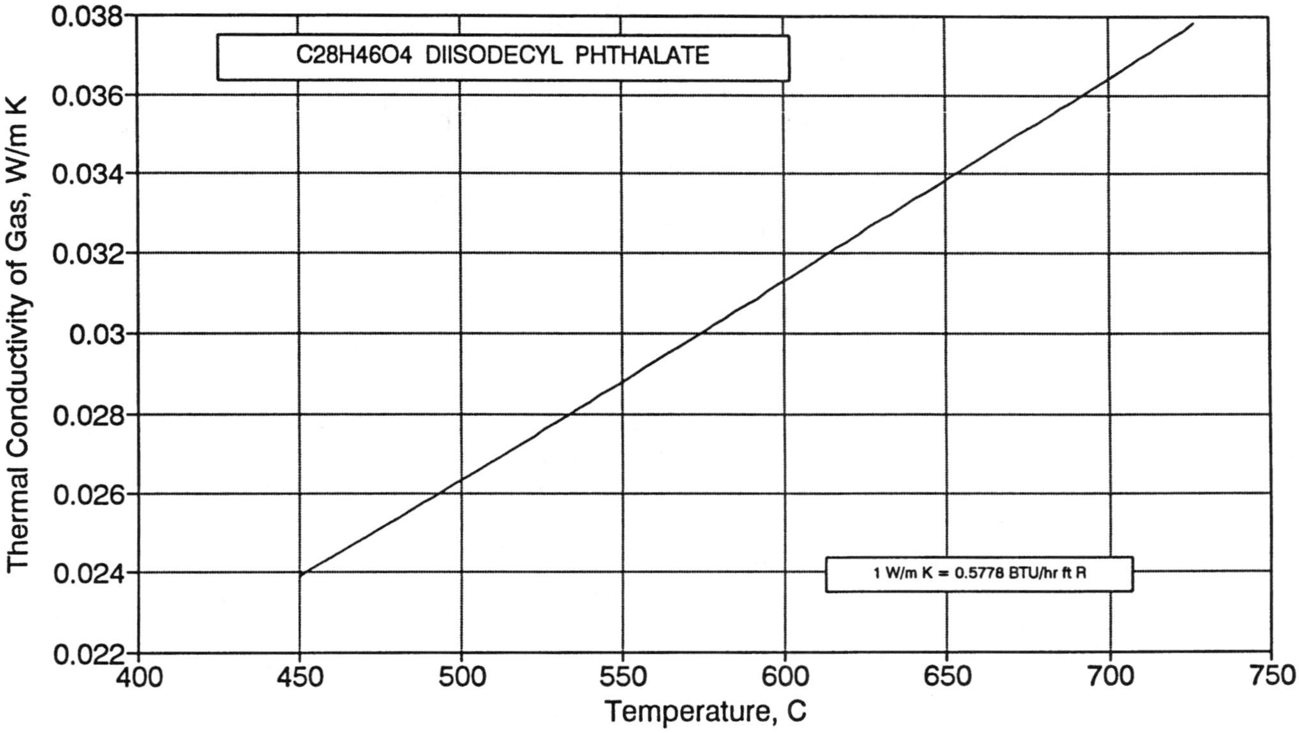
C28H46O4 DIISODECYL PHTHALATE
1 W/m K = 0.5778 BTU/hr ft R
Thermal Conductivity of Gas, W/m K
0.038
0.036
0.034
0.032
0.03
0.028
0.026
0.024
0.022
400
450
500
550
600
650
700
750
Temperature, C

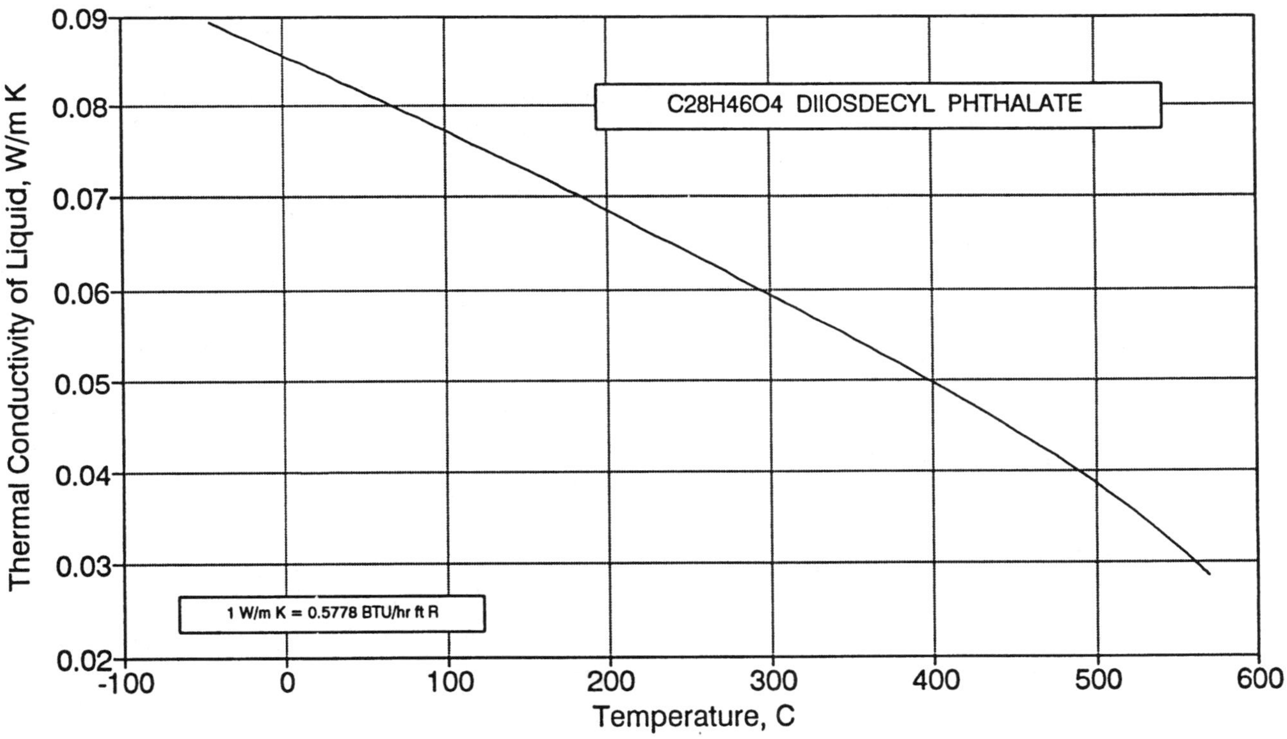
C28H46O4 DIIOSDECYL PHTHALATE
1 W/m K = 0.5778 BTU/hr ft R
Thermal Conductivity of Liquid, W/m K
0.09
0.08
0.07
0.06
0.05
0.04
0.03
0.02
-100
0
100
200
300
400
500
600
Temperature, C

REFERENCES

1. API Research Project No. 44, SELECTED VALUES OF PHYSICAL AND THERMODYNAMIC PROPERTIES OF HYDROCARBONS AND RELATED COMPOUNDS, Carnegie Press, Carnegie Institute of Technology, Pittsburgh, PA (1953).
2. SELECTED VALUES OF PROPERTIES OF HYDROCARBONS AND RELATED COMPOUNDS, Thermodynamics Research Center, TAMU, College Station, TX (1977, 1984).
3. SELECTED VALUES OF PROPERTIES OF CHEMICAL COMPOUNDS, Thermodynamics Research Center, TAMU, College Station, TX (1977, 1987).
4. TECHNICAL DATA BOOK - PETROLEUM REFINING, Vol. I and II, American Petroleum Institute, Washington, DC (1972, 1977, 1982).
5. Daubert, T. E. and R. P. Danner, DATA COMPILATION OF PROPERTIES OF PURE COMPOUNDS, Parts 1, 2, 3 and 4, Supplements 1 and 2, DIPPR Project, AIChE, New York, NY (1985-1992).
6. Ambrose, D., VAPOUR-LIQUID CRITICAL PROPERTIES, National Physical Laboratory, Teddington, England, NPL Report Chem 107 (Feb., 1980).
7. Simmrock, K. H., R. Janowsky and A. Ohnsorge, CRITICAL DATA OF PURE SUBSTANCES, Vol. II, Parts 1 and 2, Dechema Chemistry Data Series, 6000 Frankfurt/Main, Germany (1986).
8. INTERNATIONAL CRITICAL TABLES, McGraw-Hill, New York, NY (1926).
9. Maxwell, J. B., DATA BOOK ON HYDROCARBONS, D. Van Nostrand, Princeton. NJ (1958).
10. Egloff, G., PHYSICAL CONSTANTS OF HYDROCARBONS, Vols. 1-6, Reinhold Publishing Corp., New York, NY (1939-1947).
11. Braker, W. and A. L. Mossman, MATHESON GAS DATA BOOK, 6th ed., Matheson Gas Products, Secaucaus, NJ (1980).
12. CRC HANDBOOK OF CHEMISTRY AND PHYSICS, 66th - 74th eds., CRC Press, Inc., Boca Raton, FL (1985-1993).
13. LANGE'S HANDBOOK OF CHEMISTRY, 13th and 14th eds., McGraw-Hill, New York, NY (1985, 1992).
14. PERRY'S CHEMICAL ENGINEERING HANDBOOK, 6th ed., McGraw-Hill, New York, NY (1984).
15. Kaye, G. W. C. and T. H. Laby, TABLES OF PHYSICAL AND CHEMICAL CONSTANTS, Longman Group Limited, London, England (1973).
16. Raznjevic, Kuzman, HANDBOOK OF THERMODYNAMIC TABLES AND CHARTS, Hemisphere Publishing Corporation, New York, NY (1976).
17. Vargaftik, N. B., TABLES ON THE THERMOPHYSICAL PROPERTIES OF LIQUIDS AND GASES, 2nd ed., English translation, Hemisphere Publishing Corporation, New York, NY (1975, 1983).
18. Timmermans, J., PHYSICO-CHEMICAL CONSTANTS OF PURE ORGANIC COMPOUNDS, Vol. 1 and 2, Elsevier, New York, NY (1950,1965).
19. Ho, C. Y., P. E. Liley, T. Makita and Y. Tanaka, PROPERTIES OF INORGANIC AND ORGANIC FLUIDS, Hemisphere Publishing Corporation, New York, NY (1988).
20. Tsederberg, N. V., THERMAL CONDUCTIVITY OF GASES AND LIQUIDS, R. D. Cess, editor, MIT Press, Cambridge, MA (1965).
21. Vargaftik, N. B., L. P. Filippov, A. A. Tarzimanov and E. E. Totskiy, THERMAL CONDUCTIVITY OF GASES AND LIQUIDS, Standards Press, Moscow, USSR (1978).
22. Landolt-Bornstein, ZAHLENWERTE UND FUNKIONEN ANS PHYSIK, CHEMEI, ASTRONOMIE UND TECHNIK, 6th ed., Springer-Verlag, Berlin, Germany (1972).
23. Touloukian, Y. S., P. E. Liley and S. C. Saxena, THERMOPHYSICAL PROPERTIES OF MATTER, Vol. 11 - Viscosity, IFI/Plenum Press, New York, NY (1974).
24. Lyman, W. J., W. F. Reehl and D. H. Rosenblatt, HANDBOOK OF CHEMICAL PROPERTY

ESTIMATION METHODS, McGraw-Hill, New York, NY (1982).

25. Reid, R. C., J. M. Prausnitz and B. E. Poling, THE PROPERTIES OF GASES AND LIQUIDS, 3rd ed. (R. C. Reid and T. K. Sherwood), 4th ed., McGraw-Hill, New York, NY (1977, 1987).
26. Kirk, R. E. and D. F. Othmer, editors, ENCYCLOPEDIA OF CHEMICAL TECHNOLOGY, 3rd ed., Vols. 1-24, John Wiley and Sons, Inc., New York, NY (1978-1984).
27. Beaton, C. F. and G. F. Hewitt, PHYSICAL PROPERTY DATA FOR THE DESIGN ENGINEER, Hemisphere Publishing Corporation, New York, NY (1989).
28. Yaws, C. L., PHYSICAL PROPERTIES, McGraw-Hill, New York, NY (1977).
29. Yaws, C. L., THERMODYNAMIC AND PHYSICAL PROPERTY DATA, Gulf Publishing Co., Houston, TX (1992).
30. Yaws, C. L. and R. W. Gallant, PHYSICAL PROPERTIES OF HYDROCARBONS, Vols. 1 (2nd ed.), 2 (3rd ed.) and 3 (1st ed.), Gulf Publishing Co., Houston, TX (1992, 1993, 1993).

Appendix A

COEFFICIENTS FOR LIQUID THERMAL CONDUCTIVITY EQUATION*

Carl L. Yaws, Xiaoyan Lin, Li Bu and Sachin Nijhawan
Lamar University, Beaumont, Texas

$\log_{10} \lambda_{liq} = A + B\,[1-T/C]^{2/7}$ (λ_{liq} - W/m K, T - K)

NO	FORMULA	NAME	A	B	C	TMIN	TMAX	λ_{liq} @ 25 C
1	$C_8H_4Cl_2O_2$	ISOPHTHALOYL CHLORIDE	-1.7803	1.0149	768.00	317	730	---
2	$C_8H_4O_3$	PHTHALIC ANHYDRIDE	-1.6831	1.0293	791.00	404	751	---
3	$C_8H_6O_4$	ISOPHTHALIC ACID	---	---	---	---	---	---
4	$C_8H_6O_4$	PHTHALIC ACID	---	---	---	---	---	---
5	$C_8H_6O_4$	TEREPHTHALIC ACID	---	---	---	---	---	---
6	C_8H_6S	BENZOTHIOPHENE	-1.7276	1.0008	754.00	305	716	---
7	C_8H_7N	INDOLE	-1.6837	0.9961	790.00	274	751	0.1535
8	C_8H_8	STYRENE	-1.7023	1.0002	648.00	243	616	0.1369
9	C_8H_8O	ACETOPHENONE	-2.0849	1.5494	701.00	294	666	0.1728
10	C_8H_8O	p-TOLUALDEHYDE	-1.7596	1.0074	698.00	290	663	0.1258
11	$C_8H_8O_2$	METHYL BENZOATE	-1.6414	0.9769	693.00	261	658	0.1550
12	$C_8H_8O_2$	o-TOLUIC ACID	-1.6254	0.9641	751.00	377	713	---
13	$C_8H_8O_2$	p-TOLUIC ACID	-1.5980	0.9423	773.00	453	734	---
14	$C_8H_8O_3$	METHYL SALICYLATE	-1.6717	0.9822	701.00	265	666	0.1468
15	$C_8H_8O_3$	VANILLIN	-1.7080	1.0230	777.00	355	738	---
16	C_8H_9NO	ACETANILIDE	-1.7347	1.0237	825.00	387	784	---
17	C_8H_{10}	ETHYLBENZENE	-1.7498	1.0437	617.17	178	586	0.1302
18	C_8H_{10}	m-XYLENE	-1.7286	1.0193	617.05	225	586	0.1305
19	C_8H_{10}	o-XYLENE	-1.7372	1.0282	630.37	248	599	0.1315
20	C_8H_{10}	p-XYLENE	-1.7354	1.0254	616.26	286	585	0.1298
21	$C_8H_{10}O$	m-ETHYLPHENOL	---	---	---	---	---	---
22	$C_8H_{10}O$	p-ETHYLPHENOL	-1.6225	0.9776	716.45	318	681	---
23	$C_8H_{10}O$	PHENETOLE	-1.4237	0.6814	647.15	244	615	0.1404
24	$C_8H_{10}O$	2-PHENYLETHANOL	-1.6255	0.9898	684.00	247	650	0.1640
25	$C_8H_{10}O$	2,3-XYLENOL	-1.1910	0.4541	722.95	346	687	---
26	$C_8H_{10}O$	2,4-XYLENOL	-1.1954	0.4501	707.65	298	672	0.1547
27	$C_8H_{10}O$	2,5-XYLENOL	-1.1950	0.4487	707.05	348	672	---
28	$C_8H_{10}O$	2,6-XYLENOL	-1.2065	0.4624	701.05	319	666	---
29	$C_8H_{10}O$	3,4-XYLENOL	-1.6115	0.9738	729.95	338	693	---
30	$C_8H_{10}O$	3,5-XYLENOL	-1.1839	0.4551	715.65	337	680	---
31	$C_8H_{11}N$	N,N-DIMETHYLANILINE	-1.5076	0.7769	687.15	276	653	0.1422
32	$C_8H_{11}N$	o-ETHYLANILINE	-1.6969	0.9915	704.00	227	669	0.1413
33	$C_8H_{11}N$	2,4,6-TRIMETHYLPYRIDINE	-1.7344	0.9957	653.00	229	620	0.1265
34	$C_8H_{11}NO$	p-PHENETIDINE	-1.7085	1.0051	754.00	277	716	0.1452
35	C_8H_{12}	1,5-CYCLOOCTADIENE	-1.6838	0.9633	645.00	204	613	0.1327
36	C_8H_{12}	VINYLCYCLOHEXENE	-1.7094	0.9809	599.00	164	569	0.1248
37	$C_8H_{12}O_4$	1,4-CYCLOHEXANEDICARBOXYLIC ACID	-1.7081	1.0707	889.00	586	845	---
38	$C_8H_{12}O_4$	DIETHYL MALEATE	-1.7114	0.9736	680.00	264	646	0.1301
39	$C_8H_{14}O_2$	n-BUTYL METHACRYLATE	-1.7073	0.9923	616.00	223	585	0.1300
40	$C_8H_{14}O_3$	BUTYRIC ANHYDRIDE	-1.6905	0.9930	639.00	200	607	0.1378
41	$C_8H_{14}O_4$	DIETHYL SUCCINATE	-1.7317	0.9849	660.00	252	627	0.1252
42	C_8H_{16}	1,1-DIMETHYLCYCLOHEXANE	-1.7558	1.0033	591.15	240	562	0.1162
43	C_8H_{16}	cis-1,2-DIMETHYLCYCLOHEXANE	-1.7459	0.9949	606.15	223	576	0.1186
44	C_8H_{16}	trans-1,2-DIMETHYLCYCLOHEXANE	-1.2763	0.4639	596.15	185	566	0.1271
45	C_8H_{16}	cis-1,3-DIMETHYLCYCLOHEXANE	-1.7537	0.9909	591.15	198	562	0.1141
46	C_8H_{16}	trans-1,3-DIMETHYLCYCLOHEXANE	-1.7435	0.9882	598.00	183	568	0.1169
47	C_8H_{16}	cis-1,4-DIMETHYLCYCLOHEXANE	-1.7459	0.9896	598.15	186	568	0.1166
48	C_8H_{16}	trans-1,4-DIMETHYLCYCLOHEXANE	-1.7647	1.0023	590.15	236	561	0.1135
49	C_8H_{16}	ETHYLCYCLOHEXANE	-1.7405	0.9818	609.15	162	579	0.1174
50	C_8H_{16}	2-ETHYL-1-HEXENE	-1.7709	0.9929	574.00	173	545	0.1082
51	C_8H_{16}	1-METHYL-1-ETHYLCYCLOPENTANE	-1.7315	0.9795	582.00	129	553	0.1165
52	C_8H_{16}	1-OCTENE	-1.6274	0.9136	566.60	171	538	0.1290
53	C_8H_{16}	trans-2-OCTENE	-1.6935	0.9937	577.00	185	548	0.1299
54	C_8H_{16}	trans-3-OCTENE	-1.6905	0.9884	574.00	163	545	0.1292
55	C_8H_{16}	trans-4-OCTENE	-1.6921	0.9901	573.00	179	544	0.1290
56	C_8H_{16}	n-PROPYLCYCLOPENTANE	-1.7409	0.9779	603.00	156	573	0.1158
57	C_8H_{16}	2,4,4-TRIMETHYL-1-PENTENE	-1.7673	0.9910	553.00	180	525	0.1064

* A computer program, containing coefficients for all compounds, is available for a nominal fee. The computer program is in ASCII which can be accessed by other software.

$\log_{10} \lambda_{liq} = A + B\ [1-T/C]^{2/7}$ (λ_{liq} - W/m K, T - K)

NO	FORMULA	NAME	A	B	C	TMIN	TMAX	λ_{liq} @ 25 C
58	C8H16	2,4,4-TRIMETHYL-2-PENTENE	-1.7646	0.9898	558.00	167	530	0.1074
59	C8H16O	2-ETHYLHEXANAL	-1.7220	0.9981	607.00	200	577	0.1261
60	C8H16O	1-OCTANAL	-1.7252	1.0125	621.00	246	590	0.1302
61	C8H16O	2-OCTANONE	-1.6779	0.9911	624.00	253	593	0.1397
62	C8H16O2	n-BUTYL n-BUTYRATE	-1.7168	0.9953	616.00	181	585	0.1279
63	C8H16O2	n-HEXYL ACETATE	-1.3859	0.6458	618.00	192	587	0.1410
64	C8H16O2	ISOBUTYL ISOBUTYRATE	-1.7293	0.9957	602.00	192	572	0.1230
65	C8H16O2	n-OCTANOIC ACID	-1.6624	0.9819	692.00	290	657	0.1491
66	C8H16O4	DIETHYLENE GLYCOL ETHYL ETHER ACETATE	-1.7930	1.0113	660.00	248	627	0.1145
67	C8H18	2,2-DIMETHYLHEXANE	-1.7714	0.9821	549.80	152	522	0.1033
68	C8H18	2,3-DIMETHYLHEXANE	-1.7995	1.0203	563.40	272	535	0.1055
69	C8H18	2,4-DIMETHYLHEXANE	-1.8086	1.0259	553.50	272	526	0.1032
70	C8H18	2,5-DIMETHYLHEXANE	-1.7869	0.9989	550.00	182	523	0.1028
71	C8H18	3,3-DIMETHYLHEXANE	-1.7657	0.9782	562.00	147	534	0.1053
72	C8H18	3,4-DIMETHYLHEXANE	-1.7937	1.0161	568.80	272	540	0.1067
73	C8H18	3-ETHYLHEXANE	-1.7971	1.0170	565.40	272	537	0.1057
74	C8H18	3-METHYL-3-ETHYLPENTANE	-1.7687	0.9873	576.50	182	548	0.1079
75	C8H18	2-METHYLHEPTANE	-1.7035	0.9932	559.64	164	532	0.1246
76	C8H18	3-METHYLHEPTANE	-1.7146	0.9986	563.67	153	535	0.1232
77	C8H18	4-METHYLHEPTANE	-1.7083	0.9985	561.74	152	534	0.1248
78	C8H18	n-OCTANE	-1.8388	1.1699	568.83	216	540	0.1281
79	C8H18	2,2,3-TRIMETHYLPENTANE	-1.7705	0.9879	563.50	161	535	0.1062
80	C8H18	2,2,4-TRIMETHYLPENTANE	-1.6144	0.7698	543.96	166	517	0.0998
81	C8H18	2,3,3-TRIMETHYLPENTANE	-1.7681	0.9870	573.50	172	545	0.1077
82	C8H18	2,3,4-TRIMETHYLPENTANE	-1.7695	0.9879	566.30	164	538	0.1068
83	C8H18O	DI-n-BUTYL ETHER	-1.4074	0.6132	581.00	178	552	0.1235
84	C8H18O	DI-sec-BUTYL ETHER	-1.7430	0.9913	559.00	173	531	0.1133
85	C8H18O	DI-tert-BUTYL ETHER	-1.7549	0.9830	550.00	195	523	0.1075
86	C8H18O	2-ETHYL-1-HEXANOL	-1.6754	0.9952	640.25	203	608	0.1434
87	C8H18O	1-OCTANOL	-1.6972	1.0775	652.50	258	620	0.1614
88	C8H18O	2-OCTANOL	-1.2061	0.4359	637.15	242	605	0.1438
89	C8H18O2	DI-t-BUTYL PEROXIDE	-1.7630	1.0153	547.00	233	520	0.1116
90	C8H18O2S	DI-n-BUTYL SULFONE	-1.7352	1.0149	767.00	318	729	---
91	C8H18O3	DIETHYLENE GLYCOL DIETHYL ETHER	-1.4158	0.7047	624.00	229	593	0.1478
92	C8H18O3	DIETHYLENE GLYCOL MONOBUTYL EHTER	-1.3443	0.6670	654.00	205	621	0.1645
93	C8H18O4	TRIETHYLENE GLYCOL DIMETHYL ETHER	-2.2323	1.7319	651.00	229	618	0.1666
94	C8H18O5	TETRAETHYLENE GLYCOL	-1.6415	0.9871	722.00	268	686	0.1608
95	C8H18S	n-OCTYL MERCAPTAN	-1.7402	0.9950	664.00	224	631	0.1256
96	C8H18S	tert-OCTYL MERCAPTAN	-1.7613	0.9948	627.00	199	596	0.1164
97	C8H19N	DI-n-BUTYLAMINE	-1.6042	0.8870	607.50	211	577	0.1341
98	C8H19N	DIISOBUTYLAMINE	-1.7621	0.9963	580.00	203	551	0.1118
99	C8H19N	n-OCTYLAMINE	-1.0950	0.3289	627.00	273	596	0.1508
100	C8H23N5	TETRAETHYLENEPENTAMINE	-1.6786	1.0022	774.00	243	735	0.1562
101	C8H24O4Si4	OCTAMETHYLCYCLOTETRASILOXANE	-1.8019	1.0388	586.50	291	557	0.1112
102	C9H4O5	TRIMELLITIC ANHYDRIDE	-1.5385	1.0319	890.00	438	846	---
103	C9H6N2O2	TOLUENE DIISOCYANATE	-1.7078	0.9876	737.00	287	700	0.1393
104	C9H7N	ISOQUINOLINE	-1.7402	0.9953	803.15	299	763	---
105	C9H7N	QUINOLINE	-1.8878	1.2178	782.15	258	743	0.1493
106	C9H7NO	8-HYDROXYQUINOLINE	-1.7414	1.0149	788.00	346	749	---
107	C9H8	INDENE	-1.6863	1.0061	687.00	272	653	0.1475
108	C9H8O	2-METHYLBENZOFURAN	-1.7593	1.0104	698.00	290	663	0.1266
109	C9H10	INDANE	-1.7070	0.9855	684.90	222	651	0.1349
110	C9H10	alpha-METHYLSTYRENE	-1.7068	0.9915	654.00	250	621	0.1338
111	C9H10	m-METHYLSTYRENE	-1.6988	0.9833	657.00	187	624	0.1344
112	C9H10	o-METHYLSTYRENE	-1.7026	0.9871	659.00	205	626	0.1344
113	C9H10	p-METHYLSTYRENE	-1.7070	0.9953	665.00	239	632	0.1358
114	C9H10O2	BENZYL ACETATE	-1.6849	0.9991	699.00	222	664	0.1470
115	C9H10O2	ETHYL BENZOATE	-1.6750	0.9835	698.00	238	663	0.1458
116	C9H10O3	ETHYL VANILLIN	-1.6942	1.0286	748.00	351	711	---
117	C9H11NO	p-DIMETHYLAMINOBENZALDEHYDE	-1.7582	1.0148	832.00	348	790	---
118	C9H12	CUMENE	-1.1835	0.3543	631.15	177	600	0.1293
119	C9H12	m-ETHYLTOLUENE	-1.7189	0.9845	637.15	178	605	0.1268
120	C9H12	o-ETHYLTOLUENE	-1.7176	0.9874	651.15	192	619	0.1292
121	C9H12	p-ETHYLTOLUENE	-1.7291	0.9935	640.15	211	608	0.1263
122	C9H12	MESITYLENE	-1.2339	0.4698	637.36	228	605	0.1440
123	C9H12	n-PROPYLBENZENE	-1.7052	0.9688	638.38	174	606	0.1271
124	C9H12	1,2,3-TRIMETHYLBENZENE	-1.6931	0.9586	664.53	248	631	0.1305
125	C9H12	1,2,4-TRIMETHYLBENZENE	-1.7273	1.0028	649.13	229	617	0.1300

* A computer program, containing coefficients for all compounds, is available for a nominal fee. The computer program is in ASCII which can be accessed by other software.

$\log_{10} \lambda_{liq} = A + B\,[1-T/C]^{2/7}$ (λ_{liq} - W/m K, T - K)

NO	FORMULA	NAME	A	B	C	TMIN	TMAX	λ_{liq} @ 25 C
126	C9H12O	BENZYL ETHYL ETHER	-1.6898	0.9805	662.00	276	629	0.1370
127	C9H12O	2-PHENYL-2-PROPANOL	-1.6545	0.9738	660.00	309	627	---
128	C9H12O2	CUMENE HYDROPEROXIDE	-1.8736	1.0150	605.00	264	575	0.0917
129	C9H14O	ISOPHORONE	-1.6075	0.9905	715.00	265	679	0.1744
130	C9H14O6	GLYCERYL TRIACETATE	-1.7907	0.9843	704.00	277	669	0.1123
131	C9H16O4	AZELAIC ACID	-1.6152	0.9721	811.00	380	770	---
132	C9H18	ISOPROPYLCYCLOHEXANE	-1.7546	0.9867	627.00	184	596	0.1164
133	C9H18	1-NONENE	-1.8036	1.0743	593.25	192	564	0.1192
134	C9H18	n-PROPYLCYCLOHEXANE	-1.7622	0.9859	639.15	178	607	0.1152
135	C9H18O	DIISOBUTYL KETONE	-1.6430	0.9749	615.00	227	584	0.1458
136	C9H18O	1-NONANAL	-1.3939	0.6593	640.00	255	608	0.1436
137	C9H18O2	n-BUTYL VALERATE	-1.7291	0.9996	629.00	180	598	0.1267
138	C9H18O2	n-NONANOIC ACID	-1.4696	0.7282	703.00	286	668	0.1420
139	C9H18O2	n-OCTYL FORMATE	-1.3800	0.6256	645.00	234	613	0.1393
140	C9H20	3,3-DIETHYLPENTANE	-1.7892	1.0016	610.05	240	580	0.1091
141	C9H20	2,2-DIMETHYL-3-ETHYLPENTANE	-1.7833	0.9898	590.00	174	561	0.1062
142	C9H20	2,4-DIMETHYL-3-ETHYLPENTANE	-1.7726	0.9791	591.00	151	561	0.1068
143	C9H20	2,2-DIMETHYLHEPTANE	-1.7894	0.9861	576.80	160	548	0.1027
144	C9H20	2,6-DIMETHYLHEPTANE	-1.7923	0.9880	579.00	170	550	0.1026
145	C9H20	3-ETHYLHEPTANE	-1.7835	0.9841	590.00	158	561	0.1050
146	C9H20	2-METHYLOCTANE	-1.6942	0.9857	586.75	193	557	0.1290
147	C9H20	3-METHYLOCTANE	-1.7918	0.9901	590.15	166	561	0.1042
148	C9H20	4-METHYLOCTANE	-1.7871	0.9856	587.65	160	558	0.1042
149	C9H20	n-NONANE	-1.7865	1.1033	595.65	220	566	0.1313
150	C9H20	2,2,3,3-TETRAMETHYLPENTANE	-1.7988	1.0142	610.85	263	580	0.1093
151	C9H20	2,2,3,4-TETRAMETHYLPENTANE	-1.7758	0.9829	592.15	152	563	0.1069
152	C9H20	2,2,4,4-TETRAMETHYLPENTANE	-1.7906	0.9966	571.35	207	543	0.1039
153	C9H20	2,2,5-TRIMETHYLHEXANE	-2.6081	2.0588	568.05	167	540	0.1139
154	C9H20O	2,6-DIMETHYL-4-HEPTANOL	-1.6993	0.9921	603.00	208	573	0.1309
155	C9H20O	1-NONANOL	-1.4848	0.8197	673.00	268	639	0.1617
156	C9H20O	2-NONANOL	-1.6670	0.9763	623.00	238	592	0.1392
157	C9H20S	n-NONYL MERCAPTAN	-1.7464	1.0012	681.00	253	647	0.1267
158	C9H21N	n-NONYLAMINE	-1.0924	0.3278	648.00	273	616	0.1522
159	C9H21N	TRIPROPYLAMINE	-1.5650	0.7943	577.50	180	549	0.1204
160	C10H6O8	PYROMELLITIC ACID	-1.8637	1.0653	893.00	554	848	---
161	C10H7Br	1-BROMONAPHTHALENE	-1.8040	0.9932	824.00	279	783	0.1174
162	C10H7Cl	1-CHLORONAPHTHALENE	-1.7189	0.9531	785.00	269	746	0.1296
163	C10H8	NAPHTHALENE	-1.0304	0.1860	748.35	353	711	---
164	C10H9N	QUINALDINE	-1.7357	0.9944	773.00	272	734	0.1347
165	C10H10	m-DIVINYLBENZENE	-1.7132	0.9876	692.00	206	657	0.1341
166	C10H10	1-METHYLINDENE	-1.7259	1.0236	703.00	350	668	---
167	C10H10	2-METHYLINDENE	-1.7191	1.0236	684.00	353	650	---
168	C10H10O4	DIMETHYL PHTHALATE	-1.6451	0.9331	766.00	272	728	0.1463
169	C10H10O4	DIMETHYL TEREPHTHALATE	-1.7191	1.0367	772.00	414	733	---
170	C10H12	DICYCLOPENTADIENE	-1.6755	0.9683	660.00	307	627	---
171	C10H12	1,2,3,4-TETRAHYDRONAPHTHALENE	-1.1461	0.3017	720.15	237	684	0.1297
172	C10H12O	ANETHOLE	---	---	---	---	---	---
173	C10H12O4	DIALLYL MALEATE	-1.7974	1.0330	693.00	226	658	0.1208
174	C10H14	n-BUTYLBENZENE	-1.8105	1.0843	660.55	185	628	0.1267
175	C10H14	sec-BUTYLBENZENE	-1.7383	0.9837	664.54	198	631	0.1235
176	C10H14	tert-BUTYLBENZENE	-1.5954	0.7911	660.00	215	627	0.1177
177	C10H14	m-CYMENE	-1.7415	0.9895	657.00	209	624	0.1233
178	C10H14	o-CYMENE	-1.7317	0.9853	662.00	202	629	0.1255
179	C10H14	p-CYMENE	-1.7448	0.9921	653.15	205	620	0.1227
180	C10H14	m-DIETHYLBENZENE	-1.7413	0.9888	663.00	189	630	0.1237
181	C10H14	o-DIETHYLBENZENE	-1.7429	0.9984	668.00	242	635	0.1260
182	C10H14	p-DIETHYLBENZENE	-1.7474	0.9969	657.96	230	625	0.1235
183	C10H14	2-ETHYL-m-XYLENE	-1.7417	1.0025	671.00	257	637	0.1276
184	C10H14	2-ETHYL-p-XYLENE	-1.7425	0.9982	663.00	220	630	0.1256
185	C10H14	3-ETHYL-o-XYLENE	-1.7371	0.9951	680.00	224	646	0.1279
186	C10H14	4-ETHYL-m-XYLENE	-1.7409	0.9952	665.00	210	632	0.1255
187	C10H14	4-ETHYL-o-XYLENE	-1.7381	0.9903	667.00	206	634	0.1253
188	C10H14	5-ETHYL-m-XYLENE	-1.7354	0.9854	655.00	189	622	0.1239
189	C10H14	ISOBUTYLBENZENE	-1.7490	0.9961	650.15	222	618	0.1222
190	C10H14	1,2,3,5-TETRAMETHYLBENZENE	-1.7441	1.0028	679.00	249	645	0.1276
191	C10H14	1,2,4,5-TETRAMETHYLBENZENE	-1.3032	0.4911	675.15	352	641	---
192	C10H14O	p-tert-BUTYLPHENOL	-1.6377	0.9639	734.00	372	697	---
193	C10H14O2	p-tert-BUTYLCATECHOL	-1.7032	1.0160	776.00	325	737	---

* A computer program, containing coefficients for all compounds, is available for a nominal fee. The computer program is in ASCII which can be accessed by other software.

$$\log_{10} \lambda_{liq} = A + B\,[1-T/C]^{2/7} \qquad (\lambda_{liq} \text{ - W/m K, T - K})$$

NO	FORMULA	NAME	A	B	C	TMIN	TMAX	λ_{liq} @ 25 C
194	C10H15N	N,N-DIETYHLANILINE	-1.6971	0.9932	702.00	235	667	0.1416
195	C10H15N	2,6-DIETHYLANILINE	-1.7114	1.0163	678.00	277	644	0.1412
196	C10H16	CAMPHENE	-1.7932	1.0634	638.00	320	606	---
197	C10H16	D-LIMONENE	-1.7644	1.0003	660.00	199	627	0.1197
198	C10H16	alpha-PHELLANDRENE	-1.7685	1.0155	649.00	220	617	0.1212
199	C10H16	beta-PHELLANDRENE	-1.7421	0.9796	648.00	220	616	0.1200
200	C10H16	alpha-PINENE	-1.7430	0.9964	632.00	209	600	0.1223
201	C10H16	beta-PINENE	-1.7360	0.9914	643.00	212	611	0.1241
202	C10H16	alpha-TERPINENE	-1.7476	0.9807	652.00	220	619	0.1191
203	C10H16	gamma-TERPINENE	-1.7441	0.9824	661.00	220	628	0.1212
204	C10H16	TERPINOLENE	-1.7885	1.0404	672.00	220	638	0.1234
205	C10H16O	CAMPHOR	-1.5928	0.9230	709.00	453	674	---
206	C10H18	cis-DECAHYDRONANPHTALENE	-1.3879	0.5151	702.25	230	667	0.1127
207	C10H18	trans-DECAHYDRONAPHTHALENE	-1.3710	0.4977	687.05	243	653	0.1127
208	C10H18O4	SEBACIC ACID	-1.6287	0.9779	815.00	408	774	---
209	C10H20	n-BUTYLCYCLOHEXANE	-1.7790	0.9871	667.00	198	634	0.1134
210	C10H20	1-DECENE	-1.7491	1.0443	617.05	207	586	0.1305
211	C10H20O	1-DECANAL	-1.5334	0.8250	657.00	267	624	0.1448
212	C10H20O2	n-DECANOIC ACID	-1.4112	0.6591	713.00	305	677	---
213	C10H20O2	2-ETHYLHEXYL ACETATE	-1.7576	1.0042	639.00	180	607	0.1206
214	C10H20O2	ISOPENTYL ISOVALERATE	-1.7570	0.9960	637.00	215	605	0.1187
215	C10H22	n-DECANE	-1.7768	1.0839	618.45	243	588	0.1322
216	C10H22	2,2-DIMETHYLOCTANE	-1.8164	1.0045	602.00	225	572	0.1023
217	C10H22	2-METHYLNONANE	-1.6900	0.9850	610.00	199	580	0.1328
218	C10H22	3-METHYLNONANE	-1.6908	0.9873	613.00	188	582	0.1335
219	C10H22	4-METHYLNONANE	-1.6992	0.9962	610.00	174	580	0.1328
220	C10H22	5-METHYLNONANE	-1.6942	0.9883	610.00	185	580	0.1323
221	C10H22O	1-DECANOL	-1.5529	0.8957	690.00	280	656	0.1618
222	C10H22O	DI-n-PENTYL ETHER	-1.7493	0.9842	622.00	204	591	0.1168
223	C10H22O	ISODECANOL	-1.6789	0.9937	644.00	213	612	0.1423
224	C10H22O5	TETRAETHYLENE GLYCOL DIMETHYL ETHER	-1.7110	1.0058	705.00	243	670	0.1408
225	C10H22S	n-DECYL MERCAPTAN	-1.7376	1.0045	696.00	248	661	0.1314
226	C10H23N	n-DECYLAMINE	-1.1406	0.3816	663.00	289	630	0.1517
227	C11H10	1-METHYLNAPHTHALENE	-1.7031	0.9920	772.04	243	733	0.1445
228	C11H10	2-METHYLNAPHTHALENE	-1.7240	1.0073	761.00	308	723	---
229	C11H14O2	n-BUTYL BENZOATE	-1.7114	0.9826	724.00	252	688	0.1358
230	C11H16	n-PENTYLBENZENE	-1.7786	1.0126	679.90	198	646	0.1202
231	C11H16O	p-tert-AMYLPHENOL	-1.6402	0.9691	751.00	366	713	---
232	C11H20O2	2-ETHYLHEXYL ACRYLATE	-1.7687	1.0030	655.00	183	622	0.1187
233	C11H22	1-UNDECENE	-1.7475	1.0119	638.00	224	606	0.1252
234	C11H22O	1-UNDECANAL	-1.7583	1.0167	672.00	273	638	0.1263
235	C11H24	n-UNDECANE	-1.6318	0.9325	638.76	248	607	0.1404
236	C11H24O	1-UNDECANOL	-1.6798	0.9957	704.00	289	669	0.1482
237	C11H24S	UNDECYL MERCAPTAN	-1.7714	1.0137	710.00	270	675	0.1248
238	C12H8O	DIBENZOFURAN	-1.7205	0.9746	837.80	356	796	---
239	C12H9N	DIBENZOPYRROLE	-1.6626	1.0379	899.00	518	854	---
240	C12H10	ACENAPHTHENE	-1.2648	0.4524	803.15	367	763	---
241	C12H10	BIPHENYL	-1.4292	0.6647	789.26	342	750	---
242	C12H10O	DIPHENYL ETHER	-1.5285	0.7853	763.00	300	725	---
243	C12H11N	p-AMINODIPHENYL	-1.7466	1.0067	817.00	326	776	---
244	C12H11N	DIPHENYLAMINE	-1.2631	0.4669	817.00	326	776	---
245	C12H11N3	p-AMINOAZOBENZENE	-1.7397	1.0227	877.00	401	833	---
246	C12H11N3	1,3-DIPHENYLTRIAZENE	-1.7757	1.0175	845.00	372	803	---
247	C12H12	2,6-DIMETHYLNAPHTHALENE	-1.7473	1.0248	777.00	385	738	---
248	C12H12	2,7-DIMETHYLNAPHTHALENE	-1.7439	1.0203	778.00	370	739	---
249	C12H12	1-ETHYLNAPHTHALENE	-1.7225	0.9971	776.00	259	737	0.1398
250	C12H12N2	p-AMINODIPHENYLAMINE	-1.7000	1.0101	867.00	341	824	---
251	C12H12N2	HYDRAZOBENZENE	---	---	---	---	---	---
252	C12H14	1,2,3-TRIMETHYLINDENE	-1.7299	1.0200	726.00	345	690	---
253	C12H14O4	DIETHYL PHTHALATE	-1.3265	0.5567	757.00	269	719	0.1432
254	C12H16	CYCLOHEXYLBENZENE	-1.7275	0.9671	744.00	280	707	0.1282
255	C12H18	m-DIISOPROPYLBENZENE	-1.7699	0.9902	684.00	210	650	0.1177
256	C12H18	p-DIISOPROPYLBENZENE	-1.7814	1.0018	689.00	256	655	0.1177
257	C12H18	n-HEXYLBENZENE	-1.7737	0.9910	698.00	212	663	0.1179
258	C12H20O4	DIBUTYL MALEATE	-1.8319	1.0337	716.00	188	680	0.1133
259	C12H22	BICYCLOHEXYL	-1.7808	1.0044	727.00	277	691	0.1210
260	C12H23N	DICYCLOHEXYLAMINE	-1.8152	1.0039	737.00	273	700	0.1123
261	C12H24	1-DODECENE	-1.6991	0.9948	657.00	238	624	0.1373

* A computer program, containing coefficients for all compounds, is available for a nominal fee. The computer program is in ASCII which can be accessed by other software.

$\log_{10} \lambda_{liq} = A + B\,[1-T/C]^{2/7}$ (λ_{liq} - W/m K, T - K)

NO	FORMULA	NAME	A	B	C	TMIN	TMAX	λ_{liq} @ 25 C
262	C12H24O	1-DODECANAL	-1.7316	1.0818	685.00	285	651	0.1539
263	C12H24O2	n-DODECANOIC ACID	-1.4322	0.7077	734.00	317	697	---
264	C12H26	n-DODECANE	-1.7989	1.1109	658.20	264	625	0.1368
265	C12H26O	DI-n-HEXYL ETHER	-1.5209	0.7658	658.00	230	625	0.1329
266	C12H26O	1-DODECANOL	-1.3911	0.6474	721.00	297	685	0.1461
267	C12H26O3	DIETHYLENE GLYCOL DI-n-BUTYL ETHER	-1.4951	0.7795	680.00	213	646	0.1465
268	C12H26S	n-DODECYL MERCAPTAN	-1.7831	1.0100	724.00	265	688	0.1216
269	C12H27BO3	TRI-n-BUTYL BORATE	---	---	---	---	---	---
270	C12H27N	DODECYLAMINE	-1.1301	0.3850	696.00	301	661	---
271	C12H27N	TRI-n-BUTYLAMINE	-1.4764	0.6715	644.00	203	612	0.1219
272	C13H10	FLUORENE	-1.6208	1.0116	870.00	388	827	---
273	C13H10O	BENZOPHENONE	-1.5854	0.9864	816.00	321	775	---
274	C13H12	DIPHENYLMETHANE	-1.4751	0.7029	768.00	298	730	---
275	C13H20	n-HEPTYLBENZENE	-1.7909	0.9965	714.00	225	678	0.1156
276	C13H26	1-TRIDECENE	-1.7092	1.0068	675.00	250	641	0.1390
277	C13H26O	1-TRIDECANAL	-1.7559	1.0201	700.00	288	665	0.1302
278	C13H26O2	n-BUTYL NONANOATE	-1.8021	0.9790	652.00	235	619	0.1047
279	C13H26O2	METHYL DODECANOATE	-1.7783	0.9870	712.00	278	676	0.1167
280	C13H28	n-TRIDECANE	-1.6564	0.9372	675.80	268	642	0.1372
281	C13H28O	1-TRIDECANOL	-1.8051	1.1252	731.00	304	694	---
282	C14H8O2	ANTHRAQUINONE	-1.5523	0.9688	900.00	559	855	---
283	C14H10	ANTHRACENE	-1.2472	0.4397	873.00	489	829	---
284	C14H10	DIPHENYLACETYLENE	-1.7656	1.0030	832.00	336	790	---
285	C14H10	PHENANTHRENE	-1.0147	0.1435	869.25	372	826	---
286	C14H12	cis-STILBENE	-1.7315	0.9996	757.00	268	719	0.1364
287	C14H12	trans-STILBENE	-1.7801	1.0841	820.00	397	779	---
288	C14H12O2	BENZYL BENZOATE	-1.7335	0.9898	820.00	293	779	0.1369
289	C14H14	1,1-DIPHENYLETHANE	-1.7429	0.9996	775.00	255	736	0.1340
290	C14H14	1,2-DIPHENYLETHANE	-1.7400	1.0102	780.00	324	741	---
291	C14H14O	DIBENZYL ETHER	-1.7666	0.9923	777.00	277	738	0.1252
292	C14H16	1-n-BUTYLNAPHTHALENE	-1.7347	0.9739	792.00	253	752	0.1307
293	C14H22	n-OCTYLBENZENE	-1.8063	1.0017	729.00	237	693	0.1137
294	C14H22O	p-tert-OCTYLPHENOL	-1.6666	0.9739	765.00	359	727	---
295	C14H28	1-TETRADECENE	-1.6917	0.9840	692.00	260	657	0.1400
296	C14H28O2	n-TETRADECANOIC ACID	-1.3018	0.6091	756.00	328	718	---
297	C14H30	n-TETRADECANE	-1.6813	0.9584	692.40	279	658	0.1363
298	C14H30O	1-TETRADECANOL	-1.4537	0.7984	741.00	311	704	---
299	C14H31N	TETRADECYLAMINE	-1.7823	1.0245	722.30	311	686	---
300	C15H10N2O2	DIPHENYLMETHANE-4,4'-DIISOCYANATE	-1.8483	1.0108	802.00	311	762	---
301	C15H16O	p-CUMYLPHENOL	-1.7658	1.0169	834.00	346	792	---
302	C15H16O2	BISPHENOL A	-1.6286	0.9637	849.00	426	807	---
303	C15H24	n-NONYLBENZENE	-1.8155	1.0011	741.00	249	704	0.1119
304	C15H24O	2,6-DI-tert-BUTYL-p-CRESOL	-1.6963	0.9674	720.00	344	684	---
305	C15H24O	NONYLPHENOL	-1.6850	1.0015	757.00	279	719	0.1524
306	C15H30	1-PENTADECENE	-1.7166	1.0132	708.00	269	673	0.1413
307	C15H30O2	PENTADECANOIC ACID	-1.6873	0.9778	766.00	326	728	---
308	C15H32	n-PENTADECANE	-1.7650	1.0668	706.80	283	671	0.1404
309	C16H10	FLUORANTHENE	-1.7217	1.0163	905.00	383	860	---
310	C16H10	PYRENE	-1.1478	0.2832	936.00	424	889	---
311	C16H12	1-PHENYLNAPHTHALENE	-1.7308	1.0036	849.00	318	807	---
312	C16H20	1-n-HEXYLNAPHTHALENE	-1.7818	0.9943	813.00	255	772	0.1233
313	C16H22O4	DIBUTYL PHTHALATE	-1.6470	0.8980	781.00	238	742	0.1367
314	C16H26	n-DECYLBENZENE	-1.8290	1.0063	753.00	259	715	0.1102
315	C16H32	n-DECYLCYCLOHEXANE	-1.8492	1.0049	751.25	271	714	0.1048
316	C16H32	1-HEXADECENE	-1.7204	1.0172	722.00	278	686	0.1423
317	C16H32O2	n-HEXADECANOIC ACID	-1.6139	0.9634	776.00	336	737	---
318	C16H34	n-HEXADECANE	-1.8486	1.1616	720.60	291	685	0.1408
319	C16H34O	DI-n-OCTYL ETHER	-1.8375	0.9775	707.00	266	672	0.0996
320	C16H34O	1-HEXADECANOL	-1.7942	1.1033	761.00	322	723	---
321	C17H28	n-UNDECYLBENZENE	-1.8347	1.0032	764.00	268	726	0.1087
322	C17H34	1-HEPTADECENE	-1.7147	1.0094	736.00	284	699	0.1430
323	C17H36	n-HEPTADECANE	-1.5446	0.8191	733.37	295	697	0.1449
324	C17H36O	1-HEPTADECANOL	-1.7864	1.0920	770.00	327	732	---
325	C18H12	CHRYSENE	-1.7230	1.0425	979.00	531	930	---
326	C18H14	m-TERPHENYL	-1.2834	0.4800	924.85	360	879	---
327	C18H14	o-TERPHENYL	-1.4433	0.6429	890.95	329	846	---
328	C18H14	p-TERPHENYL	-1.5067	0.7626	925.95	485	880	---
329	C18H15P	TRIPHENYLPHOSPHINE	-1.7693	0.9930	1008.00	354	958	---

* A computer program, containing coefficients for all compounds, is available for a nominal fee. The computer program is in ASCII which can be accessed by other software.

$\log_{10} \lambda_{liq} = A + B\ [1-T/C]^{2/7}$ (λ_{liq} - W/m K, T - K)

NO	FORMULA	NAME	A	B	C	TMIN	TMAX	λ_{liq} @ 25 C
330	C18H15O4P	TRIPHENYL PHOSPHATE	---	---	---	---	---	---
331	C18H16N2	N,N'-DIPHENYL-p-PHENYLENEDIAMINE	-1.7812	1.0231	906.00	409	861	---
332	C18H22	2,3-DIMETHYL-2,3-DIPHENYLBUTANE	-1.7541	1.0268	805.00	392	765	---
333	C18H22O2	DICUMYL PEROXIDE	-1.7138	1.0042	884.00	311	840	---
334	C18H30	n-DODECYLBENZENE	-1.8479	1.0078	774.26	276	736	0.1070
335	C18H32O2	LINOLEIC ACID	-1.7073	0.9823	775.00	268	736	0.1405
336	C18H34O2	OLEIC ACID	-2.9905	2.6266	781.00	287	742	0.1990
337	C18H34O4	DIBUTYL SEBACATE	-1.8794	0.9921	768.00	264	730	0.0961
338	C18H34O4	DIHEXYL ADIPATE	-1.8860	0.9986	767.00	259	729	0.0958
339	C18H36	1-OCTADECENE	-1.7179	1.0127	748.00	291	711	0.1439
340	C18H36O2	STEARIC ACID	-1.8215	1.1522	799.00	343	759	---
341	C18H38	n-OCTADECANE	-1.5198	0.8067	745.26	301	708	---
342	C18H38O	DINONYL ETHER	-1.8634	0.9832	736.00	273	699	0.0964
343	C18H38O	1-OCTADECANOL	-1.6609	1.0778	777.00	331	738	---
344	C19H26	1-n-NONYLNAPHTHALENE	-1.8182	1.0024	849.00	284	807	0.1169
345	C19H32	n-TRIDECYLBENZENE	-1.8571	1.0126	783.00	283	744	0.1062
346	C19H36O2	METHYL OLEATE	-1.8537	0.9858	764.00	293	726	0.1005
347	C19H38	1-NONADECENE	-1.7206	1.0155	760.00	297	722	0.1446
348	C19H38O2	NONADECANOIC ACID	-1.6897	0.9683	810.00	341	770	---
349	C19H40	n-NONADECANE	-1.5053	0.7902	755.93	305	718	---
350	C20H16	TRIPHENYLETHYLENE	-1.7755	1.0075	908.00	342	863	---
351	C20H28	1-n-DECYLNAPHTHALENE	-1.8282	1.0039	859.00	288	816	0.1150
352	C20H30O2	ABIETIC ACID	-1.6935	0.9517	832.00	447	790	---
353	C20H31N	DEHYDROABIETYLAMINE	-2.6513	1.0122	863.00	318	820	---
354	C20H40	1-EICOSENE	-1.7233	1.0181	771.00	302	732	---
355	C20H42	n-EICOSANE	-1.8447	1.1788	767.04	310	729	---
356	C20H42O	1-EICOSANOL	-1.7565	1.0513	792.00	339	752	---
357	C21H21O4P	TRI-o-CRESYL PHOSPHATE	-1.3800	0.5756	806.00	240	680	0.1332
358	C22H44O2	n-BUTYL STEARATE	-1.4902	0.7884	764.00	299	726	---
359	C24H38O4	DIISOOCTYL PHTHALATE	-1.7906	0.9976	851.00	223	808	0.1234
360	C24H38O4	DIOCTYL PHTHALATE	-1.7804	1.0238	806.00	223	766	0.1309
361	C24H42O	DINONYLPHENOL	-1.6584	0.9739	886.00	350	842	---
362	C26H20	TETRAPHENYLETHYLENE	-1.7345	1.0370	996.00	496	946	---
363	C28H46O4	DIISODECYL PHTHALATE	-1.9715	1.0041	887.00	228	843	0.0835

* A computer program, containing coefficients for all compounds, is available for a nominal fee. The computer program is in ASCII which can be accessed by other software.

Appendix B

COEFFICIENTS FOR GAS THERMAL CONDUCTIVITY EQUATION*

Carl L. Yaws, Xiaoyan Lin, Li Bu and Sachin Nijhawan
Lamar University, Beaumont, Texas

$$\lambda_{gas} = A + B\,T + C\,T^2 \qquad (\lambda_{gas} - \mathrm{W/m\,K},\ T - \mathrm{K})$$

NO	FORMULA	NAME	A	B	C	TMIN	TMAX	λ_{gas} @ 25 C
1	C8H4Cl2O2	ISOPHTHALOYL CHLORIDE	-0.00542	3.4898E-05	3.9077E-09	549	1000	---
2	C8H4O3	PHTHALIC ANHYDRIDE	-0.00753	3.8226E-05	1.5493E-08	558	1000	---
3	C8H6O4	ISOPHTHALIC ACID	-0.00858	4.4907E-05	6.3743E-09	753	1000	---
4	C8H6O4	PHTHALIC ACID	-0.00739	4.0704E-05	8.9383E-09	598	1000	---
5	C8H6O4	TEREPHTHALIC ACID	-0.00924	4.5635E-05	6.5815E-09	832	1000	---
6	C8H6S	BENZOTHIOPHENE	-0.00886	5.6047E-05	1.0132E-08	493	1000	---
7	C8H7N	INDOLE	-0.01544	6.7583E-05	1.1291E-08	526	1000	---
8	C8H8	STYRENE	-0.00712	4.5538E-05	3.9529E-08	273	973	0.00997
9	C8H8O	ACETOPHENONE	-0.00815	4.3566E-05	2.0894E-08	475	1000	---
10	C8H8O	p-TOLUALDEHYDE	-0.00947	5.3038E-05	2.0083E-08	477	1000	---
11	C8H8O2	METHYL BENZOATE	-0.00804	4.5774E-05	1.6365E-08	473	1000	---
12	C8H8O2	o-TOLUIC ACID	-0.00835	4.5361E-05	1.4871E-08	532	1000	---
13	C8H8O2	p-TOLUIC ACID	-0.00855	4.5799E-05	1.4900E-08	548	1000	---
14	C8H8O3	METHYL SALICYLATE	-0.01382	6.2711E-05	3.3082E-10	494	1000	---
15	C8H8O3	VANILLIN	-0.00926	5.4095E-05	1.0583E-08	558	1000	---
16	C8H9NO	ACETANILIDE	-0.00954	5.0738E-05	1.4864E-08	577	1000	---
17	C8H10	ETHYLBENZENE	-0.00797	4.0572E-05	6.7289E-08	400	825	---
18	C8H10	m-XYLENE	-0.00375	2.9995E-05	7.4603E-08	400	825	---
19	C8H10	o-XYLENE	-0.00979	7.4087E-05	1.8418E-08	400	825	---
20	C8H10	p-XYLENE	-0.00870	4.7349E-05	5.8829E-08	400	825	---
21	C8H10O	m-ETHYLPHENOL	---	---	---	--	--	---
22	C8H10O	p-ETHYLPHENOL	-0.01729	7.9086E-05	-3.9212E-11	491	1000	---
23	C8H10O	PHENETOLE	-0.00831	5.0314E-05	1.5670E-08	443	1000	---
24	C8H10O	2-PHENYLETHANOL	-0.01601	7.3080E-05	4.7158E-09	492	1000	---
25	C8H10O	2,3-XYLENOL	-0.01621	8.0847E-05	4.6664E-09	490	1000	---
26	C8H10O	2,4-XYLENOL	-0.01061	6.3622E-05	1.6317E-08	484	1000	---
27	C8H10O	2,5-XYLENOL	-0.01062	6.3888E-05	1.6071E-08	484	1000	---
28	C8H10O	2,6-XYLENOL	-0.01043	6.3338E-05	1.6044E-08	474	1000	---
29	C8H10O	3,4-XYLENOL	-0.01142	6.9766E-05	1.4701E-08	500	1000	---
30	C8H10O	3,5-XYLENOL	-0.01074	6.3474E-05	1.6542E-08	495	1000	---
31	C8H11N	N,N-DIMETHYLANILINE	-0.02003	8.2062E-05	3.8364E-09	467	1000	---
32	C8H11N	o-ETHYLANILINE	-0.01014	6.1355E-05	1.5104E-08	483	1000	---
33	C8H11N	2,4,6-TRIMETHYLPYRIDINE	-0.01734	7.9239E-05	5.3151E-09	444	1000	---
34	C8H11NO	p-PHENETIDINE	-0.01005	5.8755E-05	1.3434E-08	528	1000	---
35	C8H12	1,5-CYCLOOCTADIENE	-0.02554	1.0401E-04	1.2248E-08	423	1000	---
36	C8H12	VINYLCYCLOHEXENE	-0.02497	1.0672E-04	6.0506E-09	401	1000	---
37	C8H12O4	1,4-CYCLOHEXANEDICARBOXYLIC ACID	-0.01112	5.8457E-05	1.1846E-08	669	1000	---
38	C8H12O4	DIETHYL MALEATE	0.00165	-7.1000E-06	8.6929E-08	498	1000	---
39	C8H14O2	n-BUTYL METHACRYLATE	-0.01848	8.7139E-05	6.5835E-09	434	1000	---
40	C8H14O3	BUTYRIC ANHYDRIDE	-0.00710	4.5385E-05	9.0705E-09	468	998	---
41	C8H14O4	DIETHYL SUCCINATE	-0.00826	5.1970E-05	9.8994E-09	490	1000	---
42	C8H16	1,1-DIMETHYLCYCLOHEXANE	-0.02802	1.1255E-04	1.1180E-08	393	1000	---
43	C8H16	cis-1,2-DIMETHYLCYCLOHEXANE	-0.02680	1.0843E-04	1.0955E-08	403	1000	---
44	C8H16	trans-1,2-DIMETHYLCYCLOHEXANE	-0.02699	1.1109E-04	1.0310E-08	397	1000	---
45	C8H16	cis-1,3-DIMETHYLCYCLOHEXANE	-0.02736	1.1131E-04	1.1533E-08	393	1000	---
46	C8H16	trans-1,3-DIMETHYLCYCLOHEXANE	-0.02652	1.0839E-04	1.1691E-08	398	1000	---
47	C8H16	cis-1,4-DIMETHYLCYCLOHEXANE	-0.02652	1.0841E-04	1.1697E-08	397	1000	---
48	C8H16	trans-1,4-DIMETHYLCYCLOHEXANE	-0.02750	1.1269E-04	1.0360E-08	393	1000	---
49	C8H16	ETHYLCYCLOHEXANE	-0.01194	6.6023E-05	4.1195E-08	405	995	---
50	C8H16	2-ETHYL-1-HEXENE	-0.02838	1.2204E-04	1.8230E-09	393	1000	---
51	C8H16	1-METHYL-1-ETHYLCYCLOPENTANE	-0.02598	1.1067E-04	1.0210E-08	395	1000	---
52	C8H16	1-OCTENE	-0.00463	2.8829E-05	8.4175E-08	394	500	---
53	C8H16	trans-2-OCTENE	-0.02481	1.0586E-04	2.9507E-09	398	1000	---
54	C8H16	trans-3-OCTENE	-0.02522	1.0720E-04	2.5380E-09	396	1000	---
55	C8H16	trans-4-OCTENE	-0.02519	1.0724E-04	2.6386E-09	395	1000	---
56	C8H16	n-PROPYLCYCLOPENTANE	-0.01185	6.7767E-05	3.5487E-08	404	1000	---
57	C8H16	2,4,4-TRIMETHYL-1-PENTENE	-0.02665	1.1719E-04	-5.7671E-12	375	1000	---

* A computer program, containing coefficients for all compounds, is available for a nominal fee. The computer program is in ASCII which can be accessed by other software.

NO	FORMULA	NAME	A	B	C	TMIN	TMAX	λ_{gas} @ 25 C
			$\lambda_{gas} = A + B\,T + C\,T^2$			(λ_{gas} - W/m K, T - K)		
58	C8H16	2,4,4-TRIMETHYL-2-PENTENE	-0.02613	1.1450E-04	9.0816E-10	378	1000	---
59	C8H16O	2-ETHYLHEXANAL	0.00209	-9.5110E-06	1.2283E-07	434	1000	---
60	C8H16O	1-OCTANAL	0.00204	-8.8077E-06	1.1152E-07	447	1000	---
61	C8H16O	2-OCTANONE	0.00205	-9.0803E-06	1.1603E-07	446	1000	---
62	C8H16O2	n-BUTYL n-BUTYRATE	-0.00805	5.3875E-05	1.0085E-08	438	998	---
63	C8H16O2	n-HEXYL ACETATE	0.00190	-8.2258E-06	1.0590E-07	445	1000	---
64	C8H16O2	ISOBUTYL ISOBUTYRATE	-0.01063	6.5754E-05	2.1014E-08	421	1000	---
65	C8H16O2	n-OCTANOIC ACID	-0.00981	5.8314E-05	1.3250E-08	513	1000	---
66	C8H16O4	DIETHYLENE GLYCOL ETHYL ETHER ACETATE	-0.01399	6.7839E-05	2.5453E-09	491	1000	---
67	C8H18	2,2-DIMETHYLHEXANE	-0.02694	1.1625E-04	7.4267E-09	380	1000	---
68	C8H18	2,3-DIMETHYLHEXANE	-0.02737	1.1702E-04	5.5484E-09	389	1000	---
69	C8H18	2,4-DIMETHYLHEXANE	-0.02625	1.1629E-04	5.4923E-09	383	1000	---
70	C8H18	2,5-DIMETHYLHEXANE	-0.02571	1.1255E-04	7.2788E-09	382	1000	---
71	C8H18	3,3-DIMETHYLHEXANE	-0.02788	1.2026E-04	5.8458E-09	385	1000	---
72	C8H18	3,4-DIMETHYLHEXANE	-0.02783	1.1807E-04	5.3388E-09	391	1000	---
73	C8H18	3-ETHYLHEXANE	-0.02738	1.1743E-04	3.1424E-09	392	1000	---
74	C8H18	3-METHYL-3-ETHYLPENTANE	-0.02632	1.1337E-04	1.2309E-08	391	1000	---
75	C8H18	2-METHYLHEPTANE	-0.02706	1.1601E-04	2.3365E-09	391	1000	---
76	C8H18	3-METHYLHEPTANE	-0.02729	1.1640E-04	2.9533E-09	392	1000	---
77	C8H18	4-METHYLHEPTANE	-0.02760	1.1834E-04	1.8233E-09	391	1000	---
78	C8H18	n-OCTANE	-0.00213	1.8456E-05	9.4775E-08	300	800	---
79	C8H18	2,2,3-TRIMETHYLPENTANE	-0.01237	7.2537E-05	3.9419E-08	383	993	---
80	C8H18	2,2,4-TRIMETHYLPENTANE	-0.00622	4.1978E-05	7.7251E-08	355	580	---
81	C8H18	2,3,3-TRIMETHYLPENTANE	-0.02800	1.1853E-04	1.0470E-08	388	1000	---
82	C8H18	2,3,4-TRIMETHYLPENTANE	-0.02631	1.1542E-04	9.2518E-09	387	1000	---
83	C8H18O	DI-n-BUTYL ETHER	-0.00439	4.2683E-05	3.8415E-08	323	995	---
84	C8H18O	DI-sec-BUTYL ETHER	0.00129	-7.8611E-06	1.3732E-07	394	1000	---
85	C8H18O	DI-tert-BUTYL ETHER	-0.01061	6.8838E-05	2.2520E-08	380	1000	---
86	C8H18O	2-ETHYL-1-HEXANOL	0.00206	-8.9440E-07	1.0488E-07	450	1000	---
87	C8H18O	1-OCTANOL	-0.00005	5.1932E-06	9.7714E-08	468	1000	---
88	C8H18O	2-OCTANOL	-0.01093	6.6858E-05	1.8531E-08	453	1000	---
89	C8H18O2	DI-t-BUTYL PEROXIDE	-0.01004	6.7967E-05	1.7277E-08	384	1000	---
90	C8H18O2S	DI-n-BUTYL SULFONE	-0.01331	6.2631E-05	1.8296E-08	564	1000	---
91	C8H18O3	DIETHYLENE GLYCOL DIETHYL ETHER	-0.00965	5.9025E-05	1.5504E-08	462	1000	---
92	C8H18O3	DIETHYLENE GLYCOL MONOBUTYL EHTER	-0.00965	5.6473E-05	1.4669E-08	504	1000	---
93	C8H18O4	TRIETHYLENE GLYCOL DIMETHYL ETHER	0.00189	-8.9510E-06	1.0149E-07	489	1000	---
94	C8H18O5	TETRAETHYLENE GLYCOL	-0.01018	5.5741E-05	1.2880E-08	600	990	---
95	C8H18S	n-OCTYL MERCAPTAN	-0.00576	4.7991E-05	3.6269E-08	472	1000	---
96	C8H18S	tert-OCTYL MERCAPTAN	---	---	---	--	--	---
97	C8H19N	DI-n-BUTYLAMINE	-0.01218	7.1511E-05	2.7791E-08	432	1000	---
98	C8H19N	DIISOBUTYLAMINE	-0.01109	6.8773E-05	2.2919E-08	412	1000	---
99	C8H19N	n-OCTYLAMINE	-0.00877	5.9869E-05	3.1692E-08	453	700	---
100	C8H23N5	TETRAETHYLENEPENTAMINE	-0.01101	6.0553E-05	1.2889E-08	606	1000	---
101	C8H24O4Si4	OCTAMETHYLCYCLOTETRASILOXANE	-0.01415	6.7033E-05	3.7115E-09	448	1000	---
102	C9H4O5	TRIMELLITIC ANHYDRIDE	-0.00903	4.8872E-05	8.6629E-09	663	1000	---
103	C9H6N2O2	TOLUENE DIISOCYANATE	-0.00663	4.0463E-05	7.6806E-09	523	993	---
104	C9H7N	ISOQUINOLINE	-0.01670	7.2615E-05	2.8204E-09	516	1000	---
105	C9H7N	QUINOLINE	-0.01068	6.0360E-05	1.8156E-08	511	1000	---
106	C9H7NO	8-HYDROXYQUINOLINE	-0.00821	5.0777E-05	7.8718E-09	540	1000	---
107	C9H8	INDENE	-0.00980	5.4597E-05	2.4290E-08	456	996	---
108	C9H8O	2-METHYLBENZOFURAN	-0.00952	5.8741E-05	1.4022E-08	471	1000	---
109	C9H10	INDANE	-0.02053	8.6309E-05	1.0067E-08	451	1000	---
110	C9H10	alpha-METHYLSTYRENE	-0.00992	5.5077E-05	2.7139E-08	439	999	---
111	C9H10	m-METHYLSTYRENE	-0.01828	8.0813E-05	7.8454E-09	445	1000	---
112	C9H10	o-METHYLSTYRENE	-0.01876	8.3088E-05	8.3762E-09	443	1000	---
113	C9H10	p-METHYLSTYRENE	-0.01006	5.7370E-05	2.3743E-08	446	1000	---
114	C9H10O2	BENZYL ACETATE	-0.00855	4.9702E-05	1.4755E-08	487	1000	---
115	C9H10O2	ETHYL BENZOATE	-0.00817	4.6374E-05	1.5547E-08	487	1000	---
116	C9H10O3	ETHYL VANILLIN	-0.00914	5.3952E-05	9.4382E-09	567	1000	---
117	C9H11NO	p-DIMETHYLAMINOBENZALDEHYDE	-0.01078	5.4566E-05	1.9441E-08	588	1000	---
118	C9H12	CUMENE	-0.00803	4.2071E-05	1.1791E-07	400	650	---
119	C9H12	m-ETHYLTOLUENE	-0.00984	5.5910E-05	2.5606E-08	434	994	---
120	C9H12	o-ETHYLTOLUENE	-0.01023	5.8520E-05	2.5095E-08	438	998	---
121	C9H12	p-ETHYLTOLUENE	-0.00998	5.6322E-05	2.6487E-08	435	995	---
122	C9H12	MESITYLENE	-0.01943	8.3184E-05	1.1735E-08	438	1000	---
123	C9H12	n-PROPYLBENZENE	-0.02709	1.0607E-04	-2.0873E-09	432	1000	---
124	C9H12	1,2,3-TRIMETHYLBENZENE	-0.02000	8.5193E-05	1.1559E-08	449	1000	---
125	C9H12	1,2,4-TRIMETHYLBENZENE	-0.01052	5.8378E-05	2.8098E-08	443	1000	---

* A computer program, containing coefficients for all compounds, is available for a nominal fee. The computer program is in ASCII which can be accessed by other software.

$\lambda_{gas} = A + B\,T + C\,T^2$ (λ_{gas} - W/m K, T - K)

NO	FORMULA	NAME	A	B	C	TMIN	TMAX	λ_{gas} @ 25 C
126	C9H12O	BENZYL ETHYL ETHER	0.00114	-2.9858E-06	6.6643E-08	458	998	---
127	C9H12O	2-PHENYL-2-PROPANOL	-0.01001	6.0550E-05	1.5573E-08	475	1000	---
128	C9H12O2	CUMENE HYDROPEROXIDE	---	---	---	--	--	---
129	C9H14O	ISOPHORONE	-0.01114	6.0706E-05	2.4761E-08	488	1000	---
130	C9H14O6	GLYCERYL TRIACETATE	-0.01174	5.8935E-05	-4.2145E-10	532	1000	---
131	C9H16O4	AZELAIC ACID	-0.00935	5.3205E-05	8.1020E-09	633	1000	---
132	C9H18	ISOPROPYLCYCLOHEXANE	-0.02592	1.0534E-04	1.3165E-08	428	1000	---
133	C9H18	1-NONENE	-0.01089	6.3813E-05	3.0482E-08	400	1000	---
134	C9H18	n-PROPYLCYCLOHEXANE	-0.00575	4.4891E-05	5.2081E-08	430	1000	---
135	C9H18O	DIISOBUTYL KETONE	0.00213	-1.0359E-05	1.2639E-07	441	1000	---
136	C9H18O	1-NONANAL	0.00198	-9.0458E-06	1.0749E-07	468	1000	---
137	C9H18O2	n-BUTYL VALERATE	0.00186	-8.5085E-06	1.0316E-07	460	1000	---
138	C9H18O2	n-NONANOIC ACID	-0.00971	5.8249E-05	1.1540E-08	529	1000	---
139	C9H18O2	n-OCTYL FORMATE	0.00189	-8.7169E-06	1.0248E-07	472	1000	---
140	C9H20	3,3-DIETHYLPENTANE	-0.01277	7.2251E-05	3.6371E-08	419	999	---
141	C9H20	2,2-DIMETHYL-3-ETHYLPENTANE	-0.02651	1.1035E-04	1.2089E-08	407	1000	---
142	C9H20	2,4-DIMETHYL-3-ETHYLPENTANE	-0.02585	1.0958E-04	1.0014E-08	410	1000	---
143	C9H20	2,2-DIMETHYLHEPTANE	-0.02574	1.1018E-04	7.7586E-09	406	1000	---
144	C9H20	2,6-DIMETHYLHEPTANE	-0.02543	1.0889E-04	5.8666E-09	408	1000	---
145	C9H20	3-ETHYLHEPTANE	-0.02608	1.1067E-04	3.8539E-09	416	1000	---
146	C9H20	2-METHYLOCTANE	-0.02750	1.1432E-04	-1.2209E-10	416	1000	---
147	C9H20	3-METHYLOCTANE	-0.02782	1.1521E-04	-2.2227E-11	417	1000	---
148	C9H20	4-METHYLOCTANE	-0.02793	1.1621E-04	-4.2077E-10	416	1000	---
149	C9H20	n-NONANE	-0.00655	3.2637E-05	7.7150E-08	449	678	---
150	C9H20	2,2,3,3-TETRAMETHYLPENTANE	-0.02921	1.1850E-04	1.0144E-08	413	1000	---
151	C9H20	2,2,3,4-TETRAMETHYLPENTANE	-0.02850	1.1698E-04	8.0693E-09	406	1000	---
152	C9H20	2,2,4,4-TETRAMETHYLPENTANE	-0.02637	1.1114E-04	1.1595E-08	395	1000	---
153	C9H20	2,2,5-TRIMETHYLHEXANE	-0.01211	6.9649E-05	3.7423E-08	397	997	---
154	C9H20O	2,6-DIMETHYL-4-HEPTANOL	-0.01820	8.5102E-05	3.7637E-09	451	1000	---
155	C9H20O	1-NONANOL	-0.01187	6.7488E-05	2.2469E-08	486	1000	---
156	C9H20O	2-NONANOL	-0.00147	8.6692E-06	1.0418E-07	472	1000	---
157	C9H20S	n-NONYL MERCAPTAN	---	---	---	--	--	---
158	C9H21N	n-NONYLAMINE	-0.00924	5.9820E-05	2.8719E-08	475	750	---
159	C9H21N	TRIPROPYLAMINE	-0.02328	1.0622E-04	1.3649E-08	430	1000	---
160	C10H6O8	PYROMELLITIC ACID	-0.00588	3.1496E-05	4.4893E-09	722	1000	---
161	C10H7Br	1-BROMONAPHTHALENE	-0.00686	3.9597E-05	8.4459E-09	554	1000	---
162	C10H7Cl	1-CHLORONAPHTHALENE	-0.00786	4.6122E-05	1.0111E-08	532	1000	---
163	C10H8	NAPHTHALENE	-0.02306	9.2610E-05	4.4577E-10	353	1259	---
164	C10H9N	QUINALDINE	-0.00929	5.3175E-05	1.4263E-08	520	1000	---
165	C10H10	m-DIVINYLBENZENE	-0.00988	5.7353E-05	1.8576E-08	473	1000	---
166	C10H10	1-METHYLINDENE	-0.01990	8.8360E-05	2.4675E-09	472	1000	---
167	C10H10	2-METHYLINDENE	-0.02021	9.0063E-05	3.1453E-09	458	1000	---
168	C10H10O4	DIMETHYL PHTHALATE	-0.00767	4.3884E-05	9.6831E-09	557	1000	---
169	C10H10O4	DIMETHYL TEREPHTHALATE	-0.00361	3.5675E-05	1.3239E-08	561	1000	---
170	C10H12	DICYCLOPENTADIENE	---	---	---	--	--	---
171	C10H12	1,2,3,4-TETRAHYDRONAPHTHALENE	-0.00942	5.0729E-05	2.3456E-08	481	991	---
172	C10H12O	ANETHOLE	-0.00963	5.5759E-05	1.5008E-08	508	1000	---
173	C10H12O4	DIALLYL MALEATE	-0.00806	5.3705E-05	5.7459E-09	520	1000	---
174	C10H14	n-BUTYLBENZENE	-0.02091	8.8748E-05	7.1867E-09	456	996	---
175	C10H14	sec-BUTYLBENZENE	-0.01071	6.0505E-05	2.6424E-08	446	996	---
176	C10H14	tert-BUTYLBENZENE	-0.01094	6.2733E-05	2.6379E-08	442	992	---
177	C10H14	m-CYMENE	-0.01087	6.0806E-05	2.7371E-08	448	998	---
178	C10H14	o-CYMENE	-0.01075	5.9879E-05	2.7733E-08	451	991	---
179	C10H14	p-CYMENE	-0.01029	6.7187E-05	1.3232E-08	450	1000	---
180	C10H14	m-DIETHYLBENZENE	-0.02017	8.6688E-05	8.0428E-09	454	994	---
181	C10H14	o-DIETHYLBENZENE	-0.01075	6.0773E-05	2.4600E-08	457	1000	---
182	C10H14	p-DIETHYLBENZENE	-0.01063	5.9356E-05	2.5488E-08	457	1000	---
183	C10H14	2-ETHYL-m-XYLENE	-0.01119	6.3211E-05	2.4836E-08	463	1000	---
184	C10H14	2-ETHYL-p-XYLENE	-0.01078	6.0393E-05	2.5169E-08	460	1000	---
185	C10H14	3-ETHYL-o-XYLENE	-0.01877	8.2876E-05	8.7574E-09	467	1000	---
186	C10H14	4-ETHYL-m-XYLENE	-0.01077	6.0066E-05	2.5338E-08	462	1000	---
187	C10H14	4-ETHYL-o-XYLENE	-0.01076	5.9922E-05	2.5219E-08	463	1000	---
188	C10H14	5-ETHYL-m-XYLENE	-0.01925	8.1961E-05	9.8622E-09	457	1000	---
189	C10H14	ISOBUTYLBENZENE	-0.01124	6.3606E-05	2.7317E-08	446	1000	---
190	C10H14	1,2,3,5-TETRAMETHYLBENZENE	-0.01078	5.8559E-05	2.6562E-08	471	1000	---
191	C10H14	1,2,4,5-TETRAMETHYLBENZENE	-0.00373	4.7379E-06	9.8524E-08	470	1000	---
192	C10H14O	p-tert-BUTYLPHENOL	-0.01007	5.9304E-05	1.4237E-08	513	1000	---
193	C10H14O2	p-tert-BUTYLCATECHOL	-0.00974	5.7770E-05	1.0368E-08	558	1000	---

* A computer program, containing coefficients for all compounds, is available for a nominal fee. The computer program is in ASCII which can be accessed by other software.

$\lambda_{gas} = A + B\,T + C\,T^2$ (λ_{gas} - W/m K, T - K)

NO	FORMULA	NAME	A	B	C	TMIN	TMAX	λ_{gas} @ 25 C
194	C10H15N	N,N-DIETYHLANILINE	---	---	---	--	--	---
195	C10H15N	2,6-DIETHYLANILINE	-0.00972	5.8191E-05	1.3059E-08	509	1000	---
196	C10H16	CAMPHENE	-0.02672	1.0380E-04	8.2702E-09	434	1000	---
197	C10H16	D-LIMONENE	-0.02197	9.3562E-05	7.4279E-09	450	1000	---
198	C10H16	alpha-PHELLANDRENE	-0.02222	9.6145E-05	7.8464E-09	448	1000	---
199	C10H16	beta-PHELLANDRENE	-0.02301	9.9718E-05	5.9927E-09	447	1000	---
200	C10H16	alpha-PINENE	-0.02514	1.0151E-04	-7.1878E-09	429	1000	---
201	C10H16	beta-PINENE	-0.02797	1.0908E-04	3.9401E-09	439	1000	---
202	C10H16	alpha-TERPINENE	-0.01190	7.0296E-05	2.3739E-08	450	1000	---
203	C10H16	gamma-TERPINENE	-0.01176	6.9037E-05	2.3256E-08	456	1000	---
204	C10H16	TERPINOLENE	-0.02133	9.0600E-05	8.4117E-09	458	1000	---
205	C10H16O	CAMPHOR	-0.02563	9.7273E-05	1.0104E-08	481	1000	---
206	C10H18	cis-DECAHYDRONANPHTALENE	-0.00457	-1.5454E-05	2.9327E-07	469	1000	---
207	C10H18	trans-DECAHYDRONAPHTHALENE	-0.00480	-1.2283E-05	2.7948E-07	460	1000	---
208	C10H18O4	SEBACIC ACID	-0.00925	5.2244E-05	7.9754E-09	642	1000	---
209	C10H20	n-BUTYLCYCLOHEXANE	-0.01194	6.4176E-05	3.3729E-08	454	1000	---
210	C10H20	1-DECENE	-0.01118	6.2751E-05	2.8360E-08	444	1000	---
211	C10H20O	1-DECANAL	0.00195	-9.2739E-06	1.0395E-07	488	1000	---
212	C10H20O2	n-DECANOIC ACID	-0.00965	5.5721E-05	1.2593E-08	543	1000	---
213	C10H20O2	2-ETHYLHEXYL ACETATE	-0.00816	4.9835E-05	1.2459E-08	472	1000	---
214	C10H20O2	ISOPENTYL ISOVALERATE	-0.01063	6.4970E-05	1.6592E-08	467	1000	---
215	C10H22	n-DECANE	-0.00113	8.1090E-06	9.6092E-08	470	700	---
216	C10H22	2,2-DIMETHYLOCTANE	-0.02572	1.0727E-04	5.6114E-09	430	1000	---
217	C10H22	2-METHYLNONANE	-0.02612	1.0783E-04	1.3143E-09	440	1000	---
218	C10H22	3-METHYLNONANE	-0.02629	1.0816E-04	1.7111E-09	441	1000	---
219	C10H22	4-METHYLNONANE	-0.02649	1.0941E-04	1.2206E-09	439	1000	---
220	C10H22	5-METHYLNONANE	-0.02646	1.0917E-04	1.3130E-09	438	1000	---
221	C10H22O	1-DECANOL	0.03262	-1.1884E-04	2.0994E-07	503	702	---
222	C10H22O	DI-n-PENTYL ETHER	-0.01420	7.1760E-05	1.0355E-08	460	1000	---
223	C10H22O	ISODECANOL	-0.01047	6.1440E-05	1.6754E-08	493	1000	---
224	C10H22O5	TETRAETHYLENE GLYCOL DIMETHYL ETHER	-0.00913	5.3148E-05	1.1123E-08	549	1000	---
225	C10H22S	n-DECYL MERCAPTAN	-0.01171	6.3915E-05	2.2689E-08	512	1000	---
226	C10H23N	n-DECYLAMINE	-0.00917	5.8694E-05	2.6803E-08	494	750	---
227	C11H10	1-METHYLNAPHTHALENE	-0.01703	7.1429E-05	6.7407E-09	518	998	---
228	C11H10	2-METHYLNAPHTHALENE	-0.01706	7.1614E-05	7.3389E-09	514	994	---
229	C11H14O2	n-BUTYL BENZOATE	0.00115	-4.3515E-06	6.2142E-08	523	1000	---
230	C11H16	n-PENTYLBENZENE	-0.01096	6.0321E-05	2.4626E-08	479	1000	---
231	C11H16O	p-tert-AMYLPHENOL	-0.01008	5.7404E-05	1.4538E-08	535	1000	---
232	C11H20O2	2-ETHYLHEXYL ACRYLATE	0.00187	-9.1216E-06	9.9909E-08	489	999	---
233	C11H22	1-UNDECENE	-0.02584	1.0261E-04	-2.1510E-10	466	1000	---
234	C11H22O	1-UNDECANAL	0.00191	-9.3691E-06	1.0081E-07	506	1000	---
235	C11H24	n-UNDECANE	0.01364	-4.8303E-05	1.4396E-07	470	800	---
236	C11H24O	1-UNDECANOL	-0.01150	6.4324E-05	1.9854E-08	516	1000	---
237	C11H24S	UNDECYL MERCAPTAN	-0.00988	5.5448E-05	1.5534E-08	531	1000	---
238	C12H8O	DIBENZOFURAN	-0.00911	4.9183E-05	1.4632E-08	558	1000	---
239	C12H9N	DIBENZOPYRROLE	-0.01782	7.3507E-05	2.4511E-09	628	1000	---
240	C12H10	ACENAPHTHENE	-0.00954	5.0025E-05	1.7866E-08	551	1000	---
241	C12H10	BIPHENYL	-0.00788	4.2910E-05	3.4569E-08	373	1000	---
242	C12H10O	DIPHENYL ETHER	-0.00885	5.1306E-05	1.2076E-08	531	1000	---
243	C12H11N	p-AMINODIPHENYL	-0.00918	5.2745E-05	1.0297E-08	575	1000	---
244	C12H11N	DIPHENYLAMINE	-0.00972	5.4414E-05	1.2286E-08	575	1000	---
245	C12H11N3	p-AMINOAZOBENZENE	-0.00975	5.4448E-05	9.3225E-09	633	1000	---
246	C12H11N3	1,3-DIPHENYLTRIAZENE	-0.00903	5.2512E-05	7.9461E-09	610	1000	---
247	C12H12	2,6-DIMETHYLNAPHTHALENE	-0.01677	7.1801E-05	9.5201E-09	535	1000	---
248	C12H12	2,7-DIMETHYLNAPHTHALENE	-0.01682	7.1879E-05	9.4156E-09	536	1000	---
249	C12H12	1-ETHYLNAPHTHALENE	-0.01007	5.3746E-05	1.9325E-08	531	1000	---
250	C12H12N2	p-AMINODIPHENYLAMINE	-0.01522	6.6481E-05	5.2547E-10	627	1000	---
251	C12H12N2	HYDRAZOBENZENE	-0.00927	5.3347E-05	1.0456E-08	573	1000	---
252	C12H14	1,2,3-TRIMETHYLINDENE	-0.01086	6.1319E-05	1.8696E-08	509	1000	---
253	C12H14O4	DIETHYL PHTHALATE	-0.00863	4.7444E-05	1.2248E-08	567	1000	---
254	C12H16	CYCLOHEXYLBENZENE	-0.01111	5.6012E-05	2.9171E-08	513	1000	---
255	C12H18	m-DIISOPROPYLBENZENE	-0.02095	8.8293E-05	6.1091E-09	476	1000	---
256	C12H18	p-DIISOPROPYLBENZENE	-0.02100	8.8014E-05	5.8660E-09	484	1000	---
257	C12H18	n-HEXYLBENZENE	-0.02036	8.4338E-05	6.6954E-09	499	1000	---
258	C12H20O4	DIBUTYL MALEATE	-0.00826	4.7750E-05	1.0165E-08	553	1000	---
259	C12H22	BICYCLOHEXYL	-0.01229	6.0636E-05	3.6019E-08	512	992	---
260	C12H23N	DICYCLOHEXYLAMINE	-0.01184	6.1839E-05	2.5082E-08	529	1000	---
261	C12H24	1-DODECENE	-0.02600	1.0107E-04	-1.0231E-09	487	1000	---

* A computer program, containing coefficients for all compounds, is available for a nominal fee. The computer program is in ASCII which can be accessed by other software.

			$\lambda_{gas} = A + B\,T + C\,T^2$		$(\lambda_{gas}$ - W/m K, T - K)			
NO	FORMULA	NAME	A	B	C	TMIN	TMAX	λ_{gas} @ 25 C
262	C12H24O	1-DODECANAL	0.00187	-9.4540E-06	9.7853E-08	523	1000	---
263	C12H24O2	n-DODECANOIC ACID	-0.00951	5.2199E-05	1.3328E-08	572	1000	---
264	C12H26	n-DODECANE	-0.00812	2.9150E-05	7.2085E-08	489	1000	---
265	C12H26O	DI-n-HEXYL ETHER	-0.00942	5.7682E-05	1.2128E-08	499	1000	---
266	C12H26O	1-DODECANOL	-0.01135	6.3614E-05	1.7549E-08	535	1000	---
267	C12H26O3	DIETHYLENE GLYCOL DI-n-BUTYL ETHER	-0.00930	5.3413E-05	1.3380E-08	529	1000	---
268	C12H26S	n-DODECYL MERCAPTAN	-0.00575	4.4400E-05	3.2638E-08	548	1000	---
269	C12H27BO3	TRI-n-BUTYL BORATE	---	---	---	--	--	---
270	C12H27N	DODECYLAMINE	-0.00905	5.6320E-05	2.3942E-08	532	750	---
271	C12H27N	TRI-n-BUTYLAMINE	-0.01351	7.4453E-05	2.9099E-08	487	997	---
272	C13H10	FLUORENE	-0.00904	5.6671E-05	6.6790E-09	570	1000	---
273	C13H10O	BENZOPHENONE	-0.00813	4.3772E-05	1.1875E-08	579	1000	---
274	C13H12	DIPHENYLMETHANE	-0.01004	5.3011E-05	1.9575E-08	537	1000	---
275	C13H20	n-HEPTYLBENZENE	-0.01079	5.7422E-05	2.2320E-08	519	1000	---
276	C13H26	1-TRIDECENE	-0.02607	9.9453E-05	-1.5725E-09	506	1000	---
277	C13H26O	1-TRIDECANAL	0.00184	-9.4112E-06	9.4178E-08	540	1000	---
278	C13H26O2	n-BUTYL NONANOATE	0.00202	-1.0410E-05	1.0444E-07	503	1000	---
279	C13H26O2	METHYL DODECANOATE	-0.00914	5.1369E-05	1.3643E-08	540	1000	---
280	C13H28	n-TRIDECANE	-0.00784	2.7116E-05	7.0226E-08	509	1000	---
281	C13H28O	1-TRIDECANOL	-0.00098	5.3568E-06	8.6625E-08	547	1000	---
282	C14H8O2	ANTHRAQUINONE	-0.01018	5.3038E-05	1.2081E-08	653	1000	---
283	C14H10	ANTHRACENE	-0.00508	3.4417E-05	2.3646E-08	489	615	---
284	C14H10	DIPHENYLACETYLENE	-0.00913	4.8508E-05	1.4880E-08	573	993	---
285	C14H10	PHENANTHRENE	-0.00923	4.7388E-05	1.3940E-08	613	1000	---
286	C14H12	cis-STILBENE	-0.00981	5.1510E-05	1.9775E-08	535	1000	---
287	C14H12	trans-STILBENE	-0.00980	5.1139E-05	1.6241E-08	580	1000	---
288	C14H12O2	BENZYL BENZOATE	-0.00798	4.2392E-05	1.1341E-08	597	1000	---
289	C14H14	1,1-DIPHENYLETHANE	-0.02834	9.7961E-05	-7.5061E-09	546	1000	---
290	C14H14	1,2-DIPHENYLETHANE	-0.01012	5.2760E-05	1.9062E-08	554	1000	---
291	C14H14O	DIBENZYL ETHER	-0.00858	4.5472E-05	1.5047E-08	561	991	---
292	C14H16	1-n-BUTYLNAPHTHALENE	-0.01718	7.2875E-05	9.1628E-09	563	1000	---
293	C14H22	n-OCTYLBENZENE	-0.01078	5.6575E-05	2.1438E-08	538	1000	---
294	C14H22O	p-tert-OCTYLPHENOL	-0.01018	5.8348E-05	1.2288E-08	564	1000	---
295	C14H28	1-TETRADECENE	-0.02610	9.7799E-05	-2.0865E-09	524	1000	---
296	C14H28O2	n-TETRADECANOIC ACID	-0.00936	5.0289E-05	1.2546E-08	599	1000	---
297	C14H30	n-TETRADECANE	-0.00180	1.0242E-05	7.7727E-08	527	1000	---
298	C14H30O	1-TETRADECANOL	-0.00003	2.4670E-06	7.6134E-08	560	1000	---
299	C14H31N	TETRADECYLAMINE	-0.01117	6.0842E-05	1.7026E-08	564	994	---
300	C15H10N2O2	DIPHENYLMETHANE-4,4'-DIISOCYANATE	-0.00684	3.8456E-05	7.1593E-09	609	1000	---
301	C15H16O	p-CUMYLPHENOL	-0.01833	7.2980E-05	-3.4325E-09	608	1000	---
302	C15H16O2	BISPHENOL A	-0.00886	5.0729E-05	7.4757E-09	634	1000	---
303	C15H24	n-NONYLBENZENE	-0.01084	5.6173E-05	2.0794E-08	555	1000	---
304	C15H24O	2,6-DI-tert-BUTYL-p-CRESOL	-0.01633	7.1176E-05	-1.4771E-09	538	1000	---
305	C15H24O	NONYLPHENOL	-0.00932	5.1812E-05	1.1798E-08	581	1000	---
306	C15H30	1-PENTADECENE	-0.02698	1.0116E-04	-7.5469E-09	542	1000	---
307	C15H30O2	PENTADECANOIC ACID	-0.00927	4.9669E-05	1.1808E-08	612	1000	---
308	C15H32	n-PENTADECANE	-0.00723	2.3158E-05	6.7125E-08	544	1000	---
309	C16H10	FLUORANTHENE	-0.00899	4.6986E-05	1.0387E-08	656	1000	---
310	C16H10	PYRENE	-0.00860	4.3319E-05	1.1141E-08	668	998	---
311	C16H12	1-PHENYLNAPHTHALENE	-0.00955	4.9187E-05	1.4702E-08	607	1000	---
312	C16H20	1-n-HEXYLNAPHTHALENE	-0.01676	7.0285E-05	8.6552E-09	595	1000	---
313	C16H22O4	DIBUTYL PHTHALATE	-0.00911	4.8916E-05	1.1470E-08	613	1000	---
314	C16H26	n-DECYLBENZENE	-0.01081	5.5409E-05	2.0073E-08	571	1000	---
315	C16H32	n-DECYLCYCLOHEXANE	-0.01164	5.8255E-05	2.3755E-08	571	1000	---
316	C16H32	1-HEXADECENE	-0.01088	5.5158E-05	2.2391E-08	558	1000	---
317	C16H32O2	n-HEXADECANOIC ACID	-0.00934	4.9840E-05	1.1461E-08	624	1000	---
318	C16H34	n-HEXADECANE	-0.00671	2.0080E-05	6.7235E-08	560	1000	---
319	C16H34O	DI-n-OCTYL ETHER	-0.00945	5.1927E-05	1.3988E-08	560	1000	---
320	C16H34O	1-HEXADECANOL	0.00001	2.1149E-06	8.0125E-08	585	1000	---
321	C17H28	n-UNDECYLBENZENE	-0.01083	5.5002E-05	1.9427E-08	586	1000	---
322	C17H34	1-HEPTADECENE	-0.01081	5.6622E-05	1.8218E-08	573	1000	---
323	C17H36	n-HEPTADECANE	0.00124	-6.3091E-06	8.1047E-08	575	1000	---
324	C17H36O	1-HEPTADECANOL	-0.00006	2.1078E-06	7.7468E-08	597	1000	---
325	C18H12	CHRYSENE	-0.00870	4.3518E-05	9.6092E-09	714	994	---
326	C18H14	m-TERPHENYL	-0.00741	4.0043E-05	8.0074E-09	638	1000	---
327	C18H14	o-TERPHENYL	-0.00715	3.9304E-05	8.4426E-09	605	1000	---
328	C18H14	p-TERPHENYL	-0.00748	4.0276E-05	7.8174E-09	649	1000	---
329	C18H15P	TRIPHENYLPHOSPHINE	-0.00570	4.3077E-05	2.4635E-08	650	1000	---

* A computer program, containing coefficients for all compounds, is available for a nominal fee. The computer program is in ASCII which can be accessed by other software.

NO	FORMULA	NAME	A	B	C	TMIN	TMAX	λ_{gas} @ 25 C
			$\lambda_{gas} = A + B\,T + C\,T^2$ (λ_{gas} - W/m K, T - K)					
330	C18H15O4P	TRIPHENYL PHOSPHATE	---	---	---	--	--	---
331	C18H16N2	N,N'-DIPHENYL-p-PHENYLENEDIAMINE	-0.00882	4.7619E-05	7.5718E-09	688	1000	---
332	C18H22	2,3-DIMETHYL-2,3-DIPHENYLBUTANE	-0.01060	5.4719E-05	1.7566E-08	589	1000	---
333	C18H22O2	DICUMYL PEROXIDE	---	---	---	--	--	---
334	C18H30	n-DODECYLBENZENE	-0.01081	5.2783E-05	2.1190E-08	600	1000	---
335	C18H32O2	LINOLEIC ACID	-0.00938	5.1481E-05	9.9426E-09	628	1000	---
336	C18H34O2	OLEIC ACID	-0.00917	4.8698E-05	1.1022E-08	633	1000	---
337	C18H34O4	DIBUTYL SEBACATE	-0.00826	4.5607E-05	8.7811E-09	622	1000	---
338	C18H34O4	DIHEXYL ADIPATE	-0.00727	3.9979E-05	7.9202E-09	621	1000	---
339	C18H36	1-OCTADECENE	-0.01079	5.3470E-05	2.1104E-08	588	1000	---
340	C18H36O2	STEARIC ACID	-0.00916	4.9075E-05	9.8658E-09	648	1000	---
341	C18H38	n-OCTADECANE	-0.00172	6.6775E-07	7.2881E-08	590	1000	---
342	C18H38O	DINONYL ETHER	-0.00933	5.0232E-05	1.2875E-08	591	1000	---
343	C18H38O	1-OCTADECANOL	-0.01063	5.5997E-05	1.4819E-08	608	1000	---
344	C19H26	1-n-NONYLNAPHTHALENE	-0.00998	4.9182E-05	1.5845E-08	639	1000	---
345	C19H32	n-TRIDECYLBENZENE	-0.01882	7.4211E-05	5.9657E-09	614	1000	---
346	C19H36O2	METHYL OLEATE	-0.00860	4.6444E-05	1.0352E-08	617	1000	---
347	C19H38	1-NONADECENE	-0.01069	5.3678E-05	1.8616E-08	602	1000	---
348	C19H38O2	NONADECANOIC ACID	-0.00909	4.8001E-05	9.8948E-09	659	1000	---
349	C19H40	n-NONADECANE	0.00153	-7.5609E-06	7.4184E-08	603	1000	---
350	C20H16	TRIPHENYLETHYLENE	-0.00930	4.6613E-05	1.2116E-08	669	1000	---
351	C20H28	1-n-DECYLNAPHTHALENE	-0.00994	4.8645E-05	1.5353E-08	652	1000	---
352	C20H30O2	ABIETIC ACID	-0.00981	5.1249E-05	1.1691E-08	650	1000	---
353	C20H31N	DEHYDROABIETYLAMINE	-0.00001	8.9713E-07	-6.1569E-11	660	1000	---
354	C20H40	1-EICOSENE	-0.01067	5.1723E-05	2.0085E-08	616	1000	---
355	C20H42	n-EICOSANE	0.00154	-7.5268E-06	7.0837E-08	617	1000	---
356	C20H42O	1-EICOSANOL	0.00006	1.4370E-06	7.0530E-08	629	1000	---
357	C21H21O4P	TRI-o-CRESYL PHOSPHATE	---	---	---	--	--	---
358	C22H44O2	n-BUTYL STEARATE	-0.00719	4.0258E-05	7.1090E-09	623	1000	---
359	C24H38O4	DIISOOCTYL PHTHALATE	-0.00777	4.0239E-05	7.8945E-09	694	1000	---
360	C24H38O4	DIOCTYL PHTHALATE	-0.00575	6.5471E-05	-7.3091E-09	657	1000	---
361	C24H42O	DINONYLPHENOL	-0.01218	5.3914E-05	3.2714E-09	722	1000	---
362	C26H20	TETRAPHENYLETHYLENE	-0.00895	4.3161E-05	9.4152E-09	760	1000	---
363	C28H46O4	DIISODECYL PHTHALATE	-0.00746	3.8410E-05	6.8581E-09	723	1000	---

* A computer program, containing coefficients for all compounds, is available for a nominal fee. The computer program is in ASCII which can be accessed by other software.

Appendix C

DATA CODE FOR COMPOUNDS

NO	FORMULA	NAME	LIQUID THERMAL COND.	GAS THERMAL COND.
1	C8H4Cl2O2	ISOPHTHALOYL CHLORIDE	estimate	estimate
2	C8H4O3	PHTHALIC ANHYDRIDE	estimate	estimate
3	C8H6O4	ISOPHTHALIC ACID	---	estimate
4	C8H6O4	PHTHALIC ACID	---	estimate
5	C8H6O4	TEREPHTHALIC ACID	---	estimate
6	C8H6S	BENZOTHIOPHENE	estimate	estimate
7	C8H7N	INDOLE	estimate	estimate
8	C8H8	STYRENE	estimate	experimental
9	C8H8O	ACETOPHENONE	estimate	estimate
10	C8H8O	p-TOLUALDEHYDE	estimate	estimate
11	C8H8O2	METHYL BENZOATE	estimate	estimate
12	C8H8O2	o-TOLUIC ACID	estimate	estimate
13	C8H8O2	p-TOLUIC ACID	estimate	estimate
14	C8H8O3	METHYL SALICYLATE	estimate	estimate
15	C8H8O3	VANILLIN	estimate	estimate
16	C8H9NO	ACETANILIDE	estimate	estimate
17	C8H10	ETHYLBENZENE	experimental/estimate	experimental
18	C8H10	m-XYLENE	experimental/estimate	experimental
19	C8H10	o-XYLENE	experimental/estimate	experimental
20	C8H10	p-XYLENE	experimental/estimate	experimental
21	C8H10O	m-ETHYLPHENOL	---	---
22	C8H10O	p-ETHYLPHENOL	estimate	estimate
23	C8H10O	PHENETOLE	experimental/estimate	estimate
24	C8H10O	2-PHENYLETHANOL	estimate	estimate
25	C8H10O	2,3-XYLENOL	estimate	estimate
26	C8H10O	2,4-XYLENOL	estimate	estimate
27	C8H10O	2,5-XYLENOL	estimate	estimate
28	C8H10O	2,6-XYLENOL	estimate	estimate
29	C8H10O	3,4-XYLENOL	estimate	estimate
30	C8H10O	3,5-XYLENOL	estimate	estimate
31	C8H11N	N,N-DIMETHYLANILINE	experimental/estimate	estimate
32	C8H11N	o-ETHYLANILINE	estimate	estimate
33	C8H11N	2,4,6-TRIMETHYLPYRIDINE	estimate	estimate
34	C8H11NO	p-PHENETIDINE	estimate	estimate
35	C8H12	1,5-CYCLOOCTADIENE	estimate	estimate
36	C8H12	VINYLCYCLOHEXENE	estimate	estimate
37	C8H12O4	1,4-CYCLOHEXANEDICARBOXYLIC ACID	estimate	estimate
38	C8H12O4	DIETHYL MALEATE	estimate	estimate
39	C8H14O2	n-BUTYL METHACRYLATE	estimate	estimate
40	C8H14O3	BUTYRIC ANHYDRIDE	estimate	estimate
41	C8H14O4	DIETHYL SUCCINATE	estimate	estimate
42	C8H16	1,1-DIMETHYLCYCLOHEXANE	estimate	estimate
43	C8H16	cis-1,2-DIMETHYLCYCLOHEXANE	estimate	estimate
44	C8H16	trans-1,2-DIMETHYLCYCLOHEXANE	estimate	estimate
45	C8H16	cis-1,3-DIMETHYLCYCLOHEXANE	estimate	estimate
46	C8H16	trans-1,3-DIMETHYLCYCLOHEXANE	estimate	estimate
47	C8H16	cis-1,4-DIMETHYLCYCLOHEXANE	estimate	estimate
48	C8H16	trans-1,4-DIMETHYLCYCLOHEXANE	estimate	estimate
49	C8H16	ETHYLCYCLOHEXANE	estimate	estimate
50	C8H16	2-ETHYL-1-HEXENE	estimate	estimate
51	C8H16	1-METHYL-1-ETHYLCYCLOPENTANE	estimate	estimate
52	C8H16	1-OCTENE	experimental/estimate	experimental
53	C8H16	trans-2-OCTENE	estimate	estimate
54	C8H16	trans-3-OCTENE	estimate	estimate
55	C8H16	trans-4-OCTENE	estimate	estimate
56	C8H16	n-PROPYLCYCLOPENTANE	estimate	estimate
57	C8H16	2,4,4-TRIMETHYL-1-PENTENE	estimate	estimate
58	C8H16	2,4,4-TRIMETHYL-2-PENTENE	estimate	estimate
59	C8H16O	2-ETHYLHEXANAL	estimate	estimate
60	C8H16O	1-OCTANAL	estimate	estimate
61	C8H16O	2-OCTANONE	estimate	estimate
62	C8H16O2	n-BUTYL n-BUTYRATE	estimate	estimate
63	C8H16O2	n-HEXYL ACETATE	experimental/estimate	estimate
64	C8H16O2	ISOBUTYL ISOBUTYRATE	estimate	estimate
65	C8H16O2	n-OCTANOIC ACID	estimate	estimate
66	C8H16O4	DIETHYLENE GLYCOL ETHYL ETHER ACETATE	estimate	estimate
67	C8H18	2,2-DIMETHYLHEXANE	estimate	estimate

NO	FORMULA	NAME	LIQUID THERMAL COND.	GAS THERMAL COND.
68	C8H18	2,3-DIMETHYLHEXANE	estimate	estimate
69	C8H18	2,4-DIMETHYLHEXANE	estimate	estimate
70	C8H18	2,5-DIMETHYLHEXANE	estimate	estimate
71	C8H18	3,3-DIMETHYLHEXANE	estimate	estimate
72	C8H18	3,4-DIMETHYLHEXANE	estimate	estimate
73	C8H18	3-ETHYLHEXANE	estimate	estimate
74	C8H18	3-METHYL-3-ETHYLPENTANE	estimate	estimate
75	C8H18	2-METHYLHEPTANE	estimate	estimate
76	C8H18	3-METHYLHEPTANE	estimate	estimate
77	C8H18	4-METHYLHEPTANE	estimate	estimate
78	C8H18	n-OCTANE	experimental/estimate	experimental
79	C8H18	2,2,3-TRIMETHYLPENTANE	estimate	estimate
80	C8H18	2,2,4-TRIMETHYLPENTANE	experimental/estimate	experimental
81	C8H18	2,3,3-TRIMETHYLPENTANE	estimate	estimate
82	C8H18	2,3,4-TRIMETHYLPENTANE	estimate	estimate
83	C8H18O	DI-n-BUTYL ETHER	experimental/estimate	estimate
84	C8H18O	DI-sec-BUTYL ETHER	estimate	estimate
85	C8H18O	DI-tert-BUTYL ETHER	estimate	estimate
86	C8H18O	2-ETHYL-1-HEXANOL	estimate	estimate
87	C8H18O	1-OCTANOL	experimental/estimate	estimate
88	C8H18O	2-OCTANOL	experimental/estimate	estimate
89	C8H18O2	DI-t-BUTYL PEROXIDE	estimate	estimate
90	C8H18O2S	DI-n-BUTYL SULFONE	estimate	estimate
91	C8H18O3	DIETHYLENE GLYCOL DIETHYL ETHER	experimental/estimate	estimate
92	C8H18O3	DIETHYLENE GLYCOL MONOBUTYL EHTER	experimental/estimate	estimate
93	C8H18O4	TRIETHYLENE GLYCOL DIMETHYL ETHER	experimental/estimate	estimate
94	C8H18O5	TETRAETHYLENE GLYCOL	estimate	estimate
95	C8H18S	n-OCTYL MERCAPTAN	estimate	estimate
96	C8H18S	tert-OCTYL MERCAPTAN	estimate	---
97	C8H19N	DI-n-BUTYLAMINE	experimental/estimate	estimate
98	C8H19N	DIISOBUTYLAMINE	estimate	estimate
99	C8H19N	n-OCTYLAMINE	experimental/estimate	estimate
100	C8H23N5	TETRAETHYLENEPENTAMINE	estimate	estimate
101	C8H24O4Si4	OCTAMETHYLCYCLOTETRASILOXANE	estimate	estimate
102	C9H4O5	TRIMELLITIC ANHYDRIDE	estimate	estimate
103	C9H6N2O2	TOLUENE DIISOCYANATE	estimate	estimate
104	C9H7N	ISOQUINOLINE	estimate	estimate
105	C9H7N	QUINOLINE	experimental/estimate	estimate
106	C9H7NO	8-HYDROXYQUINOLINE	estimate	estimate
107	C9H8	INDENE	estimate	estimate
108	C9H8O	2-METHYLBENZOFURAN	estimate	estimate
109	C9H10	INDANE	estimate	estimate
110	C9H10	alpha-METHYLSTYRENE	estimate	estimate
111	C9H10	m-METHYLSTYRENE	estimate	estimate
112	C9H10	o-METHYLSTYRENE	estimate	estimate
113	C9H10	p-METHYLSTYRENE	estimate	estimate
114	C9H10O2	BENZYL ACETATE	estimate	estimate
115	C9H10O2	ETHYL BENZOATE	estimate	estimate
116	C9H10O3	ETHYL VANILLIN	estimate	estimate
117	C9H11NO	p-DIMETHYLAMINOBENZALDEHYDE	estimate	estimate
118	C9H12	CUMENE	experimental/estimate	experimental
119	C9H12	m-ETHYLTOLUENE	estimate	estimate
120	C9H12	o-ETHYLTOLUENE	estimate	estimate
121	C9H12	p-ETHYLTOLUENE	estimate	estimate
122	C9H12	MESITYLENE	experimental/estimate	estimate
123	C9H12	n-PROPYLBENZENE	experimental/estimate	estimate
124	C9H12	1,2,3-TRIMETHYLBENZENE	estimate	estimate
125	C9H12	1,2,4-TRIMETHYLBENZENE	experimental/estimate	estimate
126	C9H12O	BENZYL ETHYL ETHER	estimate	estimate
127	C9H12O	2-PHENYL-2-PROPANOL	estimate	estimate
128	C9H12O2	CUMENE HYDROPEROXIDE	estimate	---
129	C9H14O	ISOPHORONE	estimate	estimate
130	C9H14O6	GLYCERYL TRIACETATE	estimate	estimate
131	C9H16O4	AZELAIC ACID	estimate	estimate
132	C9H18	ISOPROPYLCYCLOHEXANE	estimate	estimate
133	C9H18	1-NONENE	experimental/estimate	estimate
134	C9H18	n-PROPYLCYCLOHEXANE	estimate	estimate
135	C9H18O	DIISOBUTYL KETONE	estimate	estimate
136	C9H18O	1-NONANAL	experimental/estimate	estimate
137	C9H18O2	n-BUTYL VALERATE	estimate	estimate
138	C9H18O2	n-NONANOIC ACID	experimental/estimate	estimate
139	C9H18O2	n-OCTYL FORMATE	experimental/estimate	estimate
140	C9H20	3,3-DIETHYLPENTANE	estimate	estimate
141	C9H20	2,2-DIMETHYL-3-ETHYLPENTANE	estimate	estimate
142	C9H20	2,4-DIMETHYL-3-ETHYLPENTANE	estimate	estimate

NO	FORMULA	NAME	LIQUID THERMAL COND.	GAS THERMAL COND.
143	C9H20	2,2-DIMETHYLHEPTANE	estimate	estimate
144	C9H20	2,6-DIMETHYLHEPTANE	estimate	estimate
145	C9H20	3-ETHYLHEPTANE	estimate	estimate
146	C9H20	2-METHYLOCTANE	estimate	estimate
147	C9H20	3-METHYLOCTANE	estimate	estimate
148	C9H20	4-METHYLOCTANE	estimate	estimate
149	C9H20	n-NONANE	experimental/estimate	experimental
150	C9H20	2,2,3,3-TETRAMETHYLPENTANE	estimate	estimate
151	C9H20	2,2,3,4-TETRAMETHYLPENTANE	estimate	estimate
152	C9H20	2,2,4,4-TETRAMETHYLPENTANE	estimate	estimate
153	C9H20	2,2,5-TRIMETHYLHEXANE	experimental/estimate	estimate
154	C9H20O	2,6-DIMETHYL-4-HEPTANOL	estimate	estimate
155	C9H20O	1-NONANOL	experimental/estimate	estimate
156	C9H20O	2-NONANOL	estimate	estimate
157	C9H20S	n-NONYL MERCAPTAN	estimate	---
158	C9H21N	n-NONYLAMINE	experimental/estimate	estimate
159	C9H21N	TRIPROPYLAMINE	experimental/estimate	estimate
160	C10H6O8	PYROMELLITIC ACID	estimate	estimate
161	C10H7Br	1-BROMONAPHTHALENE	estimate	estimate
162	C10H7Cl	1-CHLORONAPHTHALENE	estimate	estimate
163	C10H8	NAPHTHALENE	experimental/estimate	experimental
164	C10H9N	QUINALDINE	estimate	estimate
165	C10H10	m-DIVINYLBENZENE	estimate	estimate
166	C10H10	1-METHYLINDENE	estimate	estimate
167	C10H10	2-METHYLINDENE	estimate	estimate
168	C10H10O4	DIMETHYL PHTHALATE	experimental/estimate	estimate
169	C10H10O4	DIMETHYL TEREPHTHALATE	estimate	estimate
170	C10H12	DICYCLOPENTADIENE	estimate	---
171	C10H12	1,2,3,4-TETRAHYDRONAPHTHALENE	experimental/estimate	estimate
172	C10H12O	ANETHOLE	---	estimate
173	C10H12O4	DIALLYL MALEATE	estimate	estimate
174	C10H14	n-BUTYLBENZENE	experimental/estimate	estimate
175	C10H14	sec-BUTYLBENZENE	estimate	estimate
176	C10H14	tert-BUTYLBENZENE	experimental/estimate	estimate
177	C10H14	m-CYMENE	estimate	estimate
178	C10H14	o-CYMENE	estimate	estimate
179	C10H14	p-CYMENE	estimate	estimate
180	C10H14	m-DIETHYLBENZENE	estimate	estimate
181	C10H14	o-DIETHYLBENZENE	estimate	estimate
182	C10H14	p-DIETHYLBENZENE	estimate	estimate
183	C10H14	2-ETHYL-m-XYLENE	estimate	estimate
184	C10H14	2-ETHYL-p-XYLENE	estimate	estimate
185	C10H14	3-ETHYL-o-XYLENE	estimate	estimate
186	C10H14	4-ETHYL-m-XYLENE	estimate	estimate
187	C10H14	4-ETHYL-o-XYLENE	estimate	estimate
188	C10H14	5-ETHYL-m-XYLENE	estimate	estimate
189	C10H14	ISOBUTYLBENZENE	estimate	estimate
190	C10H14	1,2,3,5-TETRAMETHYLBENZENE	estimate	estimate
191	C10H14	1,2,4,5-TETRAMETHYLBENZENE	experimental/estimate	estimate
192	C10H14O	p-tert-BUTYLPHENOL	estimate	estimate
193	C10H14O2	p-tert-BUTYLCATECHOL	estimate	estimate
194	C10H15N	N,N-DIETYHLANILINE	experimental/estimate	---
195	C10H15N	2,6-DIETHYLANILINE	estimate	estimate
196	C10H16	CAMPHENE	estimate	estimate
197	C10H16	D-LIMONENE	estimate	estimate
198	C10H16	alpha-PHELLANDRENE	estimate	estimate
199	C10H16	beta-PHELLANDRENE	estimate	estimate
200	C10H16	alpha-PINENE	estimate	estimate
201	C10H16	beta-PINENE	estimate	estimate
202	C10H16	alpha-TERPINENE	estimate	estimate
203	C10H16	gamma-TERPINENE	estimate	estimate
204	C10H16	TERPINOLENE	estimate	estimate
205	C10H16O	CAMPHOR	estimate	estimate
206	C10H18	cis-DECAHYDRONANPHTALENE	experimental/estimate	estimate
207	C10H18	trans-DECAHYDRONAPHTHALENE	experimental/estimate	estimate
208	C10H18O4	SEBACIC ACID	estimate	estimate
209	C10H20	n-BUTYLCYCLOHEXANE	estimate	estimate
210	C10H20	1-DECENE	experimental/estimate	estimate
211	C10H20O	1-DECANAL	experimental/estimate	estimate
212	C10H20O2	n-DECANOIC ACID	experimental/estimate	estimate
213	C10H20O2	2-ETHYLHEXYL ACETATE	estimate	estimate
214	C10H20O2	ISOPENTYL ISOVALERATE	estimate	estimate
215	C10H22	n-DECANE	experimental/estimate	experimental
216	C10H22	2,2-DIMETHYLOCTANE	estimate	estimate
217	C10H22	2-METHYLNONANE	estimate	estimate

NO	FORMULA	NAME	LIQUID THERMAL COND.	GAS THERMAL COND.
218	C10H22	3-METHYLNONANE	estimate	estimate
219	C10H22	4-METHYLNONANE	estimate	estimate
220	C10H22	5-METHYLNONANE	estimate	estimate
221	C10H22O	1-DECANOL	experimental/estimate	experimental
222	C10H22O	DI-n-PENTYL ETHER	estimate	estimate
223	C10H22O	ISODECANOL	estimate	estimate
224	C10H22O5	TETRAETHYLENE GLYCOL DIMETHYL ETHER	estimate	estimate
225	C10H22S	n-DECYL MERCAPTAN	estimate	estimate
226	C10H23N	n-DECYLAMINE	experimental/estimate	estimate
227	C11H10	1-METHYLNAPHTHALENE	estimate	estimate
228	C11H10	2-METHYLNAPHTHALENE	estimate	estimate
229	C11H14O2	n-BUTYL BENZOATE	estimate	estimate
230	C11H16	n-PENTYLBENZENE	estimate	estimate
231	C11H16O	p-tert-AMYLPHENOL	estimate	estimate
232	C11H20O2	2-ETHYLHEXYL ACRYLATE	estimate	estimate
233	C11H22	1-UNDECENE	experimental/estimate	estimate
234	C11H22O	1-UNDECANAL	estimate	estimate
235	C11H24	n-UNDECANE	experimental/estimate	experimental
236	C11H24O	1-UNDECANOL	estimate	estimate
237	C11H24S	UNDECYL MERCAPTAN	estimate	estimate
238	C12H8O	DIBENZOFURAN	estimate	estimate
239	C12H9N	DIBENZOPYRROLE	estimate	estimate
240	C12H10	ACENAPHTHENE	experimental/estimate	estimate
241	C12H10	BIPHENYL	experimental/estimate	estimate
242	C12H10O	DIPHENYL ETHER	experimental/estimate	estimate
243	C12H11N	p-AMINODIPHENYL	estimate	estimate
244	C12H11N	DIPHENYLAMINE	estimate	estimate
245	C12H11N3	p-AMINOAZOBENZENE	estimate	estimate
246	C12H11N3	1,3-DIPHENYLTRIAZENE	estimate	estimate
247	C12H12	2,6-DIMETHYLNAPHTHALENE	estimate	estimate
248	C12H12	2,7-DIMETHYLNAPHTHALENE	estimate	estimate
249	C12H12	1-ETHYLNAPHTHALENE	estimate	estimate
250	C12H12N2	p-AMINODIPHENYLAMINE	estimate	estimate
251	C12H12N2	HYDRAZOBENZENE	---	estimate
252	C12H14	1,2,3-TRIMETHYLINDENE	estimate	estimate
253	C12H14O4	DIETHYL PHTHALATE	experimental/estimate	estimate
254	C12H16	CYCLOHEXYLBENZENE	estimate	estimate
255	C12H18	m-DIISOPROPYLBENZENE	estimate	estimate
256	C12H18	p-DIISOPROPYLBENZENE	estimate	estimate
257	C12H18	n-HEXYLBENZENE	estimate	estimate
258	C12H20O4	DIBUTYL MALEATE	estimate	estimate
259	C12H22	BICYCLOHEXYL	estimate	estimate
260	C12H23N	DICYCLOHEXYLAMINE	estimate	estimate
261	C12H24	1-DODECENE	estimate	estimate
262	C12H24O	1-DODECANAL	experimental/estimate	estimate
263	C12H24O2	n-DODECANOIC ACID	experimental/estimate	estimate
264	C12H26	n-DODECANE	estimate	experimental
265	C12H26O	DI-n-HEXYL ETHER	experimental/estimate	estimate
266	C12H26O	1-DODECANOL	experimental/estimate	estimate
267	C12H26O3	DIETHYLENE GLYCOL DI-n-BUTYL ETHER	experimental/estimate	estimate
268	C12H26S	n-DODECYL MERCAPTAN	estimate	estimate
269	C12H27BO3	TRI-n-BUTYL BORATE	---	---
270	C12H27N	DODECYLAMINE	experimental/estimate	estimate
271	C12H27N	TRI-n-BUTYLAMINE	experimental/estimate	estimate
272	C13H10	FLUORENE	estimate	estimate
273	C13H10O	BENZOPHENONE	estimate	estimate
274	C13H12	DIPHENYLMETHANE	experimental/estimate	estimate
275	C13H20	n-HEPTYLBENZENE	estimate	estimate
276	C13H26	1-TRIDECENE	estimate	estimate
277	C13H26O	1-TRIDECANAL	estimate	estimate
278	C13H26O2	n-BUTYL NONANOATE	estimate	estimate
279	C13H26O2	METHYL DODECANOATE	estimate	estimate
280	C13H28	n-TRIDECANE	experimental/estimate	experimental
281	C13H28O	1-TRIDECANOL	estimate	estimate
282	C14H8O2	ANTHRAQUINONE	estimate	estimate
283	C14H10	ANTHRACENE	experimental/estimate	estimate
284	C14H10	DIPHENYLACETYLENE	estimate	estimate
285	C14H10	PHENANTHRENE	estimate	estimate
286	C14H12	cis-STILBENE	estimate	estimate
287	C14H12	trans-STILBENE	estimate	estimate
288	C14H12O2	BENZYL BENZOATE	estimate	estimate
289	C14H14	1,1-DIPHENYLETHANE	estimate	estimate
290	C14H14	1,2-DIPHENYLETHANE	estimate	estimate
291	C14H14O	DIBENZYL ETHER	estimate	estimate
292	C14H16	1-n-BUTYLNAPHTHALENE	estimate	estimate

NO	FORMULA	NAME	LIQUID THERMAL COND.	GAS THERMAL COND.
293	C14H22	n-OCTYLBENZENE	estimate	estimate
294	C14H220	p-tert-OCTYLPHENOL	estimate	estimate
295	C14H28	1-TETRADECENE	estimate	estimate
296	C14H2802	n-TETRADECANOIC ACID	experimental/estimate	estimate
297	C14H30	n-TETRADECANE	experimental/estimate	experimental
298	C14H300	1-TETRADECANOL	experimental/estimate	estimate
299	C14H31N	TETRADECYLAMINE	estimate	estimate
300	C15H10N2O2	DIPHENYLMETHANE-4,4'-DIISOCYANATE	estimate	estimate
301	C15H160	p-CUMYLPHENOL	estimate	estimate
302	C15H1602	BISPHENOL A	estimate	estimate
303	C15H24	n-NONYLBENZENE	estimate	estimate
304	C15H240	2,6-DI-tert-BUTYL-p-CRESOL	estimate	estimate
305	C15H240	NONYLPHENOL	estimate	estimate
306	C15H30	1-PENTADECENE	estimate	estimate
307	C15H3002	PENTADECANOIC ACID	estimate	estimate
308	C15H32	n-PENTADECANE	experimental/estimate	experimental
309	C16H10	FLUORANTHENE	estimate	estimate
310	C16H10	PYRENE	experimental/estimate	estimate
311	C16H12	1-PHENYLNAPHTHALENE	estimate	estimate
312	C16H20	1-n-HEXYLNAPHTHALENE	estimate	estimate
313	C16H2204	DIBUTYL PHTHALATE	experimental/estimate	estimate
314	C16H26	n-DECYLBENZENE	estimate	estimate
315	C16H32	n-DECYLCYCLOHEXANE	estimate	estimate
316	C16H32	1-HEXADECENE	estimate	estimate
317	C16H3202	n-HEXADECANOIC ACID	experimental/estimate	estimate
318	C16H34	n-HEXADECANE	experimental/estimate	experimental
319	C16H340	DI-n-OCTYL ETHER	estimate	estimate
320	C16H340	1-HEXADECANOL	estimate	estimate
321	C17H28	n-UNDECYLBENZENE	estimate	estimate
322	C17H34	1-HEPTADECENE	estimate	estimate
323	C17H36	n-HEPTADECANE	experimental/estimate	estimate
324	C17H360	1-HEPTADECANOL	estimate	estimate
325	C18H12	CHRYSENE	estimate	estimate
326	C18H14	m-TERPHENYL	experimental/estimate	estimate
327	C18H14	o-TERPHENYL	experimental/estimate	estimate
328	C18H14	p-TERPHENYL	experimental/estimate	estimate
329	C18H15P	TRIPHENYLPHOSPHINE	estimate	estimate
330	C18H1504P	TRIPHENYL PHOSPHATE	---	---
331	C18H16N2	N,N'-DIPHENYL-p-PHENYLENEDIAMINE	estimate	estimate
332	C18H22	2,3-DIMETHYL-2,3-DIPHENYLBUTANE	estimate	estimate
333	C18H2202	DICUMYL PEROXIDE	estimate	---
334	C18H30	n-DODECYLBENZENE	estimate	estimate
335	C18H3202	LINOLEIC ACID	estimate	estimate
336	C18H3402	OLEIC ACID	experimental/estimate	estimate
337	C18H3404	DIBUTYL SEBACATE	estimate	estimate
338	C18H3404	DIHEXYL ADIPATE	estimate	estimate
339	C18H36	1-OCTADECENE	estimate	estimate
340	C18H3602	STEARIC ACID	estimate	estimate
341	C18H38	n-OCTADECANE	experimental/estimate	estimate
342	C18H380	DINONYL ETHER	estimate	estimate
343	C18H380	1-OCTADECANOL	experimental/estimate	estimate
344	C19H26	1-n-NONYLNAPHTHALENE	estimate	estimate
345	C19H32	n-TRIDECYLBENZENE	estimate	estimate
346	C19H3602	METHYL OLEATE	estimate	estimate
347	C19H38	1-NONADECENE	estimate	estimate
348	C19H3802	NONADECANOIC ACID	estimate	estimate
349	C19H40	n-NONADECANE	experimental/estimate	estimate
350	C20H16	TRIPHENYLETHYLENE	estimate	estimate
351	C20H28	1-n-DECYLNAPHTHALENE	estimate	estimate
352	C20H3002	ABIETIC ACID	estimate	estimate
353	C20H31N	DEHYDROABIETYLAMINE	estimate	estimate
354	C20H40	1-EICOSENE	estimate	estimate
355	C20H42	n-EICOSANE	experimental/estimate	estimate
356	C20H420	1-EICOSANOL	estimate	estimate
357	C21H2104P	TRI-o-CRESYL PHOSPHATE	experimental/estimate	---
358	C22H4402	n-BUTYL STEARATE	experimental/estimate	estimate
359	C24H3804	DIISOOCTYL PHTHALATE	estimate	estimate
360	C24H3804	DIOCTYL PHTHALATE	estimate	estimate
361	C24H420	DINONYLPHENOL	estimate	estimate
362	C26H20	TETRAPHENYLETHYLENE	estimate	estimate
363	C28H4604	DIISODECYL PHTHALATE	estimate	estimate

Appendix D

COMPOUND LIST BY FORMULA

Appendix E

COMPOUND LIST BY NAME

Edwards Brothers Malloy
Ann Arbor MI. USA
December 13, 2016